Jun Ma *Editor*

Gene Expression and Regulation

高等教育出版社
HIGHER EDUCATION PRESS

Springer

Jun MA

Division of Developmental Biology
Cincinnati Children's Hospital Research Foundation
University of Cincinnati College of Medicine
3333 Burnet Avenue
Cincinnati, OH 45229
USA

E-mail: *Jun.ma@cchmc.org*

图书在版编目（CIP）数据

基因的表达与调控=Gene Expression and Regulation/马骏主编.
—北京：高等教育出版社，2006
ISBN 7-04-017675-0

Ⅰ.基… Ⅱ.马… Ⅲ.基因表达—研究—英文 Ⅳ.Q753

中国版本图书馆 CIP 数据核字（2005）第 140172 号

图字：01-2006-0366

Copyright © 2006 by
Higher Education Press
4 Dewai Dajie, Beijing 100011, P. R. China

Distributed by Springer Science+Business Media, LLC under ISBN 0-387-33208-1 worldwide except in mainland China by the arrangement of Higher Education Press.

All rights reserved. No part of this book may be reproduced or transmitted in any form or by any means, electronic or mechanical, including photocopying, recording or by any information storage and retrieval system, without permission in writing from the Publisher.

While the advice and information in this book are believed to be true and accurate at the date of going to press, neither the authors nor the editors nor the publisher can accept any legal responsibility for any errors or omissions that may be made. The publisher makes no warranty, express or implied, with respect to the material contained herein.

ISBN 7-04-017675-0

Printed in P. R. China

Contents

About the Editor
Preface

Section I   The History

Chapter 01   Transcription: The Never Ending Story   *3*

Section II   The Machinery

Chapter 02   The General Transcription Machinery and Preinitiation Complex Formation   *21*
Chapter 03   The Dynamic Association of RNA Polymerase II with Initiation, Elongation, and RNA Processing Factors during the Transcription Cycle   *49*
Chapter 04   General Cofactors: TFIID, Mediator and USA   *67*
Chapter 05   Chromatin and Regulation of Gene Expression   *95*
Chapter 06   HATs and HDACs   *111*
Chapter 07   Structure and Function of Core Promoter Elements in RNA Polymerase II Transcription   *135*

Section III   The Regulators

Chapter 08   Transcriptional Activators and Activation Mechanisms   *147*
Chapter 09   Transcriptional Repressors and Repression Mechanisms   *159*
Chapter 10   STATs in Cytokine-mediated Transcriptional Regulation   *175*
Chapter 11   Transcriptional Regulation by Smads   *185*
Chapter 12   The Rb and E2F Families of Proteins   *207*
Chapter 13   c-Jun: A Complex Tale of a Simple Transcription Factor   *219*
Chapter 14   HIV Tat and the Control of Transcriptional Elongation   *239*
Chapter 15   Post-translational Modifications of the p53 Transcription Factor   *257*
Chapter 16   Actions of Nuclear Receptors   *273*
Chapter 17   NFAT and MEF2, Two Families of Calcium-dependent Transcription Regulators   *293*
Chapter 18   Hox Genes   *309*
Chapter 19   Nuclear Factor-kappa B   *321*
Chapter 20   The ATF Transcription Factors in Cellular Adaptive Responses   *329*

Section IV   The Genome

Chapter 21   Function and Mechanism of Chromatin Boundaries   *343*
Chapter 22   Heterochromatin and X Inactivation   *365*
Chapter 23   DNA Methylation Regulates Genomic Imprinting, X Inactivation, and Gene Expression during Mammalian Development   *377*
Chapter 24   Comparative Genomics of Tissue Specific Gene Expression   *393*
Chapter 25   Transcription and Genomic Integrity   *409*
Chapter 26   Cell Death and Transcription   *431*

Section V Special Topics

Chapter 27　Pre-mRNA Splicing in Eukaryotic Cells　*447*
Chapter 28　Genome Organization: The Effects of Transcription-driven DNA Supercoiling on Gene Expression Regulation　*469*
Chapter 29　The Biogenesis and Function of MicroRNAs　*481*
Chapter 30　Transcription Factor Dynamics　*493*
Chapter 31　Actin, Actin-Related Proteins and Actin-Binding Proteins in Transcriptional Control　*503*
Chapter 32　Wnt Signaling and Transcriptional Regulation　*519*
Chapter 33　Regulatory Mechanisms for Floral Organ Identity Specification in *Arabidopsis thaliana*　*533*
Chapter 34　Transcription Control in Bacteria　*549*
Chapter 35　Gene Therapy: Back to the Basics　*565*

About the Editor

Jun Ma is an Associate Professor at the Division of Developmental Biology, Cincinnati Children's Hospital Research Foundation and University of Cincinnati College of Medicine. He graduated from Peking University in 1982, majoring in Biology. He did his graduate work with Mark Ptashne at the Department of Biochemistry and Molecular Biology in Harvard University, and was a Junior Fellow at the Harvard Society of Fellows between 1989–1992. He spent the summer of 1988 in the laboratory of Christiane Nüsslein-Volhard at the Max-Planck-Institute for Developmental Biology in Tübingen to collaborate with Wolfgang Driever. He joined the faculty of the University of Cincinnati College of Medicine in 1992 and has remained there since. Currently he also has a collaborative base at the Institute of Biophysics of the Chinese Academy of Sciences in Beijing. His earlier work on the yeast activator GAL4 helped pave the way to the development of the yeast two-hybrid system. His current research focuses on the mechanisms of transcription control and development in *Drosophila*.

Preface

All genes must be expressed to exhibit their biological activities. How genes are expressed and regulated is a central question in molecular biology and our knowledge in this area has been expanding enormously in recent years. The complexity of gene regulation is compounded by the fact that gene activities reach every corner of biology. Transcription is universally the first step toward expressing a gene. It is a highly regulated process. Understanding the molecular mechanisms of transcription regulation is of fundamental importance. For protein-coding genes, post-transcriptional steps, including pre-mRNA processing, mRNA transport and translation, can also play important roles in regulating gene expression. To contain the scope of this book, we will focus primarily on RNA polymerase II transcription and regulation. We will explore not only the biochemical basis of transcription but also the biological consequences of, and biological influences on gene transcription.

The book is composed of 35 individual review articles written by authorities in the field. The chapters are organized into five sections: The History, The Machinery, The Regulators, The Genome, and Special Topics. The History section contains one chapter, written by James Goodrich and Robert Tjian, who provide an excellent historical perspective and overview of the transcription process. The Machinery section has six chapters that cover essential topics on the transcriptional apparatus, general cofactors, chromatin structure, and core promoter structure. The Regulators section has thirteen chapters. While the first two of them investigate the mechanisms of transcriptional activation and repression, the remaining eleven chapters discuss in depth selected gene-specific transcription factors that play critical roles in a variety of biological processes, including STATs, Smads, NFκB, nuclear receptors, NFAT, Rb, p53, HIV Tat, ATFs, c-Jun and Hox proteins. The Genome section contains six chapters that examine topics relevant to transcription regulation and genome behavior, including chromatin boundaries, heterochromatin, DNA methylation, genomic analysis, genomic integrity, and cell death. Finally, the Special Topics section contains nine chapters that investigate such important issues as pre-mRNA splicing, DNA supercoiling, microRNA, transcription factor dynamics, role of actin in transcription, gene therapy, and transcription regulation in bacteria, plants and developmental signaling.

When Higher Education Press invited me to write a textbook for their Current Scientific Frontiers book series two years ago, I did not think I had the time needed to tackle such a big project. Instead, I made a proposal—endorsed quickly by HEP—to explore the possibility of editing a book (resembling a textbook style) on the topic of gene expression and regulation, with individual review articles written by experts in the field. Without the enthusiastic support and generous commitment from the contributors, this project would have never even started. I am deeply indebted to all of them. Every chapter in this book is a scholarly work reflecting numerous hours of intense efforts of the contributors. I would like to express my special thanks to Cheng-Ming Chiang for generously contributing two excellent chapters, a few contributors for kindly agreeing to write on relatively short notice, and Gordon Hager for providing the cover photo and design suggestions. I would also like to thank HEP for their flexibility and trust in this project, and the HEP and Springer editorial and design teams, in particular Li Shen at HEP, for their excellent work. Finally, I would like to thank Bingxiang Li at HEP for the countless email communications and her hard work—at every step along the way—that made this book a reality.

Jun Ma
Cincinnati, USA
November 18, 2005

Section I

The History

Chapter 01

Transcription: The Never Ending Story

James A. Goodrich[1] and Robert Tjian[2]

[1] Department of Chemistry and Biochemistry, University of Colorado, Boulder, CO 80309
[2] Molecular and Cell Biology Department, University of California, Berkeley, Berkeley, CA 94720

Key Words: transcription, promoter, activator, coactivator, general factors, chromatin, RNA polymerase II

Summary

After more than 30 years of intense and sustained activity, the field of transcriptional control in eukaryotes continues to deliver unexpected and revealing montages of the remarkably complex yet elegant consequences of evolution. Transcription research started from humble beginnings with the isolation of 3 distinct RNA polymerases. This was followed by a rich period of mapping promoters, enhancers and the isolation of the first sequence specific DNA binding regulatory factors. These studies in turn led to the unraveling of the multi-subunit pre-initiation apparatus culminating with the modern era of co-activators and chromatin remodeling complexes. Throughout this opus of biochemical discovery we have witnessed a beautiful convergence of *in vitro* biochemical tour-de-force combined with the power of molecular genetics and cell biology. In this short preamble, we offer a brief and very likely incomplete history of the maturing of eukaryotic transcription and its prospects for the future.

Fumbling in the Dark: Hoping for Simplicity

Emboldened by the inspiring successes of pioneering work in the biochemistry of DNA replication and bacterial phage transcription, early workers struggling with animal and human gene regulation followed suit by isolating not one but three distinct enzymes: RNA polymerase I, II and III each dedicated to the synthesis of rRNA, mRNA, and tRNA/5sRNA respectively (Krebs and Chambon, 1976; Sklar et al., 1975). However, due to the lack of promoter specific DNA templates or the ability to obtain sufficient quantities of "cloned" DNA, the ability of these 3 distinct enzymes to discriminate between the different classes of genes remained obscure. Nevertheless, the chromatographic separation and *in vitro* biochemical assays for detecting the RNA polymerases opened the first doors to the future development of high fidelity promoter specific and eventually activator regulated transcription in cell free systems.

Because, eukaryotic RNA polymerases behaved in a rather promiscuous and DNA template independent fashion *in vitro*, there was a brief period, (after the discovery of heterogeneous nuclear RNA) in which it was popular to posit that, unlike bacterial transcription which is temporally regulated by cascades of σ-factors, eukaryotic transcription may be "unregulated". Instead, one imagined that post transcriptional RNA processing (i.e. splicing, poly A addition, capping, etc.) would largely determine the population of mRNA's destined for gene product expression. Although this "random transcription" model fit with some early data regarding the apparent lack of promoter DNA selectivity *in vitro* of eukaryotic RNA polymerases, it soon became clear from studies of mammalian viruses (SV40, Adeno 2) that at the very least, specific DNA sequences that lie near transcription start sites (i.e. TATA elements and GC boxes) played some role in determining elements of the eukaryotic "promoter" (Fig.1.1) (Myers et al., 1981; Rio et al., 1980; Tjian, 1978).

As is often the case with biology in general but especially in the study of eukaryotic transcriptional regulation, we invariably opted for simplicity and hoped that a well defined −35/−10 like element such as the

---

Corresponding Author: Robert Tjian, Tel: (510) 642-0884, FAX: (510) 643-9547, E-mail:jmlim@Berkeley.edu

popular TATA box of Ad2 would suffice to designate the necessary cis-regulatory information of a promoter (Corden *et al.*, 1980; Hu and Manley, 1981). This rather minimalist view was, however, decisively toppled when both *in vitro* and cell based assays were developed that revealed the existence of important upstream distal as well as proximal DNA sequences in eukaryotic promoters (Banerji *et al.*, 1981; Benoist and Chambon, 1981; Fromm and Berg, 1983; Gidoni *et al.*, 1985; Myers and Tjian, 1980; Picard and Schaffner, 1984). With the emergence of cloned promoter sequences and DNA template dependent *in vitro* transcription reactions measured by run-off and primer-extension assays the combinatorial nature of multiple cis-control elements of eukaryotic gene regulatory units became firmly embedded (Mitchell and Tjian, 1989). Add to these *in vitro* assays the advent of transient transfection assays and microinjection in animal cells that revealed the existence of "orientation and distance independent" enhancer elements and we began, for the first time, to get a glimpse of the complex regulatory network of gene transcription that would follow in succeeding decades (McKnight and Tjian, 1986; McKnight, 1982; Picard and Schaffner, 1984; Treisman *et al.*, 1983). To this day, the precise mechanisms mediating "long distance" enhancer or silencer functions remain largely obscure despite many plausible models including DNA looping, scanning etc.

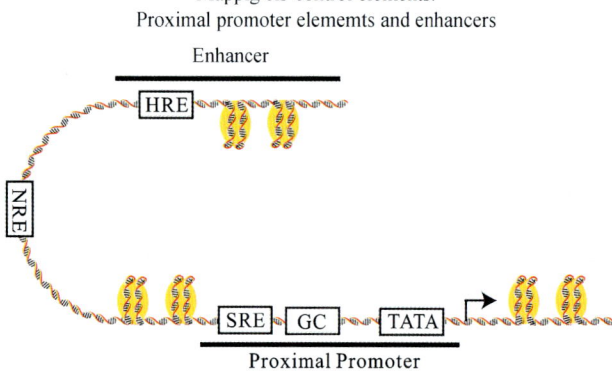

Fig.1.1 Cis-control elements in RNA polymerase II promoters can be located near the transcription start site or at great distances away. Some of the cis-control elements identified by early studies are shown. Abbreviations: TATA, TATA box; GC, GC box; SRE, sterol response element; NRE, nuclear hormone receptor response element; HRE, heat shock response element.

A Shaft of Light: Sequence Specific Transcription Factors

After a flurry of intense promoter bashing experiments with all manner of DNA templates, cell-types and gene systems, we were confronted with the daunting task of determining what was actually recognizing and keying off these composite arrays of cis-control DNA sequences to govern gene specific transcription. One important step along this pathway of discovery was the rapid deployment of various elegant *in vitro* mutagenesis techniques such as linker scanning clustered point mutations and deletions (McKnight *et al.*, 1981; McKnight and Kingsbury, 1982; Myers and Tjian, 1980). At the same time, powerful new biochemical assays such as DNase I footprint protection were being developed (Galas and Schmitz, 1978). Perhaps the single most influential strategy for those of us attempting to dissect the molecular identity of transcriptional regulatory factors was the promoter selective *in vitro* transcription assay (Manley *et al.*, 1980; Rio *et al.*, 1980; Weil *et al.*, 1979; Wu, 1978). This "bucket biochemistry" approach allowed us to use cloned DNA fragments containing well mapped and carefully defined promoters to drive accurate and factor dependent transcription by partially purified RNA polymerases. The tacit assumption in establishing such *in vitro* promoter dependent assays was that purified eukaryotic RNA polymerase II was necessary but not sufficient to direct accurate initiation of transcription. We therefore assumed that one or more additional transcription factors (whose identity and mode of action had remained unknown) was needed in order to instruct or otherwise impart upon RNA pol II the ability to discriminate one promoter from another. Indeed, since no such cellular factors in eukaryotes had yet been identified or isolated in 1980, we had little clue as to the biochemical properties of such factors (i.e. were these factors proteins, nucleic acids, carbohydrate, etc.?). The closest candidate at that time was the SV40 T-ag, a viral encoded protein that displayed many of the hallmarks of a bona fide promoter recognition factor (Rio *et al.*, 1980; Tjian, 1978). Also, whether they would directly bind RNA polymerase á la σ-factors or they would behave more like CAP in the lac operon system and bind DNA in a sequence specific manner was a big question.

Indeed, one of the unappreciated and hidden advantages of using fairly crude nuclear extracts (i.e. from Hela cells or *Drosophila* embryos) to carry out systematic biochemical "complementation" tests *in vitro* allowed us the freedom to be unbiased and simply search for whatever molecules stimulated transcription of one promoter but not another. Using this approach, factors such as Sp1 were first identified as functional transcriptional activators for RNA pol II that could discriminate, for example, between the SV40 and AdML promoters (Fig.1.2) (Carthew *et al.*, 1985; Dynan and Tjian, 1983a; Dynan and Tjian, 1983b; Sawadogo and Roeder, 1985). Similar biochemical fractionation and *in vitro* assays led to the isolation of TFIIIA for Pol III and UBF for Pol I (Engelke *et al.*, 1980; Learned *et al.*, 1985; Learned *et al.*, 1986; Pelham and Brown, 1980; Wu, 1978). However, it did not take long given the availability of various discriminating DNA binding assays available at that time to determine that these transcription factors were indeed sequence specific DNA binding "activators". And thus, there was a nice alignment of cis-regulatory elements and DNA binding transcription factors. We anticipate that a similar biochemical dissection and reconstitution of *in vitro* transcription reactions that are responsive to distal enhancers, tethering elements, silencers and boundary elements are still needed to fill-in our gaps of knowledge vis-à-vis the molecular players and mechanisms that govern "long distance" regulation so prevalent in metazoan organisms.

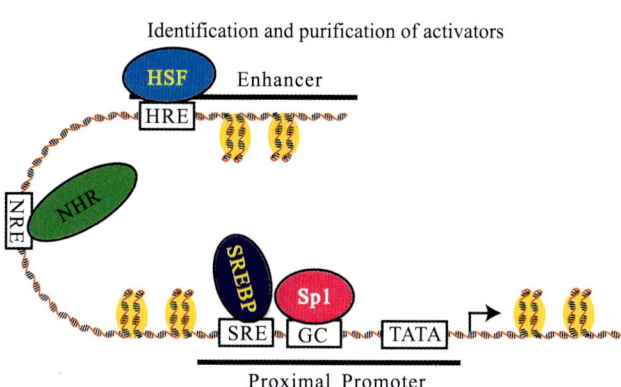

Fig.1.2 Trans-acting factors bind to RNA polymerase II promoters. Abbreviations: Sp1, specificity protein 1; SREBP, sterol response element binding protein; NHR, nuclear hormone receptor; HSF, heat shock factor.

A New Era of Transcription Biochemistry Arrives: Clone, Sequence, Express & Reconstitute

The next big hurdle was to actually purify, clone, and characterize these seemingly powerful transcriptional activators. As often happens in emerging fields, advances in concepts and techniques must go hand in hand. For the transcription field, the development of sequence specific DNA affinity chromatography and a host of affiliated techniques revolutionized our capacity to detect, purify and clone the genes encoding sequence-specific transcription factors (Briggs *et al.*, 1986; Jones *et al.*, 1985; Kadonaga *et al.*, 1987; Kadonaga and Tjian, 1986). Once the genes encoding the first few bona fide transcriptional activators (and repressors) such as Sp1, TFIIIA, CTF, AP1, GCN4, Gal4, GR, and HSF were characterized — a flood of paradigm shifting concepts emerged (Berg, 1988; Bohmann *et al.*, 1987; Courey *et al.*, 1989; Kadonaga *et al.*, 1987; Kadonaga *et al.*, 1988; Mermod *et al.*, 1989; Miller *et al.*, 1985; Mitchell *et al.*, 1987; Triezenberg *et al.*, 1988; Turner and Tjian, 1989). For instance, the remarkably modular nature of transcriptional activators was revealed (Ma and Ptashne, 1987a; Ma and Ptashne, 1987b). The subsequent cloning and sequencing of transcription factors rapidly advanced our ability to recognize DNA binding motifs (i.e. Zn finger, B-HLH, homeodomains, etc.) dimerization domains (LZ, histone folds) activation domains (gln-rich, acidic, etc.) and regulatory/ligand binding domains (AF2).

Initially, as a result of the pioneering work on transcription factor structures derived from studies of the λ-repressor and other phage and bacterial transcription factors (Anderson *et al.*, 1985; Wharton *et al.*, 1984), there was a tendency to assume that all transcription factors would utilize a helix-turn-helix DNA binding domain and an "acidic" activation domain. However, the structure/function analysis of eukaryotic transcription factors such as Sp1, TFIIIA, steroid receptors, Jun/Fos AP1, C/EBP, CTF etc. quickly dispelled the over-simplified notion that there were only one or two motifs for DNA binding and transcription activation (Gill and Ptashne, 1988). Indeed, it became clear that in eukaryotes and especially metazoan organisms, the repertoire of structural domains that had evolved to accommodate transcriptional specificity was astoundingly diverse and elaborate.

One of the most impressive accomplishments during this rich middle period (1985 — 1995) of transcription research was not only the rapid identification, cloning and characterization of hundreds of sequence specific transcription factors, but also a quantum leap in our understanding of the relationship between function and structure — particularly with regards to DNA binding motifs (Pabo and Sauer, 1992). The high resolution X ray structures of countless DNA binding domains were solved and this rich body of information continues to provide a basis for rapid genome wide functional analysis of novel gene products. The discovery of thousands of different transcriptional activators (repressors) and their pivotal role in complex biological processes such as anterior-posterior and dorsal-ventral patterning in metazoans firmly cemented the importance of this vast family of proteins. Indeed, after the first few different metazoan genomes were determined, it became apparent that between 5%~10% of the coding capacity of eukaryotes is devoted to encoding such transcriptional regulators. These findings provided another inexorable clue to the essential, universal, and yet diverse nature of transcriptional control mechanisms. However, despite this exponential growth in knowledge about transcription factors, not everything was rosy or well understood about transcriptional regulation. Indeed, although DNA binding motifs and their structures had proven to be highly informative with respect to structure/function relationships, a similar understanding of activation domains was sorely lacking and largely remains so even today.

Mix and Match: Combinatorial Control, Modularity, and Enhanceosomes

As activators and genes were being characterized in greater detail, it became apparent that the simple paradigm of a single activator or single repressor controlling transcription of a gene, as was the case in some bacterial and even yeast systems, did not apply in higher eukaryotes. The regulatory regions of mammalian and *Drosophila* genes, enhancers and silencers, contain binding sites for many transcriptional regulators. An enhancer might bind 10 or more DNA binding factors, including many different activators as well as multiple copies of a single activator. This complexity was further amplified by the presence of large activator families (Homeo-box, FOXO, AP1 etc.) in which individual members had similar DNA binding specificities, but distinct activation domains and presumably different functions (Mitchell and Tjian, 1989). Combinatorial control and the notion of cis-regulatory networks help explain observations indicating that it is the precise complement of activators and repressors present at a promoter that gives rise to gene specific activation in a spatial and temporally regulated pattern (DeFranco and Yamamoto, 1986; Diamond *et al.*, 1990). Cooperativity in DNA binding and synergy in transcriptional activation further contribute to an uncanny level of control over gene transcription. Our understanding of enhancers and activators was substantially advanced with the detailed characterization of the interferon-β and T-cell receptor α enhancers, where the correct function of the enhancers requires not only the presence of the appropriate array of transcriptional activator proteins but also the association of architectural proteins, and the proper spatial orientation of all of these factors dictates the ultimate outcome (Giese *et al.*, 1992; Giese *et al.*, 1995; Thanos and Maniatis, 1995).

Unimagined complexity: The General Transcription Factors, PIC formation, and Promoter-Specific Transcription

From early *in vitro* studies it was realized that while core RNA polymerase II was capable of synthesizing an RNA product, it required additional factors to initiate transcription at specific promoters (Weil *et al.*, 1979). The general concept of dissociable and essential transcription factors had been firmly established in bacteria, where core RNA polymerase required a sigma subunit for promoter-specific transcription. While this paradigm provided a useful framework for studying eukaryotic transcription, the requirement for a single sigma-like subunit was quickly dispelled in eukaryotic transcription systems. Employing biochemical fractionation and promoter specific DNA templates to drive *in vitro* transcription reactions, an unexpectedly large number of critical accessory factors were painstakingly teased out and characterized, initially as crude fractions eluted from columns (Matsui *et al.*, 1980). Of course, like any good biochemist, once you have an assay, next you want to purify the critical activity, characterize its biochemical properties, and identify the gene encoding the factor. After many hundreds of researcher years, all of the general factors and their genes from human, *Drosophila* and yeast eventually were isolated (Aso *et al.*, 1992; DeJong and Roeder, 1993; Eisenmann *et al.*, 1989; Finkelstein *et al.*, 1992; Fischer *et al.*, 1992; Ha *et al.*, 1991; Hahn *et al.*, 1989a; Hahn *et al.*, 1989b; Hoey *et al.*, 1990; Horikoshi *et al.*, 1989; Kao *et al.*, 1990; Ma *et al.*, 1993; Peterson *et al.*, 1991; Peterson *et al.*, 1990; Schaeffer *et al.*, 1993; Shiekhattar *et al.*, 1995; Sopta *et al.*, 1989; Yokomori *et al.*, 1993). Thus, the general or basal transcription factors TFII-A, -B, -D (TBP), -E, -F, and -H were identified (Fig.1.3).

The general transcription factors were unlike the sequence specific activators in that most of them showed little or no propensity to bind DNA in a sequence dependent manner, but instead associated with RNA polymerase II and participated in complex ways towards the assembly of the pre-initiation complex (PIC). Among this large clan of general transcription factors—one that stood out early on was the fraction originally designated TFIID which revealed a weak tendency to bind TATA elements (Reinberg *et al.*, 1987; Sawadogo and Roeder, 1985). Attempts to purify and characterize this activity proved to be particularly intransigent. After many attempts and failures on the part of several labs, through a combination of persistent biochemistry and fortuitous genetics—the all important TATA binding protein (TBP) was isolated and cloned in the late 1980s (Hahn et al., 1989b; Hoey et al., 1990; Horikoshi et al., 1989; Kao et al., 1990; Peterson et al., 1990). TBP, the central subunit of the TFIID complex, itself has the ability to bind specifically to TATA box elements found in many, but by no means most RNA polymerase II promoters. The surprising observation that the single subunit TBP could replace the crude TFIID function in directing preinitiation complex assembly and basal transcription in vitro enabled biochemical experiments to establish an order of assembly for the preinitiation complex—TBP, TFIIA, TFIIB, TFIIF/RNA polymerase II, TFIIE, and TFIIH (Buratowski et al., 1989; Buratowski et al., 1988; Flores et al., 1992). In later in vitro experiments, RNA polymerase II was also found in larger complexes containing some of the general transcription factors and provided an alternative mode of preinitiation complex assembly (Koleske and Young, 1994), in which a RNA polymerase II complex is recruited to TFIID and TFIIA pre-assembled on promoter DNA.

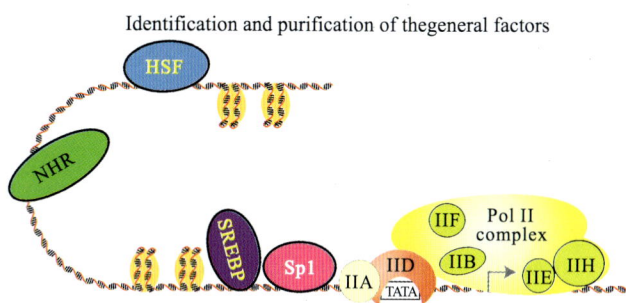

Fig.1.3 The RNA polymerase II general transcription machinery loads into preinitiation complexes encompassing the start site of transcription.

Many surprises surfaced during studies of the general transcription factors. Some of these factors functioned in multiple stages of the transcription reaction—TFIIF is required for initiation and stimulates elongation (Flores et al., 1989; Saltzman and Weinmann, 1989). Others had enzymatic activity—a subunit of TFIIH is a kinase (Feaver et al., 1994; Lu et al., 1992; Roy et al., 1994; Serizawa et al., 1995), two others are helicases (Schaeffer et al., 1994; Schaeffer et al., 1993), and the largest subunit of TFIID has kinase (Dikstein et al., 1996a), acetyltransferase (Mizzen et al., 1996), and ubiquitin-activating/conjugating (Pham and Sauer, 2000) activities. While not detected initially, some of the general transcription factors are now known to bind core promoter DNA with sequence specificity—TFIIB binds the BRE (Lagrange et al., 1998) and subunits of TFIID other than TBP bind the initiator and DPE (Burke and Kadonaga, 1997; Kaufmann and Smale, 1994; Verrijzer et al., 1994). These observations lead us to wonder what other functions will be discovered in future studies of the general transcription factors.

As the general transcription factors were discovered and some of their functions revealed, an intense interest mounted in uncovering the three dimensional structures of these critical factors. TBP was found to be "saddle shaped", with the underside making intricate contacts with and dramatically bending the TATA DNA, while the upper surface presented itself for numerous interactions with other proteins (Kim et al., 1993a; Kim et al., 1993b; Nikolov et al., 1992). This was followed by crystallography studies that unveiled the molecular architecture of complexes contain TFIIB and TFIIA along with TBP and DNA (Geiger et al., 1996; Nikolov et al., 1995; Tan et al., 1996). Larger complexes, such as TFIID and TFIIH have been envisioned using electron microscopy, which revealed the overall shape of these massive entities (Andel et al., 1999; Brand et al., 1999; Chang and Kornberg, 2000; Schultz et al., 2000). The RNA polymerase II enzyme itself was the focus of a major structural effort that began with EM and ultimately surrendered to X-ray crystallography (Cramer et al., 2000; Darst et al., 1991; Gnatt et al., 2001). The amazing structures that have so far resulted from this endeavor provide an unimaginably intricate view of the polymerase alone and bound to other molecules. The value of these structures in opening new lines of research is incredible and these initial glimpses leave us longing for high resolution structures of larger and more elaborate transcription complexes.

The Paradox of Transcriptional Activation: Insufficiency of the General Machinery

With studies rapidly progressing on transcriptional activators and the general transcription machinery a number of labs began to puzzle over the nagging question: How does a sequence specific transcription factor such as Sp1 actually promote the initiation of transcription by RNA pol II? This simple question would eventually lead to a most elaborate and unexpected molecular landscape that today dominates our thinking about how specific mechanisms of transcriptional control are executed in temporally and spatially restricted programs of gene expression. By

1989, we thought all of the molecular components necessary to form an active PIC were in hand. Thus, it seemed a simple matter to reconstitute *in vitro* transcription with purified TBP, RNA Pol II, TFII-A, -B, -E, -F, and -H. Indeed, this constellation of factors isolated either from yeast, *Drosophila* or HeLa cells efficiently produced accurately initiated transcription on any number of well defined promoters. However, to our consternation and frustration, when attempts were made to reconstitute "regulated" transcription using factors such as Sp1, VP16, NR, etc—there was no response to activators in these *in vitro* reactions—contrary to what was expected from *in vivo* studies (Fig.1.4). Thus began a new chapter in the transcription story—the hunt for the elusive co-activators, which can be categorized as the 3rd class of transcription factors, the other two being the sequence specific DNA binding factors and the general factors. Again, relying on our old standby strategy of *in vitro* biochemical complementation as well as genetics in yeast using model activators such as Sp1 or Gal4, the first evidence for a new class of transcription factors emerged — loosely named co-activators and mediators (Berger *et al.*, 1990; Kelleher III *et al.*, 1990; Pugh and Tjian, 1990). Members of this new class of factors were not required for accurate transcription initiation but provided a key function in transcriptional activation, and perhaps a link between DNA binding activators and the core transcription machinery. After much additional research, we now know that many co-activators (and co-repressors) exist, with two groups of proteins playing critical roles in activation of many if not all genes: the TAF subunits of TFIID and subunits of the Mediator complexes.

The TFIID TAF Saga: Pride and Prejudice

The first well characterized group of co-activators turned out to be subunits associated with TBP and were thus called TAFs (TBP-associated factors) (Dynlacht *et al.*, 1991; Pugh and Tjian, 1990). Although the TAF's were originally discovered in *Drosophila* and human systems, eventually it was revealed that these subunits of the TFIID complex are, in fact, universal in eukaryotes and largely conserved from yeast to man. Thus, the paradox surrounding TFIID and its relation to TBP was finally resolved: TFIID is actually composed of TBP, a subunit essential for basal transcription, while the cluster of tightly associated TAF subunits are necessary for the co-activator function of TFIID (Fig.1.5). After a great deal of structure/function analysis *in vitro* and *in vivo*, we now know that the TAF/TBP complex actually participates in several distinct aspects of transcription including recognition of composite core promoter elements (i.e. INR, TATA, DPE) by TBP and several of the TAF subunits (Burke and Kadonaga, 1997; Hahn *et al.*, 1989b; Hoey *et al.*, 1990; Horikoshi *et al.*, 1989; Kao *et al.*, 1990; Kaufmann and Smale, 1994; Peterson *et al.*, 1990; Verrijzer *et al.*, 1994). Another important co-activator function involves direct or indirect targeting of TAFs by select activation domains (Chen *et al.*, 1994; Goodrich *et al.*, 1993; Hoey *et al.*, 1993). TAFs are not limited to TFIID, but are also found in other complexes (SAGA, STAGA, and TFTC) that function in regulated transcription (Grant *et al.*, 1998; Martinez *et al.*, 1998; Wieczorek *et al.*, 1998). Moreover, some of the TAFs carry out various enzymatic functions including protein phosphorylation, acetylation, and ubiquitination (Dikstein *et al.*, 1996a; Mizzen *et al.*, 1996; Pham and Sauer, 2000). Most intriguingly, one of the TAF's bears bromo-domains that are responsible for binding and discriminating between acetylated and non-acetylated histones in the context of chromatin (Jacobson *et al.*, 2000). Thus, it appears that co-activators such as TFIID/TAFs participate in numerous functions that may serve to integrate regulatory signals from DNA bound activators (repressors) and thus help potentiate transcription activation and control.

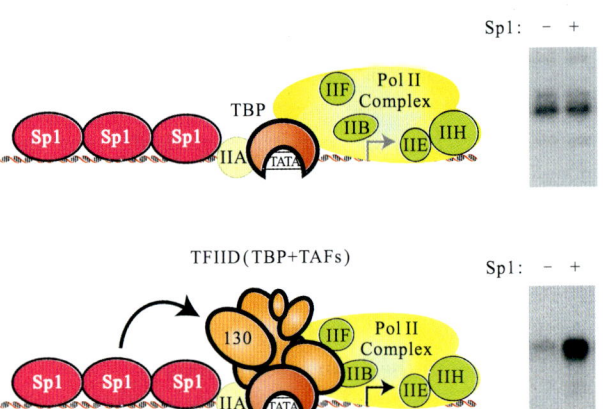

Fig.1.4 **The TAF subunits of TFIID are required for transcriptional activation *in vitro*, serving as coactivators.** The upper panel shows the lack of activation by Sp1 in transcription reactions reconstituted with TBP in place of TFIID. The lower panel shows Sp1 activation under identical conditions, with the exception that the holo-TFIID complex containing TBP and TAFs has replaced the single subunit TBP. The TAF$_{II}$130 (hTAF4) subunit of TFIID serves as a coactivator via interaction with Sp1.

Multiple functions of TFIID TAFs

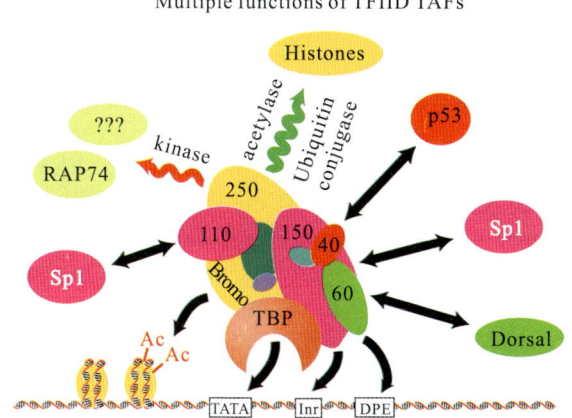

Fig.1.5 The TFIID complex plays multiple roles in the initiation and regulation of RNA polymerase II transcription. The names of the TAFs shown are the apparent molecular weights of the subunits of *Drosophila* TFIID: TAF$_{II}$250 (dTAF1), TAF$_{II}$150 (dTAF2), TAF$_{II}$110 (dTAF4), TAF$_{II}$60 (dTAF6), TAF$_{II}$40 (dTAF9).

Co-activators Abound: Mediator, CBP/p300, OCA, and Others

Of course, the TAFs and TFIID turned out to be merely the tip of the iceberg when it comes to co-activators. Using a combination of biochemistry and genetics a large number of co-factors, mediators, and co-regulators soon emerged. Among them were the yeast mediator and a series of mammalian coactivator complexes isolated in multiple labs and named CRISP, TRAP, DRIP, etc (Boyer *et al.*, 1999; Fondell *et al.*, 1996; Kim *et al.*, 1994b; Naar *et al.*, 1998; Rachez *et al.*, 1999; Ryu *et al.*, 1999; Sun *et al.*, 1998). Upon further purification and identification of the subunits, all of these complexes were found to be related and are now generally referred to as the Mediator. As a co-activator complex, the Mediator is not required for basal transcription *in vitro* and has not been found to bind DNA directly. Instead, it is thought to be recruited to promoters via interaction with promoter bound transcriptional activators where it facilitates the binding of RNA polymerase II. This class of co-activators also is able to directly bind to the CTD of RNA pol II and thus further integrate complex mechanisms of transcriptional control (Kim *et al.*, 1994a). A large and diverse group of activators have been found to bind the mediator complex, and EM studies revealed that the binding of activators can grossly alter the conformation of the co-activator complexes (Fig.1.6) (Taatjes *et al.*, 2002; Taatjes *et al.*, 2004).

Mediator complexes: Modular coregulators

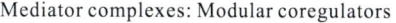
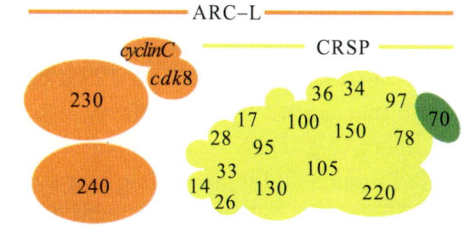
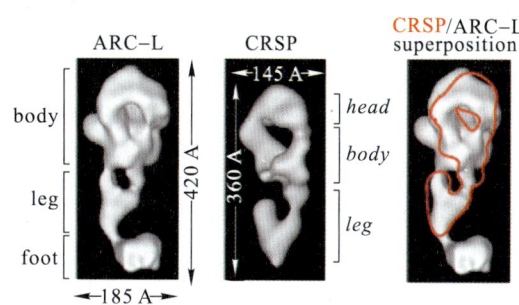

CRSP can be Converted to Alternate Conformations

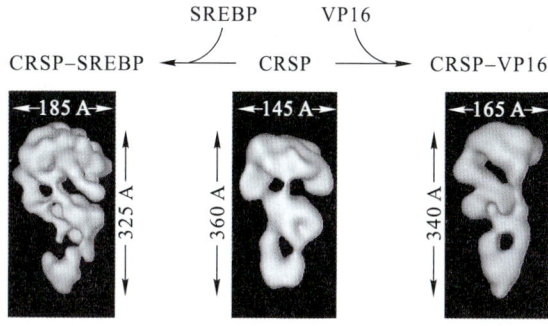

Fig.1.6 Multisubunit and Modular Co-regulatory Complexes. EM analysis revealed that ARC and CRSP, the mammalian counterpart of the yeast mediator are structurally related (yellow subunits are common) but distinct (orange and green subunits are unique) co-factors that display dramatically different functional properties. The larger ARC complex is inactive while the smaller CRSP complex is highly potent as a co-activator *in vitro*. Remarkably, the 3D structure of CRSP can undergo dramatic conformational changes dependent on the activator bound to target subunits within the CRSP assembly. Thus, the 3D structure of the unliganded, VP16-bound and SREBP-bound CRSP complexes display distinct structures as determined by negative stain EM and single particle reconstruction.

Although it may seem that the TAFs and the Mediator, which are ubiquitous transcriptional co-activators, would be sufficient for activating all genes, eukaryotic transcription once again proved to me more elaborate than imagined. Many other coactivators have now been identified. CBP, which was first identified as a co-activator for phosphorylated CREB, and p300 are two highly related proteins that are now known to function in transcriptional activation at many genes (Chrivia et al., 1993; Eckner et al., 1994; Kwok et al., 1994). The mechanism by which these two factors, as well as others (e.g. GCN5), co-activated transcription was partly illuminated by the finding that these proteins harbored histone acetyltransferase activity (Bannister and Kouzarides, 1996; Ogryzko et al., 1996). Activator-specific, cell type-specific, and developmentally regulated co-activators soon followed: for example, OCA-B co-activates Oct transcription (Luo and Roeder, 1995), $TAF_{II}105$ (TAF4b) is found in a cell type specific version of TFIID in B cells (Dikstein et al., 1996b; Freiman et al., 2002), and multiple testis-specific TAF isoforms have been found to function in spermatid development (Hiller et al., 2004; Hiller et al., 2001). Clearly, when it comes to transcriptional regulation in eukaryotes, complexity is the dominant theme. Only time and considerably more research will reveal how vast the co-activator universe is and the diverse spectrum of mechanisms they use to potentiate transcriptional activation.

Paving the Way: Remodeling Nucleosomes at Promoters

While many labs were focusing considerable effort on identifying and characterizing the transcriptional machinery, a few bold researchers had the foresight to ask how activators, co-activators, and the general machinery could possibly overcome the repressive effects of nucleosomes and higher order chromatin structures present in eukaryotic nuclei. Inroads in this area came from the integration of complementary findings from experiments in yeast, Drosophila, and human, which showed that nucleosomes could be remodeled (Cote et al., 1994; Kwon et al., 1994; Pazin et al., 1994; Tsukiyama et al., 1994). The yeast SWI/SNF complex, subunits of which had been discovered in genetic screens, turned out to be an ATP-dependent chromatin remodeling complex and effector of transcription (Cote et al., 1994). Other complexes that could assemble chromatin were found to have similar activity. These observations led to the idea that activators capable of binding native chromatin might recruit remodeling complexes to promoters, thereby opening the chromatin and allowing access to other transcriptional activators, co-activators, and the general transcription machinery. This proved to be the case (Neely et al., 1999; Yudkovsky et al., 1999), and the role of chromatin structure and its modulation was brought to the forefront of transcription research (Fig.1.7).

A Missing Link: Histone Modifications

For years, it had been known that in cells histones were differentially modified with acetyl, methyl, ubiquitin, and other post-translationally added groups. Dogma had it that histones in euchromatin, which was transcriptionally active, were hyper-acetylated, while histones in transcriptionally silenced heterochromatic regions were hypo-acetylated. Theories abounded to explain the correlation between histone acetylation and transcriptional competence, but for the most part, the transcription community paid little attention to these theories. This all changed with the identification of a nuclear histone acetyltransferase purified from Tetrahymena (Brownell et al., 1996). Surprisingly, the Tetrahymena HAT had high sequence similarity to a known yeast co-activator, Gcn5p. Instantaneously, the

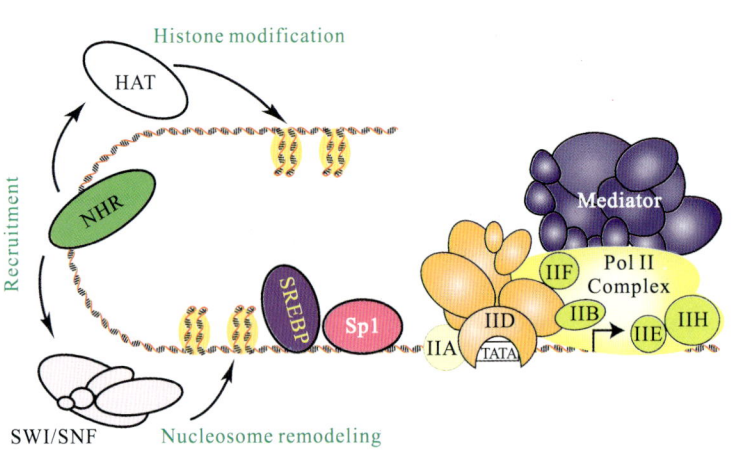

Fig.1.7 **Nucleosome remodeling and histone modifying complexes are recruited to promoters via interactions with activators.**

collective eyes of the transcription community opened to the possibility that many transcription factors might bear HAT activity. When the dust settled, multiple previously identified co-activators were found to be HATs, and ultimately it was realized that histones were not the only substrates of these acetyltransferases; indeed, activators themselves could be acetylated. With the subsequent discovery of deacetylases (Taunton *et al.*, 1996), acetylation was added to phosphorylation as a reversible post-translational modification used by intracellular signaling pathways to regulate gene expression. Ultimately, enzymes placing other modifications on histones (e.g. methylation, phosphorylation, and ubiquitination) were identified and characterized, and in some cases also found to be co-activators or co-repressors of transcription. Moreover, these enzymes can be recruited to promoters by gene specific activators and repressors to control levels of transcription.

The number of possible combinations of covalent modifications on the eight histones in any single nucleosome was dumbfounding. What was the function of all of these histone modifications? A seductive idea was posited: perhaps, specific patterns of post-translational modifications on the core histones in nucleosomes in individual promoters or regions of the genome help set the levels of transcription from those genes (Jenuwein and Allis, 2001; Strahl and Allis, 2000). For example, activation correlates with acetylation of specific lysines, while repression is observed upon acetylation or methylation of other lysines. Thus was born the Histone Code Hypothesis (Jenuwein and Allis, 2001; Strahl and Allis, 2000). While the putative histone "code" is far from understood, or even the notion of a true code accepted, it is clear that modification of histones adds another level of dynamic encoded information to the static DNA sequence present in a genome.

Escaped and On the Run: Regulation of Post-initiation Steps of Transcription

During the time that activators, co-activators, and general factors were being discovered and their roles in forming preinitiation complexes were initially characterized, some of the same labs and others embarked on understanding the mechanism and regulation of the RNA synthesis steps of the RNA polymerase II reaction. RNA synthesis is not simply the monotonous creation of phosphodiester bonds, but instead is a phase of the reaction rich in regulation (Fig.1.8). TFIIF and the TFIIH helicase function during promoter escape (Chang *et al.*, 1993; Goodrich and Tjian, 1994). The TFIIH kinase phosphorylates the CTD of RNA polymerase II as the enzyme leaves the promoter (Lu *et al.*, 1992). At the HSP70 promoter polymerase pauses after synthesis of a short (~20 nt) RNA, and is poised to fire the moment heat shock is sensed (via the Heat Shock Factor) (Gilmour and Lis, 1986; Rougvie and Lis, 1988). P-TEFb and DSIF/NELF have opposing effects on elongation (Marshall and Price, 1995; Wada *et al.*, 1998; Yamaguchi *et al.*, 1999). Elongation factors were discovered, including TFIIF, TFIIS, Elongin, etc, and indeed, the overall rate of elongation can be controlled globally and in a gene specific fashion (Aso *et al.*, 1995; Reinberg and Roeder, 1987). HIV TAT, regulates the transcription reaction by binding a TAR element in the nascent RNA, which is reminiscent of bacterial phage factors that control transcriptional termination by binding the RNA transcript (Kao *et al.*, 1987). RNA itself has recently appeared in the transcriptional regulatory picture, as a number of small noncoding RNAs have been found to control the RNA polymerase II transcription reaction via association with transcription factors and RNA polymerase II (Allen *et al.*, 2004; Espinoza *et al.*, 2004; Kwek *et al.*, 2002; Nguyen *et al.*, 2001; Yang *et al.*, 2001). It seems that evolution has taken advantage of many different regulatory mechanisms beyond simply controlling the formation of preinitiation complexes, and we have only begun to appreciate and understand the multiple layers of regulation that can come into play.

Keeping the End in Sight: Coupling RNA Processing to Transcription

As transcription factors were identified and characterized using biochemical and genetic approaches, individual pieces of data began to support the notion that the transcriptional apparatus in eukaryotic cells is tightly coupled to the RNA processing machinery, and moreover that the transcription reaction itself can be influenced by factors that add the 5′ Cap, splice the RNA, process the 3′ end of the transcript, and transport the mature transcript out of the nucleus (Fig.1.8) (Cho *et al.*, 1997; Dantonel *et al.*, 1997; Fong and Zhou, 2001; Hirose *et al.*, 1999; McCracken *et al.*, 1997a; McCracken *et al.*, 1997b; Strasser *et al.*, 2002). In hindsight, the coupling between transcription and RNA processing is logical, however, observations of splicing factors influencing transcription, and indications that RNA processing factors are recruited via interaction with the Pol II CTD were surprising, and the implications profound. We now envision that the nucleus contains mRNA synthesis/processing machines,

Promoter escape and transcript elongation: coupling transcription to mRNA processing

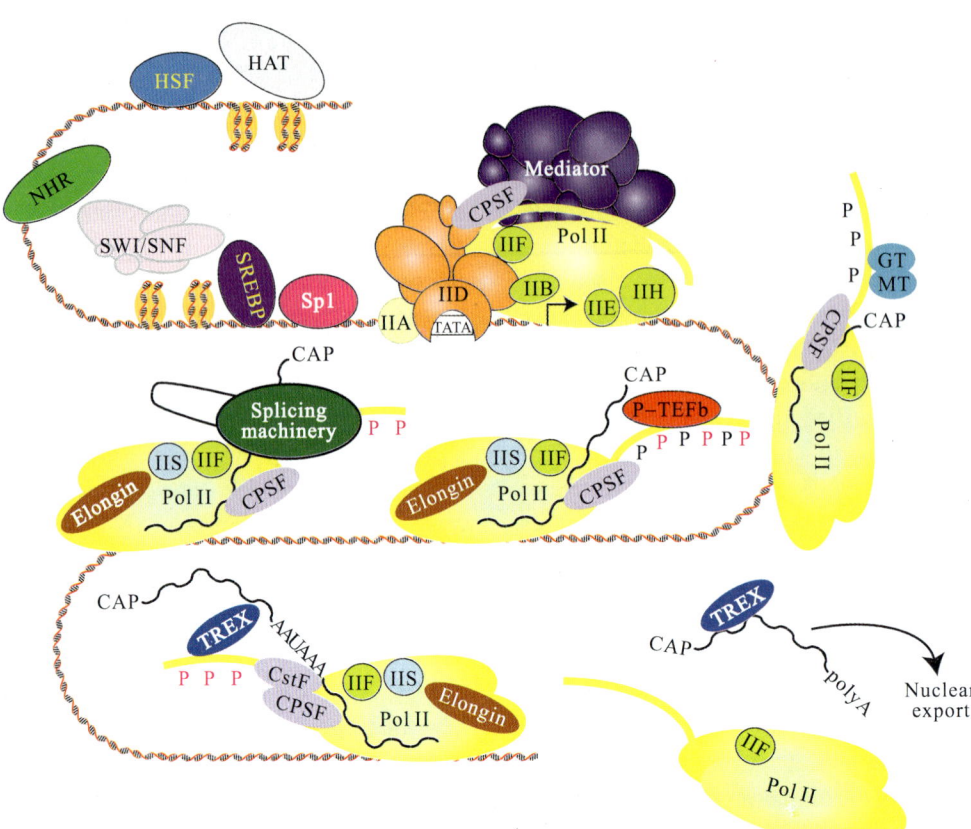

Fig.1.8 **The post-initiation steps of transcription (e.g. promoter escape and elongation) are coupled to RNA processing (e.g. Capping, splicing, termination, polyadenylation, and RNA transport).** Abbreviations: CPSF, cleavage and polyadenylaation specificity factor; GT, guanyl transferase; MT, methyl transferase; CstF, cleavage-stimulation factor; TREX, transcription/export complex.

which in response to gene specific activators coordinate the entire process from chromatin remodeling/modification through transcription, processing and export of mature mRNAs to the cytosol—perhaps akin to an assembly line where all aspects of building and refining a final product are carried out in the same location by workers in intimate contact and orchestrated in a coordinated fashion. All parts of this massive molecular machine must work collaboratively to produce a functional mRNA. Current studies of the integration and cooperativity between the transcription and mRNA processing machineries have provided glimpses of the network that exists, and future studies on the integration of transcription with other nuclear processes (repair, replication, and recombination) will undoubtedly provide many new surprises and a true appreciation for the level to which nuclear events are functionally (and physically) connected.

References

Allen, T. A., Von Kaenel, S., Goodrich, J. A., and Kugel, J. F. (2004). The SINE-encoded mouse B2 RNA represses mRNA transcription in response to heat shock. Nat Struct Mol Biol *11*, 816-821.

Andel, F., 3rd, Ladurner, A. G., Inouye, C., Tjian, R., and Nogales, E. (1999). Three-dimensional structure of the human TFIID-IIA-IIB complex. Science *286*, 2153-2156.

Anderson, J. E., Ptashne, M., and Harrison, S. C. (1985). A phage repressor-operator complex at 7 A resolution. Nature *316*, 596-601.

Aso, T., Lane, W. S., Conaway, J. W., and Conaway, R. C. (1995). Elongin (SIII): a multisubunit regulator of elongation by RNA polymerase II. Science *269*, 1439-1443.

Aso, T., Vasavada, H. A., Kawaguchi, T., Germino, F. J., Ganguly, S., Kitajima, S., Weissman, S. M., and Yasukochi, Y. (1992). Characterization of cDNA for the large subunit of the transcription initiation factor TFIIF. Nature *355*, 461-464.

Banerji, J., Rusconi, S., and Schaffner, W. (1981). Expression of a beta-globin gene is enhanced by remote SV40 DNA sequences. Cell *27*, 299-308.

Bannister, A. J., and Kouzarides, T. (1996). The CBP co-activator is a histone acetyltransferase. Nature *384*, 641-643.

Benoist, C., and Chambon, P. (1981). *In vivo* sequence requirements of the SV40 early promotor region. Nature *290*, 304-310.

Berg, J. M. (1988). Proposed structure for the zinc-binding domains from transcription factor IIIA and related proteins. Proc Natl Acad Sci USA *85*, 99-102.

Berger, S. L., Cress, W. D., Cress, A., Triezenberg, S. J., and Guarente, L. (1990). Selective Inhibition of Activated but Not Basal Transcription by the Acidic Activation Domain of VP16: Evidence for Transcriptional Adaptors. Cell *61*, 1199-1208.

Bohmann, D., Bos, T. J., Admon, A., Nishimura, T., Vogt, P. K., and Tjian, R. (1987). Human proto-oncogene c-jun encodes a DNA binding protein with structural and functional properties of transcription factor AP-1. Science *238*, 1386-1392.

Boyer, T. G., Martin, M. E., Lees, E., Ricciardi, R. P., and Berk, A. J. (1999). Mammalian Srb/Mediator complex is targeted by adenovirus E1A protein. Nature *399*, 276-279.

Brand, M., Leurent, C., Mallouh, V., Tora, L., and Schultz, P. (1999). Three-dimensional structures of the TAFII-containing complexes TFIID and TFTC. Science *286*, 2151-2153.

Briggs, M. R., Kadonaga, J. T., Bell, S. P., and Tjian, R. (1986). Purification and biochemical characterization of the promoter-specific transcription factor, Sp1. Science *234*, 47-52.

Brownell, J. E., Zhou, J., Ranalli, T., Kobayashi, R., Edmondson, D. G., Roth, S. Y., and Allis, C. D. (1996). *Tetrahymena* histone acetyltransferase A: a homolog to yeast Gcn5p linking histone acetylation to gene activation. Cell *84*, 843-851.

Buratowski, S., Hahn, S., Guarente, L., and Sharp, P. A. (1989). Five intermediate complexes in transcription initiation by RNA polymerase II. Cell *56*, 549-561.

Buratowski, S., Hahn, S., Sharp, P. A., and Guarente, L. (1988). Function of a yeast TATA element binding protein in a mammalian transcription system. Nature *334*, 37-42.

Burke, T. W., and Kadonaga, J. T. (1997). The downstream core promoter element, DPE, is conserved from *Drosophila* to humans and is recognized by TAFII60 of *Drosophila*. Genes Dev *11*, 3020-3031.

Carthew, R. W., Chodosh, L. A., and Sharp, P. A. (1985). An RNA polymerase II transcription factor binds to an upstream element in the adenovirus major late promoter. Cell *43*, 439-448.

Chang, C.-H., Kostrub, C. F., and Burton, Z. F. (1993). RAP30/74 (transcription factor IIF) is required for promoter escape by RNA polymerase II. J Biol Chem *268*, 20482-20489.

Chang, W. H., and Kornberg, R. D. (2000). Electron crystal structure of the transcription factor and DNA repair complex, core TFIIH. Cell *102*, 609-613.

Chen, J.-L., Attardi, L. D., Verrijzer, C. P., Yokomori, K., and Tjian, R. (1994). Assembly of recombinant TFIID reveals differential coactivator requirements for distinct transcriptional activators. Cell *79*, 93-105.

Cho, E. J., Takagi, T., Moore, C. R., and Buratowski, S. (1997). mRNA capping enzyme is recruited to the transcription complex by phosphorylation of the RNA polymerase II carboxy-terminal domain. Genes Dev *11*, 3319-3326.

Chrivia, J., Kwok, R., Lamb, N., Hagiwara, M., Montminy, M., and Goodman, R. (1993). Phosphorylated CREB binds specifically to the nuclear protein CBP. Nature *365*, 855-859.

Corden, J., Wasylyk, B., Buchwalder, A., Sassone-Corsi, P., Kedinger, C., and Chambon, P. (1980). Promoter sequences of eukaryotic protein-coding genes. Science *209*, 1406-1414.

Cote, J., Quinn, J., Workman, J. L., and Peterson, C. L. (1994). Stimulation of GAL4 derivative binding to nucleosomal DNA by the yeast SWI/SNF complex. Science *265*, 53-60.

Courey, A. J., Holtzman, D. A., Jackson, S. P., and Tjian, R. (1989). Synergistic activation by the glutamine-rich domains of human transcription factor Sp1. Cell *59*, 827-836.

Cramer, P., Bushnell, D. A., Fu, J., Gnatt, A. L., Maier-Davis, B., Thompson, N. E., Burgess, R. R., Edwards, A. M., David, P. R., and Kornberg, R. D. (2000). Architecture of RNA polymerase II and implications for the transcription mechanism. Science *288*, 640-649.

Dantonel, J. C., Murthy, K. G., Manley, J. L., and Tora, L. (1997). Transcription factor TFIID recruits factor CPSF for formation of 3′ end of mRNA. Nature *389*, 399-402.

Darst, S. A., Edwards, A. M., Kubalek, E. W., and Kornberg, R. D. (1991). Three-dimensional structure of yeast RNA polymerase II at 16 Å resolution. Cell *66*, 121-128.

DeFranco, D., and Yamamoto, K. R. (1986). Two different factors act separately or together to specify functionally distinct activities at a single transcriptional enhancer. Mol Cell Biol *6*, 993-1001.

DeJong, J., and Roeder, R. G. (1993). A single cDNA, hTFIIA/alpha, encodes both the p35 and p19 subunits of human TFIIA. Genes Dev *7*, 2220-2234.

Diamond, M. I., Miner, J. N., Yoshinaga, S. K., and Yamamoto, K. R. (1990). Transcription factor interactions: selectors of positive or negative regulation from a single DNA element. Science *249*, 1266-1272.

Dikstein, R., Ruppert, S., and Tjian, R. (1996a). $TAF_{II}250$ is a bipartite protein kinase that phosphorylates the basal transcription factor RAP74. Cell *84*, 781-790.

Dikstein, R., Zhou, S., and Tjian, R. (1996b). Human $TAF_{II}105$ is a cell type-specific TFIID subunit related to $hTAF_{II}130$. Cell *87*, 137-146.

Dynan, W. S., and Tjian, R. (1983a). Isolation of transcription factors that discriminate between different promoters recognized by RNA polymerase II. Cell *32*, 669-680.

Dynan, W. S., and Tjian, R. (1983b). The promoter-specific transcription factor Sp1 binds to upstream sequences in the SV40 early promoter. Cell *35*, 79-87.

Dynlacht, B. D., Hoey, T., and Tjian, R. (1991). Isolation of coactivators associated with the TATA-binding protein that mediate transcriptional activation. Cell *55*, 563-576.

Eckner, R., Ewen, M. E., Newsome, D., Gerdes, M., DeCaprio, J. A., Lawrence, J. B., and Livingston, D. M. (1994). Molecular cloning and functional analysis of the adenovirus E1A-associated 300-kD protein (p300) reveals a protein with properties of a transcriptional adaptor. Genes Dev *8*, 869-884.

Eisenmann, D. M., Dollard, C., and Winston, F. (1989). *SPT15, the Gene Encoding the Yeast TATA Binding Factor TFIID, is Required for Normal Transcription Initiation In Vivo*. Cell *58*, 1183-1191.

Engelke, D. R., Ng, S. Y., Shastry, B. S., and Roeder, R. G. (1980). Specific interaction of a purified transcription factor with an internal control region of 5S RNA genes. Cell *19*, 717-728.

Espinoza, C. A., Allen, T. A., Hieb, A. R., Kugel, J. F., and Goodrich, J. A. (2004). B2 RNA binds directly to RNA polymerase II to repress transcript synthesis. Nat Struct Mol Biol *11*, 822-829.

Feaver, W. J., Svejstrup, J. Q., Henry, N. L., and Kornberg, R. D. (1994). Relationship of CDK-activating kinase and RNA polymerase II CTD kinase TFIIH/TFIIK. Cell *79*, 1103-1109.

Finkelstein, A., Kostrub, C. F., Li, J., Chavez, D. P., Wang, B. Q., Fang, S. M., Greenblatt, J., and Burton, Z. F. (1992). A cDNA encoding RAP74, a general initiation factor for transcription by RNA polymerase II. Nature *355*, 464-467.

Fischer, L., Gerard, M., Chalut, C., Lutz, Y., Humbert, S., Kanno, M., Chambon, P., and Egly, J. M. (1992). Cloning of the 62-Kilodalton component of basic transcription factor BTF2. Science *257*, 1392-1395.

Flores, O., Lu, H., and Reinberg, D. (1992). Factors involved in specific transcription initiation by RNA polymerase II: Identification and characterization of factor IIH. J Biol Chem *267*, 2786-2793.

Flores, O., Maldonado, E., and Reinberg, D. (1989). Factors involved in specific transcription by mammalian RNA polymerase II: factors IIE and IIF independently interact with RNA polymerase II. J Biol Chem *264*, 8913-8921.

Fondell, J. D., Ge, H., and Roeder, R. G. (1996). Ligand induction of a transcriptionally active thyroid hormone receptor coactivator complex. Proc Natl Acad Sci USA *93*, 8329-8333.

Fong, Y. W., and Zhou, Q. (2001). Stimulatory effect of splicing factors on transcriptional elongation. Nature *414*, 929-933.

Freiman, R. N., Albright, S. R., Chu, L. E., Zheng, S., Liang, H. E., Sha, W. C., and Tjian, R. (2002). Redundant role of tissue-selective TAF(II)105 in B lymphocytes. Mol Cell Biol *22*, 6564-6572.

Fromm, M., and Berg, P. (1983). Simian virus 40 early- and late-region promoter functions are enhanced by the 72-base-pair repeat inserted at distant locations and inverted orientations. Mol Cell Biol *3*, 991-999.

Galas, D. J., and Schmitz, A. (1978). DNase footprinting: a simple method for the detection of protein-DNA binding specificity. Nucleic Acids Res *5*, 3157-3170.

Geiger, J. H., Hahn, S., Lee, S., and Sigler, P. B. (1996). Crystal structure of the yeast TFIIA/TBP/DNA complex. Science *272*, 830-836.

Gidoni, D., Kadonaga, J. T., Barrera-Saldana, H., Takahashi, K., Chambon, P., and Tjian, R. (1985). Bidirectional SV40 transcription mediated by tandem Sp1 binding interactions. Science *230*, 511-517.

Giese, K., Cox, J., and Grosschedl, R. (1992). The HMG domain of lymphoid enhancer factor 1 bends DNA and facilitates assembly of functional nucleoprotein structures. Cell *69*, 185-195.

Giese, K., Kingsley, C., Kirshner, J. R., and Grosschedl, R. (1995). Assembly and function of a TCR alpha enhancer complex is dependent on LEF-1-induced DNA bending and multiple protein-protein interactions. Genes Dev *9*, 995-1008.

Gill, G., and Ptashne, M. (1988). Negative effect of the transcriptional activator GAL4. Nature *334*, 721-724.

Gilmour, D. S., and Lis, J. T. (1986). RNA polymerase II interacts with the promoter region of the noninduced hsp70 gene in *Drosophila* melanogaster cells. Mol Cell Biol *6*, 3984-3989.

Gnatt, A. L., Cramer, P., Fu, J., Bushnell, D. A., and Kornberg, R. D. (2001). Structural basis of transcription: an RNA polymerase II elongation complex at 3.3 Å resolution. Science *292*, 1876-1882.

Goodrich, J. A., Hoey, T., Thut, C. J., Admon, A., and Tjian, R. (1993). *Drosophila* TAF$_{II}$40 interacts with both a VP16 activation domain and the basal transcription factor TFIIB. Cell *75*, 519-530.

Goodrich, J. A., and Tjian, R. (1994). Transcription factors IIE and IIH and ATP hydrolysis direct promoter clearance by RNA polymerase II. Cell *77*, 145-156.

Grant, P. A., Schieltz, D., Pray-Grant, M. G., Steger, D. J., Reese, J. C., Yates, J. R., 3rd, and Workman, J. L. (1998). A subset of TAF(II)s are integral components of the SAGA complex required for nucleosome acetylation and transcriptional stimulation. Cell *94*, 45-53.

Ha, I., Lane, W. S., and Reinberg, D. (1991). Cloning of a human gene encoding the general transcription factor IIB. Nature *352*, 689-695.

Hahn, S., Buratowski, S., Sharp, P. A., and Guarente, L. (1989a). Identification of a yeast protein homologous in function to the mammalian general transcription factor, TFIIA. EMBO J *11*, 3379-3382.

Hahn, S., Buratowski, S., Sharp, P. A., and Guarente, L. (1989b). Isolation of the gene encoding the yeast TATA binding protein TFIID: A gene identical to the SPT15 suppressor of Ty element insertion. Cell *58*, 1173-1181.

Hiller, M., Chen, X., Pringle, M. J., Suchorolski, M., Sancak, Y., Viswanathan, S., Bolival, B., Lin, T. Y., Marino, S., and Fuller, M. T. (2004). Testis-specific TAF homologs collaborate to control a tissue-specific transcription program. Development *131*, 5297-5308.

Hiller, M. A., Lin, T. Y., Wood, C., and Fuller, M. T. (2001). Developmental regulation of transcription by a tissue-specific TAF homolog. Genes Dev *15*, 1021-1030.

Hirose, Y., Tacke, R., and Manley, J. L. (1999). Phosphorylated RNA polymerase II stimulates pre-mRNA splicing. Genes Dev *13*, 1234-1239.

Hoey, T., Dynlacht, B. D., Peterson, M. G., Pugh, B. F., and Tjian, R. (1990). Isolation and characterization of the *Drosophila* gene encoding the TATA box binding protein, TFIID. Cell *61*, 1179-1186.

Hoey, T., Weinzierl, R. O. J., Gill, G., Chen, J.-L., Dynlacht, B. D., and Tjian, R. (1993). Molecular cloning and functional analysis of *Drosophila* TAF110 reveal properties expected of coactivators. Cell *72*, 247-270.

Horikoshi, M., Wang, C. K., Fujii, H., Cromlich, J. A., Weil, B. A., and Roeder, R. G. (1989). Cloning and structure of a yeast gene encoding a general transcription initiation factor TFIID that binds to the TATA box. Nature *341*, 299-303.

Hu, S. L., and Manley, J. L. (1981). DNA sequence required for initiation of transcription *in vitro* from the major late promoter of adenovirus 2. Proc Natl Acad Sci USA *78*, 820-824.

Jacobson, R. H., Ladurner, A. G., King, D. S., and Tjian, R. (2000). Structure and function of a human TAFII250 double bromodomain module. Science *288*, 1422-1425.

Jenuwein, T., and Allis, C. D. (2001). Translating the histone code. Science *293*, 1074-1080.

Jones, K. A., Yamamoto, K. R., and Tjian, R. (1985). Two distinct transcription factors bind to the HSV thymidine kinase promoter *in vitro*. Cell *42*, 559-572.

Kadonaga, J. T., Carner, K. R., Masiarz, F. R., and Tjian, R. (1987). Isolation of cDNA encoding transcription factor Sp1 and functional analysis of the DNA binding domain. Cell *51*, 1079-1090.

Kadonaga, J. T., Courey, A. J., Ladika, J., and Tjian, R. (1988). Distinct regions of Sp1 modulate DNA binding and transcriptional activation. Science *242*, 1566-1570.

Kadonaga, J. T., and Tjian, R. (1986). Affinity purification of sequence-specific DNA binding proteins. Proc Natl Acad Sci USA *83*, 5889-5893.

Kao, C. C., Lieberman, P. M., Schmidt, M. C., Zhou, Q., Pei, R., and Berk, A. J. (1990). Cloning of a transcriptionally active human TATA binding factor. Science *248*, 1646-1650.

Kao, S. Y., Calman, A. F., Luciw, P. A., and Peterlin, B. M. (1987). Anti-termination of transcription within the long terminal repeat of HIV-1 by tat gene product. Nature *330*, 489-493.

Kaufmann, J., and Smale, S. T. (1994). Direct recognition of initiator elements by a component of the transcription factor IID complex. Genes Dev *8*, 821-829.

Kelleher III, R. J., Flanagan, P. M., and Kornberg, R. D. (1990). A Novel Mediator between Activator Proteins and the RNA Polymerase II Transcription Apparatus. Cell *61*, 1209-1215.

Kim, J. L., Nikolov, D. B., and Burley, S. K. (1993a). Co-crystal structure of TBP recognizing the minor groove of a TATA element. Nature *365*, 520-527.

Kim, Y., Geiger, J. H., Hahn, S., and Sigler, P. B. (1993b). Crystal structure of a yeast TBP/TATA-box complex. Nature *365*, 512-520.

Kim, Y.-J., Bjoklund, S., Li, Y., Sayre, M. H., and Kornberg, R. D. (1994a). A multiprotein mediator of transcriptional activation and its interaction with the C-terminal repeat domain of RNA polymerase II. Cell *77*, 599-608.

Kim, Y. J., Bjorklund, S., Li, Y., Sayre, M. H., and Kornberg, R. D. (1994b). A multiprotein mediator of transcriptional activation and its interaction with the C-terminal repeat domain of RNA polymerase II. Cell *77*, 599-608.

Koleske, A. J., and Young, R. A. (1994). An RNA polymerase II holoenzyme responsive to activators. Nature *368*, 466-469.

Krebs, G., and Chambon, P. (1976). Animal DNA-dependent RNA polymerases. Purification and molecular structure of hen-oviduct and liver class-B RNA polymerases. Eur J Biochem *61*, 15-25.

Kwek, K. Y., Murphy, S., Furger, A., Thomas, B., O'Gorman, W., Kimura, H., Proudfoot, N. J., and Akoulitchev, A. (2002). U1 snRNA associates with TFIIH and regulates transcriptional initiation. Nat Struct Biol *9*, 800-805.

Kwok, R., Lundblad, J., Chrivia, J., Richards, J., Bachinger, H., Brennan, R., Roberts, S., Green, M., and Goodman, R. (1994). Nuclear protein CBP is a coactivator for the transcription factor CREB. Nature *370*, 223-226.

Kwon, H., Imbalzano, A. N., Khavari, P. A., Kingston, R. E., and Green, M. R. (1994). Nucleosome disruption and enhancement of activator binding by a human SWI/SNF complex. Nature *370*, 477-481.

Lagrange, T., Kapanidis, A. N., Tang, H., Reinberg, D., and Ebright, R. H. (1998). New core promoter element in RNA polymerase II-dependent transcription: sequence-specific DNA binding by transcription factor IIB. Genes Dev *12*, 34-44.

Learned, R. M., Cordes, S., and Tjian, R. (1985). Purification and characterization of a transcription factor that confers promoter specificity to human RNA polymerase I. Mol Cell Biol *5*, 1358-1369.

Learned, R. M., Learned, T. K., Haltiner, M. M., and Tjian, R. T. (1986). Human rRNA transcription is modulated by the coordinate binding of two factors to an upstream control element. Cell *45*, 847-857.

Lu, H., Zawel, L., Fisher, L., Egly, J.-M., and Reinberg, D. (1992). Human general transcription factor IIH phosphorylates the C-terminal domain of RNA polymerase II. Nature *358*, 641-645.

Luo, Y., and Roeder, R. G. (1995). Cloning, functional characterization, and mechanism of action of the B-cell-specific transcriptional coactivator OCA-B. Mol Cell Biol *15*, 4115-4124.

Ma, D., Watanabe, H., Mermelstein, F., Admon, A., Oguri, K., Sun, X., Wada, T., Imai, T., Shiroya, T., Reinberg, D., and et al. (1993). Isolation of a cDNA encoding the largest subunit of TFIIA reveals functions important for activated transcription. Genes Dev 7, 2246-2257.

Ma, J., and Ptashne, M. (1987a). Deletion analysis of GAL4 defines two transcriptional activating segments. Cell 48, 847-853.

Ma, J., and Ptashne, M. (1987b). A new class of yeast transcriptional activators. Cell 51, 113-119.

Manley, J. L., Fire, A., Cano, A., Sharp, P. A., and Gefter, M. L. (1980). DNA-dependent transcription of adenovirus genes in a soluble whole-cell extract. Proc Natl Acad Sci USA 77, 3855-3859.

Marshall, N. F., and Price, D. H. (1995). Purification of P-TEFb, a transcription factor required for the transition into productive elongation. J Biol Chem 270, 12335-12338.

Martinez, E., Kundu, T. K., Fu, J., and Roeder, R. G. (1998). A human SPT3-TAFII31-GCN5-L acetylase complex distinct from transcription factor IID. J Biol Chem 273, 23781-23785.

Matsui, T., Segall, J., Weil, P. A., and Roeder, R. G. (1980). Multiple factors required for accurate initiation of transcription by purified RNA polymerase II. J Biol Chem 255, 11992-11996.

McCracken, S., Fong, N., Rosonina, E., Yankulov, K., Brothers, G., Siderovski, D., Hessel, A., Foster, S., Shuman, S., and Bentley, D. L. (1997a). 5′-Capping enzymes are targeted to pre-mRNA by binding to the phosphorylated carboxy-terminal domain of RNA polymerase II. Genes Dev 11, 3306-3318.

McCracken, S., Fong, N., Yankulov, K., Ballantyne, S., Pan, G., Greenblatt, J., Patterson, S. D., Wickens, M., and Bentley, D. L. (1997b). The C-terminal domain of RNA polymerase II couples mRNA processing to transcription. Nature 385, 357-361.

McKnight, S., and Tjian, R. (1986). Transcriptional selectivity of viral genes in mammalian cells. Cell 46, 795-805.

McKnight, S. L. (1982). Functional relationships between transcriptional control signals of the thymidine kinase gene of herpes simplex virus. Cell 31, 355-365.

McKnight, S. L., Gavis, E. R., Kingsbury, R., and Axel, R. (1981). Analysis of transcriptional regulatory signals of the HSV thymidine kinase gene: identification of an upstream control region. Cell 25, 385-398.

McKnight, S. L., and Kingsbury, R. (1982). Transcriptional control signals of a eukaryotic protein-coding gene. Science 217, 316-324.

Mermod, N., O'Neill, E. A., Kelly, T. J., and Tjian, R. (1989). The proline-rich transcriptional activator of CTF/NF-I is distinct from the replication and DNA binding domain. Cell 58, 741-753.

Miller, J., McLachlan, A. D., and Klug, A. (1985). Repetitive zinc-binding domains in the protein transcription factor IIIA from Xenopus oocytes. EMBO J 4, 1609-1614.

Mitchell, P. J., and Tjian, R. (1989). Transcriptional regulation in mammalian cells by sequence-specific DNA binding proteins. Science 245, 371-378.

Mitchell, P. J., Wang, C., and Tjian, R. (1987). Positive and negative regulation of transcription in vitro: enhancer-binding protein AP-2 is inhibited by SV40 T antigen. Cell 50, 847-861.

Mizzen, C. A., Yang, X. J., Kokubo, T., Brownell, J. E., Bannister, A. J., Owen-Hughes, T., Workman, J., Wang, L., Berger, S. L., Kouzarides, T., et al. (1996). The $TAF_{II}250$ subunit of TFIID has histone acetyltransferase activity. Cell 87, 1261-1270.

Myers, R. M., Rio, D. C., Robbins, A. K., and Tjian, R. (1981). SV40 gene expression is modulated by the cooperative binding of T antigen to DNA. Cell 25, 373-384.

Myers, R. M., and Tjian, R. (1980). Construction and analysis of simian virus 40 origins defective in tumor antigen binding and DNA replication. Proc Natl Acad Sci USA 77, 6491-6495.

Naar, A. M., Beaurang, P. A., Robinson, K. M., Oliner, J. D., Avizonis, D., Scheek, S., Zwicker, J., Kadonaga, J. T., and Tjian, R. (1998). Chromatin, TAFs, and a novel multiprotein coactivator are required for synergistic activation by Sp1 and SREBP-1a in vitro. Genes Dev 12, 3020-3031.

Neely, K. E., Hassan, A. H., Wallberg, A. E., Steger, D. J., Cairns, B. R., Wright, A. P., and Workman, J. L. (1999). Activation domain-mediated targeting of the SWI/SNF complex to promoters stimulates transcription from nucleosome arrays. Mol Cell 4, 649-655.

Nguyen, V. T., Kiss, T., Michels, A. A., and Bensaude, O. (2001). 7SK small nuclear RNA binds to and inhibits the activity of CDK9/cyclin T complexes. Nature 414, 322-325.

Nikolov, D. B., Chen, H., Halay, E. D., Usheva, A. A., Hisatake, K., Lee, D. K., Roeder, R. G., and Burley, S. K. (1995). Crystal structure of a TFIIB-TBP-TATA-element ternary complex. Nature 377, 119-128.

Nikolov, D. B., Hu, S.-H., Lin, J., Gasch, A., Hoffman, A., Horikoshi, M., Chua, N.-H., Roeder, R. G., and Burley, S. K. (1992). Crystal structure of TFIID TATA-box binding protein. Nature 360, 40-46.

Ogryzko, V. V., Schiltz, R. L., Russanova, V., Howard, B. H., and Nakatani, Y. (1996). The transcriptional coactivators p300 and CBP are histone acetyltransferases. Cell 87, 953-959.

Pabo, C. O., and Sauer, R. T. (1992). Transcription factors: structural families and principles of DNA recognition. Annu Rev Biochem 61, 1053-1095.

Pazin, M. J., Kamakaka, R. T., and Kadonaga, J. T. (1994). ATP-dependent nucleosome reconfiguration and transcriptional activation from preassembled chromatin templates. Science 266, 2007-2011.

Pelham, H. R., and Brown, D. D. (1980). A specific transcription factor that can bind either the 5S RNA gene or 5S RNA. Proc Natl Acad Sci USA 77, 4170-4174.

Peterson, M. G., Inostroza, J., Maxon, M. E., Flores, O., Admon, A., Reinberg, D., and Tjian, R. (1991). Structure and functional properties of human general transcription factor IIE. Nature 354, 369-373.

Peterson, M. G., Tanese, N., Pugh, B. F., and Tjian, R. (1990).

Functional domains and upstream activation properties of cloned human TATA binding protein. Science *248*, 1625-1630.

Pham, A. D., and Sauer, F. (2000). Ubiquitin-activating conjugating activity of TAFII250, a mediator of activation of gene expression in *Drosophila*. Science *289*, 2357-2360.

Picard, D., and Schaffner, W. (1984). A lymphocyte-specific enhancer in the mouse immunoglobulin kappa gene. Nature *307*, 80-82.

Pugh, B. F., and Tjian, R. (1990). Mechanism of transcriptional activation by Sp1: evidence for co-activators. Cell *61*, 1187-1197.

Rachez, C., Lemon, B. D., Suldan, Z., Bromleigh, V., Gamble, M., Naar, A. M., Erdjument-Bromage, H., Tempst, P., and Freedman, L. P. (1999). Ligand-dependent transcription activation by nuclear receptors requires the DRIP complex. Nature *398*, 824-828.

Reinberg, D., Horikoshi, M., and Roeder, R. (1987). Factors involved in specific transcription by mammalian RNA Polymerase II. Functional analysis of initiation factors IIA and IID and identification of a new factor operating at sequences downstream of the initiation site. J Biol Chem *262*, 3322-3330.

Reinberg, D., and Roeder, R. G. (1987). Factors involved in specific transcription by mammalian RNA polymerase II. Transcription factor IIS stimulates elongation of RNA chains. J Biol Chem *262*, 3331-3337.

Rio, D., Robbins, A., Myers, R., and Tjian, R. (1980). Regulation of simian virus 40 early transcription *in vitro* by a purified tumor antigen. Proc Natl Acad Sci USA *77*, 5706-5710.

Rougvie, A. E., and Lis, J. T. (1988). The RNA polymerase II molecule at the 5' end of the uninduced hsp70 gene of D. melanogaster is transcriptionally engaged. Cell *54*, 795-804.

Roy, R., Adamczewski, J. P., Seroz, T., Vermeulen, W., Tassan, J. P., Schaeffer, L., Nigg, E. A., Hoeijmakers, J. H., and Egly, J. M. (1994). The MO15 cell cycle kinase is associated with the TFIIH transcription-DNA repair factor. Cell *79*, 1093-1101.

Ryu, S., Zhou, S., Ladurner, A. G., and Tjian, R. (1999). The transcriptional cofactor complex CRSP is required for activity of the enhancer-binding protein Sp1. Nature *397*, 446-450.

Saltzman, A. G., and Weinmann, R. (1989). Promoter specificity and modulation of RNA polymerase II transcription. Faseb J *3*, 1723-1733.

Sawadogo, M., and Roeder, R. G. (1985). Interaction of a gene-specific transcription factor with the adenovirus major late promoter upstream of the TATA box region. Cell *43*, 165-175.

Schaeffer, L., Moncollin, V., Roy, R., Staub, A., Mezzina, M., Sarasin, A., Weeda, G., Hoeijmakers, J. H., and Egly, J. M. (1994). The ERCC2/DNA repair protein is associated with the class II BTF2/TFIIH transcription factor. EMBO J *13*, 2388-2392.

Schaeffer, L., Roy, R., Humbert, S., Moncolli, V., Vermeulen, W., Hoeijmakers, J. H. J., Chambon, P., and Egly, J. M. (1993). DNA repair helicase: a component of BTF2 (TFIIH) basic transcription factor. Science *260*, 58-63.

Schultz, P., Fribourg, S., Poterszman, A., Mallouh, V., Moras, D., and Egly, J. M. (2000). Molecular structure of human TFIIH. Cell *102*, 599-607.

Serizawa, H., Makela, T. P., Conaway, J. W., Conaway, R. C., Weinberg, R. A., and Young, R. A. (1995). Association of Cdk-activating kinase subunits with transcription factor TFIIH. Nature *374*, 280-282.

Shiekhattar, R., Mermelstein, F., Fisher, R. P., Drapkin, R., Dynlacht, B., Wessling, H. C., Morgan, D. O., and Reinberg, D. (1995). Cdk-activating kinase complex is a component of human transcription factor TFIIH. Nature *374*, 283-287.

Sklar, V. E., Schwartz, L. B., and Roeder, R. G. (1975). Distinct molecular structures of nuclear class I, II, and III DNA-dependent RNA polymerases. Proc Natl Acad Sci USA *72*, 348-352.

Sopta, M., Burton, Z., and Greenblatt, J. (1989). Strucure and associated DNA-helicase activity of a general transcription initiation factor that binds to RNA polymerase II. Nature *341*, 410-414.

Strahl, B. D., and Allis, C. D. (2000). The language of covalent histone modifications. Nature *403*, 41-45.

Strasser, K., Masuda, S., Mason, P., Pfannstiel, J., Oppizzi, M., Rodriguez-Navarro, S., Rondon, A. G., Aguilera, A., Struhl, K., Reed, R., and Hurt, E. (2002). TREX is a conserved complex coupling transcription with messenger RNA export. Nature *417*, 304-308.

Sun, X., Zhang, Y., Cho, H., Rickert, P., Lees, E., Lane, W., and Reinberg, D. (1998). NAT, a human complex containing Srb polypeptides that functions as a negative regulator of activated transcription. Mol Cell *2*, 213-222.

Taatjes, D. J., Naar, A. M., Andel, F., 3rd, Nogales, E., and Tjian, R. (2002). Structure, function, and activator-induced conformations of the CRSP coactivator. Science *295*, 1058-1062.

Taatjes, D. J., Schneider-Poetsch, T., and Tjian, R. (2004). Distinct conformational states of nuclear receptor-bound CRSP-Med complexes. Nat Struct Mol Biol *11*, 664-671.

Tan, S., Hunziker, Y., Sargent, D. F., and Richmond, T. J. (1996). Crystal structure of a yeast TFIIA/TBP/DNA complex. Nature *381*, 127-134.

Taunton, J., Hassig, C. A., and Schreiber, S. L. (1996). A mammalian histone deacetylase related to the yeast transcriptional regulator Rpd3p. Science *272*, 408-411.

Thanos, D., and Maniatis, T. (1995). Virus induction of human IFNb gene expression requires the assembly of an enhanceosome. Cell *83*, 1091-1100.

Tjian, R. (1978). The binding site on SV40 DNA for a T antigen-related protein. Cell *13*, 165-179.

Treisman, R., Green, M. R., and Maniatis, T. (1983). cis and trans activation of globin gene transcription in transient assays. Proc Natl Acad Sci USA *80*, 7428-7432.

Triezenberg, S. J., Kingsbury, R. C., and McKnight, S. L. (1988). Functional dissection of VP16, the trans-activator of herpes simplex virus immediate early gene expression. Genes Dev *2*,

718-729.

Tsukiyama, T., Becker, P. B., and Wu, C. (1994). ATP-dependent nucleosome disruption at a heat-shock promoter mediated by binding of GAGA transcription factor. Nature *367*, 525-532.

Turner, R., and Tjian, R. (1989). Leucine repeats and an adjacent DNA binding domain mediate the formation of functional cFos-cJun heterodimers. Science *243*, 1689-1694.

Verrijzer, C. P., Yokomori, K., Chen, J.-L., and Tjian, R. (1994). *Drosophila* TAF$_{II}$150: similarity to yeast gene TSM-1 and specific binding to core promoter DNA. Science *264*, 933-941.

Wada, T., Takagi, T., Yamaguchi, Y., Ferdous, A., Imai, T., Hirose, S., Sugimoto, S., Yano, K., Hartzog, G. A., Winston, F., *et al.* (1998). DSIF, a novel transcription elongation factor that regulates RNA polymerase II processivity, is composed of human Spt4 and Spt5 homologs. Genes Dev *12*, 343-356.

Weil, P. A., Luse, D. S., Segall, J., and Roeder, R. G. (1979). Selective and accurate initiation of transcription at the Ad2 major late promoter in a soluble system dependent on purified RNA polymerase II and DNA. Cell *18*, 469-484.

Wharton, R. P., Brown, E. L., and Ptashne, M. (1984). Substituting an alpha-helix switches the sequence-specific DNA interactions of a repressor. Cell *38*, 361-369.

Wieczorek, E., Brand, M., Jacq, X., and Tora, L. (1998). Function of TAF(II)-containing complex without TBP in transcription by RNA polymerase II. Nature *393*, 187-191.

Wu, G. J. (1978). Adenovirus DNA-directed transcription of 5.5S RNA *in vitro*. Proc Natl Acad Sci USA *75*, 2175-2179.

Yamaguchi, Y., Takagi, T., Wada, T., Yano, K., Furuya, A., Sugimoto, S., Hasegawa, J., and Handa, H. (1999). NELF, a multisubunit complex containing RD, cooperates with DSIF to repress RNA polymerase II elongation. Cell *97*, 41-51.

Yang, Z., Zhu, Q., Luo, K., and Zhou, Q. (2001). The 7SK small nuclear RNA inhibits the CDK9/cyclin T1 kinase to control transcription. Nature *414*, 317-322.

Yokomori, K., Admon, A., Goodrich, J. A., Chen, J.-L., and Tjian, R. (1993). *Drosophila* TFIIA-L is processed into two subunits that are associated with the TBP/TAF complex. Genes Dev *7*, 2235-2245.

Yudkovsky, N., Logie, C., Hahn, S., and Peterson, C. L. (1999). Recruitment of the SWI/SNF chromatin remodeling complex by transcriptional activators. Genes Dev *13*, 2369-2374.

Section II

The Machinery

Chapter 02
The General Transcription Machinery and Preinitiation Complex Formation

Samuel Y. Hou and Cheng-Ming Chiang

Department of Biochemistry, Case Western Reserve University School of Medicine, 10900 Euclid Avenue, Cleveland, OH 44106-4935

Key Words: RNA polymerase II, general transcription factors, TFIIA, TFIIB, TFIID, TFIIE, TFIIF, TFIIH, preinitiation complex

Summary

Transcription of protein-coding genes in eukaryotes is regulated by RNA polymerase II (pol II). By itself, pol II is unable to direct site-specific initiation of transcription and requires a host of accessory proteins, termed general transcription factors (GTFs), to commence basal level transcription dictated by the core promoter elements. The GTFs for pol II-mediated transcription include TFIIA, TFIIB, TFIID, TFIIE, TFIIF, and TFIIH. The transcription cycle, whether basal or regulated transcription, can be divided into multiple stages: preinitiation complex (PIC) assembly, initiation, promoter clearance, elongation, termination, and reinitiation. Before initiation, TFIID binding to the core promoter marks the beginning for PIC assembly. This is followed by the entry of TFIIA, TFIIB, pol II/TFIIF, TFIIE, and TFIIH, either in a stepwise fashion or as a preassembled pol II holoenzyme complex, to form a stable PIC which is ready to make RNA when ribonucleoside triphosphates and an energy source are provided. Formation of the PIC is a critical and often rate-limiting step in transcriptional regulation. Here we discuss the properties of pol II and each GTF in relation to PIC assembly, which occurs prior to the formation of the first phosphodiester bond in transcription initiation.

Introduction

A: Discovery of RNA Polymerase Activity

The foundation of molecular biology, where information flows from DNA to RNA to protein, known as the Central Dogma, was first proposed by Francis Crick in 1958 (Crick, 1958). The discovery of RNA viruses and prions provided an expanded view of the flow of genetic information. With slight modifications, the Central Dogma as proposed by Francis Crick still holds true (Crick, 1970). The initial step of this process, the transfer of genetic information from DNA to RNA, is termed transcription. The enzymatic activity of RNA polymerase, which governs this genetic transfer, was first isolated by Weiss and Gladstone in 1959 from rat liver nuclei (Weiss and Gladstone, 1959). This enzyme could synthesize RNA in a DNA-dependent manner, as evidenced by the observation that upon the addition of DNase, incorporation of $[\alpha\text{-}^{32}P]CTP$ into RNA was severely diminished (Weiss and Gladstone, 1959). While the isolation of a single bacterial RNA polymerase from *Escherichia coli* was achieved over the next few years (Chamberlin and Berg, 1962), the biochemical purification of this eukaryotic enzyme remained elusive for an entire decade until purification of this enzyme was accomplished in 1969 by Roeder and Rutter from sea urchin embryo nuclei (Roeder and Rutter, 1969).

Surprisingly, Roeder and Rutter isolated not just one, but three different RNA polymerase activities, which were named RNA polymerase I, II, and III (or A, B, and C respectively), based upon their chromatographic fractionation on a DEAE-Sephadex column. RNA

Corresponding Author: Cheng-Ming Chiang, Tel: (216) 368-8550, Fax: (216) 368-3419, E-mail: cmc23@cwru.edu

polymerase I eluted first at the lowest salt concentration, while RNA polymerase III came off the column at the highest salt concentration (Roeder and Rutter, 1969). Although the existence of three eukaryotic RNA polymerases were reported in 1969, their functions remained undefined until the specific activities of each RNA polymerase were resolved based upon their differential sensitivities to α-amanitin (Weinmann and Roeder, 1974; Weinmann et al., 1974), a drug isolated from the death cap fungus, *Amanita phalloides*, that inhibits the activity of RNA polymerase II at low concentrations and that of RNA polymerase III at high concentrations. Using α-amanitin sensitivity assay with endogenous RNA polymerases in isolated nuclei, Roeder and colleagues discovered that RNA polymerase I is primarily involved in transcribing 18S and 28S ribosomal RNAs, while RNA polymerase II transcribes mRNAs, and RNA polymerase III is responsible for synthesis of cellular 5S rRNA, tRNAs, and adenovirus VA RNAs. These results were consistent with the finding that RNA polymerase I is localized within nucleoli, the sites for rRNA gene transcription, whereas RNA polymerase II and III are normally present in the nucleoplasm (Roeder and Rutter, 1970).

B: The General Transcription Machinery

The biochemical identification of all three eukaryotic RNA polymerases was not achieved until 1975. Analysis and comparison of these eukaryotic RNA polymerases revealed that RNA polymerase I, II and III contain multiple subunits, some of which appear to be common among all three polymerases (Sklar *et al.*, 1975). In this chapter, we concentrate on the formation of an initiation-competent RNA polymerase II (hereafter referred to as pol II) transcription complex, since pol II is responsible for transcription of all protein-coding genes in eukaryotes. Although pol II was eventually isolated and found to contain 12 subunits (Young, 1991), purified pol II was not able to recognize specific promoters and accurately initiate transcription. Site-specific initiation was only observed in whole cell lysates or nuclear extracts. This led to the hypothesis that other essential or accessory factors were required for accurate initiation of gene-specific transcription.

Biochemical evidence for necessary accessory factors became evident when crude subcellular fractions supplemented with purified pol II were able to accurately transcribe natural adenovirus DNA template *in vitro* (Weil *et al.*, 1979). This was followed by the chromatographic purification of subcellular fractions to identify the accessory factors, later termed general transcription factors (GTFs). Roeder and colleagues purified these subcellular fractions over a Whatman P11 phosphocellulose ion exchange column to isolate A, B, C, and D fractions, which correspond to the nuclear proteins sequentially eluted by 0.1, 0.3, 0.5 (or 0.6), and 0.85 (or 1.0) M KCl-containing buffer, and then proved that the A, C, and D components were necessary for accurate initiation of transcription by pol II (Matsui *et al.*, 1980). The protein factors present in the A and D fractions necessary for pol II-mediated transcription were named TFIIA and TFIID, respectively. The C fraction was later fractionated into accessory factors TFIIB, TFIIE, TFIIF, and TFIIH (Sawadogo and Roeder, 1985a; Reinberg and Roeder, 1987; Flores *et al.*, 1989; Flores *et al.*, 1992; Ge *et al.*, 1996). These accessory factors (TFIIA, TFIIB, TFIID, TFIIE, TFIIF, and TFIIH) were collectively defined as GTFs. The nomenclature for pol II GTFs soon became TFII(letter). TF represents Transcription Factor, the Roman numeral II specifies pol II transcription, and the "letter" indicates which purification fraction the specific GTF was isolated from (see Fig.2.1). Three components of the general transcription machinery are multisubunit complexes: pol II is composed of 12 subunits, TFIID is comprised of 14 subunits, and TFIIH has 9 subunits (Wu *et al.*, 1998).

C: The Sequential Assembly Pathway

Although accessory factors for accurate initiation of pol II transcription were being identified at a rapid pace, little was known about how these GTFs are assembled at the promoter region where transcription initiates. Phil Sharp and colleagues were the first to identify the hierarchical nature of GTF assembly at the promoter region using native gel electrophoresis, together with DNase I footprinting, to establish the order of addition and the relative positions of GTFs in relation to the promoter, thus suggesting a model for transcription initiation that proceeds in a stepwise manner (Buratowski *et al.*, 1989). Specifically, TFIID first recognizes the promoter, followed by TFIIA, then TFIIB, later pol II, and finally TFIIE (TFIIF and TFIIH had yet to be identified; Buratowski *et al.*, 1989). After all the GTFs were identified and purified to near homogeneity, this stepwise manner of GTF assembly was further defined as: TFIID recognition of the promoter as the first step, followed by TFIIA and TFIIB stabilizing promoter-bound TFIID, and recruiting pol II/TFIIF to the promoter. After formation of a stable TFIID-TFIIA-TFIIB-pol II/TFIIF-promoter complex, TFIIE is then recruited and followed by TFIIH (Fig. 2.2A). This stepwise manner of assembly became known as the sequential assembly pathway for the formation of the preinitiation complex (PIC).

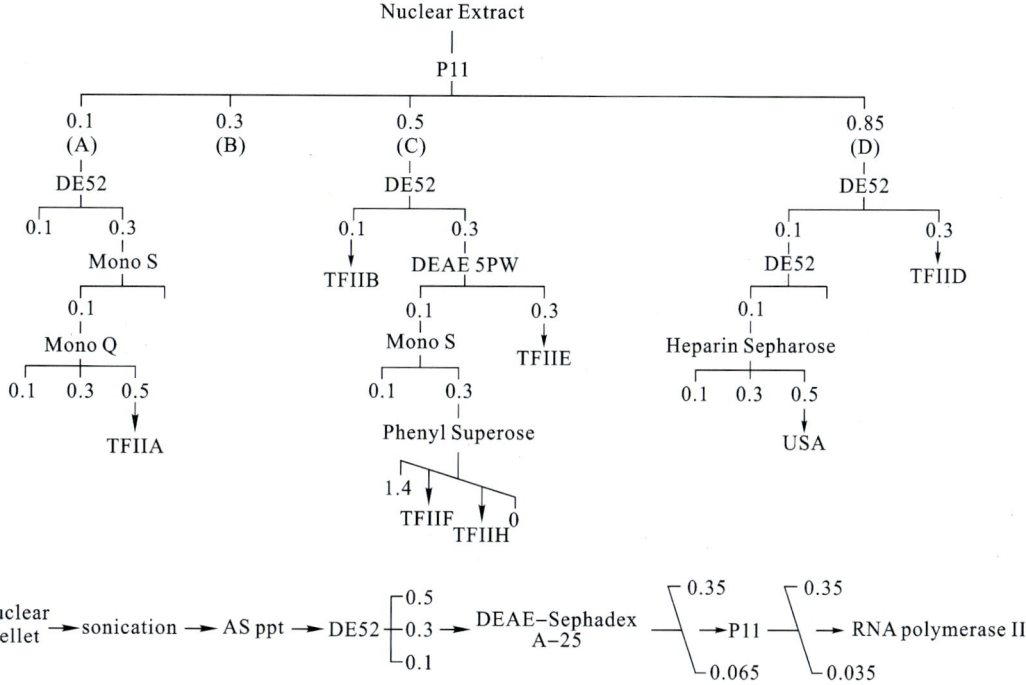

Fig.2.1 Purification scheme for partially purified general transcription factors and the USA coactivator fraction. Column chromatography fractionation of HeLa nuclear extract (upper panel) or nuclear pellet (lower panel) and the molar concentrations of KCl used for column elutions are indicated, except for the Phenyl Superose column where the molar concentrations of ammonium sulfate are shown. The purification scheme for RNA polymerase II, starting from sonication of the nuclear pellet, followed by ammonium sulfate (AS) precipitation, is also outlined. P11, DE52, Mono S, Mono Q, DEAE 5PW, Phenyl Superose, Heparin Sepharose, DEAE-Sephadex are different types of chromatography columns used for protein fractionation. A horizontal (upper panel) or vertical (lower panel) line indicates that step elutions are used for protein fractionation, whereas a slant line represents a linear gradient used for fractionation. (Figure adapted from Flores et al., 1992 and Ge et al., 1996)

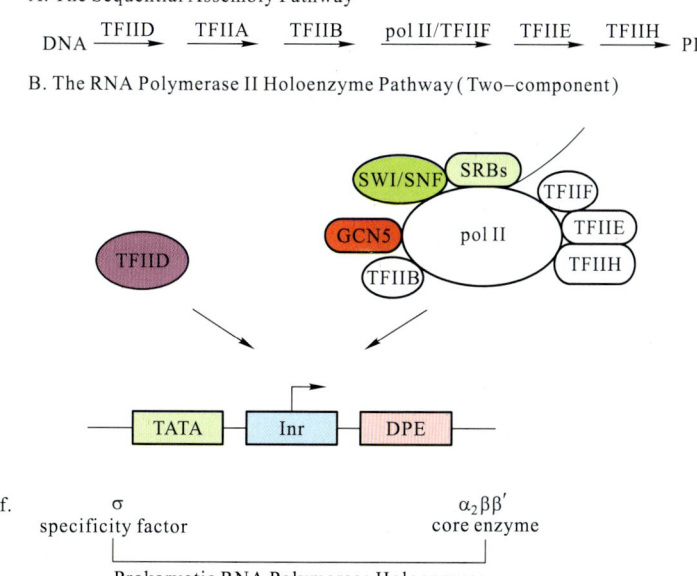

Fig.2.2 Assembly of transcription preinitiation complexes. Preinitiation complex (PIC) formation may occur via an ordered addition, as outlined in the sequential assembly pathway (A), or via a preassembled RNA polymerase II holoenzyme complex and TFIID, as depicted in the two-component pathway (B). The two-component pathway depicted here resembles the prokaryotic RNA polymerase holoenzyme system where a dissociable σ factor directs the entry of the bacterial RNA polymerase core enzyme which is comprised of $\alpha_2\beta\beta'$.

D: The Pol II Holoenzyme Pathway

An alternative pathway for PIC formation was uncovered when several laboratories found that pol II could be purified as a preassembled holoenzyme complex containing pol II, a subset of GTFs, SRBs (suppressors of RNA polymerase B mutations) (Kim *et al.*, 1994; Koleske and Young, 1994), and other proteins involved in chromatin remodeling, DNA repair, and mRNA processing (Ossipow *et al.*, 1995; Chao *et al.*, 1996; Maldonado *et al.*, 1996; Wilson *et al.*, 1996; Yuryev *et al.*, 1996; McCracken *et al.*, 1997; Nakajima *et al.*, 1997; Cho *et al.*, 1998; Wu and Chiang, 1998; Wu *et al.*, 1999). Although the composition of pol II holoenzymes isolated from different laboratories varies according to the methods of purification and the source of materials, the human pol II holoenzyme complex isolated in our laboratory contains pol II, TFIIB, TFIIE, TFIIF, TFIIH, GCN5 histone acetyltransferase, SWI/SNF chromatin remodeling factor, and SRBs, but is devoid of TFIID and TFIIA (Wu and Chiang, 1998; Wu *et al.*, 1999; see Fig. 2.2B). The identification of a TFIID-deficient pol II holoenzyme complex suggests that TFIID, as a core promoter-binding factor, may facilitate the entry of pol II holoenzyme to the promoter region, a scenario analogous to the prokaryotic RNA polymerase system where a dissociable α factor can recruit core polymerase ($\alpha_2\beta\beta'$) to the promoter region for PIC formation (Fig. 2.2B). However, there are still unresolved issues whether eukaryotic pol II indeed exists as a holoenzyme complex in the cell or whether PIC assembly occurs in a series of steps at the promoter. It is likely that both assembly pathways exist *in vivo* and, depending on signaling molecules involved and the promoter context, either pathway may be selectively used in responding to environmental cues. Indeed, evidence supporting both models has been reported for different regulatory systems (Orphanides *et al.*, 1996; Hampsey, 1998; Parvin and Young, 1998; Lee and Young, 2000; Lemon and Tjian, 2000).

TFIID

A: Identification of TATA-Binding Activity

For PIC assembly via the sequential assembly pathway or the two-component pol II holoenzyme pathway, TFIID recognition of the promoter is usually the first step to initiate the formation of a transcriptionally competent PIC. As described earlier, TFIID was originally identified as a chromatographic fraction necessary to support site-specific transcription by pol II *in vitro* (Matsui *et al.*, 1980). The TFIID chromatographic fraction that Roeder and colleagues identified was still very crude and many laboratories sought to isolate the key component(s) in TFIID. Interestingly, TFIID was identified to serve a key function in binding to the TATA box, initially from *Drosophila* (Parker and Topol, 1984), then mammals (Sawadogo and Roeder, 1985b), and yeast (Buratowski *et al.*, 1988; Cavallini *et al.*, 1988). Once TFIID bound to the TATA box, it was hypothesized to function as a scaffold upon which the PIC could assemble (Workman and Roeder, 1987; Hai *et al.*, 1988; Horikoshi *et al.*, 1988). Eventually a single polypeptide containing the TATA box-binding activity (later termed TBP for TATA box-binding protein) was purified (Horikoshi *et al.*, 1989a) and cloned from yeast (Hahn *et al.*, 1989; Horikoshi *et al.*, 1989b). The cloning of *Drosophila* and human TBP factors soon followed (Hoey *et al.*, 1990; Kao *et al.*, 1990; Peterson *et al.*, 1990).

The immediate question that followed was whether this single polypeptide TBP is the functional equivalent of TFIID. In an elegant experiment testing for transcriptional activation dependent on the transcriptional activator Sp1 (specificity protein 1), it was found that Sp1 could stimulate transcription from TATA-containing promoters only in the presence of partially purified TFIID, but not when TFIID was replaced by recombinant *Drosophila* or yeast TBP (Pugh and Tjian, 1990). Furthermore, glycerol gradient sedimentation and immunoprecipitation analyses of partially purified TFIID, which supported activator-mediated transcription, indicated that TFIID was a multiprotein complex rather than a single polypeptide (Dynlacht *et al.*, 1991; Pugh and Tjian, 1991). Thus, it was proposed that additional factors, or coactivators, in conjunction with TBP were necessary to potentiate transcriptional activation by Sp1. It was subsequently determined that TFIID is a multiprotein complex composed of TBP and TBP-associated factors (TAFs) (Dynlacht *et al.*, 1991; Tanese *et al.*, 1991). Current techniques of immunoaffinity purification using antibodies directed against TBP or an epitope tag linked to the TBP-coding sequence have pulled down TBP and TAFs (Dynlacht *et al.*, 1991; Zhou *et al.*, 1992; Chiang *et al.*, 1993; Poon and Weil, 1993; Sanders *et al.*, 2002; Auty *et al.*, 2004). This has greatly simplified the purification scheme for TFIID and further facilitated the identification and cloning of TAFs. In humans, at least 13 TAFs have been identified (Burley and Roeder, 1996; Hahn, 1998; Albright and Tjian, 2000; Green, 2000; Tora, 2002; see also the Chapter by Thomas and Chiang, 2005). A cell type-specific TAF, TAF4b (formerly named TAF$_{II}$105) has also been isolated from B cells and found to function as a coactivator for NF-κB (Dikstein et al., 1996b; Matza et al., 2001). The current concept of TFIID is that it functions as: ① a coactivator in mediating interactions between activators and GTFs to enhance PIC assembly, ② a core promoter recognition factor for both TATA-containing and TATA-less promoters, and ③ an enzyme to posttranslationally modify protein factors involved in transcriptional regulation.

B: TFIID as a Coactivator

Characterization of TAFs revealed that many gene-specific activators interact directly with specific TAFs (reviewed by Burley and Roeder, 1996; Verrijzer and Tjian, 1996). For example, the activation domain of Sp1 was found to interact with *Drosophila* TAF4 (dTAF$_{II}$110; Hoey *et al.*, 1993), whereas the DNA-binding domain of Sp1 was found to interact with human TAF7 (hTAF$_{II}$55; Chiang and Roeder, 1995). The finding that distinct domains in an activator contact different TAFs suggests that TFIID may modulate activator function through multiple domain interactions (Chiang and Roeder, 1995). While many other examples of activator-TAF interactions exist, human TAF7 (hTAF$_{II}$55) is one particular TAF shown to interact with multiple activators (Chiang and Roeder, 1995; Lavigne *et al.*, 1999). These activator-TAF interactions imply that activators may indeed function by recruiting TFIID to the promoter in order to nucleate PIC formation (Burley and Roeder, 1996; Verrijzer and Tjian, 1996; Albright and Tjian 2000; Näär *et al.*, 2001). Consistent with its role as a coactivator, only purified TFIID, but not TBP, supports activator-mediated transcription in partially purified cell-free transcription systems (Dynlacht *et al.*, 1991; Chiang *et al.*, 1993). Thus the *in vitro* biochemical concept that TFIID was a universal coactivator required for all gene transcription was formed.

Although the universal requirement for TFIID as a coactivator was proposed based on *in vitro* studies, TFIID *in vivo* appears to play a more limited role, as subsequent yeast genetic studies have revealed that TAFs are not universally required for activated transcription as previously thought (Moqtaderi *et al.*, 1996; Walker *et al.*, 1996). Rather, TAFs may mediate activator-dependent transcription from only a subset of genes. These results indicate that individual TAFs may be important for mediating distinct activator-dependent transcription *in vivo* (Walker *et al.*, 1997; Apone *et al.*, 1998; Holstege *et al.*, 1998; Mencia *et al.*, 2002). Subsequent cell-free transcription studies using highly purified and well-defined transcription systems have also shown that TBP, in the absence of TAFs but in conjunction with other GTFs, pol II and general cofactor PC4, may direct activated transcription from TATA-containing promoters in an activator-specific manner with naked (nucleosome-free) DNA templates (Oelgeschläger *et al.*, 1998; Wu and Chiang, 1998; Wu *et al.*, 1998, 1999; Fondell *et al.*, 1999).

C: TFIID as a Core Promoter Recognition Factor

In addition to mediating activator-dependent recruitment of GTFs, TFIID has been implicated as a core promoter recognition factor. Besides making contact with the TATA box through TBP binding, TFIID may contact the Initiator (Inr) via *Drosophila* TAF2 (dTAF$_{II}$150) or the downstream promoter element (DPE) via *Drosophila* TAF6 (dTAF$_{II}$60) and TAF9 (dTAF$_{II}$40; Burke and Kadonaga, 1997; Kaufmann *et al.*, 1998; Martinez *et al.*, 1998). Recent evidence has also shown that human TAF6 and TAF9 display sequence-specific binding to the core promoter of human interferon regulatory factor-1 (IRF-1) gene, which bears a functional DPE (Burke and Kadonaga, 1997; Shao *et al.*, 2005). DNase I footprinting experiments showing that TFIID gives extended protection over the TATA box, Inr, and DPE regions are consistent with a role of TFIID functioning as a core promoter recognition factor. These TFIID-DNA interactions may be especially important at promoters lacking a canonical TATA sequence, since TAFs are required for transcription from TATA-less promoters (Pugh and Tjian, 1991; Martinez *et al.*, 1994; Orphanides *et al.*, 1996; Smale, 1997). Furthermore, binding of TAFs to core promoter elements directs promoter selectivity by pol II (Hansen and Tjian, 1995; Verrijzer *et al.*, 1995). Studies in yeast also indicate that the requirement for individual TAFs may depend on core promoter sequences (Kuras *et al.*, 2000; Li *et al.*, 2000, 2002). Thus, a selective requirement of TAFs is determined by both gene-specific activators and the promoter context.

D: Histone-Like Domains in TAFs

The observation that TAFs share structural homology with histones (Hoffmann *et al.*, 1996; Xie *et al.*, 1996; reviewed by Gangloff *et al.*, 2001) suggests TAFs in TFIID may form a histone-like octamer structure. Indeed, initial crystal structures of TAF9/TAF6 (dTAF$_{II}$42/dTAF$_{II}$62) showed that there exists a heterotetramer, resembling the H3/H4 heterotetrameric core of the histone octamer. This finding indicates that TFIID may contain a histone octamer-like substructure (Xie *et al.*, 1996). Further supporting evidence for a histone-like octamer was demonstrated when it was found that yeast TAF9-TAF6-TAF12-TAF4 (yTaf17-yTaf60-yTaf61-yTaf48) may reconstitute an octamer *in vitro* (Selleck *et al.*, 2001). The immediate question that follows is whether the histone-like octamer of TAFs is able to contact DPE in the context of a nucleosome-like structure. In a study investigating protein-protein interactions among human TAF9, TAF6, TAF4b, and TAF12, which contain sequences related to histones H3, H4, H2A, and H2B, respectively, it was found that these TAFs indeed form an octamer-like complex which enhances both sequence-specific and nonspecific DNA-binding activities of TAF9-TAF6 and TAF4b-TAF12 pairs, respectively (Shao *et al.*, 2005). In contrast to the interaction studies, which suggest that TFIID may contain a histone-like octamer, low-resolution electron microscopy (EM) studies of the TFIID complex do not show an octamer-like structure within TFIID (Andel *et al.*, 1999; Brand *et al.*, 1999; Leurent *et al.*, 2002). These EM studies found TFIID to be a trilobed, horseshoe-shaped structure with TBP sitting in the central cavity while TFIIA and TFIIB bound to opposite lobes of the horseshoe-shaped structure (Andel *et al.*, 1999; Brand *et al.*, 1999; Leurent *et al.*, 2002). Each of the globular domains found in the TFIID EM structure is almost equivalent in size to the histone octamer, but none are large enough to hold all six histone motif-containing TAFs (Brand *et al.*, 1999). Therefore, a compact histone octamer-like structure incorporating all the TAFs with histone motifs is unlikely; instead, the histone folds in TAFs may form interfaces for protein-protein interactions (Brand *et al.*, 1999). Although a histone octamer containing all six histone fold motif-containing TAFs may be unlikely due to spatial constraints, the possibility that fewer TAFs may form a nucleosome-like structure remains to be explored.

E: TFIID as an Enzyme

It is well recognized that within the cell, DNA is wrapped around the histone octamer to form a nucleosome and further condensed into a higher order chromatin structure in order to compact DNA within the nucleus. Formation of this higher order chromatin structure prevents transcription factor access to the promoter region. For transcription to occur on chromatin templates, covalent modification on core histones is necessary to loosen up histone-DNA interactions at the promoter region. Therefore it was an important finding that TAF1 (formerly named $TAF_{II}250$) possesses histone acetyltransferase (HAT) activity to acetylate histones H3 and H4 (Mizzen *et al.*, 1996). Other than HAT activity, TFIID also exhibits multiple enzymatic activities via TAF1, such as kinase activity that phosphorylates the RAP74 subunit of TFIIF (Dikstein *et al.*, 1996a), the β subunit of TFIIA (Solow *et al.*, 2001), serine 33 of histone H2B (Maile *et al.*, 2004), and PC4 (Kershnar *et al.*, 1998; Malik *et al.*, 1998), and ubiquitin activating/conjugating activity that targets histone H1 (Pham and Sauer, 2000). Interestingly, TAF1-mediated phosphorylation of histone H2B correlates with gene activation (Maile *et al.*, 2004). Thus, the ability of TAF1 to modify distinct histones by acetylation, phosphorylation, or ubiquitination is consistent with the finding that TFIID, rather than TBP, is essential for chromatin transcription (Wu *et al.*, 1999), suggesting that TAFs in TFIID may play a role in modifying chromatin structure during the transcription process. The finding that a subset of TAFs are also integral components of other protein complexes, such as SAGA (Spt, Ada, Gcn5 acetyltransferase), required for nucleosome acetylation and transcriptional stimulation (Grant *et al.*, 1998) further supports the view that TAFs in TFIID are able to modify chromatin structure.

TFIIA

A: Protein Composition

Human TFIIA is composed of α, β, and γ subunits with molecular weights of 35 kDa, 19 kDa, and 12 kDa respectively. These three subunits are encoded by two genes: TFIIAαβ and TFIIAγ in higher eukaryotes, and TOA1 and TOA2 in yeast. Human TFIIAαβ encodes a 55-kDa precursor protein that is highly conserved with TOA1 in yeast within the N-terminal 54 amino acids and the C-terminal 76 amino acids, but is less conserved in the central part of TFIIA which has been shown to be dispensable for function (Ranish *et al.*, 1992; Kang *et al.*, 1995). The smallest subunit of TFIIA, TFIIAγ, is homologous to yeast TOA2 (Ozer *et al.*, 1994; Sun *et al.*, 1994). In higher eukaryotes, proteolytic cleavage of TFIIAαβ into TFIIAα and TFIIAβ subunits occurs (DeJong and Roeder, 1993; Ma *et al.*, 1993; Yokomori *et al.*, 1993). Originally it was thought that the three independent subunits (α, β, and γ) constitute TFIIA activity, since only the cleavage products were detected in cell extracts in association with one another. Recently, intact TFIIAαβ protein was detected in association with TFIIAγ along with TBP in embryonal carcinoma cells (Mitsiou and Stunnenberg, 2000, 2003). The site for proteolytic cleavage within TFIIAαβ was determined by Edman degradation; TFIIAβ was found to start at Asp 278 (Høiby *et al.*, 2004). Furthermore, cleaved TFIIAα and TFIIAβ were more efficiently degraded than the unprocessed precursor via the ubiquitin-proteasome pathway, suggesting that cleavage and degradation of TFIIA controls the level of TFIIA within the cell to adapt rapidly to the transcriptional needs in responding to environmental changes (Høiby et al., 2004). Similar to TAF4b, a cell-specific TFIIAα/β-like factor (ALF) has been identified in human testis (Upadhyaya et al., 1999; Ozer et al., 2000). ALF works in conjunction with TFIIAγ to stabilize TBP binding to the promoter (Upadhyaya et al., 2002). Further experiments indicate that ALF is also found in immature oocytes of the frog *Xenopus laevis*, in which ALF replaces TFIIA during oogenesis (Han et al., 2003).

B: Is TFIIA a Genuine GTF?

The role of TFIIA as a GTF has been controversial. As with TFIID, TFIIA was initially identified as a phosphocellulose column fraction necessary for pol II-mediated transcription *in vitro* (Matsui et al., 1980). Early *in vitro* experiments showed that TFIIA was essential for transcription (Reinberg et al., 1987), while in other *in vitro* studies TFIIA was shown to be largely dispensable for basal level transcription (Van Dyke et al., 1988; Wu et al., 1998; Pugh, 2000). It has also been suggested that TFIIA stimulates both basal and activated transcription *in vitro* 2- to 10-fold, but generally only when TFIID, instead of TBP, is used as the promoter-binding factor (Orphanides et al., 1996; Hampsey, 1998; Warfield et al., 2004). In our laboratory, we found TFIIA was not required for either basal or activator-dependent transcription in a highly purified transcription system reconstituted with recombinant GTFs (TFIIB, TFIIE, and TFIIF), recombinant PC4 coactivator, and epitope-tagged multiprotein complexes (pol II, TFIID, and TFIIH), irrespective of whether TFIID or recombinant TBP was used in the assay (Wu et al., 1998). Nevertheless, TFIIA indeed became essential for transcription in a reconstituted system containing partially purified fractions obtained according to the purification scheme outlined in Fig. 2.1. Collectively, these studies suggest that TFIIA mainly functions as an antirepressor to overcome inhibitors present in crude fractions likely by increasing the affinity of TBP or TFIID for DNA (Buratowski et al., 1989; Lee et al., 1992; Imbalzano et al., 1994; Kang et al., 1995), thereby enhancing PIC assembly. TFIIA stabilizes TBP-TATA box interactions through direct contacts with both TBP and DNA (Geiger et al., 1996; Tan et al., 1996). Binding of TFIIA to TBP dimers has also been shown to induce TBP monomer formation and accelerate the kinetics of TBP binding to DNA (Coleman et al., 1999). TFIIA is able to counteract the repressive effects of negative factors such as Dr1/NC2β, PC3/Dr2, Mot1, HMG1, and also the inhibitory activity of TAF1 on TBP binding to DNA (Meisterernst et al., 1991; Inostroza et al., 1992; Merino et al., 1993; Ge and Roeder, 1994; Ozer et al., 1994, 1998b; Orphanides et al., 1996; Auble et al., 1997; Hampsey, 1998; Kokubo et al., 1998; see also the Chapter by Thomas and Chiang, 2005). Interestingly, TAF1 interaction with TFIIA may modulate TFIIA activity, since it has been shown that TAF1 phosphorylates human TFIIAβ on serine residues important for TBP binding and transcription activity (Solow et al., 2001). Additionally, experiments have shown that the TFIIA-TBP-DNA complex may also be regulated by the transcriptional coactivator p300 through acetylation of TFIIA (Mitsiou and Stunnenberg, 2003).

C: TFIIA as a Coactivator

Many studies have demonstrated that TFIIA interacts with activators, such as Gal4-VP16, Zta, and HTLV-1 Tax (Ozer et al., 1994; Kobayashi et al., 1995; Clemens et al., 1996; Ranish et al., 1999; Dion and Coulombe, 2003), and transcriptional coactivators PC4 and HMG2 (Ge and Roeder, 1994; Sun et al., 1994). TFIIA has also been shown to interact with specific TAFs in TFIID (Yokomori et al., 1993; Kraemer et al., 2001; Robinson et al., 2005). In one proposed model of TFIIA function, interaction of TFIIA with both activators and TFIID may stimulate and stabilize binding of TFIID to DNA as part of an activator-TFIID-TFIIA-DNA complex, thereby enhancing a rate-limiting step for TFIID binding to DNA in the transcriptional process (Wang et al., 1992; Lieberman and Berk, 1994, Chi et al., 1995; Ranish et al., 1999; Dion and Coulombe, 2003). Moreover, TFIIA may upregulate PIC formation by stimulating the functions of both TFIIE and TFIIF (Langelier et al., 2001). In TAF-independent transcriptional activation experiments using a highly purified transcription system, TFIIA could potentiate TBP-mediated activation, suggesting that TFIIA may function as a coactivator especially in the absence of TAFs (Wu et al., 1998). Several studies using immunodepletion of TFIIA subunits *in vitro* and mutational studies abolishing TFIIA-TBP interactions in yeast have indicated that, similar to TAFs, TFIIA is important for transcription only from a subset of genes and does not seem to be universally required for all gene transcription (Kang et al., 1995; Ozer et al., 1998a; Liu et al., 1999; Stargell et al., 2000).

TFIIB

A: TFIIB Stabilizes TFIID Promoter Binding

Once TFIID is bound to the promoter in the

absence or presence of TFIIA, TFIIB is the next GTF to enter the PIC assembly pathway. Binding of TFIIB to promoter-bound TFIID results in a more stable ternary complex composed of TFIID-TFIIB-DNA (Orphanides et al., 1996). Besides stabilizing TFIID binding to the promoter region, TFIIB plays an important role in recruiting pol II/TFIIF to the TFIID-TFIIB-DNA ternary complex and in specifying the transcription start site (Orphanides et al., 1996; Hampsey, 1998). In humans, TFIIB exists as a single 33-kDa polypeptide (Ha et al., 1991; Malik et al., 1991), which shares sequence homology with a 38-kDa *Drosophila* TFIIB protein and also with a 38-kDa yeast protein encoded by the SUA7 gene (Pinto et al., 1992; Wampler and Kadonaga, 1992; Yamashita et al., 1992). TFIIB is conserved among humans, *Drosophila*, and yeast, and shows conservation at the N-terminal zinc-ribbon motif and the C-terminal domain containing two imperfect direct repeats (Orphanides et al., 1996; Hampsey, 1998). These two functional domains were originally identified through protease digestion, as the N-terminus is rapidly degraded leaving behind a protease-resistant C-terminal "core" that retains the two imperfect direct repeats of TFIIB (Barberis et al., 1993; Malik et al., 1993).

B: TFIIB-TBP-Promoter Structure

The C-terminal "core" of TFIIB comprises nearly two-thirds of the protein and contains two imperfect repeats each consisting of 5-helices (Bagby et al., 1995). The crystal structure of the TFIIB C-terminal "core", as revealed in a TATA-TBP-TFIIB ternary complex, indicates that TFIIB bound beneath and to one face of the TATA-TBP complex, consistent with footprinting and cross-linking experiments (Coulombe et al., 1994; Lee and Hahn, 1995; Nikolov et al., 1995; Lagrange et al., 1996). Structural evidence also exists for TFIIB competing with negative cofactor 2 (NC2) for overlapping binding sites on TBP (Kamada et al., 2001b). The conserved C-terminus of TFIIB interacts with both TBP and DNA, making contacts with DNA sequences on both sides of the TATA box (Nikolov et al., 1995; Tsai and Sigler, 2000). Experiments also indicate that TFIIB can make sequence-specific DNA contacts with a TFIIB-response element (BRE) typically located just upstream of some TATA box sequences via a helix-turn-helix motif which stabilizes the ternary TFIIB-TBP-promoter complex (Lagrange et al., 1996; Evans et al., 2001; Wolner and Gralla, 2001). Recent studies also show that TFIIB may make sequence-specific DNA contacts with a BRE-like element situated downstream of the TATA box, and that these TFIIB-DNA interactions modulate TFIIB conformation (Fairley et al., 2002). Moreover, binding of the transcriptional activator Gal4-VP16 to TFIIB induces conformational changes on TFIIB and weakens TFIIB interaction with the BRE (Evans et al., 2001). The role of activator in disrupting TFIIB-DNA interactions seems contradictory with the previously proposed role of the BRE in stabilizing the TFIIB-TBP-DNA ternary complex. It was suggested that the BRE may play an inhibitory role in PIC formation and the presence of an activator alleviates BRE-mediated repression and therefore stimulates overall transcription (Evans et al., 2001). Nevertheless, how activators, TFIIB, and BRE interactions affect PIC assembly, initiation, and promoter clearance remains to be investigated.

C: TFIIB N-Terminal Domain

The TFIIB amino terminus contains a zinc ribbon motif that interacts with components of pol II and TFIIF, and thus facilitates the recruitment of pol II/TFIIF to the TFIID-bound promoter region (Buratowski and Zhou, 1993; Ha et al., 1993; Malik and Roeder, 1993; Yamashita et al., 1993; Fang and Burton, 1996; Bangur et al., 1997). The N- and C-termini of TFIIB engage in an intramolecular interaction that undergoes an activator-induced conformational change, which frees up the N-terminal domain for the recruitment of pol II/TFIIF (Glossop et al., 2004). Immediately adjacent to the N-terminal zinc ribbon motif is a highly conserved region called the charged cluster domain (CCD) or B-finger, which contains key charged amino acid residues. It is believed that the CCD acts as a molecular switch to regulate the conformational change of TFIIB, thereby modulating the role of TFIIB in promoter recognition, start site selection, and transcriptional activation (Pinto et al., 1994; Pardee et al., 1998; Hawkes and Roberts, 1999; Wu and Hampsey, 1999; Hawkes et al., 2000; Faitar et al., 2001; Elsby and Roberts, 2004).

D: CCD-Induced Conformational Changes

As mentioned previously, TFIIB undergoes a conformational change when it interacts with DNA or activators and that the CCD plays a vital role in this conformational change (Fairley et al., 2002; Elsby and Roberts, 2004). The activator Gal4-VP16 has been shown to disrupt the intramolecular interaction between the CCD and the second repeat of the C-terminal domain (Elsby and Roberts, 2004). Corresponding evidence with mutations in the CCD, which favors N- and C-terminal intramolecular interactions, also shows defects in activator-mediated recruitment and in transcriptional activation *in vivo* and *in vitro* (Hawkes et *al.*, 2000; Glossop *et al.*, 2004; Elsby and Roberts, 2004). These CCD mutations, however, are competent in PIC assembly, indicating that mutations in the CCD do not affect TFIIB interactions with TBP/DNA and pol II/TFIIF, mediated individually by the C-terminal core and the zinc ribbon motif. Related studies in yeast TFIIB also suggest that conformational changes in TFIIB regulate the stability of the TFIIB-TBP-promoter ternary complex (Bangur *et al.*, 1999).

E: Role of CCD in Start Site Selection

The effect of yeast TFIIB (Sua7) on start site selection has been observed on CYC1, ADH1, and many other genes (Pinto *et al.*, 1992; Berroteran *et al.*, 1994). Besides its role in mediating activated transcription, the CCD also has a critical role in directing accurate initiation of transcription. The functions of the CCD in transcriptional activation and start site selection are located on different amino acid residues (Hawkes and Roberts, 1999). Similar to the activation-defective CCD mutants, distinct mutations in the CCD with aberrant transcriptional start site selection also do not show defects in PIC assembly (Hawkes and Roberts, 1999; Fairley *et al.*, 2002). It has been proposed that these distinct CCD mutations, which cannot undergo proper conformational changes, alter TFIIB's interaction with the pol II catalytic center, thus shifting the transcription start site. Recent photocrosslinking experiments have shown that the N-terminal zinc ribbon interacts with the pol II dock domain and that TFIIB helps position the path of promoter DNA across the central cleft of pol II (Chen and Hahn, 2003, 2004). Supporting crystallographic data has also shown that the CCD of TFIIB forms a finger-like structure that projects into the active center of pol II (Bushnell *et al.*, 2004). Although mutations in the CCD may not be directly involved in PIC formation, conformational changes in TFIIB does play an essential role in transcriptional activation, promoter recognition, and also start site selection.

TFIIF

A: Discovery of TFIIF

TFIIF was not initially identified when nuclear proteins were fractionated by P11 ion-exchange chromatography simply into four (A, B, C, and D) fractions (Matsui *et al.*, 1980). The discovery of TFIIF was made possible only after further purification of the C fraction and following the identification of TFIIE (Flores *et al.*, 1988). Human TFIIF was found to be composed of previously identified RNA polymerase II-associated protein 30 (RAP30) and 74 (RAP74) isolated from calf thymus extracts as well as from human and mouse cell lines (Sopta *et al.*, 1985; Flores *et al.*, 1988). Further characterization of TFIIF by size exclusion column chromatography suggests that TFIIF is a heterotetramer composed of 2 subunits each of RAP30 and RAP74 (Flores *et al.*, 1990). The structure of the RAP30-RAP74 heterodimer from respective protein-protein interaction domains was resolved by X-ray crystallography (Gaiser *et al.*, 2000). The cDNAs for RAP30 and RAP74 have already been cloned from human, *Drosophila*, and yeast (Sopta *et al.*, 1989; Aso *et al.*, 1992; Finkelstein *et al.*, 1992; Kephart *et al.*, 1993; Henry *et al.*, 1994; Frank *et al*, 1995; Gong *et al.*, 1995).

B: Subunits of TFIIF

Human RAP30 (calculated 26 kDa, apparent mass by SDS-PAGE ~30 kDa) shares two significant sequence homology regions with bacterial σ factors: a central domain that interacts with pol II and a cryptic DNA-binding domain at the C-terminus (Sopta *et al.*, 1989; McCraken and Greenblatt, 1991; Garrett *et al.*, 1992; Tan *et al.*, 1994). RAP30 also contains an N-terminal domain important for binding to RAP74 and TFIIB (Fang and Burton, 1996). Interestingly, human TFIIF prebound to *E.coli* RNA polymerase core enzyme ($\alpha_2\beta\beta'$) can be displaced by bacterial σ^{70} factor presumably via the same region in σ^{70} that shares homology with RAP30 (McCracken and Greenblatt, 1991). The solution structure of the C-terminal 86 amino acid residues of RAP30, revealed by multinuclear NMR spectroscopy, shows that the DNA-binding domain of RAP30 belongs to the eukaryotic "winged" helix-turn-helix (HTH) family of DNA-binding domains, similar to that found in histone H5 and the hepatocyte nuclear transcription factor HNF-3γ (Groft *et al.*, 1998).

The larger subunit of TFIIF, RAP74 (calculated 58 kDa, apparent mass by SDS-PAGE ~74 kDa), is highly rich in charged, in particular, acidic amino acid residues (Aso *et al.*, 1992; Finkelstein *et al.*, 1992). This highly charged central region, significantly lacking hydrophobic residues, is hypothesized to be externally exposed and unstructured within the PIC (Yong *et al*, 1998). The N-terminus of RAP74 is a globular region responsible for interaction with RAP30 (Fang and Burton, 1996). It is involved in PIC assembly, initiation and elongation (Lei *et al.*, 1998; Funk *et al.*, 2002). Similar to RAP30, RAP74 also contains a cryptic DNA-binding domain belonging to the winged HTH family, in which the structure spanning the C-terminal 155 amino acid residues of RAP74 has been revealed by X-ray crystallography (Kamada et al., 2001a). The C-terminal region of RAP74 interacts with the FCP1 phosphatase, which removes phosphorylation on serine 2 at the carboxy-terminal domain (CTD) of the pol II RPB1 subunit, and is necessary and sufficient for FCP1 phosphatase activity in vitro (Archambault et al., 1997, 1998). The interaction with FCP1 may account for the requirement of the C-terminal region of RAP74 for multiple rounds of transcription (Lei et al., 1998; see Core Pol II section). As with other GTFs, TFIIF may also be recruited to the promoter through interaction with transcriptional activators. Indeed, the androgen receptor has been shown to functionally interact with both N- and C-terminal domains of RAP74 (McEwan and Gustafsson, 1997; Reid et al., 2002).

C: Multiple Roles of TFIIF

TFIIF plays multiple roles during PIC assembly. First, TFIIF tightly associates with pol II (Sopta et al., 1985). The RAP74 subunit of yeast TFIIF has been shown to contact the dissociable RPB4/RPB7 subunits of yeast pol II, based on structural comparison between pol II-TFIIF and pol II by cryo-electron microscopy (Chung et al., 2003), and to interact with the RPB9 subunit of yeast pol II (Ziegler et al., 2003; Ghazy et al., 2004). Human RAP30 has also been shown to interact with the RPB5 subunit of human pol II (Wei et al., 2001). The interaction between TFIIF and pol II facilitates the recruitment of pol II to the promoter bound TFIID-TFIIB complex (Flores et al., 1991). Second, TFIIF serves as a stability factor to enhance the affinity of pol II for the TFIID-TFIIB-promoter complex by providing additional protein-DNA contact surfaces and also by inducing changes in DNA topology which causes the promoter to wrap around pol II (Robert et al., 1998). This TFIIF-induced conformational change creates a stable TFIID-TFIIB-pol II-TFIIF-promoter DNA complex that is likely to confer resistance to inhibition by transcriptional repressors that target PIC assembly to negatively regulate gene transcription (Hou et al., 2000). In the study of human papillomavirus (HPV) E2-mediated transcription repression, we found that E2 could still inhibit HPV transcription when added to a preformed TFIID-TFIIB-pol II-promoter DNA complex; however, E2 failed to inhibit transcription once a TFIID-TFIIB-pol II-TFIIF-DNA complex was formed, presumably a surface targeted by E2 for repression was masked by TFIIF-induced DNA wrapping on this minimal PIC (Hou et al., 2000). Third, TFIIF is necessary for subsequent recruitment of TFIIE and TFIIH (Orphanides et al., 1996) likely via direct interactions with TFIIE (Maxon et al., 1994). Fourth, TFIIF, together with pol II and TFIIB, plays a role in transcription start site selection (Fairley et al., 2002; Ghazy et al., 2004). Fifth, TFIIF has also been implicated in aiding pol II promoter escape (Yan et al., 1999). Following promoter escape, TFIIF then enhances the efficiency of pol II elongation (Shilatifard et al., 2003). Lastly, TFIIF increases the specificity and efficiency of polymerase transcription, similar to bacterial σ factors, by preventing spurious initiation through inhibiting and/or reversing the binding of pol II to nonpromoter DNA sequences (Orphanides et al., 1996; Hampsey, 1998). It is clear that TFIIF has multiple roles in the pol II transcription process.

Core Pol II

A: Subunit Composition

Pol II is the key catalytic enzyme in the PIC responsible for transcription of protein-coding genes in eukaryotes. Yeast and human pol II both contain 12 subunits, designated RPB1 to RPB12 by decreasing order of their molecular mass (Young, 1991). In general, the 12 subunits of pol II are highly conserved in sequence, architecture, and function. Indeed, seven subunits of human pol II can either partially (RPB4, RPB7, and RPB9) or completely (RPB6, RPB8, RPB10, and RPB12) substitute for the function of their yeast counterparts in complementation assays (McKune et al., 1995; Khazak et al., 1998). Of the 12 pol II subunits, five (RPB5, RPB6, RPB8, RPB10, and RPB12) are commonly shared among RNA polymerase I, II and III (Woychik et al., 1990; Carles et al., 1991; Young, 1991; Hampsey, 1998). Four pol II subunits, RPB1, RPB2, RPB3, and RPB11, have sequence-homologous counterparts in RNA polymerase I and III. Only RPB4, RPB7, RPB9 and the carboxy-terminal domain (CTD) of RPB1 are unique to pol II. In addition, RPB1, RPB2, RPB3, and RPB6 share similar primary sequences with bacterial RNA polymerase subunits β', β, α, and ω respectively (Tan et al., 2000; Minakhin et al., 2001; Mitsuzawa and Ishihama, 2004). A prokaryotic α-like sequence also exists in RPB11 (Woychik et al., 1993; Ulmasov et al., 1996). The primary sequence similarity between RPB1 and β' as well as between RPB2 and β also corresponds to functional similarity: RPB1 and β' are involved in DNA binding, while RPB2 and β bind nucleotide substrates (Hampsey, 1998). Analogous to their bacterial counterparts, RPB1 and RPB2 are responsible for most of the catalytic activity of polymerase and are essential for phosphodiester bond formation (Hampsey, 1998; Lee and Young, 2000).

B: Structure of Pol II

Recently there has been a wealth of structural information on prokaryotic and eukaryotic RNA polymerases provided by photocrosslinking, X-ray crystallography, NMR, and cryo-electron microscopy. The structures for a yeast pol II 10-subunit enzyme minus RPB4 and RPB7 in the absence of DNA (Cramer et al., 2000, 2001) and for a complete 12-subunit pol II have recently been resolved by X-ray crystallography (Armache et al., 2003, 2005; Bushnell and Kornberg, 2003). Initially, RBP4 and RBP7 were not included in the original crystals since the heterodimeric RPB4/RPB7 module is found in substoichiometric amounts in pol II and may dissociate from the "core" 10 subunits. Furthermore, the RPB4/RPB7 heterodimer, although they are required for PIC formation and initiation of transcription, are dispensable for RNA chain elongation (reviewed by Hampsey, 1998; Lee and Young, 2000; Cramer, 2004; Hahn, 2004). The structure of the 12-subunit pol II complex pinpoints the location of RPB4/RPB7 close to the RNA exit channel and also suggests a role of this heterodimer in transcriptional initiation (Armache et al., 2003, 2005; Bushnell and Kornberg, 2003). Comparison of the structures between the 10-subunit free core enzyme and a transcribing elongation complex containing the same core enzyme in complex with 9 base pairs of an RNA-DNA hybrid within a partially unwound DNA duplex has revealed detailed information regarding subunit-subunit and protein-nucleic acid contacts both within and outside the catalytic center of the enzyme (Cramer et al., 2001; Gnatt et al., 2001). Furthermore, contact residues between pol II and distinct domains of TFIIB, based on the structural information, have also been defined by photocrosslinking experiments (Chen and Hahn, 2003, 2004). The structures of pol II complexed with TFIIB (Bushnell et al., 2004), with elongation factor IIS (Kettenberger et al., 2003), and with IIS in the presence of NTPs and a transcription bubble-mimicking DNA-RNA hybrid (Kettenberger et al., 2004) have also been elucidated by X-ray crystallography. Low-resolution cryo-electron microscopy has resolved structures for a pol II-Mediator complex (Davis et al., 2002) and also for pol II interaction with TFIIF (Chung et al., 2003). The implications of these structural studies have been the focus of many recent reviews (Woychik and Hampsey, 2002; Asturias, 2004; Cramer, 2004; Hahn, 2004; Boeger et al., 2005).

C: CTD Phosphorylation

An essential feature of all pol II complexes resides in the carboxy-terminal domain (CTD) of RPB1, the largest subunit of pol II. The CTD of RPB1 contains a tandem repeat of a heptapeptide: Tyr-Ser-Pro-Thr-Ser-Pro-Ser (Lee and Young, 2000), which repeats 52 times in humans, 42 times in Drosophila, and 26 to 29 times in yeast depending upon the species (Hampsey, 1998; Lee and Young, 2000). The CTD is unstructured and tends to be degraded by proteases. Depending on the phosphorylation state and the presence or absence of the CTD, three forms of pol II (IIO, IIA, and IIB) can be easily distinguished (Kershnar et al., 1998) (Fig. 2.3). The IIA form of pol II contains a hypo- or unphosphorylated form of CTD normally implicated in PIC assembly and transcription initiation. The IIO form of pol II, involved in transcript elongation and termination, has a highly phosphorylated CTD with phosphorylation occurring primarily at serine residues 2 and 5. The IIB form of pol II does not have the CTD but remains transcriptionally active for at least the adenovirus major late promoter (Kang and Dahmus, 1993).

Several protein kinases implicated in CTD phosphorylation have been identified in humans and include cyclin-dependent kinase 7 (CDK7) associated with TFIIH, CDK8 found in general cofactor Mediator, and CDK9 present in positive transcription elongation factor b (P-TEFb). The activities of these CTD kinases are regulated by their associated cyclins that form CDK7-cyclin H, CDK8-cyclin C, and CDK9-cyclin T pairs. Similar CTD kinases have also been identified in yeast and include Cdk7/Kin28, Cdk8/Srb10, the CTD kinase 1 (CTDK-I), and Sgv1/Bur1 (Prelich, 2002). Both Bur1 and the catalytic subunit of CTDK-I, Ctk1, show sequence homology to mammalian CDK9. Phosphorylation of serine 5 by Cdk7/Kin28 following PIC assembly leads to the initiation of transcription and later recruitment of mRNA-capping enzyme guanylyltransferase (Cho et al., 1997; Komarnitsky et al., 2000; Rodriguez et al., 2000; Schroeder et al., 2000; Pei et al., 2001). Phosphorylation of serine 2 by other CTD kinases, such as CTDK-I or P-TEFb (Cho et al., 2001; Zhou et al., 2000; Shim et al., 2002), leads to transcription-coupled recruitment of splicing and 3'-end processing factors (Komarnitsky et al., 2000; Ahn et al., 2004). Two other CTD kinases, Cdk8/Srb10 (Hengartner et al., 1998) and c-Abl (Baskaran et al., 1993), has also been implicated in phosphorylation of serine 2 and tyrosine 1, respectively, although the functional effects remain to be further defined.

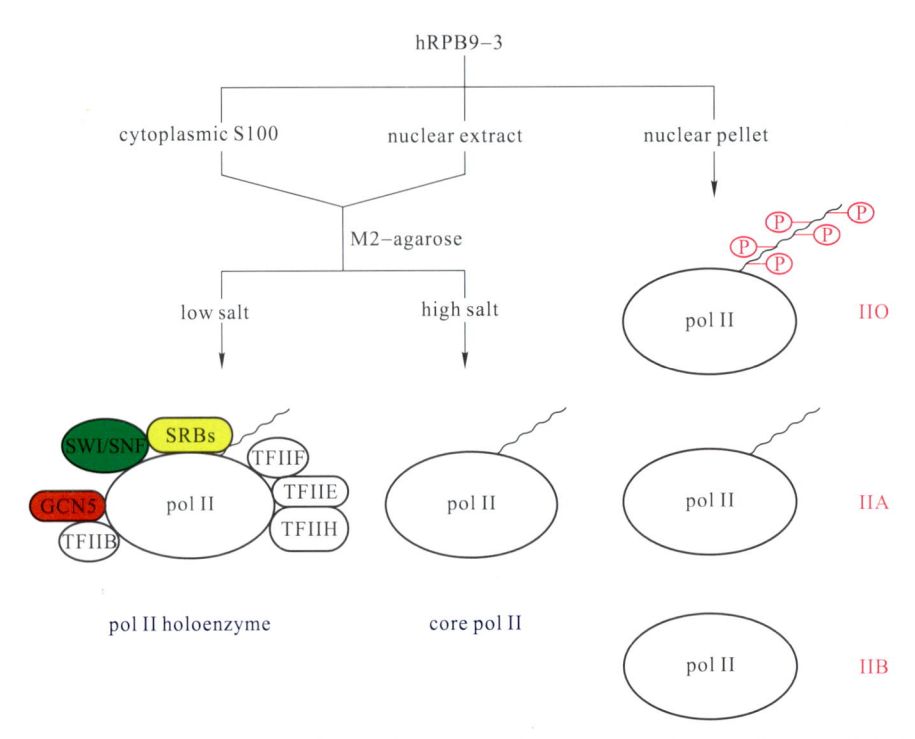

Fig.2.3 Purification of human RNA polymerase II complexes. Four different forms of RNA polymerase II (pol II) are purified from a stable cell line (hRPB9-3) conditionally expressing FLAG epitope-tagged human RPB9 (Wu and Chiang, 2001b). Pol II holoenzyme can be purified from the cytoplasmic S100 or nuclear extract fraction. Similarly, the IIA (*i.e.*, containing hypo-or unphosphorylated CTD) form of core pol II comprised of RPB1 to RPB12 subunits can also be isolated from the same S100 or nuclear extract fraction, but under high salt wash conditions (Kershnar et al., 1998). The IIO (*i.e.*, containing hyper-phosphorylated CTD) and the IIB (*i.e.*, CTD-truncated) forms of pol II can be additionally purified from the nuclear pellet. The tail represents the CTD (carboxy-terminal domain) found in the RPB1 subunit of pol II.

Protein phosphatases that remove the phosphate group on serine 2 or serine 5 have also been identified. Yeast Ssu72 (Krishnamurthy *et al.*, 2004), plant *Arabidopsis thaliana* CTD phosphatase-like proteins AtCPL1 and AtCPL2 (Koiwa *et al.*, 2004), and human small CTD phosphatase 1 (SCP1) protein (Yeo *et al.*, 2003) are able to dephosphorylate serine 5, whereas TFIIF-associated CTD phosphatase 1 (Fcp1) isolated from yeast (Archambault *et al.*, 1997; Kimura *et al.*, 2002) and humans (Archambault *et al.*, 1998; Cho *et al.*, 1999) are mainly implicated in serine 2 dephosphorylation (Cho *et al.*, 2001). Fcp1 interacts with TFIIB (Chambers *et al.*, 1995; Kobor *et al.*, 2000), the RPB4 subunit of pol II (Kimura *et al.*, 2002), and the RAP74 component of TFIIF (Chambers *et al.*, 1995; Kobor *et al.*, 2000). It has been shown that TFIIF can stimulate Fcp1 phosphatase activity and may thus accelerate reinitiation of transcription by enhancing the conversion of pol II from the elongating IIO form back to the initiating IIA form (Chambers *et al.*, 1995). Although TFIIB is able to inhibit TFIIF-stimulated Fcp1 phosphatase activity, the functional role of this inhibition is still unclear (Chambers *et al.*, 1995). Undoubtedly, the counteracting activity between Ctk1-Fcp1 and TFIIH-Ssu72 on CTD phosphorylation must play an important role in initiation, elongation, termination, and transcription-coupled mRNA processing.

TFIIE

A: Subunit Structure and Function

After formation of a TFIID-TFIIB-pol II/TFIIF-promoter complex, the next step in the sequential assembly pathway is the recruitment of TFIIE and TFIIH. TFIIE consists of 2 subunits, α and β, which form an $\alpha_2\beta_2$ heterotetramer (Ohkuma *et al.*, 1991; Sumimoto *et al.*, 1991; Peterson *et al.*, 1991). In humans, TFIIEα is the larger subunit consisting of 439 amino acids with a molecular weight of 56 kDa, while TFIIEβ is 291 amino acids with an approximate molecular weight of 34 kDa. The N-terminal half of TFIIEα is necessary for interactions with TFIIEβ and pol II via non-overlapping regions, for basal transcription, for stimulating TFIIH-mediated CTD phosphorylation, and for the transition from initiation to elongation, while the C-terminal region is involved in TFIIH interaction and appears nonessential in yeast (Ohkuma et al., 1995; Kuldell and Buratowski, 1997; Okuda et al., 2004). In analogy to TFIIF, the archaeal bacteria TFE protein, which shows sequence homology to the N-terminal half of TFIIEα but without the C-terminal domain, also has a winged HTH structure as revealed by x-ray crystallography (Meinhart et al., 2003). The NMR structure of the zinc-finger domain present in the N-terminal half of human TFIIEα different from the archaeal TFE region used for structural analysis, is comprised of one α-helix and five β-strands, which is distinct from conventional zinc finger structures (Okuda et al., 2004). In human TFIIEβ, the C-terminal basic helix loop helix region possesses single-stranded DNA-binding activity (Okamoto et al., 1998). The central core domain (amino acids 66-146) of human TFIIEβ, which contains amino acid residues homologous to the pol II-binding region of RAP30, also has a winged HTH structure (Okuda et al., 2000). Similar to the function of the TFIIEα N-terminal half, the TFIIEβ C-terminus is also implicated in the transition from transcription initiation to elongation by pol II (Watanabe et al., 2003). It remains to be determined whether the TFIIEα N-terminal region and the TFIIEβ C-terminal region function independently or cooperatively in aiding pol II transition from initiation to elongation.

B: TFIIE Function

As seen with other GTFs, TFIIE may be recruited to the promoter through direct interaction with gene-specific transcriptional activators (Sauer et al., 1995; Zhu and Kuziora, 1996). Once recruited, TFIIE interacts directly with both subunits of TFIIF, TFIIB, pol II, promoter DNA, and then recruits TFIIH (Flores et al., 1989; Maxon et al., 1994; Hampsey, 1998; Watanabe et al., 2003; Forget et al., 2004). TFIIE binds to pol II near its active center and to the promoter DNA approximately 10 bp upstream from the transcription initiation site, where promoter melting begins (Douziech et al, 2000; Forget et al., 2004). TFIIE can stimulate the ATPase, CTD kinase, and DNA helicase activities of TFIIH and thus facilitates the formation of an initiation-competent pol II complex (Ohkuma, and Roeder, 1994; Serizawa et al., 1994; Ohkuma et al., 1995; Lee and Young, 2000). Essential for the transition from PIC assembly to transcription initiation are the participation of TFIIE and TFIIH in promoter melting (Holstege et al., 1996). Consistent with TFIIE's role in promoter melting is its ability to bind single-stranded DNA (Kuldell and Buratowski, 1997). Interestingly, requirement of TFIIE and TFIIH also depends on DNA topology and the promoter sequence (Parvin and Sharp, 1993; Goodrich and Tjian, 1994; Wu et al., 1998), consistent with the observation that TFIIE and TFIIH are not necessary for transcription from premelted promoter templates (Pan and Greenblatt, 1994; Holstege et al., 1996).

TFIIH

A: Protein Composition

TFIIH is primarily recruited to the promoter complex through association with TFIIE. Historically, TFIIH is a multiprotein complex consisting of 9 subunits: p89/XPB (gene defective in xeroderma pigmentosum patients complementation group B), p80/XPD (gene defective in xeroderma pigmentosum patients complementation group D), p62, p52, p44, p40/CDK7, p38/Cyclin H, p34, and p32/MAT1. TFIIH has three enzymatic activities required for transcription: DNA-dependent ATPase, ATP-dependent helicase, and CTD kinase (Svejstrup et al., 1996; Hampsey, 1998; Lee and Young, 2000; Zurita and Merino, 2003). In addition to the enzymatic activities essential for transcription, some components of TFIIH (p89/XPD and p80/XPD) are also involved in nucleotide excision repair. Functionally, TFIIH can be separated into two subcomplexes: a cyclin-activating kinase complex (CAK) and a core complex. The CAK complex, responsible for phosphorylating pol II CTD, is consisted of CDK7, Cyclin H, and MAT1. The core complex contains XPB helicase, p62, p52, p44, and p34. CAK and core TFIIH is linked by the XPD helicase, which is essential for DNA repair activity of TFIIH but serves a structural rather than an enzymatic role in transcription (Rossignol et al., 1997; Coin et al., 1999). Mutations in either XPB or XPD lead to several human diseases, including xeroderma pigmentosum (XP), tricothiodystrophy (TTD), and Cockayne syndrome (CS; Lee and Young, 2000; Lehmann, 2001; Zurita and Merino, 2003). Recent studies have now identified a tenth subunit of TFIIH, TFB5, in yeast (Ranish et al., 2004) and humans (Giglia-Mari et al., 2004). Human TFB5 is implicated in the DNA repair syndrome TTD group A disease (Giglia-Mari et al., 2004).

B: DNA-Dependent ATPase Activity

The ATPase activity of TFIIH is required for transcription initiation and promoter clearance. Although an initial report indicated that TFIIH and ATP hydrolysis are only required for pol II promoter clearance, but not for transcription initiation and formation of the first phosphodiester bond in CpA dinucleotide-primed reactions (Goodrich and Tjian, 1994), subsequent studies found that TFIIH ATPase activity is indeed necessary for initial promoter opening and first phosphodiester bond formation when natural nucleotide substrates were used to initiate the transcription reactions (Holstege et al., 1996, 1997). In general, without TFIIH, pol II tends to stall on the promoter-proximal region, leading to abortive transcription products; the addition of TFIIH in the presence of ATP significantly reduces the amount of the promoter-stalled pol II complex, indicating a direct involvement of TFIIH in promoter clearance (Dvir et al., 1997; Kugel and Goodrich, 1998; Kumar et al., 1998).

C: ATP-Dependent Helicase Activity

TFIIH contains two helicases, XPB and XPD which unwind the DNA in a $3' \rightarrow 5'$ and $5' \rightarrow 3'$ direction, respectively, making TFIIH a bidirectional DNA helicase (Schaeffer et al., 1994). The XPB helicase activity is essential for promoter opening in the transcription process (Guzman and Lis, 1999). While XPB $3' \rightarrow 5'$ helicase activity is critical for both DNA repair and transcription, the XPD $5' \rightarrow 3'$ helicase activity is only required for DNA repair (Zurita and Merino, 2003). The ATP-dependent DNA helicase activities of TFIIH are also necessary for opening the promoter region surrounding the transcription start site and maintaining the transcription open complex (Holstege et al., 1997). This requirement for XPB helicase activity in transcription may be bypassed by the use of either supercoiled or premelted templates (Parvin and Sharp, 1993; Pan and Greenblatt, 1994; Parvin et al., 1994; Tantin and Carey, 1994), further supporting a role of TFIIH in open complex formation.

D: TFIIH and Nucleotide Excision Repair

Nucleotide excision repair (NER) is a process where damaged DNA is removed and replaced by newly synthesized DNA based on sequence information from the intact template strand. The finding that p89/XPB is identical to ERCC3, a DNA excision repair protein which is defective in patients with xeroderma pigmentosum, led to the hypothesis that transcription might be coupled to DNA repair (Schaeffer et al., 1993). Consistent with the fact that TFIIH may play a dual role in transcription and DNA repair is the observation that transcriptionally active genes are preferentially repaired (Bohr et al., 1985; Mellon and Hanawalt, 1989). For NER, the combined helicase activities of XPB and XPD seem to be required. Experiments have shown that microinjection of TFIIH into human XPD- or XPB- mutant cells led to complementation of repair-deficient phenotype (van Vuuren et al., 1994). Similarly, yeast cells with a mutation in the yeast homolog for XPD was rescued with the addition of TFIIH and not by the addition of XPD alone, suggesting that NER is only functional in the context of TFIIH (Wang et al., 1994). Subsequent studies have shown multiple components in TFIIH are required for DNA repair, including XPB, XPD, p62, p52, and p44 (Drapkin et al., 1994; Humbert et al., 1994; Schaeffer et al., 1994; Wang et al., 1995; Jawhari et al., 2002). The XPB and XPD helicase functions are required for transcription-coupled NER, as defects in helicase activity are linked to human diseases including XP, TTD, and CS. Recent studies have unraveled the mechanism by which the XPB helicase subunit of TFIIH functions in NER and transcription. Experiments showing phosphorylation of the serine 751 residue of XPB leads to inhibition of NER activity, but does not prevent TFIIH from unwinding DNA (Coin et al., 2004). Instead, phosphorylation of XPB serine 751 prevents the 5' incision triggered by the ERCC1-XPF endonuclease (Coin et al., 2004), providing convincing evidence that a separate but essential role of TFIIH is involved in both transcription and DNA repair.

E: TFIIH and CTD Phosphorylation

CDK7 is the kinase responsible for phosphorylating the serine 5 residue of the pol II CTD, whose activity is regulated by cyclin H, MAT1, TFIIE and Mediator (Svejstrup et al., 1996). The CDK7-cyclin H-MAT1 CAK complex in the context of TFIIH has higher activity in phosphorylating the CTD compared with the free form of CAK (Yankulov and Bentley, 1997). Phosphorylation of serine 5 leads to recruitment of 5' capping enzyme (Cho et al., 1997; Komarnitsky et al., 2000; Rodriguez et al., 2000; Schroeder et al., 2000; Pei et al., 2001) and promoter escape. That CTD phosphorylation regulates the transition from transcription initiation to elongation is supported by observations that pol II enters PIC assembly as the hypophosphorylated IIA form and escapes the promoter as the hyperphosphorylated IIO form (Hampsey, 1998). Besides phosphorylating CTD, TFIIH has also been shown to phosphorylate transcriptional activators, such as p53 (Lu et al., 1997), retinoic acid receptor α (Rochette-Egly et al., 1997), retinoic acid receptor γ (Bastien et al., 2000), Ets-1 (Drané et al., 2004), estrogen receptor α (Chen et al., 2000), and general cofactor PC4 (Kershnar et al., 1998). The CTD kinase activity of TFIIH can also be stimulated via interaction with transcriptional activators (Jones, 1997).

F: TFIIH-Activator Interactions

Many activators have been shown to interact with TFIIH including Gal4-VP16, E2F1, Rb, p53, ERα, RARα, RARγ, and androgen receptor (reviewed by Zurita and Merino, 2003). Consistent with TFIIH's ability to interact with many activators is the finding that TFIIH can function as a coactivator in a reconstituted cell-free transcription system (Wu *et al.*, 1998). Perhaps activators may work by enhancing the recruitment of TFIIH for PIC assembly or stimulating the enzymatic activities of TFIIH (Zurita and Merino, 2003). Conversely, TFIIH may covalently modify amino acid residues critical for activator function. For example, TFIIH has been shown to stimulate the transcriptional activity of the N-terminal activation domain (AF-1) of nuclear receptors RARα 1 and RARγ via phosphorylation of specific serine residues in AF-1 (Rochette-Egly *et al.*, 1997; Bastien *et al.*, 2000). Other than functioning through the kinase activity of CDK7, mutations in the XPD subunit of TFIIH also exhibited impaired phosphorylation of RARα (Keriel *et al.*, 2002), indicating that XPD may modulate CDK7 activity within the TFIIH complex and thereby regulate nuclear receptor phosphorylation and its transactivation activity. This finding further suggests that XPD not only participates in DNA repair but also in the transcriptional process.

PIC Assembly

A: Initiation of PIC Assembly

Formation of the PIC on the promoter is usually the rate-limiting step in transcriptional activation (Lemon and Tjian, 2000). Within eukaryotic genes, there are enhancer regions with clusters of upstream activation sequences, which allow for activator binding to regulate transcription activity. In addition to enhancers, eukaryotic genes may also contain locus control regions (LCRs) consisting of multiple transcription factor-binding sites; but unlike enhancers, which are orientation-independent and distance-independent, LCR functions are limited by position (Grosveld, 1999). Initiation of PIC formation is normally triggered by activator binding to their cognate binding sites and followed by recruitment of transcriptional coactivators or by directly contacting GTFs. A single activator may have multiple contacts with GTFs in order to regulate multiple steps of PIC formation. For example, p53 can interact with multiple GTFs, including TBP, TAFs, TFIIB, and TFIIH (Ko and Prives, 1996). These interactions may stimulate TFIID-TFIIA-promoter complex assembly (Xing *et al.*, 2001) or target different steps of PIC formation in a temporal manner in response to environmental stresses (Espinosa *et al.*, 2003). Conversely, multiple activators, as in an enhanceosome complex, may work in a combinatorial manner to initiate the assembly of a transcription-competent complex (reviewed by Carey, 1998; Merika and Thanos, 2001).

B: Chromatin Barrier to PIC Assembly

An additional level of complexity must be contemplated when considering that competent PIC formation must first overcome the inherently repressive nature of chromatin. TFIID with its ability to covalently modify histones (see TFIID section) may play a critical role in modifying chromatin structure for transcription to occur. Experiments have shown that TFIID, rather than TBP, is essential for activator-dependent transcription on chromatin templates (Wu *et al.*, 1999). In addition to TFIID, pol II holoenzyme, which also contains SWI/SNF and GCN5, is able to initiate transcription from chromatin (Wu *et al.*, 1999), implicating an important role of chromatin-modifying activity in overcoming nucleosome-mediated repression of PIC assembly. With the advent of *in vitro* chromatin assembly, initially using *Xenopus* oocyte extracts (Glikin *et al.*, 1984) and now with a completely defined recombinant chromatin assembly system coupled to transcription analysis (Fyodorov and Kadonaga, 2003; An and Roeder, 2004; Thomas and Chiang, 2005), the answers to many of these fascinating questions involving PIC formation on chromatin templates will be further uncovered in the near future.

C: Stepwise Recruitment vs. Pol II Holoenzyme Pathway

Evidence exists for both models of PIC assembly (Hampsey, 1998; Lemon and Tjian, 2000). The differences in factor recruitment and composition of the general transcription machinery assembled on distinct p53 target genes, in response to DNA-damaging agents (Espinosa *et al.*, 2003), certainly argue for the existence of the sequential assembly pathway *in vivo*. The fact that diverse transcription complexes can be detected and isolated *in vivo* and *in vitro* further supports this model. The advantage for the stepwise assembly pathway is to selectively fine-tune individual steps for different signaling events without globally inactivating the cascade leading to gene activation. It also provides an efficient way to reactivate the pathway by simply modulating the rate-limiting step. However, a *de novo* assembly of functional transcription complexes may require a significant time, considering more than 40 polypeptides must be assembled correctly in a limited time frame to respond to cellular demands. Conversely, pol II holoenzymes are preformed prior to the initiation of transcription and are thus able to respond more rapidly to the transcriptional need of the cell. A major disadvantage of the holoenzyme pathway is that a distinct set of preassembled complexes must exist for different types of transcriptional events, which is economically unfavorable for cells to generate many complexes differing in peripheral components. However, multiple pol II holoenzyme complexes with different protein compositions involved in various biological processes have indeed been isolated (Lee and Young, 2000; see also the holoenzyme section). Thus, how transcription complexes are assembled *in vivo* on distinct activator-targeted genes and its functional implications remain to be investigated both on an individual and on a genome-wide basis.

Conclusion

Tremendous progress has been achieved concerning the structure and function of GTFs and pol II. TFIID is recognized as the key promoter recognition factor, while TFIIA and TFIIB stabilize TFIID binding to DNA. In addition to core promoter recognition, TFIID also functions as a coactivator in mediating interactions between activators and GTFs. It is an enzyme with multiple activities, including kinase, acetylase, and ubiquitin activating/conjugating activities. These multiple activities of TFIID on histones may aid in PIC assembly on chromatin templates. Following the assembly of TFIID-TFIIA-TFIIB on the promoter region, pol II/TFIIF, TFIIE, and TFIIH join the complex to form a PIC. The enzymatic activities inherent to TFIIH then facilitate promoter melting and the transition from initiation to elongation following initial phosphodiester bond formation. Although PIC assembly was defined biochemically in a stepwise fashion *in vitro*, it remains unclear whether the PIC is indeed assembled in this manner *in vivo*, or as a preexisting pol II holoenzyme complex, or via other undefined pathways. Regardless of which pathway operates *in vivo*, a common theme in gene regulation is activator-mediated recruitment of GTFs and pol II to upregulate PIC assembly. Activators may target individual GTFs, such as TFIID (Burley and Roeder, 1996; Näär *et al.*, 2001; Wu and Chiang, 2001a), TFIIB (Roberts *et al.*, 1993; Colgan *et al.*, 1993; Sauer *et al.*, 1995), TFIIE (Martin *et al.*, 1996), TFIIF (Joliot *et al.*, 1995; Martin *et al.*, 1996; McEwan and Gustafsson, 1997; Reid *et al.*, 2002),

TFIIH (Zurita and Merino, 2003), TFIIA (Ozer *et al.*, 1994, 1998a; Kobayashi *et al.*, 1995; Lieberman *et al.*, 1997), and pol II (Cheong *et al.*, 1995; Wu and Chiang, 2001a). Alternatively, activators may recruit pol II holoenzyme to the promoter via interaction with Mediator, the key component in pol II holoenzyme originally identified in yeast (Näär *et al.*, 2001; Wu *et al.*, 2003). Although it is generally agreed that Mediator may function as a coactivator in mediating transcriptional activation by facilitating pol II entry to the TFIID-TFIIB-promoter complex (Wu *et al.*, 2003), whether it exists as a preassembled pol II holoenzyme complex in a cellular environment or is recruited separately to pol II by transcriptional activators remains to be further investigated.

Acknowledgment

We are grateful for Drs. Parminder Kaur, Mary C. Thomas and Shwu-Yuan Wu for their comments on this chapter. The research conducted in Dr. Chiang's laboratory is currently sponsored by grants CA103867 and GM59643 from the National Institutes of Health in the United States.

References

Ahn, S. H., Kim, M., and Buratowski, S. (2004). Phosphorylation of serine 2 within the RNA polymerase II C-terminal domain couples transcription and 3′ end processing. Mol. Cell *13*, 67-76.

Albright, S. R., and Tjian, R. (2000). TAFs revisited: more data reveal new twists and confirm old ideas. Gene *242*, 1-13.

An, W., and Roeder, R. G. (2004). Reconstitution and transcriptional analysis of chromatin *in vitro*. Methods Enzymol. *377*, 460-474.

Andel III, F., Ladurner, A. G., Inouye, C., Tjian, R., and Nogales, E. (1999). Three-dimensional structure of the human TFIID-IIA-IIB complex. Science *286*, 2153-2156.

Apone, L. M., Virbasius, C. A., Holstege, F. C. P., Wang, J., Young, R. A., and Green, M. R. (1998). Broad, but not universal, transcriptional requirement for yTAF$_{II}$17, a histone H3-like TAF$_{II}$ present in TFIID and SAGA. Mol. Cell *2*, 653-661.

Archambault, J., Chambers, R. S., Kobor, M. S., Ho, Y., Cartier, M., Bolotin, D., Andrews, B., Kane, C. M., and Greenblatt, J. (1997). An essential component of a C-terminal domain phosphatase that interacts with transcription factor IIF in *Saccharomyces cerevisiae*. Proc. Natl. Acad. Sci. USA *94*, 14300-14305.

Archambault, J., Pan, G., Dahmus, G. K., Cartier, M., Marshall, N., Zhang, S., Dahmus, M. E., and Greenblatt, J. (1998). FCP1, the RAP74-interacting subunit of a human protein phosphatase that dephosphorylates the carboxyl-terminal domain of RNA polymerase IIO. J. Biol. Chem. *273*, 27593-27601.

Armache, K.-J., Kettenberger, H., and Cramer, P. (2003). Architecture of initiation-competent 12-subunit RNA polymerase II. Proc. Natl. Acad. Sci. USA *100*, 6964-6968.

Armache, K.-J., Mitterweger, S., Meinhart, A., and Cramer, P. (2005). Structures of complete RNA polymerase II and its subcomplex, Rpb4/7. J. Biol. Chem. *280*, 7131-7134.

Aso, T., Vasavada, H. A., Kawaguchi, T., Germino, F. J., Ganguly, S., Kitajima, S., Weissman, S. M., and Yasukochi, Y. (1992). Characterization of cDNA for the large subunit of the transcription initiation factor TFIIF. Nature *355*, 461-464.

Asturias, F. J. (2004). RNA polymerase II structure, and organization of the preinitiation complex. Curr. Opin. Struct. Biol. *14*, 121-129.

Auble, D. T., Wang, D., Post, K. W., and Hahn, S. (1997). Molecular analysis of the SNF2/SWI2 protein family member MOT1, an ATP-driven enzyme that dissociates TATA-binding protein from DNA. Mol. Cell. Biol. *17*, 4842–4851.

Auty, R., Steen, H., Myers, L. C., Persinger, J., Bartholomew, B., Gygi, S. P., and Buratowski, S. (2004). Purification of active TFIID from *Saccharomyces cerevisiae*. Extensive promoter contacts and co-activator function. J. Biol. Chem. *279*, 49973-49981.

Bagby, S., Kim, S., Maldonado, E., Tong, K. I., Reinberg, D., and Ikura, M. (1995). Solution structure of the C-terminal core domain of human TFIIB: similarity to cyclin A and interaction with TATA-binding protein. Cell *82*, 857-867.

Bangur, C. S., Pardee, T. S., and Ponticelli, A. S. (1997). Mutational analysis of the D1/E1 core helices and the conserved N-terminal region of yeast transcription factor IIB (TFIIB): identification of an N-terminal mutant that stabilizes TATA-binding protein-TFIIB-DNA complexes. Mol. Cell. Biol. *17*, 6784-6793.

Bangur, C. S., Faitar, S. L., Folster, J. P., and Ponticelli, A. S. (1999). An interaction between the N-terminal region and the core domain of yeast TFIIB promotes the formation of TATA-binding protein-TFIIB-DNA complexes. J. Biol. Chem. *274*, 23203-23209.

Barberis, A., Müller, C. W., Harrison, S. C., and Ptashne, M. (1993). Delineation of two functional regions of transcription factor TFIIB. Proc. Natl. Acad. Sci. USA *90*, 5628-5632.

Baskaran, R., Dahmus, M. E., and Wang, J. Y. J. (1993). Tyrosine phosphorylation of mammalian RNA polymerase II carboxyl-terminal domain. Proc. Natl. Acad. Sci. USA *90*, 11167-11171.

Bastien, J., Adam-Stitah, S., Riedl, T., Egly, J.-M., Chambon, P., and Rochette-Egly, C. (2000). TFIIH interacts with the retinoic acid receptor γ and phosphorylates its AF-1-activating domain through cdk7. J. Biol. Chem. *275*, 21896-21904.

Berroteran, R. W., Ware, D. E., and Hampsey, M. (1994). The sua8 suppressors of *Saccharomyces cerevisiae* encode replacements of conserved residues within the largest subunit of RNA polymerase II and affect transcription start site selection similarly to sua7 (TFIIB) mutations. Mol. Cell. Biol. *14*, 226-237.

Boeger, H., Bushnell, D. A., Davis, R., Griesenbeck, J., Lorch, Y., Strattan, J. S., Westover, K. D., and Kornberg, R. D. (2005). Structural basis of eukaryotic gene transcription. FEBS Lett. *579*, 899-903.

Bohr, V. A., Smith, C. A., Okumoto, D. S., and Hanawalt, P. C. (1985). DNA repair in an active gene: removal of pyrimidine dimers from the DHFR gene of CHO cells is much more efficient than in the genome overall. Cell *40*, 359-369.

Brand, M., Leurent, C., Mallouh, V., Tora, L., and Schultz, P. (1999). Three-dimensional structures of the TAF_{II}-containing complexes TFIID and TFTC. Science *286*, 2151-2153.

Buratowski, S., Hahn, S., Sharp, P. A., and Guarente, L. (1988). Function of a yeast TATA element-binding protein in a mammalian transcription system. Nature *334*, 37-42.

Buratowski, S., Hahn, S., Guarente, L., and Sharp, P. A. (1989). Five intermediate complexes in transcription initiation by RNA polymerase II. Cell *56*, 549-561.

Buratowski, S., and Zhou, H. (1993). Functional domains of transcription factor TFIIB. Proc. Natl. Acad. Sci. USA. *90*, 5633-5637.

Burke, T. W., and Kadonaga, J. T. (1997). The downstream core promoter element, DPE, is conserved from *Drosophila* to humans and is recognized by $TAF_{II}60$ of *Drosophila*. Genes Dev. *11*, 3020-3031.

Burley, S. K., and Roeder, R. G. (1996). Biochemistry and structural biology of transcription factor IID (TFIID). Annu. Rev. Biochem. *65*, 769-799.

Bushnell, D. A., and Kornberg, R. D. (2003). Complete, 12-subunit RNA polymerase II at 4.1-A resolution: implications for the initiation of transcription. Proc. Natl. Acad. Sci. USA *100*, 6969-6973.

Bushnell, D. A., Westover, K. D., Davis, R. E., and Kornberg, R. D. (2004). Structural basis of transcription: an RNA polymerase II-TFIIB cocrystal at 4.5 Angstroms. Science *303*, 983-988.

Carey, M. (1998). The enhanceosome and transcriptional synergy. Cell *92*, 5-8.

Carles, C., Treich, I., Bouet, F., Riva, M., and Sentenac, A. (1991). Two additional common subunits, ABC10α and ABC10β, are shared by yeast RNA polymerases. J. Biol. Chem. *266*, 24092-24096.

Cavallini, B., Huet, J., Plassat, J.-L., Sentenac, A., Egly, J.-M., and Chambon, P. (1988). A yeast activity can substitute for the HeLa cell TATA box factor. Nature *334*, 77-80.

Chamberlin, M., and Berg, P. (1962). Deoxyribonucleic acid-directed synthesis of ribonucleic acid by an enzyme from *Escherichia coli*. Proc. Natl. Acad. Sci. USA *48*, 81-94.

Chambers, R. S., Wang, B. Q., Burton, Z. F., and Dahmus, M. E. (1995). The activity of COOH-terminal domain phosphatase is regulated by a docking site on RNA polymerase II and by the general transcription factors IIF and IIB. J. Biol. Chem. *270*, 14962-14969.

Chao, D. M., Gadbois, E. L., Murray, P. J., Anderson, S. F., Sonu, M. S., Parvin, J. D., and Young, R. A. (1996). A mammalian SRB protein associated with an RNA polymerase II holoenzyme. Nature *380*, 82-85.

Chen, D., Riedl, T., Washbrook, E., Pace, P. E., Coombes, R. C., Egly, J.-M., and Ali, S. (2000). Activation of estrogen receptorα by S118 phosphorylation involves a ligand-dependent interaction with TFIIH and participation of CDK7. Mol. Cell *6*, 127-137.

Chen, H.-T., and Hahn, S. (2003). Binding of TFIIB to RNA polymerase II: mapping the binding site for the TFIIB zinc ribbon domain within the preinitiation complex. Mol. Cell *12*, 437-447.

Chen, H.-T., and Hahn, S. (2004). Mapping the location of TFIIB within the RNA polymerase II transcription preinitiation complex: a model for the structure of the PIC. Cell *119*, 169-180.

Cheong, J., Yi, M., Lin, Y., and Murakami, S. (1995). Human RPB5, a subunit shared by eukaryotic nuclear RNA polymerase, binds human hepatitis B virus X protein and may play a role in X transactivation. EMBO J. *14*, 143-150.

Chi, T., Lieberman, P., Ellwood, K., and Carey, M. (1995). A general mechanism for transcriptional synergy by eukaryotic activators. Nature *377*, 254–257.

Chiang, C.-M., Ge, H., Wang, Z., Hoffmann, A., and Roeder, R. G. (1993). Unique TATA-binding protein-containing complexes and cofactors involved in transcription by RNA polymerases II and III. EMBO J. *12*, 2749-2762.

Chiang, C.-M., and Roeder, R. G. (1995). Cloning of an intrinsic human TFIID subunit that interacts with multiple transcriptional activators. Science *267*, 531-536.

Cho, E.-J., Takagi, T., Moore, C. R., and Buratowski, S. (1997). mRNA capping enzyme is recruited to the transcription complex by phosphorylation of the RNA polymerase II carboxy-terminal domain. Genes Dev. *11*, 3319-3326.

Cho, E.-J., Kobor, M. S., Kim, M., Greenblatt, J., and Buratowski, S. (2001). Opposing effects of Ctk1 kinase and Fcp1 phosphatase at Ser 2 of the RNA polymerase II C-terminal domain. Genes Dev. *15*, 3319-3329.

Cho, H., Orphanides, G., Sun, X., Yang, X.-J., Ogryzko, V., Lees, E., Nakatani, Y., and Reinberg, D. (1998). A human RNA polymerase II complex containing factors that modify chromatin structure. Mol. Cell. Biol. *18*, 5355-5363.

Cho, H., Kim, T.-K., Mancebo, H., Lane, W. S., Flores, O., and Reinberg, D. (1999). A protein phosphatase function to recycle RNA polymerase II. Genes Dev. *13*, 1540-1552.

Chung, W.-H., Craighead, J. L., Chang, W.-H., Ezeokonkwo, C., Bareket-Sarnish, A., Kornberg, R. D., and Asturias, F. J. (2003). RNA polymerase II/TFIIF structure and conserved organization of the initiation complex. Mol. Cell *12*, 1003-1013.

Clemens, K. E., Graziella, P., Radonovich, M. F., Choi, K. S., Duvall, J. F., Dejong, J., Roeder, R., and Brady, J. N. (1996). Interaction of the human T-cell lymphotropic virus type 1 Tax transactivator with transcription factor IIA. Mol. Cell. Biol. *16*, 4656-4664.

Coin, F., Bergmann, E., Tremeau-Bravard, A., and Egly, J. M. (1999). Mutations in XPB and XPD helicases found in xeroderma pigmentosum patients impair the transcription function of TFIIH. EMBO J. *18*, 1357-1366.

Coin, F., Auriol, J., Tapias, A., Clivio, P., Vermeulen, W., and Egly, J.-M. (2004). Phosphorylation of XPB helicase regulates TFIIH nucleotide excision repair activity. EMBO J. *23*, 4835-4846.

Coleman, R. A., Taggart, A. K. P., Burma, S., Chicca II, J. J., and Pugh, B. F. (1999). TFIIA regulates TBP and TFIID dimers. Mol. Cell *4*, 451-457.

Colgan, J., Wampler, S., and Manley, J. L. (1993). Interaction between a transcriptional activator and transcription factor IIB *in vivo*. Nature *362*, 549-553.

Coulombe, B., Li, J., and Greenblatt, J. (1994). Topological localization of the human transcription factors IIA, IIB, TATA box-binding protein, and RNA polymerase II-associated protein 30 on a class II promoter. J. Biol. Chem. *269*, 19962–19967.

Cramer, P., Bushnell, D. A., Fu, J., Gnatt, A. L., Maier-Davis, B., Thompson, N. E., Burgess, R. R., Edwards, A. M., David, P. R., and Kornberg, R. D. (2000). Architecture of RNA polymerase II and implications for the transcription mechanism. Science *288*, 640-649.

Cramer, P., Bushnell, D. A., and Kornberg, R. D. (2001). Structural basis of transcription: RNA polymerase II at 2.8 angstrom resolution. Science *292*, 1863-1876.

Cramer, P. (2004). RNA polymerase II structure: from core to functional complexes. Curr. Opin. Genet. Dev. *14*, 218-226.

Crick, F. H. C. (1958). Symp. Soc. Exp. Biol. The Biological Replication of Macromolecules, XII, 138.

Crick, F. H. C. (1970). Central dogma of molecular biology. Nature *227*, 561-563.

Davis, J. A., Takagi, Y., Kornberg, R. D., and Asturias, F. A. (2002). Structure of the yeast RNA polymerase II holoenzyme: mediator conformation and polymerase interaction. Mol. Cell *10*, 409-415.

DeJong, J., and Roeder, R. G. (1993). A single cDNA, hTFIIAα, encodes both the p35 and p19 subunits of human TFIIA. Genes Dev. *7*, 2220–2234.

Dikstein, R., Ruppert, S., and Tjian, R. (1996a). $TAF_{II}250$ is a bipartite protein kinase that phosphorylates the base transcription factor RAP74. Cell *84*, 781-790.

Dikstein, R., Zhou, S., and Tjian, R. (1996b). Human $TAF_{II}105$ is a cell type-specific TFIID subunit related to $hTAF_{II}130$. Cell *87*, 137-146.

Dion, V., and Coulombe, B. (2003). Interaction of a DNA-bound transcriptional activator with TBP-TFIIA-TFIIB-Promoter quaternary complex. J. Biol. Chem. *278*, 11495-11501.

Douziech, M., Coin, F., Chipoulet, J.-M., Arai, Y., Ohkuma, Y., Egly, J.-M., and Coulombe, B. (2000). Mechanism of promoter melting by the xeroderma pigmentosum complementation group B helicase of transcription factor IIH revealed by protein-DNA photo-cross-linking. Mol. Cell. Biol. *20*, 8168-8177.

Drané, P., Compe, E., Catez, P., Chymkowitch, P., and Egly, J.-M. (2004). Selective regulation of vitamin D receptor-responsive genes by TFIIH. Mol. Cell *16*, 187-197.

Drapkin, R., Reardon, J. T., Ansari, A., Huang, J. C., Zawel, L., Ahn, K., Sancar, A., and Reinberg, D. (1994). Dual role of TFIIH in DNA excision repair and in transcription by RNA polymerase II. Nature *368*, 769-772.

Dvir, A., Conaway, R. C., and Conaway, J. W. (1997). A role for TFIIH in controlling the activity of early RNA polymerase II elongation complexes. Proc. Natl. Acad. Sci. USA *94*, 9006-9010.

Dynlacht, B. D., Hoey, T., and Tjian, R. (1991). Isolation of coactivators associated with the TATA-binding protein that mediate transcriptional activation. Cell *66*, 563-576.

Elsby, L. M., and Roberts, S. G. E. (2004). The role of TFIIB conformation in transcriptional regulation. Biochem. Soc. Trans. *32*, 1098-1099.

Espinosa, J. M., Verdun, R. E., and Emerson, B. M. (2003). p53 functions through stress- and promoter-specific recruitment of transcription initiation components before and after DNA damage. Mol. Cell *12*, 1015-1027.

Evans, R., Failey, J. A., and Roberts, S. G. E. (2001). Activator-mediated disruption of sequence-specific DNA contacts by the general transcription factor TFIIB. Genes Dev. *15*, 2945-2949.

Fairley, J. A., Evans, R., Hawkes, N. A., and Roberts, S. G. E. (2002). Core promoter-dependent TFIIB conformation and a role for TFIIB conformation in transcriptional start site selection. Mol. Cell. Biol. *22*, 6697-6705.

Faitar, S. L., Brodie, S. A., and Ponticelli, A. S. (2001). Promoter-specific shifts in transcription initiation conferred by yeast TFIIB mutations are determined by the sequence in the immediate vicinity of the start sites. Mol. Cell. Biol. *21*, 4427-4440.

Fang, S. M., and Burton, Z. F. (1996). RNA polymerase II-associated protein (RAP) 74 binds transcription factor (TF) IIB and blocks TFIIB-RAP30 binding. J. Biol. Chem. *271*, 11703-11709.

Finkelstein, A., Kostrub, C. F., Li, J., Chavez, D. P., Wang, B. Q., Fang, S. M., Greenblatt, J., and Burton, Z. F. (1992). A cDNA encoding RAP74, a general initiation factor for transcription by RNA polymerase II. Nature *355*, 464-467.

Flores, O., Maldonado, E., Burton, Z., Greenblatt, J., and Reinberg, D. (1988). Factors involved in specific transcription by mammalian RNA polymerase II. RNA polymerase II-associating protein 30 is an essential component of transcription factor IIF. J. Biol. Chem. *263*, 10812-10816.

Flores, O., Maldonado, E., and Reinberg, D. (1989). Factors involved in specific transcription by mammalian RNA polymerase II. Factors IIE and IIF independently interact with RNA polymerase II. J. Biol. Chem. *264*, 8913-8921.

Flores, O., Ha, I., and Reinberg, D. (1990). Factors involved in specific transcription by mammalian RNA polymerase II. Purification and subunit composition of transcription factor IIF. J. Biol. Chem. *265*, 5629-5634.

Flores, O., Lu, H., Killeen, M., Greenblatt, J., Burton, Z. F., and Reinberg, D. (1991). The small subunit of transcription factor IIF recruits RNA polymerase II into the preinitiation complex. Proc. Natl. Acad. Sci. USA *88*, 9999-10003.

Flores, O., Lu, H., and Reinberg, D. (1992). Factors involved in specific transcription by mammalian RNA polymerase II. Identification and characterization of factor IIH. J. Biol. Chem. *267*, 2786-2793.

Fondell, J. D., Guermah, M., Malik, S., and Roeder, R. G. (1999). Thyroid hormone receptor-associated proteins and general positive cofactors mediate thyroid hormone receptor function in the absence of TATA box-binding protein-associated factors of TFIID. Proc. Natl. Acad. Sci. USA *96*, 1959-1964.

Forget, D., Langelier, M.-F., Thérien, C., Trinh, V., and Coulombe, B. (2004). Photo-cross-linking of a purified preinitiation complex reveals central roles for the RNA polymerase II mobile clamp and TFIIE in initiation mechanisms. Mol. Cell. Biol. *24*, 1122-1131.

Frank, D. J., Tyree, C. M., George, C. P., and Kadonaga, J. T. (1995). Structure and function of the small subunit of TFIIF (RAP30) from *Drosophila melanogaster*. J. Biol. Chem. *270*, 6292-6297.

Funk, J. D., Nedialkov, Y. A., Xu, D., and Burton, Z.F. (2002). A key role for the α1 helix of human RAP74 in the initiation and elongation of RNA chains. J. Biol. Chem. *277*, 46998-47003.

Fyodorov, D. V., and Kadonaga, J. T. (2003). Chromatin assembly *in vitro* with purified recombinant ACF and NAP-1. Methods Enzymol. *371*, 499-515.

Gaiser, F., Tan, S., and Richmond, T. J. (2000). Novel dimerization fold of RAP30/RAP74 in human TFIIF at 1.7 Å resolution. J. Mol. Biol. *302*, 1119-1127.

Gangloff, Y.-G., Romier, C., Thuault, S., Werten, S., and Davidson, I. (2001). The histone fold is a key structural motif of transcription factor TFIID. Trends Biochem. Sci. *26*, 250-257.

Garrett, K. P., Serizawa, H., Hanley, J. P., Bradsher, J. N., Tsuboi, A., Arai, N., Yokota, T., Ariai, K., Conaway, R. C., and Conaway, J. W. (1992). The carboxyl terminus of RAP30 is similar in sequence to region 4 of bacterial sigma factors and is require for function. J. Biol. Chem. *267*, 23942-23949.

Ge, H., and Roeder, R. G. (1994). The high mobility group protein HMG1 can reversibly inhibit class II gene transcription by interaction with the TATA-binding protein. J. Biol. Chem. *269*, 17136–17140.

Ge, H., Martinez, E., Chiang, C.-M., and Roeder, R. G. (1996).

Activator-dependent transcription by mammalian RNA polymerase II: *in vitro* reconstitution with general transcription factors and cofactors. Methods Enzymol. *274*, 57-71.

Geiger, J. H., Hahn, S., Lee, S., and Sigler, P. B. (1996). Crystal structure of the yeast TFIIA/TBP/DNA complex. Science *272*, 830–836.

Ghazy, M. A., Brodie, S. A., Ammerman, M. L., Ziegler, L. M., and Ponticelli, A. S. (2004). Amino acid substitutions in yeast TFIIF confer upstream shifts in transcription initiation and altered interaction with RNA polymerase II. Mol. Cell. Biol. *24*, 10975-10985.

Giglia-Mari, G., Coin, F., Ranish, J. A., Hoogstraten, D., Theil, A., Wijgers, N., Jaspers, N. G. J., Raams, A., Argentini, M., van der Spek, P. J., Botta, E., Stefanini, M., Egly, J.-M., Aebersold, R., Hoeijmakers, J. H. J., and Vermeulen, W. (2004). A new, tenth subunit of TFIIH is responsible for the DNA repair syndrome trichothiodystrophy group A. Nat. Genet. *36*, 714-719.

Glikin, G. C., Ruberti, I., and Worcel, A. (1984). Chromatin assembly in *Xenopus* oocytes: *in vitro* studies. Cell *37*, 33-41.

Glossop, J. A., Dafforn, T. R., and Roberts, S. G. E. (2004). A conformational change in TFIIB is required for activator-mediated assembly of the preinitiation complex. Nucleic Acids Res. *32*, 1829-1835.

Gnatt, A. L., Cramer, P., Fu, J., Bushnell, D. A., and Kornberg, R. D. (2001). Structural basis of transcription: an RNA polymerase II elongation complex at 3.3 Å resolution. Science *292*, 1876-1882.

Gong, D.-W., Mortin, M. A., Horikoshi, M., and Nakatani, Y. (1995). Molecular cloning of cDNA encoding the small subunit of *Drosophila* transcription initiation factor TFIIF. Nucleic Acids Res. *23*, 1882-1886.

Goodrich, J. A., and Tjian, R. (1994). Transcription factors IIE and IIH and ATP hydrolysis direct promoter clearance by RNA polymerase II. Cell *77*, 145-156.

Grant, P. A., Schieltz, D., Pray-Grant, M. G., Steger, D. J., Reese, J. C., Yates III, J. R., and Workman, J. L. (1998). A subset of $TAF_{II}s$ are integral components of the SAGA complex required for nucleosome acetylation and transcriptional stimulation. Cell *94*, 45-53.

Green, M. R. (2000). TBP-associated factors ($TAF_{II}s$): multiple, selective transcriptional mediators in common complexes. Trends Biochem. Sci. *25*, 59-63.

Groft, C. M., Uljon, S. N., Wang, R., and Werner, M. H. (1998). Structural homogy between the Rap30 DNA-binding domain and linker histone H5: implications for preinitiation complex assembly. Proc. Natl. Acad. Sci. USA *95*, 9117-9122.

Grosveld, F. (1999). Activation by locus control regions? Curr. Opin. Genet. Dev. *9*, 152-157.

Guzman, E., and Lis, J. T. (1999). Transcription factor TFIIH is required for promoter melting *in vivo*. Mol. Cell. Biol. *19*, 5652-5658.

Ha, I., Lane, W. S., and Reinberg, D. (1991). Cloning of a human gene encoding the general transcription initiation factor IIB. Nature *352*, 689-695.

Ha, I., Roberts, S., Maldonado, E., Sun, X., Kim, L. U., Green, M., and Reinberg, D. (1993). Multiple functional domains of human transcription factor IIB: distinct interactions with two general transcription factors and RNA polymerase II. Genes Dev. *7*, 1021-1032.

Hahn, S., Buratowski, S., Sharp, P. A., and Guarente, L. (1989). Isolation of the gene encoding the yeast TATA binding protein TFIID: a gene identical to the SPT15 suppressor of Ty element insertions. Cell *58*, 1173-1181.

Hahn, S. (1998). The role of TAFs in RNA polymerase II transcription. Cell *95*, 579-582.

Hahn, S. (2004). Structure and mechanism of the RNA polymerase II transcription machinery. Nat. Struct. Mol. Biol. *11*, 394-403.

Hai, T., Horikoshi, M., Roeder, R. G., and Green, M. R. (1988). Analysis of the role of the transcription factor ATF in the assembly of a functional preinitiation complex. Cell *54*, 1043-1051.

Hampsey, M. (1998). Molecular genetics of the RNA polymerase II general transcriptional machinery. Microbiol. Mol. Biol. Rev. *62*, 465–503.

Han, S., Xie, W., Hammes, S. R., and DeJong, J. (2003). Expression of the germ cell-specific transcription factor ALF in *Xenopus* oocytes compensates for translational inactivation of the somatic factor TFIIA. J. Biol. Chem. *278*, 45586-45593.

Hansen, S. K., and Tjian, R. (1995). TAFs and TFIIA mediate differential utilization of the tandem Adh promoters. Cell *82*, 565–575.

Hawkes, N. A., and Roberts, S. G. E. (1999). The role of human TFIIB in transcription start site selection *in vitro* and *in vivo*. J. Biol. Chem. *274*, 14337-14343.

Hawkes, N. A., Evans, R., and Roberts S. G. E. (2000). The conformation of the transcription factor TFIIB modulates the response to transcriptional activators *in vivo*. Curr. Biol. *10*, 273-276.

Hengartner, C. J., Myer, V. E., Liao, S.-M., Wilson C. J., Koh, S. S., and Young, R. A. (1998). Temporal regulation of RNA polymerase II by Srb10 and Kin28 cyclin-dependent kinases. Mol. Cell *2*, 43-53.

Henry, N. L., Campbell, A. M., Feaver, W. J., Poon, D., Weil, P. A., and Kornberg, R. D. (1994). TFIIF-TAF-RNA polymerase II connection. Genes Dev. *8*, 2868-2878.

Hoey, T., Dynlacht, B. D., Peterson, M. G., Pugh, B. F., and Tjian, R. (1990). Isolation and characterization of the *Drosophila* gene encoding the TATA box binding protein, TFIID. Cell *61*, 1179-1186.

Hoey, T., Weinzierl, R. O., Gill, G., Chen, J.-L., Dynlacht, B. D., and Tjian, R. (1993). Molecular cloning and functional analysis of *Drosophila* TAF110 reveal properties expected of coactivators. Cell *72*, 247-260.

Hoffmann, A., Chiang, C.-M., Oelgeschläger, T., Xie, X., Burley,

S. K., Nakatani, Y., and Roeder, R. G. (1996). A histone octamer-like structure within TFIID. Nature *380*, 356-359.

Høiby, T., Mitsiou, D. J., Zhou, H., Erdjument-Bromage, H., Tempst, P., and Stunnenberg, H. G. (2004). Cleavage and proteasome-mediated degradation of the basal transcription factor TFIIA. EMBO J. *23*, 3083-3091.

Holstege, F. C. P., van der Vliet, P. C., and Timmers H. T. M. (1996). Opening of an RNA polymerase II promoter occurs in two distinct steps and requires the basal transcription factors IIE and IIH. EMBO J. *15*, 1666-1677.

Holstege, F. C. P., Fiedler, U., and Timmers, H. T. M. (1997). Three transitions in the RNA polymerase II transcription complex during initiation. EMBO J. *16*, 7468-7480.

Holstege, F. C. P., Jennings, E. G., Wyrick, J. J., Lee, T. I., Hengartner, C. J., Green, M. R., Golub, T. R., Lander, E. S., and Young, R. A. (1998). Dissecting the regulatory circuitry of a eukaryotic genome. Cell *95*, 717-728.

Horikoshi, M., Hai, T., Lin, Y.-S., Green, M. R., and Roeder, R. G. (1988). Transcription factor ATF interacts with the TATA factor to facilitate establishment of a preinitiation complex. Cell *54*, 1033-1042.

Horikoshi, M., Wang, C. K., Fujii, H., Cromlish, J. A., Weil, P. A., and Roeder, R. G. (1989a). Cloning and structure of a yeast gene encoding a general transcription initiation factor TFIID that binds to the TATA box. Nature *341*, 299-303.

Horikoshi, M., Wang, C. K., Fujii, H., Cromlish, J. A., Weil, P. A., and Roeder, R. G. (1989b). Purification of a yeast TATA box-binding protein that exhibits human transcription factor IID activity. Proc. Natl. Acad. Sci. USA *86*, 4843-4847.

Hou, S. Y., Wu, S.-Y., Zhou, T., Thomas, M. C., and Chiang, C.-M. (2000). Alleviation of human papillomavirus E2-mediated transcriptional repression via formation of a TATA binding protein (or TFIID)-TFIIB-RNA polymerase II-TFIIF preinitiation complex. Mol. Cell. Biol. *20*, 113-125.

Humbert, S., van Vuuren, H., Lutz, Y., Hoeijmakers, J. H. J., Egly, J.-M., and Moncollin, V. (1994). p44 and p34 subunits of the Btf2/TFIIH transcription factor have homologies with Ssl1, a yeast protein involved in DNA repair. EMBO J. *13*, 2393-2398.

Imbalzano, A. N., Zaret, K. S., and Kingston, R. E. (1994). Transcription factor (TF) IIB and TFIIA can independently increase the affinity of the TATA-binding protein for DNA. J. Biol. Chem. *269*, 8280–8286.

Inostroza, J. A., Mermelstein, F. H., Ha, I., Lane, W. S., and Reinberg, D. (1992). Dr1, a TATA-binding protein-associated phosphoprotein and inhibitor of class II gene transcription. Cell *70*, 477–489.

Jawhari, A., Lainé, J.-P., Dubaele, S., Lamour, V., Poterszman, A., Coin, F., Moras, D., and Egly, J.-M. (2002). p52 mediates XPB function within the transcription/repair factor TFIIH. J. Biol. Chem. *277*, 31761-31767.

Joliot, V., Demma, M., and Prywes, R. (1995). Interaction with RAP74 subunit of TFIIF is required for transcriptional activation by serum response factor. Nature *373*, 632-635.

Jones, K. A. (1997). Taking a new TAK on tat transactivation. Genes Dev. *11*, 2593-2599.

Kamada, K., De Angelis, J., Roeder, R. G., and Burley, S. K. (2001a). Crystal structure of the C-terminal domain of the RAP74 subunit of human transcription factor IIF. Proc. Natl. Acad. Sci. USA. *98*, 3115-3120.

Kamada, K., Shu, F., Chen, H., Malik, S., Stelzer, G., Roeder, R. G., Meisterernst, M., and Burley, S, K. (2001b). Crystal structure of negative cofactor 2 recognizing the TBP-DNA transcription complex. Cell *106*, 71-81.

Kang, J. J., Auble, D. T., Ranish, J. A., and Hahn, S. (1995). Analysis of yeast transcription factor TFIIA: distinct functional regions and a polymerase II-specific role in basal and activated transcription. Mol. Cell. Biol. *15*, 1234–1243.

Kang, M. E., and Dahmus, M. E. (1993). RNA polymerases IIA and IIO have distinct roles during transcription from the TATA-less murine dihydrofolate reductase promoter. J. Biol. Chem. *268*, 25033-25040.

Kao, C. C., Lieberman, P. M., Schmidt, M. C., Zhou, Q., Pei, R., and Berk, A. J. (1990). Cloning of a transcriptionally active human TATA binding factor. Science *248*, 1646-1650.

Kaufmann, J., Ahrens, K., Koop, R., Smale, S. T., and Müller, R. (1998). CIF150, a human cofactor for transcription factor IID-dependent initiator function. Mol. Cell. Biol. *18*, 233-239.

Kephart, D. D., Price, M. P., Burton, Z. F., Finkelstein, A., Greenblatt, J., and Price, D. H. (1993). Cloning of a *Drosophila* cDNA with sequence similarity to human transcription factor RAP74. Nucleic Acids Res. *21*, 1319.

Keriel, A., Stary, A., Sarasin, A., Rochette-Egly, C., and Egly, J.-M. (2002). XPD mutations prevent TFIIH-dependent transactivation by nuclear receptors and phosphorylation of RAR . Cell *109*, 125-135.

Kershnar, E., Wu, S.-Y., and Chiang, C.-M. (1998). Immunoaffinity purification and functional characterization of human transcription factor IIH and RNA polymerase II from clonal cell lines that conditionally express epitope-tagged subunits of the multiprotein complexes. J. Biol. Chem. *273*, 34444-34453.

Kettenberger, H., Armache, K.-J., and Cramer, P. (2003). Architecture of the RNA polymerase II-TFIIS complex and implications for mRNA cleavage. Cell *114*, 347-357.

Kettenberger, H., Armache, K.-J., and Cramer, P. (2004). Complete RNA polymerase II elongation complex structure and its interactions with NTP and TFIIS. Mol. Cell *16*, 955-965.

Khazak, V., Estojak, J., Cho, H., Majors, J., Sonoda, G., Testa, J. R., and Golemis, E. A. (1998). Analysis of the interaction of the novel RNA polymerase II (pol II) subunit hsRPB4 with its partner hsRPB7 and with pol II. Mol. Cell. Biol. *18*, 1935-1945.

Kim, Y.-J., Björklund, S., Li, Y., Sayre, M. H., and Kornberg, R. D. (1994). A multiprotein mediator of transcriptional activation and its interaction with the C-terminal repeat domain of RNA polymerase II. Cell 77, 599-608.

Kimura, M., Suzuki, H., and Ishihama, A. (2002). Formation of a carboxy-terminal domain phosphatase (FCP1)/TFIIF/RNA polymerase II (pol II) complex in *Schizosaccharomyces pombe* involves direct interaction between Fcp1 and the Rpb4 subunit of pol II. Mol. Cell. Biol. *22*, 1577-1588.

Ko, L. J., and Prives, C. (1996). p53: puzzle and paradigm. Genes Dev. *10*, 1054-1072.

Kobayashi, N., Boyer, T. G., and Berk, A. J. (1995). A class of activation domains interacts directly with TFIIA and stimulates TFIIA-TFIID-promoter complex assembly. Mol. Cell. Biol. *15*, 6465-6473.

Kobor, M. S., Simon, L. D., Omichinski, J., Zhong, G., Archambault, J., and Greenblatt, J. (2000). A motif shared by TFIIF and TFIIB mediates their interaction with the RNA polymerase II carboxy-terminal domain phosphatase Fcp1p in *Saccharomyces cerevisiae*. Mol. Cell. Biol. *20*, 7438-7449.

Koiwa, H., Hausmann, S., Bang, W. Y., Ueda, A., Kondo, N., Hiraguri, A., Fukuhara, T., Bahk, J. D., Yun, D.-J., Bressan, R. A., Hasegawa, P. M., and Shuman, S. (2004). *Arabidopsis* C-terminal domain phosphatase-like 1 and 2 are essential Ser-5-specific C-terminal domain phosphatases. Proc. Natl. Acad. Sci. USA *101*, 14539-14544.

Kokubo, T., Swanson, M. J., Nishikawa, J.-I., Hinnebusch, A. G., and Nakatani, Y. (1998). The yeast TAF145 inhibitory domain and TFIIA competitively bind to TATA-binding protein. Mol. Cell. Biol. *18*, 1003–1012.

Koleske, A. J., and Young, R. A. (1994). An RNA polymerase II holoenzyme responsive to activators. Nature *368*, 466-469.

Komarnitsky, P., Cho, E.-J., and Buratowski, S. (2000). Different phosphorylated forms of RNA polymerase II and associated mRNA processing factors during transcription. Genes Dev. *14*, 2452-2460.

Kraemer, S. M., Ranallo, R. T., Ogg, R. C., and Stargell, L.A. (2001). TFIIA interacts with TFIID via association with TATA-binding protein and TAF40. Mol. Cell. Biol. *21*, 1737-1746.

Krishnamurthy, S., He, X., Reyes-Reyes, M., Moore, C., and Hampsey, M. (2004). Ssu72 is an RNA polymerase II CTD phosphatase. Mol. Cell *14*, 387-394.

Kugel, J. F., and Goodrich, J. A. (1998). Promoter escape limits the rate of RNA polymerase II transcription and is enhanced by TFIIE, TFIIH, and ATP on negatively supercoiled DNA. Proc. Natl. Acad. Sci. USA. *95*, 9232-9237.

Kuldell, N. H., and Buratowski, S. (1997). Genetic analysis of the large subunit of yeast transcription factor IIE reveals two regions with distinct functions. Mol. Cell. Biol. *17*, 5288-5298.

Kumar, K. P., Akoulitchev, S., and Reinberg, D. (1998). Promoter-proximal stalling results from the inability to recruit transcription factor IIH to the transcription complex and is a regulated event. Proc. Natl. Acad. Sci. USA. *95*, 9767-9772.

Kuras, L., Kosa, P., Mencia, M., and Struhl, K. (2000). TAF-containing and TAF-independent forms of transcriptionally active TBP *in vivo*. Science *288*, 1244-1248.

Lagrange, T., Kim, T. K., Orphanides, G., Ebright, Y. W., Ebright, R. H., and Reinberg, D. (1996). High-resolution mapping of nucleoprotein complexes by site-specific protein-DNA photocrosslinking: organization of the human TBP-TFIIA-TFIIB-DNA quaternary complex. Proc. Natl. Acad. Sci. USA *93*, 10620–10625.

Langelier, M.-F., Forget, D., Rojas, A., Porlier, Y., Burton, Z. F., and Coulombe, B. (2001). Structural and functional interactions of transcription factor (TF) IIA with TFIIE and TFIIF in transcription initiation by RNA polymerase II. J. Biol. Chem. *276*, 38652-38657.

Lavigne, A. C., Mengus, G., Gangloff, Y.-G., Wurtz, J.-M., and Davidson, I. (1999). Human $TAF_{II}55$ interacts with the vitamin D_3 and thyroid hormone receptors and with derivatives of the retinoid X receptor that have altered transactivation properties. Mol. Cell. Biol. *19*, 5486-5494.

Lee, D. K., De Jong, J., Hashimoto, S., Horikoshi, M., and Roeder, R. G. (1992). TFIIA induces conformational changes in TFIID via interactions with the basic repeat. Mol. Cell. Biol. *12*, 5189-5196.

Lee, S., and Hahn, S. (1995). Model for binding of transcription factor TFIIB to the TBP-DNA complex. Nature *376*, 609–612.

Lee, T. I., and Young, R. A. (2000). Transcription of eukaryotic protein-coding genes. Annu. Rev. Genet. *34*, 77-137.

Lehmann, A. R. (2001). The xeroderma pigmentosum group D (*XPD*) gene: one gene, two functions, three diseases. Genes Dev. *15*, 15-23.

Lei, L., Ren, D., Finkelstein, A., and Burton, Z. F. (1998). Functions of the N- and C-terminal domains of human RAP74 in transcription initiation, elongation, and recycling of RNA polymerase II. Mol. Cell. Biol. *18*, 2130-2142.

Lemon, B., and Tjian, R. (2000). Orchestrated response: a symphony of transcription factors for gene control. Genes Dev. *14*, 2551-2569.

Leurent, C., Sanders, S., Ruhlmann, C., Mallouh, V., Weil, P. A., Kirschner, D. B., Tora, L., and Schultz, P. (2002). Mapping histone fold TAFs within yeast TFIID. EMBO J. *21*, 3424-3433.

Li, X.-Y., Bhaumik, S. R., and Green, M. R. (2000). Distinct classes of yeast promoters revealed by differential TAF recruitment. Science *288*, 1242-1244.

Li, X.-Y., Bhaumik, S. R., Zhu, X., Li, L., Shen, W.-C., Dixit, B. L., and Green, M. R. (2002). Selective recruitment of TAFs by yeast upstream activating sequences. Implications for eukaryotic promoter structure. Curr. Biol. *12*, 1240-1244.

Lieberman, P. M., and Berk, A. J. (1994). A mechanism for TAFs in transcriptional activation: activation domain enhancement of TFIID–TFIIA–promoter DNA complex formation. Genes Dev. *9*, 995–1006.

Lieberman, P. M., Ozer, J., and Gürsel, D. B. (1997). Requirement for transcription factor IIA (TFIIA)-TFIID recruitment by an activator depends on promoter structure and template competition. Mol. Cell. Biol. *17*, 6624-6632.

Liu, Q., Gabriel, S. E., Roinick, K. L., Ward, R. D., and Arndt, K. M. (1999). Analysis of TFIIA function *in vivo*: Evidence for a role in TATA-binding protein recruitment and gene-specific activation. Mol. Cell. Biol. *19*, 8673–8685.

Lu, H., Fisher, R. P., Bailey, P., and Levine, A. J. (1997). The CDK7-cycH-p36 complex of transcription factor IIH phosphorylates p53, enhancing its sequence-specific DNA binding activity *in vitro*. Mol. Cell. Biol. *17*, 5923-5934.

Ma, D., Watanabe, H., Mermelstein, F., Admon, A., Oguri, K., Sun, X., Wada, T., Imai, T., Shiroya, T., Reinberg, D., and Handa, H. (1993). Isolation of a cDNA encoding the largest subunit of TFIIA reveals functions important for activated transcription. Genes Dev. *7*, 2246–2257.

Maile, T., Kwoczynski, S., Katzenberger, R. J., Wassarman, D. A., and Sauer, F. (2004). TAF1 activates transcription by phosphorylation of serine 33 in histone H2B. Science *304*, 1010-1014.

Maldonado, E., Shiekhattar, R., Sheldon, M., Cho, H., Drapkin, R., Rickert, P., Lees, E., Anderson, C. W., Linn, S., and Reinberg, D. (1996). A human RNA polymerase II complex associated with SRB and DNA-repair proteins. Nature *381*, 86-89.

Malik, S., Hisatake, K., Sumimoto, H., Horikoshi, M., and Roeder, R. G. (1991). Sequence of general transcription factor TFIIB and relationships to other initiation factors. Proc. Natl. Acad. Sci. USA *88*, 9553-9557.

Malik, S., Lee, D. K., and Roeder, R. G. (1993). Potential RNA polymerase II-induced interactions of transcription factor TFIIB. Mol. Cell. Biol. *13*, 6253-6259.

Malik, S., Guermah, M., and Roeder, R. G. (1998). A dynamic model for PC4 coactivator function in RNA polymerase II transcription. Proc. Natl. Acad. Sci. USA *95*, 2192-2197.

Martin, M. L., Lieberman, P. M., and Curran, T. (1996). Fos-Jun dimerization promotes interaction of the basic region with TFIIE-34 and TFIIF. Mol. Cell. Biol. *16*, 2110-2118.

Martinez, E., Chiang, C.-M., and Roeder, R. G. (1994). TATA-binding protein-associated factor(s) in TFIID function through the initiator to direct basal transcription from a TATA-less class II promoter. EMBO J. *13*, 3115-3126.

Martinez, E., Ge, H., Tao, Y., Yuan, C.-X., Palhan, V., and Roeder, R. G. (1998). Novel cofactors and TFIIA mediate functional core promoter selectivity by the human $TAF_{II}150$-containing TFIID complex. Mol. Cell. Biol. *18*, 6571-6583.

Matsui, T., Segall, J., Weil, P. A., and Roeder, R. G. (1980). Multiple factors required for accurate initiation of transcription by purified RNA polymerase II. J. Biol. Chem. *255*, 11992-11996.

Matza, D., Wolstein, O., Dikstein, R., and Shachar, I. (2001). Invariant chain induces B cell maturation by activating a $TAF_{II}105$-NF-κB-dependent transcription program. J. Biol. Chem. *276*, 27203-27206.

Maxon, M. E., Goodrich, J. A., and Tjian, R. (1994). Transcription factor IIE binds preferentially to RNA polymerase IIa and recruits TFIIH: a model for promoter clearance. Genes Dev. *8*, 515-524.

McCraken, S., and Greenblatt, J. (1991). Related RNA polymerase-binding regions in human RAP30/74 and *Escherichia coli* σ^{70}. Science *253*, 900-902.

McCracken, S., Fong, N., Yankulov, K., Ballantyne, S., Pan, G., Greenblatt, J., Patterson, S. D., Wickens, M., and Bentley, D. L. (1997). The C-terminal domain of RNA polymerase II couples mRNA processing to transcription. Nature *385*, 357-361.

McEwan, I. J., and Gustafsson, J. (1997). Interaction of the human androgen receptor transactivation function with the general transcription factor TFIIF. Proc. Natl. Acad. Sci. USA *94*, 8485-8490.

McKune, K., Moore, P. A., Hull, M. W., and Woychik, N. A. (1995). Six human RNA polymerase subunits functionally substitute for their yeast counterparts. Mol. Cell. Biol. *15*, 6895-6900.

Meinhart, A., Blobel, J., and Cramer, P. (2003). An extended winged helix domain in general transcription factor E/IIEα. J. Biol. Chem. *278*, 48267-48274.

Meisterernst, M., Roy, A. L., Lieu, H. M., and Roeder, R. G. (1991). Activation of class II gene transcription by regulatory factors is potentiated by a novel activity. Cell *66*, 981–993.

Mellon, I., and Hanawalt, P. C. (1989). Induction of the *Escherichia coli* lactose operon selectively increases repair of its transcribed DNA strand. Nature *342*, 95-98.

Mencia, M., Moqtaderi, Z., Geisberg, J. V., Kuras, L., and Struhl, K. (2002). Activator-specific recruitment of TFIID and regulation of ribosomal protein genes in yeast. Mol. Cell *9*, 823-833.

Merika, M., and Thanos, D. (2001). Enhanceosomes. Curr. Opin. Genet. Dev. *11*, 205-208.

Merino, A., Madden, K. R., Lane, W. S., Champoux, J. J., and Reinberg, D. (1993). DNA topoisomerase I is involved in both repression and activation of transcription. Nature *365*, 227–232.

Minakhin, L., Bhagat, S., Brunning, A., Campbell, E. A., Darst, S. A., Ebright, R. H., and Severinov, K. (2001). Bacterial RNA polymerase subunit ω and eukaryotic RNA polymerase subunit RPB6 are sequence, structural, and functional homologs and promote RNA polymerase assembly. Proc. Natl. Acad. Sci. USA *98*, 892-897.

Mitsiou, D. J., and Stunnenberg, H. G. (2000). TAC, a TBP-sans-TAFs complex containing the unprocessed TFIIAαβ precursor and the TFIIAγ subunit. Mol. Cell *6*, 527-537.

Mitsiou, D. J., and Stunnenberg, H. G. (2003). p300 is involved in formation of the TBP–TFIIA-containing basal transcription complex, TAC. EMBO J. *22*, 4501-4511.

Mitsuzawa, H., and Ishihama, A. (2004). RNA polymerase II transcription apparatus in *Schizosaccharomyces pombe*. Curr. Genet. *44*, 287-294.

Mizzen, C. A., Yang, X.-J., Kokubo, T., Brownell, J. E.,

Bannister, A. J., Owen-Hughes, T., Workman, J., Wang, L., Berger, S. L., Kouzarides, T., Nakatani, Y., and Allis, C. D. (1996). The TAF$_{II}$250 subunit of TFIID has histone acetyltransferase activity. Cell *87*, 1261-1270.

Moqtaderi, Z., Bai, Y., Poon, D., Weil, P. A., and Struhl, K. (1996). TBP-associated factors are not generally required for transcriptional activation in yeast. Nature *383*, 188-191.

Näär, A. M., Lemon, B. D., and Tjian, R. (2001). Transcriptional coactivator complexes. Annu. Rev. Biochem. *70*, 475-501.

Nakajima, T., Uchida, C., Anderson, S. F., Parvin, J. D., and Montminy, M. (1997). Analysis of a cAMP-responsive activator reveals a two-component mechanism for transcriptional induction via signal-dependent factors. Genes Dev. *11*, 738-747.

Nikolov, D. B., Chen, H., Halay, E. D., Usheva, A. A., Hisatake, K., Lee, D. K., Roeder, R. G., and Burley, S. K. (1995). Crystal structure of a TFIIB-TBP-TATA-element ternary complex. Nature *377*, 119-128.

Oelgeschläger, T., Tao, Y., Kang, Y. K., and Roeder, R. G. (1998). Transcription activation via enhanced preinitiation complex assembly in a human cell-free system lacking TAF$_{II}$s. Mol. Cell *1*, 925-931.

Ohkuma, Y., Sumimoto, H., Hoffmann, A., Shimasaki, S., Horikoshi, M., and Roeder, R. G. (1991). Structural motifs and potential σ homologies in the large subunit of human general transcription factor TFIIE. Nature *354*, 398-401.

Ohkuma, Y., and Roeder, R. G. (1994). Regulation of TFIIH ATPase and kinase activities by TFIIE during active initiation complex formation. Nature *368*, 160-163.

Ohkuma, Y., Hashimoto, S., Wang, C. K., Horikoshi, M., and Roeder, R. G. (1995). Analysis of the role of TFIIE in basal transcription and TFIIH-mediated carboxy-terminal domain phosphorylation through structure-function studies of TFIIE-α. Mol. Cell. Biol. *15*, 4856-4866.

Okamoto, T., Yamamoto, S., Watanabe, Y., Ohta, T., Hanaoka, F., Roeder, R. G., and Ohkuma, Y. (1998). Analysis of the role of TFIIE in transcriptional regulation through structure-function studies of the TFIIEβ subunit. J. Biol. Chem. *273*, 19866-19876.

Okuda, M., Watanabe, Y., Okamura, H., Hanaoka, F., Ohkuma, Y., and Nishimura, Y. (2000). Structure of the central core domain of TFIIEβ with a novel double-stranded DNA-binding surface. EMBO J. *19*, 1346-1356.

Okuda, M., Tanaka, A., Arai, Y., Satoh, M., Okamura, H., Nagadoi, A., Hanaoka, F., Ohkuma, Y., and Nishimura, Y. (2004). A novel zinc finger structure in the large subunit of human general transcription factor TFIIE. J. Biol. Chem. *279*, 51395-51403.

Orphanides, G., Lagrange, T., and Reinberg, D. (1996). The general transcription factors of RNA polymerase II. Genes Dev. *10*, 2657-2683.

Ossipow, V., Tassan, J.-P., Nigg, E. A., and Schibler, U. (1995). A mammalian RNA polymerase II holoenzyme containing all components required for promoter-specific transcription initiation. Cell *83*, 137-146.

Ozer, J., Moore, P. A., Bolden, A. H., Lee, A., Rosen, C. A., and Lieberman, P. M. (1994). Molecular cloning of the small γ subunit of human TFIIA reveals functions critical for activated transcription. Genes Dev. *8*, 2324-2335.

Ozer, J., Lezina, L. E., Ewing, J., Audi, S., and Lieberman, P. M. (1998a). Association of transcription factor IIA with TBP is required for transcriptional activation of a subset of promoters and cell cycle progression in *S. cerevisiae*. Mol. Cell. Biol. *18*, 2559-2570.

Ozer, J., Mitsouras, K., Zerby, D., Carey, M., and Lieberman, P.M. (1998b). Transcription factor IIA derepresses TATA binding protein (TBP)-associated factor inhibition of TBP–DNA binding. J. Biol. Chem. *273*, 14293-14300.

Ozer, J., Moore, P. A., and Lieberman, P. M. (2000). A testis-specific transcription factor IIA (TFIIAτ) stimulates TATA-binding protein-DNA binding and transcription activation. J. Biol. Chem. *275*, 122-128.

Pan, G., and Greenblatt, J. (1994). Initiation of transcription by RNA polymerase II is limited by melting of the promoter DNA in the region immediately upstream of the initiation site. J. Biol. Chem. *269*, 30101-30104.

Pardee, T. S., Bangur, C. S., and Ponticelli, A. S. (1998). The N-terminal region of yeast TFIIB contains two adjacent functional domains involved in stable RNA polymerase II binding and transcription start site selection. J. Biol. Chem. *273*, 17859-17864.

Parker, C. S., and Topol, J. (1984). A *Drosophila* RNA polymerase II transcription factor contains a promoter-region-specific DNA-binding activity. Cell *36*, 357-369.

Parvin, J. D., and Sharp, P. A. (1993). DNA topology and a minimal set of basal factors for transcription by RNA polymerase II. Cell *73*, 533-540.

Parvin, J. D., Shykind, B. M., Meyers, R. E., Kim, J., and Sharp, P. A. (1994). Multiple sets of basal factors initiate transcription by RNA polymerase II. J. Biol. Chem. *269*, 18414-18421.

Parvin, J. D., and Young, R. A. (1998). Regulatory targets in the RNA polymerase II holoenzyme. Curr. Opin. Genet. Dev. *8*, 565-570.

Pei, Y., Hausmann, S., Ho, C. K., Schwer, B., and Shuman, S. (2001). The length, phosphorylation state, and primary structure of the RNA polymerase II carboxy-terminal domain dictate interactions with mRNA capping enzymes. J. Biol. Chem. *276*, 28075-28082.

Peterson, M. G., Tanese, N., Pugh, B. F., and Tjian, R. (1990). Functional domains and upstream activation properties of cloned human TATA binding protein. Science *248*, 1625-1630.

Peterson, M. G., Inostroza, J., Maxon, M. E., Flores, O., Admon, A., Reinberg, D., and Tjian, R. (1991). Structure and functional properties of human general transcription factor IIE. Nature *354*, 369-373.

Pham, A.-D., and Sauer, F. (2000). Ubiquitin-activating/

conjugating activity of TAF$_{II}$250, a mediator of activation of gene expression in *Drosophila*. Science *289*, 2357-2360.

Pinto, I., Ware, D. E., and Hampsey, M. (1992). The yeast SUA7 gene encodes a homolog of human transcription factor TFIIB and is required for normal start site selection *in vivo*. Cell *68*, 977-988.

Pinto, I., Wu, W.-H., Na, J. G., and Hampsey, M. (1994). Characterization of sua7 mutations defines a domain of TFIIB involved in transcription start site selection in yeast. J. Biol. Chem. *269*, 30569-30573.

Poon, D., and Weil, P. A. (1993). Immunopurification of yeast TATA-binding protein and associated factors. Presence of transcription factor IIIB transcriptional activity. J. Biol. Chem. *268*, 15325-15328.

Prelich, G. (2002). RNA polymerase II carboxy-terminal domain kinases: emerging clues to their function. Eukaryot. Cell *1*, 153-162.

Pugh, B. F., and Tjian, R. (1990). Mechanism of transcriptional activation by Sp1: evidence for coactivators. Cell *61*, 1187-1197.

Pugh, B. F., and Tjian, R. (1991). Transcription from a TATA-less promoter requires a multisubunit TFIID complex. Genes Dev. *5*, 1935-1945.

Pugh, B. F. (2000). Control of gene expression through regulation of the TATA-binding protein. Gene *255*, 1-14.

Ranish, J. A., Lane, W. S., and Hahn, S. (1992). Isolation of two genes that encode subunits of the yeast transcription factor IIA. Science *255*, 1127-1229.

Ranish, J. A., Yudkovsky, N., and Hahn, S. (1999). Intermediates in formation and activity of the RNA polymerase II preinitiation complex: holoenzyme recruitment and a postrecruitment role for the TATA box and TFIIB. Genes Dev. *13*, 49-63.

Ranish, J. A., Hahn, S., Lu, Y., Yi, E. C., Li, X.-J., Eng, J., and Aebersold, R. (2004). Identification of TFB5, a new component of general transcription and DNA repair factor IIH. Nat. Genet. *36*, 707-713.

Reid, J., Murray, I., Watt, K., Betney, R., and McEwan, I. J. (2002). The androgen receptor interacts with multiple regions of the large subunit of general transcription factor TFIIF. J. Biol. Chem. *277*, 41247-41253.

Reinberg, D., and Roeder, R. G. (1987). Factors involved in specific transcription by mammalian RNA polymerase II. Purification and functional analysis of initiation factors IIB and IIE. J. Biol. Chem. *262*, 3310-3321.

Reinberg, D., Horikoshi, M., and Roeder, R. G. (1987). Factors involved in specific transcription in mammalian RNA polymerase II. Functional analysis of initiation factors IIA and IID and identification of a new factor operating at sequences downstream of the initiation site. J. Biol. Chem. *262*, 3322-3330.

Robert, F., Douziech, M., Forget, D., Egly, J.-M., Greenblatt, J., Burton, Z. F., and Coulombe, B. (1998). Wrapping of promoter DNA around the RNA polymerase II initiation complex induced by TFIIF. Mol. Cell. *2*, 341-351.

Roberts, S. G. E., Ha, I., Maldonado, E., Reinberg, D., and Green, M. R. (1993). Interaction between an acidic activator and transcription factor TFIIB is required for transcriptional activation. Nature *363*, 741-744.

Robinson, M. M., Yatherajam, G., Ranallo, R. T., Bric, A., Paule, M. R., and Stargell, L. A. (2005). Mapping and functional characterization of the TAF11 interaction with TFIIA. Mol. Cell. Biol. *25*, 945-957.

Rochette-Egly, C., Adam, S., Rossignol, M., Egly, J.-M., and Chambon, P. (1997). Stimulation of RARα activation function AF-1 through binding to the general transcription factor TFIIH and phosphorylation by CDK7. Cell *90*, 97-107.

Rodriguez, C. R., Cho, E.-J., Keogh, M.-C., Moore, C. L., Greenleaf, A. L., and Buratowski, S. (2000). Kin28, the TFIIH-associated carboxy-terminal domain kinase, facilitates the recruitment of mRNA processing machinery to RNA polymerase II. Mol. Cell. Biol. *20*, 104-112.

Roeder, R. G., and Rutter, W. J. (1969). Multiple forms of DNA-dependent RNA polymerase in eukaryotic organisms. Nature *224*, 234-237.

Roeder, R. G., and Rutter, W. J. (1970). Specific nucleolar and nucleoplasmic RNA polymerases. Proc. Natl. Acad. Sci. USA *65*, 675-682.

Rossignol, M., Kolb-Cheynel, I., and Egly, J.-M. (1997). Substrate specificity of the cdk-activating kinase (CAK) is altered upon association with TFIIH. EMBO J. *16*, 1628-1637.

Sanders, S. L., Garbett, K. A., and Weil, P. A. (2002). Molecular characterization of *Saccharomyces cerevisiae* TFIID. Mol. Cell. Biol. *22*, 6000-6013.

Sauer, F., Fondell, J. D., Ohkuma, Y., Roeder, R. G., and Jackle, H. (1995). Control of transcription by Krüppel through interactions with TFIIB and TFIIEβ. Nature *375*, 162-164.

Sawadogo, M., and Roeder, R. G. (1985a). Factors involved in specific transcription by human RNA polymerase II: analysis by a rapid and quantitative *in vitro* assay. Proc. Natl. Acad. Sci. USA *82*, 4394-4398.

Sawadogo, M., and Roeder, R. G. (1985b). Interaction of a gene-specific transcription factor with the adenovirus major late promoter upstream of the TATA box region. Cell *43*, 165-175.

Schaeffer, L., Roy, R., Humbert, S., Moncollin, V., Vermeulen, W., Hoeijmakers, J. H. J., Chambon, P., and Egly, J.-M. (1993). DNA repair helicase: a component of BTF2 (TFIIH) basic transcription factor. Science *260*, 58-63.

Schaeffer, L., Moncollin, V., Roy, R., Staub, A., Mezzina, M., Sarasin, A., Weeda, G., Hoeijmakers, J. H. J., and Egly, J.-M. (1994). The ERCC2/DNA repair protein is associated with the class II BTF2/TFIIH transcription factor. EMBO J. *13*, 2388-2392.

Schroeder, S. C., Schwer, B., Shuman, S., and Bentley, D. (2000). Dynamic association of capping enzymes with transcribing RNA polymerase II. Genes Dev. *14*, 435-2440.

Selleck, W., Howley, R., Fang, Q., Podolny, V., Fried, M. G.,

Buratowski, S., and Tan, S. (2001). A histone fold TAF octamer within the yeast TFIID transcriptional coactivator. Nat. Struct. Biol. *8*, 695-700.

Serizawa, H., Conaway, J. W., and Conaway, R. C. (1994). An oligomeric form of the large subunit of transcription factor (TF) IIE activates phosphorylation of the RNA polymerase II carboxyl-terminal domain by TFIIH. J. Biol. Chem. *269*, 20750-20756.

Shao, H., Revach, M., Moshonov, S., Tzuman, Y., Gazit, K., Albeck, S., Unger, T., and Dikstein, R. (2005). Core promoter binding by histone-like TAF complexes. Mol. Cell. Biol. *25*, 206-219.

Shilatifard, A., Conaway, R. C., and Conaway, J. W. (2003). The RNA polymerase II elongation complex. Annu. Rev. Biochem. *72*, 693-715.

Shim, E. Y., Walker, A. K., Shi, Y., and Blackwell, T. K. (2002). CDK-9/cyclin T (P-TEFb) is required in two postinitiation pathways for transcription in the *C. elegans* embryo. Genes Dev. *16*, 2135-2146.

Sklar, V. E. F., Schwartz, L. B., and Roeder, R. G. (1975). Distinct molecular structures of nuclear class I, II, and III DNA-dependent RNA polymerases. Proc. Natl. Acad. Sci. USA *72*, 348-352.

Smale, S. T. (1997). Transcription initiation from TATA-less promoters within eukaryotic protein-coding genes. Biochim. Biophys. Acta *1351*, 73-88.

Solow, S., Salunek, M., Ryan, R., and Lieberman, P. M. (2001). $TAF_{II}250$ phosphorylates human transcription factor IIA on serine residues important for TBP binding and transcription activity. J. Biol. Chem. *276*, 15886-15892.

Sopta, M., Carthew, R. W., and Greenblatt, J. (1985). Isolation of three proteins that bind to mammalian RNA polymerase II. J. Biol. Chem. *260*, 10353-10360.

Sopta, M., Burton, Z. F., and Greenblatt, J. (1989). Structure and associated DNA-helicase activity of a general transcription initiation factor that binds to RNA polymerase II. Nature *341*, 410-414.

Stargell, L. A., Moqtaderi, Z., Dorris, D. R., Ogg, R. C., and Struhl, K. (2000). TFIIA has activator-dependent and core promoter functions *in vivo*. J. Biol. Chem. *275*, 12374–12380.

Sumimoto, H., Ohkuma, Y., Sinn, E., Kato, H., Shimasaki, S., Horikoshi, M., and Roeder, R. G. (1991). Conserved sequence motifs in the small subunit of human general transcription factor TFIIE. Nature *354*, 401-404.

Sun, X., Ma, D., Sheldon, M., Yeung, K., and Reinberg, D. (1994). Reconstitution of human TFIIA activity from recombinant polypeptides: a role in TFIID-mediated transcription. Genes Dev. *8*, 2336–2348.

Svejstrup, J. Q., Vichi, P., and Egly, J.-M. (1996). The multiple roles of transcription/repair factor TFIIH. Trends Biochem. Sci. *21*, 346-350.

Tan, S., Garrett, K. P., Conaway, R. C., and Conaway, J. W. (1994). Cryptic DNA-binding domain in the C terminus of RNA polymerase II general transcription factor RAP30. Proc. Natl. Acad. Sci. USA *91*, 9808-9812.

Tan, S., Hunziker, Y., Sargent, D. F., and Richmond, T. J. (1996). Crystal structure of a yeast TFIIA/TBP/DNA complex. Nature *381*, 127–134.

Tan, Q., Linask, K. L., Ebright, R. H., and Woychik, N. A. (2000). Activation mutants in yeast RNA polymerase II subunit RPB3 provide evidence for a structurally conserved surface required for activation in eukaryotes and bacteria. Genes Dev. *14*, 339-348.

Tanese, N., Pugh, B. F., and Tjian, R. (1991). Coactivators for a proline-rich activator purified from the multisubunit human TFIID complex. Genes Dev. *5*, 2212-2224.

Tantin, D., and Carey, M. (1994). A heteroduplex template circumvents the energetic requirement for ATP during activated transcription by RNA polymerase II. J. Biol. Chem. *269*, 17397-17400.

Thomas, M. C., and Chiang, C.-M. (2005). E6 oncoprotein represses p53-dependent gene activation via inhibition of protein acetylation independently of inducing p53 degradation. Mol. Cell *17*, 251-264.

Tora, L. (2002). A unified nomenclature for TATA box binding protein (TBP)-associated factors (TAFs) involved in RNA polymerase II transcription. Genes Dev. *16*, 673-675.

Tsai, F. T. F., and Sigler, P. B. (2000). Structural basis of preinitiation complex assembly on human pol II promoters. EMBO J. *19*, 25-36.

Ulmasov, T., Larkin, R. M., and Guilfoyle, T. J. (1996). Association between 36- and 13.6-kDa α-like subunits of *Arabidopsis thaliana* RNA polymerase II. J. Biol. Chem. *271*, 5085-5094.

Upadhyaya, A. B., Lee, S. H., and DeJong, J. (1999). Identification of a general transcription factor TFIIAα/β homolog selectively expressed in testis. J. Biol. Chem. *274*, 18040-18048.

Upadhyaya, A. B., Khan, M., Mou, T.-C., Junker, M., Gray, D. M., and DeJong, J. (2002). The germ cell-specific transcription factor ALF. Structural properties and stabilization of the TATA-binding protein (TBP)-DNA complex. J. Biol. Chem. *277*, 34208-34216.

Van Dyke, M. W., Roeder, R. G., and Sawadogo, M. (1988). Physical analysis of transcriptional preinitiation complex assembly on a class II gene promoter. Science *241*, 1335-1338.

van Vuuren, A. J., Vermeulen, W., Ma, L., Weeda, G., Appeldoorn, E., Jaspers, N. G., van der Eb, A. J., Bootsma, D., Hoeijmakers, J. H. J., and Humbert, S. (1994). Correction of xeroderma pigmentosum repair defect by basal transcription factor BTF2 (TFIIH). EMBO J. *13*, 1645-1653.

Verrijzer, C. P., Chen, J.-L., Yokomori, K., and Tjian, R. (1995). Binding of TAFs to core elements directs promoter selectivity by RNA polymerase II. Cell *81*, 1115–1125.

Verrijzer, C. P., and Tjian, R. (1996). TAFs mediate transcriptional activation and promoter selectivity. Trends

Biochem. Sci. *21*, 338-342.

Walker, S. S., Reese, J. C., Apone, L. M., and Green, M. R. (1996). Transcription activation in cells lacking TAF$_{II}$S. Nature *383*, 185-188.

Walker, S. S., Shen, W.-C., Reese, J. C., Apone, L. M., and Green, M. R. (1997). Yeast TAF$_{II}$145 required for transcription of G1/S cyclin genes and regulated by the cellular growth state. Cell *90*, 607-614.

Wampler, S. L., and Kadonaga, J. T. (1992). Functional analysis of *Drosophila* transcription factor IIB. Genes Dev. *6*, 1542-1552.

Wang, W., Gralla, J. D., and Carey, M. (1992). The acidic activator GAL4-AH can stimulate polymerase II transcription by promoting assembly of a closed complex requiring TFIID and TFIIA. Genes Dev. *6*, 1716–1727.

Wang, Z., Svejstrup, J. Q., Feaver, W. J., Wu, X., Kornberg, R. D., and Friedberg, E. C. (1994). Transcription factor b (TFIIH) is required during nucleotide-excision repair in yeast. Nature *368*, 74-76.

Wang, Z., Buratowski, S., Svejstrup, J. Q., Feaver, W. J., Wu, X., Kornberg, R. D., Donahue, T. F., and Friedberg, E. C. (1995). The yeast TFB1 and SSL1 genes, which encode subunits of transcription factor IIH, are required for nucleotide excision repair and RNA polymerase II transcription. Mol. Cell. Biol. *15*, 2288-2293.

Warfield, L., Ranish, J. A., and Hahn, S. (2004). Positive and negative functions of the SAGA complex mediated through interaction of Spt8 with TBP and the N-terminal domain of TFIIA. Genes Dev. *18*, 1022-1034.

Watanabe, T., Hayashi, K., Tanaka, A., Furumoto, T., Hanaoka, F., and Ohkuma, Y. (2003). The carboxy terminus of the small subunit of TFIIE regulates the transition from transcription initiation to elongation by RNA polymerase II. Mol. Cell. Biol. *23*, 2914-2926.

Wei, W., Dorjsuren, D., Lin, Y., Qin, W., Nomura, T., Hayashi, N., and Murakami, S. (2001). Direct interaction between subunit RAP30 of the transcription factor IIF (TFIIF) and RNA polymerase subunit 5, which contributes to the association between TFIIF and RNA polymerase II. J. Biol. Chem. *276*, 12266-12273.

Weil, P. A., Luse, D. S., Segall, J., and Roeder, R. G. (1979). Selective and accurate initiation of transcription at the Ad2 major late promotor in a soluble system dependent on purified RNA polymerase II and DNA. Cell *18*, 469-484.

Weinmann, R., Raskas, H. J., and Roeder, R. G. (1974). Role of DNA-dependent RNA polymerases II and III in transcription of the adenovirus genome late in productive infection. Proc. Natl. Acad. Sci USA *71*, 3426-3439.

Weinmann, R., and Roeder, R. G. (1974). Role of DNA-dependent RNA polymerase III in the transcription of the tRNA and 5S RNA genes. Proc. Natl. Acad. Sci. USA *71*, 1790-1794.

Weiss, S., and Gladstone, L. (1959). A mammalian system for the incorporation of cytidine triphosphate into ribonucleic acid. J. Am. Chem. Soc. *81*, 4118-4119.

Wilson, C. J., Chao, D. M., Imbalzano, A. N., Schnitzler, G. R., Kingston, R. E., and Young, R. A. (1996). RNA polymerase II holoenzyme contains SWI/SNF regulators involved in chromatin remodeling. Cell *84*, 235-244.

Wolner, B. S., and Gralla, J. D. (2001). TATA-flanking sequences influence the rate and stability of TATA-binding protein and TFIIB binding. J. Biol. Chem. *276*, 6260-6266.

Workman, J. L., and Roeder, R. G. (1987). Binding of transcription factor TFIID to the major late promoter during *in vitro* nucleosome assembly potentiates subsequent initiation by RNA polymerase II. Cell *51*, 613-622.

Woychik, N. A., Liao, S. M., Kolodziej, P. A., and Young, R. A. (1990). Subunits shared by eukaryotic nuclear RNA polymerases. Genes Dev. *4*, 313-323.

Woychik, N. A., McKune, K., Lane, W. S., and Young, R. A. (1993). Yeast RNA polymerase II subunit RPB11 is related to a subunit shared by RNA polymerase I and III. Gene Expr. *3*, 77-82.

Woychik, N. A., and Hampsey, M. (2002). The RNA polymerase II machinery: structure illuminates function. Cell *108*, 453-463.

Wu, S.-Y., and Chiang, C.-M. (1998). Properties of PC4 and an RNA polymerase II complex in directing activated and basal transcription *in vitro*. J. Biol. Chem. *273*, 12492-12498.

Wu, S.-Y., Kershnar, E., and Chiang, C.-M. (1998). TAF$_{II}$-independent activation mediated by human TBP in the presence of the positive cofactor PC4. EMBO J. *17*, 4478-4490.

Wu, S.-Y., Thomas, M. C., Hou, S. Y., Likhite, V., and Chiang, C.-M. (1999). Isolation of mouse TFIID and functional characterization of TBP and TFIID in mediating estrogen receptor and chromatin transcription. J. Biol. Chem. *274*, 23480-23490.

Wu, S.-Y., and Chiang, C.-M. (2001a). TATA-binding protein-associated factors enhance the recruitment of RNA polymerase II by transcriptional activators. J. Biol. Chem. *276*, 34235-34243.

Wu, S.-Y., and Chiang, C.-M. (2001b). Expression and purification of epitope-tagged multisubunit protein complexes from mammalian cells. *Current Protocols in Molecular Biology*, Unit 16.22.1-16.22.17.

Wu, S.-Y., Zhou, T., and Chiang, C.-M. (2003). Human mediator enhances activator-facilitated recruitment of RNA polymerase II and promoter recognition by TATA-binding protein (TBP) independently of TBP-associated factors. Mol. Cell. Biol. *23*, 6229-6242.

Wu, W.-H., and Hampsey, M. (1999). An activation-specific role for transcription factor TFIIB *in vivo*. Proc. Natl. Acad. Sci. USA *96*, 2764-2769.

Xie, X., Kokubo, T., Cohen, S. L., Mirza, U. A., Hoffmann, A., Chait, B. T., Roeder, R. G., Nakatani, Y., and Burley, S. K. (1996). Structural similarity between TAFs and the heterotetrameric core of the histone octamer. Nature *380*, 316-322.

Xing, J., Sheppard, H. M., Corneillie, S. I., and Liu, X. (2001).

p53 stimulates TFIID-TFIIA-promoter complex assembly and p53-T antigen complex inhibits TATA binding protein-TATA interaction. Mol. Cell. Biol. *21*, 3652-3661.

Yamashita, S., Wada, K., Horikoshi, M., Gong, D.-W., Kokubo, T., Hisatake, K., Yokotani, N., Malik, S., Roeder, R. G., and Nakatani, Y. (1992). Isolation and characterization of a cDNA encoding *Drosophila* transcription factor TFIIB. Proc. Natl. Acad. Sci. USA *89*, 2839-2843.

Yamashita, S., Hisatake, K., Kokubo, T., Doi, K., Roeder, R. G., Horikoshi, M., and Nakatani, Y. (1993). Transcription factor TFIIB sites important for interaction with promoter-bound TFIID. Science *261*, 463-466.

Yan, Q., Moreland, R. J., Conaway, J. W., and Conaway, R. C. (1999). Dual roles for transcription factor IIF in promoter escape by RNA polymerase II. J. Biol. Chem. *274*, 25668-35675.

Yankulov, K. Y., and Bentley, D. L. (1997). Regulation of CDK7 substrate specificity by MAT1 and TFIIH. EMBO J. *16*, 1638-1646.

Yeo, M., Lin, P. S., Dahmus, M. E., and Gill, G. N. (2003). A novel RNA polymerase II C-terminal domain phosphatase that preferentially dephosphorylates serine 5. J. Biol. Chem. *278*, 26078-26085.

Yokomori, K., Admon, A., Goodrich, J. A., Chen, J.-L., and Tjian, R. (1993). *Drosophila* TFIIA-L is processed into two subunits that are associated with the TBP/TAF complex. Genes Dev. *7*, 2235–2245.

Yong, C., Mitsuyasu, H., Chun, Z., Oshiro, S., Hamasaki, N., and Kitajima, S. (1998). Structure of the human transcription factor TFIIF revealed by limited proteolysis with trypsin. FEBS Lett. *435*, 191-194.

Young, R. A. (1991). RNA polymerase II. Annu. Rev. Biochem *60*, 689-715.

Yuryev, A., Patturajan, M., Litingtung, Y., Joshi, R. V., Gentile, C., Gebara, M., and Corden, J. L. (1996). The C-terminal domain of the largest subunit of RNA polymerase II interacts with a novel set of serine/arginine-rich proteins. Proc. Natl. Acad. Sci. USA *93*, 6975-6980.

Zhou, M., Halanski, M. A., Radonovich, M. F., Kashanchi, F., Peng, J., Price, D. H., and Brady, J. N. (2000). Tat modifies the activity of CDK9 to phosphorylate serine 5 of the RNA polymerase II carboxyl-terminal domain during human immunodeficiency virus type 1 transcription. Mol. Cell. Biol. *20*, 5077-5086.

Zhou, Q., Lieberman, P. M., Boyer, T. G., and Berk, A. J. (1992). Holo-TFIID supports transcriptional stimulation by diverse activators and from a TATA-less promoter. Genes Dev. *6*, 1964-1974.

Zhu, A., and Kuziora, M. A. (1996). Homeodomain interaction with the β subunit of the general transcription factor TFIIE. J. Biol. Chem. *271*, 20993-20996.

Ziegler, L. M., Khaperskyy, D. A., Ammerman, M. L., and Ponticelli, A. S. (2003). Yeast RNA polymerase II lacking the Rpb9 subunit is impaired for interaction with transcription factor IIF. J. Biol. Chem. *278*, 48950-48956.

Zurita, M., and Merino, C. (2003). The transcriptional complexity of the TFIIH complex. Trends Genet. *19*, 578-584.

Chapter 03

The Dynamic Association of RNA Polymerase II with Initiation, Elongation, and RNA Processing Factors during the Transcription Cycle

Kristi L. Penheiter and Judith A. Jaehning

Department of Biochemistry and Molecular Genetics and Molecular Biology Program, University of Colorado Health Sciences Center, B121, 4200 East 9th Avenue, Denver, CO 80262

Key Words: transcription, RNA polymerase II, Paf1 complex, initiation, elongation, RNA processing

Summary

The transcription of protein encoding genes by RNA polymerase II (RNAPII) is a complex and highly regulated process. RNAPII and several of its associated general transcription factors (GTFs), including TBP, TFIIB, TFIIF, and TFIIH are sufficient for recognition and low levels of accurate transcription from common core promoter elements *in vitro* (Roeder, 1996 and Orphanides et al., 1996). However, in addition to these factors necessary for basal transcription initiation, accurately regulated transcription requires additional cofactors which assemble into various complexes to mediate the communication between DNA-binding activators, as well as repressors, and RNAPII. The precise manner in which the RNAPII transcription machinery is assembled upon a region of DNA determines the ability of RNAPII to initiate the synthesis of mRNA at a specific location and with a defined frequency and processivity. The dynamic nature of the RNAPII holoenzyme and the ever-increasing number of RNAPII-associated factors being identified that regulate transcription at the levels of initiation, elongation, and processing will be discussed in this chapter.

RNA Polymerase II and Transcription Initiation

In order to properly initiate transcription of eukaryotic mRNAs, transcriptional activators recruit the RNAPII-containing machinery to promoters of protein-coding genes. The assembled apparatus, or "holoenzyme" consists of the 12-subunit RNAPII, a subset of GTFs, and one or more multisubunit complexes referred to as coactivators or mediators. How this assembly and recruitment occurs is a highly complex and variable process, and a growing number of holoenzyme complexes and subcomplexes are being identified.

A: The Core RNAPII

The yeast core RNAPII is a multi-subunit enzyme comprised of 12 proteins, Rpb1-Rpb12, ranging in size from ~8kDa (Rpb10 and Rpb12) to ~200kDa (Rpb1). Of the 12 subunits, 10 are encoded by genes essential for cell viability. The extreme importance of the enzyme to all eukaryotic cells is also reflected in the high degree of structural conservation among the subunits not only between eukaryotic species, but also between the three nuclear RNAPs and the prokaryotic core RNAP. For example, six subunits of the human RNAPII can functionally substitute for their yeast homologs (McKune et al., 1995). In addition, the Rpb1, Rpb2, Rpb3, and Rpb11 subunits of RNAPII are homologous to subunits of RNAPI and RNAPIII, and Rpb5, Rpb6, Rpb8, Rpb10, and Rpb12 are shared among all three RNAPs. Only three subunits, Rpb4, Rpb7, and Rpb9 are

---

Corresponding Author: Judith A. Jaehning, Tel: (303) 315-3004, E-mail: Judith.Jaehning@UCHSC.edu

unique to RNAPII (reviewed in Hampsey, 1998).

Extensive mutational analyses of *RPB1* and *RPB2* (reviewed in Archambault and Friesen, 1993) have revealed roles for Rpb1 and Rpb2 in regulating the efficiency of transcription initiation and directing start site selection (Berroteran et al., 1994; Hekmatpanah and Young, 1991) as well as elongation (Archambault et al., 1992; Powell and Reines, 1996).

The X-ray crystallographic structure of the complete 12-subunit yeast RNAPII, which includes the reversibly-associating heterodimer of Rpb4 and Rpb7, has recently been determined (Bushnell and Kornberg, 2003; Armache et al., 2003; comment by Asturias and Craighead, 2003). The Rpb4/7 subunit was not present in previous structures of the "core" RNAPII, but it is known to be required for transcription initiation (Edwards et al., 1991), whereas the RNAPII 10-subunit core is sufficient for elongation. This structural determination reveals that the Rpb4/7 heterodimer maintains the polymerase in the conformation of a transcribing complex and is in a position to interact with general transcription factors and mediator components, thus rendering it a crucial factor for proper transcription initiation. A complete list of the RNAPII subunits and some of their features is shown in Table 3.1 (reviewed in Lee and Young, 2000; Cramer, 2002).

Table 3.1 Yeast RNA polymerase II subunit composition

RNAPII Subunit	Mass	Features
Rpb1	192 kDa	Binds DNA; involved in start site selection; contains CTD
Rpb2	139 kDa	Contains active site; involved in start site selection
Rpb3	35 kDa	Forms heterodimer with Rpb11
Rpb4	25 kDa	Forms subcomplex with Rpb7; required for initiation; involved in elongation
Rpb5	25 kDa	Target for transcriptional activators; shared with RNAPI, RNAPIII
Rpb6	18 kDa	Functions in assembly and stability of RNAPII; shared with RNAPI, RNAPIII
Rpb7	19 kDa	Forms subcomplex with Rpb4; functions in initiation and elongation
pb8	17 kDa	Has oligonucleotide binding domain; shared with RNAPI, RNAPIII
Rpb9	14 kDa	Functions in start site selection; may function in elongation Rpb10 8 kDaShared with RNAPI, RNAPIII
Rpb11	14 kDa	Forms heterodimer with Rpb3
Rpb12	8 kDa	Shared with RNAPI, RNAPIII

B: RNAPII Carboxy-Terminal Domain

The largest subunit of RNAPII, Rpb1, contains a unique C-terminal domain (CTD) consisting of tandem repeats of the highly conserved amino acid sequence Tyr-Ser-Pro-Thr-Ser-Pro-Ser (YSPTSPS) which is not found in the RNAPI or RNAPIII enzymes. This repeat sequence varies in length from 26 repeats in yeast RNAPII (Allison et al., 1985) to 52 repeats in mammalian RNAPII (Corden et al., 1985). The CTD is essential for cell viability, as partial or complete deletion of the heptapeptide repeat sequence results in lethality (Gerber et al., 1995; Nonet et al., 1987). The CTD is variably and reversibly phosphorylated at different phases throughout the transcription cycle (Laybourn and Dahmus, 1990; Dahmus, 1994), a process necessary for the recruitment of a variety of factors, including components of the Mediator complex, histone methyltransferases, as well as capping and polyadenylation enzymes, to RNAPII (reviewed in Hampsey and Reinberg, 2003; Maniatis and Reed, 2002).

C: The General Transcription Factors

Enroute to the characterization of the RNAPII holoenzyme, identification and extensive characterization of the general transcription factors (GTFs) necessary for transcription initiation has taken place. The GTFs of RNAPII, including TFIIA, TFIIB, TFIID (composed of TBP and TAFs), TFIIE, TFIIF, and TFIIH, were identified biochemically as factors required for accurate transcription initiation by RNAPII from double stranded DNA templates *in vitro* (reviewed in Conaway and Conaway, 1993; Zawel and Reinberg, 1993; Orphanides et al., 1996; Roeder, 1996). R. Kornberg and colleagues were able to reconstitute RNAPII transcription initiation from a variety of yeast and mammalian promoters in a system using only purified TBP, TFIIB, -E, -F, -H, and RNAPII (Myers et al., 1997). The addition of mediator complex components (see below) to the preparation, comprising a 44 polypeptide system, resulted in transcription from not only a minimal promoter but also that which could respond to activator proteins bound upstream of the promoter (Kim et al., 1994).

D: Coactivators

Coactivators are distinct from the general transcription factors in that they are dispensable for basal-level transcription *in vitro* but are required for regulated transcriptional activation. They differ from activators in that they generally do not bind DNA directly in a sequence-specific manner, but act as bridges between DNA binding activator proteins and the GTFs or facilitate chromatin remodeling. Several examples of coactivators that serve as bridges or mediators of transcription by acting in concert with the general transcription machinery will be discussed below.

TFIID, along with RNAPII and other GTFs is sufficient to direct promoter-specific transcription initiation. Cloning and expression of yeast TBP however, revealed that recombinant TBP could replace TFIID in RNAPII directed transcription from TATA-containing promoters *in vitro*, but these reactions were not responsive to sequence-specific transcriptional activators. In addition they could not direct transcription from TATA-less promoters (Zhou *et al.*, 1992). These results led to a model in which subunits of TFIID other than TBP are required for regulated transcriptional activation. These subunits are referred to as TBP-associated proteins or TAFs (Dynacht *et al.*, 1991, Poon and Weil, 1993; reviewed in Goodrich and Tjian 1994a; Verrijzer and Tjian, 1996). Multiple protein interactions occur between TAFs and the GTFs, as well as TAFs and the activator proteins. In addition, multiple forms of TFIID, which vary in TAF composition and function have been described, including a form of TFIID that plays a role in the repression of transcription (Wade and Jaehning, 1996). Thus, TFIID and its associated TAFs play a critical role in the regulation of transcription by relaying information from activators and repressors to the core transcriptional machinery.

E: The Mediator-Containing Holoenzyme

Earlier studies, as discussed above, centered around the identification and characterization of TAFs in association with TBP. More recent work, however, has focused on characterizing the role(s) of a coactivator component of the RNAPII holoenzyme known as the mediator complex. The first direct evidence for a yeast mediator comes from the results of transcriptional reconstitution experiments in which a mediator was required for *in vitro* transcription by the activator GAL4-VP16, but had no effect on transcription in the absence of the activator (Flanagan *et al.*, 1991). In addition, Nonet and Young discovered that a series of yeast mutants with partial deletions in the heptapeptide repeat sequence (YSPTSPS) of the largest subunit of RNAPII-the CTD- were temperature-sensitive, cold-sensitive and inositol auxotrophs. A genetic screen for suppressors of the cold-sensitive phenotype of these CTD mutants led to the identification of the Srb (**S**uppressor of **R**NA polymerase **B**) proteins (Nonet and Young,1989) whose products were shown to reside in a multiprotein complex (Thompson *et al.*,1993). A holoenzyme that supported a response to activator proteins with purified basal GTFs was identified and contained a core mediator complex consisting of Srb2, 4, 5, and 6, as well as Gal11, and Sug1 (Kim *et al.*, 1994; Koleske and Young, 1994). Other members of this complex have been purified and include Srb7, Meds1, 2, 4, 6, 7, and 8, Sin4, Rgr1, Rox3, and Pgd1 (Myers *et al.*, 1998) as well as Nut1, Nut2, Cse2, and Med11 (Gustafsson *et al.*, 1998). Genetic and biochemical studies have revealed that many of these proteins affect both positive (the Srb core proteins 2, 4, 5, 6 , and 7) and negative (Srb8, 9, 10, and 11) regulation of transcription (Hengartner, *et al.*, 1995; Holstege *et al.*, 1998; Myers *et al.*, 1999; Han *et al.*, 1999; also see review in Carlson, 1997) and, as is the case with Srb4 and Srb6, are required for transcription of most, if not all, protein-encoding genes (Holstege *et al.*, 1998; Thompson and Young, 1995). Members of this mediator complex can also exist in a large number of subcomplexes in association with the RNAPII holoenzyme, including a Sin4 subcomplex containing Sin4, Gal11, Pgd1, and Med2 (Li *et al.*, 1995; Myers *et al.*, 1999) and an Srb subcomplex comprised of Srb2, Srb4, Srb5, Srb6, Rox3, Med6, Med8, and Med11 (Lee and Kim, 1998; Kang *et al.*, 2001; Koh *et al.*, 1998) thus adding to the great complexity and broad range of regulatory response activities within transcription by RNA polymerase II (reviewed in Chang and Jaehning, 1997; Myer and Young 1998; Malik and Roeder, 2000).

A current view of the Mediator is one of an evolutionarily conserved and ubiquitously expressed complex of more than twenty subunits that connects the initiating form of RNAPII to DNA-binding regulatory factors in response to environmental signals. Indeed, much progress has been made in the characterization of several forms of a mammalian Mediator complex, and cross-species comparisons have detected metazoan homologs of nearly all yeast Mediator components. Srb/Med-containing complexes, including the thyroid hormone receptor-associated proteins/ SRB-Med containing cofactor (TRAP/SMCC), vitamin D receptor-interacting proteins (DRIP), positive cofactor 2 (PC2), the cofactor required for Sp1 activation (CRSP), and the activator-recruited factor-large (ARC-L) complexes have been isolated (reviewed in Conaway *et al.*, 2005). The complexes are highly homologous to components of the yeast Mediator. In light of these findings, a common nomenclature has recently been established under the recommendation of a large group of scientists working inside and outside of the transcription field, to clarify, across species, the conserved nature of Mediator proteins (Bourbon *et al.*, 2004). This new nomenclature has replaced the names of many of the previously identified proteins with the acronym MED, which acknowledges the discovery of Med complexes in yeast.

In addition to Mediator, some yeast holoenzyme preparations contain substoichometric levels of Swi-Snf complex components (Wilson, *et al.*, 1996). The Swi-Snf complex has been purified (Cairns *et al.*, 1994; Cote *et al.*, 1995; Treich *et al.*, 1995) and has an ATP-dependent chromatin destabilizing activity allowing for the relief of chromatin-based transcription repression (Winston and Carlson, 1992; Peterson and Tamkun, 1995). In addition, histone acetyltransferase activity, most likely carried out by the Nut1 subunit of the Mediator, has been found associated with the holoenzyme (Lorch *et al.*, 2000). Therefore, a form of the holoenzyme can be linked to transcriptional activation through chromatin remodeling.

F: The Paf1 Complex

The Paf1 complex can be distinguished from the Srb/Med-containing holoenzyme by its composition and the subset of genes it regulates. A novel collection of RNAPII associated proteins (RAPs) was isolated by immobilizing antibodies against an unphosphorylated form of the CTD of RNA polymerase on a Sepharose column. The eluted RAP fraction from a transcriptionally active yeast whole cell extract did not contain the holoenzyme's core mediator components Srb2, 4, 5, and 6, Gal11, and Sug1, but did include the known GTFs, TFIIB, and three subunits of TFIIF, as well as the transcription elongation factor TFIIS. Also isolated as RAPs were two novel factors, Paf1 and Cdc73 (Wade *et al.*, 1996). Blast searches of peptide sequences revealed that Paf1 was an uncharacterized open reading frame, and Cdc73 was the product of a previously identified gene encoding a yeast mating pheromone signaling element (Reed *et al.*, 1988). Mutations in the nonessential *PAF1* and *CDC73* genes affect cell growth and the abundance of transcripts from a subset of genes (Shi *et al.*, 1996; Chang, *et al.*, 1999; Porter *et al.*, 2002). The isolation of several proteins associated with RNAPII in a whole cell fraction that did not contain components of the Srb-containing mediator complex raised the possibility that Paf1 and Cdc73 may be associating as a second, alternative form of the RNAPII holoenzyme with distinct, but overlapping roles in transcription. Thus, in order to determine the relationship between Paf1, Cdc73 and the previously identified Srb-containing holoenzyme, studies were pursued to isolate a Paf1 complex. Glutathione S-transferase (GST)-tagged forms of Paf1, Cdc73 were used in glutathione agarose chromatography, and an RNAPII-associating complex was identified that contained Paf1 and Cdc73, as well as several components also found in the Srb-containing form of the holoenzyme, TFIIF, TFIIB, and Gal11. Srb proteins, TFIIS, TFIIH, and TBP were not isolated in this complex (Shi *et al.*, 1997). Tfg2, a subunit of TFIIF, was also GST-tagged and used to isolate proteins that may be associated with both the Paf1 complex and the Srb- mediator complex. In this case, Srb proteins, as well as Paf1 and Cdc73 were isolated, indicating that Paf1-Cdc73 and the Srbs define two separate, but partially redundant complexes (Shi *et al.*, 1997).

Further analysis also identified Hpr1 and Ccr4 as associated with the Paf1 complex but not the Srb-containing Mediator complex (Chang et. al., 1999). Both Ccr4 and Hpr1 have been found in other transcription-related complexes, have been demonstrated to have genetic interactions with the transcription machinery (Fan *et al.*, 1996; Liu *et al.*, 1998), and have been implicated in the transcription of subsets of genes (Denis and Malvar., 1990; Draper *et al.*, 1994; Zhu *et al.*, 1995). In addition, more recent studies using immunoaffinity chromatography and mass spectrometry have identified Ctr9, Rtf1, and Leo1 as components of the Paf1 complex (Koch *et al.*, 1999; Krogan, *et al.*, 2002; Mueller and Jaehning, 2002; Squazzo *et al.*, 2002). *CTR9* (*CDP1*) was isolated in a screen for mutants that failed to activate transcription of the cell cycle regulated G_1 cyclins (Koch *et al.*, 1999), and a role for Ctr9, as well as Paf1, has been established in the regulation of G_1 cyclin expression (Koch, *et al.*, 1999; Porter, *et al.*, 2002). *RTF1* was originally identified in a screen for extragenic suppressors of a TBP mutant with defects in DNA binding specificity (Stolinski *et al.*, 1997).

The continuing characterization of members of the Paf1 complex has linked the complex to a role in transcriptional regulation. While *PAF1* and *CDC73* are not essential genes in yeast, *paf1Δ* mutants are temperature-sensitive, slow-growing, and have an enlarged cell morphology. *cdc73Δ* mutants are also temperature-sensitive and grow slightly slower than wild type cells. The initial use of differential display, followed by more recent microarray analyses of a *paf1Δ* mutant demonstrated that the Paf1 complex is important in the regulation of expression of a subset of genes. Many genetic interactions between the members of the Paf1 complex and with members of the Srb-containing holoenzyme also exist: several combinations of deletions within complex members result in lethality, and *paf1Δ* and *ccr4Δ* are lethal in combination with *srb5Δ*. This indicates that factors within and between the two complexes may have overlapping essential functions.

More recent work has indicated that the Paf1 complex consists of the core components Paf1, Ctr9, Cdc73, Rtf1, and Leo1 (Mueller et al., 2002; Squazzo et al., 2002). It appears to be associated with RNAPII throughout the transcription cycle (Pokholok et al., 2002; Simic et al., 2002; Mueller et al., 2004) (Fig. 3.1) and localizes to promoter regions of active genes (Wade et al., 1996; Shi et al., 1997; Pokholok et al., 2002; Mueller et al., 2004). The Paf1 complex is also found at the 3' and 5' ends of genes with a distribution and abundance very similar to that of the elongating RNAPII (Pokholok et al., 2002; Simic et al., 2002; Mueller et al., 2004). The Paf1 complex has been found to associate with initiation factors (TFIIB, TFIIF (Wade et al., 1996, Shi et al., 1997)), elongation factors (TFIIF, Spt5, Dst1 (Mueller et al., 2002; Squazzo et al., 2002; Krogan et al., 2002b)), chromatin remodeling and modifying factors (Chd1, Set1 (Simic et al., 2003; Krogan et al., 2003a)), and factors involved in mRNA processing and export (Hpr1, Sub2 (Chang et al., 1999; Mueller et al., 2004)). The fact that the Paf1, RNAPII complex is clearly biochemically distinct from the Srb/mediator form of RNAPII establishes that it probably joins the transcription complex after the initial recruitment event (Shi et al., 1997) (see Fig. 3.1). From that point on, until RNAPII reaches the poly(A) site for 3' end processing (Kim et al., 2004), it accompanies the RNA polymerase, possibly serving as a "platform" for the dynamic association of additional factors during transcription elongation (Gerber and Shilatifard, 2003). In addition, recent evidence suggests that it may also function in the coordination of transcription and downstream events leading to proper mRNA biogenesis and maturation (Mueller et al., 2004).

RNA Polymerase II and Transcription Elongation

The transition between transcription initiation and elongation, although separated by a significant phase in transcription regulation (promoter clearance), is not well defined, and it may be difficult to distinguish between the end of one event and the beginning of the other. During promoter clearance, the appearance of a stalled RNAPII-DNA complex and the production of short RNA products, abortive initiation, occurs. At this point in the transcription cycle, a stable transcription elongation complex has not yet formed (Kireeva, et al., 2000), and RNAPII often slips (Pal and Luse, 2002). Luse and colleagues have also found that once an RNA transcript of 23 nucleotides has formed, slippage no longer occurs and a stable, elongating form of RNAPII

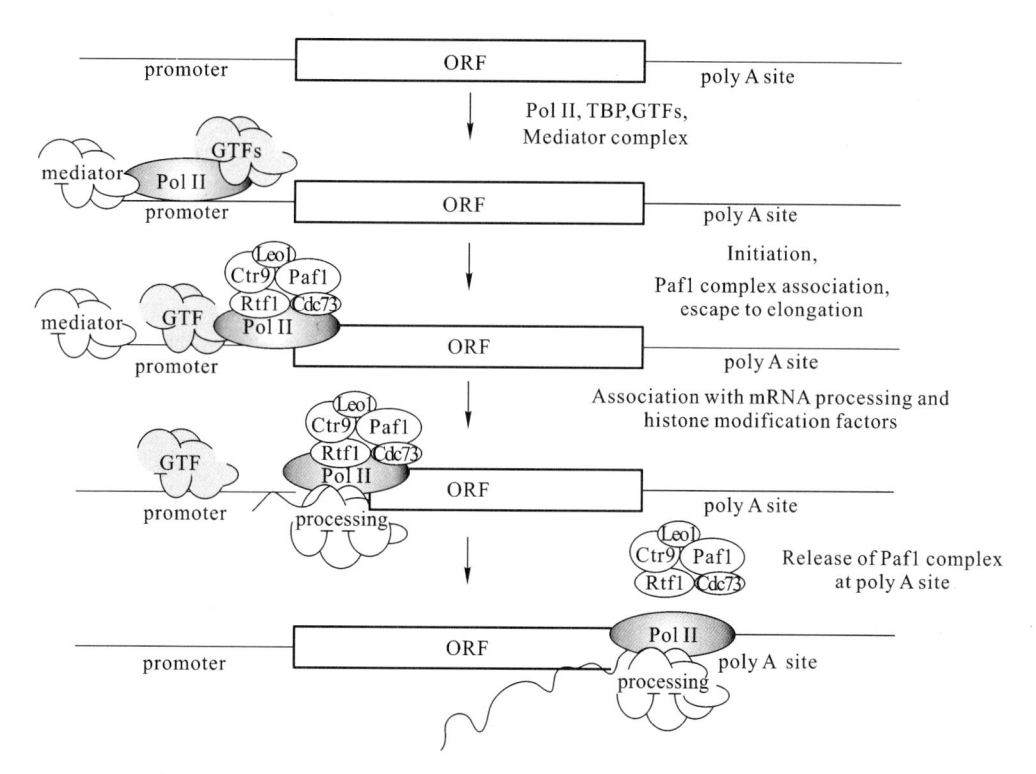

Fig.3.1 **The dynamic association of initiation, elongation and RNA processing factors with RNAPII during the transcription cycle.**

can then synthesize a complete RNA transcript (Pal and Luse, 2003). It appears that the RNAPII associated factors TFIIE and TFIIH play an important role in this transitional stage, although their specific role in the process is not yet clear (Dvir *et al.*, 1997; Goodrich and Tjian, 1994b; Kumar *et al.*, 1998; Yamamoto *et al.*, 2001; Watanabe *et al.*, 2003).

Concurrent with the promoter clearance step is the beginning of the RNAPII CTD phosphorylation cycle. RNAPII exists in a form in which the CTD of its largest subunit is either unphosphorylated, hypo- (IIA) or hyperphosphorylated (IIO) (Dahmus, 1981). The RNAPII CTD is unphosphorylated in the preinitiation complex (Laybourn and Dahmus, 1989; Lu *et al.*, 1991) and is heavily phosphorylated at specific serine residues (Ser 2 and Ser 5) during the transcription process (Christmann and Dahmus, 1981). Phosphorylation at Ser 5 in the YSPTSPS repeat of the CTD correlates with transcription initiation and early elongation, whereas phosphorylation at Ser 2 occurs further along in the elongation process (O'Brien *et al.*, 1994; Komarnitsky *et al.*, 2000) and plays a central role in recruiting elongation factors to the transcribing polymerase.

The process of transcription elongation is now viewed as a dynamic and highly regulated step in the synthesis of mRNA, and is an important factor in the coordination of downstream events as well. Recent studies have identified a multitude of factors that modulate the activity of RNAPII (active elongation factors), as well as many factors that associate with RNAPII but do not directly regulate its catalytic activity (passive elongation factors). These proteins, in association with RNAPII are often referred to as the transcription elongation complex (TEC) (reviewed in Sims, *et al.*, 2004). Factors within the TEC often associate and disassociate in a very specific manner throughout elongation, making the TEC a highly regulatable and ever-changing complex. The following sections will discuss some of the important factors, associated with RNAPII, that facilitate transcription elongation.

A: "Active" Elongation Factors That Modify RNAPII Catalytic Activity

Several impediments, such as transcriptional pause, arrest, and termination, may represent a general, intrinsic feature of RNAPII catalytic activity (Bentley, 1995). There are many elongation factors whose job it is to prevent or alleviate such blocks.

A1: TFIIF

TFIIF influences the overall rate of transcription by modulating transcriptional pause, the temporary and reversible halting of the addition of NTPs to the nascent RNA transcript. TFIIF was originally identified as a factor that directly binds immobilized RNAPII (Sopta *et al.*, 1985). During elongation, it controls the rate of transcription by decreasing the amount of time that RNAPII is paused (Price *et al.*, 1989; Bengal *et al.*, 1991; Tan *et al.*, 1994). TFIIF also has the ability to associate with both the hypo- and hyperphosphorylated forms of RNAPII (Zawel *et al.*, 1995), consistent with a role in both initiation and elongation. Mutational analyses of both subunits of TFIIF have indicated an important role for TFIIF in promoter clearance and the earliest stages of elongation, preventing arrest (Yan *et al.*, 1999). This is in agreement with chromatin immunoprecipitation experiments showing that TFIIF is primarily localized near promoter regions and not evenly distributed throughout coding regions (Krogan *et al.*, 2002a; Pokholok *et al.* 2002), and it appears that its association with RNAPII is required predominantly at times when the elongation machinery is stalled.

A2: The ELL Family

Both the ELL family and the Elongins appear to stimulate the rate of transcription in a similar manner to TFIIF, by interacting directly with RNAPII and suppressing transient pausing (Shilatifard *et al.*, 1996; Bradsher *et al.*, 1993b). The ELL gene was initially identified in humans (Thirman, *et al.*, 1994) and belongs to a family that includes ELL2 and ELL3, both of which affect the rate of elongation *in vitro* (Shilatifard *et al.*, 1997; Miller *et al.*, 2000). Consistent with its role in elongation, the single *Drosophila melanogaster* ELL homolog (dELL) associates with active sites of transcription following heat shock and colocalizes with the elongation-competent form of RNAPII on polytene chromosomes (Gerber *et al.*, 2001). In addition, mutations in dELL preferentially affect expression of long genes (Eissenberg *et al.*, 2002).

A3: Elongins

Elongin, also called SIII, is a heterotrimeric protein composed of Elongin A, B, and C (Aso, *et al.*, 1995, 1996; Garrett *et al.*, 1995; Takagi *et al.*, 1996). It was first purified to homogeneity from rat liver nuclei by its ability to stimulate the rate of RNAPII elongation *in vitro* (Bradsher *et al.*, 1993a). Although the precise mechanism by which it controls the rate of elongation is not completely understood, recent evidence suggests that it functions to properly align the 3'-OH end of a nascent transcript into the active site of RNAPII (Takagi *et al.*, 1995). There is also evidence suggesting that

Elongin is an E3 ubiquitin ligase and functions to activate ubiquitylation of RNAPII, targeting it for proteasomal degradation. However, *where* and *how* it functions in this process in unclear (reviewed in Shilatifard *et al.*, 2003).

A4: Spt4/Spt5 (DSIF)

DSIF is a heterodimeric protein complex composed of the human homologs of the yeast Spt4 and Spt5 proteins (Wada *et al.*, 1998). Early studies in yeast showed that Spt4 and Spt5 act as transcription factors that modify chromatin structure (Winston and Carlson, 1992). However, more recent genetic and biochemical evidence demonstrates that Spt4 and Spt5 are elongation factors that form a complex in which Spt5 binds RNAPII and functions in transcription elongation (Hartzog *et al.*, 1998; Wada *et al.*, 1998). Upon the discovery that DSIF functioned as an elongation factor, it became classified as a negative factor in elongation because adding it to a partially purified transcription reaction resulted in transcription inhibition. However, later experiments showed that the transcription reaction used in those studies also contained NELF, a factor which acts in concert with DSIF to inhibit elongation (Yamaguchi *et al.*, 1999). More recent studies, focusing on the yeast Spt4 protein suggest that DSIF actually promotes transcription (Rondon *et al.*, 2003). As will be discussed later in this chapter, transcription elongation and chromatin remodeling are intricately connected. Therefore, it is not surprising that DSIF has been shown to genetically and physically interact with many chromatin-related factors, including Spt6, FACT, Chd1, as well as members of the Paf1 complex (Hartzog *et al.*, 1998; Orphanides *et al.*, 1999; Costa and Arndt, 2000; Krogan *et al.*, 2002b; Mueller and Jaehning, 2002; Simic *et al.*, 2002; Squazzo *et al.*, 2002; Lindstrom *et al.*, 2002).

A5: NELF

While the previously discussed elongation factors alleviate pausing and are considered positive elongation factors, the NELF (**N**egative **E**longation **F**actor) complex actually works to *promote* pausing (Yamaguchi *et al.*, 1999). NELF binds to an assembled RNAPII/DSIF complex, but cannot bind DSIF or RNAPII alone, and this association is required for NELF to exhibit its negative effects on transcription (Yamaguchi *et al.*, 2002). When NELF dissociates from the TEC upon phosphorylation of the RNAPII CTD, as well as the Spt5 component of DSIF, pausing is overcome and elongation resumes (Ivanov *et al.*, 2000; Kim and Sharp, 2001). Once elongation resumes, it appears that the RNAPII/DSIF complex remains with the TEC, but NELF does not. *In vivo* studies in *Drosophila melanogaster* show that NELF/RNAPII/DSIF are localized to the hsp70 heat-shock promoter before induction; upon induction by heat-shock, RNAPII and DSIF localize to active sites of transcription, but NELF does not (Andrulis *et al.*, 2000; Wu *et al.*, 2003). It has been proposed that this mechanism of inhibiting elongation occurs to allow time for the assembly of other factors involved in mRNA maturation (reviewed in Sims *et al.*, 2004).

A6: TFIIS

TFIIS was originally identified by its ability to stimulate transcription *in vitro* (Natori *et al.*, 1973) and subsequently the first transcription elongation factor to be purified (Sekimizu *et al*, 1976). TFIIS promotes RNAPII read-through at transcriptional arrest sites (reviewed in Fish and Kane, 2002; Conaway *et al.*, 2003; Wind and Reines, 2000) thereby increasing the *efficiency* of transcription elongation rather than its rate (Wind-Rotolo *et al.*, 2001; reviewed in Sims *et al.*, 2004). Evidence for this comes from a series of experiments using artificial arrest sites *in vivo* in which TFIIS counteracted RNAPII arrest at a position far from the promoter (Kulish and Struhl, 2001). It is generally thought that the read-through of these transcriptional blocks is overcome by TFIIS-stimulated RNA cleavage, which creates new 3′ ends, repositions the RNA in the active site of RNAPII, and allows RNAPII to continue elongating. However, a new X–ray crystallographic model of the complete 12-subunit RNAPII in complex with TFIIS suggests that the TFIIS-induced repositioning of RNA may result in a different elongation complex conformation that is less prone to stalling. It also suggests that the acidic residues of a TFIIS hairpin structure could assist in NTP binding (Kettenberger *et al.*, 2004). Although the precise mechanism by which TFIIS alleviates transcriptional arrest is not completely understood, it does appear to be a major contributor to the efficiency of elongation.

B: "Passive" Elongation Factors and Factors That Remodel and Modify Chromatin

In addition to transcriptional pausing and arrest, another major impediment to elongation which adds a new level of complexity to its regulation is the requirement that RNAPII must transcribe through chromatin. Recent explorations into covalent histone modifications (reviewed in Strahl and Allis, 2000; Jenuwein and Allis, 2001; Gerber and Shilatifard, 2003), as well as the mechanisms by which RNAPII elongates through chromatin (including nucleosome mobilization and histone depletion models) (reviewed in Studitsky *et al.*, 2004) have provided valuable insights into regulation of elongation in a chromatin environment. Some of the key players in this process are discussed below.

B1: Swi-Snf

The Swi-Snf complex is an ATP-dependent chromatin remodeler, and, while it is generally thought of as a factor involved in initiation by acting at the promoters of active genes, it appears to also function as an elongation factor required to overcome the enhanced transcriptional pausing that occurs on chromatin templates. Evidence for this comes from the fact that while human heat shock factor 1 (HSF1) can stimulate both initiation and elongation and recruit Swi-Snf to a chromatin template, recruitment of Swi-Snf is greatly impaired when residues in HSF1 responsible for elongation are mutated (Brown *et al.*, 1996; Sullivan *et al.*, 2001). In addition, mutations in yeast Swi-Snf components (*SWI1*, *SNF5*, *SWI2*, or *SNF2*) are synthetically lethal in combination with a disruption in *PPR2* (*DST1*), the gene encoding the transcription elongation factor TFIIS (Davie and Kane, 2000). The specific mechanism by which Swi-Snf facilitates elongation is not known.

B2: Chd1

Chd1 is an ATP-dependent chromatin remodeler that appears to function in both elongation and transcription (Tran *et al.*, 2000). Much evidence suggests a role for Chd1 in elongation. Analyses of *Drosophila melanogaster* polytene chromosomes showed that Chd1 associates with highly active sites of transcription (Stokes *et al.*, 1996). Chd1 genetically interacts with Set2 and members of the IWI family, and Swi-Snf, all of which have been implicated in elongation (Tsukiyama *et al.*, 1999; Krogan *et al.*, 2003b; Tran *et al.*, 2000). In addition, Chd1 physically interacts with the elongation factors DSIF and FACT (discussed later in this chapter) (Kelley *et al.*, 1999; Simic *et al.*, 2003). Again, like Swi-Snf, little is known about *how* Chd1 facilitates elongation through chromation.

B3: CSB

Cockayne Syndrome group B (CSB) is DNA-dependent ATPase and a member of the Swi-Snf family of chromatin remodelers. CSB has been shown to enhance the rate of transcription elongation on naked DNA *in vitro* (Selby and Sancar, 1997), and can remodel chromatin *in vitro* (Citterio *et al.*, 2000). In addition, CSB can bind directly to RNAPII and affect the elongation activity of TFIIS (Tantin *et al.*, 1997). In addition to its link to elongation, CSB also plays a role in transcription-coupled nucleotide repair and base excision repair (reviewed in Licht *et al.*, 2003). Therefore, while it is possible that CSB may function as a chromatin elongation factor, its ability to remodel chromatin may simply reflect the ability of CSB to remodel protein-DNA interactions such as those between a stalled RNAPII and a DNA lesion (reviewed in Svejstrup, 2002).

B4: FACT

FACT (**fa**cilitates **c**hromatin **t**ranscription) was discovered using an assay designed to identify factors that support RNAPII transcription on chromatin templates (Orphanides *et al.*, 1998). This highly conserved heterodimer is comprised of hSpt16 and SSRP1 (Orphanides *et al.*, 1999) which are homologous to the yeast Spt16/Cdc63 and Pob3 proteins. FACT is one of the few, if only, known factors that can stimulate transcription through chromatin in a highly purified system (Orphanides *et al.*, 1998). It now appears that FACT facilitates RNAPII-induced displacement of the H2A-H2B histone dimer from the nucleosome by a direct interaction with H2A and/or H2B (Belotserkovskaya *et al.*, 2003). This is supported by experiments demonstrating that FACT activity is impaired when nucleosomes are covalently cross-linked and therefore cannot be displaced (Orphanides *et al.*, 1999; Belotserkovskaya *et al.*, 2003). Studies in yeast also suggest that not only does FACT disrupt nucleosomes to allow RNAPII access to DNA, but it also reassembles the nuclesomes afterward (Formosa *et al.*, 2002). Additional studies in yeast implicate FACT in the regulation of elongation through chromatin. Spt16 has been found to genetically and physically interact with the elongation factor DSIF, as well as the chromatin remodeler Chd1 (Orphanides *et al.*, 1999; Krogan *et al.*, 2002b; Lindstrom *et al.*, 2003; Simic *et al.*, 2003). In addition, FACT is associated with actively transcribed genes *on Drosophila melanogaster* polytene chromosomes (Saunders *et al.*, 2003). These data all support the idea that FACT functions after transcription initiation, allowing RNAPII to elongate through chromatin templates (reviewed in Belotserkovskaya *et al.*, 2004).

B5: Elongator

The Elongator complex was initially identified in yeast as a multi-subunit complex associated with the hyperphosphorylated, elongating form of RNAPII. High speed centrifugation and high salt concentrations allowed for separation of a mediator-containing, hypophosphorylated complex from a hyperphosphorylated RNAPII complex associated with chromatin. Elongator, consisting of Elp1, Elp2, and Elp3 purified with both fractions, while little, if any, mediator was found in the chromatin-associated fraction (Otero et al., 1999). Further isolation studies identified three more components of Elongator- Elp4, Elp5, and Elp6—which exist as a discrete subcomplex of the six-subunit "holo-Elongator" (Winkler et al., 2001; Krogan and Greenblatt, 2001). Characterization of Elongator components identified genetic interactions with the transcription elongation factor TFIIS and a delay in growth recovery when elpΔ cells were subjected to changes in growth medium. A similar delay in gene activation was also observed, leading to the conclusion that Elongator is a novel form of the RNAPII holoenzyme functioning in transcription elongation (Otero et al., 1999). Because Elongator was isolated from yeast chromatin, it has been speculated that it plays a role during transcription through chromatin. The discovery that Elp3 contains motifs with homology to the GNAT family of histone acetyltransferases (HATs) (Wittschieben et al., 1999; 2000) and acetylates histone H3 and H4 in vivo (Winkler et al., 2002), supports this idea. In addition, deletion of the histone H4 tail is lethal in the absence of Elp3 (Wittschieben et al., 1999). However, a role for Elongator in transcription elongation through chromatin is somewhat controversial. Subcellular localization studies demonstrated that it is primarily cytoplasmic, and chromatin immunoprecipitation experiments failed to detect Elongator recruitment to DNA (Pokholok et al., 2002). In addition, a recent proteomics approach failed to detect an interaction between Elongator and RNAPII in yeast (Krogan et al., 2002b). These data are refuted by more recent results from the Svejstrup laboratory in which an RNA immunoprecipitation (RIP) procedure was utilized to find Elongator associated with RNA along the entire coding region of genes in vivo (Gilbert et al., 2004).

B6: Histone Methyltransferases Set1 and Set2

Another histone modification that has been linked to transcription elongation is methylation (reviewed in Gerber and Shilatifard, 2003). The pattern of methylation on histone lysine residues is specific to each transcriptional state of a gene and this methylation appears to be an important transcriptional regulator (reviewed in Lachner et al., 2003; Sims et al., 2003). In yeast, Set1 is a specific histone H3 lysine 4 (H3-K4) methyltransferase that associates with the RNAPII CTD.

It was originally purified in a large complex termed COMPASS (**Com**plex of **P**roteins **As**sociated with **S**et1) (Briggs et al., 2001; Miller et al., 2001; Roguev et al., 2001; Krogan et al., 2002a; Nagy et al., 2002; Noma and Grewal, 2002; Ng et al., 2003). Set1 may play a role in the early stages of transcription elongation, as it primarily interacts with the Ser 5 phosphorylated form of RNAPII associated with promoters and early elongation complexes (Krogan et al., 2003a; Ng et al., 2003). In addition, Set1 has recently been shown to catalyze tri-methylation of H3-K4, while di-methylation occurs on a genome-wide scale, tri-methylation is present exclusively at active genes (Santos-Rosa et al., 2004). Set2 is an H3-K36-specific methyltransferase (Strahl et al., 2002) that preferentially associates with the Ser 2 phosphorylated form of RNAPII, indicating a role in transcription elongation. In addition, deletion of Ctk1, the CTD Ser 2 kinase, results in defective H3-K4 methylation (Li et al., 2003) and disrupts the interaction between Set2 and RNAPII, thus linking histone methylation to transcription elongation (Krogan et al., 2003b). The activities of Set1 and Set2 have also been found to require the Paf1 complex, although a precise role for this interaction is not completely understood. Continuing identification of RNAPII-associated factors will allow for better understanding of how transcription initiation and elongation are coordinately regulated.

RNA Polymerase II and mRNA Processing

mRNA processing, which includes 5′ end capping, splicing, and 3′ end formation by cleavage and polyadenylation is tightly coupled to transcription. Like the many protein complexes that associate with RNAPII to regulate transcription initiation and elongation, a large number of mRNA processing factors are also associated with the elongating RNAPII in a complex often referred to as the "mRNA factory" (Bentley, 2002; Zorio and Bentley, 2004). It is now becoming increasingly clear that the events comprising mRNA processing occur co-transcriptionally while the dynamic mRNA factory transcribes, processes, and packages transcripts for proper export from nucleus to cytoplasm. The coordination of processing events is directed primarily by the RNAPII CTD, which serves as a platform for many processing factors. Indeed, deletion of the CTD in many vertebrate cells inhibits capping, splicing, and poly(A) site cleavage (McCracken et al., 1997a; McCracken et al., 1997b). The phosphorylation state of the CTD is critically important in specifying where and when the array of processing factors will associate with RNAPII throughout the transcription cycle.

A: 5' End Capping

RNAPII transcripts are capped at their 5' end by a methyl guanosine cap which serves to stabilize the mRNA against 5'-to- 3' exonucleolytic degradation and promote splicing, 3' end formation, transport and translation (Lewis *et al.*, 1995; reviewed in Shuman, 2001). Nascent pre-mRNA is capped early in transcription, when only about 25-30 bases of RNA have been synthesized (Coppola *et al.*, 1983). Capping occurs via the activities of three enzymes acting in order: RNA triphosphatase (RT), guanylyltransferase (GT), and 7-methyltransferase (MT). In yeast, RT and GT are represented by two polypeptides, *Cet1* and *Ceg1*, respectively, which form a heterodimer referred to the capping enzyme (Ho *et al.*, 1998). Capping enzymes are recruited to the transcription complex shortly after transcription initiation by binding to the phosphorylated CTD (McCracken *et al.*, 1997b; Yue *et al.*, 1997; Cho *et al.*, 2001). This recruitment requires Kin28, the TFIIH-associated kinase that phosphorylates the CTD at Ser 5 (Ho and Shuman, 1999; Moteki and Price, 2002). Once a cap has been added, release of the capping enzymes correlates with the removal of Ser 5 phosphates from the elongating polymerase; however, whereas GT is rapidly released, MT may remain associated with RNAPII even at the 3' end of the transcript (Schroeder *et al.*, 2000; Komarnitsky *et al.*, 2000). The capping enzymes appear to manipulate RNAPII function, perhaps by repressing transcription reinitiation (Myers *et al.*, 2002), which serves as a checkpoint to ensure proper cap addition (reviewed in Orphanides and Reinberg, 2002).

B: 3' End Formation: Cleavage and Polyadenylation

Most protein-encoding mRNAs possess a uniform 3' end with a poly(A) tail. Prior to the addition of this tail, pre-mRNA must be cleaved. In mammalian cells, cleavage is directed by a highly conserved AAUAAA sequence and a GU- rich downstream sequence element (DPE) within the RNA. In yeast, the sequence elements that constitute the poly(A) signal are more complex and not as well defined (Graber *et al.*, 2002; also reviewed in Guo and Sherman, 1996; Keller *et al.*, 1997; Zhao *et al*, 1999). Cleavage and polyadenylation in mammalian cells is mediated by a complex of proteins including poly(A) polymerase (PAP), cleavage and polyadenylation specificity factor (CPSF), cleavage stimulation factor (CstF), and cleavage factors I and II (CFIm and CFIIm). Yeast homologs of these factors have also been identified, including Rna14, Rna15, Pcf11, Clp1, and Pap1. Like capping, 3' end formation requires factors that associate with the RNAPII CTD. More specifically, several components of the 3' end processing machinery have been shown to interact with the Ser 2 phosphorylated CTD (Barilla *et al.*, 2001; Licatalosi *et al.*, 2002). Despite the discovery of many of these interactions with RNAPII and transcription factors, the precise mechanism by which cleavage and polyadenylation occur remains somewhat of a mystery.

Conclusion

Much progress has been made in understanding the dynamic nature of initiation, elongation, and RNA processing factor associations with RNAPII. While some factors, such as the Mediator and the GTFs, remain at the promoters of transcribed genes to facilitate initiation and reinitiation, many factors, like the Paf1 complex and factors required for 5' and 3' end processing of the mature mRNA join RNAPII shortly after initiation and accompany the enzyme along the transcribed region of the gene. The presence of these factors, and the changing patterns of phosphorylation of the RNAPII CTD, serve to recruit additional elongation and processing factors to RNAPII. The dynamic association of these factors is necessary for the cycle of transcription, resulting in a mature mRNA ready for export to the cytoplasm and translation. Although it is now clear that there is extensive cross-talk between these factors important for both transcription, and post-transcriptional processing, we are still uncovering the complex associations between these factors and their effects on RNAPII.

References

Allison, L.A., Moyle, M., Shales, M., and Ingles, C.J. (1985). Extensive homology among the largest subunits of eukaryotic and prokaryotic RNA polymerases. Cell *52*, 599-610.

Andrulis, E.D., Guzman, E., Doring, P., Werner, J., and Lis, J.T. (2002). High-resolution localization of *Drosophila* Spt5 and Spt6 at heat shock genes *in vivo*: Roles in promoter proximal pausing and transcription elongation. Genes Dev. *14*, 2635-2649.

Archambault, J.F., Lacroute, F., Ruet, A., and Friesen, J.D. (1992). Genetic interactions between transcription elongation factor TFIIS and RNA polymerase II. Mol. Cell. Biol. *12*, 4142-4152.

Archambault, J., and Friesen, J.D. (1993). Genetics of eukaryotic RNA polymerases I, II, and III. Microbiol. Rev. *57*, 703-724.

Armache, K-J., Kettenberger, H., and Cramer, P. (2003). Architecture of initiation-competent 12-subunit RNA polymerase II. Proc. Natl. Acad. Sci. USA *100*, 6964-6968.

Aso, T., Conaway, J.W., and Conaway, R.C. (1994). Role of core promoter structure in assembly of the RNA polymerase II preinitiation complex. A common pathway for formation of preinitiation intermediates at many TATA and TATA-less promoters. J. Biol. Chem. *269*, 26575-26583.

Aso, T., Lane, W.S., Conaway, J.W., and Conaway, R.C. (1995). Elongin (SIII): A multisubunit regulator of elongation by RNA polymerase II. Science *269*, 1439-1443.

Aso, T., Haque, D., Barstead, R.J., Conaway, R.C., and Conaway, J.W. (1996). The inducible elongin A elongation activation domain: Structure, function and interaction with the elongin BC complex. EMBO J. *15*, 5557-5577.

Asturias, F.J., and Craighead, J.L. (2003). RNA polymerase II at initiation. Proc. Natl. Acad. Sci. USA 100, 6893-6895.

Barilla, D., Lee, B.A., and Proudfoot, N.J. (2001). Cleavage/polyadenylation factor IA associates with the carboxyl-terminal domain of RNA polymerase II in *Saccharomyces cerevisiae*. Proc. Natl. Acad. Sci. USA *98*, 445-450.

Belotserkovskaya, R., Oh, S., Bondarenko, V.A., Orphanides, G., Studitsky, V.M., and Reinberg, D. (2003). FACT facilitates transcription-dependent nucleosome alteration. Science *301*, 1090-1093.

Belotserkovskaya, R., Saunders, A., Lis, J.T., and Reinberg, D. (2004). Transcription through chromatin: understanding a complex FACT. Biochim. Biophys. Acta *1677*, 87-99

Bengal, E. Flores, O., Krauskopf, A., Reinberg, D., and Aloni, Y. (1991). Role of the mammalian transcription factors IIF, IIS, and IIX during elongation by RNA polymerase II. Mol. Cell. Biol. *11*, 1195-1206.

Bentley, D.L. (1995). Regulation of transcriptional elongation by RNA polymerase II. Curr. Opin. Genet. Dev. *5*, 210-216.

Bentley, D. (2002). The mRNA assembly line: transcription and processing machines in the same factory. Curr. Opin. Cell Biol. *14*, 3360342.

Berroteran, R.W., Ware, D.E., and Hampsey, M. (1994). The *sua8* suppressors of *Saccharomyces cerevisiae* encode replacements of conserved residues within the largest subunit of RNA polymerase II and affect transcription start site selection similarly to *sua7* (TFIIB) mutations. Mol. Cell. Biol. *14*, 226-237.

Blackwood, E., and Kadonaga, J.T. (1998). Going the distance: a current view of enhancer action. Science *281*, 60-63.

Bourbon, H-M. *et al.* (2004). A unified nomenclature for protein subunits of mediator complexes linking transcriptional regulators to RNA polymerse II. Mol. Cell. *14*, 553-557.

Bradsher, J.N., Jackson, K.W., Conaway, R.C., and Conaway, J.W. (1993a). RNA polymerase II transcription factor SIII. I. Identification, purification, and properties. J. Biol. Chem. *268*, 25587-25593.

Bradsher, J.N., Tan, S., McLaury, H.J., Conaway, J.W., and Conaway, R.C. (1993b). RNA polymerase II transcription factor SIII. II. Functional properties and role in RNA chain elongation. J. Biol. Chem. *268*, 25594-25603.

Briggs, S.D., Bryk, M., Strahl, B.D., Cheung, W.L., Davie, J.K., Dent, S.Y.R., Winston, F., and Allis, C.D. (2001). Histone H3 lysine 4 methylation is mediated by Set1 and required for cell growth and rDNA silencing in *Saccharomyces cerevisiae*. Genes Dev. *15*, 3286-3295.

Brown, S.A., Imbalzano, A.N., and Kingston, R.E. (1996). Activator-dependent regulation of transcriptional pausing on nucleosomal templates. Genes Dev. *10*, 1479-1490.

Burke, T.W., and Kadonaga, J.T. (1996) *Drosophila* TFIID binds to a conserved downstream basal promoter element that is present in many TATA-box-deficient promoters. Genes Dev. *10*, 711-724.

Bushnell, D.A., and Kornberg, R.D. (2003). Complete, 12-subunit RNA polymerase II at 4.1-A resolution: implications for the initiation of transcription. Proc. Natl. Acad. Sci. USA *100*, 6969-6973.

Cairns, B.R., Kim, Y.J., Sayre, M.H., Laurent, B.C., and Kornberg, R.D. (1994). A multisubunit complex containing the SWI1/ADR6, SWI2/SNF2. SWI3, SNF5, and SNF6 gene products isolated from yeast. Proc. Natl. Acad. Sci. USA *91*, 1950-1954.

Carcamo, J., Bucklinder, L., and Reinberg, D. (1991). The initiator directs the assembly of a transcription factor IID-dependent transcription complex. Proc. Natl. Acad. Sci. USA *88*, 8052-8056.

Carlson, M. (1997). Genetics of transcriptional regulation in yeast: connections to the RNA polymerase II CTD. Annu. Rev. Cell Dev. Biol. *13*, 1-23.

Chang, M., and Jaehning, J.A. (1997). A multiplicity of mediators: alternative forms of transcription complexes communicate with transcriptional regulators. Nucleic Acids Res. *25*, 4861-4865.

Chang, M., French-Cornay, D., Fan, H-Y., Klein, H., Denis, C.L., and Jaehning, J.A. (1999). A complex containing RNA polymerase II, Paf1p, Cdc73p, Hpr1p, and Ccr4p plays a role in protein kinase C signaling. Mol. Cell. Biol. *19*, 1056-1067.

Cho, E.J., Takagi, T., Moore, C.R., and Buratowski, S. (1997). mRNA capping enzyme is recruited to the transcription complex by phosphorylation of the RNA polymerase II carboxy-terminal domain. Genes Dev. *11,* 3319-3326.

Christmann, J.L., and Dahmus, M.E. (1981). Monoclonal antibody specific for calf thymus RNA polymerases IIO and IIA. J. Biol. Chem. *256*, 11798-11803.

Citterio, E., Van Den Boom, V., Schnitzler, G., Kanaar, R., Bonte, E., Kingston, R.E., Hoeijmakers, J.H., and Vermeulen, W. (2000). ATP-dependent chromatin remodeling by the Cockayne syndrome B DNA repair-transcription-coupling factor. Mol. Cell. Biol. *20*, 7643-7653.

Conaway, R.C., and Conaway, J.W. (1993). General initiation factors for RNA polymerase II. Annu. Rev. Biochem. *62*, 161-190.

Conaway, R.C., Kong, S.E., and Conaway, J.W. (2003). TFIIS and GreB: Two like-minded transcription elongation factors with sticky fingers. Cell *114*, 272-274.

Conaway, J.W., Florens, L., Sato, S., Tomomori-Sato, C. Parmely, T.J., Yao, T., Swanson, S.K., Banks, C.A.S., Washburn, M.P., and Conaway, R.C (2005). The mammalian Mediator complex. FEBS Letters *579*, 904-908.

Coppola, J.A., Field, A.S., and Luse, D.S. (1983). Promoter-proximal pausing by RNA polymerase II *in vitro*: transcripts shorter than 20 nucleotides are not capped. Proc. Natl. Acad. Sci. USA *80*, 1251-1255.

Corden, J.L., Cadena, D.L., Ahearn, J.M. Jr., and Dahmus, M.E. (1985). A unique structure at the carboxyl terminus of the largest subunit of eukaryotic RNA polymerase II. Proc. Natl. Acad. Sci. USA *82*, 7934-7938.

Cormack, B., and Struhl, K. (1992). The TATA-binding protein is required for transcription by all three nuclear RNA polymerases in yeast cells. Cell *69*, 685-696.

Costa, P.J., and Arndt, K.M. (2000). Synthetic lethal interactions suggest a role for the Sacharomyces cerevisiae Rtf1 protein in transcription elongation. Genetics *156*, 535-547.

Cote, J., Quinn, J., Workman, J.L., and Peterson, C.L. (1994). Stimulation of GAL4 derivative binding to nucleosomal DNA by the yeast SWI/SNF complex. Science *265*, 53-60.

Dahmus, M.E. (1981). Phosphorylation of eukaryotic DNA-dependent RNA polymerase. Identification of calf thymus RNA polymerase subunits phosphorylated by two purified protein kinases, correlation with *in vivo* sites of phosphorylation in HeLa cell RNA polymerase II. J. Biol, Chem. *256*, 3332-3339.

Dahmus, M.E. (1994). The role of multisite phosphorylation in the regulation of RNA polymerase II activity. Prog. Nucleic Acid Res. Mol. Biol. *48*, 143-179.

Davie, J.K., and Kane, C.M. (2000). Genetic interactions between TFIIS and the Swi-Snf chromatin-remodeling complex. Mol. Cell. Biol. *20*, 5960-5973.

Denis, C.L., and Malvar, T. (1990). The *CCR4* gene from *Saccharomyces cerevisiae* is required for both nonfermentative and spt-mediated gene expression. Genetics *124*, 283-291.

Draper, M.P., Liu, H., Nelsbach, A.H., Mosley, S.P., and Denis, C.L. (1994). CCR4 is a glucose-regulated transcription factor whose leucine-rich repeat binds several proteins important for placing CCR4 in its proper promoter context. Mol. Cell. Biol. *14*, 4522-4531.

Durrin, L.K., Mann, R.K., and Grunstein, M. (1992). Nucleosome loss activates CUP1 and HIS3 promoters to fully induced levels in the yeast *Saccharomyces cerevisiae*. Mol. Cell. Biol. *12*, 1621-1629.

Dvir, A., Conaway R.C., and Conaway J.W. (1997). A role for TFIIH in controlling the activity of early RNA polymerase II elongation complexes. Proc. Natl. Acad. Sci. USA *94*, 9006-9010.

Dynact, B.D., Hoey, T. and Tjian, R. (1991). Isolation of coactivators associated with the TATA binding protein that mediate transcriptional activation. Cell *66*, 563-566.

Edwards, A.M., Kane, C.M., Young, R. A., and Kornberg, R. D. (1991). Two dissociable subunits of yeast RNA polymerase II stimulate the initiation of transcription at a promoter *in vitro*. J. Biol. Chem. *266*, 71-75

Eissenberg, J.C., Ma, J., Gerber, M.A., Christensen, A., Kennison, J.A., and Shilatifard, A. (2002). dELL is an essential RNA polymerase II elongation factor with a general role in development. Proc. Natl. Acad. Sci. USA *99*, 9894-9899.

Fan, H-Y., Chang, K.K., and Klein, H.L. (1996). Mutations in the RNA polymerase II transcription machinery suppress the hyperrecombination mutant *hpr1* of *Saccharomyces cerevisiae*. Genetics *142*, 749-759.

Fish, R.N., and Kane, C.M. (2002). Promoting elongation with transcript cleavage stimulatory factors. Biochim. Biophys. Acta *1577*, 287-307.

Flanagan, P M., Kelleher R. J. III, Sayre, M. H., Tschochner, H., and Kornberg, R. D. (1991). A mediator required for activation of RNA polymerase II transcription *in vitro*. Nature *350*, 436-438.

Formosa, T., Ruone, S., Adams, M.D., Olsen, A.E., Eriksson, P., Yu, Y., Rhoades, A.R., Kaufman, P.D., and Stillman, D.J. (2002). Defects in SPT16 or POB3 (yFACT) in *Saccharomyces cerevisiae* cause dependence on the Hir/Hpc pathway: polymerase passage may degrade chromatin structure. Genetics *162*, 1447-1471.

Garrett, K.P., Aso, T., Bradsher, J.N., Foundling, S.I., Lane, W.S., Conaway, R.C., and Conaway, J.W. (1995). Positive regulation of general transcription factor SIII by a tailed ubiquitin homolog. Proc. Natl. Acad. Sci. USA *92*, 7172-7176

Gerber, H-P., Hagmann, M., Seipel, K., Georgiev, O., West, M.A., Liting Tong, Y., Schaffner, W., and Corden, J.L. (1995). RNA polymerase II C-terminal domain required for enhancer-driven transcription. Nature *374*, 660-662.

Gerber, M., Ma, J., Dean, K., Eissenberg, J.C., and Shilatifard, A. (2001). *Drosophila* ELL is associated with actively elongating RNA polymerase II on transcriptionally active sites *in vivo*. EMBO J. *20*, 6104-6114.

Gerber, M., and Shilatifard, A. (2003). Transcriptional elongation by RNA polymerase II and histone methylation. J. Biol. Chem. *278*, 26303-26306.

Gilbert, C., Kristjuhan, A., Winkler, G., and Svejstrup, J.Q. (2004). Elongator interactions with nascent mRNA revealed by RNA immunoprecipitation. Mol. Cell *14*, 457-464.

Goodrich, J.A., and Tjian, R. (1994a). TBP-TAF complexes: selectivity factors for eukaryotic transcription. Curr. Opin. Cell Biol. *6*, 403-409.

Goodrich, J.A., and Tjian, R. (1994b). Transcription factors IIE and IIH and ATP hydrolysis direct promoter clearance by RNA polymerase II. Cell *77*, 145-156.

Grunstein, M. (1990). Histone function in transcription. Annu. Rev. Cell Biol. *6*, 643-678.

Grunstein, M. (1997). Histone acetylation in chromatin structure and transcription. Nature *389*, 349-52.

Guo, Z., and Sherman, F. (1996). 3′-end-forming signals of yeast mRNA. Trends Biochem. Sci. *21*, 477-481

Gustafsson, C.M., Myers, L.C., Beve, J., Spahr, H., Lui, M., Erdjument-Bromage, P., Tempst, P., and Kornberg, R.D. (1998). Identification of new mediator subunits in the RNA polymerse II holoenzyme from *Saccharomyces cerevisieae*. J. Biol. Chem. *273*, 30851-30854.

Hahn, S., Buratowski, S., Sharp, P., and Guarente, L. (1989). Yeast TATA-binding protein TFIID binds to TATA elements with both consensus and nonconsensus DNA sequences. Proc. Natl. Acad. Sci. USA *86*, 5718-5722.

Hamspey, M. (1998). Molecular genetics of the RNA polymerase II general transcriptional machinery. Microbiol. And Mol. Biol. Rev. 465-503.

Hampsey, M., and Reinberg, D. (2003). Tails of intrigue: phosphorylation of RNA polymerase II mediates histone methylation. Cell *113*, 429-432.

Han, M., and Grunstein, M. (1988). Nucleosome loss activates yeast downstream promoters *in vivo*. Cell *55*, 1137-1145.

Han, M., Kim, U.J., Kayne, P., and Grunstein, M. (1988). Depletion of histone H4 and nucleosomes activates the PHO5 gene in *Saccharomyces cerevisiae*. EMBO J. *7*, 2221-2228.

Han, S.J., Lee, Y.C., Gim, B.S., Ryu, G-H., Park, S.J., Lane, W.S., and Kim, Y-J. (1999). Activator specific requirement of yeast mediator proteins for RNA polymerase II transcriptional activation. Mol. Cell. Biol. *19*, 979-988.

Hartzog, G.A., Wada, T., Handa, H., and Winston, F. (1998). Evidence that Spt4, Spt5, and Spt6 control transcription elongation by RNA polymerase II in *Saccharomyces cerevisiae*. Genes Dev. *12*, 357-369.

Hekmatpanah, D.S., and Young, R.A. (1991). Mutations in a conserved region of RNA polyermase II influence the accuracy of mRNA start site selection. Mol. Cell. Biol. *11*, 5781-5791.

Hengartner, C.J., Thompson, C.M., Zhang, J., Chao, D.M., Liao, S.M., Koleske, A.J., Okamura, S., and Young, R.A. (1995). Association of an activator with an RNA polyemrae II holenzyme. Genes Dev. *9*, 897-910.

Ho, C.K., Schwer, B., and Shuman S. (1998). Genetic, physical, and functional interactions between the triphosphatase and guanylyltransferase components of the yeast mRNA capping apparatus. Mol. Cell. Biol. *18*, 5189-5198.

Ho, C.K., and Shuman, S. (1999). Distinct roles for CTD Ser-2 and Ser-5 phosphorylation in the recruitment and allosteric activation of mammalian mRNA capping enzyme. Mol. Cell *3*, 405-411.

Holstege, F.C.P., Jennings, E.J., Wyrick, J.J., Lee, T.I., Hengartner, C.J., Green, M.R., Golub, T.R., Lander, E.S., and Young, R.A. (1998) Dissecting the regulatory circuitry of a eukaryotic genome. Cell *95*, 717-728.

Ivanov, D., Kwak, Y.T., Guo, J., and Gaynor, R.B. (2000). Domains in the SPT5 protein that modulate its transcriptional regulatory properties. Mol. Cell. Biol. *20*, 2970-2983.

Jenuwein, R., and Allis, C.D. (2001). Translating the histone code. Science *293*, 1074-1080.

Kang, J.S., Kim, S.H., Hwang, M.S., Han, S.J., Lee, Y.C., and Kim, Y.J. (2001). The structural and functional organization of the yeast mediator complex. J. Biol.Chem. *276*, 42003-42010.

Keller, W., and Minvielle-Sebastia, L. (1997). A comparison of mammalian and yeast pre-mRNA 3′-end processing. Curr. Opin. Cell Biol. *9*, 329-336.

Kelley, D.E., Stokes, D.G., and Perry, R.P. (1999). CHD1 Interacts with SSRP1 and depends on both its chromodomain and its ATPase/helicase-like domain for proper association with chromatin. Chromosoma *108*, 10-25.

Kettenberger, H., Armache, K-J., and Cramer, P. (2004). Complete RNA polymerase II elongation complex structure and its interactions with NTP and TFIIS. Mol. Cell *16*, 955-965.

Kim, J.L., Nikolov, D.B., and Burley, S.K. (1993a). Co-crystal structure of TBP recognizing the minor groove of a TATA element. Nature *365*, 520-527.

Kim, Y., Geiger, J H., Hahn, S., and Sigler, P.B. (1993b). Crystal structure of a yeast TBP/TATA-box complex. Nature *365*, 512-520.

Kim, Y.J., Bjorklund, S., Li, S., Sayre, M.H., and Kornberg, R D. (1994). A multiprotein mediator of transcriptional activation and its interaction with the C-terminal repeat domain of RNA polymerase II. Cell *77*, 599-608.

Kim, J.B., and Sharp, P.A. (2001). Positive transcription elongation factor B phosphorylates hSPT5 and RNA polymerase II carboxyl-terminal domain independently of cyclin-dependent kinase-activating kinase. J. Biol. Chem. *276*, 12317-12323.

Kim, M., Ahn, S. H., Krogan, N. J., Greenblatt, J. F., Buratowski, S. (2004) Transitions in RNA polymerase II elongation complexes at the 3′ ends of genes. EMBO J. 23, 354-364.

Kireeva, M.L., Komissarova, N., Waugh, D.S., and Kashlev, M. (2000). The 8-nucleotide-long RNA:DNA hybrid is a primary stability determinant of the RNA polymerase II elongation complex. J. Biol. Chem. *275*, 6530-6536.

Koch, C., Wollmann, P., Dahl, M., and Lottspeich, F. (1999). A role for Ctr9p and Paf1p in the regulation of G1 cyclin expression in yeast. Nucleic Acids Res. *27*, 2126-2134.

Koh, S.S., Ansari, A.Z., Ptashne, M., and Young, R.A. (1998). An activator target in the RNA polymerase II holoenzyme. Mol. Cell *1*, 895-904.

Komarnitsky, P., Cho, E.J., and Buratowski, S. (2000). Different phosphorylated forms of RNA polymerase II and associated mRNA processing factors during transcription. Genes Dev. *14*, 2452-2460.

Kornberg, R.D., and Lorch, Y. (1991). Irresistible force meets immovable object: Transcription and the nucleosome. Cell *67*, 833-836.

Krogan, N.J., and Greenblatt, J.F. (2001). Characterization of a six-subunit holo-elongator required for the regulated expression of a group of genes in *Saccharomyces cerevisiae*. Mol. Cell. Biol. *21*, 8203-8212.

Krogan, N.J., Dover, J., Khorrami, S., Greenblatt, J.F., Schneider, J., Johnston, M., and Shilatifard (2002a). COMPASS, a histone H3 (lysine 4) methyltransferase required for telomeric silencing of gene expression. J. Biol. Chem., *277*, 10753-10755.

Krogan, N.J., Kim, M., Ahn, S.H., Zhong, G., Kobor, M.S., Cagney, G., Emili, A., Shilatifard, A., Buratowski, S., and Greenblatt, J.F. (2002b). RNA polymerase II elongation factors of *S. cerevisiae*: a targeted proteomics approach. Mol. Cell. Biol. *22*, 6979-6992.

Krogan, N.J., Dover, J., Wood, A., Schneider, J., Heidt, J., Boateng, M.A., Dean, K., Ryan, O.W., Golshani, A., Johnston, M., Greenblat, J.F., and Shilatifard, A. (2003a). The Paf1 complex is required for histone H3 methylation by COMPASS and Dot1p: linking transcriptional elongation to histone methylation. Mol. Cell *11*, 721-729.

Krogan, N.J., Kim, M., Tong, A., Golshani, A., Cagney, G., Canadien, V., Richards, D.P., Beattie, B.K., Emili, A., Boone, C., Shilatifard, A., Buratowski, S., and Greenblatt J.F. (2003b). Methylation of histone H3 by Set2 in *Saccharomyces cerevisiae* is linked to transcriptional elongation by RNA polymerase II. Mol. Cell. Biol. 23, 4207-4218.

Kulish, D., and Struhl, K. (2001). TFIIS enhances transcriptional elongation through an artificial arrest site *in vivo*. Mol. Cell. Biol. *21*, 4162-4168.

Kumar, K.P, Akoulitchev S., and Reinberg, D. (1998). Promoter-proximal stalling results from the inability to recruit transcription factor IIH to the transcription complex and is a regulated event. Proc. Natl. Acad. Sci. USA *95*, 9767-9772.

Lachner, M., O'Sullivan, R.J., and Jenuwein, T. (2003). An epigenetic road map for histone lysine methylation. J. Cell. Sci. *116*, 2117-2124.

Lagrange, T., Kapanidis, A.N., Tang, H., Reinberg, D., and Ebright, R.H. (1998). New core promoter element in RNA polymerase II-dependent transcription: Sequence-specific DNA binding by transcription factor IIB. Genes Dev. *12*, 34-44.

Laybourn, P.J., and Dahmus, M.E. (1989). Transcription-dependent structural changes in the C-terminal doman of mammalian RNA polymerase subunit IIa/o J. Biol. Chem. *264*, 6693-6698.

Laybourn, P.J., and Dahmus, M.E. (1990). Phosphorylation of RNA polymerase IIA occurs subsequent to interaction with the promoter and before the initiation of transcription. J. Biol. Chem. *265*, 13165-13173.

Lee, C.L., and Kim, Y.J. (1998). Requirement for a functional interaction between Mediator components Med6 and Srb4 in RNA polymerase II transcription. Mol. Cell Biol. *18*, 5364-5370.

Lee, I.L., and Young, R.A. (2000). Transcription of eukaryotic protein-coding genes. Annu. Rev. Genet. *34*, 77-137.

Lewis, J.D., Gunderson, S.I., and Mattaj, I.W. (1995). The influence of 5′-end and 3′-end structures on pre-messenger-RNA metabolism. J. Cell Sci. *19*, 13-19.

Li, B., Howe, L., Anderson, S., Yates, J.R. III, and Workman, J.L. (2003). The Ser 2 histone methyltransferase functions through the phosphorylated carboxyl-terminal domain of RNA polymerase II. J. Biol. Chem. 278, 897-8903.

Li, Y., Bjorklund, S., Hiang, Y.W., Kim, Y.J., Lane, W.S., Stillman, D.J., and Kornberg, R.D. (1995). Yeast global transcriptional regulators Sin4 and Rgr1 are components of mediator complex/RNA polymerase II holoenzyme. Proc. Natl. Acad. Sci. USA *92*, 10864-10868.

Licatalosi, D.D., Geiger, G., Minet, M., Schroeder, S., Cilli, K., McNeil, J.B., and Bentley, D.L. (2002). Functional interaction of yeast pre-mRNA 3′ end processing factors with RNA polymerase II. Mol. Cell *9*, 1101-1111.

Licht, C.L., Stevnsner, T., and Bohr, V.A. (2003). Cockayne syndrome group B cellular and biochemical functions. Am. J. Hum. Genet. *73*, 1217-1239.

Lindstrom, D.L., Squazzo, S.L., Muster, N., Burckin, T.A., Wachter, K.C., Emigh, C.A., McCleery, J.A., Yates, J.R. III, and Hartzog, G.A. (2003). Dual roles for Spt5 in pre-mRNA processing and transcription elongation revealed by identification of Spt5-associated proteins. Mol. Cell. Biol. *23*, 1368-1378.

Liu, H.Y., Badarinarayana, V., Audino, D.C., Rappsilber, J., Mann, M., and Denis, C.L. (1998). The NOT proteins are part of the CCR4 transcriptional complex and affect gene expression both positively and negatively. EMBO J. *17*, 1096-1106

Lorch, Y., Beve, J., Gustafsson, C.M., Myers, L.C., and Kornberg, R.D. (2000). Mediator-nucleosome interaction. Mol. Cell *6*, 197-201.

Lu, H., Flores, O., Weinmann R., and Reinberg, D. (1991). The nonphosphoryylated form of RNA polymerase II preferentially associates with the preinitiation complex. Proc. Natl. Acad. Sci. USA *88*, 10004-10008.

Malik, S., and Roeder, R.G. (2000). Transcriptional regulation through mediator-like coactivators in yeast and metazoan cells. TIBS *25*, 277-283.

Maniatis, T., and Reed, R. (2002). An extensive network of coupling among gene expression machines. Nature *416*, 499-506.

McCracken, S., Fong, N., Yankulov, K., Ballantyne, S., Pan, G.H., Greenblatt, J., Patterson, S.D., Wickens, M., and Bentley, D.L. (1997a). The C-terminal domain of RNA polymerase II couples messenger RNA processing to transcription. Nature *385*, 357-361.

McCracken, S., Fong, N., Rosonina, R., Yankulov, K., Brothers, G., Siderovski, D., Hessel, A., Foster, S., Shuman, S., and Bentley, D.L. (1997b). 5′-capping enzymes are targeted to pre-mRNA by binding to the phosphorylated carboxy-terminal domain of RNA polymerase II. Genes Dev. *11*, 3306-3318.

McKine, K., Moore, P.A., Hull, M.W., and Woychik, N.A. (1995). Six human RNA polymerase subunits functionally substitute for their yeast counterparts. Mol. Cell. Biol. *15*, 6895-6900.

Miller, T., Williams, K., Johnstone, R.W., and Shilatifard, A. (2000). Identification, cloning, expression, and biochemical characterization of the testis-specific RNA polymerase II elongation factor ELL3 J. Biol. Chem. *275*, 32052-32056.

Miller, T., Krogan, N.J., Dover, J., Erdjument-Bromage, H., Tempst, P., Johnston, M., Greenblatt, J.F., and Shilatifard, A. (2001). COMPASS: a complex of proteins associated with a trithorax-related SET domain protein. Proc. Natl. Acad. Sci. *98*, 12902-12907.

Moteki, S., and Price, D. (2002). Functional coupling of capping and transcription of mRNA. Mol. Cell *10*, 599-609.

Mueller, C.L., and Jaehning, J.A. (2002). Ctr9, Rtf1, and Leo1 are components of the Paf1/RNApolymerase II complex. Mol. Cell. Biol. *22*, 1971-1980.

Mueller, C.L., Porter, S.E., Hoffman, M.G., and Jaehning, J.A. (2004). The Paf1 complex has functions independent of actively transcribing RNA polymerase II. Mol. Cell *14*, 447-456.

Myer, V.E., and Young, R.A. (1998). RNA polymerase II holoenzymes and subcomplexes. J. Biol. Chem. *273*, 27757-27760.

Myers, L.C., Leuther, K., Bushnell, D.A., Gustafsson, C. M., and Kornberg, R.D. (1997). Yeast RNA polymerase II transcription reconstituted with purified proteins. Methods in Enzymol. *12*, 212-216.

Myers, L.C., Gustafsson, C.M., Bushnell, D.A., Lui, M., Erdjument-Bromage, H., Tempst, P. and Kornberg, R.D. (1998). The Med proteins of yeast and their function through the RNA polymerse II carboxy-terminal domain. Genes Dev. *12*, 45-54.

Myers, L.C., Gustafsson, C.M., Hayashibara, K.C., Brown, P.O., and Kornberg, R.D. (1999). Mediator protein mutations that selectively abolish activated transcription. Proc. Natl. Acad. Sci. USA *96*, 67-72.

Myers, L.C., Lacomis, L., Erdjument-Bromage, H., and Tempst, P. (2002). The yeast capping enzyme represses RNA polymerase II transcription. Mol. Cell *10*, 883-894.

Nagy, P.L., Griesenbeck, J., Kornberg, R.D., and Cleary, M.L. (2002). A trithorax-group complex purified from *Saccharomyces cerevisiae* is required for methylation of histone H3. Proc. Natl. Acad. Sci. USA *99*, 90-94.

Natori, S., Takeuchi, K., Takahashi, K., and Mizuno, D. (1973). DNA dependent RNA polymerase from *Ehrlich ascites* tumor cells. II. Factors stimulating the activity of RNA polymerase II. J. Biochem. (Tokyo) *73*, 879-888.

Ng, H.H., Robert, F., Young, R.A., and Struhl, K. (2003). Targeted recruitment of Set1 histone methylase by elongating pol II provides a localized mark and memory of recent transcriptional activity. Mol. Cell. *11*, 709-719.

Noma, K., and Grewal, S.I. (2002). Histone H3 lysine 4 methylation is mediated by Set1 and promotes maintenance of active chromatin states in fission yeast. Proc. Natl. Acad. Sci. USA *99*, 16438-16445.

Nonet, M., Sweetser, D., and Young, R.A. (1987). Functional redundancy and structural polymorphism in the large subunit of RNA polymerase II. Cell *50*, 909-915.

Nonet, M.L., and Young, R.A. (1989). Intragenic and extragenic suppressors of mutations in the heptapeptide repeat domain *of Saccharomyces cerevisiae* RNA polymerase II. Genetics *123*, 715-724.

O'Brien, T., Hardin, S., Greenleaf, A., and Lis, J.T. (1994) Phosphorylation of RNA polymerase II C-terminal domain and transcriptional elongation. Nature *370*, 75-77.

Orphanides, G., Lagrange, T., and Reinberg, D. (1996). The general transcription factors of RNA polymerase II. Genes Dev. *10*, 2657-2683.

Orphanides, G., LeRoy, G., Chang, C.J., Luse, D.S., and Reinberg, D. (1998). FACT, a factor that facilitates transcript elongation through nucleosomes. Cell *92*, 105-116.

Orphanides, G., Wu, W.H., Lane, W.S., Hampsey, M., and Reinberg, D. (1999). The chromatin-specific transcription elongation factor FACT comprises human SPT16 and SSRP1 proteins. Nature *400*, 284-288.

Orphanides, G., and Reinberg, D. (2002). A unified theory of gene expression. Cell *108*, 439-451.

Otero, G., Fellows, J., Li, Y., de Bizemont, T., Dirac, A.M.G., Gustafsson, C.M., Erdjument-Bromage, H., Tempst, P., and Svejstrup, J.Q. (1999). Elongator, a multisubunit component of a novel RNA polymerase II holoenzyme for transcriptional elongation. Mol. Cell *3*, 109-118.

Pal, M., and Luse, D.S. (2002). Strong natural pausing by RNA polymerase II within 10 bases of transcription start may result in repeated slippage and reextension of the nascent RNA. Mol. Cell. Biol. *22*, 30-40.

Pal, M., and Luse, D.S. (2003). The initiation-elongation transition: Lateral mobility of RNA in RNA polymerase II complexes is greatly reduced at +8/+9 and absent by +23. Proc. Natl. Acad. Sci. USA *100*, 5700-5705.

Peterson, C.L., and Tamkun, J.W. (1995). The SWI-SNF complex: a chromatin remodeling machine? Trends Biochem. Sci. *20*, 143-146.

Pokholok, D.K., Hannett, N.M., and Young, R.A. (2002). Exchange of RNA polymerase II initiation and elongation factors during gene expression *in vivo*. Mol. Cell *9*, 799-809.

Poon, D., and Weil, P.A. (1993). Immunopurification of yeast TATA-binding protein and associated factors: presence of transcription factor IIB transcription activity. J. Biol. Chem. *268*, 15325-15328.

Porter, S.E., Washburn, T.M., Chang, M., and Jaehning, J.A. (2002). The yeast Paf1-RNA polymerase II complex is required for full expression of a subset of cell cycle-regulated genes. Eukaryot. Cell *1*, 830-842.

Powell, W., and Reines, D. (1996). Mutations in the second largest subunit of RNA polymerase II cause 6-azauracil sensitivity in yeast and increased transcriptional arrest *in vitro*. J. Biol. Chem. *271*, 6866-6873.

Price, D.H., Sluder, A.E., and Greenleaf, A.l. (1989). Dynamic interaction between a *Drosophila* transcription factor and RNA polymerase II. Mol. Cell. Biol. *9*, 1465-1475.

Reed, S.I., Ferguson, J., and Jahng, K.Y. (1988). Isolation and characterization of two genes encoding yeast mating pheromone elements: CDC72 and CDC73. Cold Spring Harbor Symp. on Quant. Biol. *53*, 621-627.

Roeder, R. (1996). The role of general initiation factors in transcription by RNA polymerase II. TIBS *21*, 327-333.

Roguev, A., Schaft, D., Shevchenko, A., Pijnappel, W.W.M.P., Wilm, M., Aasland, R., and Stewart, A.F. (2001). The *Saccharomyces cerevisiae* Set1 complex includes an Ash2 homologue and methylates histone 3 lysine 4. EMBO J. *20*, 7137-7148.

Rondón, A.G., Garcia-Rubio, M., Gonzaalez-Barrera, S., and Aguilera, A. (2003). Molecular evidence for a positive role of Spt4 in transcription elongation. EMBO J. *22*, 612-620.

Santos-Rosa, H., Schneider, R., Bannister, A., Sherriff, J., Bernstein, B.E., Emre, N.C.T., Schreiber, S.L., Mellor, J., and Kouzarides, T. (2002). Active genes are tri-methylated at K4 of histone H3. Nature *419*, 407-411.

Saunders, A., Werner, J., Andrulis, E.D., Nakayama, T., Hirose, S., Reinberg, D., and Lis, J.T. (2003). Tracking FACT and the RNA polymerase II elongation complex through chromatin *in vivo*. Science *301*, 1094-1096.

Schroeder, S., Schwer, B., Shuman, D., and Bentley, D.L. (2000). Dynamic association of capping enzymes with transcribing RNA polymerase II. Genes Dev. *14*, 2435-2440.

Sekimizu, K., Kobayashi, N., Mizuno, D., Natori, S. (1976). Purification of a factor from *Ehrlich ascites* tumor cells specifically stimulating RNA polymerase II. Biochemistry *15*, 5064-5070.

Selby, C.P., and Sancar, A. (1997). Cockayne syndrome group B protein enhances elongation by RNA polymerase II. Proc. Natl. Acad. Sci. USA *94*, 11205-11209.

Shi, S., Finkelstein, A., Wolf, A.J., Wade, P.A., Burton, Z.F., and Jaehning, J.A. (1996). Paf1p, an RNA polymerase II-associated factor in yeast, may have both positive and negative roles in transcription. Mol. Cell. Biol. *16*, 669-676.

Shi, X., Chang, M., Wolf, A.J., Chang, C-H., Frazer-Abel, A., Wade, P.A., Burton, Z.F., and Jaehning, J.A., (1997). Cdc73p and Paf1p are found in a novel RNA polymerase II-containing complex distinct from the Srbp-containing holoenzyme. Mol. Cell. Biol. *17*, 11160-1169.

Shilatifard, A., Lane, W.S., Jackson, K.W., Conaway, R.C., and Conaway, J.W. (1996). An RNA polymerase II elongation factor encoded by the human ELL gene. Science *271*, 1873-1876.

Shilatifard, A., Duan, D.R., Haque, D., Florence, C., Schubach, W.H., Conaway, J.W., and Conaway, R.C. (1997). ELL2, a new member of an ELL family of RNA polymerase II elongation factors. Proc. Natl. Acad. Sci. *94*, 3639-3643.

Shilatifard, A., Conaway, R.C., and Conaway, J.W. (2003). The RNA polymerase II elongation complex. Annu. Rev. Biochem. *72*, 693-715.

Shuman, S. (2001). Structure, mechanism, and evolution of the mRNA capping apparatus. Prog. Nucleic Acid Res. Mol. Biol. *66*, 1-40.

Simic, R., Lindstrom, D.L., Tran, H.G., Roinick, K.L., Costa, P.J., Johnson, A.D., Hartzog, G.A, and Arndt, K.M. (2003). Chromatin remodeling protein Chd1 interacts with transcription elongation factors and localizes to transcribed genes. EMBO J. *22*, 1846-1856.

Sims, R.J. III, Nishioka, K., and Reinberg, D. (2003). Histone lysine methylation: a signature for chromatin function. Trends Genet. *19*, 629-639.

Sims, R.J. III, Belotserkovskaya, R., and Reinberg, D. (2004). Elongation by RNA polymerase II: the short and long of it. Genes Dev. *18*, 2437-2468.

Smale, S.T., and Kadonaga, J.T. (2003). The RNA polymerase II core promoter. Annu. Rev. Biochem. *72*, 449-479.

Sopta, M., Carthew, R.W., and Greenblatt, J. F. (1985). Isolation of three proteins that bind to mammalian RNA polymerase II. J. Biol. Chem. *260*, 10353-10360.

Squazzo, S.L., Costa, P.J., Lindstrom, D.L., Kumer, K.E., Simic, R., Jennings, J.L., Link, A.J., Arndt, K.M., and Hartzog, G.A. (2002). The Paf1 complex physically and functionally associates with transcription elongation factors *in vivo*. EMBO J. *21*, 1764-1774.

Stolinski, L.A., Eisenmann, D.M., and Arndt, K.M. (1997). Identification of *RTF1*, a novel gene important for TATA site selection by TATA box-binding protein in *S. cerevisiae*. Mol. Cell. Biol. *17*, 4490-4500.

Stokes, D.G., Tartof, K.D., and Perry, R.P. (1996). CHD1 is concentrated in interbands and puffed regions of *Drosophila* polytene chromosomes. Proc.Natl. Acad. Sci. USA 93, 7137-7142.

Strahl, B.D., Grant, P.A., Briggs, S.D., Sun, Z-W., Bone, J.R., Caldwell, J.A., Mollah, S., Cook, R.G., Shabanowitz, J., Hunt, D.F., and Allis, C.D. (2002). Set2 is a nucleosomal histone H3-selective methyltransferase that mediates transcriptional repression. Mol. Cell. Biol. *22*, 1298-1306.

Strahl, B., and Allis, C.D. (2002). The language of covalent histone modifications. Nature *403*, 41-45.

Studitsky, V.S., Walter, W., Kireeva, M., Kashlev, M., and Felsenfeld, G. (2004). Chromatin remodeling by RNA polymerases. Trends Biochem. Sci. *29*, 127-135.

Sullivan, E.K., Weirich, C.S., Guyon, J.R., Sif, S., and Kingston, R.E. (2001). Transcriptional activation domains of human heat shock factor 1 recruit human SWI-SNF. Mol. Cell. Biol. 21, 5826-5837.

Svejstrup, J.Q. (2002). Chromatin elongation factors. Curr. Opin. Genet. Dev. *12*, 156-161.

Svetlov, V.V., and Cooper, T.G. (1995). Review: compilation and characterization of dedicated transcription factors in *Saccharomyces cerevisiae*. Yeast *11*, 1439-1484.

Takagi, Y., Conaway, J.W., and Conaway, R.C. (1995). A novel activity associated with RNA polymerase II elongation factor

SIII. SIII directs promoter-independent transcription initiation by RNA polymerase II in the absence of initiation factors. J. Biol. Chem. *270*, 24300-24305.

Takagi, Y., Conaway, J.W., and Conaway, R.C. (1996). Characterization of elongin C functional domains required for interaction with elongin B and activation of elongin A. J. Biol. Chem. *271*, 25562-25568.

Tan, S., Aso, T., Conaway, R.C., and Conaway, J.W. (1994). Roles for both the RAP30 and RAP74 subunits of transcription factor IIF in transcription initiation and elongation by RNA polymerase II. J. Biol. Chem. *269*, 25684-25691.

Tantin D., Kansal, A., and Carey, M. (1997). Recruitment of the putative transcription-repair coupling factor CSB/ERCC6 to RNA polymerase II elongation complexes. Mol. Cell. Biol. *17*, 6803-6814.

Thirman, M.J., Levitan, D.A., Kobayashi, H., Simon, M.C., and Rowley, J.D. (1994). Cloning of ELL, a gene that fuses to MLL in a t(11;19)(q23;p13.1) in acute myeloid leukemia. Proc. Natl. Acad. Sci. USA *91*, 12110-12114.

Thompson, C.M., Koleske, A.J., Chao, D.M., and Young, R.A. (1993). A multisubunit complex associated wit the RNA polymerase II CTD and TATA-bnding protein in yeast. Cell *73*, 1361-1375.

Thompson, C.M., and Young, R.A. (1995). General requirement for RNA polymerase II holoenzymes *in vivo*. Proc. Natl. Acad. Sci. USA *92*, 4587-4590.

Tran, H.G., Steger, D.J., Iyer, V.R., and Johnson, A.D. (2000). The chromo domain protein Chd1p from budding yeast is an ATP-dependent chromatin-modifying factor. EMBO J. *29*, 2323-2331.

Treich, I., Cairns, B.R., de los Santos, T., Brewster, E., and Carlson, M. (1995). SNF11, a new component of the yeast SNF-SWI complex that interacts with a conserved region of SNF2. Mol. Cell. Biol. *15*, 5240-4248.

Tsukiyama, T., Palmer, J., Landel, C.C., Shiloach, J., and Wu, C. (1999). Characterization of the imitation switch subfamily of ATP—dependent chromatin-remodeling factors in *Saccharomyces cerevisiae*. Genes Dev. *13*, 686-697.

Verrijzer, C.P., and Tjian, R. (1996). TAFs mediate transcriptional activation and promoter selectivity. TIBS *21*, 338-341.

Wada, T., Takagi, T., Yamaguchi, Y., Ferdous, A., Imai, T., Hirose, S., Sugimoto, S., Yano, K., Hartzog, G.A., Winston, F., Buratowski, S., and Handa, H. (1998). DSIF, a novel transcription elongation factor that regulates RNA polymerase II processivity, is composed of human Spt4 and Spt5 homologs. Genes Dev. *12*, 343-356.

Wade, P.A., and Jaehning, J.A. (1996). Transcriptional corepression *in vitro*: a Mot1p-associated form of TATA-binding protein is required for repression by Leu3p. Mol. Cell. Biol. *16*, 1641-1648.

Wade, P.A., Werel, W., Fentzke, R.C., Thompson, N.E., Leykam, J.F., Burgess, R.R., Jaehning, J.A., and Burton, Z.F. (1996). A novel collection of accessory factors associated with yeast RNA polymerase II. Protein Expr. Purif. *8*, 85-90.

Watanabe, T., Hayashi, K., Tanaka, A., Furumoto, T., Hanaoka, F., and Ohkuma, Y. (2003). The carboxy terminus of the small subunit of TFIIE regulates the transition from transcription initiation to elongation by RNA polymerase II. Mol. Cell. Biol. *23*, 2914-2926.

Wilson, C.J., Chao, D.M., Imbalzano, A.N., Schnitzler, G.R., Kingston, R.E., and Young, R.A. (1996). RNA polymerase II holoenzyme contains SWI/SNF regulators involved in chromatin remodeling. Cell *84*, 235-244.

Wind, M., and Reines, D. (2000). Transcription elongation factor SII. BioEssays *22*, 327-336.

Wind-Rotolo, M., and Reines, D. (2001). Analysis of gene induction and arrest site transcription in yeast with mutations in the transcription elongation machinery. J. Biol. Chem. *276*, 11531-11538.

Winkler, S.G., Petrakis, T.G., Ethelberg, S., Tokunaga, M., Erdjument-Bromage, H., Tempst, P., and Svejstrup J.Q. (2001). RNA polymerase II Elongator holoenzyme is composed of two discrete subcomplexes. J. Biol. Chem. *276*, 32743-32749.

Winston, F., and Carlson, M. (1992). Yeast SNF/SWI transcriptional activators and the SPT/SIN chromatin connection. Trends Genet. *8*, 387-391.

Wittschieben, B.O., Otero, G., de Bizemont, T., Fellows, J., Erdjument-Bromage, H., Ohba, Y., Li, Y., Allis, C.D., Tempst, P., and Svejstrup, J.Q. (1999). A novel histone actyltransferase is an integral subunit of elongating RNA polymerase II holoenzyme. Mol. Cell *4*, 123-128.

Wittschieben, B.O., Fellows, J., Du, W., Stillman, D.J., and Svejstrup, J.Q. (2000). Overlapping roles for the histone actyltransferase activities of SAGA and Elongator *in vivo*. EMBO J. *19*, 3060-3068.

Wu, C.H., Yamaguchi, Y., Benjamin, L.R., Horvat-Gordan, M., Washinsky, J., Enerly, E., Larsson, J., Lambertsson, A., Handa, H., and Gilmour, D. (2003). NELF and DSIF cause promoter proximal pausing on the hsp70 promoter in *Drosophila*. Genes Dev. *17*, 1402-1414.

Yamaguchi, Y., Takagi, T., Wada, T., Yano, K., Furuya, A., Sugimot, S., Hasegawa, J., and Handa, H. (1999). NELF, a multisubunit complex containing RD, cooperates with DSIF to repress RNA polymerase II elongation. Cell *97*, 41-51.

Yamaguchi, Y., Inukai, N., Narita, T., Wada, T., and Handa, H. (2002). Evidence that negative elongation factor represses transcription elongation through binding to a DRB sensitivity-inducing factor/RNA polymerase II complex and RNA. Mol. Cell. Biol. *22*, 2918-2927.

Yamamato, S., Watanabe, P.J., van der Speck, P.J., Watanabe, T., Fujimoto, H., Hanaoka, F., and Ohkuma, Y. (2001). Studies of nematode TFIIE function reveal a link between Ser-5 phosphorylation of RNA polymerse II and the transition from transcription initiation to elongation. Mol. Cell. Biol. *21*, 1-15.

Yan, Q., Moreland, R.J., Conaway, J.W., and Conaway, R.C. (1999). Dual roles for transcription factor IIF in promoter escape by RNA polymerase II. J. Biol. Chem. *274*, 35668-35675.

Yue, Z., Maldonado, E., Pillutla, R., Cho, H., Reinberg, D., and Shatkin A.J. (1997). Mammalian capping enzyme complements mutant *Saccharomyces cerevisiae* lacking mRNA guanylyltransferase and selectively binds the elongating form of RNA polymerase II. Proc. Natl. Acad, Sci. USA *94*, 12898-12903.

Zawel, L., and Reinberg, D. (1993). Initiation if transcription by RNA polymerase II: a multistep process. Prog. Nucleic Acid Res. Mol. Bio.l. *44*, 67-108.

Zawel, L., Kumar, K.P., and Reinberg, D. (1995). Recycling of the general transcription factors during RNA polymeradse II transcription. Genes Dev. *9*, 1479-1490.

Zhao, J., Hymann, L., and Moore, C. (1999). Formation of mRNA 3′ ends in eukaryotes: mechanism, regulation and interrelationships with other steps in mRNA synthesis. Microbiol. Mol. Biol. Rev. *63*, 405-445.

Zhou, Q., Lieberman, P.M., Boyer, T G., and Berk, A.J. (1992). Holo-TFIID supports transcriptional stimulation by diverse activators and from a TATA-less promoter. Genes Dev. *6*, 1964-1974.

Zhu, Y., Peterson, C.L., and Christman, M.F. (1995). *HPR1* encodes a global positive regulator of transcription in *Saccharomyces cerevisiae*. Mol. Cell. Biol. *15*, 1698-1708.

Zorio, D.A.R., and Bentley, D.L. (2004). The link between mRNA processing and transcription: communication works both ways. Exper. Cell Res. *296*, 91-97.

Chapter 04
General Cofactors: TFIID, Mediator and USA

Mary C. Thomas and Cheng-Ming Chiang

Department of Biochemistry, Case Western Reserve University School of Medicine, 10900 Euclid Avenue, Cleveland, OH 44106-4935

Key Words: TFIID, Mediator, USA, TBP, TAFs, TRF, positive cofactors, negative cofactors

Summary

Three classes of general cofactors are typically involved in gene activation to facilitate the communication between gene-specific transcription factors and components of the general transcription machinery. These general cofactors include TBP-associated factors (TAFs) found in TFIID, Mediator associated with the RNA polymerase II carboxy-terminal domain, and upstream stimulatory activity (USA)-derived positive cofactors (PC1, PC2, PC3, and PC4) and negative cofactor 1 (NC1). Among these general cofactors, only TFIID has core promoter-binding activity. While TBP and some TAF components of TFIID are able to recognize distinct core promoter elements, the largest subunit of TFIID, TAF1, also possesses enzymatic activities, which can posttranslationally modify transcriptional components and histone proteins. The enzymatic activities of TAFs constitute a unique feature of TFIID in core promoter recognition and in activator-dependent transcription, distinct from that of TBP. TFIID also interacts directly with a variety of activators and facilitates the recruitment of RNA polymerase II to the promoter region. Similar to TAFs, Mediator and USA-derived components are all capable of repressing basal transcription when activators are absent, and stimulating transcription in the presence of activators. In this chapter, we will discuss mainly in the human system how TAFs, Mediator, USA-derived cofactors, and a unique TBP-interacting negative cofactor 2 (NC2) regulate the initiation step of transcription, following the action of histone acetyltransferases and chromatin remodeling factors to expose the promoter region.

Transcription Components Modulating PIC Assembly

The nucleation pathway for preinitiation complex (PIC) formation involves TFIID binding to the core promoter, which specifies the site for the initiation of transcription, mainly through direct contact with the TATA box, initiator element (Inr), or downstream promoter element (DPE) by specific subunits of TFIID. Following promoter recognition by TFIID, the remaining general transcription factors (GTFs) and RNA polymerase II (pol II) then assemble on the TFIID-bound promoter either via a defined order of entry: TFIIA, TFIIB, pol II/TFIIF, TFIIE, and TFIIH, or via a preassembled pol II holoenzyme pathway (see Chapter by Hou and Chiang). Assembly of this promoter-bound complex is sufficient for a basal level of transcription observed in the absence of transcriptional activators. However, *in vivo*, transcriptional activators, which typically contain a DNA-binding domain (DBD) for gene targeting and an activation domain (AD) for contacting promoter-bound factors, are necessary for modulating the efficiency of PIC assembly. These activator-dependent transcription events often require general cofactors, such as TAFs, Mediator and USA-derived components, functioning as bridging molecules to transmit regulatory signals to the general transcription machinery (Fig.4.1).

Corresponding Author: Cheng-Ming Chiang, Tel: (216) 368-8550, Fax: (216) 368-3419, E-mail: cmc23@cwru.edu

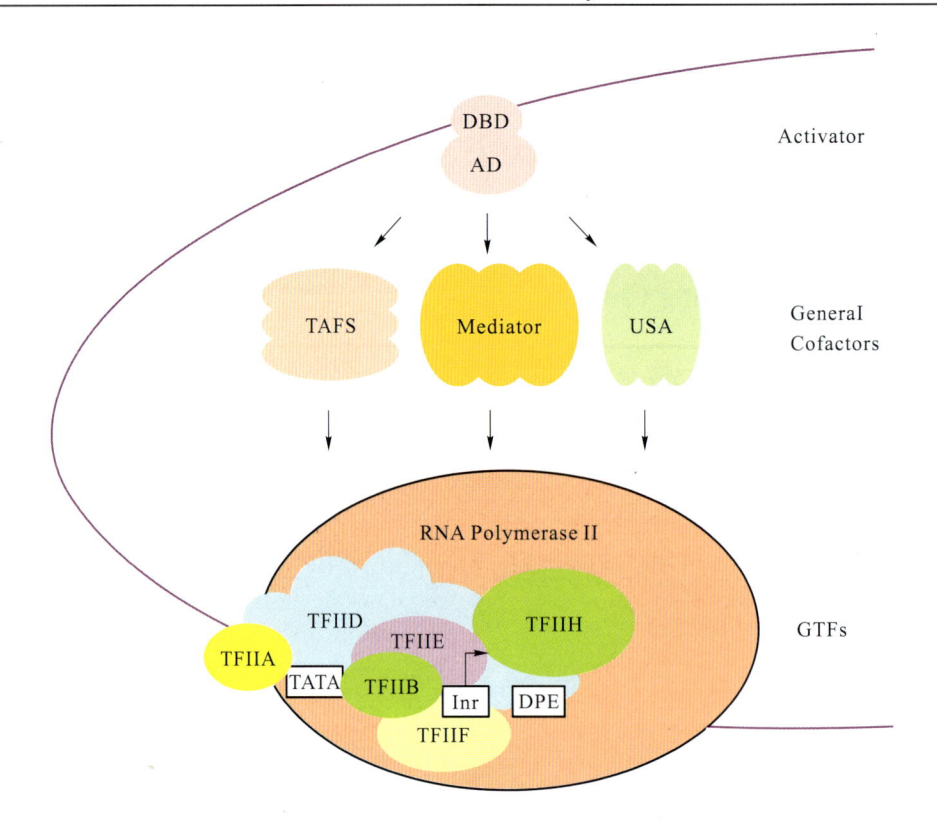

Fig.4.1 **General cofactors serve as molecular bridges in activator-dependent transcription.** General cofactors (TAFs, Mediator, USA) are required for transducing signals between gene-specific activators and components of the general transcription machinery. An activator normally contains a DNA-binding domain (DBD) contacting specific DNA sequences and an activation domain (AD) interacting with general cofactors or with components of the general transcription machinery. It is of note that TAFs normally function as an integral part of TFIID, not as a free entity in mammalian cells as drawn here.

TFIID Recognition of Core Promoter Elements

TFIID is a multiprotein complex comprised of TATA-binding protein (TBP) and approximately a dozen TBP-associated factors (TAFs). That TBP and some TAF components of TFIID bind distinct core promoter elements classifies TFIID as a core promoter-binding factor. The TBP subunit of TFIID contacts the TATA box allowing TFIID to recognize TATA-containing promoters, and the interaction between TAF-Inr and TAF-DPE also confer TFIID the ability to recognize TATA-less promoters.

A: TBP Recognition of the TATA Box

The TATA box, with a consensus sequence TATA(A/T)A(A/T), is recognized by the C-terminal region of TBP, which is phylogenetically conserved and is made up of approximately 180 amino acids (Hernandez, 1993; Nikolov and Burley, 1994; Burley and Roeder, 1996). The crystal structures of yeast TBP in complex with the TATA box of the yeast CYC1 promoter (Kim *et al.*, 1993b) as well as *Arabidopsis* TBP (Kim *et al.*, 1993a) and human TBP (Nikolov *et al.*, 1996) bound to the TATA sequence of the adenovirus major late promoter, all indicate that the DNA is severely distorted upon TBP binding. In the TATA-bound state, TBP resembles a molecular "saddle" with a pair of "stirrups" flanking the DNA-binding surface that helps bend the DNA. This saddle-shaped TBP molecule is composed of four α-helices and ten β-strands (Nikolov *et al.*, 1992), which are organized into a bipartite DNA-binding surface with each half consisting of five antiparallel β-strands located on the concave underside of TBP that straddles the DNA and one α-helix forming the side stirrup and the other α-helix situated at the convex upperside that interacts with other transcription factors (Kim *et al.*, 1993a; Kim *et al.*, 1993b; Nikolov *et al.*, 1996). This concave surface of human TBP, with two phenylalanine residues (Phe-284 and Phe-193) situated at its outermost edges, binds the TATA box in the minor groove of the DNA double helix and induces sharp kinks (more than 80°) into the DNA via intercalation of Phe-284 at the 5' end and Phe-193 at the 3' end of the TATA box (Nikolov *et al.*, 1996). Moreover, amino acid residues 190 to 194 and 281 to

285 of human TBP, which make up the two side stirrup loops, further accentuate the bending of the DNA (Juo et al., 1996; Nikolov et al., 1996). These structural studies revealing the detailed molecular interactions leading to a widening of the minor groove while compressing the major groove in order to bend the DNA helix backward are consistent with the observation that recombinant human and yeast TBP are able to induce bending of the TATA element derived from the adenovirus major late promoter in gel mobility shift assays using permuted DNA fragments (Horikoshi et al., 1992).

In contrast, the N-terminal region of human TBP, which is highly divergent among species and appears dispensable for the assembly of TFIID complexes (Zhou et al., 1993), possesses a contiguous stretch of 38 glutamine residues (amino acids 58-95) and three imperfect Pro-Met-Thr (PMT) repeats (amino acids 142-150) preceding the C-terminal core region of TBP. Although not directly involved in activator-dependent transcription by pol II (Zhou et al., 1993), this N-terminal region of human TBP is necessary for the recruitment of small nuclear RNA-activating protein complex (SNAPc) to the U6 promoter transcribed by RNA polymerase III (pol III) and seems to inhibit TATA binding by the C-terminal region of TBP on both pol III- (Mittal and Hernandez, 1997) and pol II-dependent promoters (Zhao and Herr, 2002). Since TBP binding to the TATA box is in fact a two-step process involving an initial binding of TBP to the TATA box without bending the DNA and followed by a slow transition into a more stable bent TATA-TBP complex, it is interesting to find that deletion of the glutamine-rich domain and PMT repeats in the N-terminal inhibitory region promotes formation of the stable bent TBP-DNA complex and that TFIIB's ability to enhance TBP binding to the TATA box is likely due to induced stabilization of the bent TATA-TBP complex following TFIIB binding to the solvent-exposed surface on the convex side of the TBP core (Zhao and Herr, 2002).

B: Autoregulation of TFIID-Promoter Complex Formation

As commonly observed with many DNA-binding proteins, TBP also exhibits nonspecific DNA-binding activity sometimes leading to the formation of nonproductive transcription complexes on scattered AT-rich sequences. To prevent spurious transcription events initiating at nonpromoter sequence elements, TBP may form a homodimer that is inactive in DNA binding or associate with TAFs to reduce nonspecific DNA-binding activity of TBP and concurrently increase its specificity toward TATA-containing promoters. Formation of TBP homodimers was initially observed in the crystal structure of *Arabidopsis* TBP, in the absence of DNA (Nikolov et al., 1992), and later confirmed by biochemical analysis using gel filtration/glycerol gradient profile of mouse TBP (Kato et al., 1994) and chemical cross-linking studies with human TBP (Coleman et al., 1995) and yeast TBP (Jackson-Fisher et al., 1999). Dimerization of TBP is mediated through extensive contacts between the concave surfaces of each TBP monomer, thereby masking the DNA-binding domain also located in the concave region. Clearly only the monomeric form of TBP can bind to the DNA, as evidenced by the structural analysis of the TBP-TATA complex (Kim et al., 1993a; Kim et al., 1993b; Nikolov et al., 1996). In yeast, these contacts are located in the deepest part of the concave surface, including amino acid residues N69, V71, V122, T124, N159, V161, V213, and T215, since substitutions of these amino acids individually with a bulky charged arginine residue result in destabilization of TBP dimers (Kou et al., 2003) and, as shown with the V161R mutant, also significantly reduce the half-life of TBP *in vivo* (Jackson-Fisher et al., 1999). Not surprisingly, the ability of TBP to form dimers also leads to the formation of TFIID homodimers, which can be detected by size exclusion column chromatography following chemical cross-linking and appears to be inactive in TATA recognition (Taggart and Pugh, 1996).

The DNA-binding activity of TBP is likewise subject to negative regulation by TAF1, the largest subunit of TFIID (Kokubo et al., 1993). At the N-terminus of *Drosophila* TAF1 are two separate regions able to contact TBP: amino acid residues 11-77 (TAND1, for TAF N-terminal domain 1) interacting with the concave underside of TBP to block TATA recognition (Liu et al., 1998) and amino acid residues 82-156 (TAND2) binding to the convex surface of TBP to compete with TFIIA that would otherwise facilitate TBP-TATA complex formation (Kokubo et al., 1998). Interestingly, the solution structure of *Drosophila* TAND1 bound with yeast TBP shows a remarkable resemblance to the structure of the TBP-TATA complex (Liu et al., 1998), as TAND1 exhibits an arch-shaped surface contacting the concave underside of TBP through both hydrophobic and electrostatic interactions. The negatively charged side chains of TAF1 (Asp-29, Glu-31, Glu-51, Glu-70, and Asp-73) interact with the conserved lysine and arginine residues of TBP that are also used to contact the phosphate backbone of the TATA sequence. Competition for the same binding surface underlies the mechanism for TAF1-mediated inhibition of TBP binding to the TATA box (Kokubo et al., 1998), which may also account for reduced TATA-binding and transcription activity of TFIID in comparison with that of TBP *in vivo* and *in vitro* (Ozer *et al.*, 1998; Wu and Chiang, 1998; Wu *et al.*, 1998). Likewise, TAND2-mediated inhibition of TFIIA binding to TBP is due to competition between TAF1 and TFIIA for the same conserved positive charge residues on the convex surface of TBP (Kokubo *et al.*, 1998). This TAF1-mediated effect on TBP-TFIIA and TBP-TATA interactions appear to be functionally conserved, as the same observation is also seen with experiments performed with homologous yeast and human proteins (Kokubo *et al.*, 1998; Ozer *et al.*, 1998; Banik *et al.*, 2001).

C: The Dual Function of BTAF1 in Regulating TBP-TATA Complex Formation

Other than interacting with TAFs, TBP also forms a distinct complex with a functional property similar to that of TFIID. This TFIID-like protein complex, found in the P11 0.3 M KCl (or "B") fraction (see Chapter by Hou and Chiang), is named B-TFIID which is consisted of TBP and BTAF1 (Timmers *et al.*, 1992). As observed with TAF1, human and yeast BTAF1, formerly named $TAF_{II}170$ in humans (Timmers *et al.*, 1992; van der Knapp *et al.*, 1997), Mot1 in yeast (Poon *et al.*, 1994) and 89B helicase in *Drosophila* (Goldman-Levi *et al.*, 1996), also bind the concave and convex surfaces of TBP likely in a reversible manner (Pereira *et al.*, 2001). This concave-binding region, located at the N-terminal amino acid residues 290-381 of BTAF1, not only hinders TBP-DNA complex formation, but also blocks TAF1 interaction with the concave underside of TBP (Pereira *et al.*, 2001). While the N-terminus of BTAF1 interacts with TBP, the C-terminus allows BTAF1 to induce the dissociation of TBP-TATA complexes in an ATP-dependent manner (Chicca *et al.*, 1998). The ATPase domain, which resides in the carboxy-terminus of BTAF1, has a conserved signature DEGH box within the Walker A motif, which classifies BTAF1 as a member of the DNA-dependent SWI2/SNF2 ATPase family (Pereira *et al*, 2003). This ATPase activity of human BTAF1 is strongly stimulated in the presence of both TBP and DNA (Chicca *et al.*, 1998). Moreover, while ATP is needed for TBP-TATA dissociation, it is not necessary for BTAF1 binding to TBP (Auble *et al.*, 1997). Thus, it is apparent that both N-terminal TBP interaction domain and C-terminal ATPase domain of BTAF1 are necessary for inhibiting TBP-TATA complex formation. Besides preventing the free form of TBP from binding to promoter and nonpromoter AT-rich sequences, BTAF1 also plays an active role in dissociating preformed TBP-TATA complexes, likely by acting as a molecular motor translocating across the DNA after loading through some uncharacterized DNA elements (Darst *et al.*, 2001). This nonpromoter-dissociating activity of BTAF1 may account for the coactivating activity of BTAF1 in enhancing both basal and activator-dependent transcription by redistributing TBP to correct promoter regions (Collart, 1996; Li *et al.*, 1999; Muldrow *et al.*, 1999; Andrau *et al.*, 2002; Geisberg *et al.*, 2002), consistent with the finding that the ATPase activity of BTAF1, originally implicated in transcriptional repression, also contributes to BTAF1-mediated transcriptional activation (Dasgupta *et al.*, 2002). Clearly, BTAF1 has a dual role in transcription. Under normal conditions, BTAF1-TBP seems to exist as an inactive promoter-bound complex, through nonconcave interaction, in yeast and is later activated by associating with other GTFs and pol II following environmental stress (Geisberg and Struhl, 2004). This is also in agreement with genome-wide transcriptional profiling indicating that 10% and 5% of yeast genes are respectively upregulated and downregulated by BTAF1 (Geisberg *et al.*, 2002).

D: NC2 Regulation of TBP-TATA Complex Formation

A third TBP-containing complex regulating promoter activity contains negative cofactor 2 (NC2) in association with TBP. NC2, consisting of NC2α (DRAP1) and NC2β (Dr1) that interact with each other through histone fold motifs (Goppelt *et al.*, 1996), normally functions as a repressor in inhibiting transcription from TATA-containing promoters (Meisterernst and Roeder, 1991; Inostroza *et al.*, 1992) and as a coactivator in stimulating transcription through DPE-driven TATA-less promoters (Willy *et al.*, 2000). Recent structural analysis of NC2-TBP-TATA ternary complex, revealed by X−ray crystallography at 2.6 Å resolution, shows that the N-terminal regions of both NC2 subunits bind DNA on the underside of the preformed TBP-TATA complex and the C-terminus of NC2β additionally contacts the convex surface of TBP, hence giving NC2 the appearance of a molecular "clamp" that grips both upper and lower surfaces of the TBP-TATA complex (Kamada *et al.*, 2001). Obviously, this molecular clamp is able to block PIC assembly by inhibiting TFIIA and TFIIB binding to the upper side of TBP, an observation consistent with gel-shift and *in vitro* transcription assays performed with recombinant proteins (Inostroza *et al.*, 1992; Kim *et al.*, 1995; Goppelt *et al.*, 1996). Indeed, some transcription factors, such as hypoxia-inducible factor 1α (HIF-1α), appear to inhibit transcription by inducing NC2-mediated blocking of PIC assembly (Denko *et al.*, 2003).

In contrast to our understanding regarding the repression activity of NC2, very little is known about the coactivating function of NC2. It is likely that NC2 may work in conjunction with TFIID to enhance TATA-less gene transcription, as both proteins are implicated in DPE function (Burke and Kadonaga, 1997; Willy et al., 2000). Alternatively, NC2 may positively regulate transcription by targeting events downstream of the initiation step (e.g., elongation), since NC2 association with the hyperphosphorylated form of pol II seems necessary for Gal4-VP16-mediated activation from the TATA-containing HIV-1 promoter (Castaño et al., 2000). However, the exact mechanism underlying the coactivating function of NC2 during the transcription process remains to be elucidated.

In addition to forming a heterodimeric NC2 complex, NC2α and NC2β are also individually involved in the control of specific gene transcription. It seems that free forms of NC2α and NC2β are mainly found in exponentially growing yeast cells, whereas stable NC2α and NC2β complex is only detected after glucose depletion (Creton et al., 2002). The free entity of NC2α allows it to interact with BTAF1 and in turn stimulates BTAF1 association with the convex surface of TBP on the DNA (Klejman et al., 2004). Interestingly, NC2α appears to coreside with TBP at transcriptionally active promoters and NC2β is mainly associated with TBP-bound promoters at transcriptionally repressed genes (Creton et al., 2002). This is not surprising considering the fact that NC2β binding to the upper surface of TBP blocks TFIIB association and the available NC2α structure shows only binding to the underside of the TBP-TATA complex presumably without affecting PIC assembly (Kamada et al., 2001).

E: Positive Regulators that Promote TBP-TATA Complex Formation

For TBP to nucleate the assembly of a functional PIC, it must overcome molecular impediments that prevent TBP-TATA complex formation. Some of the impediments include dimerization of TBP and inhibition of TBP binding to the TATA box exerted by TAF1, BTAF1, and NC2, as mentioned in the previous sections. Alleviation of these inhibitory activities can be achieved by transcriptional regulators, which may function by enhancing TBP binding to the TATA box, antagonizing repressor binding for the overlapping surfaces, stabilizing the bent TATA-TBP complex, or by modifying the chromatin structure surrounding the promoter region.

In general, TFIIA and TFIIB enhance TBP binding to the TATA box and further stabilize the TBP-promoter complex via distinct mechanisms. TFIIA increases the likelihood of precise promoter recognition by TBP via promoting the dissociation of TBP dimers and thus facilitating the loading of monomeric TBP onto the TATA box (Coleman et al., 1999). TFIIA also competes with the inhibitory domain of TAF1 for overlapping binding regions on TBP, hence alleviating TAF1-mediated inhibition of TATA recognition (Kokuba et al., 1998). Furthermore, incorporation of TFIIA into the TBP-TATA complex renders the ternary complex resistant to BTAF1-mediated dissociation of the TBP-TATA complex (Auble and Hahn, 1993). Similarly, TFIIB can enhance TBP binding to the TATA box (Imbalzano et al., 1994b) and also stabilize the bent TBP-TATA complex (Zhao and Herr, 2002), thereby reducing the dissociation rate of TBP from the promoter region (Wolner and Gralla, 2001). Clearly, TFIIA and TFIIB are often needed for stable TBP-TATA complex formation, yet both protein factors may not enhance TBP binding to the TATA box when the promoter is in a nucleosome configuration.

Considering that approximately 146 base pairs of DNA is wrapped around each core histone octamer *in vivo*, the promoter region may be incorporated into a nucleosome, rendering the TATA box inaccessible for TBP binding thus suppressing basal transcription (Imbalzano et al., 1994a). Undoubtedly, additional protein factors are needed to alleviate transcriptional repression from nucleosome-embedded promoters. In this regard, chromatin-modifying enzymes play a key role in enhancing TBP access to its recognition sequence by altering chromatin structure surrounding the TATA box (Martinez-Campa et al., 2004). Chromatin modifiers may reconfigure nucleosome structure by covalently modifying histones to reduce protein-DNA interactions (Fischle et al., 2003) or by using the energy of ATP hydrolysis to alter histone-DNA contacts in a process known as chromatin remodeling (Narlikar et al., 2002). Since the interaction between the TATA box and nucleosomal core histones are regulated by acetylation of lysine residues at the N-terminal tails of core histones, acetylation of nucleosomal core histones at the promoter region creates a less compact chromatin structure allowing TBP to bind the TATA box (Sewack et al., 2001). SWI/SNF is an example of an ATP-dependent chromatin remodeler that enhances access to the TATA box and subsequent binding by TBP and TFIIA *in vitro* (Imbalzano et al., 1994a). The action of SWI/SNF following the acetylation of histones by GCN5 histone acetyltransferase is seen *in vivo* as well, where SWI/SNF facilitates TBP binding to the β-interferon promoter (Agalioti et al., 2000). Interestingly, loading of TBP onto both TATA-containing and TATA-less promoters can be further enhanced by TBP association with other cellular proteins to form multiprotein complexes, such as TFIID and Spt-Ada-Gcn5 acetyltransferase (SAGA), both of which contain acetyltransferase activity (Zanton and Pugh, 2004).

F: TAF Nomenclature

By using antibodies raised against TBP, Tjian's group initially identified a set of polypeptides tightly associated with TBP, known as TAFs (Dynlacht et al., 1991). Later studies indicated that many of these polypeptides are highly conserved based on their characterization in S. cerevisiae, S. pombe, C. elegans, D. melanogaster, and H. sapiens (Burley and Roeder, 1996; Albright and Tjian, 2000). Since individual TAFs were originally named based on their apparent molecular weights, which often differ among species, a common name for each highly conserved TAF was proposed (Tora, 2002). This unified nomenclature facilitates cross-species comparisons of TAFs, designated as TAF1 to TAF15 for TFIID and a unique TBP-associated factor found in B-TFIID as BTAF1 (Table 4.1).

G: Core Promoter Recognition by Distinct TAF Components of TFIID

DNase I footprinting revealed that TFIID generates extended footprints beyond the TATA box on the adenovirus major late promoter when compared to that of TBP (Zhou et al., 1992; Chiang et al., 1993), indicating that TAFs may provide additional contacts with core promoter sequences (Fig.4.2) (Oelgeschläger et al., 1996; Verrijzer and Tjian, 1996). In a study using random DNA-binding site selection to identify TAF-targeted DNA sequences, a TAF1-TAF2 complex was shown to bind preferentially to sequences that matched the initiator consensus (Chalkley and Verrijzer, 1999). TFIID binding to TATA-less core promoter elements was also characterized by DNase I footprinting and photocrosslinking experiments which revealed that

Table 4.1 Nomenclature of TAFs involved in RNA polymerase II-mediated transcription.

New Name	H. sapiens	D. elanogaster	S. cerevisiae	S. pombe	C. elegans	
					New Name	Previous Name
TAF1	$TAF_{II}250$	$TAF_{II}230$	Taf145/130	$TAF_{II}111$	taf-1	taf-1
TAF2	$TAF_{II}150$	$TAF_{II}150$	Taf150	T38673*	taf-2	taf-2
TAF3†	$TAF_{II}140$	$TAF_{II}155$	Taf47		taf-3	C11G6.1*
TAF4†	$TAF_{II}130/135$	$TAF_{II}110$	Taf48	T50183*	taf-4	taf-5
TAF4b†	$TAF_{II}105$					
TAF5	$TAF_{II}100$	$TAF_{II}80$	Taf90	$TAF_{II}72$	taf-5	taf-4
TAF5b				$TAF_{II}73$		
TAF5L	PAF65β	Cannonball				
TAF6†	$TAF_{II}80$	$TAF_{II}60$	Taf60	CAA20756*	taf-6.1	taf-3.1
TAF6L	PAF65α	AAF52013*			taf-6.2	taf-3.2
TAF7	$TAF_{II}55$	AAF54162*	Taf67	$TAF_{II}62/PTR6$	taf-7.1	taf-8.1
TAF7L	TAF2Q				taf-7.2	taf-8.2
TAF8†	BAB71460*	Prodos	Taf65	T40895*	taf-8	ZK1320.12*
TAF9†	$TAF_{II}32/31$	$TAF_{II}40$	Taf17	S62536*	taf-9	taf-10
TAF9L	$TAF_{II}31L$					
TAF10†	$TAF_{II}30$	$TAF_{II}24$	Taf25	T39928*	taf-10	taf-11
TAF10b		$TAF_{II}16$				
TAF11†	$TAF_{II}28$	$TAF_{II}30β$	Taf40	CAA93543*	taf-11.1	taf-7.1
TAF11L					taf-11.2	taf-7.2
TAF12†	$TAF_{II}20/15$	$TAF_{II}30α$	Taf61/68	T37702*	taf-12	taf-9
TAF13†	$TAF_{II}18$	AAF53875*	Taf19	CAA19300*	taf-13	taf-6
TAF14			Taf30			
TAF15	$TAF_{II}68$					
BTAF1	$TAF_{II}170$	Hel89B	Mot1	T40642*	btaf-1	F15D4.1*

The TAF nomenclature for C. elegans is indicated on the right two columns. (Adapted from Tora, 2002)

† TAFs containing histone fold domains.

* Accession numbers for TAFs not yet published or biochemically characterized.

Drosophila TFIID subunits, TAF6 (i.e., TAF$_{II}$60) and TAF9 (i.e., TAF$_{II}$40), can be crosslinked to the DPE of the *Drosophila jockey* TATA-less promoter (Burke and Kadonaga, 1997). The DPE, with a consensus sequence (A/G)G(A/T)CGTG, is found in 46% of TATA-less and 40% of all promoters analyzed in *Drosophila* (Kutach and Kadonaga, 2000). Although the statistics for human DPE-containing promoters has not yet become available, some human promoters, such as interferon regulatory factor-1 (IRF-1) and TAF7, also possess a functional DPE similar to that defined in the *Drosophila* system (Burke and Kadonaga, 1997; Zhou and Chiang, 2001, 2002). Related to this, human TAF6 and TAF9 have recently been shown to bind the DPE of the human IRF-1 TATA-less promoter (Shao *et al.*, 2005). The finding that TBP and some TAF components of TFIID bind distinct core promoter elements (TATA, Inr and DPE) classifies TFIID as a *bona fide* core promoter-binding factor (Fig. 4.2).

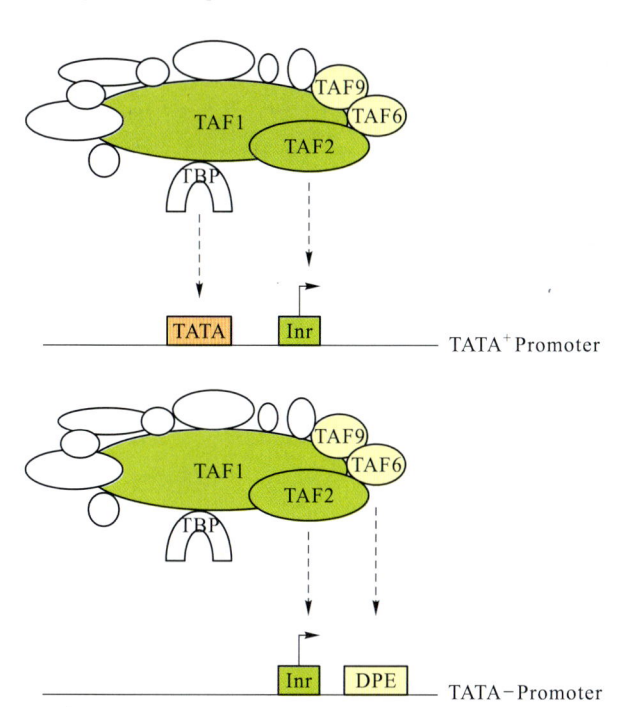

Fig.4.2 TFIID as a core promoter-binding factor for TATA-containing and TATA-less promoters. Promoter recognition is normally mediated by TBP contact with the TATA box and TAF2 interaction with the initiator element (Inr) on TATA-containing promoters (upper panel). The TAF2-Inr interaction seems to occur more efficiently and specifically in the presence of TAF1. In TATA-less promoters, an additional contact with the core promoter can be mediated by TAF6-TAF9 interactions with a downstream promoter element (DPE) or other yet-to-be characterized core promoter elements (lower panel). For simplicity, other TAFs are represented by open circles.

H: Histone-like TAFs in TFIID

Many of the TAFs present within TFIID contain histone fold domains that contribute to the recognition of core promoter elements and to the integrity of TFIID complexes. Among many histone fold-containing TAFs identified thus far (i.e., TAF3, TAF4, TAF4b, TAF6, and TAF8-13) (see Table 4.1), four of them, TAF6, TAF9, TAF12, and TAF4/TAF4b, share sequence similarity to core histones H4, H3, H2B, and H2A respectively (Gangloff *et al.*, 2001). A histone fold motif, consisting of a long α helix (α2) flanked on each side by a random coil loop (L1 or L2) and a short α helix (α1 or α3), is organized in the following order: α1L1α2L2α3 (Hoffmann *et al.*, 1996; Xie *et al.*, 1996; Birck *et al.*, 1998). This motif present in histone-like TAFs suggests that a TAF9-TAF6 heterotetramer and two of the TAF4b-TAF12 heterodimer can form a histone octamer-like structure (Selleck *et al.*, 2001), similar to H3-H4 tetramers and H2A-H2B dimers found in a nucleosome (Luger *et al.*, 1997). However, whether a histone octamer-like structure is indeed present in TFIID and contributes to its functional property remains to be investigated.

The histone fold domains of human TAF6 and TAF9 appear to be important for the DPE-binding activity of this heterotetramer (Shao *et al.*, 2005), likely due to stabilization of the protein complex. Likewise, the integrity of TFIID complexes is dependent on protein-protein interactions mediated by many of these histone fold domain-containing TAFs, as mutations in histone fold domains of these TAFs in yeast results in the disruption of TFIID complexes and reduced mRNA production (Michel *et al.*, 1998; Kirschner *et al.*, 2002).

TFIID as a Transcriptional Coactivator

Since TFIID recognition of the core promoter can be a rate-limiting step during transcription, DNA-bound activators are often necessary for contacting distinct components of TFIID in order to enhance the recruitment of TFIID to the promoter region. Because the coactivator function of TFIID has already been discussed in the chapter written by Hou and Chiang, we will address the mechanisms of TBP-mediated (i.e., TAF-independent) transcription versus TFIID-mediated (i.e., TAF-dependent) transcription in this section. This is an important issue, given the fact that TBP can be detected both as a free entity and as part of TFIID complexes in yeast (Kuras *et al.*, 2000; Li *et al.*, 2000). Although TFIID appears to be the predominant form found in mammalian cells, the free form of TBP may transiently exist under some stressed conditions (Wu *et al.*, 1998; Mitsiou and

Stunnenberg, 2000). Clearly, activator-regulated steps of transcription complex assembly are likely different, depending on whether TBP or TFIID is used as the TATA-binding factor.

A: TAF-Dependent Activation

In TATA-less promoters where TAFs are needed for promoter recognition (Burke and Kadonaga, 1997), it is plausible that the interaction between activators and TAFs are involved in the recruitment of TFIID to the Inr and DPE promoter elements, hence modulating the level of transcription. Indeed, specific DPE has been shown to be essential for the function of some transcriptional activators in *Drosophila* embryos (Butler and Kadonaga, 2001), likely reflecting unique contacts between gene-specific activators and distinct components of TFIID that allosterically modulates TAF6-TAF9 recognition of the DPE. Obviously, activators may facilitate TFIID binding to the core promoter region, which is usually the rate-limiting step for the assembly of a functional transcription complex on both TATA-containing and TATA-less promoters, as documented by many DNA-binding and transcriptional studies (see Wu and Chiang, 2001 and references therein). Normally, TFIID binds less efficiently to the TATA box compared to that of TBP, due to TAF1-mediated inhibition of TBP-promoter contacts (see the earlier section on Autoregulation of TFIID-Promoter Complex Formation). Interestingly, TAFs also contact several downstream factors, including TFIIA (Yokomori *et al.*, 1993), TFIIB (Goodrich *et al.*, 1993), TFIIE (Hisatake *et al.*, 1995), the RAP74 subunit of TFIIF (Ruppert and Tjian, 1995), and both the RPB1 and RPB2 subunits of pol II (Wu and Chiang, 2001). These TAF-mediated protein-protein interactions facilitate the entry of the remaining GTFs and pol II, whether through the sequential assembly pathway or the preassembled pol II holoenzyme pathway (see Chapter by Hou and Chiang), thereby shifting the rate-limiting step for PIC assembly from the entry of downstream factors to promoter recognition by TFIID (Fig.4.3, top panel).

B: TAF-Independent Activation

The concept that TAFs are not globally required for pol II-dependent transcription has been supported by both yeast genetic analysis (Apone *et al.*, 1996; Moqtaderi *et al.*, 1996; Walker *et al.*, 1996) and biochemical reconstitution studies (Oelgeschläger *et al.*, 1998; Wu and Chiang, 1998; Wu *et al.*, 1998; Fondell *et al.*, 1999). A dispensable role of TAFs for selective gene transcription is also consistent with genome-wide expression profiling indicating that only a subset of yeast genes, ranging from 8-20%, are affected by functional inactivation of TAF1, TAF5, TAF6, TAF10, and TAF12, individually (Green, 2000). For those genes that do not rely on TAFs, TBP appears to be the TATA-binding factor for TATA-containing gene transcription. Since TAF1-imposed inhibition of TATA binding is no longer observed with TBP, the rate-limiting step for PIC formation regulated by transcriptional activators apparently shifts from promoter recognition to the entry of downstream factors, especially the pol II recruitment step (Fig.4.3, bottom panel). This mechanism underlying TAF-independent activation is supported by order-of-addition transcription experiments and factor recruitment assays performed with individually purified general transcription factors, cofactor, pol II and activator, mimicking the assembly of transcription complexes via the sequential assembly pathway (Wu and Chiang, 2001).

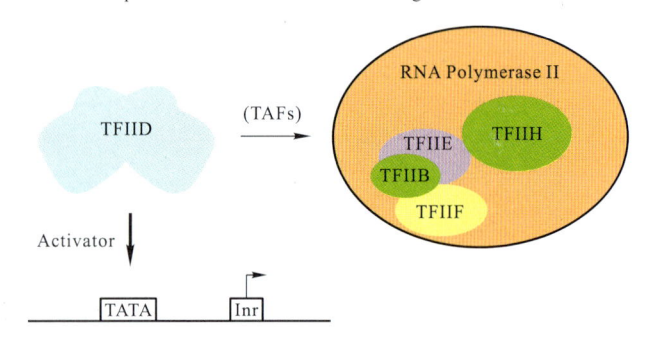

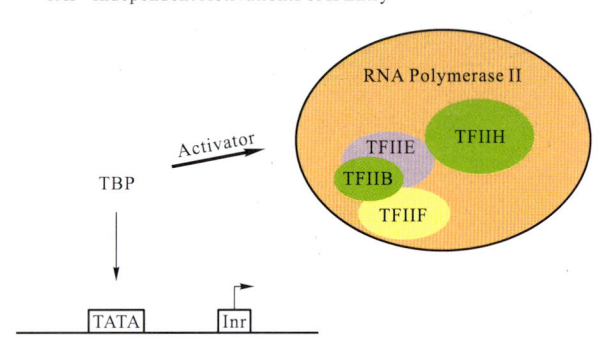

Fig.4.3 Model for TAF-dependent (*i.e.*, TFIID-mediated) and TAF-independent (*i.e.*, TBP-mediated) activation. When TFIID is present, promoter recognition by TFIID is usually the rate-limiting step (thick arrow; upper panel) facilitated by transcriptional activators. In contrast, when TBP acts as the TATA-binding factor, pol II entry is often the rate-limiting step (thick arrow; lower panel) facilitated by transcriptional activators. For simplicity, a preassembled pol II holoenzyme complex containing all the general transcription factors except TFIID is depicted. (Adapted from Wu and Chiang, 2001)

The free form of TBP, besides forming different classes of TAF complexes such as selectivity factor 1 (SL1) and TFIIIB used as core promoter-binding factors for RNA polymerase I and III, respectively, can also interact with BTAF1, NC2, or itself to downregulate the formation of TBP-TATA complexes (see the earlier sections about BTAF1, NC2, and TBP homodimers). Moreover, TBP is able to associate with the γ and unprocessed αβ subunits of TFIIA to form the TAC (TBP-TFIIA-containing) complex, which is only detectable in P19 embryonal carcinoma cells but not in differentiated cells (Mitsiou and Stunnenberg, 2000). Intriguingly, exogenous expression of p300 in monkey kidney COS-7 cells induce TAC formation from endogenous TBP and TFIIA components, correlating with the observation that acetylated TFIIA αβ is preferentially found in the TAC complex (Mitsiou and Stunnenberg, 2003). Obviously, the mechanism by which TAC regulates transcription and the identity of TAC-regulated promoters remain to be investigated.

It is important to note that, in the absence of TAFs, TBP can also function with other general cofactors, such as PC4 (Wu and Chiang, 1998; Wu et al., 1998), Mediator (Wu et al., 2003), and the coactivating activity of TFIIA and TFIIH (Wu et al., 1998), during the activation process. These general cofactors are capable of functionally replacing TAFs in conveying regulatory signals between activators and the general transcription machinery, although the precise mechanisms remain to be elucidated.

Enzymatic Activities of TFIID

The TAF1 component of TFIID possesses multiple enzymatic activities that are able to covalently modify histone tails and thus facilitate the reconfiguration of the chromatin structure in order to overcome nucleosome-mediated repression. TAF1 is known to function as an acetyltransferase in acetylating histones H3 and H4 (Mizzen et al., 1996), as a kinase in phosphorylating histone H2B (Maile et al., 2004), and as a histone-specific ubiquitin-activating/conjugating enzyme in mediating monoubiquitination of linker histone H1 (Pham and Sauer, 2000). These histone-modifying activities indicate that TFIID may be specifically needed for transcription from nucleosome-embedded promoters. Indeed, only TFIID, but not TBP, can work in conjunction with a TFIID-deficient pol II holoenzyme complex (see Fig.2.2 in the Chapter by Hou and Chiang) to facilitate activator-dependent transcription from an *in vitro*-reconstituted chromatin template, suggesting a unique involvement of TAFs in chromatin transcription (Wu et al., 1999). The importance of the acetyltransferase activity of TAF1 for gene transcription has also been demonstrated for MHC class I genes, where suppression of TAF1 enzymatic activity by TAF1-interacting protein TAF7 leads to inhibition of transcription (Gegonne et al., 2001). Similar to other histone acetyltransferases (HATs), TAF1 contains two tandem bromodomains recognizing acetylated lysine 14 (K14) of histone H3 as well as acetylated K5, K8, K12, and K16 of histone H4 (Jacobson et al., 2000; Kanno et al., 2004). This bromodomain-mediated interaction likely enhances TFIID binding to acetylated promoters previously modified by activator-recruited HATs. The structure of the human TAF1 double bromodomain spanning amino acid residues 1359-1638, resolved by X-ray crystallography at 2.1Å resolution, reveals that two acetylated lysine residues of histone H4, separated by 7 or 8 amino acids (i.e., K5/K12 or K8/K16), are preferentially recognized by the double bromodomain, where the N^ε-acetyllysine-binding pocket is situated within the amino acid residues of the two loops connecting the four antiparallel α-helices that constitute the core of each bromodomain (Jacobson et al., 2000).

Other than acting on histones, TAF1, as a free entity or part of TFIID, can also covalently modify other general transcription factors and cofactors, as demonstrated by acetylation on TFIIEβ (Imhof et al., 1997), phosphorylation on RAP74 (Dikstein et al., 1996), PC4 (Kershnar et al., 1998; Malik et al., 1998), and the β subunit of TFIIA (Solow et al., 2001), and presumably ubiquitination on TAF5 and itself as well (Auty et al., 2004).

TBP-Related Factors and TAF Variants

Some components of TFIID, including TBP and TAFs, are present in different forms often in a tissue- and development-specific manner. These TBP-related factors (TRFs) and TAF variants also play an important role in pol II-dependent transcription. In addition, multiple forms of TAF-containing complexes exhibiting HAT activity likely involved in chromatin dynamics have been identified. These protein factors will be briefly discussed in this section.

A: TBP-Related Factors

All multicellular organisms contain at least two TBP genes encoding proteins sharing sequence homology at their C-terminal 180 amino acid core DNA-binding domains. While the first gene encodes TBP, considered a universal TATA-binding transcription factor present in all eukaryotes, the second gene encodes TBP-related factor 2 (TRF2), also called TBP-related protein (TRP), TBP-like factor (TLF) or TBP-like protein (TLP), recognizing a sequence element distinct from the TATA box due to some conserved amino acid changes, including those two phenylalanine residues that kink the TATA box, within the C-terminal core that shares ~60% sequence homology and 41% identity with that of TBP (Dantonel et al., 1999; Berk, 2000; Hochheimer and Tjian, 2003). Interestingly, amino acid residues important for interactions with TFIIA and TFIIB remain mostly unaltered, allowing TRF2 to associate with TFIIA and TFIIB (Maldonado, 1999; Moore et al., 1999; Rabenstein et al., 1999; Teichmann et al., 1999) and thus assembling into a functional PIC able to transcribe some TATA-less promoters likely via undefined TRF2-binding elements (Ohbayashi et al., 2003; Chong et al., 2005). This property of TRF2 also enables it to function as a repressor preventing PIC formation initiated by TBP or TFIID on TATA-containing promoters (Moore et al., 1999; Teichmann et al., 1999; Chong et al., 2005). Not surprisingly, inactivation of TRF2 in C. elegans by RNA interference fails to support embryogenesis (Dantonel et al., 2000; Kaltenbach et al., 2000), indicating that TRF2 is essential for normal cell development.

Other than TRF2, two species-specific TRFs (TRF1 and TRF3) have also been identified. TRF1 (Crowley et al., 1993), so far only identified in neuronal and germ cells of *Drosophila* (reviewed in Berk, 2000; Hochheimer and Tjian, 2003), exhibits 63% amino acid sequence identity to TBP at its C-terminal DNA-binding core which maintains most of the conserved residues important for interactions with the TATA box, TFIIA and TFIIB (Rabenstein et al., 1999). It is thus not unexpected that TRF1 binds TFIIA and TFIIB and can partially substitute for TBP in directing pol II-dependent transcription from some TATA-containing promoters *in vitro* (Hansen et al., 1997). Interestingly, a TRF1 target gene, *Tudor*, whose expression is driven by two tandem promoters containing a TRF1-responsive upstream promoter and a TBP/TFIID-responsive downstream promoter, has been identified in *Drosophila* cells. A TC-rich sequence, located at ~25 nucleotides upstream of this TRF1-responsive *Tudor* promoter preferentially nucleates TRF1-mediated PIC assembly and transcription from the upstream *Tudor* promoter. *In vivo*, promoter selectivity by TRF1 is likely enhanced by some promoter-specific transcriptional activators or by its association with neuron-specific TRF1-associated factors (nTAFs) to form a multiprotein complex, distinct from TFIID (Holmes and Tjian, 2000). That TRF1 does not interact with TFIID-specific TAFs provides a rationale why TRF1 and TBP are not interchangeable at respective TRF1-responsive upstream and TBP-responsive downstream *Tudor* promoters (Holmes and Tjian, 2000).

TRF3, which shows 93% amino acid sequence identity to TBP at the C-terminal core region sharing the same sequence as TBP at all the conserved residues involved in TATA binding and interactions with TFIIA and TFIIB, is unique to vertebrates, ranging from fish to humans, but is not present in urochordate *Ciona intestinalis* and lower eukaryotes, such as *Drosophila* and *C. elegans* (Persengiev et al., 2003). Unlike TRF1 and TRF2, which are expressed only in selective tissues, TRF3 is ubiquitously expressed in every cell type examined, similar to that of TBP, and is present as a protein complex with a molecular size of approximately 200 kDa. Since TAF1 does not cofractionate with the TRF3 complex, it is likely that polypeptides constituting the TRF3 complex are different from TAFs defined in TFIID and nTAFs associated with TRF1. The transcriptional property of TRF3 and its regulated genes remain to be characterized.

B: TAF-Containing Complexes

Similar to TBP that forms SL1, TFIID, TFIIIB, and TAC complexes, some TAF components in TFIID are also present in distinct complexes such as TBP-free TAF_{II}-containing complex (TFTC), Spt-Ada-Gcn5-acetyltransferase (SAGA), Spt3-TAF_{II}31-GCN5L (i.e., the long form of GCN5) acetylase (STAGA), and polycomb repressive complex 1 (PRC1). These TBP-lacking TAF-containing complexes are involved in diverse aspects of pol II-dependent transcription. Except PRC1 whose role is mainly in gene silencing, the other TAF-containing complexes are mostly involved in activator-dependent transcription likely due to the HAT activity inherent to each complex.

TFTC, originally identified in HeLa cells using a monoclonal antibody against human TAF10 (Wieczorek et al., 1998), contains TAF2, TAF5, TAF5L, TAF6L, TAF7, and some histone fold-containing TAFs, including TAF4, TAF6, TAF9, TAF10, and TAF12. In addition, TFTC also has TRRAP (transformation-transactivation domain-associated protein), SAP130 (spliceosome-associated protein 130), GCN5L HAT enzyme, Ada3 adaptor protein, and histone fold-containing Spt3 (Cavusoglu et al., 2003). The three-dimensional structures of TFTC and TFIID, both resolved at 35 Å resolution by electron microscopy and single-particle image analysis, resemble a macromolecular clamp consisting of five (for TFTC) or four (for TFIID) globular domains organized around a solvent-accessible groove that may accommodate a DNA duplex (Andel et al., 1999; Brand et al., 1999a).

This configuration suggests that TFTC may adapt a DNA-binding conformation similar to that exhibited by TFIID. Indeed, like TFIID, TFTC is able to support basal and Gal4-VP16-mediated transcription *in vitro* from TATA-containing promoters and also basal transcription from a TATA-less promoter, presumably via TAF recognition of the core promoter and interaction with other components of the general transcription machinery (Wieczorek *et al.*, 1998). Moreover, TFTC has coactivator activity likely through direct protein-protein contacts between the activation domain of Gal4-VP16 and multiple subunits of TFTC including TAF5, TAF6, TAF10, SAP130, Spt3, GCN5L, and TRRAP (Hardy *et al.*, 2002). This activator-TFTC interaction may also facilitate p300-mediated transcription from a Gal4-VP16-dependent chromatin template (Hardy *et al.*, 2002), probably resulting from acetylation of nucleosomal core histones H3 and H4 by GCN5L (Brand *et al.*, 1999b). Besides the HAT component, the presence of many histone fold-containing proteins, including Spt3, TAF4, TAF6, TAF9, TAF10, and TAF12, further contributes to the integrity of the TFTC complex.

Similar to TFTC, yeast SAGA also lacks TBP but contains some common subunits such as Tra1, Ada3, TAF5, the short form of GCN5, and histone fold-containing proteins Spt3, TAF6, TAF9, TAF10, and TAF12. The unique components present in SAGA, not found in TFTC, include Tra1, Spt7, Spt8, Spt20, Sgf29, Ada1, and Ada2 (Grant *et al.*, 1998). These SAGA components are organized into five distinct domains, analogous to the structure of human TFTC, when resolved by electron microscopy at 31 Å resolution (Wu *et al.*, 2004). The largest subunit of SAGA, Tra1, is located in the outermost globular domain able to contact transcriptional activators, as evidenced by the fluorescence resonance energy transfer (FRET) technique measuring direct *in vivo* association between yeast Gal4 activator and Tra1, but not with other subunits in the SAGA complex (Bhaumik *et al.*, 2004). This is consistent with the finding that the human homologue of Tra1, TRRAP (27.3% identity and 58.9% similarity with *S. cerevisiae* Tra1 at the amino acid level), also interacts with c-Myc and E2F-1 oncoproteins implicated in transformation and transactivation (McMahon *et al.*, 1998) and then recruits GCN5 HAT activity to activator-recruited transcriptional complexes (McMahon *et al.*, 2000). However, whether recruitment of TRRAP and GCN5 occurs in the context of human SAGA-like complexes, such as TFTC, PCAF and GCN5 complexes (Ogryzko *et al.*, 1998), remains to be investigated. Interestingly, approximately 10% of yeast genes, which are mostly driven by TATA-containing promoters and are typically stress-induced, appear to be SAGA-dependent (Huisinga and Pugh, 2004). These TFIID-independent yeast promoters are recognized by the free form of TBP that in turn recruits SAGA to the targeted promoters via direct interaction between TBP and the Spt3 subunit of SAGA. Clearly, some components of the SAGA complex, such as Spt7, Spt20, Ada1, TAF5, TAF10, and TAF12 are necessary for the structural integrity of SAGA, as mutations in these subunits result in disruption of the holo complex (Grant *et al.*, 1998; Sterner *et al.*, 1999; Durso *et al.*, 2001; Kirschner *et al.*, 2002).

STAGA, the human counterpart of yeast SAGA initially isolated from HeLa nuclear extracts using polyclonal antibodies against human TAF9 (Martinez *et al.*, 1998), contains many homologues found in yeast SAGA, including TRRAP, Ada1, Ada2, Ada3, Spt3, Spt7, and histone fold-containing TAFs (TAF9, TAF10, and TAF12). While additional components of STAGA, TAF5L, TAF6L, SAP130, and GCN5L are also present in TFTC, the identities of the other subunits, such as STAF36, STAF42, STAF46, STAF55, STAF60, and STAF65γ remain to be characterized (Martinez *et al.*, 2001). The conservation of histone fold-containing Spt3, TAF9, TAF10, and TAF12 in both SAGA and STAGA suggests that these subunits are important for the structural integrity of these distinct HAT complexes. As seen with yeast SAGA, the presence of TRRAP, GCN5L and adaptor proteins Ada1, Ada2, and Ada3 in STAGA likely accounts for the coactivating activity of STAGA in supporting Gal4-VP16-mediated activation from a Gal4-driven chromatin template (Martinez *et al.*, 2001). Clearly, Ada proteins facilitate access of GCN5L to the chromatin template and thus allow acetylation on nucleosomal core histones (Balasubramanian *et al.*, 2002). That adaptor proteins are additionally required for activator-facilitated chromatin targeting of GCN5 is further supported by the observation that recombinant GCN5 protein is unable to potentiate activator-dependent transcription from preassembled chromatin templates (Thomas and Chiang, 2005). Based on our understanding of the yeast SAGA complex, it is likely that the Spt3 component of STAGA may contact TBP to facilitate transcription from STAGA-dependent promoters.

Unlike HAT-containing TFTC, SAGA and STAGA complexes involved in gene activation, the TAF-containing *Drosophila* PRC1 complex is implicated in repression of homeotic genes that govern body segmentation and the developmental process. This complex, initially isolated from *Drosophila* embryos using a monoclonal antibody against epitope-tagged polyhomeotic (PH) or posterior sex comb (PSC) protein (Shao *et al.*, 1999), contains approximately 30 subunits, including TAF1, TAF4, TAF5, TAF6, TAF9, TAF11, PH, PSC, PC (polycomb), RING1, Zeste, HSC4, SMRTER, Mi-2, Sin3A, Rpd3, p55, Sbf1, DRE4/Spt16, p90, HSC3, Modulo, Reptin, DNA Topoisomerase II, p110, Tubulin, Actin, Ribosome RS2, and Ribosome RL10 (Saurin *et al.*, 2001). Although this holo complex has histone deacetylase (HDAC) Rpd3, chromatin remodeling ATPase Mi-2, and other corepressor components (SMRTER, Sin3A, and p55) likely contributing to repression of transcription and inhibition of chromatin remodeling, it is surprising to see that a PRC1 core complex (PCC) containing only PH, PSC, PC, and RING1 is sufficient for transcriptional silencing (King *et al.*, 2002) and also blocking SWI/SNF-mediated mobilization of a nucleosomal array (Francis *et al.*, 2001). This finding plus the fact that a sequence-specific transcription factor Zeste is present in PRC1 suggests that PRC1 may use different mechanisms to target PRC1-regulated gene transcription, irrespective of the presence or absence of a Zeste-binding element (Mulholland *et al.*, 2003). The roles of TAFs, especially the enzymatic activities of TAF1, and the other subunits constituting PRC1 await further investigation.

C: TAF Variants During the Developmental Process

Some components of TFIID are present in a substoichiometric ratio relative to the other TAFs. These TAFs, such as TAF4b, TAF5L, and TAF7L, are often found in a tissue-specific manner and likely confers TFIID unique properties functioning in a specialized environment. TAF4b, a paralogue of TAF4 initially identified in TFIID purified from B cells and later found expressed at low levels in every cell types but specifically enriched in the testes and ovary, appears to function in a gonad-specific manner, as knockout of this TAF variant severely affects ovarian development in female mice (Freiman *et al.*, 2001) and also spermatogenesis in male mice (Falender *et al.*, 2005). Likewise, TAF5L and TAF7L, paralogues of TAF5 and TAF7, respectively, are implicated in male gametogenesis (Hiller *et al.*, 2001; Pointud *et al.*, 2003). The presence of these tissue-specific TAF variants likely enables TFIID to work in conjunction with germ cell-specific transcription factors, cofactors, or other components of the general transcription machinery, such as TFIIAαβ-like factor (see Chapter by Hou and Chiang). Related to this, a unique TFIID subunit, TAF8, is induced during adipocyte differentiation (Guermah *et al.*, 2003). Other than the tissue-specific expression pattern observed with these TAF variants, functional inactivation of TAF1 (Hisatake *et al.*, 1993; Ruppert *et al.*, 1993), which is present in TFIID and PRC1, and of TAF10 (Metzger *et al.*, 1999), found in TFTC, STAGA and SAGA-like complexes, causes cell cycle arrest, indicating a general role of selective mammalian TAFs in modulating cell growth. Undoubtedly, the presence of TAF variants further expands the general properties of TFIID to specialized needs in differentiated tissues.

Mediator

Mediator, which represents the second class of general cofactors that transmit the regulatory signals from gene-specific transcription factors to the general transcription machinery, was first identified in yeast and found to consist of more than 20 polypeptides (Kim *et al.*, 1994), of which 11 are essential for yeast viability (Rgr1, Rox3, Srb4, Srb6, Srb7, Med4, Med6, Med7, Med8, Med10/Nut2, and Med11; Myers and Kornberg, 2000). While nine of the Mediator components were originally defined by genetic screens as proteins interacting with the CTD of pol II (i.e., Srb2 and Srb4-11 for different dominant suppressors of RNA polymerase B mutations; Myers and Kornberg, 2000), later biochemical purification of Mediator complexes from various species has identified additional conserved as well as species-specific subunits, for which a unified nomenclature has been proposed (Table 4.2) (Bourbon *et al.*, 2004).

A: Isolation of Mediator Complexes

Human Mediator, first purified from HeLa cells as a protein complex that associates with the thyroid hormone receptor α (TR α) in a ligand-dependent manner, was able to potentiate TRα-mediated transcription *in vitro* (Fondell *et al.*, 1996). This TR-associated protein complex (TRAP) contains many protein subunits subsequently found also present in other coactivator complexes, such as SRB/MED-containing cofactor complex (SMCC; Ito *et al.*, 1999), vitamin D receptor-interacting protein complex (DRIP; Rachez *et al.*, 1998), activator-recruited complex (ARC; Näär *et al.*, 1999), positive cofactor 2 (PC2; Malik *et al.*, 2000), cofactor required for Sp1 activation (CRSP; Ryu *et al.*, 1999), and negative regulator of activated transcription (NAT; Sun *et al.*, 1998). In humans, at least two forms of Mediator complexes, Mediator-P.5 and Mediator-P.85 isolated individually from 0.5 M and 0.85 M KCl fractions of the P11 phosphocellulose ion-exchange column, have been identified and demonstrated to enhance activator-dependent and basal transcription, respectively (Wu *et al.*, 2003). Mediator-P.5 represents a class of larger Mediator complexes, including also TRAP/SMCC and ARC/DRIP, which contain a dissociable

Table 4.2 Mammalian and yeast Mediator complexes.

New Name	S. cerevisiae	TRAP/SMCC (Large)	ARC/DRIP (Large)	CRSP (Small)	PC2 (Small)	Mediator-P.5 (Large)	Mediator-P.85 (Small)
MED1	Med1	TRAP220	ARC/DRIP205	CRSP200	TRAP220	Med220	Med220
MED2	Med2†						
MED3	Med3/Pgd1/Hrs1†						
MED4	Med4	TRAP36	ARC/DRIP36		TRAP36	Med36	Med36
MED5	Nut1						
MED6	Med6	hMed6	ARC/DRIP33		hMed6	Med33	Med33
MED7	Med7	hMed7	ARC/DRIP34	CRSP33	hMed7	Med34	Med34
MED8	Med8		ARC32			Med31	Med31
MED9	Med9/Cse2†						
MED10	Med10/Nut2	hNut2	hMed10		hNut2		
MED11	Med11§						
MED12	Srb8	TRAP230	ARC/DRIP240			Med230	
MED13	Srb9/Ssn2	TRAP240	ARC/DRIP250			Med240	
MED14	Rgr1	TRAP170	ARC/DRIP150	CRSP150	TRAP170	Med150	Med150
MED15	Gal11		ARC105		PCQAP	Med105	Med105
MED16	Sin4	TRAP95	DRIP92		TRAP95	Med95	Med95
MED17	Srb4	TRAP80	ARC/DRIP77	CRSP77	TRAP80	Med78	Med78
MED18	Srb5§						
MED19	Rox3†						
MED20	Srb2	hTRFP			hTRFP		
MED21	Srb7	hSrb7			hSrb7		
MED22	Srb6§						
MED23		TRAP150β	ARC/DRIP130	CRSP130	TRAP150β		
MED24		TRAP100	ARC/DRIP100	CRSP100	TRAP100	Med100	Med100
MED25			ARC92				
MED26				CRSP70			Med70
MED27		TRAP37		CRSP34	TRAP37		
MED30		TRAP25					
MED31		hSoh1			hSoh1		
CDK8	Srb10/Ume5/Ssn3	hSrb10	CDK8			CDK8	
CycC	Srb11/Ume3/Ssn8	hSrb11	CycC			CycC	

The protein composition of human Mediator-P.5 and Mediator-P.85 complexes is described in Wu *et al*. (2003). This table is adapted from Bourbon *et al*. (2004).

† Subunits detected in *S. cerevisiae*, but not yet identified in *C. elegans*, *D. melanogaster*, or *H. sapiens* Mediator complexes.

§ Subunits also detected in *D. melanogaster*, but not yet identified in *C. elegans* or *H. sapiens* Mediator complexes.

MED12-MED13-CDK8-CycC module not found in the smaller Mediator complexes, such as Mediator-P.85, CRSP and PC2 (see Table 4.2). In contrast, the small Mediator complex has a unique polypeptide, MED26/CRSP70, normally absent in the large complex. The rest of Mediator components seem to be commonly shared between large and small Mediator complexes, although the identities of some subunits remain to be characterized. The structures of yeast and murine Mediator as well as human TRAP, ARC and CRSP complexes have been resolved by electron microscopy at 30-40Å (Asturias *et al*., 1999; Dotson *et al*., 2000; Taatjes *et al*., 2002). Comparison of these structures reveals a similarity in the overall organization of Mediator complexes. In general, three visible domains (named head, middle and tail in yeast Mediator) that may adapt to distinct conformations when complexed with activators or with the CTD are clearly distinguishable.

B: Head Module

The head module of Mediator, consisting of MED6, MED8, MED11, MED17, MED18, MED19, MED20, and MED22, forms the base of a roughly triangle-shaped Mediator complex. This triangular complex undergoes a drastic conformational change upon association with pol II, resulting in an arc-shaped structure in which the head module at the leading edge serves as a major docking site for pol II. This structural information is consistent with yeast genetic screens showing direct interaction between pol II and MED17/Srb4, MED18/Srb5, MED20/Srb2, and MED22/Srb6 respectively (Thompson et al., 1993). Moreover, MED17 and MED18 also interact with transcriptional activators p53 and Gal4-VP16, respectively (Ito et al., 1999; Lee et al., 1999), suggesting that activator-induced conformational changes (Taatjes et al., 2002) may further enhance head module-mediated recruitment of pol II.

C: Middle and Tail Modules

The middle module of Mediator contains MED1, MED4, MED5, MED7, MED9, MED10, MED21, and MED31, whereas the tail module includes MED2, MED3, MED14, MED15, and MED16 (Boube et al., 2002; Guglielmi et al., 2004). From the structures of yeast Mediator, it is obvious that, besides head module-pol II interaction, additional contacts are made between the Rpb1, Rpb3, Rpb6, and Rpb11 subunits of pol II with regions of Mediator extending from the head module to the intersection between middle and tail modules (Davis et al., 2002). Clearly, these extensive contacts may help Mediator unfold from its compact triangular shape, in which the middle and tail modules are not clearly visible in the absence of pol II, to a more extended conformation when bound to pol II. Despite the extensive contacts, the DNA-binding cleft and interaction surfaces for other components of the general transcription machinery are still accessible on pol II (Davis et al., 2002). Since MED1 in the middle module can interact with multiple nuclear receptors (Yuan et al., 1998), MED14 and MED15 in the tail module can associate with Gal4-VP16 (Lee et al., 1999; Park et al., 2000), it is likely that conformational changes induced upon activator binding to the middle and, especially the tail, module further contribute to activator-facilitated recruitment of pol II by Mediator (Taatjes et al., 2002; Wu et al., 2003).

D: MED12-MED13-CDK8-CycC Module

A dissociable module, which contains MED12/Srb8, MED13/Srb9, CDK8/Srb10, and CycC/Srb11 normally found in the large, but not small, Mediator complex (Borggrefe et al., 2002; Samuelsen et al., 2003), seems to contact mainly the middle module, via CDK8 interaction with MED1 and MED4 (Kang et al., 2001), and also the head module, through MED13 interaction with MED17 (Guglielmi et al., 2004). This CDK8 module, which could be isolated as a free entity, is able to phosphorylate serines 2 and 5 of the pol II CTD (Borggrefe et al., 2002). That transcription could be inhibited by CDK8-mediated phosphorylation of the CTD occurring prior to PIC assembly (Hengartner et al., 1998) or through phosphorylation on serines 5 and 304 of the cyclin H subunit of TFIIH by recombinant CDK8-CycC pair or by the NAT complex (Akoulitchev et al., 2000) suggests that the CDK8 module may function as a repression module in the context of Mediator. This view is supported by biochemical evidence showing that the large form of ARC (ARC-L) is transcriptionally inactive (Taatjes et al., 2002) and the coactivating activity of Mediator-P.5 was slightly enhanced when CDK8-CycC was immunodepleted from the large Mediator complex (Wu et al., 2003). Clearly, removal of CDK8 in yeast by nutrient deprivation (Holstege et al., 1998) or in mouse P19 embryonic carcinoma cells following all-*trans* retinoic acid (tRA) treatment (Pavri et al., 2005) enhances transcription from a subset of cellular genes normally suppressed by CDK8. In the latter case, it was further demonstrated that dissociation of CDK8 following tRA treatment converts Mediator from a transcriptionally suppressed state to an activated complex at the tRA-targeted RARβ2 gene promoter, indicating that CDK8 indeed functions in the context of a repression module *in vivo*.

E: Functional Properties of Mediator

In addition to the inhibitory activity conferred by the CDK8 repression module, Mediator is an authentic coactivator able to stimulate both basal and activator-dependent transcription (Kim et al., 1994). This stimulating activity of Mediator clearly relies on its ability to serve as a bridging molecule in transducing activation signals typically from activator-tail module to head module-pol II. Although Mediator can be isolated via its direct interactions with activators or with the CTD, direct biochemical evidence demonstrating that Mediator indeed functions by enhancing activator-facilitated entry of pol II to the PIC has only become possible after all the general transcription factors, cofactors, and pol II are available in purified forms devoid of any contaminating activities (Wu et al., 2003). From this *in vitro*-reconstituted transcription system, we learn that the large form of Mediator complexes, such as Mediator-P.5, has intrinsic coactivating activity able to stimulate activator-dependent transcription, whereas the small form of Mediator complexes, such as Mediator-P.85, only enhances basal transcription. Interestingly, the coactivator function of Mediator can occur in the absence of TFIID TAFs, suggesting that Mediator and TAFs may play some redundant roles in the transcriptional process. Indeed, it has been shown that TAFs can also enhance pol II entry to the PIC in the presence of transcriptional activators (Wu and Chiang, 2001). Besides targeting on pol II, Mediator has the ability to enhance TBP binding to the TATA box. This TATA-enhancing activity may help stabilize the promoter-bound scaffold complex, which contains TFIIA, TFIID, TFIIE, TFIIH, and Mediator (Yudkovsky et al., 2000), to facilitate reinitiation of transcription from the same promoter.

As defined in TFIID, Mediator exhibits multiple enzymatic activities. The kinase activity of Mediator, inherent to the CDK8 subunit, can phosphorylate the CTD of pol II (Hengartner et al., 1998; Sun et al., 1998; Borggrefe et al., 2002), the cyclin H subunit of TFIIH (Akoulitchev et al., 2000), and general cofactor PC4 (Gu et al., 1999). In addition, Mediator has been reported to exhibit HAT activity, residing in the MED5/Nut1 subunit, that preferentially acetylates histones H3 and H4 in the context of both free histones and chromatin (Lorch et al., 2000). However, no HAT activity has been reported in Mediator complexes purified from other species to date. Interestingly, some components of Mediator, such as MED8, are able to assemble into an E3 ubiquitin ligase involved in proteasome-mediated degradation (Brower et al., 2002), suggesting that Mediator may have ubiquitin ligase activity yet to be uncovered. These enzymatic activities clearly account for some roles of Mediator in the transcriptional process, including a possible stimulation of TFIIH kinase activity on the CTD (Kim et al., 1994). Many of these interesting questions remain to be addressed.

USA-Derived Cofactors

The third class of general cofactors, USA, was initially defined as a crude fraction derived from the P11 0.85 M KCl fraction of HeLa nuclear extracts (see Fig. 2.1 in the Chapter by Hou and Chiang) able to stimulate activator-dependent transcription (Meisterernst et al., 1991). Upon further fractionation of the USA fraction, several positive cofactors (PC1, PC2, PC3, and PC4) and a negative cofactor (NC1) were identified (Kaiser and Meisterernst, 1996). With ongoing research in the study of USA-derived cofactors, it is remarkable to see that each cofactor has a dual role to function as a coactivator in potentiating activator-dependent transcription and also as a repressor in inhibiting basal transcription when the activator is absent.

A: PC1

PC1 is a nuclear protein that is the functional equivalent of poly(ADP-ribose) polymerase-1 (PARP-1; Meisterernst et al., 1997), an enzyme which is well studied for its role in DNA repair by binding to damaged DNA and in nucleic acid metabolism by covalently modifying proteins involved in these pathways and also for its involvement in the maintenance of chromatin structure (Lindahl et al., 1995; D'Amours et al., 1999). These properties of PARP-1 provide an explanation for an earlier observation that TFIIC (i.e., PARP-1) is only required for site-specific initiation of transcription by pol II when nicked DNA templates were used for *in vitro* transcription assays, indicating that TFIIC facilitates transcription by binding to the nicks and suppresses nonspecific initiation from damaged DNA (Slattery et al., 1983).

Mammalian PARP-1, with a molecular size of approximately 114 kDa, catalyzes the transfer of ADP-ribose units from the donor nicotinamide adenine dinucleotide (NAD^+) to acceptor proteins, in a process known as poly(ADP-ribosyl)ation (Lindahl et al., 1995; D'Amours et al., 1999; Kraus and Lis, 2003). Not only does PARP-1 modify target proteins mediating DNA damage response, nucleic acid metabolism and chromatin dynamics, it also modifies transcription components, such as PC3 (i.e., DNA topoisomerase I), high mobility group protein 1 (HMG1), core histones (H2A, H2B, H3, and H4), pol II, TFIIF (RAP30 and RAP74), TBP, and p53 (Lindahl et al., 1995; Oei et al., 1998). Moreover, ADP-ribose units can be transferred to PARP-1 itself. Auto(ADP-ribosyl)ation of PARP-1 is mediated through its central domain containing multiple glutamic acid residues, which serve as acceptor sites for poly (ADP-ribosyl)ation. This central domain links the N-terminal zinc-finger DNA-binding domain to the C-terminal NAD^+-binding catalytic domain (D'Amours et al., 1999; Kraus and Lis, 2003). When disrupted, as observed during apoptosis where PARP-1 is cleaved by caspase death enzymes to release a 24-kDa N-terminal DNA-binding fragment and an 89-kDa C-terminal catalytic fragment, PARP-1 typically loses its enzymatic activity. For years, PARP-1 cleavage has been used as a diagnostic marker for programmed cell death.

Unlike the well-documented roles of PARP-1 in DNA repair and apoptosis, the transcriptional role of PARP-1 has not been extensively studied. Obviously, the coactivator function of PARP-1 is mediated by its direct contact with distinct transcriptional activators, including the human T-cell leukemia virus type 1 Tax protein (Anderson et al., 2000), human papillomavirus type 18 E2 (Lee et al., 2002), B-Myb (Cervellera and

Sala, 2000), E2F-1 (Simbulan-Rosenthal et al., 2003), AP2 (Kannan et al., 1999), TEF-1 (Butler and Ordahl, 1999), and NFκB (Hassa et al., 2003). Since the C-terminal domain of PARP-1 is necessary for TEF-1- and NFκB-dependent transcription (Butler and Ordahl, 1999; Hassa et al., 2003), but appears dispensable for Gal4-AH-mediated activation (Meisterernst et al., 1997), it seems that the catalytic activity of PARP-1 is differentially required for gene activation, depending on the specific activators and promoters involved. Using a cell-free transcription system performed with a Gal4-driven DNA template, it was shown that PARP-1 could stimulate PIC formation at a step post TFIID binding to the promoter region (Meisterernst et al., 1997). However, the exact step regulated by PARP-1 during PIC assembly has not been elucidated.

Other than the coactivating function, PARP-1 also possesses repressing activity able to inhibit transcription through different mechanisms. First, PARP-1 can be incorporated into chromatin via its N-terminal DNA-binding domain to promote formation of a highly condensed chromatin structure, thereby inhibiting activator-dependent transcription (Kim et al., 2004). Second, PARP-1 can be incorporated into a corepressor complex containing many protein components susceptible to modification by PARP-1, leading to disassembly of the corepressor complex on the target gene (Ju et al., 2004). Third, PARP-1 is capable of inhibiting ligand-dependent transcription by TRα in transient reporter gene assays in a catalytic domain-dependent manner (Miyamoto et al., 1999), suggesting that poly(ADP-ribosyl)ation of critical transcription components is likely involved in PARP-1-mediated transcriptional repression. This is consistent with the observation that poly(ADP-ribosyl)ation of sequence-specific DNA-binding proteins, such as p53, YY1, Sp1, and TBP, prevents binding to their cognate sequences (Malanga et al., 1998; Oei et al., 1998; Mendoza-Alvarez and Alvarez-Gonzalez, 2001). In the case of p53, the sites for poly(ADP-ribosyl)ation have been mapped to the DNA-binding and oligomerization domains of p53 (Malanga et al., 1998). Interestingly, the enzymatic activity of PARP-1 is also critical for derepression to occur on some silenced loci, as seen by the association of PARP-1 and poly(ADP-ribosyl)ated proteins with decondensed chromatin structure at transcriptionally induced *Drosophila* polytene chromosome puffs (Tulin and Spradling, 2003). Clearly, PARP-1 can function as a molecular switch to convert a silenced gene into a transcriptionally active state, by first dissociating a corepressor complex via poly(ADP-ribosyl)ation (Ju et al., 2004) or by removing a CDK8 repression module

from the larger Mediator complex (Pavri et al., 2005), and then enhancing activator-facilitated recruitment of chromatin-modifying enzymes, such as CBP or p300 HAT or ATP-dependent chromatin remodelers, to the targeted promoters.

B: PC2

PC2, originally isolated from HeLa nuclear extracts as a protein complex with a molecular size around 500 kDa, corresponds to the smaller form of human Mediator complexes described earlier in the Mediator section. This protein complex can enhance transcription mediated by Gal4-AH and HNF4 in an *in vitro*-reconstituted transcription system with DNA templates (Malik et al., 2000). The coactivating activity of PC2 requires the presence of other general cofactors, since PC2 alone only weakly stimulates activator-dependent transcription and in conjunction with TAFs, PC3, and PC4, a synergistic activation of Gal4-VP16-mediated transcription recapitulating the stimulatory activity of the USA fraction is detected (Malik et al., 2000). Likewise, PC2 is able to function together with PC3 and PC4 to regulate transcription mediated by TRα (Fondell et al., 1999), NFκB, Sp1 (Guermah et al., 1998), and the B-cell-specific Oct-1-OCA-B transcription complex (Luo et al., 1998).

C: PC3

PC3, functionally equivalent to DNA topoisomerase I (Topo I; Kretzschmar et al., 1993; Merino et al., 1993), consists of 765 amino acids with a molecular size around 91 kDa (Champoux, 2001). This protein is well known for its role in relaxing DNA by transiently introducing nicks, allowing strand passage and religation (Wang, 1996). Topo I has an unstructured N-terminal region, which contains nuclear targeting signals and available surfaces for interaction with various transcription factors, such as p53 (Gobert et al., 1999), that is dispensable for relaxing DNA *in vitro* (Stewart et al., 1996). The enzymatic activity of Topo I is regulated posttranslationally by casein kinase II- and protein kinase C-mediated phosphorylation, leading to increased relaxation activity and in responding to mitogenic stimuli (Wang, 1996). When Topo I encounters damaged DNA, it is stalled and must be removed from the lesions in order to prevent formation of irreversible single or double strand breaks that impair genomic integrity. In this regard, PARP-1 can target Topo I for poly(ADP-ribosyl)ation and facilitate Topo I to remove itself from cleaved DNA and close the resulting gap (Malanga and Althaus, 2004).

Aside from its role in modulating DNA topology,

Topo I can function as a transcriptional coactivator by contacting directly with activators, such as Gal4-AH (Kretzschmar et al., 1993) and p53 (Gobert et al., 1999), and with components of the general transcription machinery, such as TBP (Merino et al., 1993) in order to stimulate transcription. The coactivator function of Topo I is in part due to its ability to enhance TFIID-TFIIA-promoter complex formation (Shykind et al., 1997). In the absence of activator, Topo I functions as a repressor in inhibiting basal transcription. Although the transcription activity of Topo I is separated from its DNA relaxation activity (Merino et al., 1993; Kretzschmar et al., 1993), the precise mechanism by which Topo I regulates transcription remains to be elucidated. This is an area currently underexplored.

D: PC4

PC4 is a protein consisting of 127 amino acids with an N-terminal regulatory domain spanning amino acids 1 to 62 and a C-terminal single-stranded DNA-binding and dimerization domain located between amino acids 63 and 127 (Ge and Roeder, 1994a; Kretzschmar et al., 1994; Brandsen et al., 1997). The N-terminal region, important for PC4 interaction with distinct activation domains (Ge and Roeder, 1994a) and binding to double-stranded DNA in a non-sequence-specific manner (Kaiser et al., 1995), contains two serine-enriched acidic (SEAC) domains located respectively at amino acids 9 to 22 and 50 to 61, separated by a lysine-rich region lying between amino acids 23 to 41 (Kaiser et al., 1995). Both SEAC domains are susceptible to phosphorylation by several protein kinases, in particular casein kinase II (CKII), leading to inactivation of the coactivator function of PC4 likely by preventing PC4 interaction with TBP-bound TFIIA on the promoter region (Ge et al., 1994; Kretzschmar et al., 1994) and with transcriptional activators, such as the HIV Tat protein (Holloway et al., 2000). The lysine-rich region, proposed to bind nonspecific double-stranded DNA (Kaiser et al., 1995), may serve as acetylation sites for p300-enhanced PC4 binding to double-stranded DNA, consistent with the observation that CKII-mediated phosphorylation on the SEAC domains inhibits p300-dependent acetylation on PC4 (Kumar et al., 2001). The C-terminal region, in which the structure has been resolved by X-ray crystallography at 1.74 Å resolution (Brandsen et al., 1997), forming a dimer with each monomer composed of four antiparallel β-strands followed by a kinked α-helix, is able to bind with high affinity ($K_d \sim 0.07$ nM) two single strands of DNA running in opposite directions, as found in internally melted DNA duplexes. In the full-length protein, this single-stranded DNA-binding region is normally masked by intramolecular interaction with the N-terminal region and only becomes exposed after conformational changes induced, for instance, by CKII-mediated phosphorylation of the SEAC domains (Kaiser et al., 1995), thereby leading to inactivation of the coactivator function of PC4 and further inhibition of transcription mediated by the single-stranded DNA-binding activity of PC4 (Werten et al., 1998).

The inhibitory activity of PC4 allows it to function as a repressor in suppressing basal transcription when activators are absent (Malik et al., 1998; Werten et al., 1998; Wu and Chiang, 1998). This inhibition usually occurs prior to PIC assembly in the absence of TAFs (Wu and Chiang, 1998) and can be alleviated by adding increasing amounts of TFIID, TFIIH and pol II holoenzyme in the transcription reaction, correlating with the ability of these multiprotein complexes to phosphorylate PC4 (Kershnar et al., 1998; Malik et al., 1998). However, since inactivation of the ATP-binding site of ERCC3 helicase, but not ERCC2 helicase or CDK7 kinase, impairs the ability of recombinant TFIIH to overcome PC4-mediated repression (Fukuda et al., 2003), it remains unclear the precise mechanism used by components of the general transcription machinery to antagonize PC4 repressing activity.

The coactivator function of PC4 was evidenced by its ability to substitute for a crude USA fraction in mediating activator-dependent transcription in a cell-free transcription system reconstituted with recombinant general transcription factors (TFIIB, TBP, TFIIE, and TFIIF) and epitope-tagged multiprotein complexes (TFIID, TFIIH, and pol II; Wu et al., 1998). In this system, where TAFs and Mediator are not essential for activator-dependent transcription, PC4 is the only general cofactor indispensable for transcriptional activation mediated by Gal4-VP16 (Wu et al., 1998), and human papillomavirus E2 (Wu and Chiang, 2001; Hou et al., 2002; Wu et al., 2003). Not surprisingly, PC4 can interact with both transcriptional activators, such as Gal4-VP16 (Ge and Roeder, 1994a), BRCA1 (Haile and Parvin, 1999), AP2 (Kannan and Tainsky, 1999), HIV Tat (Holloway et al., 2000), human papillomavirus E2 (Wu and Chiang, 2001), and p53 (Banerjee et al., 2004), and components of the general transcription machinery, such as TFIIA (Ge and Roeder, 1994a), TFIIH (Fukuda et al., 2004), and pol II (Malik et al., 1998), thereby serving as a bridging molecule to facilitate activator-dependent transcription likely through enhancement of PIC assembly on the promoter region. In addition, PC4 may promote sequence-specific DNA-binding activity of some activators, such as p53 (Banjeree et al., 2004), stimulate promoter escape in a TFIIA- and TAF-dependent manner (Fukuda *et al.*, 2004), or enhance pol II elongation by modulating TFIIH kinase and Fcp1 phosphatase activity on CTD phosphorylation (Calvo and Manley, 2005). Clearly, PC4 can work in conjunction with other general cofactors, such as TAFs (Wu and Chiang, 1998; Wu *et al.*, 1998) and Mediator (Fondell *et al.*, 1999; Malik *et al.*, 2000; Wu *et al.*, 2003), to synergistically mediate activator-dependent transcription. Whether phosphorylation of PC4, which accounts for 95% of total PC4 in the cell (Ge *et al.*, 1994), by TFIID, TFIIH, pol II holoenzyme (Kershnar *et al.*, 1998) and Mediator (Gu *et al.*, 1999), plays a role in different steps of the transcriptional process remains to be further defined.

Besides being a transcriptional coactivator, PC4 has also been implicated in other cellular processes, such as DNA repair and DNA replication. In the aspect of DNA repair, PC4 can prevent mutagenesis arising from oxidative DNA damage caused by the interaction of reactive oxygen species (ROS) with DNA, depending upon its single-stranded DNA-binding activity (Wang *et al.*, 2004). The involvement of PC4 in DNA replication appears to be more complicated, as PC4 can interact with replication protein A (RPA) on single-stranded DNA and facilitate T-antigen-mediated unwinding of DNA containing SV40 origin of replication, while it also inhibits RNA primer synthesis and DNA polymerase δ-catalyzed DNA chain elongation (Pan *et al.*, 1996). The biological significance of these *in vitro* reactions performed in the presence of PC4 warrants further investigations.

E: NC1

NC1, also known as HMG1 or HMGB1 with a molecular size around 25 kDa, is a member of the highly conserved chromatin-associated proteins that bend DNA and bind preferentially to distorted DNA structures (Bustin, 2001; Thomas and Travers, 2001). HMGB1 is structurally divided into three domains: two homologous DNA-binding HMG-box domains A and B each containing approximately 80 amino acids, and a C-terminal tail containing a stretch of 30 acidic residues. Boxes A and B each forms an "L-shaped" structure with three α-helices constituting a minor-groove DNA-binding domain that preferentially binds distorted DNA, such as four-way junctions, cisplatin-modified DNA and bulged DNA, and induces DNA bending without sequence specificity (Thomas, 2001; Thomas and Travers, 2001). Both domains A and B contain an additional basic extension of amino acid residues that enhance the DNA-binding affinity of the HMG box. The C-terminal acidic tail modulates the DNA-binding activity of HMGB1 and seems to be inhibitory toward HMG box binding to DNA.

The transcriptional role of HMGB1 is similar to other USA-derived cofactors in that it normally functions as a repressor in the absence of an activator, but acts as a coactivator in activator-dependent transcription. The repressing activity of HMGB1 appears to work by promoting the formation of a stable HMGB1-TBP-promoter complex that prevents TFIIB entry (Ge and Roeder, 1994b), as the presence of HMGB1 increases the affinity of TBP for the TATA box by 20-fold (Das and Scovell, 2001). The interaction domains were mapped to the HMG box A of HMGB1 (Sutrias-Grau *et al.*, 1999) and the glutamine-rich region of TBP (Das and Scovell, 2001). TFIIA, as an antirepressor (see Chapter by Hou and Chiang), can displace HMGB1 from the ternary complex and overcome HMGB1-mediated inhibition of PIC formation, thereby restoring transcription activity (Ge and Roeder, 1994b). The coactivator function of HMGB1 is attributed to its direct interaction with transcriptional activators, such as p53, steroid hormone receptors, Oct-1, HOX and Rel proteins (Thomas and Travers, 2001; Agresti and Bianchi, 2003), and with components of the general transcription machinery, including TBP (Ge and Roeder, 1994b) and TAF10 (Verrier *et al.*, 1997). Undoubtedly, the architectural role of HMGB1 in bending DNA will further contribute to its coactivator function typically by enhancing sequence-specific recognition by these DNA-binding proteins.

Conclusion

Transcription in higher eukaryotes is a complex process involving a diverse set of protein factors acting through specific sequence elements surrounding the promoter region. The fact that the promoter itself can be recognized by TBP, TRFs, TFIID, and other TAF-containing complexes already lends flexibility for interaction with distinct transcriptional regulators as well as general cofactors which typically possess dual activities in repressing basal transcription and enhancing activator-dependent transcription in response to environmental changes. While tissue-specific TAFs and TRFs play an important role in regulating transcription during development, it still remains a mystery what roles positive cofactors and Mediator play in embryonic development. Without doubt, the presence of TAF variants and multiple pathways for regulating PIC assembly provide an additional way to fine-tune the transcription events occurring on individual genes. For

TAF-independent pathways, the alternative usage of other general cofactors, such as Mediator and USA-derived cofactors, help transduce regulatory signals between activators and the general transcription machinery. Considering that many of the general cofactors discussed in this chapter also exhibit multiple enzymatic activities and can function synergistically to activate transcription, it would be exciting to explore how these general cofactors communicate with one another in order to suppress their inhibitory activity during the activation process. The issue of functional redundancy and transcriptional synergy among these general cofactors warrant intensive studies for the years to come.

Acknowledgment

We are grateful for Drs. Parminder Kaur, Samuel Y. Hou and Shwu-Yuan Wu for their comments on this chapter. The research conducted in Dr. Chiang's laboratory is currently sponsored by grants CA103867 and GM59643 from the National Institutes of Health in the United States.

References

Agalioti, T., Lomvardas, S., Parekh, B., Yie, J., Maniatis, T., and Thanos, D. (2000). Ordered recruitment of chromatin modifying and general transcription factors to the IFN-β promoter. Cell *103*, 667-678.

Agresti, A., and Bianchi, M.E. (2003). HMGB proteins and gene expression. Curr. Opin. Genet. Dev. *13*, 170-178.

Akoulitchev, S., Chuikov, S., and Reinberg, D. (2000). TFIIH is negatively regulated by cdk8-containing mediator complexes. Nature *407*, 102-106.

Albright, S.R., and Tjian, R. (2000). TAFs revisited: more data reveal new twists and confirm old ideas. Gene *242*, 1-13.

Anderson, M.G., Scoggin, K.E. S., Simbulan-Rosenthal, C.M., and Steadman, J.A. (2000). Identification of poly(ADP-ribose) polymerase as a transcriptional coactivator of the human T-cell leukemia virus type 1 Tax protein. J. Virol. *74*, 2169–2177.

Andel, F., III, Ladurner, A.G., Inouye, C., Tjian, R., and Nogales, E. (1999). Three-dimensional structure of the human TFIID-IIA-IIB complex. Science *286*, 2153-2156.

Andrau, J.C., Van Oevelen, C.J., Van Teeffelen, H.A., Weil, P.A., Holstege, F.C., and Timmers, H.T. (2002). Mot1p is essential for TBP recruitment to selected promoters during *in vivo* gene activation. EMBO J. *21*, 5173-5183.

Apone, L.M., Virbasius, C.A., Holstege, F.C. P., Wang, J., Young, R.A., and Green, M.R. (1998). Broad, but not universal, transcriptional requirement for yTAF$_{II}$17, a histone H3-like TAF$_{II}$ present in TFIID and SAGA. Mol. Cell *2*, 653-661.

Apone, L.M., Virbasius, C.M., Reese, J.C., and Green, M.R. (1996). Yeast TAF$_{II}$90 is required for cell-cycle progression through G2/M but not for general transcription activation. Genes Dev. *10*, 2368-2380.

Asturias, F.J., Jiang, Y.W., Myers, L.C., Gustafsson, C.M., and Kornberg, R.D. (1999). Conserved structures of mediator and RNA polymerase II holoenzyme. Science *283*, 985-987.

Auble, D.T., and Hahn, S. (1993). An ATP-dependent inhibitor of TBP binding to DNA. Genes Dev. *7*, 844-856.

Auble, D.T., Wang, D., Post, K.W., and Hahn, S. (1997). Molecular analysis of the SNF2/SWI2 protein family member MOT1, an ATP-driven enzyme that dissociates TATA-binding protein from DNA. Mol. Cell. Biol. *17*, 4842-4851.

Auty, R., Steen, H., Myers, L.C., Persinger, J., Bartholomew, B., Gygi, S.P., and Buratowski, S. (2004). Purification of active TFIID from *Saccharomyces cerevisiae*. Extensive promoter contacts and co-activator function. J. Biol. Chem. *279*, 49973-49981.

Balasubramanian, R., Pray-Grant, M.G., Selleck, W., Grant, P.A., and Tan, S. (2002). Role of the Ada2 and Ada3 transcriptional coactivators in histone acetylation. J. Biol. Chem. *277*, 7989-7995.

Banerjee, S., Kumar, B.R., and Kundu, T.K. (2004). General transcriptional coactivator PC4 activates p53 function. Mol. Cell. Biol. *24*, 2052-2062.

Banik, U., Beechem, J.M., Klebanow, E., Schroeder, S., and Weil, P.A. (2001). Fluorescence-based analyses of the effects of full-length recombinant TAF130p on the interaction of TATA box-binding protein with TATA box DNA. J. Biol. Chem. *276*, 49100-49109.

Berk, A.J. (2000). TBP-like factors come into focus. Cell *103*, 5-8.

Bhaumik, S.R., Raha, T., Aiello, D.P., and Green, M.R. (2004). *In vivo* target of a transcriptional activator revealed by fluorescence resonance energy transfer. Genes Dev. *18*, 333-343.

Birck, C., Poch, O., Romier, C., Ruff, M., Mengus, G., Lavigne, A.-C., Davidson, I., and Moras, D. (1998). Human TAF$_{II}$28 and TAF$_{II}$18 interact through a histone fold encoded by atypical evolutionary conserved motifs also found in the SPT3 family. Cell *94*, 239-249.

Borggrefe, T., Davis, R., Erdjument-Bromage, H., Tempst, P., and Kornberg, R. D. (2002). A complex of the Srb8, -9, -10, and -11 transcriptional regulatory proteins from yeast. J. Biol. Chem. *277*, 44202-44207.

Boube, M., Joulia, L., Cribbs, D.L., and Bourbon, H.-M. (2002). Evidence for a mediator of RNA polymerase II transcriptional regulation conserved from yeast to man. Cell *110*, 143-151.

Bourbon, H.M., Aguilera, A., Ansari, A.Z., Asturias, F.J., Berk, A.J., Bjorklund, S., Blackwell, T.K., Borggrefe, T., Carey, M., Carlson, M., Conaway, J.W., Conaway, R.C., Emmons, S.W., Fondell, J.D., Freedman, L.P., Fukasawa, T., Gustafsson, C.M., Han, M., He, X., Herman, P.K., Hinnebusch, A.G., Holmberg, S.,

Holstege, F.C., Jaehning, J.A., Kim, Y.J., Kuras, L., Leutz, A., Lis, J.T., Meisterernest, M., Näär, A.M., Nasmyth, K., Parvin, J.D., Ptashne, M., Reinberg, D., Ronne, H., Sadowski, I., Sakurai, H., Sipiczki, M., Sternberg, P.W., Stillman, D.J., Strich, R., Struhl, K., Svejstrup, J.Q., Tuck, S., Winston, F., Roeder, R.G., and Kornberg, R.D. (2004). A unified nomenclature for protein subunits of mediator complexes linking transcriptional regulators to RNA polymerase II. Mol. Cell *14*, 553-557.

Brand, M., Leurent, C., Mallouh, V., Tora, L., and Schultz, P. (1999a). Three-dimensional structures of the TAF$_{II}$-containing complexes TFIID and TFTC. Science *286*, 2151-2153.

Brand, M., Yamamoto, K., Staub, A., and Tora, L. (1999b). Identification of TATA-binding protein-free TAF$_{II}$-containing complex subunits suggests a role in nucleosome acetylation and signal transduction. J. Biol. Chem. *274*, 18285-18289.

Brandsen, J., Werten, S., van der Vliet, P.C., Meisterernst, M., Kroon, J., and Gros, P. (1997). C-terminal domain of transcription cofactor PC4 reveals dimeric ssDNA binding site. Nat. Struct. Biol. *4*, 900-903.

Brower, C.S., Sato, S., Tomomori-Sato, C., Kamura, T., Pause, A., Stearman, R., Klausner, R.D., Malik, S., Lane, W.S., Sorokina, I., Roeder, R.G., Conaway, J.W., Conaway, R.C. (2002). Mammalian mediator subunit mMED8 is an Elongin BC-interacting protein that can assemble with Cul2 and Rbx1 to reconstitute a ubiquitin ligase. Proc. Natl. Acad. Sci. USA *99*, 10353-10358.

Burke, T.W., and Kadonaga, J.T. (1997). The downstream core promoter element, DPE, is conserved from *Drosophila* to humans and is recognized by TAF$_{II}$60 of *Drosophila*. Genes Dev. *11*, 3020-3031.

Burley, S.K., and Roeder, R.G. (1998). TATA box mimicry by TFIID: autoinhibition of pol II transcription. Cell *94*, 551-553.

Burley, S. K., and Roeder, R. G. (1996). Biochemistry and structural biology of transcription factor IID (TFIID). Annu. Rev. Biochem. *65*, 769-799.

Bustin, M. (2001). Revised nomenclature for high mobility group (HMG) chromosomal proteins. Trends Biochem. Sci. *26*, 152-153.

Butler, A.J., and Ordahl, C.P. (1999). Poly(ADP-ribose) polymerase binds with transcription enhancer factor 1 to MCAT1 elements to regulate muscle-specific transcription. Mol. Cell. Biol. *19*, 296–306.

Butler, J.E., and Kadonaga, J.T. (2001). Enhancer-promoter specificity mediated by DPE or TATA core promoter motifs. Genes Dev. *15*, 2515-2519.

Calvo, O., and Manley, J.L. (2005). The transcriptional coactivator PC4/Sub1 has multiple functions in RNA polymerase II transcription. EMBO J. *24*, 1009-1020.

Castaño, E., Gross, P., Wang, Z., Roeder, R.G., and Oelgeschläger, T. (2000). The C-terminal domain-phosphorylated IIO form of RNA polymerase II is associated with the transcription repressor NC2 (Dr1/DRAP1) and is required for transcription activation in human nuclear extracts. Proc. Natl. Acad. Sci. USA *97*, 7184-7189.

Cavusoglu, N., Brand, M., Tora, L., and Dorsselaer, A.V. (2003). Novel subunits of the TATA binding protein free TAF$_{II}$-containing transcription complex identified by matrix-assisted laser desorption/ionization-time of flight mass spectrometry following one-dimensional gel electrophoresis. Proteomics *3*, 217-223.

Cervellera, M.N. and Sala, A. (2000). Poly(ADP-ribose) polymerase is a B-MYB coactivator. J. Biol. Chem. *275*, 10692–10696.

Chalkley, G.E., and Verrijzer, C.P. (1999). DNA binding site selection by RNA polymerase II TAFs: a TAF$_{II}$250-TAF$_{II}$150 complex recognizes the initiator. EMBO J. *18*, 4835-4845.

Champoux, J.J. (2001). DNA topoisomerases: structure, function, and mechanism. Annu Rev. Bioch. *70*, 369-413.

Chiang, C.-M., Ge, H., Wang, Z., Hoffmann, A., and Roeder, R.G. (1993). Unique TATA-binding protein-containing complexes and cofactors involved in transcription by RNA polymerases II and III. EMBO J. *12*, 2749-2762.

Chiang, C.-M., and Roeder, R.G. (1995). Cloning of an intrinsic human TFIID subunit that interacts with multiple transcriptional activators. Science *267*, 531-536.

Chicca, J.J., II, Auble, D.T., and Pugh, B.F. (1998). Cloning and biochemical characterization of TAF-172, a human homolog of yeast Mot1. Mol. Cell. Biol. *18*, 1701-1710.

Chong, J. A., Moran, M.M., Teichmann, M., Kaczmarek, J.S., Roeder, R., and Clapham, D.E. (2005). TATA-binding protein (TBP)-like factor (TLF) is a functional regulator of transcription: reciprocal regulation of the neurofibromatosis type 1 and c-fos genes by TLF/TRF2 and TBP. Mol. Cell. Biol. *25*, 2632 - 2643.

Coleman, R.A., Taggart, A.K.P., Benjamin, L.R., and Pugh, B.F. (1995). Dimerization of the TATA binding protein. J. Biol. Chem. *270*, 13842-13849.

Coleman, R.A., Taggart, A.K.P., Burma, S., Chicca, J.J., II, and Pugh, B.F. (1999). TFIIA regulates TBP and TFIID dimers. Mol. Cell *4*, 451-457.

Collart, M.A. (1996). The NOT, SPT3, and MOT1 genes functionally interact to regulate transcription at core promoters. Mol. Cell. Biol. *16*, 6668-6676.

Creton, S., Svejstrup, J.Q., and Collart, M.A. (2002). The NC2 α and β subunits play different roles *in vivo*. Genes Dev. *16*, 3265-3276.

Crowley, T.E., Hoey, T., Liu, J.K., Jan, Y.N., Jan, L.Y., and Tjian, R. (1993). A new factor related to TATA-binding protein has highly restricted expression patterns in *Drosophila*. Nature *361*, 557-561.

D'Amours, D., Desnoyers, S., D'Silva, I., and Poierier, G.G. (1999). Poly(ADP-ribosyl)ation reactions in the regulation of nuclear functions. Biochem. J. *342*, 249-268.

Dantonel, J.C., Wurtz, J.M., Poch, O., Moras, D., and Tora, L. (1999). The TBP-like factor: an alternative transcription factor in metazoan? Trends Biochem. Sci. *24*, 335-339.

Dantonel, J.C., Quintin, S., Lakatos, L., Labouesse, M., and Tora, L. (2000). TBP-like factor is required for embryonic RNA polymerase II transcription in *C. elegans*. Mol. Cell *6*, 715-722.

Darst, R.P., Wang, D., and Auble, D.T. (2001). MOT1-catalyzed TBP-DNA disruption: uncoupling DNA conformational change and role of upstream DNA. EMBO J. *20*, 2028-2040.

Das, D., and Scovell, W.M. (2001). The binding interaction of HMG-1 with the TATA-binding protein/TATA complex. J. Biol. Chem. *276*, 32597-32605.

Dasgupta, A., Darst, R.P., Martin, K.J., Afshari, C.A., and Auble, D.T. (2002). Mot1 activates and represses transcription by direct, ATPase-dependent mechanisms. Proc. Natl. Acad. Sci. USA *99*, 2666-2671.

Davis, J.A., Takagi, Y., Kornberg, R.D., and Asturias, F.A. (2002). Structure of the yeast RNA polymerase II holoenzyme: Mediator conformation and polymerase interaction. Mol. Cell *10*, 409-415.

Denko, N., Wernke-Dollries, K., Johnson, A.B., Hammond, E., Chiang, C.-M., and Barton, M.C. (2003). Hypoxia actively represses transcription by inducing negative cofactor 2 (Dr1/DrAP1) and blocking preinitiation complex assembly. J. Biol. Chem. *278*, 5744-5749.

Dikstein, R., Ruppert, S., and Tjian, R. (1996). $TAF_{II}250$ is a bipartite protein kinase that phosphorylates the base transcription factor RAP74. Cell *84*, 781-790.

Dotson, M.R., Yuan, C.X., Roeder, R.G., Myers, L.C., Gustafsson, C.M., Jiang, Y.W., Li, Y., Kornberg, R.D., and Asturias, F.J. (2000). Structural organization of yeast and mammalian mediator complexes. Proc. Natl. Acad. Sci. USA *97*, 14307-14310.

Durso, R.J., Fisher, A.K., Albright-Frey, T.J., and Reese, J.C. (2001). Analysis of TAF90 mutants displaying allele-specific and broad defects in transcription. Mol. Cell. Biol. *21*, 7331-7344.

Dynlacht, B.D., Hoey, T., and Tjian, R. (1991). Isolation of coactivators associated with the TATA-binding protein that mediate transcriptional activation. Cell *66*, 563-576.

Falender, A.E., Freiman, R.N., Geles, K.G., Lo, K.C., Hwang, K., Lamb, D.J., Morris, P.L., Tjian, R., and Richards, J.S. (2005). Maintenance of spermatogenesis requires TAF4b, a gonad-specific subunit of TFIID. Genes Dev. *19*, 794-803.

Fischle, W., Wang, Y., and Allis, C.D. (2003). Histone and chromatin cross-talk. Curr. Opin. Cell. Biol. *15*, 172-183.

Fondell, J.D., Ge, H., and Roeder, R.G. (1996). Ligand induction of a transcriptionally active thyroid hormone receptor coactivator complex. Proc. Natl. Acad. Sci. USA *93*, 8329-8333.

Fondell, J.D., Guermah, M., Malik, S., and Roeder, R.G. (1999). Thyroid hormone receptor-associated proteins and general positive cofactors mediate thyroid hormone receptor function in the absence of the TATA box-binding protein-associated factors of TFIID. Proc. Natl. Acad. Sci. USA *96*, 1959-1964.

Francis, N.J., Saurin, A.J., Shao, Z., and Kingston, R.E. (2001). Reconstitution of a functional core polycomb repressive complex. Mol. Cell *8*, 545-556.

Freiman, R.N., Albright, S.R., Zheng, S., Sha, W.C., Hammer, R.E., and Tjian, R. (2001). Requirement of tissue-selective TBP-associated factor $TAF_{II}105$ in ovarian development. Science *293*, 2084-2087.

Fukuda, A., Tokonabe, S., Hamada, M., Matsumoto, M., Tsukui, T., Nogi, Y., and Hisatake, K. (2003). Alleviation of PC4-mediated transcriptional repression by the ERCC3 helicase activity of general transcription factor TFIIH. J. Biol. Chem. *278*, 14827-14831.

Fukuda, A., Nakadai, T., Shimada, M., Tsukui, T., Matsumoto, M., Nogi, Y., Meisterernst, M., and Hisatake, K. (2004). Transcriptional coactivator PC4 stimulates promoter escape and facilitates transcriptional synergy by GAL4-VP16. Mol. Cell. Biol. *24*, 6525-6535.

Gangloff, Y.-G., Romier, C., Thuault, S., Werten, S., and Davidson, I. (2001). The histone fold is a key structural motif of transcription factor TFIID. Trends Biochem. Sci. *26*, 250-257.

Ge, H., and Roeder, R.G. (1994a). Purification, cloning, and characterization of a human coactivator, PC4, that mediates transcriptional activation of class II genes. Cell *78*, 513-523.

Ge, H., and Roeder, R.G. (1994b). The high mobility group protein HMG1 can reversibly inhibit class II gene transcription by interaction with the TATA-binding protein. J. Biol. Chem. *269*, 17136-17140.

Ge, H., Zhao, Y., Chait, B.T., and Roeder, R.G. (1994). Phosphorylation negatively regulates the function of coactivator PC4. Proc. Natl. Acad. Sci. USA *91*, 12691-12695.

Gegonne, A., Weissman, J.D., and Singer, D.S. (2001). $TAF_{II}55$ binding to $TAF_{II}250$ inhibits its acetyltransferase activity. Proc. Natl. Acad. Sci. USA *98*, 12432-12437.

Geisberg, J.V., Moqtaderi, Z., Kuras, L., and Struhl, K. (2002). Mot1 associates with transcriptionally active promoters and inhibits the association of NC2 in yeast. Mol. Cell. Biol. *22*, 8122–8134.

Geisberg, J.V., and Struhl, K. (2004). Cellular stress alters the transcriptional properties of promoter-bound Mot1-TBP complexes. Mol. Cell *14*, 479-489.

Gobert, C., Skladanowski, A., and Larsen, A.K. (1999). The interaction between p53 and DNA topoisomerase I is regulated differently in cells with wild-type and mutant p53. Proc. Natl. Acad. Sci. USA *96*, 10355-10360.

Goldman-Levi, R., Miller, C., Bogoch, J., and Zak, N.B. (1996). Expanding the Mot1 subfamily: 89B helicase encodes a new *Drosophila* melanogaster SNF2-related protein which binds to multiple sites on polytene chromosomes. Nucleic Acids Res. *24*, 3121-3128.

Goodrich, J.A., Hoey, T., Thut, C.J., Admon, A., and Tjian, R. (1993). *Drosophila* $TAF_{II}40$ interacts with both a VP16 activation domain and the basal transcription factor TFIIB. Cell *75*, 519-530.

Goppelt, A., Stelzer, G., Lottspeich, F., and Meisterernst, M.

(1996). A mechanism for repression of class II gene transcription through specific binding of NC2 to TBP-promoter complexes via heterodimeric histone fold domains. EMBO J. *15*, 3105-3116.

Grant, P.A., Schieltz, D., Pray-Grant, M.G., Steger, D.J., Reese, J.C., Yates, J.R., III, and Workman, J.L. (1998). A subset of TAFs are integral components of the SAGA complex required for nucleosome acetylation and transcriptional stimulation. Cell *94*, 45-53.

Green, M.R. (2000). TBP-associated factors (TAF$_{II}$s): multiple, selective transcriptional mediators in common complexes. Trends Biochem. Sci. *25*, 59-63.

Gu, W., Malik, S., Ito, M., Yuan, C.X., Fondell, J.D., Zhang, X., Martinez, E., Qin, J., and Roeder, R.G. (1999). A novel human SRB/MED-containing cofactor complex SMCC, involved in transcriptional regulation. Mol. Cell *3*, 97-108.

Guermah, M., Ge, K., Chiang, C.-M., and Roeder, R.G. (2003). The TBN protein, which is essential for early embryonic mouse development, is an inducible TAF$_{II}$ implicated in adipogenesis. Mol. Cell *12*, 991-1001.

Guglielmi, B., van Berkum, N.L., Klapholz, B., Bijma, T., Boube, M., Boschiero, C., Bourbon, H.M., Holstege, F.C., and Werner, M. (2004). A high resolution protein interaction map of the yeast Mediator complex. Nucleic Acids Res. *32*, 5379-5391.

Haile, D.T., and Parvin, J.D. (1999). Activation of transcription *in vitro* by the BRCA1 carboxyl-terminal domain. J. Biol. Chem. *274*, 2113-2117.

Hansen, S.K., Takada, S., Jacobson, R.H., Lis, J.T., and Tjian, R. (1997). Transcription properties of a cell type-specific TATA-binding protein, TRF. Cell *91*, 71-83.

Hardy, S., Brand, M., Mittler, G., Yanagisawa, J., Kato, S., Meisterernst, M., and Tora, L. (2002). TATA-binding protein-free TAF-containing complex (TFTC) and p300 are both required for efficient transcriptional activation. J. Biol. Chem. *277*, 32875-32882.

Hassa, P.O., Buerki, C., Lombardi, C., Imhof, R., and Hottiger, M.O. (2003). Transcriptional coactivation of nuclear factor-κB-dependent gene expression by p300 is regulated by poly(ADP)-ribose polymerase-1. J. Biol. Chem. *278*, 45145-45153.

Hernandez, N. (1993). TBP, a universal eukaryotic transcription factor? Genes Dev. *7*, 1291-1308.

Hengartner, C.J., Myer, V.E., Liao, S.M., Wilson, C.J., Koh, S.S., and Young, R.A. (1998). Temporal regulation of RNA polymerase II by Srb 10 and Kin28 cyclin-dependent kinases. Mol. Cell *2*, 43-53.

Hengartner, C..J., Thompson, C..M., Zhang, J., Chao, D.M., Liao, S.-M., Koleske, A.J., Okamura, S., and Young, R.A. (1995). Association of an activator with an RNA polymerase II holoenzyme. Genes Dev. *9*, 897-910.

Hiller, M.A., Lin, T.Y., Wood, C., and Fuller, M.T. (2001). Developmental regulation of transcription by a tissue-specific TAF homolog. Genes Dev. *15*, 1021-1030.

Hisatake, K., Hasegawa, S., Takada, R., Nakatani, Y., Horikoshi, M., and Roeder, R.G. (1993). The p250 subunit of native TATA box-binding factor TFIID is the cell-cycle regulatory protein CCG1. Nature *362*, 179-181.

Hisatake, K., Ohta, T., Takada, R., Guermah, M., Horikoshi, M., Nakatani, Y., and Roeder, R.G. (1995). Evolutionary conservation of human TATA-binding-polypeptide-associated factors TAF$_{II}$31 and TAF$_{II}$80 and interactions of TAF$_{II}$80 with other TAFs and with general transcription factors. Proc. Natl. Acad. Sci. USA *92*, 8195-8199.

Hochheimer, A., and Tjian, R. (2003). Diversified transcription initiation complexes expand promoter selectivity and tissue-specific gene expression. Genes Dev. *17*, 1309-1320.

Hoffmann, A., Chiang, C.-M., Oelgeschläger, T., Xie, X., Burley, S.K., Nakatani, Y., and Roeder, R.G. (1996). A histone octamer-like structure within TFIID. Nature *380*, 356-359.

Holloway, A.F., Occhiodoro, F., Mittler, G., Meisterernst, M., and Shannon, M.F. (2000). Functional interaction between the HIV transactivator Tat and the transcriptional coactivator PC4 in T cells. J. Biol. Chem. *275*, 21668-21677.

Holstege, F.C., Jennings, E.G., Wyrick, J.J., Lee, T.I., Hengartner, C.J., Green, M.R., Golub, T.R., Lander, E.S., and Young, R.A. (1998). Dissecting the regulatory circuitry of a eukaryotic genome. Cell *95*, 717-728.

Holmes, M.C., and Tjian, R. (2000). Promoter-selective properties of the TBP-related factor TRF1. Science *288*, 867-70.

Horikoshi, M., Bertuccioli, C., Takada, R., Wang, J., Yamamoto, T., and Roeder, R.G. (1992). Transcription factor TFIID induces DNA bending upon binding to the TATA element. Proc. Natl. Acad. Sci. USA *89*, 1060-1064.

Hou, S.Y., Wu, S.-Y., and Chiang, C.-M. (2002). Transcriptional activity among high and low risk human papillomavirus E2 proteins correlates with E2 DNA binding. J. Biol. Chem. *277*, 45619-45629.

Huisinga, K.L., and Pugh, B.F. (2004). A genome-wide housekeeping role for TFIID and a highly regulated stress-related role for SAGA in *Saccharomyces cerevisiae*. Mol. Cell *13*, 573-585.

Imbalzano, A.N., Kwon, H., Green, M.R., and Kingston, R.E. (1994a). Facilitated binding of TATA-binding protein to nucleosomal DNA. Nature *370*, 481-485.

Imbalzano, A.N., Zaret, K.S., and Kingston, R.E. (1994b). Transcription factor (TF) IIB and TFIIA can independently increase the affinity of the TATA-binding protein for DNA. J. Biol. Chem. *269*, 8280-8286.

Imhof, A., Yang, X.J., Ogryzko, V.V., Nakatani, Y., Wolffe, A.P., and Ge, H. (1997). Acetylation of general transcription factors by histone acetyltransferases. Curr. Biol. *7*, 689-692.

Inostroza, J.A., Mermelstein, F.H., Ha, I., Lane, W.S., and Reinberg, D. (1992). Dr1, a TATA-binding protein-associated phosphoprotein and inhibitor of class II gene transcription. Cell *70*, 477-489.

Ito, M., Yuan, C.X., Malik, S., Gu, W., Fondell, J.D., Yamamura,

S., Fu, Z.Y., Zhang, X., Qin, J., and Roeder RG. (1999). Identity between TRAP and SMCC complexes indicates novel pathways for the function of nuclear receptors and diverse mammalian activators. Mol. Cell *3*, 361-370.

Jackson-Fisher, A.J., Chitikila, C., Mitra, M., and Pugh, B.F. (1999). A role for TBP dimerization in preventing unregulated gene expression. Mol. Cell *3*, 717-727.

Jacobson, R.H., Ladurner, A.G., King, D.S., and Tjian, R. (2000). Structure and function of a human $TAF_{II}250$ double bromodomain module. Science *288*, 1422-1425.

Ju, B.G., Solum, D., Song, E.J., Lee, K.J., Rose, D.W., Glass, C.K., and Rosenfeld, M.G. (2004). Activating the PARP-1 sensor component of the groucho/ TLE1 corepressor complex mediates a CaMKinase IIδ-dependent neurogenic gene activation pathway. Cell *119*, 815-829.

Juo, Z.S., Chiu, T.K., Leiberman, P.M., Baikalov, I., Berk, A.J., and Dickerson, R.E. (1996). How proteins recognize the TATA box. J. Mol. Biol. *261*, 239-254.

Kaiser, K., Stelzer, G., and Meisterernst, M. (1995). The coactivator p15 (PC4) initiates transcriptional activation during TFIIA-TFIID-promoter complex formation. EMBO J. *14*, 3520-3527.

Kaiser, K., and Meisterernst, M. (1996). The human general co-factors. Trends Biochem. Sci. *21*, 342-345.

Kaltenbach, L., Horner, M.A., Rothman, J.H., and Mango, S.E. (2000). The TBP-like factor CeTLF is required to activate RNA polymerase II transcription during *C. elegans* embryogenesis. Mol. Cell *6*, 705-713.

Kamada, K., Shu, F., Chen, H., Malik, S., Stelzer, G., Roeder, R.G., Meisterernst, M., and Burley, S.K. (2001). Crystal structure of negative cofactor 2 recognizing the TBP-DNA transcription complex. Cell *106*, 71-81.

Kannan, P., and Tainsky, M.A. (1999). Coactivator PC4 mediates AP-2 transcriptional activity and suppresses ras-induced transformation dependent on AP-2 transcriptional interference. Mol. Cell. Biol. *19*, 899-908.

Kannan, P., Yu, Y., Wankhade, S., and Tainsky, M.A. (1999). PolyADP-ribose polymerase is a coactivator for AP-2-mediated transcriptional activation. Nucleic Acids Res. *27*, 866-874.

Kanno, T., Kanno, Y., Siegel, R.M., Jang, M.K., Lenardo, M.J., and Ozato, K. (2004). Selective recognition of acetylated histones by bromodomain proteins visualized in living cells. Mol. Cell *13*, 33-43.

Kato, K., Makino, Y., Kishimoto, T., Yamauchi, J., Kato, S., Muramatsu, M., and Tamura, T. (1994). Multimerization of the mouse TATA-binding protein (TBP) driven by its C-terminal conserved domain. Nucleic Acids Res. *22*, 1179-1185.

Kershnar, E., Wu, S.-Y., and Chiang, C.-M. (1998). Immunoaffinity purification and functional characterization of human transcription factor IIH and RNA polymerase II from clonal cell lines that conditionally express epitope-tagged subunits of the multiprotein complexes. J. Biol. Chem. *273*, 34444-34453.

Kim, J.L., Nikolov, D.B., and Burley, S.K. (1993a). Co-crystal structure of TBP recognizing the minor groove of a TATA element. Nature *365*, 520-527.

Kim, Y., Geiger, J.H., Hahn, S., and Sigler, P.B. (1993b). Crystal structure of a yeast TBP/TATA-box complex. Nature *365*, 512-520.

Kim, Y.J., Bjorklund, S., Li, Y., Sayre, M.H., and Kornberg, R.D. (1994). A multiprotein mediator of transcriptional activation and its interaction with the C-terminal repeat domain of RNA polymerase II. Cell *77*, 599-608.

Kim, J.L., and Burley, S.K. (1994). 1.9 Å resolution refined structure of TBP recognizing the minor groove of TATAAAAG. Nat. Struct. Biol. *1*, 638-653.

Kim, T.K., Zhao, Y., Ge, H., Bernstein, R., and Roeder, R.G. (1995). TATA-binding protein residues implicated in a functional interplay between negative cofactor NC2 (Dr1) and general factors TFIIA and TFIIB. J. Biol. Chem. *270*, 10976-10981.

Kim, M.Y., Mauro, S., Gevry, N., Lis, J.T., and Kraus, W.L. (2004). NAD^+-dependent modulation of chromatin structure and transcription by nucleosome binding properties of PARP-1. Cell *119*, 803-814.

King, I.F., Francis, N.J., and Kingston, R.E. (2002). Native and recombinant polycomb group complexes establish a selective block to template accessibility to repress transcription *in vitro*. Mol. Cell. Biol. *22*, 7919-7928.

Kirschner, D.B., vom Baur, E., Thibault, C., Sanders, S.L., Gangloff, Y.G., Davidson, I., Weil, P.A., and Tora, L. (2002). Distinct mutations in yeast $TAF_{II}25$ differentially affect the composition of TFIID and SAGA complexes as well as global gene expression patterns. Mol. Cell. Biol. *22*, 3178-3193.

Klejman, M.P., Pereira, L.A., van Zeeburg, H.J., Gilfillan, S., Meisterernst, M., Timmers, H.T. (2004). NC2α interacts with BTAF1 and stimulates its ATP-dependent association with TATA-binding protein. Mol. Cell. Biol. *24*, 10072-10082.

Kokubo, T., Gong, D.W., Yamashita, S., Horikoshi, M., Roeder, R.G., and Nakatani, Y. (1993). *Drosophila* 230-kD TFIID subunit, a functional homolog of the human cell cycle gene product, negatively regulates DNA binding of the TATA box-binding subunit of TFIID. Genes Dev. *7*, 1033-1046.

Kokubo, T., Swanson, M.J., Nishikawa, J.I., Hinnebusch, A.G., and Nakatani, Y. (1998). The yeast TAF145 inhibitory domain and TFIIA competitively bind to TATA-binding protein. Mol. Cell. Biol. *18*, 1003-1012.

Kou, H., Irvin, J.D., Huisinga, K.L., Mitra, M., and Pugh B.F. (2003). Structural and functional analysis of mutations along the crystallographic dimer interface of the yeast TATA binding protein. Mol. Cell. Biol. *23*, 3186-3201.

Kraus, W.L., and Lis, J.T. (2003). PARP goes transcription. Cell *113*, 677-683.

Kretzschmar, M., Kaiser, K., Lottspeich, F., and Meisterernst, M. (1994). A novel mediator of class II gene transcription with homology to viral immediate-early transcriptional regulators. Cell *78*, 525-534.

Kretzschmar, M., Meisterernst, M., and Roeder, R.G. (1993). Identification of human DNA topoisomerase I as a cofactor for activator-dependent transcription by RNA polymerase II. Proc. Natl. Acad. Sci. USA *90*, 11508-11512.

Kumar, B.R.P., Swaminathan, V., Banerjee, S., and Kundu, T. (2001). p300-mediated acetylation of human transcriptional coactivator PC4 is inhibited by phosphorylation. J. Biol. Chem. *276*, 16804-16809.

Kuras, L., Kosa, P., Mencia, M., and Struhl, K. (2000). TAF-containing and TAF-independent forms of transcriptionally active TBP *in vivo*. Science *288*, 1244-1248.

Kutach, A., and Kadonaga, J. (2000). The downstream promoter element DPE appears to be as widely used as the TATA box in *Drosophila* core promoters. Mol. Cell. Biol. *20*, 4754-4764.

Lee, Y.C., Park, J.M., Min, S., Han, S.J., and Kim, Y.J. (1999). An activator binding module of yeast RNA polymerase II holoenzyme. Mol. Cell. Biol. *19*, 2967-2976.

Lee, D., Kim, J.W., Kim, K., Joel, C.O., Schreiber,V., Ménissier-de Murcia, J., and Choe, J. (2002). Functional interaction between human papillomavirus type 18 E2 and poly(ADP-ribose) polymerase 1. Oncogene *21*, 5877-5885.

Li, X. Y., Virbasius, A., Zhu, X., and Green, M.R. (1999). Enhancement of TBP binding by activators and general transcription factors. Nature *399*, 605-609.

Li, X-Y., Bhaumik, S.R., and Green, M. (2000). Distinct classes of yeast promoters revealed by differential TAF recruitment. Science *288*, 1242-1244.

Lindahl, T., Satoh, M.S., Poirier, G.G., and Klungland, A. (1995). Post-translational modification of poly(ADP-ribose) polymerase induced by DNA strand breaks. Trends Biochem. Sci. *20*, 405-411.

Liu, D., Ishima, R., Tong, K.I., Bagby, S., Kokubo, T., Muhandiram, D.R., Kay, L.E., Nakatani, Y., and Ikura, M. (1998). Solution structure of a TBP-TAF$_{II}$230 complex: protein mimicry of the minor groove surface of the TATA box unwound by TBP. Cell *94*, 573-583.

Lorch, Y., Beve, J., Gustafsson, C.M., Myers, L.C., and Kornberg, R.D. (2000). Mediator-nucleosome interaction. Mol. Cell *6*, 197-201.

Luger, K., Mader, A.W., Richmond, R.K., Sargent, D.F., and Richmond, T.J. (1997). Crystal structure of the nucleosome core particle at 2.8 Å resolution. Nature *389*, 251-260.

Luo, Y., Ge, H., Stevens, S., Xiao, H., and Roeder, R.G. (1998). Coactivation by OCA-B: definition of critical regions and synergism with general cofactors. Mol. Cell. Biol. *18*, 3803-3810.

Maile, T., Kwoczynski, S., Katzenberger, R.J., Wassarman, D.A., and Sauer, F. (2004). TAF1 activates transcription by phosphorylation of serine 33 in histone H2B. Science *304*, 1010-1014.

Malanga, M., Pleschke, J.M., Kleczkowska, H.E., and Althaus, F.R. (1998). Poly(ADP-ribose) binds to specific domains of p53 and alters its DNA binding functions. J. Biol. Chem. *273*, 11839-11843.

Malanga, M., and Althaus, F.R. (2004). Poly(ADP-ribose) reactivates stalled DNA topoisomerase I and induces DNA strand break resealing. J. Biol. Chem. *279*, 5244-5248.

Maldonado, E. (1999). Transcriptional functions of a new mammalian TATA-binding protein-related factor. J. Biol. Chem. *274*, 12963-12966.

Malik, S., Guermah, M., and Roeder, R.G. (1998). A dynamic model for PC4 coactivator function in RNA polymerase II transcription. Proc. Natl. Acad. Sci. USA *95*, 2192-2197.

Malik, S., Gu, W., Wu, W., Qin, J., and Roeder, R.G. (2000). The USA-derived transcriptional coactivator PC2 is a submodule of TRAP/SMCC and acts synergistically with other PCs. Mol. Cell *5*, 753-760.

Martinez-Campa, C., Politis, P., Moreau, J.-L., Kent, N., Goodall, J., Mellor, J., and Goding, C.R. (2004). Precise nucleosome positioning and the TATA box dictate requirements for the histone H4 tail and the bromodomain factor Bdf1. Mol. Cell *15*, 69-81.

Martinez, E., Kundu, T.K., Fu, J., and Roeder, R.G. (1998). A human SPT3-TAF$_{II}$31-GCN5-L acetylase complex distinct from transcription factor IID. J. Biol. Chem. *273*, 23781-23785.

Martinez, E., Palhan, V.B., Tjernberg, A., Lymar, E.S., Gamper, A.M., Kundu, T.K., Chait, B.T., and Roeder, R.G. (2001). Human STAGA complex is a chromatin-acetylating transcription coactivator that interacts with pre-mRNA splicing and DNA damage-binding factors *in vivo*. Mol. Cell. Biol. *21*, 6782-6795.

McMahon, S.B., Van Buskirk, H.A., Dugan, K.A., Copeland, T.D., and Cole, M.D. (1998). The novel ATM-related protein TRRAP is an essential cofactor for the c-Myc and E2F oncoproteins. Cell *94*, 363-74.

McMahon, S.B., Wood, M.A., and Cole, M.D. (2000). The essential cofactor TRRAP recruits the histone acetyltransferase hGCN5 to c-Myc. Mol. Cell. Biol. *20*, 556-562.

Meisterernst, M., and Roeder, R.G. (1991). Family of proteins that interact with TFIID and regulate promoter activity. Cell *67*, 557-567.

Meisterernst, M., Roy, A.L., Lieu, H.M., Roeder, R.G. (1991). Activation of class II gene transcription by regulatory factors is potentiated by a novel activity. Cell *66*, 981-993.

Meisterernst, M., Stelzer, G., and Roeder, R.G. (1997). Poly(ADP-ribose) polymerase enhances activator-dependent transcription *in vitro*. Proc. Natl. Acad. Sci. USA *94*, 2261-2265.

Mendoza-Alvarez, H., and Alvarez-Gonzalez, R. (2001). Regulation of p53 sequence-specific DNA-binding by covalent poly(ADP-ribosyl)ation. J. Biol. Chem. *276*, 36425-36430.

Merino, A., Madden, K.R., Lane, W.S., Champoux, J.J., and Reinberg, D. (1993). DNA topoisomerase I is involved in both repression and activation of transcription. Nature *365*, 227-232.

Metzger, D., Scheer, E., Soldatov, A., and Tora, L. (1999).

Mammalian TAF$_{II}$30 is required for cell cycle progression and specific cellular differentiation programmes. EMBO J. *17*, 4823-4834.

Michel, B., Komarnitsky, P., and Buratowski, S. (1998). Histone-like TAFs are essential for transcription *in vivo*. Mol. Cell *2*, 663-673.

Mitsiou, D.J., and Stunnenberg, H.G. (2000). TAC, a TBP-sans-TAFs complex containing the unprocessed TFIIA precursor and the TFIIA subunit. Mol. Cell *6*, 527-537.

Mitsiou, D.J., and Stunnenberg, H.G. (2003). p300 is involved in formation of the TBP-TFIIA-containing basal transcription complex, TAC. EMBO J. *22*, 4501-4511.

Mittal, V., and Hernandez, N. (1997). Role for the amino-terminal region of human TBP in U6 snRNA transcription. Science *275*, 1136-1140.

Miyamoto, T., Kakizawa, T., and Hashizume, K. (1999). Inhibition of nuclear receptor signalling by poly(ADP-ribose) polymerase. Mol. Cell. Biol. *19*, 2644–2649.

Mizzen, C.A., Yang, X.J., Kokubo, T., Brownell, J.E., Bannister, A.J., Owen-Hughes, T., Workman, J., Wang, L., Berger, S.L., Kouzarides, T., Nakatani, Y., and Allis, C.D. (1996). The TAF$_{II}$250 subunit of TFIID has histone acetyltransferase activity. Cell *87*, 1261-1270.

Moore, P.A., Ozer, J., Salunek, M., Jan, G., Zerby, D., Campbell, S., and Lieberman, P.M. (1999). A human TATA binding protein-related protein with altered DNA binding specificity inhibits transcription from multiple promoters and activators. Mol. Cell. Biol. *19*, 7610-7620.

Moqtaderi, Z., Bai, Y., Poon, D., Weil, P.A., and Struhl, K. (1996). TBP-associated factors are not generally required for transcriptional activation in yeast. Nature *383*, 188–191.

Moqtaderi, Z., Keaveney, M., and Struhl, K. (1998). The histone H3-like TAF is broadly required for transcription in yeast. Mol. Cell *2*, 675-682.

Muldrow, T.A., Campbell, A.M., Weil, P.A., and Auble, D.T. (1999). MOT1 can activate basal transcription *in vitro* by regulating the distribution of TATA binding protein between promoter and nonpromoter sites. Mol. Cell. Biol. *19*, 2835-2845.

Mulholland, N.M., King, I.F., and Kingston, R.E. (2003). Regulation of Polycomb group complexes by the sequence-specific DNA binding proteins Zeste and GAGA. Genes Dev. *17*, 2741-2746.

Myers, L.C., and Kornberg, R.D. (2000). Mediator of transcriptional regulation. Annu. Rev. Biochem. *69*, 729-749.

Näär, A.M., Beaurang, P.A., Zhou, S., Abraham, S., Solomon, W., and Tjian, R. (1999). Composite co-activator ARC mediates chromatin-directed transcriptional activation. Nature *398*, 828-832.

Narlikar, G.J., Fan, H.Y., and Kingston, R.E. (2002). Cooperation between complexes that regulate chromatin structure and transcription. Cell *108*, 475-487.

Nikolov, D.B., Hu, S.-H., Lin, J., Gasch, A., Hoffman, A., Horikoshi, M., Chua, N.-H., Roeder, R.G., and Burley, S.K. (1992). Crystal structure of TFIID TATA-box binding protein. Nature *360*, 40-46.

Nikolov, D.B., and Burley S.K. (1994). 2.1 Å resolution refined structure of a TATA box-binding protein (TBP). Nat. Struct. Biol. *1*, 621-637.

Nikolov, D.B., Chen, H., Halay, E.D., Hoffman, A., Roeder, R.G., and Burley, S.K. (1996). Crystal structure of a human TATA box-binding protein/TATA element complex. Proc. Natl. Acad. Sci. USA *93*, 4862-4867.

Oei, S.L., Griesenbeck, J., Schweiger, M., and Ziegler, M. (1998). Regulation of RNA polymerase II-dependent transcription by poly(ADP-ribosyl)ation of transcription factors. J. Biol. Chem. *273*, 31644-31647.

Oelgeschläger, T., Chiang, C.-M., and Roeder RG. (1996). Topology and reorganization of a human TFIID-promoter complex. Nature *382*, 735-738.

Oelgeschläger, T., Tao, Y., Kang, Y.K., and Roeder, R.G. (1998). Transcription activation via enhanced preinitiation complex assembly in a human cell-free system lacking TAF$_{II}$s. Mol. Cell *1*, 925-931.

Ogryzko, V.V., Kotani, T., Zhang, X., Schiltz, R.L., Howard, T., Yang, X.J., Howard, B.H., Qin, J., and Nakatani, Y. (1998). Histone-like TAFs within the PCAF histone acetylase complex. Cell *94*, 35-44.

Ohbayashi, T., Makino, Y., and Tamura, T. (1999). Identification of a mouse TBP-like protein (TLP) distantly related to the *drosophila* TBP-related factor. Nucleic Acids Res. *27*, 3750-3755.

Ohbayashi, T., Shimada, M., Nakadai, T., Wada, T., Handa, H., and Tamura, T. (2003). Vertebrate TBP-like protein (TLP/TRF2/TLF) stimulates TATA-less terminal deoxynucleotidyl transferase promoters in a transient reporter assay, and TFIIA-binding capacity of TLP is required for this function. Nucleic Acids Res. *31*, 2127-2133.

Ozer, J., Mitsouras, K., Zerby, D., Carey, M., and Lieberman, P.M. (1998). Transcription factor IIA derepresses TATA-binding protein (TBP)-associated factor inhibition of TBP-DNA binding. J. Biol. Chem. *273*, 14293-14300.

Pan, Z.Q., Ge, H., Amin, A.A., and Hurwitz, J. (1996). Transcription-positive cofactor 4 forms complexes with HSSB (RPA) on single-stranded DNA and influences HSSB-dependent enzymatic synthesis of simian virus 40 DNA. J. Biol. Chem. *271*, 22111-22116.

Park, J.M., Kim, H.S., Han, S.J., Hwang, M.S., Lee, Y.C., and Kim, Y.J. (2000). *In vivo* requirement of activator-specific binding targets of mediator. Mol. Cell. Biol. *20*, 8709-8719.

Pavri, R., Lewis, B., Kim, T.K., Dilworth, F.J., Erdjument-Bromage, H., Tempst, P., de Murcia, G., Evans, R., Chambon, P., and Reinberg, D. (2005). PARP-1 determines specificity in a retinoid signaling pathway via direct modulation of Mediator. Mol. Cell *18*, 83-96.

Pereira, L.A., van der Knaap, J.A., van den Boom, V., van den Heuvel, F.A.J., and Timmers, H.T.M. (2001). TAF$_{II}$170 interacts with the concave surface of TATA-binding protein to inhibit its DNA binding activity. Mol. Cell. Biol. *21*, 7523-7534.

Pereira, L.A., Klejman, M.P., and Timmers, H.T. (2003). Roles for BTAF1 and Mot1p in dynamics of TATA-binding protein and regulation of RNA polymerase II transcription. Gene *315*, 1-13.

Persengiev, S.P., Zhu, X., Dixit, B.L., Maston, G.A., Kittler, E.L., and Green, M.R. (2003). TRF3, a TATA-box-binding protein-related factor, is vertebrate-specific and widely expressed. Proc. Natl. Acad. Sci. USA *100*, 14887-14891.

Pham, A.D., and Sauer, F. (2000). Ubiquitin-activating/conjugating activity of TAF$_{II}$250, a mediator of activation of gene expression in *Drosophila*. Science *289*, 2357-2360.

Pointud, J.-C., Mengus, G., Brancorsini, S., Monaco, L., Parvinen, M., Sassone-Corsi, P., and Davidson, I. (2003). The intracellular localization of TAF7L, a paralogue of transcription factor TFIID subunit TAF7, is developmentally regulated during male germ-cell differentiation. J. Cell Science *116*, 1847-1858.

Poon, D., Campbell, A.M., Bai, Y., and Weil, P.A. (1994). Yeast Taf170 is encoded by MOT1 and exists in a TATA box-binding protein (TBP)-TBP-associated factor complex distinct from transcription factor IID. J. Biol. Chem. *269*, 23135-23140.

Rabenstein, M.D., Zhou, S., Lis, J.T., and Tjian, R. (1999). TATA box-binding protein (TBP)-related factor 2 (TRF2), a third member of the TBP family. Proc. Natl. Acad. Sci. USA *96*, 4791-4796.

Rachez, C., Suldan, Z., Ward, J., Chang, C.P., Burakov, D., Erdjument-Bromage, H., Tempst, P., and Freedman, L.P. (1998). A novel protein complex that interacts with the vitamin D$_3$ receptor in a ligand-dependent manner and enhances VDR transactivation in a cell-free system. Genes Dev. *12*, 1787-1800.

Ruppert, S., Wang, E.H., and Tjian, R. (1993). Cloning and expression of human TAF$_{II}$250: a TBP-associated factor implicated in cell-cycle regulation. Nature *362*, 175-179.

Ruppert, S. and Tjian, R. (1995). Human TAF$_{II}$250 interacts with RAP74: implications for RNA polymerase II initiation. Genes Dev. *9*, 2747-2755.

Ryu, S., Zhou, S., Ladurner, A.G., and Tjian, R. (1999). The transcriptional cofactor complex CRSP is required for activity of the enhancer-binding protein Sp1. Nature *397*, 446-450.

Saurin, A.J., Shao, Z., Erdjument-Bromage, H., Tempst, P., and Kingston, R.E. (2001). A *Drosophila* Polycomb group complex includes Zeste and dTAF$_{II}$ proteins. Nature *412*, 655-660.

Selleck, W., Howley, R., Fang, Q., Podolny, V., Fried, M.G., Buratowski, S., and Tan, S. (2001). A histone fold TAF octamer within the yeast TFIID transcriptional coactivator. Nat. Struct. Biol. *8*, 695-700.

Sewack, G. F., Ellis, T. W., and Hansen, U. (2001). Binding of TATA binding protein to a naturally positioned nucleosome is facilitated by histone acetylation. Mol. Cell. Biol. *21*, 1404-1415.

Shao, Z., Raible, F., Mollaaghababa, R., Guyon, J.R., Wu, C.T., Bender, W., and Kingston, R.E. (1999) Stabilization of chromatin structure by PRC1, a Polycomb complex. Cell *98*, 37-46.

Shao, H., Revach, M., Moshonov, S., Tzuman, Y., Gazit, K., Albeck, S., Unger, T., and Dikstein, R. (2005). Core promoter binding by histone-like TAF complexes. Mol. Cell. Biol. *25*, 206-219.

Shykind, B.M., Kim, J., Stewart, L., Champoux, J.J., and Sharp, P.A. (1997). Topoisomerase I enhances TFIID-TFIIA complex assembly during activation of transcription. Genes Dev. *11*, 397-407.

Simbulan-Rosenthal, C.M., Rosenthal, D.S., Luo, R., Samara, R., Espinoza, L.A., Hassa, P.O., Hottiger, M.O., and Smulson, M.E. (2003). PARP-1 binds E2F-1 independently of its DNA binding and catalytic domains, and acts as a novel coactivator of E2F-1-mediated transcription during re-entry of quiescent cells into S phase. Oncogene. *22*, 8460-8471.

Slattery, E., Dignam, J.D., Matsui, T., and Roeder, R. G. (1983). Purification and analysis of a factor which suppresses nick-induced transcription by RNA polymerase II and its identity with Poly(ADP-ribose) polymerase. J. Biol. Chem. *258*, 5955-5959.

Solow, S., Salunek, M., Ryan, R., and Lieberman, P.M. (2001). TAF$_{II}$250 phosphorylates human transcription factor IIA on serine residues important for TBP binding and transcription activity. J. Biol. Chem. *276*, 15886–15892.

Sterner, D.E., Grant, P.A., Roberts, S.M., Duggan, L.J., Belotserkovskaya, R., Pacella, L.A., Winston, F., Workman, J.L., and Berger, S.L. (1999). Functional organization of the yeast SAGA complex: distinct components involved in structural integrity, nucleosome acetylation, and TATA-binding protein interaction. Mol. Cell. Biol. *19*, 86-98.

Stewart, L., Ireton, G.C., and Champouxx, J.J. (1996). The domain organization of human topoisomerase I. J. Biol. Chem. *271*, 7602-7608.

Sun, X., Zhang, Y., Cho, H., Rickert, P., Lees, E., Lane, W., and Reinberg, D. (1998). NAT, a human complex containing Srb polypeptides that functions as a negative regulator of activated transcription. Mol. Cell *2*, 213-222.

Sutrias-Grau, M., Bianchi, M.E., and Bernues, J. (1999). High mobility group protein 1 interacts specifically with the core domain of human TATA box-binding protein and interferes with transcription factor IIB within the preinitiation complex. J. Biol. Chem. *274*, 1628-1634.

Taatjes, D.J., Näär, A.M., Andel, F., III, Nogales, E., and Tjian, R. (2002). Structure, function, and activator-induced conformations of the CRSP coactivator. Science *295*, 1058-1062.

Taggart, A.K.P., and Pugh, B.F. (1996). Dimerization of TFIID when not bound to DNA. Science *272*, 1331-1333.

Teichmann, M., Wang, Z., Martinez, E., Tjernberg, A., Zhang, D., Vollmer, F., Chait, B.T., and Roeder, R.G. (1999). Human TATA-binding protein-related factor-2 (hTRF2) stably associates with hTFIIA in HeLa cells. Proc. Natl. Acad. Sci. USA *96*, 13720-13725.

Thomas, J.O. (2001). HMG1 and 2: architectural DNA-binding proteins. Biochem. Soc. Trans. *29*, 395-401.

Thomas, J.O., and Travers, A.A. (2001). HMG1 and 2, and related 'architectural' DNA-binding proteins. Trends Biochem. Sci. *26*, 167-174.

Thomas, M.C., and Chiang, C.-M. (2005). E6 oncoprotein represses p53-dependent gene activation via inhibition of protein acetylation independently of inducing p53 degradation. Mol. Cell *17*, 251-264.

Thompson, C.M., Koleske, A.J., Chao, D.M. and Young, R.A. (1993). A multisubunit complex associated with the RNA polymerase II CTD and TATA-binding protein in yeast. Cell *73*, 1361-1375.

Timmers, H.T., Meyers, R.E., and Sharp, P.A. (1992). Composition of transcription factor B-TFIID. Proc. Natl. Acad. Sci. USA *89*, 8140-8144.

Tora, L. (2002). A unified nomenclature for TATA box binding protein (TBP)-associated factors (TAFs) involved in RNA polymerase II transcription. Genes Dev. *16*, 673-675.

Tulin, A., and Spradling, A. (2003). Chromatin loosening by poly(ADP)-ribose polymerase at *Drosophila* puff loci. Science *299*, 560-562.

van der Knaap, J.A., Borst, J.W., van der Vliet, P.C., Gentz, R., and Timmers, H.T.M. (1997). Cloning of the cDNA for the TATA-binding protein-associated factor$_{II}$170 subunit of transcription factor B-TFIID reveals homology to global transcription regulators in yeast and *Drosophila*. Proc. Natl. Acad. Sci. USA *94*, 11827-11832.

Verrier, C.S., Roodi, N., Yee, C.J., Bailey, L.R., Jensen, R.A., Bustin, M., and Parl, F.F. (1997). High-mobility group (HMG) protein and TATA-binding protein-associated factor TAF$_{II}$30 affect estrogen receptor-mediated transcriptional activation. Mol. Endocrinol. *11*, 1009-1019.

Verrijzer, C. P., and Tjian, R. (1996). TAFs mediate transcriptional activation and promoter selectivity. Trends Bioch. Sci. *21*, 338-342.

Walker, S.S., Reese, J.C., Apone, L.M., and Green, M.R. (1996). Transcription activation in cells lacking TAF$_{II}$s. Nature *383*, 185–188.

Walker, A.K., Shi, Y., and Blackwell, T.K. (2004). An extensive requirement for transcription factor IID-specific TAF-1 in *Caenorhabditis elegans* embryonic transcription. J. Biol. Chem. *279*, 15339-15347.

Wang, J.C. (1996). DNA topoisomerases. Annu. Rev. Biochem. *65*, 635-692.

Wang, J.C. (2002). Cellular roles of DNA topoisomerases: a molecular perspective. Nature Rev. Mol. Cell. Biol. *3*, 430-440.

Wang, J.Y., Sarker, A.H., Cooper, P.K., and Volkert, M.R.. (2004).The single-strand DNA binding activity of human PC4 prevents mutagenesis and killing by oxidative DNA damage. Mol. Cell. Biol. *24*, 6084-6093.

Werten, S., Stelzer, G., Goppelt, A., Langen, F.M., Gros, P., Timmers, H.T.M., Van der Vliet, P.C., and Meisterernst, M. (1998). Interaction of PC4 with melted DNA inhibits transcription. EMBO J. *17*, 5103-5111.

Wieczorek, E., Brand, M., Jacq, X., and Tora, L. (1998). Function of TAF$_{II}$-containing complex without TBP in transcription by RNA polymerase II. Nature *393*, 187-191.

Willy, P.J., Kobayashi, R., and Kadonaga, J.T. (2000). A basal transcription factor that activates or represses transcription. Science *290*, 982-985.

Wolner, B.S., and Gralla, J.D. (2001). TATA-flanking sequences influence the rate and stability of TATA-binding protein and TFIIB binding. J. Biol. Chem. *276*, 6260-6266.

Wu, S.-Y., and Chiang, C.-M. (1998). Properties of PC4 and an RNA polymerase II complex in directing activated and basal transcription *in vitro*. J. Biol. Chem. *273*, 12492-12498.

Wu, S.-Y., Kershnar, E., and Chiang, C.-M. (1998). TAF$_{II}$-independent activation mediated by human TBP in the presence of the positive cofactor PC4. EMBO J. *17*, 4478-4490.

Wu, S.-Y., Thomas, M.C., Hou, S.Y., Likhite, V., and Chiang, C.-M. (1999). Isolation of mouse TFIID and functional characterization of TBP and TFIID in mediating estrogen receptor and chromatin transcription. J. Biol. Chem. *274*, 23480-23490.

Wu, S.-Y., and Chiang, C.-M. (2001). TATA-binding protein-associated factors enhance the recruitment of RNA polymerase II by transcriptional activators. J. Biol. Chem. *276*, 34235-34243.

Wu, S.-Y., Zhou, T., and Chiang, C.-M. (2003). Human Mediator enhances activator-facilitated recruitment of RNA polymerase II and promoter recognition by TATA-binding protein (TBP) independently of TBP-associated factors. Mol. Cell. Biol. *23*, 6229-6242.

Wu, P.-Y.J., Ruhlmann, C., Winston, F., and Schultz, P. (2004). Molecular architecture of the S. cerevisiae SAGA complex. Mol. Cell *15*, 199-208.

Xie, X., Kokubo, T., Cohen, S.L., Mirza, U.A., Hoffmann, A., Chait, B.T., Roeder, R.G., Nakatani, Y., and Burley, S.K. (1996). Structural similarity between TAFs and the heterotetrameric core of the histone octamer. Nature *380*, 316-322.

Yudkovsky, N., Ranish, J.A., and Hahn, S. (2000). A transcription reinitiation intermediate that is stabilized by activator. Nature *408*, 225-229.

Yuan, C.X., Ito, M., Fondell, J.D., Fu, Z.Y., and Roeder, R.G. (1998). The TRAP220 component of a thyroid hormone receptor-associated protein (TRAP) coactivator complex interacts directly with nuclear receptors in a ligand-dependent fashion. Proc. Natl. Acad. Sci. USA *95*, 7939-7944.

Zanton, S.J., and Pugh, B.F. (2004). Changes in genomewide occupancy of core transcriptional regulators during heat stress. Proc. Natl. Acad. Sci. USA *101*,16843-16848.

Zhao, X., and Herr, W. (2002). A regulated two-step mechanism of TBP binding to DNA: a solvent-exposed surface of TBP inhibits TATA box recognition. Cell *108*, 615-627.

Zhou, Q., Lieberman, P.M., Boyer, T.G., and Berk, A.J. (1992). Holo-TFIID supports transcriptional stimulation by diverse activators and from a TATA-less promoter. Genes Dev. *6*, 1964-1974.

Zhou, Q., Boyer, T.G., and Berk, A.J. (1993). Factors (TAFs) required for activated transcription interact with TATA box-binding protein conserved core domain. Genes Dev. *7*, 180-187.

Zhou, T., and Chiang, C.-M. (2001). The intronless and TATA-less human $TAF_{II}55$ gene contains a functional initiator and a downstream promoter element. J. Biol. Chem. *276*, 25503-25511.

Zhou, T., and Chiang, C.-M. (2002). Sp1 and AP2 regulate but do not constitute TATA-less human $TAF_{II}55$ core promoter activity. Nucleic Acids Res. *30*, 4145-4157.

Chapter 05
Chromatin and Regulation of Gene Expression

Joseph H. Taube and Michelle Craig Barton

Department of Biochemistry and Molecular Biology, Program in Genes and Development, Graduate School of Biological Sciences, University of Texas MD Anderson Cancer Center, Houston, TX 77030

Key Words: nucleosome, histones, histone modifications, histone acetylase, histone deacetylase, histone methyltransferase, chromatin remodeling complex

Summary

The human genome is approximately three billion nucleotides in composition and, if stretched along a single, continuous strand, would span almost two meters. The organization of this entity within the confines of a nucleus is an essential challenge for every eukaryotic cell. Darkly staining nuclear material was observed by early microscopists and called chromatin, which later proved to be the structural answer to this problem. Chromatin structure and its place in regulated gene expression, cellular homeostasis and loss of function in disease form ever-expanding and fundamental interests for basic scientists, as well as clinicians. This chapter addresses the basics of chromatin structure, how regulation by chromatin structure has been studied, the specifics of activated, repressed and silenced chromatin, identification of chromatin remodeling complexes and histone modifiers, and their functions in regulation of gene expression.

Introduction

Chromatin is composed of both DNA and histone proteins, H1, H2A, H2B, H3, and H4. The repeating unit of chromatin is the nucleosome, defined as two of each of the core histones, H2A, H2B, H3, and H4 wrapped almost twice by approximately 146 base pairs of DNA. The nucleosome establishes the first level of chromatin organization. Nucleosomes themselves can be stacked and coiled to form higher order structures; one example of which is the 30 nm chromatin fiber. Greater levels of compaction result in the looped and organized fibers of the metaphase chromosome. Necessarily, wrapping of DNA as nucleosomes and compaction into higher order structures inhibits DNA-dependent nuclear processes. Accessing DNA, which is structured as chromatin, is a major challenge for transcription, replication and repair in all cells. Several mechanisms have evolved to allow access including chromatin remodeling by ATP-dependent protein complexes and enzymes that modify histone N-terminal tails (discussed below and in the next chapter).

Histone Proteins

The histone proteins are highly conserved, essential proteins found in every eukaryotic organism. The properties of these proteins are fundamental for the structure and function of chromatin overall. The core histones, H2A, H2B, H3, and H4 are low molecular weight (10-14 kDa) proteins (Johns, 1967). Histone H1, the linker histone, is slightly larger than the core histones (21 kDa) and is found at half the molar concentration of H2A, H2B, H3, and H4 (Hayashi *et al.*, 1978). Each core histone contains a characteristic histone-fold domain, which consists of a central α-helix flanked by two smaller α-helices and loops (Arents and Moudrianakis, 1995). This domain mediates antiparallel dimerization of H2A and H2B as well as tetramerization of H3 and H4. Interestingly, the histone fold domain is also found in the general transcription factor TFIID and other TBP-associated factors (Xie *et al.*, 1996).

Corresponding Author: Michelle Craig Barton, Dept. of Biochemistry and Molecular Biology, The University of Texas MD Anderson Cancer Center, 1515 Holcombe Blvd., Box 1000, Houston, TX 77030. E-mail: mbarton@odin.mdacc.tmc.edu.

Extending from the central histone fold core domain of histones are the N-terminal tails (Fig. 5.1). The crystal structure of the nucleosome lacks definition of the N-terminal tails outside of the nucleosome, leading to the conclusion that the tails are highly dynamic, exist in a random configuration and extend from the nucleosomal core (Luger *et al.*, 1997).

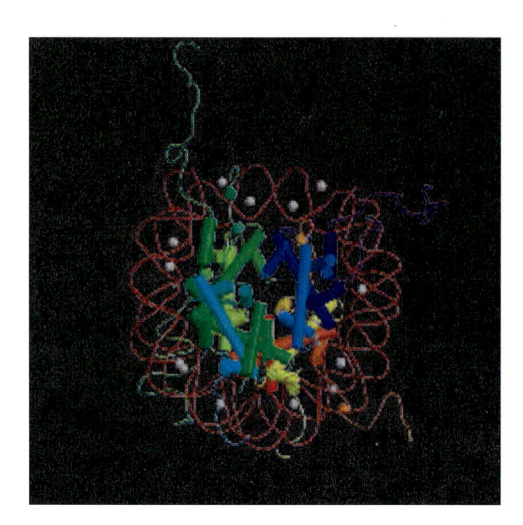

Fig.5.1 The structure of the nucleosome. Shown is a cylindrical representation of the nucleosome, an X-ray structure solved at 1.9 angstrom resolution by T. Richmond and colleagues (Davey *et al.*, 2002). Recombinant core histone proteins of *Xenopus laevis* were used in assembly of a mononucleosome with 147 bp of DNA. The DNA helical strands are shown in red. The histone proteins in the foreground of the whole nucleosome shown are as follows: H3, dark blue, H4, medium blue, H2B, green, H2A, and light blue. The histone fold domains interact and are encircled by the DNA; the histone tails project from the core nucleosome particle. This image is available on the Protein Data Bank web site.

The N-terminal tails are subject to post-translational protein modifications such as acetylation, phosphorylation, ubiquitylation, methylation, and others, which influence chromatin structure and transcription factor binding. Combinatorial patterns of these modifications are correlated with specific regulatory states and, as an informational mechanism for gene regulation, are called the "histone code" (Fischle *et al.*, 2003). Histones, which are the most abundant nuclear proteins, oligomerize to function as a scaffold for DNA; how they do so is the story of the nucleosome.

Nucleosome Structure

The histone proteins interact with each other, to form a histone octamer, wrapped by approximately 146 base pairs of DNA, 1.7-1.8 times in a left-handed superhelix, to build a nucleosome (Kornberg and Lorch, 1999). Additional linker DNA between neighboring nucleosomes varies in length depending on cell type (McGhee and Felsenfeld, 1980). Properties of octamer formation were determined by ion concentration-dependent interactions. In the absence of DNA or salt the core histones form dimers of H2A and H2B, and tetramers of two H3/H4 dimers. With addition of DNA or NaCl, the core histone proteins combine to form the histone octamer. DNA-histone octamer interactions occur through the neutralization of negatively charged backbone phosphate groups by direct interactions and water-mediated hydrogen bonds with lysines and arginines. Additionally, hydrophobic interactions occur between residues such as threonine, proline, valine, and isoleucine and the deoxyribose groups of DNA (Luger *et al.*, 1997). The lack of specific interactions with the nitrogenous base pairs corresponds with the lack of nucleotide-sequence specificity of histone-DNA interactions. Also of note, the major grooves of the two turns of DNA on each side of the nucleosome line up, forming a channel, allowing the tails of H2B and H3 to escape while the tails of H2A and H4 pass through the minor groove. Finally, the nucleosome can accommodate DNA with a slightly altered twist, allowing for potential mechanisms to shift nucleosome positions along the DNA sequence (Luger *et al.*, 1997).

Nucleosome Dynamics

DNA wrapped around a histone octamer can be blocked from interactions with proteins essential for transcription, replication and other nuclear processes. Exposure of regulatory elements within DNA can occur either by sliding of nucleosomes, to position these sequences within linker DNA, or by transitionally lifting the DNA molecule off the nucleosome (Fig. 5.2). The first process is actively regulated by chromatin remodeling complexes, while the second process occurs passively in solution and is termed nucleosome dynamics. Sliding of the nucleosome across DNA does not occur without the aid of enzymes because the energy-of-activation barrier is prohibitively high, due to numerous hydrogen and electrostatic interactions that anchor the nucleosome in place. DNA positioned near the border of the nucleosome could partially unwrap and then rebind further upstream, leading to a bulge at the opposite side of the nucleosome (Kassabov *et al.*, 2003). Recently, the method of fluorescence resonance energy transfer (FRET) analysis was used to assign an equilibrium constant of approximately .05 for unwrapping the first few base pairs of DNA (at the entry/exit sites) from the histones octamer. Thus, at any given time ~5% of nucleosomes on average are partially unwrapped or breathing (Li and Widom, 2004). Additional kinetic experiments reveal that the rate constant for unwrapping is 4 s^{-1}, while rewrapping occurs much faster at 20-90 s^{-1} (Li et al., 2005). Integration of these data with equilibrium constants and rate constants of DNA-binding proteins reveals a model where proteins have a window of opportunity to bind their sequences during its brief exposure thus preventing rewrapping of the DNA back onto the nucleosome. Physiologically, active cellular processes, such as DNA replication and chromatin remodeling, may widen or close this window of opportunity.

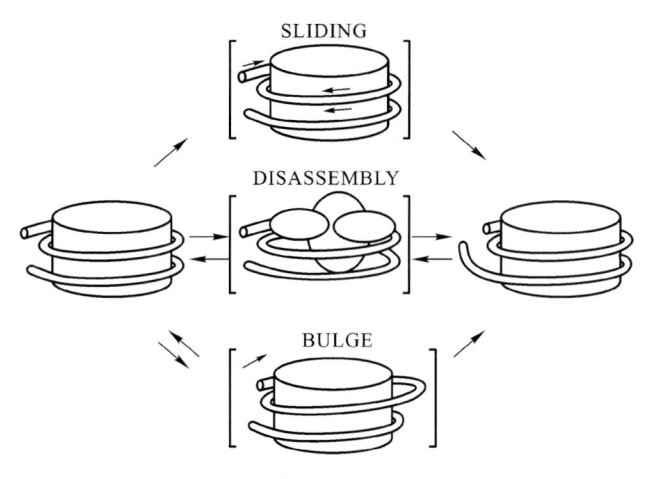

Fig.5.2 **Potential mechanisms of chromatin remodeling by SWI/SNF and ISWI complexes.** Sliding of DNA around the histone octamer is thermodynamically prohibited but could be catalyzed by a chromatin remodeling, ATPase complex (top). Disassembly of the nucleosome into an H3/H4 tetramer and two H2A/H2B dimmers could account for remodeling of tightly spaced nucleosomes (center). Bulge propagation involves twisting or bending of the DNA to force a loop of exposed DNA around the nucleosome (bottom). In all cases, DNA incorporation into the nucleosome is maintained in the product of remodeling. Small arrows indicate direction of DNA movement.

Histone H1

Linker histone H1 is found at half the molar concentration of the other histones, and it was long believed that every nucleosome also contained a single H1 molecule in order to compact chromatin. However, X-ray diffraction experiments required only the four core histones and DNA to reconstitute the diffraction pattern obtained with in vivo chromatin, indicating that H1 is not essential (Kornberg and Thomas, 1974). Nevertheless, the H1 protein, which consists of N-terminal, C-terminal and central winged-helix domains, aids chromatin compaction in vitro. Recent studies identified sub-domains within the unstructured C-terminal domain, which induce chromatin fiber folding (Lu and Hansen, 2004). H1 binds to DNA through its winged-helix domain, but cannot bind to the histone octamer on its own. Instead, H1 binds to the nucleosome close to where the DNA enters and exits.

Linker histones likely play a role in regulation rather than in structure alone, as yeast lacking the homolog of H1 survive (Shen et al., 1995). Additionally, H1 knockouts in *Tetrahymena* and mouse 3T3 cells resulted in down-regulation, as well as up-regulation of certain genes (Brown et al., 1997). As with most regulatory proteins, H1 is evolutionarily divergent and several isoforms exist. Mammals contain six somatic H1 subtypes, which are expressed at various levels in different tissues, cell-cycle phases and developmental stages (Wang et al., 1997). Specifically, histone H1.2 translocates from the nucleus to the mitochondria and signals cytochrome c release in apoptosis and H1b is recruited by a homeobox transcription factor Msx1 to repress transcription of the muscle-specific determinant MyoD (Konishi et al., 2003; Lee et al., 2004). Examples such as these highlight H1's role as a transcription factor.

Histone Variants

In addition to several loci encoding core histones, eukaryotes also contain genes that encode variants of the core histones. These variants often impart special properties to the chromatin where they are incorporated and may, in part, provide a mechanism for inheritance of chromatin transcription states (Henikoff et al., 2004). Variant histones are incorporated into previously assembled nucleosomes by specific histone chaperones during DNA replication and/or by a process that may be coupled to transcription, independently of DNA replication. In this manner, specific variants that denote transcriptional domains can be maintained through multiple rounds of replication.

Specialized functions of the variant histones illustrate the breadth of influence that chromatin has on nuclear processes. CENP-A is an H3-like, histone variant that is found only in nucleosomes at the centromere, a chromatin, structural organization that plays an essential role during cell division. H3.3, another H3-like histone, is incorporated into chromatin throughout the cell cycle, unlike canonical H3.1, which enters chromatin during active DNA replication in S-phase of the cell cycle. H3.3 is enriched in transcribed chromatin. Chaperone-mediated swapping of H3.3 for existing H3.1, within a nucleosome, serves as a means of erasing repressive H3.1 post-translation modifications even in the absence of DNA replication (Tagami et al., 2004). Likewise, variants of H2A perform specific roles. In *S. pombe*, H2A.Z is incorporated near the edges of silenced chromatin and inhibits spreading of higher-order structures into active chromatin by promoting formation of the 30 nm chromatin fiber (Fan et al., 2002). Finally, H2A.X is localized to the foci of DNA repair proteins at the sites of double-strand breaks and is phosphorylated, either to mark the damage site or as a result of the process of repair.

Higher Order Structures

Electron microscopy studies of extracted chromatin reveal a 30 nm chromatin fiber, which can be further compacted, ultimately forming the mitotic chromosome (Adolph, 1981). For decades intense research has not yet resolved the molecular structure of this fiber, nor of any higher order chromatin structure. Chromatin reconstituted *in vitro* can undergo salt-dependent compaction from the nucleosomal array (at 1-10 mM salt concentration) to the 30 nm fiber (100-200 mM salt concentration) and addition of H1 increases the rate of compaction (Carruthers et al., 1998). X−ray and neutron-scattering experiments led to numerous models of chromatin structure beyond repeated nucleosomes or "beads-on-a-string", including the solenoid, continuous superhelix, twisted-ribbon, helical-ribbon, and cross-linker (Horn and Peterson, 2002). These can be divided into one-start and two-start models based on whether one strand wraps around itself (solenoid, continuous superhelix, and twisted-ribbon) or two strands wrap around each other (helical-ribbon and crossed-linker). Current, *in vitro* analysis of a twelve- nucleosome array favors the two-start class of models due to the ability of an endonuclease to separate the array into two six-nucleosome halves (Dorigo et al., 2004). Deletion experiments have shown that the histone H4 tail is essential for compaction and that it interacts with neighboring H2A/H2B (Davey et al., 2002). Substitution of H2A with variant histone H2A.Z results in a higher rate of compaction (Fan et al., 2004).

In order to view higher order chromatin structures in a physiological setting, microscopy studies were performed to visualize expressed Green Fluorescent Protein (GFP)-LacI fusion proteins interacting with LacI-binding sites, inserted as tandem, multi-copy arrays within an endogenous, genomic locus in cells. Exogenous expression of GFP-LacI proteins showed GFP-LacI fluorescence at a single focus of compacted chromatin.

However, when a GFP-LacI-VP16 fusion protein was used to simulate binding of a transcription-activating protein to chromatin, the single focus dispersed and spread into a ribbon of GFP-fluorescence, 80 to 100 nm in diameter (Tumbar et al., 1999). This type of chromatin structure, which likely contains three 30 nm chromatin fibers, may be the basic unit of higher order chromatin structure that is competent for activated gene expression.

Histones are Required for Life

While biochemists were analyzing the structure of chromatin and the nucleosome, and developing means of assembling chromatin *in vitro* (see below), geneticists were studying the impact that chromatin structure had on gene expression *in vivo*. These studies verified that histones, and thus chromatin, were essential for life itself. Deletion of the genes encoding histone proteins, using the yeast *S. cerevisiae* as a model organism, led to death of the yeast cells. Histone proteins are encoded by multiple copies of each gene; cells are viable with just one copy of each, but die in the complete absence of any one histone. Plasmid shuffle techniques were used where exogenously introduced plasmid DNA, encoding a single copy of a histone gene, could replace the endogenous copy. This methodology allows the regulated expression of specific histone proteins, as well as introduction of mutations or deletions within a histone gene, and analysis of their effect on gene expression *in vivo*. This approach, employing a genetically tractable model organism, revealed that individual regions of histones have specific functions in gene regulation (Grunstein, 1990). The amino-terminal tails of histones play major roles in transcription activation and repression. Deletions of these tails do not kill cells but the life cycle of yeast and expression of many genes, both activation and repression, are dramatically affected. Further deletions more C-terminal, into the hydrophobic cores of the histones, led to cell death. These findings show that histones perform essential functions in gene expression, both activating and repressing, and in maintenance of viability.

Heterochromatin and Euchromatin

Arguably, the most important function of chromatin may be repression and silencing of gene expression. While repressed genes may share some of the characteristics of continuously silenced chromatin, defined as heterochromatin, they retain the potential for active transcription and are placed in the subdivision of chromatin known as euchromatin. The term "silenced" is used here with regards to a maintained absence of transcriptional activity, and "repression" for short-term or active down-regulation of transcription. Silenced chromatin (versus repressed chromatin) can be further subdivided into constitutive, primarily at centromeric and telomeric regions, and facultative heterochromatin, such as the inactivated X-chromosome (Maison et al., 2002; Richards and Elgin, 2002). Briefly, both types of heterochromatin are present within highly condensed and pericentric regions of chromosomes, often consisting of long stretches of repeated elements. They also share the properties of widespread histone hypoacetylation, extended areas of methylated histone H3 at lysine 9 (metH3K9) and DNA methylation. Additionally, recent results support the participation of unique, non-coding siRNAs and dedicated enzymatic complexes in the silencing of both constitutive and facultative heterochromatin (Grewal and Elgin, 2002), as will be discussed further in a later chapter.

Each heterochromatin subgroup may have its own specific, associated histone methyltransferases (HMT) and variant histones. The lack of acetylation at H3K9, which permits its modification by HMT's to form metH3K9, is characteristic of histone tails present in heterochromatin. The association of proteins such as Heterochromatin Protein 1 (HP1), which is usually concentrated in pericentric heterochromatin and telomeres, promotes spreading of silenced chromatin along the repetitive DNA characteristic of constitutive heterochromatin. HP1 interacts with metH3K9 via HP1's N-terminal chromo domain and further with HMT's through its C-terminal chromo-shadow domain. Thus, by interacting with both the enzyme and its substrate, HP1 ensures establishment and maintenance of a stable silenced state (Kouzarides, 2002; Richards and Elgin, 2002).

Histone deacetylation, metH3K9 modification and HP1 binding are associated with gene inactivation but are not unique to heterochromatin, as they also occur at repressed euchromatin (Nguyen et al., 2005; Nielsen et al., 2001; Schultz et al., 2002). Hypoacetylation, induced by histone deacetylases (HDAC's), and histone methylation, by HMT's, can spread along a region of silenced chromatin. The best-studied example of histone deacetylation/methylation linked to silencing is its propagation over a large region at the silent mating type HML/HMR loci of S. cerevisiae. Although not cytologically distinct as heterochromatin, silent mating type chromatin is never expressed and is heterochromatin-like. Sequence-specific, silencer-element binding proteins, such as Rap1, along with Ku70 and Ku80, recruit the multi-component, nucleosome-binding Sir (1-4) complex in yeast. Sir2, a unique histone deacetylase, is dependent on the cofactor nicotinamide adenine dinucleotide (NAD) for its enzymatic activity (Tanny et al., 1999), catalyzes deacetylation of histone H3/H4 tails, which then facilitates the binding of Sir3 and Sir4. Interactions between the Sir proteins enable propagation of the silenced state (Gottschling, 2000; Shore, 2000). Interestingly, silencing and Sir2 activity, in particular, are linked to the process of ageing, which has generated considerable interest in this facet of chromatin function (Denu, 2003; Guarente, 2000).

Heterochromatin of higher eukaryotes is generally marked by methylation of DNA. Methylated DNA can serve as a recruiting center for enzyme complexes that promote chromatin inactivation, which can further support additional DNA methylation. DNA methylases, such as Dnmt, can methylate DNA following directions from histone modification codes rather than the DNA sequence itself. Recent evidence links histone methylation and DNA methylation through specific HMT-mediated methylation of H3K9 and recruitment of DNA methyltransferases. These HMT's, e.g, Neurospora crassa Dim-5 and Krypotonite from A. thaliana, through their enzymatic function at K9 of H3, recruit HP1-like proteins that in turn attract specific DNA methyltransferases (Lindroth et al., 2004; Naumann et al., 2005; Tamaru and Selker, 2001). Cytosine-methylated DNA is bound by MeCP/MBD (methyl cytosine-binding protein/methyl binding domain) proteins, which act as adaptors between DNA methylases and histone deacetylases. There is also evidence for direct interaction between DNA methylase Dnmt and HDACs, which is independent of Dnmt enzymatic activity. Again, similar to histone methylation, DNA methylation-mediated recruitment of HDACs via MBDs may not be limited to heterochromatin silencing (Fuks et al., 2003; Jaenisch and Bird, 2003).

Silencing of gene expression must be tightly regulated, just as activation of gene expression is. Aberrant methylation of DNA at gene sequences enriched in C-G nucleotide pairs, CpG islands, has been reported in a number of tumor-derived cells. In this example, proximal promoter sequences of tumor suppressor genes are methylated at CpG sites. This DNA methylation and chromatin silencing disrupts regulatory protein binding, and the tumor suppressor gene cannot be activated to perform its protective functions when cellular homeostasis is threatened. Tumor cells may express abnormally high levels of active Dnmt's or other factors, normally protective of the CpG regulatory sequences found enriched in the promoters of many genes, are not active. The link between increased HMT activity, by a number of family members bearing conserved SET domains, and aberrant DNA methylation (Marmorstein, 2003) has led to combined therapeutic approaches whereby 5-azacytidine, an inhibitor of Dnmt's, and HDAC inhibitors, which can maintain acetylated H3K9 and inhibit its methylation, are used in an attempt to re-activate aberrantly silenced tumor suppressor function (Baylin *et al.*, 2001; Jones, 2002).

Questions of accessibility

As structural knowledge of chromatin increased, the natural question arose of how DNA, wrapped around a nucleosome and condensed into fibers, can be accessed for transcription. It was apparent to microscopists that chromatin structure was different somehow when actively transcribed. Stained images of polytene chromosomes revealed an unevenly spaced banding pattern and physical distortions along the length of the chromosomal fibers. Polytene chromosomes, consist of thousands of copies of the same chromosomal DNA, which are amplified, align one alongside the other and are highly enriched in the giant cells of *Drosophila* salivary glands. These amplified chromosomes are visible by light microscopy and can be stained or probed for specific RNA expression, by *in situ* hybridization, or the presence of specific proteins, using immunological or histochemical methods (Fig.5.3). Regions of polytene chromosomes that are actively transcribing RNA are expanded or "puffed", while regions that are silent and not transcribing are more condensed. These patterns of RNA expression accounted for the distorted shape of the chromosome fiber, observed under the microscope. What causes the expanded regions to puff? Is it the act of RNA polymerase moving along the chromatin and pushing its way through nucleosome after nucleosome? Are there mechanisms at work that open up the chromatin structure to allow passage to a polymerase, and others that close up the chromatin behind it? What marks a certain section of a chromosome for activation at a specific time in a particular cell? How does RNA polymerase find this specific site when chromatin is so highly structured?

Biochemists attempted to study these problems much the same way as reconstruction of basal transcription proceeded, by *in vitro* assembly of the process using purified components. The additional challenge, beyond the essential need to purify and reconstitute active RNA polymerase and basal transcription factors, as discussed in the previous chapter, was how to build chromatin *in vitro* that bore any resemblance to chromatin in real life. Bringing pieces of chromatin from living cells into the test tube and attempting to transcribe them was not a successful approach. Most of the chromatin in any given cell is silent or heterochromatic, and isolation of chromatin in the act of gene expression is difficult. The problem of how to build a nucleosomal substrate for RNA polymerase to copy needed to be addressed first.

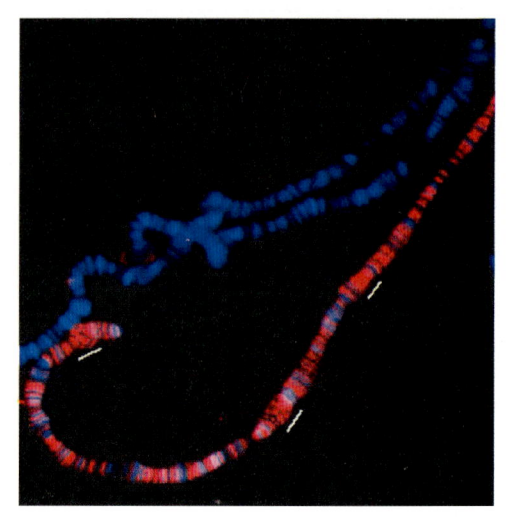

Fig.5.3 The polytene chromosomes of *Drosophila* salivary glands. Light microscopy can be used to visualize areas of expansion (puffs, some marked by white bars) and constriction of amplified, polytene chromosomes, associated with active and silent transcription, respectively. DAPI (blue) stains DNA and is brightest when DNA is heterochromatic. Immunofluorescence is used to probe for the presence of a protein complex involved in *Drosophila* dosage compensation (red). Image kindly provided by Dr. R. Kelley, Baylor College of Medicine, Houston, TX, USA.

The components of chromatin seem fairly simple: two copies each of four proteins, histones H2A, H2B, H3, and H4, 146 base pairs of DNA and a linker histone, H1, if chromatin is more condensed. However, addition of these constituent parts together in a test tube does not lead to self-assembly of the highly organized nucleosome structure. Histones are positively charged due to their enrichment in basic amino acids; DNA is negatively charged. Placing them together in a test tube will cause aggregation and precipitation. An ordered, stepwise assembly process is needed to bring all of the pieces together and form a biologically active, nucleosomal substrate. This challenge must be faced within a cell every time a DNA replication fork passes through chromatin. The component parts must be re-assembled without disruption of overall nuclear structure and with complete integrity and reproducibility to maintain cellular "memory".

Assembly of Chromatin: Chemical Means

The simplest way to avoid nonspecific, histone aggregation and DNA precipitation is to slow down the process of chromatin assembly and give the natural, histone partners, which interact by virtue of their histone fold domains, time to associate with each other: H2A with H2B to form dimers, H3 with H4 to form a tetramer. Two dimers and a tetramer then interact to form an octamer, and DNA wraps around the octamer core approximately 1.8 turns. Slowing the process down can be accomplished *in vitro* by starting the chromatin assembly process in molar concentration salt solutions with all of the components present, and then diluting the salt over a very long period of time, as much as 48 hours in a cold room. This slow approach toward physiological salt levels gives the histone partners time to associate in a stable, ordered way. This is especially effective with shorter pieces of DNA, approximately 150-200 base pairs in length, to form mononucleosomes, and was employed to assemble the mononucleosomes used in determination of the crystallographic structure (Luger et al., 1997).

Polymers of nucleosomes will form, using the high-salt approach and longer pieces of DNA, but they lack the repetitive spacing of true chromatin. Random clustering of nucleosomes and stretches of bare DNA may result, as there is no specific, DNA binding sequence for nucleosomes. Certain, naturally occurring sequences are considered to have high affinity for nucleosomal assembly, but these enrich 1000-fold or less for a specific nucleosome position. Compared to DNA sequence-specific binding of a transcription regulatory factor, this is a relatively moderate probability and usually nucleosomes assume even these "specific" positions with some random clustering. The enrichment sites, as well as artificial sequences of DNA designed for the specific purpose of higher affinity binding for a nucleosome, are often referred to as nucleosome-spacer sequences (Lowary and Widom, 1998; Thastrom et al., 2004). Plasmids carrying nucleosome-spacer sequences, set approximately 200 base pairs apart, have been developed to use with the simple, high-salt approach to assemble nucleosomes and form repetitive, multi-nucleosome structures. Stretches of chromatin, whether formed *in vivo* or *in vitro*, that carry a number of evenly spaced nucleosomes, which form in an ordered manner at specific sites, are referred to as nucleosome arrays.

Well-defined arrays of nucleosomes can be assembled *in vitro* but these are not truly representative of chromatin found *in vivo*. The vast majority of chromatin is assembled in a more random, or stochastic, manner to yield a highly repetitive structure but one not generally defined by specific DNA-binding sequences. Specific stretches of chromatin formed by arrays of ordered nucleosomes can occur *in vivo*, due to some sequence specificity or as a result of DNA binding by transcription factors to set or "phase" nucleosomes at specific positions. Some well-studied examples of positioned nucleosomes, which play an integral role in gene regulation, are the phosphate-responsive *Pho5* gene of *S. cerevisiae* and the glucocorticoid-regulated mouse mammary tumor virus promoter (Archer et al., 1991; Schmid et al., 1992). Positions of nucleosomes within the promoters of these genes are altered in response to environmental factors, which then renders DNA sequences, normally blocked by nucleosomes, accessible to transcription factors and basal transcription machinery. In other cases, positioned nucleosomes enable communication between activating regulatory elements and basal transcription machinery at the core promoter, providing a platform for looping and long-distance interactions in chromatin (Wolffe, 1994).

Assembly of Chromatin: Biological Choices

Within a nucleus there is no option to assemble chromatin in a high-salt solution over a number of hours, nor are protein and enzymatic complexes generally stable under these conditions. Specific proteins, which establish a stepwise order to the process of nucleosome assembly, have evolved to form chromatin *in vivo*. Histones are found in association with specific chaperone proteins, which escort them to the site of nucleosome assembly and prevent aggregation. For example, histones H2A and H2B are chaperoned by proteins Nap-1/2 and H3/H4 associate with the CAF-1 and ASF-1 proteins. Chaperoned histones are met by protein complexes, which perform the energy-requiring, enzymatic function of nucleosome assembly and spacing. Examples of these assembly factors are ACF1 and its catalytic, ATPase-subunit, Iswi1p (discussed below with chromatin remodeling enzymes). As might be expected, these factors come into play especially during DNA replication *in vivo*, as chromatin assembly occurs efficiently and most frequently in the wake of the processive replication machinery. Energy in the form of ATP is hydrolyzed by specific chromatin remodeling complexes to work with the histones, chaperones and nucleosome assembly factors and form physiological chromatin *in vivo* (Tyler, 2002).

Biochemical isolation and identification of these factors stemmed from *in vitro* chromatin assembly using extracts from embryonic sources: *Xenopus laevis* eggs or oocytes and *Drosophila* embryos. The building blocks of early development in these, and other, organisms are provided by maternal stores of proteins, membrane vesicles and other components, as supplies for rapid cell division and growth to a point where the developing embryo can rely on its own metabolic processes. These maternal stores can be harvested in the form of cell-free extracts to use in assembly of chromatin *in vitro*. Any cloned DNA can be added to these extracts and, with the addition of ATP and the right physiological buffers, assembly of the DNA into nucleosomes proceeds over time. Once assembled, the *in vitro* chromatin must be validated for its physiological relevance by comparison to *in vivo* chromatin.

The primary point of validation is comparison to *in vivo* chromatin at the level of first-order structure, meaning the repetitive spacing of nucleosomes along the DNA. The enzyme micrococcal nuclease (MNase) from *Staphylococcus aureus* digests DNA into nucleotides, in a nearly, but not entirely, sequence-independent way, wherever DNA is not protected by a nucleosome. When limited amounts of MNase are added to chromatin, whether it is assembled *in vivo* or *in vitro*, digestion of DNA occurs only within the linker regions lying between adjacent nucleosomes. Complete digestion should yield roughly the mononucleosomal length of DNA, 146 base pairs, plus a few, additional nucleotides due to restricted accessibility of the enzyme. The length of the protected fragment can vary due to the presence of linker histones, such as H1, and sometimes even tightly bound, non-histone proteins. The DNA protected from MNase digestion can be visualized by electrophoretic separation on an agarose gel, which is ethidium-stained or probed as a Southern blot. By titrating the degree of MNase digestion among individual reactions, either by varying enzyme concentration or time of digestion, a "ladder" of DNA, representing protection by mono-, di-, tri-nucleosomes and increasing numbers of nucleosomes, can be seen on these gels with rungs or bands of DNA in increments of 150-160 base pairs (Fig. 5.4). With this standard of effective chromatin assembly, biochemists could then proceed with their attempts to study chromatin regulation *in vitro*.

How to Activate Chromatin

Regulation of chromatin structure, whether the outcome is repression or activation of gene expression, is due to the combined actions of chromatin remodeling factories, histone modifying enzyme complexes and gene-specific transcription factors, acting in a temporally ordered series of steps. The ground state of any activation process is repression, and this was the first property of chromatin-mediated gene expression that was reconstituted *in vitro*. Critical DNA elements, including the sites of preinitiation complex assembly and binding sites for activators of transcription, are often occluded or blocked by assembled nucleosomes. Addition of regulatory proteins, such as TBP or specific trans-activators, during nucleosomal assembly allowed "open" or active chromatin formation. This chromatin substrate could be transcribed in a reconstituted transcription system *in vitro* by addition of basal transcription factors and RNA polymerase II. An "order-of-addition" approach, by which individual general transcription factors, e.g. TFIID, TFIIB, TFIIE, TFIIH, etc, were added before, during or after chromatin assembly, established that TBP or TFIID addition was critical to maintain an open core promoter region and that the general rules of preinitiation complex assembly were followed with chromatin templates as with chromatin-free templates (Knezetic and Luse, 1986; Workman and Roeder, 1987).

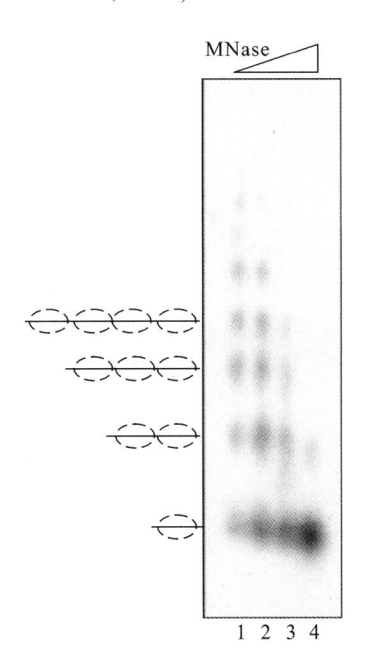

Fig.5.4 Micrococcal nuclease analysis of chromatin structure. MNase digestion of liver nuclei was followed by DNA purification and separation on an agarose gel by electrophoresis. A Southern blot of the gel was probed for the presence of the alpha-fetoprotein gene (S. Stratton and M.C. Barton, unpublished results). Increasing amounts of MNase (denoted by the triangle) used in each digestion reaction leads to progressive degradation of DNA in the linker regions, unprotected by nucleosomes (lanes 1-4). The diagram to the left illustrates the amount of DNA (solid line) at each position of the blot, protected by nucleosome(s) (dashed oval) from MNase cleavage.

Using embryonic-origin, cell-free extracts that promote nucleosomal assembly, the process of establishing activated chromatin could be dissected. Interestingly, now that repression of transcription could be recapitulated *in vitro*, regulated transcription approaching the levels of gene activation observed *in vivo* could be achieved. *In vitro* transcription of chromatin-free templates suffered from the high background signal of basal transcription and addition of most trans-activator proteins had little or no impact on expression. Since basal transcription is essentially nonexistent *in vivo*, and now was repressed *in vitro*, regulation of transcription by multiple activators, such as Gal4-VP16 and HIV Tat protein, as examples, was possible in the test tube (Kamakaka et al., 1993; Steger et al., 1998). Biochemical analyses of chromatin structure and function suggested that multiple enzymatic complexes, working at the level of chromatin structure alteration, were required for regulation of gene expression. This work interfaced with the findings of geneticists, who found that suppression of specific mutations, which caused loss of regulated gene expression, was sometimes achieved by genes encoding modifiers or components of chromatin structure. These studies laid the groundwork, linking structural modification of chromatin and regulation of gene expression *in vivo*.

Chromatin Remodeling Enzymes

The ability to remodel nucleosomes is vital for accurate and regulated transcription and efficient DNA replication. Nucleosome remodeling is defined as an enzymatic activity capable of altering the position or stability of the nucleosome. The flagship of the ATP-dependent, enzyme complexes that remodel chromatin is the SWI/SNF complex. The SWI/SNF class of chromatin remodeling complexes is named for the genetic screens used to identify sucrose non-fermenting (Snf) or mating-type switching defective (Swi) mutants. The SWI/SNF chromatin remodeling complexes contain an enzymatic ATPase protein subunit, which on its own is capable of remodeling chromatin *in vitro* (Cote et al., 1994). In yeast this protein is called Snf2/Swi2 and in humans there are two homologous proteins: Brg1 and Brm. Each of these proteins contains a structural motif called a bromodomain, which binds to acetylated lysines. Acetylation of histones by HATs promotes recruitment of SWI/SNF remodeling complexes through bromodomain interaction and stabilizes template binding by SWI/SNF (Hassan et al., 2002). Chromatin remodeling by the stabilized SWI/SNF complex can then facilitate binding of additional factors required for activation.

The second class of chromatin remodeling enzymes is defined by the presence of the ATPase protein ISWI. This protein was discovered due to its homology to SWI/SNF and named imitation switch (ISWI). The ISWI protein subunit is present in at least three different complexes in *Drosophila* (Tsukiyama and Wu, 1995). Each was purified biochemically, based on the complex's ability to promote restriction enzyme accessibility of a mononucleosomal template. NURF (nucleosome remodeling factor) contains four proteins, which disrupt nucleosomes at specific sites in the genome, whereas another complex CHRAC (chromatin-accessibility complex), appears to affect chromatin condensation globally (Badenhorst et al., 2002; Fyodorov et al., 2004; Varga-Weisz et al., 1997). In addition to remodeling activity, the third complex, ACF (ATP-utilizing chromatin assembly and remodeling factor), functions in nucleosome assembly and spacing and is used for assembly of chromatin *in vitro* (Fyodorov and Kadonaga, 2002). Two other classes of chromatin remodelers are the CHD class, including mammalian Mi-2, which interacts with both HDACs and methylated DNA binding proteins, and the INO-80 class, which has been implicated in DNA repair (Saito and Ishikawa, 2002; van Attikum et al., 2004).

The mechanisms by which these protein complexes remodel chromatin are under intense investigation. Unlike "passive" nucleosome dynamics, all of these enzymes utilize the energy of ATP to catalyze a reaction with energy-of-activation barriers and transition states. Defining the transition states has been pursued through biochemical assays that demonstrate exposure of buried DNA due to movement of the nucleosome. Comparison of the ATPase subunits from both SWI/SNF complexes and ISWI complexes highlight some of the differences in the mechanisms of each remodeler. For instance, histone tails are not required for SWI/SNF to remodel but are required for ISWI activity (Guyon et al., 1999). Also, unlike ISWI, SWI/SNF is capable of exposing restriction enzyme sites within a nucleosome even when there is no neighboring DNA onto which the nucleosome could move (Aalfs et al., 2001). This may indicate that SWI/SNF operates either by disassembling the nucleosome and repositioning it elsewhere or by exposing portions of the DNA opposite to the DNA entry/exit site (Fig.5.2). Additional evidence for a chromatin remodeling mechanism other than sliding comes from experiments demonstrating exposure of DNA in the center of a tightly packed polynucleosome template. If sliding were the only remodeling mechanism then tightly packed nucleosomes could not be removed. Indeed, Brg1, the human SWI/SNF ATPase, generates DNA loops without sliding away the histone octamer (Fan et al., 2003).

Functionally, SWI/SNF and ISWI-containing complexes regulate different sets of genes and are involved in many other processes (Fyodorov and Kadonaga, 2001). Efficient incorporation of nucleosomes onto newly replicated DNA can be mediated by the SWI/SNF complex member, Ini1, and the ISWI-containing complex, CHRAC/ACF (Alexiadis et al., 1998); both chromatin remodelers stimulate initiation of viral DNA replication *in vitro* (Lee et al., 1999). DNA repair proteins also require access to the DNA template, and the chromatin remodeler INO80 is recruited to phosphorylated histone H2A.X found at double strand break sites (Morrison et al., 2004; van Attikum et al., 2004). The human SWI/SNF complex ATPase, Brg1, has also been implicated as a tumor suppressor with a role in cell cycle control. Cyclin E/cdk2 phosphorylates the protein during the G_1-phase of the cell cycle and Rb requires SWI/SNF to induce cell cycle arrest (Zhang et al., 2000). Additionally, mutations in the SWI/SNF complex member Ini1 are found in many rhabdoid tumors and some chronic myeloid leukemia (Versteege et al., 1998).

Chromatin Structure Alteration and Transcription

The end result of chromatin remodeling and post-translational modification of histones can either be repression of transcription or activation of transcription (Tyler and Kadonaga, 1999). The choice is primarily dictated by the specific co-regulators associated with transcription factors, which target their enzyme complexes to effect alterations in chromatin structure. Co-regulatory complexes, whether co-activators, such as CBP/p300, P/CAF, CARM1, PRMT1, and others, or co-repressors, such as mSin3A/HDAC's, NCoR/SMRT, HMT/SET protein complexes, and others, may interact with Mediator components to translate the appropriate regulatory signal into action at the core promoter. In the case of chromatin activation, accessibility of the core promoter and assembly of a preinitiation complex is facilitated. The methodology of chromatin immunoprecipitation (ChIP) has been applied extensively to define a temporal order of interactions between regulatory centers of DNA and chromatin remodeling complexes, histone modifiers and transcription factors, and has provided a much-needed bridge between *in vitro* and *in vivo* experimentation (Fig. 5.5). ChIP analysis is an essential tool in further understanding of the histone code, the regulatory proteins that establish or erase it, and correlation of the final readout in terms of nuclear functions. Step-wise ratcheting toward fully activated or fully repressed gene expression can take different paths, depending on the target gene, activation signal and cell-specific background, as well as other variables (Fry and Peterson, 2002; Narlikar et al., 2002).

A number of histone acetyltransferases (HATs) such as Gcn5, CBP/p300, P/CAF, and TAF1 also function as transcriptional co-regulators, to regulate gene expression positively or negatively in concert with other modifications. These proteins, as well as BRM, BRG1, and SWI2/SNF2, can associate with chromatin through structural motifs, called bromodomains, which interact directly with acetylated histone tails (Jacobson et al., 2000). Specific interpretation of histone tail modifications, with regards to activation or repression, depends on the amino acid residue modified and the presence of other specific modifications, within the same N-terminal histone tail, such as H3S10 phosphorylation, or in *trans* between histones, e.g. ubiquitylation of H2B and increased HMT-mediated methylation of H3K4 (Sims et al., 2004).

Histone methylation can have opposing, downstream effects on chromatin transcription, depending on the specific residue that is a substrate for various HMT's (Kouzarides, 2002; Sanders et al., 2004; Santos-Rosa et al., 2004). For example, histone methylation by arginine methyltransferases, such as CARM1 and PMRT, is targeted by activated nuclear hormone receptors, which promote active transcription. In contrast, we have already discussed methylation of H3K9 and its role in gene repression and silencing. Methylation of a lysine residue at a different position in the H3 N-terminal tail, such as H3K4, opposes silencing and correlates with gene activation. H3K4 methylation occurs not only at regulatory elements, but may be even more important during the process of transcription elongation. Chromatin acts as a formidable barrier to RNA polymerase during elongation (introduced in the preceding chapter). HMT's that mediate methylation of H3K4 have been found in association with RNA polymerase II during its passage along chromatin. This enzymatic activity, along with other protein complexes that promote nucleosomal movement, such as FACT and Spt6, facilitate expression of the physiological, chromatin substrate (Sims et al., 2004).

Phosphorylation of histones is essential for condensation of chromosomes during meiosis and mitosis. There are a few, specific examples of histone phosphorylation correlated with activated gene expression, but there is no clear relationship established as a general case. In yeast, the genes regulated by Gcn5

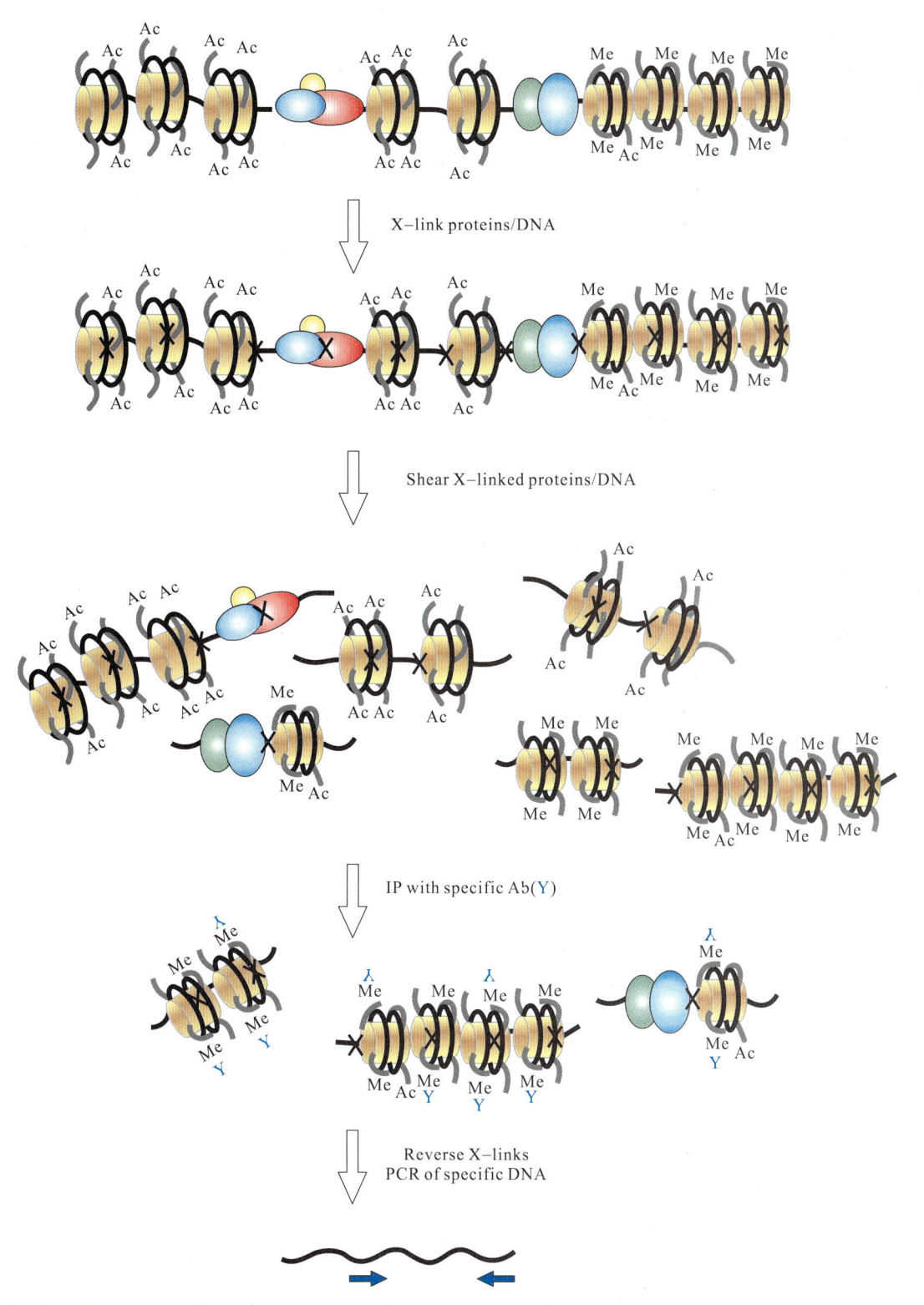

Fig.5.5 ChIP analysis. Intact cells or tissue are exposed to a cross-linking chemical agent, e.g. formaldehyde, which covalently links (X) DNA and proteins within Angstroms of each other. A lysate is prepared and the chromatin/DNA is randomly sheared into fragments, generally smaller than 500 bp. The chromatin/DNA fragments can be immunoprecipitated with antibodies (Y) that recognize histones, specific post-translational modifications of histones, transcription factors, co-regulatory proteins or other chromatin-interacting proteins. Following immunoprecipitation, the cross-links are reversed by heating and DNA is extracted. The presence of a specific region of DNA within the immunoprecipitated materials is analyzed by PCR with gene-specific primers. Symbols are as follows: nucleosomes (brown, shaded barrels), DNA (black line), histone modifications: Ac (acetylated), Me (methylated), chromatin-associated proteins (green, red, orange and blue ovoid shapes) and colored arrows (PCR primers).

HAT activity are also positively affected by H3S10 phosphorylation, but inhibition of this modification by mutation of H3S10 has no direct effect on gene expression (Lo *et al.*, 2000). However, H3S10 is needed to determine if the majority of effects ascribed to H3S10 phosphorylation are due to direct effects on transcription, its function in modulating acetylation or methylation of neighboring residues and/or its integral role in mitotic and meiotic condensation.

Chromatin and the Future

This chapter has addressed the basics of chromatin structure, how specific regulatory patterns are put into place, mechanisms of altering chromatin structure and the complex interplay of chromatin-interacting factors and regulators of transcription. We have emphasized that regulated gene expression is the outcome of an intricate choreography of transcription factors, chromatin structure alterations and modifications. The impact of chromatin as a key regulatory player is being felt across an amazing variety of endeavors in basic and clinical sciences. The study of chromatin structure and function has greatly expanded in recent years and continues to grow in excitement and implications for regulation of gene expression and numerous nuclear processes. New modifications of histone residues and their specific roles in regulation of gene expression and/or other cellular processes are being reported. The enzyme complexes that promote these modifications, the ones that erase them and their mechanisms of action continue to be identified. Perhaps the most exciting areas of chromatin research are these studies, as well as explorations of higher order chromatin structures and the basis for epigenetic/cellular memory. The study of chromatin is likely to continue as a growth industry for some time.

Acknowledgement

The authors are grateful for the contributions of R. Kori, early in development of this chapter, and to D. Wilkinson, for critical reading. R. Kelley and S. Stratton graciously provided unpublished data.

References

Aalfs, J. D., Narlikar, G. J., and Kingston, R. E. (2001). Functional differences between the human ATP-dependent nucleosome remodeling proteins BRG1 and SNF2H. J Biol Chem *276*, 34270-34278.

Adolph, K. W. (1981). A serial sectioning study of the structure of human mitotic chromosomes. Eur J Cell Biol *24*, 146-153.

phosphorylation, associated with active gene expression, is linked to specific core promoters present within *Drosophila* (Labrador and Corces, 2003). Further study

Alexiadis, V., Varga-Weisz, P. D., Bonte, E., Becker, P. B., and Gruss, C. (1998). *In vitro* chromatin remodelling by chromatin accessibility complex (CHRAC) at the SV40 origin of DNA replication. Embo J *17*, 3428-3438.

Archer, T. K., Cordingley, M. G., Wolford, R. G., and Hager, G. L. (1991). Transcription factor access is mediated by accurately positioned nucleosomes on the mouse mammary tumor virus promoter. Molec Cell Biol *11*, 688-698.

Arents, G., and Moudrianakis, E. N. (1995). The histone fold: a ubiquitous architectural motif utilized in DNA compaction and protein dimerization. Proc Natl Acad Sci USA *92*, 11170-11174.

Badenhorst, P., Voas, M., Rebay, I., and Wu, C. (2002). Biological functions of the ISWI chromatin remodeling complex NURF. Genes Dev *16*, 3186-3198.

Baylin, S. B., Esteller, M., Rountree, M. R., Bachman, K. E., Schuebel, K., and Herman, J. G. (2001). Aberrant patterns of DNA methylation, chromatin formation and gene expression in cancer. Hum Mol Genet *10*, 687-692.

Brown, D. T., Gunjan, A., Alexander, B. T., and Sittman, D. B. (1997). Differential effect of H1 variant overproduction on gene expression is due to differences in the central globular domain. Nucleic Acids Res *25*, 5003-5009.

Carruthers, L. M., Bednar, J., Woodcock, C. L., and Hansen, J. C. (1998). Linker histones stabilize the intrinsic salt-dependent folding of nucleosomal arrays: mechanistic ramifications for higher-order chromatin folding. Biochemistry *37*, 14776-14787.

Cote, J., Quinn, J., Workman, J. L., and Peterson, C. L. (1994). Stimulation of GAL4 derivative binding to nucleosomal DNA by the yeast SWI/SNF complex. Science *265*, 53-69.

Davey, C. A., Sargent, D. F., Luger, K., Maeder, A. W., and Richmond, T. J. (2002). Solvent mediated interactions in the structure of the nucleosome core particle at 1.9 a resolution. J Mol Biol *319*, 1097-1113.

Denu, J. M. (2003). Linking chromatin function with metabolic networks: Sir2 family of NAD(+)-dependent deacetylases. Trends Biochem Sci *28*, 41-48.

Dorigo, B., Schalch, T., Kulangara, A., Duda, S., Schroeder, R., and Richmond, T. J. (2004). Nucleosome arrays reveal the two-start organization of the chromatin fiber. Science *306*, 1571-1573.

Fan, H. Y., He, X., Kingston, R. E., and Narlikar, G. J. (2003). Distinct strategies to make nucleosomal DNA accessible. Mol Cell *11*, 1311-1322.

Fan, J. Y., Gordon, F., Luger, K., Hansen, J. C., and Tremethick, D. J. (2002). The essential histone variant H2A.Z regulates the equilibrium between different chromatin conformational states. Nat Struct Biol *9*, 172-176.

Fan, J. Y., Rangasamy, D., Luger, K., and Tremethick, D. J.

(2004). H2A.Z alters the nucleosome surface to promote HP1alpha-mediated chromatin fiber folding. Mol Cell *16*, 655-661.

Fischle, W., Wang, Y., and Allis, C. D. (2003). Histone and chromatin cross-talk. Curr Opin Cell Biol *15*, 172-183.

Fry, C. J., and Peterson, C. L. (2002). Unlocking the gates to gene expression. Science *295*, 1847-1848.

Fuks, F., Hurd, P. J., Wolf, D., Nan, X., Bird, A. P., and Kouzarides, T. (2003). The methyl-CpG-binding protein MeCP2 links DNA methylation to histone methylation. J Biol Chem *278*, 4035-4040.

Fyodorov, D. V., Blower, M. D., Karpen, G. H., and Kadonaga, J. T. (2004). Acf1 confers unique activities to ACF/CHRAC and promotes the formation rather than disruption of chromatin *in vivo*. Genes Dev *18*, 170-183.

Fyodorov, D. V., and Kadonaga, J. T. (2001). The many faces of chromatin remodeling: SWItching beyond transcription. Cell *106*, 523-525.

Fyodorov, D. V., and Kadonaga, J. T. (2002). Dynamics of ATP-dependent chromatin assembly by ACF. Nature *418*, 897-900.

Gottschling, D. E. (2000). Gene silencing: two faces of SIR2. Curr Biol *10*, R708-711.

Grewal, S. I., and Elgin, S. C. (2002). Heterochromatin: new possibilities for the inheritance of structure. Curr Opin Genet Dev *12*, 178-187.

Grunstein, M. (1990). Histone function in transcription. Annu Rev Cell Biol *6*, 643-678.

Guarente, L. (2000). Sir2 links chromatin silencing, metabolism, and aging. Genes Dev *14*, 1021-1026.

Guyon, J. R., Narlikar, G. J., Sif, S., and Kingston, R. E. (1999). Stable remodeling of tailless nucleosomes by the human SWI-SNF complex. Mol Cell Biol *19*, 2088-2097.

Hassan, A. H., Prochasson, P., Neely, K. E., Galasinski, S. C., Chandy, M., Carrozza, M. J., and Workman, J. L. (2002). Function and selectivity of bromodomains in anchoring chromatin-modifying complexes to promoter nucleosomes. Cell *111*, 369-379.

Hayashi, K., Hofstaetter, T., and Yakuwa, N. (1978). Asymmetry of chromatin subunits probed with histone H1 in an H1-DNA complex. Biochemistry *17*, 1880-1883.

Henikoff, S., Furuyama, T., and Ahmad, K. (2004). Histone variants, nucleosome assembly and epigenetic inheritance. Trends Genet *20*, 320-326.

Horn, P. J., and Peterson, C. L. (2002). Molecular biology. Chromatin higher order folding--wrapping up transcription. Science *297*, 1824-1827.

Jacobson, R. H., Ladurner, A. G., King, D. S., and Tjian, R. (2000). Structure and function of a human TAFII250 double bromodomain module. Science *288*, 1422-1425.

Jaenisch, R., and Bird, A. (2003). Epigenetic regulation of gene expression: how the genome integrates intrinsic and environmental signals. Nat Genet *33 Suppl*, 245-254.

Johns, E. W. (1967). The electrophoresis of histones in polyacrylamide gel and their quantitative determination. Biochem J *104*, 78-82.

Jones, P. A. (2002). DNA methylation and cancer. Oncogene *21*, 5358-5360.

Kamakaka, R. T., Bulger, M., and Kadonaga, J. T. (1993). Potentiation of RNA polymerase II transcription by Gal4-VP16 during but not after DNA replication and chromatin assembly. Genes Dev *7*, 1779-1795.

Kassabov, S. R., Zhang, B., Persinger, J., and Bartholomew, B. (2003). SWI/SNF unwraps, slides, and rewraps the nucleosome. Mol Cell *11*, 391-403.

Knezetic, J. A., and Luse, D. S. (1986). The presence of nucleosomes on a DNA template prevents initiation by RNA polymerase II *in vitro*. Cell *45*, 95-104.

Konishi, A., Shimizu, S., Hirota, J., Takao, T., Fan, Y., Matsuoka, Y., Zhang, L., Yoneda, Y., Fujii, Y., Skoultchi, A. I., and Tsujimoto, Y. (2003). Involvement of histone H1.2 in apoptosis induced by DNA double-strand breaks. Cell *114*, 673-688.

Kornberg, R. D., and Lorch, Y. (1999). Twenty-five years of the nucleosome, fundamental particle of the eukaryote chromosome. Cell *98*, 285-294.

Kornberg, R. D., and Thomas, J. O. (1974). Chromatin structure; oligomers of the histones. Science *184*, 865-868.

Kouzarides, T. (2002). Histone methylation in transcriptional control. Curr Opin Genet Dev *12*, 198-209.

Labrador, M., and Corces, V. G. (2003). Phosphorylation of histone H3 during transcriptional activation depends on promoter structure. Genes Dev *17*, 43-48.

Lee, D., Sohn, H., Kalpana, G. V., and Choe, J. (1999). Interaction of E1 and hSNF5 proteins stimulates replication of human papillomavirus DNA. Nature *399*, 487-491.

Lee, H., Habas, R., and Abate-Shen, C. (2004). MSX1 cooperates with histone H1b for inhibition of transcription and myogenesis. In Science, pp. 1675-1678.

Li, G., Levitus, M., Bustamante, C., and Widom, J. (2005). Rapid spontaneous accessibility of nucleosomal DNA. Nat Struct Mol Biol *12*, 46-53.

Li, G., and Widom, J. (2004). Nucleosomes facilitate their own invasion. Nat Struct Mol Biol *11*, 763-769.

Lindroth, A. M., Shultis, D., Jasencakova, Z., Fuchs, J., Johnson, L., Schubert, D., Patnaik, D., Pradhan, S., Goodrich, J., Schubert, I., *et al.* (2004). Dual histone H3 methylation marks at lysines 9 and 27 required for interaction with CHROMOMETHYLASE3. Embo J *23*, 4286-4296.

Lo, W. S., Trievel, R. C., Rojas, J. R., Duggan, L., Hsu, J. Y., Allis, C. D., Marmorstein, R., and Berger, S. L. (2000). Phosphorylation of serine 10 in histone H3 is functionally linked *in vitro* and *in vivo* to Gcn5-mediated acetylation at lysine 14. Mol Cell *5*, 917-926.

Lowary, P. T., and Widom, J. (1998). New DNA sequence rules

for high affinity binding to histone octamer and sequence-directed nucleosome positioning. J Mol Biol *276*, 19-42.

Lu, X., and Hansen, J. C. (2004). Identification of specific functional subdomains within the linker histone H10 C-terminal domain. J Biol Chem *279*, 8701-8707.

Luger, K., Mader, A. W., Richmond, R. K., Sargent, D. F., and Richmond, T. J. (1997). Crystal structure of the nucleosome core particle at 2.8 A resolution. Nature *389*, 251-260.

Maison, C., Bailly, D., Peters, A. H., Quivy, J. P., Roche, D., Taddei, A., Lachner, M., Jenuwein, T., and Almouzni, G. (2002). Higher-order structure in pericentric heterochromatin involves a distinct pattern of histone modification and an RNA component. Nat Genet *30*, 329-334.

Marmorstein, R. (2003). Structure of SET domain proteins: a new twist on histone methylation. Trends Biochem Sci *28*, 59-62.

McGhee, J. D., and Felsenfeld, G. (1980). Nucleosome structure. Annu Rev Biochem *49*, 1115-1156.

Morrison, A. J., Highland, J., Krogan, N. J., Arbel-Eden, A., Greenblatt, J. F., Haber, J. E., and Shen, X. (2004). INO80 and gamma-H2AX interaction links ATP-dependent chromatin remodeling to DNA damage repair. Cell *119*, 767-775.

Narlikar, G. J., Fan, H. Y., and Kingston, R. E. (2002). Cooperation between complexes that regulate chromatin structure and transcription. Cell *108*, 475-487.

Naumann, K., Fischer, A., Hofmann, I., Krauss, V., Phalke, S., Irmler, K., Hause, G., Aurich, A. C., Dorn, R., Jenuwein, T., and Reuter, G. (2005). Pivotal role of AtSUVH2 in heterochromatic histone methylation and gene silencing in *Arabidopsis*. Embo J.

Nguyen, T. T., Cho, K., Stratton, S. A., and Barton, M. C. (2005). Transcription factor interactions and chromatin modifications associated with p53-mediated, developmental repression of the alpha-fetoprotein gene. Mol Cell Biol *25*, 2147-2157.

Nielsen, S. J., Schneider, R., Bauer, U. M., Bannister, A. J., Morrison, A., O'Carroll, D., Firestein, R., Cleary, M., Jenuwein, T., Herrera, R. E., and Kouzarides, T. (2001). Rb targets histone H3 methylation and HP1 to promoters. Nature *412*, 561-565.

Richards, E. J., and Elgin, S. C. (2002). Epigenetic codes for heterochromatin formation and silencing: rounding up the usual suspects. Cell *108*, 489-500.

Saito, M., and Ishikawa, F. (2002). The mCpG-binding domain of human MBD3 does not bind to mCpG but interacts with NuRD/Mi2 components HDAC1 and MTA2. J Biol Chem *277*, 35434-35439.

Sanders, S. L., Portoso, M., Mata, J., Bahler, J., Allshire, R. C., and Kouzarides, T. (2004). Methylation of histone H4 lysine 20 controls recruitment of Crb2 to sites of DNA damage. Cell *119*, 603-614.

Santos-Rosa, H., Bannister, A. J., Dehe, P. M., Geli, V., and Kouzarides, T. (2004). Methylation of H3 lysine 4 at euchromatin promotes Sir3p association with heterochromatin. J Biol Chem *279*, 47506-47512.

Schmid, A., Fascher, K.-D., and Horz, W. (1992). Nucleosome disruption at the yeast PHO5 promoter on PHO5 induction occurs in the absence of DNA replication. Cell *71*, 853-864.

Schultz, D. C., Ayyanathan, K., Negorev, D., Maul, G. G., and Rauscher, F. J., 3rd (2002). SETDB1: a novel KAP-1-associated histone H3, lysine 9-specific methyltransferase that contributes to HP1-mediated silencing of euchromatic genes by KRAB zinc-finger proteins. Genes Dev *16*, 919-932.

Shen, X., Yu, L., Weir, J. W., and Gorovsky, M. A. (1995). Linker histones are not essential and affect chromatin condensation *in vivo*. Cell *82*, 47-56.

Shore, D. (2000). The Sir2 protein family: A novel deacetylase for gene silencing and more. Proc Natl Acad Sci USA *97*, 14030-14032.

Sims, R. J., 3rd, Belotserkovskaya, R., and Reinberg, D. (2004). Elongation by RNA polymerase II: the short and long of it. Genes Dev *18*, 2437-2468.

Steger, D. J., Eberharter, A., John, S., Grant, P. A., and Workman, J. L. (1998). Purified histone acetyltransferase complexes stimulate HIV-1 transcription from preassembled nucleosomal arrays. Proc Natl Acad Sci *95*, 12924-12929.

Tagami, H., Ray-Gallet, D., Almouzni, G., and Nakatani, Y. (2004). Histone H3.1 and H3.3 complexes mediate nucleosome assembly pathways dependent or independent of DNA synthesis. Cell *116*, 51-61.

Tamaru, H., and Selker, E. U. (2001). A histone H3 methyltransferase controls DNA methylation in Neurospora crassa. Nature *414*, 277-283.

Tanny, J. C., Dowd, G. J., Huang, J., Hilz, H., and Moazed, D. (1999). An enzymatic activity in the yeast Sir2 protein that is essential for gene silencing. Cell *99*, 735-745.

Thastrom, A., Bingham, L. M., and Widom, J. (2004). Nucleosomal locations of dominant DNA sequence motifs for histone-DNA interactions and nucleosome positioning. J Mol Biol *338*, 695-709.

Tsukiyama, T., and Wu, C. (1995). Purification and protperties of an ATP-dependent nucleosome remodeling factor. Cell *83*, 1011-1020.

Tumbar, T., Sudlow, G., and Belmont, A. S. (1999). Large-scale chromatin unfolding and remodeling induced by VP16 acidic activation domain. J Cell Biol *145*, 1341-1354.

Tyler, J. K. (2002). Chromatin assembly. Cooperation between histone chaperones and ATP-dependent nucleosome remodeling machines. Eur J Biochem *269*, 2268-2274.

van Attikum, H., Fritsch, O., Hohn, B., and Gasser, S. M. (2004). Recruitment of the INO80 complex by H2A phosphorylation links ATP-dependent chromatin remodeling with DNA double-strand break repair. Cell *119*, 777-788.

Varga-Weisz, P. D., Wilm, M., Bonte, E., Dumas, I., Mann, M., and Becker, P. B. (1997). Chromatin-remodeling factor CHRAC contains the ATPases ISWI and topoisomerase II. Nature *388*, 598-602.

Versteege, I., Sevenet, N., Lange, J., Rousseau-Merck, M. F.,

Ambros, P., Handgretinger, R., Aurias, A., and Delattre, O. (1998). Truncating mutations of hSNF5/INI1 in aggressive paediatric cancer. Nature 394, 203-206.

Wang, Z. F., Sirotkin, A. M., Buchold, G. M., Skoultchi, A. I., and Marzluff, W. F. (1997). The mouse histone H1 genes: gene organization and differential regulation. J Mol Biol 271, 124-138.

Wolffe, A. P. (1994). Nucleosome positioning and modification: chromatin structures that potentiate transcription. TIBS 19, 240-244.

Workman, J. L., and Roeder, R. G. (1987). Binding of transcription factor TFIID to the major late promoter during in vitro nucleosome assembly potentiates subsequent initiation by RNA polymerase II. Cell 51, 613-622.

Xie, X., Kokubo, T., Cohen, S. L., Mirza, U. A., Hoffmann, A., Chait, B. T., Roeder, R. G., Nakatani, Y., and Burley, S. K. (1996). Structural similarity between TAFs and the heterotetrameric core of the histone octamer. Nature 380, 316-322.

Zhang, H. S., Gavin, M., Dahiya, A., Postigo, A. A., Ma, D., Luo, R. X., Harbour, J. W., and Dean, D. C. (2000). Exit from G1 and S phase of the cell cycle is regulated by repressor complexes containing HDAC-Rb-hSWI/SNF and Rb-hSWI/SNF. Cell 101, 79-89.

Chapter 06
HATs and HDACs

Timothy A. Bolger, Todd Cohen and Tso-Pang Yao

Department of Pharmacology and Cancer Biology, Duke University, Durham, North Carolina 27710, USA

Key Words: acetylation, histone, transcription, acetyltransferase, deacetylase, HAT, HDAC, Sir2, CBP/p300, PCAF, MYST

Abstract

DNA in eukaryotes is arranged into higher order structures termed chromatin. Chromatin poses an obstacle for gene transcription because it is not readily accessible to DNA binding transcription factors. Chromatin also provides the structural basis for gene regulation, keeping most genes in a default state of repression. The regulation of gene expression involves dynamic chromatin remodeling associated with reversible histone acetylation. The enzymes that catalyze histone acetylation (HAT) and deacetylation (HDAC) are critical elements that regulate chromatin dynamics and gene expression. These chromatin-modifying enzymes are recruited to the genome by DNA binding transcription factors and are the key effectors that control chromatin structure. HAT and HDAC proteins are important in both normal cellular physiology and disease states. The recent identification of a large number of non-histone substrates for HATs and HDACs suggest even broader biological functions for these versatile enzymes.

Introduction

Although genetic information is stored in a linear fashion in DNA, the genetic material in eukaryotes is arranged into higher order structures for efficient packaging in the nucleus (reviewed in Horn and Peterson, 2002). It is this DNA-protein complex, termed chromatin, rather than naked DNA, where transcription factors operate. The formation of chromatin is repressive for gene transcription because it reduces access to DNA by the various DNA binding transcription factors. This configuration therefore allows for efficient regulation of gene transcription under specific physiological conditions through dynamic remodeling of the chromatin. The mechanism that dictates chromatin structure and the machinery that controls chromatin remodeling are the central elements in gene regulation.

Chromatin is made of repeated units of DNA and specialized DNA associated proteins called histones. Approximately 146 base pairs of DNA are wrapped twice around an octamer of core histones consisting of two subunits each of H2A, H2B, H3, and H4. This structure is known as a nucleosome and is the basic building block of the chromatin. Each of the histones has a loosely structured amino-terminal tail that extends out of the octamer core and wraps around the outside of the DNA. The tails have numerous positively-charged side chains that interact with the negatively-charged DNA backbone, and help to bind the DNA to the nucleosome more tightly. The amino-terminal tails are also subject to many post-translational modifications, including acetylation, methylation, phosphorylation, and ubiquitination. These groups modulate histone-DNA and histone-protein interactions, as well as the chromatin structure important for gene transcription and other chromatin-dependent processes, such as DNA replication and repair (reviewed in Jenuwein and Allis, 2001; Kurdistani and Grunstein, 2003). Here we will discuss the regulation of histone acetylation and the enzymes that are responsible for this process.

Histone acetylation was first discovered more than 40 years ago. Very quickly it was noticed that acetylated histones are associated with active RNA transcription (Allfrey *et al.*, 1964). Over the years, our understanding of this association has gradually increased as experimental proof has revealed much of the mechanism and function

Corresponding Author: Tso-Pang Yao, E-mail: yao00001@mc.duke.edu of histone acetylation. As mentioned above, chromatin is a naturally repressive state for transcription as it is difficult for transcription factors to access the DNA in that highly organized structure. Acetylation of histones, which occurs on ε-amino group of specific lysines, neutralizes the positive charge of the lysine residues within histones. This causes a loosened association between the DNA and the histones and allows greater availability of the DNA to transcription factors. This is believed to assist in the formation of the large transcription initiation complexes, as well as increasing the processivity of RNA polymerase (Kouzarides, 2000). Elegant genetic studies on the histone variant H2A.Z in *Tetrahymena* have provided strong evidence supporting the idea that the main function of histone acetylation is to neutralize the positive charge of lysine (Ren and Gorovsky, 2001). However, it should be noted that acetylation of lysines has an additional function in the recruitment of specific transcriptional co-activators by acting as a specific structural moiety recognized by an acetyllysine-binding motif termed a bromodomain (see below). Histone acetylation can also influence other histone modifications, such as arginine methylation, which is important for gene transcription (Wang *et al.*, 2001).

Acetylation of histones is a reversible and dynamically regulated process controlled by a large family of enzymes termed histone acetyltransferases (HATs) and histone deacetylases (HDACs). In this review, we will first discuss the function of HATs and HDACs in histone acetylation and gene transcription. We will then discuss the role of HATs and HDACs in the regulation of non-histone protein acetylation, expanding the functional repertoire of HAT and HDAC activity. Lastly, we will discuss therapeutic approaches that target the functions of these highly versatile enzymes in human disease.

An Overview of Histone Acetyltransferases and Histone Deacetylases

Although a tight correlation between the level of histone acetylation and gene transcription activity has been observed for decades, the definitive proof of this relationship could not be established until the enzymes that catalyze histone acetylation and deacetylation were discovered. The first histone acetyltransferase (HAT) identified was yeast Hat1 in 1995 (Kleff *et al.*, 1995). However, Hat1 was found to be a cytoplasmic protein that acetylates newly-synthesized histones for deposition on DNA, a housekeeping function that does not play a critical role in dynamic histone acetylation linked to gene transcription. Soon after, a HAT specifically involved in gene transcription was identified by Brownell and Allis (Brownell and Allis, 1995). Using an ingenious in-gel histone acetylation assay, the acetyltransferase, termed HAT-A (for type A), was purified and cloned from *Tetrahymena thermophila*. The sequence of HAT-A revealed a surprising homology to Gcn5, a well-known transcriptional coactivator in yeast (Brownell *et al.*, 1996). Transcriptional coactivators, in contrast to DNA binding transcription factors, generally do not have sequence-specific DNA binding activity. Instead, they are recruited to specific gene regulatory regions by associating with DNA binding transcription factors in order to influence gene transcription. The striking discovery that HAT-A and GCN5 are related genes immediately converted the long-known phenomenological correlation of histone acetylation and gene transcription into a mechanistic understanding. In this understanding, transcriptional coactivators such as GCN5 promote gene transcription by catalyzing histone acetylation. Indeed, subsequent genetic studies verified that the acetyltransferase activity of GCN5 is required for regulating gene transcription (Kuo *et al.*, 1998).

With mounting excitement, researchers soon showed that several other known transcriptional coactivators such as CBP (Bannister and Kouzarides, 1996), p300 (Ogryzko *et al.*, 1996), TAF250 (Mizzen *et al.*, 1996), and SRC1 (Spencer *et al.*, 1997) all possess HAT activity. These identifications directly linked the regulation of histone acetylation to the transcription machinery and provided a unifying model of transcriptional co-activators as histone-modifying enzymes. Thus, HATs such as GCN5 and CBP associate directly or indirectly with DNA binding transcriptional activators and acetylate histones in nearby nucleosomes (Fig.6.1A). The acetylation allows greater access to the DNA by the general transcription machinery, helping it to initiate transcription. Interestingly, many of the HATs also contain bromodomains, which have acetyllysine binding activity (Dhalluin *et al.*, 1999). Bromodomains bind acetylated lysines on the histones, thus allowing the HATs to further acetylate their targets, creating a positive feedback system to efficiently activate transcription (Yang, 2004b).

Soon after the identification of the first histone acetyltransferases, the first histone deacetylase, HDAC1, was purified by an affinity column made of the HDAC inhibitor trapoxin B (Taunton et al., 1996). The cloning of HDAC1 once again revealed surprising homology to a yeast transcriptional regulator. This time the yeast gene, Rpd3, was known to be involved in gene repression. Thus, the identification of Gcn5 and Rpd3 as HAT and

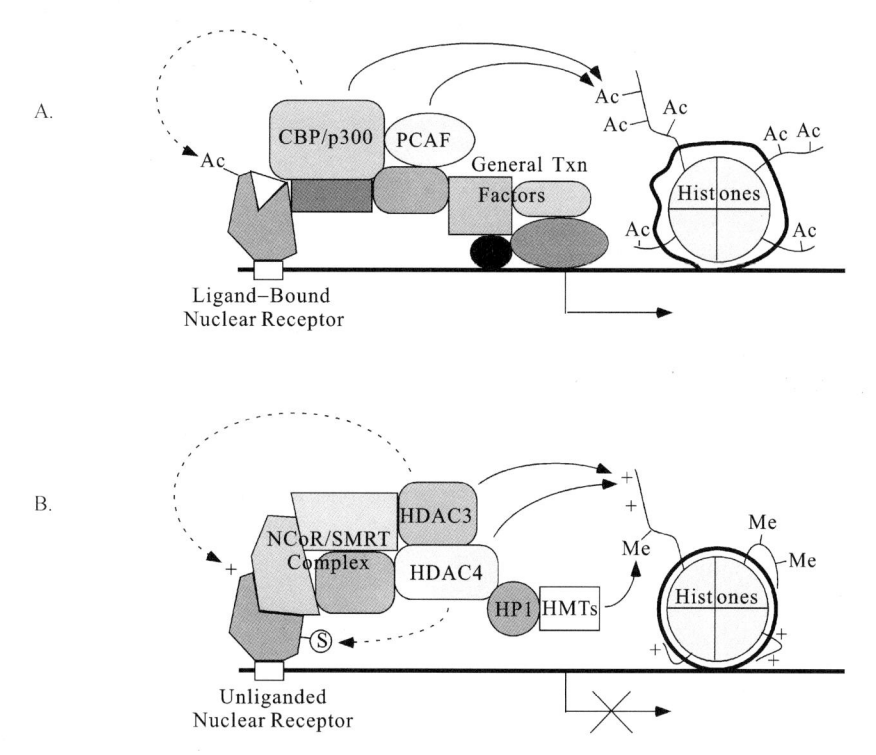

Fig.6.1 HAT and HDAC complexes in transcription. In (A.), a hypothetical example of a histone acetyltransferase (HAT) complex activating transcription is shown. A DNA-bound nuclear receptor binds its ligand, recruiting various transcription coactivators, including HATs CBP/p300 and PCAF. The HATs acetylate histone tails, neutralizing their positive charge and loosening the histone-DNA interaction. The acetylated tails also provide a platform for additional factors to bind (not pictured). The HATs may also acetylate the transcription activator itself, altering its activity. The combined action of HATs and other coactivators allow the general transcription factors to bind DNA and initiate transcription. In (B.), a hypothetical example of a histone deacetylase (HDAC) complex that is repressing transcription is shown. The unliganded nuclear receptor recruits a corepressor complex such as N-CoR or SMRT, which recruits class I HDACs like HDAC3. HDAC4 or other class IIA HDACs can bind both the corepressor and HDAC3 (and sometimes the transcription factor - not shown). The HDACs can catalyze histone deacetylation, allowing the histone tails to bind DNA more tightly and leading to repression. HDACs recruit histone methyltransferases (HMTs) through heterochromatin protein-1 (HP1), which can methylate histones for transcriptional silencing. HDACs may also deacetylate the transcription activator itself, altering its activity. HDACs can also promote other lysine modifications on transcription factors such as sumoylation.

HDAC respectively provides a simple paradigm to understand how gene transcription and histone acetylation are regulated. Namely, HATs and HDACs are recruited through DNA-binding transcription factors to specific gene regulatory regions. Thus, the prototypical function of HATs and HDACs are as critical regulators of gene transcription. However, as discussed in later sections, HATs and HDACs are multi-functional enzymes, and they regulate biological processes through substrates other than histones.

The Histone Acetyltransferase Family

In less than 10 years since the discovery of the first HAT, a large number of proteins with acetyltransferase activity have been identified. Most, though not all, of these HAT proteins have also been shown to be transcription activators or coactivators. The most well-characterized ones include GCN5, PCAF, CBP/p300, and SRC1 family members as well as a family of proteins with MYST homology. In addition, several subunits of the general transcription machinery (TFIIB and TFIIIC) (Choi et al., 2003; Kundu et al., 1999) as well as DNA-binding transcription factors such as ATF2 also possess intrinsic HAT activity (Kawasaki et al., 2000); however, the significance of HAT activity in these latter proteins are not as well characterized. Here we will focus on GCN5/PCAF, CBP/p300, and the MYST acetyltransferases.

A: GCN5 and PCAF

Yeast GCN5 is the founding member of a large superfamily of acetyltransferases called Gcn5-related N-acetyltransferases (GNATs). GNAT members share several structural motifs, including an acetyl co-A binding motif (Marmorstein and Roth, 2001). Among other known HAT proteins, PCAF, Hat1, Hpa2, and Elp3 belong to the GNATs family. The GNATs also include enzymes that catalyze acetylation at the N-terminus of many proteins (N-acetylation) and more distantly related acyltransferases that modify small molecules (e.g. serotonin and antibiotics) (Marmorstein and Roth, 2001; Vetting *et al.*, 2005).

Unlike yeast GCN5, in higher eukaryotes GCN5 is alternatively spliced, existing as both a short and long form (sometimes called GCN5L) (Xu *et al.*, 1998). In addition, PCAF is also closely related to GCN5, to the extent that mammalian GCN5 and PCAF are both considered homologs of yeast GCN5 (Sterner and Berger, 2000). Both GCN5 and PCAF contain an acetyllysine binding bromodomain in addition to the HAT domain, and they share many biochemical features. Isolated GCN5 acetylates histones, but it is much more effective against complete nucleosomes in the context of multisubunit complexes (Grant *et al.*, 1997). HAT complexes are often necessary both for the regulation of activity and for conferring substrate specificity. Yeast GCN5, for example, is the catalytic HAT in a large transcriptional complex termed SAGA (Grant *et al.*, 1997). The SAGA complex contains transcription adaptor proteins, TAFs, the transcription regulator TRRAP, and others (Grant *et al.*, 1998; Ogryzko *et al.*, 1998). TRRAP, a member of the ATM superfamily (McMahon *et al.*, 1998), has been shown to recruit GCN5 complexes to the adenoviral E1A protein, which contributes to oncogenic transformation and possibly apoptosis, (Lang and Hearing, 2003; Deleu *et al.*, 2001; Samuelson *et al.*, 2005). GCN5 was also found to be part of different complexes, such as STAGA and ADA, further increasing the possible regulation of its function (Martinez *et al.*, 1998; Grant *et al.*, 1997). As with SAGA, these complexes typically contain a number of different transcription-related proteins. The exact composition of the complexes probably differs from cell to cell and from gene to gene. This would provide yet another level of regulation of the HAT activity (Yang, 2004a). Strikingly, like GCN5, mammalian PCAF also exists as part of a large protein complex, containing many of the same components as yeast SAGA (Ogryzko *et al.*, 1998). Thus GCN5 family members appear to utilize an evolutionarily conserved mechanism in order to regulate histone acetylation and gene expression.

GCN5 is often viewed as a general transcription activator; however, it may be more important in specific biological functions. GCN5 has been implicated in cell cycle progression in both yeast and higher eukaryotes (Howe *et al.*, 2001; Kikuchi *et al.*, 2005). In GCN5 mutants, cells accumulate in G2/M, suggesting that GCN5 is important for normal progression through mitosis (Zhang *et al.*, 1998a). GCN5 complexes activate transcription mediated by the E2F transcription factors (Lang *et al.*, 2001) and regulate E2F expression as well (Kikuchi *et al.*, 2005), which could be mechanisms of GCN5 control of the cell cycle.

B: CBP and p300

CBP (CREB binding protein) and p300 were probably the most extensively characterized transcriptional coactivators even before their HAT activity was first identified (Goodman and Smolik, 2000). CBP was initially identified as a protein that selectively binds the phosphorylated active form of the CREB transcription factor (thus the name CBP) (Chrivia *et al.*, 1993) while p300 was purified as a cellular protein targeted by the E1A oncoprotein (Yee and Branton, 1985; Harlow *et al.*, 1986). Despite being identified in different ways, CBP and p300 are closely related proteins and are often referred to as CBP/p300 (Arany *et al.*, 1994). Although both p300 and CBP have potent histone acetyltransferase activity, their relation to other HATs is a distant one, having only very limited homology with the GNAT superfamily (Martinez-Balbas *et al.*, 1998). CBP and p300 interact with a large number of other proteins through at least four protein-protein interaction domains and, similar to the GCN5 HAT, a bromodomain (Sterner and Berger, 2000). In fact, they possess the ability to act as physical scaffolds for the transcription machinery as another mechanism of activation in addition to their HAT activity. As such, they are essential cofactors for more than 40 different transcription factors (Kalkhoven, 2004).

Unlike GCN5 and PCAF, there is no evidence that p300 or CBP must stably exist in a large protein complex in order to function. However, CBP and p300 are regulated by several posttranslational modifications. CBP and p300 are both phosphoproteins and subject to regulation by several kinases. For example, CBP can be activated by calcium/calmodulin-dependent kinases in neuronal signaling and survival (Chawla *et al.*, 1998; Hu *et al.*, 1999). In addition, protein kinase A, cdk2, and p42/44 MAP kinase have also been reported to upregulate CBP (Kalkhoven, 2004; Ait-Si-Ali *et al.*, 1998; Ait-Si-Ali *et al.*, 1999) and p300 HAT activity while protein kinase C(δ) has been shown to reduce it (Yuan *et al.*, 2002). CBP and p300 are also regulated at the level of protein stability. They are subject to ubiquitination and degraded by the proteasome (Poizat *et al.*, 2000; Lonard *et al.*, 2000). p300 is progressively ubiquitinated and degraded during retinoic acid-induced differentiation (Iwao *et al.*, 1999), suggesting that the ubiquitination of CBP and p300 may play a regulatory role during specific cellular conditions. Interestingly, degradation of CBP accompanies cell death triggered by the accumulation of mutant huntingtin, an expanded poly-glutamine protein (Jiang et al., 2003). CBP is also cleaved by caspases under apoptotic conditions in neurons (Rouaux et al., 2003), and p300 is subject to sumoylation, which converts p300 into a transcriptional repressor (Girdwood et al., 2003).

Given its functional interaction with a large set of important transcriptional factors, it is not surprising that ablation of CBP or p300 leads to embryonic lethality (Yao et al., 1998). Interestingly, mice with heterozygous mutations for both CBP and p300 also die as embryos, supporting the idea that CBP and p300 share certain functions during development (Yao et al., 1998). However, further genetic studies reveal that CBP and p300 are not identical and have distinct functions *in vivo*. For example, heterozygous mutation of CBP in humans is the cause of Rubinstein-Taybi syndrome (RTS), a genetic disorder characterized by mental retardation and skeletal malformations (Rouaux et al., 2004). A similar phenotype is seen in mice heterozygous for CBP mutations (Tanaka et al., 1997; Oike et al., 1999). Although mice with heterozygous mutations of p300 show reduced viability similar to the CBP heterozygous mutant mice, they do not develop an RTS-like syndrome (Yao et al., 1998).

In addition to developmental defects, both p300 and CBP appear to be involved in tumor formation. CBP heterozygous mutant mice are predisposed to hematological malignancy (Kung et al., 2000). In those tumors, the wild type CBP allele is lost, supporting a function of CBP as a tumor suppressor. Indeed, a recent study showed that inactivation of both CBP alleles by conditional knockout leads to T cell lymphoma in mice (Kang-Decker et al., 2004). In fact, human RTS patients are also prone to developing tumors (Goodman and Smolik, 2000). In contrast, heterozygous mutation of p300 did not predispose mice to tumor formation. However, rare p300 genetic alterations have been reported in solid tumors in humans (Muraoka et al., 1996; Gayther et al., 2000). Interestingly, in one gastric tumor, the wild type p300 allele is lost while the other allele harbors a point mutation that inactivates p300 HAT activity. These results suggest that HAT activity is important for the function of p300 and probably CBP as tumor suppressors.

Both CBP and p300 are also involved in hematological malignancy through aberrant chromosomal translocation. Chromosomal translocations of both CBP and p300 are common in acute myeloid leukemias (AML) (reviewed in (Yang, 2004a)). Fusion proteins are formed with the MYST acetyltransferases MOZ and MORF, as well as with the methyltransferase MLL (mixed lineage leukemia). These fusions retain the HAT domain of CBP or p300 and often the HAT domain of MOZ and MORF as well. These fusions most likely lead to mistargeting and misregulation of HAT activity leading to aberrant gene expression that contributes to the leukemic phenotype. It is also worth noting both CBP and p300 appear to have a role in hematopoietic stem cells, though self-renewal requires the former and differentiation the latter (Rebel et al., 2002). Although the exact mechanism underlying the role of CBP and p300 in hematological malignancy remains to be established, it is clear that CBP and p300 can function both as tumor suppressors and oncogenes depending on the molecular and cellular context.

CBP is also implicated in the complex pathogenesis of neurodegenerative diseases. CBP loss of function has been seen in both models and patient samples from several such diseases, including Alzheimer's disease, Huntington's disease, amyotrophic lateral sclerosis, spinocerebellar ataxias, and other poly-glutamine (polyQ) diseases (Rouaux et al., 2003; Steffan et al., 2000; Takahashi et al., 2002; Hughes, 2002). PolyQ diseases typically involve aggregation of an expanded poly-glutamine protein into insoluble masses that can also sequester other native polyQ-containing proteins. CBP has a tract of 18 glutamines, and it has been shown to colocalize with several different polyQ aggregates (McCampbell et al., 2000; Yamada et al., 2001). In addition, a mutant huntingtin protein can also repress CBP and p300 HAT activity (Nucifora et al., 2001). CBP is also cleaved and degraded by caspase-6 in a cellular model of neurodegeneration (Rouaux et al., 2004). Strikingly, overexpression of CBP can reduce cell death in both culture models and a fly model of neurodegeneration (McCampbell et al., 2000; Taylor et al., 2003). This loss of CBP function can also be rescued by inhibition of HDACs in *Drosophila* and mice (see below). Therefore, the regulation of acetylation plays an important role in neurodegeneration, and this data suggests a new method of therapy for these diseases.

C: The MYST Proteins

MYST is an acronym from the four founding members of the HAT family: MOZ, Ybf2 (also called Sas3), Sas2, and TIP60 (Reifsnyder et al., 1996). Additional members include Esa1, MOF, HBO1, and MORF (Sterner and Berger, 2000). While not as well-characterized as GCN5 or CBP, the MYST proteins have several intriguing properties. They all share a highly homologous MYST domain that is required for HAT activity. This domain contains a zinc finger (except Esa1 which has a zinc finger-like structure) (Yang, 2004a) and an acetyl co-A binding motif similar to the GNAT motif (Neuwald and Landsman, 1997). Despite the shared domains, MYST proteins are quite divergent in their association. Most of the MYST members are part of their own distinct protein complexes. Yeast Esa1, for example, forms complexes called NuA3 and NuA4, which among other proteins contain TRRAP, a component of the SAGA complex (Allard et al., 1999; Grant et al., 1997). The Tip60 complex has been purified and also found to contain TRRAP and other shared proteins as well as unique components such as p400, a SWI2/SNF2 subunit (Ikura et al., 2000). Tip60 may also be able to translocate between the nucleus and cytoplasm, where it associates with the endothelin receptor (Lee et al., 2001).

Given the range of complexes, it is perhaps not surprising that the MYST proteins have divergent functions as well. Esa1 was first shown to be required for proper cell cycle progression, and later it was implicated in DNA repair as well (Clarke et al., 1999; Bird et al., 2002). The original MYST member Sas2 is involved in transcriptional silencing and opposes the action of Sir2 (see below). Sas2 helps establish the boundary between euchromatin and heterochromatin (Kimura et al., 2002; Suka et al., 2002). Interestingly, it seems to have opposite effects on silencing at different loci. Sas2 inhibits silencing at the HMR yeast mating type locus while promoting silencing at the HML locus (Ehrenhofer-Murray et al., 1997), indicating the complexity of silencing regulation. Sas3 was originally found to be involved in silencing as a weaker version of Sas2 (Sterner and Berger, 2000), however, it also regulates transcription elongation (John et al., 2000). MOZ and the related MORF appear to be linked to specific developmental processes since they act as coactivators for the Runx1 and Runx2 transcription factors (Pelletier et al., 2002; Bristow and Shore, 2003). This agrees with data that reduction in levels of Querkopf, the murine MORF, leads to defects in osteogenesis and neurogenesis (Thomas et al., 2000). MOZ and MORF also contain transcription repression as well as activation domains, suggesting that they may have repressive activity in certain conditions (Champagne et al., 1999). In the same manner, the functional consequence of Tip60 activity appears to depend on the context. For example, it acts as a coactivator for NF-κB (Baek et al., 2002) and c-Myc (Frank et al., 2003), and as a corepressor of STAT3

(Xiao et al., 2003). Further, it has been implicated in DNA repair, specifically chromatin remodeling involving H2AX (Kusch et al., 2004).

MOF is a *Drosophila* MYST protein that plays a role in dosage compensation (Sterner and Berger, 2000). It provides a definitive picture of how histone acetylation can be important in a specific biological process. In flies, dosage compensation is achieved by doubling the expression of X-linked genes in males, rather than by female X-inactivation as is generally the case in humans. It was first noted that the X chromosomes of male flies were hyperacetylated compared to autosomes and female X chromosomes, and that this hyperacetylation occurred specifically on lysine 16 of histone H4 (Turner et al., 1992). Later, MOF (males absent on the first) was identified, and MOF mutant flies were characterized as lacking this acetylation on K-16 (Hilfiker et al., 1997). A complex called MSL (male-specific lethality) had already been observed to associate at the hyperacetylated sites (Bone et al., 1994), and MOF was shown to be a part of that complex (Smith et al., 2000). *In vitro*, MOF strongly acetylates H4, and a partially purified MSL complex specifically acetylated K-16 on H4 *in vitro* (Smith et al., 2000), thus completing the link between dosage compensation and MOF.

Dysregulation of MYST proteins is also implicated in disease pathways. MOZ and MORF are frequently involved in chromosomal rearrangements in leukemia. As discussed above, CBP and p300 form fusion proteins with MOZ and MORF, which usually retain the HAT domains of both proteins (Yang, 2004a). These fusions have been shown to leukemogenic (e.g. (Lavau et al., 2000)). MOZ has also been observed in two similar fusions with TIF2, a transcription-related protein (Carapeti et al., 1998; Carapeti et al., 1998). TIF2 can bind CBP, and it appears that the recruitment of CBP to the fusion protein is responsible for its oncogenic potential (Deguchi et al., 2003). MYST proteins can be involved in other disease processes as well. Tip60, which stands for Tat interacting protein of 60kD, specifically interacts with the HIV transactivator protein Tat (Kamine et al., 1996). Tat is able to inhibit the HAT activity of Tip60, and it is postulated that Tat thus blocks Tip60 function at gene targets, such as superoxide dismutase (Mn-SOD) (Creaven et al., 1999). MYST proteins are less characterized than PCAF, CBP, and p300; however, what is known about this HAT family suggests that they have many important biological functions.

A number of different HATs have been identified. An interesting question is why it was necessary for nature to evolve so many different types of HATs. Several specific HATs, most prominently p300 and CBP, appear only in higher organisms and are not present in yeast. One possibility is that the demand is much greater in higher eukaryotes for processing and integrating multiple signals for regulation of transcription. The fact that p300 and CBP can interact with a plethora of transcription factors have led to the proposal that they can function as transcriptional integrators that allow communication among many transcriptional networks. It is also possible that different HATs could have different substrate specificities that dictate their functions. For instance, in the context of the SAGA complex, GCN5 primarily acetylates histone H3 and not others (Sterner and Berger, 2000). Acetyltransferases can also target proteins other than histones (see below).

The Histone Deacetylase Family

By its nature, reversible acetylation of a target implies that the acetyl group can also be removed. For every acetyltransferase that acetylates a substrate, there must be a deacetylase that does the reverse. Although prominent exceptions do exist (see below), generally speaking, HDACs are anti-HATs: they are associated with transcription repression complexes and repressive chromatin states (Fig. 6.1B). Dynamic control of gene expression is critical in biology, and shutting off transcription is as important as turning it on. Therefore HDACs have as wide a range of biological roles as HATs. Based on sequence homology, there are eighteen mammalian HDACs known. They are divided into three classes (class I, II, and III) based on the similarity of their enzymatic (HDAC) domains to the yeast prototypes Rpd3 (Class I) and Hda1 (Class II) as well as their requirement for NAD^+ for enzymatic activity (Class III). We will consider each of the classes in turn.

A: Class I HDACs

The HDAC domains of class I are homologous to the yeast protein Rpd3 (Taunton et al., 1996). In humans, the class I HDACs consist of HDAC1, 2, 3, and 8. HDAC11 is also nominally a class I HDAC, but it is more divergent from the others and has some similarities with class II HDACs (Gao et al., 2002). Class I HDACs are significantly smaller than class II HDACs and are predominantly nuclear proteins, although a recent report has indicated that HDAC8 may be cytosolic in smooth muscle cells (Waltregny et al., 2004).

Like some HATs, the class I HDACs generally function as part of protein complexes. HDAC1 and 2 are in fact often found in the same complexes. The complexes include Sin3, NuRD, and CoREST (Heinzel et al., 1997; Xue et al., 1998; Humphrey et al., 2001). In addition to other components, the Sin3 and NuRD complexes each contain a "core complex" of HDAC1/2 and the histone binding proteins RbAp46 and 48, which may help target HDACs to histones (Zhang et al., 1999). The NuRD complex also includes CH3 and 4 (Mi-2α and β), which contain DNA helicase/ATPase domains found in the SWI/SNF chromatin remodeling proteins (Tong et al., 1998; Zhang et al., 1998b). This suggests that histone deacetylation and chromatin remodeling are coupled events. Further supporting this idea, the SWI/SNF complexes themselves have also been reported to be associated with class I HDACs (Zhang et al., 2000). Histone deacetylation may also be linked to DNA methylation, another epigenetic mechanism associated with gene repression, as proteins with methyl-CpG binding motifs are also a component of HDAC1 complexes (Zhang et al., 1999; Ng et al., 1999). However, the mechanistic details about the interplay of these proteins in the regulation of HDAC1 function and in the repression of gene transcription remain unclear.

Similar to HDAC1, HDAC3 was found to reside in a large complex. In this case, endogenous HDAC3 is present in complexes with the corepressors N-CoR and SMRT that also contain several other components including transducin (beta)-like I (TBL1) protein (Yoon et al., 2003). Using RNA interference to specifically inactivate these components, it was demonstrated that the HDAC3-containing N-CoR and SMRT complexes are important for mediating the transcriptional repression activity of unliganded thyroid hormone receptor (Yoon et al., 2003). Interestingly, HDAC3 appears to be the only HDAC present in these complexes. Sin3, a co-repressor in HDAC1 complexes, is also not found in SMRT or N-CoR complexes. Therefore HDAC1 and HDAC3 clearly exist in different complexes and probably regulate chromatin and transcription through distinct mechanisms. Using RNAi knockdown approaches, it should be possible to delineate the specific functions of HDAC1 and HDAC3 deacetylase complexes in gene transcription.

Recruitment of class I HDACs to their targets takes place through at least a couple of mechanisms. Usually, class I HDACs are recruited via other members of their repressor complex. For example, the Sin3 and N-CoR complexes can both bind unliganded nuclear hormone receptors (Heinzel et al., 1997; Horlein et al., 1995). The silencing repressors polycomb and hunchback can recruit NuRD components (Kehle et al., 1998). Other factors like Ikaros and REST can also recruit HDAC complexes (Koipally *et al.*, 1999; Naruse *et al.*, 1999). Class I HDACs can also sometimes bind DNA-binding factors directly. This appears to be the case with HDAC1 and the transcription factor YY1 (Yang *et al.*, 1996). This is more common for the class IIA HDACs, however (e.g. Miska *et al.*, 1999 and see below).

B: Class II HDACs

The HDAC domain of class II HDACs is homologous to the yeast protein Hda1 (Grozinger *et al.*, 1999). In contrast to class I, class II HDACs are relatively large proteins (140kD or more) and have multiple domains. Unlike class I, which are mostly localized to the nucleus, many of the class II HDACs have a significant cytoplasmic component. Class II deacetylases can be further divided into two subfamilies, IIA and IIB.

B1: Class IIA

The class IIA HDACs consist of HDAC4, 5, 7, and 9 (Verdin *et al.*, 2003). This subfamily of HDACs shares several unique properties. First, they all contain the N-terminal non-catalytic MITR (MEF2-interacting transcription repressor) homology domain (Zhou *et al.*, 2000). This domain serves critical functions as both a protein-protein interaction domain and a regulatory domain subject to phosphorylation (Miska *et al.*, 1999; Grozinger and Schreiber, 2000). Second, they are all critical regulators of MEF2, a family of transcription factors important in muscle differentiation and neuronal apoptosis (Miska *et al.*, 1999). Third and most importantly, they are all regulated by phosphorylation-dependent subcellular trafficking (McKinsey *et al.*, 2000).

All of the class IIa HDACs are subject to nuclear export via a nuclear export signal at the C-terminus (Wang and Yang, 2001; McKinsey *et al.*, 2001). For example, HDAC4 is often found in the cytoplasm; however treatment with nuclear export inhibitor leptomycin B causes its nuclear accumulation (Zhao *et al.*, 2001). The subcellular localization of these HDAC members is cell type-dependent and tightly regulated by specific signaling events. For example, in C2C12 myoblasts, HDAC5 is normally a nuclear protein and it translocates from the nucleus to the cytoplasm upon differentiation, presumably to allow MEF2 to execute the transcription program for muscle differentiation (McKinsey *et al.*, 2000). In contrast, HDAC5 and HDAC4 move from the cytoplasm into the nucleus in neurons in response to apoptotic stimuli (Bolger and Yao, 2005). Thus, the subcellular localization of HDAC4 and HDAC5 is dynamically regulated, and this principle has generally held true for HDAC7 and 9 as well.

It is now known that HDAC4 and HDAC5 intracellular trafficking is regulated by phosphorylation. Upon appropriate signaling, HDAC4 or 5 become phosphorylated. The calcium/calmodulin dependent kinases (CaMKs) and more recently, protein kinase D (PKD or PKCμ) have been shown to be the kinases responsible (McKinsey *et al.*, 2000; Zhao *et al.*, 2001; Vega *et al.*, 2004a). The phosphorylated HDAC then binds the phospho-binding protein 14-3-3 (Grozinger and Schreiber, 2000), and this interaction activates HDAC4 and HDAC5 nuclear export through the Crm-1 dependent machinery (Wang and Yang, 2001). It is thought that cytoplasmic HDAC4 and 5 are then sequestered by 14-3-3 until they are dephosphorylated. Thus, by regulating the nuclear export activity of HDAC4 and HDAC5, specific cell signaling events can then induce the transcriptional program that is normally controlled by HDAC4 and HDAC5.

The most well-characterized target for class IIA deacetylases is the myocyte enhancer factor 2 (MEF2) (Miska *et al.*, 1999). MEF2 is one of the master activators of the regulatory cascade required for muscle differentiation, and HDAC4 and 5 were first shown to repress MEF2 in that context (McKinsey *et al.*, 2000). MEF2 is also involved in several other biological pathways, such as cardiac hypertrophy and T cell apoptosis (reviewed in McKinsey *et al.*, 2002), most of which have now been shown to involve one or more class IIA HDACs as well. The class IIa HDACs bind MEF2 directly and repress MEF2-dependent transcription (Miska *et al.*, 1999). The transcriptional repression by HDAC4 family members on MEF2 has been naturally attributed to its ability to induce histone deacetylation. Indeed, deacetylase-deficient HDAC4 or HDAC5 mutants cannot inhibit C2C12 muscle differentiation (Lu *et al.*, 2000). However, recent studies reveal that HDAC4 and HDAC5 use additional mechanisms to repress transcription. For example HDAC4 and HDAC5 associate with the transcriptional co-repressor CtBP and the heterochromatin-associated protein HP1 (Zhang *et al.*, 2001; Zhang *et al.*, 2002b). HP1 is an essential gene for establishing transcriptional silencing that binds histone H3 methylated at lysine-9 (MethH3-K9), a common hallmark of heterochromatin (Bannister *et al.*, 2001; Lachner *et al.*, 2001). The interaction of HP1 and MethH3-K9 is apparently important for establishing the chromatin structure required for various type of gene silencing. The interaction of HP1 with HDAC4 as well as HDAC11 (K. Wei and T.P.Y., unpublished result) suggests that in addition to catalyzing local histone deacetylation, HDACs can contribute to transcriptional repression through direct interaction with components of the histone methylation network (Zhang et al., 2002b). Indeed, pharmacological inhibition of HDACs by trichostatin A disrupted the spatial distribution of HP1 without affecting the total amounting of H3-K9 methylation (Maison et al., 2002).

Surprisingly, HDAC4-related deacetylases can also act to promote transcription factor sumoylation. In characterizing HDAC4, we have discovered that HDAC4 can function as a SUMO E3 ligase for MEF2 (Zhao et al., 2005). We and others have found that HDAC4 promotes MEF2 sumoylation and leads an apparent loss of MEF2 transcriptional activity (Gregoire and Yang, 2005). Thus the class IIA HDACs are able to repress transcription in at least three different ways: by deacetylating histones, by interacting with other corepressors and by inducing sumoylation of target activators.

Although they were initially identified as MEF2-interacting HDACs, the class IIA HDACs target multiple transcription activators for repression. HDAC4 and/or 5 repress and interact with Runx2, GATA-1, and SRF (serum response factor) (Vega et al., 2004b; Watamoto et al., 2003; Davis et al., 2003). It is thought that the recruitment of HDAC converts these transcriptional activators into repressors. Class IIA HDACs also interact with known DNA-binding transcriptional repressors including BCL6, and PLZF (Lemercier et al., 2002; Chauchereau et al., 2004). In these cases, HDACs function more as classic transcriptional co-repressors. Consistent with the idea of multiple mechanisms of action for class IIA HDACs, recent gene ablation and other functional studies of HDAC4, 5, 7 and 9 have revealed much broader roles for class IIA HDACs in various developmental processes as summarized below.

Mice with mutations in either HDAC5 or HDAC9 are viable and grossly normal. However, they are prone to developing cardiac hypertrophy (reviewed in Metzger, 2002). Hypertrophic growth of the heart muscle is a stress response that increases myocyte size and activates a fetal cardiac gene program dependent on MEF2 transcription factors. Loss of HDAC5 and HDAC9 lead to ectopic activation of this gene program and result in enlarged hearts that are acutely sensitive to hypertrophic stimuli (Zhang et al., 2002a; Chang et al., 2004). These observations indicate that HDAC5 and HDAC9 have overlapping functions and play a major role in preventing cardiac hypertrophy. The HDAC5/9 double knockouts are prone to lethal hemorrhages and display proportional hypertrophy as well (Chang et al., 2004). It is also of great interest to note that transgenic expression of a HDAC5 mutant that is resistant to phosphorylation-induced export causes lethality associated with severe defects in mitochondria (Czubryt et al., 2003). This result substantiates the importance of intracellular trafficking of HDAC5 and reveals a potential role for class IIA HDACs in metabolic regulation.

The HDAC4 knockout mouse has a different phenotype. This mouse displays significant skeletal defects (Vega et al., 2004b). These result from defects in chrondrocyte hypertrophy. In normal development, most bones form as a cartilage template first, then chondrocytes in the cartilage undergo hypertrophy, secreting a matrix that allows vascular invasion and subsequent ossification. HDAC4 represses the hypertrophy, at least in part by inhibiting Runx2-mediated transcription. Without HDAC4, the chondrocytes undergo early and improper hypertrophy and premature ossification, leading to the skeletal defects. HDAC4 has also been implicated in DNA damage response (Kao et al., 2003). It associates with p53-binding protein-1 (53BP1) and localizes to nuclear foci induced by DNA binding agents. HDAC4 can also be cleaved by caspases after DNA damage (Liu et al., 2004; Paroni et al., 2004); however, the physiological significance of this observation is not clear. Since the HAT Esa1 is also involved in DNA repair (Bird et al., 2002), these findings suggest both acetylation and deacetylation are required for chromatin remodeling during DNA repair.

While no HDAC7 knockout has been reported, HDAC7 has been implicated in T-cell maturation in the thymus (Dequiedt et al., 2003). Nur77 is a steroid receptor that induces apoptosis of T-cells during negative selection, and its expression is controlled by MEF2 (Youn et al., 1999). HDAC7 represses Nur77 transcription and inhibits the activity of MEF2. It thus plays a critical role in controlling negative selection. MEF2 has also been shown to be an important factor in neuronal survival (Mao et al., 1999). It functions to inhibit neuronal cell death, at least partly through activation of neurotrophin-3 (Shalizi et al., 2003). In fact, many of the pathways involved in muscle differentiation, like CaMK signaling, function similarly to promote neuronal survival. Therefore, it is likely that class IIA HDACs have a significant role in neuronal survival and death as well. This is supported by evidence that overexpression of HDAC5 can induce cell death in cerebellar granule neurons (Linseman et al., 2003). It will be interesting to see how HDACs are involved, given that HDAC4 is highly expressed in the brain (Grozinger et al., 1999) and that HDAC inhibitors are being explored as therapy for neurodegenerative diseases (see below).

B2: Class IIb

Class IIb deacetylases include HDAC6 and HDAC10. They are characterized by a tandem repeat of complete (HDAC6) or partial (HDAC10) catalytic domains (Grozinger *et al.*, 1999; Guardiola and Yao, 2002). HDAC6 and HDAC10 are also unique in their resistance to select deacetylase inhibitors, such as trapoxin B, which can potently inhibit the deacetylase activity of both class I and IIa HDACs (Guardiola and Yao, 2002). Unlike other HDAC members, HDAC6 does not appear to play a direct role in histone acetylation and transcriptional regulation. In fact, HDAC6 is almost always cytoplasmic and has functions other than histone acetylation and transcription (Hubbert *et al.*, 2002; Kawaguchi *et al.*, 2003). The HDAC6-specific inhibitor tubacin does not induce histone acetylation and has little effect on gene transcription (Haggarty *et al.*, 2003). HDAC6, however, has potent tubulin deacetylase activity which may regulate microtubule function (Hubbert *et al.*, 2002). HDAC6 also helps to clear misfolded protein aggregates from the cell, which may or may not be related to its tubulin deacetylase activity (Kawaguchi *et al.*, 2003). The biological function of HDAC10 remains unclear. However, it does have a nuclear component, and when attached to GAL4 DNA binding domain, it is capable of repressing transcription (Guardiola and Yao, 2002).

C: Class III HDACs

The prototype of the class III deacetylase is yeast SIR2 (silencing information regulator 2). Within the last ten years, Sir2 has gained notoriety as being essential for lifespan determination. In yeast, additional copies of this gene can enhance lifespan, while eliminating this gene altogether can reduce lifespan (Kaeberlein *et al.*, 1999). Sir2, along with two other adaptor proteins, Sir3 and Sir4, can transcriptionally silence mating-type loci, telomeres, and rDNA, the last of which is thought to be the molecular basis for lifespan extension in yeast (Rine and Herskowitz, 1987; Smith and Boeke, 1997). The silencing of rDNA occurs via specific histone deacetylation events, which are thought to prevent the accumulation of extrachromosomal rDNA circles (ERCs), a molecular sign of ageing in a yeast mother cell (Sinclair and Guarente, 1997).

These processes can now be understood after SIR2 was shown to be a histone deacetylase (Imai *et al.*, 2000). However, SIR2 and class I and II HDAC family members are structurally unrelated. Unlike HDACs, SIR2 deacetylase activity uniquely requires NAD^+, a cofactor whose levels are regulated by cellular respiration rates (Imai *et al.*, 2000). The enzymatic reaction is completely different from that of HDAC-catalyzed deacetylation, and it produces a deacetylated peptide substrate as well as two reaction byproducts: nicotinamide, which can act to inhibit Sir2 activity through a negative feedback mechanism, and O-acetyl-ADP-ribose (OAAR), a novel metabolite whose function remains unknown (Tanner *et al.*, 2000). The requirement of NAD^+ for Sir2 activity suggests that class III HDACs could be regulated by changes in the cellular NAD^+/NADH ratio (see below), providing a logical link to metabolic regulation and ageing processes.

Genome-wide searches for homologs in other species have been fruitful in identifying Sir2 homologs in bacteria, *C.elegans*, *Drosophila*, and higher eukaryotes including humans. Although Sir2-related proteins in worms and flies also seem to regulate the ageing process, the molecular mechanisms appear to be quite different and more complex than the yeast Sir2 protein (Guarente and Picard, 2005). There are seven class III HDACs in humans: SIRT1 to SIRT7. SIRT1 is the closest homolog to yeast Sir2 and is most well-characterized so far. It has been shown to repress a handful of different activators, and it interacts with several repressors including BCL6, as well as HES1 and HEY2, homologs of the Drosphila repressor Hairy (Bereshchenko *et al.*, 2002; Takata and Ishikawa, 2003). This evidence supports the idea that SIRT1 can modulate histone acetylation directly. Surprisingly, however, SIRT1 also appears to regulate transcription factors themselves through direct deacetylation events. Most notably, SIRT1 can regulate the acetylation status of the tumor suppressor p53 (Vaziri *et al.*, 2001; Luo *et al.*, 2001). Acetylation of p53 correlates with p53-dependent transcriptional activity, while deacetylated p53 abrogates this transcriptional activity (Gu and Roeder, 1997). Under conditions favoring cell survival and increased lifespan, SIRT1 activity would deacetylate and inactivate p53, a master regulator of apoptosis. However, during harsh conditions or prolonged environmental stress, when p53 activity is required to mount an apoptotic response, SIRT1 may become inactivated leading to enhanced p53-dependent transcription (Smith, 2002).

SIRT1 appears to modulate a suite of additional transcription factors including the pro-survival factor, NF-κB (Yeung *et al.*, 2004). Similar to p53, regulation of NF-κB occurs through direct acetylation of the RelA/p65 subunit of NF-κB. Deacetylation of this subunit is thought to prevent NF-κB-dependent transcription and therefore sensitize cells to apoptosis induced by TNFα. In this specific instance, it appears that SIRT1 possesses anti-proliferative capabilities and may instead act as a fine-tuning response regulator at the transcriptional level. It is unclear why SIRT1 would promote apoptosis through NF-κB-dependent pathways. The Foxo family of transcription factors are also involved in cellular decisions between survival and apoptosis during the cellular response to stress (i.e oxidative stress, heat shock) (Giannakou and Partridge, 2004). Accordingly, Foxo1, Foxo3a and Foxo4 are all subject to reversible acetylation and SIRT1 appears to modulate their transcriptional activity, in this case favoring cellular survival and increased lifespan in the presence of cellular stress (van der Horst et al., 2004; Brunet et al., 2004; Yang et al., 2005). Lastly, SIRT1 also regulates transcription during muscle cell differentiation. Similarly to the class IIA HDACs, SIRT1 acts as a negative regulator of the muscle differentiation program (Fulco et al., 2003). SIRT1 has been reported to directly interact with PCAF in muscle cells and can deacetylate both PCAF and the muscle transcription activator MyoD. Furthermore, the NAD^+/NADH ratio decreases during muscle differentiation, possibly regulating SIRT1 activity.

In light of the profound effects of Sir2 on yeast lifespan, one may ask whether the mammalian Sir2 homologues have similar life-extending benefits. As stated above, SIRT1 appears to promote cell survival through direct histone interactions, or through regulation of transcription factor activity. But how might SIRT1 regulate organismal ageing in higher eukaryotes? Since SIRT1 is an NAD^+-dependent deacetylase, it has the ability to couple the energy status of the cell to various cellular processes including the regulation of pro-and anti-survival factors mentioned above. As a NAD^+-responsive protein, one might speculate that SIRT1 is regulated by nutritional cues at the whole-body level. Consistent with this hypothesis, calorie restriction (CR), which is known to increase longevity in many organisms, induces SIRT1 expression in mammalian cells (Cohen et al., 2004). It has been proposed that CR increases NAD^+ levels and consequently enhances Sir2 activity; however, the connection between CR, Sir2, and the NAD^+/NADH ratio is currently under debate (Anderson et al., 2003).

Recently, Rodgers et al. reported another link between SIRT1 and nutritional status (Rodgers et al., 2005). SIRT1 was shown to interact with and regulate PGC-1 (PPARγ co-activator 1), which is a master regulator of gluconeogenesis. As part of the response to fasting, hormonal cues impinge on the transcriptional machinery including PGC-1, and as a consequence PGC-1 transcriptionally activates the necessary gluconeogenic enzymes and precursors required for glucose production under starvation conditions. Remarkably, SIRT1 was shown to deacetylate PGC-1 and enhance its transcriptional ability under fasting conditions. These results support the idea that the Sir2 class of proteins can participate in nutritional responses and regulate whole-body metabolism and glucose homeostasis. The ability of this class of proteins to utilize NAD^+ as a cofactor allows a fine-tuned response to nutritional conditions. Since energy status could potentially be a factor that dictates the ageing process, SIRT1 could represent a link to ageing in higher organisms (Rodgers et al., 2005; Nemoto et al., 2005). In light of this fact, there has been an effort to identify pharmacological tools that can activate the Sir2 class of proteins in order to harness their potential health benefits. In a search for Sir2 activators, it was shown that resveratrol, a phenol derivative found in red wine, is able to activate Sir2 and increase lifespan in yeast (Howitz et al., 2003). This study and follow-ups have shown that resveratrol and calorie restriction operate in the same pathway to promote lifespan extension, suggesting they both work through similar mechanisms involving Sir2 activation. However, more detailed structural analysis is required, however, to understand the molecular details behind this compound's potential anti-ageing effects.

Although roles for SIRT1 in ageing have begun to be elucidated, much less is known about its six mammalian homologs, SIRT2 to 7. Like HDAC6, SIRT2 has been reported to be a tubulin deacetylase, although the biological function of this is not known (North et al., 2003). SIRT2 also is regulated by the cell cycle, and the levels are dramatically increased during mitosis (Dryden et al., 2003). The SIRT3 protein is localized to the mitochondrial matrix (Onyango et al., 2002; Schwer et al., 2002). It appears to have a role in adaptive thermogenesis in brown adipocytes (Shi et al., 2005). Very little is known about the other members of this family, leaving open the possibility that other Sir2 family members can regulate transcription or other biological processes by coupling energy in the form of NAD^+ to various cellular functions. Indeed, much remains unknown about this class of proteins including any potential roles in the ageing process, a field that is certain to receive more attention in the future.

Cross Talk between HDAC Family Members

It is important to recognize that the 18 HDAC

members probably do not function independently. In fact, studies have presented ample evidence that several of these deacetylases might work in conjunction and regulate common biological targets and processes. For example, HDAC4 and HDAC5 can associate with HDAC3 (Grozinger et al., 1999; Fischle et al., 2002). These observations led to the hypothesis that HDAC3 might be the catalytic subunit for the HDAC4 deacetylase (Fischle et al., 2002). However, it is important to note that although endogenous HDAC3 is abundantly present in the NcoR and Smrt co-repressor complex, HDAC4 is not part of the complex (Yoon et al., 2003). Thus, it is unlikely that endogenous HDAC4 (or another class IIA HDAC) forms an exclusive complex with HDAC3 in vivo. We have found that HDAC4 and SIRT1 can also form a complex and this complex could be important in regulating MEF2 transcriptional activity (Zhao et al., 2005). As discussed, both HDAC1 and SIRT1 have been shown to promote p53 deacetylation (see above). Lastly, the cytoplasmically localized HDAC6 and SIRT2 were reported to interact, and both can function as tubulin deacetylases (North et al., 2003). Although the exact functional significance of this obvious cross-talk between HDAC family members remains to be defined, these observations clearly show that the various HDAC members can operate in coordination with one another.

HAT, HDAC and Histone Acetylation-a More Complicated Picture

Although it is generally thought that histone deacetylation causes gene repression, several studies now reveal a much more complex picture. For example, in a genome-wide CHIP analysis in yeast, it was revealed that the recruitment of the histone deacetylase HOS2 to the coding region of specific genes is correlated with gene activation, rather than repression (Wang et al., 2002). Since there is concurrent deacetylation of histone H3 and H4, these observations demonstrate that histone deacetylation by a deacetylase does not necessarily cause gene repression. How HOS2-catalyzed deacetylation causes transcriptional activation is not yet clear. Wang et al. propose that deacetylation by HOS2 might be required to restore a permissive transcription state after one round of acetylation associated with active transcription (Wang et al., 2002). It is of interest to note that although Rpd3 is also present in the genomic regions occupied by HOS2, Rpd3 mutation gives rise to the opposite phenotype (Robyr et al., 2002). On the other hand, Rpd3 is also able to behave as a transcriptional activator in conditions of osmotic stress (De Nadal et al., 2004). Upon activation of osmotic stress signaling, Rpd3 is required for the deacetylation of promoters from stress response genes, including the heat shock factor HSP12 among others. It was concluded that Rpd3 responds to various stimuli, including osmo-stress and heat shock, by positively regulating target gene expression. These results clearly illustrate the complexity of HDAC function. HOS2 was found to associate with SET3, which belongs to a family of histone methyltransferases, although such an activity has not been demonstrated for SET3 (Pijnappel et al., 2001). One interesting possibility is that deacetylation of specific histone residues in conjunction with specific methylation might confer gene activation instead of repression. Thus, although all HDAC family members possess histone deacetylase activity, their functions could be vastly different depending on many other factors.

It was initially thought that HATs and HDACs are recruited to specific promoter regions and affect histone acetylation spanning as little as two nucleosomes (Kadosh and Struhl, 1998; Wu et al., 2001). However, genome-wide surveys by chromatin-immunoprecipitation (ChIP) in yeast also reveal non-targeted and broad distributions of both HATs and HDACs throughout the yeast genome. This configuration is proposed to allow for rapid reversal of the hyper- or hypo-acetylation of chromatin induced by the targeted recruitment of HAT and HDAC, thereby restoring the chromatin to a "ground" state (Vogelauer et al., 2000; Katan-Khaykovich and Struhl, 2002). This model is somewhat analogous to that proposed for HOS2-dependent gene activation.

The last important issue concerns the complexity of histone acetylation. Different HATs and HDACs appear to have distinct preferences toward different lysine residues in histones. For example, in yeast cells, Rpd3 is capable of deacetylating all lysines examined in the core histones, whereas Hda1 specifically deacetylates histone H2B and H3, and HOS2 deacetylates histone H3 and H4 (for review, see Kurdistani and Grunstein, 2003). The specificity of HDAC family members in higher eukaryotes is less well defined; however, it is likely it exists there as well. HATs also show specificity. GCN5 acetylates mostly H3 and H2B while CBP and p300 appear to acetylate all histones (for review, see Sterner and Berger, 2000). Whether this apparent specificity toward different histones is important for the chromatin remodeling activity of these enzymes remains to be established.

Acetylation of Non-histone Proteins

Histones are the prototypical target of acetylation, but they are far from the only ones. It has become clear in recent years that acetylation has a much wider range than simply histones. As already mentioned in previous sections, a growing number of non-histone acetylated proteins have been identified (reviewed in Cohen and Yao, 2004; Yang, 2004b). These include several different types of proteins. First, other non-histone chromatin proteins can be acetylated, such as several members of the High Mobility Group proteins (HMG1, HMG2, HMG14, and HMG17) and cohesin subunits (Sterner et al., 1978; Sterner et al., 1981; Ivanov et al., 2002). Secondly, a number of transcription factors can be acetylated, including such well-known proteins as p53, E2F1, CREB, MyoD, and more than 30 other activators (For a list of acetylated TFs, see (Yang, 2004b)). Several general transcription factors are also acetylated, including TFIIB, E, and F (Choi et al., 2003; Imhof et al., 1997). Acetylation of proteins involved in DNA replication and repair, such as PCNA, and the nuclear transport factor importin-α have also been reported (Naryzhny and Lee, 2004; Bannister et al., 2000). In the cytoplasm, α-tubulin has long been known to be acetylated (L'Hernault and Rosenbaum, 1985). A recent study also demonstrates that the molecular chaperone Hsp90 is also regulated by acetylation (Yu et al., 2002). The recent explosion of acetylated proteins suggests that acetylation is a very common modification, and that many more can and will be identified in the future. Furthermore, these observations strongly support the proposition that reversible protein acetylation plays a broad biological role beyond histone modification.

Given the wide array of targets, it is not surprising that acetylation can have diverse functional consequences. Since many acetylated substrates are DNA-binding proteins, a common effect of acetylation is to alter the DNA-binding ability of the protein. Acetylation can stimulate DNA binding, as shown for GATA1 (Boyes et al., 1998), or it can disrupt it, as is the case with histones and HMGI(Y) (Munshi et al., 1998). Acetylation can affect protein transport. The deacetylation of RelA subunit of NF-κB by HDAC3 leads to NF-κB nuclear export, attenuating its transcriptional activity (Chen et al., 2001). Acetylation has also been reported to affect protein stability, possibly by interfering with protein ubiquitination. In fact, acetylation, ubiquitination, sumoylation, methylation, and neddylation all target lysine residues. An emerging concept is that acetylation can compete with other modifications for a specific lysine residue and block its effects. If acetylation blocks ubiquitin binding of a lysine, for example, it may prevent its degradation by the proteasome, suggesting a mechanism for the stability increase mentioned above. It appears p53 acetylation can function in this manner (Ito et al., 2002). Importantly, these acetylation events appear to be controlled by the same enzymes that were initially thought to be dedicated to histone modification. Thus, the classically defined HAT and HDAC are likely to have functions other than chromatin-dependent processes. In fact in at least one case, HDAC6, it does not appear to function as a histone deacetylase at all (see above), further splitting the "H" from the "AT" and "DAC".

HDAC Inhibitors in Diseases and Therapy

HDAC inhibitors are currently being examined for clinical use in several different areas. So far, these approaches focus on class I and class II inhibitors, although class III HDACs may soon be considered as therapeutic targets for ageing or metabolic disorders (see above). Generally, the inhibitors lack specificity, inhibiting all of class I and II, although several do not inhibit HDAC6 or 10 very efficiently (Furumai et al., 2001). One major area of research is the use of HDAC inhibitors as anti-cancer chemotherapeutic agents. Sodium butyrate is one inhibitor that has been approved for clinical use (Thiagalingam et al., 2003). It is able to reduce proliferation of tumor cells, induce apoptosis, and increase differentiation (Sawa et al., 2001; Chopin et al., 2002). It has, however, effects on several different enzymes besides HDACs. Trichostatin A (TSA), originally discovered as an antifungal agent, is another widely used HDAC inhibitor (Tsuji et al., 1976). It causes cell cycle arrest and apoptosis in a number of cell lines (Kim et al., 2000b; Sawa et al., 2001; Herold et al., 2002). TSA is a reversible inhibitor that binds to the active site of HDACs (Finnin et al., 1999). In contrast, trapoxin is an irreversible inhibitor that was originally identified as a compound capable of reversing transformation of NIH3T3 cells (Itazaki et al., 1990). It was subsequently shown to be an HDAC inhibitor. Unfortunately TSA and trapoxin do not have high antitumor activity in vivo, probably due to stability issues (Yoshida et al., 2001). On the other hand, suberoylanilide hydroxamic acid (SAHA), which is similar to TSA, appears to be more effective (Richon et al., 1998). Like other inhibitors, it stimulates apoptosis and differentiation while inhibiting growth of cancer cell lines (Butler et al., 2002; Munster et al., 2001), and it also shows substantial antitumor activity in animal models (Cohen et al., 2002). It reduced both tumor number and size in rats with mammary carcinomas as well as inhibiting tumor growth in at least two different mouse models (Cohen *et al.*, 2002; Marks *et al.*, 2000). SAHA is currently undergoing clinical trials for both solid tumors and hematological malignancies (Kelly *et al.*, 2003). Several other HDAC inhibitors have also been identified, such as FK228, MS-275, and apicidin, which have also shown promise as chemotherapy drugs (Nakajima *et al.*, 1998; Jaboin *et al.*, 2002; Kim *et al.*, 2000a).

Although HDAC inhibitors are effective against many cell lines and tumors, surprisingly, they appear to have the opposite effect in a different context. Another major area of study for clinical use of HDAC inhibitors is neurodegenerative diseases. HDAC inhibitors were first shown to reduce neurodegeneration in *Drosophila* in models of polyglutamine disease (Steffan *et al.*, 2001). Since then, SAHA and sodium butyrate have both been shown to ameliorate symptoms in mouse models of Huntington's disease (Ferrante *et al.*, 2003; Hockly *et al.*, 2003). Sodium butyrate has also been shown to ameliorate the phenotypes of another polyQ disease, spinal and bulbar muscular atrophy (SBMA) (Minamiyama *et al.*, 2004). As mentioned above, polyQ proteins can sequester CBP, thus reducing HAT activity and leading to histone hypoacetylation (Taylor *et al.*, 2003). HDAC inhibition may be a way of restoring the proper balance between HATs and HDACs in these diseases (Rouaux *et al.*, 2004). However, the fact that HDAC inhibition does work in these animal models suggests that HDACs do play an active role in this system. This is supported by the high level of expression of several HDACs in the brain (Grozinger *et al.*, 1999), and the observation that HDAC5 can induce neuronal cell death when overexpressed (Linseman *et al.*, 2003). The advantage of developing inhibitors specific for individual HDAC members is apparent, though, given the important role for HDAC5 and 9 in preventing cardiac hypertrophy (see above).

The reasons for the difference in response to HDAC inhibitors between tumor and neurodegenerative models are not clear. It may be due to the relative expression levels of various HDACs in the tissues or cell lines. There are also likely to be differences in associated factors and targets in different tissues and even in the same tissues under different conditions. This may cause completely different responses to the same HDAC inhibitors. For instance, HDAC inhibitors such as TSA have been shown to arrest the cell cycle (Kim *et al.*, 2000b). However, neurons are largely postmitotic in the adult brain, so they might be insensitive to effects caused by the arrest. These recent studies underline the potential therapeutic utility of HDAC inhibitors and further highlight the importance of protein acetylation in human disease.

Concluding Remarks

Through more than forty years of research, reversible acetylation of histones catalyzed by a large number of HATs and HDACs has led us to an unprecedented understanding of transcriptional regulation and chromatin remodeling. The realization that reversible acetylation is not restricted to histones and that HATs and HDACs can regulate non-histone protein acetylation further expands the biological functions and significance of protein acetylation. As more evidence suggests a broad involvement of HAT and HDAC proteins in critical biological processes and disease (Fig.6.2), the study of histone acetylation and general protein acetylation promises many important discoveries to come.

References

Ait-Si-Ali, S., Carlisi, D., Ramirez, S., Upegui-Gonzalez, L. C., Duquet, A., Robin, P., Rudkin, B., Harel-Bellan, A., and Trouche, D. (1999). Phosphorylation by p44 MAP Kinase/ERK1 stimulates CBP histone acetyl transferase activity *in vitro*. Biochem Biophys Res Commun *262*, 157-162.

Ait-Si-Ali, S., Ramirez, S., Barre, F. X., Dkhissi, F., Magnaghi-Jaulin, L., Girault, J. A., Robin, P., Knibiehler, M., Pritchard, L. L., Ducommun, B., *et al.* (1998). Histone acetyltransferase activity of CBP is controlled by cycle-dependent kinases and oncoprotein E1A. Nature *396*, 184-186.

Allard, S., Utley, R. T., Savard, J., Clarke, A., Grant, P., Brandl, C. J., Pillus, L., Workman, J. L., and Cote, J. (1999). NuA4, an essential transcription adaptor/histone H4 acetyltransferase complex containing Esa1p and the ATM-related cofactor Tra1p. Embo J *18*, 5108-5119.

Allfrey, V. G., Faulkner, R., and Mirsky, A. E. (1964). Acetylation and Methylation of Histones and Their Possible Role in the Regulation of Rna Synthesis. Proc Natl Acad Sci USA *51*, 786-794.

Anderson, R. M., Latorre-Esteves, M., Neves, A. R., Lavu, S., Medvedik, O., Taylor, C., Howitz, K. T., Santos, H., and Sinclair, D. A. (2003). Yeast life-span extension by calorie restriction is independent of NAD fluctuation. Science *302*, 2124-2126.

Arany, Z., Sellers, W. R., Livingston, D. M., and Eckner, R. (1994). E1A-associated p300 and CREB-associated CBP belong to a conserved family of coactivators. Cell *77*, 799-800.

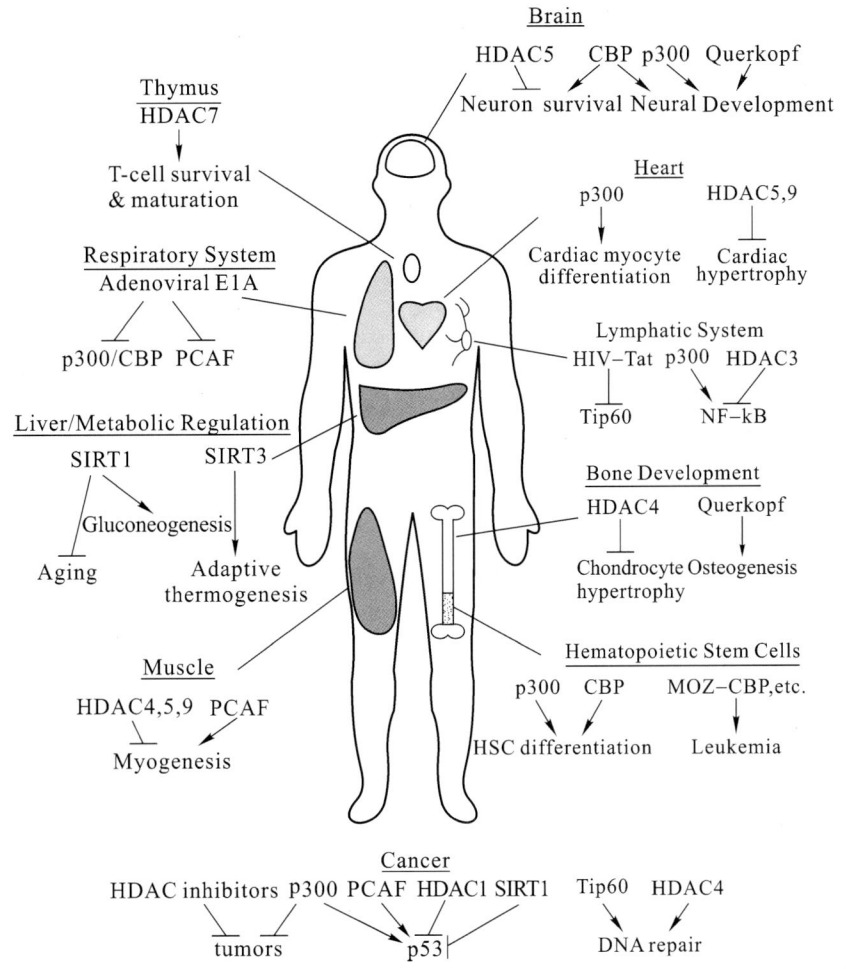

Fig 6.2 Biological pathways involving HATs and HDACs. This figure illustrates the wide array of biological functions in which HATs and HDACs have been specifically implicated. Many of these pathways are developmental, as in skeletal and cardiac myogenesis, neural development, osteogenesis, and hematopoietic stem cell differentiation. HDACs are also involved in metabolic regulation. HATs and HDACs have been implicated in disease or resistance pathways, including adenoviral and HIV infection, inflammation responses, T-cell maturation, and neurodegenerative diseases. A number of cancer-related pathways also include HATs and HDACs, such as DNA repair, p53, and leukemias. HDAC inhibitors and p300 act as tumor suppressors.

Baek, S. H., Ohgi, K. A., Rose, D. W., Koo, E. H., Glass, C. K., and Rosenfeld, M. G. (2002). Exchange of N-CoR corepressor and Tip60 coactivator complexes links gene expression by NF-kappaB and beta-amyloid precursor protein. Cell *110*, 55-67.

Bannister, A. J., and Kouzarides, T. (1996). The CBP co-activator is a histone acetyltransferase. Nature *384*, 641-643.

Bannister, A. J., Miska, E. A., Gorlich, D., and Kouzarides, T. (2000). Acetylation of importin-alpha nuclear import factors by CBP/p300. Curr Biol *10*, 467-470.

Bannister, A. J., Zegerman, P., Partridge, J. F., Miska, E. A., Thomas, J. O., Allshire, R. C., and Kouzarides, T. (2001). Selective recognition of methylated lysine 9 on histone H3 by the HP1 chromo domain. Nature *410*, 120-124.

Bereshchenko, O. R., Gu, W., and Dalla-Favera, R. (2002). Acetylation inactivates the transcriptional repressor BCL6. Nat Genet *32*, 606-613.

Bird, A. W., Yu, D. Y., Pray-Grant, M. G., Qiu, Q., Harmon, K. E., Megee, P. C., Grant, P. A., Smith, M. M., and Christman, M. F. (2002). Acetylation of histone H4 by Esa1 is required for DNA double-strand break repair. Nature *419*, 411-415.

Bolger, T. A., and Yao, T. P. (2005). Intracellular trafficking of histone deacetylase 4 regulates neuronal cell death. J Neurosci. *25(41)*, 9544-53.

Bone, J. R., Lavender, J., Richman, R., Palmer, M. J., Turner, B. M., and Kuroda, M. I. (1994). Acetylated histone H4 on the male X chromosome is associated with dosage compensation in *Drosophila*. Genes Dev *8*, 96-104.

Boyes, J., Byfield, P., Nakatani, Y., and Ogryzko, V. (1998). Regulation of activity of the transcription factor GATA-1 by acetylation. Nature *396*, 594-598.

Bristow, C. A., and Shore, P. (2003). Transcriptional regulation of the human MIP-1alpha promoter by RUNX1 and MOZ. Nucleic Acids Res *31*, 2735-2744.

Brownell, J. E., and Allis, C. D. (1995). An activity gel assay detects a single, catalytically active histone acetyltransferase subunit in *Tetrahymena* macronuclei. Proc Natl Acad Sci USA 92, 6364-6368.

Brownell, J. E., Zhou, J., Ranalli, T., Kobayashi, R., Edmondson, D. G., Roth, S. Y., and Allis, C. D. (1996). *Tetrahymena* histone acetyltransferase A: a homolog to yeast Gcn5p linking histone acetylation to gene activation. Cell 84, 843-851.

Brunet, A., Sweeney, L. B., Sturgill, J. F., Chua, K. F., Greer, P. L., Lin, Y., Tran, H., Ross, S. E., Mostoslavsky, R., Cohen, H. Y., et al. (2004). Stress-dependent regulation of FOXO transcription factors by the SIRT1 deacetylase. Science 303, 2011-2015.

Butler, L. M., Zhou, X., Xu, W. S., Scher, H. I., Rifkind, R. A., Marks, P. A., and Richon, V. M. (2002). The histone deacetylase inhibitor SAHA arrests cancer cell growth, up-regulates thioredoxin-binding protein-2, and down-regulates thioredoxin. Proc Natl Acad Sci USA 99, 11700-11705.

Carapeti, M., Aguiar, R. C., Goldman, J. M., and Cross, N. C. (1998). A novel fusion between MOZ and the nuclear receptor coactivator TIF2 in acute myeloid leukemia. Blood 91, 3127-3133.

Champagne, N., Bertos, N. R., Pelletier, N., Wang, A. H., Vezmar, M., Yang, Y., Heng, H. H., and Yang, X. J. (1999). Identification of a human histone acetyltransferase related to monocytic leukemia zinc finger protein. J Biol Chem 274, 28528-28536.

Chang, S., McKinsey, T. A., Zhang, C. L., Richardson, J. A., Hill, J. A., and Olson, E. N. (2004). Histone deacetylases 5 and 9 govern responsiveness of the heart to a subset of stress signals and play redundant roles in heart development. Mol Cell Biol 24, 8467-8476.

Chauchereau, A., Mathieu, M., de Saintignon, J., Ferreira, R., Pritchard, L. L., Mishal, Z., Dejean, A., and Harel-Bellan, A. (2004). HDAC4 mediates transcriptional repression by the acute promyelocytic leukaemia-associated protein PLZF. Oncogene 23, 8777-8784.

Chawla, S., Hardingham, G. E., Quinn, D. R., and Bading, H. (1998). CBP: a signal-regulated transcriptional coactivator controlled by nuclear calcium and CaM kinase IV. Science 281, 1505-1509.

Chen, L., Fischle, W., Verdin, E., and Greene, W. C. (2001). Duration of nuclear NF-kappaB action regulated by reversible acetylation. Science 293, 1653-1657.

Choi, C. H., Hiromura, M., and Usheva, A. (2003). Transcription factor IIB acetylates itself to regulate transcription. Nature 424, 965-969.

Chopin, V., Toillon, R. A., Jouy, N., and Le Bourhis, X. (2002). Sodium butyrate induces P53-independent, Fas-mediated apoptosis in MCF-7 human breast cancer cells. Br J Pharmacol 135, 79-86.

Chrivia, J. C., Kwok, R. P., Lamb, N., Hagiwara, M., Montminy, M. R., and Goodman, R. H. (1993). Phosphorylated CREB binds specifically to the nuclear protein CBP. Nature 365, 855-859.

Clarke, A. S., Lowell, J. E., Jacobson, S. J., and Pillus, L. (1999). Esa1p is an essential histone acetyltransferase required for cell cycle progression. Mol Cell Biol 19, 2515-2526.

Cohen, H. Y., Miller, C., Bitterman, K. J., Wall, N. R., Hekking, B., Kessler, B., Howitz, K. T., Gorospe, M., de Cabo, R., and Sinclair, D. A. (2004). Calorie restriction promotes mammalian cell survival by inducing the SIRT1 deacetylase. Science 305, 390-392.

Cohen, L. A., Marks, P. A., Rifkind, R. A., Amin, S., Desai, D., Pittman, B., and Richon, V. M. (2002). Suberoylanilide hydroxamic acid (SAHA), a histone deacetylase inhibitor, suppresses the growth of carcinogen-induced mammary tumors. Anticancer Res 22, 1497-1504.

Cohen, T., and Yao, T. P. (2004). AcK-knowledge reversible acetylation. Sci STKE 2004, pe42.

Creaven, M., Hans, F., Mutskov, V., Col, E., Caron, C., Dimitrov, S., and Khochbin, S. (1999). Control of the histone-acetyltransferase activity of Tip60 by the HIV-1 transactivator protein, Tat. Biochemistry 38, 8826-8830.

Czubryt, M. P., McAnally, J., Fishman, G. I., and Olson, E. N. (2003). Regulation of peroxisome proliferator-activated receptor gamma coactivator 1 alpha (PGC-1 alpha) and mitochondrial function by MEF2 and HDAC5. Proc Natl Acad Sci USA 100, 1711-1716.

Davis, F. J., Gupta, M., Camoretti-Mercado, B., Schwartz, R. J., and Gupta, M. P. (2003). Calcium/calmodulin-dependent protein kinase activates serum response factor transcription activity by its dissociation from histone deacetylase, HDAC4. Implications in cardiac muscle gene regulation during hypertrophy. J Biol Chem 278, 20047-20058.

De Nadal, E., Zapater, M., Alepuz, P. M., Sumoy, L., Mas, G., and Posas, F. (2004). The MAPK Hog1 recruits Rpd3 histone deacetylase to activate osmoresponsive genes. Nature 427, 370-374.

Deguchi, K., Ayton, P. M., Carapeti, M., Kutok, J. L., Snyder, C. S., Williams, I. R., Cross, N. C., Glass, C. K., Cleary, M. L., and Gilliland, D. G. (2003). MOZ-TIF2-induced acute myeloid leukemia requires the MOZ nucleosome binding motif and TIF2-mediated recruitment of CBP. Cancer Cell 3, 259-271.

Deleu, L., Shellard, S., Alevizopoulos, K., Amati, B., and Land, H. (2001). Recruitment of TRRAP required for oncogenic transformation by E1A. Oncogene 20, 8270-8275.

Dequiedt, F., Kasler, H., Fischle, W., Kiermer, V., Weinstein, M., Herndier, B. G., and Verdin, E. (2003). HDAC7, a thymus-specific class II histone deacetylase, regulates Nur77 transcription and TCR-mediated apoptosis. Immunity 18, 687-698.

Dhalluin, C., Carlson, J. E., Zeng, L., He, C., Aggarwal, A. K., and Zhou, M. M. (1999). Structure and ligand of a histone acetyltransferase bromodomain. Nature 399, 491-496.

Dryden, S. C., Nahhas, F. A., Nowak, J. E., Goustin, A. S., and Tainsky, M. A. (2003). Role for human SIRT2 NAD-dependent deacetylase activity in control of mitotic exit in the cell cycle.

Mol Cell Biol *23*, 3173-3185.

Ehrenhofer-Murray, A. E., Rivier, D. H., and Rine, J. (1997). The role of Sas2, an acetyltransferase homologue of *Saccharomyces cerevisiae*, in silencing and ORC function. Genetics *145*, 923-934.

Ferrante, R. J., Kubilus, J. K., Lee, J., Ryu, H., Beesen, A., Zucker, B., Smith, K., Kowall, N. W., Ratan, R. R., Luthi-Carter, R., and Hersch, S. M. (2003). Histone deacetylase inhibition by sodium butyrate chemotherapy ameliorates the neurodegenerative phenotype in Huntington's disease mice. J Neurosci *23*, 9418-9427.

Finnin, M. S., Donigian, J. R., Cohen, A., Richon, V. M., Rifkind, R. A., Marks, P. A., Breslow, R., and Pavletich, N. P. (1999). Structures of a histone deacetylase homologue bound to the TSA and SAHA inhibitors. Nature *401*, 188-193.

Fischle, W., Dequiedt, F., Hendzel, M. J., Guenther, M. G., Lazar, M. A., Voelter, W., and Verdin, E. (2002). Enzymatic activity associated with class II HDACs is dependent on a multiprotein complex containing HDAC3 and SMRT/N-CoR. Mol Cell *9*, 45-57.

Frank, S. R., Parisi, T., Taubert, S., Fernandez, P., Fuchs, M., Chan, H. M., Livingston, D. M., and Amati, B. (2003). MYC recruits the TIP60 histone acetyltransferase complex to chromatin. EMBO Rep *4*, 575-580.

Fulco, M., Schiltz, R. L., Iezzi, S., King, M. T., Zhao, P., Kashiwaya, Y., Hoffman, E., Veech, R. L., and Sartorelli, V. (2003). Sir2 regulates skeletal muscle differentiation as a potential sensor of the redox state. Mol Cell *12*, 51-62.

Furumai, R., Komatsu, Y., Nishino, N., Khochbin, S., Yoshida, M., and Horinouchi, S. (2001). Potent histone deacetylase inhibitors built from trichostatin A and cyclic tetrapeptide antibiotics including trapoxin. Proc Natl Acad Sci USA *98*, 87-92.

Gao, L., Cueto, M. A., Asselbergs, F., and Atadja, P. (2002). Cloning and functional characterization of HDAC11, a novel member of the human histone deacetylase family. J Biol Chem *277*, 25748-25755.

Gayther, S. A., Batley, S. J., Linger, L., Bannister, A., Thorpe, K., Chin, S. F., Daigo, Y., Russell, P., Wilson, A., Sowter, H. M., et al. (2000). Mutations truncating the EP300 acetylase in human cancers. Nat Genet *24*, 300-303.

Giannakou, M. E., and Partridge, L. (2004). The interaction between FOXO and SIRT1: tipping the balance towards survival. Trends Cell Biol *14*, 408-412.

Girdwood, D., Bumpass, D., Vaughan, O. A., Thain, A., Anderson, L. A., Snowden, A. W., Garcia-Wilson, E., Perkins, N. D., and Hay, R. T. (2003). P300 transcriptional repression is mediated by SUMO modification. Mol Cell *11*, 1043-1054.

Goodman, R. H., and Smolik, S. (2000). CBP/p300 in cell growth, transformation, and development. Genes Dev *14*, 1553-1577.

Grant, P. A., Duggan, L., Cote, J., Roberts, S. M., Brownell, J. E., Candau, R., Ohba, R., Owen-Hughes, T., Allis, C. D., Winston, F., et al. (1997). Yeast Gcn5 functions in two multisubunit complexes to acetylate nucleosomal histones: characterization of an Ada complex and the SAGA (Spt/Ada) complex. Genes Dev *11*, 1640-1650.

Grant, P. A., Schieltz, D., Pray-Grant, M. G., Yates, J. R., 3rd, and Workman, J. L. (1998). The ATM-related cofactor Tra1 is a component of the purified SAGA complex. Mol Cell *2*, 863-867.

Gregoire, S., and Yang, X. J. (2005). Association with class IIa histone deacetylases upregulates the sumoylation of MEF2 transcription factors. Mol Cell Biol *25*, 2273-2287.

Grozinger, C. M., Hassig, C. A., and Schreiber, S. L. (1999). Three proteins define a class of human histone deacetylases related to yeast Hda1p. Proc Natl Acad Sci USA *96*, 4868-4873.

Grozinger, C. M., and Schreiber, S. L. (2000). Regulation of histone deacetylase 4 and 5 and transcriptional activity by 14-3-3-dependent cellular localization. Proc Natl Acad Sci USA *97*, 7835-7840.

Gu, W., and Roeder, R. G. (1997). Activation of p53 sequence-specific DNA binding by acetylation of the p53 C-terminal domain. Cell *90*, 595-606.

Guardiola, A. R., and Yao, T. P. (2002). Molecular cloning and characterization of a novel histone deacetylase HDAC10. J Biol Chem *277*, 3350-3356.

Guarente, L., and Picard, F. (2005). Calorie restriction--the SIR2 connection. Cell *120*, 473-482.

Haggarty, S. J., Koeller, K. M., Wong, J. C., Grozinger, C. M., and Schreiber, S. L. (2003). Domain-selective small-molecule inhibitor of histone deacetylase 6 (HDAC6)-mediated tubulin deacetylation. Proc Natl Acad Sci USA *100*, 4389-4394.

Harlow, E., Whyte, P., Franza, B. R., Jr., and Schley, C. (1986). Association of adenovirus early-region 1A proteins with cellular polypeptides. Mol Cell Biol *6*, 1579-1589.

Heinzel, T., Lavinsky, R. M., Mullen, T. M., Soderstrom, M., Laherty, C. D., Torchia, J., Yang, W. M., Brard, G., Ngo, S. D., Davie, J. R., et al. (1997). A complex containing N-CoR, mSin3 and histone deacetylase mediates transcriptional repression. Nature *387*, 43-48.

Herold, C., Ganslmayer, M., Ocker, M., Hermann, M., Geerts, A., Hahn, E. G., and Schuppan, D. (2002). The histone-deacetylase inhibitor Trichostatin A blocks proliferation and triggers apoptotic programs in hepatoma cells. J Hepatol *36*, 233-240.

Hilfiker, A., Hilfiker-Kleiner, D., Pannuti, A., and Lucchesi, J. C. (1997). mof, a putative acetyl transferase gene related to the Tip60 and MOZ human genes and to the SAS genes of yeast, is required for dosage compensation in *Drosophila*. Embo J *16*, 2054-2060.

Hockly, E., Richon, V. M., Woodman, B., Smith, D. L., Zhou, X., Rosa, E., Sathasivam, K., Ghazi-Noori, S., Mahal, A., Lowden, P. A., et al. (2003). Suberoylanilide hydroxamic acid, a histone deacetylase inhibitor, ameliorates motor deficits in a mouse model of Huntington's disease. Proc Natl Acad Sci USA *100*, 2041-2046.

Horlein, A. J., Naar, A. M., Heinzel, T., Torchia, J., Gloss, B., Kurokawa, R., Ryan, A., Kamei, Y., Soderstrom, M., Glass, C. K., and et al. (1995). Ligand-independent repression by the thyroid hormone receptor mediated by a nuclear receptor co-repressor. Nature 377, 397-404.

Horn, P. J., and Peterson, C. L. (2002). Molecular biology. Chromatin higher order folding--wrapping up transcription. Science 297, 1824-1827.

Howe, L., Auston, D., Grant, P., John, S., Cook, R. G., Workman, J. L., and Pillus, L. (2001). Histone H3 specific acetyltransferases are essential for cell cycle progression. Genes Dev 15, 3144-3154.

Howitz, K. T., Bitterman, K. J., Cohen, H. Y., Lamming, D. W., Lavu, S., Wood, J. G., Zipkin, R. E., Chung, P., Kisielewski, A., Zhang, L. L., et al. (2003). Small molecule activators of sirtuins extend Saccharomyces cerevisiae lifespan. Nature 425, 191-196.

Hu, S. C., Chrivia, J., and Ghosh, A. (1999). Regulation of CBP-mediated transcription by neuronal calcium signaling. Neuron 22, 799-808.

Hubbert, C., Guardiola, A., Shao, R., Kawaguchi, Y., Ito, A., Nixon, A., Yoshida, M., Wang, X. F., and Yao, T. P. (2002). HDAC6 is a microtubule-associated deacetylase. Nature 417, 455-458.

Hughes, R. E. (2002). Polyglutamine disease: acetyltransferases awry. Curr Biol 12, R141-143.

Humphrey, G. W., Wang, Y., Russanova, V. R., Hirai, T., Qin, J., Nakatani, Y., and Howard, B. H. (2001). Stable histone deacetylase complexes distinguished by the presence of SANT domain proteins CoREST/kiaa0071 and Mta-L1. J Biol Chem 276, 6817-6824.

Ikura, T., Ogryzko, V. V., Grigoriev, M., Groisman, R., Wang, J., Horikoshi, M., Scully, R., Qin, J., and Nakatani, Y. (2000). Involvement of the TIP60 histone acetylase complex in DNA repair and apoptosis. Cell 102, 463-473.

Imai, S., Armstrong, C. M., Kaeberlein, M., and Guarente, L. (2000). Transcriptional silencing and longevity protein Sir2 is an NAD-dependent histone deacetylase. Nature 403, 795-800.

Imhof, A., Yang, X. J., Ogryzko, V. V., Nakatani, Y., Wolffe, A. P., and Ge, H. (1997). Acetylation of general transcription factors by histone acetyltransferases. Curr Biol 7, 689-692.

Itazaki, H., Nagashima, K., Sugita, K., Yoshida, H., Kawamura, Y., Yasuda, Y., Matsumoto, K., Ishii, K., Uotani, N., Nakai, H., and et al. (1990). Isolation and structural elucidation of new cyclotetrapeptides, trapoxins A and B, having detransformation activities as antitumor agents. J Antibiot (Tokyo) 43, 1524-1532.

Ito, A., Kawaguchi, Y., Lai, C. H., Kovacs, J. J., Higashimoto, Y., Appella, E., and Yao, T. P. (2002). MDM2-HDAC1-mediated deacetylation of p53 is required for its degradation. Embo J 21, 6236-6245.

Ivanov, D., Schleiffer, A., Eisenhaber, F., Mechtler, K., Haering, C. H., and Nasmyth, K. (2002). Eco1 is a novel acetyltransferase that can acetylate proteins involved in cohesion. Curr Biol 12, 323-328.

Iwao, K., Kawasaki, H., Taira, K., and Yokoyama, K. K. (1999). Ubiquitination of the transcriptional coactivator p300 during retinic acid induced differentiation. Nucleic Acids Symp Ser, 207-208.

Jaboin, J., Wild, J., Hamidi, H., Khanna, C., Kim, C. J., Robey, R., Bates, S. E., and Thiele, C. J. (2002). MS-27-275, an inhibitor of histone deacetylase, has marked in vitro and in vivo antitumor activity against pediatric solid tumors. Cancer Res 62, 6108-6115.

Jenuwein, T., and Allis, C. D. (2001). Translating the histone code. Science 293, 1074-1080.

Jiang, H., Nucifora, F. C., Jr., Ross, C. A., and DeFranco, D. B. (2003). Cell death triggered by polyglutamine-expanded huntingtin in a neuronal cell line is associated with degradation of CREB-binding protein. Hum Mol Genet 12, 1-12.

John, S., Howe, L., Tafrov, S. T., Grant, P. A., Sternglanz, R., and Workman, J. L. (2000). The something about silencing protein, Sas3, is the catalytic subunit of NuA3, a yTAF(II)30-containing HAT complex that interacts with the Spt16 subunit of the yeast CP (Cdc68/Pob3)-FACT complex. Genes Dev 14, 1196-1208.

Kadosh, D., and Struhl, K. (1998). Targeted recruitment of the Sin3-Rpd3 histone deacetylase complex generates a highly localized domain of repressed chromatin in vivo. Mol Cell Biol 18, 5121-5127.

Kaeberlein, M., McVey, M., and Guarente, L. (1999). The SIR2/3/4 complex and SIR2 alone promote longevity in Saccharomyces cerevisiae by two different mechanisms. Genes Dev 13, 2570-2580.

Kalkhoven, E. (2004). CBP and p300: HATs for different occasions. Biochem Pharmacol 68, 1145-1155.

Kamine, J., Elangovan, B., Subramanian, T., Coleman, D., and Chinnadurai, G. (1996). Identification of a cellular protein that specifically interacts with the essential cysteine region of the HIV-1 Tat transactivator. Virology 216, 357-366.

Kang-Decker, N., Tong, C., Boussouar, F., Baker, D. J., Xu, W., Leontovich, A. A., Taylor, W. R., Brindle, P. K., and van Deursen, J. M. (2004). Loss of CBP causes T cell lymphomagenesis in synergy with p27Kip1 insufficiency. Cancer Cell 5, 177-189.

Kao, G. D., McKenna, W. G., Guenther, M. G., Muschel, R. J., Lazar, M. A., and Yen, T. J. (2003). Histone deacetylase 4 interacts with 53BP1 to mediate the DNA damage response. J Cell Biol 160, 1017-1027.

Katan-Khaykovich, Y., and Struhl, K. (2002). Dynamics of global histone acetylation and deacetylation in vivo: rapid restoration of normal histone acetylation status upon removal of activators and repressors. Genes Dev 16, 743-752.

Kawaguchi, Y., Kovacs, J. J., McLaurin, A., Vance, J. M., Ito, A., and Yao, T. P. (2003). The deacetylase HDAC6 regulates aggresome formation and cell viability in response to misfolded protein stress. Cell 115, 727-738.

Kawasaki, H., Schiltz, L., Chiu, R., Itakura, K., Taira, K.,

Nakatani, Y., and Yokoyama, K. K. (2000). ATF-2 has intrinsic histone acetyltransferase activity which is modulated by phosphorylation. Nature *405*, 195-200.

Kehle, J., Beuchle, D., Treuheit, S., Christen, B., Kennison, J. A., Bienz, M., and Muller, J. (1998). dMi-2, a hunchback-interacting protein that functions in polycomb repression. Science *282*, 1897-1900.

Kelly, W. K., Richon, V. M., O'Connor, O., Curley, T., MacGregor-Curtelli, B., Tong, W., Klang, M., Schwartz, L., Richardson, S., Rosa, E., et al. (2003). Phase I clinical trial of histone deacetylase inhibitor: suberoylanilide hydroxamic acid administered intravenously. Clin Cancer Res *9*, 3578-3588.

Kikuchi, H., Takami, Y., and Nakayama, T. (2005). GCN5: a supervisor in all-inclusive control of vertebrate cell cycle progression through transcription regulation of various cell cycle-related genes. Gene.

Kim, M. S., Son, M. W., Kim, W. B., In Park, Y., and Moon, A. (2000a). Apicidin, an inhibitor of histone deacetylase, prevents H-ras-induced invasive phenotype. Cancer Lett *157*, 23-30.

Kim, Y. B., Ki, S. W., Yoshida, M., and Horinouchi, S. (2000b). Mechanism of cell cycle arrest caused by histone deacetylase inhibitors in human carcinoma cells. J Antibiot (Tokyo) *53*, 1191-1200.

Kimura, A., Umehara, T., and Horikoshi, M. (2002). Chromosomal gradient of histone acetylation established by Sas2p and Sir2p functions as a shield against gene silencing. Nat Genet *32*, 370-377.

Kleff, S., Andrulis, E. D., Anderson, C. W., and Sternglanz, R. (1995). Identification of a gene encoding a yeast histone H4 acetyltransferase. J Biol Chem *270*, 24674-24677.

Koipally, J., Renold, A., Kim, J., and Georgopoulos, K. (1999). Repression by Ikaros and Aiolos is mediated through histone deacetylase complexes. Embo J *18*, 3090-3100.

Kouzarides, T. (2000). Acetylation: a regulatory modification to rival phosphorylation? Embo J *19*, 1176-1179.

Kundu, T. K., Wang, Z., and Roeder, R. G. (1999). Human TFIIIC relieves chromatin-mediated repression of RNA polymerase III transcription and contains an intrinsic histone acetyltransferase activity. Mol Cell Biol *19*, 1605-1615.

Kung, A. L., Rebel, V. I., Bronson, R. T., Ch'ng, L. E., Sieff, C. A., Livingston, D. M., and Yao, T. P. (2000). Gene dose-dependent control of hematopoiesis and hematologic tumor suppression by CBP. Genes Dev *14*, 272-277.

Kuo, M. H., Zhou, J., Jambeck, P., Churchill, M. E., and Allis, C. D. (1998). Histone acetyltransferase activity of yeast Gcn5p is required for the activation of target genes in vivo. Genes Dev *12*, 627-639.

Kurdistani, S. K., and Grunstein, M. (2003). Histone acetylation and deacetylation in yeast. Nat Rev Mol Cell Biol *4*, 276-284.

Kusch, T., Florens, L., Macdonald, W. H., Swanson, S. K., Glaser, R. L., Yates, J. R., 3rd, Abmayr, S. M., Washburn, M. P., and Workman, J. L. (2004). Acetylation by Tip60 is required for selective histone variant exchange at DNA lesions. Science *306*, 2084-2087.

L'Hernault, S. W., and Rosenbaum, J. L. (1985). Chlamydomonas alpha-tubulin is posttranslationally modified by acetylation on the epsilon-amino group of a lysine. Biochemistry *24*, 473-478.

Lachner, M., O'Carroll, D., Rea, S., Mechtler, K., and Jenuwein, T. (2001). Methylation of histone H3 lysine 9 creates a binding site for HP1 proteins. Nature *410*, 116-120.

Lang, S. E., and Hearing, P. (2003). The adenovirus E1A oncoprotein recruits the cellular TRRAP/GCN5 histone acetyltransferase complex. Oncogene *22*, 2836-2841.

Lang, S. E., McMahon, S. B., Cole, M. D., and Hearing, P. (2001). E2F transcriptional activation requires TRRAP and GCN5 cofactors. J Biol Chem *276*, 32627-32634.

Lavau, C., Du, C., Thirman, M., and Zeleznik-Le, N. (2000). Chromatin-related properties of CBP fused to MLL generate a myelodysplastic-like syndrome that evolves into myeloid leukemia. Embo J *19*, 4655-4664.

Lee, H. J., Chun, M., and Kandror, K. V. (2001). Tip60 and HDAC7 interact with the endothelin receptor a and may be involved in downstream signaling. J Biol Chem *276*, 16597-16600.

Lemercier, C., Brocard, M. P., Puvion-Dutilleul, F., Kao, H. Y., Albagli, O., and Khochbin, S. (2002). Class II histone deacetylases are directly recruited by BCL6 transcriptional repressor. J Biol Chem *277*, 22045-22052.

Linseman, D. A., Bartley, C. M., Le, S. S., Laessig, T. A., Bouchard, R. J., Meintzer, M. K., Li, M., and Heidenreich, K. A. (2003). Inactivation of the myocyte enhancer factor-2 repressor histone deacetylase-5 by endogenous Ca(2+) //calmodulin-dependent kinase II promotes depolarization-mediated cerebellar granule neuron survival. J Biol Chem *278*, 41472-41481.

Liu, F., Dowling, M., Yang, X. J., and Kao, G. D. (2004). Caspase-mediated specific cleavage of human histone deacetylase 4. J Biol Chem *279*, 34537-34546.

Lonard, D. M., Nawaz, Z., Smith, C. L., and O'Malley, B. W. (2000). The 26S proteasome is required for estrogen receptor-alpha and coactivator turnover and for efficient estrogen receptor-alpha transactivation. Mol Cell *5*, 939-948.

Lu, J., McKinsey, T. A., Zhang, C. L., and Olson, E. N. (2000). Regulation of skeletal myogenesis by association of the MEF2 transcription factor with class II histone deacetylases. Mol Cell *6*, 233-244.

Luo, J., Nikolaev, A. Y., Imai, S., Chen, D., Su, F., Shiloh, A., Guarente, L., and Gu, W. (2001). Negative control of p53 by Sir2alpha promotes cell survival under stress. Cell *107*, 137-148.

Maison, C., Bailly, D., Peters, A. H., Quivy, J. P., Roche, D., Taddei, A., Lachner, M., Jenuwein, T., and Almouzni, G. (2002). Higher-order structure in pericentric heterochromatin involves a distinct pattern of histone modification and an RNA component. Nat Genet *30*, 329-334.

Mao, Z., Bonni, A., Xia, F., Nadal-Vicens, M., and Greenberg, M.

E. (1999). Neuronal activity-dependent cell survival mediated by transcription factor MEF2. Science 286, 785-790.

Marks, P. A., Richon, V. M., and Rifkind, R. A. (2000). Histone deacetylase inhibitors: inducers of differentiation or apoptosis of transformed cells. J Natl Cancer Inst 92, 1210-1216.

Marmorstein, R., and Roth, S. Y. (2001). Histone acetyltransferases: function, structure, and catalysis. Curr Opin Genet Dev 11, 155-161.

Martinez, E., Kundu, T. K., Fu, J., and Roeder, R. G. (1998). A human SPT3-TAFII31-GCN5-L acetylase complex distinct from transcription factor IID. J Biol Chem 273, 23781-23785.

Martinez-Balbas, M. A., Bannister, A. J., Martin, K., Haus-Seuffert, P., Meisterernst, M., and Kouzarides, T. (1998). The acetyltransferase activity of CBP stimulates transcription. Embo J 17, 2886-2893.

McCampbell, A., Taylor, J. P., Taye, A. A., Robitschek, J., Li, M., Walcott, J., Merry, D., Chai, Y., Paulson, H., Sobue, G., and Fischbeck, K. H. (2000). CREB-binding protein sequestration by expanded polyglutamine. Hum Mol Genet 9, 2197-2202.

McKinsey, T. A., Zhang, C. L., Lu, J., and Olson, E. N. (2000). Signal-dependent nuclear export of a histone deacetylase regulates muscle differentiation. Nature 408, 106-111.

McKinsey, T. A., Zhang, C. L., and Olson, E. N. (2001). Identification of a signal-responsive nuclear export sequence in class II histone deacetylases. Mol Cell Biol 21, 6312-6321.

McKinsey, T. A., Zhang, C. L., and Olson, E. N. (2002). MEF2: a calcium-dependent regulator of cell division, differentiation and death. Trends Biochem Sci 27, 40-47.

McMahon, S. B., Van Buskirk, H. A., Dugan, K. A., Copeland, T. D., and Cole, M. D. (1998). The novel ATM-related protein TRRAP is an essential cofactor for the c-Myc and E2F oncoproteins. Cell 94, 363-374.

Metzger, J. M. (2002). HDAC lightens a heavy heart. Nat Med 8, 1078-1079.

Minamiyama, M., Katsuno, M., Adachi, H., Waza, M., Sang, C., Kobayashi, Y., Tanaka, F., Doyu, M., Inukai, A., and Sobue, G. (2004). Sodium butyrate ameliorates phenotypic expression in a transgenic mouse model of spinal and bulbar muscular atrophy. Hum Mol Genet 13, 1183-1192.

Miska, E. A., Karlsson, C., Langley, E., Nielsen, S. J., Pines, J., and Kouzarides, T. (1999). HDAC4 deacetylase associates with and represses the MEF2 transcription factor. Embo J 18, 5099-5107.

Mizzen, C. A., Yang, X. J., Kokubo, T., Brownell, J. E., Bannister, A. J., Owen-Hughes, T., Workman, J., Wang, L., Berger, S. L., Kouzarides, T., et al. (1996). The TAF(II)250 subunit of TFIID has histone acetyltransferase activity. Cell 87, 1261-1270.

Munshi, N., Merika, M., Yie, J., Senger, K., Chen, G., and Thanos, D. (1998). Acetylation of HMG I(Y) by CBP turns off IFN beta expression by disrupting the enhanceosome. Mol Cell 2, 457-467.

Munster, P. N., Troso-Sandoval, T., Rosen, N., Rifkind, R., Marks, P. A., and Richon, V. M. (2001). The histone deacetylase inhibitor suberoylanilide hydroxamic acid induces differentiation of human breast cancer cells. Cancer Res 61, 8492-8497.

Muraoka, M., Konishi, M., Kikuchi-Yanoshita, R., Tanaka, K., Shitara, N., Chong, J. M., Iwama, T., and Miyaki, M. (1996). p300 gene alterations in colorectal and gastric carcinomas. Oncogene 12, 1565-1569.

Nakajima, H., Kim, Y. B., Terano, H., Yoshida, M., and Horinouchi, S. (1998). FR901228, a potent antitumor antibiotic, is a novel histone deacetylase inhibitor. Exp Cell Res 241, 126-133.

Naruse, Y., Aoki, T., Kojima, T., and Mori, N. (1999). Neural restrictive silencer factor recruits mSin3 and histone deacetylase complex to repress neuron-specific target genes. Proc Natl Acad Sci USA 96, 13691-13696.

Naryzhny, S. N., and Lee, H. (2004). The post-translational modifications of proliferating cell nuclear antigen: acetylation, not phosphorylation, plays an important role in the regulation of its function. J Biol Chem 279, 20194-20199.

Nemoto, S., Fergusson, M. M., and Finkel, T. (2005). SIRT1 functionally interacts with the metabolic regulator and transcriptional coactivator PGC-1alpha. J Biol Chem.

Neuwald, A. F., and Landsman, D. (1997). GCN5-related histone N-acetyltransferases belong to a diverse superfamily that includes the yeast SPT10 protein. Trends Biochem Sci 22, 154-155.

Ng, H. H., Zhang, Y., Hendrich, B., Johnson, C. A., Turner, B. M., Erdjument-Bromage, H., Tempst, P., Reinberg, D., and Bird, A. (1999). MBD2 is a transcriptional repressor belonging to the MeCP1 histone deacetylase complex. Nat Genet 23, 58-61.

North, B. J., Marshall, B. L., Borra, M. T., Denu, J. M., and Verdin, E. (2003). The human Sir2 ortholog, SIRT2, is an NAD+-dependent tubulin deacetylase. Mol Cell 11, 437-444.

Nucifora, F. C., Jr., Sasaki, M., Peters, M. F., Huang, H., Cooper, J. K., Yamada, M., Takahashi, H., Tsuji, S., Troncoso, J., Dawson, V. L., et al. (2001). Interference by huntingtin and atrophin-1 with cbp-mediated transcription leading to cellular toxicity. Science 291, 2423-2428.

Ogryzko, V. V., Kotani, T., Zhang, X., Schiltz, R. L., Howard, T., Yang, X. J., Howard, B. H., Qin, J., and Nakatani, Y. (1998). Histone-like TAFs within the PCAF histone acetylase complex. Cell 94, 35-44.

Ogryzko, V. V., Schiltz, R. L., Russanova, V., Howard, B. H., and Nakatani, Y. (1996). The transcriptional coactivators p300 and CBP are histone acetyltransferases. Cell 87, 953-959.

Oike, Y., Hata, A., Mamiya, T., Kaname, T., Noda, Y., Suzuki, M., Yasue, H., Nabeshima, T., Araki, K., and Yamamura, K. (1999). Truncated CBP protein leads to classical Rubinstein-Taybi syndrome phenotypes in mice: implications for a dominant-negative mechanism. Hum Mol Genet 8, 387-396.

Onyango, P., Celic, I., McCaffery, J. M., Boeke, J. D., and Feinberg, A. P. (2002). SIRT3, a human SIR2 homologue, is an NAD-dependent deacetylase localized to mitochondria. Proc Natl

Acad Sci USA *99*, 13653-13658.

Paroni, G., Mizzau, M., Henderson, C., Del Sal, G., Schneider, C., and Brancolini, C. (2004). Caspase-dependent regulation of histone deacetylase 4 nuclear-cytoplasmic shuttling promotes apoptosis. Mol Biol Cell *15*, 2804-2818.

Pelletier, N., Champagne, N., Stifani, S., and Yang, X. J. (2002). MOZ and MORF histone acetyltransferases interact with the Runt-domain transcription factor Runx2. Oncogene *21*, 2729-2740.

Pijnappel, W. W., Schaft, D., Roguev, A., Shevchenko, A., Tekotte, H., Wilm, M., Rigaut, G., Seraphin, B., Aasland, R., and Stewart, A. F. (2001). The *S. cerevisiae* SET3 complex includes two histone deacetylases, Hos2 and Hst1, and is a meiotic-specific repressor of the sporulation gene program. Genes Dev *15*, 2991-3004.

Poizat, C., Sartorelli, V., Chung, G., Kloner, R. A., and Kedes, L. (2000). Proteasome-mediated degradation of the coactivator p300 impairs cardiac transcription. Mol Cell Biol *20*, 8643-8654.

Rebel, V. I., Kung, A. L., Tanner, E. A., Yang, H., Bronson, R. T., and Livingston, D. M. (2002). Distinct roles for CREB-binding protein and p300 in hematopoietic stem cell self-renewal. Proc Natl Acad Sci USA *99*, 14789-14794.

Reifsnyder, C., Lowell, J., Clarke, A., and Pillus, L. (1996). Yeast SAS silencing genes and human genes associated with AML and HIV-1 Tat interactions are homologous with acetyltransferases. Nat Genet *14*, 42-49.

Ren, Q., and Gorovsky, M. A. (2001). Histone H2A.Z acetylation modulates an essential charge patch. Mol Cell *7*, 1329-1335.

Richon, V. M., Emiliani, S., Verdin, E., Webb, Y., Breslow, R., Rifkind, R. A., and Marks, P. A. (1998). A class of hybrid polar inducers of transformed cell differentiation inhibits histone deacetylases. Proc Natl Acad Sci USA *95*, 3003-3007.

Rine, J., and Herskowitz, I. (1987). Four genes responsible for a position effect on expression from HML and HMR in *Saccharomyces cerevisiae*. Genetics *116*, 9-22.

Robyr, D., Suka, Y., Xenarios, I., Kurdistani, S. K., Wang, A., Suka, N., and Grunstein, M. (2002). Microarray deacetylation maps determine genome-wide functions for yeast histone deacetylases. Cell *109*, 437-446.

Rodgers, J. T., Lerin, C., Haas, W., Gygi, S. P., Spiegelman, B. M., and Puigserver, P. (2005). Nutrient control of glucose homeostasis through a complex of PGC-1alpha and SIRT1. Nature *434*, 113-118.

Rouaux, C., Jokic, N., Mbebi, C., Boutillier, S., Loeffler, J. P., and Boutillier, A. L. (2003). Critical loss of CBP/p300 histone acetylase activity by caspase-6 during neurodegeneration. Embo J *22*, 6537-6549.

Rouaux, C., Loeffler, J. P., and Boutillier, A. L. (2004). Targeting CREB-binding protein (CBP) loss of function as a therapeutic strategy in neurological disorders. Biochem Pharmacol *68*, 1157-1164.

Samuelson, A. V., Narita, M., Chan, H. M., Jin, J., de Stanchina, E., McCurrach, M. E., Fuchs, M., Livingston, D. M., and Lowe, S. W. (2005). p400 is required for E1A to promote apoptosis. J Biol Chem.

Sawa, H., Murakami, H., Ohshima, Y., Sugino, T., Nakajyo, T., Kisanuki, T., Tamura, Y., Satone, A., Ide, W., Hashimoto, I., and Kamada, H. (2001). Histone deacetylase inhibitors such as sodium butyrate and trichostatin A induce apoptosis through an increase of the bcl-2-related protein Bad. Brain Tumor Pathol *18*, 109-114.

Schwer, B., North, B. J., Frye, R. A., Ott, M., and Verdin, E. (2002). The human silent information regulator (Sir)2 homologue hSIRT3 is a mitochondrial nicotinamide adenine dinucleotide-dependent deacetylase. J Cell Biol *158*, 647-657.

Shalizi, A., Lehtinen, M., Gaudilliere, B., Donovan, N., Han, J., Konishi, Y., and Bonni, A. (2003). Characterization of a neurotrophin signaling mechanism that mediates neuron survival in a temporally specific pattern. J Neurosci *23*, 7326-7336.

Shi, T., Wang, F., Stieren, E., and Tong, Q. (2005). SIRT3, a mitochondrial Sirtuin deacetylase, regulates mitochondrial function and thermogenesis in Brown adipocytes. J Biol Chem.

Sinclair, D. A., and Guarente, L. (1997). Extrachromosomal rDNA circles--a cause of aging in yeast. Cell *91*, 1033-1042.

Smith, E. R., Pannuti, A., Gu, W., Steurnagel, A., Cook, R. G., Allis, C. D., and Lucchesi, J. C. (2000). The *drosophila* MSL complex acetylates histone H4 at lysine 16, a chromatin modification linked to dosage compensation. Mol Cell Biol *20*, 312-318.

Smith, J. (2002). Human Sir2 and the 'silencing' of p53 activity. Trends Cell Biol *12*, 404-406.

Smith, J. S., and Boeke, J. D. (1997). An unusual form of transcriptional silencing in yeast ribosomal DNA. Genes Dev *11*, 241-254.

Spencer, T. E., Jenster, G., Burcin, M. M., Allis, C. D., Zhou, J., Mizzen, C. A., McKenna, N. J., Onate, S. A., Tsai, S. Y., Tsai, M. J., and O'Malley, B. W. (1997). Steroid receptor coactivator-1 is a histone acetyltransferase. Nature *389*, 194-198.

Steffan, J. S., Bodai, L., Pallos, J., Poelman, M., McCampbell, A., Apostol, B. L., Kazantsev, A., Schmidt, E., Zhu, Y. Z., Greenwald, M., *et al.* (2001). Histone deacetylase inhibitors arrest polyglutamine-dependent neurodegeneration in *Drosophila*. Nature *413*, 739-743.

Steffan, J. S., Kazantsev, A., Spasic-Boskovic, O., Greenwald, M., Zhu, Y. Z., Gohler, H., Wanker, E. E., Bates, G. P., Housman, D. E., and Thompson, L. M. (2000). The Huntington's disease protein interacts with p53 and CREB-binding protein and represses transcription. Proc Natl Acad Sci USA *97*, 6763-6768.

Sterner, D. E., and Berger, S. L. (2000). Acetylation of histones and transcription-related factors. Microbiol Mol Biol Rev *64*, 435-459.

Sterner, R., Vidali, G., and Allfrey, V. G. (1981). Studies of acetylation and deacetylation in high mobility group proteins. Identification of the sites of acetylation in high mobility group proteins 14 and 17. J Biol Chem *256*, 8892-8895.

Sterner, R., Vidali, G., Heinrikson, R. L., and Allfrey, V. G. (1978). Postsynthetic modification of high mobility group proteins. Evidence that high mobility group proteins are acetylated. J Biol Chem *253*, 7601-7604.

Suka, N., Luo, K., and Grunstein, M. (2002). Sir2p and Sas2p opposingly regulate acetylation of yeast histone H4 lysine16 and spreading of heterochromatin. Nat Genet *32*, 378-383.

Takahashi, J., Fujigasaki, H., Zander, C., El Hachimi, K. H., Stevanin, G., Durr, A., Lebre, A. S., Yvert, G., Trottier, Y., The, H., et al. (2002). Two populations of neuronal intranuclear inclusions in SCA7 differ in size and promyelocytic leukaemia protein content. Brain *125*, 1534-1543.

Takata, T., and Ishikawa, F. (2003). Human Sir2-related protein SIRT1 associates with the bHLH repressors HES1 and HEY2 and is involved in HES1- and HEY2-mediated transcriptional repression. Biochem Biophys Res Commun *301*, 250-257.

Tanaka, Y., Naruse, I., Maekawa, T., Masuya, H., Shiroishi, T., and Ishii, S. (1997). Abnormal skeletal patterning in embryos lacking a single Cbp allele: a partial similarity with Rubinstein-Taybi syndrome. Proc Natl Acad Sci USA *94*, 10215-10220.

Tanner, K. G., Landry, J., Sternglanz, R., and Denu, J. M. (2000). Silent information regulator 2 family of NAD- dependent histone/protein deacetylases generates a unique product, 1-O-acetyl-ADP-ribose. Proc Natl Acad Sci USA *97*, 14178-14182.

Taunton, J., Hassig, C. A., and Schreiber, S. L. (1996). A mammalian histone deacetylase related to the yeast transcriptional regulator Rpd3p. Science *272*, 408-411.

Taylor, J. P., Taye, A. A., Campbell, C., Kazemi-Esfarjani, P., Fischbeck, K. H., and Min, K. T. (2003). Aberrant histone acetylation, altered transcription, and retinal degeneration in a *Drosophila* model of polyglutamine disease are rescued by CREB-binding protein. Genes Dev *17*, 1463-1468.

Thiagalingam, S., Cheng, K. H., Lee, H. J., Mineva, N., Thiagalingam, A., and Ponte, J. F. (2003). Histone deacetylases: unique players in shaping the epigenetic histone code. Ann N Y Acad Sci *983*, 84-100.

Thomas, T., Voss, A. K., Chowdhury, K., and Gruss, P. (2000). Querkopf, a MYST family histone acetyltransferase, is required for normal cerebral cortex development. Development *127*, 2537-2548.

Tong, J. K., Hassig, C. A., Schnitzler, G. R., Kingston, R. E., and Schreiber, S. L. (1998). Chromatin deacetylation by an ATP-dependent nucleosome remodelling complex. Nature *395*, 917-921.

Tsuji, N., Kobayashi, M., Nagashima, K., Wakisaka, Y., and Koizumi, K. (1976). A new antifungal antibiotic, trichostatin. J Antibiot (Tokyo) *29*, 1-6.

Turner, B. M., Birley, A. J., and Lavender, J. (1992). Histone H4 isoforms acetylated at specific lysine residues define individual chromosomes and chromatin domains in *Drosophila* polytene nuclei. Cell *69*, 375-384.

van der Horst, A., Tertoolen, L. G., de Vries-Smits, L. M., Frye, R. A., Medema, R. H., and Burgering, B. M. (2004). FOXO4 is acetylated upon peroxide stress and deacetylated by the longevity protein hSir2(SIRT1). J Biol Chem *279*, 28873-28879.

Vaziri, H., Dessain, S. K., Ng Eaton, E., Imai, S. I., Frye, R. A., Pandita, T. K., Guarente, L., and Weinberg, R. A. (2001). hSIR2(SIRT1) functions as an NAD-dependent p53 deacetylase. Cell *107*, 149-159.

Vega, R. B., Harrison, B. C., Meadows, E., Roberts, C. R., Papst, P. J., Olson, E. N., and McKinsey, T. A. (2004a). Protein kinases C and D mediate agonist-dependent cardiac hypertrophy through nuclear export of histone deacetylase 5. Mol Cell Biol *24*, 8374-8385.

Vega, R. B., Matsuda, K., Oh, J., Barbosa, A. C., Yang, X., Meadows, E., McAnally, J., Pomajzl, C., Shelton, J. M., Richardson, J. A., et al. (2004b). Histone deacetylase 4 controls chondrocyte hypertrophy during skeletogenesis. Cell *119*, 555-566.

Verdin, E., Dequiedt, F., and Kasler, H. G. (2003). Class II histone deacetylases: versatile regulators. Trends Genet *19*, 286-293.

Vetting, M. W., LP, S. d. C., Yu, M., Hegde, S. S., Magnet, S., Roderick, S. L., and Blanchard, J. S. (2005). Structure and functions of the GNAT superfamily of acetyltransferases. Arch Biochem Biophys *433*, 212-226.

Vogelauer, M., Wu, J., Suka, N., and Grunstein, M. (2000). Global histone acetylation and deacetylation in yeast. Nature *408*, 495-498.

Waltregny, D., De Leval, L., Glenisson, W., Ly Tran, S., North, B. J., Bellahcene, A., Weidle, U., Verdin, E., and Castronovo, V. (2004). Expression of histone deacetylase 8, a class I histone deacetylase, is restricted to cells showing smooth muscle differentiation in normal human tissues. Am J Pathol *165*, 553-564.

Wang, A., Kurdistani, S. K., and Grunstein, M. (2002). Requirement of Hos2 histone deacetylase for gene activity in yeast. Science *298*, 1412-1414.

Wang, A. H., and Yang, X. J. (2001). Histone deacetylase 4 possesses intrinsic nuclear import and export signals. Mol Cell Biol *21*, 5992-6005.

Wang, H., Huang, Z. Q., Xia, L., Feng, Q., Erdjument-Bromage, H., Strahl, B. D., Briggs, S. D., Allis, C. D., Wong, J., Tempst, P., and Zhang, Y. (2001). Methylation of histone H4 at arginine 3 facilitating transcriptional activation by nuclear hormone receptor. Science *293*, 853-857.

Watamoto, K., Towatari, M., Ozawa, Y., Miyata, Y., Okamoto, M., Abe, A., Naoe, T., and Saito, H. (2003). Altered interaction of HDAC5 with GATA-1 during MEL cell differentiation. Oncogene *22*, 9176-9184.

Wu, J., Suka, N., Carlson, M., and Grunstein, M. (2001). TUP1 utilizes histone H3/H2B-specific HDA1 deacetylase to repress gene activity in yeast. Mol Cell 7, 117-126.

Xiao, H., Chung, J., Kao, H. Y., and Yang, Y. C. (2003). Tip60 is a co-repressor for STAT3. J Biol Chem 278, 11197-11204.

Xu, W., Edmondson, D. G., and Roth, S. Y. (1998). Mammalian GCN5 and P/CAF acetyltransferases have homologous amino-terminal domains important for recognition of nucleosomal substrates. Mol Cell Biol 18, 5659-5669.

Xue, Y., Wong, J., Moreno, G. T., Young, M. K., Cote, J., and Wang, W. (1998). NURD, a novel complex with both ATP-dependent chromatin-remodeling and histone deacetylase activities. Mol Cell 2, 851-861.

Yamada, M., Wood, J. D., Shimohata, T., Hayashi, S., Tsuji, S., Ross, C. A., and Takahashi, H. (2001). Widespread occurrence of intranuclear atrophin-1 accumulation in the central nervous system neurons of patients with dentatorubral-pallidoluysian atrophy. Ann Neurol 49, 14-23.

Yang, W. M., Inouye, C., Zeng, Y., Bearss, D., and Seto, E. (1996). Transcriptional repression by YY1 is mediated by interaction with a mammalian homolog of the yeast global regulator RPD3. Proc Natl Acad Sci USA 93, 12845-12850.

Yang, X. J. (2004a). The diverse superfamily of lysine acetyltransferases and their roles in leukemia and other diseases. Nucleic Acids Res 32, 959-976.

Yang, X. J. (2004b). Lysine acetylation and the bromodomain: a new partnership for signaling. Bioessays 26, 1076-1087.

Yang, Y., Hou, H., Haller, E. M., Nicosia, S. V., and Bai, W. (2005). Suppression of FOXO1 activity by FHL2 through SIRT1-mediated deacetylation. Embo J 24, 1021-1032.

Yao, T. P., Oh, S. P., Fuchs, M., Zhou, N. D., Ch'ng, L. E., Newsome, D., Bronson, R. T., Li, E., Livingston, D. M., and Eckner, R. (1998). Gene dosage-dependent embryonic development and proliferation defects in mice lacking the transcriptional integrator p300. Cell 93, 361-372.

Yee, S. P., and Branton, P. E. (1985). Detection of cellular proteins associated with human adenovirus type 5 early region 1A polypeptides. Virology 147, 142-153.

Yeung, F., Hoberg, J. E., Ramsey, C. S., Keller, M. D., Jones, D. R., Frye, R. A., and Mayo, M. W. (2004). Modulation of NF-kappaB-dependent transcription and cell survival by the SIRT1 deacetylase. Embo J 23, 2369-2380.

Yoon, H. G., Chan, D. W., Huang, Z. Q., Li, J., Fondell, J. D., Qin, J., and Wong, J. (2003). Purification and functional characterization of the human N-CoR complex: the roles of HDAC3, TBL1 and TBLR1. Embo J 22, 1336-1346.

Yoshida, M., Furumai, R., Nishiyama, M., Komatsu, Y., Nishino, N., and Horinouchi, S. (2001). Histone deacetylase as a new target for cancer chemotherapy. Cancer Chemother Pharmacol 48 Suppl 1, S20-26.

Youn, H. D., Sun, L., Prywes, R., and Liu, J. O. (1999). Apoptosis of T cells mediated by Ca2+-induced release of the transcription factor MEF2. Science 286, 790-793.

Yu, X., Guo, Z. S., Marcu, M. G., Neckers, L., Nguyen, D. M., Chen, G. A., and Schrump, D. S. (2002). Modulation of p53, ErbB1, ErbB2, and Raf-1 expression in lung cancer cells by depsipeptide FR901228. J Natl Cancer Inst 94, 504-513.

Yuan, L. W., Soh, J. W., and Weinstein, I. B. (2002). Inhibition of histone acetyltransferase function of p300 by PKCdelta. Biochim Biophys Acta 1592, 205-211.

Zhang, C. L., McKinsey, T. A., Chang, S., Antos, C. L., Hill, J. A., and Olson, E. N. (2002a). Class II histone deacetylases act as signal-responsive repressors of cardiac hypertrophy. Cell 110, 479-488.

Zhang, C. L., McKinsey, T. A., Lu, J. R., and Olson, E. N. (2001). Association of COOH-terminal-binding protein (CtBP) and MEF2-interacting transcription repressor (MITR) contributes to transcriptional repression of the MEF2 transcription factor. J Biol Chem 276, 35-39.

Zhang, C. L., McKinsey, T. A., and Olson, E. N. (2002b). Association of class II histone deacetylases with heterochromatin protein 1: potential role for histone methylation in control of muscle differentiation. Mol Cell Biol 22, 7302-7312.

Zhang, H. S., Gavin, M., Dahiya, A., Postigo, A. A., Ma, D., Luo, R. X., Harbour, J. W., and Dean, D. C. (2000). Exit from G1 and S phase of the cell cycle is regulated by repressor complexes containing HDAC-Rb-hSWI/SNF and Rb-hSWI/SNF. Cell 101, 79-89.

Zhang, W., Bone, J. R., Edmondson, D. G., Turner, B. M., and Roth, S. Y. (1998a). Essential and redundant functions of histone acetylation revealed by mutation of target lysines and loss of the Gcn5p acetyltransferase. Embo J 17, 3155-3167.

Zhang, Y., LeRoy, G., Seelig, H. P., Lane, W. S., and Reinberg, D. (1998b). The dermatomyositis-specific autoantigen Mi2 is a component of a complex containing histone deacetylase and nucleosome remodeling activities. Cell 95, 279-289.

Zhang, Y., Ng, H. H., Erdjument-Bromage, H., Tempst, P., Bird, A., and Reinberg, D. (1999). Analysis of the NuRD subunits reveals a histone deacetylase core complex and a connection with DNA methylation. Genes Dev 13, 1924-1935.

Zhao, X., Ito, A., Kane, C. D., Liao, T. S., Bolger, T. A., Lemrow, S. M., Means, A. R., and Yao, T. P. (2001). The modular nature of histone deacetylase HDAC4 confers phosphorylation-dependent intracellular trafficking. J Biol Chem 276, 35042-35048.

Zhao, X., Sternsdorf, T., Bolger, T. A., Evans, R. M., Yao, T. P. (2005). Regulation of MEF2 by histone deacetylase 4- and SIRT1 deacetylase-mediated lysine modifications. Mol Cell Biol. 25(19), 8456-64.

Zhou, X., Richon, V. M., Rifkind, R. A., and Marks, P. A. (2000). Identification of a transcriptional repressor related to the noncatalytic domain of histone deacetylases 4 and 5. Proc Natl Acad Sci USA 97, 1056-1061.

Chapter 07

Structure and Function of Core Promoter Elements in RNA Polymerase II Transcription

Edith H. Wang

Department of Pharmacology, University of Washington, 1959 NE Pacific Street, Health Sciences Bldg, Box 357280, Seattle, WA 98195

Key Words: core promoter, basal transcription, TATA box, initiator, downstream promoter elements, TFIID, enhancer selectivity

Abstract

The initiation of messenger RNA synthesis by RNA polymerase II is regulated by a collection of cis-acting DNA elements. The core promoter, which is typically found ~40 bp upstream and downstream of the transcription start site, plays a crucial role in the recruitment and positioning of the basic transcription machinery. DNA elements within the core promoter can also control the levels of gene transcription and specify enhancer selectivity. These regulatory functions are consistent with the large degree of diversity that has become apparent within core promoters. In this article, we will discuss our current understanding of the structure and function of core promoter elements in the regulation of transcription by RNA polymerase II.

Introduction

The synthesis of messenger RNA (mRNA) from eukaryotic protein encoding genes, a process known as transcription, is essential for all biological processes including cell proliferation, differentiation, and survival. Transcription is carried out by RNA polymerase II and a set of general transcription factors (for review see Orphanides *et al.*, 1996). The ability of these proteins to support mRNA synthesis is subject to regulation by cis-acting DNA elements. These DNA elements can be found both upstream and downstream of the transcription start site of a gene. They serve as recognition sites for proteins that facilitate or hinder the recruitment, binding and/or assembly of the transcription apparatus. There are distinct types of transcription control modules for RNA polymerase II transcribed genes, including enhancers, silencers, proximal promoter elements, the core promoter, and boundary/insulator elements. A schematic diagram of the transcription control modules of a typical eukaryotic protein encoding gene is provided in Fig.7.1.

Enhancers and silencers are distal control elements that can be located many thousands of base pairs away from the transcription start site (designated +1) and activate or repress transcription respectively (Bondarenko *et al.*, 2003; Ogbourne and Antalis, 1998). A unique feature of enhancer elements is their ability to function in a position and orientation independent manner. Boundary/insulator sequences limit the domain within the genome at which enhancers and silencers exert their effect (Kuhn and Geyer, 2003). The core promoter, typically restricted to a region centered on the transcription start site, is where the transcription apparatus is assembled and the site of RNA polymerase II action. DNA sequences located near the immediate vicinity of the core promoter (from approximately -250 to +250) are often referred to as proximal promoter elements. Proximal promoter elements sometimes are considered part of a gene's promoter and include binding sites for trans-acting factors that affect transcription levels. The structure and function of cis-acting DNA elements within the core promoter will be the focus of

---

Corresponding Author: Tel: (206) 616-5376, E-mail: ehwang@u.washington.edu

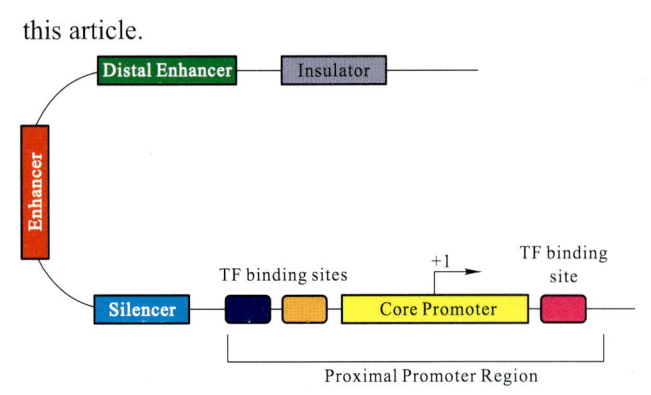

Fig.7.1 **Transcription control modules of a eukaryotic protein encoding-gene**. Cis-acting DNA elements for an RNA polymerase II transcribed gene are shown. The core promoter is recognized by the general transcriptional machinery, which includes RNA polymerase II. The start site of transcription maps to the core promoter and is indicated by the arrow. Enhancer and silencer elements, shown here upstream of the core promoter, contain binding sites for gene-specific regulatory factors that stimulate or repress mRNA synthesis. Insulators limit the distance over which enhancer and silencer sequences exert their stimulatory and inhibitory effects. The proximal promoter region encompasses the core promoter and proximal promoter elements that contain binding sites for other transcription factors (TF binding sites) that affect core promoter activity. The arrangement of these regulatory elements is commonly dispersed over a large region of genomic DNA and can extend both upstream and downstream of the core promoter.

Core Promoter

The core promoter was originally defined as the minimal length of contiguous DNA sufficient to support the accurate initiation of transcription in the absence of other cis-acting regulatory elements. The low level of activity detected *in vitro* from core promoters was termed basal transcription. Sequence analysis of eukaryotic core promoters has led to the identification of a number of prevalent DNA sequence motifs (Fig. 7.2). These core promoter elements are the TATA box, initiator (Inr), downstream promoter element (DPE), TFIIB recognition element (BRE), and motif ten element (MTE), each of which can be found in some but not all eukaryotic core promoter regions. The absence of a universal core promoter element has made it difficult to effectively map the transcriptional control modules for RNA polymerase II transcribed genes. The arrival of the genomic era will greatly aid the development of computational algorithms for predicting the core promoter region of a given gene.

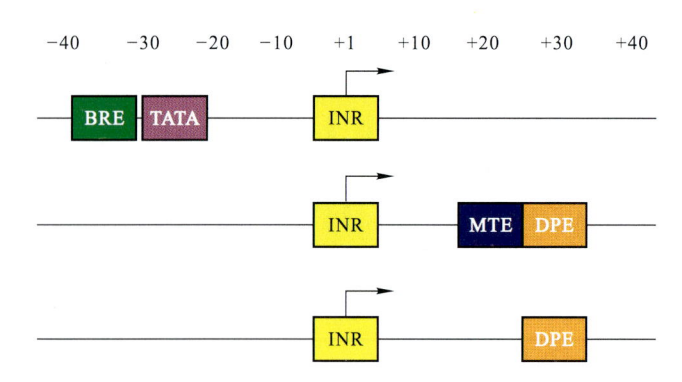

Fig.7.2 **RNA polymerase II core promoter elements.** DNA sequences prevalent within the core promoters of RNA polymerase II transcribed genes have been identified. These sequence motifs include the TATA box, initiator (INR), downstream promoter element (DPE), and motif ten element (MTE), all of which serve as binding sites for subunits of the TFIID complex. The BRE, located immediately upstream of the TATA box in a subset of TATA- containing promoters, is a binding site for the general transcription factor TFIIB. Any given promoter will contain one or more of these core promoter motifs with the preferred combinations illustrated. The diversity of core promoter regions provides an additional level of complexity to transcriptional regulation in higher eukaryotes.

TAtA Box Element

The first eukaryotic core promoter element identified in protein encoding genes was the TATA box (Goldberg, 1979). The A-T rich sequence was discovered by comparing the 5′ flanking regions of a number of eukaryotic and viral genes. In these early studies, the sequence TATAAA, now considered the canonical TATA sequence, was present in nearly all the RNA polymerase II transcribed genes that were examined. Other studies later revealed that a wide variety of A-T rich sequences could function as a TATA element (Singer et al., 1990). These results lead to a modified TATA consensus sequence, 5′-TATA(A/T)AA(G/A)-3′.

In metazoans, the TATA box is typically found around 25 to 30 nucleotides upstream of the transcription start site. However, the position of TATA elements in yeast promoters is more variable, ranging from nucleotide −40 to −100 relative to the start site of transcription (Struhl, 1989). Mutation or removal of the TATA box has been shown to reduce or abolish the transcriptional activity of eukaryotic promoters in transient transfection experiments and in *in vitro* transcription assays (Grosveld et al., 1982; Grosveld et al., 1981; Hu and Manley, 1981; Wasylyk et al., 1980). These findings suggested that the TATA box is essential for transcription and would be present in all RNA polymerase II transcribed genes.

With the sequencing of eukaryotic genomes and the identification of more and more core promoters, it has become clear that the prevalence of the TATA box is much lower than originally predicted. Computational analysis of 1941 *Drosophila* genes indicated that, between nucleotides –45 and –15, a match to the TATA consensus motif was found in 28.3% of the genes examined (Ohler *et al.*, 2002). Expanding the region of analysis to nucleotides –60 to –15 only increased the percentage of TATA containing promoters to 33.9%. A similar database analysis of 1031 potential human core promoters revealed that a TATA box was present in only 32% of the genes (Suzuki *et al.*, 2001). Therefore computational algorithms that use the presence of a TATA sequence as one of their criteria for promoter prediction will be ineffective for the majority of eukaryotic protein encoding genes.

A: TATA Box Binding

The TATA box is a binding site for the TATA binding protein (TBP), a subunit of the transcription factor IID (TFIID) complex (Hahn *et al.*, 1989; Hoey *et al.*, 1990; Peterson *et al.*, 1990). TFIID was biochemically defined by the Roeder lab as an activity present in human HeLa cell nuclear extracts that is essential for transcription from TATA-containing promoters (Matsui T, 1980). DNase I footprinting experiments demonstrated that TFIID directly contacted the TATA box within the adenovirus major late promoter (Parker and Topol, 1984; Sawadogo and Roeder, 1985). Unexpectedly, it was very difficult to purify and clone TFIID from human or *Drosophila* cells. The activity appeared to reside in a large protein complex (Nakajima *et al.*, 1988). Cloning of TBP was facilitated by the discovery that in *Saccharomyces cerevisiae* TATA binding activity resided in a single polypeptide (Buratowski *et al.*, 1988; Hahn *et al.*, 1989). The yeast TBP sequence was subsequently used to isolate homologues from other eukaryotes (Hoey *et al.*, 1990; Peterson *et al.*, 1990). Antibodies against TBP revealed that in human and *Drosophila* cells TBP is associated with other proteins ranging in size from ~20 kD to 250 kD, subunits of the TFIID complex (Dynlacht *et al.*, 1991; Tanese *et al.*, 1991). A detailed discussion of TBP and the TBP associated factors (TAFs) is provided in a subsequent article.

The binding of TFIID to promoter DNA is required to nucleate the assembly of the basic transcription machinery. It is now widely accepted that the majority of RNA polymerase II promoters lack a binding site for TBP. Therefore the recruitment of TFIID to TATA-less promoters cannot be driven by TBP-TATA box interactions and must involve other DNA elements within the core promoter, allowing for greater diversity in the mechanisms regulating transcription initiation.

Intiator Element (Inr)

The initiator is a core promoter element that encompasses the start site of mRNA synthesis (Smale and Baltimore, 1989). A comparison of efficiently transcribed protein encoding genes revealed a conserved adenosine at the +1 position and a conserved cytosine at nucleotide –1, surrounded by pyrimidines (Corden *et al.*, 1980). Removal of the putative initiator element (Inr) within a number of eukaryotic TATA-containing promoters reduced transcription levels and increased the heterogeneity of the position of transcription initiation (Concino *et al.*, 1984). These findings suggested that sequences within the vicinity of the transcription start site can direct the location and level of transcription mediated by RNA polymerase II from the core promoter.

The importance of sequences around the transcription start site was further strengthened by a comprehensive analysis of the murine terminal deoxynucleotidyltransferase (TdT) promoter, a TATA-less promoter that initiates transcription from a single well-characterized start site. Mutations within the TdT core promoter demonstrated that the sequence between –3 and +5 functioned as a distinct DNA element that was necessary and sufficient for accurate transcription in the absence of a TATA sequence (Javahery *et al.*, 1994; Smale and Baltimore, 1989). This region of the TdT promoter matched the conserved motif proposed in earlier studies as an Inr element (Corden *et al.*, 1980).

A functional consensus sequence for an Inr was subsequently defined as 5′-(C/T)(C/T)A$_{+1}$N(T/A)(C/T)(C/T)-3′ by examining the ability of initiator mutants and randomly generated sequences to promote transcription in mammalian cells (Javahery *et al.*, 1994; Lo and Smale, 1996). Insertion of the Inr consensus into a synthetic promoter that lacked a TATA box but contained upstream binding sites for the transcription factor Sp1 was sufficient to support high levels of transcription that initiated from a specific site in the Inr sequence (Smale *et al.*, 1990). Although not all flanking pyrimidine nucleotides (-2, +4 and +5 positions) were essential for Inr function, more activity was associated with motifs that contained more pyrimidines in the flanking positions (Javahery *et al.*, 1994; Lo and Smale, 1996).

In *Drosophila*, a bioinformatics approach produced

the Inr consensus 5'-TCA$_{+1}$(G/T)T(C/T)-3', a sequence that is very similar but not identical to the functional Inr consensus determined for mammalian cells (Arkhipova, 1995; Hultmark et al., 1986). The A$_{+1}$ position in the Inr is most commonly the site of transcription initiation. However, under conditions transcription does not begin at the A$_{+1}$ nucleotide, the Inr element retained the ability to stimulate the efficiency of transcriptional initiation from the alternative start sites (O'Shea-Greenfield and Smale, 1992). Therefore, the Inr carries out two distinct and separable functions in RNA polymerase II dependent gene transcription.

A: Initiator-binding Proteins
A1: TAF Subunits of TFIID

In DNase I footprinting experiments, TFIID protected sequences downstream of the TATA box including the Inr (Martinez et al., 1994; Parker and Topol, 1984; Purnell and Gilmour, 1993; Sawadogo and Roeder, 1985; Zhou et al., 1992). UV crosslinking and electrophoretic mobility shift assays with recombinant TAF subunits of TFIID revealed that TAF1 and TAF2 directly contacted the Inr (Verrijzer et al., 1995; Verrijzer et al., 1994). Moreover, a TAF1-TAF2 dimeric complex enriched for the Inr consensus in a random double-stranded oligonucleotide site selection screen (Chalkley and Verrijzer, 1999). A temperature-sensitive mutation in TAF1 was also shown to disrupt the ability of TFIID to bind to the Inr of TATA-less promoters (Dehm et al., 2004; Hilton and Wang, 2003). These findings suggest that in the absence of a TATA sequence, the TAF2 and TAF1 subunits of TFIID recognize the Inr element leading to the stable binding of TFIID to the core promoter.

A2: Transcription Factor II-I

Another model that has been proposed for TFIID recruitment to TATA-less promoters involves Inr binding by a cellular protein, which subsequently interacts with TFIID to bring the transcription factor complex to the core promoter. In support of this model, TFII-I was purified as a protein that binds to the initiator sequence within the adenovirus major late promoter (Roy et al., 1991). The addition of TFII-I to in vitro transcription assays stimulated transcription from a naturally occurring TATA-less, Inr-containing murine V β 5.2 promoter (Cheriyath et al., 1998; Manzano-Winker et al., 1996). The immunodepletion of TFII-I from nuclear extracts with an anti-TFII-I antibody also abolished transcription from the Vβ promoter. These results supported the hypothesis that TFII-I is a general transcription factor involved in initiator recognition. To complicate matters, TFII-I was subsequently found to be identical to proteins that interacted with distal promoter elements in several inducible genes and stabilized the DNA binding of other transcription factors (Grueneberg et al., 1997; Kim et al., 1998; Parker et al., 2001). These findings suggest that TFII-I is a multifunctional transcription factor. Whether the protein acts at the Inr or at more distal control elements may be determined by the cis-elements within the promoter region of interest.

A3: Ying Yang 1

Ying Yang 1 (YY1) is a zinc finger protein that can bind to the Inr of the adeno-associated adenovirus (AAV) core promoter and stimulate Inr dependent transcription in an in vitro transcription system reconstituted with pure proteins (Seto et al., 1991). Mutations that disrupt YY1 binding, however, had no effect when transcription levels were monitored in crude nuclear extracts or in vivo (Javahery et al., 1994; Lo and Smale, 1996). Likewise, Inr mutations that abolished transcriptional activity had little effect on YY1 binding to core promoter DNA. YY1 was originally isolated as a protein that bound to a distal element in the AAV P5 promoter and repressed transcription (Shi et al., 1991). More recently, YY1 has been shown to belong to a family of mammalian polycomb group proteins that function in high molecular weight complexes to repress transcription (Atchison L, 2003). YY1 clearly has an important role in the regulation of gene transcription, but whether its primary site of action is at the Inr remains unclear.

B: Inr Function in TATA-Containing Core promoters

The Inr and TATA box, found together in many core promoters, are binding sites for different components of the TFIID complex. These core promoter elements act synergistically when spaced 25-30 nucleotides apart but function as independent DNA elements when separated by greater than 30 nucleotides (O'Shea-Greenfield and Smale, 1992). The extended footprint reported for TFIID covered approximately 40 bp downstream of the transcription start site, suggesting that the observed synergy could be attributed to the TATA box and Inr cooperatively acting as contact points for the TFIID complex. When these two core promoter elements are greater than 30 nucleotides apart, the binding of a single TFIID molecule would no longer be stabilized by two sites of DNA-protein interaction. Interestingly, if the length of the intervening sequence is reduced to 15 or 20 nucleotides, the TATA box and Inr remained synergistic in supporting transcription (O'Shea-Greenfield and Smale, 1992). However the start site of transcription was now dictated by the position of the TATA box, with transcription initiation occurring 25 bp downstream of the TATA sequence and not within the Inr motif. Therefore, the distance between the TATA box and Inr determines not only how these two core promoter elements functionally interact to influence the level of transcription but also the position of transcription initiation.

Downstream Promoter Element (DPE)

The DPE is a downstream core promoter element located at +28 to +32 relative to the +1 nucleotide of the Inr (Kutach and Kadonaga, 2000). It was originally identified in *Drosophila* as a DNA recognition site for purified TFIID (Burke and Kadonaga, 1996; Purnell *et al.*, 1994). Using different experimental approaches, a variety of sequences was found to function as a DPE in *Drosophila* cells. The analysis of natural *Drosophila* promoters led to the consensus sequence 5'-(A/G)G(A/T)(C/T)GT-3' between +28 and +33, and a preference for a G nucleotide at position +24 (Kutach and Kadonaga, 2000). A comparison of eukaryotic promoters revealed that the DPE is conserved from *Drosophila* to human and is most commonly, but not exclusively, found in promoters lacking a TATA sequence (Burke and Kadonaga, 1997). However, a homologue of the DPE in *Saccharomyces cerevisiae* has yet to be discovered.

A: Inr Function in DPE Dependent Transcription

Studies analyzing the *Drosophila* genome suggest that approximately 8% of *Drosophila* core promoters contain a DPE (Ohler *et al.*, 2002). Almost 50% of the DPE-containing promoters also contained an Inr while only 11% contained a TATA sequence, indicating a strong bias for the coexistence of the DPE and Inr motifs in core promoters. In TATA-less promoters, TFIID has been shown to contact both the DPE and Inr elements by DNase I footprinting (Burke and Kadonaga, 1996; Purnell *et al.*, 1994). Mutations within either sequence motif severely reduced promoter activity (Burke and Kadonaga, 1996; Burke and Kadonaga, 1997; Kutach and Kadonaga, 2000). The functional relationship between the DPE and Inr is also illustrated by the maintenance of a strict spacing requirement between the two core promoter elements in confirmed DPE-dependent *Drosophila* promoters (Kutach and Kadonaga, 2000). Removal or insertion of nucleotides between the DPE and Inr led to a significant reduction in TFIID binding and basal transcription levels. These findings suggest that the DPE and Inr cooperate to mediate the interaction of TFIID with core promoter DNA when a binding site for TBP, specifically a TATA box, is absent.

B: TAFs Bind the DPE

TFIID is a multi-subunit complex. Which subunit(s) of TFIID directly makes contact with the DPE? Photo-crosslinking experiments indicated that TAF6 and TAF9 were likely candidates (Burke and Kadonaga, 1997). TAF6 and TAF9 each contain a histone fold domain (HFD), which has been proposed to play a role in direct DNA binding (Gangloff *et al.*, 2001). Experiments with purified proteins revealed that TAF9 but not TAF6 was able to bind to the DPE in a 60 bp promoter fragment from the human IRF-1 gene in electrophoretic mobility shift assays (Shao *et al.*, 2005). A TAF6-TAF9 complex displayed greater DPE binding activity but only when both proteins retained their respective HFDs. The sequence specificity of the heterodimer for DPE binding was also increased compared to the individual TAFs. In competition assays, unlabeled double-stranded oligonucleotides containing wild-type or mutant DPE were equally effective at competing for TAF9 binding to the radio-labeled DPE fragment. With the TAF6-TAF9 complex, wild-type DPE was a significantly better competitor than the mutant DPE sequence. While the HFDs of TAF6 and TAF9 do not directly bind to promoter DNA, the interaction of these domains contributed to the binding affinity and sequence specificity of the TAF-DNA interactions at the DPE.

C: DPE- vs. TATA-Dependent Promoters

The DPE is required for the efficient binding of TFIID to a subset of TATA-less promoter. Therefore, it could be considered a downstream TATA box, as it is serving as a binding site for TFIID. As predicted, the insertion of the DPE at its proper position restored transcriptional activity to promoters that contained a mutant TATA sequence (Burke and Kadonaga, 1996). These findings suggested that transcription at DPE and TATA dependent promoter utilizes similar molecular mechanisms. The identification of NC2 (negative cofactor 2, also known as Dr1-Drap1), an activity that stimulated transcription from DPE-dependent promoters but repressed TATA-dependent transcription, indicates otherwise (Willy *et al.*, 2000). In addition, a mutation in NC2 that disrupted activation of DPE-dependent transcription had no effect on the ability of the mutant NC2 to repress TATA-dependent transcription. These results suggest that NC2 carries out distinct and separable functions at DPE- and TATA-driven promoters and that NC2 is able to recognize the fundamental differences between these two types of core promoter elements.

TFIIB Recognition Element (BRE)

The BRE is the fourth core promoter element identified in eukaryotes, after the TATA box, Inr and DPE. The BRE functions as a DNA binding site for the general transcription factor TFIIB (Lagrange *et al.*, 1998). It is found in only a subset of TATA-containing core promoters. The importance of the BRE in transcriptional regulation was first suggested by X−ray crystallography data. The structure of a TBP-TFIIB-TATA box ternary complex showed that TFIIB interacted with the major groove of DNA upstream of the TATA box (Nikolov DB, 1995). In these studies the DNA fragment used to assemble the ternary complex contained only 3 base pairs upstream of the TATA sequence. Subsequent photo-crosslinking experiments confirmed the interaction of TFIIB with core promoter DNA and suggested that the site of contact spanned 7-9 base pairs (Lagrange *et al.*, 1998; Lee and Hahn, 1995). A consensus sequence for the BRE was subsequently determined by binding-site selection using a library of DNA fragments containing 12 randomized nucleotide pairs followed by the TATA element from the adenovirus major late promoter (Lagrange *et al.*, 1998). After two rounds of selection, the 7 bp consensus sequence 5′-(G/C)(G/C)(G/A)CGCC-3′ was defined, with the 3′ C nucleotide of the BRE immediately followed by the 5′ T nucleotide of the TATA motif. The ability of TFIIB to bind to the BRE consensus in the absence of TBP was confirmed by fluorescence anisotropy DNA-binding and site-specific protein-DNA photo-crosslinking. Interestingly, the BRE motif does not appear to be conserved in yeast or plants, suggesting that its function in transcriptional regulation may be restricted to higher eukaryotes.

In vitro transcription studies carried out with purified proteins demonstrated that the BRE stimulated the initiation of transcription by RNA polymerase II. The sequence specific interaction of TFIIB with the core promoter led to the hypothesis that the BRE facilitates the entry of TFIIB into the transcription complex. However, Evans *et al* later reported a repressor function for BRE when examining basal transcription levels *in vitro* using crude nuclear extracts and *in vivo* by transient transfection (Evans *et al.*, 2001). The inhibitory effect of the BRE was overcome by the addition of the transcriptional activator Gal4-VP16 such that comparable levels of activated transcription was detected from wild-type and mutant BRE containing promoters. The net result was that the magnitude of transcriptional activation mediated by DNA bound activator proteins from BRE containing promoters was significantly increased. It will be interest to test if the core promoter content and/or surrounding environment are factors that determine whether the BRE will have a positive or negative role in transcriptional regulation.

Motif Ten Element (MTE)

The most recently defined core promoter element involved in RNA polymerase II transcription is the motif ten element or MTE (Lim *et al.*, 2004). The MTE was recognized as a potential core promoter element when the sequence was found overrepresented near the start site of transcription in nearly 2000 *Drosophila* genes (Ohler *et al.*, 2002). The identification of known core promoter elements, including the TATA box, Inr, and DPE, at their predicted positions relative to +1, validated the algorithm that led to the discovery of the MTE.

The MTE is conserved from *Drosophila* to humans and supports transcription by RNA polymerase II in the absence of a TATA motif. MTE activity requires a properly positioned Inr, similar to what has been reported for Inr-DPE dependent transcription. The spacing requirement, however, is more forgiving, as removal of one nucleotide only modestly reduced transcription from Inr-MTE promoters. The identical change in Inr-DPE spacing results in a several-fold decrease in transcriptional activity.

The consensus sequence for the MTE is 5′-C(G/C)A(A/G)C(G/C)(G/C)AACGC-3′ and can be found between nucleotides +18 and +29, thereby overlapping the DPE by two nucleotides. Deletion of nucleotides unique to the MTE, in the context of an inactive DPE, compromised transcription levels, with nucleotides +18 to +22 being essential for MTE activity. A search of the *Drosophila* genome uncovered ten putative MTE dependent core promoters, none of which contained a TATA box. Primer extension analysis using poly(A)+ *Drosophila* mRNA mapped the start site of transcription to the C_{-1} position of the Inr in 9 out of 10 of the promoters. Using hybrid promoters in which the TATA box and Inr region from the *hb*P2 core promoter were fused to the downstream promoter region of MTE-containing genes, the MTE restored transcriptional activity to core promoters that contained a TATA box deletion. The MTE could also compensate for the loss of DPE activity. These findings indicate that the MTE is a bonafide core promoter element that is distinct from the DPE.

A: MTE Binding

Core promoter elements represent points of contact for transcription factors that regulate the process of transcription initiation. The ability of the MTE to compensate for loss of the TATA box or the DPE, both binding sites for TFIID subunits, suggested that the MTE may represent another entry point for TFIID. Mutations within the MTE that disrupted transcriptional activity also reduced the interaction of TFIID with the Inr region, as determined by DNase I protection (Lim *et al.*, 2004). The binding of TFIID to the MTE-containing Tollo promoter was quite weak, suggesting that other factors, yet to be determined, may be contributing to the efficient binding of TFIID to TATA-less MTE-Inr core promoters.

Core Promoter Selectivity of Transcription Regulators

The core promoter is the assembly site for the transcription machinery. However, enhancer-binding proteins ultimately target the core promoter in order to activate transcription. A body of evidence exists indicating that transcriptional enhancers display core promoter selectivity. The ability of such regulatory proteins to activate transcription can be dictated by the DNA elements within a given core promoter. The diversification of core promoters provides an addition level of combinatorial gene regulation.

A: Distinct TATA Box Sequences

The first example of this type of regulation was observed in studies on the his3 promoter in *S. cerevisiae*. The his3 promoter contains two TATA boxes, named T_C and T_R, each initiating transcription from a distinct site (Struhl, 1986). The downstream T_C, is a canonical TATA box (TATAAAA) whereas the upstream T_R is a non-canonical TATA that matches the looser consensus sequence 5'-TATA(A/T)AA(G/A)-3' (Mahadevan and Struhl, 1990). The low level of his3 transcription observed under normal growth conditions is supported by T_C (Iyer and Struhl, 1995). Upon his3 induction, the activator proteins Gal4 and Gcn5 specifically increase transcription from T_R but not from T_C. These experiments demonstrated that transcriptional regulators can distinguish between core promoter elements, in the case of the his3 gene, between different TATA sequences.

The ability of enhancer sequences to discriminate between canonical and non-canonical TATA sequences has also been observed in mammalian cells. For example, the myoglobin enhancer is capable of activating transcription from the myoglobin promoter, which contains a canonical TATA sequence, but not from the SV40 early promoter (Wefald *et al.*, 1990). When the SV40 TATA sequence of TATTTAT was changed to the canonical motif, the resulting promoter responded to the myoglobin enhancer. A similar finding was reported for transcription from the hsp70 promoter. Activation of the hsp70 promoter by E1A is dictated by the TATA sequence within the core promoter. When the canonical hsp70 TATA element was replaced with the SV40 early promoter TATA sequence, the promoter was no longer activated by E1A (Simon *et al.*, 1988). Both the myoglobin enhancer and the transcriptional activator E1A specifically preferred the TATAAAA sequence compared to the TATTTAT sequence within the core promoter for proper function. The ability of enhancers and activator proteins to distinguish between TATA sequences in the core promoter represents one way of targeting their actions to a specific gene.

B: TATA versus DPE Core Promoters

The analysis of gene transcription in *Drosophila* has revealed that enhancer elements can also distinguish between promoters that are driven by a TATA element versus the DPE. These studies examined the transcriptional activity of enhancers that were located either between or within close proximity of two distinct core promoters. One promoter, from the even-skipped gene, contained a TATA element. The other promoter was the TATA-less, DPE-dependent white promoter. The two enhancers analyzed, AE1 and IAB5, preferentially activated transcription from the TATA containing even-skipped core promoter (Ohtsuki *et al.*, 1998). In the absence of the even-skipped promoter, AE1 and IAB5 activated transcription from the white promoter. Therefore, both enhancer elements possessed the ability to function with the core promoter of the white gene, but, when given a choice, displayed a strong preference for the TATA-containing even-skipped core promoter. Enhancer elements that can discriminate between different core promoter elements is extremely useful when the gene to be expressed is situated within a cluster of other genes or is located many kb away from the enhancer element. Therefore, core promoter elements can play a key role in directing enhancer function to the proper core promoter region.

Conclusion

The identification of core promoter motifs has expanded our understanding of the mechanisms regulating

RNA polymerase II transcription. The TATA box, Inr, DPE, BRE, and MTE have unique and overlapping functions in the process of transcription initiation. However, a significant number of genes do not contain any of these predominant core promoter motifs within their transcriptional regulatory region. Computational genome analysis has revealed additional novel, uncharacterized sequence motifs that are prevalent from –60 to +40 in the core promoter of *Drosophila* genes (Ohler *et al.*, 2002). Therefore, it is very likely that many more core promoter elements still remain to be discovered and characterized. The long-term goal will be to determine how different combinations of core promoter elements contribute to the regulation of RNA polymerase II dependent transcription in eukaryotic cells.

References

Arkhipova, I. R. (1995). Promoter elements in *Drosophila melanogaster* revealed by sequence analysis. Genetics *139*, 1359-1369.

Atchison, L. G. A., Wilkinson, F., Bonini, N., and Atchison, M.L. (2003). Transcription factor YY1 functions as a PcG protein *in vivo*. EMBO J *22*, 1347-1358.

Bondarenko, V. A., Liu, Y. V., Jiang, Y. I., and Studitsky, V. M. (2003). Communication over a large distance: enhancers and insulators. Biochem Cell Biol *81*, 241-251.

Buratowski, S., Hahn, S., Sharp, P. A., and Guarente, L. (1988). Function of a yeast TATA element binding protein in a mammalian transcription system. Nature *334*, 37-42.

Burke, T. W., and Kadonaga, J. T. (1996). *Drosophila* TFIID binds to a conserved downstream basal promoter element that is present in many TATA-box-deficient promoters. Genes Dev *10*, 711-724.

Burke, T. W., and Kadonaga, J. T. (1997). The downstream core promoter element, DPE, is conserved from *Drosophila* to humans and is recognized by $TAF_{II}60$ of *Drosophila*. Genes Dev *11*, 3020-3031.

Chalkley, G. E., and Verrijzer, C. P. (1999). DNA binding site selection by RNA polymerase II TAFs: a $TAF_{II}250$-$TAF_{II}150$ complex recognizes the initiator. EMBO J *18*, 4835-4845.

Cheriyath, V., Novina, C. D., and Roy, A. L. (1998). TFII-I regulates Vbeta promoter activity through an initiator element. Mol Cell Biol *18*, 4444-4454.

Concino, M. F., Lee, R. F., Merryweather, J. P., and Weinmann, R. (1984). The adenovirus major late promoter TATA box and initiation site are both necessary for transcription *in vitro*. Nucleic Acids Res *12*, 7423-7433.

Corden, J., Wasylyk, B., Buchwalder, A., Sassone-Corsi, P., Kedinger, C., and Chambon, P. (1980). Promoter sequences of eukaryotic protein-coding genes. Science *209*, 1406-1414.

Dehm, S. M., Hilton, T. L., Wang, E. H., and Bonham, K. (2004). SRC proximal and core promoter elements dictate TAF1 dependence and transcriptional repression by histone deacetylase inhibitors. Mol Cell Biol *24*, 2296-2307.

Dynlacht, B. D., Hoey, T., and Tjian, R. (1991). Isolation of coactivators associated with the TATA-binding protein that mediate transcriptional activation. Cell *55*, 563-576.

Evans, R., Fairley, J. A., and Roberts, S. G. (2001). Activator-mediated disruption of sequence-specific DNA contacts by the general transcription factor TFIIB. Genes Dev *15*, 2945-2949.

Gangloff, Y. G., Romier, C., Thuault, S., Werten, S., and Davidson, I. (2001). The histone fold is a key structural motif of transcription factor TFIID. Trends Biochem Sci *26*, 250-257.

Goldberg, M. L. (1979) Sequence analysis of *Drosophila* histone genes, Ph.D. thesis, Stanford University.

Grosveld, G. C., de Boer, E., Shewmaker, C. K., and Flavell, R. A. (1982). DNA sequences necessary for transcription of the rabbit beta-globin gene *in vivo*. Nature *295*, 120-126.

Grosveld, G. C., Shewmaker, C. K., Jat, P., and Flavell, R. A. (1981). Localization of DNA sequences necessary for transcription of the rabbit beta-globin gene *in vitro*. Cell *25*, 215-226.

Grueneberg, D. A., Henry, R. W., Brauer, A., Novina, C. D., Cheriyath, V., Roy, A. L., and Gilman, M. (1997). A multifunctional DNA-binding protein that promotes the formation of serum response factor/homeodomain complexes: identity to TFII-I. Genes Dev *11*, 2482-2493.

Hahn, S., Buratowski, S., Sharp, P. A., and Guarente, L. (1989). Isolation of the gene encoding the yeast TATA binding protein TFIID: A gene identical to the SPT15 suppressor of Ty element insertion. Cell *58*, 1173-1181.

Hilton, T. L., and Wang, E. H. (2003). Transcription factor IID recruitment and Sp1 activation. Dual function of TAF1 in cyclin D1 transcription. Mol Cell Biol *278*, 12992-13002.

Hoey, T., Dynlacht, B. D., Peterson, M. G., Pugh, B. F., and Tjian, R. (1990). Isolation and characterization of the *Drosophila* gene encoding the TATA box binding protein, TFIID. Cell *61*, 1179-1186.

Hu, S. L., and Manley, J. L. (1981). DNA sequence required for initiation of transcription *in vitro* from the major late promoter of adenovirus 2. Proc Natl Acad Sci USA *78*, 820-824.

Hultmark, D., Klemenz, R., and Gehring, W. J. (1986). Translational and transcriptional control elements in the untranslated leader of the heat-shock gene hsp22. Cell *44*, 429-438.

Iyer, V., and Struhl, K. (1995). Mechanism of differential utilization of the his3 TR and TC TATA elements. Mol Cell Biol *15*, 7059-7066.

Javahery, R., Khachi, A., Lo, K., Zenzie-Gregory, B., and Smale, S. T. (1994). DNA sequence requirements for transcriptional initiator activity in mammalian cells. Mol Cell Biol *14*, 116-127.

Kim, D. W., Cheriyath, V., Roy, A. L., and Cochran, B. H. (1998). TFII-I enhances activation of the c-fos promoter through interactions with upstream elements. Mol Cell Biol *18*, 3310-3320.

Kuhn, E. J., and Geyer, P. K. (2003). Genomic insulators: connecting properties to mechanism. Curr Opin Cell Biol *15*, 259-265.

Kutach, A. K., and Kadonaga, J. T. (2000). The downstream promoter element DPE appears to be as widely used as the TATA box in *Drosophila* core promoters. Mol Cell Biol *20*, 4754-4764.

Lagrange, T., Kapanidis, A. N., Tang, H., Reinberg, D., and Ebright, R. (1998). New core promoter element in RNA polymerase II-dependent transcription: sequence-specific DNA binding by transcription factor IIB. Genes Dev *12*, 34-44.

Lee, S., and Hahn, S. (1995). Model for binding of transcription factor TFIIB to the TBP-DNA complex. Nature *376*, 609-612.

Lim, C. Y., Santoso, B., Boulay, T., Dong, E., Ohler, U., and Kadonaga, J. T. (2004). The MTE, a new core promoter element for transcription by RNA polymerase II. Genes Dev *18*, 1606-1617.

Lo, K., and Smale, S. T. (1996). Generality of a functional initiator consensus sequence. Gene *182*, 13-22.

Mahadevan, S., and Struhl, K. (1990). Tc, an unusual promoter element required for constitutive transcription of the yeast HIS3 gene. Mol Cell Biol *10*, 4447-4455.

Manzano-Winker, B., Novina, C. D., and Roy, A. L. (1996). TFII is required for transcription of the naturally TATA-less but initiator-containing Vbeta promoter. J Biol Chem *271*, 12076-12081.

Martinez, E., Chiang, C. M., Ge, H., and Roeder, R. G. (1994). TATA-binding protein-associated factor(s) in TFIID function through the initiator to direct basal transcription from a TATA-less class II promoter. EMBO J *13*, 3115-3126.

Matsui, T. S. J., Weil, P.A., and Roeder, R.G. (1980). Multiple factors required for accurate initiation of transcription by purified RNA polymerase II. J Biol Chem *255*, 11992-11996.

Nakajima, N., Horikoshi, M., and Roeder, R. G. (1988). Factors involved in specific transcription by mammalian RNA polymerase II: Purification, genetic specificity, and TATA box-promoter interactions of TFIID. Mol Cell Biol *8*, 4028-4040.

Nikolov DB, C. H., Halay ED, Usheva AA, Hisatake K, Lee DK, Roeder RG, and Burley, S.K. (1995). Crystal structure of a TFIIB-TBP-TATA-element ternary complex. Nature *377*, 119-128.

O'Shea-Greenfield, A., and Smale, S. T. (1992). Roles of TATA and initiator elements in determining the start site location and direction of RNA polymerase II transcription. J Biol Chem *267*, 1391-1402.

Ogbourne, S., and Antalis, T. M. (1998). Transcriptional control and the role of silencers in transcriptional regulation in eukaryotes. Biochem J *331*, 1-14.

Ohler, U., Laio, G. C., Niemann, H., and Rubin, G. M. (2002). Computational analysis of core promoters in the *Drosophila* genome. Geonome Biol *3*, research0087.0081-0087.0012.

Ohtsuki, S., Levine, M., and Cai, H. N. (1998). Different core promoters possess distinct regulatory activities in the *Drosophila* embryo. Genes Dev *12*, 547-556.

Orphanides, G., Lagrange, T., and Reinberg, D. (1996). The general transcription factor of RNA polymerase II. Genes Dev *10*, 2657-2683.

Parker, C. S., and Topol, J. (1984). A *Drosophila* RNA polymerase II transcription factor contains a promoter-region-specific DNA-binding activity. Cell *36*, 357-369.

Parker, R., Phan, T., Baumeister, P., Roy, B., Cheriyath, V., Roy, A. L., and Lee, A. S. (2001). Identification of TFII-I as the endoplasmic reticulum stress response element binding factor ERSF: its autoregulation by stress and interaction with ATF6. Mol Cell Biol *21*, 3220-3233.

Peterson, M. G., Tanese, N., Pugh, B. F., and Tjian, R. (1990). Functional domains and upstream activation properties of cloned human TATA binding protein. Science *248*, 1625-1630.

Purnell, B. A., Emanuel, P. A., and Gilmour, D. S. (1994). TFIID sequence recognition of the initiator and sequences farther downstream in *Drosophila* class II genes. Genes Dev *8*, 83-842.

Purnell, B. A., and Gilmour, D. S. (1993). Contribution of sequences downstream of the TATA element to a protein-DNA complex containing the TATA-binding protein. Mol Cell Biol *13*, 2593-2603.

Roy, A. L., Meisterernst, M., Pognonec, P., and Roeder, R. G. (1991). Cooperative interaction of an initiator-binding transcription initiation factor and the helix-loop-helix activator USF. Nature *354*, 245-248.

Sawadogo, M., and Roeder, R. G. (1985). Interaction of a gene-specific transcription factor with the adenovirus major late promoter upstream of the TATA box region. Cell *43*, 165-175.

Seto, E., Shi, Y., and Shenk, T. (1991). YY1 is an initiator sequence-binding protein that directs and activates transcription *in vitro*. Nature *354*, 241-245.

Shao, H., Revach, M., Moshonov, S., Tzuman, Y., Gazit, K., Albeck, S., Unger, T., and Dikstein, R. (2005). Core promoter binding by histone-like TAF complexes. Mol Cell Biol *25*, 206-219.

Shi, Y., Seto, E., Chang, L. S., and Shenk, T. (1991). Transcriptional repression by YY1, a human GLI-Kruppel-related protein, and relief of repression by adenovirus E1A protein. Cell *67*, 377-388.

Simon, M. C., Fisch, T. M., Benecke, B. J., Nevins, J. R., and Heintz, N. (1988). Definition of multiple, functionally distinct TATA elements, one of which is a target in the hsp70 promoter for E1A regulation. Cell *52*, 723-729.

Singer, V. L., Wobbe, C. R., and Struhl, K. (1990`). A wide variety of DNA sequences can functionally replace a yeast TATA element for transcriptional activation. Genes Dev *4*, 636-645.

Smale, S. T., and Baltimore, D. (1989). The "initiator" as a transcription control element. Cell *57*, 103-113.

Smale, S. T., Schmidt, M. C., Berk, A. J., and Baltimore, D. (1990). Transcriptional activation by Sp1 as directed through TATA or initiator: specific requirement for mammalian transcription factor IID. Proc Natl Acad Sci USA *87*, 4509-4513.

Struhl, K. (1986). Constitutive and inducible *Saccharomyces cerevisiae* promoters: evidence for two distinct molecular mechanisms. Mol Cell Biol *6*, 3847-3853.

Struhl, K. (1989). Molecular mechanisms of transcriptional regulation in yeast. Annu Rev Biochem *58*, 1051-1077.

Suzuki, Y., Tsunoda, T., Sese, J., Taira, H., Mizushima-Sugano, J., Hata, H., Ota, T., Isogai, T., Tanaka, T., Nakamura, Y.*, et al.* (2001). Identification and characterization of the potential promoter regions of 1031 kinds of human genes. Genome Res *11*, 677-684.

Tanese, N., Pugh, B. F., and Tjian, R. (1991). Coactivators for a proline-rich activator purified from the multisubunit human TFIID complex. Genes & Dev *5*, 2212-2224.

Verrijzer, C. P., Chen, J.-L., Yokomori, K., and Tjian, R. (1995). Binding of TAFs to core elements directs promoter selectivity by RNA polymerase II. Cell *81*, 1115-1128.

Verrijzer, C. P., Yokomori, K., Chen, J.-L., and Tjian, R. (1994). *Drosophila* $TAF_{II}150$: similarity to yeast gene TSM-1 and specific binding to core promoter DNA. Science *264*, 933-941.

Wasylyk, B., Derbyshire, R., Guy, A., Molko, D., Roget, A., Teoule, R., and Chambon, P. (1980). Specific *in vitro* transcription of conalbumin gene is drastically decreased by single-point mutation in T-A-T-A box homology sequence. Proc Natl Acad Sci USA *77*, 7024-7028.

Wefald, F. C., Devlin, B. H., and Williams, R. S. (1990). Functional heterogeneity of mammalian TATA-box sequences revealed by interaction with a cell-specific enhancer. Nature *344*, 260-262.

Willy, P. J., Kobayashi, R., and Kadonaga, J. T. (2000). A basal transcription factor that activates or represses transcription. Science *290*, 982-985.

Zhou, Q., Lieberman, P. M., Boyer, T. G., and Berk, A. J. (1992). Holo-TFIID supports transcriptional stimulation by diverse activators and from a TATA-less promoter. Genes & Dev *6*, 1964-1974.

Section III

The Regulators

Chapter 08

Transcriptional Activators and Activation Mechanisms

Jun Ma

Division of Developmental Biology, Cincinnati Children's Hospital Research Foundation, University of Cincinnati College of Medicine, 3333 Burnet Avenue, Cincinnati, OH 45229, USA

Key Words: activator, transcription, co-activator, enhancer, promoter, signal transduction, development.

Summary

Transcriptional activators are required to turn on the expression of genes in a eukaryotic cell. Activators bound to enhancers stimulate the assembly and activity of the transcription machinery at gene promoters. This article examines selected issues in understanding activator functions and activation mechanisms.

Introduction

Transcription is the process of copying (transcribing) the information from one strand of DNA into RNA by the enzyme called RNA polymerase (RNAP). In bacteria there is only one RNAP, but in eukaryotes there are three different RNAPs that transcribe different classes of genes (Hahn, 2004). RNAPII is responsible for transcribing protein-coding genes, whereas RNAPI and III are responsible for synthesizing rRNA and tRNA respectively. This article deals with transcription by RNAPII (referred to as RNAP from now on), which has been subject to intensive investigation over the past decades (Kadonaga, 2004; Sims *et al.*, 2004b). A major focus of this article is to discuss mechanisms leading to increased levels of transcription, a process called activation.

In addition to the coding sequence, a typical class II gene contains at least two other types of DNA sequences that are required for initiating transcription. The first such elements are called promoters (also referred to as core promoters). These are specific DNA sequences located upstream of the coding regions of the genes. Promoters help orient RNAP so that it "knows" where on DNA to start transcribing and in which direction. RNAP itself does not have the ability to recognize specific DNA sequences such as promoters. Instead, a group of proteins, called general transcription factors (GTFs), help RNAP to find promoter sequences (Hampsey, 1998; Orphanides *et al.*, 1996). One of these GTFs is the TATA-box binding protein (TBP), which directly binds to the TATA element of a promoter. The protein complex assembled at the promoter is often referred to as the preinitiation complex or transcription machinery (or apparatus). This complex contains GTFs and RNAP. It also contains co-factors and chromatin modifying/remodeling factors (or complexes) that are part of the RNAP holoenzyme. Many of these additional factors play important roles in mediating transcription regulation by responding to regulatory proteins (Levine and Tjian, 2003; Malik and Roeder, 2000; Naar *et al.*, 2001; Narlikar *et al.*, 2002).

The second type of DNA elements required for initiating gene transcription is the regulatory elements, to which regulatory proteins bind. Those elements that play positive roles in transcription are called upstream activation sequences (UASs) in yeast and enhancers in higher eukaryotes such as humans. These sequences provide the binding sites for transcriptional activators that increase the levels of gene transcription.[1] Enhancers (and proximal-promoter elements) play particularly important roles in gene expression: genes in eukaryotic cells tend to stay silent (off) unless they are

---

[1] For our discussion in this chapter, we treat proximal-promoter elements, which are located immediately upstream of the core promoters, as part of the regulatory elements.

stimulated (turned on) by activators bound to enhancers. This is obvious for genes that need to be specifically turned on at precise times and locations in response to environmental or developmental signals. This is even true for housekeeping genes that appear to be transcribed at all times; for these genes, their transcription is also dependent on activators bound to regulatory sequences. Many enhancers are located upstream of the genes, but they have also been found in introns or even downstream of the genes. A special feature of enhancers is that they can stimulate transcription in an orientation- and distance- independent manner (Blackwood and Kadonaga, 1998). There are also regulatory elements that play negative roles in transcription; these elements contain binding sites for transcriptional repressors. This article primarily deals with activator functions and mechanisms of activation.

There are several other types of DNA sequences that also play important roles in transcription but are not further discussed in this article. For example, the polyadenylation site located at the end of a gene instructs RNAP to terminate transcription (Ares and Proudfoot, 2005; Tollervey, 2004). In this article we will first discuss how a typical activator looks like and how it might activate transcription. We will then expand our discussion by examining selected issues to further explore activator functions and activation mechanisms.

A Typical Transcriptional Activator

A typical activator has two essential functions: DNA binding and transcriptional activation (Ptashne, 1988). Many activators have separate protein domains to confer these two functions. The DNA binding domain of an activator enables the protein to recognize specific DNA sites located within enhancers. There are different families of DNA binding domains that form distinct structures to recognize DNA (Garvie and Wolberger, 2001). These domains tend to bear names that depict their structural and/or functional properties or follow their founding member's names. For example, a zinc-finger DNA binding domain uses zinc to maintain its three-dimensional structure required for DNA recognition. A basic region-leucine zipper (bZIP) domain contains a basic region (that contacts DNA) and a leucine zipper (that forms dimers). A homeodomain is a conserved 60-aa DNA binding domain initially identified in proteins encoded by *Drosophila* homeotic genes, which play critical roles in specifying segment identity. A Rel homology domain is a DNA binding domain that bears the name of its founding member Rel.

Most DNA binding domains, including all the examples mentioned above, recognize short, specific DNA sequences by making elaborate contacts with the bases in the major groove of the DNA double helix (Garvie and Wolberger, 2001; Patikoglou and Burley, 1997). Others, e.g., the high-mobility-group (HMG) domain, recognize DNA sequences by interacting with the minor groove (Travers, 2000).[②] While most DNA binding domains recognize DNA sites as dimers (e.g., bZIP and Rel family members), others can bind as monomers (e.g., some homeodomain proteins). For proteins that bind DNA as dimers, many can form both homodimers and heterodimers with other family members. Heterodimer formation can increase the repertoire of DNA sequences recognized by a given family of transcription factors. Many activators can bind DNA cooperatively with one another, which can increase the stability of the protein complexes formed at the enhancers (Adams and Workman, 1995; Ma *et al.*, 1996).

The activation domain of an activator plays critical roles in stimulating transcription. Unlike DNA binding domains that require elaborate structures for DNA recognition, activation domains tend to be short sequences often with very limited sequence complexity. There are different types of activation domains, which are named after their sequence characteristics, such as acidic, glutamine-rich, proline-rich, and alanine-rich. For the acidic class of activating sequences, it was estimated that 1% of the peptides encoded by random DNA sequences (from the *E.coli* genome) can activate transcription when fused to a DNA binding domain (Ma and Ptashne, 1987c). This finding further highlights the "relaxed" specificity between activation sequences and their targets (Ma, 2004), a feature that contrasts the interaction mode between DNA binding domains and DNA sites.

One important finding in understanding activator functions was the demonstration of the modular nature of activators, i.e., the DNA binding and activation functions are provided by separable domains (Brent and Ptashne, 1985; Keegan *et al.*, 1986). This finding suggested that DNA binding *per se* was insufficient for activation in eukaryotes (Brent, 2004; Ptashne, 2004). Subsequent demonstration that activation sequences are short, simple peptides (Hope and Struhl, 1986; Ma and Ptashne, 1987b; Ma and Ptashne, 1987c) further supported the notion that activation domains achieved their functions by touching other proteins (also see below). The demonstration of activators' modular nature has also enabled researchers to determine easily whether a

---

[②] TBP, a GTF that can bind specific DNA sequences, also makes contacts with the minor groove (Kim *et al.*, 1993; Nikolov *et al.*, 1992).

particular transcription factor has an activation function, through experiments of assaying the activity of hybrid proteins containing the factor's fragments fused to a heterologous DNA binding domain.

The Recruitment Model

What does an activator do to stimulate transcription? As discussed above, an essential domain of an activator is its DNA binding domain, which brings the activator to DNA sites in an enhancer. But the DNA binding domain itself is insufficient to activate transcription; an activation domain is required for activation. Activation domains have been shown to have the ability to interact with a wide array of proteins, many of which are components of the transcription machinery, including the GTFs (e.g., TBP, TFIIB, TFIIE, and TAFs), co-factors and chromatin modifying/remodeling complexes (Malik and Roeder, 2000; Naar et al., 2001; Narlikar et al., 2002; Orphanides et al., 1996; Peterson and Workman, 2000; Ptashne and Gann, 1990). All these (and other) interactions lead to a unified final outcome: increased level of transcription. According to a well-established recruitment model, the ultimate and only goal of these interactions is to bring the transcription machinery, in particular RNAP, to the promoter (Ptashne and Gann, 1997; Stargell and Struhl, 1996). During the activation process, a DNA loop may be formed as a result of the interaction between the activator bound at the enhancer and the transcription machinery at the core promoter (Ptashne, 1986).

Several lines of evidence support the recruitment model. First, for many genes, the GTFs and RNAP are absent from their promoters unless the genes are turned on by activators (Chatterjee and Struhl, 1995; Klein and Struhl, 1994; Li et al., 1999). Second, activators are known to interact with components of the transcription machinery; as noted above, one property common to the activation domains is that they tend to have the ability to interact with multiple target proteins (Bryant and Ptashne, 2003; Ma, 2004; Ptashne and Gann, 1990). Finally, in a set of "artificial recruitment" experiments that provided pivotal support to this model, it was shown that transcription can be elicited by artificially attaching components of the transcription machinery to a DNA binding domain (Chatterjee and Struhl, 1995; Farrell et al., 1996; Gonzalez-Gouto et al., 1997; Nevado et al., 1999; Xiao et al., 1995). In these artificial recruitment experiments, the requirement of a classical activator is bypassed, i.e., the activator is no longer needed for transcription. This suggested that, at least for the promoters tested, all the functions that are provided by the activators could be substituted by physically bringing the RNAP holoenzyme to the promoter. It should be noted that, since the eukaryotic DNA is wrapped in nucleosomes, the recruitment model may also cover situations in which the chromatin modifying or remodeling factors recruited by activators facilitate the assembly of the transcription machinery by increasing the accessibility of promoter DNA. As discussed below, the recruitment model, though attractive due to its simplicity and experimental support, does not exclude other possibilities of how activators may stimulate transcription.

Now with this broad description of what a typical activator looks like and how it may activation transcription according to one model, we will discuss several additional issues to further our understanding of activator functions and activation mechanisms. Readers should refer to other chapters in this volume that discuss specific examples of activators in greater details.

Composite Activators

Although a typical activator contains both an activation domain and a DNA binding domain, sometimes these two domains can reside on separate proteins. For example, the herpes simplex virus (HSV) activator VP16 dos not bind to DNA, but rather, it is brought to DNA by interacting with other DNA-binding proteins (Triezenberg et al., 1988). The activation domain of VP16 can also activate transcription when directly linked to a DNA binding domain (Sadowski et al., 1988). This finding demonstrated that an activation domain can be brought to DNA by distinct, but interchangeable, means, either directly binding to DNA (through its linked DNA binding domain) or interacting with other DNA-binding proteins. This concept was further demonstrated by the creation of an artificial composite activator (Ma and Ptashne, 1988). The yeast repressor protein GAL80 inhibits the activation function of GAL4 by interacting with and masking its activation domain (Johnston et al., 1987; Lue et al., 1987; Ma and Ptashne, 1987a). When GAL80 was attached to an activation domain, the hybrid GAL80 protein, which itself cannot bind DNA, gained an ability to activate transcription, but only through a GAL4 derivative that could interact with both DNA and GAL80 (Ma and Ptashne, 1988). The concept that an activation domain can be brought to DNA through protein-protein interactions led to the proposal of the yeast two-hybrid system (Fields and Song, 1989). This powerful genetic system has allowed researchers to dissect protein-protein interactions and to identify proteins' interacting partners (Bai and Elledge, 1996; Fields and Sternglanz, 1994; Ma, 2000).

Transcriptional activators that do not bind DNA but interact with other DNA-binding proteins are sometimes also referred to as co-activators. But it may be useful to make a distinction between these non-DNA binding activators and the "true" co-activators that play more general roles in transcription. Unlike non-DNA binding activators, which are gene-specific, co-activators of the latter class (e.g., CBP and Swi-Snf complexes) play important roles in facilitating the actions of many activators. Some of these general co-activators are components of the RNAP holoenzyme (Myers and Kornberg, 2000; Ranish and Hahn, 1996).

The concept that an activation domain can be brought to its action site, the vicinity of a gene's promoter, through multiple means can be further extended to activators that bind to RNA sequences. One such example is the HIV activator Tat, which is discussed in further detail in another chapter of this volume. Relevant to this discussion is the finding that the RNA-binding activator Tat can also activate transcription from DNA sites when fused to a DNA binding domain (Southgate and Green, 1991), further illustrating that an activation domain can be brought to the vicinity of a promoter through distinct, but interchangeable, mechanisms.

Conformational Changes

The artificial recruitment experiments mentioned above support the notion that the ultimate and only function of activators is to bring RNAP to the promoter. It is known that the preinitiation complex undergoes several conformational changes before RNAP actually initiates transcription (Carey and Smale, 2000). For example, the promoter DNA is significantly bent and unwound upon TBP binding (Kim *et al.*, 1993; Nikolov *et al.*, 1992). In addition, the DNA double helix at the transcription start site becomes unpaired, or melted, to form a bubble prior to transcription initiation by RNAP (Giardina and Lis, 1993; Wang *et al.*, 1992). In one study, it was shown that activators can change the conformation of the TFIIA-TFIID-TATA complex and such a conformational change is necessary and sufficient for activation in an *in vitro* system (Chi and Carey, 1996). Thus, conformational changes of the transcription machinery represent potential steps that can also be targeted by transcriptional activators.

Initiation vs. Elongation

Although transcription initiation is a critical step that can be stimulated by many activators, other steps of transcription, such as elongation, can also be activated. For example, the *Drosophila* heat shock gene *hsp70* already has the transcription machinery loaded at its promoter even before heat shock (induction) (Rougvie and Lis, 1988). In fact, RNAP is able to transcribe the 5′ region of the gene prior to induction, but it fails to transcribe through the gene (Rasmussen and Lis, 1993; Rasmussen and Lis, 1995). Upon induction, the transcriptional activator HSF stimulates elongation by RNAP, enabling it to complete transcription through the gene.

Many proteins (or complexes) have been identified that play important roles in facilitating transcription elongation, and some of these factors represent targets for activators (Sims *et al.*, 2004a). For example, experiments in *Drosophila* suggested that the elongation factor P-TEFb is recruited (likely by the activator HSF) to the heat shock loci to facilitate transcription elongation upon heat shock induction (Lis *et al.*, 2000). In addition, *in vitro* experiments using the human *hsp70* gene demonstrated that the Swi-Snf complex was recruited by the human activator HSF1 to facilitate transcription elongation through the chromatin template of the gene (Brown *et al.*, 1996). As discussed elsewhere in this volume, the HIV Tat activator stimulates transcription elongation by recruiting the elongation factor P-TEFb (Mancebo *et al.*, 1997; Zhou *et al.*, 1998; Zhu *et al.*, 1997). Together, these examples highlight the importance of the elongation step in transcriptional activation.

In this context, it should be noted that recent studies in both yeast and *Drosophila* suggest that transcription silencing can work at a step after the assembly of the transcription machinery (Breiling *et al.*, 2001; Dellino *et al.*, 2004; Sekinger and Gross, 2001). In other words, the mere presence of the transcription machinery (including the RNAP itself) assembled at the promoter does not necessarily equate to productive transcription of the gene.

Synergism

One of the characteristic features of transcriptional activation is synergism. Synergy refers to the situation where the transcription level achieved by multiple activators is higher than the sum of the levels by individual factors separately. Synergy can arise from different mechanisms. In the simplest case, it can be due to cooperative binding of activators to multiple sites in the enhancer. This is obvious particularly if the activators are at limiting (sub-saturating) concentrations. An enhanceosome model has been proposed that further emphasizes the role of multiple activators for activation (Merika and Thanos, 2001; Thanos and Maniatis, 1995). According to this model, different activators, including those that play architectural roles, are together required to form a stable complex at the enhancer for efficient transcriptional activation.

Synergy can also be achieved even when activators are at saturating levels. This particular form of synergy, which can be demonstrated readily under *in vitro* conditions, provides useful insights into how activators work. It suggests that activators can contact multiple targets in the transcription machinery (Ptashne and Gann, 1998). If all activator molecules contacted the same target through the same surface, then increasing the number of activator molecules should only increase transcription level in an additive, rather than synergistic, manner.

Studies to compare the roles of different activators suggest that synergy may represent a consequence of combinatorial actions of activators that work on distinct steps of transcription (Blau *et al.*, 1996). By comparing the RNAP density along a gene, it is possible to gain information about which step, initiation or elongation, an activator may stimulate. Using this and other analyses, Blau *et al.* concluded that, while some activators (e.g., Sp1 and CTF) work primarily on the initiation step, others (e.g., Tat) work primarily on the elongation step (Blau *et al.*, 1996). Another class of activators (e.g., VP16, p53 and E2F1) can work on both initiation and elongation steps. An analysis of these activators revealed that synergy was only achieved between those that work on different steps of transcription (Blau *et al.*, 1996).

Interplay Between DNA Binding and Activation Functions

It is well established that, in general, the DNA binding and transcriptional activation domains of an activator are physically separable. It should be emphasized, however, that the functions provided by these two domains are interconnected. First, an activation domain cannot exert its activating effect unless it is brought to DNA, either through a physical link to a DNA binding domain or through other means such as interacting with another DNA-binding protein. Second, some activators, e.g., the glucocorticoid receptor and MyoD, have DNA binding and activation functions conferred by single protein domains (Davis *et al.*, 1990; Schena *et al.*, 1989). Furthermore, several studies have suggested that the DNA binding properties of an activator can be influenced by its activation function (Bunker and Kingston, 1996; Tanaka, 1996). In particular, it was observed that activators that stimulated transcription strongly bound DNA better than those that activated weakly. It has been proposed that the interaction between the activation domain and the transcription machinery can help the binding of not only the transcription machinery to the promoter but also the activator to its DNA sites. For activators that work by recruiting chromatin remodeling or modifying complexes, the increased DNA accessibility is beneficial to not only GTFs but also activators themselves. The connection between the activation and DNA binding functions may also contribute to how activator gradients, such as *Drosophila* Bicoid, stimulate transcription in a concentration-dependent manner (Driever *et al.*, 1989; Fu *et al.*, 2003; Zhao *et al.*, 2003; Zhao *et al.*, 2002).

Activation vs. De-repression

One of the major differences between eukaryotes and prokaryotes is that eukaryotic genomes are packaged into nucleosomes. Nucleosomes can impede DNA binding of transcription factors and GTFs, thus repressing transcription (Narlikar *et al.*, 2002; Peterson and Workman, 2000; Wu, 1997; Wu and Grunstein, 2000). One of the questions regarding mechanisms of eukaryotic transcription activation is how much is due to de-repression and how much is due to "real" activation. It is well established that genes can be de-repressed when histones are depleted from cells (Han and Grunstein, 1988). To further obtain insights into the role of histones in gene activation, Wyrick *et al.* used the microarray strategy to determine the profiles of gene expression upon histone H4 depletion (Wyrick *et al.*, 1999). The authors found that 15% of the genes exhibited de-repressed (increased) expression in response to the removal of histone H4. Genes that are located near telomeres tend to be more sensitive to histone depletion than genes located elsewhere. These results show that depletion of histone can lead to gene-specific de-repression. The authors also found that histones did not play a generally repressive role for all genes, since the majority (75%) of genes appeared to be insensitive to histone depletion. Interestingly, 10% of the genes in yeast had reduced expression upon histone depletion, suggesting that histones and nucleosomes may also play positive roles in transcription (Wyrick *et al.*, 1999).

Activation and Cellular Memory

In some cases the effect of transcriptional activators can be maintained or inherited even after the activators themselves are no longer present. One such example is homeotic gene expression in *Drosophila* (Levine *et al.*, 2004; Orlando, 2003). During early embryogenesis the homeotic genes respond to transcription factors that are encoded by gap and pair rule genes. The active and silent states of these genes are subsequently maintained by proteins encoded by the *trithorax* group (*trxG*) and *Polycomb* group (*PcG*) genes, respectively. trxG and PcG proteins form co-factor complexes that work through DNA elements called Polycomb response elements (PREs). In an elegant study, an isolated PRE, *Fab-7*, was shown to be able to maintain the active state of a linked reporter construct that had been activated by a transiently expressed activator (Cavalli and Paro, 1998). In other words, the reporter gene remained on even after the activator itself was no longer present. Intriguingly, such memory can be transmitted in an activator-independent manner to subsequent generations through female (but not male) germline. Recent studies show that components in both PcG and trxG complexes contain histone methyltransferase (HMT) activities with different specificities/preferences for different lysine residues in histone tails (Levine *et al.*, 2004; Lund and van Lohuizen, 2004; Sims *et al.*, 2003). It is thought that distinct histone methylation patterns represent cell memory systems to maintain the active and silent states of homeotic genes (Orlando, 2003; Sims *et al.*, 2003).

In yeast, genes that are transcribed recently are also marked by a specific pattern of histone methylation (Hampsey and Reinberg, 2003). This is achieved by the recruitment of the HMT Set1 to the genes by the elongating RNAP (Ng *et al.*, 2003). Interestingly, the Set1-mediated histone methylation pattern persists for some time even after the genes are no longer transcribed. Unlike the long-term memory of active genes mediated by trxG proteins in *Drosophila* (which can last for several generations), Set1-mediated marking of recently active genes in yeast is only short term (up to several hours). In addition, while the consequence of the trxG-mediated marking is to maintain the genes on, the yeast Set1-mediated system only marks the recently transcribed genes without actually keeping them on. Interestingly, yeast Set1 is also involved in the long-term memory of gene silencing (Bryk *et al.*, 2002; Krogan *et al.*, 2002).

Another case of activator-induced memory is noteworthy in this context. This is an extremely short-term memory, which lasts only through the initial activation process itself. Under some *in vitro* conditions, transcriptional activators can induce conformational changes of the preinitiation complex. Interestingly, one study demonstrated that an activator-induced conformational change persisted, and led to the completion of the transcription process, even after the activator itself was removed (Chi and Carey, 1996). This result further illustrates the importance of conformational changes in transcriptional activation.

Activator-repressor Switches

Although this chapter deals primarily with activators and activation mechanisms, it should be noted that many transcription factors can often work as either activators or repressors in a context-dependent manner (Ma, 2005). For example, many transcription factors that mediate signal transduction processes work as repressors in the absence of the signals but as activators in the presence of the signals. In addition, the concentrations and posttranslational modifications of a transcription factor can affect its ability to either activate or repress transcription. The presence of other nearby DNA binding proteins on DNA, as well as the availability and concentration of co-factors, can also influence the behavior of a transcription factor. See a recent review article for further details (Ma, 2005).

Short Distance and Long Distance Actions

One of the questions in eukaryotic gene activation concerns actions at long distances. Enhancers in higher eukaryotes have the ability to exert their effects even when they are located many kilobases away from the promoters (Blackwood and Kadonaga, 1998). There are no specific definitions of short distance vs. long distance, but for our discussion we can consider short distance as anything up to a few hundred base pairs and long distance greater than one kilobase (Blackwood and Kadonaga, 1998; Dorsett, 1999). The mechanisms for activation at short or long distances may be fundamentally similar in that they are both achieved through a network of protein-protein interactions and alterations of chromatin structure. But activation at a long distance (e.g., 50-60 kilobases) faces two additional challenges that are less relevant to activation at a short distance (e.g., 100-200 bp). First, how can promoters and enhancers communicate through such long distances? Second, how does an enhancer "choose" to activate one promoter, but not another one that is also within its reach?

Proteins called facilitators have been proposed to promote the interaction between enhancers and promoters that are separated by long distances (Bulger and Groudine, 1999; Dorsett, 1999). One such example is a *Drosophila* protein called Chip (Morcillo *et al.*, 1997; Torigoi *et al.*, 2000). It is thought Chip can interact with proteins that may bind throughout the genome, such as homeodomain proteins, thus bringing enhancers closer to promoters through the formation of a series of loops (Dorsett, 1999).

Recent studies suggest that the efficiency (and specificity) of the communication between enhancers and promoters can also be augmented by DNA elements located near the core promoters. These elements have been called tethering elements (Bertolino and Singh, 2002; Calhoun *et al.*, 2002). In one study, it was shown that the POU domain of Oct-1 bound to DNA sites near a promoter enables the promoter to respond to a distant enhancer (Bertolino and Singh, 2002). Interestingly, the POU domain itself does not work as a classical activator because it cannot activate transcription. It was suggested that the POU domain of Oct-1 recruits the TFIID complex to the promoter, so that the promoter becomes poised for activation by an enhancer at a distance (Bertolino and Singh, 2002).

The specificity of long-distance communication between enhancers and promoters can be regulated by different mechanisms (Blackwood and Kadonaga, 1998). First, the tethering elements mentioned above can selectively facilitate the communication between a promoter and one, but not another, enhancer (Calhoun *et al.*, 2002). Second, in some cases promoters can compete with each other for an enhancer, thus the enhancer preferentially communicates with the strong promoter, while ignoring the weak promoter (Foley and Engel, 1992; Sharpe *et al.*, 1998). Finally, insulator elements can prevent "unwanted" communications between enhancers and promoters thus encouraging "wanted" interactions; an insulator is a DNA element that can block the communication between an enhancer (or a silencer) and a promoter when the insulator is located between them, but not when it is located outside the enhancer-promoter unit (Kuhn and Geyer, 2003; West *et al.*, 2002).

Recent studies reveal that the *Drosophila* genome contains organized domains—some as large as 200 kilobases—that contain many genes with similar expression profiles (Spellman and Rubin, 2002). How genes within these large domains are coordinately regulated is currently unclear. It is proposed that each of these domains may contain some higher order control elements (Calhoun and Levine, 2003), such as the recently discovered global control region (GCR) for the mouse *HoxD* complex (Spitz *et al.*, 2003; Zuniga *et al.*, 2004). In this context it is noted that UASs in yeast generally do not work at distances greater than several hundred base pairs (also see de Bruin *et al.*, 2001). It is evident metazoans have evolved mechanisms to facilitate long-distance enhancer-promoter communications and accommodate the increased complexity of gene regulation.

Modifications of Activators

Many activators are subject to posttranslational modifications, such as phosphorylation (Brivanlou and Darnell, 2002), acetylation (Brooks and Gu, 2003), and glycosylation (Jackson and Tjian, 1988; Kamemura and Hart, 2003). In many cases the modifications can have positive roles in transcriptional activation. For example, phosphorylation of STAT is responsible for mediating the JAK/STAT signal transduction pathway (Brivanlou and Darnell, 2002; Darnell *et al.*, 1994). Acetylation of p53 can increase its ability to bind DNA (Brooks and Gu, 2003; Gu and Roeder, 1997; Prives and Manley, 2001). Recent studies suggest that ubiquitination and SUMOylation also play important roles in regulating the activity of transcription factors. In several cases, the transcriptional activation functions of activators are dependent on ubiquitination. Due to space limitation, readers should refer to several recent review articles on this topic for further details (Conaway *et al.*, 2002; Freiman and Tjian, 2003; Gill, 2004; Herrera and Triezenberg, 2004).

Activators with Enzymatic Activities

Although eukaryotic activators themselves generally contain no enzymatic activities, recent studies challenge this generalization. A group of activators, which belong to the family of Eyes absent (Eya), play important roles in the development of multiple tissues and organs including the eye, kidney and muscle (Epstein and Neel, 2003; Rebay *et al.*, 2005). Recent studies show that Eya proteins, which are non-DNA binding activators, contain phosphatase activities (Li *et al.*, 2003; Rayapureddi *et al.*, 2003; Tootle *et al.*, 2003). It is currently not fully understood what substrates these phosphatases work on and how they specifically contribute to the transcription activation process. Numerous co-factors and components in the transcription machinery contain various enzymatic activities that play critical roles in transcription regulation (Shi and Shi, 2004; Sims *et al.*, 2004b). Therefore, the presence of enzymatic activities in activators may not fundamentally change our way of thinking about transcription. Nevertheless, the identification of enzyme-containing activators establishes a new paradigm of increased complexity in transcription regulation.

Concluding Remarks

I would like to end our discussion by returning to the issue introduced at the beginning of this chapter, i.e., a typical activator contains two important functions, DNA binding and activation. Why, then, do activators have to bind DNA, or for non-DNA binding activators, interact with other DNA-binding proteins? This question touches the very heart of the activation process. DNA binding brings an activator closer to the promoter, its action site, thus effectively increasing its local concentration for the promoter. This in turn leads to more efficient, localized interactions between the activator bound at the enhancer and the transcription machinery bound at the promoter. According to the recruitment model, such localized interactions help recruit the transcription machinery to the promoter. Activator-recruited chromatin remodeling/modifying complexes also exert greater, local effects on DNA accessibility (than untargeted complexes do) to facilitate the assembly of the transcription machinery at the promoter. For genes that are activated through other mechanisms, e.g., elongation, the rate-limiting steps also respond to local stimulation more favorably than untargeted signals.

The relatively weak interactions between activators and components of the transcription machinery represent a critical means to achieve specificity in activation: these interactions (or their effects) may only occur efficiently when the enhancer (to which activators bind) and the promoter (to which the transcription machinery binds) are linked. As discussed already, there are mechanisms that can facilitate long-distance communications between enhancers and promoters. Interestingly, some of these and/or additional mechanisms may also play roles in facilitating a rare class of communications that occur between enhancers and promoters on separate chromosomes (Dorsett, 1999; Duncan, 2002; Muller and Schaffner, 1990). It should be reminded that transcription regulation is not restricted to activation. Genes are also subject to repression and silencing. Similar to activation, repression is also facilitated by targeted, local (or regional) interactions, only to achieve the opposite outcome of reducing transcription levels. Understanding gene regulation requires considerations of the integration of transcriptional activation and repression.

Acknowledgement

I would like to thank members of my lab for discussions. Research in this lab is supported by grants from the NSF, NIH, AHA and DOD. Support from the NSFC is also acknowledged.

References

Adams, C. C., and Workman, J. L. (1995). Binding of disparate transcriptional activators to nucleosomal DNA is inherently cooperative. Mol Cell Biol *15*, 1405-1421.

Ares, M., Jr., and Proudfoot, N. J. (2005). The spanish connection; transcription and mRNA processing get even closer. Cell *120*, 163-166.

Bai, C., and Elledge, S. J. (1996). Gene identification using the yeast two-hybrid system. Meth in Enzymol *273*, 331-347.

Bertolino, E., and Singh, H. (2002). POU/TBP cooperativity: a mechanism for enhancer action from a distance. Mol Cell *10*, 397-407.

Blackwood, E. M., and Kadonaga, J. T. (1998). Going the distance: a current view of enhancer action. Science *281*, 61-63.

Blau, J., Xiao, H., McCracken, S., O'Hare, P., Greenblatt, J., and Bentley, D. (1996). Three functional classes of transcriptional activation domains. Mol Cell Biol *16*, 2044-2055.

Breiling, A., Turner, B. M., Bianchi, M. E., and Orlando, V. (2001). General transcription factors bind promoters repressed by Polycomb group proteins. Nature *412*, 651-655.

Brent, R. (2004). Building an artificial regulatory system to understand a natural one. Cell *116*, S73-74, 71 p following S76.

Brent, R., and Ptashne, M. (1985). A eukaryotic transcriptional activator bearing the DNA specificity of a prokaryotic repressor. Cell *43*, 729-736.

Brivanlou, A. H., and Darnell, J. E., Jr. (2002). Signal transduction and the control of gene expression. Science *295*, 813-818.

Brooks, C. L., and Gu, W. (2003). Ubiquitination, phosphorylation and acetylation: the molecular basis for p53 regulation. Curr Opin Cell Biol *15*, 164-171.

Brown, S. A., Imbalzano, A. N., and Kingston, R. E. (1996). Activator-dependent regulation of transcriptional pausing on nucleosomal templates. Genes Dev *10*, 1479-1490.

Bryant, G. O., and Ptashne, M. (2003). Independent recruitment *in vivo* by Gal4 of two complexes required for transcription. Mol Cell *11*, 1301-1309.

Bryk, M., Briggs, S. D., Strahl, B. D., Curcio, M. J., Allis, C. D., and Winston, F. (2002). Evidence that Set1, a factor required for methylation of histone H3, regulates rDNA silencing in *S. cerevisiae* by a Sir2-independent mechanism. Curr Biol *12*, 165-170.

Bulger, M., and Groudine, M. (1999). Looping versus linking: toward a model for long-distance gene activation. Genes Dev *13*,

2465-2477.

Bunker, C. A., and Kingston, R. E. (1996). Activation domain-mediated enhancement of activator binding to chromatin in mammalian cells. Proc Natl Acad Sci USA *93*, 10820-10825.

Calhoun, V. C., and Levine, M. (2003). Coordinate regulation of an extended chromosome domain. Cell *113*, 278-280.

Calhoun, V. C., Stathopoulos, A., and Levine, M. (2002). Promoter-proximal tethering elements regulate enhancer-promoter specificity in the *Drosophila* Antennapedia complex. Proc Natl Acad Sci USA *99*, 9243-9247.

Carey, M., and Smale, S. T. (2000). Transcriptional regulation in eukaryotes: Concepts, Strategies, and Techniques (Cold Spring Harbor, New York, Cold Spring Harbor Laboratory Press).

Cavalli, G., and Paro, R. (1998). The *Drosophila* Fab-7 chromosomal element conveys epigenetic inheritance during mitosis and meiosis. Cell *93*, 505-518.

Chatterjee, S., and Struhl, K. (1995). Connecting a promoter-bound protein to TBP bypasses the need for a transcriptional activation domain. Nature *374*, 820-822.

Chi, T., and Carey, M. (1996). Assembly of the isomerized TFIIA-TFIID-TATA ternary complex is necessary and sufficient for gene activation. Gen & Dev *10*, 2540-2550.

Conaway, R. C., Brower, C. S., and Conaway, J. W. (2002). Emerging roles of ubiquitin in transcription regulation. Science *296*, 1254-1258.

Darnell, J. E., Jr., Kerr, I. M., and Stark, G. R. (1994). Jak-STAT pathways and transcriptional activation in response to IFNs and other extracellular signaling proteins. Science *264*, 1415-1421.

Davis, R. L., Cheng, P. F., Lassar, A. B., and Weintraub, H. (1990). The MyoD DNA binding domain contains a recognition code for muscle-specific gene activation. Cell *60*, 733-746.

de Bruin, D., Zaman, Z., Liberatore, R. A., and Ptashne, M. (2001). Telomere looping permits gene activation by a downstream UAS in yeast. Nature *409*, 109-113.

Dellino, G. I., Schwartz, Y. B., Farkas, G., McCabe, D., Elgin, S. C., and Pirrotta, V. (2004). Polycomb silencing blocks transcription initiation. Mol Cell *13*, 887-893.

Dorsett, D. (1999). Distant liaisons: long-range enhancer-promoter interactions in *Drosophila*. Curr Opin Genet Dev *9*, 505-514.

Driever, W., Ma, J., Nusslein-Volhard, C., and Ptashne, M. (1989). Rescue of bicoid mutant *Drosophila* embryos by Bicoid fusion proteins containing heterologous activating sequences. Nature *342*, 149-154.

Duncan, I. W. (2002). Transvection effects in *Drosophila*. Annu Rev Genet *36*, 521-556.

Epstein, J. A., and Neel, B. G. (2003). Signal transduction: an eye on organ development. Nature *426*, 238-239.

Farrell, S., Simkovich, N., Wu, Y., Barberis, A., and Ptashne, M. (1996). Gene activation by recruitment of the RNA polymerase II holoenzyme. Gen & Dev *10*, 2359-2367.

Fields, S., and Song, O. (1989). A novel genetic system to detect protein-protein interactions. Nature *340*, 245-246.

Fields, S., and Sternglanz, R. (1994). The two-hybrid system: an assay for protein-protein interactions. TIG *10*, 286-291.

Foley, K. P., and Engel, J. D. (1992). Individual stage selector element mutations lead to reciprocal changes in beta- vs. epsilon-globin gene transcription: genetic confirmation of promoter competition during globin gene switching. Genes Dev *6*, 730-744.

Freiman, R. N., and Tjian, R. (2003). Regulating the regulators: lysine modifications make their mark. Cell *112*, 11-17.

Fu, D., Zhao, C., and Ma, J. (2003). Enhancer sequences influence the role of the amino terminal domain of Bicoid in transcription. Mol Cell Biol *23*, 4439-4448.

Garvie, C. W., and Wolberger, C. (2001). Recognition of specific DNA sequences. Mol Cell *8*, 937-946.

Giardina, C., and Lis, J. T. (1993). DNA melting on yeast RNA polymerase II promoters. Science *261*, 759-762.

Gill, G. (2004). SUMO and ubiquitin in the nucleus: different functions, similar mechanisms? Genes Dev *18*, 2046-2059.

Gonzalez-Gouto, E., Klages, N., and Strubin, M. (1997). Synergistic and promoter-selective activation of transcription by recruitment of transcription factors TFIID and TFIIB. Proc Natl Acad Sci USA *94*, 8036-8041.

Gu, W., and Roeder, R. G. (1997). Activation of p53 sequence-specific DNA binding by acetylation of the p53 C-terminal domain. Cell *90*, 595-606.

Hahn, S. (2004). Structure and mechanism of the RNA polymerase II transcription machinery. Nat Struct Mol Biol *11*, 394-403.

Hampsey, M. (1998). Molecular Genetics of the RNA polymerase II general transcription machinery. Microbiol & Mol Biol Rev *62*, 465-503.

Hampsey, M., and Reinberg, D. (2003). Tails of intrigue: phosphorylation of RNA polymerase II mediates histone methylation. Cell *113*, 429-432.

Han, M., and Grunstein, M. (1988). Nucleosome loss activates yeast downstream promoters *in vivo*. Cell *55*, 1137-1145.

Herrera, F. J., and Triezenberg, S. J. (2004). Molecular biology: what ubiquitin can do for transcription. Curr Biol *14*, R622-624.

Hope, I. A., and Struhl, K. (1986). Functional dissection of a eukaryotic transcriptional activator protein, GCN4 of yeast. Cell *46*, 885-894.

Jackson, S. P., and Tjian, R. (1988). O-glycosylation of eukaryotic transcription factors: implications for mechanisms of transcriptional regulation. Cell *55*, 125-133.

Johnston, S. A., Salmeron, J. M., and Dincher, S. S. (1987). Interaction of positive and negative regulatory proteins in the galactose regulon of yeast. Cell *50*, 143-146.

Kadonaga, J. T. (2004). Regulation of RNA polymerase II transcription by sequence-specific DNA binding factors. Cell *116*, 247-257.

Kamemura, K., and Hart, G. W. (2003). Dynamic interplay between

O-glycosylation and O-phosphorylation of nucleocytoplasmic proteins: a new paradigm for metabolic control of signal transduction and transcription. Prog Nucleic Acid Res Mol Biol 73, 107-136.

Keegan, L., Gill, G., and Ptashne, M. (1986). Separation of DNA binding from the transcriptional-activating function of a eukaryotic regulatory protein. Science 231, 699-704.

Kim, Y., Geiger, J. H., Hahn, S., and Sigler, P. B. (1993). Crystal structure of a yeast TBP/TATA-box complex. Nature 365, 512-520.

Klein, C., and Struhl, K. (1994). Increased recruitment of TATA-binding protein to the promoter by transcriptional activation domains in vivo. Science 266, 280-282.

Krogan, N. J., Dover, J., Khorrami, S., Greenblatt, J. F., Schneider, J., Johnston, M., and Shilatifard, A. (2002). COMPASS, a histone H3 (Lysine 4) methyltransferase required for telomeric silencing of gene expression. J Biol Chem 277, 10753-10755.

Kuhn, E. J., and Geyer, P. K. (2003). Genomic insulators: connecting properties to mechanism. Curr Opin Cell Biol 15, 259-265.

Levine, M., and Tjian, R. (2003). Transcription regulation and animal diversity. Nature 424, 147-151.

Levine, S. S., King, I. F., and Kingston, R. E. (2004). Division of labor in polycomb group repression. Trends Biochem Sci 29, 478-485.

Li, X., Oghi, K. A., Zhang, J., Krones, A., Bush, K. T., Glass, C. K., Nigam, S. K., Aggarwal, A. K., Maas, R., Rose, D. W., and Rosenfeld, M. G. (2003). Eya protein phosphatase activity regulates Six1-Dach-Eya transcriptional effects in mammalian organogenesis. Nature 426, 247-254.

Li, X. Y., Virbasius, A., Zhu, X., and Green, M. R. (1999). Enhancement of TBP binding by activators and general transcription factors. Nature 399, 605-609.

Lis, J. T., Mason, P., Peng, J., Price, D. H., and Werner, J. (2000). P-TEFb kinase recruitment and function at heat shock loci. Genes Dev 14, 792-803.

Lue, N. F., Chasman, D. I., Buchman, A. R., and Kornberg, R. D. (1987). Interaction of GAL4 and GAL80 gene regulatory proteins in vitro. Mol Cell Biol 7, 3446-3451.

Lund, A. H., and van Lohuizen, M. (2004). Polycomb complexes and silencing mechanisms. Curr Opin Cell Biol 16, 239-246.

Ma, J. (2000). Yeast transcriptional activation and the two-hybrid system. In Yeast Hybrid Technologies, L. Zhu, and G. J. Hannon, eds. (Natick, MA, BioTechniques/Eaton Publishing), pp. 3-12.

Ma, J. (2004). Actively seeking activating sequences. Cell S116, S75-S76.

Ma, J. (2005). Crossing the line between activation and repression. Trends in Genetics 21, 54-59.

Ma, J., and Ptashne, M. (1987a). The carboxy-terminal 30 amino acids of GAL4 are recognized by GAL80. Cell 50, 137-142.

Ma, J., and Ptashne, M. (1987b). Deletion analysis of GAL4 defines two transcriptional activating segments. Cell 48, 847-853.

Ma, J., and Ptashne, M. (1987c). A new class of yeast transcriptional activators. Cell 51, 113-119.

Ma, J., and Ptashne, M. (1988). Converting a eukaryotic transcriptional inhibitor into an activator. Cell 55, 443-446.

Ma, X., Yuan, D., Diepold, K., Scarborough, T., and Ma, J. (1996). The Drosophila morphogenetic protein Bicoid binds DNA cooperatively. Development 122, 1195-1206.

Malik, S., and Roeder, R. G. (2000). Transcriptional regulation through Mediator-like coactivators in yeast and metazoan cells. Trends Biochem Sci 25, 277-283.

Mancebo, H. S., Lee, G., Flygare, J., Tomassini, J., Luu, P., Zhu, Y., Peng, J., Blau, C., Hazuda, D., Price, D., and Flores, O. (1997). P-TEFb kinase is required for HIV Tat transcriptional activation in vivo and in vitro. Genes Dev 11, 2633-2644.

Merika, M., and Thanos, D. (2001). Enhanceosomes. Curr Opin Genet Dev 11, 205-208.

Morcillo, P., Rosen, C., Baylies, M. K., and Dorsett, D. (1997). Chip, a widely expressed chromosomal protein required for segmentation and activity of a remote wing margin enhancer in Drosophila. Genes Dev 11, 2729-2740.

Muller, H. P., and Schaffner, W. (1990). Transcriptional enhancers can act in trans. Trends Genet 6, 300-304.

Myers, L. C., and Kornberg, R. D. (2000). Mediator of transcriptional regulation. Annu Rev Biochem 69, 729-749.

Naar, A. M., Lemon, B. D., and Tjian, R. (2001). Transcriptional coactivator complexes. Annu Rev Biochem 70, 475-501.

Narlikar, G. J., Fan, H. Y., and Kingston, R. E. (2002). Cooperation between complexes that regulate chromatin structure and transcription. Cell 108, 475-487.

Nevado, J., Gaudreau, L., Adam, M., and Ptashne, M. (1999). Transcriptional activation by artificial recruitment in mammalian cells. Proc Natl Acad Sci USA 96, 2674-2677.

Ng, H. H., Robert, F., Young, R. A., and Struhl, K. (2003). Targeted recruitment of Set1 histone methylase by elongating Pol II provides a localized mark and memory of recent transcriptional activity. Mol Cell 11, 709-719.

Nikolov, D. B., Hu, S.-H., Lin, J., Gasch, A., Hoffmann, A., Horikoshi, M., Chua, N.-H., Roeder, R. G., and Burley, S. K. (1992). Crystal structure of TFIID TATA-box binding protein. Nature 360, 40-46.

Orlando, V. (2003). Polycomb, epigenomes, and control of cell identity. Cell 112, 599-606.

Orphanides, G., Lagrange, T., and Reinberg, D. (1996). The general transcription factors of RNA polymerase II. Genes & Dev 10, 2657-2683.

Patikoglou, G., and Burley, S. K. (1997). Eukaryotic transcription factor-DNA complexes. Annu Rev Biophys Biomol Struct 26, 289-325.

Peterson, C. L., and Workman, J. L. (2000). Promoter targeting and chromatin remodeling by the SWI/SNF complex. Curr Opin Genet Dev 10, 187-192.

Prives, C., and Manley, J. L. (2001). Why is p53 acetylated? Cell

107, 815-818.

Ptashne, M. (1986). Gene regulation by proteins acting nearby and at a distance. Nature *322*, 697-701.

Ptashne, M. (1988). How eukaryotic transcriptional activators work. Nature *335*, 683-689.

Ptashne, M. (2004). Two "what if" experiments. Cell *S116*, S71-S72.

Ptashne, M., and Gann, A. (1997). Transcriptional activation by recruitment. Nature *386*, 569-577.

Ptashne, M., and Gann, A. (1998). Imposing specificity by localization: mechanism and evolution. Curr Biol *8*, R812-R822.

Ptashne, M., and Gann, A. A. F. (1990). Activators and targets. Nature *346*, 329-331.

Ranish, J. A., and Hahn, S. (1996). Transcription: basal factors and activation. Curr Opin Genet Dev *6*, 151-158.

Rasmussen, E. B., and Lis, J. T. (1993). *In vivo* transcriptional pausing and cap formation on three *Drosophila* heat shock genes. Proc Natl Acad Sci USA *90*, 7923-7927.

Rasmussen, E. B., and Lis, J. T. (1995). Short transcripts of the ternary complex provide insight into RNA polymerase II elongational pausing. J Mol Biol *252*, 522-535.

Rayapureddi, J. P., Kattamuri, C., Steinmetz, B. D., Frankfort, B. J., Ostrin, E. J., Mardon, G., and Hegde, R. S. (2003). Eyes absent represents a class of protein tyrosine phosphatases. Nature *426*, 295-298.

Rebay, I., Silver, S. J., and Tootle, T. L. (2005). New vision from Eyes absent: transcription factors as enzymes. Trends Genet *21*, 163-171.

Rougvie, A. E., and Lis, J. T. (1988). The RNA polymerase II molecule at the 5′-end of the uninduced hsp70 genes of *D. melanogaster* is transcriptionally engaged. Cell *54*, 795-804.

Sadowski, I., Ma, J., Triezenberg, S., and Ptashne, M. (1988). GAL4-VP16 is an unusually potent transcriptional activator. Nature *335*, 563-564.

Schena, M., Freedman, L. P., and Yamamoto, K. R. (1989). Mutations in the glucocorticoid receptor zinc finger region that distinguish interdigitated DNA binding and transcriptional enhancement activities. Genes Dev *3*, 1590-1601.

Sekinger, E. A., and Gross, D. S. (2001). Silenced chromatin is permissive to activator binding and PIC recruitment. Cell *105*, 403-414.

Sharpe, J., Nonchev, S., Gould, A., Whiting, J., and Krumlauf, R. (1998). Selectivity, sharing and competitive interactions in the regulation of Hoxb genes. Embo J *17*, 1788-1798.

Shi, Y., and Shi, Y. (2004). Metabolic enzymes and coenzymes in transcription--a direct link between metabolism and transcription? Trends Genet *20*, 445-452.

Sims, R. J., 3rd, Belotserkovskaya, R., and Reinberg, D. (2004a). Elongation by RNA polymerase II: the short and long of it. Genes Dev *18*, 2437-2468.

Sims, R. J., 3rd, Mandal, S. S., and Reinberg, D. (2004b). Recent highlights of RNA-polymerase-II-mediated transcription. Curr Opin Cell Biol *16*, 263-271.

Sims, R. J., 3rd, Nishioka, K., and Reinberg, D. (2003). Histone lysine methylation: a signature for chromatin function. Trends Genet *19*, 629-639.

Southgate, C. D., and Green, M. R. (1991). The HIV-1 Tat protein activates transcription from an upstream DNA-binding site: implications for Tat function. Genes Dev *5*, 2496-2507.

Spellman, P. T., and Rubin, G. M. (2002). Evidence for large domains of similarly expressed genes in the *Drosophila* genome. J Biol *1*, 5.

Spitz, F., Gonzalez, F., and Duboule, D. (2003). A global control region defines a chromosomal regulatory landscape containing the HoxD cluster. Cell *113*, 405-417.

Stargell, L. A., and Struhl, K. (1996). Mechanisms of transcriptional activation *in vivo*: two steps forward. Trends Genet *12*, 311-315.

Tanaka, M. (1996). Modulation of promoter occupancy by cooperative DNA binding and activation-domain function is a major determinant of transcriptional regulation by activators *in vivo*. Proc Natl Acad Sci USA *93*, 4311-4315.

Thanos, D., and Maniatis, T. (1995). Virus induction of human INFb gene expression requires the assembly of an enhanceosome. Cell *83*, 1091-1100.

Tollervey, D. (2004). Molecular biology: termination by torpedo. Nature *432*, 456-457.

Tootle, T. L., Silver, S. J., Davies, E. L., Newman, V., Latek, R. R., Mills, I. A., Selengut, J. D., Parlikar, B. E., and Rebay, I. (2003). The transcription factor Eyes absent is a protein tyrosine phosphatase. Nature *426*, 299-302.

Torigoi, E., Bennani-Baiti, I. M., Rosen, C., Gonzalez, K., Morcillo, P., Ptashne, M., and Dorsett, D. (2000). Chip interacts with diverse homeodomain proteins and potentiates bicoid activity *in vivo*. Proc Natl Acad Sci USA *97*, 2686-2691.

Travers, A. (2000). Recognition of distorted DNA structures by HMG domains. Curr Opin Struct Biol *10*, 102-109.

Triezenberg, S. J., Kingsbury, R. C., and McKnight, S. L. (1988). Functional dissection of VP16, the trans-activator of herpes simplex virus immediate early gene expression. Genes & Dev *2*, 718-729.

Wang, W., Carey, M., and Gralla, J. D. (1992). Polymerase II promoter activation: Closed complex formation and ATP-driven start-site opening. Science *255*, 450-453.

West, A. G., Gaszner, M., and Felsenfeld, G. (2002). Insulators: many functions, many mechanisms. Genes Dev *16*, 271-288.

Wu, C. (1997). Chromatin remodeling and the control of gene expression. J Biol Chem *272*, 28171-28174.

Wu, J., and Grunstein, M. (2000). 25 years after the nucleosome model: chromatin modifications. Trends Biochem Sci *25*, 619-623.

Wyrick, J. J., Holstege, F. C., Jennings, E. G., Causton, H. C., Shore, D., Grunstein, M., Lander, E. S., and Young, R. A. (1999). Chromosomal landscape of nucleosome-dependent gene expression and silencing in yeast. Nature *402*, 418-421.

Xiao, H., Friesen, J. D., and Lis, J. T. (1995). Recruiting TATA-binding protein to a promoter: transcriptional activation without an upstream activator. Mol Cell Biol *15*, 5757-5761.

Zhao, C., Dave, V., Fu, D., York, A., and Ma, J. (2003). Insights into the molecular functions of the *Drosophila* morphogenetic protein Bicoid. Recent Res Devel Mol Cell Biol *4*, 115-126.

Zhao, C., York, A., Yang, F., Forsthoefel, D. J., Dave, V., Fu, D., Zhang, D., Corado, M. S., Small, S., Seeger, M. A., and Ma, J. (2002). The activity of the *Drosophila* morphogenetic protein Bicoid is inhibited by a domain located outside its homeodomain. Development *129*, 1669-1680.

Zhou, Q., Chen, D., Pierstorff, E., and Luo, K. (1998). Transcription elongation factor P-TEFb mediates Tat activation of HIV-1 transcription at multiple stages. Embo J *17*, 3681-3691.

Zhu, Y., Pe'ery, T., Peng, J., Ramanathan, Y., Marshall, N., Marshall, T., Amendt, B., Mathews, M. B., and Price, D. H. (1997). Transcription elongation factor P-TEFb is required for HIV-1 tat transactivation *in vitro*. Genes Dev *11*, 2622-2632.

Zuniga, A., Michos, O., Spitz, F., Haramis, A. P., Panman, L., Galli, A., Vintersten, K., Klasen, C., Mansfield, W., Kuc, S., *et al.* (2004). Mouse limb deformity mutations disrupt a global control region within the large regulatory landscape required for Gremlin expression. Genes Dev *18*, 1553-1564.

Chapter 09

Transcriptional Repressors and Repression Mechanisms

Lorena Perrone, Hitoshi Aihara and Yutaka Nibu

Department of Cell and Developmental Biology, Weill Medical College of Cornell University, 1300 York Avenue, Box 60, A308, New York, NY 10021, USA

Key Words: transcriptional repressors, transcriptional repression, corepressor, short-range repression, long-range repression

Summary

A harmonious balance between transcriptional activation and repression in eukaryotes is necessary for a variety of biological phenomena, such as pattern formation, tissue differentiation, and normal development. In this chapter, we will use well-understood cases to provide an overview of the molecular mechanisms by which transcription factors mediate repression.

Introduction

Drosophila melanogaster (fruit fly) has 13,379 protein-coding genes and of which 700 encodings transcription factors (Adams *et al.*, 2000; Celniker and Rubin, 2003; Misra *et al.*, 2002). Of the 15,832 protein-coding genes identified in the primitive chordate, *Ciona intestinalis* (ascidian), 400 genes are transcription factors (Dehal *et al.*, 2002; Imai *et al.*, 2004). In genomes of other species, such as *Arabidopsis thaliana* (plant), *Caenorhabditis elegans* (worm), and *Saccharomyces cerevisiae* (yeast), 3-6% of the protein-coding genes are transcription factors (Riechmann *et al.*, 2000). Thus, genes encoding transcription factors are a relevant fraction of the genome, perhaps reflecting their crucial role in several biological processes.

Obviously, not all of the protein-coding genes are transcribed in any specific cell. For instance, 70% of the *Drosophila* genes tested are transcribed tissue-specifically during a wide range of embryonic stages (Tomancak *et al.*, 2002). Conversely, these genes are not transcribed in some tissues. In addition, 20% of the fly genes are maternally expressed, while 15% are not expressed during the entire embryogenesis. In *Ciona intestinalis* tailbud embryos, 30% of the genes tested are specifically expressed in only one tissue, such as epidermis, nervous system, endoderm, mesenchyme, notochord, and muscle (Satou *et al.*, 2001). Another 30% of the genes are expressed in multiple tissues and expression of 12% of the genes is not detected by *in situ* hybridization. How is such a large number of genes expressed (or not) at the right moment and at the right place? A well-balanced program of gene expression is mostly regulated at the level of transcription. Ultimately, the coordinate expression of different sets of genes directs normal development, differentiation, and morphogenesis.

For example, cooperation of transcriptional repression and activation is essential for establishing localized stripes, bands, and tissue-specific patterns of gene expression in the early *Drosophila* embryo (Gray and Levine, 1996b; Ip and Hemavathy, 1997; Jackle *et al.*, 1992; Mannervik *et al.*, 1999). The initial formation of both the anterior-posterior and dorsal-ventral axes during early *Drosophila* embryogenesis depends on broadly distributed activators that are maternally expressed as well as localized sequence-specific repressors. In brief, after fertilization, the maternal activators begin turning on a set of zygotic genes called "gap genes" that mostly encode DNA-binding repressors. The same maternal activators also activate expression of a second set of downstream genes, called "pair-rule genes" that are expressed in patterns of seven stripes. The seven stripes are evenly spaced along the anterior-posterior axis. The borders of individual stripes are formed by the localized repressors (gap gene products)

Corresponding Author: Yutaka Nibu, Tel: +1-(212) 746-6184, Fax: +1-(212) 746-8175, E-mail: yun2001@med.cornell.edu

which turn off transcription via a concentration threshold mechanism. Thus, the pair-rule genes define the initial metameric layout of the body plan in *Drosophila*.

Transcription of protein-coding genes driven by RNA polymerase II can be repressed or modulated at a local or at a genomic level. For instance, expression of some genes is locally repressed by DNA-binding transcriptional repressors. This process is the major focus of this chapter. On the other hand, some genes are silenced along with large regions of the genome, genes located near centromeres and telomeres, nearly all genes on one of the two mammalian female X chromosomes, and Hox genes, are repressed.

Transcriptional Repressors

In general, transcriptional repressors affecting RNA polymerase II dependent transcription can be classified into two categories, DNA-binding and non-DNA-binding factors (Burke and Baniahmad, 2000; Gaston and Jayaraman, 2003; Hanna-Rose and Hansen, 1996). DNA-binding repressors bind sequence-specifically to DNA via DNA-binding domains, such as zinc finger, homeodomain, basic helix-loop-helix (bHLH), basic leucine zipper (bZip), and others. In addition to the DNA-binding domain, these factors usually have repression domains that mediate "active" repression. It has been shown in the early *Drosophila* embryo that sequence-specific DNA-binding repressors can be classified as either short-range or long-range repressors, depending upon their range of action (Courey and Jia, 2001; Gray and Levine, 1996b; Mannervik *et al.*, 1999; Nibu *et al.*, 1998a). Non-DNA-binding factors are typically termed corepressors and can either be recruited to DNA by DNA-binding repressors or directly interact with the components of the preinitiation complex.

Mechanisms of Repression

In the past two decades, by the virtue of the dramatic advancement of experimental techniques and molecular tools, several molecular mechanisms have been proposed to explain how transcriptional repression is achieved (Burke and Baniahmad, 2000; Gaston and Jayaraman, 2003; Hanna-Rose and Hansen, 1996; Johnson, 1995; Levine and Manley, 1989). Here we describe the molecular mechanisms of how eukaryotic protein-coding genes are repressed.

A: Inhibition of TBP and General Transcription Factors (GTFs)

Recruitment of TATA-binding protein (TBP) to eukaryotic promoters is an essential step for the initiation of transcription. Extensive analyses of TBP have shown that the function of TBP can be inhibited by direct binding of several factors, such as human BTAF1 (TAF-172) and its yeast ortholog Mot1p, the Dr1/Drap1 complex (also known as NC2), and TAF1 (*Drosophila* TAFII230) (Burley and Roeder, 1998; Lee and Young, 1998; Pugh, 2000).

Human BTAF1 and yeast Mot1p are members of the evolutionarily conserved SWI/SNF family, a DNA-dependent ATPase (Pereira *et al.*, 2003). BTAF1/Mot1p are large proteins (210-kDa) and Mot1p is essential for yeast cell viability. BTAF1/Mot1p directly interact with TBP and are able to remove TBP from the TATA box using ATPase activity (Fig. 9.1A) (Auble *et al.*, 1994; Chicca *et al.*, 1998). Hence, Mot1p negatively regulates transcription by impeding the binding of TBP to DNA. In contrast, it has been shown that, in some cases, Mot1p can also activate a few genes (Andrau *et al.*, 2002; Dasgupta *et al.*, 2002; Geisberg *et al.*, 2002; Prelich, 1997). Microarray analysis revealed that 178 genes (3% of yeast genes) are repressed by Mot1p, while 6 genes are activated (Dasgupta *et al.*, 2002). Both northern blot and microarray analyses have demonstrated that transcription of *BNA1*, *URA1*, and *YDR539W* genes is decreased in *mot1* mutant yeast, while the *INO1* gene is activated. An ATPase-defective Mot1p introduced in the *mot1* mutant yeast could not rescue their expression levels. Hence, the ATPase activity of Mot1p is essential for both repression and activation.

The Dr1/Drap1 complex consists of two subunits, Drap1 (NC2alpha) and Dr1 (NC2beta), and is conserved among eukaryotes and yeast (Lee and Young, 1998). *In vitro* studies suggested that mammalian Drap1 heterodimerizes with Dr1 through the histone fold motifs at their amino-terminal ends, and increases the repression activity of Dr1 (Fig. 9.1B) (Goppelt *et al.*, 1996; Mermelstein *et al.*, 1996). The Dr1 subunit directly interacts with the basic repeat domain of TBP bound to the TATA box. This interaction blocks the subsequent recruitment of TFIIA and TFIIB to the core promoter, thereby resulting in repression. X−ray structural studies also support this mechanism (Kamada *et al.*, 2001). However, additional studies on Dr1 and Drap1 indicate alternative modes of action. For example, mice lacking Drap1 exhibit severe gastrulation defects, likely due to an increased expression of Nodal (Iratni *et al.*, 2002). In this case, mouse Drap1, but not Dr1, is sufficient to prevent DNA binding of the FoxH1 activator that regulates the *Nodal* gene. Moreover, the Dr1/Drap1 complex can be purified from postdiauxic yeast, however, Drap1 does not associate with Dr1 in growing yeast (Creton et al., 2002). In chromatin immunoprecipitation assays, both yeast Drap1 and TBP associate with active promoters, while the Dr1/Drap1 complex is found on repressed promoters. Finally, in vitro transcription assays, the Drosophila Dr1/Drap1 complex purified from embryonic nuclear extract represses TATA-box containing promoters, but it activates promoters containing the downstream promoter element (DPE) (Willy et al., 2000).

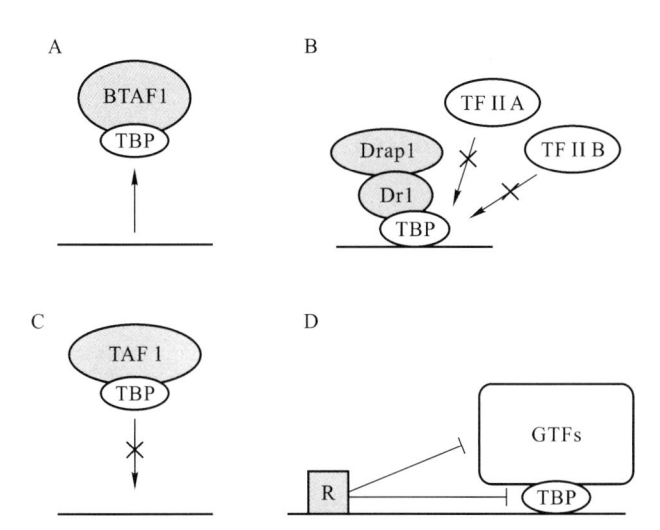

Fig. 9.1 Inhibition of TBP and general transcription factors (GTFs). TBP function can be inhibited by BTAF1/Mot1p **(A)**, Drap1/Dr1 **(B)**, and TAF1 **(C)**. **(D)** DNA-binding repressors (R) inhibit TBP and GTFs. The black bars represent double-stranded DNA.

Recent studies suggest that human Drap1 interacts with the central region of BTAF1 in vitro and in yeast (Klejman et al., 2004). The physical interaction between Drap1 and BTAF1 may account for the regulation of the same genes by these two factors in yeast (Lemaire et al., 2000; Prelich, 1997).

TAF1 (formerly named TAF250 in human and TAF230 in Drosophila) is the largest subunit of TFIID complex (Pugh, 2000). TAF1 contains multiple enzymatic activities, a histone acetyltransferase, a serine/threonine kinase, and a histone-specific ubiquitin-activating/conjugating enzyme (Dikstein et al., 1996; Mizzen et al., 1996; Pham and Sauer, 2000). TAF1 has bromodomains that bind to acetylated histone H4 (Jacobson et al., 2000). In vitro studies displayed that the 80-residue N-terminal region of Drosophila TAF1, containing three alpha helices and a beta hairpin, binds directly to TBP and inhibits the TBP function (Fig. 9.1C) (Kokubo et al., 1994; Nishikawa et al., 1997). Subsequent NMR structural studies revealed that the structure of the

N-terminal region of Drosophila TAF1 is similar to the minor groove surface of the TATA box DNA sequence and that the TATA box-binding domain of TBP binds to this region of TAF1 (Liu et al., 1998). Thus, TAF1, by mimicking the TATA box, can prevent TBP from binding to DNA. Additional studies in yeast suggested the following two-step hand off model (Kotani et al., 2000). The interaction between TAF1 and TBP can be compromised by activators containing acidic activation domains that also interact with the TATA box-binding domain of TBP. The intermediate complex containing TBP and the activator can not bind the TATA box. Finally, TBP is released from the activator and can bind the TATA box to initiate transcription.

It is conceivable that TBP may need to be in an inactive state before activators turn on transcription, and/or that TBP may need to be removed from DNA to turn off transcription after the genes receive the signal leading to transcriptional repression.

In addition to the factors mentioned above, DNA-binding repressors also directly interact with TBP or GTFs to inhibit transcription (Fig. 9.1D). This mechanism is called "direct repression". For instance, Even-skipped (Eve), a Drosophila homeodomain protein, interacts with TBP in vitro and in tissue culture (Austin and Biggin, 1995; Li and Manley, 1998; Um et al., 1995). A repression domain of Eve directly contacts the C-terminal region of TBP, leading to inhibition of TFIID binding to the promoter. This repression is independent of the distance between the promoter and the Eve binding sites in vitro transcription assays. Unliganded thyroid hormone receptor (TR) also interacts with TBP and inhibits the formation of preinitiation complex in vitro, however TR bound to its ligand activates transcription by interacting with TFIIB (Baniahmad et al., 1993; Fondell et al., 1996; Fondell et al., 1993). Similarly, TFIIEbeta binds to a dimeric form of Krüppel in vitro, a Drosophila zinc finger protein, through its C-terminal domain (Sauer et al., 1995). This direct association is sufficient to cause repression and hence this is a corepressor-independent mechanism. On the contrary, a monomer form of Krüppel acts as an activator that interacts with TFIIB in vitro (Sauer et al., 1995). It should be noted that both Krüppel and Eve interact with corepressors: dCtBP, dRpd3, Groucho, and Atrophin, in vitro and/or in yeast, and that the corepressors are required for their repression activities in Drosophila genetic assays (Kobayashi et al., 2001; Mannervik and Levine, 1999; Nibu et al., 2003; Nibu et al., 1998a; Zhang et al., 2002b).

B: Short-range Repression

The DNA-binding repressors expressed in the

Drosophila embryo appear to fall into two categories, short-range or long-range repressors, depending upon their range of action (Courey and Jia, 2001; Gray and Levine, 1996b; Mannervik et al., 1999). Short-range repression takes place through three mechanisms, quenching (B1), direct repression (B2), and competition (B3).

B1: Quenching

An activator bound to DNA can be inactivated by an adjacent repressor and this mechanism is called quenching (Fig. 9.2B). Short-range repressors, such as Krüppel (Cys2His2 zinc fingers), Knirps (nuclear receptor), Snail (Cys2His2 zinc fingers), and Giant (bZip), work over distances of less than 100 bp to inhibit adjacent activators in the *Drosophila* embryo (Arnosti et al., 1996b; Gray and Levine, 1996a; Gray et al., 1994; Hewitt et al., 1999; Nibu and Levine, 2001). When binding sites for these short-range repressors are located within 100bp from activator binding sites, these repressors inhibit these adjacent activators in transgenic *Drosophila* embryos. However, repression is lost when the repressor sites are moved more than 100 bp away from the activator sites.

A short-range repressor bound to one enhancer is not able to interfere with activators bound to a neighboring enhancer. For instance, the pair-rule *eve* gene carries five enhancers located 5′ and 3′ of the transcription unit and each of them controls one or two stripes (Clyde et al., 2003; Fujioka et al., 1999; Small et al., 1996; Small et al., 1991). These enhancers are typically 300 bp to 1 kb in length and contain clustered binding sites for both activators and repressors (Berman et al., 2002). Even though expression of *Krüppel* and *eve* stripe 3 overlap, the binding of Krüppel to the *eve* stripe 2 enhancer to form the posterior border does not interfere with stripe 3 expression (Gray and Levine, 1996b; Small et al., 1993; Small et al., 1992; Small et al., 1991). Similarly, the anterior border of stripe 2 is defined by the binding of Giant repressor to the stripe 2 enhancer, while Giant expression overlaps with stripe 1 (Arnosti et al., 1996a; Small et al., 1992; Small et al., 1991). The maternal D-Stat activator turns on the expression of *eve* stripe 3 and 7, while Knirps establishes the borders of *eve* stripe 3/7 and 4/6 in a concentration dependent manner (Clyde et al., 2003; Struffi et al., 2004). However, binding of Knirps to the *eve* stripe 3/7 and 4/6 enhancers does not inhibit expression of *eve* stripe 5. Quenching is also widely employed for establishment of the dorsal-ventral axis. The maternal Dorsal (rel domain) nuclear gradient activates *rhomboid, short gastrulation, singleminded,* and *ventral nervous system defective* genes, in both ventral and lateral regions of early embryos, but the Snail repressor keeps these genes off in the ventral mesoderm (Cowden and Levine, 2003; Ip et al., 1992; Nibu et al., 1998a; Stathopoulos and Levine, 2002).

Interestingly, removal of the corepressor *Drosophila* CtBP (dCtBP) that interacts with Krüppel, Knirps, and Snail, through the conserved PxDLS motif, impairs the activity of these repressors in the *Drosophila* embryo (Nibu et al., 1998a; Nibu et al., 1998b). Thus, quenching is a corepressor-dependent repression.

What is the molecular mechanism by which dCtBP operates? dCtBP is similar to NAD^+- dependent D-isomer-specific 2-hydroxy acid dehydrogenases, which are metabolic enzymes such as pyruvate dehydrogenase (Chinnadurai, 2002; Turner and Crossley, 2001). Human CtBP1 (hCtBP1) has been shown to be a functional dehydrogenase (Balasubramanian et al., 2003; Kumar et al., 2002; Shi et al., 2003). It is known that NAD^+/NADH binding to hCtBP1 enhances the interaction with the adenovirus E1A oncoprotein containing the PxDLS motif and then facilitates the oligomerization of hCtBP1 itself (Balasubramanian et al., 2003; Kumar et al., 2002; Zhang et al., 2002a). Additional studies indicated that the dehydrogenase domain of hCtBP1 is essential for its repression activity in tissue culture (Kumar et al., 2002). However, it has been argued that a mutation of the catalytic histidine residue does not alter the repression activity of mouse CtBP2, dCtBP, or hCtBP1 in transient reporter assays (Grooteclaes et al., 2003; Phippen et al., 2000; Sutrias-Grau and Arnosti, 2004; Turner and Crossley, 1998). A complex containing hCtBP1 tagged with both the Flag and haemagglutinin (HA) epitopes has been purified from HeLa cells. The purified fraction contains histone methyltransferases (HMT) and histone deacetylases (HDAC) (Shi et al., 2003). These results suggest that the CtBP complex represses transcription by directing deacetylation and methylation of histones through HDACs and HMTs. However, it should be pointed out that, in *Drosophila* S2 cells, the HDAC inhibitor TSA did not inhibit dCtBP-mediated repression, but blocked the repression activity of Groucho which interacts with the histone deacetylase dRpd3 (Ryu and Arnosti, 2003).

B2:Short-range Direct Repression

When the binding sites for the short-range repressors (Krüppel, Knirps, Snail, and Giant) are located within 100 bp from the core promoter, these factors can dominantly shut down the promoter activity regulated by multiple enhancers in the transgenic *Drosophila* embryo (Fig. 9.2C) (Arnosti *et al.*, 1996b; Gray and Levine, 1996a; Gray *et al.*, 1994; Hewitt *et al.*, 1999). In this case, these repressors do not directly inhibit the enhancers, since the enhancers can actually activate the other linked promoter which lacks the repressor sites. In addition, when these sites are moved away from the core promoter, repression is lost. The direct repression activity of Krüppel and Knirps, as monitored by transgenes, is diminished in the *dCtBP* mutant embryo, indicating that short-range direct repression is corepressor-dependent (Nibu *et al.*, 2003). In contrast, Krüppel-mediated repression detected *in vitro* is corepressor-independent, as mentioned earlier.

B3: Competition

Some repressors compete with activators for identical or overlapping DNA sequences (Fig. 9.2D). This is often termed "passive" repression, in contrast to the corepressor-dependent repression, which is called "active" repression.

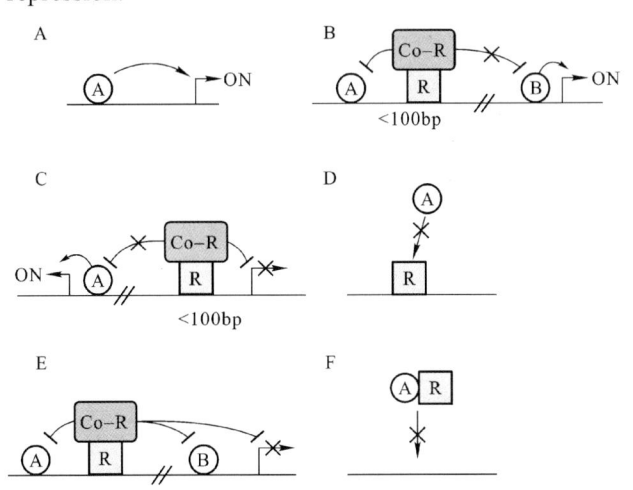

Fig.9.2 Transcriptional repression by DNA-binding repressors (R) and corepressors (Co-R). (A) Distal DNA-binding activators (A) can turn on transcription by contacting the preinitiation complex formed on the core promoter. (B) Quenching: a complex containing a short-range repressor (R) and a corepressors (Co-R) inhibits an activator-A (A) bound to an enhancer within 100 bp away from the repressor, while an activator-B (B), located more than 100 bp away from the repressor, can initiate transcription. (C) Direct repression: the repressor/corepressor complex binds within 100 bp from the transcription start site and directly blocks the promoter activity. (D) Competition: activators and repressors can compete for common binding sites. This repression does not require corepressors. (E) Long-range repression. The repressor/corepressor complex blocks transcription over the distances of more than 1 kb from its binding site. (F) Inhibition of activators by their heterodimeric partners.

Many liver specific genes are activated by a nuclear receptor HNF-4 (Hayashi *et al.*, 1999). The homodimeric form of HNF-4 binds a direct repeat of hexamer sequence, typically AGGTCA or related sites (Jiang and Sladek, 1997). Other nuclear receptors, COUP-TFI and COUP-TFII (also known as ARP-1), wich are ubiquitously expressed, can recognize the HNF-4 binding sites (Kimura *et al.*, 1993; Ladias *et al.*, 1992; Mietus-Snyder *et al.*, 1992). COUP-TFs inhibit HNF-4 dependent activation by competing for binding to the common sites in cotransfection assays. COUP-TFs can also occupy binding sites for other nuclear receptors, PPAR, VDR, TR, RAR, and SF1, leading to repression of their target genes (Park *et al.*, 2003; Pereira *et al.*, 2000).

Sp1 family proteins, Sp1, Sp3, and Sp4, regulate gene expression through GC and GT boxes in cellular and viral promoters (Lania *et al.*, 1997; Suske, 1999). These factors contain three highly conserved Cys2His2 zinc fingers that bind DNA with similar affinities. Sp1 and Sp3 are ubiquitously expressed, while Sp4 is a brain-specific factor. Sp1 and Sp4 were originally characterized as activators, while Sp3 acts as both a repressor and a weak activator. Cotransfection assays in *Drosophila* Schneider SL2 cells lacking endogenous Sp activity demonstrated that Sp3 can repress expression of a reporter gene activated by Sp1 via competition (Majello *et al.*, 1997; Yu *et al.*, 2003).

Brinker has a helix-turn-helix (HTH) DNA-binding domain and negatively regulates TGFbeta (Dpp) signaling pathway in *Drosophila* embryos and wing imaginal discs (Campbell and Tomlinson, 1999; Jazwinska *et al.*, 1999; Minami *et al.*, 1999). This signaling pathway triggers phosphorylation of the DNA-binding factor Mad (a member of the Smad family) and the activated Mad then turns on downstream target genes. Several lines of evidence indicate that Mad binding sites in the enhancers controlling *zerknüllt* (*zen*) and *Ultrabitorax* (*Ubx*) genes are also recognized by Brinker which therefore competes with Mad (Kirkpatrick *et al.*, 2001; Rushlow *et al.*, 2001; Saller and Bienz, 2001).

Bicoid, a homeodomain protein, governs development of the anterior region of the early *Drosophila* embryo by activating gap genes, pair-rule genes, and other target genes (Ochoa-Espinosa *et al.*, 2005). Some of the Bicoid binding sites located within embryonic enhancers overlap with binding sites for the gap gene products, Knirps, Giant, and Krüppel. Cotransfection assays show that increasing amounts of Knirps gradually eliminates Bicoid-mediated activation in a 16-bp element found within a *Krüppel* enhancer which contains overlapping sites for Bicoid and Knirps

(Hoch et al., 1992). In transgenic embryos, the 16-bp element fused to the *lacZ* reporter is repressed by overexpression of Knirps and is not expressed in the absence of Bicoid activity, consistent with a mechanism of repression by competition. Furthermore, *in vitro* gel shift assays, cotransfection assays, and reporter assays in transgenic embryos show that transcriptional activation by some Bicoid sites within the *eve* stripe 2 enhancer is repressed by Krüppel and Giant through competition (Arnosti et al., 1996a; Small et al., 1992; Small et al., 1991; Stanojevic et al., 1991). For the competition activity of Krüppel, the dCtBP corepressor is not required in transgenic embryos (Nibu et al., 2003).

It should be noted that most of the repressors mentioned in this section have active repression domains that are known to interact with corepressors. For example, N-CoR/SMRT is a corepressor for COUP-TFs, dCtBP is a corepressor for Knirps, Krüppel, and Brinker, and Groucho is a corepressor for Brinker.

C: Long-range Repression

Long-range repressors, such as Hairy (bHLH) and Dorsal, can function over distances of more than 1 kb to silence the transcription complex in the *Drosophila* embryo, thereby resulting in simple on/off patterns of gene expression (Fig. 9.2E) (Barolo and Levine, 1997; Cai et al., 1996). Dorsal is the major activator for genes expressed along the dorsal-ventral axis, but it also mediates long-range repression on the 600 bp *zen* ventral repression element together with additional DNA-binding factors, Cut, Dead Ringer, and Capicua (Dubnicoff et al., 1997; Jimenez et al., 2000; Valentine et al., 1998). A corepressor for long-range repressors is Groucho, a WD-repeat containing protein (Paroush et al., 1994; Dubnicoff et al., 1997), which interacts genetically and physically with dRpd3 (HDAC1) (Chen et al., 1999), although a genetic interaction between *dRpd3* and *hairy* has not been observed (Rosenberg and Parkhurst, 2002). The Courey group has proposed a spreading model to explain how Groucho mediates repression (Courey and Jia, 2001; Song et al., 2004). A ternary complex containing repressors, Groucho, and dRpd3 removes acetyl residues from nearby histones. Subsequently, Groucho is polymerized and covers the chromatin template through the deacetylated histones, as it has been shown that Groucho predominantly binds to the deacetylated histones *in vitro*. The corepressor polymer ultimately leads chromatin to a repressed state.

dCtBP-dependent short-range and Groucho-dependent long-range repression are qualitatively different. Dorsal/Groucho mediated long-range repression is clearly dCtBP-independent (Nibu et al., 1998a). Repression of *zen* expression is normal in the *dCtBP* mutant embryo. Hairy/Groucho mediated long-range repression and Krüppel/dCtBP mediated short-range repression are also different (Nibu et al., 2001). For example, when Krüppel is misexpressed in the ventral region it represses the *hairy* 6 stripe enhancer in a dCtBP-dependent manner. In fact, disruption of the PEDLSMH motif, which is necessary for interaction with dCtBP, abolishes this repression. Addition of the Hairy repression domain including the Groucho interaction motif (WRPW) to the Krüppel protein leads to the repression of both stripes 5 and 6, indicating long-range repression. Thus, it is clear that Groucho and dCtBP mediate two separate modes of repression, long-range and short-range, respectively.

Recently, it has been clearly demonstrated that *Drosophila* Heterochromatin Protein 1 (HP1) mediates long-range repression (silencing) within euchromatin by two mechanisms (Danzer and Wallrath, 2004). HP1 spreads along DNA bidirectionally from its entry site and alters chromatin structure. The chromo domain of HP1 binds a methylated lysine 9 of histone H3 generated by the histone methyltransferase SU(VAR)3-9. This study also demonstrates that repression by HP1 up to 1.9 kb is likely due to self-propagation, but repression over longer distances is SU(VAR)3-9-dependent.

D: Activator Inhibition is Dependent on Heterodimeric Partners

Heterodimeric partners lacking functional DNA-binding domains can make DNA-binding defective dimers with activators (Fig. 9.2F). For example, a heterodimer of two bHLH proteins, MyoD and E12/E47, promotes myogenesis (Lassar et al., 1991), however, dimerization of Id with MyoD inhibits both DNA binding of MyoD and muscle differentiation (Benezra et al., 1990; Jen et al., 1992). The Id protein lacks the basic domain that binds to DNA and is usually adjacent to the HLH dimerization motif. Similarly, CHOP (also known as C/EBPzeta), a member of C/EBP family, contains the leucine zipper dimerization domain but a defective basic domain (Ramji and Foka, 2002; Ron and Habener, 1992). When heterodimerized with CHOP, other members of the C/EBP family that have the basic leucine zipper domain fail to activate transcription due to their inability to bind DNA.

Heterodimeric partners lacking activation domains can repress transcription. NF-E2 p45 hetrodimerizes with small Mafs (MafF, MafG, and MafK) through the bZip domain (Igarashi et al., 1994; Motohashi et al.,

1997). This heterodimer is able to activate expression of erythroid-specific genes. The small Mafs lack a canonical activation domain and their homodimers bind the p45/small Mafs activator site. As a result, the small Maf homodimers act as repressors as they compete for the same binding site. Similarly, dimerization of bHLH-Zip proteins, Myc, Max and Mad, regulate transcription through their DNA binding site called E-box (Grandori *et al.*, 2000; Laherty *et al.*, 1997; Luscher, 2001). A Max/Myc heterodimer that recruits SWI/SNF complex works as an activator, however a Max/Max homodimer bound to the E-box keeps transcription off passively. Furthermore, Max can heterodimerize with Mad, which in turn interacts with the HDAC corepressor complex and hence the Max/Mad heterodimer bound to the E-box actively represses transcription.

E: Repression at the Chromatin Level

As discussed in other chapters, both histone and DNA modifications are involved in repression (Fig. 9.3A). In brief, deacetylation of histones by HDACs might establish repressive chromatin (Thiel *et al.*, 2004; Yang and Seto, 2003). In addition, the lysine-specific methylation of histones by SU(VAR)3-9 is essential for the formation of heterochromatin (Jenuwein, 2001). DNA methylation is also correlated with transcriptional repression (Wade, 2001). DNA methylation changes the chromatin structure by affecting nucleosome position and stability. In particular, a STAT3 activator binding site containing the dinucleotide sequence CpG in the GFAP promoter (an astrocyte marker gene) is highly methylated in astrocyte precursor cells (Takizawa *et al.*, 2001). In differentiated astrocytes, demethylation of the STAT3 binding site allows binding of STAT3 to the GFAP promoter, resulting in the activation of the GFAP gene.

In this section, we will describe epigenetic transcriptional repression involving the Polycomb group proteins (PcG). PcG proteins maintain Hox genes stably and heritably silenced during development of *Drosophila* and vertebrates (Gould, 1997; Orlando, 2003; Orlando and Paro, 1995; Otte and Kwaks, 2003; Pirrotta, 1999). *Drosophila* PcG complexes inhibit transcription upon binding to specialized DNA elements, known as Polycomb response elements (PREs), while PREs have not been identified in mammalian genomes (Orlando, 2003). Chromatin immunoprecipitation assays show that *Drosophila* PcG is not only associated to the PRE sequences near the *Ubx* gene, but also to adjacent DNA regions over several thousand base pairs (Orlando *et al.*, 1998; Orlando and Paro, 1993). In contrast, PcG binding is not detected near active homeotic genes. Spreading of the PcG proteins along chromatin fibers perhaps prevents activators from binding to DNA (Fig.9.3C) (Fitzgerald and Bender, 2001; Zink and Paro, 1995). *Drosophila* PcG proteins consist of up to 15 genes and can be separated into two classes biochemically. The 2-6 MDa Polycomb repressive complex 1 (PRC1) contains Polycomb (Pc), Polyhomeotic (Ph), Posterior sex combs (Psc), Zeste, dSbf1, and Ring1, while the 400-600 kDa Polycomb repressive complex 2 (PRC2) is composed of Extra sex combs (Esc), Enhancer of zeste (E(z)), Pleiohomeotic (Pho), Suppressor 12 of zeste (Su(z)12), and NURF-55 (Lund and van Lohuizen, 2004; Otte and Kwaks, 2003). Two evolutionarily conserved complexes are found also in mammals: PRC1 contains HPC1, HPC2, HPC3, HPH1, HPH2, HPH3, RNF110, and RING1, while PRC2 contains EZH1, EZH2, EED, YY1, and SUZ12 (Lund and van Lohuizen, 2004; Otte and Kwaks, 2003). Pho and YY1 are the only sequence-specific DNA binding factors in the PcG complexes. E(z) contains a SET domain and is a methyltransferase that preferentially facilitates the methylation of histone H3 at lysine 27 (Cao and Zhang, 2004). Both *Drosophila* E(z) and mammalian EZH2 interact with HDACs (Chang *et al.*, 2001; Tie *et al.*, 2001; van der Vlag and Otte, 1999). The E(z)/EZH2-HDAC interaction may function as a "silencing core" in which deacetylation and subsequent methylation of histones induce transcriptional repression. Thus, histone H3 is epigenetically marked by the methylation through the PRC2 complex. The methylated histone H3 then serves as a recognition site for the chromodomain of Pc in the PRC1 complex (Cao *et al.*, 2002; Czermin *et al.*, 2002). The PRC1 complex bound to DNA inhibits chromatin remodeling by the SWI/SNF complex *in vitro* (Shao *et al.*, 1999).

Spreading of the PcG complex along DNA was thought to convert the chromatin to a repressed state, thus blocking the access of activators. In contrast, recent reports suggest an alternative "looping" model (Fig. 9.3D) (Orlando, 2003; Pirrotta *et al.*, 2003; Wang *et al.*, 2004). This model is supported by the evidence that the *Drosophila* PcG complex can be purified together with TBP and TAFs, and that the PcG complex is found in the core promoter regions together with components of preinitiation complex, TBP, TFIIB, and TFIIF (Breiling *et al.*, 2001; Saurin *et al.*, 2001). Perhaps, the methylation of histone H3 by the PRC2 complex defines the landing zone for the PRC1 complex to PREs. Subsequently, PRC1 bound to PRE can contact TFIID and other GTFs on the core promoter by looping, thereby inhibiting transcription. In fact, it has been recently reported that PcG-mediated repression blocks the formation of the preinitiation complex (Dellino *et al.*, 2004).

It should be noted that expression of each component of human PcG is different between tissues, cell types, and developmental stages. These observations suggest that the PcG complexes containing different components may regulate distinct target genes (Gunster *et al.*, 2001).

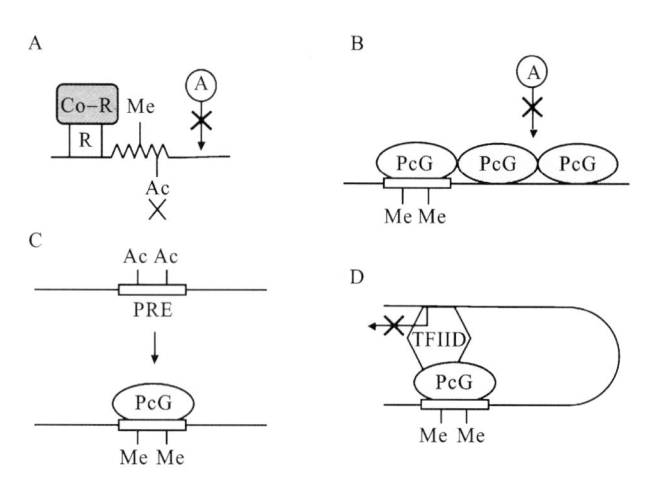

Fig. 9.3 **Repression by chromatin-remodeling factors.** (A) Levels of histone acetylation are reduced by HDACs (crossed "Ac"), while levels of histone methylation (Me) are increased by histone methyl transferases. DNA can be methylated by DNA methyltransferases (indicated by a zig-zag). These modifications change the chromatin structure, possibly preventing activators from binding DNA. (B) Repression mediated by Polycomb group proteins (PcG). The PcG complex represses transcription by either spreading to cover DNA (C) or inhibiting the promoter activity via a looping mechanism (D).

F: Role of Multiple Repression Domains

It is known that repressors often have multiple repression domains. These domains function qualitatively and/or quantitatively different, or act in a tissue/cell-type specific manner.

Two evolutionarily conserved repression domains of Krüppel have been characterized using *Drosophila* S2 cells, non *Drosophila* cell lines CV-1 and U2OS, and *Drosophila* transgenic embryos (Hanna-Rose *et al.*, 1997; Licht *et al.*, 1994; Nibu *et al.*, 2003; Nibu *et al.*, 1998a; Nibu *et al.*, 2001; Sauer *et al.*, 1995). The C-terminal repression domain (402-502aa), which is dCtBP-dependent *in vivo* and TFIIbeta-dependent *in vitro*, can function in all of the cell types tested. On the other hand, the N-terminal repression domain (62-92aa) is active in transfected CV-1 cells (Hanna-Rose *et al.*, 1997; Licht *et al.*, 1994), but is dispensable in the *Drosophila* blastoderm embryo (Nibu *et al.*, 2003), suggesting a tissue/cell-type specific role.

Two repression domains of Knirps, one dCtBP-dependent and the other dCtBP-independent domains, function in transgenic embryos and in S2 cells (Keller *et al.*, 2000; Struffi *et al.*, 2004). In *dCtBP* mutant embryos, expression of *eve* stripes 4/6 is derepressed due to the loss of Knirps-mediated repression, but Knirps continues to repress the *eve* stripes 3/7 enhancer. In contrast, the mutant form of Knirps lacking the dCtBP-dependent repression domain is able to repress not only the *eve* stripes 3/7 enhancer but also the *eve* stripes 4/6 enhancer in a concentration-dependent manner. These results indicate that the two domains contribute to the full repression activity of Knirps quantitatively rather than qualitatively.

Brinker is known to interact with both Groucho and dCtBP. Brinker carries at least three repression domains, a dCtBP interacting domain, a Groucho interacting domain, and a newly identified region called 3R (Hasson *et al.*, 2001; Winter and Campbell, 2004). These domains enable to repress different target genes, both quantitatively and qualitatively.

G: Molecular Switches: Interchanging Coactivators and Corepressors

Some DNA-binding transcription factors responding to signaling pathways play dual roles in transcription (Barolo and Posakony, 2002). They can become either activators or repressors by interchanging coactivators and corepressors in response to signals.

When cells are stimulated by the Wnt ligand through Frizzled receptors, beta-Catenin (Armadillo) is activated and translocated into the nucleus, where it serves as a coactivator for a DNA-binding factor TCF (Pangolin) to induce its target genes (Roose and Clevers, 1999). In cells lacking a Wnt signal, however, TCF keeps its target genes off by recruiting Groucho (Cavallo *et al.*, 1998). In addition, *Drosophila* CBP acetylates a lysine residue in the beta-Catenin binding domain of TCF and this modification dissociates beta-Catenin from TCF, resulting in repression (Waltzer and Bienz, 1998).

Similarly, a *Drosophila* DNA-binding factor Su(H) acts as an activator by recruiting coactivators in the presence of Notch signaling (Lai, 2002). In contrast, binding of Hairless, Groucho, and dCtBP convert Su(H) into a repressor that inhibits its target genes in unstimulated cells (Barolo *et al.*, 2002).

Finally, nuclear hormone receptors that sequence-specifically bind DNA can change their roles in transcription (Baek and Rosenfeld, 2004; Glass and Rosenfeld, 2000). In the absence of ligands, a heterodimer containing TR and retinoic acid receptor (TR/RXR) bound to its binding site interacts with the N-CoR corepressor and with two other corepressors, Sin3A and HDAC1 (Heinzel et al., 1997; Horlein et al., 1995). However, ligand binding alters the conformation of the receptors and this conformational change allows TR to switch its interacting factors from these corepressors to the coactivators containing SRC-1, PCAF, and CBP. These histone acetylases and histones deacetylase ultimately define the fate of transcription.

H: Protein Modification

The activities of transcription factors are often modulated by post-translational modifications including ubiquitination, sumoylation, acetylation, methylation, and phosphorylation (Baek and Rosenfeld, 2004; Gill, 2004). Such modifications are known to regulate the DNA binding activity, repression activity, protein-protein interactions (such as the interaction between repressors and corepressors), and the subcellular localization of factors. Activators are often inactivated by protein modifications, resulting in loss of activation and hence repression. For example, phosphorylation inhibits the DNA-binding activity of several activators, such as c-Jun, Oct1, HNF-4, and others (Hunter and Karin, 1992; Viollet et al., 1997). Phosphorylation of *Drosophila* Groucho by MAPK attenuates its repression activity (Hasson et al., 2005). Acetylation of Sp3 by p300 converts it from a repressor to an activator (Ammanamanchi et al., 2003). When sumoylation of BKLF, a zinc finger repressor, is blocked, its repression activity is reduced, but its DNA-binding activity is not affected (Perdomo et al., 2005). The tumor-suppressor Retinoblastoma protein (Rb) interacts with both the DNA binding factor E2F and HDACs in G0 and in early G1 phases of the cell cycle (Cress and Seto, 2000; Ferreira et al., 2001). The association with these factors represses E2F-regulated promoters through the deacetylation of nearby histones. Rb is then phosphorylated by cyclin-CDK kinase complexes during the transition from G1 to S phase. The phosphorylated Rb can dissociate from E2F and HDAC (Takaki et al., 2004), and as a consequence, E2F can recruit CBP to activate transcription. As mentioned above, the acetylation of TCF facilitates separation from its coactivator beta-Catenin (Waltzer and Bienz, 1998). Similarly, the acetylation of E1A by p300 and P/CAF disrupts the interaction with CtBP (Zhang et al., 2000). Finally, when sumoylation of hCtBP1 is blocked, this factor is translocated from the nucleus into the cytoplasm (Lin et al., 2003).

Concluding Remarks

We have discussed a variety of mechanisms by which repressors/corepressors inhibit transcription. To activate transcription, it is essential that activators, bound to an enhancer located far from the core promoter, contact the preinitiation complex containing RNA polymerase II directly and/or indirectly through multiple coactivator complexes, such as the Mediator complex (Lemon and Tjian, 2000). In principle, disruption of these interactions at any level could lead to repression, so there might be more molecular mechanisms yet to be discovered. Some corepressors are known to interact with specific amino acid sequences (motifs or domains). For instance, dCtBP and Groucho interact with specific motifs, PxDLS and WRPW/FKPY/FxIxxIL, respectively. Specific amino acids of the DNA-binding factors/coregulators can be modified, although the consensus sequences allowing such modifications are not often tight. A profound deeper knowledge of such "sequence features" will enable us to easily predict whether a factor of interest can work as an activator or a repressor.

Acknowledgment

We thank Dr. Mark Stern and Dr. Anna Di Gregorio for critically reading the manuscript.

References

Adams, M. D., Celniker, S. E., Holt, R. A., Evans, C. A., Gocayne, J. D., Amanatides, P. G., Scherer, S. E., Li, P. W., Hoskins, R. A., Galle, R. F., et al. (2000). The genome sequence of *Drosophila* melanogaster. Science *287*, 2185-2195.

Ammanamanchi, S., Freeman, J. W., and Brattain, M. G. (2003). Acetylated sp3 is a transcriptional activator. J Biol Chem *278*, 35775-35780.

Andrau, J. C., Van Oevelen, C. J., Van Teeffelen, H. A., Weil, P. A., Holstege, F. C., and Timmers, H. T. (2002). Mot1p is essential for TBP recruitment to selected promoters during *in vivo* gene activation. Embo J *21*, 5173-5183.

Arnosti, D. N., Barolo, S., Levine, M., and Small, S. (1996a). The eve stripe 2 enhancer employs multiple modes of transcriptional synergy. Development *122*, 205-214.

Arnosti, D. N., Gray, S., Barolo, S., Zhou, J., and Levine, M. (1996b). The gap protein knirps mediates both quenching and direct repression in the *Drosophila* embryo. Embo J *15*, 3659-3666.

Auble, D. T., Hansen, K. E., Mueller, C. G., Lane, W. S., Thorner, J., and Hahn, S. (1994). Mot1, a global repressor of RNA polymerase II transcription, inhibits TBP binding to DNA by an ATP-dependent mechanism. Genes Dev *8*, 1920-1934.

Austin, R. J., and Biggin, M. D. (1995). A domain of the even-skipped protein represses transcription by preventing TFIID binding to a promoter: repression by cooperative blocking. Mol

Cell Biol *15*, 4683-4693.

Baek, S. H., and Rosenfeld, M. G. (2004). Nuclear receptor coregulators: their modification codes and regulatory mechanism by translocation. Biochem Biophys Res Commun *319*, 707-714.

Balasubramanian, P., Zhao, L. J., and Chinnadurai, G. (2003). Nicotinamide adenine dinucleotide stimulates oligomerization, interaction with adenovirus E1A and an intrinsic dehydrogenase activity of CtBP. FEBS Lett *537*, 157-160.

Baniahmad, A., Ha, I., Reinberg, D., Tsai, S., Tsai, M. J., and O'Malley, B. W. (1993). Interaction of human thyroid hormone receptor beta with transcription factor TFIIB may mediate target gene derepression and activation by thyroid hormone. Proc Natl Acad Sci USA *90*, 8832-8836.

Barolo, S., and Levine, M. (1997). hairy mediates dominant repression in the *Drosophila* embryo. Embo J *16*, 2883-2891.

Barolo, S., and Posakony, J. W. (2002). Three habits of highly effective signaling pathways: principles of transcriptional control by developmental cell signaling. Genes Dev *16*, 1167-1181.

Barolo, S., Stone, T., Bang, A. G., and Posakony, J. W. (2002). Default repression and Notch signaling: Hairless acts as an adaptor to recruit the corepressors Groucho and dCtBP to Suppressor of Hairless. Genes Dev *16*, 1964-1976.

Benezra, R., Davis, R. L., Lockshon, D., Turner, D. L., and Weintraub, H. (1990). The protein Id: a negative regulator of helix-loop-helix DNA binding proteins. Cell *61*, 49-59.

Berman, B. P., Nibu, Y., Pfeiffer, B. D., Tomancak, P., Celniker, S. E., Levine, M., Rubin, G. M., and Eisen, M. B. (2002). Exploiting transcription factor binding site clustering to identify cis-regulatory modules involved in pattern formation in the *Drosophila* genome. Proc Natl Acad Sci USA *99*, 757-762.

Breiling, A., Turner, B. M., Bianchi, M. E., and Orlando, V. (2001). General transcription factors bind promoters repressed by Polycomb group proteins. Nature *412*, 651-655.

Burke, L. J., and Baniahmad, A. (2000). Co-repressors 2000. Faseb J *14*, 1876-1888.

Burley, S. K., and Roeder, R. G. (1998). TATA box mimicry by TFIID: autoinhibition of pol II transcription. Cell *94*, 551-553.

Cai, H. N., Arnosti, D. N., and Levine, M. (1996). Long-range repression in the *Drosophila* embryo. Proc Natl Acad Sci USA *93*, 9309-9314.

Campbell, G., and Tomlinson, A. (1999). Transducing the Dpp morphogen gradient in the wing of *Drosophila*: regulation of Dpp targets by brinker. Cell *96*, 553-562.

Cao, R., Wang, L., Wang, H., Xia, L., Erdjument-Bromage, H., Tempst, P., Jones, R. S., and Zhang, Y. (2002). Role of histone H3 lysine 27 methylation in Polycomb-group silencing. Science *298*, 1039-1043.

Cao, R., and Zhang, Y. (2004). The functions of E(Z)/EZH2-mediated methylation of lysine 27 in histone H3. Curr Opin Genet Dev *14*, 155-164.

Cavallo, R. A., Cox, R. T., Moline, M. M., Roose, J., Polevoy, G. A., Clevers, H., Peifer, M., and Bejsovec, A. (1998). *Drosophila* Tcf and Groucho interact to repress Wingless signalling activity. Nature *395*, 604-608.

Celniker, S. E., and Rubin, G. M. (2003). The *Drosophila melanogaster* genome. Annu Rev Genomics Hum Genet *4*, 89-117.

Chang, Y. L., Peng, Y. H., Pan, I. C., Sun, D. S., King, B., and Huang, D. H. (2001). Essential role of *Drosophila* Hdac1 in homeotic gene silencing. Proc Natl Acad Sci USA *98*, 9730-9735.

Chen, G., Fernandez, J., Mische, S., and Courey, A. J. (1999). A functional interaction between the histone deacetylase Rpd3 and the corepressor groucho in *Drosophila* development. Genes Dev *13*, 2218-2230.

Chicca, J. J., 2nd, Auble, D. T., and Pugh, B. F. (1998). Cloning and biochemical characterization of TAF-172, a human homolog of yeast Mot1. Mol Cell Biol *18*, 1701-1710.

Chinnadurai, G. (2002). CtBP, an unconventional transcriptional corepressor in development and oncogenesis. Mol Cell *9*, 213-224.

Clyde, D. E., Corado, M. S., Wu, X., Pare, A., Papatsenko, D., and Small, S. (2003). A self-organizing system of repressor gradients establishes segmental complexity in *Drosophila*. Nature *426*, 849-853.

Courey, A. J., and Jia, S. (2001). Transcriptional repression: the long and the short of it. Genes Dev *15*, 2786-2796.

Cowden, J., and Levine, M. (2003). Ventral dominance governs sequential patterns of gene expression across the dorsal-ventral axis of the neuroectoderm in the *Drosophila* embryo. Dev Biol *262*, 335-349.

Cress, W. D., and Seto, E. (2000). Histone deacetylases, transcriptional control, and cancer. J Cell Physiol *184*, 1-16.

Creton, S., Svejstrup, J. Q., and Collart, M. A. (2002). The NC2 alpha and beta subunits play different roles *in vivo*. Genes Dev *16*, 3265-3276.

Czermin, B., Melfi, R., McCabe, D., Seitz, V., Imhof, A., and Pirrotta, V. (2002). *Drosophila* enhancer of Zeste/ESC complexes have a histone H3 methyltransferase activity that marks chromosomal Polycomb sites. Cell *111*, 185-196.

Danzer, J. R., and Wallrath, L. L. (2004). Mechanisms of HP1-mediated gene silencing in *Drosophila*. Development *131*, 3571-3580.

Dasgupta, A., Darst, R. P., Martin, K. J., Afshari, C. A., and Auble, D. T. (2002). Mot1 activates and represses transcription by direct, ATPase-dependent mechanisms. Proc Natl Acad Sci USA *99*, 2666-2671.

Dehal, P., Satou, Y., Campbell, R. K., Chapman, J., Degnan, B., De Tomaso, A., Davidson, B., Di Gregorio, A., Gelpke, M., Goodstein, D. M., et al. (2002). The draft genome of Ciona intestinalis: insights into chordate and vertebrate origins. Science *298*, 2157-2167.

Dellino, G. I., Schwartz, Y. B., Farkas, G., McCabe, D., Elgin, S. C., and Pirrotta, V. (2004). Polycomb silencing blocks transcription initiation. Mol Cell *13*, 887-893.

Dikstein, R., Ruppert, S., and Tjian, R. (1996). TAFII250 is a bipartite protein kinase that phosphorylates the base transcription factor RAP74. Cell *84*, 781-790.

Dubnicoff, T., Valentine, S. A., Chen, G., Shi, T., Lengyel, J. A., Paroush, Z., and Courey, A. J. (1997). Conversion of dorsal from an activator to a repressor by the global corepressor Groucho. Genes Dev *11*, 2952-2957.

Ferreira, R., Naguibneva, I., Pritchard, L. L., Ait-Si-Ali, S., and Harel-Bellan, A. (2001). The Rb/chromatin connection and epigenetic control: opinion. Oncogene *20*, 3128-3133.

Fitzgerald, D. P., and Bender, W. (2001). Polycomb group repression reduces DNA accessibility. Mol Cell Biol *21*, 6585-6597.

Fondell, J. D., Brunel, F., Hisatake, K., and Roeder, R. G. (1996). Unliganded thyroid hormone receptor alpha can target TATA-binding protein for transcriptional repression. Mol Cell Biol *16*, 281-287.

Fondell, J. D., Roy, A. L., and Roeder, R. G. (1993). Unliganded thyroid hormone receptor inhibits formation of a functional preinitiation complex: implications for active repression. Genes Dev *7*, 1400-1410.

Fujioka, M., Emi-Sarker, Y., Yusibova, G. L., Goto, T., and Jaynes, J. B. (1999). Analysis of an even-skipped rescue transgene reveals both composite and discrete neuronal and early blastoderm enhancers, and multi-stripe positioning by gap gene repressor gradients. Development *126*, 2527-2538.

Gaston, K., and Jayaraman, P. S. (2003). Transcriptional repression in eukaryotes: repressors and repression mechanisms. Cell Mol Life Sci *60*, 721-741.

Geisberg, J. V., Moqtaderi, Z., Kuras, L., and Struhl, K. (2002). Mot1 associates with transcriptionally active promoters and inhibits association of NC2 in *Saccharomyces cerevisiae*. Mol Cell Biol *22*, 8122-8134.

Gill, G. (2004). SUMO and ubiquitin in the nucleus: different functions, similar mechanisms? Genes Dev *18*, 2046-2059.

Glass, C. K., and Rosenfeld, M. G. (2000). The coregulator exchange in transcriptional functions of nuclear receptors. Genes Dev *14*, 121-141.

Goppelt, A., Stelzer, G., Lottspeich, F., and Meisterernst, M. (1996). A mechanism for repression of class II gene transcription through specific binding of NC2 to TBP-promoter complexes via heterodimeric histone fold domains. Embo J *15*, 3105-3116.

Gould, A. (1997). Functions of mammalian Polycomb group and trithorax group related genes. Curr Opin Genet Dev *7*, 488-494.

Grandori, C., Cowley, S. M., James, L. P., and Eisenman, R. N. (2000). The Myc/Max/Mad network and the transcriptional control of cell behavior. Annu Rev Cell Dev Biol *16*, 653-699.

Gray, S., and Levine, M. (1996a). Short-range transcriptional repressors mediate both quenching and direct repression within complex loci in *Drosophila*. Genes Dev *10*, 700-710.

Gray, S., and Levine, M. (1996b). Transcriptional repression in development. Curr Opin Cell Biol *8*, 358-364.

Gray, S., Szymanski, P., and Levine, M. (1994). Short-range repression permits multiple enhancers to function autonomously within a complex promoter. Genes Dev *8*, 1829-1838.

Grooteclaes, M., Deveraux, Q., Hildebrand, J., Zhang, Q., Goodman, R. H., and Frisch, S. M. (2003). C-terminal-binding protein corepresses epithelial and proapoptotic gene expression programs. Proc Natl Acad Sci USA *100*, 4568-4573.

Gunster, M., raaphorst, F., Hamer, K., den Blaauwen, J., Fieret, E., Meijer, C., and Otte, A. (2001). Differential expression of human Polycomb group proteins in various tissues and cell types. J Cell Biochem *81(S36)*, 129-143.

Hanna-Rose, W., and Hansen, U. (1996). Active repression mechanisms of eukaryotic transcription repressors. Trends Genet *12*, 229-234.

Hanna-Rose, W., Licht, J. D., and Hansen, U. (1997). Two evolutionarily conserved repression domains in the *Drosophila* Kruppel protein differ in activator specificity. Mol Cell Biol *17*, 4820-4829.

Hasson, P., Egoz, N., Winkler, C., Volohonsky, G., Jia, S., Dinur, T., Volk, T., Courey, A. J., and Paroush, Z. (2005). EGFR signaling attenuates Groucho-dependent repression to antagonize Notch transcriptional output. Nat Genet *37*, 101-105.

Hasson, P., Muller, B., Basler, K., and Paroush, Z. (2001). Brinker requires two corepressors for maximal and versatile repression in Dpp signalling. Embo J *20*, 5725-5736.

Hayashi, Y., Wang, W., Ninomiya, T., Nagano, H., Ohta, K., and Itoh, H. (1999). Liver enriched transcription factors and differentiation of hepatocellular carcinoma. Mol Pathol *52*, 19-24.

Heinzel, T., Lavinsky, R. M., Mullen, T. M., Soderstrom, M., Laherty, C. D., Torchia, J., Yang, W. M., Brard, G., Ngo, S. D., Davie, J. R., *et al.* (1997). A complex containing N-CoR, mSin3 and histone deacetylase mediates transcriptional repression. Nature *387*, 43-48.

Hewitt, G. F., Strunk, B. S., Margulies, C., Priputin, T., Wang, X. D., Amey, R., Pabst, B. A., Kosman, D., Reinitz, J., and Arnosti, D. N. (1999). Transcriptional repression by the *Drosophila* giant protein: cis element positioning provides an alternative means of interpreting an effector gradient. Development *126*, 1201-1210.

Hoch, M., Gerwin, N., Taubert, H., and Jackle, H. (1992). Competition for overlapping sites in the regulatory region of the *Drosophila* gene Kruppel. Science *256*, 94-97.

Horlein, A. J., Naar, A. M., Heinzel, T., Torchia, J., Gloss, B., Kurokawa, R., Ryan, A., Kamei, Y., Soderstrom, M., Glass, C. K., and *et al.* (1995). Ligand-independent repression by the thyroid hormone receptor mediated by a nuclear receptor co-repressor. Nature *377*, 397-404.

Hunter, T., and Karin, M. (1992). The regulation of transcription by phosphorylation. Cell *70*, 375-387.

Igarashi, K., Kataoka, K., Itoh, K., Hayashi, N., Nishizawa, M., and Yamamoto, M. (1994). Regulation of transcription by dimerization of erythroid factor NF-E2 p45 with small Maf proteins. Nature *367*, 568-572.

Imai, K. S., Hino, K., Yagi, K., Satoh, N., and Satou, Y. (2004). Gene expression profiles of transcription factors and signaling molecules in the ascidian embryo: towards a comprehensive understanding of gene networks. Development *131*, 4047-4058.

Ip, Y. T., and Hemavathy, K. (1997). *Drosophila* development. Delimiting patterns by repression. Curr Biol *7*, R216-218.

Ip, Y. T., Park, R. E., Kosman, D., Bier, E., and Levine, M. (1992). The dorsal gradient morphogen regulates stripes of rhomboid expression in the presumptive neuroectoderm of the *Drosophila* embryo. Genes Dev *6*, 1728-1739.

Iratni, R., Yan, Y. T., Chen, C., Ding, J., Zhang, Y., Price, S. M., Reinberg, D., and Shen, M. M. (2002). Inhibition of excess nodal signaling during mouse gastrulation by the transcriptional corepressor DRAP1. Science *298*, 1996-1999.

Jackle, H., Hoch, M., Pankratz, M. J., Gerwin, N., Sauer, F., and Bronner, G. (1992). Transcriptional control by *Drosophila* gap genes. J Cell Sci Suppl *16*, 39-51.

Jacobson, R. H., Ladurner, A. G., King, D. S., and Tjian, R. (2000). Structure and function of a human TAFII250 double bromodomain module. Science *288*, 1422-1425.

Jazwinska, A., Kirov, N., Wieschaus, E., Roth, S., and Rushlow, C. (1999). The *Drosophila* gene brinker reveals a novel mechanism of Dpp target gene regulation. Cell *96*, 563-573.

Jen, Y., Weintraub, H., and Benezra, R. (1992). Overexpression of Id protein inhibits the muscle differentiation program: *in vivo* association of Id with E2A proteins. Genes Dev *6*, 1466-1479.

Jenuwein, T. (2001). Re-SET-ting heterochromatin by histone methyltransferases. Trends Cell Biol *11*, 266-273.

Jiang, G., and Sladek, F. M. (1997). The DNA binding domain of hepatocyte nuclear factor 4 mediates cooperative, specific binding to DNA and heterodimerization with the retinoid X receptor alpha. J Biol Chem *272*, 1218-1225.

Jimenez, G., Guichet, A., Ephrussi, A., and Casanova, J. (2000). Relief of gene repression by torso RTK signaling: role of capicua in *Drosophila* terminal and dorsoventral patterning. Genes Dev *14*, 224-231.

Johnson, A. D. (1995). The price of repression. Cell *81*, 655-658.

Kamada, K., Shu, F., Chen, H., Malik, S., Stelzer, G., Roeder, R. G., Meisterernst, M., and Burley, S. K. (2001). Crystal structure of negative cofactor 2 recognizing the TBP-DNA transcription complex. Cell *106*, 71-81.

Keller, S. A., Mao, Y., Struffi, P., Margulies, C., Yurk, C. E., Anderson, A. R., Amey, R. L., Moore, S., Ebels, J. M., Foley, K., et al. (2000). dCtBP-dependent and -independent repression activities of the *Drosophila* Knirps protein. Mol Cell Biol *20*, 7247-7258.

Kimura, A., Nishiyori, A., Murakami, T., Tsukamoto, T., Hata, S., Osumi, T., Okamura, R., Mori, M., and Takiguchi, M. (1993). Chicken ovalbumin upstream promoter-transcription factor (COUP-TF) represses transcription from the promoter of the gene for ornithine transcarbamylase in a manner antagonistic to hepatocyte nuclear factor-4 (HNF-4). J Biol Chem *268*, 11125-11133.

Kirkpatrick, H., Johnson, K., and Laughon, A. (2001). Repression of dpp targets by binding of brinker to mad sites. J Biol Chem *276*, 18216-18222.

Klejman, M. P., Pereira, L. A., van Zeeburg, H. J., Gilfillan, S., Meisterernst, M., and Timmers, H. T. (2004). NC2alpha interacts with BTAF1 and stimulates its ATP-dependent association with TATA-binding protein. Mol Cell Biol *24*, 10072-10082.

Kobayashi, M., Goldstein, R. E., Fujioka, M., Paroush, Z., and Jaynes, J. B. (2001). Groucho augments the repression of multiple Even skipped target genes in establishing parasegment boundaries. Development *128*, 1805-1815.

Kokubo, T., Yamashita, S., Horikoshi, M., Roeder, R. G., and Nakatani, Y. (1994). Interaction between the N-terminal domain of the 230-kDa subunit and the TATA box-binding subunit of TFIID negatively regulates TATA-box binding. Proc Natl Acad Sci USA *91*, 3520-3524.

Kotani, T., Banno, K., Ikura, M., Hinnebusch, A. G., Nakatani, Y., Kawaichi, M., and Kokubo, T. (2000). A role of transcriptional activators as antirepressors for the autoinhibitory activity of TATA box binding of transcription factor IID. Proc Natl Acad Sci USA *97*, 7178-7183.

Kumar, V., Carlson, J. E., Ohgi, K. A., Edwards, T. A., Rose, D. W., Escalante, C. R., Rosenfeld, M. G., and Aggarwal, A. K. (2002). Transcription corepressor CtBP is an NAD(+)-regulated dehydrogenase. Mol Cell *10*, 857-869.

Ladias, J. A., Hadzopoulou-Cladaras, M., Kardassis, D., Cardot, P., Cheng, J., Zannis, V., and Cladaras, C. (1992). Transcriptional regulation of human apolipoprotein genes ApoB, ApoCIII, and ApoAII by members of the steroid hormone receptor superfamily HNF-4, ARP-1, EAR-2, and EAR-3. J Biol Chem *267*, 15849-15860.

Laherty, C. D., Yang, W. M., Sun, J. M., Davie, J. R., Seto, E., and Eisenman, R. N. (1997). Histone deacetylases associated with the mSin3 corepressor mediate mad transcriptional repression. Cell *89*, 349-356.

Lai, E. C. (2002). Keeping a good pathway down: transcriptional repression of Notch pathway target genes by CSL proteins. EMBO Rep *3*, 840-845.

Lania, L., Majello, B., and De Luca, P. (1997). Transcriptional regulation by the Sp family proteins. Int J Biochem Cell Biol *29*, 1313-1323.

Lassar, A. B., Davis, R. L., Wright, W. E., Kadesch, T., Murre, C., Voronova, A., Baltimore, D., and Weintraub, H. (1991). Functional activity of myogenic HLH proteins requires hetero-oligomerization with E12/E47-like proteins *in vivo*. Cell *66*, 305-315.

Lee, T. I., and Young, R. A. (1998). Regulation of gene expression by TBP-associated proteins. Genes Dev *12*, 1398-1408.

Lemaire, M., Xie, J., Meisterernst, M., and Collart, M. A. (2000). The NC2 repressor is dispensable in yeast mutated for the Sin4p component of the holoenzyme and plays roles similar to Mot1p *in vivo*. Mol Microbiol *36*, 163-173.

Lemon, B., and Tjian, R. (2000). Orchestrated response: a symphony of transcription factors for gene control. Genes Dev *14*, 2551-2569.

Levine, M., and Manley, J. L. (1989). Transcriptional repression of eukaryotic promoters. Cell *59*, 405-408.

Li, C., and Manley, J. L. (1998). Even-skipped represses transcription by binding TATA binding protein and blocking the TFIID-TATA box interaction. Mol Cell Biol *18*, 3771-3781.

Licht, J. D., Hanna-Rose, W., Reddy, J. C., English, M. A., Ro, M., Grossel, M., Shaknovich, R., and Hansen, U. (1994). Mapping and mutagenesis of the amino-terminal transcriptional repression domain of the *Drosophila* Kruppel protein. Mol Cell Biol *14*, 4057-4066.

Lin, X., Sun, B., Liang, M., Liang, Y. Y., Gast, A., Hildebrand, J., Brunicardi, F. C., Melchior, F., and Feng, X. H. (2003). Opposed regulation of corepressor CtBP by SUMOylation and PDZ binding. Mol Cell *11*, 1389-1396.

Liu, D., Ishima, R., Tong, K. I., Bagby, S., Kokubo, T., Muhandiram, D. R., Kay, L. E., Nakatani, Y., and Ikura, M. (1998). Solution structure of a TBP-TAF(II)230 complex: protein mimicry of the minor groove surface of the TATA box unwound by TBP. Cell *94*, 573-583.

Lund, A. H., and van Lohuizen, M. (2004). Polycomb complexes and silencing mechanisms. Curr Opin Cell Biol *16*, 239-246.

Luscher, B. (2001). Function and regulation of the transcription factors of the Myc/Max/Mad network. Gene *277*, 1-14.

Majello, B., De Luca, P., and Lania, L. (1997). Sp3 is a bifunctional transcription regulator with modular independent activation and repression domains. J Biol Chem *272*, 4021-4026.

Mannervik, M., and Levine, M. (1999). The Rpd3 histone deacetylase is required for segmentation of the *Drosophila* embryo. Proc Natl Acad Sci USA *96*, 6797-6801.

Mannervik, M., Nibu, Y., Zhang, H., and Levine, M. (1999). Transcriptional coregulators in development. Science *284*, 606-609.

Mermelstein, F., Yeung, K., Cao, J., Inostroza, J. A., Erdjument-Bromage, H., Eagelson, K., Landsman, D., Levitt, P., Tempst, P., and Reinberg, D. (1996). Requirement of a corepressor for Dr1-mediated repression of transcription. Genes Dev *10*, 1033-1048.

Mietus-Snyder, M., Sladek, F. M., Ginsburg, G. S., Kuo, C. F., Ladias, J. A., Darnell, J. E., Jr., and Karathanasis, S. K. (1992). Antagonism between apolipoprotein AI regulatory protein 1, Ear3/COUP-TF, and hepatocyte nuclear factor 4 modulates apolipoprotein CIII gene expression in liver and intestinal cells. Mol Cell Biol *12*, 1708-1718.

Minami, M., Kinoshita, N., Kamoshida, Y., Tanimoto, H., and Tabata, T. (1999). brinker is a target of Dpp in *Drosophila* that negatively regulates Dpp-dependent genes. Nature *398*, 242-246.

Misra, S., Crosby, M. A., Mungall, C. J., Matthews, B. B., Campbell, K. S., Hradecky, P., Huang, Y., Kaminker, J. S., Millburn, G. H., Prochnik, S. E., *et al.* (2002). Annotation of the *Drosophila* melanogaster euchromatic genome: a systematic review. Genome Biol *3*, RESEARCH0083.

Mizzen, C. A., Yang, X. J., Kokubo, T., Brownell, J. E., Bannister, A. J., Owen-Hughes, T., Workman, J., Wang, L., Berger, S. L., Kouzarides, T., *et al.* (1996). The TAF(II)250 subunit of TFIID has histone acetyltransferase activity. Cell *87*, 1261-1270.

Motohashi, H., Shavit, J. A., Igarashi, K., Yamamoto, M., and Engel, J. D. (1997). The world according to Maf. Nucleic Acids Res *25*, 2953-2959.

Nibu, Y., and Levine, M. S. (2001). CtBP-dependent activities of the short-range Giant repressor in the *Drosophila* embryo. Proc Natl Acad Sci USA *98*, 6204-6208.

Nibu, Y., Senger, K., and Levine, M. (2003). CtBP-independent repression in the *Drosophila* embryo. Mol Cell Biol *23*, 3990-3999.

Nibu, Y., Zhang, H., Bajor, E., Barolo, S., Small, S., and Levine, M. (1998a). dCtBP mediates transcriptional repression by Knirps, Kruppel and Snail in the *Drosophila* embryo. Embo J *17*, 7009-7020.

Nibu, Y., Zhang, H., and Levine, M. (1998b). Interaction of short-range repressors with *Drosophila* CtBP in the embryo. Science *280*, 101-104.

Nibu, Y., Zhang, H., and Levine, M. (2001). Local action of long-range repressors in the *Drosophila* embryo. Embo J *20*, 2246-2253.

Nishikawa, J., Kokubo, T., Horikoshi, M., Roeder, R. G., and Nakatani, Y. (1997). *Drosophila* TAF(II)230 and the transcriptional activator VP16 bind competitively to the TATA box-binding domain of the TATA box-binding protein. Proc Natl Acad Sci USA *94*, 85-90.

Ochoa-Espinosa, A., Yucel, G., Kaplan, L., Pare, A., Pura, N., Oberstein, A., Papatsenko, D., and Small, S. (2005). The role of binding site cluster strength in Bicoid-dependent patterning in *Drosophila*. Proc Natl Acad Sci USA.

Orlando, V. (2003). Polycomb, epigenomes, and control of cell identity. Cell *112*, 599-606.

Orlando, V., Jane, E. P., Chinwalla, V., Harte, P. J., and Paro, R. (1998). Binding of trithorax and Polycomb proteins to the bithorax complex: dynamic changes during early *Drosophila* embryogenesis. Embo J *17*, 5141-5150.

Orlando, V., and Paro, R. (1995). Chromatin multiprotein complexes involved in the maintenance of transcription patterns. Curr Opin Genet Dev *5*, 174-179.

Orlando, V., and Paro, V. (1993). Mapping Polycomb-repressed domains in the bithorax complex using *in vivo* formaldehyde cross-linked chromatin. Cell *75*, 1187-1198.

Otte, A. P., and Kwaks, T. H. (2003). Gene repression by Polycomb group protein complexes: a distinct complex for every occasion? Curr Opin Genet Dev *13*, 448-454.

Park, J. I., Tsai, S. Y., and Tsai, M. J. (2003). Molecular mechanism of chicken ovalbumin upstream promoter-transcription factor (COUP-TF) actions. Keio J Med *52*, 174-181.

Perdomo, J., Verger, A., Turner, J., and Crossley, M. (2005). Role for SUMO modification in facilitating transcriptional repression by BKLF. Mol Cell Biol *25*, 1549-1559.

Pereira, F. A., Tsai, M. J., and Tsai, S. Y. (2000). COUP-TF orphan nuclear receptors in development and differentiation. Cell Mol Life Sci *57*, 1388-1398.

Pereira, L. A., Klejman, M. P., and Timmers, H. T. (2003). Roles for BTAF1 and Mot1p in dynamics of TATA-binding protein and regulation of RNA polymerase II transcription. Gene *315*, 1-13.

Pham, A. D., and Sauer, F. (2000). Ubiquitin-activating/conjugating activity of TAFII250, a mediator of activation of gene expression in *Drosophila*. Science *289*, 2357-2360.

Phippen, T. M., Sweigart, A. L., Moniwa, M., Krumm, A., Davie, J. R., and Parkhurst, S. M. (2000). *Drosophila* C-terminal binding protein functions as a context-dependent transcriptional co-factor and interferes with both mad and groucho transcriptional repression. J Biol Chem *275*, 37628-37637.

Pirrotta, V. (1999). Polycomb silencing and the maintenance of stable chromatin states. In Results and problems in cell differentiation: genomic imprinting., R. Ohlsson, ed. (Berlin, Germany, Springer-Verlag), pp. 205-228.

Pirrotta, V., Poux, S., Melfi, R., and Pilyugin, M. (2003). Assembly of Polycomb complexes and silencing mechanisms. Genetica *117*, 191-197.

Prelich, G. (1997). *Saccharomyces cerevisiae* BUR6 encodes a DRAP1/NC2alpha homolog that has both positive and negative roles in transcription *in vivo*. Mol Cell Biol *17*, 2057-2065.

Pugh, B. F. (2000). Control of gene expression through regulation of the TATA-binding protein. Gene *255*, 1-14.

Ramji, D. P., and Foka, P. (2002). CCAAT/enhancer-binding proteins: structure, function and regulation. Biochem J *365*, 561-575.

Riechmann, J. L., Heard, J., Martin, G., Reuber, L., Jiang, C., Keddie, J., Adam, L., Pineda, O., Ratcliffe, O. J., Samaha, R. R., *et al.* (2000). *Arabidopsis* transcription factors: genome-wide comparative analysis among eukaryotes. Science *290*, 2105-2110.

Ron, D., and Habener, J. F. (1992). CHOP, a novel developmentally regulated nuclear protein that dimerizes with transcription factors C/EBP and LAP and functions as a dominant-negative inhibitor of gene transcription. Genes Dev *6*, 439-453.

Roose, J., and Clevers, H. (1999). TCF transcription factors: molecular switches in carcinogenesis. Biochim Biophys Acta *1424*, M23-37.

Rosenberg, M. I., and Parkhurst, S. M. (2002). *Drosophila* Sir2 is required for heterochromatic silencing and by euchromatic Hairy/E(Spl) bHLH repressors in segmentation and sex determination. Cell *109*, 447-458.

Rushlow, C., Colosimo, P. F., Lin, M. C., Xu, M., and Kirov, N. (2001). Transcriptional regulation of the *Drosophila* gene zen by competing Smad and Brinker inputs. Genes Dev *15*, 340-351.

Ryu, J. R., and Arnosti, D. N. (2003). Functional similarity of Knirps CtBP-dependent and CtBP-independent transcriptional repressor activities. Nucleic Acids Res *31*, 4654-4662.

Saller, E., and Bienz, M. (2001). Direct competition between Brinker and *Drosophila* Mad in Dpp target gene transcription. EMBO Rep *2*, 298-305.

Satou, Y., Takatori, N., Yamada, L., Mochizuki, Y., Hamaguchi, M., Ishikawa, H., Chiba, S., Imai, K., Kano, S., Murakami, S. D., *et al.* (2001). Gene expression profiles in Ciona intestinalis tailbud embryos. Development *128*, 2893-2904.

Sauer, F., Fondell, J. D., Ohkuma, Y., Roeder, R. G., and Jackle, H. (1995). Control of transcription by Kruppel through interactions with TFIIB and TFIIE beta. Nature *375*, 162-164.

Saurin, A. J., Shao, Z., Erdjument-Bromage, H., Tempst, P., and Kingston, R. E. (2001). A *Drosophila* Polycomb group complex includes Zeste and dTAFII proteins. Nature *412*, 655-660.

Shao, Z., Raible, F., Mollaaghababa, R., Guyon, J. R., Wu, C. T., Bender, W., and Kingston, R. E. (1999). Stabilization of chromatin structure by PRC1, a Polycomb complex. Cell *98*, 37-46.

Shi, Y., Sawada, J., Sui, G., Affar el, B., Whetstine, J. R., Lan, F., Ogawa, H., Luke, M. P., and Nakatani, Y. (2003). Coordinated histone modifications mediated by a CtBP co-repressor complex. Nature *422*, 735-738.

Small, S., Arnosti, D. N., and Levine, M. (1993). Spacing ensures autonomous expression of different stripe enhancers in the even-skipped promoter. Development *119*, 762-772.

Small, S., Blair, A., and Levine, M. (1992). Regulation of even-skipped stripe 2 in the *Drosophila* embryo. Embo J *11*, 4047-4057.

Small, S., Blair, A., and Levine, M. (1996). Regulation of two pair-rule stripes by a single enhancer in the *Drosophila* embryo. Dev Biol *175*, 314-324.

Small, S., Kraut, R., Hoey, T., Warrior, R., and Levine, M. (1991). Transcriptional regulation of a pair-rule stripe in *Drosophila*. Genes Dev *5*, 827-839.

Song, H., Hasson, P., Paroush, Z., and Courey, A. J. (2004). Groucho oligomerization is required for repression *in vivo*. Mol Cell Biol *24*, 4341-4350.

Stanojevic, D., Small, S., and Levine, M. (1991). Regulation of a segmentation stripe by overlapping activators and repressors in the *Drosophila* embryo. Science *254*, 1385-1387.

Stathopoulos, A., and Levine, M. (2002). Dorsal gradient networks in the *Drosophila* embryo. Dev Biol *246*, 57-67.

Struffi, P., Corado, M., Kulkarni, M., and Arnosti, D. N. (2004). Quantitative contributions of CtBP-dependent and -independent repression activities of Knirps. Development *131*, 2419-2429.

Suske, G. (1999). The Sp-family of transcription factors. Gene *238*, 291-300.

Sutrias-Grau, M., and Arnosti, D. N. (2004). CtBP contributes quantitatively to Knirps repression activity in an NAD binding-dependent manner. Mol Cell Biol *24*, 5953-5966.

Takaki, T., Fukasawa, K., Suzuki-Takahashi, I., and Hirai, H. (2004). Cdk-mediated phosphorylation of pRB regulates HDAC binding *in vitro*. Biochem Biophys Res Commun *316*, 252-255.

Takizawa, T., Nakashima, K., Namihira, M., Ochiai, W., Uemura, A., Yanagisawa, M., Fujita, N., Nakao, M., and Taga, T. (2001). DNA methylation is a critical cell-intrinsic determinant of astrocyte differentiation in the fetal brain. Dev Cell *1*, 749-758.

Thiel, G., Lietz, M., and Hohl, M. (2004). How mammalian transcriptional repressors work. Eur J Biochem *271*, 2855-2862.

Tie, F., Furuyama, T., Prasad-Sinha, J., Jane, E., and Harte, P. (2001). The *Drosophila* polycom group proteins ESC and E(Z) are present in a complex containing the histone-binding protein p55 and the histone deacetylase RPD3. Development *128*, 275-286.

Tomancak, P., Beaton, A., Weiszmann, R., Kwan, E., Shu, S., Lewis, S. E., Richards, S., Ashburner, M., Hartenstein, V., Celniker, S. E., and Rubin, G. M. (2002). Systematic determination of patterns of gene expression during *Drosophila* embryogenesis. Genome Biol *3*, RESEARCH0088.

Turner, J., and Crossley, M. (1998). Cloning and characterization of mCtBP2, a co-repressor that associates with basic Kruppel-like factor and other mammalian transcriptional regulators. Embo J *17*, 5129-5140.

Turner, J., and Crossley, M. (2001). The CtBP family: enigmatic and enzymatic transcriptional co-repressors. Bioessays *23*, 683-690.

Um, M., Li, C., and Manley, J. L. (1995). The transcriptional repressor even-skipped interacts directly with TATA-binding protein. Mol Cell Biol *15*, 5007-5016.

Valentine, S. A., Chen, G., Shandala, T., Fernandez, J., Mische, S., Saint, R., and Courey, A. J. (1998). Dorsal-mediated repression requires the formation of a multiprotein repression complex at the ventral silencer. Mol Cell Biol *18*, 6584-6594.

van der Vlag, J., and Otte, A. (1999). Transcriptional repression mediated by the human polycomb-group protein EED involves histone deacetylation. Nat Genet *23*, 474-478.

Viollet, B., Kahn, A., and Raymondjean, M. (1997). Protein kinase A-dependent phosphorylation modulates DNA-binding activity of hepatocyte nuclear factor 4. Mol Cell Biol *17*, 4208-4219.

Wade, P. (2001). Methyl CpG-binding proteins and transcriptional repression. BioEssays *23*, 1131-1137.

Waltzer, L., and Bienz, M. (1998). *Drosophila* CBP represses the transcription factor TCF to antagonize Wingless signalling. Nature *395*, 521-525.

Wang, L., Brown, J. L., Cao, R., Zhang, Y., Kassis, J. A., and Jones, R. S. (2004). Hierarchical recruitment of polycomb group silencing complexes. Mol Cell *14*, 637-646.

Willy, P. J., Kobayashi, R., and Kadonaga, J. T. (2000). A basal transcription factor that activates or represses transcription. Science *290*, 982-985.

Winter, S. E., and Campbell, G. (2004). Repression of Dpp targets in the *Drosophila* wing by Brinker. Development *131*, 6071-6081.

Yang, X.-J., and Seto, E. (2003). Collaborative spirit of histone deacetylases in regulating chromatin structure and gene expression. Curr Opin Genet Dev *13*, 143-153.

Yu, B., Datta, P. K., and Bagchi, S. (2003). Stability of the Sp3-DNA complex is promoter-specific: Sp3 efficiently competes with Sp1 for binding to promoters containing multiple Sp-sites. Nucleic Acids Res *31*, 5368-5376.

Zhang, Q., Piston, D. W., and Goodman, R. H. (2002a). Regulation of corepressor function by nuclear NADH. Science *295*, 1895-1897.

Zhang, Q., Yao, H., Vo, N., and Goodman, R. H. (2000). Acetylation of adenovirus E1A regulates binding of the transcriptional corepressor CtBP. Proc Natl Acad Sci USA *97*, 14323-14328.

Zhang, S., Xu, L., Lee, J., and Xu, T. (2002b). *Drosophila* atrophin homolog functions as a transcriptional corepressor in multiple developmental processes. Cell *108*, 45-56.

Zink, D., and Paro, R. (1995). *Drosophila* Polycomb-group regulated chromatin inhibits the accessibility of a trans-activator to its target DNA. Embo J *14*, 5660-5671.

Chapter 10
STATs in Cytokine-mediated Transcriptional Regulation

Ke Shuai

Division of Hematology-Oncology, Department of Medicine; Department of Biological Chemistry, University of California, Los Angeles, Los Angeles, California 90095

Key Words: STAT, gene activation, cytokine, signal transduction, protein modification, PIAS, immune regulation

Summary

The signal transducer and activator of transcription (STAT) proteins are latent cytoplasmic transcription factors that are activated by a variety of cytokines and growth factors to regulate transcription. Upon stimulation, STATs become tyrosine phosphorylated to form functional dimers. Activated STAT dimers then translocate into the nucleus where they directly bind to the promoters of downstream genes to activate transcription. Posttranslational modifications such as phosphorylation and acetylation, the protein inhibitor of activate STAT (PIAS) proteins, and several protein tyrosine phosphatases (PTPases) have been shown to regulate the activity of STATs. This article reviews the current understanding of the functional domains, the regulation, and the physiological roles of STATs in cytokine signaling.

Introduction

Through modifying the transcription profile of the cell, extracellular ligands such as cytokines and growth factors play important roles in regulating numerous fundamental cellular processes, including cell growth, differentiation, apoptosis, and survival. One important research area in modern cell biology is to understand how an extacellular ligand signals to the nucleus for gene regulation.

Interferons (IFNs) have antiviral, antiproliferative, and immuregulatory functions. Type I (IFN-α and IFN-β) and type II (IFN-γ) IFNs bind to distinct receptors and they activate overlapping but different sets of genes. Earlier biochemical and genetic studies aimed at understanding how IFNs activate transcription lead to the discovery of the JAK-STAT signaling pathway (Darnell, 1997; Darnell *et al.*, 1994). Subsequently, the JAK-STAT signaling has been shown to be a common gene-activation pathway utilized by many cytokines and growth factors. Although the receptors for many cytokines such as IFNs do not contain intrinsic enzymatic activity, they are constitutively associated with a family of protein tyrosine kinases named as JAKs (Janus kinases) through non-covalent protein-protein interactions (Fig. 10.1). The binding of a cytokine to its cell surface receptor leads to receptor dimerization and the subsequent activation of receptor- associated JAKs. Specific tyrosine residues present in the cytoplasmic region of receptors become phosphorylated by activated JAKs and serve as docking sites to recruit STATs, resulting in the tyrosine phosphorylation and dimerization of STATs. Activated STAT dimers then translocate into the nucleus, where they directly bind to the promoters of target genes to activate transcription (Fig. 10.2). The STAT signaling pathway plays an important role in the control of many biological processes, including immune regulation, cell growth and differentiation, cell survival and apoptosis (Levy and Darnell, 2002; Shuai, 2000; Shuai and Liu, 2003). This review focuses on the current understanding of the domain structures, the regulation, and the physiological functions of STATs.

Corresponding Author: Division of Hematology/Oncology, 11-934 Factor Bldg., 10833 Le Conte Avenue, Los Angeles, California 90095-1678, Tel: (310) 206-9168, Fax: (310) 825-2493, E-mail: kshuai@mednet.ucla.edu

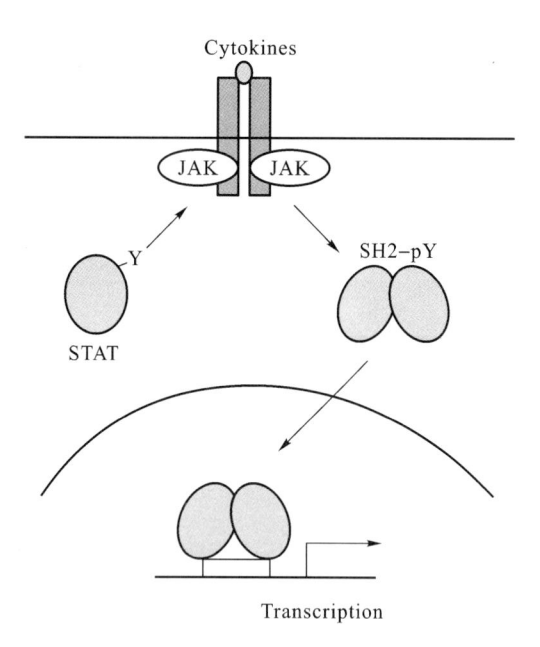

Fig.10.1 **The JAK-STAT signaling pathway.** A schematic representation of the Janus kinase (JAK)-signal transducer and activator of transcription (STAT) pathway. JAKs are constitutively associated with cytokine receptors. Upon ligand stimulation, JAKs are activated which then phosphorylate STATs. Tyrosine phosphorylated STATs form dimers and translocate into the nucleus, where they directly bind to DNA to activate transcription.

Domain Structures of STATs

Seven members of the mammalian STAT family have been isolated: STAT1, STAT2, STAT3, STAT4, STAT5A, STAT5B, and STAT6. Sequence analysis and functional studies indicate that STAT proteins contain several conserved domains (Fig. 10.2).

A: STAT SH2 Domain

All STAT proteins contain a Src homology 2 (SH2) domain, which is often found in many signaling proteins (Fu et al., 1992; Fu and Zhang, 1993). Two functional roles of the STAT SH2 domain have been demonstrated. The SH2 domain is required for the cytokine-induced tyrosine phosphorylation of STATs and it also mediates STAT dimerization by forming intermolecular SH2-phosphotyrosyl peptide interactions.

The replacement of a conserved Arg602 with Leu in the SH2 binding pocket abolished the tyrosine phosphorylation of STAT1 in response to IFN stimulation (Shuai et al., 1993b). The SH2 domain of a STAT protein is important for STAT activation since it targets STAT to the receptor by binding to the tyrosine phosphorylated residue on the receptor. For example, STAT1 binds to the phosphotyrosine residue 440 of the IFNγR1 subunit of the IFN-γ receptor (Greenlund et al., 1995), while the SH2 domain of STAT2 binds to the phosphotyrosine residue 466 on the IFNAR1 chain of the IFN-α receptor (Uddin et al., 1995; Yan et al., 1996).

It has been documented that different SH2 domains display specificity toward different tyrosine-phosphorylated peptide sequences (Pawson and Gish, 2004). Indeed, SH2 domain swapping analysis indicates that the STAT SH2 domain confers specificity in the selective activation of STATs (Heim et al., 1995). For example, while STAT1 is activated by both IFN-α and IFN-γ, STAT2 is activated only by IFN-α, but not IFN-γ (Schindler et al., 1992; Shuai et al., 1992). A chimeric STAT2 protein, in which the SH2 domain of STAT2 was replaced by that of STAT1, became tyrosine phosphorylated in response to both IFN-α and IFN-γ (Heim et al., 1995). These results support an important role of the STAT SH2 domain in the specific activation of STATs in response to cytokine stimulation.

A role of the STAT SH2 domain in the formation of STAT dimers was first documented by biochemical studies. It was found that STAT1 existed as a monomer in untreated cells and formed a homodimer upon IFN stimulation (Shuai et al., 1994). A peptide containing the phosphotyrosine residue Tyr701 from the COOH-terminal region of STAT1 or a phosphotyrosyl peptide from the SH2 domain of STAT1 disrupted the dimerization of STAT1, suggesting that the dimerization of STAT1 is mediated through SH2-phosphotyrosyl peptide interactions (Shuai et al., 1994). Recent studies indicate that a fraction

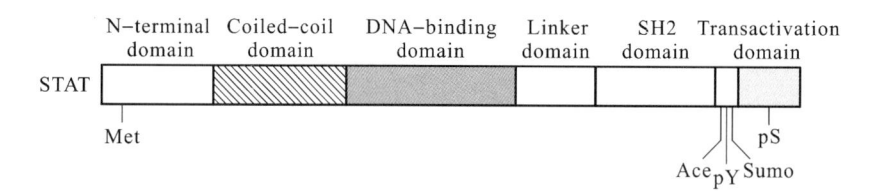

Fig.10.2 **The domain structure of STAT proteins.** STATs contain several conserved domains. The activity of STATs may be regulated by several posttranslational modifications, including tyrosine phosphorylation (pY), serine phosphorylation (pS), acetylation (Ace), sumoylation (Sumo), and methylation (Met). Contradictory reports on the roles of sumoylation and methylation in the regulation of STAT activity have been reported. See text for details.

of STATs in unstimulated cells also existed as dimers (Braunstein et al., 2003; Ota et al., 2004; Yuan et al., 2005; Zhong et al., 2005), although the exact amount of non-phosphorylated STATs existing as dimers under physiological conditions is not known.

The crystal structures of tyrosine phosphorylated STAT1 or STAT3 dimers binding to DNA have been determined (Becker et al., 1998; Chen et al., 1998). The reciprocal and highly specific SH2-phosphotyrosyl peptide interactions are solely responsible for holding two STAT monomers together and may be involved in stabilizing the STAT-DNA binding (Chen et al., 1998).

B: DNA Binding Domain

The first evidence that suggests the direct binding of a STAT protein to DNA comes from studies on IFN-γ signaling. Using two-dimensional mobility gel shift and SDS-PAGE analysis and UV cross-linking assays, STAT1 was found to directly bind to DNA (Shuai et al., 1992).

Sequence analysis of STAT proteins failed to reveal significant homology to any known DNA binding domains. The DNA binding domain of STATs was identified by domain-swapping studies (Horvath et al., 1995). It was shown that the region between residues 400 to 500 of STAT might contact DNA. The DNA binding domain has been further defined from the crystal structures of STATs binding to DNA (Becker et al., 1998; Chen et al., 1998). The region between residues 317 to 488 of STAT1 is in contact with DNA and this region contains an immunoglobulin-like fold, which resembles the DNA binding domains of the p50 subunit of NF-κB and the tumor suppressor p53.

All STAT proteins, except STAT2, can bind to DNA alone. The consensus DNA binding sequence for STAT1, 3, 4, 5 is TTN5AA (where N represents any nucleotide). The optimal DNA binding sequence for STAT6 is TTN6AA (Darnell, 1997). However, it should be noted that the natural DNA binding sequences present in the promoters of endogenous genes have different affinity toward STAT binding, which may provide additional specificity in transcriptional regulation.

C: Transcriptional Activation Domain

STAT1β, a differentially spliced product of *Stat1* gene lacking the COOH-terminal 38 amino acids present in STAT1α, failed to activate IFN-γ-induced genes, although it was tyrosine phosphorylated, translocated to the nucleus, and bound to DNA (Muller et al., 1993; Shuai et al., 1993a). These findings suggest that the COOH-terminal 38 amino acid residues may act as the transcriptional activation domain of STAT1. Indeed, when fused to the GAL4 DNA-binding domain, the COOH-terminal regions of STATs can act as transcriptional activation domains (Bhattacharya et al., 1996).

The COOH-terminal regions of STATs have been shown to interact with several transcriptional coactivators, including the histone acetyltransferases p300/CREB-binding protein (CBP) (Bhattacharya et al., 1996; Zhang et al., 1996) and the acetyltransferase general control non-repressed 5 (GCN5) (Paulson et al., 2002), resulting in the enhanced transcriptional activity of STAT1 and STAT2.

D: The NH2-Terminal Region

The NH2-terminal region of STATs contains a N-terminal domain and a coil-coil domain (Fig. 10.2). Tandem arrays of multiple copies of weak STAT-binding sites are present in some genes such as MIG (monokine induced by IFN-γ) and IFN-γ (Xu et al., 1996). The binding of STAT dimers to two adjacent DNA sites is cooperative and involves the formation of a tetrameric STAT-DNA binding complex. The NH2-terminal regions of STAT1 and STAT4, although not required for the binding of a dimer to a single high-affinity binding site, were required for a cooperative interaction between two STAT dimers (Vinkemeier et al., 1996; Xu et al., 1996).

Recently, a portion of STATs in untreated cells was found to exist as dimers (Braunstein et al., 2003; Ota et al., 2004; Yuan et al., 2005; Zhong et al., 2005). The crystal structure of unphosphorylated STAT1 has been resolved (Yuan et al., 2005). The NH2-terminal region of STAT1 was shown to mediate the dimerization of unphosphorylated STAT1 molecules. However, the NH2-region appears to play different roles in the activation of STATs. For example, mutations in the NH2-terminal region of STAT4 abolished the tyrosine phosphorylation of STAT4 (Ota et al., 2004). In contrast, the mutation of several conserved residues at the NH2-terminal region of STAT1 increased the tyrosine phosphorylation of STAT1, which probably resulted from the defects in the tyrosine dephophorylation of STAT1 caused by these mutations (Zhong et al., 2005). Thus, the NH2-terminal regions of different STATs may play different roles in the regulation of STAT activity.

Regulation of STATs

The activity of STATs is tightly regulated by multiple mechanisms, including posttranslational modifications of STATs, inhibition of STAT activity by PIAS proteins, and the dephosphorylation of STATs by protein tyrosine phosphatases (Shuai and Liu, 2003).

A: Regulation by Posttranslational Modifications

In quiescent cells, STAT proteins exist as inactive monomers. One of the key signals that triggers the activation of STATs is through tyrosine phosphorylation. A single conserved tyrosine residue, which is located immediately after the SH2 domain of STAT proteins, becomes rapidly phosphorylated after ligand stimulation. For example, IFN treatment of cells triggers the phosphorylation of Tyr701 residue of STAT1 (Shuai et al., 1993a). Mutation of Tyr701 to Phe abolishes the dimerization, DNA binding, and nuclear translocation of STAT1 (Shuai et al., 1993a). Thus, tyrosine phosphorylation plays a critical role in the activation of STATs.

In addition to tyrosine phosphorylation, STAT1, STAT3, STAT4, STAT5A, and STAT5B have also been shown to be modified by phosphorylation at a serine residue located in the COOH-terminal transcriptional activation domain (Decker and Kovarik, 2000). Serine phosphorylation and tyrosine phosphorylation of STATs are independent events. STATs are constitutively serine phosphorylated, which can be further enhanced by cytokine stimulation. Mutational analysis indicates that serine phosphorylation is required for the maximum transcriptional activity of STATs (Wen and Darnell, 1997; Wen et al., 1995). Several serine kinases, including extracellular signal-regulated protein kinase (ERK), p38, JUN N-terminal kinase (JNK), and protein kinase Cδ (PKCδ) have been suggested to participate in STAT serine phosphorylation under different conditions (Decker and Kovarik, 2000). It has been suggested that serine phosphorylation of STAT1 is required for the interaction of STAT1 with the co-activator CREB-binding protein (CBP), a histone acetyltransferase (Varinou et al., 2003).

Recently, the physiological importance of STAT serine phosphorylation has been investigated. The STAT1S727A mutant mice, in which the serine phosphorylation site of STAT1 (Ser727) was substituted with alanine, had been generated (Varinou et al., 2003). The transcription of only a subset of IFN-γ-responsive genes was reduced in STAT1S727A macrophages. Furthermore, when infected with lower doses of bacteria, the Stat1S727A and wild-type control mice exhibited similar survival rates. However, the mutant STAT1S727A mice displayed increased mortality when challenged with higher doses of bacteria (Varinou et al., 2003). Thus, these studies demonstrate that serine phosphorylation of STAT1 plays a critical role in IFN-mediated innate immunity to highly pathogenic infection, but serine phosphorylation of STAT1 is dispensable for host defense to a milder pathogenic infection.

The physiological importance to STAT3 serine phosphorylation has also been examined by gene targeting studies. Mice carrying a mutant STAT3 gene with the serine727 residue substituted to alanine (STAT3S727A) have been created (Shen et al., 2004). The transcriptional activation of STAT3-dependent genes in response to IL-6 and oncostatin M (OSM) was significantly reduced in embryonic fibroblasts derived from the STAT3S727A mice (SA/SA). Surprisingly, in contrast to the STAT3-null (-/-) mice which were embryonic lethal, the SA/SA mice showed no defect in development. However, STAT3 SA/null (SA/-) mice displayed perinatal lethality and growth retardation (Shen et al., 2004), while STAT3 wild type/null (+/-) were normal in animal development. These results support the importance of STAT3 serine phosphorylation under certain physiological conditions.

STATs can be modified by protein acetylation. Recently, it has been reported that STAT3 becomes acetylated on a single lysine residue, Lys685, in response to cytokine stimulation (Yuan et al., 2005). The histone acetyltransferase p300 catalyzes the acetylation of Lys685, which can be removed by type I histone deacetylase (HDAC). Mutational analysis suggests that Lys685 acetylation is important for the formation of STAT3 dimer as well as the DNA binding and transcriptional activity of STAT3 (Yuan et al., 2005).

Protein methylation has been suggested to regulate the activity of STATs (Mowen et al., 2001). STAT1 was reported to be methylated on Arg31 by protein arginine methyl-transferase I (PRMT1). It was proposed that arginine methylation of STAT1 increases the DNA binding activity of STAT1 (Mowen et al., 2001). However, other studies have also been reported that argue against the possible modification of STAT1 by protein methylation (Meissner et al., 2004). Thus, the role of protein methylation in the regulation of STAT activity remains to be clarified.

The ubiquitin-proteasome pathway, which targets protein for degradation in various fundamental cellular processes such as transcriptional control, apoptosis, and cell cycle regulation (Pickart, 2001), has been suggested to regulate STAT signaling. For example, polyubiquitination of STAT1 has been reported (Kim and Maniatis, 1996). However, the newly synthesized STAT1 protein was found to be rather stable by pulse-chase analysis (Haspel et al., 1996). Thus, the physiological importance of protein ubiquitination in the regulation of STAT protein stability is still unclear.

Protein sumoylation has recently been implicated in the regulation of STAT signaling. SUMO conjugation occurs through a pathway that is distinct from, but analogous to, protein ubiquitination (Johnson, 2004). Protein sumoylation has been suggested to regulate a wide variety of biological processes, including protein stability, protein-protein interaction, protein localization, and modulation of transcription factors (Johnson, 2004). The Lys703 residue of STAT1 was found to be sumoylated (Rogers et al., 2003; Ungureanu et al., 2003). As will be discussed in the next section, PIAS proteins, which possess SUMO E3 ligase activity, have been shown to promote STAT1 sumoylation. However, contradictory results have been reported on the role of Lys703 sumoylation in the regulation of STAT1 activity (Rogers et al., 2003; Ungureanu et al., 2005; Ungureanu et al., 2003). Further studies are needed to clarify the physiological role of protein sumoylation in STAT signaling.

B: Regulation by PIAS Proteins

The activity of STAT proteins can be regulated by the protein inhibitor of activated STAT (PIAS) family (Shuai and Liu, 2003). The involvement of PIAS proteins in the regulation of STATs was first revealed through yeast two-hybrid assays. Using STAT1β as the bait, PIAS1 was isolated as a STAT1-interacting protein (Chung et al., 1997; Liu et al., 1998). Subsequent cDNA library screening and sequence analysis have identified additional members of the PIAS family. The mammalian PIAS family consists of four members: PIAS1, PIAS3, PIASx, and PIASy (Shuai and Liu, 2003). Except for PIAS1, each member of the PIAS protein family has two splice isoforms. Recent studies suggest that PIAS proteins possess SUMO E3 ligase activity (Jackson, 2001; Schmidt and Muller, 2003).

The PIAS protein family contains several conserved domains (Shuai and Liu, 2003). A domain named as SAP (SAF-A/B, Acinus and PIAS) is present in the NH2-terminal region of all PIAS proteins (Aravind and Koonin, 2000). The SAP domain binds to non-specific AT-rich DNA sequences in scaffold/matrix attachment regions (S/MARs) of the chromatin (Kipp et al., 2000). PIAS proteins contain a C3HC4-type RING-finger-like zinc-binding domain (RLD) that is required for the SUMO E3 ligase activity of PIAS proteins. An LXXLL signature motif, which is known to mediate interactions between nuclear receptors and their co-regulators (Glass and Rosenfeld, 2000), is located within the SAP domain (Liu et al., 2001). The PINIT motif, located within a highly conserved region of PIAS proteins, may be involved in the nuclear retention of PIAS proteins (Duval et al., 2003). The carboxyl-terminal regions of PIAS proteins are the most diversified, which contain a highly acidic region (AD), a serine/threonine rich (S/T) region, and a putative SUMO1 interaction motif (SIM). Interestingly, the S/T region and SIM motif are absent in PIASy. The functional roles of AD, S/T, or SIM of PIAS proteins remain to be defined.

PIAS proteins, normally expressing in the nucleus, do not interact with STATs in unstimulated cells. Upon cytokine stimulation, tyrosine phosphorylated STATs translocate into the nucleus where they interact with PIAS proteins. *In vivo* co-immunoprecipitation studies using specific antibodies against PIAS proteins suggest that there is specificity as well as redundancy in PIAS-STAT interactions (Shuai and Liu, 2003). PIAS1, PIAS3, and PIASx interact with STAT1, STAT3, and STAT4 respectively, in response to cytokine stimulation (Arora et al., 2003; Chung et al., 1997; Liu et al., 1998). In addition, PIASy also interacts with tyrosine phosphorylated STAT1 (Liu et al., 2001). PIAS1 binds to the dimeric, but not the monomeric form of STAT1, which may explain why PIAS-STAT interaction is cytokine-dependent (Liao et al., 2000). Members of the PIAS family have been shown to inhibit STAT-mediated gene activation through several distinct mechanisms. First, PIAS proteins can block the DNA binding activity of STAT. For example, PIAS1 and PIAS3 can inhibit the DNA binding activity of STAT1 and STAT3 *in vitro* respectively (Chung et al., 1997; Liu et al., 1998). Second, PIAS proteins may inhibit STAT-dependent transcription by recruiting other transcriptional co-repressors such as histone deacetyl transferases (HDACs). For example, PIASy and PIASx inhibit STAT1- and STAT4-dependent transcription without affecting their DNA binding activities (Arora et al., 2003; Liu et al., 2001). PIASx and PIASy act as transcriptional co-repressors of STATs, possibly by recruiting HDACs (Arora et al., 2003; Liu et al., 2001). Finally, PIAS proteins may inhibit the transcriptional activity of STATs by promoting SUMO modification of STATs. However, contradictory results have been reported on the role sumoylation in the regulation of STAT activity (Rogers et al., 2003; Ungureanu et al., 2005; Ungureanu et al., 2003). Further studies are needed to clarify the physiological role of PIAS SUMO ligase activity in STAT signaling.

Recently, gene-targeting analysis has been performed to understand the physiological role of PIAS proteins in cytokine signaling. *Pias1* null mice were runted and showed perinatal lethality (Liu et al., 2004). In PIAS1-deficient cells, the induction of a subgroup of IFN-responsive genes was enhanced, suggesting that PIAS1 has specificity in regulating IFN signaling. Consistently, *Pias1* null mice displayed increased protection against viral and bacterial infection. In addition, *Pias1* null mice were hypersensitive to LPS-induced endotoxic shock (Liu *et al.*, 2004). These studies reveal an important role of PIAS1 in the regulation of innate immune responses and demonstrate that PIAS1 is a physiological negative regulator of STAT1.

C: Regulation by Protein Tyrosine Phosphatases (PTPs)

The activity of STATs is regulated by PTPs in both the cytoplasm and nucleus (Shuai and Liu, 2003). Since tyrosine phosphorylation is required to form the active STAT dimmer structure, the dephosphorylation of STATs in the nucleus is believed to be important in terminating the transcriptional activity of STATs. The inactivation of STAT1 in the nucleus has been extensively investigated. Through biochemical purification, TC45, the nuclear isoform of T-cell protein tyrosine phosphatase (TC-PTP), has been identified as a PTP responsible for the nuclear dephosphorylation of STAT1 (ten Hoeve *et al.*, 2002). *In vitro*, TC45 can directly dephosphorylate STAT1. In *Tc-ptp* null mouse embryonic fibroblasts (MEFs) and primary lymphocytes, the dephosphorylation of STAT1 in the nucleus is defective. Interestingly, TC45 is also found to be involved in the nuclear dephosphorylation of STAT3, but not STAT5 or STAT6 (ten Hoeve *et al.*, 2002). These results suggest that there exists specificity in the nuclear dephosphorylation of STATs. Future studies are needed to identify the PTPs involved in the dephosphorylation of other STATs in the nucleus. In addition to TC45, SHP-2, an SH2-containing PTP, is also involved in the nuclear dephosphorylation of STAT1 (Wu *et al.*, 2002).

In the cytoplasm, SHP-2 has been suggested in the dephosphorylation of STAT5. SHP-2 interacts with STAT5 and can directly dephosphorylate STAT5 (Chen *et al.*, 2003). In addition, the dephosphorylation of STAT5 in the cytoplasm is inhibited in *Shp-2* null cells (Chen *et al.*, 2003).

Biological Functions of STATs

Gene targeting studies have identified the physiological functions of STATs. The phenotypes of STAT knockout mice are summarized in this section.

A: STAT1

STAT1 was first shown to be activated by both type I and type II IFNs (Schindler *et al.*, 1992; Shuai *et al.*, 1992). Subsequently, it was found that many cytokines and growth factors, such as IL-2 and EGF, could also activate STAT1 (Fu and Zhang, 1993; Sadowski *et al.*, 1993). However, gene-targeting studies have demonstrated that the physiological function of STAT1 in cytokine signaling is rather specific. *Stat1* null mice showed no apparent defects in animal development. However, *Stat1* null mice were hypersensitive to viral and microbial infections (Durbin *et al.*, 1996; Meraz *et al.*, 1996). These findings are consistent with an essential role of STAT1 in IFN signaling.

B: STAT2

Biochemical studies indicate that STAT2 is solely activated by IFN-α/β (Schindler *et al.*, 1992). Consistently, *Stat2* null cells showed specific defects in IFN-α/β signaling, but were normal in response to IFN-γ stimulation (Park *et al.*, 2000). *Stat2*-deficient mice displayed increased susceptibility to viral infections, otherwise were viable and developmentally normal. These results have established a specific function of STAT2 in IFN-α/β signaling.

C: STAT3

STAT3 is activated by many cytokines, including the IL-6 family of cytokines (Darnell, 1997). *Stat3* null mice were embryonic lethal (Takeda *et al.*, 1997). Conditional *Stat3* knockouts using the Cre-loxP system have uncovered physiological roles of STAT3 in various tissues and cell types, including liver (Alonzi *et al.*, 2001), T lymphocytes (Takeda *et al.*, 1998), macrophages and neutrophils (Takeda *et al.*, 1999), and bone marrows (Lee *et al.*, 2002; Welte *et al.*, 2003). These studies demonstrate that STAT3 participates in various cytokine signaling pathways and that STAT3 is important in cell proliferation, differentiation, migration, and apoptosis (Levy and Lee, 2002). For example, mice with *Stat3* deletion in macrophages and neutrophils showed increased inflammatory responses and developed chronic colitis (Takeda *et al.*, 1999). The deletion of *Stat3* in T lymphocytes caused impaired IL-6-dependent cell survival (Takeda *et al.*, 1998).

D: STAT4

STAT4 is largely activated by IL-12. Consistently, *Stat4* deficient mice displayed impaired TH1 differentiation (Kaplan *et al.*, 1996; Thierfelder *et al.*, 1996), decreased nature killer (NK) cell cytotoxicity, and abrogated IFN-γ production upon IL-12 treatment. These data support a physiological role of STAT4 in IL-12 signaling.

E: STAT5A and STAT5B

STAT5A and STAT5B are encoded by two different

genes but they share over 90% sequence homology (Ihle, 2001b). STAT5A and STAT5B are activated by a variety of cytokines, including growth hormone (GH), prolactin (PRL), and IL-2 (Ihle, 2001a). Gene targeting studies suggest that STAT5A and STAT5B have both redundant and non-redundant roles in cytokine signaling. *Stat5a* deficient mice showed defects in mammary gland development and lactation, consistent with the essential role of STAT5A in PRL signaling (Liu *et al.*, 1997). *Stat5b* null mice were defective in growth hormone (GH)-mediated sexual dimorphic growth, suggesting an important role of STAT5B in GH signaling (Udy *et al.*, 1997). Although *Stat5a* or *Stat5b* null mice showed defects in IL-2-mediated T cell proliferation and NK cell functions (Imada *et al.*, 1998; Nakajima *et al.*, 1997), *Stat5a/5b* double knockout mice displayed even more severe phenotypes. *Stat5a/5b* double knockout mice had no NK cells and T cells from these mice were defective in anti-CD3-mediated cell proliferation (Teglund *et al.*, 1998).

F: STAT6

Biochemical studies indicate that STAT6 is mainly activated by IL-4 and IL-13. Consistently, *Stat6* null mice showed defective TH2 responses and enhanced TH1 differentiation (Shimoda *et al.*, 1996; Takeda *et al.*, 1996). Furthermore, the IL-4/IL-13-mediated expulsion of the gastrointestingal parasite was defective in *Stat6* null mice (Urban *et al.*, 1998). These results support a critical role of STAT6 in IL-4/IL-13 signaling.

Conclusion

The STAT signaling pathway plays an important role in the control of many fundamental biological processes. Great progress has been made in the understanding of the molecular basis of STAT-mediated gene regulation. Since abnormal STAT signaling is associated with immune disorders and cancers, one of the great challenges in the STAT field is to understand the regulation of STAT signaling under both physiological and pathological conditions. These studies will lead to the design of rationale therapeutic strategies for the treatment of human diseases.

Acknowledgement

Supported by grants from The National Institutes of Health (K.S.).

References

Alonzi, T., Maritano, D., Gorgoni, B., Rizzuto, G., Libert, C., and Poli, V. (2001). Essential role of STAT3 in the control of the acute-phase response as revealed by inducible gene inactivation (correction of activation) in the liver. Mol Cell Biol *21*, 1621-1632.

Aravind, L., and Koonin, E. V. (2000). SAP - a putative DNA-binding motif involved in chromosomal organization. Trends Biochem Sci *25*, 112-114.

Arora, T., Liu, B., He, H., Kim, J., Murphy, T. L., Murphy, K. M., Modlin, R. L., and Shuai, K. (2003). PIASx Is a transcriptional co-repressor of signal transducer and activator of transcription 4. J Biol Chem *278*, 21327-21330.

Becker, S., Groner, B., and Muller, C. W. (1998). Three-dimensional structure of the Stat3beta homodimer bound to DNA. Nature *394*, 145-151.

Bhattacharya, S., Eckner, R., Grossman, S., Oldread, E., Arany, Z., D'Andrea, A., and Livingston, D. M. (1996). Cooperation of Stat2 and p300/CBP in signalling induced by interferon-alpha. Nature *383*, 344-347.

Braunstein, J., Brutsaert, S., Olson, R., and Schindler, C. (2003). STATs dimerize in the absence of phosphorylation. J Biol Chem *278*, 34133-34140.

Chen, X., Vinkemeier, U., Zhao, Y., Jeruzalmi, D., Darnell, J. E., Jr., and Kuriyan, J. (1998). Crystal structure of a tyrosine phosphorylated STAT-1 dimer bound to DNA. Cell *93*, 827-839.

Chen, Y., Wen, R., Yang, S., Schuman, J., Zhang, E. E., Yi, T., Feng, G. S., and Wang, D. (2003). Identification of Shp-2 as a Stat5A phosphatase. J Biol Chem *278*, 16520-16527.

Chung, C. D., Liao, J., Liu, B., Rao, X., Jay, P., Berta, P., and Shuai, K. (1997). Specific inhibition of Stat3 signal transduction by PIAS3. Science *278*, 1803-1805.

Darnell, J. E., Jr. (1997). STATs and gene regulation. Science *277*, 1630-1635.

Darnell, J. E., Jr., Kerr, I. M., and Stark, G. R. (1994). Jak-STAT pathways and transcriptional activation in response to IFNs and other extracellular signaling proteins. Science *264*, 1415-1421.

Decker, T., and Kovarik, P. (2000). Serine phosphorylation of STATs. Oncogene *19*, 2628-2637.

Durbin, J. E., Hackenmiller, R., Simon, M. C., and Levy, D. E. (1996). Targeted disruption of the mouse Stat1 gene results in compromised innate immunity to viral disease. Cell *84*, 443-450.

Duval, D., Duval, G., Kedinger, C., Poch, O., and Boeuf, H. (2003). The 'PINIT' motif, of a newly identified conserved domain of the PIAS protein family, is essential for nuclear retention of PIAS3L. FEBS Lett *554*, 111-118.

Fu, X. Y., Schindler, C., Improta, T., Aebersold, R., and Darnell, J. E., Jr. (1992). The proteins of ISGF-3, the interferon alpha-induced transcriptional activator, define a gene family involved in signal transduction. Proc Natl Acad Sci USA *89*, 7840-7843.

Fu, X. Y., and Zhang, J. J. (1993). Transcription factor p91 interacts with the epidermal growth factor receptor and mediates activation of the c-fos gene promoter. Cell *74*, 1135-1145.

Glass, C. K., and Rosenfeld, M. G. (2000). The coregulator exchange in transcriptional functions of nuclear receptors. Genes and Development *14*, 121-141.

Greenlund, A. C., Morales, M. O., Viviano, B. L., Yan, H., Krolewski, J., and Schreiber, R. D. (1995). Stat recruitment by tyrosine-phosphorylated cytokine receptors: an ordered reversible affinity-driven process. Immunity *2*, 677-687.

Haspel, R. L., Salditt-Georgieff, M., and Darnell, J. E., Jr. (1996). The rapid inactivation of nuclear tyrosine phosphorylated Stat1 depends upon a protein tyrosine phosphatase. Embo J *15*, 6262-6268.

Heim, M. H., Kerr, I. M., Stark, G. R., and Darnell, J. E., Jr. (1995). Contribution of STAT SH2 groups to specific interferon signaling by the Jak-STAT pathway. Science *267*, 1347-1349.

Horvath, C. M., Wen, Z., and Darnell, J. E., Jr. (1995). A STAT protein domain that determines DNA sequence recognition suggests a novel DNA-binding domain. Genes and Development *9*, 984-994.

Ihle, J. (2001a). Pathways in cytokine regulation of hematopoiesis. Ann N Y Acad Sci *938*, 129-130.

Ihle, J. N. (2001b). The Stat family in cytokine signaling. Curr Opin Cell Biol *13*, 211-217.

Imada, K., Bloom, E. T., Nakajima, H., Horvath-Arcidiacono, J. A., Udy, G. B., Davey, H. W., and Leonard, W. J. (1998). Stat5b is essential for natural killer cell-mediated proliferation and cytolytic activity. J Exp Med *188*, 2067-2074.

Jackson, P. K. (2001). A new RING for SUMO: wrestling transcriptional responses into nuclear bodies with PIAS family E3 SUMO ligases. Genes Dev *15*, 3053-3058.

Johnson, E. S. (2004). Protein modification by SUMO. Annu Rev Biochem *73*, 355-382.

Kaplan, M. H., Sun, Y. L., Hoey, T., and Grusby, M. J. (1996). Impaired IL-12 responses and enhanced development of Th2 cells in Stat4-deficient mice. Nature *382*, 174-177.

Kim, T. K., and Maniatis, T. (1996). Regulation of interferon-gamma-activated STAT1 by the ubiquitin-proteasome pathway. Science *273*, 1717-1719.

Kipp, M., Gohring, F., Ostendorp, T., van Drunen, C. M., van Driel, R., Przybylski, M., and Fackelmayer, F. O. (2000). SAF-Box, a conserved protein domain that specifically recognizes scaffold attachment region DNA. Mol Cell Biol *20*, 7480-7489.

Lee, C. K., Raz, R., Gimeno, R., Gertner, R., Wistinghausen, B., Takeshita, K., DePinho, R. A., and Levy, D. E. (2002). STAT3 is a negative regulator of granulopoiesis but is not required for G-CSF-dependent differentiation. Immunity *17*, 63-72.

Levy, D. E., and Darnell, J. E. (2002). Signalling: Stats: transcriptional control and biological impact. Nat Rev Mol Cell Biol *3*, 651-662.

Levy, D. E., and Lee, C. K. (2002). What does Stat3 do? J Clin Invest *109*, 1143-1148.

Liao, J., Fu, Y., and Shuai, K. (2000). Distinct roles of the NH2- and COOH-terminal domains of the protein inhibitor of activated signal transducer and activator of transcription (STAT)1 (PIAS1) in cytokine-induced PIAS1-Stat1 interaction. Proc Natl Acad Sci USA *97*, 5267-5272.

Liu, B., Gross, M., ten Hoeve, J., and Shuai, K. (2001). A transcriptional corepressor of Stat1 with an essential LXXLL signature motif. Proc Natl Acad Sci USA *98*, 3203-3207.

Liu, B., Liao, J., Rao, X., Kushner, S. A., Chung, C. D., Chang, D. D., and Shuai, K. (1998). Inhibition of Stat1-mediated gene activation by PIAS1. Proc Natl Acad Sci USA *95*, 10626-10631.

Liu, B., Mink, S., Wong, K. A., Stein, N., Getman, C., Dempsey, P. W., Wu, H., and Shuai, K. (2004). PIAS1 selectively inhibits interferon-inducible genes and is important in innate immunity. Nat Immunol *5*, 891-898.

Liu, X., Robinson, G. W., Wagner, K. U., Garrett, L., Wynshaw-Boris, A., and Hennighausen, L. (1997). Stat5a is mandatory for adult mammary gland development and lactogenesis. Genes Dev *11*, 179-186.

Meissner, T., Krause, E., Lodige, I., and Vinkemeier, U. (2004). Arginine methylation of STAT1: a reassessment. Cell *119*, 587-589; discussion 589-590.

Meraz, M. A., White, J. M., Sheehan, K. C., Bach, E. A., Rodig, S. J., Dighe, A. S., Kaplan, D. H., Riley, J. K., Greenlund, A. C., Campbell, D., et al. (1996). Targeted disruption of the Stat1 gene in mice reveals unexpected physiologic specificity in the JAK-STAT signaling pathway. Cell *84*, 431-442.

Mowen, K. A., Tang, J., Zhu, W., Schurter, B., Shuai, K., Herschman, H. R., and David, M. (2001). Arginine methylation of STAT1 modulates IFNa/b induced transcription. Cell *104*, 731-741.

Muller, M., Laxton, C., Briscoe, J., Schindler, C., Improta, T., Darnell, J. E., Jr., Stark, G. R., and Kerr, I. M. (1993). Complementation of a mutant cell line: central role of the 91 kDa polypeptide of ISGF3 in the interferon-alpha and -gamma signal transduction pathways. Embo J *12*, 4221-4228.

Nakajima, H., Liu, X. W., Wynshaw-Boris, A., Rosenthal, L. A., Imada, K., Finbloom, D. S., Hennighausen, L., and Leonard, W. J. (1997). An indirect effect of Stat5a in IL-2-induced proliferation: a critical role for Stat5a in IL-2-mediated IL-2 receptor alpha chain induction. Immunity *7*, 691-701.

Ota, N., Brett, T. J., Murphy, T. L., Fremont, D. H., and Murphy, K. M. (2004). N-domain-dependent nonphosphorylated STAT4 dimers required for cytokine-driven activation. Nat Immunol *5*, 208-215.

Park, C., Li, S., Cha, E., and Schindler, C. (2000). Immune response in Stat2 knockout mice. Immunity *13*, 795-804.

Paulson, M., Press, C., Smith, E., Tanese, N., and Levy, D. E. (2002). IFN-Stimulated transcription through a TBP-free acetyltransferase complex escapes viral shutoff. Nat Cell Biol *4*, 140-147.

Pawson, T., and Gish, G. D. (2004). Specificity in signal

transduction: from phosphotyrosine-SH2 domain interactions to complex cellular systems SH2 and SH3 domains: from structure to function. Cell *116*, 191-203.

Pickart, C. M. (2001). Mechanisms underlying ubiquitination. Annu Rev Biochem *70*, 503-533.

Rogers, R. S., Horvath, C. M., and Matunis, M. J. (2003). SUMO modification of STAT1 and its role in PIAS-mediated inhibition of gene activation. J Biol Chem *278*, 30091-30097.

Sadowski, H. B., Shuai, K., Darnell, J. E., Jr., and Gilman, M. Z. (1993). A common nuclear signal transduction pathway activated by growth factor and cytokine receptors (see comments). Science *261*, 1739-1744.

Schindler, C., Shuai, K., Prezioso, V. R., and Darnell, J. E., Jr. (1992). Interferon-dependent tyrosine phosphorylation of a latent cytoplasmic transcription factor [see comments]. Science *257*, 809-813.

Schmidt, D., and Muller, S. (2003). PIAS/SUMO: new partners in transcriptional regulation. Cell Mol Life Sci *60*, 2561-2574.

Shen, Y., Schlessinger, K., Zhu, X., Meffre, E., Quimby, F., Levy, D. E., and Darnell, J. E., Jr. (2004). Essential role of STAT3 in postnatal survival and growth revealed by mice lacking STAT3 serine 727 phosphorylation. Mol Cell Biol *24*, 407-419.

Shimoda, K., van Deursen, J., Sangster, M. Y., Sarawar, S. R., Carson, R. T., Tripp, R. A., Chu, C., Quelle, F. W., Nosaka, T., Vignali, D. A., *et al.* (1996). Lack of IL-4-induced Th2 response and IgE class switching in mice with disrupted Stat6 gene. Nature *380*, 630-633.

Shuai, K. (2000). Modulation of STAT signaling by STAT-interacting proteins. Oncogene *19*, 2638-2644.

Shuai, K., Horvath, C. M., Huang, L. H., Qureshi, S. A., Cowburn, D., and Darnell, J. E., Jr. (1994). Interferon activation of the transcription factor Stat91 involves dimerization through SH2-phosphotyrosyl peptide interactions. Cell *76*, 821-828.

Shuai, K., and Liu, B. (2003). Regulation of JAK-STAT signalling in the immune system. Nat Rev Immunol *3*, 900-911.

Shuai, K., Schindler, C., Prezioso, V. R., and Darnell, J. E., Jr. (1992). Activation of transcription by IFN-gamma: tyrosine phosphorylation of a 91-kD DNA binding protein. Science *258*, 1808-1812.

Shuai, K., Stark, G. R., Kerr, I. M., and Darnell, J. E., Jr. (1993a). A single phosphotyrosine residue of Stat91 required for gene activation by interferon-gamma [see comments]. Science *261*, 1744-1746.

Shuai, K., Ziemiecki, A., Wilks, A. F., Harpur, A. G., Sadowski, H. B., Gilman, M. Z., and Darnell, J. E. (1993b). Polypeptide signalling to the nucleus through tyrosine phosphorylation of Jak and Stat proteins. Nature *366*, 580-583.

Takeda, K., Clausen, B. E., Kaisho, T., Tsujimura, T., Terada, N., Forster, I., and Akira, S. (1999). Enhanced Th1 activity and development of chronic enterocolitis in mice devoid of Stat3 in macrophages and neutrophils. Immunity *10*, 39-49.

Takeda, K., Kaisho, T., Yoshida, N., Takeda, J., Kishimoto, T., and Akira, S. (1998). Stat3 activation is responsible for IL-6-dependent T cell proliferation through preventing apoptosis: generation and characterization of T cell-specific Stat3-deficient mice. J Immunol *161*, 4652-4660.

Takeda, K., Noguchi, K., Shi, W., Tanaka, T., Matsumoto, M., Yoshida, N., Kishimoto, T., and Akira, S. (1997). Targeted disruption of the mouse Stat3 gene leads to early embryonic lethality. Proc Natl Acad Sci USA *94*, 3801-3804.

Takeda, K., Tanaka, T., Shi, W., Matsumoto, M., Minami, M., Kashiwamura, S., Nakanishi, K., Yoshida, N., Kishimoto, T., and Akira, S. (1996). Essential role of Stat6 in IL-4 signalling. Nature *380*, 627-630.

Teglund, S., McKay, C., Schuetz, E., van Deursen, J. M., Stravopodis, D., Wang, D., Brown, M., Bodner, S., Grosveld, G., and Ihle, J. N. (1998). Stat5a and Stat5b proteins have essential and nonessential, or redundant, roles in cytokine responses. Cell *93*, 841-850.

ten Hoeve, J., de Jesus Ibarra-Sanchez, M., Fu, Y., Zhu, W., Tremblay, M., David, M., and Shuai, K. (2002). Identification of a nuclear Stat1 protein tyrosine phosphatase. Mol Cell Biol *22*, 5662-5668.

Thierfelder, W. E., van Deursen, J. M., Yamamoto, K., Tripp, R. A., Sarawar, S. R., Carson, R. T., Sangster, M. Y., Vignali, D. A., Doherty, P. C., Grosveld, G. C., and Ihle, J. N. (1996). Requirement for Stat4 in interleukin-12-mediated responses of natural killer and T cells. Nature *382*, 171-174.

Uddin, S., Chamdin, A., and Platanias, L. C. (1995). Interaction of the transcriptional activator Stat-2 with the type I interferon receptor. Journal of Biological Chemistry *270*, 24627-24630.

Udy, G. B., Towers, R. P., Snell, R. G., Wilkins, R. J., Park, S. H., Ram, P. A., Waxman, D. J., and Davey, H. W. (1997). Requirement of STAT5b for sexual dimorphism of body growth rates and liver gene expression. Proc Natl Acad Sci USA *94*, 7239-7244.

Ungureanu, D., Vanhatupa, S., Gronholm, J., Palvimo, J. J., and Silvennoinen, O. (2005). SUMO-1 conjugation selectively modulates STAT1-mediated gene responses. Blood.

Ungureanu, D., Vanhatupa, S., Kotaja, N., Yang, J., Aittomaki, S., Janne, O. A., Palvimo, J. J., and Silvennoinen, O. (2003). PIAS proteins promote SUMO-1 conjugation to STAT1. Blood *102*, 3311-3313.

Urban, J. F., Jr., Noben-Trauth, N., Donaldson, D. D., Madden, K. B., Morris, S. C., Collins, M., and Finkelman, F. D. (1998). IL-13, IL-4Ralpha, and Stat6 are required for the expulsion of the gastrointestinal nematode parasite Nippostrongylus brasiliensis. Immunity *8*, 255-264.

Varinou, L., Ramsauer, K., Karaghiosoff, M., Kolbe, T., Pfeffer, K., Muller, M., and Decker, T. (2003). Phosphorylation of the Stat1 transactivation domain is required for full-fledged IFN-gamma-dependent innate immunity. Immunity *19*, 793-802.

Vinkemeier, U., Cohen, S. L., Moarefi, I., Chait, B. T., Kuriyan, J., and Darnell, J. E., Jr. (1996). DNA binding of *in vitro* activated Stat1 alpha, Stat1 beta and truncated Stat1: interaction between NH2-terminal domains stabilizes binding of two dimers to tandem DNA sites. Embo J *15*, 5616-5626.

Welte, T., Zhang, S. S., Wang, T., Zhang, Z., Hesslein, D. G., Yin, Z., Kano, A., Iwamoto, Y., Li, E., Craft, J. E., *et al.* (2003). STAT3 deletion during hematopoiesis causes Crohn's disease-like pathogenesis and lethality: a critical role of STAT3 in innate immunity. Proc Natl Acad Sci USA *100*, 1879-1884.

Wen, Z., and Darnell, J. E., Jr. (1997). Mapping of Stat3 serine phosphorylation to a single residue (727) and evidence that serine phosphorylation has no influence on DNA binding of Stat1 and Stat3. Nucleic Acids Research *25*, 2062-2067.

Wen, Z., Zhong, Z., and Darnell, J. E., Jr. (1995). Maximal activation of transcription by Stat1 and Stat3 requires both tyrosine and serine phosphorylation. Cell *82*, 241-250.

Wu, T. R., Hong, Y. K., Wang, X. D., Ling, M. Y., Dragoi, A. M., Chung, A. S., Campbell, A. G., Han, Z. Y., Feng, G. S., and Chin, Y. E. (2002). SHP-2 is a dual-specificity phosphatase involved in Stat1 dephosphorylation at both tyrosine and serine residues in nuclei. J Biol Chem *277*, 47572-47580.

Xu, X., Sun, Y. L., and Hoey, T. (1996). Cooperative DNA binding and sequence-selective recognition conferred by the STAT amino-terminal domain [see comments]. Science *273*, 794-797.

Yan, H., Krishnan, K., Greenlund, A. C., Gupta, S., Lim, J. T., Schreiber, R. D., Schindler, C. W., and Krolewski, J. J. (1996). Phosphorylated interferon-alpha receptor 1 subunit (IFNaR1) acts as a docking site for the latent form of the 113 kDa STAT2 protein. Embo J. *15*, 1064-1074.

Yuan, Z. L., Guan, Y. J., Chatterjee, D., and Chin, Y. E. (2005). Stat3 dimerization regulated by reversible acetylation of a single lysine residue. Science *307*, 269-273.

Zhang, J. J., Vinkemeier, U., Gu, W., Chakravarti, D., Horvath, C. M., and Darnell, J. E., Jr. (1996). Two contact regions between Stat1 and CBP/p300 in interferon gamma signaling. Proc Natl Acad Sci USA *93*, 15092-15096.

Zhong, M., Henriksen, M. A., Takeuchi, K., Schaefer, O., Liu, B., ten Hoeve, J., Ren, Z., Mao, X., Chen, X., Shuai, K., and Darnell, J. E., Jr. (2005). Implications of an antiparallel dimeric structure of nonphosphorylated STAT1 for the activation-inactivation cycle. Proc Natl Acad Sci USA *102*, 3966-3971.

Chapter 11

Transcriptional Regulation by Smads

Fang Liu

Center for Advanced Biotechnology and Medicine, Susan Lehman Cullman Laboratory for Cancer Research, Department of Chemical Biology, Ernest Mario School of Pharmacy, Rutgers, The State University of New Jersey, Cancer Institute of New Jersey, 679 Hoes Lane, Piscataway, NJ 08854

Key Words: TGF-β, BMP, Smad, signal transduction, transcriptional regulation

Summary

Transforming growth factor-β (TGF-β) and related polypeptides, including activins and bone morphogenetic proteins (BMPs) regulate a wide variety of biological activities, such as cell proliferation, differentiation, adhesion, migration, and apoptosis. The TGF-β family members signal through transmembrane serine/threonine kinase receptors. Smads are major intracellular mediators that transduce the TGF-β signal from the cell surface to the nucleus. Upon TGF-β binding, the TGF-β type I receptor phosphorylates Smad2 and Smad3, which then form complexes with Smad4 and together accumulate in the nucleus to regulate transcription of a variety of genes. Microarray studies indicate that Smad3 and Smad4 are essential for regulation of many TGF-β-responsive genes. Smad3 and Smad4 possess DNA binding activities. For high affinity binding sites, Smad3-Smad4 can bind on its own. For low affinity binding sites, Smads are recruited to target promoters through interaction with DNA-binding partners. Smads can activate transcription by recruiting transcriptional coactivators, and they can also inhibit transcription by recruiting transcriptional corepressors. In addition, Smad3 can also repress transcription by several other distinct mechanisms. Through regulation of different target genes in a cell context dependent manner, TGF-β/Smads elicit distinct biological responses.

Introduction

TGF-β superfamily members are a group of secreted polypeptides that regulate various biological and developmental processes (Massague, 1990; Roberts and Sporn, 1990). TGF-β is multifunctional. It potently inhibits cell proliferation, regulates differentiation, enhances extracellular matrix production, induces apoptosis, suppresses immune function, and promotes angiogenesis. Alterations of the components of the TGF-β signaling pathways are associated with a number of diseases, in particular, with cancer and fibrosis (Massague *et al.*, 2000; Derynck *et al.*, 2001; Miyazono *et al.*, 2003; Roberts *et al.*, 2003; Roberts and Wakefield 2003).

TGF-β signals through binding and bringing together two classes of transmembrane receptors, the type I and type II receptors (Heldin *et al.*, 1997; Massague, 1998; Massague and Chen, 2000). The TGF-β type II receptor is constitutively active. Upon TGF-β binding, the type II receptor trans-phosphorylates the type I receptor, which then plays a major role in specifying downstream events (Heldin *et al.*, 1997; Massague, 1998; Massague and Chen, 2000). Smads are direct substrates of the TGF-β family type I receptors and can transduce the signal at the cell surface into transcriptional regulation in the nucleus, leading to various biological responses (Fig.11.1, Heldin *et al.*, 1997; Attisano and Wrana, 2000; Miyazono *et al.*, 2001; Derynck and Zhang, 2003; Shi and Massague, 2003; ten Dijke and Hill, 2004).

Corresponding Author: Tel: (732) 235-5372, Fax: (732) 235-4850, E-mail: fangliu@cabm.rutgers.edu

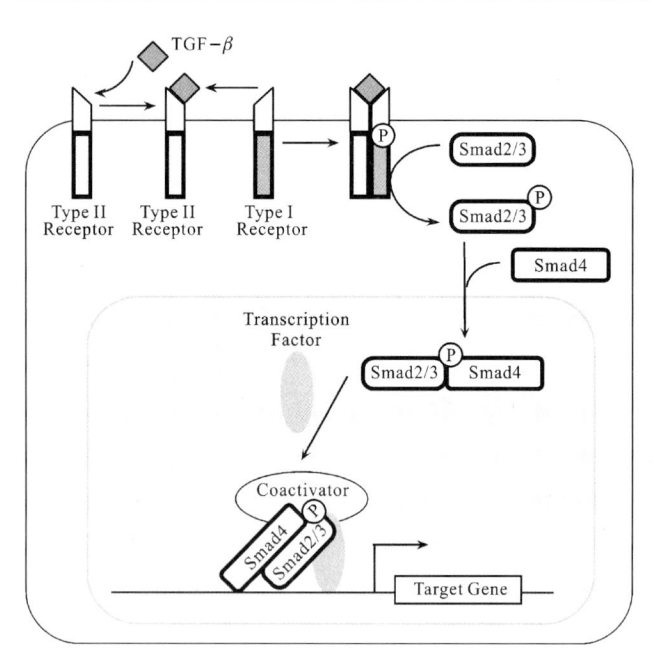

Fig.11.1 A simplified TGF-β/Smad signal transduction and transcriptional activation pathway. TGF-β type I and type II receptors are transmembrane serine/threonine kinase receptors. TGF-β binds directly to its type II receptor. Binding of TGF-β to the type II receptor results in the recruitment of type I receptor into the ligand-receptor complex. In this complex, the constitutively active type II receptor phosphorylates and activates the type I receptor, which then plays a major role in specifying downstream signaling events. The activated type I receptor directly phosphorylates Smad2 and Smad3 at their C-terminal tail. Smad4 cannot be phosphorylated by the receptor, but can form complexes with receptor phosphorylated Smad2/3. The Smad complexes then accumulate in the nucleus. Smad3 and Smad4 possess DNA binding activities, whereas Smad2 cannot bind to DNA. Many TGF-β-responsive promoters contain low affinity binding sites for Smads, and Smads are recruited to these promoters through interaction with other transcription factors. Smads can activate transcription through recruitment of coactivators. Smads can repress transcripiotn through multiple distinct mechanisms (not shown).

The Smad family can be structurally and functionally divided into three groups (Fig.11.2). One group includes those receptor-regulated Smads (R-Smads, also termed pathway-specific Smads) that are phosphorylated by receptor kinases. Smad1, Smad5 and Smad8 are phosphorylated by BMP receptor kinases (Hoodless et al., 1996; Kretzschmar et al., 1997), whereas Smad2 and Smad3 are phosphorylated by the homologous TGF-β and activin receptor kinases (Macias-Silva et al., 1996; Zhang et al., 1996; Abdollah et al., 1997; Souchelnytskyi et al., 1997). The second group includes common Smads (Co-Smads), which are not phosphorylated by receptors but are essential for TGF-β, activin and BMP signaling by associating with a receptor activated Smad (Lagna et al., 1996). The only known member of this group in mammalian cells is Smad4. Smad4 is necessary for transcriptional activation (Liu et al., 1997; Zhou et al., 1998a). The third group includes inhibitory Smads (I-Smads) that antagonize the function of receptor-activated Smads. Smad6 binds and inhibits BMP receptor phosphorylation of Smad1 (Imamura et al., 1997). In addition, Smad6 antagonizes BMP signaling by interacting with Smad1, thus preventing Smad1 from forming a complex with Smad4 (Hata et al., 1998). Smad7 antagonizes TGF-β, activin, and BMP signaling pathways. For example, Smad7 antagonizes TGF-β signaling by binding to the receptor and thus inhibiting its capacity to phosphorylate Smad2 and Smad3 (Hayashi et al., 1997; Nakao et al., 1997). Interestingly, Smad6 and Smad7 are themselves direct target genes of Smad proteins (Nagarajan et al., 1999; Brodin et al., 2000; Denissova et al., 2000; Ishida et al., 2000; von Gersdorff et al., 2000). Their transcription is upregulated by treatment with BMP and TGF-β, respectively, thus providing a negative feedback loop control of TGF-β family signaling (Hayashi et al., 1997; Nakao et al., 1997; Nagarajan et al., 1999; Brodin et al., 2000; Denissova et al., 2000; Ishida et al., 2000; von Gersdorff et al., 2000).

The TGF-β/Smad-mediated transcriptional responses are subjected to regulation at multiple levels (Derynck and Zhang, 2003; Shi and Massague, 2003; ten Dijke and Hill, 2004). For instance, regulation of receptor activity and activation of Smad2 and Smad3 by TGF-β receptor are controlled by a variety of TGF-β receptor and/or Smad interacting proteins and by Smad2 and Smad3 homo-trimerization and hetero-oligomerization with Smad4. Other controls are mediated by Smads nucleocytoplasmic shuttling, DNA binding properties, interacting partners for transcriptional activation or repression, and by post-translational modifications including phosphorylation, ubiquitination and sumoylation. This chapter is mostly focused on Smads DNA binding, interaction with transcription factors, transcriptional activation and transcriptional repression.

Structure-Function Relationship of Smads

Smads are evolutionarily conserved. The name Smad was coined to reflect it being the vertebrate homologue of the *Drosophila* Mad and the *C. elegans* Sma. The R-Smads and Smad4 contain conserved N- and C-terminal regions, also designated as the MH1 (Mad Homology 1) and MH2 (Mad Homology 2) domains, respectively, separated by a divergent proline-rich linker region (Fig.11.3). Smad2 and Smad3 are highly homologous with each other, sharing over 90% homology. They are divergent in the proline-rich linker region. In addition, Smad2 contains two stretches of amino acids that are not present in Smad3. The first stretch is 10 amino acids. The second stretch is 30 amino acids that is encoded by a separate exon (exon 3) (Takenoshita et al., 1998). The presence of these 30 amino acids interferes with Smad2 binding to DNA (Dennler et al., 1998; Shi et al., 1998; Yagi et al., 1999; Shi, 2001). Smad2 and Smad3 have overlapping as well as distinct properties and functions (Liu, 2003; Roberts et al., 2003). Smad1, Smad5 and Smad8 share over 90% homology. The R-Smads are more similar with each other than with Smad4. The C-terminal domains of Smad6 and Smad7 are similar to those of the R-Smads and Smad4 but do not contain the SSXS motif for receptor

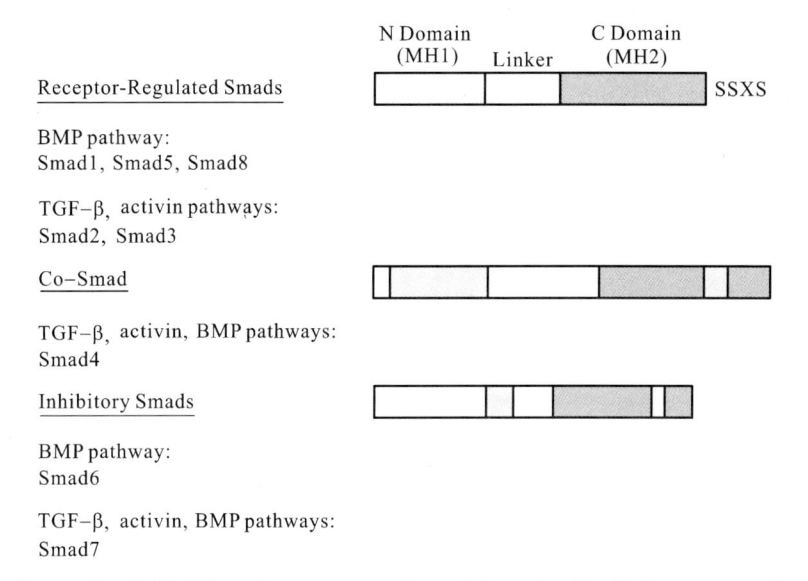

Fig.11.2 **The Smad family can be divided into three structurally and functionally distinct groups.** One group includes receptor regulated Smads (R-Smads). These Smads contain the SSXS motif at the C-tail for phosphorylation by receptors. Smad1, Smad5 and Smad8 are phosphorylated by BMP receptors, whereas Smad2 and Smad3 are phosphorylated by the homologous TGF-β and activin receptors, Smad4 is a common Smad (Co-Smad) that participates in TGF-β, activin and BMP signaling by associating with a receptor-phosphorylated Smad. In mammalian cells, Smad4 is the only Co-Smad. The third group includes those inhibitory Smads (I-Smads) that antagonize TGF-β family signaling. Smad6 specifically inhibits BMP signaling. Smad7 inhibits TGF-β, activin and BMP signaling. The N-terminal and C-terminal domains are also termed as MH1 (Mad Homology 1) and MH2 (Mad Homology 2) domains, respectively. The MH1 and MH2 domains are conserved among R-Smads and Smad4. The MH2 domain is also similar in Smad6 and Smad7. The MH1 domains of Smad6 and Smad7 are not conserved with those of the R-Smads and Smad4.

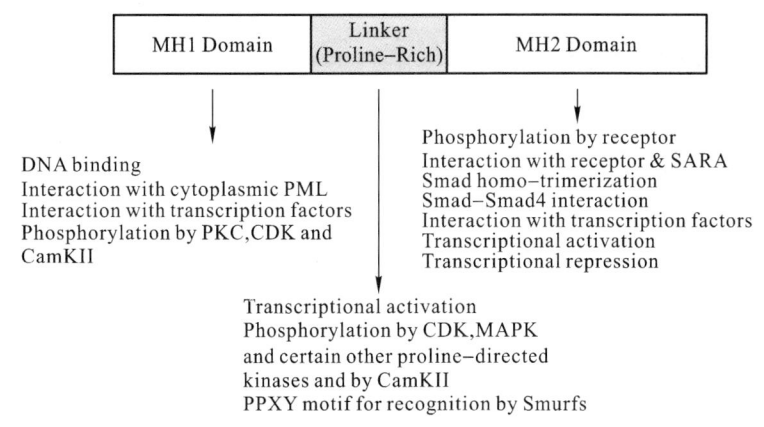

Fig.11.3 **Structure-function relationship of Smads.** SARA (Smad anchor for receptor activation), PML (promyelocytic leukaemia protein), CDK (cyclin-dependent kinase), MAPK (mitogen-activated protein kinase), PKC (protein kinase C), CamKII (Ca^{2+}-calmodulin-dependent kinase II), Smurf (Smad ubiqutin regulatory factor).

phosphorylation. The N-terminal domains of Smad6 and Smad7 share very limited homology with those of the R-Smads and Smad4 (Fig.11.3). A structure-function relationship of Smads is shown in Fig.11.3 and is further discussed in subsequent sections.

Smads Activation by the TGF-β Receptor

Both Smad2 and Smad3 are phosphorylated by TGF-β receptor in the C-terminal SSXS motif (Abdollah et al., 1997; Souchelnytskyi et al., 1997). At the basal state, the MH1 and MH2 domains interact intramolecularly, thus inhibiting each other's activities (Hata et al., 1997). Smad phosphorylation by receptor disrupts the intramolecular interaction (Hata et al., 1997). Smad2/3-receptor interactions are mediated by the L3 loop in the MH2 domain of Smad2/3 and L45 loop in the TGF-β type I receptor (Feng and Derynck, 1997; Chen et al., 1998; Lo et al., 1998; Huse et al., 2001).

SARA (Smad anchor for receptor activation) functions to recruit Smad2 and Smad3 to the TGF-β receptor (Tsukazaki et al., 1998). SARA contains a FYVE domain for membrane localization, a Smad binding domain (SBD) for binding to Smad2/3, and a carboxyl-terminal domain for interacting with the receptor kinase (Tsukazaki et al., 1998). At the basal state, Smad2 and Smad3 are bound by SARA (Wu et al., 2000; Moustakas and Heldin, 2002; Qin et al., 2002). Structural studies indicate that SARA binds and stabilizes monomeric forms of Smad2 and Smad3, and inhibit Smad2 and Smad3 trimerization (Moustakas and Heldin, 2002; Qin et al. 2002). This helps explain, at least in part, why overexpression of Smad3 or Smad2, which can saturate SARA, leads to transcriptional activation in the absence of TGF-β signaling. Recent studies have shown that the cytoplasmic PML (promyelocytic leukaemia) protein physically interacts with both Smad2/3 and SARA and is required for association of Smad2/3 with SARA (Lin et al., 2004). Cytoplasmic PML can also bind to the receptors. Upon TGF-β binding and activation of the receptor, Cytoplasmic PML can promote the transfer of the complex containing TGF-β receptors, SARA, Smad2/3, and probably cytoplasmic PML itself into early endosome (Lin et al., 2004). Cytoplasmic PML then seems to dissociate from the complex, followed by SARA presenting Smad2 and Smad3 for receptor recognition, and precisely positioning the phosphorylation sites of Smad2 and Smad3 in the kinase catalytic center (Tsukazaki et al., 1998; Wu et al., 2000; Moustakas and Heldin, 2002; Qin et al., 2002; Shi and Massague, 2003; Lin et al., 2004). Phosphorylation of Smad2/3 by TGF-β receptor allows Smad2/3 to dissociate from SARA and the receptor, then associate with Smad4 and accumulate in the nucleus (Wu et al., 2000; Moustakas and Heldin, 2002; Qin et al., 2002).

In addition to SARA and cytoplasmic PML, a number of other proteins with anchoring, scaffolding or chaperone activity, such as Hgs, Disabled-2, Axin, Caveolin-1, ARIP1, GIPC, STRAP, Filamin, TRAP-1, and microtubules, can regulate the recruitment of Smad2 and Smad3 to the TGF-β receptor complex or bring Smad4 into the proximity of the receptor complex and aid in the formation of heteromeric complexes between Smad2/3 and Smad4 (Wrana, 2000; Miyazono, et al. 2001; ten Dijke, et al. 2002; Liu, 2003)

Stoichiometry State of Smads

As described above, structural studies and computer modeling indicate that Smad2 and Smad3 are present as monomers at basal state through interacton with SARA (Moustakas and Heldin, 2002; Qin et al., 2002). Biochemical studies suggest that Smads are present as monomers or a mixture of monomers and oligomers at basal state (Hata et al., 1997; Kawabata et al., 1998). The crystal structure of the Smad4 C-terminal domain is a trimer (Shi et al., 1997). The crystal structure of phosphorylated Smad2 reveals that it is a trimer with the phosphoserine being recognized by the MH2 domain of Smad2 as well as Smad4 (Wu et al., 2001b). Similar conclusions were also made for Smad1 (Qin et al., 2001). The crystal structure of pseudo-phosphorylated Smad3 indicates that it has an increased propensity to homo-trimerize and recruits Smad4 to form a hetero-trimer (Chacko et al., 2001; Qin et al., 2002). Thus, phosphorylation of Smad2 and Smad3 by the receptor drives their homo-trimerization as well as hetero-oligomerization with the common mediator Smad4 (Chacko et al., 2001; Wu et al., 2001b; Moustakas and Heldin, 2002; Qin et al., 2002; Wrana, 2002). This was also supported by tissue culture studies (Kawabata et al., 1998). Recently, the crystal structures of C-tail phosphorylated Smad3 or Smad2 in complex with Smad4 have been solved. The crystal structures show trimeric Smad3-Smad4 and Smad2-Smad4 complexes: containing 2 copies of Smad3 C-terminal domain (or Smad2 C-terminal domain) in complex with the Smad4 C-terminal domain (Chacko et al., 2004). This supports the previous notion that the Smad3-Smad4 complex exists as a heterotrimer containing two Smad3 and one Smad4 molecule (Chacko et al., 2001; Moustakas and Heldin, 2002; Qin et al., 2002). The trimeric interaction is mediated through conserved interfaces where tumorigenic mutations map (Shi et al., 1997; Chacko et al., 2004). Some biochemical studies suggest that Smad2-Smad4 exists as a heterodimer (Jayaraman and Massague, 2000; Wu et al., 2001a). The basis for the discrepancy remains to be elucidated.

To add another level of complexity, it is also possible that the stoichiometry of active Smad complexes on DNA may be different from those in solution or in crystal structure. It was shown that the FAST-1-Smad2-Smad4 complex contains one copy of FAST-1, 2 copies of Smad2, and one copy of Smad4 on the ARE of the Mix.2 promoter (Inman and Hill, 2002). In contrast, Smad3-Smad4 complex that binds to the Smad binding element of the c-Jun promoter is dimeric. Future studies are necessary to further investigate the stoichiometry of Smads in various states.

Smads Nucleocytoplasm Shuttling

Smads does not reside statically in the cytoplasm at the basal state or in the nucleus after TGF-β treatment. Instead, they continuously shuttle between the cytoplasm and the nucleus (Shi and Massague, 2003; ten Dijke and Hill, 2004). At basal state, Smad2 is more localized in the cytoplasm in a steady state. Significant proportion of Smad3 is present in the nucleus even at the basal state. It is possible that the DNA binding activity of Smad3 may prevent it from being exported to the cytoplasm. Nevertheless, Smad3, and especially Smad2, are nuclear-cytoplasm shuttling proteins (Shi and Massague, 2003; ten Dijke and Hill, 2004). Smad4 constantly shuttles between the cytoplasm and the nucleus (Shi and Massague, 2003; ten Dijke and Hill, 2004). Smad4 moving into the nucleus is autonomous (Chen et al., 2005). However, Smad4 is exported into the cytoplasm at the basal state. In the presence of TGF-β signaling, Smad4 forms complexes with Smad2 and Smad3. Through interaction with Smad2 and Smad3, Smad4 is retained in the nucleus (Chen et al., 2005). The activated Smad2 and Smad3 is thought to be continuously dephosphorylated by a yet to be identified phosphatase, dissociated from Smad4 and exported back to the cytoplasm by a CRM1-independent mechanism (Shi and Massague, 2003; ten Dijke and Hill, 2004). Smad4 is exported separately into the cytoplasm by a CRM1-dependent mechanism (Shi and Massague, 2003; ten Dijke and Hill, 2004). If the receptor is still active, Smad2 and Smad3 will be rephosphorylated, form complexes with Smad4 and return to the nucleus. If the receptor is inactive after the TGF-β signal is shut off, Smad2 and Smad3 will remain in the cytoplasm and interact with SARA again. Thus, for the duration of active signaling, Smad2 and Smad3 constantly monitor the activity of the receptors (ten Dijke and Hill, 2004).

Smads DNA Binding Activities and DNA-Binding Cofactors

Smad3 and Smad4 possess intrinsic DNA binding activities through their MH1 domains (Figs.11.3 and 11.4) (Yingling et al., 1997; Dennler et al., 1998; Shi et al., 1998; Zawel et al., 1998). *In vitro*, full length recombinant Smad4 can bind to DNA, yet full length recombinant Smad3 has only weak DNA binding activity even when present at high amount (Zawel et al., 1998). The MH1 domain of recombinant Smad3 can bind to DNA much more efficiently than the intact protein. Using GST-Smad3 (MH1 domain) and GST-Smad4 together with a random oligonucleotide pool in a PCR-based SELEX approach, an 8 base pair palindromic sequence GTCTAGAC was identified as the SMAD binding element (SBE) (Zawel et al., 1998).

The crystal structure of the Smad3 MH1 domain binding to the 8 base pair palindromic SBE has been solved (Fig.11.4) (Shi et al., 1998; Chai et al., 2003). An 11 amino acid β-hairpin that is conserved among receptor-regulated Smads and Smad4 is embedded in the major groove of DNA (Fig.11.4) (Shi et al., 1998; Chai et al., 2003). Water molecules play an important role in buttressing the asymmetric placement of the DNA-binding motif in the major groove of DNA (Chai et al., 2003). In addition, the MH1 domain is found to contain a bound zinc atom using three conserved cysteines and one histidine among Smads (Fig.11.4) (Chai et al., 2003). Removal of the zinc atom results in compromised DNA binding activity (Chai et al., 2003). Two molecules of Smad3 MH1 bind to the 8 base pair SBE, with each molecule contacting a single half site, 5′-GTCT-3′ (also termed as Smad box or SBE) (Shi et al., 1998; Chai et al., 2003). In the GTCT sequence, the G at position 1, the G at position 3 on the complemetary strand, and the A at position 4 on the complementary strand form hydrogen bonds with amino acids of the β-hairpin (Shi et al., 1998; Chai et al., 2003). Thus, it is predicted from the crystal structure that the second base in the GTCT motif can tolerate substitutions (Shi et al., 1998). Indeed, during the selection of the consensus binding site for Smads, it was found that substitution of the second base only modestly reduced DNA binding by Smad3 and Smad4 (Zawel et al., 1998).

A number of TGF-β/Smad responsive promoters, such as the PAI-1, collagenase I, c-Jun, IgA, and Jun B promoters, contain one or multiple copies of the sequence GTCT or AGAC, and Smad3/Smad4 complex can bind to these elements (Yingling et al., 1996; Dennler et al., 1998; Hua et al., 1998; Jonk et al., 1998; Song et al., 1998; Zhang et al., 1998; Hanai et al., 1999; Hua et al., 1999; Stroschein et al., 1999a; Wong et al., 1999; Yeo et al., 1999; Massague and Wotton, 2000; Pardali et al., 2000a; Zhang and Derynck, 2000). Moreover, the GTCT or AGAC sequences have been shown to be critical for TGF-β inducibility of a number of responsive genes. Tandem repeats of GTCT, AGAC, or the 8 bp SBE can confer TGF-β inducibility to heterologous promoters (Dennler et al., 1998; Jonk et al., 1998; Zawel et al., 1998; Johnson et al., 1999; Massague and Wotton, 2000). Since the 11 amino acid β-hairpin is conserved also in BMP pathway Smads, a natural question is whether the GTCT and/or AGAC elements can also constitue as BMP responsive elements and bound by BMP Smads for certain natural BMP target genes. Indeed, the BMP responsive element of the *Xvent-2* promoter contains the AGAC motif and can be bound by the Smad1-Smad4 complex (Hata et al., 2000).

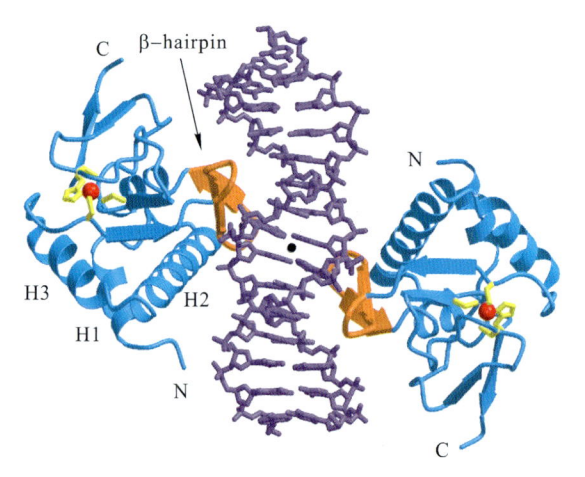

Fig.11.4 Overall structure of the Smad3 MH1 domain bound to the SBE. The palindromic DNA and the MH1 domain are colored purple and cyan, respectively. The DNA-binding motif is highlighted in orange. The bound zinc atom is shown in red, and its coordinating residues are colored yellow (adapted from Chai et al., 2003).

Smads possess relatively flexible DNA binding properties. In addition to the GTCT and AGAC elements, Smad3, Smad4, and BMP receptor regulated Smads can also recognize a GC-rich sequence (Kim et al., 1997; Labbe et al., 1998; Ishida et al., 2000; Kusanagi et al., 2000). For example, the MH1 domains of Smad3 and Smad4 have been shown to bind to a GC-rich sequence of the *goosecoid* promoter (Labbe et al., 1998), and the MH1 domain of the *Drosophila* Mad is implicated in binding to a GC-rich sequence of the DPP-responsive *vestigial* enhancer (Kim et al., 1997). A GC-rich sequence has also been shown to be essential for BMP-induced activation of the Smad6 promoter (Ishida et al., 2000). The crystal structure of the Smad3 MH1 domain revealed a highly positively charged and solvent-exposed helix 2 (H2) that contains several conserved Lys residues (Fig.11.4) (Shi et al., 1998; Shi, 2001; Chai et al., 2003). Two Lys residues that are invariant among R-Smads and Smad4 are completely solvent-exposed (Shi et al., 1998; Shi, 2001; Chai et al., 2003). The C-terminal half of the helix contains a cluster of Lys residues and can be readily modeled into the major groove of the DNA. Because Lys residues are known to prefer GC-rich sequences, the helix 2 (H2) may bind to some GC-rich sequence independent of the DNA binding β-hairpin. This likely provides, at least in part, a structural basis for Smads binding to GC-rich sequences. Nevertheless, it appears that TGF-β responsive promoters contain the GTCT or AGAC sequence more frequently, whereas the BMP responsive promoters prefer the GC-rich binding sites.

The palindromic SBE represents an optimal binding site for Smad, which is the basis of its being selected from a pool of random oligonucleotides. A study examined Smad3/Smad4 complex binding to two or three copies of abutting sequences GTCT and AGAC in different combinations (Johnson et al., 1999). Interestingly, Smad3/Smad4 has little or no capacity to bind to two or three copies of the GTCT sequence, or the AGAC sequence followed by one or two copies of the GTCT sequence. These observations further indicate that the 8 bp palindromic SBE is a high affinity-binding site for Smads. Smad7 promoter is the only natural TGF-β responsive promoter in vertebrates that has been shown to contain the 8 bp palindromic SBE (Nagarajan et al., 1999; Brodin et al., 2000; Denissova et al., 2000; von Gersdorff et al., 2000). Upon TGF-β induction, endogenous Smad complex can bind to a Smad7 promoter DNA as short as 14 or 16 base pairs containing the 8 bp palindromic SBE with only 3 or 4 base pairs adjacent sequences on each side, suggesting that Smad complex can bind to the 8 bp palindromic SBE on its own *in vivo* (Denissova et al., 2000).

Through comparing TGF-β-regulated gene expression in wild type versus Smad2 or Smad3 knockout mouse embryonic fibroblasts, it has been shown that deletion of Smad3 abolishes TGF-β induction of vast majority of TGF-β-inducible genes (Yang et al., 2003). These genes can be classified into distinct groups: immediate-early target genes, intermediate-induced genes, and intermediate-repressed genes. The SBE box repeats with short spacer lengths (0-3 bp) is present specifically in the proximal promoters of majority (~80%) of Smad3-dependent immediate-early target genes (Yang et al., 2003). It is likely that at least some of the SBE box repeats constitute high-affinity binding sites for the Smad3-Smad4 complex. Interestingly, no statistically significant co-occurance of the SBE box with other transcription factor binding sites were observed among the immediate-early target genes (Yang et al., 2003). This led to the conclusion that Smad3-Smad4 complex binds to these immediate-early genes on its own (Yang et al., 2003). Taken together, these observations support the notion that Smad3-Smad4 complex can bind directly to high affinity DNA binding sites.

For many TGF-β/activin responsive promoters that do not contain high affinity DNA binding sites for Smads, Smads are usually recruited to the promoters through interaction with DNA binding cofactors (Massague and Wotton, 2000). The first example is from studies of the activin responsive gene Mix.2 (Chen et al., 1996). The Mix.2 promoter contains a 51 base pair activin responsive element (ARE), which is upregulated by activin and also by TGF-β in the same manner (Chen et al., 1996; Liu et al., 1997). The ARE contains a binding site for FAST-1, a winged helix transcription factor, and a GTCT sequence for binding to Smad proteins (Chen et al., 1996; Chen et al., 1997; Yeo et al., 1999). FAST-1 can bind directly to Smad2, and the interaction is significantly increased in the presence of TGF-β or activin. Moreover, FAST-1, Smad2 and Smad4 form a stable complex, which binds to the ARE and activates transcription of the Mix.2 gene (Chen et al., 1996; Chen et al., 1997; Liu et al., 1997; Yeo et al., 1999). In addition to FAST-1, a large number of transcription factors have been found to interact with Smads to regulate transcription of diverse genes (Tables 11.1 and 11.2). Some of them are discussed in detail below.

Table 11.1 Smad-interacting transcription factors for activation.

Interacting Proteins	Properties and Functions	Selected References
ARC105	A component of the activator-recruited cofactor (ARC) complex; Smad coactivator	Kato et al., 2002
ATF2	ATF/CREB family member; cooperates with Smad3	Hanafusa et al., 1999; Sano et al., 1999
c-Jun, JunB, JunD c-fos	AP-1 family member; cooperate with Smads to activate c-Jun and collagenase promoters	Zhang et al., 1998; Liberati et al., 1999; Wong et al., 1999; Qing et al., 2000
FAST-1, FAST-3	Winged-helix factor; cooperate with Smad2,3	Chen et al., 1996, 1997; Liu et al., 1997
FAST-2	Winged-helix factor; cooperates with Smad2	Labbé et al., 1998; Zhou et al., 1998b
FoxO	Forkhead factor; cooperates with Smad3/4	Seoane et al., 2004
HNF4	Hepatocyte nuclear factor 4; cooperates with Smad3/4	Kardassis et al. 2000
GCN5	Smads coactivator	Kahata et al., 2004
IRF-7	Cooperates with Smad3 to activate INF-β transcription	Qing et al., 2004
Lef1/Tcf	HMG box factor; cooperates with Smad2/3/4	Labbé et al., 2000; Nishita et al., 2000
Menin	Tumor suppressor; necessary for TGF-β signaling	Kaji et al., 2001
Milk, Mixer	Paired-like homeodomain factor; cooperate with Smad2	Germain et al., 2000; Randall et al., 2002
Miz-1	Zinc finger protein; cooperates with Smads and Sp1 to activate p15 and p21 promoters	Staller et al., 2001; Seoane et al., 2001, 2002
MSG-1	Smad4 coactivator	Shioda et al., 1998; Yahata et al., 2000
p52 (NFκB)	NFκB/Rel family factor; cooperates with Smad3	Lopez-Rovira et al., 2000
p300/CBP	Smads coactivator	Feng et al., 1998; Janknecht et al., 1998
P/CAF	Smad3 and Smad2 coactivator	Itoh et al., 2000
PIAS3	Acting as a coactivator to recruit p300 to Smad3	Long et al., 2004
Runx/PEBP2/CBFA /AML	Runt-domain protein; cooperate with Smads to activate germline Ig C-α promoter	Hanai et al., 1999; Pardali et al., 2000a; Zhang et al., 2000
SKIP	Ski-interacting protein; nuclear hormone receptor coactivator; enhances TGF-β-dependent transcription	Leong et al., 2001
SMIF	EVH1/WH1 protein; Smad4 coactivator	Bai et al., 2002; Callebaut et al., 2002
Sp1, Sp3	Cooperate with Smads and Miz-1 to activate p15 and p21 promoters	Moustakas and Kardassis, 1998; Pardali et al., 2000; Feng et al., 2000; Lai et al., 2000

continued

Interacting Proteins	Properties and Functions	Selected References
Swift	BRCT domain factor; Smad2 coactivator	Shimizu *et al.*, 2001
TFE3 (μE3)	Helix-loop-helix leucine zipper factor; cooperates with Smads to activate PAI-I promoter	Hua *et al.*, 1998, 1999; Grinberg and Kerppola, 2003
VDR	Vitamin D receptor; cooperates with Smad3	Yanagi *et al.*, 1999; Yanagisawa *et al.*, 1999
ZEB1 (δEF1)	Two-handed zinc finger protein; synergizes with Smads	Postigo, 2003; Postigo *et al.*, 2003

Table 11.2 Smad-interacting transcription factors for repression.

Interacting Proteins	Properties and Functions	Selected References
Androgen receptor	AR inhibits TGF-β signaling; whether Smad3 activates or inhibits AR activity has conflict results	Hayes *et al.*, 2001; Chipuk *et al.*, 2001; Kang *et al.*, 2001, 2002
ATF3	Smad3 binds ATF3 to inhibit Id1 gene expression	Kang *et al.*, 2003
BF-1	Winged-helix brain factor-1; proto-oncogene; binds to Smad DNA binding cofactor, such as FAST-2, thereby inhibiting Smads; also inhibits Smad3 binding to DNA	Dou *et al.*, 2000; Rodriguez *et al.*, 2001
CBFA1	Smad3 inhibits osteoblast differentiation through interacting with CBFA1 and repressing its activity	Alliston *et al.*, 2001
CBF-Cb	Splice variant of CCAAT-binding factor C subunit (CBF-C); binds Smad2,3 to inhibit Smad-mediated transcriptional activation	Chen *et al.*, 2002b
C/EBP	Smad3 inhibits adipocyte differentiation by binding to C/EBP and inhibiting C/EBP transcriptional activation	Choy and Derynck, 2003
c-Jun	Inhibits Smad2,3 transcriptional activity	Dennler *et al.*, 2000; Pessah *et al.*, 2002
Dach1	Interacts with the Smad complex and the Sin3A corepressor to inhibit BMP-mediated transcription	Kida *et al.*, 2004
E1a	Adenoviral oncoprotein; competes with p300 to bind Smad3	Nishihara *et al.*, 1999
E2F4/5	Interact with Smad3 to downregulate c-myc	Chen *et al.*, 2002a; Yagi *et al.*, 2002
Estrogen receptor-α	ER inhibits Smad3 activity; TGF-β enhances ER transcriptional activity	Matsuda *et al.*, 2001
Evi-1	Zinc finger factor; inhibits Smad3 DNA binding; recruits corepressor CtBP	Kurokawa *et al.*, 1998a, 1998b; Izutsu *et al.*, 2001; Alliston *et al.*, 2005
Gli3ΔC	C-terminally truncated Gli3 zinc finger factor; repressor form	Liu *et al.*, 1998
Glucocorticoid receptor	Inhibits Smad3 transcriptional activation	Song *et al.*, 1999
Max	Both Max and TFE3 cooperate with Smads to bind to the PAI-1 promoter; Max inhibits TGF-β/Smad3 activation of the PAI-1 promoter	Grinberg and Kerppola, 2003
MEF2	Myocyte enhancer factor 2; Smad3 binds and inhibits MEF2 transactivation	Liu *et al.*, 2004
Nkx3.2	Homeo domain factor; Nkx3.2, Smad1/4 and HDAC/Sin3A form a complex to inhibit transcription	Kim and Lassar, 2003
Myc	Interacts with Smad2,3 to inhibit p15 induction	Feng *et al.*, 2002
MyoD	Smad3 inhibits myogenesis by interacting with MyoD and MEF2 and inhibit their activity	Liu *et al.*, 2001
PIASy	Acting as a corepressor by recruiting HDAC to Smads	Imoto *et al.*, 2003; Long *et al.*, 2003
SIP1	Two-handed zinc finger repressor; represses E-cadherin and other genes; represseses *Xenopusbrachyury*	Verschueren *et al.*, 1999; Comijn *et al.*, 2001

Interacting Proteins	Properties and Functions	Selected References
Ski	Proto-oncoprotein; Smads corepressor; recruits HDAC; binds Smad4 L3 loop to inhibit Smad4 association with Smad2/3; stabilizes inactive Smads on SBE	Akiyoshi et al., 1999; Luo et al., 1999; Nomura et al., 1999; Sun et al., 1999a; Wu et al., 2002; Suzuki et al., 2004
SNIP1	Forkhead-associated nuclear protein; competes with p300 to bind Smad4; upregulates cyclin D expression	Kim et al., 2000
SnoN	Proto-oncoprotein and potential tumor suppressor; Smads corepressor; recruits HDAC	Stroschein et al., 1999; Sun et al., 1999
SREBP-2	Binds and inhibits Smad3 transcriptional activity	Grimsby et al., 2004
TGIF	Homeo-domain factor; Smads corepressor; recruits HDAC, mSin3A and CtBP	Wotton et al., 1999a; 1999b; Melhuish and Wotton, 2000; Wotton et al., 2001
YB-1	Y box-binding protein; represses TGF-β-stimulated α2(I) procollagen gene transcription by interfering with Smad3 binding to DNA and disrupting the Smad3-p300 interaction	Higashi et al., 2003
YY1	Interacts with Smads; inhibits TGF-β- and BMP-induced cell differentiation by repressing Smad activity in a gene-specific manner	Kurisaki et al., 2003

Despite being highly homologous with Smad3, Smad2 is unable to bind to DNA due to the insertion of the exon 3-encoded extra 30 amino acids present immediately before the DNA binding hairpin (Dennler et al., 1998; Shi et al., 1998; Yagi et al., 1999; Shi, 2001). If exon 3 is removed, as found in an alternatively spliced Smad2, it can then bind to the SBE sequence (Yagi et al., 1999). The transcript of this alternatively spliced variant Smad2 is present in certain cells and tissues at a level of about 1/10 of full length Smad2 (Yagi et al., 1999). Recent studies have shown that the full length Smad2 and this short form of Smad2 are coexpressed throughout mouse development (Dunn et al., 2005). Further studies in mouse indicate that the short isoform of Smad2 or Smad3, but not full-length Smad2, activates all essential target genes downstream of TGF-β-related ligands (Dunn et al., 2005). Similarly, as described above, genome-wide microarray screening of TGF-β responsive genes by comparing wild type with Smad2 or Smad3 abalated fibroblasts have indicated that Smad3 is an essential mediator of TGF-β signal transduction and transcriptional regulation (Yang et al., 2003). Ongoing studies by RNAi approach in a few laboratories reach the same conclusion for certain other cell types analyzed. Thus, Smad3 plays a critical physiological role in mediating TGF-β responses.

The DNA binding activity of Smad3 is regulated by protein kinase C. PKC can phosphorylate Smad3 at Ser 37 and Ser 70 both in vivo and in vitro (Fig.11.3) (Yakymovych et al., 2001). Phosphorylation of Smad3 by PKC abrogates its DNA-binding activity and thus inhibits the transcriptional responses (Yakymovych et al., 2001). PKC can also phosphorylate Smad2 at the analogous Ser 47 and Ser 110 (Yakymovych et al., 2001). The physiological significance of PKC phosphorylation of Smad2 remains to be elucidated.

Smads Transcriptional Activation

A: Interaction with p300/CBP, P/CAF and GCN5 Histon Acetyltransferases

A1: Overview

GAL4 fusion studies provided the first evidence that Smads are TGF-β family-regulated transcription factors (Liu et al., 1996). The C terminal domain together with a segment of the linker region from a receptor regulated Smad or Smad4 can activate transcription when fused to the GAL4 DNA binding domain (Liu et al., 1996). Full length Smads have very little activity in the GAL4 fusion assay, but their transcription activities are greatly increased by treatment with the corresponding agonists, BMP or TGF-β (Liu et al., 1996; Liu et al., 1997). Subsequent studies have shown that transcriptional activation by Smad3 and Smad2 occurs in part by their ability to recruit general transcriptional coactivator p300/CBP (Feng et al., 1998; Janknecht et al., 1998; Nishihara et al., 1998; Pouponnot et al., 1998; Shen et al., 1998; Topper et al., 1998). p300/CBP have intrinsic histone acetyltransferase activity (HAT) that facilitates transcription by altering nucleosome structure through histone acetylation and thereby remodeling the chromatin template (Roth et al., 2001). This interaction occurs through the MH2 and a segment of the linker region of Smad3 or Smad2 and the C-terminal domain of p300/CBP. In addition, P/CAF, another HAT-containing transcriptional co-activator, has been shown to associate with Smad3 and Smad2 upon TGF-β receptor activation and to enhance Smad3 and Smad2 transcriptional activity (Itoh et al., 2000). GCN5, another co-activator containing the HAT, enhances transcriptional activation of a variety of genes. GCN5 is structurally related to P/CAF, and therefore is also designated P/CAF-B. Unlike P/CAF, which binds to only TGF-β activated Smad3 and Smad2, GCN5 binds to TGF-β activated Smad3 and Smad2 as well as BMP activated Smad1 and Smad5, and enhancing both TGF-β- and BMP-mediated transcriptional responses (Kahata et al., 2004).

The transcriptional activity of receptor-regulated Smads is greatly increased in the presence of Smad4. This was shown by studies using Smad4 deficient colon cancer cells (Liu et al., 1997; Zhou et al., 1998a). In such cells, Smad2 and FAST-1 together have minimal ability to stimulate a typical activin/TGF-β reporter gene. In addition, GAL4-Smad1, GAL4-Smad2 have little transcriptional activities in the Smad4 deficient cells compared to the same cells with transfected Smad4 (Liu et al., 1997). Thus, Smad4 plays an important role in Smad-mediated transcriptional activation for many TGF-β-responsive genes. This is partly due to the presence of a unique Smad activation domain (SAD), a 48 amino acid proline-rich regulatory element in the linker region of Smad4 (de Caestecker et al., 2000). The crystal structure of a Smad4 fragment containing the SAD and the MH2 domain has been solved (Qin et al., 1999). The MH2 domain of Smad4 is highly homologous with that of Smad2 and Smad3 (50% identity), except that Smad4 has a unique insert of ~35 amino acids which interacts with the C-terminal tail to form a TOWER-like structural extension from the core. The crystal structure suggests that SAD provides transcriptional capability by reinforcing the structural core and coordinating with the TOWER to present the proline rich surface and a glutamine-rich surface in the TOWER for interaction with transcription partners (Qin et al., 1999). It has been shown that the SAD domain physically interacts with the N-terminal domain of p300/CBP (de Caestecker et al., 2000).

The Smad3 linker region also contains an activation domain (Wang et al., 2005). When the linker region is fused to the GAL4 DNA binding domain, it has constitutive transcriptional activity, comparable to that of SAD in Smad4. In the context of full length Smad3, deletion of the linker region renders Smad3 unable to support TGF-β transcriptional activation, although it can still be phosphorylated by the TGF-β receptor at the C-tail and has a markedly increased capacity to form a heteromeric complex with Smad4 (Wang et al., 2005). Further experiments using the GAL4 system indicate that the linker region and the C-terminal domain of Smad3 synergize for transcriptional activation in the presence of TGF-β (Wang et al., 2005).

The transcriptional activity of the Smad3 linker region can be blocked by wild type E1a, but not by an E1a mutant that cannot bind to p300. Overexpression of p300 can partially but not completely rescue E1a-meidated repression. Immunoprecipitation analyses indicate that the linker region is capable of recruiting the p300 coactivator (Wang et al., 2005). Thus, these observations suggest that in addition to p300, other regulatory components may participate with the Smad3 linker region for transcriptional activation.

The Smad3 and Smad2 linker region contains demonstrated as well as suspected phosphorylation sites for multiple kinases, such as the cyclin-dependent kinase (CDK), ERK mitogen activated protein (MAP) kinase, c-Jun N-terminal kinase, p38 MAP kinase, and Ca^{2+}-calmodulin-dependent kinase II (CamKII) (Fig.11.3) (Kretzschmar et al. 1999; Wicks et al. 2000; Derynck and Zhang 2003; Matsuura et al. 2004; Mori et al. 2004). For example, it has been shown that CDK phosphoryaltion of Smad3 inhibits its transcriptional activity and antiproliferative function (Matsuura et al., 2004). It will be very interesting to elucidate the exact mechanism by which CDK phosphorylation of Smad3 inhibits its transcriptional activity. Phosphorylation of the linker region by the various kinases may differentially influence Smad3 transcriptional activity in a context-dependent manner. The C-terminal domain of Smad3 is regulated by TGF-β receptor phosphorylation. The C-terminal domain of Smad3 is also a protein-protein interaction domain, responsible for homo-trimerization, hetero-trimerization with Smad4, and also interaction with a number of DNA binding proteins (Fig.11.3). Under different conditions, the linker region and the C-terminal domain may functionally interact differentially and therefore display varying transcriptional activities, leading to distinct biological responses in a cell-context dependent manner.

A2: Recruitment of p300 by Smad4 Coactivator MSG1

MSG1 is a potent transcriptional activator (Shioda et al., 1997). It was originally identified as a candidate pigmentation-related gene in melanocytes (Shioda et al., 1996). Its possible involvement in differentiation and development was suggested based on its restricted and developmentally regulated expression (Shioda et al., 1996). MSG1 lacks an intrinsic DNA binding activity

(Shioda et al., 1997). In a yeast two-hybrid screen using MSG1 as bait, it was found that MSG1 interacted with Smad4. Subsequent studies indicate that MSG1 associates with p300, recruits p300 to Smad4, and enhance Smad-mediated transcriptional activiation in the presence of TGF-β. Thus, MSG1 functions as a co-activator of Smad4 (Shioda et al., 1998; Yahata et al., 2000).

A3: Requirement of p300/CBP for the Effect of Smad4 Coactivator SMIF

SMIF is another coactivor of Smad4. SMIF is a ubiquitously expressed protein that contains an EVH1/WH1 (enabled VASP homology 1 /WASP homology 1) domain (Bai et al., 2002). The EVH1/WH1 domain is a protein interaction module for binding to proline-rich regions (Callebaut, 2002). SMIF interacts with Smad4 but not with other Smads. The interaction occurs through the EVH1/WH1 domain of SMIF and the SAD domain of Smad4. TGF-β induces the formation of a SMIF-Smad4 complex, which translocates to the nucleus. Whereas SMIF does not directly bind to p300/CBP, SMIF possesses strong TGF-β-inducible Smad4 and p300-dependent transcriptional activity (Bai et al., 2002).

A4: Recruitment of p300 by PIAS3

The family of protein inhibitor of activated STAT (PIAS) represents a group of proteins that play an important role in regulating a variety of signaling pathways and transcriptional responses (Schmidt and Muller, 2003). The acronym PIAS stems from the initial observation that members of the PIAS family inhibit the DNA binding activity of activated STATs (see Chapter 10). Subsequent studies indicate that PIAS proteins interact with a variety of transcription factors and regulate diverse cellular processes. The mammalian PIAS family includes five members: PIAS1, PIAS3, PIASxα, PIASxβ, and PIASy. The PIAS proteins contain a SAP domain in the N-terminal domain. The term SAP refers to three of the founding members of SAP-containing proteins: scaffold attachment factor (SAF), acinus, and PIAS. A common feature of SAP-containing proteins is their ability to bind to chromatin. PIAS proteins contain a RING domain. All PIAS proteins possess SUMO E3 ligase activity, and the RING domain is essential for this activity (Schmidt and Muller, 2003).

PIAS3 can activate transcription of a Smad-dependent reporter gene, and TGF-β treatment markedly increases the transcriptional response (Long et al., 2004b). PIAS3 interacts with Smad proteins. The strongest interaction is with Smad3 and it occurred through the C-terminal domain of Smad3. The interaction can be detected at endogenous protein levels. PIAS3 can also interact with p300/CBP. Interestingly, the RING domain of PIAS3, which is essential for the SUMO E3 ligase activity of PIAS3, is also necessary for interaction with p300/CBP and for activation of Smad-dependent transcription. Importantly, PIAS3, Smad3 and p300 can form a ternary complex. This complex formation is significantly increased in the presence of TGF-β, which help explain that PIAS3 activation of Smad-dependent transcription is significantly increased in the presence of TGF-β (Long et al., 2004b), p300 does not interact with all PIAS proteins. For example, PIASy, which inhibits Smad3 transcriptional activity and other transcriptional responses, cannot bind to p300/CBP (Long et al., 2004b). Since PIAS3 has SUMO E3 ligase activity, whether sumoylation of certain target is necessary for the stimulatory effect of PIAS3 remains to be elucidated. It is possible that PIAS3 regulates transcription through both sumoylation-dependent and sumoylation-independent mechanisms.

B: Interaction with ARC105

Activator-recruited co-factor (ARC) and the related or identical metazoan Mediator play an important role in transcriptional regulation (Taatjes et al., 2004). These complexes contain multiple polypeptides, and can associate with many transcription factors and activate transcription *in vitro*. ARC105, a component of the ARC/mediator complex, is essential for signaling by TGF-β, activin and nodal, a member of the TGF-β superfamily (Kato et al., 2002). ARC105 interacts with the C-terminal domains of Smad2, Smad3, and Smad4, and the interaction is induced or increased in the presence of ligand. Moreover, ARC105 is recruited to the responsive promoters as shown by chromatin immunoprecipitation assay. Overexpression of ARC105 enhances TGF-β/activin/nodal-inducible transcription. Conversely, knockdown ARC105 by siRNA inhibited the transcription. Thus, ARC105 links the ARC/Mediator complex with Smad-mediated transcriptional control (Kato et al., 2002).

Smads Transcriptional Repression

A: Multiple Repression Mechanisms

Smads can inhibit transcription through recruitment of corepressors, which is described in detail below. In addition, Smads, in particular Smad3, can repress transcription through other modes. For instances, TGF-β inhibits osteoblast, skeletal muscle, and adipocyte differentiation. Smad3 mediates the TGF-β inhibitory effects. Smad3 inhibits osteoblast differentiation by physically interacts with CBFA1, which is a key transcription factor for osteoblast differentiation, and prevents CBFA1 from activating osteocalcin and the CBFA1 promoter itself (Alliston et al., 2001). This inhibition is both cell type- and promoter-dependent. It occurs in mesenchymal cells but not in epithelial cells (Alliston et al., 2001). For myogenesis, Smad3 inhibits myogenic differentiation through interaction with MyoD and MEF2 (Liu et al., 2001a; Liu et al., 2004). Smad3 physically interacts with the HLH domain of MyoD, thus inhibiting MyoD heterodimerization with an E-box binding protein (such as E12 and E47) and subsequent binding of the heterodimer to the E-box (Liu et al., 2001a). Smad3 interacts with MEF2 and inhibits its transcriptional activity (Liu et al., 2004). Similarly, Smad3 inhibits adipocyte differentiation through interaction with C/EBP and inhibits its transactivation function (Choy and Derynck, 2003). Thus, Smad3 is a key regulator for TGF-β-mediated differentiation control.

Smad3 also plays an important role to inhibit the expression of genes that are necessary for the TGF-β cytostatic effects. c-Myc downregulation is essential for the TGF-β-mediated growth inhibitory responses (Pietenpol et al., 1990; Chen et al., 2001; Shi and Massague, 2003). TGF-β treatment induces a preassembled complex containing Smad3, E2F4/5 and DP1, and the Rb-related factor p107 to move into the nucleus, associate with Smad4, bind to a compository Smad-E2F site for repression (Chen et al., 2002a; Yagi et al., 2002; Frederick et al., 2004). Inhibition of Id1 expression is also a general feature of the TGF-β-induced growth inhibition. TGF-β-activated Smad3 directly induces the expression of ATF3, a transcriptional repressor. ATF3, Smad3 and Smad4 then form a complex that directly mediates Id1 repression (Kang et al., 2003).

Smad3 can also compete for DNA binding. Smad3 can inhibit the expression of the *goosecoid* gene (Labbe et al., 1998). This is thought to occur through Smad3 competing with Smad4 binding to a GC-rich sequence. While binding of Smad4 in complex with Smad2 and FAST-2 to this GC-rich sequence activates transcription, binding of Smad3 to this sequence may alter the conformation of the DNA binding complex, thus leading to inhibition of transcription (Labbe et al., 1998).

Smad3 has been reported to interact with HDAC through its MH1 domain (Liberati et al., 2001). Although it is not clear whether the interaction is direct, one possibility is that direct interaction of Smad3 with HDAC may contribute to some of the repression examples described above. Future studies are necessary to explore this interesting possibility.

B: Interaction with Corepressors

Three Smad corepressors have been identified: TGIF (Massague and Wotton 2000), Ski and the related SnoN (Massague and Wotton, 2000; Liu et al., 2001b; Frederick and Wang, 2002; Luo, 2004). In addition, PIASy can also act as a corepressor for Smads (Long et al., 2003).

B1:TGIF

TGIF is a member of the evolutionarily conserved three-amino-acid loop extension (TALE) family of atypical homeodomain proteins. TGIF exhibits several modes of repression. TGIF can repress transcription by competing with retinoid receptors for the DNA binding sites, the retinoid X receptor (RXR) responsive element, in the promoter regions of the regulated genes (Massague and Wotton, 2000). TGIF can also repress transcription through recruiting HDAC (Wotton et al., 1999a). TGIF can interact directly with the paired amphipathic α-helix 2 domain of the mSin3 corepressor (Wotton et al., 1999b; Wotton et al., 2001). TGIF can also recruit CtBP (Melhuish and Wotton, 2000).

TGIF is highly conserved in mammals. In contrast, *Drosophila* TGIF proteins, *achintya* and *vismay*, share homology with human TGIF only in the TALE homeodomain. The *Drosophila* TGIF proteins are transcriptional activators (Hyman et al., 2003). Obviously, regions outside of the TALE domain specify activation or repression of transcription.

TGIF interacts with the C-terminal domain of Smad2 and Smad3, and the interactions are increased in the presence of TGF-β (Wotton et al., 1999a). TGIF represses Smads-mediated transcriptional activation in part by interacting with histone deacetylases (HDACs) (Wotton et al., 1999a). In addition, TGIF recruits mSin3 to a TGF-β activated Smad complex to inhibit transcriptional responses (Wotton et al., 1999b; Wotton et al., 2001). Efficient repression of TGF-β transcriptional responses by TGIF is also dependent on the interaction with CtBP (Melhuish and Wotton, 2000). Thus, TGIF uses several modes to repress Smads-mediated transcription.

TGIF is a short-lived protein. Small changes in the physiological levels of TGIF can result in profound effects on human development, as shown by the devastating brain and craniofacial developmental defects in heterozygotes carrying a hypomorphic TGIF mutant allele (Gripp et al., 2000). EGF signaling can lead to the phosphorylation of TGIF at two Erk MAP kinase sites. This stabilizes TGIF and increases Smad2–TGIF complex in response to TGF-β (Lo et al., 2001).

Recent studies have shown that TGIF knockout mice are viable and fertile (Shen and Walsh, 2005). In addition, there were no discernible derangements in all the major organs (Shen and Walsh, 2005). TGIF2 shows an expression pattern very similar to that of TGIF. TGIF2 and TGIF share 77% identity in the homeodomain and 49% similarity outside of the homeodomain. It is possible that TGIF and TGIF2 have redundant or at least overlapping roles, and that TGIF2 can compansate for the loss of TGIF. Germline knockout or conditional knockout both TGIF and TGIF2 will help to identify genes that are repressed by TGIF under natural setting.

B2:Ski/SnoN

Ski was first identified as a viral oncogene (v-Ski) from the Sloan-Kettering avian retrovirus that transforms chicken embryonic fibroblasts (Liu et al., 2001b; Luo, 2004). SnoN is a member of the Ski proto-oncogene family. In addition to being an oncogene, SnoN appears to act as a tumor suppressor, at least in certain cells (Liu et al., 2001b; Luo, 2004). Ski and SnoN directly binds to the N-CoR and mSin3A that form a complex with HDAC (Luo et al., 1999; Nomura et al., 1999). In addition, Ski has been shown to be able to bind directly to the corepressors HIPK2 and MeCP2 (Kokura et al., 2001; Harada et al., 2003). Ski is also required for transcriptional repression by several other proteins, including the Mad, the thyroid hormone receptor-β, Rb protein and the Gli3 repressor (Liu et al., 2001b; Luo, 2004). Thus, Ski appears to be an integral part of the transcriptional repression machinery.

Ski and SnoN interact with Smad2 and Smad3 in a TGF-β dependent manner, whereas Ski and SnoN associate with Smad4 constitutively (Akiyoshi et al., 1999; Luo et al., 1999; Stroschein et al., 1999b; Sun et al., 1999a; Sun et al., 1999b; Xu et al., 2000). The MH2 domains of Smad2, 3, 4 are essential for these interactions (Akiyoshi et al., 1999; Luo et al., 1999; Stroschein et al., 1999b; Sun et al., 1999a; Sun et al., 1999b; Xu et al., 2000), and Ski has been shown to recognize trimeric Smad3 (Moustakas and Heldin, 2002; Qin et al., 2002). Ski and SnoN can inhibit TGF-β transcriptional responses through multiple mechanims. Ski and SnoN can recruit HDAC to TGF-β activated Smad complexes through direct interaction with N-CoR and mSin3A (Luo et al., 1999; Liu et al., 2001b; Luo, 2004). Ski also competes with the coactivator p300/CBP for binding to the activated Smad3 (Akiyoshi et al., 1999). In addition, Ski can inhibit Smad2 and Smad3 binding to the L3 loop of the Smad4 MH2 domain (Frederick and Wang, 2002; Wu et al., 2002). Ski also stabilizes inactive Smad complexes on SBE (Suzuki et al., 2004). Finally, Ski has been reported to inhibit TGF-β signaling through inhibition of Smad2 phosphorylation by the TGF-β receptor (Prunier et al., 2003). The transforming activity of Ski and SnoN is dependent on their interaction with Smads (He et al., 2003). In addition to inhibition of the TGF-β pathway, Ski but not SnoN, also associates with BMP specific Smad complex in a ligand-dependent manner and blocks BMP transcriptional responses (Luo, 2004).

SnoN, and to a lesser extent, Ski, are degraded upon TGF-β treatment (Stroschein et al., 1999b; Sun et al., 1999b; Liu et al., 2001b; Luo, 2004). Therefore, SnoN and Ski have been proposed as nuclear corepressors for Smad4 to maintain TGF-β responsive genes in a repressed state in the absence of ligand. TGF-β also induces the expression of SnoN, which likely to function in a negative feedback control to turn off TGF-β signaling at late stages (Stroschein et al., 1999b; Liu et al., 2001b; Luo, 2004).

Studies on the natural Smad7 promoter have provided evidence that Ski is indeed a corepressor for Smad4 at basal state. As described in the DNA binding section, the Smad7 promoter contains the perfect 8 bp palindromic SBE. Smad4 binds to this SBE. Ski is recruited to the Smad7 promoter through interaction with Smad4 and represses Smad7 expression at basal state (Denissova and Liu, 2004).

B3:PIASy

PIASy also inhibits TGF-β-inducible transcriptional responses (Imoto et al., 2003; Long et al., 2003). It interacts most strongly with Smad3 and also associates with other receptor-regulated Smads and Smad4. Smad3, Smad4 and PIASy can form a complex. PIASy can associate with HDAC1, and the inhibitory effect of PIASy can be disrupted by treatment with TSA, a HDAC inhibitor (Long et al., 2003). Thus, PIASy can inhibit TGF-β-mediated transcription by recruiting HDAC to Smads. Similarly, PIASxβ can also interact with HDAC3 (Tussie-Luna et al., 2002). PIASxα and PIASxβ have also been implicated to recruit HDAC to inhibit IL12-mediated and STAT4-dependent gene activation (Arora et al., 2003). Thus, some PIAS proteins can act as corepressors through association with HDAC.

PIAS family members have SUMO E3 ligase activities (Schmidt and Muller, 2003). HDAC1 itself may be a target for sumoylation by PIAS proteins. A study has demonstrated that HDAC1 is sumoylated both *in vivo* and *in vitro* (David *et al.*, 2002). Moreover, mutation of the sumoylation sites in HDAC1 reduces its transcriptional repression activity in reporter gene assays (David *et al.*, 2002). Further studies are necessary to fully explore this interesting possibility.

Related to the issue of sumoylation targets of PIASy, it is worth pointing out that Smad4 is modified by SUMO and sumoylation represses its transcriptional activity (Long *et al.*, 2004a). This occurs through Daxx (Chang *et al.*, 2005), which regulates apoptosis and represses transcription through its interaction with various cytoplasmic and nuclear proteins. Daxx interacts with Smad4 and represses its transcriptional activity. Binding of Daxx to Smad4 is dependent on Smad4 sumoylation. Mutation of the major Smad4 sumoylation site not only disrupted Smad4-Daxx interaction but also relieved Daxx-mediated repression of Smad4 transcriptional activity. Thus, Daxx represses Smad4-mediated transcriptional activity by direct interaction with the sumoylated Smad4 (Chang *et al.*, 2005). Interestingly, sumoylation of Smad4 also increases its stability and nuclear accumulation (Lee *et al.*, 2003; Lin *et al.*, 2003a; Lin *et al.*, 2003b). Thus, the net effect of sumoylation of Smad4 can be either stimulatory or inhibitory, depending on the target promoter that is analyzed.

B4: Why Multiple Corepressors?

With the presence of TGIF, Ski/SnoN and PIASy, a natural question is whether each one can inhibit all TGF-β-responisve genes. Several studies indicate that TGIF, Ski/SnoN and PIASy have distinct promoter specificities. For example, TGF-β induction of p15 is greatly inhibited by overexpression of PIASy (Long *et al.*, 2003), partially inhibited by overexpression of TGIF (Lo *et al.*, 2001), but not inhibited by overexpression of Ski (Sun *et al.*, 1999a). Conversely, the early responsiveness of JunB to TGF-β is only moderately inhibited by overexpression of PIASy (Long *et al.*, 2003) but is significantly inhibited by overexpression of Ski (Luo *et al.*, 1999). One possibility is that TGIF, Ski/SnoN and PIASy are present in different corepressor complexes that inhibit distinct sets of promoters. Both TGIF and PIASy may inhibit TGF-β activated Smad complex, whereas Ski/SnoN may function as corepressors for Smad4 to maintain TGF-β responsive genes in a repressed state in the absence of TGF-β and contribute to TGF-β signal termination. The presence of multiple regulators that share overlapping functions in the same cells confers ample flexibility under complex circumstances. In addition, TGIF, Ski/SnoN and PIASy are differentially expressed in various tissues, which may confer cell-type specific functions.

Perspectives

TGF-β regulates a wide variety of biological activities through transcriptional regulation of distinct target genes. Extensive research on Smads in the past nine years has provided many fundamental insights into the basic mechanisms of Smads-mediated transcriptional control as well as the correlation with developmetal status or disease conditions. Still, many interesting questions remain to be answered. For instance, Smads have been shown to recruit two classes of coactivators: histone acetyltransferases and the ARC complex. The role of SWI/SNF complex, a distinct class of coactivator that contains ATP-dependent DNA unwinding activities, remains to be addressed for Smads-mediated transcriptional activation. Post-translational modifications, such as phosphorylation and sumoylation, regulate Smad DNA binding and transcriptional activities. Do other types of modification, such as methylation, affect Smad activity? Genome-wide gene expression profiling coupled with conditional knockout, knockin and RNAi approaches has emerged, and will continue to be extremely useful in delineating cell specific TGF-β responsive genes with disease states. The findings may lead to the development of effective drugs for therapeutic applications.

Acknowledgement

I apologize to those colleagues whose work were not cited due to space constrain. I thank Y. Shi, Y. E. Zhang, and many colleagues for stimulating discussions. I also thank the American Association for Cancer Research-National Foundation for Cancer Research, the Burroughs Wellcome Fund, the Sidney Kimmel Foundation for Cancer Research, the Pharmaceutical Research and Manufacturer of America Foundation, the Emerald Foundation, the New Jersey Commission on Cancer Research, the Department of Defense Breast Cancer Research Program, and the National Institutes of Health for support of our research.

References

Abdollah, S., Macias-Silva, M., Tsukazaki, T., Hayashi, H., Attisano, L., and Wrana, J. L. (1997). TbetaRI phosphorylation of Smad2 on Ser465 and Ser467 is required for Smad2-Smad4 complex formation and signaling. J. Biol. Chem. *272*, 27678-27685.

Akiyoshi, S., Inoue, H., Hanai, J., Kusanagi, K., Nemoto, N., Miyazono, K., and Kawabata, M. (1999). c-Ski acts as a transcriptional co-repressor in transforming growth factor-beta signaling through interaction with smads. J. Biol. Chem. *274*, 35269-35277.

Alliston, T., Choy, L., Ducy, P., Karsenty, G., and Derynck, R. (2001). TGF-beta-induced repression of CBFA1 by Smad3 decreases cbfa1 and osteocalcin expression and inhibits osteoblast differentiation. EMBO J. *20*, 2254-2272.

Alliston, T., Ko, T. C., Cao, Y., Liang, Y. Y., Feng, X. H., Chang, C., and Derynck, R. (2005). Repression of BMP and activin-inducible transcription by Evi-1. J. Biol. Chem. E-pub ahead of print.

Arora, T., Liu, B., He, H., Kim, J., Murphy, T. L., Murphy, K. M., Modlin, R. L., and Shuai, K. (2003). PIASx is a transcriptional co-repressor of signal transducer and activator of transcription 4. J. Biol. Chem. *278*, 21327-21330.

Attisano, L., and Wrana, J. L. (2000). Smads as transcriptional co-modulators. Curr. Opin. Cell Biol. *12*, 235-243.

Bai, R. Y., Koester, C., Ouyang, T., Hahn, S. A., Hammerschmidt, M., Peschel, C., and Duyster, J. (2002). SMIF, a Smad4-interacting protein that functions as a co-activator in TGFbeta signalling. Nat. Cell Biol. *4*, 181-190.

Brodin, G., Ahgren, A., ten Dijke, P., Heldin, C. H., and Heuchel, R. (2000). Efficient TGF-beta induction of the Smad7 gene requires cooperation between AP-1, Sp1, and Smad proteins on the mouse Smad7 promoter. J. Biol. Chem. *275*, 29023-29030.

Callebaut, I. (2002). An EVH1/WH1 domain as a key actor in TGFbeta signalling. FEBS Lett. *519*, 178-180.

Chacko, B. M., Qin, B., Correia, J. J., Lam, S. S., de Caestecker, M. P., and Lin, K. (2001). The L3 loop and C-terminal phosphorylation jointly define Smad protein trimerization. Nat. Struct. Biol. *8*, 248-253.

Chacko, B. M., Qin, B. Y., Tiwari, A., Shi, G., Lam, S., Hayward, L. J., De Caestecker, M., and Lin, K. (2004). Structural basis of heteromeric smad protein assembly in TGF-beta signaling. Mol. Cell *15*, 813-823.

Chai, J., Wu, J. W., Yan, N., Massague, J., Pavletich, N. P., and Shi, Y. (2003). Features of a Smad3 MH1-DNA complex. Roles of water and zinc in DNA binding. J. Biol. Chem. *278*, 20327-20331.

Chang, C. C., Lin, D. Y., Fang, H. I., Chen, R. H., and Shih, H. M. (2005). Daxx mediates the small ubiquitin-like modifier-dependent transcriptional repression of Smad4. J. Biol. Chem. *280*, 10164-10173.

Chen, X., Rubock, M. J., and Whitman, M. (1996). A transcriptional partner for MAD proteins in TGF-beta signalling. Nature *383*, 691-696.

Chen, X., Weisberg, E., Fridmacher, V., Watanabe, M., Naco, G., and Whitman, M. (1997). Smad4 and FAST-1 in the assembly of activin-responsive factor. Nature *389*, 85-89.

Chen, Y. G., Hata, A., Lo, R. S., Wotton, D., Shi, Y., Pavletich, N., and Massague, J. (1998). Determinants of specificity in TGF-beta signal transduction. Genes Dev. *12*, 2144-2152.

Chen, C. R., Kang, Y., and Massague, J. (2001). Defective repression of c-myc in breast cancer cells: A loss at the core of the transforming growth factor beta growth arrest program. Proc. Natl. Acad. Sci. USA *98*, 992-999.

Chen, C. R., Kang, Y., Siegel, P. M., and Massague, J. (2002a). E2F4/5 and p107 as Smad cofactors linking the TGFbeta receptor to c-myc repression. Cell *110*, 19-32.

Chen, F., Ogawa, K., Liu, X., Stringfield, T. M., and Chen, Y. (2002b). Repression of Smad2 and Smad3 transactivating activity by association with a novel splice variant of CCAAT-binding factor C subunit. Biochem. J. *364*, 571-577.

Chen, H. B., Rud, J. G., Lin, K., and Xu, L. (2005). Nuclear targeting of TGF-beta -activated Smad complexes. J. Biol. Chem. E-pub ahead of print

Chipuk, J. E., Cornelius, S. C., Pultz, N. J., Jorgensen, J. S., Bonham, M. J., Kim, S. J., and Danielpour, D. (2002). The androgen receptor represses transforming growth factor-beta signaling through interaction with Smad3. J. Biol. Chem. *277*, 1240-1248.

Choy, L., and Derynck, R. (2003). Transforming growth factor-beta inhibits adipocyte differentiation by Smad3 interacting with CCAAT/enhancer-binding protein (C/EBP) and repressing C/EBP transactivation function. J. Biol. Chem. *278*, 9609-9619.

Comijn, J., Berx, G., Vermassen, P., Verschueren, K., van Grunsven, L., Bruyneel, E., Mareel, M., Huylebroeck, D., and van Roy, F. (2001). The two-handed E box binding zinc finger protein SIP1 downregulates E-cadherin and induces invasion. Mol. Cell *7*, 1267-1278.

David, G., Neptune, M. A., and DePinho, R. A. (2002). SUMO-1 modification of histone deacetylase 1 (HDAC1) modulates its biological activities. J. Biol. Chem. *277*, 23658-23663.

de Caestecker, M. P., Yahata, T., Wang, D., Parks, W. T., Huang, S., Hill, C. S., Shioda, T., Roberts, A. B., and Lechleider, R. J. (2000). The Smad4 activation domain (SAD) is a proline-rich, p300-dependent transcriptional activation domain. J. Biol. Chem. *275*, 2115-2122.

Denissova, N. G., Pouponnot, C., Long, J., He, D., and Liu, F. (2000). Transforming growth factor beta-inducible independent binding of SMAD to the Smad7 promoter. Proc. Natl. Acad. Sci. USA *97*, 6397-6402.

Denissova, N. G., and Liu, F. (2004). Repression of endogenous Smad7 by Ski. J. Biol. Chem. *279*, 28143-28148.

Dennler, S., Itoh, S., Vivien, D., ten Dijke, P., Huet, S., and Gauthier, J. M. (1998). Direct binding of Smad3 and Smad4 to critical TGF beta-inducible elements in the promoter of human plasminogen activator inhibitor-type 1 gene. EMBO J. *17*, 3091-3100.

Dennler, S., Prunier, C., Ferrand, N., Gauthier, J. M., and Atfi, A. (2000). c-Jun inhibits transforming growth factor beta-mediated transcription by repressing Smad3 transcriptional activity. J. Biol.

Chem. *275*, 28858-28865.

Derynck, R., Akhurst, R. J., and Balmain, A. (2001). TGF-beta signaling in tumor suppression and cancer progression. Nat. Genet. *29*, 117-129.

Derynck, R., and Zhang, Y. E. (2003). Smad-dependent and Smad-independent pathways in TGF-beta family signalling. Nature *425*, 577-584.

Dou, C., Lee, J., Liu, B., Liu, F., Massague, J., Xuan, S., and Lai, E. (2000). BF-1 interferes with transforming growth factor beta signaling by associating with Smad partners. Mol. Cell. Biol. *20*, 6201-6211.

Dunn, N. R., Koonce, C. H., Anderson, D. C., Islam, A., Bikoff, E. K., and Robertson, E. J. (2005). Mice exclusively expressing the short isoform of Smad2 develop normally and are viable and fertile. Genes Dev. *19*, 152-163.

Feng, X. H., and Derynck, R. (1997). A kinase subdomain of transforming growth factor-beta (TGF-beta) type I receptor determines the TGF-beta intracellular signaling specificity. EMBO J. *16*, 3912-3923.

Feng, X. H., Zhang, Y., Wu, R. Y., and Derynck, R. (1998). The tumor suppressor Smad4/DPC4 and transcriptional adaptor CBP/p300 are coactivators for smad3 in TGF-beta-induced transcriptional activation. Genes Dev. *12*, 2153-2163.

Feng, X. H., Lin, X., and Derynck, R. (2000). Smad2, Smad3 and Smad4 cooperate with Sp1 to induce p15(Ink4B) transcription in response to TGF-beta. EMBO J. *19*, 5178-5193.

Feng, X. H., Liang, Y. Y., Liang, M., Zhai, W., and Lin, X. (2002). Direct interaction of c-Myc with Smad2 and Smad3 to inhibit TGF-beta-mediated induction of the CDK inhibitor p15(Ink4B). Mol. Cell *9*, 133-143.

Frederick, J. P., and Wang, X. F. (2002). Smads "freeze" when they ski. Structure (Camb) *10*, 1607-1611.

Frederick, J. P., Liberati, N. T., Waddell, D. S., Shi, Y., and Wang, X. F. (2004). Transforming growth factor beta-mediated transcriptional repression of c-myc is dependent on direct binding of Smad3 to a novel repressive Smad binding element. Mol. Cell. Biol. *24*, 2546-2559.

Germain, S., Howell, M., Esslemont, G. M., and Hill, C. S. (2000). Homeodomain and winged-helix transcription factors recruit activated Smads to distinct promoter elements via a common Smad interaction motif. Genes Dev. *14*, 435-451.

Grimsby, S., Jaensson, H., Dubrovska, A., Lomnytska, M., Hellman, U., and Souchelnytskyi, S. (2004). Proteomics-based identification of proteins interacting with Smad3: SREBP-2 forms a complex with Smad3 and inhibits its transcriptional activity. FEBS Lett. *577*, 93-100.

Grinberg, A. V., and Kerppola, T. (2003). Both Max and TFE3 cooperate with Smad proteins to bind the plasminogen activator inhibitor-1 promoter, but they have opposite effects on transcriptional activity. J. Biol. Chem. *278*, 11227-11236.

Gripp, K. W., Wotton, D., Edwards, M. C., Roessler, E., Ades, L., Meinecke, P., Richieri-Costa, A., Zackai, E. H., Massague, J., Muenke, M., and Elledge, S. J. (2000). Mutations in TGIF cause holoprosencephaly and link NODAL signalling to human neural axis determination. Nat. Genet. *25*, 205-208.

Hanafusa, h., Ninomiya-Tsuji, J., Masuyama, N., Nishita, M., Fujisawa, J., Shibuya, H., Matsumoto, K., and Nishida, E. (1999). Involvement of the p38 mitogen-activated protein kinase pathway in transforming growth factor-beta-induced gene expression. J. Biol. Chem. *274*, 27161-27167.

Hanai, J., Chen, L. F., Kanno, T., Ohtani-Fujita, N., Kim, W. Y., Guo, W. H., Imamura, T., Ishidou, Y., Fukuchi, M., Shi, M. J., *et al.* (1999). Interaction and functional cooperation of PEBP2/CBF with Smads. Synergistic induction of the immunoglobulin germline Calpha promoter. J. Biol. Chem. *274*, 31577-31582.

Harada, J., Kokura, K., Kanei-Ishii, C., Nomura, T., Khan, M. M., Kim, Y., and Ishii, S. (2003). Requirement of the co-repressor homeodomain-interacting protein kinase 2 for ski-mediated inhibition of bone morphogenetic protein-induced transcriptional activation. J. Biol. Chem. *278*, 38998-39005.

Hata, A., Lo, R. S., Wotton, D., Lagna, G., and Massague, J. (1997). Mutations increasing autoinhibition inactivate tumour suppressors Smad2 and Smad4. Nature *388*, 82-87.

Hata, A., Lagna, G., Massague, J., and Hemmati-Brivanlou, A. (1998). Smad6 inhibits BMP/Smad1 signaling by specifically competing with the Smad4 tumor suppressor. Genes Dev. *12*, 186-197.

Hata, A., Seoane, J., Lagna, G., Montalvo, E., Hemmati-Brivanlou, A., and Massague, J. (2000). OAZ uses distinct DNA- and protein-binding zinc fingers in separate BMP-Smad and Olf signaling pathways. Cell *100*, 229-240.

Hayashi, H., Abdollah, S., Qiu, Y., Cai, J., Xu, Y. Y., Grinnell, B. W., Richardson, M. A., Topper, J. N., Gimbrone, M. A., Jr., Wrana, J. L., and Falb, D. (1997). The MAD-related protein Smad7 associates with the TGFbeta receptor and functions as an antagonist of TGFbeta signaling. Cell *89*, 1165-1173.

Hayes, S. A., Zarnegar, M., Sharma, M., Yang, F., Peehl, D. M., ten Dijke, P., and Sun, Z. (2001). SMAD3 represses androgen receptor-mediated transcription. Cancer Res. *61*, 2112-2118.

He, J., Tegen, S. B., Krawitz, A. R., Martin, G. S., and Luo, K. (2003). The transforming activity of Ski and SnoN is dependent on their ability to repress the activity of Smad proteins. J. Biol. Chem. *278*, 30540-30547.

Heldin, C. H., Miyazono, K., and ten Dijke, P. (1997). TGF-beta signalling from cell membrane to nucleus through SMAD proteins. Nature *390*, 465-471.

Higashi, K., Inagaki, Y., Fujimori, K., Nakao, A., Kaneko, H., and Nakatsuka, I. (2003). Interferon-gamma interferes with transforming growth factor-beta signaling through direct interaction of YB-1 with Smad3. J. Biol. Chem. *278*, 43470-43479.

Hoodless, P. A., Haerry, T., Abdollah, S., Stapleton, M., O'Connor, M. B., Attisano, L., and Wrana, J. L. (1996). MADR1, a MAD-related protein that functions in BMP2 signaling pathways. Cell *85*, 489-500.

Hua, X., Liu, X., Ansari, D. O., and Lodish, H. F. (1998). Synergistic cooperation of TFE3 and smad proteins in TGF-beta-induced transcription of the plasminogen activator inhibitor-1 gene. Genes Dev. *12*, 3084-3095.

Hua, X., Miller, Z. A., Wu, G., Shi, Y., and Lodish, H. F. (1999). Specificity in transforming growth factor beta-induced transcription of the plasminogen activator inhibitor-1 gene: interactions of promoter DNA, transcription factor muE3, and Smad proteins. Proc. Natl. Acad. Sci. USA *96*, 13130-13135.

Huse, M., Muir, T. W., Xu, L., Chen, Y. G., Kuriyan, J., and Massague, J. (2001). The TGF beta receptor activation process: an inhibitor- to substrate-binding switch. Mol. Cell *8*, 671-682.

Hyman, C. A., Bartholin, L., Newfeld, S. J., and Wotton, D. (2003). *Drosophila* TGIF proteins are transcriptional activators. Mol. Cell. Biol. *23*, 9262-9274.

Imamura, T., Takase, M., Nishihara, A., Oeda, E., Hanai, J., Kawabata, M., and Miyazono, K. (1997). Smad6 inhibits signalling by the TGF-beta superfamily. Nature *389*, 622-626.

Imoto, S., Sugiyama, K., Muromoto, R., Sato, N., Yamamoto, T., and Matsuda, T. (2003). Regulation of transforming growth factor-beta signaling by protein inhibitor of activated STAT, PIASy through Smad3. J. Biol. Chem. *278*, 34253-34258.

Inman, G. J., and Hill, C. S. (2002). Stoichiometry of active smad-transcription factor complexes on DNA. J. Biol. Chem. *277*, 51008-51016.

Ishida, W., Hamamoto, T., Kusanagi, K., Yagi, K., Kawabata, M., Takehara, K., Sampath, T. K., Kato, M., and Miyazono, K. (2000). Smad6 is a Smad1/5-induced smad inhibitor. Characterization of bone morphogenetic protein-responsive element in the mouse Smad6 promoter. J. Biol. Chem. *275*, 6075-6079.

Itoh, S., Ericsson, J., Nishikawa, J., Heldin, C. H., and ten Dijke, P. (2000). The transcriptional co-activator P/CAF potentiates TGF-beta/Smad signaling. Nucleic Acids Res. *28*, 4291-4298.

Izutsu, K., Kurokawa, M., Imai, Y., Maki, K., Mitani, K., and Hirai, H. (2001). The corepressor CtBP interacts with Evi-1 to repress transforming growth factor beta signaling. Blood *97*, 2815-2822.

Janknecht, R., Wells, N. J., and Hunter, T. (1998). TGF-beta-stimulated cooperation of smad proteins with the coactivators CBP/p300. Genes Dev. *12*, 2114-2119.

Jayaraman, L., and Massague, J. (2000). Distinct oligomeric states of SMAD proteins in the transforming growth factor-beta pathway. J. Biol. Chem. *275*, 40710-40717.

Johnson, K., Kirkpatrick, H., Comer, A., Hoffmann, F. M., and Laughon, A. (1999). Interaction of Smad complexes with tripartite DNA-binding sites. J. Biol. Chem. *274*, 20709-20716.

Jonk, L. J., Itoh, S., Heldin, C. H., ten Dijke, P., and Kruijer, W. (1998). Identification and functional characterization of a Smad binding element (SBE) in the JunB promoter that acts as a transforming growth factor-beta, activin, and bone morphogenetic protein-inducible enhancer. J. Biol. Chem. *273*, 21145-21152.

Kahata, K., Hayashi, M., Asaka, M., Hellman, U., Kitagawa, H., Yanagisawa, J., Kato, S., Imamura, T., and Miyazono, K. (2004). Regulation of transforming growth factor-beta and bone morphogenetic protein signalling by transcriptional coactivator GCN5. Genes Cells *9*, 143-151.

Kaji, H., Canaff, L., Lebrun, J. J., Goltzman, D., and Hendy, G. N. (2001). Inactivation of menin, a Smad3-interacting protein, blocks transforming growth factor type beta signaling. Proc. Natl. Acad. Sci. USA *98*, 3837-3842.

Kang, H. Y., Lin, H. K., Hu, Y. C., Yeh, S., Huang, K. E., and Chang, C. (2001). From transforming growth factor-beta signaling to androgen action: identification of Smad3 as an androgen receptor coregulator in prostate cancer cells. Proc. Natl. Acad. Sci. USA *98*, 3018-3023.

Kang, H. Y., Huang, K. E., Chang, S. Y., Ma, W. L., Lin, W. J., and Chang, C. (2002). Differential modulation of androgen receptor-mediated transactivation by Smad3 and tumor suppressor Smad4. J. Biol. Chem. *277*, 43749-43756.

Kang, Y., Chen, C. R., and Massague, J. (2003). A self-enabling TGFbeta response coupled to stress signaling: Smad engages stress response factor ATF3 for Id1 repression in epithelial cells. Mol. Cell *11*, 915-926.

Kardassis, D., Pardali, K., and Zannis, V. I. (2000). SMAD proteins transactivate the human ApoCIII promoter by interacting physically and functionally with hepatocyte nuclear factor 4. J. Biol. Chem. *275*, 41405-41414.

Kato, Y., Habas, R., Katsuyama, Y., Naar, A. M., and He, X. (2002). A component of the ARC/Mediator complex required for TGF beta/Nodal signalling. Nature *418*, 641-646.

Kawabata, M., Inoue, H., Hanyu, A., Imamura, T., and Miyazono, K. (1998). Smad proteins exist as monomers *in vivo* and undergo homo- and hetero-oligomerization upon activation by serine/threonine kinase receptors. EMBO J. *17*, 4056-4065.

Kida, Y., Maeda, Y., Shiraishi, T., Suzuki, T., and Ogura, T. (2004). Chick Dach1 interacts with the Smad complex and Sin3a to control AER formation and limb development along the proximodistal axis. Development *131*, 4179-4187.

Kim, J., Johnson, K., Chen, H. J., Carroll, S., and Laughon, A. (1997). *Drosophila* Mad binds to DNA and directly mediates activation of vestigial by Decapentaplegic. Nature *388*, 304-308.

Kim, R. H., Wang, D., Tsang, M., Martin, J., Huff, C., de Caestecker, M. P., Parks, W. T., Meng, X., Lechleider, R. J., Wang, T., and Roberts, A. B. (2000). A novel smad nuclear interacting protein, SNIP1, suppresses p300-dependent TGF-beta signal transduction. Genes Dev. *14*, 1605-1616.

Kim, D. W., and Lassar, A. B. (2003). Smad-dependent recruitment of a histone deacetylase/Sin3A complex modulates the bone morphogenetic protein-dependent transcriptional repressor activity of Nkx3.2. Mol. Cell. Biol. *23*, 8704-8717.

Kokura, K., Kaul, S. C., Wadhwa, R., Nomura, T., Khan, M. M.,

Shinagawa, T., Yasukawa, T., Colmenares, C., and Ishii, S. (2001). The Ski protein family is required for MeCP2-mediated transcriptional repression. J. Biol. Chem. *276*, 34115-34121.

Kretzschmar, M., Liu, F., Hata, A., Doody, J., and Massague, J. (1997). The TGF-beta family mediator Smad1 is phosphorylated directly and activated functionally by the BMP receptor kinase. Genes Dev. *11*, 984-995.

Kretzschmar, M., Doody, J., Timokhina, I., and Massague, J. (1999). A mechanism of repression of TGFbeta/ Smad signaling by oncogenic Ras. Genes Dev. *13*, 804-816.

Kurisaki, K., Kurisaki, A., Valcourt, U., Terentiev, A. A., Pardali, K., Ten Dijke, P., Heldin, C. H., Ericsson, J., and Moustakas, A. (2003). Nuclear factor YY1 inhibits transforming growth factor beta- and bone morphogenetic protein-induced cell differentiation. Mol. Cell. Biol. *23*, 4494-4510.

Kurokawa, M., Mitani, K., Imai, Y., Ogawa, S., Yazaki, Y., and Hirai, H. (1998a). The t(3;21) fusion product, AML1/Evi-1, interacts with Smad3 and blocks transforming growth factor-beta-mediated growth inhibition of myeloid cells. Blood *92*, 4003-4012.

Kurokawa, M., Mitani, K., Irie, K., Matsuyama, T., Takahashi, T., Chiba, S., Yazaki, Y., Matsumoto, K., and Hirai, H. (1998b). The oncoprotein Evi-1 represses TGF-beta signalling by inhibiting Smad3. Nature *394*, 92-96.

Kusanagi, K., Inoue, H., Ishidou, Y., Mishima, H. K., Kawabata, M., and Miyazono, K. (2000). Characterization of a bone morphogenetic protein-responsive Smad-binding element. Mol. Biol. Cell *11*, 555-565.

Labbe, E., Silvestri, C., Hoodless, P. A., Wrana, J. L., and Attisano, L. (1998). Smad2 and Smad3 positively and negatively regulate TGF beta-dependent transcription through the forkhead DNA-binding protein FAST2. Mol. Cell *2*, 109-120.

Labbe, E., Letamendia, A., and Attisano, L. (2000). Association of Smads with lymphoid enhancer binding factor 1/T cell-specific factor mediates cooperative signaling by the transforming growth factor-beta and wnt pathways. Proc. Natl. Acad. Sci. USA *97*, 8358-8363.

Lagna, G., Hata, A., Hemmati-Brivanlou, A., and Massague, J. (1996). Partnership between DPC4 and SMAD proteins in TGF-beta signalling pathways. Nature *383*, 832-836.

Lai, C. F., Feng, X., Nishimura, R., Teitelbaum, S. L., Avioli, L. V., Ross, F. P., and Cheng, S. L. (2000). Transforming growth factor-beta up-regulates the beta 5 integrin subunit expression via Sp1 and Smad signaling. J. Biol. Chem. *275*, 36400-36406.

Lee, P. S., Chang, C., Liu, D., and Derynck, R. (2003). Sumoylation of Smad4, the common Smad mediator of transforming growth factor-beta family signaling. J. Biol. Chem. *278*, 27853-27863.

Leong, G. M., Subramaniam, N., Figueroa, J., Flanagan, J. L., Hayman, M. J., Eisman, J. A., and Kouzmenko, A. P. (2001). Ski-interacting protein interacts with Smad proteins to augment transforming growth factor-beta-dependent transcription. J. Biol. Chem. *276*, 18243-18248.

Liberati, N. T., Datto, M. B., Frederick, J. P., Shen, X., Wong, C., Rougier-Chapman, E. M., and Wang, X. F. (1999). Smads bind directly to the Jun family of AP-1 transcription factors. Proc. Natl. Acad. Sci. USA *96*, 4844-4849.

Liberati, N. T., Moniwa, M., Borton, A. J., Davie, J. R., and Wang, X. F. (2001). An essential role for Mad homology domain 1 in the association of Smad3 with histone deacetylase activity. J. Biol. Chem. *276*, 22595-22603.

Lin, X., Liang, M., Liang, Y. Y., Brunicardi, F. C., and Feng, X. H. (2003a). SUMO-1/Ubc9 promotes nuclear accumulation and metabolic stability of tumor suppressor Smad4. J. Biol. Chem. *278*, 31043-31048.

Lin, X., Liang, M., Liang, Y. Y., Brunicardi, F. C., Melchior, F., and Feng, X. H. (2003b). Activation of transforming growth factor-beta signaling by SUMO-1 modification of tumor suppressor Smad4/DPC4. J. Biol. Chem. *278*, 18714-18719.

Lin, H. K., Bergmann, S., and Pandolfi, P. P. (2004). Cytoplasmic PML function in TGF-beta signalling. Nature *431*, 205-211.

Liu, F., Hata, A., Baker, J. C., Doody, J., Carcamo, J., Harland, R. M., and Massague, J. (1996). A human Mad protein acting as a BMP-regulated transcriptional activator. Nature *381*, 620-623.

Liu, F., Pouponnot, C., and Massague, J. (1997). Dual role of the Smad4/DPC4 tumor suppressor in TGFbeta-inducible transcriptional complexes. Genes Dev. *11*, 3157-3167.

Liu, F., Massague, J., and Ruiz i Altaba, A. (1998). Carboxy-terminally truncated Gli3 proteins associate with Smads. Nat. Genet. *20*, 325-326.

Liu, D., Black, B. L., and Derynck, R. (2001a). TGF-beta inhibits muscle differentiation through functional repression of myogenic transcription factors by Smad3. Genes Dev. *15*, 2950-2966.

Liu, X., Sun, Y., Weinberg, R. A., and Lodish, H. F. (2001b). Ski/Sno and TGF-beta signaling. Cytokine Growth Factor Rev. *12*, 1-8.

Liu, F. (2003). Receptor-regulated Smads in TGF-beta signaling. Front. Biosci. *8*, s1280-1303.

Liu, D., Kang, J. S., and Derynck, R. (2004). TGF-beta-activated Smad3 represses MEF2-dependent transcription in myogenic differentiation. EMBO J. *23*, 1557-1566.

Lo, R. S., Chen, Y. G., Shi, Y., Pavletich, N. P., and Massague, J. (1998). The L3 loop: a structural motif determining specific interactions between SMAD proteins and TGF-beta receptors. EMBO J. *17*, 996-1005.

Lo, R. S., Wotton, D., and Massague, J. (2001). Epidermal growth factor signaling via Ras controls the Smad transcriptional co-repressor TGIF. EMBO J. *20*, 128-136.

Long, J., Matsuura, I., He, D., Wang, G., Shuai, K., and Liu, F. (2003). Repression of Smad transcriptional activity by PIASy, an inhibitor of activated STAT. Proc. Natl. Acad. Sci. USA *100*, 9791-9796.

Long, J., Wang, G., He, D., and Liu, F. (2004a). Repression of

Smad4 transcriptional activity by SUMO modification. Biochem. J. *379*, 23-29.

Long, J., Wang, G., Matsuura, I., He, D., and Liu, F. (2004b). Activation of Smad transcriptional activity by protein inhibitor of activated STAT3 (PIAS3). Proc. Natl. Acad. Sci. USA *101*, 99-104.

Lopez-Rovira, T., Chalaux, E., Rosa, J. L., Bartrons, R., and Ventura, F. (2000). Interaction and functional cooperation of NF-kappa B with Smads. Transcriptional regulation of the junB promoter. J. Biol. Chem. *275*, 28937-28946.

Luo, K., Stroschein, S. L., Wang, W., Chen, D., Martens, E., Zhou, S., and Zhou, Q. (1999). The Ski oncoprotein interacts with the Smad proteins to repress TGFbeta signaling. Genes Dev. *13*, 2196-2206.

Luo, K. (2004). Ski and SnoN: negative regulators of TGF-beta signaling. Curr. Opin. Genet. Dev. *14*, 65-70.

Macias-Silva, M., Abdollah, S., Hoodless, P. A., Pirone, R., Attisano, L., and Wrana, J. L. (1996). MADR2 is a substrate of the TGFbeta receptor and its phosphorylation is required for nuclear accumulation and signaling. Cell *87*, 1215-1224.

Massague, J. (1990). The transforming growth factor-beta family. Annu Rev Cell Biol *6*, 597-641.

Massague, J. (1998). TGF-beta signal transduction. Annu. Rev. Biochem. *67*, 753-791.

Massague, J., Blain, S. W., and Lo, R. S. (2000). TGFbeta signaling in growth control, cancer, and heritable disorders. Cell *103*, 295-309.

Massague, J., and Chen, Y. G. (2000). Controlling TGF-beta signaling. Genes Dev. *14*, 627-644.

Massague, J., and Wotton, D. (2000). Transcriptional control by the TGF-beta/Smad signaling system. EMBO J. *19*, 1745-1754.

Matsuda, T., Yamamoto, T., Muraguchi, A., and Saatcioglu, F. (2001). Cross-talk between transforming growth factor-beta and estrogen receptor signaling through Smad3. J. Biol. Chem. *276*, 42908-42914.

Matsuura, I., Denissova, N. G., Wang, G., He, D., Long, J., and Liu, F. (2004). Cyclin-dependent kinases regulate the antiproliferative function of Smads. Nature *430*, 226-231.

Melhuish, T. A., and Wotton, D. (2000). The interaction of the carboxyl terminus-binding protein with the Smad corepressor TGIF is disrupted by a holoprosencephaly mutation in TGIF. J. Biol. Chem. *275*, 39762-39766.

Miyazono, K., Kusanagi, K., and Inoue, H. (2001). Divergence and convergence of TGF-beta/BMP signaling. J. Cell Physiol. *187*, 265-276.

Miyazono, K., Suzuki, H., and Imamura, T. (2003). Regulation of TGF-beta signaling and its roles in progression of tumors. Cancer Sci. *94*, 230-234.

Mori, S., Matsuzaki, K., Yoshida, K., Furukawa, F., Tahashi, Y., Yamagata, H., Sekimoto, G., Seki, T., Matsui, H., Nishizawa, M., et al. (2004). TGF-beta and HGF transmit the signals through JNK-dependent Smad2/3 phosphorylation at the linker regions.

Oncogene *23*, 7416-7429.

Moustakas, A., and Kardassis, D. (1998). Regulation of the human p21/WAF1/Cip1 promoter in hepatic cells by functional interactions between Sp1 and Smad family members. Proc. Natl. Acad. Sci. USA *95*, 6733-6738.

Moustakas, A., and Heldin, C. H. (2002). From mono- to oligo-Smads: the heart of the matter in TGF-beta signal transduction. Genes Dev. *16*, 1867-1871.

Nagarajan, R. P., Zhang, J., Li, W., and Chen, Y. (1999). Regulation of Smad7 promoter by direct association with Smad3 and Smad4. J. Biol. Chem. *274*, 33412-33418.

Nakao, A., Afrakhte, M., Moren, A., Nakayama, T., Christian, J. L., Heuchel, R., Itoh, S., Kawabata, M., Heldin, N. E., Heldin, C. H., and ten Dijke, P. (1997). Identification of Smad7, a TGFbeta-inducible antagonist of TGF-beta signalling. Nature *389*, 631-635.

Nishihara, A., Hanai, J. I., Okamoto, N., Yanagisawa, J., Kato, S., Miyazono, K., and Kawabata, M. (1998). Role of p300, a transcriptional coactivator, in signalling of TGF-beta. Genes Cells *3*, 613-623.

Nishihara, A., Hanai, J., Imamura, T., Miyazono, K., and Kawabata, M. (1999). E1A inhibits transforming growth factor-beta signaling through binding to Smad proteins. J. Biol. Chem. *274*, 28716-28723.

Nishita, M., Hashimoto, M. K., Ogata, S., Laurent, M. N., Ueno, N., Shibuya, H., and Cho, K. W. (2000). Interaction between Wnt and TGF-beta signalling pathways during formation of Spemann's organizer. Nature *403*, 781-785.

Nomura, T., Khan, M. M., Kaul, S. C., Dong, H. D., Wadhwa, R., Colmenares, C., Kohno, I., and Ishii, S. (1999). Ski is a component of the histone deacetylase complex required for transcriptional repression by Mad and thyroid hormone receptor. Genes Dev. *13*, 412-423.

Pardali, E., Xie, X. Q., Tsapogas, P., Itoh, S., Arvanitidis, K., Heldin, C. H., ten Dijke, P., Grundstrom, T., and Sideras, P. (2000a). Smad and AML proteins synergistically confer transforming growth factor beta1 responsiveness to human germ-line IgA genes. J. Biol. Chem. *275*, 3552-3560.

Pardali, K., Kurisaki, A., Moren, A., ten Dijke, P., Kardassis, D., and Moustakas, A. (2000b). Role of Smad proteins and transcription factor Sp1 in p21(Waf1/Cip1) regulation by transforming growth factor-beta. J. Biol. Chem. *275*, 29244-29256.

Pessah, M., Marais, J., Prunier, C., Ferrand, N., Lallemand, F., Mauviel, A., and Atfi, A. (2002). c-Jun associates with the oncoprotein Ski and suppresses Smad2 transcriptional activity. J. Biol. Chem. *277*, 29094-29100.

Pietenpol, J. A., Stein, R. W., Moran, E., Yaciuk, P., Schlegel, R., Lyons, R. M., Pittelkow, M. R., Munger, K., Howley, P. M., and Moses, H. L. (1990). TGF-beta 1 inhibition of c-myc transcription and growth in keratinocytes is abrogated by viral transforming proteins with pRB binding domains. Cell *61*,

777-785.

Postigo, A. A. (2003). Opposing functions of ZEB proteins in the regulation of the TGFbeta/BMP signaling pathway. EMBO J. *22*, 2443-2452.

Postigo, A. A., Depp, J. L., Taylor, J. J., and Kroll, K. L. (2003). Regulation of Smad signaling through a differential recruitment of coactivators and corepressors by ZEB proteins. EMBO J. *22*, 2453-2462.

Pouponnot, C., Jayaraman, L., and Massague, J. (1998). Physical and functional interaction of SMADs and p300/CBP. J. Biol. Chem. *273*, 22865-22868.

Prunier, C., Pessah, M., Ferrand, N., Seo, S. R., Howe, P., and Atfi, A. (2003). The oncoprotein Ski acts as an antagonist of transforming growth factor-beta signaling by suppressing Smad2 phosphorylation. J. Biol. Chem. *278*, 26249-26257.

Qin, B., Lam, S. S., and Lin, K. (1999). Crystal structure of a transcriptionally active Smad4 fragment. Structure Fold Des. *7*, 1493-1503.

Qin, B. Y., Chacko, B. M., Lam, S. S., de Caestecker, M. P., Correia, J. J., and Lin, K. (2001). Structural basis of Smad1 activation by receptor kinase phosphorylation. Mol. Cell *8*, 1303-1312.

Qin, B. Y., Lam, S. S., Correia, J. J., and Lin, K. (2002). Smad3 allostery links TGF-beta receptor kinase activation to transcriptional control. Genes Dev. *16*, 1950-1963.

Qing, J., Zhang, Y., and Derynck, R. (2000). Structural and functional characterization of the transforming growth factor-beta -induced Smad3/c-Jun transcriptional cooperativity. J. Biol. Chem. *275*, 38802-38812.

Qing, J., Liu, C., Choy, L., Wu, R. Y., Pagano, J. S., and Derynck, R. (2004). Transforming growth factor beta/Smad3 signaling regulates IRF-7 function and transcriptional activation of the beta interferon promoter. Mol. Cell. Biol. *24*, 1411-1425.

Randall, R. A., Germain, S., Inman, G. J., Bates, P. A., and Hill, C. S. (2002). Different Smad2 partners bind a common hydrophobic pocket in Smad2 via a defined proline-rich motif. EMBO J. *21*, 145-156.

Roberts A. B., and Sporn, M. B. (1990) The transforming growth factor-betas. In *Peptide growth factors and their receptors*, Eds Sporn M. B., and Roberts, A. B. (Springer-Verlag, Heidelberg) 419-472.

Roberts, A. B., Russo, A., Felici, A., and Flanders, K. C. (2003). Smad3: a key player in pathogenetic mechanisms dependent on TGF-beta. Ann. NY. Acad. Sci. *995*, 1-10.

Roberts, A. B., and Wakefield, L. M. (2003). The two faces of transforming growth factor beta in carcinogenesis. Proc. Natl. Acad. Sci. USA *100*, 8621-8623.

Rodriguez, C., Huang, L. J., Son, J. K., McKee, A., Xiao, Z., and Lodish, H. F. (2001). Functional cloning of the proto-oncogene brain factor-1 (BF-1) as a Smad-binding antagonist of transforming growth factor-beta signaling. J. Biol. Chem. *276*, 30224-30230.

Roth, S. Y., Denu, J. M., and Allis, C. D. (2001). Histone acetyltransferases. Annu. Rev. Biochem. *70*, 81-120.

Sano, Y., Harada, J., Tashiro, S., Gotoh-Mandeville, R., Maekawa, T., and Ishii, S. (1999). ATF-2 is a common nuclear target of Smad and TAK1 pathways in transforming growth factor-beta signaling. J. Biol. Chem. *274*, 8949-8957.

Schmidt, D., and Muller, S. (2003). PIAS/SUMO: new partners in transcriptional regulation. Cell. Mol. Life Sci. *60*, 2561-2574.

Seoane, J., Pouponnot, C., Staller, P., Schader, M., Eilers, M., and Massague, J. (2001). TGFbeta influences Myc, Miz-1 and Smad to control the CDK inhibitor p15INK4b. Nat. Cell Biol. *3*, 400-408.

Seoane, J., Le, H. V., and Massague, J. (2002). Myc suppression of the p21(Cip1) Cdk inhibitor influences the outcome of the p53 response to DNA damage. Nature *419*, 729-734.

Seoane, J., Le, H. V., Shen, L., Anderson, S. A., and Massague, J. (2004). Integration of Smad and forkhead pathways in the control of neuroepithelial and glioblastoma cell proliferation. Cell *117*, 211-223.

Shen, X., Hu, P. P., Liberati, N. T., Datto, M. B., Frederick, J. P., and Wang, X. F. (1998). TGF-beta-induced phosphorylation of Smad3 regulates its interaction with coactivator p300/CREB-binding protein. Mol. Biol. Cell *9*, 3309-3319.

Shen, J., and Walsh, C. A. (2005). Targeted disruption of tgif, the mouse ortholog of a human holoprosencephaly gene, does not result in holoprosencephaly in mice. Mol. Cell. Biol. *25*, 3639-3647.

Shi, Y., Hata, A., Lo, R. S., Massague, J., and Pavletich, N. P. (1997). A structural basis for mutational inactivation of the tumour suppressor Smad4. Nature *388*, 87-93.

Shi, Y., Wang, Y. F., Jayaraman, L., Yang, H., Massague, J., and Pavletich, N. P. (1998). Crystal structure of a Smad MH1 domain bound to DNA: insights on DNA binding in TGF-beta signaling. Cell *94*, 585-594.

Shi, Y. (2001). Structural insights on Smad function in TGFbeta signaling. Bioessays *23*, 223-232.

Shi, Y., and Massague, J. (2003). Mechanisms of TGF-beta signaling from cell membrane to the nucleus. Cell *113*, 685-700.

Shimizu, K., Bourillot, P. Y., Nielsen, S. J., Zorn, A. M., and Gurdon, J. B. (2001). Swift is a novel BRCT domain coactivator of Smad2 in transforming growth factor beta signaling. Mol. Cell. Biol. *21*, 3901-3912.

Shioda, T., Fenner, M. H., and Isselbacher, K. J. (1996). msg1, a novel melanocyte-specific gene, encodes a nuclear protein and is associated with pigmentation. Proc. Natl. Acad. Sci. USA *93*, 12298-12303.

Shioda, T., Fenner, M. H., and Isselbacher, K. J. (1997). MSG1 and its related protein MRG1 share a transcription activating domain. Gene *204*, 235-241.

Shioda, T., Lechleider, R. J., Dunwoodie, S. L., Li, H., Yahata, T., de Caestecker, M. P., Fenner, M. H., Roberts, A. B., and Isselbacher, K. J. (1998). Transcriptional activating activity of

Smad4: roles of SMAD hetero-oligomerization and enhancement by an associating transactivator. Proc. Natl. Acad. Sci. USA 95, 9785-9790.

Song, C. Z., Siok, T. E., and Gelehrter, T. D. (1998). Smad4/DPC4 and Smad3 mediate transforming growth factor-beta (TGF-beta) signaling through direct binding to a novel TGF-beta-responsive element in the human plasminogen activator inhibitor-1 promoter. J. Biol. Chem. 273, 29287-29290.

Song, C. Z., Tian, X., and Gelehrter, T. D. (1999). Glucocorticoid receptor inhibits transforming growth factor-beta signaling by directly targeting the transcriptional activation function of Smad3. Proc. Natl. Acad. Sci. USA 96, 11776-11781.

Souchelnytskyi, S., Tamaki, K., Engstrom, U., Wernstedt, C., ten Dijke, P., and Heldin, C. H. (1997). Phosphorylation of Ser465 and Ser467 in the C terminus of Smad2 mediates interaction with Smad4 and is required for transforming growth factor-beta signaling. J. Biol. Chem. 272, 28107-28115.

Staller, P., Peukert, K., Kiermaier, A., Seoane, J., Lukas, J., Karsunky, H., Moroy, T., Bartek, J., Massague, J., Hanel, F., and Eilers, M. (2001). Repression of p15INK4b expression by Myc through association with Miz-1. Nat. Cell Biol. 3, 392-399.

Stroschein, S. L., Wang, W., and Luo, K. (1999a). Cooperative binding of Smad proteins to two adjacent DNA elements in the plasminogen activator inhibitor-1 promoter mediates transforming growth factor beta-induced smad-dependent transcriptional activation. J. Biol. Chem. 274, 9431-9441.

Stroschein, S. L., Wang, W., Zhou, S., Zhou, Q., and Luo, K. (1999b). Negative feedback regulation of TGF-beta signaling by the SnoN oncoprotein. Science 286, 771-774.

Sun, Y., Liu, X., Eaton, E. N., Lane, W. S., Lodish, H. F., and Weinberg, R. A. (1999a). Interaction of the Ski oncoprotein with Smad3 regulates TGF-beta signaling. Mol. Cell 4, 499-509.

Sun, Y., Liu, X., Ng-Eaton, E., Lodish, H. F., and Weinberg, R. A. (1999b). SnoN and Ski protooncoproteins are rapidly degraded in response to transforming growth factor beta signaling. Proc. Natl. Acad. Sci. USA 96, 12442-12447.

Suzuki, H., Yagi, K., Kondo, M., Kato, M., Miyazono, K., and Miyazawa, K. (2004). c-Ski inhibits the TGF-beta signaling pathway through stabilization of inactive Smad complexes on Smad-binding elements. Oncogene 23, 5068-5076.

Taatjes, D. J., Marr, M. T., and Tjian, R. (2004). Regulatory diversity among metazoan co-activator complexes. Nat. Rev. Mol. Cell. Biol. 5, 403-410.

Takenoshita, S., Mogi, A., Nagashima, M., Yang, K., Yagi, K., Hanyu, A., Nagamachi, Y., Miyazono, K., and Hagiwara, K. (1998). Characterization of the MADH2/Smad2 gene, a human Mad homolog responsible for the transforming growth factor-beta and activin signal transduction pathway. Genomics 48, 1-11.

ten Dijke, P., Goumans, M. J., Itoh, F., and Itoh, S. (2002). Regulation of cell proliferation by Smad proteins. J. Cell Physiol. 191, 1-16.

ten Dijke, P., and Hill, C. S. (2004). New insights into TGF-beta-Smad signalling. Trends Biochem. Sci. 29, 265-273.

Topper, J. N., DiChiara, M. R., Brown, J. D., Williams, A. J., Falb, D., Collins, T., and Gimbrone, M. A., Jr. (1998). CREB binding protein is a required coactivator for Smad-dependent, transforming growth factor beta transcriptional responses in endothelial cells. Proc. Natl. Acad. Sci. USA 95, 9506-9511.

Tsukazaki, T., Chiang, T. A., Davison, A. F., Attisano, L., and Wrana, J. L. (1998). SARA, a FYVE domain protein that recruits Smad2 to the TGFbeta receptor. Cell 95, 779-791.

Tussie-Luna, M. I., Bayarsaihan, D., Seto, E., Ruddle, F. H., and Roy, A. L. (2002). Physical and functional interactions of histone deacetylase 3 with TFII-I family proteins and PIASxbeta. Proc. Natl. Acad. Sci. USA 99, 12807-12812.

Verschueren, K., Remacle, J. E., Collart, C., Kraft, H., Baker, B. S., Tylzanowski, P., Nelles, L., Wuytens, G., Su, M. T., Bodmer, R., et al. (1999). SIP1, a novel zinc finger/homeodomain repressor, interacts with Smad proteins and binds to 5'-CACCT sequences in candidate target genes. J. Biol. Chem. 274, 20489-20498.

von Gersdorff, G., Susztak, K., Rezvani, F., Bitzer, M., Liang, D., and Bottinger, E. P. (2000). Smad3 and Smad4 mediate transcriptional activation of the human Smad7 promoter by transforming growth factor beta. J. Biol. Chem. 275, 11320-11326.

Wang, G., Long, J., Matsuura, I., He, D., and Liu, F. (2005). The Smad3 linker region contains a transcriptional activation domain. Biochem. J. 386, 29-34.

Wicks, S. J., Lui, S., Abdel-Wahab, N., Mason, R. M., and Chantry, A. (2000). Inactivation of smad-transforming growth factor beta signaling by Ca(2+)-calmodulin-dependent protein kinase II. Mol. Cell. Biol. 20, 8103-8111.

Wong, C., Rougier-Chapman, E. M., Frederick, J. P., Datto, M. B., Liberati, N. T., Li, J. M., and Wang, X. F. (1999). Smad3-Smad4 and AP-1 complexes synergize in transcriptional activation of the c-Jun promoter by transforming growth factor beta. Mol. Cell. Biol. 19, 1821-1830.

Wotton, D., Lo, R. S., Lee, S., and Massague, J. (1999a). A Smad transcriptional corepressor. Cell 97, 29-39.

Wotton, D., Lo, R. S., Swaby, L. A., and Massague, J. (1999b). Multiple modes of repression by the Smad transcriptional corepressor TGIF. J. Biol. Chem. 274, 37105-37110.

Wotton, D., Knoepfler, P. S., Laherty, C. D., Eisenman, R. N., and Massague, J. (2001). The Smad transcriptional corepressor TGIF recruits mSin3. Cell Growth Differ. 12, 457-463.

Wrana, J. L. (2000). Regulation of Smad activity. Cell 100, 189-192.

Wrana, J. L. (2002). Phosphoserine-dependent regulation of protein-protein interactions in the Smad pathway. Structure (Camb) 10, 5-7.

Wu, G., Chen, Y. G., Ozdamar, B., Gyuricza, C. A., Chong, P. A., Wrana, J. L., Massague, J., and Shi, Y. (2000). Structural basis of

Smad2 recognition by the Smad anchor for receptor activation. Science *287*, 92-97.

Wu, J. W., Fairman, R., Penry, J., and Shi, Y. (2001a). Formation of a stable heterodimer between Smad2 and Smad4. J. Biol. Chem. *276*, 20688-20694.

Wu, J. W., Hu, M., Chai, J., Seoane, J., Huse, M., Li, C., Rigotti, D. J., Kyin, S., Muir, T. W., Fairman, R., *et al.* (2001b). Crystal structure of a phosphorylated Smad2. Recognition of phosphoserine by the MH2 domain and insights on Smad function in TGF-beta signaling. Mol. Cell *8*, 1277-1289.

Wu, J. W., Krawitz, A. R., Chai, J., Li, W., Zhang, F., Luo, K., and Shi, Y. (2002). Structural mechanism of Smad4 recognition by the nuclear oncoprotein Ski: insights on Ski-mediated repression of TGF-beta signaling. Cell *111*, 357-367.

Xu, W., Angelis, K., Danielpour, D., Haddad, M. M., Bischof, O., Campisi, J., Stavnezer, E., and Medrano, E. E. (2000). Ski acts as a co-repressor with Smad2 and Smad3 to regulate the response to type beta transforming growth factor. Proc. Natl. Acad. Sci. USA *97*, 5924-5929.

Yagi, K., Goto, D., Hamamoto, T., Takenoshita, S., Kato, M., and Miyazono, K. (1999). Alternatively spliced variant of Smad2 lacking exon 3. Comparison with wild-type Smad2 and Smad3. J. Biol. Chem. *274*, 703-709.

Yagi, K., Furuhashi, M., Aoki, H., Goto, D., Kuwano, H., Sugamura, K., Miyazono, K., and Kato, M. (2002). c-myc is a downstream target of the Smad pathway. J. Biol. Chem. *277*, 854-861.

Yahata, T., de Caestecker, M. P., Lechleider, R. J., Andriole, S., Roberts, A. B., Isselbacher, K. J., and Shioda, T. (2000). The MSG1 non-DNA-binding transactivator binds to the p300/CBP coactivators, enhancing their functional link to the Smad transcription factors. J. Biol. Chem. *275*, 8825-8834.

Yakymovych, I., Ten Dijke, P., Heldin, C. H., and Souchelnytskyi, S. (2001). Regulation of Smad signaling by protein kinase C. Faseb J. *15*, 553-555.

Yanagi, Y., Suzawa, M., Kawabata, M., Miyazono, K., Yanagisawa, J., and Kato, S. (1999). Positive and negative modulation of vitamin D receptor function by transforming growth factor-beta signaling through smad proteins. J. Biol. Chem. *274*, 12971-12974.

Yanagisawa, J., Yanagi, Y., Masuhiro, Y., Suzawa, M., Watanabe, M., Kashiwagi, K., Toriyabe, T., Kawabata, M., Miyazono, K., and Kato, S. (1999). Convergence of transforming growth factor-beta and vitamin D signaling pathways on SMAD transcriptional coactivators. Science *283*, 1317-1321.

Yang, Y. C., Piek, E., Zavadil, J., Liang, D., Xie, D., Heyer, J., Pavlidis, P., Kucherlapati, R., Roberts, A. B., and Bottinger, E. P. (2003). Hierarchical model of gene regulation by transforming growth factor beta. Proc. Natl. Acad. Sci. USA *100*, 10269-10274.

Yeo, C. Y., Chen, X., and Whitman, M. (1999). The role of FAST-1 and Smads in transcriptional regulation by activin during early *Xenopus* embryogenesis. J. Biol. Chem. *274*, 26584-26590.

Yingling, J. M., Datto, M. B., Wong, C., Frederick, J. P., Liberati, N. T., and Wang, X. F. (1997). Tumor suppressor Smad4 is a transforming growth factor beta-inducible DNA binding protein. Mol. Cell. Biol. *17*, 7019-7028.

Zawel, L., Dai, J. L., Buckhaults, P., Zhou, S., Kinzler, K. W., Vogelstein, B., and Kern, S. E. (1998). Human Smad3 and Smad4 are sequence-specific transcription activators. Mol. Cell *1*, 611-617.

Zhang, Y., Feng, X., We, R., and Derynck, R. (1996). Receptor-associated Mad homologues synergize as effectors of the TGF-beta response. Nature *383*, 168-172.

Zhang, Y., Feng, X. H., and Derynck, R. (1998). Smad3 and Smad4 cooperate with c-Jun/c-Fos to mediate TGF-beta-induced transcription. Nature *394*, 909-913.

Zhang, Y., and Derynck, R. (2000). Transcriptional regulation of the transforming growth factor-beta -inducible mouse germ line Ig alpha constant region gene by functional cooperation of Smad, CREB, and AML family members. J Biol Chem *275*, 16979-16985.

Zhou, S., Buckhaults, P., Zawel, L., Bunz, F., Riggins, G., Dai, J. L., Kern, S. E., Kinzler, K. W., and Vogelstein, B. (1998a). Targeted deletion of Smad4 shows it is required for transforming growth factor beta and activin signaling in colorectal cancer cells. Proc Natl Acad Sci USA *95*, 2412-2416.

Zhou, S., Zawel, L., Lengauer, C., Kinzler, K. W., and Vogelstein, B. (1998b). Characterization of human FAST-1, a TGF-beta and activin signal transducer. Mol Cell. *2*, 121-127.

Chapter 12

The Rb and E2F Families of Proteins

Wei Du and Jennifer Pogoriler

Ben May Institute for Cancer Research and Center for Molecular Oncology, University of Chicago, 924 E. 57th Street, Chicago, IL 60637

Key Words: Rb, E2F, cell cycle, transcription.

Summary

The Retinoblastoma gene Rb was the first tumor suppressor gene cloned, and it is well known as a negative cell cycle regulator. A simplified model for Rb function is that it blocks cell cycle progression by binding to the E2F transcription factors and recruiting co-repressor complexes to repress the expression of E2F target genes. Both E2F and Rb have multiple family members, with specific members of the Rb family capable of binding only to a specific subset of the E2F family. In addition, different E2F proteins exhibit distinct capacities to regulate natural E2F targets due to their different ability to interact with other transcription factors which are also required for the activation of particular E2F target genes. Although Rb and E2F are best known for their cell cycle roles, recent genome wide analysis of E2F targets has suggested that E2F and Rb control not only genes important in the cell cycle but also genes involved in apoptosis, differentiation, and development. Consistent with these observations, analysis of mice and *Drosophila* with mutations in the Rb and E2F genes has revealed important roles of E2F and Rb proteins in all these areas.

Cloning and Early Characterization of Rb

Rb is a prototypical tumor suppressor gene. Retinoblastoma, a childhood tumor of the eye, is found in either an inherited or a sporadic pattern, and it was hypothesized that this observation could be accounted for by a model requiring two mutations in a tumor suppressor gene (Knudson, 1971). In the inherited form, the existence of one germline mutation would predispose to multiple tumors because only a single additional mutation in the remaining wild type allele would need to occur in the initial tumor cell. In contrast, in sporadic cases only single tumors were found because each tumor required two independent mutational events in the same precursor cell. Analysis of chromosomal deletions in inherited cases of retinoblastoma allowed the chromosomal position of Rb to be determined, and the Rb gene was then identified by chromosomal walking to find genes that were deleted, mis-expressed or contained point mutations in either hereditary or sporadic cases of retinoblastoma but not in normal tissues or other tumor types (Dunn *et al.*, 1988; Friend *et al.*, 1986; Lee *et al.*, 1987). Although Rb was shown to be a nuclear protein that could associate with DNA, the sequence itself did not reveal a biological function of Rb that could underlie its tumor suppressor function.

Insight into this tumor suppressor function was initially obtained from studies of the viral oncogene E1a, an adenovirus immediate early protein. Expression of E1a alone had been shown to immortalize tissue culture cells, while expression of E1a and either E1b or Ras together could transform cells. To determine the mechanism by which E1a immortalized cells, cellular proteins that co-precipitated with E1a were analyzed. Among the major proteins was a 105 KD band that was identified as the tumor suppressor, Rb (Whyte *et al.*, 1988). Mutations of E1a that abolished binding to Rb also eliminated its ability to transform cells in conjunction with Ras. Deletion mapping showed that the region of E1a required to induce DNA synthesis and to immortalize cells was also required to bind to both Rb and p107, another E1a associated protein later found to be a

---

Corresponding Author: Wei Du, Tel: (773) 834-1949, Fax: (773) 702-4394, E-mail: wdu@huggins.bsd.uchicago.edu

member of the Rb family (Whyte *et al.*, 1989). These studies connected the ability of viral oncogenes to transform cells with their ability to interact with a cellular tumor suppressor protein and raised the possibility that the tumor suppressor function of Rb could be mediated by its interaction with other cellular targets.

Subsequent experiments led to the identification of numerous cellular targets of Rb over the years. Among these, the first and the best studied are the E2F transcription factors, which were originally identified among the cellular factors that bound to the adenovirus E2 promoter. In uninfected cells, a large complex was found bound to the E2F binding site, while in cells expressing E1a a smaller, "free" E2F was present. E1a was shown to be capable of dissociating the larger E2F complex, and this depended on a conserved domain showing homology to other viral oncoproteins (Bagchi *et al.*, 1990). Given E1a's interaction with Rb and its ability to release "free" E2F, it was hypothesized that Rb might be part of the larger E2F complex, and this was confirmed by purification of that complex (Bagchi *et al.*, 1991). Furthermore, it was shown that E2F only interacted with a hypophosphorylated form of Rb (the active form) and that Rb was part of the complex dissociated by E1a (Chellappan *et al.*, 1991).

The E2F Family of Proteins

Cloning of the genes that gave rise to the E2F activities revealed that E2F transcription factors are heterodimers composed of a subunit of the E2F gene family and a subunit of the DP gene family. All the E2F and DP proteins, with the exception of the newly identified E2F 7, have both a conserved DNA binding domain (DBD) and a dimerization domain (Dim, see Fig.12.1). In mammalian systems, there are seven E2F family members and two DP family members (for reviews, see (Attwooll *et al.*, 2004a; Dyson, 1998)). The E2F family members can be divided into four subgroups. The first subgroup includes E2Fs 1-3 and is often referred to as the "activating E2Fs". In addition to the DNA binding and dimerization sequences, these E2F proteins also contain a cdk binding domain and a nuclear localization sequence in the N terminus, and a transcriptional activation domain and an Rb binding sequence in the C terminus (Fig. 12.1). This subgroup of E2F proteins is required for the activation of E2F target genes at the G1/S transition and is important for proper cell cycle progression. Compound knock-out of E2Fs 1-3 leads to cell cycle arrest of mouse embryonic fibroblasts (Wu *et al.*, 2001). The second E2F subgroup includes E2Fs 4 and 5 and is often referred to as the "repressive E2Fs". This subgroup of E2F proteins does not have the cdk binding or the nuclear localization sequences in the N terminus but still contains the Rb binding sequence at the C terminus. Consistent with these structural features, this subgroup of E2Fs functions mainly as repressors of E2F target gene expression and can get into the nucleus only when in complex with the

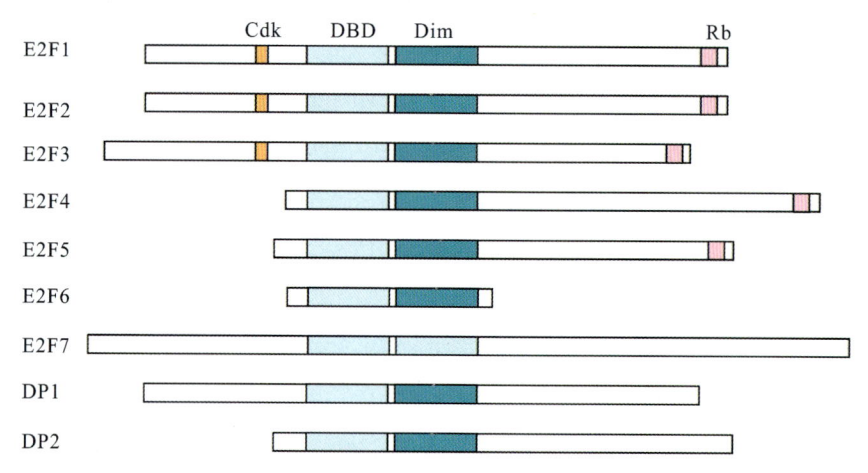

Fig.12.1 The mammalian E2F protein family. The E2F family is composed of E2Fs 1-7, DP1 and DP2. All E2Fs have a conserved DNA binding domain (DBD) and (except for E2F 7) a dimerization domain for binding to DP (Dim). The activating E2Fs 1-3 contain a cyclin/cdk binding domain (cdk) at the N terminus and a strong activation domain at the C terminus. This activation domain overlaps with the Rb binding domain (Rb) so that binding by Rb masks the activation domain. The repressive E2Fs 4 and 5 also bind the pocket proteins but lack the cyclin/cdk binding domain and have weak transactivation domains. E2F 6 shares only the DNA binding domain and dimerization domain with the rest of the family, while E2F 7 has two conserved DNA binding domains, allowing it to function without DP to form a heterodimer. The DP proteins are distantly related members of the family that share the DNA binding domain and dimerization domain, allowing them to bind to E2Fs.

Rb family of proteins during the G0/G1 phases of the cell cycle (Muller et al., 1997; Verona et al., 1997). E2Fs 6 and 7, each representing one of the remaining two subgroups, also function as transcriptional repressors (Attwooll et al., 2004b; de Bruin et al., 2003; Di Stefano et al., 2003; Trimarchi et al., 1998; Trimarchi et al., 2001). However, in contrast to E2Fs 1-5, these E2F proteins do not have the Rb binding sequence at the C terminus, and therefore their function and regulation are independent of the Rb family of proteins.

The Rb Family of Proteins

Rb, p107, and p130 are members of a family of closely related proteins (Fig. 12.2). Together, these proteins are often referred to as the "pocket proteins" because their main sequence similarity resides in a domain, the pocket domain, which mediates interactions with viral oncoproteins such as E1a. A spacer region that is not conserved separates the pocket domain into the A and B pockets. The spacers of p107 and p130 but not Rb contain binding sites for cyclin/cdk complexes. There are many phosphorylation sites on Rb which play critical roles in regulating the function of Rb (see below). Currently, over 100 proteins have been reported to interact with the Rb protein (Morris and Dyson, 2001), and most, if not all, of these interactions also involve the pocket domain. Interestingly, despite the sequence similarities between the Rb family members, specific members of the Rb family preferentially interact with specific members of the E2F family. As shown in Fig.12.3, while Rb preferentially binds to E2Fs 1-4, p107 and p130 predominantly bind to E2F 4 and E2F 5 (Classon and Harlow, 2002). The preferential binding of activating E2Fs by Rb but not by p107 or p130 potentially underlies the observation that only Rb mutations are detected in cancers.

RB and E2F Proteins in *Drosophila*

There are two E2F (*dE2F1* and *dE2F2*), one DP (*dDP*), and two Rb family (*RBF* and *RBF2*) genes in the *Drosophila* genome (Du et al., 1996a; Dynlacht et al., 1994; Ohtani and Nevins, 1994; Sawado et al., 1998; Stevaux et al., 2002). The *Drosophila* E2F proteins share many of the functional properties of their mammalian counterparts: they can dimerize with dDP and bind to E2F binding sites to activate or repress transcription. Similar to their mammalian counterparts, *Drosophila* E2F proteins regulate a set of genes coordinately expressed during S phase such as RNR2, PCNA, cyclin E, and DNA polα (Duronio and O'Farrell, 1994; Duronio et al., 1995). Interestingly, the two *Drosophila* E2F proteins behave like the first two subgroups of the mammalian E2F proteins: dE2F1 mainly functions as a transcription activator (Du, 2000), comparable to the mammalian activating E2Fs 1-3, while dE2F2 mainly functions to mediate active repression, similar to the mammalian repressive E2Fs 4 and 5 (Frolov et al., 2001). Furthermore, similar to the mammalian Rb protein that can bind to both the activating and the repressive E2F proteins, RBF can bind to both dE2F1 and dE2F2 proteins in *Drosophila* (see Fig. 12.3). In contrast, RBF2 can only bind dE2F2 (Stevaux et al., 2002) analogous to the mammalian p107/p130 proteins that bind specifically to the repressive E2F proteins (see Fig.12.3). Thus the Rb-E2F pathway is well conserved and is much simpler in *Drosophila* than in the mammalian systems. The advantages of this simplified model system have been exploited in examining the relative importance of the E2F and Rb family members as seen in many of the experiments described below.

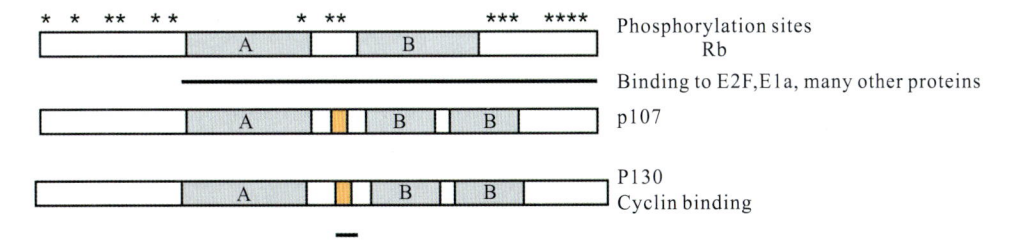

Fig.12.2 The mammalian pocket protein family. The pocket protein family in mammals consists of Rb, p107 and p130. The pocket domain, responsible for most protein-protein interactions, consists of two conserved sequences, A and B. The spacer region between A and B is conserved between p107 and p130 (yellow box), and can bind to cyclin/cdk complexes. The activity of Rb protein is controlled by phosphorylation at numerous phosphorylation sites (*).

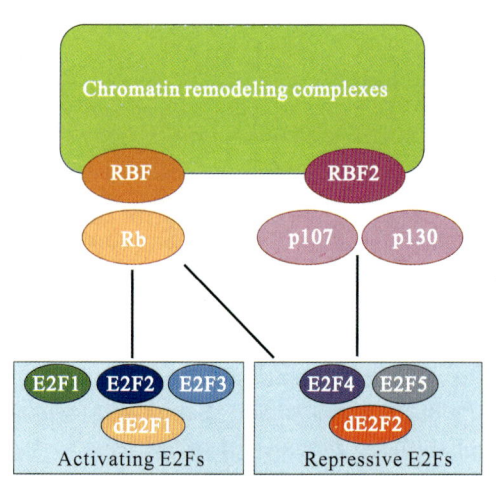

Fig.12.3 Interactions between the Rb and E2F proteins in mammals and *Drosophila*. The mammalian activating E2Fs, E2F 1-3, interact only with Rb. Similarly, the *Drosophila* activating E2F, dE2F1, interacts only with RBF. The mammalian repressive E2Fs, E2F 4-5, interact with p107 and p130. In addition, E2F4 can also interact with Rb. Similarly, the *Drosophila* repressive E2F, dE2F2, interacts with both RBF and RBF2.

Transcriptional Targets of Rb and E2F Proteins

Since Rb and E2F proteins are transcriptional regulators, identification of their targets is critical to our understanding of their biological functions. Rb and E2F proteins were originally known for their ability to regulate the G1/S transition, so E2F target genes were initially identified by examining the promoters of genes known to be turned on at the G1/S transition for E2F binding sites. These early studies identified E2F targets genes that are either cell cycle regulators (such as cyclin E, cyclin A, and cyclin B) or DNA replication factors (such as PCNA, DHFR, and thymidine kinase).

Recently, microarray technology has allowed identification of E2F target genes at the genome level. Several different approaches have been carried out. Initial microarray screens focused on the over-expression of E2F or Rb family members (Muller *et al*., 2001; Polager *et al*., 2002). While these screens identified large arrays of altered genes, it is difficult to distinguish genes directly regulated by Rb or E2F from genes that changed expression due to secondary effects of Rb or E2F expression (such as alterations in the cell cycle). Furthermore, over-expression of E2Fs or Rb might also result in non-physiological binding and activation or repression.

To identify potential direct targets of Rb and E2F proteins, a second generation of screens used the so called ChIP (Chromosomal immunoprecipitation) on CHIP approach to identify genes which have promoters bound to endogenous E2F/Rb (Ren *et al*., 2002). Specifically, small DNA fragments that are bound by the endogenous Rb and E2F proteins are enriched by Chromosomal IP and are then amplified and hybridized to a promoter array. However, as the complete human promoter microarray is not currently available, the current ChIP on CHIP screens have used either a promoter library of genes known to be regulated in a cell-cycle dependent manner (Ren *et al*., 2002) or a library of CpG islands (Wells *et al*., 2003), as GC rich regions are often found at promoters and origins of replication. The ChIP on CHIP screen with the cell-cycle regulated promoter array found that in quiescent cells, antibodies against the repressive E2F 4 precipitated the promoters of genes involved not only in the cell cycle but also in DNA replication, DNA damage and repair, chromosome condensation and separation, the G2/M checkpoint, and mitosis. Many components of multimeric complexes appeared in the screen, suggesting that they are transcriptionally co-regulated. It was confirmed that in p107-/-;p130-/- mouse embryonic fibroblasts that many of these genes were deregulated, as expected if repression by E2F 4 complexes controlled their expression. In S-phase cells, the activating E2F 1 precipitated primarily genes involved in DNA replication and repair, suggesting that these genes require not only de-repression but also activation. In the ChIP on CHIP screen of CpG islands, targets other than cell cycle regulated genes were found, including genes involved in embryogenesis, differentiation and development. As the binding of E2F or Rb to a given promoter does not necessary mean that E2F or Rb will be the rate-limiting determinant of expression for that gene, combining the microarray gene expression data and the CHIP on CHIP data should allow a better identification of Rb/E2F target genes.

A third type of screen has been used in *Drosophila* cell lines by depleting individual Rb/E2F proteins. Specifically, RNAi was used to deplete the activating E2F (dE2F1), the inhibitory E2F (dE2F2), dDP, RBF or RBF2. (Dimova *et al*., 2003). Besides the benefits of avoiding the non-physiological consequences of over-expression, this type of screen also allowed a greater range of genes to be analyzed than in the ChIP on ChIP screens since it was not dependent on a promoter array. In addition, although it was possible that secondary changes in gene expression could be responsible, it did detect actual changes in expression, suggesting that the E2F/Rb interactions were rate limiting under these conditions. Furthermore, the presence of a single activating E2F and a single repressive E2F in *Drosophila* removed the complication of redundant functions. A surprising result of this screen was that very few genes showed *both* a decrease in expression in the absence of dE2F1 *and* an increase in expression in the absence of dE2F2, suggesting that most genes are more influenced either by repression or by activation rather than by both activities equally. Most of the genes found to be regulated largely by dE2F1 were involved in processes similar to those found in the ChIP on CHIP screens, including the cell cycle, DNA replication and repair, mitosis, chromosome segregation, and checkpoints. However, the genes regulated primarily by dE2F2 were not involved in DNA repair or S-phase processes. While the functions of many genes in this category were unknown, a surprising number appear to be unrelated to the cell cycle and instead were involved in development, including male and female specific genes. The importance of dE2F2 repression was confirmed by finding that expression of many of these genes was deregulated in *de2f2* null flies (Dimova *et al.*, 2003).

Biological Functions of Rb and E2F

Because of the simplicity of the *Drosophila* E2F/Rb protein families, significant insights into the biological functions of the E2F/Rb proteins have been derived from studies of this model system. Analysis of the phenotype of flies with mutations of *de2f1* showed that dE2F1 plays a critical role in the expression of a set of S phase genes such as RNR2 and PCNA, which are two replication factors regulated by E2F (Duronio *et al.*, 1995). However, in *de2f1* mutants, although the rate of BrdU incorporation was significantly reduced, it was not completely blocked (Royzman *et al.*, 1997). This effect of *de2f1* mutation on BrdU incorporation is in contrast to that of one of its targets, the cyclin E mutant, which showed a complete block to BrdU incorporation (Royzman *et al.*, 1997). Therefore, during normal development dE2F1 is limiting for the expression of replication factors but is not limiting for the expression of cell cycle regulators such as cyclin E. Unexpectedly, reducing the gene dosage of *rbf* or *de2f2*, or *dDp* all substantially suppress the phenotype of *de2f1* null mutants (Du, 2000; Frolov *et al.*, 2001), suggesting that the phenotype of *de2f1* mutant is at least in part due to the presence of an RBF/dE2F2 repressor complex.

Interestingly, not all of the E2F target genes are coordinately deregulated by *rbf* mutation. For example, the epidermal cells of the developing embryo normally arrest at G1 after the 16th cell division and have low levels of the E2F targets *PCNA*, *RNR2*, and *cyclin E* expression. In embryos devoid of RBF, although all the epidermal cells initially still arrest in G1, they exhibit strongly deregulated PCNA and RNR2 expression. In contrast, cyclin E is not immediately upregulated, but later a subset of epidermal cells accumulate cyclin E and enter S phase (Du and Dyson, 1999). The initial G1 cell cycle arrest of the epidermal cells of the developing embryos requires the cdk inhibitor dacapo (de Nooij *et al.*, 1996; Lane *et al.*, 1996), and sufficient *cyclin E* expression in the absence of RBF can overcome this arrest. RBF therefore functions to maintain the G1 arrest by inhibiting the expression of key cell cycle regulators such as *cyclin E* (Du and Dyson, 1999).

In summary, it appears that while E2F activity is not absolutely required for S phase entry, lack of E2F activation, either due to mutation of *de2f1* or due to overexpression of RBF, slows S phase progression (Royzman *et al.*, 1997; Xin *et al.*, 2002). On the other hand, inappropriate activation of E2F, either by ectopic expression of dE2F1 or by mutation of *rbf*, does lead to ectopic S phase entry and increased apoptosis (Du and Dyson, 1999; Du *et al.*, 1996b). These effects of E2F on the cell cycle are likely due to the fact that the expression of the key regulator of S phase, cyclin E, does not absolutely require E2F but ectopic expression of E2F does activate cyclin E.

Biochemical studies using mammalian tissue culture systems have provided significant insight into the different E2F/Rb complexes that function in different phases of the cell cycle. Although simplified models suggest that the Rb/E2F complexes present during quiescence or G0 are replaced with free E2Fs at the G1/S transition, the actual composition of these complexes appears to vary with the cell cycle (reviewed in (Bracken *et al.*, 2004; Classon and Harlow, 2002). In quiescent cells, p130 is the main pocket protein complexed to inhibitory E2Fs; however, in cycling cells it is replaced by p107/E2F 4 in G0/G1 phases and by Rb/E2F 1-3 complexes in S phase. Although mouse embryonic fibroblasts (MEFs) deficient in any one of the inhibitory E2Fs or in both p107 and p130 proliferate normally, they show defects in response to growth inhibitory signals. While pocket protein members may compensate for one another to some extent, Rb appears to be particularly important for senescence, and MEFs that are acutely inactivated for Rb are able to re-enter the cell cycle even from a quiescent state (Sage *et al.*, 2003).

Given this important role of E2F and Rb in the cell cycle, it is not surprising that alterations in the Rb pathway are found in most tumors. Although mutation of Rb itself does not necessarily occur, inactivation may occur through overexpression of cyclin D or through loss of cdk inhibitors such as p16, both leading to phosphorylation of Rb. In addition to adenovirus E1a,

E2F can also be released from E2F/Rb complexes by viral oncogenes such as the SV40 large T antigen or by the human papilloma virus E7 protein. The strains of HPV that are associated with cervical cancer encode E7 proteins with the highest affinity and the ability to degrade Rb (Munger, 2002).

In addition to the cell cycle, Rb and E2F are involved in functions such as apoptosis, differentiation, and development (reviewed in Bracken et al., 2004; Classon and Harlow, 2002; Liu et al., 2004). Besides altering proliferation, loss of Rb may contribute to cancer by affecting differentiation. It was initially believed that loss of Rb would indirectly contribute to loss of differentiation because cells would be unable to terminally exit the cell cycle. However, screens of E2F-dependent genes have demonstrated that tissue-specific genes are among the targets of Rb/E2F. The importance of E2F and pocket protein activity has been shown in confluent (non-dividing) MEFs that are induced to differentiate into adipocytes. Under these conditions, the loss of E2F 4 results in spontaneous differentiation even though the cell cycle is not affected, suggesting that E2F 4 suppresses differentiation independently of proliferation (Landsberg et al., 2003). Moreover, mouse models of pocket protein or E2F family members show that loss of various E2Fs is associated with specific developmental abnormalities. For example, in some hematopoetic cell types Rb null cells cannot reach the final stage of differentiation and, the mice are prone to myeloproliferative disorders. It has been speculated that this is due to the increase in the pool of precursor cells that can give rise to tumors (Spike et al., 2004).

In addition to their roles in the cell cycle and development, Rb and E2F have also been found to affect apoptosis. Overexpression of E2F 1 has been shown to induce apoptosis, and E2F 1-/- thymocytes are defective in undergoing apoptosis (Field et al., 1996). Whether this ability to induce apoptosis is restricted to E2F 1 or is shared by other activating E2Fs is still unclear. E2F targets related to apoptosis include APAF1, caspases 3 and 7, and p73. Although some apoptosis appears to be independent of p53, E2F can activate the p53 pathway through upregulation of Arf and Pin-1, both of which contribute to p53 stabilization. There is also evidence that Rb itself may contribute to inhibiting apoptosis by stabilizing the anti-apoptotic protein Bcl-xL (reviewed in Liu et al., 2004). The ability of Rb and E2F to affect apoptosis would be expected to *inhibit* tumor formation by leading to apoptosis of cells with excessive E2F activity. Consistent with a role for E2F activity in tumor suppression, E2F 1-/- mice are surprisingly tumor prone (Cloud et al., 2002), although they show a different tumor spectrum from that of Rb+/- mice. Tumor incidence of E2F 1-/- mice can be further increased by removing E2F 2, but not E2F 3, suggesting that tumor suppression is a characteristic only of E2Fs 1 and 2.

Besides apoptosis, mouse models confirm many of the additional biological functions ascribed to Rb and E2F (reviewed in Classon and Harlow, 2002; Liu et al., 2004). The best characterized of the pocket protein knockout animals are Rb deficient mice. Rb heterozygous mice show a similar phenotype to heterozygous humans except that instead of developing retinoblastomas, Rb+/- mice develop pituitary and thyroid tumors (Jacks et al., 1992). The incidence of these tumors can be decreased by eliminating E2F 1, suggesting that the ability of Rb to suppress tumors is due to its inhibition of activating E2Fs. Rb null mice die at embryonic day 13.5 and show ectopic S phases, extensive neuronal apoptosis, and defects in the differentiation of muscle and red blood cells, among other tissues. However, many of these defects appear not to be cell autonomous but may instead be due to hypoxia since the Rb-/- placenta and red blood cell development are abnormal (Wu et al., 2003). In contrast, p107 or p130 null animals appear to develop normally, although p107-/-;p130-/- animals have defects in chondrocytes leading to bone abnormalities (Cobrinik et al., 1996), suggesting some essential, tissue-specific roles for these proteins. The ability of E2F 1-/- or E2F 3-/- to decrease the extent of neuronal apoptosis in Rb-/- embryos has suggested that it is excessive E2F activity that leads to apoptosis; however, restoration of Rb in the placenta also rescued apoptosis in the central nervous system, so it is not clear whether the apoptotic rescue by E2Fs 1 and 3 is cell autonomous or whether loss of the E2Fs corrected the placental phenotypes.

Regulation of E2F and Rb Activity

Rb/E2F complexes have been shown to be central targets of factors that influence the cell cycle. Rb contains numerous phosphorylation sites which can by phosphorylated by the G1 phase cyclinD/cdk4 complexes and by the G1/S phase cyclinE/cdk2 and cyclinA/cdk2 complexes. Hyper-phosphorylation of Rb by cyclin/cdk complexes results in a decreased ability to interact both with its target E2F and with co-repressors such as histone deacetylase enzymes. Growth stimulating factors and growth inhibitory factors generally affect the transcription, translation and stabilities of D and E type of cyclins as well as cdk inhibitors. The net effect of these growth signaling pathways controls the E2F/Rb/

corepressor complex formation through regulating the phosphorylation of the Rb family of proteins, with the hypo-phosphorylated Rb being active and the hyper-phosphorylated forms of Rb being inactive.

In addition to the phosphorylation status of Rb, E2F activity can also be regulated by a large number of other factors directly. Regulation of E2F activity has been primarily studied for E2F 1. The histone acetylase enzymes P300/CBP and P/CAF can directly bind to E2F 1 and can acetylate E2F 1 *in vitro* (Martinez-Balbas *et al.*, 2000). *In vivo,* the N-terminal lysines of E2F 1 are found to be acetylated, and this acetylation increases the DNA binding activity of E2F as well as its half life. The increased stability and DNA binding contribute to increased transcriptional activity of acetylated E2F 1. This acetylation can be reversed by histone deacetylase enzymes recruited by Rb. In addition, E2F 1 stability can also be increased specifically following activation of the DNA damage signaling response. Treatment of cells with DNA damaging agents leads to activation of the ATM kinase, which can specifically phosphorylate E2F 1, resulting in increased protein accumulation (Lin *et al.*, 2001). In contrast, E2F activity was found to be negatively regulated by the kinase activity of cyclinA/cdk2 (Krek *et al.*, 1994). Cyclin A binds to the N terminus of E2F 1 and can phosphorylate both E2F 1 and DP *in vitro*, resulting in decreased DNA binding activity, and *in vivo* phosphorylation of DP is dependent on cyclin A binding to E2F 1.

Mechanisms of Transcriptional Control

E2F and Rb control transcription through a variety of mechanisms. While E2F alone can activate transcription, binding by Rb not only blocks transcriptional activation but also leads to active repression. Inhibition of Rb function would be predicted to both remove active repression and to allow transactivation by E2F. The relative importance of de-repression versus activation may depend on the specific target. For genes at which active repression is important, mutation of the E2F binding site would be predicted to increase the basal level of transcription. Such target genes include B-myb and cyclin E, which show premature expression when the E2F binding site is mutated. In contrast, for genes at which activation is important, mutation of the E2F binding site would be expected to result in loss of appropriate expression. Examples of these genes include DHFR and thymidine kinase (reviewed in Mundle and Saberwal, 2003).

The effects of Rb and E2F on gene expression have been primarily evaluated using reporter assays of transiently transfected plasmids. Although a great deal has been learned from these experiments, they may not necessarily be an accurate reflection of events in a chromosomal setting. Additional experiments have used chromatin immunoprecipitation (ChIP) to examine the proteins localized at endogenous sites on chromatin or have examined patterns of endogenous gene expression to further understand the mechanisms by which E2F and Rb control transcription.

The ability of Rb to actively repress transcription was suggested by the observation that E2F binding sites conferred increased reporter activity in Rb negative cells but decreased reporter activity in Rb positive cells (Weintraub *et al.*, 1992). Fusion of Rb to the Gal4 DNA binding domain demonstrated that it could actively repress transcription of a reporter gene independently of E2F. Moreover, this repression was inhibited by cyclin E expression, indicating that hyperphosphorylated Rb could not repress transcription (Weintraub *et al.*, 1995).

Recruitment of co-repressors contributes to Rb-dependent repression. Rb was first shown to interact with histone deacetylase (HDAC) 1 through its pocket domain. The HDAC inhibitor trichostatin can abolish Rb-dependent repression from a promoter containing E2F binding sites, and over-expression of HDAC and Rb results in a cooperative decrease in expression of the E2F target cyclin E, providing genetic evidence for the role of histone deacetylation in Rb's repression of normal chromatin (Brehm *et al.*, 1998). Although HDAC and E2F both bind to Rb's pocket domain, they can simultaneously interact with hypophosphorylated Rb, allowing HDAC to be recruited to E2F sites. Rb's pocket domain has been shown to interact with histone deacetylase (HDAC) enzymes 1-3 through their LXCXE sequences. Interestingly, Rb that has been phosphorylated by cyclinD/cdk4 cannot bind to HDACs, which allows the expression of targets such as cyclin E (Zhang *et al.*, 2000).

This mode of repression by Rb is in direct contrast to activation by E2F, which can bind to the histone acetylases (HATs) p300/CBP and P/CAF, making the DNA more accessible to transcription factors. The coordination of RB/E2F binding, the acetylation status of E2F-dependent promoters and induction of transcription has been shown in a number of studies examining the proteins localized to the promoters of E2F-dependent genes at different stages of the cell cycle. As compared to quiescent cells, in late G1 there is a loss of the pocket proteins and repressive E2Fs (E2F 4/p130), which are replaced by activating E2Fs. At the same time, there is a switch from HDACs to HATs. This is accompanied by hyperacetylation of histones H3 and

H4 and the initiation of E2F-dependent gene expression (reviewed in Frolov and Dyson, 2004).

A role for chromatin remodeling by Rb as a part of its repressive activity was first identified by the interaction of Rb with BRG1, a component of the human SWI/SNF complex. BRG1 interacts with the unphosphorylated pocket domain, and in BRG1 deficient cells its re-expression is capable of conferring a growth arrested, flat cell phenotype (Dunaief et al., 1994). However, it has since been shown that the Rb binding site of BRG1 is not conserved in the Drosophila BRG homolog, and mutation of human BRG1 so that it could no longer bind Rb did not affect its ability to arrest cells in G1, suggesting that the genetic interaction between BRG1 and Rb does not reflect a direct biochemical interaction. Instead, it appears that BRG1 induces expression of a cyclinE/cdk2 inhibitor, resulting in decreased phosphorylation of Rb, thus allowing it to remain in an active state (Kang et al., 2004).

Rb may also repress transcription through its ability to recruit histone methylase activity. Endogenous Rb associates with SUV39H1, an enzyme which methylates lysine 9 on histone H3. In reporter assays SUV39H1 can repress transcription when Rb is targeted to the promoter, and it can cooperate with Rb to repress endogenous cyclin E expression. Furthermore, HP1, a protein which binds methylated lysine 9 and is associated with transcriptionally silent regions of chromatin, can bind Rb. The nucleosome at the cyclin E promoter is associated with methylated H3 and HP1, but only in Rb positive cells, suggesting that Rb is responsible for this chromatin modification. (Nielsen et al., 2001)

A further mechanism of Rb repression may be its ability to recruit the DNA methyltransferase enzyme DNMT1. Rb physically directly interacts with DNMT1, and DNMT1 shows cooperative effects on repression of reporter genes. However, DNA was not methylated at these sites, suggesting that the effect of DNMT1 may not be through its enzymatic activity but instead might help recruit other co-repressors.

Although a large number of general mechanisms of repression and activation have been studied, more recently effort has been directed towards elucidating the mechanism by which various E2F and Rb family members carry out their distinct functions. Since the crystallization of an E2F 4/DP DNA-bound heterodimer suggested that all the DNA-contacting residues are conserved among E2F family members (Zheng et al., 1999), it is unlikely that specificity is conferred by variations in the DNA sequence of the E2F binding sites. An alternative hypothesis is that specificity of E2F family members is determined by their interaction with other transcription factors. A number of papers have examined the role of the marked box domain of E2F family members. This domain, directly downstream of the DNA binding domain, was first found to be important in the interaction of E2F 1 with the adenoviral protein E4, and it allowed for a synergistic effect of these two proteins on the activation of the E2 promoter. It was hypothesized that this region might also bind to other cellular transcription factors (Jost et al., 1996). The importance of this domain for specificity was initially suggested in studies examining why E2F 1, but not the other activating E2Fs, was capable of inducing apoptosis. Construction of chimaeric proteins containing E2F 1 and E2F 3 sequences indicated that the marked box domain and adjacent regions of E2F 1 were critically involved in the ability of activating E2Fs to induce apoptosis (Hallstrom and Nevins, 2003).

To further examine this region, the marked box domain of E2F 3 was used as bait in a yeast 2-hybrid screen to identifying interacting proteins (Giangrande et al., 2003; Giangrande et al., 2004). Although this domain shows 55% homology between E2F 1 and E2F 3, one of the proteins from the screen, the ubiquitous, E box binding transcription factor TFE-3, specifically bound only to E2F 3. The promoter of p68, a polymerase α subunit with both an E-box and E2F sites proximal to the promoter, was used to examine the interaction between these proteins. E2F 3 and TFE-3 were shown to synergistically activate p68 and to directly bind to one another through the E2F 3 marked box domain. In fact, the synergistic activation required this interaction, since an E2F 3 chimera with the E2F 1 marked box domain could not activate transcription. In the absence of E2F 3, TFE-3 was unable to bind to the p68 promoter. Conversely, in the absence of TFE-3, E2F 3 showed delayed binding, which was due to the presence of another weakly compensating E-box binding protein, USF1. USF1 was also capable of binding specifically to the E2F 3 marked box domain and could synergistically activate p68 expression. These findings could be extended to some, but not all, E2F 3 specific promoters that contained both E2F and E box sites proximal to the promoter, suggesting that additional factors also contribute to control of expression.

In a separate approach, another group found that the repressive E2Fs, E2F 4 and 5, could specifically interact with the SMAD transcription factors (Chen et al., 2002). In response to TGFβ signaling, phosphorylation of SMAD 2 and 3 allows them to bind to SMAD4, and this complex can either activate or repress transcription of target genes. In studying TGFβ-mediated repression, it was found that the TGFβ response element of c-myc contained a consensus E2F site adjacent to the SMAD binding site. Mutation of either site led to decreased SMAD binding and loss of TGFβ responsiveness. Chromatin immunoprecipitation experiments demonstrated that SMADs 2, 3, and 4, as well as E2Fs 4 and 5 and p107 were found at the TGFβ response element, but that the other RB family members and activating E2Fs were absent. In fact, SMAD3 could directly bind to both E2F 4 and 5 as well as p107. This interaction was required for repression of c-myc in response to TGFβ since c-myc expression did not decrease in E2F 4-/-; E2F 5-/- or p107-/- cells. This repressive ability of E2Fs 4 and 5 may therefore in some cases be due to their ability to be recruited to the promoter by their specific interaction with other transcription factors.

Finally, recent work in *Drosophila* has also contributed to further understanding of the differences between activating and repressive E2Fs. In an attempt to understand the specific function of dE2F2, the repressive E2F, native complexes were purified from embryo extracts (Korenjak et al., 2004). In addition to either RBF or RBF2, these complexes contained components of the dMyb complex, a known transcriptional regulator. The dMyb complex showed no co-staining with actively transcribed regions of chromatin, consistent with its association with repressive E2Fs. Depletion of these dMyb subunits by RNAi resulted in up-regulation of genes normally controlled exclusively by dE2F2 repression, demonstrating that dMyb is required for dE2F2 repression. Rather than having a role in the cell cycle, these E2F target genes are involved in differentiation, often in expression of sex-specific genes. The Myb/Rb/E2F interaction also appears to be evolutionarily conserved: Rb and components of the Myb complex form part of the synMuv class of genes in *C. elegans* that control vulval development, and, although less studied, the human Myb homologs also appear to bind to Rb directly.

Concluding Remarks

Although Rb has been extensively studied since its identification as the first tumor suppressor gene, much remains to be discovered about both Rb and E2F's physiological roles and the specific mechanisms of action that account for those roles. Recent molecular discoveries have begun to provide a mechanistic explanation for the observation that some target genes are controlled by specific E2F family members, and further understanding of the specific interactions between E2F/Rb family members and other transcription factors may also explain complex patterns of timing and levels of induction. The mechanisms and targets found in biochemical and microarray experiments will need to be examined *in vivo* to determine which interactions and targets are crucial for determining the physiological outcome. This information can be combined with advances in mouse targeting constructs and microarray techniques which have already expanded the arenas in which Rb and E2F are known to play key roles.

References

Attwooll, C., Denchi, E. L., and Helin, K. (2004a). The E2F family: specific functions and overlapping interests. Embo J *23*, 4709-4716.

Attwooll, C., Oddi, S., Cartwright, P., Prosperini, E., Agger, K., Steensgaard, P., Wagener, C., Sardet, C., Moroni, M. C., and Helin, K. (2004b). A novel repressive E2F6 complex containing the polycomb group protein, EPC1, that interacts with EZH2 in a proliferation-specific manner. J Biol Chem.

Bagchi, S., Raychaudhuri, P., and Nevins, J. R. (1990). Adenovirus E1A proteins can dissociate heteromeric complexes involving the E2F transcription factor: a novel mechanism for E1A trans-activation. Cell *62*, 659-669.

Bagchi, S., Weinmann, R., and Raychaudhuri, P. (1991). The retinoblastoma protein copurifies with E2F-I, an E1A-regulated inhibitor of the transcription factor E2F. Cell *65*, 1063-1072.

Bracken, A. P., Ciro, M., Cocito, A., and Helin, K. (2004). E2F target genes: unraveling the biology. Trends Biochem Sci *29*, 409-417.

Brehm, A., Miska, E. A., McCance, D. J., Reid, J. L., Bannister, A. J., and Kouzarides, T. (1998). Retinoblastoma protein recruits histone deacetylase to repress transcription. Nature *391*, 597-601.

Chellappan, S. P., Hiebert, S., Mudryj, M., Horowitz, J. M., and Nevins, J. R. (1991). The E2F transcription factor is a cellular target for the RB protein. Cell *65*, 1053-1061.

Chen, C. R., Kang, Y., Siegel, P. M., and Massague, J. (2002). E2F4/5 and p107 as Smad cofactors linking the TGFbeta receptor to c-myc repression. Cell *110*, 19-32.

Classon, M., and Harlow, E. (2002). The retinoblastoma tumour suppressor in development and cancer. Nat Rev Cancer *2*, 910-917.

Cloud, J. E., Rogers, C., Reza, T. L., Ziebold, U., Stone, J. R., Picard, M. H., Caron, A. M., Bronson, R. T., and Lees, J. A. (2002). Mutant mouse models reveal the relative roles of E2F1 and E2F3 *in vivo*. Mol Cell Biol *22*, 2663-2672.

Cobrinik, D., Lee, M. H., Hannon, G., Mulligan, G., Bronson, R. T., Dyson, N., Harlow, E., Beach, D., Weinberg, R. A., and Jacks, T. (1996). Shared role of the pRB-related p130 and p107 proteins in limb development. Genes Dev *10*, 1633-1644.

de Bruin, A., Maiti, B., Jakoi, L., Timmers, C., Buerki, R., and

Leone, G. (2003). Identification and characterization of E2F7, a novel mammalian E2F family member capable of blocking cellular proliferation. J Biol Chem 278, 42041-42049.

de Nooij, J. C., Letendre, M. A., and Hariharan, I. K. (1996). A cyclin-dependent kinase inhibitor, Dacapo, is necessary for timely exit from the cell cycle during Drosophila embryogenesis. Cell 87, 1237-1247.

Di Stefano, L., Jensen, M. R., and Helin, K. (2003). E2F7, a novel E2F featuring DP-independent repression of a subset of E2F-regulated genes. Embo J 22, 6289-6298.

Dimova, D. K., Stevaux, O., Frolov, M. V., and Dyson, N. J. (2003). Cell cycle-dependent and cell cycle-independent control of transcription by the Drosophila E2F/RB pathway. Genes Dev 17, 2308-2320.

Du, W. (2000). Suppression of the rbf null mutants by a de2f1 allele that lacks transactivation domain. Development 127, 367-379.

Du, W., and Dyson, N. (1999). The role of RBF in the introduction of G1 regulation during Drosophila embryogenesis. Embo J 18, 916-925.

Du, W., Vidal, M., Xie, J. E., and Dyson, N. (1996a). RBF, a novel RB-related gene that regulates E2F activity and interacts with cyclin E in Drosophila. Genes Dev 10, 1206-1218.

Du, W., Xie, J. E., and Dyson, N. (1996b). Ectopic expression of dE2F and dDP induces cell proliferation and death in the Drosophila eye. Embo J 15, 3684-3692.

Dunaief, J. L., Strober, B. E., Guha, S., Khavari, P. A., Alin, K., Luban, J., Begemann, M., Crabtree, G. R., and Goff, S. P. (1994). The retinoblastoma protein and BRG1 form a complex and cooperate to induce cell cycle arrest. Cell 79, 119-130.

Dunn, J. M., Phillips, R. A., Becker, A. J., and Gallie, B. L. (1988). Identification of germline and somatic mutations affecting the retinoblastoma gene. Science 241, 1797-1800.

Duronio, R. J., and O'Farrell, P. H. (1994). Developmental control of a G1-S transcriptional program in Drosophila. Development 120, 1503-1515.

Duronio, R. J., O'Farrell, P. H., Xie, J. E., Brook, A., and Dyson, N. (1995). The transcription factor E2F is required for S phase during Drosophila embryogenesis. Genes Dev 9, 1445-1455.

Dynlacht, B. D., Brook, A., Dembski, M., Yenush, L., and Dyson, N. (1994). DNA-binding and trans-activation properties of Drosophila E2F and DP proteins. Proc Natl Acad Sci USA 91, 6359-6363.

Dyson, N. (1998). The regulation of E2F by pRB-family proteins. Genes Dev 12, 2245-2262.

Field, S. J., Tsai, F. Y., Kuo, F., Zubiaga, A. M., Kaelin, W. G., Jr., Livingston, D. M., Orkin, S. H., and Greenberg, M. E. (1996). E2F-1 functions in mice to promote apoptosis and suppress proliferation. Cell 85, 549-561.

Friend, S. H., Bernards, R., Rogelj, S., Weinberg, R. A., Rapaport, J. M., Albert, D. M., and Dryja, T. P. (1986). A human DNA segment with properties of the gene that predisposes to retinoblastoma and osteosarcoma. Nature 323, 643-646.

Frolov, M. V., and Dyson, N. J. (2004). Molecular mechanisms of E2F-dependent activation and pRB-mediated repression. J Cell Sci 117, 2173-2181.

Frolov, M. V., Huen, D. S., Stevaux, O., Dimova, D., Balczarek-Strang, K., Elsdon, M., and Dyson, N. J. (2001). Functional antagonism between E2F family members. Genes Dev 15, 2146-2160.

Giangrande, P. H., Hallstrom, T. C., Tunyaplin, C., Calame, K., and Nevins, J. R. (2003). Identification of E-box factor TFE3 as a functional partner for the E2F3 transcription factor. Mol Cell Biol 23, 3707-3720.

Giangrande, P. H., Zhu, W., Rempel, R. E., Laakso, N., and Nevins, J. R. (2004). Combinatorial gene control involving E2F and E Box family members. Embo J 23, 1336-1347.

Hallstrom, T. C., and Nevins, J. R. (2003). Specificity in the activation and control of transcription factor E2F-dependent apoptosis. Proc Natl Acad Sci USA 100, 10848-10853.

Jacks, T., Fazeli, A., Schmitt, E. M., Bronson, R. T., Goodell, M. A., and Weinberg, R. A. (1992). Effects of an Rb mutation in the mouse. Nature 359, 295-300.

Jost, C. A., Ginsberg, D., and Kaelin, W. G., Jr. (1996). A conserved region of unknown function participates in the recognition of E2F family members by the adenovirus E4 ORF 6/7 protein. Virology 220, 78-90.

Kang, H., Cui, K., and Zhao, K. (2004). BRG1 controls the activity of the retinoblastoma protein via regulation of p21CIP1/WAF1/SDI. Mol Cell Biol 24, 1188-1199.

Knudson, A. G., Jr. (1971). Mutation and cancer: statistical study of retinoblastoma. Proc Natl Acad Sci USA 68, 820-823.

Korenjak, M., Taylor-Harding, B., Binne, U. K., Satterlee, J. S., Stevaux, O., Aasland, R., White-Cooper, H., Dyson, N., and Brehm, A. (2004). Native E2F/RBF complexes contain Myb-interacting proteins and repress transcription of developmentally controlled E2F target genes. Cell 119, 181-193.

Krek, W., Ewen, M. E., Shirodkar, S., Arany, Z., Kaelin, W. G., Jr., and Livingston, D. M. (1994). Negative regulation of the growth-promoting transcription factor E2F-1 by a stably bound cyclin A-dependent protein kinase. Cell 78, 161-172.

Landsberg, R. L., Sero, J. E., Danielian, P. S., Yuan, T. L., Lee, E. Y., and Lees, J. A. (2003). The role of E2F4 in adipogenesis is independent of its cell cycle regulatory activity. Proc Natl Acad Sci USA 100, 2456-2461.

Lane, M. E., Sauer, K., Wallace, K., Jan, Y. N., Lehner, C. F., and Vaessin, H. (1996). Dacapo, a cyclin-dependent kinase inhibitor, stops cell proliferation during Drosophila development. Cell 87, 1225-1235.

Lee, W. H., Bookstein, R., Hong, F., Young, L. J., Shew, J. Y., and Lee, E. Y. (1987). Human retinoblastoma susceptibility gene: cloning, identification, and sequence. Science 235, 1394-1399.

Lin, W. C., Lin, F. T., and Nevins, J. R. (2001). Selective induction of E2F1 in response to DNA damage, mediated by

ATM-dependent phosphorylation. Genes Dev *15*, 1833-1844.

Liu, H., Dibling, B., Spike, B., Dirlam, A., and Macleod, K. (2004). New roles for the RB tumor suppressor protein. Curr Opin Genet Dev *14*, 55-64.

Martinez-Balbas, M. A., Bauer, U. M., Nielsen, S. J., Brehm, A., and Kouzarides, T. (2000). Regulation of E2F1 activity by acetylation. Embo J *19*, 662-671.

Morris, E. J., and Dyson, N. J. (2001). Retinoblastoma protein partners. Adv Cancer Res *82*, 1-54.

Muller, H., Bracken, A. P., Vernell, R., Moroni, M. C., Christians, F., Grassilli, E., Prosperini, E., Vigo, E., Oliner, J. D., and Helin, K. (2001). E2Fs regulate the expression of genes involved in differentiation, development, proliferation, and apoptosis. Genes Dev *15*, 267-285.

Muller, H., Moroni, M. C., Vigo, E., Petersen, B. O., Bartek, J., and Helin, K. (1997). Induction of S-phase entry by E2F transcription factors depends on their nuclear localization. Mol Cell Biol *17*, 5508-5520.

Mundle, S. D., and Saberwal, G. (2003). Evolving intricacies and implications of E2F1 regulation. Faseb J *17*, 569-574.

Munger, K. (2002). The role of human papillomaviruses in human cancers. Front Biosci *7*, d641-649.

Nielsen, S. J., Schneider, R., Bauer, U. M., Bannister, A. J., Morrison, A., O'Carroll, D., Firestein, R., Cleary, M., Jenuwein, T., Herrera, R. E., and Kouzarides, T. (2001). Rb targets histone H3 methylation and HP1 to promoters. Nature *412*, 561-565.

Ohtani, K., and Nevins, J. R. (1994). Functional properties of a *Drosophila* homolog of the E2F1 gene. Mol Cell Biol *14*, 1603-1612.

Polager, S., Kalma, Y., Berkovich, E., and Ginsberg, D. (2002). E2Fs up-regulate expression of genes involved in DNA replication, DNA repair and mitosis. Oncogene *21*, 437-446.

Ren, B., Cam, H., Takahashi, Y., Volkert, T., Terragni, J., Young, R. A., and Dynlacht, B. D. (2002). E2F integrates cell cycle progression with DNA repair, replication, and G(2)/M checkpoints. Genes Dev *16*, 245-256.

Royzman, I., Whittaker, A. J., and Orr-Weaver, T. L. (1997). Mutations in *Drosophila* DP and E2F distinguish G1-S progression from an associated transcriptional program. Genes Dev *11*, 1999-2011.

Sage, J., Miller, A. L., Perez-Mancera, P. A., Wysocki, J. M., and Jacks, T. (2003). Acute mutation of retinoblastoma gene function is sufficient for cell cycle re-entry. Nature *424*, 223-228.

Sawado, T., Yamaguchi, M., Nishimoto, Y., Ohno, K., Sakaguchi, K., and Matsukage, A. (1998). dE2F2, a novel E2F-family transcription factor in *Drosophila* melanogaster. Biochem Biophys Res Commun *251*, 409-415.

Spike, B. T., Dirlam, A., Dibling, B. C., Marvin, J., Williams, B. O., Jacks, T., and Macleod, K. F. (2004). The Rb tumor suppressor is required for stress erythropoiesis. Embo J *23*, 4319-4329.

Stevaux, O., Dimova, D., Frolov, M. V., Taylor-Harding, B., Morris, E., and Dyson, N. (2002). Distinct mechanisms of E2F regulation by *Drosophila* RBF1 and RBF2. Embo J *21*, 4927-4937.

Trimarchi, J. M., Fairchild, B., Verona, R., Moberg, K., Andon, N., and Lees, J. A. (1998). E2F-6, a member of the E2F family that can behave as a transcriptional repressor. Proc Natl Acad Sci USA *95*, 2850-2855.

Trimarchi, J. M., Fairchild, B., Wen, J., and Lees, J. A. (2001). The E2F6 transcription factor is a component of the mammalian Bmi1-containing polycomb complex. Proc Natl Acad Sci USA *98*, 1519-1524.

Verona, R., Moberg, K., Estes, S., Starz, M., Vernon, J. P., and Lees, J. A. (1997). E2F activity is regulated by cell cycle-dependent changes in subcellular localization. Mol Cell Biol *17*, 7268-7282.

Weintraub, S. J., Chow, K. N., Luo, R. X., Zhang, S. H., He, S., and Dean, D. C. (1995). Mechanism of active transcriptional repression by the retinoblastoma protein. Nature *375*, 812-815.

Weintraub, S. J., Prater, C. A., and Dean, D. C. (1992). Retinoblastoma protein switches the E2F site from positive to negative element. Nature *358*, 259-261.

Wells, J., Yan, P. S., Cechvala, M., Huang, T., and Farnham, P. J. (2003). Identification of novel pRb binding sites using CpG microarrays suggests that E2F recruits pRb to specific genomic sites during S phase. Oncogene *22*, 1445-1460.

Whyte, P., Buchkovich, K. J., Horowitz, J. M., Friend, S. H., Raybuck, M., Weinberg, R. A., and Harlow, E. (1988). Association between an oncogene and an anti-oncogene: the adenovirus E1A proteins bind to the retinoblastoma gene product. Nature *334*, 124-129.

Whyte, P., Williamson, N. M., and Harlow, E. (1989). Cellular targets for transformation by the adenovirus E1A proteins. Cell *56*, 67-75.

Wu, L., de Bruin, A., Saavedra, H. I., Starovic, M., Trimboli, A., Yang, Y., Opavska, J., Wilson, P., Thompson, J. C., Ostrowski, M. C., et al. (2003). Extra-embryonic function of Rb is essential for embryonic development and viability. Nature *421*, 942-947.

Wu, L., Timmers, C., Maiti, B., Saavedra, H. I., Sang, L., Chong, G. T., Nuckolls, F., Giangrande, P., Wright, F. A., Field, S. J., et al. (2001). The E2F1-3 transcription factors are essential for cellular proliferation. Nature *414*, 457-462.

Xin, S., Weng, L., Xu, J., and Du, W. (2002). The role of RBF in developmentally regulated cell proliferation in the eye disc and in Cyclin D/Cdk4 induced cellular growth. Development *129*, 1345-1356.

Zhang, H. S., Gavin, M., Dahiya, A., Postigo, A. A., Ma, D., Luo, R. X., Harbour, J. W., and Dean, D. C. (2000). Exit from G1 and S phase of the cell cycle is regulated by repressor complexes containing HDAC-Rb-hSWI/SNF and Rb-hSWI/SNF. Cell *101*, 79-89.

Zheng, N., Fraenkel, E., Pabo, C. O., and Pavletich, N. P. (1999). Structural basis of DNA recognition by the heterodimeric cell cycle transcription factor E2F-DP. Genes Dev *13*, 666-674.

Chapter 13

c-Jun: A Complex Tale of a Simple Transcription Factor

Ying Xia

Department of Environmental Health, University of Cincinnati, College of Medicine, 123 East Shields Street, Cincinnati, Ohio 45267-0056

Key Words: c-Jun, AP-1 complex, gene transcription, phosphorylation dimerization, DNA-binding

Summary

Since the discovery of its first member more than two decades ago, the AP-1 transcription complex has evolved into a paradigm for the transcription factors. The AP-1 regulate many aspects of cell physiology in response to growth factor signals and environmental insults, acting as master regulators of cell activities. c-Jun, the most extensively studied AP-1 protein, is involved in numerous cell functions, such as proliferation, apoptosis, survival, tumorigenesis, and tissue morphogenesis. Most of these activities rely on c-Jun function as a transcription factor, but its specificity in target gene selection and the intensity of its activity appear to be decided in a cell type or context-dependent manner. This article will focus on the mechanisms involved in c-Jun regulation, target gene expression and physiological functions, and will address the complex nature of this transcription factor.

The AP-1 Transcription Complex

AP-1 is a collective term for dimmers formed by proteins of the Jun (c-Jun, JunB and JunD), Fos (c-Fos, FosB, Fra1 and Fra2), activating transcription factor (ATF) (ATF2, ATF3/LRF1, B-ATF, JDP1 and JDP2) and musculoaponeurotic fibrosarcoma (MAF) (c-Maf, MafB, MafA, MafG/F/K and Nrl) families (Angel and Karin, 1991; Hai et al., 1988). These are structurally similar and functionally related basic leucine zipper (bZIP) proteins, forming homo- and/or hetero-dimers through the heptad repeat of leucine residues. Dimerization is essential for AP-1 function, because it brings together the two basic regions, constituting a contiguous DNA-contact interface that interacts with specific sequences in the major groove of one half-site on DNA (Ellenberger et al., 1992; Schumacher et al., 2000). Binding to the target sites allows the AP-1 to activate or repress genes in the nucleus and to function as fundamental transcription regulators.

This seemingly simple regulatory scheme, however, is entangled due to the complexity involved in the structure and regulation of each AP-1 protein. The relative abundance of a given AP-1 component in the cell is largely determined by its living environment, which controls its promoter activity, mRNA turnover and protein stability. Once expressed, the AP-1 component selectively binds to other available AP-1 proteins in its surroundings to form various protein dimers. The Jun proteins form homo- or hetero-dimers with members of Fos and ATF families. ATF, but not Fos, also form stable homodimers. While c-Maf and Nrl heterodimerize with c-Jun and c-Fos, other Maf-related proteins, including MafB, MafF, MafG and MafK, heterodimerizes with only Fos, but not Jun (Landschulz et al., 1988). Each AP-1 dimer may vary in affinity for binding to DNA with the consensus 5′-TGA(C/G)TCA-3′ sequence. The most well known AP-1 binding sites are the phorbol 12 O-tetradecanoate-13-acetate (TPA)-responsive element (TRE, 5′-TGAG/CTCA-3′) and the cAMP-responsive element (CRE, 5′-TGACGTCA-3′), while some dimers possess weak affinity for non-consensus sequences deviated from the traditional AP-1 sites. The Jun/ATF

Corresponding Author: Tel: (513) 558-0371, Fax: (513) 558-0974, E-mail: Xiay@email.uc.edu

dimers bind to the CRE sequence, Jun/Fos binds to TRE, while the Jun/Maf binds to both. The AP-1 dimers regulate diverse gene expression through binding to the AP-1 sites that are widely spread in gene promoter or enhancer regions (Karin *et al.*, 1997).

The rapid induction of AP-1 activity by a vast number of factors, including growth factors, cytokines, neurotransmitters, oncoproteins, cell-matrix interactions and a variety of physiological and chemical stresses is mainly attributed to the post-translational modification that alters the activities of AP-1 proteins in dimerization, DNA-binding and transactivation. This property allows the AP-1 to convert extracellular signals into gene expression events, which in turn control a plethora of cellular processes, including proliferation, survival, apoptosis and transformation. Although the complexity of AP-1 regulation and function is well appreciated, its functions rely largely on the specific roles for individual AP-1 protein.

c-Jun, A Component of AP-1

The Jun protein was originally identified in transformed cells carrying the genome of a replication defective avian sarcoma virus 17 (ASV 17) that directs the expression of a 65-kDa gag-jun fusion product, designated as v-Jun (Maki *et al.*, 1987; Bos *et al.*, 1988). Soon after, its homolog, the proto-oncogene c-Jun, was isolated from human and murine tissues (Bohmann *et al.*, 1987; Ryder *et al.*, 1988). c-Jun shares sequence similarity not only with v-jun, but also with the yeast transcription regulatory protein GCN4. GCN4 binds to a DNA element that is bound also by the mammalian AP-1, known at that time only as a transcription factor interacting with specific enhancer sequences (Haluska *et al.*, 1988; Short, 1987). The co-purification of c-Jun as the Fos-binding protein p39 from an AP-1 oligonucleotide affinity chromatography was the first indication that AP-1 was a heterodimeric transcription factor complex of c-Jun and Fos (Rauscher, III *et al.*, 1988b; Rauscher, III *et al.*, 1988a; Angel *et al.*, 1988a; Harshman *et al.*, 1988).

c-Jun and Fos are both basic-leucine zipper proteins. Unlike c-Jun, which can bind to AP-1 DNA site as a homodimer, Fos and other Fos family proteins only heterodimerize with Jun. Dimerization is determined by the "leucine zipper" domain; exchanging the leucine zipper in Fos with that of c-Jun generates a protein that forms a complex with Fos (Neuberg *et al.*, 1989; O'Shea *et al.*, 1989; Verma *et al.*, 1989). Comparing to Jun/Jun homodimer, the Jun/Fos heterodimer is a much more stable complex with enhanced DNA-binding and increased transcriptional activity (Chiu *et al.*, 1988; Halazonetis *et al.*, 1988; Kouzarides and Ziff, 1988; Cohen *et al.*, 1989; Zerial *et al.*, 1989; Angel *et al.*, 1989). Hence, although different dimers bind common target sites on DNA, they do not necessarily result in equivalent transcriptional responses. In fact, of the multiple protein complexes that form in a given cellular environment, some activate transcription while others that bind to the same site may repress it. For example, Jun B may dimerizes with Fos and prevent c-Jun/Fos dimer formation. Because JunB differs from c-Jun in amino acid sequences within the DNA-binding and dimerization motif, the JunB-Fos dimer has a much decreased DNA binding activity. In the presence of JunB, c-Jun mediated trans-activation and transformation are significantly attenuated (Chiu *et al.*, 1989; Schutte *et al.*, 1989; Deng and Karin, 1993). c-Jun can form dimers as well with other members of the Fos family, such as Fra-1 and Fra-2, and can dimerize with several bZIP proteins of the AP-1 family, including the ATF and Maf proteins (Fig.13.1) (Cohen *et al.*, 1989). Since different dimers have unique DNA binding specificity, dimerization with diverse partners greatly expands the regulatory potential of c-Jun (Chatton *et al.*, 1993; Hai and Curran, 1991; Kataoka *et al.*, 1994).

Transcriptional and Post-transcriptional Regulation of c-Jun Expression

When cells are exposed to the tumor promoter phorbol-ester 12-o-tetradecanoyl phorbol 13-acetate (TPA), growth factors or stress stimuli, there is a considerable enhancement in c-Jun/AP-1 DNA binding and transcription activity (Lee *et al.*, 1987; Sherman *et al.*, 1990; Lamph *et al.*, 1988; Ryder and Nathans, 1988; Quantin and Breathnach, 1988; Devary *et al.*, 1991; Pertovaara *et al.*, 1989). Enhanced AP-1 activity is accounted for in part by a rapid induction of *c-Jun* transcription. The newly synthesized *c-Jun* mRNA, however, decays quickly, with a short half-life that lasts for just 20-25 minutes. The *c-Jun* mRNA turnover is govern by an AU-rich RNA-destabilizing elements in the 3′ untranslated region, an intrinsic activity that does not appear to be tightly coupled to ongoing translation by ribosomes and is insensitive to blockage of transcription (Peng *et al.*, 1996). Exposure of cells to UV, MCSF and photodynamic therapy increases *c-Jun* mRNA stability considerably, possibly due to the activation of RNA binding proteins by these stimuli (Kick *et al.*, 1996; Blattner *et al.*, 2000; Nakamura *et al.*, 1991).

The stability of the c-Jun protein depends on its

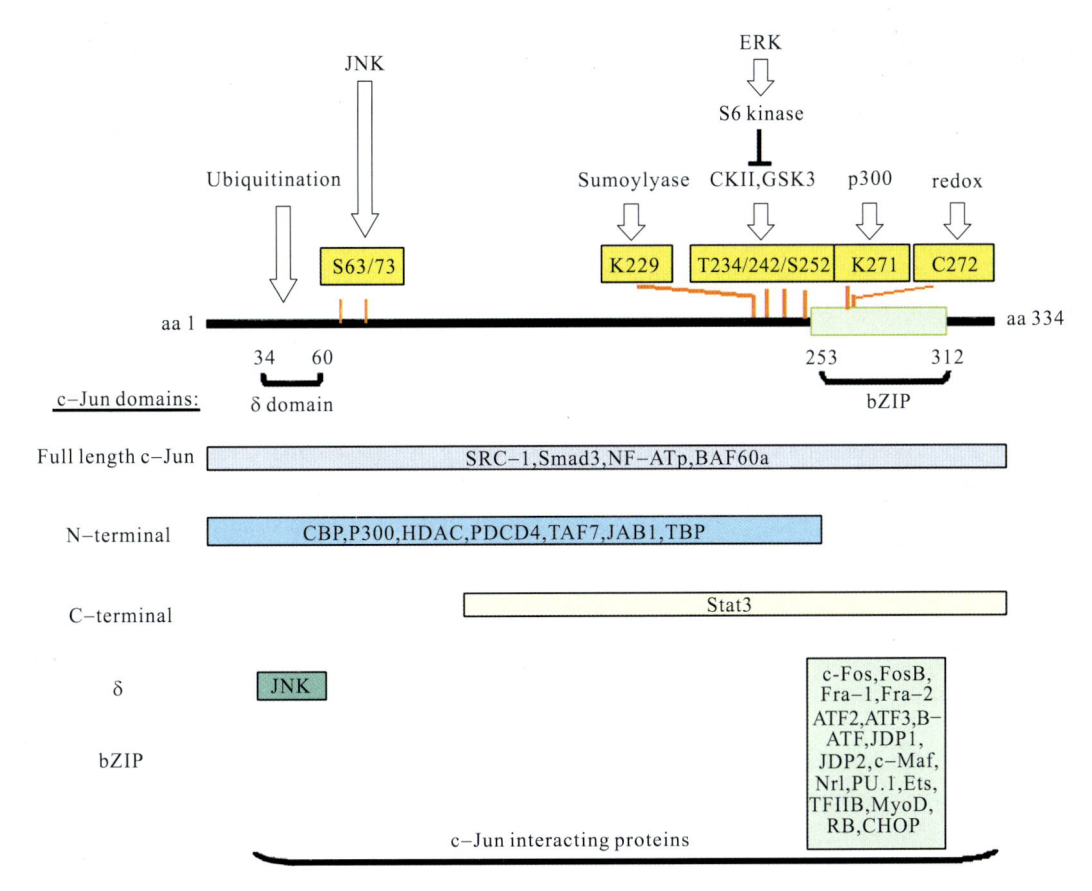

Fig.13.1 **c-Jun activity is affected by post-translational modification and protein-protein interactions.** Protein modification (red lines and yellow boxes) takes place at amino-terminal serines 63 and 73 phosphorylated by JNK; carboxy-terminal threonines 234 and 242, and serine 252 phosphorylated by CKII and GSK-3, whose activities are negatively regulated by the ERK-S6 kinase pathway; sumoylation at lysine 229; acetylation at lysine 271 by p300 and redox-mediated modification of cysteine 272. c-Jun contains two established domains: domain (aa 34-60) and bZIP domain (aa 253-312) (brackets). The domain mediates JNK binding and ubiquitination. A large number of proteins interact with c-Jun bZIP domain, while some are only known to interact with N-terminal, C-terminal, or the full-length of c-Jun.

modification by ubiquitination and sumoylation. While sumoylation targets lysine-229 (Muller et al., 2000), ubiquitination is determined by the cis-acting signal provided in the delta domain (Fig. 13.1). Lacking this domain in v-Jun, allows escape from ubiquitin-dependent degradation and is responsible for the extended stability and longer half-life of v-Jun than c-Jun (Treier et al., 1994). The fact that c-Jun expression is superinduced in the presence of protein synthesis inhibitors suggests that *de novo* synthesized proteins are involved in its degradation (Lamph et al., 1988).

Induction of *c-Jun* mRNA by extracellular stimuli is attributed mostly to the activation of its promoter. Besides the binding elements for NF-κB, Sp1 and CCAAT-binding transcription factors, the *c-Jun* promoter contains a high-affinity AP-1 binding site that is essential for promoter activation. Site-specific mutagenesis of this binding site prevents induction of *c-Jun* transcription, suggesting the existence of a positive regulatory loop, in which *c-Jun* transcription is directly stimulated by its own gene products (Lamph et al., 1988; Angel et al., 1988b; Nakamura et al., 1991). Such autocrine regulation explains how transient TPA treatment or growth factor signals can trigger the prolonged activation of the *c-Jun* promoter for a biological effect to take place (Angel et al., 1988b). The initiation of the activation loop is independent of *de novo* protein synthesis; rather it is originated from the pre-existing c-Jun that is activated by signal-induced protein modifications that is most commonly the result of changes in the phosphorylation state of c-Jun (Lamph et al., 1988; Bohmann, 1990; Papavassiliou et al., 1992). Phosphorylation affects c-Jun DNA-binding affinity and modulates the activities of its transcriptional activation domain. In resting cells, c-Jun is phosphorylated on serine and threonine at five sites (Boyle et al., 1991; Baker et al., 1992). Three phosphorylation sites are located just upstream of the basic region in the DNA-binding domain (residues 227-252) and two sites are within the N-terminal transcription activation domain at serines 63 and 73 (Fig. 13.1). When cells are exposed to AP-1 activators, such as TPA, c-Jun undergoes C-terminal domain dephosphorylation that increases in DNA binding activity and a significant increase in N-terminal phosphorylation that enhances its transcriptional activity (Smeal et al., 1991; Smeal et al., 1992; Lin et al., 1992; Nikolakaki et al., 1993). Lack of serine-63 and 73 in Jun B accounts for its irresponsiveness to TPA induced transcriptional activation (Franklin et al., 1992).

Less well-understood modifications are induced by changes of the cellular oxidative status. These changes lead to modulation of the redox state of a conserved cysteine residue in the DNA-binding domains of the Fos and Jun proteins that affect the AP-1 heterodimer DNA binding activity (Fig. 13.1) (Frame et al., 1991; Abate et al., 1990). The redox state of the cell represents the precise balance between the levels of oxidizing and reducing equivalents (Matsuzawa and Ichijo, 2005). Because redox signaling is commonly modulated in response to alterations of both external and internal environments, the resulting c-Jun modification may actually have quite a broad implication in modulation of cell function.

The Transcriptional Activity of c-Jun

Transcription initiation requires the assembly of a preinitiation complex, involving a sequential recruitment of the basal transcriptional machinery, TFIIB-TFIID-TFIIA-UpSA complex, co-activators and the RNA polymerase II holoenzymes (Kim et al., 1998; Edelstein et al., 2003). N-terminal phosphorylation of c-Jun at serines 63 and 73 allows it to interact with CREB-binding protein (CBP), a histone acetyltransferase (HAT) known to hyperacetylate N-terminal histone tails to facilitate chromatin opening. Mutation of c-Jun serines 63 and 73 to alanines (Jun 63/73 AA) prevents CBP binding and severely reduces its transcriptional ability (Bannister et al., 1995; Goldman et al., 1997). Other co-activators that act in a similar way are BAF60a, a component of the SWI/SNF chromatin remodeling complex, p300 protein, a homolog of CBP, and steroid receptor coactivator-1 (SRC-1), all of which interact with c-Jun and potentiate AP-1-mediated transactivaton (Fig.13.1) (Ito et al., 2001; Lee et al., 1996; Lee et al., 1998). These co-activators, however, do not necessarily operate alone and are not mutually exclusive, because SRC-1-stimulated AP-1 activity can be further enhanced by p300 (Lee et al., 1998). Conversely, c-Jun phosphorylation can trigger its dissociation from an inhibitory complex of histone deacetylase 3 (HDAC3), suggesting that c-Jun-mediated transcription activity can also be induced by a de-repression mechanism (Weiss et al., 2003).

The p300 protein, originally cloned by virtue of its interaction with the adenovirus E1A protein, causes acetylation not only of histones, but also of c-Jun at Lys271 (Eckner et al., 1994; Vries et al., 2001). By binding to and sequestrating p300 from c-Jun, E1A represses AP1-mediated promoter activity in adenovirus infected cells (Lee et al., 1996; Hagmeyer et al., 1995). Correspondingly, the tumor suppressor PDCD4 suppresses AP-1 activity by binding to c-Jun and preventing its subsequent association with p300 (Bitomsky et al., 2004).

Once the chromatin structure is opened, c-Jun helps recruit the proteins involved in basal transcriptional machinery, of which both TATA-binding protein (TBP) and TBP-associated factors (TAFs), TAF7, bind to the c-Jun N-terminal activation domain, whereas TFIIB binds to the c-Jun bZIP domain (Fig.13.1) (Franklin et al., 1995; Munz et al., 2003). TAF7 preferentially binds to the DNA-bound and phosphorylated c-Jun, explaining the transcription activation directed only by the activated c-Jun. What likely to happen is that extracellular stimuli-induced signal transduction pathways lead to N-terminal phosphorylation of c-Jun and that only the hyperphosphorylated c-Jun is recognized by co-activators with histone acetyltransferase activity, like p300/CBP, for the task of opening the local chromatin structure. Subsequent interaction of c-Jun with TBP, TAF7 and TFIIB stabilizes the transcription complex and enables more efficient initiation of transcription processes. Such inducible functional interactions with co-factors allow c-Jun to function as a bridge between extracellular signals and the basal transcription machinery to activate specific promoters (Munz et al., 2003).

Crosstalk of c-Jun with Sequence Specific Transcription Factors

c-Jun does not usually act alone, rather, it physically interacts with other transcription factors, allowing signal integration on promoter DNA for a combinatorial transcriptional regulation of gene expression. Interaction of this sort expands the diversity of c-Jun effects on gene regulation. For instance, it can bring c-Jun to a non-AP-1 site to regulate gene expression. c-Jun activates the monocyte-specific macrophage colony-stimulating factor (M-CSF) receptor promoter, which contains DNA binding sites for only PU.1, but not AP-1. c-Jun is recruited to the PU.1 binding site in the M-CSF receptor promoter through binding to the Ets domain of PU.1 for a transactivation effect (Behre et al., 1999). Binding to c-Jun does not always lead to transcription activation, as the c-Jun/MyoD complex suppresses the MyoD promoter and inhibits myogenesis (Bengal et al., 1992) and the c-Jun/Smad complex acts as a co-repressor for Smad3-mediated transactivation of TGF β responsive genes (Verrecchia et al., 2001; Dennler et al., 2000; Atfi et al., 1997). The tumor necrosis factor α (TNFα) and the human T-cell leukemia viral oncoprotein Tax agonize TGF β signaling, an effect likely resulting from the induction of c-Jun expression by TNFα and the increase of c-Jun N-terminal phosphorylation by Tax, both leading to c-Jun/Smad complex formation and abrogation of Smad binding to its target site for gene transcription (Arnulf et al., 2002).

Conversely, recruitment of Smad by c-Jun to the AP-1 binding sites synergistically activates AP-1-dependent promoters (Verrecchia et al., 2001). Apparently only some c-Jun binding proteins, including retinoblastoma (RB) protein and transcription factor CHOP, act as transcription co-activators on the AP-1 sites (Nead et al., 1998; Ubeda et al., 1999), while others act as co-repressors, such as MyoD and c-Jun dimerization protein 2 (JDP2) (Bengal et al., 1992; Heinrich et al., 2004). The c-Jun activation domain binding protein, JAB1, was originally identified as a c-Jun co-activator (Claret et al., 1996), as well as a component of the constitutive photomorphogenesis (COP9) signalosome. A fraction of JAB1 is associated with the LFA-1 integrin at the plasma membrane. Induction of JAB1 nuclear translocation by the LFA-1 signal allows its association with c-Jun to potentiate AP-1-dependent promoter activity (Bianchi et al., 2000). JAB1 also brings to c-Jun the COP9 signalosome, consisting of protein kinase CK2 (CK2) and protein kinase D (PKD) that can phosphorylate c-Jun and block c-Jun-Ub conjugates and degradation (Uhle et al., 2003; Muller et al., 2000). Hence, proteins like JAB1 may act through multiple channels to activate c-Jun-mediated transcription.

Gene promoters usually contain binding sequences for several transcription factors, such that promoter activity is determined by these factors acting in synergy. This appears to be the case for the *tissue inhibitor of metalloproteinases-1* (*Timp-1*) promoter, containing adjacent Ets and AP-1 binding sites. The optimal promoter activity requires the binding of a trimolecular complex, consisting c-Jun, Ets and Fos, which is particularly efficient in DNA binding and synergistically activating transcription (Bassuk and Leiden, 1995; Logan et al., 1996). The optimal activation of the *osteopontin* (*Opn*) promoter derives from the interaction and cooperation of beta-catenin/Lef-1, Ets, and AP-1 transcription factors (El Tanani et al., 2004). Likewise, interactions of c-Jun and NFATp or c-Jun and Stat3 are responsible for the maximal transcriptional activity of specific promoters containing closely adjacent binding sequences for respective transcription factors (Zhang et al., 1999a; Alroy et al., 1995). Hence, association with other transcription factors provides ample explanation for the diversity, the intensity and the specificity exerted by c-Jun that sometimes activates and others suppresses gene transcription.

Signaling Mechanisms Involved in c-Jun Phosphorylation

Transmission of the extracellular signals through the cytoplasm is mediated by cascades of protein kinases, leading to changes of the transcription factor phosphorylation state and modulation of transcription activity (Karin, 1991; Bohmann, 1990). Glycogen synthase kinase 3 (GSK-3) and CKII can phosphorylate c-Jun at its C-termini, keeping c-Jun in a non DNA-binding state (Boyle et al., 1991), while extracellular signal regulated kinase (ERK)-mediated activation of p70 S6 kinase phosphorylates GSK-3 at serine-21 and inactivates it (Sutherland et al., 1994). It is therefore possible that ERK activation leads to c-Jun DNA binding activity through p70 S6 kinase-mediated GSK-3 inactivation and c-Jun C-terminal dephosphorylation (Fig.13.1). A connection between ERK and c-Jun has been suggested by mutation studies in fission yeast, of which homologs of mammalian ERK and Jun act in concert to control yeast cell elongation immediately after division (Toda et al., 1991). The ERK, however, does not seem to be involved in modulating the c-Jun N-terminal phosphorylation, which must be catalyzed by other protein kinases (Westwick et al., 1994).

The c-Jun amino-terminal kinases (JNKs) have been identified based on their activation by UV and oncoproteins. Like the ERKs, the JNKs are proline-directed kinases with optimal sequence Pro-Xaa-Ser/Thr-Pro for phosphorylation; while unlike the ERKs, the JNKs have distinct substrate specificity, being unable to phosphorylate pp90rsk but more efficient in phosphorylating the c-Jun transactivation domain (Fig.13.1). JNK phosphorylates c-Jun at serines 63 and 73 and phosphorylation requires binding of JNK to a specific region within the c-Jun transactivation domain (Hibi *et al.*, 1993; Derijard *et al.*, 1994; Kyriakis *et al.*, 1994). This feature explains why v-Jun, having completely conserved Ser-63 to Ser-73, but lacking the JNK binding domain (amino acids 34-60), is resistant to TPA-induced N-terminal phosphorylation (Adler *et al.*, 1992).

The MAP Kinase Cascade

The JNKs and the ERKs, together with the later discovered p38s, constitute three separate groups of Mitogen-Activated Protein Kinases (MAPKs), which are themselves activated through concomitant phosphorylation on tyrosine and threonine residues in the Thr-Xxx-Tyr motif (Fig.13.2). For each MAPK group, the phosphorylation is catalyzed by the designated MAP kinase kinases (MAPKKs), a novel class of dual specific protein kinases (Mordret, 1993). The connection between MAPKK to the MAPK is fairly specific, as the MEK4 and MEK7 activate the JNKs, the MEK1 and MEK2 activate the ERKs, and the MEK3 and MEK6 activate the p38s (Johnson and Lapadat, 2002). The activities of the MAPKKs are turned on by serine/threonine phosphorylation catalyzed by their immediate upstream kinases, the MAPKK kinases (MAPKKKs) (Davis, 1994). A rapidly increasing number of protein kinases are categorized as MAPKKKs, including Rafs, MEK kinase 1-4 (MEKKs), germinal center kinase (GCK), mixed lineage kinases (MLK), apoptosis-stimulated kinase 1(ASK1), tumor progression locus 2 (TPL2), and TGF-beta-activated kinase (TAK). Having rather distinct regulatory motifs for connection to unique input signals, but relatively conserved kinase domains for activating the MAPKK-MAPK cascade, the MAPKKKs appear to be exactly what is needed for conferring numerous signals to MAPK activation (Schlesinger *et al.*, 1998). The complex role for MAPKKK in cell signaling is best illustrated by studies on the regulation and function of MEKK1 (Xia and Karin, 2004).

MEKK1 was originally identified as a mammalian homolog of the yeast MEK kinases, Byr2 and Ste11, involved in pheromone-induced mating (Lange-Carter *et al.*, 1993). Although initially considered as an upstream activator of ERK, it soon becomes clear that MEKK1 preferentially activate the JNK pathway, by interacting with and phosphorylating the JNK activator, MEK4 (Minden *et al.*, 1994; Yan *et al.*, 1994; Lin *et al.*, 1995; Xia *et al.*, 1998). MEKK1 is in turn activated by various upstream signals, and intriguingly, each signal appears to be coupled to MEKK1 by a distinct mechanism. The diversity of MEKK1 in cell signaling is attributed, in part, to its N-terminal regulatory domain, which can interact with various upstream regulators to establish a connection (Fig. 13.3). Binding to the SH3 domain of Grb2 through this region allows the recruitment of the MEKK1/Grb2 complex by the Shc proteins to the activated EGF receptor and a transient plasma membrane localization of MEKK1 (Pomerance *et al.*,1998). The membrane localized MEKK1 undergoes conformational changes to initiate autoactivation, by which MEKK1 phosphorylates its own threonine residues between the kinase subdomains VII and VIII (Deak and Templeton, 1997; Siow *et al.*, 1997). The N-terminal domain of MEKK1 also offers binding sites for RhoA, JNK and alpha-actinin, while its C-terminal kinase domain binds MEK4 (Gallagher *et al.*, 2004; Christerson *et al.*, 1999; Xu *et al.*, 1996; Xia *et al.*, 1998).

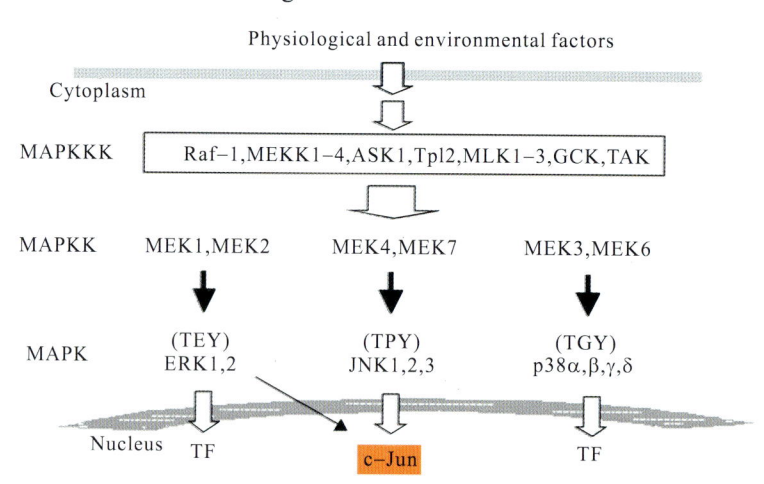

Fig. 13.2 Signal transduction through the MAPK cascades that lead to c-Jun activation. Extracellular signals are transduced by various MAPKKKs that lead to the phosphorylation and activation of the MAPKKs. The MAPKKs in turn phosphorylate the MAPKs at threonine and tyrosine residues in the TXY motif, as TEY for ERK, TPY for JNK and TGY for p38. Phosphorylation results in MAPK activation. Through different mechanisms, JNK and ERK can modulate c-Jun phosphorylation and activity.

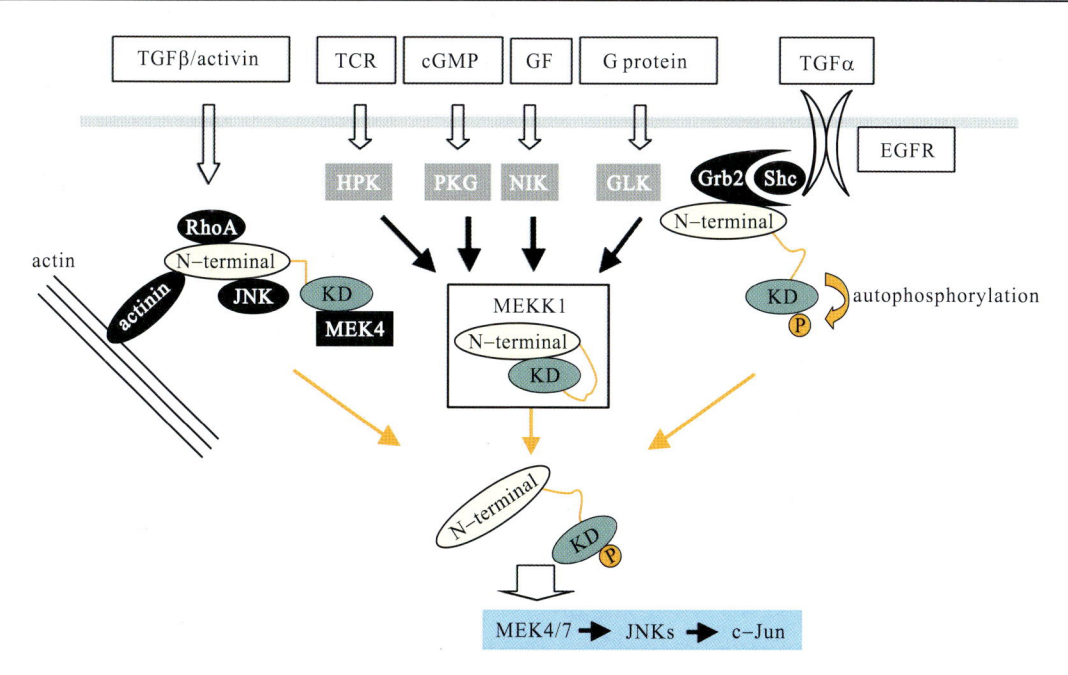

Fig. 13.3 **Molecular regulation of MEKK1.** MEKK1 exists in the cytoplasm in an inactive form. Its activity is induced by various extra- or intracellular signals that activate the MEKK1 N-terminal domain-associated upstream regulators, including HPK, PKG, NIK, GLK, leading to MEKK1 phosphorylation and activation. Alternatively, the N-terminal domain of MEKK1 can interact with Grb2 upon EGF stimulation, which causes MEKK1 membrane localization, conformational change and autophosphorylation. Simultaneous interaction of MEKK1 with RhoA, JNK, actinin and MEK4 allows efficient signal transduction to regulate actin cytoskeleton, a likely pathway used by TGFβ to induce actin stress fiber formation.

Such scaffolding function of MEKK1 organizes a signaling complex for the RhoA signal efficiently transmitting to MEKK1 and to the downstream MEK4-JNK pathway that is linked to the actin cytoskeleton. This pathway is likely to be utilized by TGFβ, which activates RhoA, to control c-Jun phosphorylation and the organization of actin cytoskeleton (Gallagher *et al.*, 2004; Atfi *et al.*, 1997; Zhang *et al.*, 2005). Through interaction with axin, MEKK1 may mediate Wnt signal in JNK activation and planar polarity determination (Zhang *et al.*, 1999b). Alternatively, several protein kinases, including HPK, NIK, PKG and GLK, have been shown to directly associate with and phosphorylate MEKK1 *in vitro*, acting as the possible upstream activators for MEKK1 (Su *et al.*, 1997; Diener *et al.*, 1997; Soh *et al.*, 2001). The complexity of MEKK1 regulation provides a promising angle to explain signal integration and segregation that takes place on MAPK activation. As far as the c-Jun phosphorylation is concerned, each MAPKKK may be responsible for transducing a subset of signals to the activation of the MAPK-c-Jun, whereas, a particular signal-mediated by MAPKKKs can lead to the activation of not only JNK-c-Jun, but also other downstream effectors.

With this idea in mind, it is not surprising to find that each MAPKKK knockout mice have rather distinct phenotypes. Like the *Jnk1(-/-)Jnk2(-/-)* and *c-Jun(-/-)* mice that are embryonic lethal, some MAPKKK knockout mice, including *Raf-1(-/-)*, *Mekk3(-/-)*, and *Mekk4(-/-)*, die in embryogenesis, but by different reasons (Kuan CY, 1999; Johnson, R.S, 1993). The *Raf-1(-/-)* fetuses show vascular defects in the yolk sac and placenta as well as increased apoptosis of embryonic tissues (Huser *et al.*, 2001), the *Mekk3(-/-)* embryos die due to impaired blood vessels development (Yang *et al.*, 2000), while the *Mekk4(-/-)* mice die from neural tube defects that are associated with massively elevated apoptosis before and during neural tube closure (Chi *et al.*, 2005). Conversely, some MAPKKKs, such as ASK1, Tpl2 and MEKK2, are dispensable for embryonic development, as their knockout mice are all born alive with no overt developmental defects. Only under certain pathological or environmental conditions, these MAPKKK are required for a cell type specific response. The *Ask1(-/-)* embryonic fibroblasts are resistant to apoptosis induced by TNFα and H_2O_2 (Tobiume *et al.*, 2001), the *Tpl2 (-/-)* mice produce low levels of TNFα when exposed to lipopolysaccharide (LPS) (Dumitru *et al.*, 2000), and the *Mekk2(-/-)* T cells show increased proliferation and susceptibility to apoptosis induced by T cell receptor cross-linking (Guo *et al.*, 2002). The MEKK1-deficient mice also survive embryonic development, but show a specific defect on embryonic eyelid closure, a developmental process resulting from eyelid epidermal cell migration and movement (Xia et al., 2000; Yujiri et al., 2000). Correspondingly, the MEKK1-deficient epithelial cells are impaired in TGFβ/activin-induced migration. What makes it more interesting is that the activin βB knockout mice are also impaired in embryonic eyelid closure (Schrewe et al., 1994). Hence, MEKK1 plays a specific role in transducing the activin signals that lead to epithelial cell migration and eyelid morphogenesis (Zhang et al., 2003).

The activation of MEKK1 by activins in turn activates JNK and p38, leading to c-Jun N-terminal phosphorylation. The c-Jun phosphorylation, however, does not appear to be necessary for eyelid closure, because the c-Jun 63/73 AA transgenic mice display normal eyelid development (Behrens et al., 2000). It turns out that besides c-Jun activation, the MEKK1-JNK pathway also regulates actin cytoskeleton re-organization, crucial for epithelial cell migration (Zhang et al., 2003; Zhang et al., 2005)(Fig. 13.4). Besides the TGFβ/activin signals, eyelid closure is controlled by the TGFα signal that activates the EGR receptor, leading to a MEKK1-independent activation of the ERK pathway (Berkowitz et al., 1996; Zhang et al., 2003). This pathway is controlled by the transcription factor c-Jun, which regulates the expression of EGFR or its ligand HB-EGF (Grose, 2003) (Fig.13.4). For a developmental process as simple as embryonic eyelid closure, complex signaling pathways are involved in its regulation. Not only does it require the TGFβ/activin signal that activates the MEKK1-JNK pathway, leading to c-Jun activation, but it also needs a c-Jun-dependent activation of the TGFα/EGFR signal that activates the ERK pathway. c-Jun is apparently playing a crucial role in integrating the signals derived from these two pathways for eyelid closure. c-Jun activity is controlled by a complex regulatory machinery involving the integration of various signals, the regulation of c-Jun expression and activity and the interaction with other specific and general transcription factors. It is therefore not surprising that c-Jun is involved in the control of diverse cell activities that are not only critical for tissue development, but also for a wide variety of biological outcomes, such as proliferation, apoptosis and tumorigenesis.

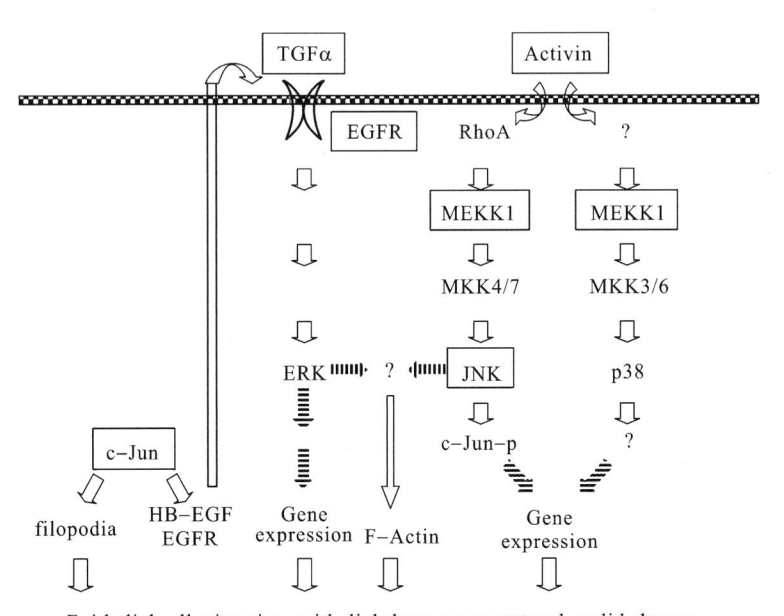

Fig. 13.4 **The molecular pathways in the control of mouse epithelial sheet movement and eyelid closure.** In the eyelid epithelium, at least two pathways are critical for actin polymerization and epithelial sheet movement: (1) the c-Jun controlled expression of HB-EGF and EGFR, leading to activation of the ERK pathway; and (2) the TGFβ/activin-induced MEKK1-JNK pathway, which induces actin polymerization and c-Jun phosphorylation. Both pathways are essential for eyelid closure and they may be connected through transcription factor c-Jun. c-Jun regulated filopodia might be of importance for eyelid fusion. The signaling factors whose ablation or perturbation lead to EOB in mice are boxed, the unknown factors in these pathways are denoted as question marks, the open arrows indicate the pathways or cell functions demonstrated by experimental data and the shaded arrows represent hypothetic pathways that are yet to be established.

Cell Proliferation

The very first clue for the involvement of c-Jun in cell growth stems from the observation that c-Jun expression takes place in quiescent cells at the G0 to G1 transition (Ryseck *et al.*, 1988) and upon exposure to mitogenic signals at the G1 to S transition (Carter *et al.*, 1994; Mayo *et al.*, 1994). A more clear indication derives from studies in c-Jun-null fibroblasts and in erythroleukemia cells expressing c-Jun antisense sequences. Both cells display greatly reduced growth rates and blockage of G1-to-S-phase cell cycle progression (Johnson *et al.*, 1993; Smith and Prochownik, 1992; Schreiber *et al.*, 1999). In addition to its expression, c-Jun activity also fluctuates in a cell cycle dependent manner, possibly subjected to regulation by cyclin-dependent kinase (cdk) inhibitor, p27(Kip1). In quiescent cells, p27(Kip1) is phosphorylated at serine 10 and the phosphorylated p27(Kip) binds to Jab1, preventing Jab1-c-Jun interaction. Upon quiescence exit, p27(Kip1) dephosphorylates, releasing Jab1 from its suppression state to activate c-Jun (Chopra *et al.*, 2002). Once activated, c-Jun must turn on gene expression programs crucial for cell progression through the G1-S checkpoint.

One such gene regulated by c-Jun is *cyclin D1*. Having an AP-1 binding site in its promoter, *cyclin D1* transcripton is directly induced by c-Jun, which recruits p300 for transcription activation, but is suppressed by adenovirus E1A, which competes for p300 (Wisdom *et al.*, 1999; Albanese *et al.*, 1999). Cyclin D1 in turn forms complex with cdks for RB phosphorylation. The phosphorylated RB releases from E2F, allowing E2F to activate genes that enable cells to advance into late G_1 and S phases (Grana and Reddy, 1995; Zetterberg *et al.*, 1995). Another cell cycle regulator is the tumor suppressor p53, which is known to inhibit cell cycle progression by transcriptional activation of p21, the inhibitor of cyclin/cdks (Levine, 1997; Agarwal *et al.*, 1995). p53 null fibroblasts have accelerated proliferation rate (Harvey *et al.*, 1993), in contrast, c-Jun null cells show growth defects. It turns out that c-Jun is a negative regulator for the transcription of p53 and p21, an effect mediated by direct binding of c-Jun to a variant AP-1 site in the *p53* promoter or indirectly through a SP-1 site in the *p21* promoter (Wang *et al.*, 2000). In the absence of c-Jun, an elevated expression of p53 and its target gene *p21* leads to growth arrest (Schreiber *et al.*, 1999).

Despite the crucial role of c-Jun in cell proliferation, c-Jun null embryos survive until mid-gestation without overt abnormalities before death and c-Jun-null embryonic stem cells grow normally in culture and contribute to most tissues in chimeric mice (Hilberg *et al.*, 1993; Hilberg and Wagner, 1992; Johnson *et al.*, 1993). Hence, the function of c-Jun in cell proliferation is likely cell type specific, and it is required only at late developmental stages or under certain environmental conditions, such as following UV irradiation, to trigger cell cycle re-entry (Shaulian and Karin, 2002).

Apoptotic Cell Death

The pro-apoptotic function of c-Jun was first observed in neuronal cells, from which nerve growth factor (NGF) withdrawal caused cell death, correlating with an induction of c-Jun expression and an increase of AP-1 DNA binding activity (Ham *et al.*, 1995). Expression of a dominant negative c-Jun mutant protects these cells from death. Apoptosis is an intrinsic program to activate caspase-mediated proteolysis pathways to eliminate cells that have suffered serious damage, a process required for normal development and homeostasis maintenance. Apoptotic cells have a unique morphological pattern characterized by chromatin condensation, membrane blebbing and DNA fragmentation. The property of c-Jun of being involved in both cell proliferation and apoptosis is in fact common to a number of oncogene products, including Ras, c-Myc and E2F (Sears and Nevins, 2002). These proteins, originally identified as positive regulators of cell growth, were subsequently found to be involved in apoptosis. It actually makes sense for cells to employ an overlapping system to regulate cell growth and death, as it provides the most convenient way to switch cell functions in response to environmental changes. c-Jun is apparently part of the program for life-death decisions of neuronal cells depending on the presence or absence of NGF.

Neurotoxicity is mediated by JNK phosphorylation of c-Jun (Xia *et al.*, 1995). Supporting this view is the finding that mice deficient in the neural specific JNK3 and mice of which the endogenous Jun is replaced by c-Jun 63/73 AA are protected from excitatory amino acid kainite induced neuronal cell apoptosis (Yang *et al.*, 1997; Behrens *et al.*, 1999). We now know that the JNK-c-Jun pathway is widely implicated in apoptosis induced by many stress stimuli, including DNA damaging agents, microtubule inhibitors, protein synthesis inhibitors, cytokines and lipid mediators; its effect is not limited to the neuronal cells, but also applies to other cell types, such as T cells, macrophages and epithelial cells (Chen *et al.*, 1996; Park *et al.*, 1997; Mosieniak *et al.*, 1997; Verheij *et al.*, 1996).

Apoptosis requires the participation of both the transactivation domain and the bZIP domain of c-Jun and is prevented by a c-Jun N-terminal truncation mutant TAM67, c-Jun antisense oligonucleotides, and c-Jun genetic knockout, suggesting that c-Jun transcriptional activities are needed for the expression of apoptosis effectors (Fan et al., 2001; Sawai et al., 1995; Shaulian et al., 2000). Candidate effectors include FasL, CD95L and the death receptor 4(DR4), all having AP-1 binding sites in their gene promoters and being induced by a number of stress stimuli in a JNK dependent manner (Eichhorst et al., 2000; Faris et al., 1998; Guan et al., 2002). Their induction initiates apoptosis by the death-receptor mediated pathways. Another effector is BIM, a member of the proapoptotic BCL-2 family, whose expression is induced by NGF withdrawal in a c-Jun dependent manner (Toh et al., 2004). BIM can in turn be phosphorylated by JNK for a further enhancement of the proapoptotic activity, allowing signal amplification that activates the mitochondria death pathway (Putcha et al., 2003).

Not all death response requires c-Jun and gene transcription. JNK can potentiate apoptosis by transcription independent modulation of the 14-3-3 protein, whose unphosporylated form serves as a cytoplasmic anchor for c-Abl and BAX (Yoshida et al., 2005; Tsuruta et al., 2004). Phosphorylation of 14-3-3 by JNK triggers the release of c-Abl to the nucleus to activate apoptotic genes and of BAX to mitochondria to induce cytochrome c release (Tournier et al., 2000; Koo et al., 2002). The multiple facets of JNK in apoptosis through both c-Jun dependent and independent mechanisms explain its unique role in the apoptotic response during tissue development. Developmental defects associated with JNK inactivation, including dysregulation of brain cell apoptosis in Jnk1(-/-)Jnk2(-/-) mice and reduction of double positive thymocyte death in dominant negative JNK transgenic mice, are not observed in mice of which c-Jun activity is compromised, such as TAM-67 or c-Jun 63/73 AA transgenic mice (King et al., 1999; Behrens et al., 1999; Kuan et al., 1999; Rincon et al., 1998).

Interestingly, c-Jun appears also to have anti-apoptotic functions, probably by the induction of the anti-apoptotic factor BCL3 (Rebollo et al., 2000). c-Jun mutant fetuses harbor increased numbers of apoptotic cells in liver (Eferl et al., 1999) and c-Jun-null primary embryonic fibroblasts show enhanced sensitivity to UV-induced apoptosis (Wisdom et al., 1999). Correspondingly, suppression of AP-1 activity by the glucocorticoid receptor induces leukemia cell apoptosis (Helmberg et al., 1995; Chen et al., 1996). The observations that AP-1 is involved in Ras-induced survival of only RB-null fibroblasts (Young and Longmore, 2004) and that JNK potentiates the growth of only p53-deficient tumor cells (Potapova et al., 2000), suggest that the cross talk to the tumor suppressor pathway may have a lot to do with life and death decision made by the JNK-c-Jun pathway. An example of such cross-talk is the rapid and sustained activation of ASK1 and JNK, which leads to c-Jun phosphorylation, when rhabdomyosarcoma cells are subjected to rapamycin or amino acid deprivation. In the absence of p53, this pathway leads to apoptosis, while in the presence of wild type p53 or p21(Cip), cells arrest in G1 and remain viable (Huang et al., 2003). The apparent contradictory roles for c-Jun in p53-dependent survival and apoptosis, together with the previously discussed role for c-Jun in the regulation of p53 expression and cell cycle progression, strongly indicate that c-Jun does not act alone and that its function in cell fate determination must be subjected to the regulation by a complex system in a cell type and stimuli specific fashion (Potapova et al., 2000; Huang et al., 2003; Schreiber et al., 1999; Shaulian and Karin, 2002).

How are the apoptotic signals transduced to JNK? NGF deprivation signals are known to be organized by a protein called POSH, which stands for plenty of SH3s. POSH acts as a scaffold for the formation of a complex consisting of the activated Rac1/Cdc42, MLKs, MEK4/7, JNK and c-Jun, allowing the sequential activation of the signaling cascade that leads to neuronal death (Xu et al., 2003). Although it has been suggested that the TNFα signal is mediated by ASK1 and the DNA-damaging signals are transduced by BRCA1 (Ichijo et al., 1997; Harkin et al., 1999), the precise mechanisms involved in JNK activation during these signaling processes are still unknown.

Tumorigenesis

Tumorigenesis is a multistep process involving cell transformation, invasive growth, angiogenesis and tumor spread to distant sites (Hanahan and Weinberg, 2000). The ability of the retroviral oncogene v-Jun to transform quail embryo fibroblasts lead to the discovery of the oncogenic potential of its cellular homologue c-Jun (Bader et al., 2000; Vogt and Bos, 1989). Retroviral-mediated chronic expression of c-Jun in fibroblasts causes phenotypic changes characteristic of transformation, including sustained growth in low serum medium, ability to develop colonies in agar and to form tumors in nude mice (Castellazzi et al., 1990; Schutte et al., 1989). Cell transformation by c-Jun requires its transactivation domain, because expression of c-Jun TAM67 suppresses cancer cell growth in vitro and in nude mice and reverts the transformed phenotype of keratinocyte and mammary epithelial cells (Neyns et al., 1999; Li et al., 2000). Although c-Jun overexpression in transgenic mice does not result in the development of tumors, c-Jun ablation or its inhibition suppresses tumor progression (Grigoriadis et al., 1993). c-Jun knockout mice have drastically reduced number and size of hepatic tumors (Eferl et al., 2003). c-Jun ablation in the skin of the tumor-prone K5-SOS-F transgenic mice results in less EGF receptor expression in basal keratinocytes and smaller papillomas, and transgenic mice expressing c-Jun TAM67 in the basal keratinocytes escape chemically induced papilloma development (Zenz et al., 2003; Young et al., 1999).

Ras proteins are mutationally activated in many human cancers and are responsible for malignant transformation (Coleman et al., 2004). Compelling evidence suggests that c-Jun expression and activity are required for Ras-induced cell transformation, because c-Jun-null fibroblasts are resistant to Ras-induced transformation and c-Jun 63/73 AA transgenic mice are impaired in skin tumor development induced by expressing of SOS K5-SOS-F that activates the Ras pathway (Schutte et al., 1989; Johnson et al., 1996; Behrens et al., 2000). Since the constitutively active Ras activates MAPKs that lead to c-Jun phosphorylation, it is possible that c-Jun is an essential downstream mediator for Ras to induce cell transformation (Sistonen et al., 1989; Pertovaara et al., 1989; Mansour et al., 1994).

In transformed cells, c-Jun is expressed constitutively at high levels, which may contribute to tumor progression (Carter et al., 1994; Piette et al., 1988). High c-Jun expression possibly co-operates with tumor specific proteins, such as the prolyl isomerase Pin1, which is distinctly overexpressed in breast cancer cells. Pin1 binds to the N-terminal phosphorylated c-Jun to activate the *cyclin D1* promoter, thereby promoting tumor cell growth (Wulf et al., 2001). Conversely, the tumor suppressors Pdcd4 and JDP2 bind to and inactivate c-Jun to suppress the constitutive AP-1 activity in tumor cells and to inhibit tumor phenotype (Yang et al., 2003; Heinrich et al., 2004).

cDNA microarray studies have been employed to search for c-Jun-regulated genes involved in cell transformation and tumorigenesis. Differential gene expression profiles have been established by comparing mRNAs from TPA-treated epidermal cells that are either transformed by wild type c-Jun or by non-transformed c-Jun TAM-67. The high-mobility group protein A1 (HMGA1) has been identified as a TPA-induced and c-Jun transactivation domain dependent protein. HMGA1 expression is essential for epidermal cell transformation; however, its expression alone is insufficient, suggesting that c-Jun must control the expression of a group of genes for a combined effect on the transformation phenotype (Dhar et al., 2004). In transformed tumor cells, profiling the AP-1 target genes has revealed tumor suppressor TSCL-1 and gelsolin-like actin capping protein CapG, demonstrating a role for AP-1 in tumor cell motility and cytoskeletal dynamics (Bahassi et al., 2004). The apparent differences in gene expression profile revealed by these studies support the idea that c-Jun regulated gene expression is cell type and cell transformation status dependent. Finding the c-Jun target genes responsible for cell transformation and tumor promotion will uncover pertinent targets for cancer prevention. In this regard, this is area will be worthwhile of more extensive investigation.

Concluding Remarks

After close to two decades of extensive efforts, we have become painfully aware of the complex nature of c-Jun regulation and function. Still, as originally believed, c-Jun is a transcription factor that binds to a relatively simple sequence in the regulatory domain of genes, but just its ability to bind DNA and activate AP-1-driven promoters is insufficient to explain its role in diverse and sometimes opposing physiological processes, including proliferation, apoptosis, survival, tumorigenesis and tissue morphogenesis. c-Jun-mediated cell responses can be affected by many factors and are dependent on the abundance of c-Jun protein, its dimerization partners, and its interactions with other transcription factors, co-activators and co-repressors. All these parameters may be influenced by subtle alterations of the signaling properties within a cell that modulate c-Jun or its co-factors. There are still too many unknowns involved in c-Jun regulation and function. For instance, we do not know how various signals cross-talk in the regulation of c-Jun activities and how c-Jun selects its target genes in a tissue or cell type specific manner. In this regard, genetic knockout mice or cells that are deficient in c-Jun regulation pathways are instrumental in elucidating the signaling and genetic programs controlled by c-Jun.

Acknowledgement

The author would like to thank Dr. Alvaro Puga for proof reading of this article.

References

Abate,C., Patel,L., Rauscher,F.J., III, and Curran,T. (1990).

Redox regulation of fos and jun DNA-binding activity *in vitro*. Science *249*, 1157-1161.

Adler,V., Polotskaya,A., Wagner,F., and Kraft,A.S. (1992). Affinity-purified c-Jun amino-terminal protein kinase requires serine/threonine phosphorylation for activity. J. Biol. Chem. *267*, 17001-17005.

Agarwal,M.L., Agarwal,A., Taylor,W.R., and Stark,G.R. (1995). p53 controls both the G2/M and the G1 cell cycle checkpoints and mediates reversible growth arrest in human fibroblasts. Proc. Natl. Acad. Sci. U. S. A *92*, 8493-8497.

Albanese,C., D'Amico,M., Reutens,A.T., Fu,M., Watanabe,G., Lee,R.J., Kitsis,R.N., Henglein,B., Avantaggiati,M., Somasundaram, K., Thimmapaya,B., and Pestell,R.G. (1999). Activation of the cyclin D1 gene by the E1A-associated protein p300 through AP-1 inhibits cellular apoptosis. J. Biol. Chem. *274*, 34186-34195.

Alroy,I., Towers,T.L., and Freedman,L.P. (1995). Transcriptional repression of the interleukin-2 gene by vitamin D3: direct inhibition of NFATp/AP-1 complex formation by a nuclear hormone receptor. Mol. Cell Biol. *15*, 5789-5799.

Angel,P., Allegretto,E.A., Okino,S.T., Hattori,K., Boyle,W.J., Hunter,T., and Karin,M. (1988a). Oncogene jun encodes a sequence-specific trans-activator similar to AP-1. Nature *332*, 166-171.

Angel,P., Hattori,K., Smeal,T., and Karin,M. (1988b). The jun proto-oncogene is positively autoregulated by its product, Jun/AP-1. Cell *55*, 875-885.

Angel,P. and Karin,M. (1991). The role of Jun, Fos and the AP-1 complex in cell-proliferation and transformation. Biochim. Biophys. Acta *1072*, 129-157.

Angel,P., Smeal,T., Meek,J., and Karin,M. (1989). Jun and v-jun contain multiple regions that participate in transcriptional activation in an interdependent manner. New Biol. *1*, 35-43.

Arnulf,B., Villemain,A., Nicot,C., Mordelet,E., Charneau,P., Kersual,J., Zermati,Y., Mauviel,A., Bazarbachi,A., and Hermine,O. (2002). Human T-cell lymphotropic virus oncoprotein Tax represses TGF-beta 1 signaling in human T cells via c-Jun activation: a potential mechanism of HTLV-I leukemogenesis. Blood *100*, 4129-4138.

Atfi,A., Djelloul,S., Chastre,E., Davis,R., and Gespach,C. (1997). Evidence for a role of Rho-like GTPases and stress-activated protein kinase/c-Jun N-terminal kinase (SAPK/JNK) in transforming growth factor beta-mediated signaling. J Biol Chem *272*, 1429-32.

Bader,A.G., Hartl,M., and Bister,K. (2000). Conditional cell transformation by doxycycline-controlled expression of the ASV17 v-jun allele. Virology *270*, 98-110.

Bahassi,e.M., Karyala,S., Tomlinson,C.R., Sartor,M.A., Medvedovic, M., and Hennigan,R.F. (2004). Critical regulation of genes for tumor cell migration by AP-1. Clin. Exp. Metastasis *21*, 293-304.

Baker,S.J., Kerppola,T.K., Luk,D., Vandenberg,M.T., Marshak, D.R., Curran,T., and Abate,C. (1992). Jun is phosphorylated by several protein kinases at the same sites that are modified in serum-stimulated fibroblasts. Mol. Cell Biol. *12*, 4694-4705.

Bannister,A.J., Oehler,T., Wilhelm,D., Angel,P., and Kouzarides, T. (1995). Stimulation of c-Jun activity by CBP: c-Jun residues Ser63/73 are required for CBP induced stimulation *in vivo* and CBP binding *in vitro*. Oncogene *11*, 2509-2514.

Bassuk,A.G. and Leiden,J.M. (1995). A direct physical association between ETS and AP-1 transcription factors in normal human T cells. Immunity. *3*, 223-237.

Behre,G., Whitmarsh,A.J., Coghlan,M.P., Hoang,T., Carpenter, C.L., Zhang,D.E., Davis,R.J., and Tenen,D.G. (1999). c-Jun is a JNK-independent coactivator of the PU.1 transcription factor. J. Biol. Chem. *274*, 4939-4946.

Behrens,A., Jochum,W., Sibilia,M., and Wagner,E.F. (2000). Oncogenic transformation by ras and fos is mediated by c-Jun N-terminal phosphorylation. Oncogene *19*, 2657-2663.

Behrens,A., Sibilia,M., and Wagner,E.F. (1999). Amino-terminal phosphorylation of c-Jun regulates stress-induced apoptosis and cellular proliferation. Nat. Genet. *21*, 326-329.

Bengal,E., Ransone,L., Scharfmann,R., Dwarki,V.J., Tapscott,S.J., Weintraub,H., and Verma,I.M. (1992). Functional antagonism between c-Jun and MyoD proteins: a direct physical association. Cell *68*, 507-519.

Berkowitz,E.A., Seroogy,K.B., Schroeder,J.A., Russell,W.E., Evans,E.P., Riedel,R.F., Phillips,H.K., Harrison,C.A., Lee,D.C., and Luetteke,N.C. (1996). Characterization of the mouse transforming growth factor alpha gene: its expression during eyelid development and in waved 1 tissues. Cell Growth Differ *7*, 1271-82.

Bianchi,E., Denti,S., Granata,A., Bossi,G., Geginat,J., Villa,A., Rogge,L., and Pardi,R. (2000). Integrin LFA-1 interacts with the transcriptional co-activator JAB1 to modulate AP-1 activity. Nature *404*, 617-621.

Bitomsky,N., Bohm,M., and Klempnauer,K.H. (2004). Transformation suppressor protein Pdcd4 interferes with JNK-mediated phosphorylation of c-Jun and recruitment of the coactivator p300 by c-Jun. Oncogene *23*, 7484-7493.

Blattner,C., Kannouche,P., Litfin,M., Bender,K., Rahmsdorf,H.J., Angulo,J.F., and Herrlich,P. (2000). UV-Induced stabilization of c-fos and other short-lived mRNAs. Mol. Cell Biol. *20*, 3616-3625.

Bohmann,D. (1990). Transcription factor phosphorylation: a link between signal transduction and the regulation of gene expression. Cancer Cells *2*, 337-344.

Bohmann,D., Bos,T.J., Admon,A., Nishimura,T., Vogt,P.K., and Tjian,R. (1987). Human proto-oncogene c-jun encodes a DNA binding protein with structural and functional properties of transcription factor AP-1. Science *238*, 1386-1392.

Bos,T.J., Bohmann,D., Tsuchie,H., Tjian,R., and Vogt,P.K. (1988). v-Jun encodes a nuclear protein with enhancer binding properties of AP-1. Cell *52*, 705-712.

Boyle,W.J., Smeal,T., Defize,L.H., Angel,P., Woodgett,J.R., Karin,M., and Hunter,T. (1991). Activation of protein kinase C decreases phosphorylation of c-Jun at sites that negatively regulate its DNA-binding activity. Cell 64, 573-84.

Carter,R., Yumet,G., Pena,A., Soprano,D.R., and Soprano,K.J. (1994). Transcriptional regulation of c-Jun expression during late G1/S in normal human cells is lost in human tumor cells. Oncogene 9, 2675-2682.

Castellazzi,M., Dangy,J.P., Mechta,F., Hirai,S., Yaniv,M., Samarut,J., Lassailly,A., and Brun,G. (1990). Overexpression of avian or mouse c-jun in primary chick embryo fibroblasts confers a partially transformed phenotype. Oncogene 5, 1541-1547.

Chatton,B., Bocco,J.L., Gaire,M., Hauss,C., Reimund,B., Goetz,J., and Kedinger,C. (1993). Transcriptional activation by the adenovirus larger E1a product is mediated by members of the cellular transcription factor ATF family which can directly associate with E1a. Mol. Cell Biol. 13, 561-570.

Chen,Y.R., Meyer,C.F., and Tan,T.H. (1996). Persistent activation of c-Jun N-terminal kinase 1 (JNK1) in gamma radiation-induced apoptosis. J. Biol. Chem. 271, 631-634.

Chi,H., Sarkisian,M.R., Rakic,P., and Flavell,R.A. (2005). Loss of mitogen-activated protein kinase kinase kinase 4 (MEKK4) results in enhanced apoptosis and defective neural tube development. Proc. Natl. Acad. Sci. U. S. A 102, 3846-3851.

Chiu,R., Angel,P., and Karin,M. (1989). Jun-B differs in its biological properties from, and is a negative regulator of, c-Jun. Cell 59, 979-986.

Chiu,R., Boyle,W.J., Meek,J., Smeal,T., Hunter,T., and Karin,M. (1988). The c-Fos protein interacts with c-Jun/AP-1 to stimulate transcription of AP-1 responsive genes. Cell 54, 541-552.

Chopra,S., Fernandez,D.M., Lam,E.W., and Mann,D.J. (2002). Jab1 co-activation of c-Jun is abrogated by the serine 10-phosphorylated form of p27Kip1. J. Biol. Chem. 277, 32413-32416.

Christerson,L.B., Vanderbilt,C.A., and Cobb,M.H. (1999). MEKK1 interacts with alpha-actinin and localizes to stress fibers and focal adhesions. Cell Motil Cytoskeleton 43, 186-98.

Claret,F.X., Hibi,M., Dhut,S., Toda,T., and Karin,M. (1996). A new group of conserved coactivators that increase the specificity of AP-1 transcription factors. Nature 383, 453-457.

Cohen,D.R., Ferreira,P.C., Gentz,R., Franza,B.R., Jr., and Curran,T. (1989). The product of a fos-related gene, fra-1, binds cooperatively to the AP-1 site with Jun: transcription factor AP-1 is comprised of multiple protein complexes. Genes Dev. 3, 173-184.

Coleman,M.L., Marshall,C.J., and Olson,M.F. (2004). RAS and RHO GTPases in G1-phase cell-cycle regulation. Nat. Rev. Mol. Cell Biol. 5, 355-366.

Davis,R.J. (1994). MAPKs: new JNK expands the group. Trends Biochem Sci 19, 470-3.

Deak,J.C. and Templeton,D.J. (1997). Regulation of the activity of MEK kinase 1 (MEKK1) by autophosphorylation within the kinase activation domain. Biochem. J. 322 (Pt 1), 185-192.

Deng,T. and Karin,M. (1993). JunB differs from c-Jun in its DNA-binding and dimerization domains, and represses c-Jun by formation of inactive heterodimers. Genes Dev. 7, 479-490.

Dennler,S., Prunier,C., Ferrand,N., Gauthier,J.M., and Atfi,A. (2000). c-Jun inhibits transforming growth factor beta-mediated transcription by repressing Smad3 transcriptional activity. J. Biol. Chem. 275, 28858-28865.

Derijard,B., Hibi,M., Wu,I.H., Barrett,T., Su,B., Deng,T., Karin,M., and Davis,R.J. (1994). JNK1: a protein kinase stimulated by UV light and Ha-Ras that binds and phosphorylates the c-Jun activation domain. Cell 76, 1025-1037.

Devary,Y., Gottlieb,R.A., Lau,L.F., and Karin,M. (1991). Rapid and preferential activation of the c-jun gene during the mammalian UV response. Mol. Cell Biol. 11, 2804-2811.

Dhar,A., Hu,J., Reeves,R., Resar,L.M., and Colburn,N.H. (2004). Dominant-negative c-Jun (TAM67) target genes: HMGA1 is required for tumor promoter-induced transformation. Oncogene 23, 4466-4476.

Diener,K., Wang,X.S., Chen,C., Meyer,C.F., Keesler,G., Zukowski,M., Tan,T.H., and Yao,Z. (1997). Activation of the c-Jun N-terminal kinase pathway by a novel protein kinase related to human germinal center kinase. Proc. Natl. Acad. Sci. U. S. A 94, 9687-9692.

Dumitru,C.D., Ceci,J.D., Tsatsanis,C., Kontoyiannis,D., Stamatakis,K., Lin,J.H., Patriotis,C., Jenkins,N.A., Copeland,N.G., Kollias,G., and Tsichlis,P.N. (2000). TNF-alpha induction by LPS is regulated posttranscriptionally via a Tpl2/ERK-dependent pathway. Cell 103, 1071-1083.

Eckner,R., Arany,Z., Ewen,M., Sellers,W., and Livingston,D.M. (1994). The adenovirus E1A-associated 300-kD protein exhibits properties of a transcriptional coactivator and belongs to an evolutionarily conserved family. Cold Spring Harb. Symp. Quant. Biol. 59, 85-95.

Edelstein,L.C., Lagos,L., Simmons,M., Tirumalai,H., and Gelinas,C. (2003). NF-kappa B-dependent assembly of an enhanceosome-like complex on the promoter region of apoptosis inhibitor Bfl-1/A1. Mol. Cell Biol. 23, 2749-2761.

Eferl,R., Ricci,R., Kenner,L., Zenz,R., David,J.P., Rath,M., and Wagner,E.F. (2003). Liver tumor development. c-Jun antagonizes the proapoptotic activity of p53. Cell 112, 181-192.

Eferl,R., Sibilia,M., Hilberg,F., Fuchsbichler,A., Kufferath,I., Guertl,B., Zenz,R., Wagner,E.F., and Zatloukal,K. (1999). Functions of c-Jun in liver and heart development. J. Cell Biol. 145, 1049-1061.

Eichhorst,S.T., Muller,M., Li-Weber,M., Schulze-Bergkamen,H., Angel,P., and Krammer,P.H. (2000). A novel AP-1 element in the CD95 ligand promoter is required for induction of apoptosis in hepatocellular carcinoma cells upon treatment with anticancer drugs. Mol. Cell Biol. 20, 7826-7837.

El Tanani,M., Platt-Higgins,A., Rudland,P.S., and Campbell,F.C. (2004). Ets gene PEA3 cooperates with beta-catenin-Lef-1 and c-Jun in regulation of osteopontin transcription. J. Biol. Chem. 279, 20794-20806.

Ellenberger,T.E., Brandl,C.J., Struhl,K., and Harrison,S.C. (1992). The GCN4 basic region leucine zipper binds DNA as a dimer of uninterrupted alpha helices: crystal structure of the protein-DNA complex. Cell *71*, 1223-1237.

Fan,M., Goodwin,M.E., Birrer,M.J., and Chambers,T.C. (2001). The c-Jun NH(2)-terminal protein kinase/AP-1 pathway is required for efficient apoptosis induced by vinblastine. Cancer Res. *61*, 4450-4458.

Faris,M., Latinis,K.M., Kempiak,S.J., Koretzky,G.A., and Nel,A. (1998). Stress-induced Fas ligand expression in T cells is mediated through a MEK kinase 1-regulated response element in the Fas ligand promoter. Mol Cell Biol *18*, 5414-24.

Frame,M.C., Wilkie,N.M., Darling,A.J., Chudleigh,A., Pintzas,A., Lang,J.C., and Gillespie,D.A. (1991). Regulation of AP-1/DNA complex formation *in vitro*. Oncogene *6*, 205-209.

Franklin,C.C., McCulloch,A.V., and Kraft,A.S. (1995). *In vitro* association between the Jun protein family and the general transcription factors, TBP and TFIIB. Biochem. J. *305 (Pt 3)*, 967-974.

Franklin,C.C., Sanchez,V., Wagner,F., Woodgett,J.R., and Kraft,A.S. (1992). Phorbol ester-induced amino-terminal phosphorylation of human JUN but not JUNB regulates transcriptional activation. Proc. Natl. Acad. Sci. U. S. A *89*, 7247-7251.

Gallagher,E.D., Gutowski,S., Sternweis,P.C., and Cobb,M.H. (2004). RhoA binds to the amino terminus of MEKK1 and regulates its kinase activity. J. Biol. Chem. *279*, 1872-1877.

Goldman,P.S., Tran,V.K., and Goodman,R.H. (1997). The multifunctional role of the co-activator CBP in transcriptional regulation. Recent Prog. Horm. Res. *52*, 103-119.

Grana,X. and Reddy,E.P. (1995). Cell cycle control in mammalian cells: role of cyclins, cyclin dependent kinases (CDKs), growth suppressor genes and cyclin-dependent kinase inhibitors (CKIs). Oncogene *11*, 211-219.

Grigoriadis,A.E., Schellander,K., Wang,Z.Q., and Wagner,E.F. (1993). Osteoblasts are target cells for transformation in c-fos transgenic mice. J. Cell Biol. *122*, 685-701.

Grose,R. (2003). Epithelial migration: open your eyes to c-Jun. Curr. Biol. *13*, R678-R680.

Guan,B., Yue,P., Lotan,R., and Sun,S.Y. (2002). Evidence that the human death receptor 4 is regulated by activator protein 1. Oncogene *21*, 3121-3129.

Guo,Z., Clydesdale,G., Cheng,J., Kim,K., Gan,L., McConkey,D.J., Ullrich,S.E., Zhuang,Y., and Su,B. (2002). Disruption of Mekk2 in mice reveals an unexpected role for MEKK2 in modulating T-cell receptor signal transduction. Mol. Cell Biol. *22*, 5761-5768.

Hagmeyer,B.M., Angel,P., and van Dam,H. (1995). Modulation of AP-1/ATF transcription factor activity by the adenovirus-E1A oncogene products. Bioessays *17*, 621-629.

Hai,T. and Curran,T. (1991). Cross-family dimerization of transcription factors Fos/Jun and ATF/CREB alters DNA binding specificity. Proc Natl Acad Sci USA *88*, 3720-4.

Hai,T.W., Liu,F., Allegretto,E.A., Karin,M., and Green,M.R. (1988). A family of immunologically related transcription factors that includes multiple forms of ATF and AP-1. Genes Dev. *2*, 1216-1226.

Halazonetis,T.D., Georgopoulos,K., Greenberg,M.E., and Leder,P. (1988). c-Jun dimerizes with itself and with c-Fos, forming complexes of different DNA binding affinities. Cell *55*, 917-924.

Haluska,F.G., Huebner,K., Isobe,M., Nishimura,T., Croce,C.M., and Vogt,P.K. (1988). Localization of the human JUN protooncogene to chromosome region 1p31-32. Proc. Natl. Acad. Sci. U. S. A *85*, 2215-2218.

Ham,J., Babij,C., Whitfield,J., Pfarr,C.M., Lallemand,D., Yaniv,M., and Rubin,L.L. (1995). A c-Jun dominant negative mutant protects sympathetic neurons against programmed cell death. Neuron *14*, 927-939.

Hanahan,D. and Weinberg,R.A. (2000). The hallmarks of cancer. Cell *100*, 57-70.

Harkin,D.P., Bean,J.M., Miklos,D., Song,Y.H., Truong,V.B., Englert,C., Christians,F.C., Ellisen,L.W., Maheswaran,S., Oliner,J.D., and Haber,D.A. (1999). Induction of GADD45 and JNK/SAPK-dependent apoptosis following inducible expression of BRCA1. Cell *97*, 575-586.

Harshman,K.D., Moye-Rowley,W.S., and Parker,C.S. (1988). Transcriptional activation by the SV40 AP-1 recognition element in yeast is mediated by a factor similar to AP-1 that is distinct from GCN4. Cell *53*, 321-330.

Harvey,M., McArthur,M.J., Montgomery,C.A., Jr., Butel,J.S., Bradley,A., and Donehower,L.A. (1993). Spontaneous and carcinogen-induced tumorigenesis in p53-deficient mice. Nat. Genet. *5*, 225-229.

Heinrich,R., Livne,E., Ben Izhak,O., and Aronheim,A. (2004). The c-Jun dimerization protein 2 inhibits cell transformation and acts as a tumor suppressor gene. J. Biol. Chem. *279*, 5708-5715.

Helmberg,A., Auphan,N., Caelles,C., and Karin,M. (1995). Glucocorticoid-induced apoptosis of human leukemic cells is caused by the repressive function of the glucocorticoid receptor. EMBO J. *14*, 452-460.

Hibi,M., Lin,A., Smeal,T., Minden,A., and Karin,M. (1993). Identification of an oncoprotein- and UV-responsive protein kinase that binds and potentiates the c-Jun activation domain. Genes Dev *7*, 2135-48.

Hilberg,F., Aguzzi,A., Howells,N., and Wagner,E.F. (1993). c-jun is essential for normal mouse development and hepatogenesis. Nature *365*, 179-181.

Hilberg,F. and Wagner,E.F. (1992). Embryonic stem (ES) cells lacking functional c-Jun: consequences for growth and differentiation, AP-1 activity and tumorigenicity. Oncogene *7*, 2371-2380.

Huang,S., Shu,L., Dilling,M.B., Easton,J., Harwood,F.C., Ichijo,H., and Houghton,P.J. (2003). Sustained activation of the JNK cascade and rapamycin-induced apoptosis are suppressed by p53/p21(Cip1). Mol. Cell *11*, 1491-1501.

Huser,M., Luckett,J., Chiloeches,A., Mercer,K., Iwobi,M., Giblett,S., Sun,X.M., Brown,J., Marais,R., and Pritchard,C. (2001). MEK kinase activity is not necessary for Raf-1 function. EMBO J. *20*, 1940-1951.

Ichijo,H., Nishida,E., Irie,K., ten Dijke,P., Saitoh,M., Moriguchi,T., Takagi,M., Matsumoto,K., Miyazono,K., and Gotoh,Y. (1997). Induction of apoptosis by ASK1, a mammalian MAPKKK that activates SAPK/JNK and p38 signaling pathways. Science *275*, 90-4.

Ito,T., Yamauchi,M., Nishina,M., Yamamichi,N., Mizutani,T., Ui,M., Murakami,M., and Iba,H. (2001). Identification of SWI.SNF complex subunit BAF60a as a determinant of the transactivation potential of Fos/Jun dimers. J. Biol. Chem. *276*, 2852-2857.

Johnson,G.L. and Lapadat,R. (2002). Mitogen-activated protein kinase pathways mediated by ERK, JNK, and p38 protein kinases. Science *298*, 1911-1912.

Johnson,R., Spiegelman,B., Hanahan,D., and Wisdom,R. (1996). Cellular transformation and malignancy induced by ras require c-Jun. Mol. Cell Biol. *16*, 4504-4511.

Johnson,R.S., van Lingen,B., Papaioannou,V.E., and Spiegelman,B.M. (1993). A null mutation at the c-Jun locus causes embryonic lethality and retarded cell growth in culture. Genes Dev. *7*, 1309-1317.

Karin,M. (1991). Signal transduction and gene control. Curr. Opin. Cell Biol. *3*, 467-473.

Karin,M., Liu,Z., and Zandi,E. (1997). AP-1 function and regulation. Curr Opin Cell Biol *9*, 240-6.

Kataoka,K., Noda,M., and Nishizawa,M. (1994). Maf nuclear oncoprotein recognizes sequences related to an AP-1 site and forms heterodimers with both Fos and Jun. Mol. Cell Biol. *14*, 700-712.

Kick,G., Messer,G., Plewig,G., Kind,P., and Goetz,A.E. (1996). Strong and prolonged induction of c-jun and c-fos proto-oncogenes by photodynamic therapy. Br. J. Cancer *74*, 30-36.

Kim,T.K., Kim,T.H., and Maniatis,T. (1998). Efficient recruitment of TFIIB and CBP-RNA polymerase II holoenzyme by an interferon-beta enhanceosome *in vitro*. Proc. Natl. Acad. Sci. U. S. A *95*, 12191-12196.

King,L.B., Tolosa,E., Lenczowski,J.M., Lu,F., Lind,E.F., Hunziker,R., Petrie,H.T., and Ashwell,J.D. (1999). A dominant-negative mutant of c-Jun inhibits cell cycle progression during the transition of CD4(-)CD8(-) to CD4(+)CD8(+) thymocytes. Int. Immunol. *11*, 1203-1216.

Koo,M.S., Kwo,Y.G., Park,J.H., Choi,W.J., Billiar,T.R., and Kim,Y.M. (2002). Signaling and function of caspase and c-jun N-terminal kinase in cisplatin-induced apoptosis. Mol. Cells *13*, 194-201.

Kouzarides,T. and Ziff,E. (1988). The role of the leucine zipper in the fos-jun interaction. Nature *336*, 646-651.

Kuan,C.Y., Yang,D.D., Samanta Roy,D.R., Davis,R.J., Rakic,P., and Flavell,R.A. (1999). The Jnk1 and Jnk2 protein kinases are required for regional specific apoptosis during early brain development. Neuron *22*, 667-76.

Kyriakis,J.M., Banerjee,P., Nikolakaki,E., Dai,T., Rubie,E.A., Ahmad,M.F., Avruch,J., and Woodgett,J.R. (1994). The stress-activated protein kinase subfamily of c-Jun kinases. Nature *369*, 156-160.

Lamph,W.W., Wamsley,P., Sassone-Corsi,P., and Verma,I.M. (1988). Induction of proto-oncogene JUN/AP-1 by serum and TPA. Nature *334*, 629-631.

Landschulz,W.H., Johnson,P.F., and McKnight,S.L. (1988). The leucine zipper: a hypothetical structure common to a new class of DNA binding proteins. Science *240*, 1759-1764.

Lange-Carter,C.A., Pleiman,C.M., Gardner,A.M., Blumer,K.J., and Johnson,G.L. (1993). A divergence in the MAP kinase regulatory network defined by MEK kinase and Raf. Science *260*, 315-9.

Lee,J.S., See,R.H., Deng,T., and Shi,Y. (1996). Adenovirus E1A downregulates cJun- and JunB-mediated transcription by targeting their coactivator p300. Mol. Cell Biol. *16*, 4312-4326.

Lee,S.K., Kim,H.J., Na,S.Y., Kim,T.S., Choi,H.S., Im,S.Y., and Lee,J.W. (1998). Steroid receptor coactivator-1 coactivates activating protein-1-mediated transactivations through interaction with the c-Jun and c-Fos subunits. J. Biol. Chem. *273*, 16651-16654.

Lee,W., Mitchell,P., and Tjian,R. (1987). Purified transcription factor AP-1 interacts with TPA-inducible enhancer elements. Cell *49*, 741-752.

Levine,A.J. (1997). p53, the cellular gatekeeper for growth and division. Cell *88*, 323-331.

Li,J.J., Cao,Y., Young,M.R., and Colburn,N.H. (2000). Induced expression of dominant-negative c-jun downregulates NFkappaB and AP-1 target genes and suppresses tumor phenotype in human keratinocytes. Mol. Carcinog. *29*, 159-169.

Lin,A., Frost,J., Deng,T., Smeal,T., al-Alawi,N., Kikkawa,U., Hunter,T., Brenner,D., and Karin,M. (1992). Casein kinase II is a negative regulator of c-Jun DNA binding and AP-1 activity. Cell *70*, 777-89.

Lin,A., Minden,A., Martinetto,H., Claret,F.X., Lange-Carter,C., Mercurio,F., Johnson,G.L., and Karin,M. (1995). Identification of a dual specificity kinase that activates the Jun kinases and p38-Mpk2. Science *268*, 286-90.

Logan,S.K., Garabedian,M.J., Campbell,C.E., and Werb,Z. (1996). Synergistic transcriptional activation of the tissue inhibitor of metalloproteinases-1 promoter via functional interaction of AP-1 and Ets-1 transcription factors. J. Biol. Chem. *271*, 774-782.

Maki,Y., Bos,T.J., Davis,C., Starbuck,M., and Vogt,P.K. (1987). Avian sarcoma virus 17 carries the jun oncogene. Proc. Natl. Acad. Sci. U. S. A *84*, 2848-2852.

Mansour,S.J., Matten,W.T., Hermann,A.S., Candia,J.M., Rong,S., Fukasawa,K., Vande Woude,G.F., and Ahn,N.G. (1994).

Transformation of mammalian cells by constitutively active MAP kinase kinase. Science 265, 966-970.

Matsuzawa,A. and Ichijo,H. (2005). Stress-responsive protein kinases in redox-regulated apoptosis signaling. Antioxid. Redox. Signal. 7, 472-481.

Mayo,M.W., Steelman,L.S., and McCubrey,J.A. (1994). Phorbol esters support the proliferation of a hematopoietic cell line by upregulating c-jun expression. Oncogene 9, 1999-2008.

Minden,A., Lin,A., McMahon,M., Lange-Carter,C., Derijard,B., Davis,R.J., Johnson,G.L., and Karin,M. (1994). Differential activation of ERK and JNK mitogen-activated protein kinases by Raf-1 and MEKK. Science 266, 1719-23.

Mordret,G. (1993). MAP kinase kinase: a node connecting multiple pathways. Biol. Cell 79, 193-207.

Mosieniak,G., Figiel,I., and Kaminska,B. (1997). Cyclosporin A, an immunosuppressive drug, induces programmed cell death in rat C6 glioma cells by a mechanism that involves the AP-1 transcription factor. J. Neurochem. 68, 1142-1149.

Muller,S., Berger,M., Lehembre,F., Seeler,J.S., Haupt,Y., and Dejean,A. (2000). c-Jun and p53 activity is modulated by SUMO-1 modification. J. Biol. Chem. 275, 13321-13329.

Munz,C., Psichari,E., Mandilis,D., Lavigne,A.C., Spiliotaki,M., Oehler,T., Davidson,I., Tora,L., Angel,P., and Pintzas,A. (2003). TAF7 (TAFII55) plays a role in the transcription activation by c-Jun. J. Biol. Chem. 278, 21510-21516.

Nakamura,T., Datta,R., Kharbanda,S., and Kufe,D. (1991). Regulation of jun and fos gene expression in human monocytes by the macrophage colony-stimulating factor. Cell Growth Differ. 2, 267-272.

Nead,M.A., Baglia,L.A., Antinore,M.J., Ludlow,J.W., and McCance,D.J. (1998). Rb binds c-Jun and activates transcription. EMBO J. 17, 2342-2352.

Neuberg,M., Adamkiewicz,J., Hunter,J.B., and Muller,R. (1989). A Fos protein containing the Jun leucine zipper forms a homodimer which binds to the AP1 binding site. Nature 341, 243-245.

Neyns,B., Teugels,E., Bourgain,C., Birrerand,M., and De Greve,J. (1999). Alteration of jun proto-oncogene status by plasmid transfection affects growth of human ovarian cancer cells. Int. J. Cancer 82, 687-693.

Nikolakaki,E., Coffer,P.J., Hemelsoet,R., Woodgett,J.R., and Defize,L.H. (1993). Glycogen synthase kinase 3 phosphorylates Jun family members in vitro and negatively regulates their transactivating potential in intact cells. Oncogene 8, 833-840.

O'Shea,E.K., Rutkowski,R., Stafford,W.F., III, and Kim,P.S. (1989). Preferential heterodimer formation by isolated leucine zippers from fos and jun. Science 245, 646-648.

Papavassiliou,A.G., Chavrier,C., and Bohmann,D. (1992). Phosphorylation state and DNA-binding activity of c-Jun depend on the intracellular concentration of binding sites. Proc. Natl. Acad. Sci. U. S. A 89, 11562-11565.

Park,J., Kim,I., Oh,Y.J., Lee,K., Han,P.L., and Choi,E.J. (1997). Activation of c-Jun N-terminal kinase antagonizes an anti-apoptotic action of Bcl-2. J. Biol. Chem. 272, 16725-16728.

Peng,S.S., Chen,C.Y., and Shyu,A.B. (1996). Functional characterization of a non-AUUUA AU-rich element from the c-jun proto-oncogene mRNA: evidence for a novel class of AU-rich elements. Mol. Cell Biol. 16, 1490-1499.

Pertovaara,L., Sistonen,L., Bos,T.J., Vogt,P.K., Keski-Oja,J., and Alitalo,K. (1989). Enhanced jun gene expression is an early genomic response to transforming growth factor beta stimulation. Mol. Cell Biol. 9, 1255-1262.

Piette,J., Hirai,S., and Yaniv,M. (1988). Constitutive synthesis of activator protein 1 transcription factor after viral transformation of mouse fibroblasts. Proc. Natl. Acad. Sci. U. S. A 85, 3401-3405.

Pomerance,M., Multon,M.C., Parker,F., Venot,C., Blondeau,J.P., Tocque,B., and Schweighoffer,F. (1998). Grb2 interaction with MEK-kinase 1 is involved in regulation of Jun-kinase activities in response to epidermal growth factor. J Biol Chem 273, 24301-4.

Potapova,O., Gorospe,M., Dougherty,R.H., Dean,N.M., Gaarde,W.A., and Holbrook,N.J. (2000). Inhibition of c-Jun N-terminal kinase 2 expression suppresses growth and induces apoptosis of human tumor cells in a p53-dependent manner. Mol Cell Biol 20, 1713-22.

Putcha,G.V., Le,S., Frank,S., Besirli,C.G., Clark,K., Chu,B., Alix,S., Youle,R.J., LaMarche,A., Maroney,A.C., and Johnson,E.M., Jr. (2003). JNK-mediated BIM phosphorylation potentiates BAX-dependent apoptosis. Neuron 38, 899-914.

Quantin,B. and Breathnach,R. (1988). Epidermal growth factor stimulates transcription of the c-Jun proto-oncogene in rat fibroblasts. Nature 334, 538-539.

Rauscher,F.J., III, Cohen,D.R., Curran,T., Bos,T.J., Vogt,P.K., Bohmann,D., Tjian,R., and Franza,B.R., Jr. (1988a). Fos-associated protein p39 is the product of the jun proto-oncogene. Science 240, 1010-1016.

Rauscher,F.J., III, Voulalas,P.J., Franza,B.R., Jr., and Curran,T. (1988b). Fos and Jun bind cooperatively to the AP-1 site: reconstitution in vitro. Genes Dev. 2, 1687-1699.

Rebollo,A., Dumoutier,L., Renauld,J.C., Zaballos,A., Ayllon,V., and Martinez,A. (2000). Bcl-3 expression promotes cell survival following interleukin-4 deprivation and is controlled by AP1 and AP1-like transcription factors. Mol. Cell Biol. 20, 3407-3416.

Rincon,M., Whitmarsh,A., Yang,D.D., Weiss,L., Derijard,B., Jayaraj,P., Davis,R.J., and Flavell,R.A. (1998). The JNK pathway regulates the In vivo deletion of immature CD4(+)CD8(+) thymocytes. J. Exp. Med. 188, 1817-1830.

Ryder,K., Lau,L.F., and Nathans,D. (1988). A gene activated by growth factors is related to the oncogene v-jun. Proc. Natl. Acad. Sci. U. S. A 85, 1487-1491.

Ryder,K. and Nathans,D. (1988). Induction of protooncogene c-jun by serum growth factors. Proc. Natl. Acad. Sci. U. S. A 85, 8464-8467.

Ryseck,R.P., Hirai,S.I., Yaniv,M., and Bravo,R. (1988).

Transcriptional activation of c-Jun during the G0/G1 transition in mouse fibroblasts. Nature *334*, 535-537.

Sawai,H., Okazaki,T., Yamamoto,H., Okano,H., Takeda,Y., Tashima,M., Sawada,H., Okuma,M., Ishikura,H., and Umehara,H. (1995). Requirement of AP-1 for ceramide-induced apoptosis in human leukemia HL-60 cells. J. Biol. Chem. *270*, 27326-27331.

Schlesinger,T.K., Fanger,G.R., Yujiri,T., and Johnson,G.L. (1998). The TAO of MEKK. Front Biosci *3*, 1181-6.

Schreiber,M., Kolbus,A., Piu,F., Szabowski,A., Mohle-Steinlein,U., Tian,J., Karin,M., Angel,P., and Wagner,E.F. (1999). Control of cell cycle progression by c-Jun is p53 dependent. Genes Dev. *13*, 607-619.

Schrewe,H., Gendron-Maguire,M., Harbison,M.L., and Gridley,T. (1994). Mice homozygous for a null mutation of activin beta B are viable and fertile. Mech. Dev. *47*, 43-51.

Schumacher,M.A., Goodman,R.H., and Brennan,R.G. (2000). The structure of a CREB bZIP.somatostatin CRE complex reveals the basis for selective dimerization and divalent cation-enhanced DNA binding. J. Biol. Chem. *275*, 35242-35247.

Schutte,J., Viallet,J., Nau,M., Segal,S., Fedorko,J., and Minna,J. (1989). Jun-B inhibits and c-fos stimulates the transforming and trans-activating activities of c-Jun. Cell *59*, 987-997.

Sears,R.C. and Nevins,J.R. (2002). Signaling networks that link cell proliferation and cell fate. J. Biol. Chem. *277*, 11617-11620.

Shaulian,E. and Karin,M. (2002). AP-1 as a regulator of cell life and death. Nat. Cell Biol. *4*, E131-E136.

Shaulian,E., Schreiber,M., Piu,F., Beeche,M., Wagner,E.F., and Karin,M. (2000). The mammalian UV response: c-Jun induction is required for exit from p53-imposed growth arrest. Cell *103*, 897-907.

Sherman,M.L., Stone,R.M., Datta,R., Bernstein,S.H., and Kufe,D.W. (1990). Transcriptional and post-transcriptional regulation of c-Jun expression during monocytic differentiation of human myeloid leukemic cells. J. Biol. Chem. *265*, 3320-3323.

Short,N.J. (1987). Regulation of transcription. Are some controlling factors more equal than others? Nature *326*, 740-741.

Siow,Y.L., Kalmar,G.B., Sanghera,J.S., Tai,G., Oh,S.S., and Pelech,S.L. (1997). Identification of two essential phosphorylated threonine residues in the catalytic domain of Mekk1. Indirect activation by Pak3 and protein kinase C. J. Biol. Chem. *272*, 7586-7594.

Sistonen,L., Holtta,E., Lehvaslaiho,H., Lehtola,L., and Alitalo,K. (1989). Activation of the neu tyrosine kinase induces the fos/jun transcription factor complex, the glucose transporter and ornithine decarboxylase. J. Cell Biol. *109*, 1911-1919.

Smeal,T., Binetruy,B., Mercola,D., Grover-Bardwick,A., Heidecker,G., Rapp,U.R., and Karin,M. (1992). Oncoprotein-mediated signalling cascade stimulates c-Jun activity by phosphorylation of serines 63 and 73. Mol. Cell Biol. *12*, 3507-3513.

Smeal,T., Binetruy,B., Mercola,D.A., Birrer,M., and Karin,M. (1991). Oncogenic and transcriptional cooperation with Ha-Ras requires phosphorylation of c-Jun on serines 63 and 73. Nature *354*, 494-496.

Smith,M.J. and Prochownik,E.V. (1992). Inhibition of c-Jun causes reversible proliferative arrest and withdrawal from the cell cycle. Blood *79*, 2107-2115.

Soh,J.W., Mao,Y., Liu,L., Thompson,W.J., Pamukcu,R., and Weinstein,I.B. (2001). Protein kinase G activates the JNK1 pathway via phosphorylation of MEKK1. J Biol Chem *276*, 16406-10.

Su,Y.C., Han,J., Xu,S., Cobb,M., and Skolnik,E.Y. (1997). NIK is a new Ste20-related kinase that binds NCK and MEKK1 and activates the SAPK/JNK cascade via a conserved regulatory domain. Embo J *16*, 1279-90.

Sutherland,C., Renaux,B.S., McKay,D.J., and Walsh,M.P. (1994). Phosphorylation of caldesmon by smooth-muscle casein kinase II. J. Muscle Res. Cell Motil. *15*, 440-456.

Tobiume,K., Matsuzawa,A., Takahashi,T., Nishitoh,H., Morita,K., Takeda,K., Minowa,O., Miyazono,K., Noda,T., and Ichijo,H. (2001). ASK1 is required for sustained activations of JNK/p38 MAP kinases and apoptosis. EMBO Rep. *2*, 222-228.

Toda,T., Shimanuki,M., and Yanagida,M. (1991). Fission yeast genes that confer resistance to staurosporine encode an AP-1-like transcription factor and a protein kinase related to the mammalian ERK1/MAP2 and budding yeast FUS3 and KSS1 kinases. Genes Dev. *5*, 60-73.

Toh,W.H., Siddique,M.M., Boominathan,L., Lin,K.W., and Sabapathy,K. (2004). c-Jun regulates the stability and activity of the p53 homologue, p73. J. Biol. Chem. *279*, 44713-44722.

Tournier,C., Hess,P., Yang,D.D., Xu,J., Turner,T.K., Nimnual,A., Bar-Sagi,D., Jones,S.N., Flavell,R.A., and Davis,R.J. (2000). Requirement of JNK for stress-induced activation of the cytochrome c-mediated death pathway. Science *288*, 870-4.

Treier,M., Staszewski,L.M., and Bohmann,D. (1994). Ubiquitin-dependent c-Jun degradation *in vivo* is mediated by the delta domain. Cell *78*, 787-798.

Tsuruta,F., Sunayama,J., Mori,Y., Hattori,S., Shimizu,S., Tsujimoto,Y., Yoshioka,K., Masuyama,N., and Gotoh,Y. (2004). JNK promotes Bax translocation to mitochondria through phosphorylation of 14-3-3 proteins. EMBO J. *23*, 1889-1899.

Ubeda,M., Vallejo,M., and Habener,J.F. (1999). CHOP enhancement of gene transcription by interactions with Jun/Fos AP-1 complex proteins. Mol. Cell Biol. *19*, 7589-7599.

Uhle,S., Medalia,O., Waldron,R., Dumdey,R., Henklein,P., Bech-Otschir,D., Huang,X., Berse,M., Sperling,J., Schade,R., and Dubiel,W. (2003). Protein kinase CK2 and protein kinase D are associated with the COP9 signalosome. EMBO J. *22*, 1302-1312.

Verheij,M., Bose,R., Lin,X.H., Yao,B., Jarvis,W.D., Grant,S., Birrer,M.J., Szabo,E., Zon,L.I., Kyriakis,J.M., Haimovitz-Friedman,A., Fuks,Z., and Kolesnick,R.N. (1996). Requirement for ceramide-initiated SAPK/JNK signalling in stress-induced apoptosis. Nature *380*, 75-79.

Verma,I.M., Ransone,L.J., Visvader,J., Sassone-Corsi,P., and

Lamph,W.W. (1989). fos-jun Conspiracy: implications for the cell. Princess Takamatsu Symp. *20*, 119-126.

Verrecchia,F., Vindevoghel,L., Lechleider,R.J., Uitto,J., Roberts,A.B., and Mauviel,A. (2001). Smad3/AP-1 interactions control transcriptional responses to TGF-beta in a promoter-specific manner. Oncogene *20*, 3332-3340.

Vogt,P.K. and Bos,T.J. (1989). The oncogene jun and nuclear signalling. Trends Biochem. Sci. *14*, 172-175.

Vries,R.G., Prudenziati,M., Zwartjes,C., Verlaan,M., Kalkhoven,E., and Zantema,A. (2001). A specific lysine in c-Jun is required for transcriptional repression by E1A and is acetylated by p300. EMBO J. *20*, 6095-6103.

Wang,C.H., Tsao,Y.P., Chen,H.J., Chen,H.L., Wang,H.W., and Chen,S.L. (2000). Transcriptional repression of p21((Waf1/Cip1/Sdi1)) gene by c-Jun through Sp1 site. Biochem. Biophys. Res. Commun. *270*, 303-310.

Weiss,C., Schneider,S., Wagner,E.F., Zhang,X., Seto,E., and Bohmann,D. (2003). JNK phosphorylation relieves HDAC3-dependent suppression of the transcriptional activity of c-Jun. EMBO J. *22*, 3686-3695.

Westwick,J.K., Cox,A.D., Der,C.J., Cobb,M.H., Hibi,M., Karin,M., and Brenner,D.A. (1994). Oncogenic Ras activates c-Jun via a separate pathway from the activation of extracellular signal-regulated kinases. Proc. Natl. Acad. Sci. U. S. A *91*, 6030-6034.

Wisdom,R., Johnson,R.S., and Moore,C. (1999). c-Jun regulates cell cycle progression and apoptosis by distinct mechanisms. EMBO J. *18*, 188-197.

Wulf,G.M., Ryo,A., Wulf,G.G., Lee,S.W., Niu,T., Petkova,V., and Lu,K.P. (2001). Pin1 is overexpressed in breast cancer and cooperates with Ras signaling in increasing the transcriptional activity of c-Jun towards cyclin D1. EMBO J. *20*, 3459-3472.

Xia,Y. and Karin,M. (2004). The control of cell motility and epithelial morphogenesis by Jun kinases. Trends Cell Biol. *14*, 94-101.

Xia,Y., Makris,C., Su,B., Li,E., Yang,J., Nemerow,G.R., and Karin,M. (2000). MEK kinase 1 is critically required for c-Jun N-terminal kinase activation by proinflammatory stimuli and growth factor-induced cell migration. Proc Natl Acad Sci USA *97*, 5243-8.

Xia,Y., Wu,Z., Su,B., Murray,B., and Karin,M. (1998). JNKK1 organizes a MAP kinase module through specific and sequential interactions with upstream and downstream components mediated by its amino-terminal extension. Genes Dev *12*, 3369-81.

Xia,Z., Dickens,M., Raingeaud,J., Davis,R.J., and Greenberg,M.E. (1995). Opposing effects of ERK and JNK-p38 MAP kinases on apoptosis. Science *270*, 1326-1331.

Xu,S., Robbins,D.J., Christerson,L.B., English,J.M., Vanderbilt,C.A., and Cobb,M.H. (1996). Cloning of rat MEK kinase 1 cDNA reveals an endogenous membrane-associated 195-kDa protein with a large regulatory domain. Proc Natl Acad Sci USA *93*, 5291-5.

Xu,Z., Kukekov,N.V., and Greene,L.A. (2003). POSH acts as a scaffold for a multiprotein complex that mediates JNK activation in apoptosis. EMBO J. *22*, 252-261.

Yan,M., Dai,T., Deak,J.C., Kyriakis,J.M., Zon,L.I., Woodgett,J.R., and Templeton,D.J. (1994). Activation of stress-activated protein kinase by MEKK1 phosphorylation of its activator SEK1. Nature *372*, 798-800.

Yang,D.D., Kuan,C.Y., Whitmarsh,A.J., Rincon,M., Zheng,T.S., Davis,R.J., Rakic,P., and Flavell,R.A. (1997). Absence of excitotoxicity-induced apoptosis in the hippocampus of mice lacking the Jnk3 gene. Nature *389*, 865-870.

Yang,H.S., Knies,J.L., Stark,C., and Colburn,N.H. (2003). Pdcd4 suppresses tumor phenotype in JB6 cells by inhibiting AP-1 transactivation. Oncogene *22*, 3712-3720.

Yang,J., Boerm,M., McCarty,M., Bucana,C., Fidler,I.J., Zhuang,Y., and Su,B. (2000). Mekk3 is essential for early embryonic cardiovascular development. Nat Genet *24*, 309-13.

Yoshida,K., Yamaguchi,T., Natsume,T., Kufe,D., and Miki,Y. (2005). JNK phosphorylation of 14-3-3 proteins regulates nuclear targeting of c-Abl in the apoptotic response to DNA damage. Nat. Cell Biol. *7*, 278-285.

Young,A.P. and Longmore,G.D. (2004). Ras protects Rb family null fibroblasts from cell death: a role for AP-1. J. Biol. Chem. *279*, 10931-10938.

Young,M.R., Li,J.J., Rincon,M., Flavell,R.A., Sathyanarayana,B.K., Hunziker,R., and Colburn,N. (1999). Transgenic mice demonstrate AP-1 (activator protein-1) transactivation is required for tumor promotion. Proc. Natl. Acad. Sci. U. S. A *96*, 9827-9832.

Yujiri,T., Ware,M., Widmann,C., Oyer,R., Russell,D., Chan,E., Zaitsu,Y., Clarke,P., Tyler,K., Oka,Y., Fanger,G.R., Henson,P., and Johnson,G.L. (2000). MEK kinase 1 gene disruption alters cell migration and c-Jun NH2-terminal kinase regulation but does not cause a measurable defect in NF-kappa B activation. Proc Natl Acad Sci USA *97*, 7272-7.

Zenz,R., Scheuch,H., Martin,P., Frank,C., Eferl,R., Kenner,L., Sibilia,M., and Wagner,E.F. (2003). c-Jun Regulates Eyelid Closure and Skin Tumor Development through EGFR Signaling. Dev. Cell *4*, 879-889.

Zerial,M., Toschi,L., Ryseck,R.P., Schuermann,M., Muller,R., and Bravo,R. (1989). The product of a novel growth factor activated gene, fos B, interacts with JUN proteins enhancing their DNA binding activity. EMBO J. *8*, 805-813.

Zetterberg,A., Larsson,O., and Wiman,K.G. (1995). What is the restriction point? Curr. Opin. Cell Biol. *7*, 835-842.

Zhang,L., Deng,M., Parthasarathy,R., Wang,L., Mongan,M., Molkentin,J.D., Zheng,Y., and Xia,Y. (2005). MEKK1 transduces activin signals in keratinocytes to induce actin stress fiber formation and migration. Mol. Cell Biol. *25*, 60-65.

Zhang,L., Wang,W., Hayashi,Y., Jester,J.V., Birk,D.E., Gao,M., Liu,C.Y., Kao,W.W., Karin,M., and Xia,Y. (2003). A role for

MEK kinase 1 in TGF-beta/activin-induced epithelium movement and embryonic eyelid closure. Embo J 22, 4443-4454.

Zhang,X., Wrzeszczynska,M.H., Horvath,C.M., and Darnell,J.E., Jr. (1999a). Interacting regions in Stat3 and c-Jun that participate in cooperative transcriptional activation. Mol. Cell Biol. 19, 7138-7146.

Zhang,Y., Neo,S.Y., Wang,X., Han,J., and Lin,S.C. (1999b). Axin forms a complex with MEKK1 and activates c-Jun NH(2)-terminal kinase/stress-activated protein kinase through domains distinct from Wnt signaling. J. Biol. Chem. 274, 35247-35254.

Chapter 14

HIV Tat and the Control of Transcriptional Elongation

Ruichuan Chen[1] and Qiang Zhou[1,2]

[1]School of Life Sciences, Xiamen University, Xiamen 361005, China
[2]Department of Molecular and Cell Biology, University of California, Berkeley, CA 94720

Key Words: transcriptional elongation, P-TEFb, HIV-1 Tat, Pol II CTD

Summary

The elongation stage of eukaryotic transcription has only recently been recognized as a highly dynamic and regulated process that couples transcription with other major gene expression events. Our appreciation of its role in controlling gene expression has benefited tremendously from the analysis of a single model system that centers on the mechanism regulating HIV transcriptional elongation. Tat is an essential regulatory protein encoded by the HIV virus. It cooperates with the host cellular cofactor P-TEFb to stimulate the production of the full-length HIV transcripts. Consisting of CDK9 and cyclin T, P-TEFb also functions as a general transcriptional elongation factor globally required for eukaryotic gene expression. It activates transcription by phosphorylating RNA polymerase II, leading to the formation of a highly processive elongation complex. Here, we review recent advances concerning Tat, P-TEFb and their control of HIV elongation. Additionally, we discuss how the studies of P-TEFb impact our understanding of the general control of eukaryotic gene expression at the stage of transcriptional elongation.

Introduction

In eukaryotic cells, the transcription of messenger RNA (mRNA) from protein-coding genes is performed by RNA polymerase II in a cyclic process consisting of several tightly connected stages designated as pre-initiation, initiation, promoter clearance, elongation and termination (Shilatifard, 2004; Sims et al., 2004). Over the past two decades, efforts to understand the molecular mechanisms of transcriptional regulation have mainly focused on the assembly of pre-initiation complexes and the initiation process, resulting in the identification of many factors that are required for these early steps of mRNA transcription (Shilatifard, 2004; Sims et al., 2004). In contrast, studies analyzing the elongation stage of transcription have lagged behind, largely due to the lack of an appropriate model system as well as gene-specific elongation factors. For many years, transcriptional elongation has been thought of as repetitive, unregulated additions of ribonucleoside triphosphates to the growing mRNA chain. In addition, the processing of pre-mRNA has been considered as a series of post-transcriptional events separable from the transcription process. Only until recently, it has become increasingly clear that the elongation stage of eukaryotic transcription is a highly regulated process that is capable of not only generating full-length RNA transcripts but also coupling transcription with other major gene expression events such as pre-mRNA capping, splicing, and polyadenylation (Bentley, 1999; Howe, 2002; Kim et al., 2001; Proudfoot et al., 2002; Reines et al., 1999; Shilatifard et al., 2003; Steinmetz, 1997). Moreover, the recent years have also seen great advances in the biochemical analysis of the mechanisms and factors that control the elongation process. Some of the most significant advances have come from studies of one particular HIV-encoded regulatory protein called Tat and its host cellular cofactor P-TEFb (positive transcription elongation factor b). Today, Tat remains the best-characterized gene-specific transcriptional

Corresponding Author: Qiang Zhou, Tel: (510) 643-1697 (Office), (510) 643-0494 (Lab), Fax: (510) 643-6334, E-mail: qzhou@uclink4.berkeley.edu

elongation factor and P-TEFb a Tat-specific cofactor as well as a general elongation factor. These two factors work together to stimulate HIV elongation by modulating the activity of Pol II, resulting in the critical transition from abortive to productive elongation (Garriga and Grana, 2004; Jones, 1997; Price, 2000). In this review, we will focus our attention on novel insights into the mechanisms and factors controlling general and HIV-specific transcriptional elongation with a special emphasis on P-TEFb and Tat. Furthermore, we will examine the contributions of P-TEFb and another elongation factor Tat-SF1 in coupling transcription with pre-mRNA splicing. Finally, we will discuss how the activity of P-TEFb as a general elongation factor is regulated and how this regulation may lead to the global control of cell growth and differentiation.

RNA Pol II in Transcriptional Elongation

A: RNA Pol II CTD

Two major forms of RNA Pol II exist in eukaryotic cells: the hypophosphorylated Pol IIa or the hyperphosphorylated RNA Pol IIo. In Pol IIo, the carboxy-terminal domain (CTD) of the largest subunit (Rpb1) of Pol II is extensively phosphorylated (Lin et al., 2002; Svejstrup, 2004). The CTD consists of heptapeptide repeats with a consensus sequence $Y_1S_2P_3T_4S_5P_6S_7$, which is conserved throughout the eukaryotic kingdom. However, the number of the repeats varies among different species. The CTD of S. cerevisiae Pol II has 26 repeats, while C. elegans has 32, Drosophila 42, and mammals contain 52 repeats (Allison et al., 1985; Corden et al., 1985; Dahmus, 1995; Kobor and Greenblatt, 2002; Oelgeschlager, 2002). For a given species, there is a set threshold number of heptapeptide repeats that is critical for Pol II to exert its full function. For example, a minimal length of 9 repeats in yeast or 28 repeats in human cells is necessary for cell survival. Yeast strains with a mutant Pol II containing between 9 and 20 repeats are cold-sensitive and display defects in the transcription of a number of genes (Allison et al., 1988; Nonet et al., 1987). On the other hand, mice with a Pol II that has only 39 repeats are either growth retarded or display an increased neonatal lethality (Litingtung et al., 1999). The reason for the requirement for a critical number of heptapeptide repeats within the CTD is because during transcription the CTD interacts with a variety of accessory factors functioning at different stages of the transcription cycle and also serves as a platform for the operation of various machineries involved in co-transcriptional processing. This point is illustrated by the observations that the truncation of the CTD causes defects not only in transcription but also in pre-mRNA capping, splicing as well as cleavage/polyadenylation (Kobor and Greenblatt, 2002; McCracken et al., 1997; McCracken et al., 1997; West and Corden, 1995).

B: Phosphorylation Cycle of the Pol II CTD

During the transcription cycle, the CTD of RNA Pol II undergoes a cycle of phosphorylation and dephosphorylation (Dahmus, 1996). Hypophosphorylated Pol II is the form that is recruited to promoters for formation of the pre-initiation complex. Shortly after transcription begins, the CTD becomes heavily phosphorylated (Lin et al., 2002). In eukaryotes, three major cyclin-dependent kinases, CDK7, CDK8, and CDK9 specifically target and phosphorylate the CTD of RNA Pol II (Prelich, 2002). Although these three kinases are part of different protein complexes, they themselves are evolutionarily conserved in all eukaryotes. CDK7 (also called MO15 or Kin28 in yeast) is a component of the general transcription factor TFIIH, which contains several core subunits and an enzymatically active CAK sub-complex consisting of CDK7, cyclin H and Mat1 (Orphanides et al., 1996). Whereas the TFIIH-free CAK can exist as a free, independent complex and function as a cell cycle regulator by targeting and activating a set of CDKs that are involved in cell cycle progression, the association with the TFIIH core subunits changes CAK's substrate specificity and enables it to phosphorylate the CTD at the Ser5 position within the heptapeptide repeats (Akoulitchev et al., 1995). The phosphorylation of Ser5 is not only essential for the formation of the first phosphodiester bond in the nascent RNA chain, but also important for Pol II to clear the promoter and shift into an early elongation mode (Orphanides and Reinberg, 2002; Rodriguez et al., 2000; Sims et al., 2004). Moreover, it also allows the CTD to recruit capping activities to the 5' end of the nascent pre-mRNA (also see below) (Komarnitsky et al., 2000; Rodriguez et al., 2000; Schroeder et al., 2000).

The second CTD kinase, CDK8, is a component of the Mediator complex, which plays an intermediary role in bridging the interactions of transcriptional activators with the basal transcriptional apparatus. CDK8 has been reported to function in transcriptional events prior to the elongation step owing to its ability to phosphorylate the Pol II CTD (Leclerc et al., 1996; Nelson et al., 2003; Rickert et al., 1996). However, its precise role in regulating transcription remains largely unknown.

Soon after transcription initiation, the transcription elongation complex is arrested at a "checkpoint" (Fig.14.1) to ensure proper pre-mRNA capping. Release from this checkpoint requires the enzymatic action of the third CTD kinase CDK9 (Bur1 or Ctk1 in yeast), which targets Ser2 in the CTD. From now on till the end of the transcription cycle, phosphorylation of Ser2 predominates (Price, 2000; Zhu et al., 1997). This particular phosphorylation event enables Pol II to resist pausing caused by the concerted actions of DSIF and NELF (see below for details) (Wada et al., 1998; Yamaguchi et al., 1999; Yamaguchi et al., 2002). The phosphorylation of Ser2 in the CTD, the Spt5 subunit of DSIF (Ivanov et al. 2000) and the RD subunit of NELF (Fujinaga et al. 2004) by CDK9 results in heavily phosphorylated Pol IIo, converts DSIF and NELF into positive elongation factors, and enables a shift from the abortive to the productive phase of transcriptional elongation (Renner et al., 2001; Wada et al., 1998; Yamaguchi et al., 1999). In addition, the Ser2 phosphorylation functions in part by recruiting RNA processing factors involved in pre-mRNA splicing and 3′ end polyadenylation to facilitate the cotranscriptional processing (see below for details).

At the end of the transcription cycle and upon the generation of a mature mRNA, Pol II must dissociate from DNA and undergo dephosphorylation in order to be recycled into the hypophosphorylated IIa form for subsequent rounds of transcription (Cho et al., 2001; Licciardo et al., 2001; Palancade et al., 2001). A protein phosphatase named FCP1 is the best-characterized enzyme that acts on the Pol II CTD (Fig. 14.1). Human FCP1 is able to dephosphorylate both Ser2 and Ser5 in the CTD, whereas the S. pombe FCP1 displays a preference for Ser2 over Ser5 (Cho et al., 2001; Hausmann et al., 2004; Mandal et al., 2002). Interestingly, besides participating in the recycling of Pol II, FCP1 was found to associate with the elongating Pol II both in vivo and in vitro and exhibit elongation stimulatory activity in a manner independent of its phosphatase catalytic activity (Cho et al., 2001; Mandal et al., 2002). In addition to FCP1, a few other protein phosphatases (e.g. Ssu72 and protein phosphatase 1) have also been shown to target the CTD and implicated in recycling of the RNA Pol II (Krishnamurthy et al., 2004; Washington et al., 2002), although the mechanistic details of their actions remain very sketchy at this moment.

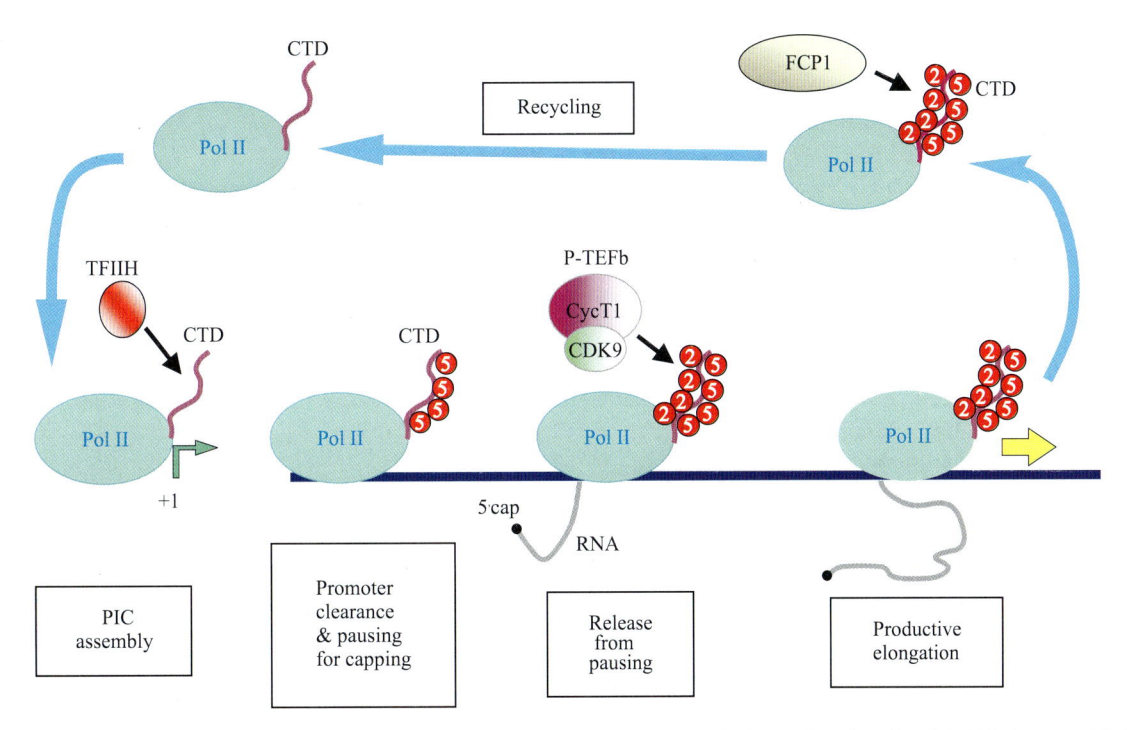

Fig. 14.1 Phosphorylation cycle of the Pol II CTD. During the transcription cycle, hypophosphorylated Pol IIa is assembled into the pre-initiation complex (PIC). Shortly after initiation, Ser5 of the CTD is phosphorylated by CDK7, a component of the TFIIH complex. This phosphorylation facilitates both the promoter clearance to shift into an early elongation mode and the formation of the 5′ cap structure in the nascent RNA chain. Subsequently, the phosphorylation on Ser2 of the CTD by P-TEFb enables Pol II to shift into a productive elongation mode. Upon the generation of a mature mRNA, the heavily phosphorylated Pol II undergoes dephosphorylation by the specific phosphatase FCP1 in order to be recycled back into the hypophosphorylated IIa form for subsequent rounds of transcription.

The Role of P-TEFb in Controlling Transcriptional Elongation and the Coupling of Transcription with RNA Processing

A: Discovery, Composition and Biogenesis of P-TEFb

Like all other cyclin-dependent kinases (CDKs), CDK9 does not have any catalytic activity unless it complexes with its regulatory partner cyclin T1 (CycT1) to form the CDK9/CycT1 heterodimer, which is termed P-TEFb (positive transcription elongation factor b). The 42-kDa CDK9 in complex with CycT1 account for ~80% of the total P-TEFb in human HeLa cells (Peng et al., 1998; Price, 2000; Wei et al., 1998). Besides CDK9(42), there is also a second isoform of CDK9 called CDK9(55) with a 55-kDa molecular mass and a 117-residue amino terminal extension not present in the 42-kDa form of CDK9 (Shore et al., 2005; Shore et al., 2003). In addition to CycT1, minor CDK9-associated CycT2a, T2b, and K molecules are also present at much lower concentrations than CycT1 in many cell types (Price, 2000; Shore et al., 2003). CDK9 was first identified as the cell division cycle 2 (CDC2)-related protein kinase during a cDNA screening intended to isolate novel regulators of the mammalian cell cycle (Grana et al., 1994). Subsequently, it was rediscovered as a component of the P-TEFb complex purified from *Drosophila* and mammalian cell nuclear extracts (Gold et al., 1998; Herrmann and Rice, 1995; Marshall et al., 1996; Marshall and Price, 1995; Yang et al., 1996; Zhu et al., 1997). Unlike other CDKs with known cell cycle regulatory functions, the expression levels of CDK9 and CycT1 as well as the kinase activity of the isolated CDK9/CycT1 heterodimer in the absence of any associated factors are fairly constant throughout the cell cycles and also during cell cycle entry from a quiescent state (Garriga et al., 2003; Grana et al., 1994). These observations suggest that the CDK9/CycT1 heterodimer of P-TEFb is unlikely to function as a conventional cell cycle regulator, although its activity could still be regulated by the associated factors (see below) for cell cycle progression.

Although a connection of P-TEFb with the cell cycle control is yet to be established, a role for P-TEFb in regulating RNA Pol II transcriptional elongation has become crystal clear. In fact, P-TEFb is required for the expression of most genes in *C. elegans* and mammals (Chao and Price, 2001; Garriga and Grana, 2004; Shim et al., 2002). When CDK9 in *C. elegans* early embryo was depleted by RNAi, a technique used to specifically suppress a target gene expression, the CTD phosphorylation on Ser2 but not on Ser5 was found to be dramatically reduced (Shim et al., 2002). These RNAi experiments further demonstrate that P-TEFb is broadly essential for the expression of early embryonic genes. This result agrees with an earlier observation, which uses a pharmacological inhibitor of CDK9 to show that P-TEFb is required for the expression of most protein-coding genes in mammalian cells (Chao and Price, 2001), indicating that P-TEFb is globally required for the transcription of cellular genes.

In mammalian cells, the formation of the kinase-active CDK9/CycT1 heterodimer of P-TEFb involves more than just the combining of CDK9 and CycT1 together. Rather, a kinase-specific chaperone pathway is required for the production of the mature CDK9/CycT1 heterodimer responsible for P-TEFb-mediated Tat stimulation of HIV-1 transcription (O'Keeffe et al., 2000). Before interacting with CycT1 in the nucleus, CDK9 binds to the molecular chaperone Hsp70 and then a kinase-specific chaperone complex, Hsp90/Cdc37, to form two separate chaperone-CDK9 complexes in the cytoplasm. Pharmacological inactivation of Hsp90/Cdc37 function by geldanamycin revealed that these two complexes act sequentially to facilitate CDK9 folding/stabilization and the subsequent formation of a mature and active CDK9/CycT1 heterodimer (O'Keeffe et al., 2000).

B: Targets of P-TEFb during Early Transcriptional Elongation

In species ranging from *Drosophila*, *C. elegans* to mammals, phosphorylation of Ser2 of the Pol II CTD is mediated primarily by CDK9 of P-TEFb (Garriga and Grana, 2004; Price, 2000). This phosphorylation is crucial for the transition from the abortive to the productive phase of transcriptional elongation, leading to the generation of full-length RNA transcripts (Jones, 1997; Price, 2000). In support of a model in which P-TEFb acts to stimulate the elongation potential (namely processivity) of the paused Pol II in the vicinity of promoters, the hypophosphorylated Pol IIa is found to co-localize with P-TEFb at promoter-proximal pausing sites (Andrulis et al., 2000; Boehm et al., 2003). Additional experiments suggest that P-TEFb appears to track along with the transcribing Pol II during productive elongation (Boehm et al., 2003). Besides the phosphorylation of the Pol II CTD on Ser2, P-TEFb can also phosphorylate the subunits of DSIF and NELF, removing the coordinated inhibition of Pol II elongation exerted by these two regulatory factors and leading to the stimulation of early elongation (Fig.14.2) (Fujinaga et al., 2004; Ivanov et al., 2000; Kim and Sharp, 2001; Renner et al., 2001; Wada et al., 1998; Yamaguchi et al., 1999; Zhu et al., 1997).

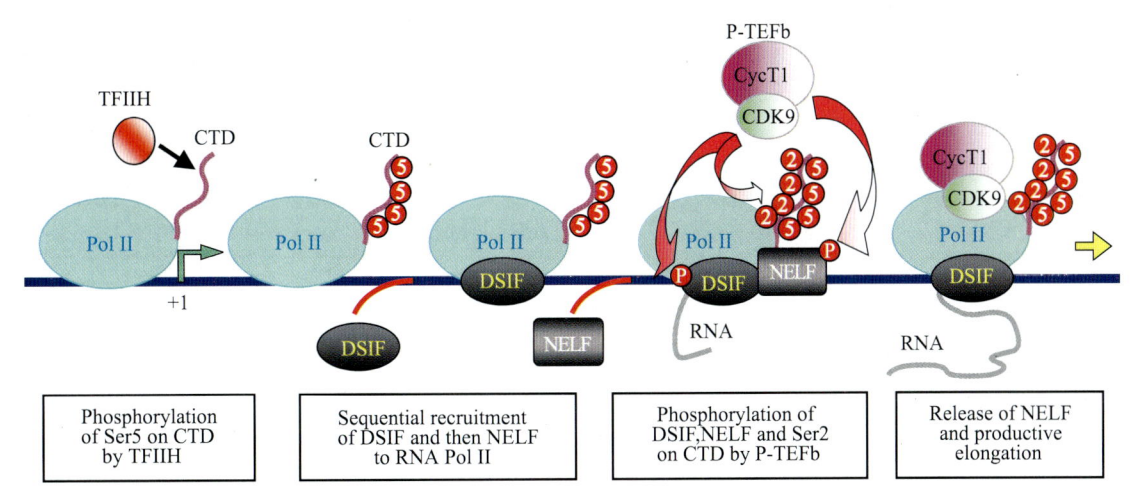

Fig. 14.2 Negative and positive control of eukaryotic transcriptional elongation. RNA Pol II, which is phosphorylated on Ser5 in the CTD and thus in an early elongation mode, is paused by the concerted actions of DSIF and NELF, which are recruited sequentially to Pol II. This pausing establishes a natural and necessary "checkpoint" to ensure the proper capping of the nascent pre-mRNA. Release from this "checkpoint" requires P-TEFb, which is recruited to the elongation complex and which phosphorylates Ser2 in the Pol II CTD, the Spt5 subunit of DSIF and the RD subunit of NELF. These phosphorylation events presumably result in the disassociation of NELF and resumption of elongation. During productive elongation, P-TEFb and DSIF track along with the elongating Pol II, mostly through direct interactions with Pol II.

DSIF (for DRB-sensitivity-inducing factor), a heterodimer consisting of Spt4 and Spt5, was initially identified based on its ability to confer sensitivity to DRB (5,6-dichloro-1-β-D-ribofuranosylbenzimidazole), a purine nucleoside analogue and an elongation inhibitor, to transcription reactions in vitro (Fujinaga et al., 2004; Ivanov et al., 2000; Kim and Sharp, 2001; Renner et al., 2001; Wada et al., 1998; Yamaguchi et al., 1999; Zhu et al., 1997). For a long time, DSIF has been thought as a pure and constitutively negative elongation factor. However, the recent findings have revealed a more complicated picture, implicating DSIF in the promotion of elongation (Sims et al., 2004). It now appears that in the absence of NELF, DSIF can bind to RNA Pol II directly to promote elongation and also interact with factors implicated in mRNA maturation and surveillance (Andrulis et al., 2002; Lindstrom et al., 2003; Pei and Shuman, 2002; Wen and Shatkin, 1999). The suppression of RNA Pol II elongation is in fact due to the recruitment of NELF, a true negative elongation factor, to the assembled DSIF/Pol II complex (Yamaguchi et al., 2002). It should be noted that the induction of RNA Pol II pausing during early elongation through the coordinative actions of DSIF and NELF is a natural and necessary step to provide a window of opportunity for the assembly of pre-mRNA processing activities. A "checkpoint" model has thus been proposed based on the pertinent information relevant to this step. According to this model (Fig.14.2), DSIF binds to Pol II during or shortly after initiation through a direct interaction with the phosphorylated CTD. The DSIF/Pol II complex then recruits NELF to induce polymerase pausing and establishes a "checkpoint" for the recruitment of capping enzymes to the 5'-end of the nascent RNA. P-TEFb, which is recruited either prior to or precisely at the "checkpoint", phosphorylates Ser2 of the CTD, the Spt5 subunit of DSIF and the RD subunit of NELF, presumably resulting in the disassociation of NELF and the resumption of elongation (Sims et al., 2004). In addition to P-TEFb, a recent study suggests that the capping enzyme itself can also relieve the NELF-mediated repression (Mandal et al., 2004).

C: Role of P-TEFb in Coupling Transcriptional Elongation with Pre-mRNA Splicing

Although the major events in gene expression have traditionally been analyzed individually, recent evidence indicates that they are in fact highly interconnected inside the cell. The interconnections described thus far, however, have all been one directional, involving one upstream event (such as splicing) affecting one or more downstream events (e.g. mRNA export and mRNA decay). We have recently observed that in a process that involves P-TEFb, another human elongation factor Tat-SF1 and the spliceosomal U snRNP complexes, transcriptional elongation and splicing can actually exhibit reciprocal synergism, suggesting that at least some of the interconnections can be mutually beneficial

(Fong and Zhou, 2001).

Previously, it has been shown that the C-terminal domain of the CycT1 subunit of P-TEFb is key to P-TEFb's elongation activity and it interacts with Tat-SF1 (Fong and Zhou, 2000), a factor also implicated in HIV-1 transcription (Kim et al., 1999; Li and Green, 1998; Parada and Roeder, 1999; Zhou et al., 1998; Zhou and Sharp, 1996). The observed importance of the CycT1 C-terminal domain in transcription prompted us to identify and investigate the roles of transcription factors including Tat-SF1 that may be targeted by this domain. Our studies have led to the demonstration that human Tat-SF1, like its yeast counterpart CUS2, interacts with the spliceosomal U snRNPs to form a multi-subunit complex. This complex strongly stimulates Pol II elongation when directed to an intron-free DNA template. This effect is mediated through the binding of Tat-SF1 to the CycT1 C-terminal domain. Notably, inclusion of splicing signals in the nascent transcript further stimulates transcription, supporting the notion that the recruitment of U snRNPs near the elongating polymerase is important for transcription. Since the Tat-SF1-snRNP complex also stimulates splicing *in vitro*, it may serve as a dual-function factor to mediate efficient coupling of transcription and splicing.

Transcription and pre-mRNA splicing are tightly coupled gene expression events in eukaryotic cells. An interaction between the Pol II CTD and components of the splicing machinery has been postulated to mediate this coupling. Our studies have uncovered a direct role of splicing factors in promoting transcriptional elongation, rendering transcription more intimately coupled to splicing than previously thought. Furthermore, besides the previously known function of P-TEFb in phosphorylating the CTD, DSIF and NELF, our studies have also revealed a novel role of P-TEFb in delivering the Tat-SF1:U snRNP complex to Pol II for stimulation of the coupled transcription and splicing (Fig.14.3).

In addition to playing a role in coupling splicing with transcription, recent reports have also implicated a role for *Drosophila* P-TEFb or *S. cerevisiae* Ctk1, a kinase similar to the mammalian CDK9, in coupling transcription with mRNA 3'-end polyadenylation, as the absence of *Drosophila* P-TEFb or yeast Ctk1 activity results in defects in 3' end processing (Ahn et al., 2004; Ni et al., 2004). It remains to be determined whether mammalian CDK9 plays a similar role in this process.

D: The Recruitment of P-TEFb to Transcriptional Templates

Neither CDK9 nor CycT1 is known to have any sequence-specific DNA-binding activity. How then is P-TEFb recruited to the transcriptional templates for stimulation of Pol II elongation? In fact, several DNA

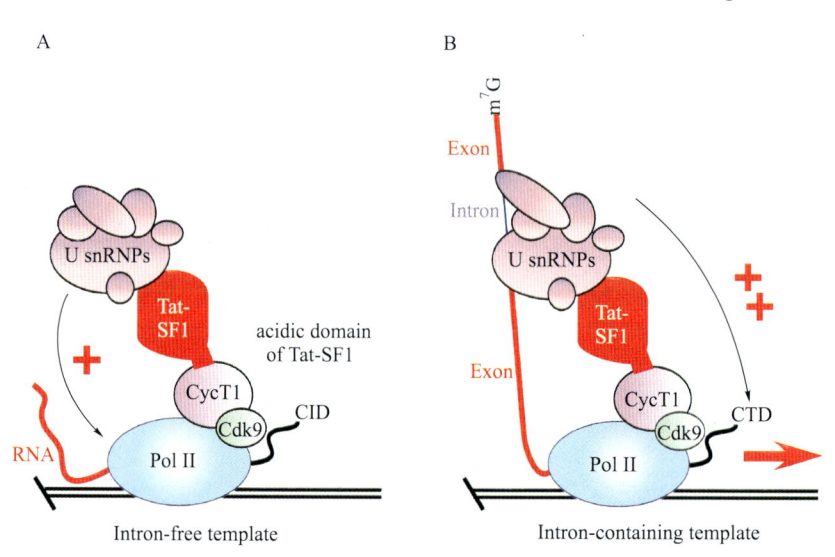

Fig.14.3 **A role for P-TEFb in coupling transcriptional elongation with pre-mRNA splicing. (A)** The Tat-SF1:U snRNP complex is recruited by P-TEFb to elongating Pol II to stimulate elongation on intron-free DNA templates. The spliceosomal U snRNPs interact with human transcription elongation factor Tat-SF1, which associates with the elongation complex through a direct interaction with CycT1 of P-TEFb. The formation of this multi-protein complex strongly stimulates polymerase elongation when directed to an intron-free DNA template. **(B)** The presence of splice sites in nascent transcripts may help recruit/stabilize U snRNPs to further stimulate Pol II transcriptional activity. Inclusion of splicing signals in the nascent transcript can further stimulate transcriptional elongation, as these signals contribute to the recruitment of the Tat-SF1:U snRNP complex to the elongating Pol II through an interaction between U snRNPs and the splice sites in RNA. On the other hand, the Tat-SF1:U snRNP complex also stimulates splicing *in vitro*, suggesting that it may serve as a dual-function factor to mediate efficient coupling of transcription and splicing.

sequence-specific transcription factors, such as CIITA (Kanazawa *et al.*, 2000), NF-κB (Barboric *et al.*, 2001), Myc (Eberhardy and Farnham, 2001), STAT3 (Giraud *et al.*, 2004), the androgen receptor (Lee *et al.*, 2001), the aryl hydrocarbon receptor (Tian *et al.*, 2003), MyoD (Simone *et al.*, 2002), HIC (Young *et al.*, 2003), B-Myb (De Falco *et al.*, 2000), GRIP1 (Kino *et al.*, 2002) and MCEF (Estable *et al.*, 2002) have been identified as P-TEFb-associated factors that can potentially recruit P-TEFb to their respective promoter targets (Simone and Giordano, 2001). Because these are gene-specific transcription factors, their recruitment of P-TEFb would be expected to affect only a limited group of genes that bear their specific recognition sequences.

However, given that P-TEFb is globally required for transcription of a vast majority of cellular genes (Chao and Price, 2001; Lam *et al.*, 2001; Shim *et al.*, 2002), it is conceivable that there should exist a common mechanism to recruit P-TEFb to generic cellular promoters that do not contain any binding sites for the above-mentioned P-TEFb-associated transcription factors. The recent identification of Brd4 as a major P-TEFb-associated factor (Jang *et al.*, 2005) has prompted us to investigate a potential role of Brd4 in the recruitment of P-TEFb to transcriptional templates. Brd4 (also known as MCAP) is an ubiquitously expressed protein belonging to the conserved BET family of proteins that carry two tandem bromodomains and an ET (extra terminal) domain (Jeanmougin *et al.*, 1997; Shang *et al.*, 2004). The bromodomain has been recognized as a functional module in helping decipher the histone code through interacting with acetylated histone tails (Zeng and Zhou, 2002). Consistent with this view, Brd4 has been shown to bind to acetylated euchromatin through acetylated histone H3 and H4 (Dey *et al.*, 2003). Using immuno-affinity purification, both CDK9 and CycT1 have been identified as Brd4-associated factors. Subsequent studies indicate that the association with Brd4 is strictly required for forming the transcriptionally active P-TEFb for stimulation of RNA Pol II elongation. Furthermore, Brd4 contributes to general transcriptional elongation through its recruitment of P-TEFb to transcriptional templates *in vivo* and *in vitro* (see Fig.14.6 below) (Jang *et al.*, 2005; Yang *et al.*, 2005). What remains to be determined is the target of the Brd4-mediated recruitment of P-TEFb to the transcriptional templates, although the acetylated histones are an obvious candidate because of the bromodomains in Brd4. However, since the stimulatory effect of Brd4 on P-TEFb's transcription and template-binding activities can also be observed *in vitro*, where non-chromatinized DNA template is used, it is possible that other factor such as the human Mediator complex may also be targeted by Brd4 for the recruitment of P-TEFb (Yang *et al.*, 2005).

Upon the recruitment of P-TEFb by Brd4, it is not known whether the Brd4:P-TEFb complex can track along with the elongating Pol II or it has to be disrupted to leave P-TEFb alone with Pol II. In a series of experiment, we found that the P-TEFb-recruitment role of Brd4 can be functionally substituted by that of the HIV Tat protein (Yang *et al.*, 2005), which recruits P-TEFb to the HIV LTR for activated HIV transcription (see below). In addition, the overexpressed Brd4 proves to be inhibitory for Tat-transactivation, due the competition between Tat and Brd4 for binding to the same P-TEFb complex. These observations strongly suggest that the Brd4:P-TEFb complex may be disrupted to leave P-TEFb traveling with Pol II along at least the HIV proviral DNA template. A previous study has shown that the C-terminus of CycT1 can directly interact with the Pol II CTD (Taube *et al.*, 2002), suggesting that P-TEFb may not require other factors to mediate its interaction with the elongating Pol II once it has been brought to the DNA template by Brd4. The recent observation that the PIE-1 transcriptional repressor of *C. elegans* can inhibit transcriptional elongation by blocking the binding of CycT1 to the Pol II CTD through an alanine-heptapeptide repeat (Zhang *et al.*, 2003) is consistent with the notion that Brd4 may not be needed for the P-TEFb:CTD binding after the recruitment step.

Stimulation of HIV Transcriptional Elongation by Tat and P-TEFb

A: HIV Tat Protein and TAR RNA

Unlike simpler retroviruses that rely exclusively on host cellular machinery for gene expression, HIV, a lentivirus that causes the acquired immunodeficiency syndrome (AIDS), encodes additional regulatory proteins that work together with cellular machineries to further control viral replication (Barboric and Peterlin, 2005). Among these regulatory proteins, Tat, a small polypeptide of 101 amino acids in most clinical HIV-1 isolates, is recognized as a key factor essential for activating transcriptional elongation from the HIV LTR (long terminal repeat) and for productive viral replication (Brigati *et al.*, 2003; Karn, 1999). Because the proteins that regulate elongation for many cellular and viral genes in a gene-specific fashion have not been identified, HIV Tat has been used as a model system to study this process. As a result, our understanding of not only the regulation of HIIV gene expression but also the general mechanism controlling eukaryotic elongation has benefited tremendously from the analysis of Tat and its host cellular cofactor P-TEFb.

In the absence of Tat, the HIV LTR generates short or so-called nonprocessive transcripts. The presence of Tat, however, results in a large increase in the level of transcripts that extend through the more than 9-kb HIV genome. Tat stimulation of the efficiency of transcriptional elongation is primarily responsible for this dramatic increase in the level of full-length HIV transcripts (Barboric and Peterlin, 2005). Tat stimulates elongation by recognizing the trans-acting-response (TAR) RNA element. Located at the 5′ end of the nascent viral transcript (nucleotides +1 to +59), TAR forms a stem-loop structure (Fig. 14.4). The specific binding of Tat to TAR is primarily dependent on the 3-nt bulge and immediately flanking sequences in the double-stranded RNA region just below the apical loop. This binding enables the recruitment of Tat and through Tat the key cellular cofactor P-TEFb to the paused Pol II on the viral promoter (Barboric and Peterlin, 2005; Jones, 1997; Karn, 1999).

B: P-TEFb Stimulation of HIV Transcriptional Elongation through Tat and TAR

Besides functioning as a general elongation factor capable of overcoming the pausing of Pol II during early elongation (Price, 2000), P-TEFb was also identified as a Tat-associated kinase (so-called TAK) specifically required for HIV transcription and viral replication (Herrmann and Rice, 1995; Yang et al., 1997). RNA Pol II transcribing the integrated HIV proviral DNA has a particularly strong tendency to pause and then terminate close to the start site, producing only short transcripts. Preventing Pol II from stalling is essential for HIV transcription, during which P-TEFb is recruited to the nascent mRNA by Tat (Garriga and Grana, 2004; Jones, 1997; Price, 2000). Unlike most DNA sequence-specific transcription activators, Tat stimulates polymerase elongation by recognizing the HIV TAR RNA structure and recruiting P-TEFb to the HIV promoter through the Tat:CycT1 interaction and the formation of a ternary complex containing P-TEFb, Tat and TAR (Garber et al., 1998; Wei et al., 1998; Zhang et al., 2000). Once recruited, P-TEFb phosphorylates the Pol II CTD and stimulates transcriptional elongation to produce the full-length HIV-1 transcripts (Fig. 14.4).

A definitive proof that P-TEFb is the specific cellular cofactor for Tat function came from studies of HIV transcription in rodent cells. Tat activity has been shown to be species-specific, as it trans-activates the HIV LTR efficiently in many human and primate cell types but not in cells of other species (e.g., yeast, Drosophila, and rodent cells) (Jones and Peterlin, 1994), suggesting that there exists a species-restricted cellular cofactor for Tat function (Alonso et al., 1994; Hart et al., 1993). To investigate whether human P-TEFb contributes

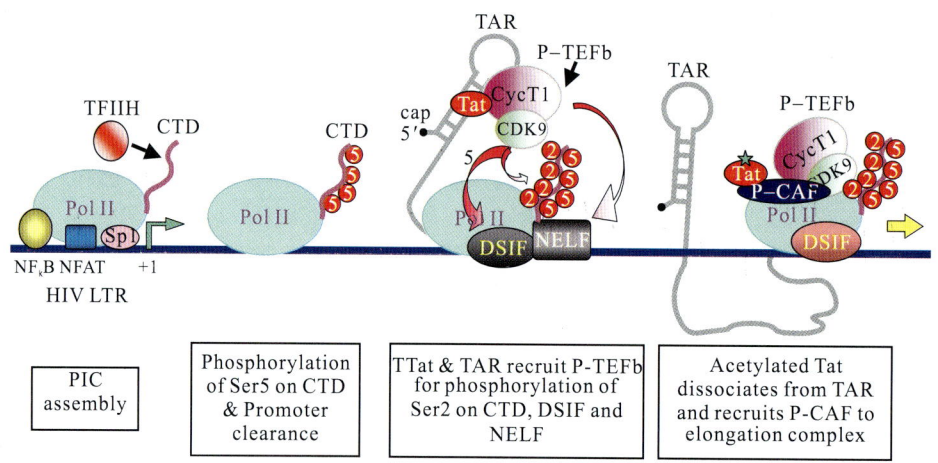

Fig. 14.4 Tat activation of HIV transcription. Transcription factors (e.g. NF-κB, NF-AT, Sp1, and etc.) bound to the HIV LTR promoter promote the formation of the PIC and recruitment of Pol IIa to the start site of transcription. After being phosphorylated on Ser5 in the CTD by TFIIH, Pol II clears the promoter and transcribes a short RNA transcript that forms the TAR RNA stem-loop structure. However, due to the concerted actions of DISF and NELF, Pol II pauses to form the 5′ cap structure. Tat, an HIV-encoded regulatory protein that recognizes the TAR RNA structure, then interacts with the CycT1 subunit of P-TEFb and recruits P-TEFb to the paused Pol II through formation of a ternary complex containing P-TEFb, Tat and TAR. Once recruited, P-TEFb phosphorylates the Pol II CTD on Ser2, DSIF and NELF to releases Pol II from pausing. Upon shifting into the productive elongation stage, Tat is acetylated by p300 and dissociated from the TAR RNA. The released Tat protein can recruit the P-CAF factor to the elongating Pol II, possibly to facilitate chromatin remodeling.

to the species-specific Tat function, we generated and examined the activities of stable human-rodent "hybrid" P-TEFb complexes that consist of CDK9 and CycT1 derived from these two different species. Although all P-TEFb complexes were found to phosphorylate the CTD and support basal level HIV transcription to a similar extent, only the complexes containing human CycT1 were capable of interacting with Tat/TAR and mediating Tat trans-activation (Chen et al., 1999). Moreover, we showed that the cyclin-box of human CycT1 and its immediate flanking region are responsible for the specific P-TEFb:Tat interaction. These studies identify the interaction of Tat with human CycT1 as essential for the species-specific Tat function. Concurrent to these studies, published results from several other laboratories largely came to the same conclusion (Garber et al., 1998; Ivanov et al., 1999; Kwak et al., 1999). They further showed that overexpression of human CycT1 in non-permissive rodent cells rescued Tat activation of HIV transcription. This activity was attributed to a critical cysteine residue at position 261 that is present in human CycT1 but absent in the rodent homologue. This cysteine residue is proposed to interact with the cysteine-rich activation domain in Tat coordinated by a zinc atom (Garber et al., 1998; Ivanov et al., 1999; Kwak et al., 1999).

C: The Assembly and Disassembly of the Complex Containing P-TEFb, Tat, and TAR

The assembly and disassembly of the ternary complex containing P-TEFb, Tat, and TAR is a highly regulated process *in vivo*. In fact, P-TEFb is intrinsically incapable of forming a stable complex with Tat and TAR due to two built-in autoinhibitory mechanisms in P-TEFb. The first arises from the lack of phosphorylation on several key serine and threonine residues near the C-terminus of CDK9, which renders P-TEFb to adopt a conformation unfavorable for TAR recognition. The second is caused by the intramolecular interaction between the N- and C-terminal regions of CycT1, which sterically blocks the P-TEFb:TAR interaction (Fong and Zhou, 2000; Garber et al., 2000). While the autophosphorylation of CDK9 (at least *in vitro*, could be mediated by other kinases *in vivo*) can overcome the first inhibition by inducing conformational changes in P-TEFb and thereby exposing a region in CycT1 for possible TAR binding, the second is relieved by the binding of the C-terminal region of CycT1 to Tat-SF1 and perhaps other cellular factors.

Whereas the phosphorylation of CDK9 stabilizes the P-TEFb:Tat:TAR ternary complex (Fong and Zhou, 2000; Garber et al., 2000), the acetylation of Tat by p300 on Lys50, which is located in the TAR RNA-binding domain of Tat, has been shown to dissociate Tat from the TAR RNA. Once liberated, the acetylated Tat can recruit the P-CAF factor to the elongating Pol II (Fig. 14.4), possibly to facilitate chromatin remodeling (Bres et al., 2002; Bres et al., 2002; Kiernan et al., 1999). Recently, it has been shown that Tat can be deacetylated by SIRT1, which recycles Tat to its unacetylated form and acts as a transcriptional co-activator during Tat trans-activation (Pagans et al., 2005). Taken together, these results have revealed novel control mechanisms for the assembly/disassembly of a multi-component transcription elongation complex at the HIV promoter through reversible modifications of the components within this complex.

Regulation of P-TEFb Activity

A: HEXIM1 and 7SK Cooperate to Inhibit P-TEFb

The activities of CDKs involved in cell cycle regulation are negatively regulated by proteins that associate with either CDKs or CDK/cyclin complexes to inhibit their kinase activities (Sherr and Roberts, 1999). Similarly, in the nucleus, not every CDK9/CycT1 heterodimer displays the P-TEFb kinase and transcriptional activities. In human HeLa cells, only about of half of cellular P-TEFb are in the active state and the regulation of P-TEFb activity is a dynamic and tightly controlled process involving the P-TEFb-associated factors. In an effort to isolate nuclear factors that can bind to and control the activity of human P-TEFb, we and others have identified the 7SK snRNA as a specific P-TEFb-associated factor (Nguyen et al., 2001; Yang et al., 2001). Transcribed by RNA Pol III, the human 7SK snRNA is an abundant (2×10^5 copies/cell) and evolutionarily conserved small nuclear RNA (snRNA) of 331 nucleotides (Murphy et al., 1987; Zieve et al., 1977). Despite the first description of 7SK in the 1970's, little is known about its function for the next quarter of century (Murphy et al., 1987; Zieve et al., 1977). Our data indicate that there exist two major forms of P-TEFb with distinct 7SK-binding abilities in human cells. The 7SK-free form, which accounts for about half of total P-TEFb in the cell, can function as both a general and HIV-specific transcription factor. In contrast, the 7SK-bound P-TEFb is inactive as its kinase and transcriptional activities are suppressed. Moreover, when associated with 7SK, P-TEFb cannot even be recruited to the HIV promoter *in vivo* and *in vitro* (Nguyen et al., 2001; Yang et al., 2001).

While investigating the functional significance of a reconstituted interaction between 7SK and P-TEFb, we found that the association with 7SK is necessary but not sufficient to inactivate P-TEFb, implicating the presence of another factor in the 7SK snRNP for P-TEFb inactivation (Yik et al., 2003). Indeed, through affinity-purification, a protein factor called HEXIM1 has been identified as the third protein component of the 7SK:P-TEFb snRNP formed *in vivo* (Michels et al., 2003; Yik et al., 2003). Importantly, HEXIM1 can potently and specifically inhibit the kinase and transcriptional activities of P-TEFb in a 7SK-dependent manner (Yik et al., 2003). HEXIM1 has previously been identified as a nuclear protein whose expression is induced in many transformed cell types treated with hexamethylene bisacetamide (HMBA) (Ouchida et al., 2003), a potent inducer of cell differentiation. The 7SK-dependent inhibition of P-TEFb by HEXIM1 can be explained by the fact that 7SK plays a scaffolding role in mediating the interaction between HEXIM1 and P-TEFb, enabling HEXIM1 to exert its inhibitory effect on P-TEFb *in vivo* and *in vitro* (Yik et al., 2003).

B: A Similar Architectural Plan for Forming the Tat:TAR: P-TEFb and HEXIM1:7SK:P-TEFb Complexes

In an effort to define the sequence requirements for HEXIM1 to interact with 7SK and inactivate P-TEFb, an arginine-rich motif that overlaps with the nuclear localization signal (NLS) near the center of HEXIM1 (Fig. 14.5) has been shown to mediate a direct and specific interaction of HEXIM1 with 7SK (Yik et al., 2004). This motif, together with the HEXIM1 C-terminal domain that is required for the interaction with and inhibition of P-TEFb, allow HEXIM1 to inhibit Pol II transcription through the 7SK-mediated inactivation of P-TEFb. Interestingly, the 7SK-binding motif in HEXIM1 is highly homologous to and functionally interchangeable with the arginine-rich TAR-binding motif in HIV Tat (Fig.14.5) (Yik et al., 2004), suggesting that a similar architectural plan may exist to form both the Tat:TAR:P-TEFb and the HEXIM1:7SK:P-TEFb ternary complexes. This hypothesis, while yet to be confirmed through further structural analysis, raises an intriguing possibility that the nuclear level of HEXIM1 can be therapeutically manipulated to effectively modulate the amount of P-TEFb in the Tat:TAR:P-TEFb complex, which in turn would affect HIV transcription.

C: Dynamic Exchange of Partners between Active and Inactive P-TEFb Complexes

In HeLa cell, about half of nuclear P-TEFb are sequestered in the inactive 7SK:HEXIM1:P-TEFb snRNP. Importantly, the amount of the HEXIM1/7SK-bound P-TEFb does not remain static in the cell but rather undergoes dynamic changes in response to various stimuli. For example, treatment of cells with several stress-inducing agents, such as the global transcription inhibitors actinomycin D and DRB or the DNA-damaging agent UV-irradiation, rapidly dissociates 7SK and HEXIM1 from P-TEFb without affecting the expressions of CDK9 and CycT1 or the CDK9/CycT1 dimer formation (Fig. 14.6) (Nguyen et al., 2001; Yang et al., 2001). Although these agents generally cause a global inhibition of transcription, earlier observations have indicated that at least for actinomycin D and UV, a low dosage can induce the phosphorylation of the Pol II CTD and activation of HIV transcription (Casse et al., 1999; Valerie et al., 1988). Moreover, UV irradiation of human T-cells prior to HIV infection also significantly shortens the viral growth cycle (Valerie et al., 1988). The ability of these agents to dissociate HEXIM1 and 7SK and activate P-TEFb provides a mechanistic explanation for their effects on HIV transcription and replication. In contrast to the stress treatment that causes the disruption of the 7SK:HEXIM1:P-TEFb snRNP, HEXIM1 expression is known to be elevated in cells treated with HMBA (Ouchida et al., 2003). The induced HEXIM1 expression could potentially sequester more P-TEFb into the inactive 7SK snRNP (Fig.14.6), although this notion is yet to be tested experimentally.

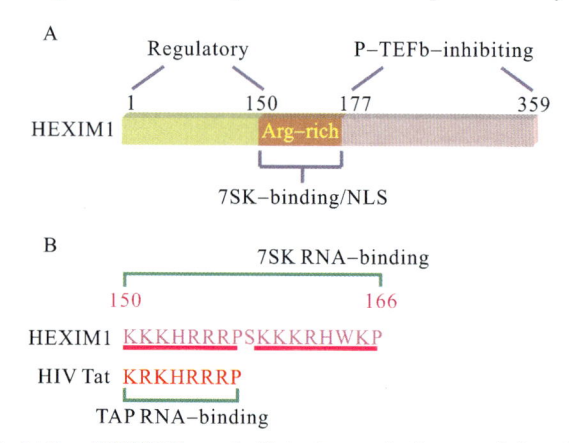

Fig.14.5 HEXIM1 and Tat share similar arginine-rich RNA-binding domains. (**A**) The molecule anatomy of the HEXIM1 protein. An arginine-rich motif that overlaps with the nuclear localization signal (NLS) near the center of HEXIM1 is essential for the direct and specific interaction of HEXIM1 with the 7SK snRNA. The C-terminal domain of HEXIM1 is required for the interaction with and inhibition of P-TEFb, whereas the N-terminal domain presumably functions as a regulatory domain. (**B**) The arginine-rich RNA-binding domains of HEXIM1 and Tat are highly homologous to each other. The 7SK-binding motif in HEXIM1 has a bipartite structure with the first half showing a near perfect match with the arginine-rich TAR-binding motif in HIV Tat.

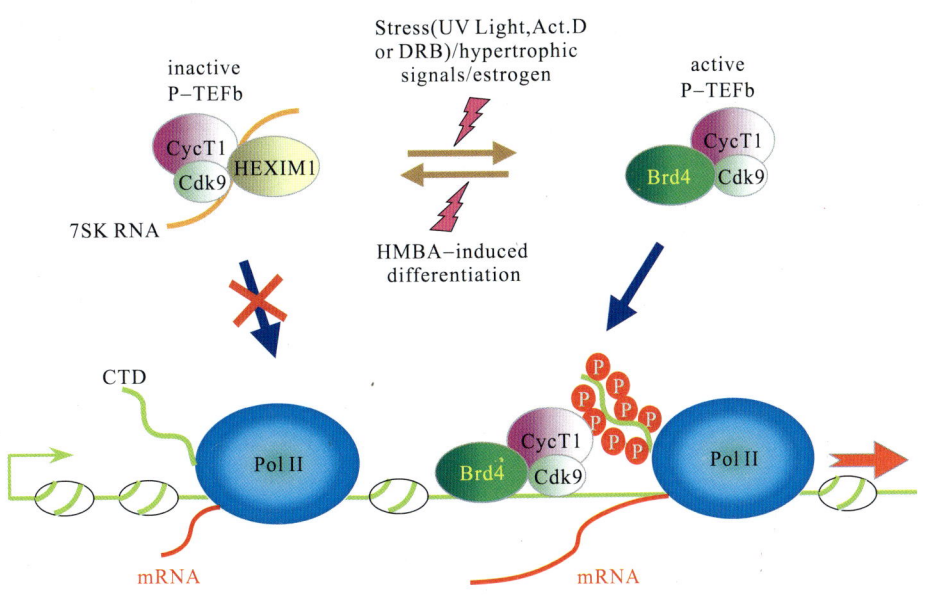

Fig.14.6 Cellular P-TEFb heterodimers exist in two distinct states that are kept in a functional equilibrium. In HeLa cell, about half of nuclear P-TEFb are sequestered in the inactive 7SK:HEXIM1:P-TEFb snRNP, where HEXIM1 binds to and inhibits P-TEFb's kinase activity through a 7SK-dependent process. The 7SK snRNP is unable to phosphorylate Pol II or associate with a promoter. Treatment of cells with certain stress-inducing agents or hypertrophic signals can cause rapidly dissociation of 7SK and HEXIM1 from P-TEFb and quantitatively convert the released P-TEFb into a Brd4-bound form. In contract, the cell differentiation inducer HMBA could presumably sequester more P-TEFb into the inactive 7SK snRNP by inducing the expression of HEXIM1. Although the exact target of the Brd4-mediated recruitment of P-TEFb to promoters remains to be characterized, Brd4 has been shown to contribute to general transcriptional elongation through its recruitment of P-TEFb to transcriptional templates.

Besides stress-induction in HeLa cells, treatment of cardiac myocytes with conditions that cause cardiac hypertrophy has also been shown to induce the disruption of the 7SK snRNP and activation of P-TEFb (Sano et al., 2002). Because P-TEFb activity is limiting in normal cardiac myocytes, the activation of P-TEFb by hypertrophic signals can lead to a global increase in cellular RNA and protein contents and consequently the enlargement of heart cells, which is the cause of hypertrophy (Sano et al., 2002). Consistent with the notion that HEXIM1 has a growth-suppressing function in normal heart growth and development through its inactivation of P-TEFb, ablation of the HEXIM1 (called CLP-1 in mouse) gene in mice has been shown to cause the same disease of cardiac hypertrophy in developing embryos, which results in fetal death (Huang et al., 2004).

As mentioned above, in contrast to HEXIM1 and 7SK that are inhibitory to P-TEFb's function, the bromodomain protein Brd4 has recently been identified as a P-TEFb-associated factor, which stimulates transcription by recruiting P-TEFb to DNA templates (Jang et al., 2005; Yang et al., 2005).It now appears that there is no free P-TEFb in HeLa cells. Brd4 in fact associates with the other half of P-TEFb that are free of HEXIM1 and 7SK. More importantly, in stress-treated cells, the 7SK/HEXIM1-bound P-TEFb can be efficiently and quantitatively converted into the Brd4-associated form for stress-induced transcription (Jang et al., 2005; Yang et al., 2005). Taken together, these results have revealed a highly controlled and dynamic process in which P-TEFb is kept in a delicate balance between two distinctive states: the HEXIM1/7SK-bound inactive state and the Brd4-associated active state ready for transcriptional elongation (Fig.14.6). Perturbation of this balance will result in dynamic exchange of binding partners between the two P-TEFb sub-populations and may produce serious physiological consequences.

D: Compensatory Contributions of HEXIM1 and HEXIM2 in Maintaining the Balance of Active and Inactive P-TEFb Complexes

Sequence analysis reveals a high degree of homology between a hypothetical protein and HEXIM1. However, the expression of this protein, renamed HEXIM2, has not been confirmed and whether it displays a function similar to that of HEXIM1 has yet to be established. Recently, published data from us and others indicate that like HEXIM1, HEXIM2 also possesses ability to inactivate P-TEFb to suppress transcription through a 7SK-mediated interaction with

P-TEFb (Yik *et al.*, 2005). Furthermore, HEXIM1 and HEXIM2 can form stable homo- and hetero-oligomers (most likely dimers), which may nucleate the formation of the 7SK snRNP (Yik *et al.*, 2005). Despite their similar functions, HEXIM1 and HEXIM2 exhibit distinct expression patterns in various human tissues and established cell lines. In HEXIM1-knocked down cells, HEXIM2 can functionally and quantitatively compensate for the loss of HEXIM1 to maintain a constant level of the 7SK/HEXIM-bound P-TEFb (Yik *et al.*, 2005). These results demonstrate that there is a tightly regulated cellular process to maintain the balance between active and inactive P-TEFb complexes.

E: Targeting P-TEFb for Global Control of Cell Growth and Differentiation

What could be the reason for cells striving to maintain a delicate balance between the active and inactive P-TEFb subpopulations? We believe the growth-regulatory functions of the P-TEFb-associated factors hold the key to this important question. As a major negative regulator of P-TEFb activity, HEXIM1 expression is elevated in cells that are induced to differentiate by HMBA (Ouchida *et al.*, 2003), suggesting that this protein may play an important role during the critical cellular decision to transit from proliferative growth to terminal differentiation. Consistent with this view, an anti-growth function of HEXIM1 has also been demonstrated in cardiac myocytes, where the absence of HEXIM1 causes the enlargement of heart cells in a pathological condition known as hypertrophy (Huang *et al.*, 2004). In breast epithelial cells, HEXIM1 has also been recognized as an inhibitor of cell proliferation, as its expression is down-regulated by estrogens and decreased in breast tumors (Wittmann *et al.*, 2003). In contrast to HEXIM1, Brd4, the positive regulator of P-TEFb, has been implicated to play a growth-stimulatory role as suggested by the demonstration that the Brd4-heterozygotic mice display pre- and post-natal growth defects associated with a reduced proliferation rate (Houzelstein *et al.*, 2002). The opposing effects on cellular growth exerted by Brd4 and HEXIM1, both of which target the same P-TEFb but produce antagonizing results, supports the idea that controlling the activity of the general transcription factor P-TEFb, which affects the expression of a vast majority of cellular genes, is central to the global regulation of cell growth and differentiation. With this notion in mind, it is now easy to understand that for normal cell growth, a well-controlled equilibrium has to be maintained between the two P-TEFb subpopulations, so that cells would not accidentally slip into either an over-proliferative state or terminal differentiation.

Conclusion and Perspectives

The intense biochemical characterization of the factors and mechanisms regulating HIV transcription over the past decade has established a brand new paradigm in the field of eukaryotic gene expression control. Central to this new paradigm is the intricate interplay between the viral encoded Tat protein and its host cellular cofactor P-TEFb, which act together to control HIV transcriptional elongation, a previously neglected but extremely important control step for eukaryotic gene expression. Besides HIV replication, P-TEFb is globally required for transcriptional elongation and is targeted by both positive and negative regulators for control of cell growth, proliferation and differentiation. The exciting findings described in this chapter suggest numerous future experiments. The ones that demand our immediate attention are structure-function analyses of the 7SK:HEXIM1:P-TEFb snRNP to elucidate the precise mechanism by which HEXIM1 functions as an RNA-dependent cyclin-dependent kinase inhibitor (CKI) to inhibit CDK9. Another area worth exploring is the unraveling of the signaling pathway controlling the formation and induced disruption of the 7SK:HEXIM1: P-TEFb snRNP and the involvement of this pathway in diverse disease processes ranging from cardiac hypertrophy, breast cancer to HIV infection. So far, our preliminary data indicate that the phosphorylation of CDK9 at the tip of the flexible T-loop (Thr186) by a yet-to-be identified cellular kinase may serve as a key molecular switch for controlling the formation of the 7SK snRNP (Chen *et al.*, 2004). Future work will attempt to identify this kinase and map out the details of the entire signaling network that sequesters P-TEFb for inactivation. On the Brd4 front, future studies will be crucial to elucidate the mechanism by which Brd4 recruits P-TEFb to cellular promoters and how this action may involve chromatin modification and/or other components of the transcriptional machinery. In addition, it will be interesting to determine whether through Brd4, the activity of P-TEFb can be regulated throughout the cell cycle, as Brd4 has been implicated to play an important role during the cell cycle progression. As for HIV, a long-term goal of the Tat field is to apply our knowledge of P-TEFb and Tat to the development of interventions to modulate HIV gene expression, thereby addressing the major problem of viral latency. Our studies of the mechanism and factors required for Tat-activation of HIV gene expression have been extremely rewarding and we can certainly count on this deadly virus, despite its devastating impact on human health worldwide, to teach us a few new lessons down the road.

Acknowledgment

This work is supported by grants from the National Institute of Health (AI41757), American Cancer Society (RSG-01-171-01-MBC) and the National Natural Science Foundation of China (30428004) to Q. Zhou, and the National Natural Science Foundation of China (30470371) to R. Chen.

References

Ahn, S. H., Kim, M., and Buratowski, S. (2004). Phosphorylation of serine 2 within the RNA polymerase II C-terminal domain couples transcription and 3' end processing. Mol Cell 13, 67-76.

Akoulitchev, S., Makela, T. P., Weinberg, R. A., and Reinberg, D. (1995). Requirement for TFIIH kinase activity in transcription by RNA polymerase II. Nature 377, 557-560.

Allison, L. A., Moyle, M., Shales, M., and Ingles, C. J. (1985). Extensive homology among the largest subunits of eukaryotic and prokaryotic RNA polymerases. Cell 42, 599-610.

Allison, L. A., Wong, J. K., Fitzpatrick, V. D., Moyle, M., and Ingles, C. J. (1988). The C-terminal domain of the largest subunit of RNA polymerase II of Saccharomyces cerevisiae, Drosophila melanogaster, and mammals: a conserved structure with an essential function. Mol Cell Biol 8, 321-329.

Alonso, A., Cujec, T. P., and Peterlin, B. M. (1994). Effects of human chromosome 12 on interactions between Tat and TAR of human immunodeficiency virus type 1. J Virol 68, 6505-6513.

Andrulis, E. D., Guzman, E., Doring, P., Werner, J., and Lis, J. T. (2000). High-resolution localization of Drosophila Spt5 and Spt6 at heat shock genes in vivo: roles in promoter proximal pausing and transcription elongation. Genes Dev 14, 2635-2649.

Andrulis, E. D., Werner, J., Nazarian, A., Erdjument-Bromage, H., Tempst, P., and Lis, J. T. (2002). The RNA processing exosome is linked to elongating RNA polymerase II in Drosophila. Nature 420, 837-841.

Barboric, M., Nissen, R. M., Kanazawa, S., Jabrane-Ferrat, N., and Peterlin, B. M. (2001). NF-kappaB binds P-TEFb to stimulate transcriptional elongation by RNA polymerase II. Molecular Cell 8, 327-337.

Barboric, M., and Peterlin, B. M. (2005). A new paradigm in eukaryotic biology: HIV Tat and the control of transcriptional elongation. PLoS Biol 3, e76.

Bentley, D. (1999). Coupling RNA polymerase II transcription with pre-mRNA processing. Curr Opin Cell Biol 11, 347-351.

Boehm, A. K., Saunders, A., Werner, J., and Lis, J. T. (2003). Transcription factor and polymerase recruitment, modification, and movement on dhsp70 in vivo in the minutes following heat shock. Mol Cell Biol 23, 7628-7637.

Bres, V., Kiernan, R., Emiliani, S., and Benkirane, M. (2002). Tat acetyl-acceptor lysines are important for human immunodeficiency virus type-1 replication. J Biol Chem 277, 22215-22221.

Bres, V., Tagami, H., Peloponese, J. M., Loret, E., Jeang, K. T., Nakatani, Y., Emiliani, S., Benkirane, M., and Kiernan, R. E. (2002). Differential acetylation of Tat coordinates its interaction with the co-activators cyclin T1 and PCAF. Embo J 21, 6811-6819.

Brigati, C., Giacca, M., Noonan, D. M., and Albini, A. (2003). HIV Tat, its TARgets and the control of viral gene expression. FEMS Microbiol Lett 220, 57-65.

Casse, C., Giannoni, F., Nguyen, V. T., Dubois, M.-F., and Bensaude, O. (1999). The transcriptional inhibitors, actinomycin D and alpha-amanitin, activate the HIV-1 promoter and favor phosphorylation of the RNA polymerase II C-terminal domain. Journal of Biological Chemistry 274, 16097-16106.

Chao, S.-H., and Price, D. H. (2001). Flavopiridol inactivates P-TEFb and blocks most RNA polymerase II transcription in vivo. Journal of Biological Chemistry 276, 31793-31799.

Chao, S. H., and Price, D. H. (2001). Flavopiridol inactivates P-TEFb and blocks most RNA polymerase II transcription in vivo. J Biol Chem 276, 31793-31799.

Chen, D., Fong, Y., and Zhou, Q. (1999). Specific interaction of Tat with the human but not rodent P-TEFb complex mediates the species-specific Tat activation of HIV-1 transcription. Proc Natl Acad Sci USA 96, 2728-2733.

Chen, R., Yang, Z., and Zhou, Q. (2004). Phosphorylated positive transcription elongation factor b (P-TEFb) is tagged for inhibition through association with 7SK snRNA. J Biol Chem 279, 4153-4160.

Cho, E. J., Kobor, M. S., Kim, M., Greenblatt, J., and Buratowski, S. (2001). Opposing effects of Ctk1 kinase and Fcp1 phosphatase at Ser 2 of the RNA polymerase II C-terminal domain. Genes Dev 15, 3319-3329.

Corden, J. L., Cadena, D. L., Ahearn, J. M., Jr., and Dahmus, M. E. (1985). A unique structure at the carboxyl terminus of the largest subunit of eukaryotic RNA polymerase II. Proc Natl Acad Sci USA 82, 7934-7938.

Dahmus, M. E. (1995). Phosphorylation of the C-terminal domain of RNA polymerase II. Biochim Biophys Acta 1261, 171-182.

Dahmus, M. E. (1996). Phosphorylation of mammalian RNA polymerase II. Methods Enzymol 273, 185-193.

De Falco, G., Bagella, L., Claudio, P. P., De Luca, A., Fu, Y., Calabretta, B., Sala, A., and Giordano, A. (2000). Physical interaction between CDK9 and B-Myb results in suppression of B-Myb gene autoregulation. Oncogene 19, 373-379.

Dey, A., Chitsaz, F., Abbasi, A., Misteli, T., and Ozato, K. (2003).

The double bromodomain protein Brd4 binds to acetylated chromatin during interphase and mitosis. Proc Natl Acad Sci USA *100*, 8758-8763.

Eberhardy, S. R., and Farnham, P. J. (2001). C-Myc mediates activation of the cad promoter via a post-RNA polymerase II recruitment mechanism. J Biol Chem *276*, 48562-48571.

Estable, M. C., Naghavi, M. H., Kato, H., Xiao, H., Qin, J., Vahlne, A., and Roeder, R. G. (2002). MCEF, the newest member of the AF4 family of transcription factors involved in leukemia, is a positive transcription elongation factor-b-associated protein. J Biomed Sci *9*, 234-245.

Fong, Y. W., and Zhou, Q. (2000). Relief of two built-In autoinhibitory mechanisms in P-TEFb is required for assembly of a multicomponent transcription elongation complex at the human immunodeficiency virus type 1 promoter. Mol Cell Biol *20*, 5897-5907.

Fong, Y. W., and Zhou, Q. (2001). Stimulatory effect of splicing factors on transcriptional elongation. Nature *414*, 929-933.

Fujinaga, K., Irwin, D., Huang, Y., Taube, R., Kurosu, T., and Peterlin, B. M. (2004). Dynamics of human immunodeficiency virus transcription: P-TEFb phosphorylates RD and dissociates negative effectors from the transactivation response element. Mol Cell Biol *24*, 787-795.

Garber, M. E., Mayall, T. P., Suess, E. M., Meisenhelder, J., Thompson, N. E., and Jones, K. A. (2000). CDK9 autophosphorylation regulates high-affinity binding of the human immunodeficiency virus type 1 tat-P-TEFb complex to TAR RNA. Mol Cell Biol *20*, 6958-6969.

Garber, M. E., Wei, P., and Jones, K. A. (1998). HIV-1 Tat interacts with cyclin T1 to direct the P-TEFb CTD kinase complex to TAR RNA. Cold Spring Harb Symp Quant Biol *63*, 371-380.

Garber, M. E., Wei, P., KewalRamani, V. N., Mayall, T. P., Herrmann, C. H., Rice, A. P., Littman, D. R., and Jones, K. A. (1998). The interaction between HIV-1 Tat and human cyclin T1 requires zinc and a critical cysteine residue that is not conserved in the murine CycT1 protein. Genes Dev *12*, 3512-3527.

Garriga, J., Bhattacharya, S., Calbo, J., Marshall, R. M., Truongcao, M., Haines, D. S., and Grana, X. (2003). CDK9 is constitutively expressed throughout the cell cycle, and its steady-state expression is independent of SKP2. Mol Cell Biol *23*, 5165-5173.

Garriga, J., and Grana, X. (2004). Cellular control of gene expression by T-type cyclin/CDK9 complexes. Gene *337*, 15-23.

Giraud, S., Hurlstone, A., Avril, S., and Coqueret, O. (2004). Implication of BRG1 and cdk9 in the STAT3-mediated activation of the p21waf1 gene. Oncogene *23*, 7391-7398.

Gold, M. O., Yang, X., Herrmann, C. H., and Rice, A. P. (1998). PITALRE, the catalytic subunit of TAK, is required for human immunodeficiency virus Tat transactivation *in vivo*. Journal of Virology *72*, 4448-4453.

Grana, X., De Luca, A., Sang, N., Fu, Y., Claudio, P. P., Rosenblatt, J., Morgan, D. O., and Giordano, A. (1994). PITALRE, a nuclear CDC2-related protein kinase that phosphorylates the retinoblastoma protein *in vitro*. Proc Natl Acad Sci USA *91*, 3834-3838.

Hart, C. E., Galphin, J. C., Westhafer, M. A., and Schochetman, G. (1993). TAR loop-dependent human immunodeficiency virus trans activation requires factors encoded on human chromosome 12. J Virol *67*, 5020-5024.

Hausmann, S., Schwer, B., and Shuman, S. (2004). An encephalitozoon cuniculi ortholog of the RNA polymerase II carboxyl-terminal domain (CTD) serine phosphatase Fcp1. Biochemistry *43*, 7111-7120.

Herrmann, C. H., and Rice, A. P. (1995). Lentivirus Tat proteins specifically associate with a cellular protein kinase, TAK, that hyperphosphorylates the carboxyl-terminal domain of the large subunit of RNA polymerase II: candidate for a Tat cofactor. J Virol *69*, 1612-1620.

Herrmann, C. H., and Rice, A. P. (1995). Lentivirus Tat proteins specifically associate with a cellular protein kinase, TAK, that hyperphosphorylates the carboxyl-terminal domain of the large subunit of RNA polymerase II: Candidate for a Tat cofactor. Journal of Virology *69*, 1612-1620.

Houzelstein, D., Bullock, S. L., Lynch, D. E., Grigorieva, E. F., Wilson, V. A., and Beddington, R. S. (2002). Growth and early postimplantation defects in mice deficient for the bromodomain-containing protein Brd4. Mol Cell Biol *22*, 3794-3802.

Howe, K. J. (2002). RNA polymerase II conducts a symphony of pre-mRNA processing activities. Biochim Biophys Acta *1577*, 308-324.

Huang, F., Wagner, M., and Siddiqui, M. A. (2004). Ablation of the CLP-1 gene leads to down-regulation of the HAND1 gene and abnormality of the left ventricle of the heart and fetal death. Mech Dev *121*, 559-572.

Ivanov, D., Kwak, Y. T., Guo, J., and Gaynor, R. B. (2000). Domains in the SPT5 protein that modulate its transcriptional regulatory properties. Mol Cell Biol *20*, 2970-2983.

Ivanov, D., Kwak, Y. T., Nee, E., Guo, J., Garcia-Martinez, L. F., and Gaynor, R. B. (1999). Cyclin T1 domains involved in complex formation with Tat and TAR RNA are critical for tat-activation. J Mol Biol *288*, 41-56.

Jang, M., Mochizuki, K., Zhou, M., Jeong, H., Brady, J., and Ozato, K. (2005). Bromodomain protein Brd4 is a positive regulatory component of P-TEFb and stimulates RNA polymerase II dependent transcription. Mol Cell *19*, 523-534.

Jeanmougin, F., Wurtz, J. M., Le Douarin, B., Chambon, P., and Losson, R. (1997). The bromodomain revisited. Trends Biochem Sci *22*, 151-153.

Jones, K. A. (1997). Taking a new TAK on tat transactivation. Genes Dev *11*, 2593-2599.

Jones, K. A., and Peterlin, B. M. (1994). Control of RNA initiation and elongation at the HIV-1 promoter. Annu Rev

Biochem *63*, 717-743.

Kanazawa, S., Okamoto, T., and Peterlin, B. M. (2000). Tat competes with CIITA for the binding to P-TEFb and blocks the expression of MHC class II genes in HIV infection. Immunity *12*, 61-70.

Karn, J. (1999). Tackling Tat. J Mol Biol *293*, 235-254.

Kiernan, R. E., Vanhulle, C., Schiltz, L., Adam, E., Xiao, H., Maudoux, F., Calomme, C., Burny, A., Nakatani, Y., Jeang, K. T., *et al.* (1999). HIV-1 tat transcriptional activity is regulated by acetylation. Embo J *18*, 6106-6118.

Kim, D. K., Yamaguchi, Y., Wada, T., and Handa, H. (2001). The regulation of elongation by eukaryotic RNA polymerase II: a recent view. Mol Cells *11*, 267-274.

Kim, J. B., and Sharp, P. A. (2001). Positive transcription elongation factor B phosphorylates hSPT5 and RNA polymerase II carboxyl-terminal domain independently of cyclin-dependent kinase-activating kinase. J Biol Chem *276*, 12317-12323.

Kim, J. B., Yamaguchi, Y., Wada, T., Handa, H., and Sharp, P. A. (1999). Tat-SF1 protein associates with RAP30 and human SPT5 proteins. Mol Cell Biol *19*, 5960-5968.

Kino, T., Slobodskaya, O., Pavlakis, G. N., and Chrousos, G. P. (2002). Nuclear receptor coactivator p160 proteins enhance the HIV-1 long terminal repeat promoter by bridging promoter-bound factors and the Tat-P-TEFb complex. J Biol Chem *277*, 2396-2405.

Kobor, M. S., and Greenblatt, J. (2002). Regulation of transcription elongation by phosphorylation. Biochim Biophys Acta *1577*, 261-275.

Komarnitsky, P., Cho, E. J., and Buratowski, S. (2000). Different phosphorylated forms of RNA polymerase II and associated mRNA processing factors during transcription. Genes Dev *14*, 2452-2460.

Krishnamurthy, S., He, X., Reyes-Reyes, M., Moore, C., and Hampsey, M. (2004). Ssu72 Is an RNA polymerase II CTD phosphatase. Mol Cell *14*, 387-394.

Kwak, Y. T., Ivanov, D., Guo, J., Nee, E., and Gaynor, R. B. (1999). Role of the human and murine cyclin T proteins in regulating HIV-1 tat-activation. J Mol Biol *288*, 57-69.

Lam, L. T., Pickeral, O. K., Peng, A. C., Rosenwald, A., Hurt, E. M., Giltnane, J. M., Averett, L. M., Zhao, H., Davis, R. E., Sathyamoorthy, M., *et al.* (2001). Genomic-scale measurement of mRNA turnover and the mechanisms of action of the anti-cancer drug flavopiridol. Genome Biol *2*, RESEARCH0041.

Leclerc, V., Tassan, J. P., O'Farrell, P. H., Nigg, E. A., and Leopold, P. (1996). *Drosophila* Cdk8, a kinase partner of cyclin C that interacts with the large subunit of RNA polymerase II. Mol Biol Cell *7*, 505-513.

Lee, D. K., Duan, H. O., and Chang, C. (2001). Androgen receptor interacts with the positive elongation factor P-TEFb and enhances the efficiency of transcriptional elongation. Journal of Biological Chemistry *276*, 9978-9984.

Li, X. Y., and Green, M. R. (1998). The HIV-1 Tat cellular coactivator Tat-SF1 is a general transcription elongation factor. Genes Dev *12*, 2992-2996.

Licciardo, P., Ruggiero, L., Lania, L., and Majello, B. (2001). Transcription activation by targeted recruitment of the RNA polymerase II CTD phosphatase FCP1. Nucleic Acids Res *29*, 3539-3545.

Lin, P. S., Marshall, N. F., and Dahmus, M. E. (2002). CTD phosphatase: role in RNA polymerase II cycling and the regulation of transcript elongation. Prog Nucleic Acid Res Mol Biol *72*, 333-365.

Lindstrom, D. L., Squazzo, S. L., Muster, N., Burckin, T. A., Wachter, K. C., Emigh, C. A., McCleery, J. A., Yates, J. R., 3rd, and Hartzog, G. A. (2003). Dual roles for Spt5 in pre-mRNA processing and transcription elongation revealed by identification of Spt5-associated proteins. Mol Cell Biol *23*, 1368-1378.

Litingtung, Y., Lawler, A. M., Sebald, S. M., Lee, E., Gearhart, J. D., Westphal, H., and Corden, J. L. (1999). Growth retardation and neonatal lethality in mice with a homozygous deletion in the C-terminal domain of RNA polymerase II. Mol Gen Genet *261*, 100-105.

Mandal, S. S., Cho, H., Kim, S., Cabane, K., and Reinberg, D. (2002). FCP1, a phosphatase specific for the heptapeptide repeat of the largest subunit of RNA polymerase II, stimulates transcription elongation. Mol Cell Biol *22*, 7543-7552.

Mandal, S. S., Chu, C., Wada, T., Handa, H., Shatkin, A. J., and Reinberg, D. (2004). Functional interactions of RNA-capping enzyme with factors that positively and negatively regulate promoter escape by RNA polymerase II. Proc Natl Acad Sci USA *101*, 7572-7577.

Marshall, N. F., Peng, J., Xie, Z., and Price, D. H. (1996). Control of RNA polymerase II elongation potential by a novel carboxyl-terminal domain kinase. Journal of Biological Chemistry *271*, 27176-27183.

Marshall, N. F., and Price, D. H. (1995). Purification of P-TEFb, a transcription factor required for the transition into productive elongation. Journal of Biological Chemistry *270*, 12335-12338.

McCracken, S., Fong, N., Rosonina, E., Yankulov, K., Brothers, G., Siderovski, D., Hessel, A., Foster, S., Shuman, S., and Bentley, D. L. (1997). 5′-Capping enzymes are targeted to pre-mRNA by binding to the phosphorylated carboxy-terminal domain of RNA polymerase II. Genes Dev *11*, 3306-3318.

McCracken, S., Fong, N., Yankulov, K., Ballantyne, S., Pan, G., Greenblatt, J., Patterson, S. D., Wickens, M., and Bentley, D. L. (1997). The C-terminal domain of RNA polymerase II couples mRNA processing to transcription. Nature *385*, 357-361.

Michels, A. A., Nguyen, V. T., Fraldi, A., Labas, V., Edwards, M., Bonnet, F., Lania, L., and Bensaude, O. (2003). MAQ1 and 7SK RNA interact with CDK9/cyclin T complexes in a transcription-dependent manner. Mol Cell Biol *23*, 4859-4869.

Murphy, S., Di Liegro, C., and Melli, M. (1987). The *in vitro* transcription of the 7SK RNA gene by RNA polymerase III is dependent only on the presence of an upstream promoter. Cell *51*,

81-87.

Nelson, C., Goto, S., Lund, K., Hung, W., and Sadowski, I. (2003). Srb10/Cdk8 regulates yeast filamentous growth by phosphorylating the transcription factor Ste12. Nature *421*, 187-190.

Nguyen, V. T., Kiss, T., Michels, A. A., and Bensaude, O. (2001). 7SK small nuclear RNA binds to and inhibits the activity of CDK9/cyclin T complexes. Nature *414*, 322-325.

Ni, Z., Schwartz, B. E., Werner, J., Suarez, J. R., and Lis, J. T. (2004). Coordination of transcription, RNA processing, and surveillance by P-TEFb kinase on heat shock genes. Mol Cell *13*, 55-65.

Nonet, M., Sweetser, D., and Young, R. A. (1987). Functional redundancy and structural polymorphism in the large subunit of RNA polymerase II. Cell *50*, 909-915.

O'Keeffe, B., Fong, Y., Chen, D., Zhou, S., and Zhou, Q. (2000). Requirement for a kinase-specific chaperone pathway in the production of a Cdk9/cyclin T1 heterodimer responsible for P-TEFb-mediated tat stimulation of HIV-1 transcription. J Biol Chem *275*, 279-287.

Oelgeschlager, T. (2002). Regulation of RNA polymerase II activity by CTD phosphorylation and cell cycle control. J Cell Physiol *190*, 160-169.

Orphanides, G., Lagrange, T., and Reinberg, D. (1996). The general transcription factors of RNA polymerase II. Genes Dev *10*, 2657-2683.

Orphanides, G., and Reinberg, D. (2002). A unified theory of gene expression. Cell *108*, 439-451.

Ouchida, R., Kusuhara, M., Shimizu, N., Hisada, T., Makino, Y., Morimoto, C., Handa, H., Ohsuzu, F., and Tanaka, H. (2003). Suppression of NF-kappaB-dependent gene expression by a hexamethylene bisacetamide-inducible protein HEXIM1 in human vascular smooth muscle cells. Genes Cells *8*, 95-107.

Pagans, S., Pedal, A., North, B. J., Kaehlcke, K., Marshall, B. L., Dorr, A., Hetzer-Egger, C., Henklein, P., Frye, R., McBurney, M. W., *et al.* (2005). SIRT1 regulates HIV transcription via Tat deacetylation. PLoS Biol *3*, e41.

Palancade, B., Dubois, M. F., Dahmus, M. E., and Bensaude, O. (2001). Transcription-independent RNA polymerase II dephosphorylation by the FCP1 carboxy-terminal domain phosphatase in *Xenopus* laevis early embryos. Mol Cell Biol *21*, 6359-6368.

Parada, C. A., and Roeder, R. G. (1999). A novel RNA polymerase II-containing complex potentiates Tat-enhanced HIV-1 transcription. Embo J *18*, 3688-3701.

Pei, Y., and Shuman, S. (2002). Interactions between fission yeast mRNA capping enzymes and elongation factor Spt5. J Biol Chem *277*, 19639-19648.

Peng, J., Zhu, Y., Milton, J. T., and Price, D. H. (1998). Identification of multiple cyclin subunits of human P-TEFb. Genes & Development *12*, 755-762.

Prelich, G. (2002). RNA polymerase II carboxy-terminal domain kinases: emerging clues to their function. Eukaryot Cell *1*, 153-162.

Price, D. H. (2000). P-TEFb, a cyclin-dependent kinase controlling elongation by RNA polymerase II. Mol Cell Biol *20*, 2629-2634.

Price, D. H. (2000). P-TEFb, a cyclin-dependent kinase controlling elongation by RNA polymerase II. Molecular and Cellular Biology *20*, 2629-2634.

Proudfoot, N. J., Furger, A., and Dye, M. J. (2002). Integrating mRNA processing with transcription. Cell *108*, 501-512.

Reines, D., Conaway, R. C., and Conaway, J. W. (1999). Mechanism and regulation of transcriptional elongation by RNA polymerase II. Curr Opin Cell Biol *11*, 342-346.

Renner, D. B., Yamaguchi, Y., Wada, T., Handa, H., and Price, D. H. (2001). A highly purified RNA polymerase II elongation control system. J Biol Chem *276*, 42601-42609.

Rickert, P., Seghezzi, W., Shanahan, F., Cho, H., and Lees, E. (1996). Cyclin C/CDK8 is a novel CTD kinase associated with RNA polymerase II. Oncogene *12*, 2631-2640.

Rodriguez, C. R., Cho, E. J., Keogh, M. C., Moore, C. L., Greenleaf, A. L., and Buratowski, S. (2000). Kin28, the TFIIH-associated carboxy-terminal domain kinase, facilitates the recruitment of mRNA processing machinery to RNA polymerase II. Mol Cell Biol *20*, 104-112.

Sano, M., Abdellatif, M., Oh, H., Xie, M., Bagella, L., Giordano, A., Michael, L. H., DeMayo, F. J., and Schneider, M. D. (2002). Activation and function of cyclin T-Cdk9 (positive transcription elongation factor-b) in cardiac muscle-cell hypertrophy. Nat Med *8*, 1310-1317.

Schroeder, S. C., Schwer, B., Shuman, S., and Bentley, D. (2000). Dynamic association of capping enzymes with transcribing RNA polymerase II. Genes Dev *14*, 2435-2440.

Shang, E., Salazar, G., Crowley, T. E., Wang, X., Lopez, R. A., and Wolgemuth, D. J. (2004). Identification of unique, differentiation stage-specific patterns of expression of the bromodomain-containing genes Brd2, Brd3, Brd4, and Brdt in the mouse testis. Gene Expr Patterns *4*, 513-519.

Sherr, C. J., and Roberts, J. M. (1999). CDK inhibitors: positive and negative regulators of G1-phase progression. Genes Dev *13*, 1501-1512.

Shilatifard, A. (2004). Transcriptional elongation control by RNA polymerase II: a new frontier. Biochim Biophys Acta *1677*, 79-86.

Shilatifard, A., Conaway, R. C., and Conaway, J. W. (2003). The RNA polymerase II elongation complex. Annu Rev Biochem *72*, 693-715.

Shim, E. Y., Walker, A. K., Shi, Y., and Blackwell, T. K. (2002). CDK-9/cyclin T (P-TEFb) is required in two postinitiation pathways for transcription in the *C. elegans* embryo. Genes Dev *16*, 2135-2146.

Shore, S. M., Byers, S. A., Dent, P., and Price, D. H. (2005). Characterization of Cdk9(55) and differential regulation of two

Cdk9 isoforms. Gene *350*, 51-58.

Shore, S. M., Byers, S. A., Maury, W., and Price, D. H. (2003). Identification of a novel isoform of Cdk9. Gene *307*, 175-182.

Simone, C., and Giordano, A. (2001). New insight in cdk9 function: from Tat to MyoD. Front Biosci *6*, D1073-1082.

Simone, C., Stiegler, P., Bagella, L., Pucci, B., Bellan, C., De Falco, G., De Luca, A., Guanti, G., Puri, P. L., and Giordano, A. (2002). Activation of MyoD-dependent transcription by cdk9/cyclin T2. Oncogene *21*, 4137-4148.

Sims, R. J., 3rd, Belotserkovskaya, R., and Reinberg, D. (2004). Elongation by RNA polymerase II: the short and long of it. Genes Dev *18*, 2437-2468.

Steinmetz, E. J. (1997). Pre-mRNA processing and the CTD of RNA polymerase II: the tail that wags the dog? Cell *89*, 491-494.

Svejstrup, J. Q. (2004). The RNA polymerase II transcription cycle: cycling through chromatin. Biochim Biophys Acta *1677*, 64-73.

Taube, R., Lin, X., Irwin, D., Fujinaga, K., and Peterlin, B. M. (2002). Interaction between P-TEFb and the C-terminal domain of RNA polymerase II activates transcriptional elongation from sites upstream or downstream of target genes. Mol Cell Biol *22*, 321-331.

Tian, Y., Ke, S., Chen, M., and Sheng, T. (2003). Interactions between the aryl hydrocarbon receptor and P-TEFb. Sequential recruitment of transcription factors and differential phosphorylation of C-terminal domain of RNA polymerase II at cyp1a1 promoter. J Biol Chem *278*, 44041-44048.

Valerie, K., Delers, A., Bruck, C., Thiriart, C., Rosenberg, H., Debouck, C., and Rosenberg, M. (1988). Activation of human immunodeficiency virus type 1 by DNA damage in human cells. Nature *333*, 78-81.

Wada, T., Takagi, T., Yamaguchi, Y., Watanabe, D., and Handa, H. (1998). Evidence that P-TEFb alleviates the negative effect of DSIF on RNA polymerase II-dependent transcription *in vitro*. Embo J *17*, 7395-7403.

Washington, K., Ammosova, T., Beullens, M., Jerebtsova, M., Kumar, A., Bollen, M., and Nekhai, S. (2002). Protein phosphatase-1 dephosphorylates the C-terminal domain of RNA polymerase-II. J Biol Chem *277*, 40442-40448.

Wei, P., Garber, M. E., Fang, S. M., Fischer, W. H., and Jones, K. A. (1998). A novel CDK9-associated C-type cyclin interacts directly with HIV-1 Tat and mediates its high-affinity, loop-specific binding to TAR RNA. Cell *92*, 451-462.

Wen, Y., and Shatkin, A. J. (1999). Transcription elongation factor hSPT5 stimulates mRNA capping. Genes Dev *13*, 1774-1779.

West, M. L., and Corden, J. L. (1995). Construction and analysis of yeast RNA polymerase II CTD deletion and substitution mutations. Genetics *140*, 1223-1233.

Wittmann, B. M., Wang, N., and Montano, M. M. (2003). Identification of a novel inhibitor of breast cell growth that is down-regulated by estrogens and decreased in breast tumors. Cancer Res *63*, 5151-5158.

Yamaguchi, Y., Inukai, N., Narita, T., Wada, T., and Handa, H. (2002). Evidence that negative elongation factor represses transcription elongation through binding to a DRB sensitivity-inducing factor/RNA polymerase II complex and RNA. Mol Cell Biol *22*, 2918-2927.

Yamaguchi, Y., Takagi, T., Wada, T., Yano, K., Furuya, A., Sugimoto, S., Hasegawa, J., and Handa, H. (1999). NELF, a multisubunit complex containing RD, cooperates with DSIF to repress RNA polymerase II elongation. Cell *97*, 41-51.

Yang, X., Gold, M. O., Tang, D. N., Lewis, D. E., Aguilar-Cordova, E., Rice, A. P., and Herrmann, C. H. (1997). TAK, an HIV Tat-associated kinase, is a member of the cyclin-dependent family of protein kinases and is induced by activation of peripheral blood lymphocytes and differentiation of promonocytic cell lines. Proc Natl Acad Sci USA *94*, 12331-12336.

Yang, X., Herrmann, C. H., and Rice, A. P. (1996). The human immunodeficiency virus Tat proteins specifically associated with TAK *in vivo* and require the carboxyl-terminal domain of RNA polymerase II for function. Journal of Virology *70*, 4576-4584.

Yang, Z., Yik, J. H., Chen, R., Jang, M. K., Ozato, K., and Zhou, Q. (2005). Recruitment of P-TEFb for stimulation of transcriptional elongation by bromodomain protein Brd4. Mol Cell *19*, 535-545.

Yang, Z., Zhu, Q., Luo, K., and Zhou, Q. (2001). The 7SK small nuclear RNA inhibits the CDK9/cyclin T1 kinase to control transcription. Nature *414*, 317-322.

Yik, J. H., Chen, R., Nishimura, R., Jennings, J. L., Link, A. J., and Zhou, Q. (2003). Inhibition of P-TEFb (CDK9/Cyclin T) kinase and RNA polymerase II transcription by the coordinated actions of HEXIM1 and 7SK snRNA. Mol Cell *12*, 971-982.

Yik, J. H., Chen, R., Pezda, A. C., Samford, C. S., and Zhou, Q. (2004). A human immunodeficiency virus type 1 Tat-like arginine-rich RNA-binding domain is essential for HEXIM1 to inhibit RNA polymerase II transcription through 7SK snRNA-mediated inactivation of P-TEFb. Mol Cell Biol *24*, 5094-5105.

Yik, J. H., Chen, R., Pezda, A. C., and Zhou, Q. (2005). Compensatory contributions of HEXIM1 and HEXIM2 in maintaining the balance of active and inactive positive transcription elongation factor b complexes for control of transcription. J Biol Chem *280*, 16368-16376.

Young, T. M., Wang, Q., Pe'ery, T., and Mathews, M. B. (2003). The human I-mfa domain-containing protein, HIC, interacts with cyclin T1 and modulates P-TEFb-dependent transcription. Mol Cell Biol *23*, 6373-6384.

Zeng, L., and Zhou, M. M. (2002). Bromodomain: an acetyl-lysine binding domain. FEBS Lett *513*, 124-128.

Zhang, F., Barboric, M., Blackwell, T. K., and Peterlin, B. M. (2003). A model of repression: CTD analogs and PIE-1 inhibit transcriptional elongation by P-TEFb. Genes Dev *17*, 748-758.

Zhang, J., Tamilarasu, N., Hwang, S., Garber, M. E., Huq, I., Jones, K. A., and Rana, T. M. (2000). HIV-1 TAR RNA enhances the interaction between Tat and cyclin T1. J Biol Chem *275*, 34314-34319.

Zhou, Q., Chen, D., Pierstorff, E., and Luo, K. (1998). Transcription elongation factor P-TEFb mediates Tat activation of HIV-1 transcription at multiple stages. Embo J *17*, 3681-3691.

Zhou, Q., and Sharp, P. A. (1996). Tat-SF1: cofactor for stimulation of transcriptional elongation by HIV-1 Tat. Science *274*, 605-610.

Zhu, Y., Pe'ery, T., Peng, J., Ramanathan, Y., Marshall, N., Marshall, T., Amendt, B., Mathews, M. B., and Price, D. H. (1997). Transcription elongation factor P-TEFb is required for HIV-1 tat transactivation *in vitro*. Genes Dev *11*, 2622-2632.

Zhu, Y., Pe'ery, T., Peng, J., Ramanathan, Y., Marshall, N., Marshall, T., Amendt, B., Mathews, M. B., and Price, D. H. (1997). Transcription elongation factor P-TEFb is required for HIV-1 tat transactivation *in vitro*. Genes and Development *11*, 2622-2632.

Zieve, G., Benecke, B. J., and Penman, S. (1977). Synthesis of two classes of small RNA species *in vivo* and *in vitro*. Biochemistry *16*, 4520-4525.

Chapter 15

Post-translational Modifications of the p53 Transcription Factor

Christopher L. Brooks and Wei Gu

Institute for Cancer Genetics, and Department of Pathology, College of Physicians & Surgeons, Columbia University, 1150 St. Nicholas Ave, New York, NY 10032, USA

Key Words: p53, mdm2, acetylation, ubiquitination, phosphorylation, deubiquitination, deacetylation, apoptosis.

Summary

The p53 tumor suppressor acts as a potent transcription factor in response to bombardment by a variety of cellular stresses. Its importance in maintaining genomic stability and exerting anit-proliferative effects is underscored by the fact that it is found mutated in approximately 50% of all human tumors. p53 is tightly regulated at the post-translational level though multiple modifications including ubiquitination, phosphorylation, acetylation, and neddylation. Mdm2 is a key regulator of p53 by acting as a specific E3 ligase on the protein and targeting it to the ubiquitin-proteasome pathway. Together, these two proteins provide a critical node in the cellular circuitry for countless signaling pathways to converge.

Introduction

The p53 tumor suppressor exerts its anti-proliferative effects, including growth arrest, apoptosis, and cell senescence in response to various types of cellular stress (Levine, 1997). The activity of p53 as a sequence-specific transcription factor is highly regulated by post-translational modifications, protein-protein interactions, and protein stabilization (Brooks and Gu, 2003). Tight regulation of p53 is critical for maintaining cellular homeostasis. Its imperative function is underscored by the discovery of p53 mutations in over 50% of all human tumors, and those harboring wild-type p53 have strong evidence of other p53 pathway disruptions (Vousden, 2000). Indeed, Mdm2, a powerful regulator of p53, is found over expressed in approximately 5-10% of all human tumors (Juven-Gershon and Oren, 1999).

Normally at low levels in unstressed cells, p53 activity as a potent transcription factor is highly dependent on rapid and effective stabilization of the protein. The key regulator of p53 is the E3 ligase Mdm2, which ubiquitinates p53 under normal conditions and maintains the protein at low levels (Freedman *et al.*, 1999). Upon DNA damage, however, the cell strategically targets the intricate p53-Mdm2 relationship for disruption through a variety of mechanisms. Blocking this interaction leads to quick stabilization of the protein, though the precise sequence of events leading to p53 activation remains ambiguous (Michael and Oren, 2002).

Post-translational modifications are critical events for the regulation of p53. While ubiquitination of p53 is important for regulation through the ubiquitin-proteosome pathway, other post-translational modifications such as acetylation, phosphorylation, and neddylation are important for its activity as a sequence-specific transcription factor. Several residues within p53 have been shown to be differentially phosphorylated or dephosphorylated depending on the type of DNA damage that occurs (Appella and Anderson, 2001). Further, p53 is specifically acetylated at multiple lysine residues (Lys370, Lys371, Lys372, Lys381, and Lys382) of the carboxy-terminal regulatory domain by CBP/p300 and, to a lesser extent, Lys320 by PCAF. While elucidation

---

Corresponding Author: Wei Gu, Berrie Research Pavilion Rm 412C, Institute for Cancer Genetics, Columbia University, 1150 St. Nicholas Avenue, New York, NY 10032, Tel: (212) 851-5282 (Office), (212) 851-5285/5286 (Lab), Fax: (212) 851-5284, E-mail: wg8@columbia.edu

of their precise physiological role continues, these modifications could alter the p53 protein structure to a more energetically favorable and active structure. Additionally, these modifications could serve as docking sites for other co-factors and effectors involved in p53 function.

Post-translational Modifications

Post-translational modifications that occur on p53 include ubiquitination, sumoylation, phosphorylation, acetylation, and neddylation all of which are covalent linkages that serve to regulate either protein levels or protein activity (Fig.15.1). They provide layers of regulation that, when combined, offer virtually endless combinations of nuanced regulatory possibilities. The life of p53 exists in a sea of regulatory enzymes all competing for interaction time with the protein upon appropriate activation. The fact that p53 is so highly regulated at the post-translational level underscores its central importance in cellular homeostasis.

A: Ubiquitination

p53 was first shown to be ubiquitinated by a cellular factor that associated with the viral E6 protein in papilloma-virus-infected cells (Scheffner *et al.*, 1993). It was later determined to be predominantly degraded through the ubiquitin-proteosomal pathway (Chowdary *et al.*, 1994; Maki *et al.*, 1996). However, in normal cells Mdm2 (Hdm2 in humans; hereafter referred to as Mdm2) acts as the predominant regulator of p53 protein levels by implementing its E3 ligase activity on p53 (Haupt *et al.*, 1997; Honda *et al.*, 1997; Kubbutat *et al.*, 1997). Additionally, Mdm2 can inhibit the transactivation ability of p53 (Momand *et al.*, 1992). Maintaining p53 at low levels is critical for cellular homeostasis in unstressed conditions. p53 assists in its own regulation by driving the gene expression of *mdm2* in a negative feedback loop, effectively guaranteeing minimal effects when not needed (Wu *et al.*, 1993). Mdm2 was first discovered as an extrachromsomal amplification present in several mouse tumors and is itself a highly regulated protein.

The quick response needed from p53 upon cellular stress requires this autoregulatory feedback loop to be blocked efficiently and effectively for sufficient amounts of the protein to accumulate. Logistically, this can occur either by a) disruptive modifications occurring on p53, b) disruptive modifications on Mdm2, c) protein-protein interactions, or d) inhibition of mdm2 gene expression. Although the latter has been proposed as one mechanism for p53 stability, the fact that p53 continues to drive Mdm2 expression, coupled with the abundance of information supporting the first two hypotheses, suggest post-translational events as the predominant mechanisms. The vast number of mechanisms used by the cell to regulate both p53 and Mdm2 suggests the true physiologic importance they have in maintaining genomic integrity in resting cells. Indeed, p53 homozygous null mutant mice develop tumor formation within 6 months of age and the removal of *mdm2* gene expression leads to embryonic lethality in mice (Donehower *et al.*, 1992; Jones *et al.*, 1995; Montes de Oca Luna *et al.*, 1995). Importantly, crossing both nullizygous mice yields a normal, viable *p53/mdm2* double-knockout mouse. These findings indicate that Mdm2 is not only in the same genetic pathway as p53, but that it can also be viewed as one of the most important regulators of p53 function. Because of its intimate association with p53, Mdm2 provides an important regulatory level and point of attack for quickly stabilizing and activating the transactivation activity of p53.

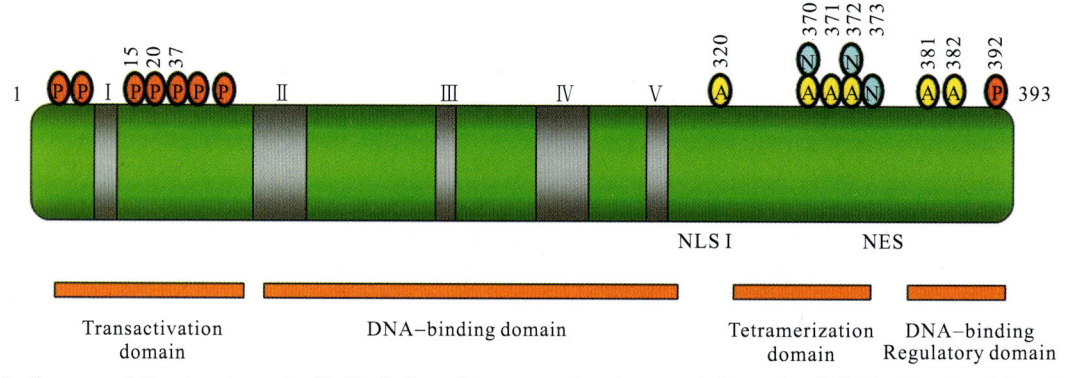

Fig.15.1 A diagram of the structure of p53. Evolutionarily conserved regions are indicated as I-V. The functional domains are listed as: transactivation domain, DNA-binding domain, tetramerization domain, and DNA-binding regulatory domain. Phosphorylation, Acetylation, and Neddylation residues are indicated.

Ubiquitin ligases are divided into two classes depending on whether they contain a HECT domain or a RING domain (Pickart, 2004). Mdm2 falls within the family of RING E3 ligases possessing the characteristic CX2CX(9-39)CX(1-3)HX(2-3)C/HX2CX(4-48)CX2C motif at its extreme C-terminus (Joazeiro and Weissman, 2000). Its predominant substrate remains p53, though Mdm2 has been shown to be capable of ubiquitinating substrates such as β-arrestin, PCAF, and the insulin-like growth factor 1 receptor (Girnita et al., 2003; Jin et al., 2004; Shenoy et al., 2001). Mdm2 can also ubiquitinate itself, and while this may be the predominant mechanism for maintaining its short half-life, it is clearly not the only answer. Mdm2 constructs expressed in cells that are either missing their RING domain (Mdm2 1-440) or contain a point mutation in the active site (Mdm2 C464A) are still capable of being degraded, suggesting other mechanisms for Mdm2 regulation are at play (Brooks and Gu, unpublished data).

In addition to Mdm2, other E3 ligases have been shown to impart specificity toward p53 and promote its proteasome-mediated degradation (Fig.15.2). Pirh2, a RING-H2 domain containing protein, interacts with p53 and promotes Mdm2-independent p53 ubiquitination and degradation (Leng et al., 2003). Reminiscent of Mdm2, Pirh2 is a p53 transcriptional target gene and participates in a similar autoregulatory negative feedback loop. Another E3 ligase, COP1, has also been described recently as a direct ubiquitin ligase for p53. Using an epitope-tagging strategy, Dornan et al. purified a mammalian COP1 protein complex from the U2OS cell line and found p53 as a strong interacting cellular factor (Dornan et al., 2004). COP1 is also a p53-inducible gene, and can ubiquitinate and degrade p53. Further, COP1 depletion by siRNA enhances p53-mediated G1 arrest and can sensitize cells to ionizing radiation. Together, Mdm2, COP1, and Pirh2 represent an army of E3 ligases the cell can call upon to regulate and maintain p53 levels (Fig. 15.1). They suggest that both Mdm2-dependent and independent mechanisms are used cooperatively by the cell for tight p53 regulation. It is yet uncertain exactly how these proteins are specifically regulated and under what situations they may be differentially activated. However, the redundancy of three ubiquitin ligases for p53 emphasizes the importance of keeping this tumor suppressor under tight lock and key.

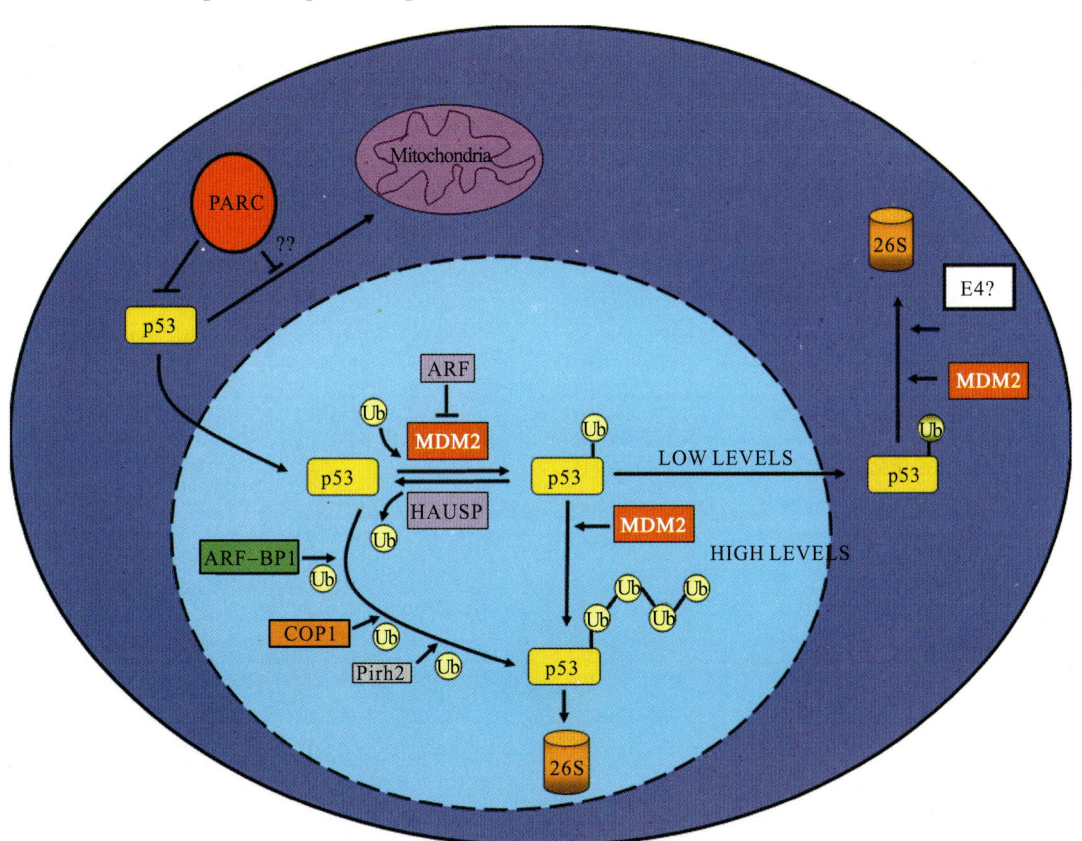

Fig.15.2 A model for p53 regulation with both Mdm2-dependent and Mdm2-independent mechanisms. Mdm2, Pirh2, and COP1 all function as ubiquitin E3 ligases for p53 leading to its 26S proteasome dependent regulation. The protein level of Mdm2 is important for p53 mono- and polyubiquitination. When the level of Mdm2 is high it induces polyubiquitination of p53 and causes its efficient degradation. When the level is low, p53 is monoubiquitinated and moves into the cytoplasm where it can potentially be modified further.

B: Deubiquitination

The strong effect that ubiquitination has on p53 regulation suggests that its reversal may have an equal and opposite effect on protein stability. The first deubiquitinase specific for p53, HAUSP (Herpesvirus-associated ubiquitin-specific protease), was recently shown to have a profound effect on p53 stability (Li et al., 2002). HAUSP is capable of removing ubiquitin moieties for p53 both *in vitro* and *in vivo* and can stabilize p53 when overexpressesd. HAUSP was originally identified as a cellular protein that associated with the Herpesvirus protein ICP0 and also been shown to associate with the Ebstein-Barr virus protein EBNA1 (Everett et al., 1999; Holowaty et al., 2003). The intimate relationship between HAUSP and p53 may be an attractive target for both viruses and oncogenes to exploit as a way of inhibiting p53's role in cell cycle arrest.

The role of HAUSP in p53 regulation became more obscure with the discovery that removal of the gene in somatic HCT116 cells caused profound p53 stabilization. This result was counterintuitive to what was expected, and a detailed biochemical analsysis of this pathway revealed a provocative model. Two independent systems, namely the HeLa cell line and transient partial reduction of HAUSP gene expression by siRNA, both show p53 destabilization, indicating that HAUSP does in part function as a stabilizing effector of p53. However, full reduction or removal of HAUSP expression leads to destabilized Mdm2 and the subsequent stabilization of p53. HAUSP was also shown to act as a deubiquitinase for Mdm2, and this intricate interplay between p53, Mdm2, and HAUSP yields a cellular system that is delicately responsive to changes in HAUSP level (Brooks and Gu, 2004). The physiology underlying this pathway remains to be elucidated. One possibility is that HAUSP is differentially regulated depending on the cellular signaling involved, so that at times it preferentially stabilizes p53 (i.e. DNA damage) and at others it stabilizes Mdm2 (i.e. normal cellular conditions). It is also quite likely that there is a substrate balance in this pathway and that HAUSP does not exhibit an all-or-nothing response. Nevertheless, future studies into HAUSP regulation may yield more insight into the p53-Mdm2 pathway as well as some viral commandeering strategies.

C: Neddylation

The Nedd8 protein is in the ubiquitin-like family of proteins and uses E1 activating and E2 transfer enzymes in a similar fashion as ubiquitin. The Nedd8 modifying pathway is important for yeast growth and viability (Lammer et al., 1998; Osaka et al., 1998). In mammalian systems, it has been shown to be an important modification for the cullin family of proteins. In the case of the SCF complexes, neddylation of cullins seems to regulate their activity. Neddylation can promote the recruitment of an E2 ubiquitin-conjugating enzyme to the SCF complex (Kawakami et al., 2001; Wu et al., 2002). It has also been shown to promote the disassociation of the SCF ligase and cullin inhibitor p120 ^{Cand1}(Liu et al., 2002). In this context, a component of the SCF complex, Roc1/Rbx1/Hrt1, has been proposed to possess E3 Nedd8 ligase activity toward the complex (Gray et al., 2002; Kamura et al., 1999; Morimoto et al., 2003).

Neddylation was recently shown to occur on p53 as well, and this modification inhibited its transactivation activity (Xirodimas et al., 2004). Mdm2 acts as a specific E3 Nedd8 ligase for p53 and can undergo self-neddylation as well. Lysines 370, 372, and 373 seem to be important for the Nedd8 conjugation pathway as p53 KR mutants of this region are no longer inhibited in their transcriptional activity. Together, the data suggests yet another layer of regulation of p53 as a transcription factor and an increase in complexity of Mdm2 function. Further work on this newly evolving aspect of p53 regulation will certainly yield more answers in the future.

D: Phosphorylation

Several residues at both the amino terminus and the carboxyl terminus of p53 are phosphorylated or dephosphorylated in response to genotoxic stress (Appella and Anderson, 2001). Several of these phosphorylation events occur within the N-terminal Mdm2 binding domain of p53 and abrogate the p53-Mdm2 interaction. The DNA-damage induced kinases Chk1 and Chk2 both phosphorylate Ser20 after DNA damage while ATM and ATR both phosphorylate Ser15 and Ser37 (Bode and Dong, 2004). Other kinases including DNA-PK, CK1, CAK, and p38 also phosphorylate p53 at specific residues within this region with unclear roles in the DNA damage response. Despite these findings, it has been difficult to show strong phenotypic changes when these sites are mutated, suggesting a more complex and yet undetermined physiologic role for these events. It is also possible that specific combinations of phosphorylation events are differentially mediating specific p53 responses or regulating effector interactions.

The role of phosphorylation sites outside of the Mdm2 binding domain is less clear. A protein complex consisting of CK2, hSpt16, and SSRP1 specifically phosphorylates Ser392 upon UV irradiation and enhances p53 transactivation activity on the p21 promoter (Keller et al., 2001). JNK also phosphorylates p53 at residue Thr81 and stabilizes the protein (Buschmann et al., 2001). However, when not active, JNK seems to promote p53 degradation by the 26S proteasome through an ill-defined mechanism. In addition, CKII phosphorylates Ser392 and PK-C phosphorylates Ser371, 376, and 378 (Pluquet and Hainaut, 2001). These modifications have been proposed to enhance DNA binding as they occur within the often alluded to C-terminal regulatory domain (see acetylation below).

The missing mechanistic link between phosphorylation status and p53 transactivation was characterized with the finding of the peptidyl-prolyl isomerase Pin1 (Zacchi et al., 2002; Zheng et al., 2002). This protein specifically binds to phosphorylated Ser/Thr-Pro motifs on p53 and induces a conformational change in the protein thereby enhancing its transactivation activity. Importantly, cells deficient in Pin1 show a marked decrease in p53 transactivation of some pro-apoptotic genes, indicating a role for Pin1 in linking p53 phosphorylation with transactivation. Pin1 also interacts with the family member p73 and induces a similar conformational change as with p53 (Mantovani et al., 2004). Here, c-Abl-dependent phosphorylation of p73 enhances the Pin1-p73 interaction and this in turn enhances p300-mediated acetylation of p73. Acetylated p73 is then able to bind to and transactivate pro-apoptotic genes such as *bax*, *p53AIP*, and *pig3*. Pin1 therefore may represent a common translator of phosphorylation status for the p53 family of transcription factors. It also might prove to be a more general mechanism for linking phosphorylation with transactivation function particularly in the case of p53.

E: Acetylation

The covalent linkage of an acetyl group to lysine residues located on histone tails, the enzymatic process of acetylation (Jenuwein and Allis, 2001), is soundly believed to be involved with transcriptional regulation as it was first shown decades ago to correlate with an increase in transcriptional activity (Kouzarides, 2000). The significance of histone acetylation in transcriptional regulation seems indisputable; however, the precise role of this event is still not completely understood. Histones are not the only proteins that can be acetylated however, and the discovery that acetylation of a transcription factor could increase its transactivation activity has opened up an entirely new field of research.

F: HATS/FATS

CBP/p300, a protein possessing histone acetyl-transferase (HAT) activity, acts as a coactivator of p53 and augments its transcriptional activity as well as its biological function *in vivo* (Avantaggiati et al., 1997; Gu et al., 1997; Lill et al., 1997). The observation that p53 could have functional synergism with CBP/p300, together with the intrinsic HAT activity of CBP/p300, led to the discovery of a novel transcriptional factor acetyl-transferase (FAT) function of CBP/p300 on p53 (Gu and Roeder, 1997). The significance of CBP/p300, and arguably its interaction with p53, is emphasized by the presence of p300 mutations in several types of tumors (Goodman and Smolik, 2000). Additionally, mutations of CBP in human Rubeinstein-Taybi syndrome as well as CBP knockout mice lead to a higher risk of tumorigenesis.

Many transcription factors have been demonstrated as bona fide substrates for acetyl-transferases such as GATA-1, MyoD, HMG-1, E2F-1, ACTR, EKLF and Smad7 (Sterner and Berger, 2000). The functional consequences of acetylation are diverse and include increased DNA binding, enhancement of stability, and changes in protein-protein interactions. p53 is specifically acetylated at multiple lysine residues (Lys370, 371, 372, 381, 382) of the C-terminal regulatory domain by CBP/p300, and to a lesser extent Lys320 by PCAF (Appella and Anderson, 2001). The acetylation levels of p53 are significantly enhanced *in vivo* in response to almost every type of stress, well correlated with its activation and stabilization induced by stress (Ito et al., 2001). These acetylation sites of p53 are essential for its ubiquitination and subsequent degradation by Mdm2. Acetylation may even have a more direct role in p53 stabilization, as it was shown that C-terminal acetylation of p53 can inhibit Mdm2-mediated ubiquitination and prolong the half-life of p53. They may also have a significant impact on protein-protein interactions between p53 and transcriptional co-activators such as CBP/p300 and PCAF. Indeed, p53 acetylation has been shown to be critically important for the efficient recruitment of these complexes to promoter regions and the activation of p53 target genes *in vivo* (Barlev et al., 2001).

Numerous studies indicate that the C-terminus of p53 acts as a critical regulator of p53 and negatively modulates its transcriptional activation. Deletion of the C-terminus, injection of antibodies specific for the C-terminus (PAb421), single-strand DNA, protein-protein interactions such as HMG-1, and post-translational modifications at this region all induce profound p53 transactivation abilities (Hupp et al., 1992; Jayaraman and Prives, 1995; Jayaraman *et al.*, 1998). Consistent with this model, the acetylation of p53 can dramatically stimulate its sequence-specific DNA binding activity *in vitro*, possibly as a result of an acetylation-induced conformational change (Espinosa and Emerson, 2001; Gu and Roeder, 1997; Liu *et al.*, 1999; Sakaguchi *et al.*, 1998). The model has also been confirmed *in vitro* and *in vivo* using purified acetylated p53, which augments its site-specific DNA binding to both short and long DNA fragments.

An opposing model suggests that the C-terminus of p53 has a regulatory effect on short pieces of DNA but has no effect on longer DNA templates (Espinosa and Emerson, 2001). This is based on the observation that unpurified acetylated p53 significantly inhibits its activity in an *in vitro* chromatin assay (Espinosa and Emerson, 2001). However, this notion is flawed when considering the data set forth. First, Espinosa and Emerson show that acetylation on the C-terminus of p53 has no effect on DNA binding when compared to unacetylated p53 in an *in vitro* DNase footprinting assay. However, the enzyme used in the assay was subsequently supershifted by an antibody that specifically recognizes the C-terminus of p53, pAb421. If the enzyme used in this assay was indeed completely acetylated, then pAb421 would not recognize p53 and the protein would not be supershifted. Indeed, when purified acetylated p53 is used in a similar assay, there is a profound increase in the DNA binding ability of p53 (Luo *et al.*, 2004). This holds true for the *in vivo* setting as well. The concept of the C-terminus of p53 acting as a positive regulator of transcriptional activation is also unlikely fit for the endogenous p21 promoter, since the p53 C-terminal mutant exhibits very strong cell growth repression and transactivation of this promoter *in vivo* (Crook *et al.*, 1994; Hupp *et al.*, 1992; Jayaraman *et al.*, 1998; Marston *et al.*, 1994). Taken together, a substantial amount of data obtained by several researchers still indicates that acetylation of the C-terminus of p53 is involved in transcriptional regulation, protein stability, and protein-protein interactions.

G: Deacetylases

Histone deacetylase complexes (HDACs) have emerged as notable components in regulating transcriptional activation as well. HDACs are often associated with corepressor complexes and can exert their repressive effects on both histone and non-histone proteins by removing acetyl groups (Smith, 2002). In contrast, much less is known about HDAC activity on p53 function and the general role of p53 deacetylation. Normal resting cells have a very low level of acetylated p53. Treatment of cells with the HDAC inhibitor trichostatin A (TSA) increases levels of acetylated p53 and led to the identification of the adaptor protein PID/MTA2, a component of the HDAC1 complex that can enhance HDAC1-mediated deacetylation of p53 (Juan *et al.*, 2000; Luo *et al.*, 2000). Subsequent work has identified Sir2α (SIRT1), a TSA-resistant, NAD-dependent histone deacetylase that can both deacetylate p53 and attenuate its transcriptional activity (Luo *et al.*, 2001; Vaziri *et al.*, 2001). Sir2α co-localizes in PML nuclear bodies with p53 (Langley *et al.*, 2002) and was structurally shown to undergo a conformational change when bound to acetylated p53 (Avalos *et al.*, 2002). Further, PML and oncogenic Ras can upregulate acetylated p53 levels in primary fibroblasts (Pearson *et al.*, 2000). The novelty of the Sir2α family of HDACs suggests an interesting link between nicotinamide (vitamin B_3), cellular metabolism, and p53-mediated cellular responses to genotoxic stress. Transgenic mice harboring an N-terminus p53 deletion mutant exhibit an early-ageing phenotype (Tyner *et al.*, 2002), and Sir2α is involved in gene silencing and extension of life span in yeast and *C. elegans* (Guarente, 2000). Taken together, Sir2α may provide a possible link between p53 and mammalian longevity.

The prominence of deacetylase activity on p53 certainly raises the defining question of its physiological purpose. One possibility is that deacetylation provides a quick acting mechanism to stop p53 function once transcriptional activation of target genes is no longer needed (Fig.15.3). Targeted deacetylation has been shown to occur very quickly amidst a global equilibrium of genomic acetylation and deacetylation (Katan-Khaykovich and Struhl, 2002). Restoration of this steady-state level at p53 target genes is crucial for cellular homeostasis once DNA repair is complete. Deacetylation could also serve as an important step in MDM2-mediated p53 degradation.

Protein-protein Interactions

p53 is highly dependent on its interaction with specific cellular proteins for rapid stabilization during times of cellular stress. These proteins tend to function by binding to either p53 (e.g. ING1b) or binding to Mdm2 (e.g. pRb, p19ARF, and MdmX) (Hsieh *et al.*, 1999; Kamijo *et al.*, 1998; Leung *et al.*, 2002; Sharp *et al.*, 1999). Presumably, interactions in this regard block Mdm2-mediated ubiquitination of p53 and thereby stabilize the protein.

p14^{ARF} was first discovered as an alternative reading frame gene product from the *INK4a/ARF* gene

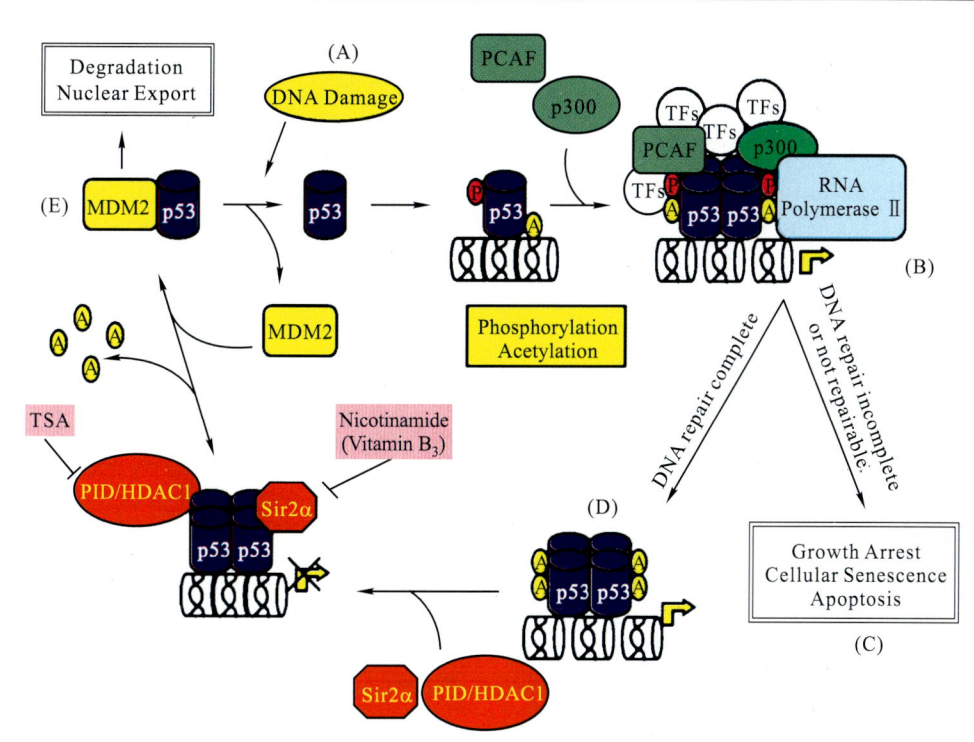

Fig.15.3 A model for p53 transcriptional activation. (A) DNA damage signaling causes Mdm2-p53 disassociation and post-translational events leading to p53 activation. (B) The cell chooses a particular cell fate through known and unknown mechanisms. (C) If the DNA repair is not complete or is not repairable, the cell may choose G1 arrest, apoptosis, or cellular senescence as a fate. (D) On the other hand, if DNA repair is complete the cell may choose to re-enter the cell cycle through recruitment of Sir2α and/or PID/HDAC1 to deacetylate p53 and shut down p53-dependent transcription. (E) p53 is then regulated at low levels by Mdm2 until the next DNA damage event.

locus, a chromosomal region frequently mutated in cancer (Ruas and Peters, 1998). Today the locus is known to code for two genes: p16^{INK4a} and p14^{ARF}, which function to activate the Rb and p53 pathways, respectively. ARF functions upstream of Mdm2 and blocks its activity on p53, a hypothesis strengthened by the tumorigenicity of ARF knockout mice (Lowe and Sherr, 2003). Still, its exact mechanistic function remains unclear. ARF is capable of sequestering Mdm2 in the nucleolus, though more recent data show that ARF can stabilize p53 independent of Mdm2 relocalization (Llanos et al., 2001). ARF can also directly inhibit the enzymatic function of Mdm2 in vitro, however in vivo it can only reduce polyubiquitinated forms of p53 and shows no effect on Mdm2 self-ubiquitination (Honda and Yasuda, 1999; Midgley et al., 2000; Xirodimas et al., 2001). Based on these observations, as well as other data, several working hypotheses of ARF-mediated p53 stabilization can be drawn. ARF may physically sequester Mdm2 away from p53, ARF can directly inhibit Mdm2 enzymatic activity, or ARF possesses distinct and unique functions aside from those proposed that are Mdm2 independent. It is quite likely that all are correct, and ARF has the capability of multi-tasking several functions that have direct or indirect implications on p53 stability.

The nucleolus, possessing ill-defined structural features and entry requirements, is nevertheless an up-and-coming star of complexity contained within the nucleoplasm. In addition to its role as a required exit port for mRNA transcripts ready for translation, it is also becoming an important link between protein synthesis and the p53-Mdm2 pathway (Horn and Vousden, 2004). ARF resides within the nucleolus, and its recently documented interaction with the nucleolar protein B23 has implicated ARF as a link between ribosomal biogenesis and p53 (Bertwistle et al., 2004; Itahana et al., 2003). ARF promotes the polyubiquitination and degradation of B23, and by doing so, may prevent it from acting as a required rRNA processing enzyme needed for proper cellular proliferation. In this regard, ARF seems to serve a dual role as a tumor suppressor, acting on both B23 and the Mdm2-p53 pathway to stop cell growth and proliferation. ARF is clearly not the only player in this novel offshoot of the pathway, and recent evidence has shown Mdm2 may be involved in ribosomal biogenesis. Mdm2 was previously shown to interact with ribosomal protein L5 suggesting that

Mdm2 was involved in some aspect of protein translation (Marechal et al., 1994). More recently it has been shown the Mdm2 can also interact with ribosomal protein L11 and that this protein can inhibit Mdm2's activity on p53 (Lohrum et al., 2003; Zhang et al., 2003). L11, as well as L5, may then act as sensors of nucleolar stress that can inhibit Mdm2 activity so that p53 can be efficiently stabilized. While it appears that no individual interaction is an exclusive mechanism for p53 stabilization and activation, it is quite likely that the cell utilizes a balance of pathways under various stress responses to sufficiently activate the protein (Ashcroft et al., 2000).

MdmX is yet another protein that has an intricate and poorly understood involvement in p53 regulation. The embryonic lethal phenotype of MdmX null embryos and the failure to rescue this phenotype when crossed with the p53 null mice clearly places it as an important negative regulator of p53 during embryonic development. Still, its physiologic function and the question of whether they hold true in all cell types remains to be seen. MdmX possesses structural similarities with Mdm2, and though it has a C-terminal RING domain, does not possess an *in vivo* ability to ubiquitinate and degrade p53. MdmX can stabilize p53, as polyubiquitinated forms of p53 readily accumulate within the nucleus (Jackson and Berberich, 2000; Stad et al., 2001). However, when the ratio of MdmX:Mdm2 is low, these proteins cooperatively decrease p53 levels (Gu et al., 2002; Iwakuma and Lozano, 2003). It has been shown that MdmX can act as a transcriptional repressor suggesting another possible physiologic role for MdmX (Kadakia et al., 2002; Wunderlich et al., 2004; Yam et al., 1999) MdmX imparts a negative affect on p53 acetylation, possibly through inhibition of p300/CBP. This observation has also been supported by an increased level of acetylated p53 in *mdmx*-mutant cells. Regardless of the mechanisms, MdmX may rival the functional importance of Mdm2, considering that it is found upregulated in many tumors expressing wildtype p53.

In addition, some proteins can stabilize p53 in an Mdm2-independent manner. Calpain 1, β-catenin, and JNK have all been shown to stabilize p53 independently of Mdm2 (Damalas et al., 1999; Fuchs et al., 1998; Kubbutat and Vousden, 1997). NQO1, an NADH quinone oxidoreductase, stabilizes p53 by regulating its interaction with the 20S proteosome and represents a ubiquitin-independent pathway for p53 regulation (Asher et al., 2005). The protein Sin3a can also bind to and stabilize p53 on promoters of genes targeted for transcriptional repression in response to DNA damage (Zilfou et al., 2001). This spatial interaction is thought to extend p53 promoter association.

A: Cytoplasmic p53

Compartmentalization of the cell into distinct structural components provides physical isolation of proteins from one another and adds an additional layer of protein regulation. In the case of p53, its activity as a potent transcription factor is in part regulated by sequestering it away from its targets in the nucleus during times of homeostasis. The importance of this mechanism is underscored by its exploitation by several viruses. The cytomegalovirus and adenovirus type 12 both maintain p53 in the cytoplasm, presumably as a way to ensure continued cell proliferation (Kovacs et al., 1996; Zhao and Liao, 2003).

The cellular protein PARC (p53-associated, Parkin-like cytoplasmic protein) has recently been isolated from stable p53 cytoplasmic protein complexes in unstressed cells (Nikolaev et al., 2003). PARC acts as a cytoplasmic anchor for p53 by physically binding to and sequestering it in the cytoplasm. Indeed, during unstressed conditions p53 is found diffusely dispersed throughout the cytoplasm of cells. Several neuroblastoma cell lines also show high expression of PARC and abnormally distributed cytoplasmic wildtype p53 that fails to respond to DNA damage signals. When PARC was specifically ablated using siRNA in these tumor cells, the p53-mediated DNA damage response was restored. These findings are hopeful when considering novel chemotherapeutic targets, as neuroblastomas represent the most common extracranial malignancy in children, often with poor prognosis (Kastan and Zambetti, 2003).

Moving p53 away from its transcriptional targets seems to be a critical requirement for inhibiting its function, and Mdm2 may play an important role in this regard. Interestingly, recent evidence suggests monoubiquitination is a critical signal for nuclear export, and Mdm2 is capable of inducing both monoubiquitination and polyubiquitination on p53 (Li et al., 2003). When Mdm2 is low, it catalyzes monoubiquitination of p53 that is effectively exported out of the nucleus for degradation and/or further modifications in the cytoplasm. When Mdm2 levels are high, p53 is quickly polyubiquitinated within the nucleus and degraded by nuclear proteasomes. One question that is quickly raised is why this type of regulation is needed. It has been known for some time that proteasomes exist in both the nucleus and cytoplasm, and therefore some type of regulation is clearly needed to prevent rapid degradation of proteins in close contact. It also has been shown that during the late stages of DNA damage, a time where p53 has accomplished its jobs and needs to be quickly shut down, that Mdm2 protein levels are very high. While p53 drives Mdm2 expression through a negative feedback loop, the Mdm2-p53 interaction is known to be blocked during cellular stress and DNA damage. A high level of Mdm2 at the end of the DNA damage response would allow the cell to quickly degrade p53 through polyubiquitination as soon as the mechanisms for p53 stabilization are reversed.

The reasons for moving monoubiquitinated p53 into the cytoplasm are less clear. Perhaps this mechanism removes p53 from a region where its actions have large consequences on the fate of the cell. Another intriguing possibility is that monoubiquitinated p53 is acted upon by cellular factors present in the cytoplasm for diverse functions other than transcriptional activation.

Commanding Cellular Fate

As a leader of the cell's proteome, p53 possesses transactivation abilities that can lead the cell down particular pathways including cell cycle arrest, cellular senescence, and apoptosis. The molecular mechanisms behind how p53 functions in each of these pathways are becoming clearer; how p53 *chooses* a particular pathway is not. Originally, it seemed to be cell type specific as γ-irradiation of thymocytes induced p53-mediated apoptosis while in normal human fibroblasts it induced a p53-mediated G1 arrest (Clarke *et al.*, 1993; Kuerbitz *et al.*, 1992; Lowe *et al.*, 1993). However, further findings suggested a more complicated scenario involving other factors. Evidence detailed below begins to paint a global picture of p53 function at a post-stabilization point whereby p53 is guided toward a particular pathway by covalent post-translational modifications, co-activators, and promoter selectivity. Though different types of genotoxic events trigger different responses of p53, the net result appears to be a more general guidance toward particular cell fates. Understanding the specifics of this aspect of p53 biology is of utmost interest, as it may yield new ideas for cancer therapy in tumors retaining wildtype p53.

Cell-cycle arrest mediated by p53 occurs largely through the induction of the CDK inhibitor $p21^{Waf1}$. p21 halts cells in G1 of the cell cycle by inhibiting several cyclin dependent kinases and can also inhibit DNA replication through interactions with PCNA (Fotedar *et al.*, 2004). p53 induces p21 after the occurrence of several types of genotoxic stress, including UV and ionizing irradiation, and may represent the first wave of cellular arsenal p53 utilizes to protect the cell. In addition to *p21*, *GADD45* and *14-3-3σ* are other genes involved with inhibiting the cell cycle that are capable of being transactivated by p53 (Taylor and Stark, 2001).

Such a robust induction of p21 raises at least two possibilities of explanation. The first is that by evoking a cell cycle arrest, p53 buys time for the cell to assess damaged DNA, attempt DNA repair, transactivate other genes, and decide to revert back to the cell cycle or initiate apoptosis. Indeed, both BRCA1 and WT1 tumor suppressors can selectively coax p53 to induce cell cycle arrest and DNA repair genes (MacLachlan *et al.*, 2002; Maheswaran *et al.*, 1995). In addition, p53 itself has roles in nucleotide excision repair and base excision repair through the induction of $p48^{DDB2}$, *XPC*, and *MSH2* (Sengupta and Harris, 2005). p53 also enhances mismatch repair through associations with APE1/REF1 and DNA polymerase β. The decision of apoptosis isn't to be taken lightly, as this represents an irreversible and terminal effort the cell uses as a last option. By putting the cell in a cell cycle arrest, p53 gives the cell time to assess its options. Alternatively, p21 induction may just be the first wave of defense in a long list of options culminating in apoptosis if everything prior fails.

The first indication that p53 may possess some selectivity toward choosing a particular cell fate came from the analysis of a naturally occurring p53 tumor mutant, p53175P, which retained its ability to induce cell cycle arrest but lost its apoptotic function (Rowan *et al.*, 1996). Though a majority of tumor derived mutations in p53 cause a loss of transactivation function, this mutant was shown to retain its ability to transactivate the cell cycle arrest promoting gene $p21^{WAF1}$, but not the pro-apoptotic gene *bax*. Others have shown similar phenotypes for an increasing number of these types of mutants (Friedlander *et al.*, 1996; Ludwig *et al.*, 1996; Ryan and Vousden, 1998; Smith *et al.*, 1999). In this regard, p53 may possess the inherent ability, through yet undetermined mechanisms, to selectively choose subsets of genes involved in either cell cycle regulation or apoptosis.

Further evidence of covalent post-translational modifications on p53 driving particular responses supports the idea that p53 has selectivity for choosing a cell fate. Homeodomain-interacting protein kinase 2 (HIPK2) phosphorylates Ser46 after UV irradiation and drives an apoptotic response (D'Orazi *et al.*, 2002; Hofmann *et al.*, 2002). HIPK2-mediated Ser46 phosphorylation, in conjunction with the *p53DINP1* gene product, is important for the induction of the pro-apoptotic gene *p53AIP1* (Oda *et al.*, 2000). Ser20 was previously shown to be phosphorylated by the checkpoint kinases

Chk1 and Chk2 in response to IR causing the abrogation of the p53-Mdm2 interaction (Chehab et al., 2000; Hirao et al., 2000; Shieh et al., 2000). More recently, however, it has been shown that Chk2 in not needed for p53-mediated cell cycle arrest but is required for an apoptotic response (Jack et al., 2002). The exact mechanism for this is currently unknown, but it does support the notion of p53 having selectivity in determining cell fate. Co-factors, such as JMY and the ASPP family of proteins, also enhance a p53-mediated apoptotic response (Samuels-Lev et al., 2001; Shikama et al., 1999). In the case of JMY, the result may be an indirect effect through interaction with p300 and alteration of p53 acetylation status. The p53 family members p63 and p73 are also required for p53-mediated induction of apoptotic genes, as DNA damaged-induced p53 failed to evoke apoptosis in cells deficient in p63 and p73 (Flores et al., 2002).

Another possibility is that p53 responsive elements exist in promoters throughout the genome that have inherently different binding affinities for p53. Binding and transactivaton could again depend on medications and co-factors of p53, or it could conceivably depend on quantitative levels of p53 (Fridman and Lowe, 2003). Engineered systems with conditional p53 expression do indeed show a shift toward G1 arrest when p53 levels are low and an apoptotic response when p53 levels are high. p53 protein levels are in a delicate flux at all times, governed by Mdm2, promoter binding affinity could be one mechanism of several that exists for p53-mediated cell fate selectivity.

A: Apoptosis

Once the apoptotic response is determined and initiated, p53 has been shown to be capable of inducing several apoptotic genes in both the so-called extrinsic and intrinsic apoptotic pathways (Haupt et al., 2003). The role of p53 in the extrinsic pathway includes transactivation of *DR5* and *Fas/CD95* that encode transmembrane proteins in the TNF-R family (Muller et al., 1998; Owen-Schaub et al., 1995; Wu et al., 1997). These proteins mediate apoptosis upon activation through the downstream activation of several caspases, though they appear to function in a tissue specific manner (Bouvard et al., 2000; Burns et al., 2001). PERP, another transmembrane protein, is induced by p53 after DNA damage and thought to play an as yet ill-defined role in the apoptotic response as well (Attardi et al., 2000). Similarly, *PIDD* is yet another gene containing a p53 response element that is induced upon shifting to the permissive temperature in a p53 conditional erythroleukemia cell line (Lin et al., 2000).

p53 has several genetic targets within the intrinsic apoptotic pathway, and they represent a majority of the genes mediating the p53-dependent apoptotic response. p53 transactivates several Bcl-2 pro-apoptotic family members including *bax*, *puma*, *noxa*, and *bid* (Nakano and Vousden, 2001; Oda et al., 2000; Sax et al., 2002). p53 can also induce the expression *apaf-1*, an essential component of the apoptotic effector machinery and co-activator of caspase 9 (Kannan et al., 2001; Moroni et al., 2001; Robles et al., 2001; Rozenfeld-Granot et al., 2002). Interestingly, p53-mediated induction of *apaf-1* is required for Myc-induced apoptosis in mouse embryonic fibroblasts (Soengas et al., 1999). The activation of p53 and subsequent induction of *apaf-1* may represent one failsafe mechanism the cell has evolved for inducing apoptosis in response to aberrant oncogenic signals.

Transcription independent mechanisms also exist for the induction of an apoptotic response by p53. In certain contexts or cell types, p53 has been shown to induce apoptosis independently of its transactivation ability. p53 can also directly interact with mitochondrial proteins and permeabilize the outer membrane releasing cytochrome c (Chipuk et al., 2004; Mihara et al., 2003). Further, p53 can directly activate both pro-apoptotic proteins Bax and Bak, with the latter by disruption of the Bak-Mcl1 complex (Leu et al., 2004). This shift in the balance of pro- and anti-apoptotic Bcl-2 family members may be enough to enhance the apoptotic response in a transcription-independent manner. Under certain conditions, p53 accumulation also induces a biphasic apoptotic response that is first transcriptionally independent followed by one that is transcriptionally dependent (Erster et al., 2004).

p53 also functions as a transrepressor of gene transcription, generally thought to occur through the formation of an mSin3a repressor complex that recruits histone deacetylases to the promoters (Zilfou et al., 2001). The genes to date shown to be repressed by p53 include *bcl-2*, *bcl-X*, and *survivin*, all of which are anti-apoptotic genes (Haldar et al., 1994; Hoffman et al., 2002; Mirza et al., 2002; Miyashita et al., 1994; Sugars et al., 2001). Repression of these genes may therefore be an additional mechanism for p53 to enhance apoptosis, as the net result of inhibition of anti-apoptotic genes is apoptosis.

Taken together, the data emerging suggest an important and centralized role of p53 in the apoptosis pathway that is highly, but not exclusively, dependent on its transactivation function. p53 has been described as a critical 'node' in the cellular circuitry, and as such, has many downstream effector circuits. It is no surprise then that there is no one critical downstream component that governs the apoptotic response; rather, p53 has its "hand" in several death circuit "cookie jars" as to increase the likelihood that apoptosis will occur, as so eloquently put by others (Fridman and Lowe, 2003).

Conclusion

The study of p53 biology and its function as a transcription factor has grown exponentially since it was first discovered in 1982. Still, the precise regulation of this critical transcription factor and the precise sequence of events leading up to its potent activity in the cell cycle remain unclear. Regardless, p53 has a clear implication in tumorigenesis and remains a sought after target for chemotherapeutics. Though little can be done for endogenous p53 that is inactive in tumors, introducing wildtype p53 in this setting may restore critical genetic connections enough to induce an apoptotic response. Understanding p53 function and the pathways that regulate it is also extremely important, as they may yield novel interventions for lowering the likelihood that p53 is inactivated. The future of p53 is open to endless possibilities. Despite continuing complexity, it still remains "the cellular gatekeeper for growth and division" (Levine, 1997).

References

Appella, E., and Anderson, C. W. (2001). Post-translational modifications and activation of p53 by genotoxic stresses. Eur J Biochem *268*, 2764-2772.

Ashcroft, M., Taya, Y., and Vousden, K. H. (2000). Stress signals utilize multiple pathways to stabilize p53. Mol Cell Biol *20*, 3224-3233.

Asher, G., Tsvetkov, P., Kahana, C., and Shaul, Y. (2005). A mechanism of ubiquitin-independent proteasomal degradation of the tumor suppressors p53 and p73. Genes Dev *19*, 316-321.

Attardi, L. D., Reczek, E. E., Cosmas, C., Demicco, E. G., McCurrach, M. E., Lowe, S. W., and Jacks, T. (2000). PERP, an apoptosis-associated target of p53, is a novel member of the PMP-22/gas3 family. Genes Dev *14*, 704-718.

Avalos, J. L., Celic, I., Muhammad, S., Cosgrove, M. S., Boeke, J. D., and Wolberger, C. (2002). Structure of a Sir2 Enzyme Bound to an Acetylated p53 Peptide. Mol Cell *10*, 523-535.

Avantaggiati, M. L., Ogryzko, V., Gardner, K., Giordano, A., Levine, A. S., and Kelly, K. (1997). Recruitment of p300/CBP in p53-dependent signal pathways. Cell *89*, 1175-1184.

Barlev, N. A., Liu, L., Chehab, N. H., Mansfield, K., Harris, K. G., Halazonetis, T. D., and Berger, S. L. (2001). Acetylation of p53 activates transcription through recruitment of coactivators/histone acetyltransferases. Mol Cell *8*, 1243-1254.

Bertwistle, D., Sugimoto, M., and Sherr, C. J. (2004). Physical and functional interactions of the Arf tumor suppressor protein with nucleophosmin/B23. Mol Cell Biol *24*, 985-996.

Bode, A. M., and Dong, Z. (2004). Post-translational modification of p53 in tumorigenesis. Nat Rev Cancer *4*, 793-805.

Bouvard, V., Zaitchouk, T., Vacher, M., Duthu, A., Canivet, M., Choisy-Rossi, C., Nieruchalski, M., and May, E. (2000). Tissue and cell-specific expression of the p53-target genes: bax, fas, mdm2 and waf1/p21, before and following ionising irradiation in mice. Oncogene *19*, 649-660.

Brooks, C. L., and Gu, W. (2003). Ubiquitination, phosphorylation and acetylation: the molecular basis for p53 regulation. Curr Opin Cell Biol *15*, 164-171.

Brooks, C. L., and Gu, W. (2004). Dynamics in the p53-Mdm2 ubiquitination pathway. Cell Cycle *3*, 895-899.

Burns, T. F., Bernhard, E. J., and El-Deiry, W. S. (2001). Tissue specific expression of p53 target genes suggests a key role for KILLER/DR5 in p53-dependent apoptosis *in vivo*. Oncogene *20*, 4601-4612.

Buschmann, T., Potapova, O., Bar-Shira, A., Ivanov, V. N., Fuchs, S. Y., Henderson, S., Fried, V. A., Minamoto, T., Alarcon-Vargas, D., Pincus, M. R., *et al.* (2001). Jun NH2-terminal kinase phosphorylation of p53 on Thr-81 is important for p53 stabilization and transcriptional activities in response to stress. Mol Cell Biol *21*, 2743-2754.

Chehab, N. H., Malikzay, A., Appel, M., and Halazonetis, T. D. (2000). Chk2/hCds1 functions as a DNA damage checkpoint in G(1) by stabilizing p53. Genes Dev *14*, 278-288.

Chipuk, J. E., Kuwana, T., Bouchier-Hayes, L., Droin, N. M., Newmeyer, D. D., Schuler, M., and Green, D. R. (2004). Direct activation of Bax by p53 mediates mitochondrial membrane permeabilization and apoptosis. Science *303*, 1010-1014.

Chowdary, D. R., Dermody, J. J., Jha, K. K., and Ozer, H. L. (1994). Accumulation of p53 in a mutant cell line defective in the ubiquitin pathway. Mol Cell Biol *14*, 1997-2003.

Clarke, A. R., Purdie, C. A., Harrison, D. J., Morris, R. G., Bird, C. C., Hooper, M. L., and Wyllie, A. H. (1993). Thymocyte apoptosis induced by p53-dependent and independent pathways. Nature *362*, 849-852.

Crook, T., Marston, N. J., Sara, E. A., and Vousden, K. H. (1994). Transcriptional activation by p53 correlates with suppression of growth but not transformation. Cell *79*, 817-827.

D'Orazi, G., Cecchinelli, B., Bruno, T., Manni, I., Higashimoto, Y., Saito, S., Gostissa, M., Coen, S., Marchetti, A., Del Sal, G., *et al.* (2002). Homeodomain-interacting protein kinase-2 phosphorylates p53 at Ser 46 and mediates apoptosis. Nat Cell Biol *4*, 11-19.

Damalas, A., Ben-Ze'ev, A., Simcha, I., Shtutman, M., Leal, J. F., Zhurinsky, J., Geiger, B., and Oren, M. (1999). Excess beta-catenin promotes accumulation of transcriptionally active p53. Embo J *18*, 3054-3063.

Donehower, L. A., Harvey, M., Slagle, B. L., McArthur, M. J., Montgomery, C. A., Jr., Butel, J. S., and Bradley, A. (1992). Mice deficient for p53 are developmentally normal but susceptible to spontaneous tumours. Nature *356*, 215-221.

Dornan, D., Wertz, I., Shimizu, H., Arnott, D., Frantz, G. D., Dowd, P., K, O. R., Koeppen, H., and Dixit, V. M. (2004). The ubiquitin ligase COP1 is a critical negative regulator of p53. Nature.

Erster, S., Mihara, M., Kim, R. H., Petrenko, O., and Moll, U. M. (2004). *In vivo* mitochondrial p53 translocation triggers a rapid first wave of cell death in response to DNA damage that can precede p53 target gene activation. Mol Cell Biol *24*, 6728-6741.

Espinosa, J. M., and Emerson, B. M. (2001). Transcriptional regulation by p53 through intrinsic DNA/chromatin binding and site-directed cofactor recruitment. Mol Cell *8*, 57-69.

Everett, R. D., Meredith, M., and Orr, A. (1999). The ability of herpes simplex virus type 1 immediate-early protein Vmw110 to bind to a ubiquitin-specific protease contributes to its roles in the activation of gene expression and stimulation of virus replication. J Virol *73*, 417-426.

Flores, E. R., Tsai, K. Y., Crowley, D., Sengupta, S., Yang, A., McKeon, F., and Jacks, T. (2002). p63 and p73 are required for p53-dependent apoptosis in response to DNA damage. Nature *416*, 560-564.

Fotedar, R., Bendjennat, M., and Fotedar, A. (2004). Role of p21WAF1 in the cellular response to UV. Cell Cycle *3*, 134-137.

Freedman, D. A., Wu, L., and Levine, A. J. (1999). Functions of the MDM2 oncoprotein. Cell Mol Life Sci *55*, 96-107.

Fridman, J. S., and Lowe, S. W. (2003). Control of apoptosis by p53. Oncogene *22*, 9030-9040.

Friedlander, P., Haupt, Y., Prives, C., and Oren, M. (1996). A mutant p53 that discriminates between p53-responsive genes cannot induce apoptosis. Mol Cell Biol *16*, 4961-4971.

Fuchs, S. Y., Adler, V., Buschmann, T., Yin, Z., Wu, X., Jones, S. N., and Ronai, Z. (1998). JNK targets p53 ubiquitination and degradation in nonstressed cells. Genes Dev *12*, 2658-2663.

Girnita, L., Girnita, A., and Larsson, O. (2003). Mdm2-dependent ubiquitination and degradation of the insulin-like growth factor 1 receptor. Proc Natl Acad Sci USA *100*, 8247-8252.

Goodman, R. H., and Smolik, S. (2000). CBP/p300 in cell growth, transformation, and development. Genes Dev *14*, 1553-1577.

Gray, W. M., Hellmann, H., Dharmasiri, S., and Estelle, M. (2002). Role of the *Arabidopsis* RING-H2 protein RBX1 in RUB modification and SCF function. Plant Cell *14*, 2137-2144.

Gu, J., Kawai, H., Nie, L., Kitao, H., Wiederschain, D., Jochemsen, A. G., Parant, J., Lozano, G., and Yuan, Z. M. (2002). Mutual dependence of MDM2 and MDMX in their functional inactivation of p53. J Biol Chem *277*, 19251-19254.

Gu, W., and Roeder, R. G. (1997). Activation of p53 sequence-specific DNA binding by acetylation of the p53 C-terminal domain. Cell *90*, 595-606.

Gu, W., Shi, X. L., and Roeder, R. G. (1997). Synergistic activation of transcription by CBP and p53. Nature *387*, 819-823.

Guarente, L. (2000). Sir2 links chromatin silencing, metabolism, and aging. Genes Dev *14*, 1021-1026.

Haldar, S., Negrini, M., Monne, M., Sabbioni, S., and Croce, C. M. (1994). Down-regulation of bcl-2 by p53 in breast cancer cells. Cancer Res *54*, 2095-2097.

Haupt, S., Berger, M., Goldberg, Z., and Haupt, Y. (2003). Apoptosis - the p53 network. J Cell Sci *116*, 4077-4085.

Haupt, Y., Maya, R., Kazaz, A., and Oren, M. (1997). Mdm2 promotes the rapid degradation of p53. Nature *387*, 296-299.

Hirao, A., Kong, Y. Y., Matsuoka, S., Wakeham, A., Ruland, J., Yoshida, H., Liu, D., Elledge, S. J., and Mak, T. W. (2000). DNA damage-induced activation of p53 by the checkpoint kinase Chk2. Science *287*, 1824-1827.

Hoffman, W. H., Biade, S., Zilfou, J. T., Chen, J., and Murphy, M. (2002). Transcriptional repression of the anti-apoptotic survivin gene by wild type p53. J Biol Chem *277*, 3247-3257.

Hofmann, T. G., Moller, A., Sirma, H., Zentgraf, H., Taya, Y., Droge, W., Will, H., and Schmitz, M. L. (2002). Regulation of p53 activity by its interaction with homeodomain-interacting protein kinase-2. Nat Cell Biol *4*, 1-10.

Holowaty, M. N., Zeghouf, M., Wu, H., Tellam, J., Athanasopoulos, V., Greenblatt, J., and Frappier, L. (2003). Protein profiling with Epstein-Barr nuclear antigen-1 reveals an interaction with the herpesvirus-associated ubiquitin-specific protease HAUSP/USP7. J Biol Chem *278*, 29987-29994.

Honda, R., Tanaka, H., and Yasuda, H. (1997). Oncoprotein MDM2 is a ubiquitin ligase E3 for tumor suppressor p53. FEBS Lett *420*, 25-27.

Honda, R., and Yasuda, H. (1999). Association of p19(ARF) with Mdm2 inhibits ubiquitin ligase activity of Mdm2 for tumor suppressor p53. Embo J *18*, 22-27.

Horn, H. F., and Vousden, K. H. (2004). Cancer: guarding the guardian? Nature *427*, 110-111.

Hsieh, J. K., Chan, F. S., O'Connor, D. J., Mittnacht, S., Zhong, S., and Lu, X. (1999). RB regulates the stability and the apoptotic function of p53 via MDM2. Mol Cell *3*, 181-193.

Hupp, T. R., Meek, D. W., Midgley, C. A., and Lane, D. P. (1992). Regulation of the specific DNA binding function of p53. Cell *71*, 875-886.

Itahana, K., Bhat, K. P., Jin, A., Itahana, Y., Hawke, D., Kobayashi, R., and Zhang, Y. (2003). Tumor suppressor ARF degrades B23, a nucleolar protein involved in ribosome biogenesis and cell proliferation. Mol Cell *12*, 1151-1164.

Ito, A., Lai, C. H., Zhao, X., Saito, S., Hamilton, M. H., Appella, E., and Yao, T. P. (2001). p300/CBP-mediated p53 acetylation is commonly induced by p53-activating agents and inhibited by MDM2. Embo J *20*, 1331-1340.

Iwakuma, T., and Lozano, G. (2003). MDM2, an introduction. Mol Cancer Res *1*, 993-1000.

Jack, M. T., Woo, R. A., Hirao, A., Cheung, A., Mak, T. W., and

Lee, P. W. (2002). Chk2 is dispensable for p53-mediated G1 arrest but is required for a latent p53-mediated apoptotic response. Proc Natl Acad Sci USA 99, 9825-9829.

Jackson, M. W., and Berberich, S. J. (2000). MdmX protects p53 from Mdm2-mediated degradation. Mol Cell Biol 20, 1001-1007.

Jayaraman, J., and Prives, C. (1995). Activation of p53 sequence-specific DNA binding by short single strands of DNA requires the p53 C-terminus. Cell 81, 1021-1029.

Jayaraman, L., Moorthy, N. C., Murthy, K. G., Manley, J. L., Bustin, M., and Prives, C. (1998). High mobility group protein-1 (HMG-1) is a unique activator of p53. Genes Dev 12, 462-472.

Jenuwein, T., and Allis, C. D. (2001). Translating the histone code. Science 293, 1074-1080.

Jin, Y., Zeng, S. X., Lee, H., and Lu, H. (2004). MDM2 mediates PCAF ubiquitination and degradation. J Biol Chem.

Joazeiro, C. A., and Weissman, A. M. (2000). RING finger proteins: mediators of ubiquitin ligase activity. Cell 102, 549-552.

Jones, S. N., Roe, A. E., Donehower, L. A., and Bradley, A. (1995). Rescue of embryonic lethality in Mdm2-deficient mice by absence of p53. Nature 378, 206-208.

Juan, L. J., Shia, W. J., Chen, M. H., Yang, W. M., Seto, E., Lin, Y. S., and Wu, C. W. (2000). Histone deacetylases specifically down-regulate p53-dependent gene activation. J Biol Chem 275, 20436-20443.

Juven-Gershon, T., and Oren, M. (1999). Mdm2: the ups and downs. Mol Med 5, 71-83.

Kadakia, M., Brown, T. L., McGorry, M. M., and Berberich, S. J. (2002). MdmX inhibits Smad transactivation. Oncogene 21, 8776-8785.

Kamijo, T., Weber, J. D., Zambetti, G., Zindy, F., Roussel, M. F., and Sherr, C. J. (1998). Functional and physical interactions of the ARF tumor suppressor with p53 and Mdm2. Proc Natl Acad Sci USA 95, 8292-8297.

Kamura, T., Conrad, M. N., Yan, Q., Conaway, R. C., and Conaway, J. W. (1999). The Rbx1 subunit of SCF and VHL E3 ubiquitin ligase activates Rub1 modification of cullins Cdc53 and Cul2. Genes Dev 13, 2928-2933.

Kannan, K., Kaminski, N., Rechavi, G., Jakob-Hirsch, J., Amariglio, N., and Givol, D. (2001). DNA microarray analysis of genes involved in p53 mediated apoptosis: activation of Apaf-1. Oncogene 20, 3449-3455.

Kastan, M. B., and Zambetti, G. P. (2003). Parc-ing p53 in the cytoplasm. Cell 112, 1-2.

Katan-Khaykovich, Y., and Struhl, K. (2002). Dynamics of global histone acetylation and deacetylation in vivo: rapid restoration of normal histone acetylation status upon removal of activators and repressors. Genes Dev 16, 743-752.

Kawakami, T., Chiba, T., Suzuki, T., Iwai, K., Yamanaka, K., Minato, N., Suzuki, H., Shimbara, N., Hidaka, Y., Osaka, F., et al. (2001). NEDD8 recruits E2-ubiquitin to SCF E3 ligase. Embo J 20, 4003-4012.

Keller, D. M., Zeng, X., Wang, Y., Zhang, Q. H., Kapoor, M., Shu, H., Goodman, R., Lozano, G., Zhao, Y., and Lu, H. (2001). A DNA damage-induced p53 serine 392 kinase complex contains CK2, hSpt16, and SSRP1. Mol Cell 7, 283-292.

Kouzarides, T. (2000). Acetylation: a regulatory modification to rival phosphorylation? Embo J 19, 1176-1179.

Kovacs, A., Weber, M. L., Burns, L. J., Jacob, H. S., and Vercellotti, G. M. (1996). Cytoplasmic sequestration of p53 in cytomegalovirus-infected human endothelial cells. Am J Pathol 149, 1531-1539.

Kubbutat, M. H., Jones, S. N., and Vousden, K. H. (1997). Regulation of p53 stability by Mdm2. Nature 387, 299-303.

Kubbutat, M. H., and Vousden, K. H. (1997). Proteolytic cleavage of human p53 by calpain: a potential regulator of protein stability. Mol Cell Biol 17, 460-468.

Kuerbitz, S. J., Plunkett, B. S., Walsh, W. V., and Kastan, M. B. (1992). Wild-type p53 is a cell cycle checkpoint determinant following irradiation. Proc Natl Acad Sci USA 89, 7491-7495.

Lammer, D., Mathias, N., Laplaza, J. M., Jiang, W., Liu, Y., Callis, J., Goebl, M., and Estelle, M. (1998). Modification of yeast Cdc53p by the ubiquitin-related protein rub1p affects function of the SCFCdc4 complex. Genes Dev 12, 914-926.

Langley, E., Pearson, M., Faretta, M., Bauer, U. M., Frye, R. A., Minucci, S., Pelicci, P. G., and Kouzarides, T. (2002). Human SIR2 deacetylates p53 and antagonizes PML/p53-induced cellular senescence. Embo J 21, 2383-2396.

Leng, R. P., Lin, Y., Ma, W., Wu, H., Lemmers, B., Chung, S., Parant, J. M., Lozano, G., Hakem, R., and Benchimol, S. (2003). Pirh2, a p53-induced ubiquitin-protein ligase, promotes p53 degradation. Cell 112, 779-791.

Leu, J. I., Dumont, P., Hafey, M., Murphy, M. E., and George, D. L. (2004). Mitochondrial p53 activates Bak and causes disruption of a Bak-Mcl1 complex. Nat Cell Biol 6, 443-450.

Leung, K. M., Po, L. S., Tsang, F. C., Siu, W. Y., Lau, A., Ho, H. T., and Poon, R. Y. (2002). The candidate tumor suppressor ING1b can stabilize p53 by disrupting the regulation of p53 by MDM2. Cancer Res 62, 4890-4893.

Levine, A. J. (1997). p53, the cellular gatekeeper for growth and division. Cell 88, 323-331.

Li, M., Brooks, C. L., Wu-Baer, F., Chen, D., Baer, R., and Gu, W. (2003). Mono- versus polyubiquitination: differential control of p53 fate by Mdm2. Science 302, 1972-1975.

Li, M., Chen, D., Shiloh, A., Luo, J., Nikolaev, A. Y., Qin, J., and Gu, W. (2002). Deubiquitination of p53 by HAUSP is an important pathway for p53 stabilization. Nature 416, 648-653.

Lill, N. L., Grossman, S. R., Ginsberg, D., DeCaprio, J., and Livingston, D. M. (1997). Binding and modulation of p53 by p300/CBP coactivators. Nature 387, 823-827.

Lin, Y., Ma, W., and Benchimol, S. (2000). Pidd, a new death-domain-containing protein, is induced by p53 and promotes apoptosis. Nat Genet 26, 122-127.

Liu, J., Furukawa, M., Matsumoto, T., and Xiong, Y. (2002).

NEDD8 modification of CUL1 dissociates p120(CAND1), an inhibitor of CUL1-SKP1 binding and SCF ligases. Mol Cell 10, 1511-1518.

Liu, L., Scolnick, D. M., Trievel, R. C., Zhang, H. B., Marmorstein, R., Halazonetis, T. D., and Berger, S. L. (1999). p53 sites acetylated in vitro by PCAF and p300 are acetylated in vivo in response to DNA damage. Mol Cell Biol 19, 1202-1209.

Llanos, S., Clark, P. A., Rowe, J., and Peters, G. (2001). Stabilization of p53 by p14ARF without relocation of MDM2 to the nucleolus. Nat Cell Biol 3, 445-452.

Lohrum, M. A., Ludwig, R. L., Kubbutat, M. H., Hanlon, M., and Vousden, K. H. (2003). Regulation of HDM2 activity by the ribosomal protein L11. Cancer Cell 3, 577-587.

Lowe, S. W., Schmitt, E. M., Smith, S. W., Osborne, B. A., and Jacks, T. (1993). p53 is required for radiation-induced apoptosis in mouse thymocytes. Nature 362, 847-849.

Lowe, S. W., and Sherr, C. J. (2003). Tumor suppression by Ink4a-Arf: progress and puzzles. Curr Opin Genet Dev 13, 77-83.

Ludwig, R. L., Bates, S., and Vousden, K. H. (1996). Differential activation of target cellular promoters by p53 mutants with impaired apoptotic function. Mol Cell Biol 16, 4952-4960.

Luo, J., Li, M., Tang, Y., Laszkowska, M., Roeder, R. G., and Gu, W. (2004). Acetylation of p53 augments its site-specific DNA binding both in vitro and in vivo. Proc Natl Acad Sci USA 101, 2259-2264.

Luo, J., Nikolaev, A. Y., Imai, S., Chen, D., Su, F., Shiloh, A., Guarente, L., and Gu, W. (2001). Negative control of p53 by Sir2alpha promotes cell survival under stress. Cell 107, 137-148.

Luo, J., Su, F., Chen, D., Shiloh, A., and Gu, W. (2000). Deacetylation of p53 modulates its effect on cell growth and apoptosis. Nature 408, 377-381.

MacLachlan, T. K., Takimoto, R., and El-Deiry, W. S. (2002). BRCA1 directs a selective p53-dependent transcriptional response towards growth arrest and DNA repair targets. Mol Cell Biol 22, 4280-4292.

Maheswaran, S., Englert, C., Bennett, P., Heinrich, G., and Haber, D. A. (1995). The WT1 gene product stabilizes p53 and inhibits p53-mediated apoptosis. Genes Dev 9, 2143-2156.

Maki, C. G., Huibregtse, J. M., and Howley, P. M. (1996). In vivo ubiquitination and proteasome-mediated degradation of p53(1). Cancer Res 56, 2649-2654.

Mantovani, F., Piazza, S., Gostissa, M., Strano, S., Zacchi, P., Mantovani, R., Blandino, G., and Del Sal, G. (2004). Pin1 links the activities of c-Abl and p300 in regulating p73 function. Mol Cell 14, 625-636.

Marechal, V., Elenbaas, B., Piette, J., Nicolas, J. C., and Levine, A. J. (1994). The ribosomal L5 protein is associated with mdm-2 and mdm-2-p53 complexes. Mol Cell Biol 14, 7414-7420.

Marston, N. J., Crook, T., and Vousden, K. H. (1994). Interaction of p53 with MDM2 is independent of E6 and does not mediate wild type transformation suppressor function. Oncogene 9, 2707-2716.

Michael, D., and Oren, M. (2002). The p53 and Mdm2 families in cancer. Curr Opin Genet Dev 12, 53-59.

Midgley, C. A., Desterro, J. M., Saville, M. K., Howard, S., Sparks, A., Hay, R. T., and Lane, D. P. (2000). An N-terminal p14ARF peptide blocks Mdm2-dependent ubiquitination in vitro and can activate p53 in vivo. Oncogene 19, 2312-2323.

Mihara, M., Erster, S., Zaika, A., Petrenko, O., Chittenden, T., Pancoska, P., and Moll, U. M. (2003). p53 has a direct apoptogenic role at the mitochondria. Mol Cell 11, 577-590.

Mirza, A., McGuirk, M., Hockenberry, T. N., Wu, Q., Ashar, H., Black, S., Wen, S. F., Wang, L., Kirschmeier, P., Bishop, W. R., et al. (2002). Human survivin is negatively regulated by wild-type p53 and participates in p53-dependent apoptotic pathway. Oncogene 21, 2613-2622.

Miyashita, T., Harigai, M., Hanada, M., and Reed, J. C. (1994). Identification of a p53-dependent negative response element in the bcl-2 gene. Cancer Res 54, 3131-3135.

Momand, J., Zambetti, G. P., Olson, D. C., George, D., and Levine, A. J. (1992). The mdm-2 oncogene product forms a complex with the p53 protein and inhibits p53-mediated transactivation. Cell 69, 1237-1245.

Montes de Oca Luna, R., Wagner, D. S., and Lozano, G. (1995). Rescue of early embryonic lethality in mdm2-deficient mice by deletion of p53. Nature 378, 203-206.

Morimoto, M., Nishida, T., Nagayama, Y., and Yasuda, H. (2003). Nedd8-modification of Cul1 is promoted by Roc1 as a Nedd8-E3 ligase and regulates its stability. Biochem Biophys Res Commun 301, 392-398.

Moroni, M. C., Hickman, E. S., Denchi, E. L., Caprara, G., Colli, E., Cecconi, F., Muller, H., and Helin, K. (2001). Apaf-1 is a transcriptional target for E2F and p53. Nat Cell Biol 3, 552-558.

Muller, M., Wilder, S., Bannasch, D., Israeli, D., Lehlbach, K., Li-Weber, M., Friedman, S. L., Galle, P. R., Stremmel, W., Oren, M., and Krammer, P. H. (1998). p53 activates the CD95 (APO-1/Fas) gene in response to DNA damage by anticancer drugs. J Exp Med 188, 2033-2045.

Nakano, K., and Vousden, K. H. (2001). PUMA, a novel proapoptotic gene, is induced by p53. Mol Cell 7, 683-694.

Nikolaev, A. Y., Li, M., Puskas, N., Qin, J., and Gu, W. (2003). Parc: a cytoplasmic anchor for p53. Cell 112, 29-40.

Oda, E., Ohki, R., Murasawa, H., Nemoto, J., Shibue, T., Yamashita, T., Tokino, T., Taniguchi, T., and Tanaka, N. (2000). Noxa, a BH3-only member of the Bcl-2 family and candidate mediator of p53-induced apoptosis. Science 288, 1053-1058.

Oda, K., Arakawa, H., Tanaka, T., Matsuda, K., Tanikawa, C., Mori, T., Nishimori, H., Tamai, K., Tokino, T., Nakamura, Y., and Taya, Y. (2000). P53AIP1, a potential mediator of p53-dependent apoptosis, and its regulation by Ser-46-phosphorylated p53. Cell 102, 849-862.

Osaka, F., Kawasaki, H., Aida, N., Saeki, M., Chiba, T., Kawashima, S., Tanaka, K., and Kato, S. (1998). A new NEDD8-ligating system for cullin-4A. Genes Dev 12,

2263-2268.

Owen-Schaub, L. B., Zhang, W., Cusack, J. C., Angelo, L. S., Santee, S. M., Fujiwara, T., Roth, J. A., Deisseroth, A. B., Zhang, W. W., Kruzel, E., and et al. (1995). Wild-type human p53 and a temperature-sensitive mutant induce Fas/APO-1 expression. Mol Cell Biol *15*, 3032-3040.

Pearson, M., Carbone, R., Sebastiani, C., Cioce, M., Fagioli, M., Saito, S., Higashimoto, Y., Appella, E., Minucci, S., Pandolfi, P. P., and Pelicci, P. G. (2000). PML regulates p53 acetylation and premature senescence induced by oncogenic Ras. Nature *406*, 207-210.

Pickart, C. M. (2004). Back to the future with ubiquitin. Cell *116*, 181-190.

Pluquet, O., and Hainaut, P. (2001). Genotoxic and non-genotoxic pathways of p53 induction. Cancer Lett *174*, 1-15.

Robles, A. I., Bemmels, N. A., Foraker, A. B., and Harris, C. C. (2001). APAF-1 is a transcriptional target of p53 in DNA damage-induced apoptosis. Cancer Res *61*, 6660-6664.

Rowan, S., Ludwig, R. L., Haupt, Y., Bates, S., Lu, X., Oren, M., and Vousden, K. H. (1996). Specific loss of apoptotic but not cell-cycle arrest function in a human tumor derived p53 mutant. Embo J *15*, 827-838.

Rozenfeld-Granot, G., Krishnamurthy, J., Kannan, K., Toren, A., Amariglio, N., Givol, D., and Rechavi, G. (2002). A positive feedback mechanism in the transcriptional activation of Apaf-1 by p53 and the coactivator Zac-1. Oncogene *21*, 1469-1476.

Ruas, M., and Peters, G. (1998). The p16INK4a/CDKN2A tumor suppressor and its relatives. Biochim Biophys Acta *1378*, F115-177.

Ryan, K. M., and Vousden, K. H. (1998). Characterization of structural p53 mutants which show selective defects in apoptosis but not cell cycle arrest. Mol Cell Biol *18*, 3692-3698.

Sakaguchi, K., Herrera, J. E., Saito, S., Miki, T., Bustin, M., Vassilev, A., Anderson, C. W., and Appella, E. (1998). DNA damage activates p53 through a phosphorylation-acetylation cascade. Genes Dev *12*, 2831-2841.

Samuels-Lev, Y., O'Connor, D. J., Bergamaschi, D., Trigiante, G., Hsieh, J. K., Zhong, S., Campargue, I., Naumovski, L., Crook, T., and Lu, X. (2001). ASPP proteins specifically stimulate the apoptotic function of p53. Mol Cell *8*, 781-794.

Sax, J. K., Fei, P., Murphy, M. E., Bernhard, E., Korsmeyer, S. J., and El-Deiry, W. S. (2002). BID regulation by p53 contributes to chemosensitivity. Nat Cell Biol *4*, 842-849.

Scheffner, M., Huibregtse, J. M., Vierstra, R. D., and Howley, P. M. (1993). The HPV-16 E6 and E6-AP complex functions as a ubiquitin-protein ligase in the ubiquitination of p53. Cell *75*, 495-505.

Sengupta, S., and Harris, C. C. (2005). p53: traffic cop at the crossroads of DNA repair and recombination. Nat Rev Mol Cell Biol *6*, 44-55.

Sharp, D. A., Kratowicz, S. A., Sank, M. J., and George, D. L. (1999). Stabilization of the MDM2 oncoprotein by interaction with the structurally related MDMX protein. J Biol Chem *274*, 38189-38196.

Shenoy, S. K., McDonald, P. H., Kohout, T. A., and Lefkowitz, R. J. (2001). Regulation of receptor fate by ubiquitination of activated beta 2-adrenergic receptor and beta-arrestin. Science *294*, 1307-1313.

Shieh, S. Y., Ahn, J., Tamai, K., Taya, Y., and Prives, C. (2000). The human homologs of checkpoint kinases Chk1 and Cds1 (Chk2) phosphorylate p53 at multiple DNA damage-inducible sites. Genes Dev *14*, 289-300.

Shikama, N., Lee, C. W., France, S., Delavaine, L., Lyon, J., Krstic-Demonacos, M., and La Thangue, N. B. (1999). A novel cofactor for p300 that regulates the p53 response. Mol Cell *4*, 365-376.

Smith, J. (2002). Human Sir2 and the 'silencing' of p53 activity. Trends Cell Biol *12*, 404.

Smith, P. D., Crossland, S., Parker, G., Osin, P., Brooks, L., Waller, J., Philp, E., Crompton, M. R., Gusterson, B. A., Allday, M. J., and Crook, T. (1999). Novel p53 mutants selected in BRCA-associated tumours which dissociate transformation suppression from other wild-type p53 functions. Oncogene *18*, 2451-2459.

Soengas, M. S., Alarcon, R. M., Yoshida, H., Giaccia, A. J., Hakem, R., Mak, T. W., and Lowe, S. W. (1999). Apaf-1 and caspase-9 in p53-dependent apoptosis and tumor inhibition. Science *284*, 156-159.

Stad, R., Little, N. A., Xirodimas, D. P., Frenk, R., van der Eb, A. J., Lane, D. P., Saville, M. K., and Jochemsen, A. G. (2001). Mdmx stabilizes p53 and Mdm2 via two distinct mechanisms. EMBO Rep *2*, 1029-1034.

Sterner, D. E., and Berger, S. L. (2000). Acetylation of histones and transcription-related factors. Microbiol Mol Biol Rev *64*, 435-459.

Sugars, K. L., Budhram-Mahadeo, V., Packham, G., and Latchman, D. S. (2001). A minimal Bcl-x promoter is activated by Brn-3a and repressed by p53. Nucleic Acids Res *29*, 4530-4540.

Taylor, W. R., and Stark, G. R. (2001). Regulation of the G2/M transition by p53. Oncogene *20*, 1803-1815.

Tyner, S. D., Venkatachalam, S., Choi, J., Jones, S., Ghebranious, N., Igelmann, H., Lu, X., Soron, G., Cooper, B., Brayton, C., et al. (2002). p53 mutant mice that display early ageing-associated phenotypes. Nature *415*, 45-53.

Vaziri, H., Dessain, S. K., Ng Eaton, E., Imai, S. I., Frye, R. A., Pandita, T. K., Guarente, L., and Weinberg, R. A. (2001). hSIR2(SIRT1) functions as an NAD-dependent p53 deacetylase. Cell *107*, 149-159.

Vousden, K. H. (2000). p53: death star. Cell *103*, 691-694.

Wu, G. S., Burns, T. F., McDonald, E. R., 3rd, Jiang, W., Meng, R., Krantz, I. D., Kao, G., Gan, D. D., Zhou, J. Y., Muschel, R., et al. (1997). KILLER/DR5 is a DNA damage-inducible p53-regulated death receptor gene. Nat Genet *17*, 141-143.

Wu, K., Chen, A., Tan, P., and Pan, Z. Q. (2002). The Nedd8-conjugated ROC1-CUL1 core ubiquitin ligase utilizes Nedd8 charged surface residues for efficient polyubiquitin chain assembly catalyzed by Cdc34. J Biol Chem *277*, 516-527.

Wu, X., Bayle, J. H., Olson, D., and Levine, A. J. (1993). The p53-mdm-2 autoregulatory feedback loop. Genes Dev *7*, 1126-1132.

Wunderlich, M., Ghosh, M., Weghorst, K., and Berberich, S. J. (2004). MdmX Represses E2F1 Transactivation. Cell Cycle *3*, 472-478.

Xirodimas, D., Saville, M. K., Edling, C., Lane, D. P., and Lain, S. (2001). Different effects of p14ARF on the levels of ubiquitinated p53 and Mdm2 *in vivo*. Oncogene *20*, 4972-4983.

Xirodimas, D. P., Saville, M. K., Bourdon, J. C., Hay, R. T., and Lane, D. P. (2004). Mdm2-mediated NEDD8 conjugation of p53 inhibits its transcriptional activity. Cell *118*, 83-97.

Yam, C. H., Siu, W. Y., Arooz, T., Chiu, C. H., Lau, A., Wang, X. Q., and Poon, R. Y. (1999). MDM2 and MDMX inhibit the transcriptional activity of ectopically expressed SMAD proteins. Cancer Res *59*, 5075-5078.

Zacchi, P., Gostissa, M., Uchida, T., Salvagno, C., Avolio, F., Volinia, S., Ronai, Z., Blandino, G., Schneider, C., and Del Sal, G. (2002). The prolyl isomerase Pin1 reveals a mechanism to control p53 functions after genotoxic insults. Nature *419*, 853-857.

Zhang, Y., Wolf, G. W., Bhat, K., Jin, A., Allio, T., Burkhart, W. A., and Xiong, Y. (2003). Ribosomal protein L11 negatively regulates oncoprotein MDM2 and mediates a p53-dependent ribosomal-stress checkpoint pathway. Mol Cell Biol *23*, 8902-8912.

Zhao, L. Y., and Liao, D. (2003). Sequestration of p53 in the cytoplasm by adenovirus type 12 E1B 55-kilodalton oncoprotein is required for inhibition of p53-mediated apoptosis. J Virol *77*, 13171-13181.

Zheng, H., You, H., Zhou, X. Z., Murray, S. A., Uchida, T., Wulf, G., Gu, L., Tang, X., Lu, K. P., and Xiao, Z. X. (2002). The prolyl isomerase Pin1 is a regulator of p53 in genotoxic response. Nature *419*, 849-853.

Zilfou, J. T., Hoffman, W. H., Sank, M., George, D. L., and Murphy, M. (2001). The corepressor mSin3a interacts with the proline-rich domain of p53 and protects p53 from proteasome-mediated degradation. Mol Cell Biol *21*, 3974-3985.

Chapter 16

Actions of Nuclear Receptors

Kurt Schillinger[1], Sophia Y. Tsai[1,2] and Ming-Jer Tsai[1,2]

[1]*Department of Molecular and Cellular Biology and*
[2]*Program of Development Biology, Baylor College of Medicine, Houston Texas, 77030*

Key Words: nuclear receptor, COUP-TF, GCNF, TR, PPAR, AR, VDR, RAR, CAR, LXR, PXR, FXR, RXR, HNF4, GR, PR, ER

Summary

The Nuclear Hormone Receptor Superfamily is a group of proteins which function as transcriptional regulators in a vast array of diverse processes. In mammals, these processes include embryonic development, maintenance of body fluid and electrolyte composition, regulation of energy sources and metabolism, and protection of organ systems from endogenous and exogenous toxic compounds. The nuclear hormone receptors may also be involved in the pathophysiology of diseases such as diabetes, cancer, and atherosclerosis. This chapter begins with a brief overview of general nuclear hormone receptor structure. It then introduces the basic mechanistic concepts underlying transcriptional activation of gene expression by these proteins. In closing, this chapter looks in more detail at specific members of the Nuclear Hormone Receptor Superfamily, and describes their individual roles in the processes introduced above.

Introduction

Over two decades have passed since the first nuclear hormone receptor (NR) was cloned. Since that time, spectacular growth in the number of recognized NRs has occurred, leading to the identification and characterization of over 40 NRs in vertebrates. Among the many consequences of this rapid growth in characterization of NRs has been the development of a novel concept known as "reverse endocrinology" in which the characterization of a putative receptor precedes identification of its ligand or study of its physiological function. This approach has succeeded in identifying a number of "orphan receptors" for which a corresponding ligand has not been found. It has also succeeded in underscoring the tremendous number of physiological processes in which NRs play a role. Far from being involved in simply mediating the effects of the steroid hormones, NRs have been shown to be involved in processes ranging from axis patterning and organ morphogenesis in the embryo to lipid homeostasis and xenobiotic metabolism in the adult. NRs are also involved in pathophysiological processes, as roles for NRs have been demonstrated in diseases such as diabetes, cancer, and atherosclerosis. Not surprisingly, given the involvement of this superfamily in processes ranging from development to disease, considerable effort has been spent in attempting to elucidate the mechanisms by which NRs regulate gene expression. However, while a basic understanding of the function of NRs at the promoters of target genes has been garnered through a host of studies, much remains to be understood about NR function.

General Stucture of Nuclear Receptors

Like other transcriptional regulators, NRs possess a modular structure with autonomous functional domains. In many instances, these domains can be interchanged between related receptors with little or no loss of function. Typically, a NR consists of a variable N-terminal domain (Domain A/B), a conserved DNA-

---

Corresponding Author: Ming-Jer Tsai, Department of Molecular and Cellular Biology, Baylor College of Medicine, One Baylor Plaza, Houston Texas 77030, Tel: (713) 798-6253, Fax: (713) 798-8227, E-mail: mtsai@bcm.tmc.edu

binding domain (Domain C), a linker or hinge region (Domain D), and a conserved ligand binding domain (Domain E). Many receptors also contain a C-terminal region (Domain F) with unknown function, and it is not uncommon to see this region described as part of a C-terminal E/F domain in combination with the ligand binding domain (Aranda and Pascual, 2001). In addition to distinct domains, the NRs also contain regions within the tertiary structure of the individual receptors which confer functions such as transcriptional activation, dimerization, and nuclear localization upon the fully folded NR polypeptide (see Fig.16.1) (Aranda and Pascual, 2001).

A: The Hypervariable Region: Domain A/B

The N-terminal A/B region is the most variable domain, both in size and sequence, between all members of the NR superfamily. In some instances, it is also the most variable region between members of a single group of NRs, as many receptor isoforms generated from a single gene by alternative splicing or alternative promoter usage diverge in their A/B domains (Aranda and Pascual, 2001). Complicating the study of the A/B domain of NRs is the fact that almost no three-dimensional structural information is available for the A/B domain from any member of the NR superfamily (Kumar and Thompson, 1999). Still, molecular genetic analysis has revealed that the A/B domain is important for transcriptional activity of NRs and serves as a target for modulation through phosphorylation (Aranda and Pascual, 2001; Kumar and Thompson, 1999).

Transcriptional activity attributed to the A/B domain is due in large part to a powerful transactivation region within this domain called Activation Function-1 (AF-1). Circular dichroism and nuclear magnetic resonance studies of AF-1 have shown that this region is rich in acidic amino acids, and may be composed of as many as three α-helices (Dahlman-Wright et al., 1995; Folkers et al., 1995). Mutational studies of the AF-1 regions from the Glucocorticoid Receptor have suggested that the ability of AF-1 to transactivate a reporter gene in vivo correlates with the ability of this region to form α-helices in vitro (Dahlman-Wright et al., 1995; McEwan et al., 1993). However, it is not clear how AF-1, or the A/B domain in general, interacts with other components of the transcriptional apparatus to initiate transcription (Kumar and Thompson, 1999).

The A/B domain also shows promoter and cell-specific activity, suggesting that this domain is the target of many NR modulatory mechanisms (Aranda and Pascual, 2001). One such mechanism is phosphorylation. Experiments conducted with a number of different NRs have indicated that phosphorylation may be mediated by kinases such as cyclin-dependent kinases and mitogen-activated protein kinase (MAPK), among others (Juge-Aubry et al., 1999; Shao et al., 1998).

B: The DNA-Binding Domain: Domain C

The DNA-binding domain (DBD) is the most highly conserved domain among members of the NR superfamily (Aranda and Pascual, 2001). This domain confers the ability to recognize specific target sequences and, consequently, the ability to interact with specific promoters to transactivate gene expression. The DBD consists of two interdependent subdomains which are both required for high-affinity DNA binding (Kumar and Thompson, 1999).

The first subdomain of the DBD is composed of an α-helix, designated Helix I, and a zinc-finger motif. In the zinc-finger motif, four invariable cysteines coordinate tetrahedrically with a single zinc ion (Kumar and Thompson, 1999). Helix I, which lies at the base of this first zinc-finger, is involved in the site-specific recognition of the DNA to which the entire NR is bound. This recognition is accomplished by a portion of Helix I composed of 3-4 amino acids and termed the "P-box" (Luisi et al., 1991). During NR binding to DNA, Helix I of the DBD fits into the major groove of the DNA helix, and the amino acids of the P-box make critical contacts with very specific bases in the major groove (Luisi et al., 1991).

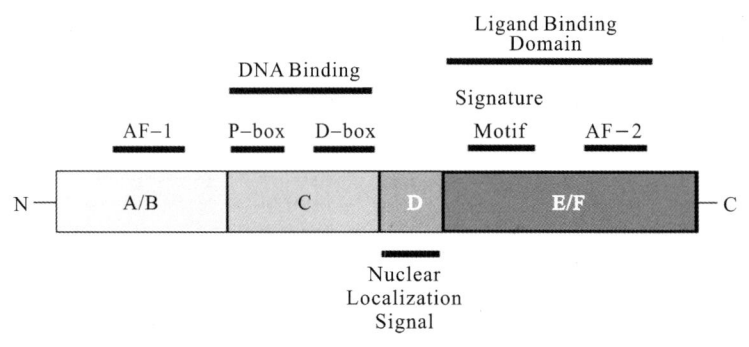

Fig.16.1 General stucture of nuclear hormone receptors.

The second subdomain of the NR DBD also contains a zinc-finger motif and an α-helix, termed Helix III. As with the zinc-finger motif in the first subdomain, a single zinc ion is tetrahedrically complexed to four highly conserved cysteine residues in the second subdomain (Luisi et al., 1991). The existence of two zinc-finger motifs in such close proximity allows these motifs to fold together, giving the NR DBD its characteristic tertiary structure. In this tertiary structure, Helix III is oriented perpendicular to Helix I. The second subdomain of the DBD contains one other important region, termed the "D-box". This region is composed of amino acids in the zinc-finger of the second subdomain, and is partly responsible for the dimerization of the NRs (Aranda and Pascual, 2001; Kumar and Thompson, 1999).

C: The Hinge Domain: Domain D

The hinge domain is not well conserved among members of the NR superfamily. This domain serves as a linker between the DBD and the ligand binding domain (LBD, Domain E), and allows for rotation between these two domains. The hinge domain also contains a nuclear localization signal which is critical for the nuclear translocation of NRs (Aranda and Pascual, 2001).

D: The Ligand Binding Domain and C-Terminal Extension: Domain E/F

The ligand binding domain (LBD) of NRs is a multifunctional domain which contains regions mediating binding of ligand, homo- and heterodimerization, and ligand-dependent transcriptional activity. The function of Domain F, by contrast, is not currently known. This domain is mentioned in general NR superfamily structure out of recognition for its existence in some members of the superfamily. However, given the general lack of understanding of its function, it will not be discussed further here (Aranda and Pascual, 2001).

Crystal structures of the LBD of multiple NRs have been solved and have demonstrated some level of similarity between LBDs from different NRs. Based on these studies, a canonical three-dimensional structure of NRs has been proposed (Moras and Gronemeyer, 1998). In this structure, the NR LBD is composed of a series of 12α -helical regions, termed helices 1-12, closely folded into a three-layer, anti-parallel helical "sandwich" (Bourguet et al., 1995). A highly conserved α-turn is situated between helices 5 and 6. In this conformation, a central core layer of three helices is packed between 2 additional layers of helices, creating a cavity which can accommodate a ligand (Aranda and Pascual, 2001).

Contacts between amino acids in the ligand binding pocket and the ligand can be extensive and varies among different receptors. In some NRs, the ligand binding pocket of the LBD is quite large, allowing the binding of several differently-sized ligands. In other NRs, a smaller ligand binding pocket accommodates only a single, specific ligand (Uppenberg et al., 1998). Regardless of the ligand pocket size, interaction between residues in the LBD and the ligand are hypothesized to result in a conformational change which may have important consequences for NR function (Aranda and Pascual, 2001).

Within the ligand binding domain are two important regions, a "signature motif" and an Activation Function-2 (AF-2) motif. The AF-2 motif, located on helix 12 of the LBD, is a region that has been shown to be essential for ligand-induced transcriptional activation. However, the AF-2 motif does not act alone in this capacity, and must interact with a region called the "signature motif" located on and between helices 3 and 4. It has been hypothesized that the conformational change induced by ligand binding to the LBD results in physical approximation between AF-2 and the signature motif. This results in the formation of a surface over the liganded NR which is capable of interacting with other protein mediators of transcription (Danielian et al., 1992; McInerney et al., 1996). This interaction is described in more detail below.

Binding of Nuclear Receptors to DNA

The transcriptional effects of NRs are largely dependent upon the ability of these proteins to bind specific sequences in the 5′-flanking regions of target genes. These sequences are termed hormone response elements, or HREs. Often, HREs are located close to the transcriptional initiation site in the promoters of target genes. However, HREs have been shown to be present in enhancer regions several thousand basepairs upstream of the transcriptional initiation site (Aranda and Pascual, 2001). Regardless of location, a sequence of six nucleotides typically constitutes the core recognition motif for the DNA-binding domain of most NRs. In the case of Class III NRs (described later), the consensus sequence AGAACA is preferentially recognized. The remainder of the members of the NR superfamily, however, recognize the consensus sequence AGG/ TTCA (Beato et al., 1995). It is worth noting, however, that these consensus sequences represent idealized sequences, and naturally occurring HREs can show marked deviation from their corresponding consensus sequence.

While some NRs have the capacity to bind to a single hexameric motif as a monomer, most NRs must bind as homo- or heterodimers to HREs composed of 2 adjacent hexameric motifs. In HREs designed to bind NR dimers, the hexameric motifs which are termed "half-sites" can be configured as palindromes, inverted palindromes, or direct repeats. Furthermore, half-sites may be separated by as few as 0 nucleotides, or as many as 9 nucleotides (Glass, 1994). Careful analysis of natural and synthetic HREs has revealed that Class III NRs typically bind to palindromic HREs with 3 nucleotides between half-sites while all other NRs preferentially bind to HREs configured as direct repeats separated by 3, 4, or 5 nucleotides (Glass, 1994; Mangelsdorf and Evans, 1995). While such statements can be made about NR-HRE interaction in general, in the context of specific genes it is also true that small differences in half-site sequence and the sequences flanking an HRE are important parameters in determining individual NR binding efficiency (Mader *et al.*, 1993).

Several NRs bind to DNA with high affinity as monomers (Giguere, 1999). As described above, this binding entails interaction between the DNA-binding domain of the NR and a hexameric consensus sequence, or half-site. However, sequence diversity 5′ of the monomeric HRE also plays a role in the binding of NRs to DNA. Typically, monomeric HREs are preceded by an A/T rich sequence which interacts with C-terminal portions of the DBD of the monomeric NR (Giguere *et al.*, 1994). Portions of the C-terminal DBD can make extensive contacts with the minor groove of DNA and, in so doing, extend the surface contact of the receptor DBD beyond the original half-site. As a result, binding affinity of the NR for the DNA is markedly increased (Aranda and Pascual, 2001).

NRs can also bind DNA as homodimers, and this activity has been best characterized in the Class III NRs. Class III NRs bind to palindromic HREs, which impose a symmetrical structure resulting in "head-to-head" arrangements of individual NR monomers. In this arrangement, each of the two Class III NR monomers comprising the homodimers make analogous contacts with one half-site through their DBDs (Aranda and Pascual, 2001). Simultaneously, dimerization of the monomers is maintained through interfaces between regions in both the DBD and LBD of the NR monomers. The interfaces in the LBD are likely to involve amino acids from helices 7, 8, and 9 as well as a conserved hydrophobic region at the N-terminus of helices 10/11 (Tanenbaum *et al.*, 1998). In the DBD, the dimerization interface is generated by amino acids of the D-box (Aranda and Pascual, 2001).

Certain NRs can also bind to HREs as heterodimers. In these cases, an individual NR monomers is paired with a retinoid X receptor (RXR) monomers. Typically, binding of such heterodimers to DNA involves HREs composed of two half-sites in a direct repeat configuration separated by 3, 4, or 5 nucleotides (Kurokawa *et al.*, 1993). However, heterodimers formed by a combination of the retinoid A receptor and RXR have also been shown to bind to HREs organized as direct repeats separated by 1 or 2 nucleotides (Kurokawa *et al.*, 1994). This direct repeat organization of HREs for heterodimers has important consequences for the mechanism of dimerization between the NRs involved. In contrast to the head-to-head configuration observed with homodimers, heterodimerized NRs are configured in a "head-to-tail" organization. While the LBD of heterodimers interface in a manner identical to that of homodimers, the change in orientation of the DBDs of the NR monomers as a heterodimer results in a second, different dimerization interface. Typically, this second interface involves interaction of the D-box of the upstream monomer (typically RXR) with the first zinc-finger of the downstream monomer (Zechel *et al.*, 1994).

Based on ligand binding, two-types of heterodimers can be identified. "Non-permissive heterodimers" are heterodimers that cannot be activated by the ligand for RXR, 9-cis retinol, but can be activated by the ligand of the partner monomer (Forman *et al.*, 1995). "Permissive heterodimers" are heterodimers that can be activated by either 9-cis retinol or the ligand of the partner monomer. In addition, permissive heterodimers are synergistically activated in the presence of both ligands (Janowski *et al.*, 1996; Kliewer *et al.*, 1992). Examples of non-permissive heterodimers include RXR complexed with Thyroid Hormone Receptor (TR), Vitamin D Receptor (VDR), or Retinoic Acid Receptor (RAR). Examples of permissive heterodimers include RXR complexed with Peroxisome Proliferator Activated Receptor (PPAR), Farnesoid X Receptor (FXR), or NGF-Induced Clone B (NGFI-B) (Aranda and Pascual, 2001).

Mechanisms of Nuclear Receptor Mediated Transcriptional Activation

The transcription of mammalian genes by RNA polymerase II requires the interaction of multiple protein complexes with the promoter region immediately adjacent to the gene. A typical core promoter region contains a TATA box close to a transcriptional start site where RNA polymerase II binds (Wilson *et al.*, 1996). In the initial stages of gene transcription, this TATA box is bound by a complex of basal transcriptional factors collectively referred to as the TFIID complex. One of the proteins in the TFIID complex is TATA binding protein (TBP), a highly conserved protein which binds the minor groove of DNA over the TATA box and generates a drastic bend in the DNA of the promoter (Beato and Sanchez-Pacheco, 1996). Binding of TBP to the TATA box is aided by the other members of the TFIID complex, designated TBP-associated factors (TAF$_{II}$s). Examples of common TAF$_{II}$s include TAF$_{II}$250, TAF$_{II}$135, TAF$_{II}$100, and TAF$_{II}$28 (Beato and Sanchez-Pacheco, 1996; Roeder, 1996). Subsequent to TFIID binding, a second basal transcription factor, TFIIB, is recruited to the promoter. TFIIB contacts DNA upstream and downstream of the TATA box on the concave side of the bend induced by TBP binding (Beato and Sanchez-Pacheco, 1996). Recruitment of TFIIB to the promoter is critical, as it leads to the formation of a multi-protein structure called the preinitiation complex. This complex eventually recruits RNA polymerase to the promoter and creates a microenvironment at the promoter which favors initiation of gene transcription by RNA polymerase II.

It has been hypothesized that NRs regulate gene expression by influencing the rate at which the preinitiation complex forms in the promoters of NR-responsive genes (Aranda and Pascual, 2001). Regulation of preinitiation complex formation may involve direct protein-protein interaction between NRs and components of the transcription preinitiation complex. Supporting this possibility is the observation that TBP has the ability to interact with several different NRs. As well, the overexpression of TBP in transient transfection assays can enhance ligand-dependent transactivation from certain NRs (Sadovsky *et al.*, 1995b; Schulman *et al.*, 1995). However, TBP is not the only preinitiation complex component thought to interact directly with NRs. Several TAF$_{II}$s have also been identified as potential targets for protein-protein interaction with NRs. For example, TAF$_{II}$135 has been shown to strongly potentiate transcriptional stimulation by RAR, TR, and VDR (Mengus *et al.*, 1997). However, while it is tempting to speculate on the role of interaction between NRs and the basal transcriptional machinery, it is important to note that the functionality of these interactions has yet to be determined (Aranda and Pascual, 2001).

A second mechanism by which NRs may modulate the assembly of the preinitiation complex involves interaction between NRs and proteins called coactivators and corepressors. These bridging molecules bind to DNA-bound NRs and mediate the interaction of NRs with components of the basal transcriptional machinery. In addition, these coactivators change chromatin to a transcriptionally active state through histone modifications such as acetylation, methylation, and phosphorylation. To date, a large number of proteins have been found to be coactivators for NRs (McKenna *et al.*, 1999; Robyr *et al.*, 2000). While an extensive description of these proteins is beyond the scope of this chapter, all coactivators are similar in the sense that their interaction with NRs results in enhanced assembly of the preinitiation complex. Examples of these proteins include the members of the p160 family of coactivators SRC-1, SRC-2, and SRC-3 (Onate *et al.*, 1995; Torchia *et al.*, 1998; Voegel *et al.*, 1996). NR corepressors exist in far fewer numbers, and in a manner which strongly contrasts that of NR coactivators, bind to NRs and reduce preinitiation complex assembly. Two of the best-studied examples of NR corepressors are NcoR and SMRT. These polypeptides, which bind to DNA-bound TR and RAR, serve as potent silencers of gene expression in the absence of ligands for the NRs (Aranda and Pascual, 2001; Chen *et al.*, 1999; Horlein *et al.*, 1995).

At a molecular level, the interaction between NRs and coactivators is mediated predominantly by the most C-terminal portion of the LBD of NRs, termed the AF-2 domain (Barettino *et al.*, 1994). This region possesses high homology over a very short region with the consensus sequence nnXEnn, where n represents a hydrophobic amino acid. This conserved sequence adopts an amphipathic α-helical conformation with the two well-conserved paris of hydrophobic residues pointing towards the core of the LBD and negatively charged residues on its surface (Zenke *et al.*, 1990). While this α-helix contains the core activity of the NR AF-2 domain, AF-2 activity is greatly enhanced through approximation with a portion of the NR called the NR "signature motif". This motif is formed by the C-terminal half of NR LBD helix 3, all of NR LBD helix 4, and the loop of amino acids between them. In liganded NRs, a conformational change in the LBD places the AF-2 domain of helix 12 and the signature motif in a contiguous conformation at the surface of the receptor. This generates a hydrophilic surface on the liganded NR that can interact with a coactivator (Moras and Gronemeyer, 1998). Interestingly, this model also explains how coactivator-mediated preinitiation complex formation can be effectively linked to ligand binding by a NR.

Classification and Specific Examples of Nuclear Hormone Receptors

The most current classification scheme for vertebrate NRs defines 6 NR subfamilies based on evolutionary alignment of sequences from the C, D, and E domains (Aranda and Pascual, 2001; Laudet, 1997). This classification scheme, and the phylogenetic tree derived from it, group all NRs on the basis of their behavior with respect to DNA binding and dimerization. It does not, however, classify NRs on the basis of the ligands they bind. As a result, each NR Class in this scheme contains NRs with defined ligands, as well as NRs with currently unknown ligands. The following sections are descriptions of the NR classes defined by this evolutionary approach. While not an exhaustive characterization of all known NRs, each section does describe relevant NR subtype examples for each class.

A: Class I

With very few exceptions, the Class I NRs bind to their corresponding HRE as a heterodimer(Aranda and Pascual, 2001; Laudet, 1997). Class I NRs include Thyroid Hormone Receptor (TR), Retinoic Acid Receptor (RAR), Vitamin D Receptor (VDR), Peroxisome Proliferator Activated Receptor (PPAR), Constitutive Androstane Receptor (CAR), Pregnane X Receptor (PXR), Liver X Receptor (LXR), and Farnesoid X Receptor (FXR).

A1: Thyroid Hormone Receptor (TR)

Triiodothronine (T3), also called thyroid hormone, plays an important role in the normal development and homeostasis of adult mammals. The effects of this hormone are mediated by the Thyroid Hormone Receptor (TR), a Class I NR. Developmentally, TR mediates the effects of T3 in tissues such as the brain, intestine, and long bones. In the adult, TR is critical for control of basal metabolism, heart rate, and temperature. In the adult liver, a major target of T3 activity, TR regulates genes involved in gluconeogenesis, lipogenesis, insulin signaling, adenylate cyclase signaling, cell proliferation, and apoptosis (Eckey *et al.*, 2003; Feng *et al.*, 2000; Feng *et al.*, 2002).

Eight TR isoforms are encoded by 2 different genes designated TRα and TRβ. Five of these isoforms, TRα1, TRβ1, TRβ2, TRβ3, and TRΔβ3 can bind T3 (Eckey *et al.*, 2003). Three other isoforms derived from the TRα locus, TRα2, TRΔα1, and TRΔα2 are not able to bind T3 and act as TR inhibitors (O'Shea and Williams, 2002). All of these isoforms are derived from alternative splicing or alternative promoter usage at the TRα and TRβ genes (Eckey *et al.*, 2003). The structure of the vast majority of these isoforms is similar to that of other NRs, and consists of a variable A/B region, as well as C, D, and E/F domains (Eckey *et al.*, 2003).

In a fashion that is characteristic of Class I NRs, TR prefers heterodimerization with a Class II NR called retinoid X receptor (RXR) when binding to DNA. In the absence of T3, TR-RXR heterodimers are bound to DNA at specific sequences called thyroid hormone response elements (TREs) (Eckey *et al.*, 2003; Lazar, 2003). The classic TRE consists of two half-sites directly repeated with a spacing of 4 nucleotides. In one subset of TRE-containing target genes, binding of a TR-RXR heterodimer to the TRE in the absence of T3 leads to gene repression. Upon binding T3, the TR-RXR complex undergoes a conformational change that leads to release of repression and active transcription of target genes. In a second subset of TRE-containing target genes, however, binding of a liganded TR-RXR complex to DNA leads to target gene repression. The mechanisms by which binding of TR-RXR to this subset of TREs results in repression are poorly understood, but have been hypothesized to involve DNA-mediated changes in TR conformation or binding of additional protein complexes to the region immediately adjacent to the TRE (Lazar, 2003).

A2: Retinoic Acid Receptor (RAR)

Vitamin A and its derivatives, known collectively as retinoids, have been known to be essential throughout the lifecycle for decades. These compounds play a critical role in the development and homeostasis of virtually every vertebrate tissue through their regulatory effects on cell differentiation, proliferation, and apoptosis (Bastien and Rochette-Egly, 2004). Vitamin A deficiency during embryogenesis can lead to significant abnormalities in heart development, including thin-walled and dilated cardiac ventricles, and defects in cardiac outflow tracts. In the central nervous system, retinoids are known to be crucial for development of the visual system and the retina, as well as the inner ear primordia and the spinal cord. The retinoids are also known to play important roles in limb morphogenesis, as well as fetal lung and urogenital development (Zile, 2001).

Two derivatives of Vitamin A, all-trans retinoic acid and 9-cis retinoid acid, exert their morphogenic effects by binding to the Class I NR Retinoic Acid Receptor (RAR) and Retinoid X Receptor (RXR) (Bastien and Rochette-Egly, 2004). RAR has 3 isotypes, RARα, RARβ, and RARγ, which are encoded by three separate genes. Each isotype also has at least two isoforms that are generated by differential promoter usage and alternative splicing (Mangelsdorf and Evans, 1995). Regardless of isotype and isoform, all RARs are composed of the 6 domains, A/B through E/F, that are conserved throughout the NR superfamily (Bastien and Rochette-Egly, 2004).

In the absence of ligand, RARs are found predominantly in the nucleus bound as RAR-RXR heterodimers to DNA sequences called retinoic acid receptor response elements (RAREs). These RAREs are typically composed of two direct repeats of a core motif separated by 5 nucleotides (Leid et al., 1992; Mangelsdorf and Evans, 1995). However, RAR-RXR heterodimers have also been shown to bind to direct repeats of the same core motif separated by 1 or 2 nucleotides (Bastien and Rochette-Egly, 2004). Binding of ligand to the LBD of RAR induces a conformational change that simultaneously increases the affinity of this NR for the RARE and initiates assembly of the pre-initiation machinery at the promoter (Bastien and Rochette-Egly, 2004). In this way, all-trans retinoic acid and 9-cis retinoid acid upregulate the transcription and expression of specific target genes.

A3: Vitamin D Receptor (VDR)

Vitamin D_3 was recognized in 1936 to be necessary for the prevention of the disease rickets, and was later identified as a critical micronutrient responsible for regulating calcium homeostasis (Lin and White, 2004). This secosteroid acts as a ligand for the Vitamin D Receptor (VDR), which mediates the effects of this compound through activation of gene expression. Recently, vitamin D_3, acting through VDR, has been shown to play roles in the inhibition of cellular proliferation and the modulation of cellular differentiation (Gurlek et al., 2002). Given these properties, it is not surprising that VDR activated gene expression has recently been shown to play a role in vitamin D_3-mediated inhibition of cancer cell proliferation (Gurlek et al., 2002). VDR has also been shown to be widely expressed throughout the central nervous system and most cells of the immune system, including T-lymphocytes and antigen presenting cells (Prufer et al., 1999). It is believed that VDR-mediated transcriptional activation plays an important role in immune system function and nerve growth factor expression, two processes that are likely associated with the ability of vitamin D_3 to promote cellular differentiation (Lin and White, 2004).

The structure of VDR is similar to other NRs, with the six conserved domains A/B through E/F (Lin and White, 2004). Like most Class I NRs, VDR prefers to form heterodimers with RXR to enable DNA binding. Upon forming a heterodimer, the VDR-RXR complex binds DNA at vitamin D_3 response elements (VDREs) composed of direct repeats of a core motif with consensus sequence PuG(G/T)TCA separated by three nucleotides. Interestingly, VDR-RXR heterodimers have also been shown to bind to inverted palindromic sequences separated by nine nucleotides. Subsequent to binding to a VDRE, ligand bound VDR serves as an activator of gene expression (Schrader et al., 1995).

A4: Peroxisome Proliferator Activated Receptor (PPAR)

The Peroxisome Proliferator Activated Receptors (PPARs) form a group of Class I NRs with 3 isoforms called PPARα, PPARβ, and PPARγ. These three isoforms are encoded by separate genes, and have been implicated in regulating both normal cellular differentiation and the pathophyiology of carcinogenesis (Berger and Moller, 2002). PPARα has been shown to play an important role in the regulation of uptake and β-oxidation of fatty acids by individual cells, and may have a role in atherosclerosis. PPARα has also been hypothesized to be a modulator of inflammation. The role of PPARβ is less well-defined, but may involve blastocyte implantation during early stages of pregnancy, and differentiation of cells within the central nervous system. In addition, it may be involved in lipid metabolism and transport. Finally, PPARγ, the most well-studied isoform of this group, has been demonstrated to be necessary and sufficient for adipocyte differentiation. PPARγ may also play an important inhibitory role in monocyte/macrophage-mediated inflammation as well as colon cancer cell-cycle progression (Berger and Moller, 2002; Desvergne et al., 2004).

The stucture and function of PPAR isoforms mirror that of other Class I NRs. As such, PPARs contain all six functional domains A/B through E/F associated with members of the NR superfamily (Laudet et al., 1992). Also, heterodimerization with RXR is a prerequisite for binding of PPARs to peroxisome proliferator response elements (PPREs) (Miyata et al., 1994). These PPREs are direct repeat elements composed of two hexanucleotide AGGTCA consensus sequences separated by a single nucleotide (A et al., 1997). An important structural difference does distinguish PPARs from other Class I NRs, though. The LBD of PPARs is quite large in comparison to other NRs. This may allow the PPARs to interact with a broader than usual range of structurally distinct ligands (Nolte et al., 1998; Xu et al., 1999).

The ligands for the PPAR NRs have been shown to be numerous fatty acids and their derivatives, including a variety of prostaglandins and eicosanoids (Xu et al.,

1999). PPARα can be activated by a wide variety of saturated and unsaturated fatty acids, including palmitic acid, oleic acid, linoleic acid, and arachidonic acid (Kliewer et al., 1997). PPARγ has also been shown to interact with components called hypolipidemic fibrates that induce hepatic peroxisome proliferation and hepatocarcinogenesis in rodents (Issemann and Green, 1990). Like PPARα, PPARβ interacts with saturated and unsaturated fatty acids such as dihomo-g-linoleic acid and eicosapentanoic acid, but has also been shown to bind to eicosanoids such as prostaglandin A1 and prostaglandin D2 as well (Forman et al., 1997). Studies of PPARα ligand binding have indicated that, unlike PPARα and PPARβ, this NR clearly prefers polyunsaturated fatty acids such as arachidonic acid for binding (Xu et al., 1999). Perhaps most interestingly, however, is the observation that thiazolidinedione (TZD) antidiabetic agents have been shown to interact with the PPAR NRs. This fact, taken together with the propensity of these NRs for fatty acids, has led some to speculate about important roles for the PPAR receptors in insulin-resistant diabetes (Willson et al., 1996).

A5: Constitutive Androstane Receptor (CAR), Pregnane X Receptor (PXR), Liver X Receptor (LXR), and Farnesoid X Receptor (FXR)

Overlap in function, ligand binding, and structural characteristics make the discussion of Constitutive Androstane Receptor (CAR), Pregnane X Receptor (PXR), Liver X Receptor (LXR), and Farnesoid X Receptor (FXR) relevant only in the context of their interactions with each other. Like other Class I NRs, CAR, PXR, LXR, and FXR heterodimerize with the Class II NR RXR upon binding DNA. In a manner reminiscent of the PPARs, these NRs also have large ligand binding pockets in their LBDs which enable them to bind a large number of structurally related ligands. This property of their LBDs, in conjunction with similarities in their DNA binding specificities, have allowed CAR, PXR, LXR, and FXR to develop a lipid-sensing NR network with important physiological consequences in the liver (Handschin and Meyer, 2005).

CAR, first cloned in the mouse in 1997, has become known as a "xenobiotic sensing receptor" because of its ability to regulate the induction of the CYP2B gene family, the gene products of which are known for metabolism of a variety of compounds (Swales and Negishi, 2004). This CAR-mediated induction typically occurs after exposure of hepatocytes to exogenous compounds such as phenobarbital, or endogenous compounds such as bilirubin (Guo et al., 2003). However, it is important to note that these compounds do not function as ligands for CAR. Instead, these compounds increase CAR activity by promoting translocation of CAR from the cytoplasm to the nucleus in a process that is though to involve a leucine-rich peptide domain in the C-terminal portion of CAR (Zelko et al., 2001). Two androstane metabolites, androstenol and androstanol, have been identified as true CAR ligands, but to date have only been shown to repress the constitutive activity of CAR (Forman et al., 1998). These observations have led to the conclusion that CAR is a constitutively active orphan NR that is active in the absence of ligand binding and has the unique capability to be further upregulated by activators. Because these activators consist predominantly of toxic exogenous and endogenous compounds, CAR activity is consistently associated with protection of the liver against insults through its ability to upregulate genes involved in the metabolism of these compounds (Swales and Negishi, 2004).

Closely related to CAR, and often functioning in an overlapping capacity, is the Class I NR Pregnane X Receptor (PXR). PXR coordinately regulates a program of genes involved in the metabolism, transport, and elimination of molecules such as various xenobiotics, natural and synthetic steroids, and bile acids from the body (Moore et al., 2003). As with CAR, genes upregulated by PXR include those in the CYP family (Kliewer et al., 1998). However, PXR has also been shown to regulate expression of genes such as the organic anion transporter protein-2 and the multidrug resistance related protein-2 (Geick et al., 2001; Staudinger et al., 2001).

While PXR and CAR are very similar structurally, significant differences exist in the molecular basis of their respective functions. Unlike CAR, which does not bind a majority of its activators through its LBD, PXR demonstrates direct interaction between a vast majority of its activating compounds and its LBD (Jones et al., 2000). Furthermore, while CAR is predominantly cytoplasmic, constitutively active, and undergoes nuclear translocation upon indirect interaction with an activator, PXR is located predominantly in the nucleus, exhibits low basal activity, and is highly activated through binding of activators to its LBD. However, in a manner analogous to CAR, PXR does require heterodimerization with RXR for binding to DNA (Kliewer et al., 2002).

Although bile acids have been shown to activate both CAR and PXR, these compounds also interact directly with another Class I NR called Farnesoid X Receptor (FXR). In fact, given the 10-fold higher affinity of bile acids for FXR, it is likely that activation of FXR-regulated genes occurs before activation of CAR or PXR-regulated genes under physiological conditions (Handschin and Meyer, 2005). Binding of four of the five human bile acids to FXR-RXR heterodimers leads to upregulation of target genes through transcriptional activation. The remaining known bile salt, lithocholate, acts as an antagonist to FXR and down-regulates gene expression when bound to the LBD of FXR (Lew et al., 2004). Interestingly, the products of genes upregulated upon bile acid binding to FXR are responsible predominantly for increasing bile acid export from the liver into the bile duct. One such gene product is bile salt export pump (BSEP) (Ananthanarayanan et al., 2001). Consequently, functionally redundant pathways exist for protection of mammals from elevated plasma levels of bile acids. With small elevations in serum bile acid levels, FXR-mediated gene expression increases normal bile-acid excretion. At potentially toxic bile acid levels, CAR and PXR can reduce the levels of endogenous toxins such as bile acids through inducing genes involved in bile acid hydroxylation, conjugation, and excretion (Handschin and Meyer, 2005).

Maintaining lipid homeostasis, a related but mechanistically separate means by which the liver is protected from the accumulation of bile acids, is accomplished to a large degree by the Liver X Receptor (LXR) isoforms LXRα and LXRβ (Svensson et al., 2003). Direct binding of specific oxysterols in the cholesterol biosynthetic pathway to the LBD of LXRs promotes binding of LXR-RXR heterodimers to LXR response elements (LXREs). LXREs are typically composed of direct repeats of a central core motif separated by four nucleotides (Janowski et al., 1996; Svensson et al., 2003). These response elements are located in the promoters of a large number of genes involved in several aspects of cholesterol biology, including reverse cholesterol transport, intestinal absorption, and lipoprotein remodeling (Costet et al., 2000; Zhang et al., 2001). In the liver, activation of LXRs by oxysterols also upregulates genes involved in *de novo* fatty acid synthesis (Svensson et al., 2003). It has also been observed that activated LXRs reduce the binding of PPARα to the promoters of genes whose products are responsible for fatty acid metabolism. This observation has led some to speculate that LXRs may serve as sensors of the balance between cholesterol synthesis and fatty acid metabolism (Handschin and Meyer, 2005; Ide et al., 2003). By doing so, LXRs act as the final component of the lipid-sensing NR network, also involving CAR, PXR, and FXR. This network is designed to act as the primary defense against accumulation of potentially toxic lipophilic compounds in the mammalian body (Handschin and Meyer, 2005).

B: Class II

In contrast to the Class I NRs, Class II NRs possess the ability to bind to HREs as homodimers(Aranda and Pascual, 2001; Laudet, 1997). However, at least 2, RXR and COUP-TF, also retain the ability to heterodimerize (Aranda and Pascual, 2001; Giguere, 1999; Laudet, 1997). The Class II NRs include Retinoid X Receptor (RXR), Chicken Ovalbumin Upstream Promoter Transcription Factor (COUP-TF), and Hepatocyte Nuclear Factor-4 (HNF-4).

B1: Retinoid X Receptor (RXR)

Given its necessity as a heterodimerization partner for Class I NRs, perhaps no Class II NR is more important than Retinoid X Receptor (RXR). The functions of RXR throughout the vertebrate life cycle are intrinsically linked to the various Class I NRs with which it heterodimerizes. However there is also a hypothesized role for the homodimerization of RXR in the control of specific target genes (Vivat-Hannah et al., 2003).

Like RAR, there are three isotypes of RXR, each derived from a separate gene, and each with at least two isoforms (Bastien and Rochette-Egly, 2004). Also, RXR is composed of the six domains that are conserved throughout the NR superfamily, and is found primarily in the nucleus in its unliganded state. However, while RAR has the capacity to bind both all-trans retinoid acid and 9-cis retinoic acid, RXR can bind only 9-cis retinoic acid (Bastien and Rochette-Egly, 2004). Furthermore, while NR-RXR heterodimers have the capacity to bind to response elements containing direct repeats separated by 1, 2, or 5 nucleotides, RXR homodimers can bind only direct repeats separated by a single nucleotide. To date, the only natural response element which binds RXR homodimers has been found in the rat CRBPII promoter (Mangelsdorf et al., 1991).

B2: Chicken Ovalbumin Upstream Promoter Transcription Factor (COUP-TF)

The Chicken Ovalbumin Upstream Promoter Transcription Factor (COUP-TF) group of Class II NRs is composed of 2 members in mammals, COUP-TFI and COUP-TFII. Both COUP-TFI and COUP-TFII are highly expressed at embryonic stages and play a key role in embryonic development. Studies conducted in mice have revealed that COUP-TFI is more highly expressed in neuronal tissues of the central and peripheral nervous systems while COUP-TFII is localized predominantly in developing organs (Jonk et al., 1994; Pereira et al., 1995; Qiu et al., 1994). To date, COUP-TFI has been shown to be important in processes such as axon guidance, axon myelination, regionalization of the forebrain and neocortex, and inner ear development. COUP-TFII is thought to play a role in general regulation of mesenchymal-endothelial interactions during embryonic vascular development as well as heart and stomach development (Pereira et al., 1999; Qiu et al., 1997). Recently, it was also shown to be a master regulator of arterial-venous specification. Interestingly, a role for COUP-TFII has also been hypothesized in the progression of cancer, a disease which is highly dependent on neovascularization. This hypothesis is based largely on observed COUP-TFII expression in several tumor cell lines (Shibata et al., 1998).

The COUP-TFs are true orphan receptors, as a ligand for these NRs has never been identified. COUP-TFs exist in solution as homodimers, but have also been shown to form heterodimeric complexes with RXR on DNA. COUP-TFs are also capable of binding to a wide variety of response elements, including direct, inverted, and everted repeats of the AGGTCA core motif (Cooney et al., 1992; Kadowaki et al., 1992). Not surprisingly, given this range of potential binding modalities, these NRs have been identified as potential regulators in the expression of a large number of genes (Giguere, 1999). Specifically, COUP-TFs act as potent transcriptional repressors which antagonize the transcriptional activation mediated by other NRs such as PPAR, HNF4, ER, RAR, VDR, and TR (Giguere, 1999). The precise mechanism by which COUP-TFs behave as transcriptional repressors is currently unknown, but is thought to involve both passive and active mechanisms in a promoter-dependent context. These mechanisms may include competition for DNA binding sites, competition for RXR through formation of inactive heterodimeric complexes, and direct repression mediated by N- and C-terminal repressor domains on COUP-TFs (Achatz et al., 1997; Cooney et al., 1993; Shibata et al., 1997).

B3: Hepatocyte Nuclear Factor 4 (HNF4)

First identified as a transcription factor required for liver-specific gene expression, Hepatocyte Nuclear Factor 4 (HNF4) has recently been shown to be a Class II NR which modulates the expression of genes involved in cholesterol, xenobiotic, amino-acid, carbohydrate, and lipid metabolism (Giguere, 1999). Not surprisingly, the expression pattern of HNF4 in mammals reflects its profound effects on multiple pathways involved in metabolism. High levels of HNF4 subtypes are found predominantly in the liver, kidney, intestine, and pancreas (Drewes et al., 1996). Also, recent evidence has linked mutations in the human HNF4α gene to a locus linked to maturity onset diabetes of the young (MODY) (Yamagata et al., 1996). These observations have led to insight into the potential role of this NR in human energy homeostasis.

Two mammalian subtypes of HNF4, HNF4α and HNF4γ, have been identified and are encoded by separate genes. Both isotypes contain the six conserved domains shared by almost all NRs, and bind as homodimers to response elements composed of direct repeats of a core motif separated by a single nucleotide (Jiang et al., 1995). Like Constitutive Androstane Receptor (CAR), HNF4 subtypes exhibit constitutive transactivation function (Wang et al., 1998). This constitutive activity of HNF4α and HNF4γ is regulated by binding of long-chain fatty acyl-CoA thioesters to the LBD of the NRs (Hertz et al., 1998). These long-chain fatty acyl-CoA thioesters demonstrate considerable differences in their ability to modulate HNF4γ transcriptional activity based on their molecular composition. Specifically, poly and mono-unsaturated acyl-CoA thioesters inhibit the constitutive activity of HNF4α, while differentially saturated acyl-CoA thioesters can either increase or decrease the constitutive activity of HNF4α (Hertz et al.,1998). This ability of fatty acyl-CoA thioesters to regulate gene expression through modulating the activity of HNF4 may be one of the mechanisms by which elevated intracellular fatty acyl-CoA levels switch the primary cellular energy source from glycolysis to fatty acid oxidation (Kruszynska et al., 1990).

C: Class III

The Class III NRs consist of the classically defined steroid hormone receptors, and constitute the most thoroughly studied group of NRs (Aranda and Pascual, 2001; Laudet, 1997). These NRs bind to their corresponding HREs exclusively as homodimers. The Class III NRs include Glucocorticoid Receptor (GR), Androgen Receptor (AR), Progesterone Receptor (PR), and Estrogen Receptor (ER) and Mineralocorticoid Receptor (MR).

C1: Glucocorticoid Receptor (GR)

Glucocorticoids are lipophilic compounds produced by the adrenal cortex under the regulatory influence of ACTH. Lack of glucocorticoids is incompatible with life in primates as these compounds play an indispensable role in maintaining basal and stress-related homeostasis. In the resting state, basal levels of glucocorticoids sustain normoglycemia and prevent arterial hypotension (Bamberger *et al.*, 1996). However, physical and emotional stress activate the hypothalamic-pituitary axis (HPA), elevating plasma ACTH levels and, ultimately, plasma glucocorticoid levels. As a result, plasma glucose levels are elevated, as is mean blood pressure (Chrousos, 1995; Chrousos and Gold, 1992). Concurrently, glucocorticoids restrain inflammatory and immune reactions which could potentially lead to tissue damage in the stressed state (Bamberger *et al.*, 1996). These effects of glucocorticoids are driven at a genetic level by the up-regulation of genes whose products are responsible for gluconeogenesis, lipolysis, and proteolysis while simultaneously down-regulating genes whose products participate in the inflammatory process (Schoneveld *et al.*, 2004).

At a molecular level, the genetic effects of glucocorticoids are mediated by the NR Glucocorticoid Receptor (GR). Although only one gene for GR has been identified, several GR isoforms have been identified that exist as a result of alternative splicing and use of multiple promoters (Yudt and Cidlowski, 2002). By far, however, GRα is the form most often associated with glucocorticoid-driven effects in target tissues, and analysis of GR mRNA levels in different tissues have revealed that this NR is ubiquitously expressed in mammals (Nordeen *et al.*, 1990). Like other NRs, GR is composed of the six functional domains A/B through E/F. In its unliganded state, GR exists as a monomeric form complexed with heat shock proteins 90 (hsp90), 70 (hsp70), and 56 (hsp56). Upon interaction with naturally-occurring and synthetic glucocorticoids with its LBD, GR undergoes a conformational change which causes dissociation of hsps, hyperphosphorylation, and translocation to the nucleus (Wright *et al.*, 1993).

Ligand-bound, nuclear GR can influence gene expression through four different types of binding sites (Schoneveld *et al.*, 2004). The first type of binding site, a simple glucocorticoid response element (GRE) has been identified as a pentadecameric imperfect palindrome with consensus sequence GGTACAnnnTGTTCT (Nordeen *et al.*, 1990). Liganded GR binds such simple GREs as a homodimer and activates expression of adjacent target genes. The second type of binding site is a GRE half-site (GRE1/2) (Segard-Maurel *et al.*, 1996). These sites are bound by monomeric, liganded GR, and typically require additional accessory elements or must be present in multiple copies in order to mediate a glucocorticoid response (Bristeau *et al.*, 2001; Schuetz *et al.*, 1996). The third type of binding site enables GR to down-regulate expression of responsive genes and is called a negative GRE (nGRE). NGREs are bound by liganded GR homodimers and have a consensus sequence which is more variable than that of GREs (Truss and Beato, 1993). The fourth and final binding site is termed a "tethering GRE" and is a sequence in a promoter that is not directly bound by GR, but rather recruits GR to the promoter using other DNA-bound transcription factors. An example occurs in the b-casein gene, where promoter-bound Stat5 recruits ligand-bound GR (Stocklin *et al.*, 1996). These tethering GREs allow either up-regulation or down-regulation of adjacent target gene expression.

C2: Androgen Receptor (AR)

Testosterone, and its more potent 5a-reduced metabolite dihyroxytestosterone (DHT), have been known for decades to be responsible for development of male genital organs *in utero*. These compounds are also known to be critical for the development of secondary sexual characteristics and male reproductive function and fertility at puberty and in adult life (McEwan, 2004). Analysis of murine gene ablation models has also suggested roles for testosterone and DHT in maintenance of normal trabecular and cortical bone volumes, as well as maintenance of body fat composition. These models also confirmed the role of testosterone and DHT in promoting sexual behavior in both males and females (Kato *et al.*, 2004). Finally, the roles of these compounds has been of significant interest recently, as they have been demonstrated to play direct roles in the progression of some forms of prostate cancer (Gelmann, 2002).

Testosterone and DHT, collectively referred to as androgens, have been shown to exert their effects through a Class III NR called Androgen Receptor (AR). AR is encoded by a single gene located on the long arm of the X-chromosome (McEwan, 2004; Quigley *et al.*, 1995). Like all conventional NRs, AR contains six functional domains designated A/B through E/F. However, AR is significantly longer than most NRs, due in large part to an extended A/B domain (Glass and Rosenfeld, 2000). While androgen activity is associated predominantly with male secondary sexual characteristic development, recent results from DNA microarrays and proteomic analysis have led to identification of a large number of genes regulated by androgens (Eder *et al.*, 2003; Nelson *et al.*, 1999; Umar *et al.*, 2003). These include genes coding for proteins involved in protein folding, trafficking, secretion, metabolism, the cytoskeleton, cell-cycle regulation, and signal transduction (McEwan, 2004).

Multiple mechanisms, alone or in combination, are

likely to be involved in ensuring AR-dependent gene regulation. In the initial steps of AR-mediated transactivation, binding of androgen to the LBD of AR causes a conformational change that results in the shedding of molecular chaperones by cytoplasmic AR and nuclear translocation of ligand-bound AR. Binding of ligand-bound AR homodimers to the palindromic consensus sequence AGA/TACA/TgcagT/AGTTCT, also called an Androgen Response Element (ARE), is widely recognized as a major mechanism by which AR transactivates adjacent target genes (McEwan, 2004; Roche *et al.*, 1992). However, recent studies have argued compellingly that ligand-bound AR homodimers, organized in a head-to-tail configuration, may be capable of binding direct repeat ARE sequences to activate gene expression (Schoenmakers *et al.*, 2000). Furthermore, multiple studies have revealed that the activity of AR in specific promoters depends not only on the DNA architecture of the response element, but also on the presence or absence of non-receptor accessory proteins (Gonzalez and Robins, 2001).

C3: Progesterone Receptor (PR)

The ovarian steroid hormone progesterone plays a pivotal role in normal female reproduction. All tissues of the female reproductive system, including the ovary, uterus, and mammary gland, are affected by this molecule. In the ovary, progesterone is essential for ovulation, while in the uterus progesterone directs glandular differentiation, stromal proliferation, and development of predecidual cells (Graham and Clarke, 1997). In the mammary gland, progesterone is required for appropriate ductal branching morphogenesis and lobular-alveolar differentiation (Graham and Clarke, 2002; Lydon *et al.*, 1995). Progesterone has also been implicated in neuroendocrine gonadotropin regulation, bone density maintenance, and the regulation of thymic involution (Conneely *et al.*, 2002; Lydon *et al.*, 1995). Recently, a great deal of research has been focused on understanding the role of progesterone after menopause, when estradiol levels are reduced. These studies have been driven largely by the epidemiological observation that breast cancer incidence increases with advancing age and through time points where progesterone levels are higher than estradiol levels (Bernstein, 2002; Li *et al.*, 2004).

The biological actions of progesterone are mediated through the action of the Progesterone Receptor (PR), a Class III NR. Two natural isoforms of PR exist, and are generated by alternative promoter usage from a single PR gene (Kastner *et al.*, 1990). The two isoforms, PR-A and PR-B, differ only in that PR-B contains an additional N-terminal 164 amino acids. Both PR-A and PR-B resemble the prototypic NR, and contain the six conserved functional domains characteristic of the NR superfamily (Leonhardt *et al.*, 2003).

The complex transcriptional activity of PR-A and PR-B is best described as a combination of cooperative action and distinct activity (Graham and Clarke, 2002). Newly transcribed cytoplasmic PR is maintained in an inactive multiprotein chaperone complex. This complex dissociates on ligand binding, and PR undergoes nuclear translocation. In the nucleus, ligand-bound PR-A or PR-B homodimers bind to palindromic progesterone response elements (PREs) in the promoters of target genes. PR-B exhibits hormone-dependent transactivation in all cell types irrespective of the complexity of the response elements (Graham and Clarke, 2002). PR-A, by contrast, displays transactivation activity that is cell-specific and reporter-specific.

C4: Estrogen Receptor (ER)

Estrogens are steroid hormones that regulate a large number of physiological processes involving cell growth and development. Estrogens regulate the reproductive organs of females, where the effects of these hormones drive sexual maturation of the uterus, vagina, and mammary gland while maintaining the ovulatory capacity of the ovaries. In the male, estrogens may also influence reproductive organ development and may have a profound influence on sperm motility. In both sexes, recent evidence from murine models has pointed to a role for estrogen in skeletal remodeling as well as plasma cholesterol and blood pressure homeostasis (Couse and Korach, 1999). Finally, estrogens also influence a number of pathological processes in hormone-dependent diseases such as breast, endometrial, and ovarian cancer (Matthews and Gustafsson, 2003).

The biological actions of estrogens are mediated by the Class III NR called Estrogen Receptor (ER). Two distinct ERs exist, and are called ERα and ERβ. These separate proteins are the product of different genes on different chromosomes, and each has several isoforms derived from splice variants (Enmark *et al.*, 1997; Menasce *et al.*, 1993). ERα and EPβ hare the evolutionarily conserved structural and functional domains of all NRs but exhibit different affinities for natural compounds and novel receptor-specific ligands (Kuiper *et al.*, 1998; Nilsson *et al.*, 2001; Sun *et al.*, 1999). Additionally, while both ERs are widely distributed throughout the mammalian body, the ERs demonstrate distinct, but overlapping expression patterns in a variety of tissues (Couse and Korach, 1999). ERα is expressed primarily in the uterus, liver, kidney, and heart, while ERβ is expressed primarily in the ovary, prostate, lung, gastrointestinal tract, bladder, bone marrow, and central nervous system (Couse and Korach, 1999).

ERs have the capacity to regulate gene expression through several different modes. In a manner similar to that seen with the Glucocorticoid Receptor (GR), ligand bound ER homodimers can bind palindromic estrogen response elements (EREs) to activate gene expression. However, because two distinct types of ER exist, it is also possible for ligand bound ERα-ERβ heterodimers to bind EREs. While ERα homodimers have been shown to have greater transactivation potential than ERβ homodimers, ERα-ERβ heterodimers have a transactivation potential which is midway between that of ERα homodimers and ERβ homodimers (Matthews and Gustafsson, 2003). Also similar to GR, ER-mediated gene transactivation has been associated with "tethering EREs". Such DNA sequences promote the binding of transcription factors such as activating protein-1 (AP-1) and stimulating protein-1 (SP-1) which then recruit ligand-bound ER (Matthews and Gustafsson, 2003). In general, Erα and ERβ exhibit opposing actions in the regulation of promoters from specific response elements (Paech et al., 1997). In such instances, ERβ appears to act as a dominant regulator of estrogen signaling, and when coexpressed with ERα, causes a concentration-dependent reduction in ERα-mediated transcriptional activation (Liu et al., 2002; Pettersson et al., 2000).

D: Class IV

The Class IV NR subfamily is composed off one group of NRs designated NGFI-B (Aranda and Pascual, 2001; Laudet, 1997). There are three members of this group, NGFI-Bα (Nurr77), NGFI-Bγ (Nurr1), and NGFI-Bβ (Nor-1), which are all true orphan receptors (Giguere, 1999). The members of the NGFI-B group of NRs are highly expressed in the adult nervous system, where they are induced as part of an immediate early response to stimuli such as nerve growth factor and membrane depolarization (Law et al., 1992; Ryseck et al., 1989; Watson and Milbrandt, 1990; Williams and Lau, 1993). In addition, these NRs demonstrate a pattern of expression outside the nervous system which is broad, and includes the liver, testis, ovary, thymus, muscle, lung, prostate, adrenal gland, thyroid gland, and pituitary gland (Bandoh et al., 1997; Ryseck et al., 1989; Zetterstrom et al., 1996). Most evidence has pointed to a role for the NGFI-B group of NRs in the regulation of signaling functions in both the hypothalamic-pituitary axis and during T-cell development (Giguere, 1999). In addition, NGFI-B has recently been shown to be an important mediator for the development of midbrain neurons with a dopaminergic phenotype, an observation which may have important implications in the pathogenesis of Parkinson's disease (Zetterstrom et al., 1997).

Like all members of the NR superfamily, the members of the NGFI-B group of orphan receptors retain all six conserved functional domains from A/B to E/F(Wansa et al., 2002). The NGFI-B members have also been shown to bind to DNA as monomers, homodimers, or heterodimers with RXR, and are potent activators of target gene expression. Monomeric binding occurs specifically at monomeric response elements (NBREs) which contain a special 5′-extended core motif with the sequence AAAGGTCA(Wilson et al., 1991). Homodimerization occurs at binding sites consisting of two inverted NBREs which are spaced six basepairs apart. Finally, while no ligand has been identified for the members of the NGFI-B group, both NGFI-Bα and NGFI-Bβ have been shown to bind to DR5 response elements as heterodimers with RXR (Perlmann and Jansson, 1995; Zetterstrom et al., 1996). On these elements, heterodimer complex formation can be efficiently induced by rexinoids, suggesting that rexinoids may be able to enhance responses to growth factors mediated by the NGFI-B members(Giguere, 1999).

E: Class V

The Class V NR subfamily is composed of only one group of NRs, called the Fushi Tarazu-Factor 1 (FTZ-F1 also called SF-1) group (Aranda and Pascual, 2001; Laudet, 1997). This group was initially composed entirely of a single gene product, FTZ-F1α, which was characterized as an adrenal gland-specific factor capable of binding to the proximal promoter region of the steroid hydroxylase genes CYP11A, CYP11B2, and CYP21 (Parker and Schimmer, 1997). However, a closely related gene, FTZ-F1β, has recently been cloned and may play a role in regulation of theα-fetoprotein locus and the CYP7A gene (Galarneau et al., 1996). Gene ablation models in mice have provided strong evidence for a role of FTZ-F1α in regulating mammalian sexual development as well as the differentiation of steroidogenic factors (Luo et al., 1994; Sadovsky et al., 1995a; Shinoda et al., 1995).

FTZ-F1α and FTZ-F1β are orphan receptors and, as such, have no known ligands. However, FTZ-F1α constitutively activates gene expression, and its activity can be regulated by phosphorylation (Giguere, 1999; Zhang and Mellon, 1996). FTZ-F1α functions as a monomeric receptor, and binds to HREs with the consensus sequence TCAAGGTCA (Wilson et al., 1993). Based on this sequence, FTZ-F1α is thought to regulate transcription of genes such as Mullerian inhibiting substance, Mullerian inhibiting substance receptor, the β-subunit of leutinizing hormone, and the ACTH receptor, among others (Parker and Schimmer, 1997). Interestingly, recent studies have indicated that certain oxysterols, distinct from those that regulate LXRα, can increase FTZ-F1α transcriptional activity (Lala et al., 1997). However, direct binding of these oxysterols to FTZ-F1α has not been demonstrated.

F: Class VI

The sixth subfamily of NRs is composed of only one NR called Germ Cell Nuclear Factor (GCNF) (Aranda and Pascual, 2001; Laudet, 1997). This NR was originally cloned by low-stringency screening and is not closely related to any other members of the NR superfamily(Chen et al., 1994; Chung and Cooney, 2001). GCNF is expressed at very high levels during embryogenesis and in the developing oocytes and spermatogenic cells of adults (Chung and Cooney, 2001). Recent studies in mouse models have suggested that GCNF plays an important role in development of a normal anterior-posterior axis during embryonic development. However, the target genes through which GCNF exerts its effects have yet to be identified. Moreover, the physiological role of GCNF in gametogenesis in adults remains incompletely understood, but may involve regulation of post-meiotically expressed protamine genes(Cooney et al., 2001).

GCNF is a true orphan receptor with no ligand identified for its activity to date. Although GCNF is not closely related to other members of the NR superfamily, it does retain the six conserved functional domains A/B through E/F that are the hallmark of all NRs (Chung and Cooney, 2001). Like other members of the NR superfamily, GCNF is also capable of binding to DNA at response elements, and it has been shown that GCNF binds a novel response element composed of a direct repeat of AGGTCA separated by no nucleotides (DR0) (Chung and Cooney, 2001). GCNF binds to these response elements as a homodimer, and in the absence of it putative ligand, serves as a repressor of gene function (Chung and Cooney, 2001; Cooney et al., 2001). COUP-TFs can also bind to these response elements with good binding affinity.

Conclusion

Over the past two decades, our understanding of the molecular mechanisms governing NR activity has grown considerably. Concomitant with this growth in mechanistic understanding has been a rapid delineation of the various physiological effects of the NRs. This has led to recognition of the tremendous importance of the members of this protein superfamily in both normal human physiology as well as pathophysiological processes. However, despite the astounding amount of information acquired recently, a great deal remains unknown about the NR superfamily. Specifically, the exact mechanisms by which these proteins stimulate or repress transcription remain unclear. Likewise, the roles that direct interaction between NRs and basal transcription factors, or interactions between NRs and coactivators/corepressors, play in pre-initiation complex formation remain incompletely understood. Undoubtedly, the future will provide new insights into the nature of these complex interactions between NRs and the promoters of the target genes which they regulate.

References

A, I. J., Jeannin, E., Wahli, W., and Desvergne, B. (1997). Polarity and specific sequence requirements of peroxisome proliferator-activated receptor (PPAR)/retinoid X receptor heterodimer binding to DNA. A functional analysis of the malic enzyme gene PPAR response element. J Biol Chem *272*, 20108-20117.

Achatz, G., Holzl, B., Speckmayer, R., Hauser, C., Sandhofer, F., and Paulweber, B. (1997). Functional domains of the human orphan receptor ARP-1/COUP-TFII involved in active repression and transrepression. Mol Cell Biol *17*, 4914-4932.

Ananthanarayanan, M., Balasubramanian, N., Makishima, M., Mangelsdorf, D. J., and Suchy, F. J. (2001). Human bile salt export pump promoter is transactivated by the farnesoid X receptor/bile acid receptor. J Biol Chem *276*, 28857-28865.

Aranda, A., and Pascual, A. (2001). Nuclear hormone receptors and gene expression. Physiol Rev *81*, 1269-1304.

Bamberger, C. M., Schulte, H. M., and Chrousos, G. P. (1996). Molecular determinants of glucocorticoid receptor function and tissue sensitivity to glucocorticoids. Endocr Rev *17*, 245-261.

Bandoh, S., Tsukada, T., Maruyama, K., Ohkura, N., and Yamaguchi, K. (1997). Differential expression of NGFI-B and RNR-1 genes in various tissues and developing brain of the rat: comparative study by quantitative reverse transcription-polymerase chain reaction. J Neuroendocrinol *9*, 3-8.

Barettino, D., Vivanco Ruiz, M. M., and Stunnenberg, H. G. (1994). Characterization of the ligand-dependent transactivation domain of thyroid hormone receptor. Embo J *13*, 3039-3049.

Bastien, J., and Rochette-Egly, C. (2004). Nuclear retinoid receptors and the transcription of retinoid-target genes. Gene *328*, 1-16.

Beato, M., Herrlich, P., and Schutz, G. (1995). Steroid hormone receptors: many actors in search of a plot. Cell *83*, 851-857.

Beato, M., and Sanchez-Pacheco, A. (1996). Interaction of steroid hormone receptors with the transcription initiation complex. Endocr Rev *17*, 587-609.

Berger, J., and Moller, D. E. (2002). The mechanisms of action of PPARs. Annu Rev Med *53*, 409-435.

Bernstein, L. (2002). Epidemiology of endocrine-related risk factors for breast cancer. J Mammary Gland Biol Neoplasia *7*, 3-15.

Bourguet, W., Ruff, M., Chambon, P., Gronemeyer, H., and Moras, D. (1995). Crystal structure of the ligand-binding domain of the human nuclear receptor RXR-alpha. Nature *375*, 377-382.

Bristeau, A., Catherin, A. M., Weiss, M. C., and Faust, D. M. (2001). Hormone response of rodent phenylalanine hydroxylase requires HNF1 and the glucocorticoid receptor. Biochem Biophys Res Commun *287*, 852-858.

Chen, F., Cooney, A. J., Wang, Y., Law, S. W., and O'Malley, B. W. (1994). Cloning of a novel orphan receptor (GCNF) expressed during germ cell development. Mol Endocrinol *8*, 1434-1444.

Chen, H., Lin, R. J., Xie, W., Wilpitz, D., and Evans, R. M. (1999). Regulation of hormone-induced histone hyperacetylation and gene activation via acetylation of an acetylase. Cell *98*, 675-686.

Chrousos, G. P. (1995). The hypothalamic-pituitary-adrenal axis and immune-mediated inflammation. N Engl J Med *332*, 1351-1362.

Chrousos, G. P., and Gold, P. W. (1992). The concepts of stress and stress system disorders. Overview of physical and behavioral homeostasis. Jama *267*, 1244-1252.

Chung, A. C., and Cooney, A. J. (2001). Germ cell nuclear factor. Int J Biochem Cell Biol *33*, 1141-1146.

Conneely, O. M., Mulac-Jericevic, B., DeMayo, F., Lydon, J. P., and O'Malley, B. W. (2002). Reproductive functions of progesterone receptors. Recent Prog Horm Res *57*, 339-355.

Cooney, A. J., Lee, C. T., Lin, S. C., Tsai, S. Y., and Tsai, M. J. (2001). Physiological function of the orphans GCNF and COUP-TF. Trends Endocrinol Metab *12*, 247-251.

Cooney, A. J., Leng, X., Tsai, S. Y., O'Malley, B. W., and Tsai, M. J. (1993). Multiple mechanisms of chicken ovalbumin upstream promoter transcription factor-dependent repression of transactivation by the vitamin D, thyroid hormone, and retinoic acid receptors. J Biol Chem *268*, 4152-4160.

Cooney, A. J., Tsai, S. Y., O'Malley, B. W., and Tsai, M. J. (1992). Chicken ovalbumin upstream promoter transcription factor (COUP-TF) dimers bind to different GGTCA response elements, allowing COUP-TF to repress hormonal induction of the vitamin D3, thyroid hormone, and retinoic acid receptors. Mol Cell Biol *12*, 4153-4163.

Costet, P., Luo, Y., Wang, N., and Tall, A. R. (2000). Sterol-dependent transactivation of the ABC1 promoter by the liver X receptor/retinoid X receptor. J Biol Chem *275*, 28240-28245.

Couse, J. F., and Korach, K. S. (1999). Estrogen receptor null mice: what have we learned and where will they lead us? Endocr Rev *20*, 358-417.

Dahlman-Wright, K., Baumann, H., McEwan, I. J., Almlof, T., Wright, A. P., Gustafsson, J. A., and Hard, T. (1995). Structural characterization of a minimal functional transactivation domain from the human glucocorticoid receptor. Proc Natl Acad Sci USA *92*, 1699-1703.

Danielian, P. S., White, R., Lees, J. A., and Parker, M. G. (1992). Identification of a conserved region required for hormone dependent transcriptional activation by steroid hormone receptors. Embo J *11*, 1025-1033.

Desvergne, B., Michalik, L., and Wahli, W. (2004). Be fit or be sick: peroxisome proliferator-activated receptors are down the road. Mol Endocrinol *18*, 1321-1332.

Drewes, T., Senkel, S., Holewa, B., and Ryffel, G. U. (1996). Human hepatocyte nuclear factor 4 isoforms are encoded by distinct and differentially expressed genes. Mol Cell Biol *16*, 925-931.

Eckey, M., Moehren, U., and Baniahmad, A. (2003). Gene silencing by the thyroid hormone receptor. Mol Cell Endocrinol *213*, 13-22.

Eder, I. E., Haag, P., Basik, M., Mousses, S., Bektic, J., Bartsch, G., and Klocker, H. (2003). Gene expression changes following androgen receptor elimination in LNCaP prostate cancer cells. Mol Carcinog *37*, 181-191.

Enmark, E., Pelto-Huikko, M., Grandien, K., Lagercrantz, S., Lagercrantz, J., Fried, G., Nordenskjold, M., and Gustafsson, J. A. (1997). Human estrogen receptor beta-gene structure, chromosomal localization, and expression pattern. J Clin Endocrinol Metab *82*, 4258-4265.

Feng, X., Jiang, Y., Meltzer, P., and Yen, P. M. (2000). Thyroid hormone regulation of hepatic genes *in vivo* detected by complementary DNA microarray. Mol Endocrinol *14*, 947-955.

Feng, X., Meltzer, P., and Yen, P. M. (2002). Transgenic targeting of a dominant negative corepressor to liver and analyses by cDNA microarray. Methods Mol Biol *202*, 31-54.

Folkers, G. E., van Heerde, E. C., and van der Saag, P. T. (1995). Activation function 1 of retinoic acid receptor beta 2 is an acidic activator resembling VP16. J Biol Chem *270*, 23552-23559.

Forman, B. M., Chen, J., and Evans, R. M. (1997). Hypolipidemic drugs, polyunsaturated fatty acids, and eicosanoids are ligands for peroxisome proliferator-activated receptors alpha and delta. Proc Natl Acad Sci USA *94*, 4312-4317.

Forman, B. M., Tzameli, I., Choi, H. S., Chen, J., Simha, D., Seol, W., Evans, R. M., and Moore, D. D. (1998). Androstane metabolites bind to and deactivate the nuclear receptor CAR-beta. Nature *395*, 612-615.

Forman, B. M., Umesono, K., Chen, J., and Evans, R. M. (1995). Unique response pathways are established by allosteric interactions among nuclear hormone receptors. Cell *81*, 541-550.

Galarneau, L., Pare, J. F., Allard, D., Hamel, D., Levesque, L., Tugwood, J. D., Green, S., and Belanger, L. (1996). The alpha1-fetoprotein locus is activated by a nuclear receptor of the *Drosophila* FTZ-F1 family. Mol Cell Biol *16*, 3853-3865.

Geick, A., Eichelbaum, M., and Burk, O. (2001). Nuclear receptor response elements mediate induction of intestinal MDR1 by rifampin. J Biol Chem *276*, 14581-14587.

Gelmann, E. P. (2002). Molecular biology of the androgen receptor. J Clin Oncol *20*, 3001-3015.

Giguere, V. (1999). Orphan nuclear receptors: from gene to function. Endocr Rev *20*, 689-725.

Giguere, V., Tini, M., Flock, G., Ong, E., Evans, R. M., and Otulakowski, G. (1994). Isoform-specific amino-terminal domains dictate DNA-binding properties of ROR alpha, a novel family of orphan hormone nuclear receptors. Genes Dev *8*, 538-553.

Glass, C. K. (1994). Differential recognition of target genes by nuclear receptor monomers, dimers, and heterodimers. Endocr Rev *15*, 391-407.

Glass, C. K., and Rosenfeld, M. G. (2000). The coregulator exchange in transcriptional functions of nuclear receptors. Genes Dev *14*, 121-141.

Gonzalez, M. I., and Robins, D. M. (2001). Oct-1 preferentially interacts with androgen receptor in a DNA-dependent manner that facilitates recruitment of SRC-1. J Biol Chem *276*, 6420-6428.

Graham, J. D., and Clarke, C. L. (1997). Physiological action of progesterone in target tissues. Endocr Rev *18*, 502-519.

Graham, J. D., and Clarke, C. L. (2002). Expression and transcriptional activity of progesterone receptor A and progesterone receptor B in mammalian cells. Breast Cancer Res *4*, 187-190.

Guo, G. L., Lambert, G., Negishi, M., Ward, J. M., Brewer, H. B., Jr., Kliewer, S. A., Gonzalez, F. J., and Sinal, C. J. (2003). Complementary roles of farnesoid X receptor, pregnane X receptor, and constitutive androstane receptor in protection against bile acid toxicity. J Biol Chem *278*, 45062-45071.

Gurlek, A., Pittelkow, M. R., and Kumar, R. (2002). Modulation of growth factor/cytokine synthesis and signaling by 1alpha,25-dihydroxyvitamin D(3): implications in cell growth and differentiation. Endocr Rev *23*, 763-786.

Handschin, C., and Meyer, U. A. (2005). Regulatory network of lipid-sensing nuclear receptors: roles for CAR, PXR, LXR, and FXR. Arch Biochem Biophys *433*, 387-396.

Hertz, R., Magenheim, J., Berman, I., and Bar-Tana, J. (1998). Fatty acyl-CoA thioesters are ligands of hepatic nuclear factor-4alpha. Nature *392*, 512-516.

Horlein, A. J., Naar, A. M., Heinzel, T., Torchia, J., Gloss, B., Kurokawa, R., Ryan, A., Kamei, Y., Soderstrom, M., Glass, C. K., and *et al.* (1995). Ligand-independent repression by the thyroid hormone receptor mediated by a nuclear receptor co-repressor. Nature *377*, 397-404.

Ide, T., Shimano, H., Yoshikawa, T., Yahagi, N., Amemiya-Kudo, M., Matsuzaka, T., Nakakuki, M., Yatoh, S., Iizuka, Y., Tomita, S., *et al.* (2003). Cross-talk between peroxisome proliferator-activated receptor (PPAR) alpha and liver X receptor (LXR) in nutritional regulation of fatty acid metabolism. II. LXRs suppress lipid degradation gene promoters through inhibition of PPAR signaling. Mol Endocrinol *17*, 1255-1267.

Issemann, I., and Green, S. (1990). Activation of a member of the steroid hormone receptor superfamily by peroxisome proliferators. Nature *347*, 645-650.

Janowski, B. A., Willy, P. J., Devi, T. R., Falck, J. R., and Mangelsdorf, D. J. (1996). An oxysterol signalling pathway mediated by the nuclear receptor LXR alpha. Nature *383*, 728-731.

Jiang, G., Nepomuceno, L., Hopkins, K., and Sladek, F. M. (1995). Exclusive homodimerization of the orphan receptor hepatocyte nuclear factor 4 defines a new subclass of nuclear receptors. Mol Cell Biol *15*, 5131-5143.

Jones, S. A., Moore, L. B., Shenk, J. L., Wisely, G. B., Hamilton, G. A., McKee, D. D., Tomkinson, N. C., LeCluyse, E. L., Lambert, M. H., Willson, T. M., *et al.* (2000). The pregnane X receptor: a promiscuous xenobiotic receptor that has diverged during evolution. Mol Endocrinol *14*, 27-39.

Jonk, L. J., de Jonge, M. E., Pals, C. E., Wissink, S., Vervaart, J. M., Schoorlemmer, J., and Kruijer, W. (1994). Cloning and expression during development of three murine members of the COUP family of nuclear orphan receptors. Mech Dev *47*, 81-97.

Juge-Aubry, C. E., Hammar, E., Siegrist-Kaiser, C., Pernin, A., Takeshita, A., Chin, W. W., Burger, A. G., and Meier, C. A. (1999). Regulation of the transcriptional activity of the peroxisome proliferator-activated receptor alpha by phosphorylation of a ligand-independent trans-activating domain. J Biol Chem *274*, 10505-10510.

Kadowaki, Y., Toyoshima, K., and Yamamoto, T. (1992). Ear3/COUP-TF binds most tightly to a response element with tandem repeat separated by one nucleotide. Biochem Biophys Res Commun *183*, 492-498.

Kastner, P., Krust, A., Turcotte, B., Stropp, U., Tora, L., Gronemeyer, H., and Chambon, P. (1990). Two distinct estrogen-regulated promoters generate transcripts encoding the two functionally different human progesterone receptor forms A and B. Embo J *9*, 1603-1614.

Kato, S., Matsumoto, T., Kawano, H., Sato, T., and Takeyama, K. (2004). Function of androgen receptor in gene regulations. J Steroid Biochem Mol Biol *89-90*, 627-633.

Kliewer, S. A., Goodwin, B., and Willson, T. M. (2002). The nuclear pregnane X receptor: a key regulator of xenobiotic metabolism. Endocr Rev *23*, 687-702.

Kliewer, S. A., Moore, J. T., Wade, L., Staudinger, J. L., Watson, M. A., Jones, S. A., McKee, D. D., Oliver, B. B., Willson, T. M., Zetterstrom, R. H., *et al.* (1998). An orphan nuclear receptor activated by pregnanes defines a novel steroid signaling pathway. Cell 92, 73-82.

Kliewer, S. A., Sundseth, S. S., Jones, S. A., Brown, P. J., Wisely, G. B., Koble, C. S., Devchand, P., Wahli, W., Willson, T. M., Lenhard, J. M., and Lehmann, J. M. (1997). Fatty acids and eicosanoids regulate gene expression through direct interactions with peroxisome proliferator-activated receptors alpha and gamma. Proc Natl Acad Sci USA 94, 4318-4323.

Kliewer, S. A., Umesono, K., Mangelsdorf, D. J., and Evans, R. M. (1992). Retinoid X receptor interacts with nuclear receptors in retinoic acid, thyroid hormone and vitamin D3 signalling. Nature 355, 446-449.

Kruszynska, Y. T., McCormack, J. G., and McIntyre, N. (1990). Effects of non-esterified fatty acid availability on insulin stimulated glucose utilisation and tissue pyruvate dehydrogenase activity in the rat. Diabetologia 33, 396-402.

Kuiper, G. G., Lemmen, J. G., Carlsson, B., Corton, J. C., Safe, S. H., van der Saag, P. T., van der Burg, B., and Gustafsson, J. A. (1998). Interaction of estrogenic chemicals and phytoestrogens with estrogen receptor beta. Endocrinology 139, 4252-4263.

Kumar, R., and Thompson, E. B. (1999). The structure of the nuclear hormone receptors. Steroids 64, 310-319.

Kurokawa, R., DiRenzo, J., Boehm, M., Sugarman, J., Gloss, B., Rosenfeld, M. G., Heyman, R. A., and Glass, C. K. (1994). Regulation of retinoid signalling by receptor polarity and allosteric control of ligand binding. Nature 371, 528-531.

Kurokawa, R., Yu, V. C., Naar, A., Kyakumoto, S., Han, Z., Silverman, S., Rosenfeld, M. G., and Glass, C. K. (1993). Differential orientations of the DNA-binding domain and carboxy-terminal dimerization interface regulate binding site selection by nuclear receptor heterodimers. Genes Dev 7, 1423-1435.

Lala, D. S., Syka, P. M., Lazarchik, S. B., Mangelsdorf, D. J., Parker, K. L., and Heyman, R. A. (1997). Activation of the orphan nuclear receptor steroidogenic factor 1 by oxysterols. Proc Natl Acad Sci USA 94, 4895-4900.

Laudet, V. (1997). Evolution of the nuclear receptor superfamily: early diversification from an ancestral orphan receptor. J Mol Endocrinol 19, 207-226.

Laudet, V., Hanni, C., Coll, J., Catzeflis, F., and Stehelin, D. (1992). Evolution of the nuclear receptor gene superfamily. Embo J 11, 1003-1013.

Law, S. W., Conneely, O. M., DeMayo, F. J., and O'Malley, B. W. (1992). Identification of a new brain-specific transcription factor, NURR1. Mol Endocrinol 6, 2129-2135.

Lazar, M. A. (2003). Thyroid hormone action: a binding contract. J Clin Invest 112, 497-499.

Leid, M., Kastner, P., and Chambon, P. (1992). Multiplicity generates diversity in the retinoic acid signalling pathways. Trends Biochem Sci 17, 427-433.

Leonhardt, S. A., Boonyaratanakornkit, V., and Edwards, D. P. (2003). Progesterone receptor transcription and non-transcription signaling mechanisms. Steroids 68, 761-770.

Lew, J. L., Zhao, A., Yu, J., Huang, L., De Pedro, N., Pelaez, F., Wright, S. D., and Cui, J. (2004). The farnesoid X receptor controls gene expression in a ligand- and promoter-selective fashion. J Biol Chem 279, 8856-8861.

Li, X., Lonard, D. M., and O'Malley, B. W. (2004). A contemporary understanding of progesterone receptor function. Mech Ageing Dev 125, 669-678.

Lin, R., and White, J. H. (2004). The pleiotropic actions of vitamin D. Bioessays 26, 21-28.

Liu, M. M., Albanese, C., Anderson, C. M., Hilty, K., Webb, P., Uht, R. M., Price, R. H., Jr., Pestell, R. G., and Kushner, P. J. (2002). Opposing action of estrogen receptors alpha and beta on cyclin D1 gene expression. J Biol Chem 277, 24353-24360.

Luisi, B. F., Xu, W. X., Otwinowski, Z., Freedman, L. P., Yamamoto, K. R., and Sigler, P. B. (1991). Crystallographic analysis of the interaction of the glucocorticoid receptor with DNA. Nature 352, 497-505.

Luo, X., Ikeda, Y., and Parker, K. L. (1994). A cell-specific nuclear receptor is essential for adrenal and gonadal development and sexual differentiation. Cell 77, 481-490.

Lydon, J. P., DeMayo, F. J., Funk, C. R., Mani, S. K., Hughes, A. R., Montgomery, C. A., Jr., Shyamala, G., Conneely, O. M., and O'Malley, B. W. (1995). Mice lacking progesterone receptor exhibit pleiotropic reproductive abnormalities. Genes Dev 9, 2266-2278.

Mader, S., Leroy, P., Chen, J. Y., and Chambon, P. (1993). Multiple parameters control the selectivity of nuclear receptors for their response elements. Selectivity and promiscuity in response element recognition by retinoic acid receptors and retinoid X receptors. J Biol Chem 268, 591-600.

Mangelsdorf, D. J., and Evans, R. M. (1995). The RXR heterodimers and orphan receptors. Cell 83, 841-850.

Mangelsdorf, D. J., Umesono, K., Kliewer, S. A., Borgmeyer, U., Ong, E. S., and Evans, R. M. (1991). A direct repeat in the cellular retinol-binding protein type II gene confers differential regulation by RXR and RAR. Cell 66, 555-561.

Matthews, J., and Gustafsson, J. A. (2003). Estrogen signaling: a subtle balance between ER alpha and ER beta. Mol Interv 3, 281-292.

McEwan, I. J. (2004). Molecular mechanisms of androgen receptor-mediated gene regulation: structure-function analysis of the AF-1 domain. Endocr Relat Cancer 11, 281-293.

McEwan, I. J., Wright, A. P., Dahlman-Wright, K., Carlstedt-Duke, J., and Gustafsson, J. A. (1993). Direct interaction of the tau 1 transactivation domain of the human glucocorticoid receptor with the basal transcriptional machinery. Mol Cell Biol 13, 399-407.

McInerney, E. M., Tsai, M. J., O'Malley, B. W., and Katzenellenbogen, B. S. (1996). Analysis of estrogen receptor transcriptional enhancement by a nuclear hormone receptor coactivator. Proc Natl Acad Sci USA 93, 10069-10073.

McKenna, N. J., Lanz, R. B., and O'Malley, B. W. (1999). Nuclear receptor coregulators: cellular and molecular biology. Endocr Rev 20, 321-344.

Menasce, L. P., White, G. R., Harrison, C. J., and Boyle, J. M. (1993). Localization of the estrogen receptor locus (ESR) to chromosome 6q25.1 by FISH and a simple post-FISH banding technique. Genomics 17, 263-265.

Mengus, G., May, M., Carre, L., Chambon, P., and Davidson, I. (1997). Human TAF(II)135 potentiates transcriptional activation by the AF-2s of the retinoic acid, vitamin D3, and thyroid hormone receptors in mammalian cells. Genes Dev 11, 1381-1395.

Miyata, K. S., McCaw, S. E., Marcus, S. L., Rachubinski, R. A., and Capone, J. P. (1994). The peroxisome proliferator-activated receptor interacts with the retinoid X receptor in vivo. Gene 148, 327-330.

Moore, J. T., Moore, L. B., Maglich, J. M., and Kliewer, S. A. (2003). Functional and structural comparison of PXR and CAR. Biochim Biophys Acta 1619, 235-238.

Moras, D., and Gronemeyer, H. (1998). The nuclear receptor ligand-binding domain: structure and function. Curr Opin Cell Biol 10, 384-391.

Nelson, C. C., Hendy, S. C., Shukin, R. J., Cheng, H., Bruchovsky, N., Koop, B. F., and Rennie, P. S. (1999). Determinants of DNA sequence specificity of the androgen, progesterone, and glucocorticoid receptors: evidence for differential steroid receptor response elements. Mol Endocrinol 13, 2090-2107.

Nilsson, S., Makela, S., Treuter, E., Tujague, M., Thomsen, J., Andersson, G., Enmark, E., Pettersson, K., Warner, M., and Gustafsson, J. A. (2001). Mechanisms of estrogen action. Physiol Rev 81, 1535-1565.

Nolte, R. T., Wisely, G. B., Westin, S., Cobb, J. E., Lambert, M. H., Kurokawa, R., Rosenfeld, M. G., Willson, T. M., Glass, C. K., and Milburn, M. V. (1998). Ligand binding and co-activator assembly of the peroxisome proliferator-activated receptor-gamma. Nature 395, 137-143.

Nordeen, S. K., Suh, B. J., Kuhnel, B., and Hutchison, C. D. (1990). Structural determinants of a glucocorticoid receptor recognition element. Mol Endocrinol 4, 1866-1873.

Onate, S. A., Tsai, S. Y., Tsai, M. J., and O'Malley, B. W. (1995). Sequence and characterization of a coactivator for the steroid hormone receptor superfamily. Science 270, 1354-1357.

O'Shea, P. J., and Williams, G. R. (2002). Insight into the physiological actions of thyroid hormone receptors from genetically modified mice. J Endocrinol 175, 553-570.

Paech, K., Webb, P., Kuiper, G. G., Nilsson, S., Gustafsson, J., Kushner, P. J., and Scanlan, T. S. (1997). Differential ligand activation of estrogen receptors ERalpha and ERbeta at AP1 sites. Science 277, 1508-1510.

Parker, K. L., and Schimmer, B. P. (1997). Steroidogenic factor 1: a key determinant of endocrine development and function. Endocr Rev 18, 361-377.

Pereira, F. A., Qiu, Y., Tsai, M. J., and Tsai, S. Y. (1995). Chicken ovalbumin upstream promoter transcription factor (COUP-TF): expression during mouse embryogenesis. J Steroid Biochem Mol Biol 53, 503-508.

Pereira, F. A., Qiu, Y., Zhou, G., Tsai, M. J., and Tsai, S. Y. (1999). The orphan nuclear receptor COUP-TFII is required for angiogenesis and heart development. Genes Dev 13, 1037-1049.

Perlmann, T., and Jansson, L. (1995). A novel pathway for vitamin A signaling mediated by RXR heterodimerization with NGFI-B and NURR1. Genes Dev 9, 769-782.

Pettersson, K., Delaunay, F., and Gustafsson, J. A. (2000). Estrogen receptor beta acts as a dominant regulator of estrogen signaling. Oncogene 19, 4970-4978.

Prufer, K., Veenstra, T. D., Jirikowski, G. F., and Kumar, R. (1999). Distribution of 1,25-dihydroxyvitamin D3 receptor immunoreactivity in the rat brain and spinal cord. J Chem Neuroanat 16, 135-145.

Qiu, Y., Cooney, A. J., Kuratani, S., DeMayo, F. J., Tsai, S. Y., and Tsai, M. J. (1994). Spatiotemporal expression patterns of chicken ovalbumin upstream promoter-transcription factors in the developing mouse central nervous system: evidence for a role in segmental patterning of the diencephalon. Proc Natl Acad Sci USA 91, 4451-4455.

Qiu, Y., Pereira, F. A., DeMayo, F. J., Lydon, J. P., Tsai, S. Y., and Tsai, M. J. (1997). Null mutation of mCOUP-TFI results in defects in morphogenesis of the glossopharyngeal ganglion, axonal projection, and arborization. Genes Dev 11, 1925-1937.

Quigley, C. A., De Bellis, A., Marschke, K. B., el-Awady, M. K., Wilson, E. M., and French, F. S. (1995). Androgen receptor defects: historical, clinical, and molecular perspectives. Endocr Rev 16, 271-321.

Robyr, D., Wolffe, A. P., and Wahli, W. (2000). Nuclear hormone receptor coregulators in action: diversity for shared tasks. Mol Endocrinol 14, 329-347.

Roche, P. J., Hoare, S. A., and Parker, M. G. (1992). A consensus DNA-binding site for the androgen receptor. Mol Endocrinol 6, 2229-2235.

Roeder, R. G. (1996). The role of general initiation factors in transcription by RNA polymerase II. Trends Biochem Sci 21, 327-335.

Ryseck, R. P., Macdonald-Bravo, H., Mattei, M. G., Ruppert, S., and Bravo, R. (1989). Structure, mapping and expression of a growth factor inducible gene encoding a putative nuclear hormonal binding receptor. Embo J 8, 3327-3335.

Sadovsky, Y., Crawford, P. A., Woodson, K. G., Polish, J. A., Clements, M. A., Tourtellotte, L. M., Simburger, K., and Milbrandt, J. (1995a). Mice deficient in the orphan receptor steroidogenic factor 1 lack adrenal glands and gonads but express P450 side-chain-cleavage enzyme in the placenta and have normal embryonic serum levels of corticosteroids. Proc Natl Acad Sci USA 92, 10939-10943.

Sadovsky, Y., Webb, P., Lopez, G., Baxter, J. D., Fitzpatrick, P. M., Gizang-Ginsberg, E., Cavailles, V., Parker, M. G., and Kushner, P. J. (1995b). Transcriptional activators differ in their responses to overexpression of TATA-box-binding protein. Mol Cell Biol *15*, 1554-1563.

Schoenmakers, E., Verrijdt, G., Peeters, B., Verhoeven, G., Rombauts, W., and Claessens, F. (2000). Differences in DNA binding characteristics of the androgen and glucocorticoid receptors can determine hormone-specific responses. J Biol Chem *275*, 12290-12297.

Schoneveld, O. J., Gaemers, I. C., and Lamers, W. H. (2004). Mechanisms of glucocorticoid signalling. Biochim Biophys Acta *1680*, 114-128.

Schrader, M., Nayeri, S., Kahlen, J. P., Muller, K. M., and Carlberg, C. (1995). Natural vitamin D3 response elements formed by inverted palindromes: polarity-directed ligand sensitivity of vitamin D3 receptor-retinoid X receptor heterodimer-mediated transactivation. Mol Cell Biol *15*, 1154-1161.

Schuetz, J. D., Schuetz, E. G., Thottassery, J. V., Guzelian, P. S., Strom, S., and Sun, D. (1996). Identification of a novel dexamethasone responsive enhancer in the human CYP3A5 gene and its activation in human and rat liver cells. Mol Pharmacol *49*, 63-72.

Schulman, I. G., Chakravarti, D., Juguilon, H., Romo, A., and Evans, R. M. (1995). Interactions between the retinoid X receptor and a conserved region of the TATA-binding protein mediate hormone-dependent transactivation. Proc Natl Acad Sci USA *92*, 8288-8292.

Segard-Maurel, I., Rajkowski, K., Jibard, N., Schweizer-Groyer, G., Baulieu, E. E., and Cadepond, F. (1996). Glucocorticosteroid receptor dimerization investigated by analysis of receptor binding to glucocorticosteroid responsive elements using a monomer-dimer equilibrium model. Biochemistry *35*, 1634-1642.

Shao, D., Rangwala, S. M., Bailey, S. T., Krakow, S. L., Reginato, M. J., and Lazar, M. A. (1998). Interdomain communication regulating ligand binding by PPAR-gamma. Nature *396*, 377-380.

Shibata, H., Ando, T., Suzuki, T., Kurihara, I., Hayashi, K., Hayashi, M., Saito, I., Kawabe, H., Tsujioka, M., Mural, M., and Saruta, T. (1998). Differential expression of an orphan receptor COUP-TFI and corepressors in adrenal tumors. Endocr Res *24*, 881-885.

Shibata, H., Nawaz, Z., Tsai, S. Y., O'Malley, B. W., and Tsai, M. J. (1997). Gene silencing by chicken ovalbumin upstream promoter-transcription factor I (COUP-TFI) is mediated by transcriptional corepressors, nuclear receptor-corepressor (N-CoR) and silencing mediator for retinoic acid receptor and thyroid hormone receptor (SMRT). Mol Endocrinol *11*, 714-724.

Shinoda, K., Lei, H., Yoshii, H., Nomura, M., Nagano, M., Shiba, H., Sasaki, H., Osawa, Y., Ninomiya, Y., Niwa, O., and *et al.* (1995). Developmental defects of the ventromedial hypothalamic nucleus and pituitary gonadotroph in the Ftz-F1 disrupted mice. Dev Dyn *204*, 22-29.

Staudinger, J. L., Goodwin, B., Jones, S. A., Hawkins-Brown, D., MacKenzie, K. I., LaTour, A., Liu, Y., Klaassen, C. D., Brown, K. K., Reinhard, J., *et al.* (2001). The nuclear receptor PXR is a lithocholic acid sensor that protects against liver toxicity. Proc Natl Acad Sci USA *98*, 3369-3374.

Stocklin, E., Wissler, M., Gouilleux, F., and Groner, B. (1996). Functional interactions between Stat5 and the glucocorticoid receptor. Nature *383*, 726-728.

Sun, J., Meyers, M. J., Fink, B. E., Rajendran, R., Katzenellenbogen, J. A., and Katzenellenbogen, B. S. (1999). Novel ligands that function as selective estrogens or antiestrogens for estrogen receptor-alpha or estrogen receptor-beta. Endocrinology *140*, 800-804.

Svensson, S., Ostberg, T., Jacobsson, M., Norstrom, C., Stefansson, K., Hallen, D., Johansson, I. C., Zachrisson, K., Ogg, D., and Jendeberg, L. (2003). Crystal structure of the heterodimeric complex of LXRalpha and RXRbeta ligand-binding domains in a fully agonistic conformation. Embo J *22*, 4625-4633.

Swales, K., and Negishi, M. (2004). CAR, driving into the future. Mol Endocrinol *18*, 1589-1598.

Tanenbaum, D. M., Wang, Y., Williams, S. P., and Sigler, P. B. (1998). Crystallographic comparison of the estrogen and progesterone receptor's ligand binding domains. Proc Natl Acad Sci USA *95*, 5998-6003.

Torchia, J., Glass, C., and Rosenfeld, M. G. (1998). Co-activators and co-repressors in the integration of transcriptional responses. Curr Opin Cell Biol *10*, 373-383.

Truss, M., and Beato, M. (1993). Steroid hormone receptors: interaction with deoxyribonucleic acid and transcription factors. Endocr Rev *14*, 459-479.

Umar, A., Luider, T. M., Berrevoets, C. A., Grootegoed, J. A., and Brinkmann, A. O. (2003). Proteomic analysis of androgen-regulated protein expression in a mouse fetal vas deferens cell line. Endocrinology *144*, 1147-1154.

Uppenberg, J., Svensson, C., Jaki, M., Bertilsson, G., Jendeberg, L., and Berkenstam, A. (1998). Crystal structure of the ligand binding domain of the human nuclear receptor PPARgamma. J Biol Chem *273*, 31108-31112.

Vivat-Hannah, V., Bourguet, W., Gottardis, M., and Gronemeyer, H. (2003). Separation of retinoid X receptor homo- and heterodimerization functions. Mol Cell Biol *23*, 7678-7688.

Voegel, J. J., Heine, M. J., Zechel, C., Chambon, P., and Gronemeyer, H. (1996). TIF2, a 160 kDa transcriptional mediator for the ligand-dependent activation function AF-2 of nuclear receptors. Embo J *15*, 3667-3675.

Wang, J. C., Stafford, J. M., and Granner, D. K. (1998). SRC-1 and GRIP1 coactivate transcription with hepatocyte nuclear factor 4. J Biol Chem *273*, 30847-30850.

Wansa, K. D., Harris, J. M., and Muscat, G. E. (2002). The activation function-1 domain of Nur77/NR4A1 mediates trans-activation, cell specificity, and coactivator recruitment. J Biol Chem 277, 33001-33011.

Watson, M. A., and Milbrandt, J. (1990). Expression of the nerve growth factor-regulated NGFI-A and NGFI-B genes in the developing rat. Development 110, 173-183.

Williams, G. T., and Lau, L. F. (1993). Activation of the inducible orphan receptor gene nur77 by serum growth factors: dissociation of immediate-early and delayed-early responses. Mol Cell Biol 13, 6124-6136.

Willson, T. M., Cobb, J. E., Cowan, D. J., Wiethe, R. W., Correa, I. D., Prakash, S. R., Beck, K. D., Moore, L. B., Kliewer, S. A., and Lehmann, J. M. (1996). The structure-activity relationship between peroxisome proliferator-activated receptor gamma agonism and the antihyperglycemic activity of thiazolidinediones. J Med Chem 39, 665-668.

Wilson, C. J., Chao, D. M., Imbalzano, A. N., Schnitzler, G. R., Kingston, R. E., and Young, R. A. (1996). RNA polymerase II holoenzyme contains SWI/SNF regulators involved in chromatin remodeling. Cell 84, 235-244.

Wilson, T. E., Fahrner, T. J., Johnston, M., and Milbrandt, J. (1991). Identification of the DNA binding site for NGFI-B by genetic selection in yeast. Science 252, 1296-1300.

Wilson, T. E., Fahrner, T. J., and Milbrandt, J. (1993). The orphan receptors NGFI-B and steroidogenic factor 1 establish monomer binding as a third paradigm of nuclear receptor-DNA interaction. Mol Cell Biol 13, 5794-5804.

Wright, A. P., Zilliacus, J., McEwan, I. J., Dahlman-Wright, K., Almlof, T., Carlstedt-Duke, J., and Gustafsson, J. A. (1993). Structure and function of the glucocorticoid receptor. J Steroid Biochem Mol Biol 47, 11-19.

Xu, H. E., Lambert, M. H., Montana, V. G., Parks, D. J., Blanchard, S. G., Brown, P. J., Sternbach, D. D., Lehmann, J. M., Wisely, G. B., Willson, T. M., et al. (1999). Molecular recognition of fatty acids by peroxisome proliferator-activated receptors. Mol Cell 3, 397-403.

Yamagata, K., Furuta, H., Oda, N., Kaisaki, P. J., Menzel, S., Cox, N. J., Fajans, S. S., Signorini, S., Stoffel, M., and Bell, G. I. (1996). Mutations in the hepatocyte nuclear factor-4alpha gene in maturity-onset diabetes of the young (MODY1). Nature 384, 458-460.

Yudt, M. R., and Cidlowski, J. A. (2002). The glucocorticoid receptor: coding a diversity of proteins and responses through a single gene. Mol Endocrinol 16, 1719-1726.

Zechel, C., Shen, X. Q., Chambon, P., and Gronemeyer, H. (1994). Dimerization interfaces formed between the DNA binding domains determine the cooperative binding of RXR/RAR and RXR/TR heterodimers to DR5 and DR4 elements. Embo J 13, 1414-1424.

Zelko, I., Sueyoshi, T., Kawamoto, T., Moore, R., and Negishi, M. (2001). The peptide near the C terminus regulates receptor CAR nuclear translocation induced by xenochemicals in mouse liver. Mol Cell Biol 21, 2838-2846.

Zenke, M., Munoz, A., Sap, J., Vennstrom, B., and Beug, H. (1990). V-erbA oncogene activation entails the loss of hormone-dependent regulator activity of c-erbA. Cell 61, 1035-1049.

Zetterstrom, R. H., Solomin, L., Jansson, L., Hoffer, B. J., Olson, L., and Perlmann, T. (1997). Dopamine neuron agenesis in Nurr1-deficient mice. Science 276, 248-250.

Zetterstrom, R. H., Solomin, L., Mitsiadis, T., Olson, L., and Perlmann, T. (1996). Retinoid X receptor heterodimerization and developmental expression distinguish the orphan nuclear receptors NGFI-B, Nurr1, and Nor1. Mol Endocrinol 10, 1656-1666.

Zhang, P., and Mellon, S. H. (1996). The orphan nuclear receptor steroidogenic factor-1 regulates the cyclic adenosine 3',5'-monophosphate-mediated transcriptional activation of rat cytochrome P450c17 (17 alpha-hydroxylase/c17-20 lyase). Mol Endocrinol 10, 147-158.

Zhang, Y., Repa, J. J., Gauthier, K., and Mangelsdorf, D. J. (2001). Regulation of lipoprotein lipase by the oxysterol receptors, LXRalpha and LXRbeta. J Biol Chem 276, 43018-43024.

Zile, M. H. (2001). Function of vitamin A in vertebrate embryonic development. J Nutr 131, 705-708.

Chapter 17

NFAT and MEF2, Two Families of Calcium-dependent Transcription Regulators

Jun O. Liu[1], Lin Chen[2], Fan Pan[1] and James C. Stroud[2]

[1]Department of Pharmacology, Johns Hopkins School of Medicine, 725 North Wolfe Street, Baltimore, MD 21205
[2]Department of Chemistry and Biochemistry, University of Colorado, Boulder, CO 80309

Key Words: NFAT, MEF2, calcium signaling, T cells, chromatin remodeling

The inorganic ion calcium plays a pivotal role in signal transduction throughout biology. How a small metal ion causes such profound changes in chromatin structures and patterns of gene expression in the nucleus has fascinated biologists and chemists alike for decades. A universal sensor protein for intracellular calcium is a small protein known as calmodulin or its homologs. Calmodulin binds to four calcium ions in a cooperative manner, much like hemoglobin senses oxygen. Upon binding to calcium, calmodulin undergoes significant conformational changes and gain the ability to interact with diverse cellular target proteins to transmit the calcium signal. Two common classes of calmodulin-dependent signaling proteins are calmodulin-dependent kinases and a calmodulin-dependent phosphatase known as calcineurin. Upon activation by calmodulin, the kinases and phosphatase will act on their respective substrates, often transcription factors, to propagate the signal from the cytosol into the nucleus. A surprising finding in recent years was the existence of a direct signaling pathway from calmodulin to nuclear chromatin remodeling proteins to modulate the activity of a family of transcription factors. In this chapter, we will focus on two families of calcium-dependent transcription regulators, the nuclear factor of activated T cells (NFAT) family and the myocyte enhancer factor (MEF2) family. We will focus on the structure and activity of these transcription factors at the molecular and cellular levels without getting into their functions in whole organisms. By comparing and contrasting these two distinct families of transcription factors which also interact with each other in some cellular processes, the differences in both design and activities of these two families shall become apparent.

General Features of NFAT

NFAT was named as such, as it was first identified in T cells as a rapidly inducible nuclear transcription factor bound to the distal antigen receptor responsive element of the human IL-2 promoter (Shaw et al., 1988). As we know it today, NFAT is expressed in a number of cell types and is involved in numerous cellular processes. Among all known transcription factors, NFAT is unique in that it is sequestered in the cytoplasm of cells in the absence of calcium signaling and it undergoes nuclear translocation to access the DNA elements of its target genes upon the onset of calcium signaling. To date, NFAT has been shown to play critical roles not only in the immune system, the nervous system for both development and the cognitive functions of adult neurons, but also during embryo development, particularly heart valve formation, and in muscle cell differentiation (Crabtree and Olson, 2002; Graef et al., 2001; Rao et al., 1997).

NFAT consists of a family of distinct isoforms encoded by distinct genes; most of the NFAT isoforms exist in multiple splice variants. For historic reasons, the different isoforms of NFAT have been given different names. Under one nomenclature system, NFAT is named NFAT1-5 whereas under another nomenclature system, it is named NFATc1-5. In this review, we will adopt the original names accorded to each isoform of NFAT, with

---

Corresponding Author: Jun O. Liu, Tel: (410) 955-4619, Fax: (410) 955-4620, E-mail: joliu@jhu.edu

NFATc (NFAT2 or NFATc1), NFATp (NFAT1 or NFATc2), NFAT3 (NFATc4), NFAT4 (NFATc3) and NFAT5 (TonEBP) (Hoey et al., 1995; McCaffrey et al., 1993; Northrop et al., 1994). Most tissues express a subset of NFAT isoforms, some of which appear to play redundant functions. It is also noteworthy that of the five known members of the NFAT family, NFAT5 is distantly related to the other four family members in that it is not responsive to calcium signaling (Lopez-Rodriguez et al., 1999) and it will not be included in ensuing discussions.

NFAT can be divided into several distinct domains, each of which plays a role in its function (Fig. 17.1). Its DNA binding domain, also known as Rel Homology Region (RHR), lies in the middle of the protein. It has two transactivation domains at both the N- and the C-termini, respectively. A large regulatory domain separates the N-terminal transactivation domain and the RHR, which mediates its regulation by the calcium and calmodulin-dependent protein phosphatase calcineurin and its subcellular localization. The rich regulatory domain can be further subdivided into multiple motifs: calcineurin-binding regions mediating the interaction between NFAT and calcineurin, two serine rich regions (SRR1 and SRR2) and three SPXX repeats that undergo dephosphorylation by calcineurin; and a nuclear localization sequence that is responsible for nuclear translocation of NFAT upon calcium signaling. The regulatory domain confers to NFAT its most distinct feature—calcium-dependent dephosphorylation by the protein phosphatase calcineurin and its subsequent nuclear import to serve its function as a transcription regulator.

The Regulatory Domain-1: Regulation of NFAT by Calcineurin

In resting T cells (as well as other cell types), NFAT resides in the cytosol as a hyerphosphorylated protein. There are an estimated total of 13 serine residues that are phosphorylated, five in SRR-1 region, three and four in each of the SP-2 and SP-3 motif and one in SRR-2 that is in close proximity to the nuclear localization region (Okamura et al., 2000). Upon T cell activation and the accompanying calcium influx, first from intracellular stores via the IP3 receptor and subsequently from the extracellular space through the cacium-release activated calcium channels (CRAC) on the plasma membrane, the cytosolic protein phosphatase calcineurin is activated by calmodulin. Activated calcineurin then dephosphorylates NFAT, exposing its nuclear localization sequence and causing it to translocate into the nucleus (Winslow and Crabtree, 2005) (Fig.17.2).

Unlike other serine/threonine phosphatases such as PP1 and PP2A, calcineurin is unique in that it has a built-in calcium switch to turn on its phosphatase activity (Klee et al., 1998). Calcineurin contains two subunits, a catalytic A subunit and a regulatory B subunit. The regulatory B subunit has been shown to confer structural stability to calcineurin and is likely to participate in substrate recognition, responsible for the relatively narrow substrate specificity of calcineurin (Klee et al., 1998). The catalytic subunit of calcinurin contains two regulatory domains toward its C-terminus, that together form the on-and-off switch for calcineurin. At the very C-terminus of the catalytic subunit of calcineurin is an autoinhibitory domain that serves as a pseudosubstrate and binds to the active site of calcineurin, keeping it inactive in the absence of a calcium signal. Next to the autoinhibitory domain is a calmodulin-binding domain that mediates the binding of calcium-bound calmodulin, which leads to the dissociation of the autoinhibitory domain from the active site and access of hyperphosphorylated NFAT to the active site of calcineurin.

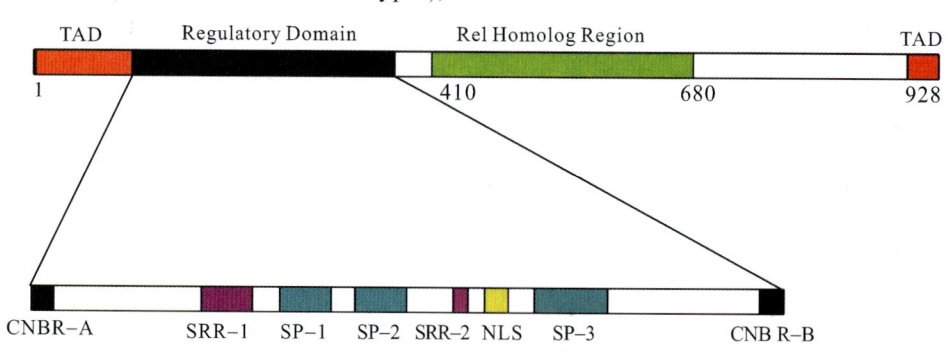

Fig.17.1 NFAT domain structures. Different functional domains of NFAT are highlighted with different colors, with the regulatory region expanded to show different regulatory features. Abbreviations: TAD, transactivation domain; CNBR, calcineurin-binding region, SRR, serine-rich region; SP, serine-proline repeat motif; NLS, nuclear localization signal.

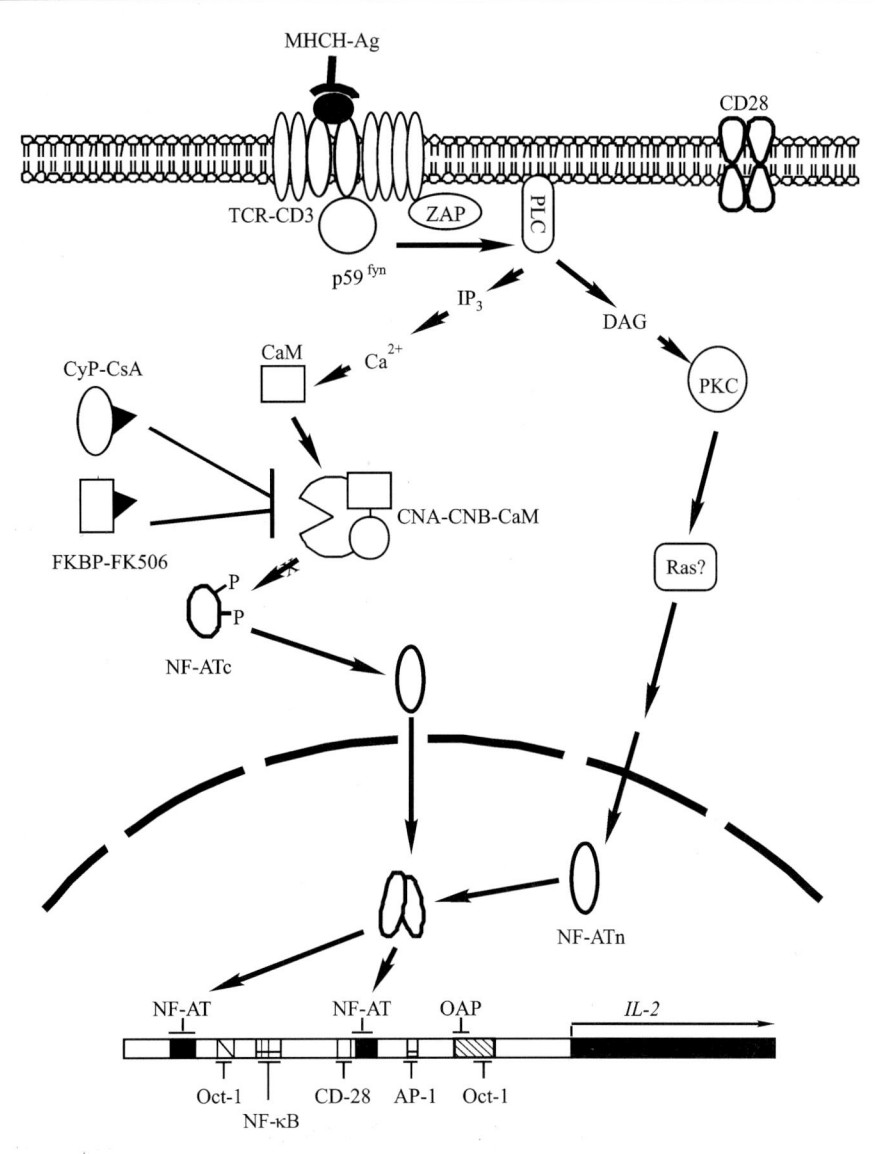

Fig.17.2 **Calcium signaling pathway leading to the activation of NFAT.** NFAT exists as a hyperphosphorylated form in the cytosol in resting T cells. Upon engagement of the TCR complex with MHCII-antigen complex on antigen presenting cells, membrane proximal protein tyrosine kinases including Fyn and Lck are activated, which leads to activation of phospholipase C (PLC)γ that generates the second messenger inositol-1, 4, 5-triphosphate (IP$_3$). IP$_3$ in turn causes release of calcium from intracellular stores that leads to further calcium influx from extracellular space. The second messenger calcium activates calmodulin (CaM) that then activates calcineurin (CAN-CNB). Activated calcineurin dephosphorylates NFAT to enable it to translocate into the nucleus to cause the transcriptional activation of such cytokine genes as IL-2.

To calcineurin, NFAT is not an ordinary substrate. There are a total of 13 phosphoserine residues to be dephosphorylated spanning a region of over 200 amino acids, as opposed to one or a few phosphoserine or phosphothreonine in other known calcineurin substrates such as inhibitor of PP1 or the regulatory subunit of protein kinase A (Okamura et al., 2000). The multiplicity of phosphoserines that require dephosphorylation by calcineurin necessitates that NFAT is associated with calcineurin for relatively longer period of time so that phosphoserines at different regions of NFAT can be efficiently and cooperatively dephosphorylated upon a single encounter with calcineurin. It is likely for this reason that NFAT also contains two calcineurin-binding regions (CNBR) or docking sites flanking the phosphoserine-containing segment. The N-terminal calcineurin docking site contains the consensus sequence PxIxIT, in which x can be any amino acids. While the N-terminal calcineurin docking site share homology with another calcineurin binding protein Cabin1/Cain, the C-terminal calcineurin binding region is similar to the DSCR/MCIP/calcipressin family of endogenous calcineurin inhibitors (Fuentes *et al.*, 2000; Jiang *et al.*, 1997; Kingsbury and Cunningham, 2000; Lai *et al.*, 1998; Liu, 2003; Rothermel *et al.*, 2003; Sun *et al.*, 1998).

The PxIxIT calcineurin-docking motif has been studied in great detail (Aramburu *et al.*, 1998; Aramburu *et al.*, 1999; Li *et al.*, 2004; Liu *et al.*, 2001) whereas the role of the C-terminal calcineurin binding region in NFAT remains to be elucidated. It has been shown that PxIxIT motif of NFAT is likely to associate with the catalytic domain of calcineurin at a site that is conserved between calcineurin and PP1 or PP2A and that are used frequently by other PP1 regulatory proteins for association (Li *et al.*, 2004). This site, sandwitched by the loops connecting β strands, lie in proximity to the metal ion-containing active site of calcineurin, rendering it possible for the adjacent phosphoserine residues to gain access to the active site of calcineurin in a pheudo-intramolecular fashion once NFAT is bound through this site to calcineurin. This is consistent with the observation that synthesized oligopeptide containing PxIxIT or its high affinity variants, specifically inhibits dephosphorylation of NFAT, but not other substrates, by calcineurin *in vitro* (Aramburu *et al.*, 1999), raising the possibility that small molecules bound to this docking site in calcineurin will become substrate-specific inhibitors of calcineurin. Indeed, small molecule inhibitors were identified through a high-throughput screen that selectively blocked NFAT dephosphorylation by calcineurin *in vivo* and TCR-mediated IL-2 transcription (Roehrl *et al.*, 2004). It is conceivable that these inhibitors, if improved to possess efficacy *in vivo*, will be have less toxicity than the currently used inhibitors of calcineurin such as cyclosproin A and FK506 in the clinic.

The dephosphorylation of NFAT by calcineurin results in a major conformational change that masks the nuclear export sequence while revealing the nuclear localization sequence (Zhu and McKeon, 1999). Dephosphorylated NFAT will translocate from the cytosol into the nucleus where it binds, either alone or more often, in complex with other DNA-binding transcription factors, to its cognate recognition sequences. In response to calcium signaling, the nuclear translocation of NFAT is complete within minutes, underlying the relatively rapid conversion of calcium signal in the cytosol into a nuclear transactivation output. The nuclear localization of NFAT requires sustained activation of calcineurin, as inhibition of calcineurin by CsA or FK506, leads to rapid rephosphorylation of NFAT and its return from the nucleus into the cytosol (Zhu and McKeon, 1999). Indeed, NFAT is under dynamic control of cellular kinases that oppose the activity of calcineurin.

The Regulatory Domain-2: Regulation of NFAT by Kinases

In the lifetime of NFAT, it has to undergo phosphorylation on two distinct occasions. First, upon translation, NFAT has to be phosphorylated by cellular kinases to prevent its entry into the nuclear compartment. Second, upon termination of calcium signaling, NFAT is subject to phosphorylation to exit the nucleus. It is unclear whether the same cellular kinases are responsible for post-translational phosphorylation and post-activation phosphorylation of NFAT.

A number of candidate enzymes have been identified that are capable of phosphorylating NFAT (Hogan *et al.*, 2003; Neilson *et al.*, 2001; Zhu and McKeon, 2000). This is not surprising, as the multiple phosphorylated residues in NFAT reside within completely distinct sequence contexts, suggesting that they are targets for different classes of kinases. It is likely that phosphorylation of NFAT is accomplished by the collaborative action of different kinases. Wherease the kinases responsible for posttranslational phosphorylation of NFAT are likely to be constitutively active enzymes, those responsible for post-activation phosphorylation of NFAT may include signal-activated kinases.

Among the putative NFAT kinases identified to date, some, such as casein kinase (CK)1 (Zhu *et al.*, 1998), are constitutive, while others, such as GSK3 (Beals *et al.*, 1997), cAMP-dependent kinase (Chow and Davis, 2000), MEKK (Zhu *et al.*, 1998), p38 (Gomez del Arco *et al.*, 2000; Yang *et al.*, 2002) and JNK (Chow *et al.*, 1997), are subject to signal dependent regulation. It has been shown that for GSK3 to phosphorylate NFAT, a priming phosphorylation by PKA is required (Sheridan *et al.*, 2002), necessitating a collaborative action of those two kinases to phosphorylate NFAT. Due to the difference in the sequences in the regulatory region, different NFAT isoforms are substrates for different, though overlapping, subsets of the kinases. For example, JNK1 was shown to phosphorylate NFATc and NFAT4 (Chow *et al.*, 1997). In contrast, p38 is selective for NFATp and NFAT4 (Gomez del Arco *et al.*, 2000; Yang *et al.*, 2002). It is possible that there may exist redundant NFAT kinases in the same cell type and the primary kinase responsible for NFAT phosphorylation will be dependent on which signaling pathways are operative at a given time.

It is worth noting that it takes energy to keep NFAT in a multiply phosphorylated state in resting cells. It is more common and economical for cells to keep the basal states of signaling proteins in unphosphorylated form and use phosphorylation cascades to transmit cellular signals. Although calcium and calmodulin-dependent kinases are expressed in T cells and participate in T cell receptor signaling, eventually a dephophosphorylation-dependent nuclear translocation was chosen over a phosphorylation-dependent nuclear translocation of NFAT. This may be due to the use of calmodulin-dependent kinase cascades in regulating other transcription regulators such as CREB and Cabin1 (F. P., Means, A., and J.O.L., unpublished results), making it difficult to "rewire" calmodulin-dependent kinase cascade to the activation of NFAT.

The Rel-homology Domain: Interaction with DNA and other Transcription Factors

The NFAT family of proteins shares a highly conserved DNA binding domain that is similar to the Rel homology region/domain (RHR) initially identified in the Rel/NF-κB proteins (Chen, 1998; Ho, 1994). The canonical mode of DNA binding by the RHR has been well characterized in the structures of several NFκB-DNA complexes (Chen, 1998; Ghosh, 1995; Muller, 1995): the RHR contains two functionally distinct domains, a N-terminal specificity domain (RHR-N) that makes base-specific DNA contacts, and a C-terminal domain involved in dimer formation and IκB binding (RHR-C) (Huxford, 1998; Jacobs, 1998). Of the five NFAT proteins, NFAT5, also known as TonEBP, shows the highest degree of structural similarity to NFκB, forming a symmetric NFAT5 dimer with a striking resemblance to the NFκB-DNA complex (Lopez-Rodriguez, 2001; Miyakawa, 1999; Stroud, 2003). Unlike the classical Rel proteins and the related NFAT5/TonEBP, NFAT1-4 exists as monomer in solution (Chen, 1995; Chytil, 1996; Hoey et al., 1995), and binds DNA as a monomer at its cognate site (Stroud, 2003), as a dimer at κB like sites (Giffin, 2003; Jin, 2003), or in complexes with other transcription factors at composite sites (Chen, 1998). The structures of many of these NFAT complexes have been characterized in the last few years.

A: DNA Binding by a NFAT Monomer

The crystal structure of the human NFAT1 RHR bound to DNA as a monomer has recently been solved (Stroud, 2003). In this complex, a DNA recognition loop from the RHR-N binds the core NFAT site GGAA through mechanisms highly conserved in other Rel/DNA complexes (Chen, 1998; Ghosh, 1995; Muller, 1995). The NFAT1/DNA binary complex also reveals a novel DNA-binding mode by the NFAT RHR wherein the RHR-C wraps around the DNA and makes extensive contacts to the phosphate backbone (Fig. 17.3A). This mode of protein-DNA interaction is consistent with the DNA methylation and ethylation footprints of the binary NFAT/DNA complex in solution (Stroud, 2003). Surprisingly, there are four independent NFAT/DNA complexes in the asymmetric unit of the crystal. In all four complexes, the structure of the RHR-N and its interaction with DNA are almost identical, suggesting that the RHR-N/DNA interaction is a conserved feature of the NFAT/DNA complexes. Indeed, in all NFAT complexes characterized so far, including the NFAT monomer/DNA complexes, the NFAT dimer/DNA complex (see below) (Giffin, 2003; Jin, 2003) and the NFAT/Fos-Jun/DNA ternary complex (see below), the interaction between RHR-N and its cognate DNA site is kept constant. However, the RHR-C of each NFAT/DNA complex in the NFAT1 monomer/DNA complex adopts a different orientation. Thus, it seems that when NFAT binds DNA, its RHR-C constantly cast in space to look for interaction partner. This is reminiscent of DNA binding of Fos-Jun, which rotates between two distinct orientations until NFAT joins in to bind the ARRE2 site as a stereo-specific NFAT/Fos-Jun/DNA complex (Chen, 1995). The conformational flexibility of the RHR-C is therefore likely a key structural feature of NFAT to allow the assembly of higher-order transcription complexes between NFAT and a variety of partners in different promoter contexts.

B: Binding of κB-like DNA Sites by NFAT Dimes

DNA binding by an NFAT dimer has been implicated in the activation of specific subsets of host and viral genes. Two NFAT dimer complexes, bound to κB sites from the IL-8 promoter and the HIV-1 LTR, have been characterized at the structural level (Giffin, 2003; Jin, 2003) (Fig. 17.3B, C). As in the NFAT5 and Rel dimer complexes, the NFAT1 dimerization is mediated by the RHR-C (Fig. 17.3B, C). However, unlike the symmetric and hydrophobic RHR-C dimer interface seen in NFAT5 and NFκB, the RHR-C dimer interface in NFAT1 is asymmetric and largely hydrophilic, and involves different residues on the RHR-C. The RHR-C dimer interface is identical in the NFAT1 dimers bound to the IL-8 and HIV-1 LTR κB sites (compare Fig. 17.3B, C), and mutational and biochemical evidence suggests that the interface also occurs in solution and contributes to cooperative DNA binding by NFAT1 dimers to the κB sites. Many of the RHR-C interface residues observed in the NFAT1 dimer are conserved in NFAT2 and NFAT4, suggesting that at least some members of the NFAT family could form homo- or heterodimers on κB-like DNA sites. These studies provide a direct structural explanation for the long-standing puzzle of how NFAT recognizes the κB sites as a dimer, which are usually bound by preformed dimers of the NF-κB family of proteins. These studies also support a potential mechanism by which HIV-1 exploit host transcription factor (NFAT and NF-κB) for its transcription and replication.

Comparison of the NFAT dimer bound to κB sites from the IL-8 promoter and the HIV-1 LTR has also provided a dramatic example of how subtle sequence variations in the DNA binding site can affect the conformation of bound transcription factor. Since this finding has major implications for how cis-regulatory elements in human genome may gain diverse functions (Leung, 2004), we will discuss this phenomenon further here. The two NFAT1 dimer complexes on the IL-8 promoter and the HIV-1 LTR differ significantly in their RHR-N interactions and hence their overall conformation (compare Fig. 17.3B, C). The IL-8 and HIV-1 LTR κB

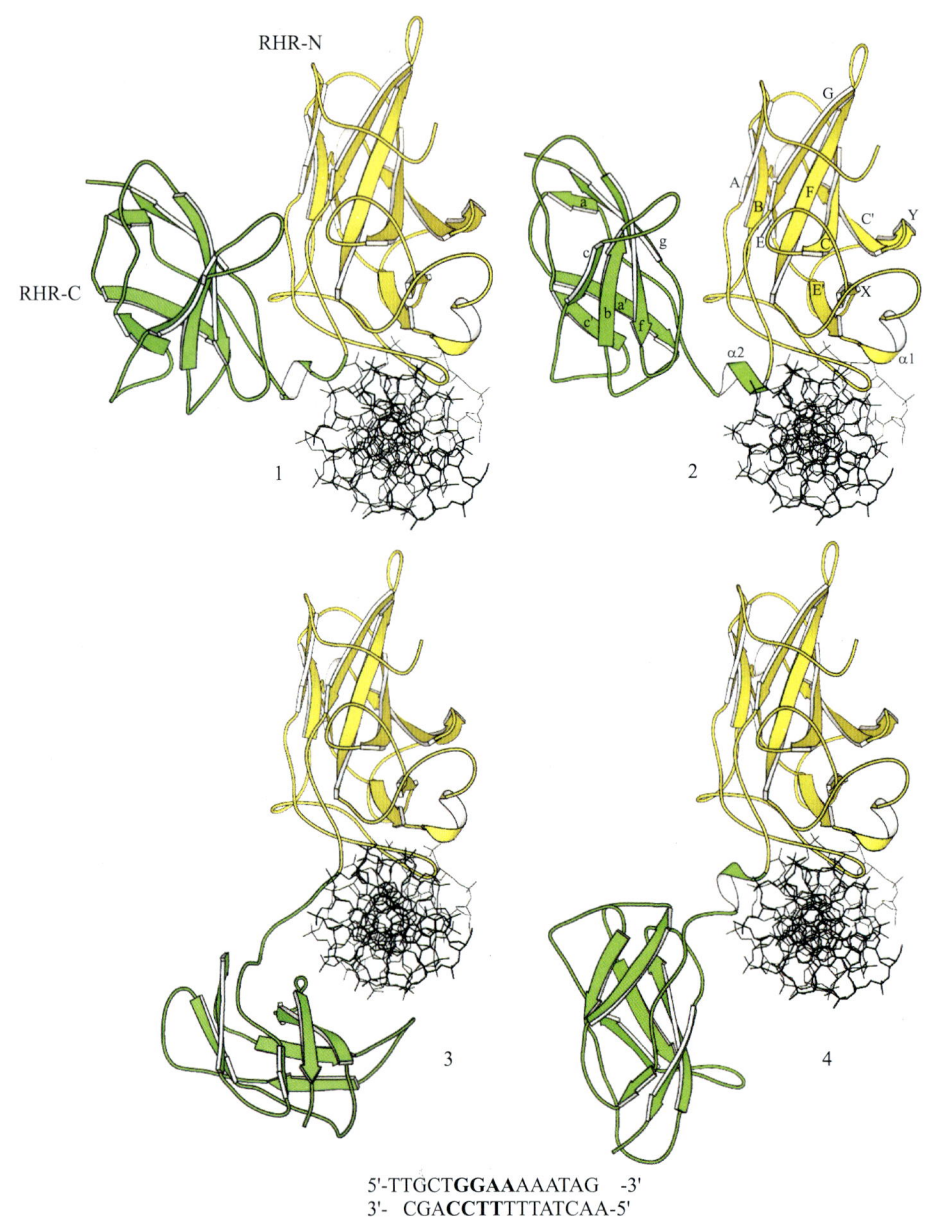

5'-TTGCT**GGAA**AAATAG -3'
3'- CGA**CCTT**TTTATCAA-5'

Fig.17.3A Systematic structural studies of NFAT complexes. The proteins are shown in ribbon style, with the RHR-N domain in green and the RHR-C domain in yellow. The DNA is drawn in stick model with its axis perpendicular to the paper. The DNA sequences used in the crystallographic studies are listed under each structure. The NFAT binding sites are bold and the AP-1 site is underlined in *d*. Also in panel *d*, Fos and Jun are colored in red and blue, respectively. (*a*) NFAT monomer bound to a cognate NFAT site. The four independent NFAT/DNA complexes are shown; (*b*) The NFAT1 dimer bound to the κB site from the HIV-1 LTR; (*c*) The NFAT1 dimer bound to a κB site similar to that in the human *IL-8* promoter. (*d*) The NFAT1/Fos-Jun complex bound to the ARRE2 site from the murine IL-2 promoter.

5'-AATG**GGGACTTTCC**A -3'
AC**CCCTGAAAGG**TT -5'

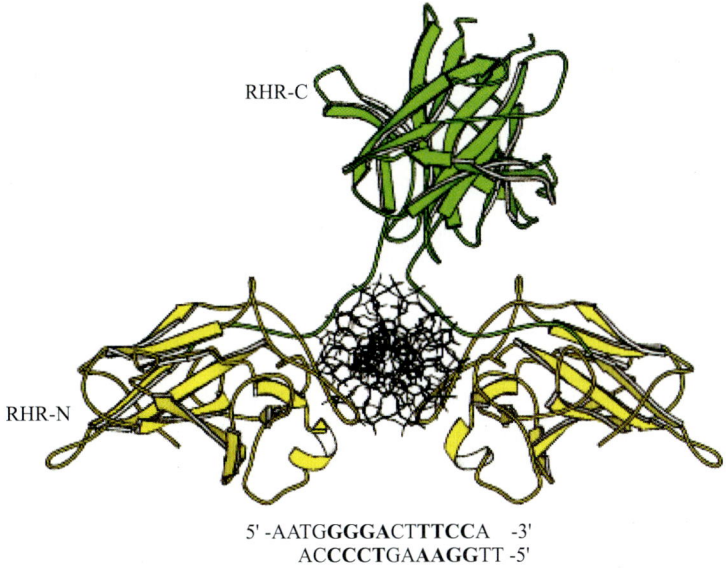

Fig.17.3B

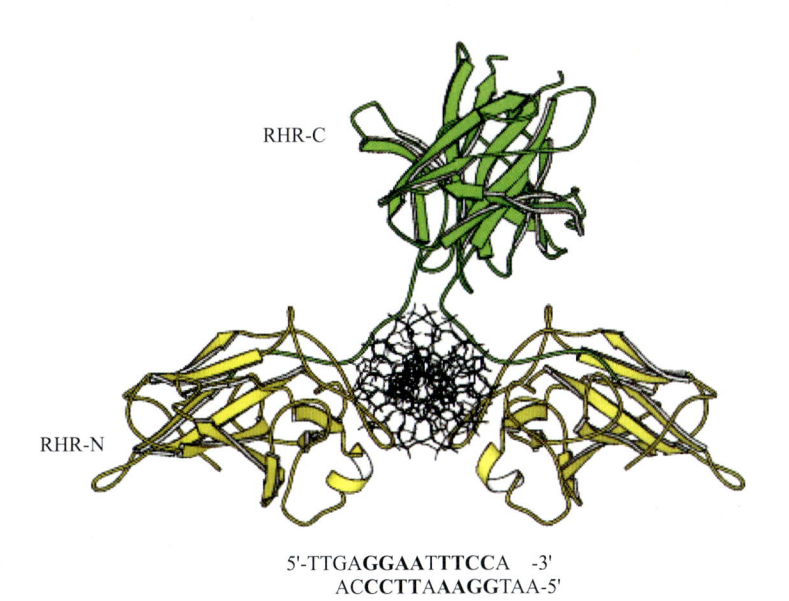

5'-TTGA**GGAATTTCC**A -3'
AC**CCTTAAAGG**TAA-5'

Fig.17.3C

sequences can be viewed as two NFAT sites arranged in a dyad orientation, but separated by 9 or 10 bp respectively (IL-8 κB, G̲G̲A̲A̲A̲T̲T̲C̲C̲; HIV-1 LTR κB, G̲G̲G̲A̲C̲T̲T̲T̲C̲C̲). In the IL-8 κB complex, each consensus NFAT site is bound to the RHR-N of one NFAT1 monomer; while in the HIV-1 LTR κB complex, the NFAT1 monomer bound to the 5′ end of the site surprisingly binds the lower consensus sequence (G̲G̲G̲A̲CTTTCC) with a 10 bp spacing, instead of binding to the more consensus sequence (GGGAC̲T̲T̲T̲C̲C̲) with a 9 bp spacing. As a result, the NFAT1 dimer on the HIV-1 LTR κB site completely encircles the DNA through E′F loop interactions, in a manner similar to that seen in the NFAT5/DNA complex. The extra conserved guanine residue at the 5′ end of this sequence (G̲GGACTTTTCC) allows the NFAT dimer to "slip" into the 10 bp binding mode, in which it gains additional stability through E′F loop interactions without significantly losing protein-DNA interactions. By contrast, on the IL-8 κB site, the two NFAT monomers are strictly anchored to the two consensus NFAT sites in a 9 bp spacing mode, resulting in an open conformation that does not permit E′F loop interactions. Notably, the E′F loop is also used in the ternary NFAT/Fos-Jun/DNA complex (Chen, 1998), where it constitutes the major binding site for Fos-Jun (see below), emphasizing the versatility of this protein surface in promoting assembly of distinct NFAT transcription complexes.

C: The NFAT/Fos-Jun/DNA Ternary Complex

NFAT cooperates with a myriad of partner transcription factors to regulate distinct transcription programs. This cooperation occurs at composite elements that generally have a consensus site for one member of the complex and a non-consensus site for the other. For instance, the ARRE2 element in the human IL-2 promoter has a consensus NFAT site (GGAA) and a non-consensus AP-1 site (TGTTTCA) that differs considerably from the consensus (TGACTCA). NFAT1-4 proteins form cooperative complexes on such composite DNA elements with the unrelated transcription factor AP-1 (Fos-Jun dimers) (Fig. 17.3D), thereby integrating two different signaling pathways, the calcium/ calcineurin pathway that activates NFAT and the phorbol ester-responsive MAP kinase pathway that promotes the synthesis and activation of Fos and Jun family proteins (Chen, 1999). Since the NFAT/AP-1 complex is not preformed before binding to DNA, the complex must undergo step-wise assembly at the promoter where the conformation of NFAT and the orientation of Fos-Jun on DNA become fixed (see above). The crystal structure of the NFAT/Fos-Jun/DNA complex formed on the ARRE2 site was solved (Chen, 1998). The NFAT/Fos:Jun complexes contact an ~15 basepair stretch of DNA, in which the NFAT and AP-1 elements are precisely apposed to each other. The residues involved in Fos-Jun contact are located largely in the N-terminal RHR domains of NFAT, and form an extensive network of mostly polar interactions with residues in the basic-leucine zipper regions of Fos and Jun (Chen, 1998). The interacting residues are almost completely conserved in NFAT1-4 but are absent from NFAT5 (Lopez-Rodriguez *et al.*, 1999), indicating that the ability to cooperate with Fos and Jun was a restricted to the calcium-responsive members of the NFAT family (NFAT1-4).

D: Complexes with Other Transcription Factors

Although NFAT is known to be essential for its target genes, including cytokine genes in T cells in response to antigen receptor activation, its activation alone does not seems to efficient for driving the expression of those genes. As such, NFAT does not belong to the category of master control genes. Consistent with this notion, NFAT often works in concert with other transcription factors to activate gene transcription, which allows for the integration of multiple signals that act on different transcription factors. NFAT engages in direct protein-protein interactions and/ or influences transcription synergistically with several families of transcription factors. These include the AP-1- like bZIP proteins such as Maf, ICER, (Bodor, 2000; Bower, 2002; Ho, 1996); the zinc finger proteins GATA and EGR (Decker, 2003; Decker, 1998; Molkentin, 1998); the helix-turn-helix domain proteins Oct, HNF3 and IRF-4 (Bert, 2000; Furstenau, 1999; Hu, 2002); the MADS-box protein MEF2 (Olson, 2000). For transcriptional partners other than AP-1, it is not known whether the synergy with NFAT occurs in the context of true "composite" regulatory elements that have a defined geometry and spatial orientation for cooperative binding of NFAT and these partner proteins. In the future, it will be interesting to analyze the structure and function of distinct NFAT transcription complexes to gain insight into the structural and functional versatility of NFAT.

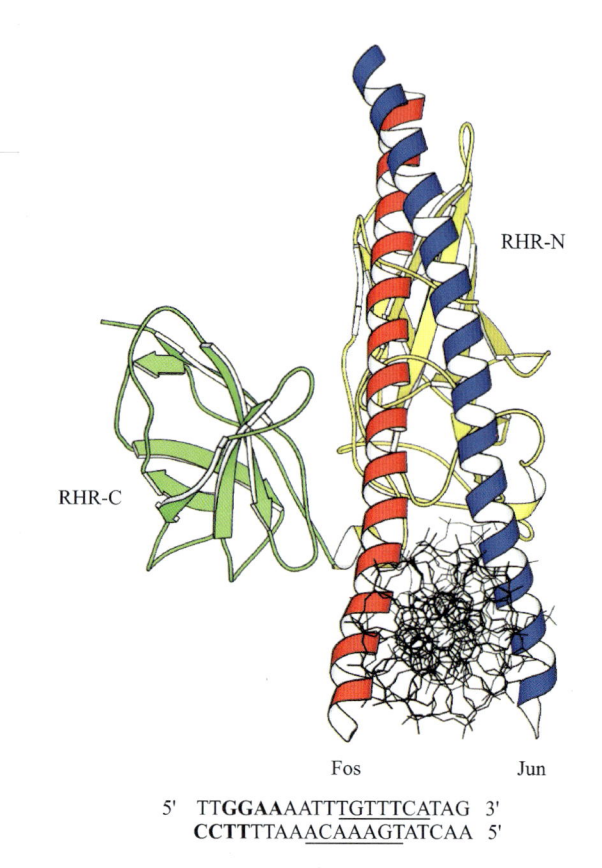

5' TTGGAAAATTTGTTTCATAG 3'
CCTTTTAAACAAAGTATCAA 5'

Fig.17.3D

In summary, systematic structural studies of NFAT complexes reveal that NFAT can bind DNA as monomers at cognate (GGAA) sites, as dimers at κB-like response elements, and as cooperative complexes with Fos-Jun at NFAT:AP-1 composite sites. In different complexes, NFAT adopts distinct conformations and its protein surface mediates distinct protein-protein interactions. This diversity of binding modes arises, at least in part, from the fact that NFAT is not an obligate dimer. The remarkable conformational flexibility of NFAT on different DNA sites is likely to facilitate the assembly of NFAT into distinct higher-order complexes containing diverse DNA-binding partners. Studies in the past a few years have expanded the function of the calcineurin/NFAT pathway to many medically important processes. These studies not only broaden the potential clinical application of drugs targeting the calcineurin/NFAT pathway, but also raise the question as to whether specific branches of the calcineurin/NFAT signaling pathway may be targeted for therapeutic benefits. Detailed studies of NFAT complexes to identify unique protein-protein interface in distinct NFAT complexes will help address this important question.

The Transactivation Domains: Interaction of NFAT with p300 and CBP

A common transactivation domain among all four isoforms of NFAT lies at the N-terminus, consisting of about 100 amino acids. Despite the lack of sequence similarity, contiguous stretches of acidic amino acids are found in the transactivation domains of different NFAT isoforms, which are likely to be involved in the recruitment of the basal transcription machinery to initiate transcription as has been shown for other acidic transactivation domains (Ruden et al., 1991). The C-terminus of NFAT has also been shown to be capable of activating transcription when fused to DNA binding domains of other transcription factors (Luo et al., 1996). In addition, the N-terminal transactivation domain of NFATp has been shown to interact with p300 and CBP for further enhance transcription (Garcia-Rodriguez and Rao, 1998). Thus, NFAT is likely to activate gene transcription by interacting with both chromatin remodeling enzymes and the basal transcription machinery.

Unconventional Modes of Transcription Regulation by NFAT as Repressors and Coactivators

Although NFAT has been shown to serve as a transcription activator for a number of genes from T cells to muscle cells, several lines of evidence suggest that it may also be involved in repressing gene expression. First, DNA microarray analyses revealed that the activation of the calcineurin-NFAT pathway led to both activation of certain genes and suppression of others, indicating that NFAT is capable of repressing gene expression (Diehn et al., 2002; Feske et al., 2001). Second, it has been shown that knockout of several isoforms of NFAT led to an enhanced, rather than decreased, production of cytokines genes, consistent with a repressive role of NFAT for gene expression (Hogan et al., 2003). Third, based on analyses of several NFAT-DNA complexes, it has been shown that NFAT is capable of binding to DNA in dramatically different conformations, making it possible for NFAT to switch partners that may be associated with either transcriptional corepressors or coactivators. It has been unambiguously shown that NFAT is responsible for the repression of cyclin-dependent kinase 4 expression in T cells (Baksh et al., 2002). NFAT binding sites were found in the promoter region of Cdk4. Activation of the calcineurin-NFAT pathway was found to inhibit Cdk4 gene activation, suggesting that NFAT plays a central role in controlling cell cycle during T cell activation. Consistent with this finding, Cdk4 expression was found to be upregulated in mouse embryofibroblasts derived from NFATp and calcineurinAα null animals.

That NFAT may serve as a pure coactivator independent of its DNA binding domain was found through the analysis of the activation of the Nur77 promoter during thymocyte apoptosis (Youn and Liu, 2000; Youn et al., 1999). TCR-dependent Nur77 expression requires calcineurin-NFAT pathway as judged by the sensitivity of Nur77 promoter to FK506 and cyclosporin A, but not rapamycin. Although the Nur77 promoter contains two consensus NFAT binding sites next to each of the two MEF2 binding sites, mutagenesis of the two NFAT-binding sequences had no effect on calcium-dependent Nur77-luciferase reporter gene activation (Youn et al., 2000a). Moreover, mutants of NFAT defective in DNA binding were equally active in causing Nur77 gene activation. These observations strongly suggest that in the context of Nur77 gene activation in thymocytes, NFAT serves as a pure coactivator. Indeed, NFATp was shown to bind to both MEF2, via its C-terminal transactivation domain, and p300, via its N-terminal transactivation domain (vide infra), to stabilize the MEF2-p300 complex, causing full activation of the Nur77 gene to cause thymocyte apoptosis upon calcium signaling.

General Features of MEF2

MEF2 is composed of a family of four isoforms encoded by four distinct genes, MEF2A-D (Olson et al., 1995). Similar to NFAT, each MEF2 gene can be expressed as multiple splicing variants. Although MEF2 was originally identified in muscle cells and was shown to play an important role in muscle cell differentiation, it was subsequently found to be ubiquitously expressed with each cell type expressing a subset of MEF2 isoforms. To date, MEF2 has been shown to regulate cell growth, differentiation and death. Like NFAT, MEF2 is also activated upon calcium signaling. Unlike NFAT, however, MEF2 can also be stimulated by a wide variety of other signals through phosphorylation or ubiquitination. Therefore, MEF2 is not a dedicated to respond to calcium signaling as NFAT.

MEF2 belongs to the superfamily of MADS box proteins, MADS being derived from MCM1, Agamous, Deficiens and Serum-response Factor, transcription factors involved in diverse cellular process from yeast to mammals. MEF2 can be divided into two structural domains, an N-terminal domain (MADS/MEF2S) domain and a C-terminal transactivation and regulatory domain. The MADS/MEF2S domain is responsible for DNA binding, dimerization and association with either HDACs or HATs. What distinguish MEF2 from other MADS superfamily members such as serum response factor is that MEF2 contains, immediately next to the MADS domain, a MEF2S extension that is unique and characteristic of MEF2 family members. This MEF2S extension not only eliminates potential interactions of MEF2 from other MADS domain partner proteins but also endowed MEF2 with the ability to interact with both HATs and HDACs, setting up the calcium-responsiveness within the MEF2 family members. Among the four different isoforms of MEF2, the N-terminal MADS/MEF2S domain is highly conserved. As MEF2 exists as obligatory dimers, this sequence conservation in the MADS/MEF2S domain allows different members of MEF2 expressed within the same cell type to form various combinations of homo- and heterodimers. In contrast to the highly conserved N-terminal MADS/MEF2S domain, the C-terminal domains are distinct between the different isoforms of MEF2. As the C-terminal domains are recipients of non-calcium signals for MEF2 activation, this allows for integration of other stimulation signals with calcium signaling by MEF2.

Interaction of MEF2 with HATs and HDACs

It has been shown that p300 binds to the N-terminal MADS/MEF2S domain, enhancing the transcriptional activity of MEF2. Interestingly, a superfamily of transcription corepressors was found to bind the MADS/MEF2S domain of MEF2 as well. They include Cabin1/cain (Lai et al., 1998; Sun et al., 1998), originally identified as a calcineurin-binding protein, and the entire family of class II HDACs (McKinsey et al., 2002). That both HATs and HDACs are capable of binding to the relatively small MADS/MEF2S domain of MEF2 raised the question of whether they bind to the same site in MEF2 and if so, how their binding to MEF2 is regulated under different physiological conditions. Using thymocytes as a model system, it was shown the binding of p300 and Cabin1 to MEF2 is mutually exclusive (Youn et al., 2000a). This is consistent with the model that Cabin1 as well as class II HDACs are associated with MEF2 when its activity is suppressed, and p300 replaces corepressors when MEF2 is activated.

Although Cabin1 does not possess intrinsic HDAC activity, it is capable of recruiting HDAC1 and 2 via another corepressor mSin3 (Youn et al., 2000b). Thus, Cabin1 and different members of the class II HDAC family are functionally redundant for suppression of MEF2 activity. Limited genetic and cell biological evidence suggests a cell-type specific role for different members of the MEF2 corepressor family. For example, the expression of Cabin1 is not seen in muscle cells, where there is abundant expression of class II HDACs, such as HDAC4 and 5. In thymocytes, it appears that Cabin1 plays negligible role in regulating MEF2-dependent expression of Nur77 while it seems to be critical for controlling MEF2 in peripheral T cells (Esau et al., 2001).

Regulation of MEF2/HDAC or HAT Interaction by Calcium Signaling

Different from NFAT, which undergoes calcium dependent cytosol-to-nucleus translocation upon calcium signaling, MEF2 is constitutively bound to its target DNA element in the nucleus regardless of activation status as judged by the results of chromatin immunoprecipitation-PCR assay (Pan et al., 2004; Youn and Liu, 2000). The calcium-dependent activation of MEF2 is mediated through the modulation of association of MEF2 corepressors and the calcium switches are built into the corepressors, including Cabin1 and class II HDACs.

There are at least two complementary and distinct mechanisms by which calcium signaling regulates the association of MEF2 with its corepressors. The first mechanism involves the binding of activated calmodulin to Cabin1 or class II HDACs and the consequent dissociation of these corepressors from MEF2 as a result of competition between active calmodulin and MEF2 for the corepressors (Liu, 2005) (Fig.17.4). The mutually exclusive binding of MEF2 and calmodulin to Cabin1 or class II HDACs are made possible by conferring to the same overlapping fragment of the corepressors the ability to bind to either MEF2 or calmodulin. The calcium-dependent dissociation of corepressors from MEF2 was first demonstrated for Cabin1 and was subsequently extended to HDAC4. By sequence comparison, it seems that this mechanism applied to all other members of class II HDAC (Berger et al., 2003; Youn et al., 2000b). This represents an unprecedented mechanism of direct signaling from

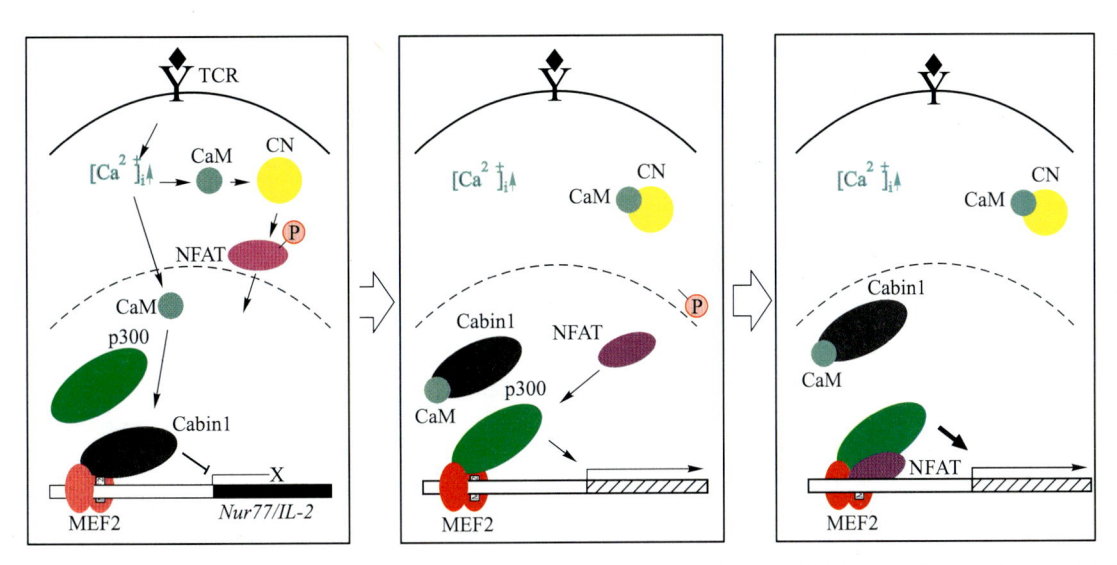

Fig.17.4 Regulation of MEF2 by calcium signal and integration of the calcineurin-NFAT signaling pathway with MEF2. In resting T cells, MEF2 recruits HDAC-containing repressors such as Cabin1 in the context of the IL2 and Nur77 promoter to repress the expression of the target genes. Upon calcium signaling, nuclear calmodulin directly bind to Cabin1 or class II HDACs (not shown) to dissociate them from MEF2 and allow for the association of p300. The MEF2-p300 complex can be further stabilized by NFAT, which forms a ternary complex with MEF2 and p300.

calcium to a nuclear chromatin remodeling complex to activate transcription.

A second mechanism of calcium-dependent dissociation of corepressors from MEF2 was through the CaM kinase-mediated and 14-3-3-dependent nuclear export of the corepressors. This was originally demonstrated for HDAC4 and HDAC5 in muscle cells (Lu et al., 2000; McKinsey et al., 2000). Thus, upon calcium signaling, calmudulin-dependent protein kinase II and IV are activated. They are capable of phosphorylating class II HDACs, creating docking sites for 14-3-3. Upon binding of 14-3-3 to class II HDACs, they undergo translocation from the nucleus to the cytosol. More recently, it was shown that Cabin1, like class II HDACs, also becomes phosphorylated by CaM kinase IV in human T cells upon activation, associates with 14-3-3 and exit the nucleus (F. P, J.O.L., and A. Means, unpublished). Thus, it appears that two independent mechanisms exist that ensures the dissociation of repressors from MEF2 and further exclusion of them away from MEF2 upon egression into the cytosol. In muscle cells, it has been shown that the kinase(s) responsible for class II HDAC nuclear export may not be a calmodulin-dependent kinase, thus allowing for the integration of other cellular signals into MEF2 for its full activation.

Once corepressors are removed from MEF2 upon calcium signaling, the vacant binding site on MEF2 for the corepressors will become available for the binding of coactivators such as p300. The switch from corepressors to coactivators is a dynamic yet ordered process, so that there is no aberrant p300 binding to MEF2 in the absence of a stimulation signal. It is extraordinary that the relatively small N-terminal SMADS/MEF2S domain could be responsible for both DNA binding and for interaction with both HATs and HDACs. The secret was revealed upon the attainment of a high-resolution crystal structure of MEF2 in complex with both DNA and the MEF2 binding peptide derived from Cabin1.

The crystal structure of the MADS-box/MEF2S domain of human MEF2B bound to the MEF2-binding motif of Cabin1 and DNA has been published (Han et al., 2003) (Fig. 17.5A). This structure provides the first view of co-repressor recruitment by MEF2. The crystal structure reveals a stably folded MEF2S domain (helices H2, H3 and strand S3) on top of the MADS-box (H1, S1 and S2). Cabin1 adopts an amphipathic-helix (red) that binds a hydrophobic groove on the MEF2S domain. The protein surface of MEF2 is made up primarily by the MEF2S domain, which forms a concave hydrophobic pocket resembling the antigen peptide-binding site of the Major Histocompatibility Complex (MHC). This hydrophobic pocket may act a signaling module to bind a variety of protein factors including both co-repressors and co-activators that function in complex with MEF2.

More recently, the complex between the MEF2-binding motif of class II HDACs and the MADS-box/MEF2S domain of MEF2B has been characterized by structural and biochemical methods (Han et al., 2005). The crystal structure of an HDAC9/MEF2/DNA

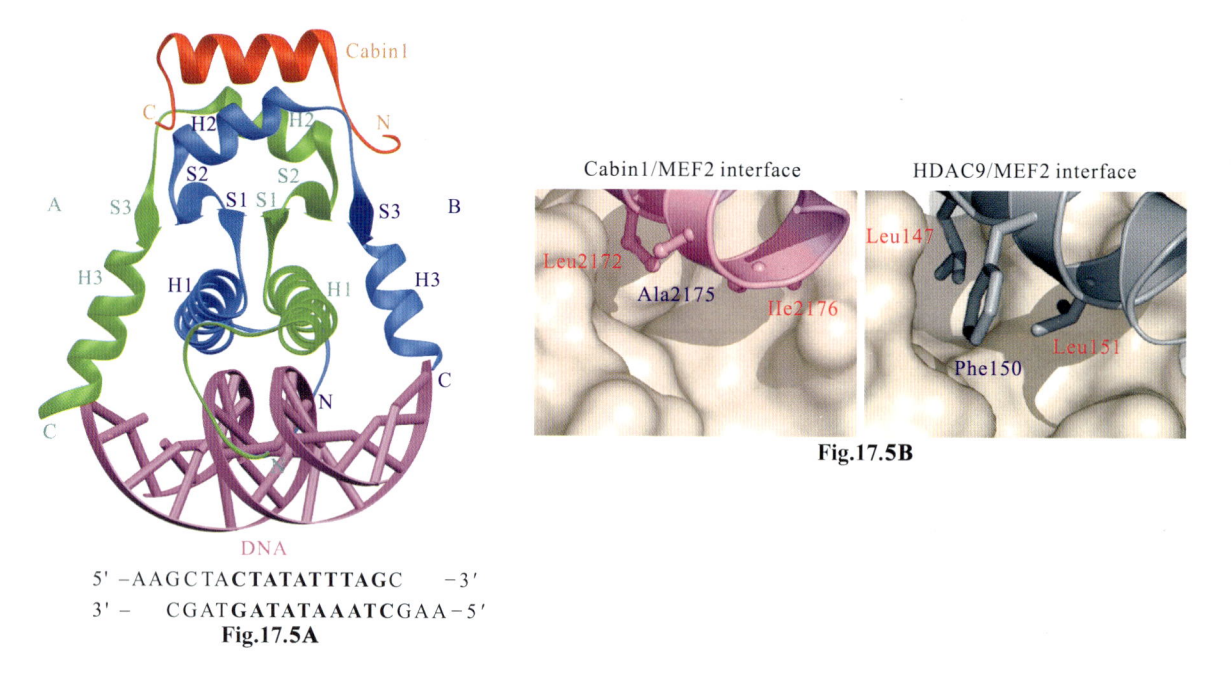

Fig. 17.5 The crystal structures of the co-repressor/ MEF2/DNA complex. (A) The overall structure of the Cabin1/MEF2/DNA complex. The MEF2-binding motif of Cabin1 (red), MEF2B (monomer A in green, monomer B in blue) and DNA (magenta) in the complex are colored differently. The DNA sequence used in the crystal is listed below, with the MEF2 binding site in Bold. (B) A key difference between the Cabin1/MEF2 (left panel) and the HDAC9/MEF2 (right panel) interface. At the Cabin1/MEF2 interface, Ala2175 does not fill up a hydrophobic pocket of the MEF2 surface, whereas Phe150 in HDAC9 fit snugly into this pocket at the HDAC9/MEF2 interface.

complex reveals that HDAC9 binds to a hydrophobic groove of the MEF2 dimer. The overall binding mode is similar to that seen in the Cabin1/MEF2/DNA complex. The detailed binding interactions at the HDAC9/MEF2 interface, however, show marked differences from those at the Cabin1/MEF2 interface (Fig.17.5B). These studies further support a general mechanism by which class II HDACs and possibly other transcriptional co-regulators are recruited by MEF2. On the other hand, the differential binding between MEF2 and its various partners may confer specific regulatory and functional properties to MEF2 in distinct cellular processes.

Interaction between MEF2 and other Transcription Factors

Similar to NFAT, MEF2 is capable of interacting with a number of unrelated transcription factors to cooperatively drive target gene expression. In muscle cells, it has been shown that MEF2 interacts with myogenic bHLH proteins such as MyoD to induce muscle cell differentiation (Black *et al.*, 1998). As discussed earlier in this chapter, MEF2 also binds NFAT and form a ternary complex with NFAT and p300 on the Nur77 promoter to cause full activation of Nur77 expression during thymocyte apoptosis (Youn *et al.*, 2000a) (Fig. 17.4). Surprisingly, the NFAT-interacting domain was confined to the N-terminal MADS/MEF2S domain, which also interact with both DNA and p300. It will be interesting to see how structurally this relatively small domain accommodates both NFAT and p300 for the formation of a ternary complex on DNA.

Function and Regulation of the Transactivation Domain of MEF2

Although the four isoforms of MEF2 diverge in sequence in the C-terminal transactivation domains, some motifs are conserved. Contiguous acidic amino acid stretches characteristic of eukaryotic transactivation domains are found in all forms of MEF2, although they are sometimes spliced out in certain splice isoforms. These acidic blobs are likely to play a part in the transactivation of target genes by MEF2. In addition, the transactivation domains of different isoforms of MEF2 contain the consensus phophsorylation sites for the MAP kinases including p38 and ERK5. It has been shown that phosphorylation by these kinases further enhances the transactivation activity of MEF2. It is interesting to note that the MADS/MEF2S domain contains the binding sites for p300. Therefore, it is likely that MEF2-bound p300 will work in concert with the transactivation domain to cause transcriptional activation.

Perspectives

How the second messenger calcium signals from the cytosol into the nucleus to cause a transcriptional output can be achieved in different ways. Although both NFAT and MEF2 are calcium-responsive transcription factors, the ways they respond to calcium signaling are significantly different, even though both are mediated through the immediate calcium sensor calmodulin. For NFAT, the calcium switch is built into its upstream phosphatase calcineurin. It is the C-terminal autoinhibitory domain and calmodulin-binding domain, which together forms the on-and-off switch for calcium signaling. Once calcineurin is activated, it dephosphorylates NFAT to enable it to translocate into the nucleus. It is through compartmental segregation that NFAT is kept inactive in the absence of calcium signaling. For MEF2, however, the calcium switch is embedded in its associated transcriptional corepressors, be it Cabin1 or class II HDACs. It is through calcium and calmodulin-dependent dissociation of the HDAC-containing corepressors and the consequent association of HATs that calcium regulates the transcription of MEF2. Although it took different paths to design the two calcium-sensing transcription/chromatin remodeling modules, the same endpoint—calcium-dependent transcriptional activation—was achieved.

References

Aramburu, J., Garcia-Cozar, F., Raghavan, A., Okamura, H., Rao, A., and Hogan, P. G. (1998). Selective inhibition of NFAT activation by a peptide spanning the calcineurin targeting site of NFAT. Mol Cell *1*, 627-637.

Aramburu, J., Yaffe, M. B., Lopez-Rodriguez, C., Cantley, L. C., Hogan, P. G., and Rao, A. (1999). Affinity-driven peptide selection of an NFAT inhibitor more selective than cyclosporin A. Science *285*, 2129-2133.

Baksh, S., Widlund, H. R., Frazer-Abel, A. A., Du, J., Fosmire, S., Fisher, D. E., DeCaprio, J. A., Modiano, J. F., and Burakoff, S. J. (2002). NFATc2-mediated repression of cyclin-dependent kinase 4 expression. Mol Cell *10*, 1071-1081.

Beals, C. R., Sheridan, C. M., Turck, C. W., Gardner, P., and Crabtree, G. R. (1997). Nuclear export of NF-ATc enhanced by glycogen synthase kinase-3. Science *275*, 1930-1934.

Berger, I., Bieniossek, C., Schaffitzel, C., Hassler, M., Santelli, E., and Richmond, T. J. (2003). Direct interaction of Ca2+/calmodulin inhibits histone deacetylase 5 repressor core binding to myocyte enhancer factor 2. J Biol Chem *278*, 17625-17635.

Bert, A. G., Burrows, J., Hawwari, A., Vadas, M. A., and Cockerill, P. N. (2000). Reconstitution of T cell-specific transcription directed by composite NFAT/Oct elements. J Immunol *165*, 5646-5655.

Black, B. L., Molkentin, J. D., and Olson, E. N. (1998). Multiple roles for the MyoD basic region in transmission of transcriptional activation signals and interaction with MEF2. Mol Cell Biol *18*, 69-77.

Bodor, J., Bodorova, J., and Gress, R. E. (2000). Suppression of T cell function: a potential role for transcriptional repressor ICER. J Leukoc Biol *67*, 774-779.

Bower, K. E., Zeller, R. W., Wachsman, W., Martinez, T., and McGuire, K. L. (2002). Correlation of transcriptional repression by p21(SNFT) with changes in DNA.NF-AT complex interactions. J Biol Chem *277*, 34967-34977.

Chen, F. E., Huang, D. B., Chen, Y. Q., and Ghosh, G. (1998). Crystal structure of p50/p65 heterodimer of transcription factor NF-kappaB bound to DNA. Nature *391*, 410-413.

Chen, L., Oakley, M. G., Glover, J. N., Jain, J., Dervan, P. B., Hogan, P. G., Rao, A., and Verdine, G. L. (1995). Only one of the two DNA-bound orientations of AP-1 found in solution cooperates with NFATp. Curr Biol *5*, 882-889.

Chen, L., Rao, A., and Harrison, S. C. (1999). Signal integration by transcription-factor assemblies: interactions of NF-AT1 and AP-1 on the IL-2 promoter. Cold Spring Harb Symp Quant Biol *64*, 527-531.

Chow, C. W., and Davis, R. J. (2000). Integration of calcium and cyclic AMP signaling pathways by 14-3-3. Mol Cell Biol *20*, 702-712.

Chow, C. W., Rincon, M., Cavanagh, J., Dickens, M., and Davis, R. J. (1997). Nuclear accumulation of NFAT4 opposed by the JNK signal transduction pathway. Science *278*, 1638-1641.

Chytil, M., and Verdine, G. L. (1996). The Rel family of eukaryotic transcription factors. Curr Opin Struct Biol *6*, 91-100.

Crabtree, G. R., and Olson, E. N. (2002). NFAT signaling: choreographing the social lives of cells. Cell *109*, S67-79.

Decker, E. L., Nehmann, N., Kampen, E., Eibel, H., Zipfel, P. F., and Skerka, C. (2003). Early growth response proteins (EGR) and nuclear factors of activated T cells (NFAT) form heterodimers and regulate proinflammatory cytokine gene expression. Nucleic Acids Res *31*, 911-921.

Decker, E. L., Skerka, C., and Zipfel, P. F. (1998). The early growth response protein (EGR-1) regulates interleukin-2 transcription by synergistic interaction with the nuclear factor of activated T cells. J Biol Chem *273*, 26923-26930.

Diehn, M., Alizadeh, A. A., Rando, O. J., Liu, C. L., Stankunas, K., Botstein, D., Crabtree, G. R., and Brown, P. O. (2002). Genomic expression programs and the integration of the CD28 costimulatory signal in T cell activation. Proc Natl Acad Sci USA *99*, 11796-11801.

Esau, C., Boes, M., Youn, H. D., Tatterson, L., Liu, J. O., and Chen, J. (2001). Deletion of calcineurin and myocyte enhancer factor 2 (MEF2) binding domain of Cabin1 results in enhanced cytokine gene expression in T cells. J Exp Med *194*, 1449-1459.

Feske, S., Giltnane, J., Dolmetsch, R., Staudt, L. M., and Rao, A. (2001). Gene regulation mediated by calcium signals in T lymphocytes. Nat Immunol *2*, 316-324.

Fuentes, J. J., Genesca, L., Kingsbury, T. J., Cunningham, K. W., Perez-Riba, M., Estivill, X., and de la Luna, S. (2000). DSCR1, overexpressed in Down syndrome, is an inhibitor of calcineurin-mediated signaling pathways. Hum Mol Genet *9*, 1681-1690.

Furstenau, U., Schwaninger, M., Blume, R., Jendrusch, E. M., and Knepel, W. (1999). Characterization of a novel calcium response element in the glucagon gene. J Biol Chem *274*, 5851-5860.

Garcia-Rodriguez, C., and Rao, A. (1998). Nuclear factor of activated T cells (NFAT)-dependent transactivation regulated by the coactivators p300/CREB-binding protein (CBP). J Exp Med *187*, 2031-2036.

Ghosh, G., van Duyne, G., Ghosh, S., and Sigler, P. B. (1995). Structure of NF-kappa B p50 homodimer bound to a kappa B site. Nature *373*, 303-310.

Giffin, M. J., Stroud, J. C., Bates, D. L., von Koenig, K. D., Hardin, J., and Chen, L. (2003). Structure of NFAT1 bound as a dimer to the HIV-1 LTR kappa B element. Nat Struct Biol *10*, 800-806.

Gomez del Arco, P., Martinez-Martinez, S., Maldonado, J. L., Ortega-Perez, I., and Redondo, J. M. (2000). A role for the p38 MAP kinase pathway in the nuclear shuttling of NFATp. J Biol Chem *275*, 13872-13878.

Graef, I. A., Chen, F., and Crabtree, G. R. (2001). NFAT signaling in vertebrate development. Curr Opin Genet Dev *11*, 505-512.

Han, A., He, J., Wu, Y., Liu, J. O., and Chen, L. (2005). Mechanism of recruitment of class II histone deacetylases by myocyte enhancer factor-2. J Mol Biol *345*, 91-102.

Han, A., Pan, F., Stroud, J. C., Youn, H. D., Liu, J. O., and Chen, L. (2003). Sequence-specific recruitment of transcriptional co-repressor Cabin1 by myocyte enhancer factor-2. Nature *422*, 730-734.

Ho, I. C., Hodge, M. R., Rooney, J. W., and Glimcher, L. H. (1996). The proto-oncogene c-maf is responsible for tissue-specific expression of interleukin-4. Cell *85*, 973-983.

Ho, S., Timmerman, L., Northrop, J., and Crabtree, G. R. (1994). Cloning and characterization of NF-ATc and NF-ATp: the cytoplasmic components of NF-AT. Adv Exp Med Biol *365*, 167-173.

Hoey, T., Sun, Y. L., Williamson, K., and Xu, X. (1995). Isolation of two new members of the NF-AT gene family and functional characterization of the NF-AT proteins. Immunity *2*, 461-472.

Hogan, P. G., Chen, L., Nardone, J., and Rao, A. (2003). Transcriptional regulation by calcium, calcineurin, and NFAT. Genes Dev *17*, 2205-2232.

Hu, C. M., Jang, S. Y., Fanzo, J. C., and Pernis, A. B. (2002). Modulation of T cell cytokine production by interferon regulatory factor-4. J Biol Chem *277*, 49238-49246.

Huxford, T., Huang, D. B., Malek, S., and Ghosh, G. (1998). The crystal structure of the IkappaBalpha/NF-kappaB complex reveals mechanisms of NF-kappaB inactivation. Cell *95*, 759-770.

Jacobs, M. D., and Harrison, S. C. (1998). Structure of an IkappaBalpha/NF-kappaB complex. Cell *95*, 749-758.

Jiang, H., Xiong, F., Kong, S., Ogawa, T., Kobayashi, M., and Liu, J. O. (1997). Distinct tissue and cellular distribution of two major isoforms of calcineurin. Mol Immunol *34*, 663-669.

Jin, L., Sliz, P., Chen, L., Macian, F., Rao, A., Hogan, P. G., and Harrison, S. C. (2003). An asymmetric NFAT1 dimer on a pseudo-palindromic kappa B-like DNA site. Nat Struct Biol *10*, 807-811.

Kingsbury, T. J., and Cunningham, K. W. (2000). A conserved family of calcineurin regulators. Genes Dev *14*, 1595-1604.

Klee, C. B., Ren, H., and Wang, X. (1998). Regulation of the calmodulin-stimulated protein phosphatase, calcineurin. J Biol Chem *273*, 13367-13370.

Lai, M. M., Burnett, P. E., Wolosker, H., Blackshaw, S., and Snyder, S. H. (1998). Cain, a novel physiologic protein inhibitor of calcineurin. J Biol Chem *273*, 18325-18331.

Leung, T. H., Hoffmann, A., and Baltimore, D. (2004). One nucleotide in a kappaB site can determine cofactor specificity for NF-kappaB dimers. Cell *118*, 453-464.

Li, H., Rao, A., and Hogan, P. G. (2004). Structural delineation of the calcineurin-NFAT interaction and its parallels to PP1 targeting interactions. J Mol Biol *342*, 1659-1674.

Liu, J., Arai, K., and Arai, N. (2001). Inhibition of NFATx activation by an oligopeptide: disrupting the interaction of NFATx with calcineurin. J Immunol *167*, 2677-2687.

Liu, J. O. (2003). Endogenous protein inhibitors of calcineurin. Biochem Biophys Res Commun *311*, 1103-1109.

Liu, J. O. (2005). The yins of T cell activation. Sci STKE *2005*, re1.

Lopez-Rodriguez, C., Aramburu, J., Rakeman, A. S., and Rao, A. (1999). NFAT5, a constitutively nuclear NFAT protein that does not cooperate with Fos and Jun. Proc Natl Acad Sci USA *96*, 7214-7219.

Lopez-Rodriguez, C., Aramburu, J., Jin, L., Rakeman, A. S., Michino, M., and Rao, A. (2001). Bridging the NFAT and NF-kappaB families: NFAT5 dimerization regulates cytokine gene transcription in response to osmotic stress. Immunity *15*, 47-58.

Lu, J., McKinsey, T. A., Nicol, R. L., and Olson, E. N. (2000). Signal-dependent activation of the MEF2 transcription factor by dissociation from histone deacetylases. Proc Natl Acad Sci USA *97*, 4070-4075.

Luo, C., Burgeon, E., and Rao, A. (1996). Mechanisms of transactivation by nuclear factor of activated T cells-1. J Exp Med *184*, 141-147.

Macian, F., Lopez-Rodriguez, C., and Rao, A. (2001). Partners in transcription: NFAT and AP-1. Oncogene *20*, 2476-2489.

McCaffrey, P. G., Luo, C., Kerppola, T. K., Jain, J., Badalian, T. M., Ho, A. M., Burgeon, E., Lane, W. S., Lambert, J. N., Curran, T., and et al. (1993). Isolation of the cyclosporin-sensitive T cell transcription factor NFATp. Science *262*, 750-754.

McKinsey, T. A., Zhang, C. L., Lu, J., and Olson, E. N. (2000). Signal-dependent nuclear export of a histone deacetylase regulates muscle differentiation. Nature *408*, 106-111.

McKinsey, T. A., Zhang, C. L., and Olson, E. N. (2002). MEF2: a calcium-dependent regulator of cell division, differentiation and death. Trends Biochem Sci *27*, 40-47.

Miyakawa, H., Rim, J. S., Handler, J. S., and Kwon, H. M. (1999). Identification of the second tonicity-responsive enhancer for the betaine transporter (BGT1) gene. Biochim Biophys Acta *1446*, 359-364.

Molkentin, J. D., Lu, J. R., Antos, C. L., Markham, B., Richardson, J., Robbins, J., Grant, S. R., and Olson, E. N. (1998). A calcineurin-dependent transcriptional pathway for cardiac hypertrophy. Cell *93*, 215-228.

Muller, C. W., Rey, F. A., Sodeoka, M., Verdine, G. L., and Harrison, S. C. (1995). Structure of the NF-kappa B p50 homodimer bound to DNA. Nature *373*, 311-317.

Neilson, J., Stankunas, K., and Crabtree, G. R. (2001). Monitoring the duration of antigen-receptor occupancy by calcineurin/glycogen-synthase-kinase-3 control of NF-AT nuclear shuttling. Curr Opin Immunol *13*, 346-350.

Northrop, J. P., Ho, S. N., Chen, L., Thomas, D. J., Timmerman, L. A., Nolan, G. P., Admon, A., and Crabtree, G. R. (1994). NF-AT components define a family of transcription factors targeted in T-cell activation. Nature *369*, 497-502.

Okamura, H., Aramburu, J., Garcia-Rodriguez, C., Viola, J. P., Raghavan, A., Tahiliani, M., Zhang, X., Qin, J., Hogan, P. G., and Rao, A. (2000). Concerted dephosphorylation of the transcription factor NFAT1 induces a conformational switch that regulates transcriptional activity. Mol Cell *6*, 539-550.

Olson, E. N., and Williams, R. S. (2000). Calcineurin signaling and muscle remodeling. Cell *101*, 689-692.

Olson, E. N., Perry, M., and Schulz, R. A. (1995). Regulation of muscle differentiation by the MEF2 family of MADS box transcription factors. Dev Biol *172*, 2-14.

Pan, F., Ye, Z., Cheng, L., and Liu, J. O. (2004). Myocyte enhancer factor 2 mediates calcium-dependent transcription of the interleukin-2 gene in T lymphocytes: a calcium signaling module that is distinct from but collaborates with the nuclear factor of activated T cells (NFAT). J Biol Chem *279*, 14477-14480.

Rao, A., Luo, C., and Hogan, P. G. (1997). Transcription factors of the NFAT family: regulation and function. Annu Rev Immunol *15*, 707-747.

Roehrl, M. H., Kang, S., Aramburu, J., Wagner, G., Rao, A., and Hogan, P. G. (2004). Selective inhibition of calcineurin-NFAT signaling by blocking protein-protein interaction with small organic molecules. Proc Natl Acad Sci USA *101*, 7554-7559.

Rothermel, B. A., Vega, R. B., and Williams, R. S. (2003). The role of modulatory calcineurin-interacting proteins in calcineurin signaling. Trends Cardiovasc Med *13*, 15-21.

Ruden, D. M., Ma, J., Li, Y., Wood, K., and Ptashne, M. (1991). Generating yeast transcriptional activators containing no yeast protein sequences. Nature *350*, 250-252.

Shaw, J. P., Utz, P. J., Durand, D. B., Toole, J. J., Emmel, E. A., and Crabtree, G. R. (1988). Identification of a putative regulator of early T cell activation genes. Science *241*, 202-205.

Sheridan, C. M., Heist, E. K., Beals, C. R., Crabtree, G. R., and Gardner, P. (2002). Protein kinase A negatively modulates the nuclear accumulation of NF-ATc1 by priming for subsequent phosphorylation by glycogen synthase kinase-3. J Biol Chem *277*, 48664-48676.

Stroud, J. C., and Chen, L. (2003). Structure of NFAT bound to DNA as a monomer. J Mol Biol *334*, 1009-1022.

Sun, L., Youn, H. D., Loh, C., Stolow, M., He, W., and Liu, J. O. (1998). Cabin 1, a negative regulator for calcineurin signaling in T lymphocytes. Immunity *8*, 703-711.

Winslow, M. M., and Crabtree, G. R. (2005). Immunology. Decoding calcium signaling. Science *307*, 56-57.

Yang, T. T., Xiong, Q., Enslen, H., Davis, R. J., and Chow, C. W. (2002). Phosphorylation of NFATc4 by p38 mitogen-activated protein kinases. Mol Cell Biol *22*, 3892-3904.

Youn, H. D., Chatila, T. A., and Liu, J. O. (2000a). Integration of calcineurin and MEF2 signals by the coactivator p300 during T-cell apoptosis. EMBO J *19*, 4323-4331.

Youn, H. D., Grozinger, C. M., and Liu, J. O. (2000b). Calcium regulates transcriptional repression of myocyte enhancer factor 2 by histone deacetylase 4. J Biol Chem *275*, 22563-22567.

Youn, H. D., and Liu, J. O. (2000). Cabin1 represses MEF2-dependent Nur77 expression and T cell apoptosis by controlling association of histone deacetylases and acetylases with MEF2. Immunity *13*, 85-94.

Youn, H. D., Sun, L., Prywes, R., and Liu, J. O. (1999). Apoptosis of T cells mediated by Ca^{2+}-induced release of the transcription factor MEF2. Science *286*, 790-793.

Zhu, J., and McKeon, F. (1999). NF-AT activation requires suppression of Crm1-dependent export by calcineurin. Nature *398*, 256-260.

Zhu, J., and McKeon, F. (2000). Nucleocytoplasmic shuttling and the control of NF-AT signaling. Cell Mol Life Sci *57*, 411-420.

Zhu, J., Shibasaki, F., Price, R., Guillemot, J. C., Yano, T., Dotsch, V., Wagner, G., Ferrara, P., and McKeon, F. (1998). Intramolecular masking of nuclear import signal on NF-AT4 by casein kinase I and MEKK1. Cell *93*, 851-861.

Chapter 18
Hox Genes

S. Steven Potter

Division of Developmental Biology, Children's Hospital Medical Center, 3333 Burnet Ave., Cincinnati OH, 45229

Key Words: homeotic transformations, Hox clusters, Hox cofactors, Hox protein functional specificity, Hox gene targets, homeobox

Summary

The Hox genes were first found in *Drosophila*, where mutations in these genes often resulting in dramatic transformations of one body part into another. The Hox genes encode transcription factors with a DNA binding homeodomain carrying a helix-turn-helix motif. Several features of Hox genes are well conserved during evolution, including their clustered organization, their encoded homeodomain amino acid sequences, and their colinear expression, with the position of a gene in a cluster reflecting its domain of expression in the embryo. In *Drosophila* one important function of the Hox genes is to determine segment identity. The Hox encoded proteins, with their similar DNA binding preferences, achieve functional specificity at least in part through interactions with multiple cofactors. Hox genes appear to occupy high level positions in the genetic hierarchy of development.

A Brief History of the Homeobox

The homeobox genes are among the master switch genetic regulators of development. This was apparent from their initial discovery, when *Drosophila* geneticists found mutant flies with quite remarkable phenotypes. For example, mutation of the Antennapedia gene can result in a fly with legs found on the head in place of the antennae. In this case the imaginal discs that would normally give rise to antennae go down the wrong developmental pathway and form legs, which now protrude from the head. Other mutations, which clustered near Antennapedia, also gave dramatic transformations of body parts. These changes in developmental destiny were termed homeotic transformations, because they result in the conversion of the normal structure into a distinct, yet evolutionarily homologous structure. For example, one appendage, the antenna, is transformed into another appendage, the leg.

The cloning, and eventual sequencing of multiple homeotic genes from *Drosophila* showed that they carried a highly conserved 180 bp block of sequence, which was named the homeobox (McGinnis *et al.*, 1984a). This sequence was conserved among *Drosophila* homeotic genes, and also phylogenetically. It was used as a probe to rapidly clone orthologous (corresponding) homeobox genes from many species, including frog, mouse and man (McGinnis *et al.*, 1984b). The homeobox was observed to encode the 60 amino acid homeodomain, with a helix-turn-helix DNA binding motif previously observed in prokaryotic transcription factors. These findings began to explain the ability of these genes to control developmental fates. These genes encoded proteins that could regulate other genes, some of which might represent additional regulators. The homeobox genes therefore appeared to occupy high-level positions in the genetic hierarchy of development. At least in some cases they were capable of initiating genetic cascades that could drive the developmental destinies of groups of cells.

Groupings of the Homeobox Genes

The complete sequence of the mouse genome reveals the presence of several hundred genes with a homeobox. The 39 Antennapedia-like mammalian homeobox genes orthologous to the *Drosophila* homeotic genes are arranged in four clusters, (A, B, C and D) and are

Corresponding Author: Tel: (513) 636-4850, Fax: (513) 636-4317, E-mail: steve.potter@cchmc.org

termed the Hox genes. In addition there are other clusters of homeobox genes. The Rhox cluster of 12 homeobox genes is located on the X chromosome and is expressed in reproductive tissues, including testis, ovary and epididymis, with these genes therefore named reproductive homeobox, or Rhox (Maclean *et al.*, 2005). The Rhox genes encode proteins with divergent DNA binding homeodomains, suggesting that they recognize downstream target genes that are distinct from other homeobox genes. The ParaHox cluster includes Cdx, Xlox and Gsx genes (Pollard and Holland, 2000), while the 93D/E (NKL) cluster has additional homeobox genes. These non-Hox clustered homeobox genes perform diverse functions during development.

The homeodomain is often found associated with other functional domains in proteins. For example the POU genes encode proteins with both a homeodomain and a POU domain. Many of the Pax gene encoded proteins have both a Paired domain and a homeodomain. Other proteins carry both a homeodomain and a Lim domain, or a homeodomain and a zinc finger domain. Genes encoding these proteins are most often dispersed in chromosomal position, and the specific genes are often named after their *Drosophila* orthologs. This chapter is focused on the Hox genes.

The Organization and Evolutionary Conservation of the Hox Genes

The clustered chromosomal arrangement of the homeotic genes suggested early on that they evolved by tandem duplication of a single original gene, likely by unequal crossing over. In *Drosophila* the resulting gene cluster has been split, making the Antennapedia and Bithorax complexes. In mammals, however, the full cluster has been quadruplicated to give a total of 39 genes in four clusters (A-D) at four different chromosomal locations (Fig.18.1).

The quadruplication of a single initial cluster results in interesting sequence relationships among the Hox genes. A single gene on the original cluster can give rise to four progeny genes on the resulting four clusters. Such genes, derived from a single ancestor, show the most closely related sequences and are termed paralogs. These genes are shown vertically aligned in Fig. 18.1. By examining the sequences of the Hox genes it is possible to divide them into 13 such paralogous groups.

The mammalian Hox genes are sensibly named in matrix fashion according to the cluster, A-D, and paralogous group, 1-13. For example, the Hoxa 11 gene is from paralogous group 11 on cluster A. The *Drosophila* orthologs, however, from a single split cluster, are all given individual names, labial (lab), proboscopoedia (Pb), deformed (dfd), sex combs reduced (Scr), Antennapedia (Antp), ultrabithorax (Ubx), abdominal A (Abd-A), and abdominal B (Abd-B).

While paralogs are most closely related at the sequence level, there is also a strong tendency for genes flanking each other on a cluster to be more closely related than genes more distant. For example, paralogous

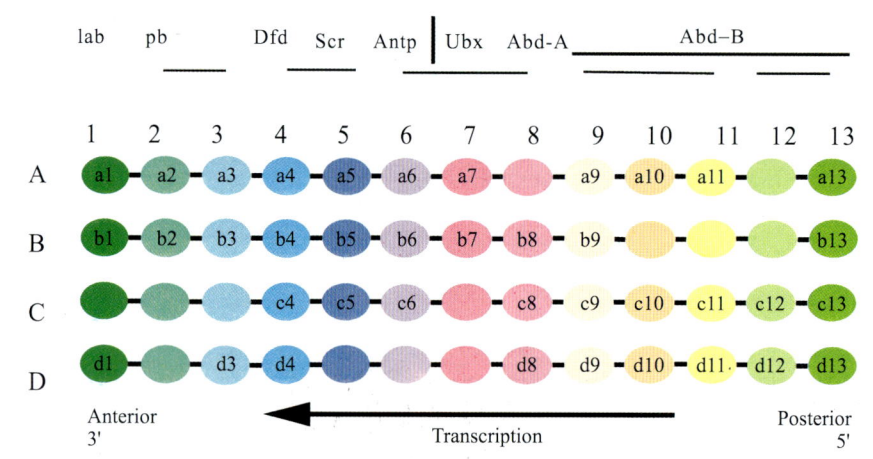

Fig.18.1 The mammalian Hox genes. The four clusters are marked A, B, C and D. The 39 Hox genes are divided into 13 paralogous groups, with the numbers shown above the clusters. Genes within each paralogous groups are given a single color, to indicate their close relationship. Individual genes are named according to cluster and paralogous group. Blank ovals represent genes apparently lost during the duplication process. The 5′ genes are shown at the right and the 3′ cluster genes are shown at the left, according to custom, but reverse of the standard for most genes. The genes are transcribed right to left. The *drosophila* orthologs are shown at the top. These are divided into two clusters by the vertical line. Mammalian paralogous groups 9-13 are considered Abd-B like. Sequence comparisons of the encoded homeodomains allows the paralogous groups to be subdivided into the more closely related 2,3 and 4,5 and 6-8 and 9-11 and 12,13, as marked by the thin horizontal lines. The figure shown is correct for both mouse and man Hox clusters.

groups 9-13 are all classified as Abd-B type, as they are most closely related to the single Abd-B *Drosophila* gene. There has been an interesting expansion of this group from a single gene in *Drosophila* to 15 genes in mammals. Hierarchical cluster analysis of sequence relationships, however, shows that within this group the paralogs 9, 10 and 11 are much more closely related to each other than to 12 and 13. Similarly, paralogs 6, 7 and 8 form another closely related subgroup, as do 2 and 3 and also 4 and 5.

The most conserved regions of the Hox encoded proteins are the homeodomains. Within paralogous groups this homology is quite strong. For example, the mouse Hoxa 4 homeodomain differs from that of Hoxd 4, Hoxc 4 and Hoxb 4 by only two, three and four amino acids respectively, and the orthologous *Drosophila* Deformed homeodomain is different at only seven amino acids out of 60.

There is also significant conservation of sequence flanking the homeodomain within a paralogous group. For example, the Hoxa 4 and Hoxb 4 encoded proteins show amino acid sequence identity at 19/22 residues immediately amino terminal and at 15/21 residues immediately carboxy terminal of the homeodomain. Outside of these regions the genes show relatively little homology.

The mammalian Hox genes are generally quite simple in structure, most often consisting of only two exons separated by a small intron. An entire cluster of approximately 10 Hox genes may comprise only about 150 Kb of DNA. Interestingly, the Hox genes in *Drosophila* are much larger and more complex in their organization, This is the reverse of what is seen for most genes, where simpler organization in *Drosophila* is the general rule.

There are several other interesting features of the mammalian homeobox clusters to consider, although their importance remains a matter of conjecture. First, transcripts spanning multiple Hox genes have been observed. The presence of such cross-Hox mRNAs suggests that at least in some cases alternate processing can combine exons of different homeobox genes for unique functions. In addition there is a conspicuous absence of repetitive sequences in the homeobox clusters. It appears that almost no insertions of mobile DNA sequences are tolerated in these clusters. This argues for strong function for both their intergenic and intronic sequences. This is also suggested by the evolutionary conservation of these noncoding sequences among the mammalian clusters.

These observations provide a possible explanation for the conserved clustered organization of the Hox genes. The relative absence of repetitive elements argues that the integrity of even noncoding regions is important, and the break up of a Hox cluster would destroy this integrity. A common interpretation of the data is that Hox genes extensively share cis-regulatory elements dispersed throughout the clusters, and that this results in strong selective pressure to conserve the clustered arrangement. This topic is touched again later in the chapter.

It should be noted that in *Drosophila* the Hox cluster has been split in two, and that in other simpler organisms the Hox clusters have experienced even more severe fragmentation. It is clear that principles that apply to the mammalian Hox genes are not always universally valid on a wider evolutionary scale.

Colinearity

One of the most interesting evolutionarily conserved features of Hox genes is referred to as colinearity. In *Drosophila*, mouse and man the position of a Hox gene within a cluster shows a striking colinear relationship to the anterior-posterior expression pattern in the developing embryo. Genes at the more 3' positions in the clusters are expressed at earlier developmental times and at more anterior (rostral) positions. In contrast, genes at more 5' positions in clusters are expressed later in development and show more posterior (caudal) domains of expression. For example, the Abd-B type Hox genes, from paralog groups 9-13, are at the extreme 5' ends of the clusters, and show the most posterior restricted expression. It is also interesting to note that the clustered Hox genes are, with few exceptions, transcribed from the same DNA strand, so they are colinear in this sense as well.

Functions of Hox Genes

The functions of Hox genes are perhaps best understood in *Drosophila*. In this organism the actions of genes expressed earlier in development create an anterior-posterior axis and create segments. The Hox genes then act to define segment identities. As noted earlier, mutations in *Drosophila* Hox genes often result in the development of a segment into the incorrect identity, causing a homeotic transformation of one body part into another.

A typical segment will express several of the eight *Drosophila* Hox genes, and, clearly, more than eight segment and structural identities are specified by the Hox genes. There is not, therefore, a simple one to one relationship between identity and Hox gene expression, with each Hox gene specifying a single structure. Instead, there is a combinatorial Hox code system at work, where a particular combination of expression of Hox genes determines segment identity (Lewis, 1978).

During embryogenesis the *Drosophila* Hox genes are generally expressed with each having a distinct anterior boundary of robust expression, which then trails off posteriorly. As mentioned, Hox genes more 5' on the clusters have more posterior restricted domains of expression. As a result there is a pattern of overlapping Hox expression with more Hox genes expressed in posterior segments and fewer in anterior segments.

A null mutation in a Hox gene removes its function from the most anterior segment of its expression, and thereby typically converts the code of that segment to a pattern normally found more anteriorly. Null Hox mutations therefore normally result in homeotic transformations towards more anterior identities.

Interestingly, Hox genes located more 5', with more posterior expression domains, are generally dominant over the more 3' Hox genes. This is referred to as posterior prevalence, or phenotypic suppression. The result is that ectopic anterior expression of the 5' Hox genes will often convert a segment into a more posterior identity. For example, the Antennapedia mutation that causes the formations of legs on the head in place of antennae is the result of a chromosomal rearrangement that brings the Antennapedia coding sequences under the control of a heterologous promoter, driving expression in a more anterior domain than normal, the antennal imaginal discs, and producing the resulting homeotic transformation of antenna to leg.

It is interesting to note that in some cases mammalian Hox genes can functionally substitute for their *Drosophila* orthologues. For example, the mammalian ortholog of the Antennapedia gene can be placed in a transgenic fly under the control of the heat shock promoter, allowing induced ectopic expression (Malicki *et al.*, 1990). The amazing result is that the mammalian Antennapedia like gene can also drive the homeotic transformation of antennae into legs. The legs that form are fly legs and not mammalian legs, because the Hox gene simply initiates the genetic program, which in the context of the fly would be a fly program. It is also interesting to note that even though the heat shock promoter drives near ubiquitous expression of the transgene, only the antennal imaginal disc is mis-directed to form legs. It appears that a correct combination of other Hox gene and co-factor gene expression is required for this transformation to take place.

The ability of some mammalian Hox genes to substitute for their *Drosophila* orthologs strongly suggests an evolutionarily conserved function. Nevertheless, this possible conservation has been an area of controversy. The functional equivalence model of Duboule, for example, proposes that *Drosophila* and Mammalian Hox genes have acquired quite distinct functions (Duboule, 1995).

Functional Specificity of Hox Proteins

The evolutionary history of the mammalian Hox genes includes the repeated tandem duplication of a single gene to create a cluster, followed by a quadruplication of this ancestral cluster to generate four clusters. The result is a set of 39 Hox genes with closely related homeoboxes and overlapping expression domains. It is therefore not surprising that the Hox genes would show a complex pattern of overlapping functions.

Immediately following a gene duplication event the resulting twin copies would have the same sequence and would share identical or very closely related functions. Over evolutionary time the two copies would experience sequence divergence, effecting both expression pattern and encoded protein, and their degree of functional overlap would diminish. The tandem duplication events creating the original cluster preceded the quadruplication to generate the four clusters. The nonparoalogous genes have therefore had more evolutionary time for functional divergence to take place, compared to paralogs. This explains why their sequences and functions are more distinct.

How functionally distinct are the Hox proteins? They all carry a 60-aa homeodomain with three α-helices. Helix two and three make up the helix-turn-helix motif, with the third helix termed the recognition helix because of its important role in interaction with the major groove of the target DNA. In addition there are several amino acids throughout the homeodomain that make contact with the phosphate backbone of DNA, and amino acids upstream of helix 1 interact with the minor groove of DNA. The well-conserved nature of the Hox homeodomains suggests they would recognize similar or identical target sequences and the results of *in vitro* DNA binding assays indicate that this is indeed the case. Most Hox homeodomains appear to recognize a six base sequence with a well-conserved TAAT core. This short target sequence is found randomly approximately once every Kb of DNA throughout the genome, including the cis regulatory regions of most genes, suggesting that additional factors must be involved in generating Hox functional specificity.

Despite the apparently common target binding sequences it is nevertheless abundantly clear that the

Drosophila Hox genes do have distinct functions. This conclusion derives from a host of both recessive null and dominant ectopic expression studies, with discrete phenotypes usually resulting for different Hox genes. There are, however, a few interesting exceptions, where genetic assays show functional equivalence for the different *Drosophila* Hox genes. For example, the *Drosophila* Hox gene labial normally drives the specification of the tritocerebral neuromere. In labial null mutants the cells that would normally make this structure do not acquire neuronal identity, consistent with the idea that labial initiates this developmental genetic program. Hirth *et al* (Hirth *et al.*, 2001) showed that a null labial mutant could be rescued by a labial transgene using labial central nervous system specific cis-regulatory elements. Surprisingly, they went on to show that using these same cis regulatory elements it was possible to rescue tritocerebral neuromere development using any of the *Drosophila* Hox genes except for Abd-B. They concluded "…most Hox proteins are functionally equivalent in their ability to replace Labial in the specification of neuronal identity." It is interesting to note that the efficiency of the rescue varied among the Hox genes, with genes closer to Labial on the cluster showing higher efficiency, consistent with their closer sequence relationship. A similar functional overlap has been observed in the development of the haltere, where the Abd-A and Abd-B can substitute for UBX (Casares *et al.*, 1996), as well as in development of the gonads (Greig and Akam, 1993).

It must be emphasized, however, that these particular studies illustrating common Hox protein function in *Drosophila* are the exception and not the rule. Most work has indicated that the Hox genes specify distinct identities along the A-P axis by acting as selector genes, initiating alternative developmental genetic programs (Gellon and McGinnis, 1998; Lawrence and Morata, 1994; Mann and Morata, 2000).

The genetic analysis of the functional relationships of the mammalian Hox genes has been much more complex, because the greater number of Hox genes and the weaker genetic tools available. The phenotypes resulting from Hox gene mutations in mammals have been a great deal less dramatic than those seen in *Drosophila*, with no resulting mice having legs protruding from their heads, for example. Indeed, many of the observed phenotypes have involved structures that are absent or reduced in size. This is consistent with Hox function in initiating a genetic program of development. In the absence of the Hox gene the program would not execute and the resulting structure would not form. It is also consistent with a function in driving cell proliferation, with reduced division giving smaller size.

The milder phenotypes observed for mammalian Hox mutations has resulted in the proposal that mammalian Hox genes have functions that are distinct from their counterparts in *Drosophila*. Indeed, the functional equivalence model of Duboule proposes that the mammalian Hox proteins all recognize the same, or functionally equivalent downstream targets, which are involved in the regulation of cell proliferation (Duboule, 1995). This model states that the quantity of Hox expression is more important than the quality of Hox expression. That is, the clusters of Hox genes function to provide the precisely correct spatio-temporal Hox dosage during development, but that one Hox protein is functionally interchangeable with another. According to this model apparent homeotic transformations of segment identity that had been observed in mammalian Hox gene mutants are interpreted as changes in the sculpted shapes of the bones resulting from changes in cell proliferation rates (Duboule, 1995).

A series of mice with genetically engineered homeobox swaps were made in order to test this model. This represents a fairly mild test of the model, which predicts that the entire coding sequences of the Hox genes, and not just their most highly conserved homeobox regions, should be functionally interchangeable. The results of the homeobox swap experiments were interesting, in that they showed surprising tolerance of homeobox exchange in some aspects of development, as predicted by the functional equivalence model, but in other cases the homeoboxes were shown to be very functionally distinct (Zhao and Potter, 2001; Zhao and Potter, 2002). The axial skeleton, including the bones of the vertebrae and ribs, formed normally when any of three distinct homeoboxes were substituted into the Hoxa 11 gene, as would be predicted by this model. In the case of the female reproductive tract, however, the results were quite different. The Hoxa 11 gene is normally expressed in the developing uterus, while the Hoxa 13 gene is expressed in the vagina. A mouse with the Hoxa 11 gene carrying a Hoxa 13 homeobox showed a striking homeotic transformation of the uterus into a vagina. That is, the expression of the Hoxa 11 gene in the developing uterus, only now with the heterologous Hoxa13 homeobox, converted the uterus into vagina, as measured by altered histology as well as gene expression patterns examined globally by microarray. It is also interesting that this was a dominant effect, with one copy of the homeobox swap gene giving this phenotype, even in the presence of a normal Hoxa 11 gene. It is interesting to note that the Hoxa 13 gene is located 5′ of the Hoxa 11 gene, and according to the posterior prevalence principle of Hox genes in *Drosophila*, would be expected to have dominant effects. These results illustrate that Hox proteins are not all functionally equivalent, at least not in all developing structures, and that indeed functionally specificity can track with the homeobox, even though it is the most conserved part of the Hox protein.

It now appears that the relatively milder phenotypes of the mammalian Hox mutants are simply the result of the presence of four clusters in mammals instead of the single split cluster in flies. This greater number of genes provides much more opportunity for functional overlap in mammals, which can result in a less dramatic phenotype when a single Hox gene is mutated.

There is considerable genetic evidence indicating a very strong functional overlap of Hox genes within a single paralogous group. For example, mutation of either the Hoxa 11 or Hoxd 11 gene gives a relatively minor limb phenotype, with relatively minor malformations of some bones. Mice homozygous for mutations in both genes, however, show a much more dramatic limb phenotype, with an almost complete absence of the zeugopod, or forelimb, with the paw now essentially attached to the elbow (Davis *et al.*, 1995). Similarly, the kidneys of single mutants are relatively normal, while the kidneys of double mutants are dramatically reduced in size, and sometimes absent. Comparable results have been obtained in the genetic analysis of other paralogous groups. For example, mutation of both Hoxa 13 and Hoxd 13 result in almost complete absence of the autopod, or paw (Fromental-Ramain *et al.*, 1996). In another study it was shown that the coding sequences of the genes of paralogous group 3 were functionally interchangeable (Greer *et al.*, 2000), again illustrating the close functional relationships of paralogs. When only one paralog is removed, the remaining genes can often provide adequate, although not complete function. Therefore, in order to define the function of a paralogous group of Hox genes in mammals it is necessary to mutate multiple members of the gene group.

Genetic studies have also demonstrated functional overlap of flanking Hox genes, although much less dramatic than observed for paralogs. For example, the relationships of the 10 and 11 paralogous groups have been examined by making Hoxa 10, Hoxa 11 transheterozygotes, with one mutant copy of each gene. The result was a synergistic phenotype not seen in either single mutant (Branford *et al.*, 2000). In addition Hoxd 10, Hoxa 11 double homozygotes show significant evidence of functional overlap between these genes (Favier *et al.*, 1996), and the use of clusters carrying double neo insertions into the 10 and 11 paralogous groups allowed further dissection of the relationships of these genes (Wellik *et al.*, 2002).

The picture that emerges from these genetic studies is that genes within a paralogous group have very strong functional overlap. Nevertheless, they are not identical in function, as mutation of one Hox gene does produce a phenotype, although often mild, showing that its absence cannot be completely compensated for by the remaining members of the group. This non-identity is presumably the result of differences in expression, and in some groups perhaps the divergence of encoded protein function. Nevertheless, the evolutionarily recent duplication of the Hox cluster has left us with paralogs that are still very closely related in function. There is also some functional overlap of nonparalogous Hox genes, in particular among flanking Hox genes with greater sequence similarity, as discussed above.

Mechanisms of Functional Specificity

How do the Hox proteins achieve their functional specificity? As previously mentioned, *in vitro* binding assays indicate that their homeodomains almost all recognize the same core TAAT sequence, which is found by chance in the cis regulatory elements of most genes. Further, since almost all Hox proteins bind to the same DNA sequences this suggests that they recognize the same downstream targets. These observations do not explain the Hox functional specificity observed in genetic assays.

Perhaps *in vivo* there is more target sequence specificity than suggested by *in vitro* DNA binding assays. Indeed, even in the *in vitro* assays some variations in target preferences have been observed. In addition it is possible that the different Hox proteins bind to similar sequences, but with different affinities, or strengths (Lamka *et al.*, 1992). This has been proposed as one possible mechanism for posterior prevalence, with more 5′ Hox proteins binding to targets more tightly, and thereby able to displace more 3′ Hox proteins. It is also possible that different Hox proteins would carry different transcription activation and repression domains, and would therefore have different effects on their targets. These considerations begin to explain Hox functional specificity.

Hox Cofactors

One main source of Hox functional specificity is thought to reside in their interacting cofactors. The specific milieu of Hox interacting proteins present in a particular cell is thought to play an important role in driving target gene specificity as well as determining the activation or repression effect on target genes. That is, a single Hox protein can regulate different target genes in different cell types. Moreover, a single Hox protein can activate some targets and repress others. Various combinations of cofactors are likely responsible for mediating these differential effects.

The best studied binding partners for Hox proteins are the TALE (three amino acid loop extension) proteins Extradenticle (Exd) (Peifer and Wieschaus, 1990) and homothorax (Hth) (Phelan et al., 1995; Ryoo et al., 1999) in *Drosophila*, with the mammalian homologs referred to as Pbx (Phelan et al., 1995) and Meis (Moskow et al., 1995) respectively. These cofactors also include homeodomains which are able to bind DNA. The Pbx interaction with Hox proteins is often mediated through a YPWM motif present on most Hox proteins N-terminal of the homeodomain.

There is good evidence that Hox proteins can bind to DNA as monomers, as a duplex with one of these cofactors, or as a trimer consisting of one Hox, one Pbx and one Meis protein. When binding as a multimer the target sequence specificity is greatly increased. In addition to the Hox binding sequence, there must reside a Pbx binding sequence in close proximity, for example, and perhaps a Meis site as well. It is easy to envision subtle differences in target specificities for the different Hox proteins, and the different proteins of the Pbx and Meis families. In addition, distinct interactions of these proteins in forming multimers could result in discrete spacing preferences for the different target sequences. Hence, the interaction of Hox proteins with Pbx and Meis proteins is considered an important mechanism for increasing Hox target specificity (Fig.18.2). Indeed, several studies have shown that different Hox complexes bind to distinct target DNA sequences, giving rise to the Hox binding selectivity model (Chan and Mann, 1996; Chang et al., 1996; Gebelein et al., 2002; Mann and Chan, 1996).

Nevertheless, the number of Pbx and Meis family cofactors available is quite limited (four Pbx and three Meis genes in the mouse), and strong target sequence binding similarities among family members place limits on the specificity attainable.

It is interesting to note, however, that the number of possible protein interactions is ever expanding, offering more opportunities for achieving specificity. In *Drosophila*, for example, it has been shown that the Exd and Hth proteins can interact not only with Hox proteins, but also with a dispersed, non-clustered homeobox gene encoded protein, Engrailed, and thereby repress target genes (Kobayashi et al., 2003). This shows that the Exd and Hom cofactors can interact with other non-Hox transcription factors. It is also true that Hox proteins can interact with non-Exd, non-Hth cofactors.

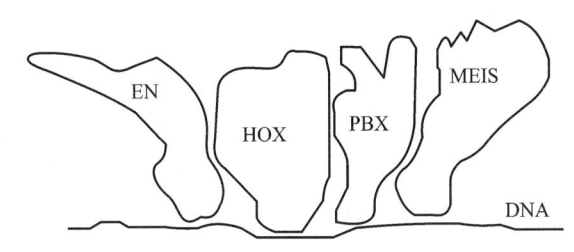

Fig.18.2 Hox-Cofactor interactions. This diagram shows a HOX protein binding DNA as a part of a complex that includes PBX and MEIS protein. The combined DNA binding domains can contribute to a much more specific target binding sequence than could be achieved by any of the proteins alone. Further specificity can result from differences in preferred interactions of the multiple members of the Hox, Pbx and Meis families, resulting in different combinations of DNA binding domains and distinct spacing. Other factors, such as the ENGRAILED protein shown, can also bind, as a function of the Hox complex, and/or the flanking DNA sequence, and contribute gene activator or repressor function.

In a particularly informative example, the Hox regulation of the target gene Distalless (Dll) was carefully dissected (Gebelein et al., 2004). It was shown that the Ubx Hox protein binds to the Dll promoter as a tetramer, consisting of two Ubx proteins, one Exd and one Hth protein. The Dll promoter carries one binding site for each of these four proteins, in close proximity. If the Antp Hox protein is substituted for Ubx then the resulting tetramer binds with ten fold lower affinity, clearly demonstrating the presence of Hox specificity although not revealing its mechanistic source. Of interest the Engrailed (En) protein was found to bind to the Dll promoter poorly on its own, but in a strongly cooperative fashion with either the AbdA or Ubx Hox proteins. Since En is a powerful transcriptional repressor, this interaction of Hox, Exd, Hth and En on the Dll promoter can explain how the Hox protein achieves repression of Dll. Further, this effect can be compartment specific, since En and other possible cofactors show compartment specific expression. This begins to explain how a single Hox protein can activate some targets and repress others, or even effect the activation and repression of a single target in different compartments.

Another interesting study showed how Hox proteins can directly influence growth factor signaling pathways. In particular, the 5′ HoxD cluster genes in the mouse were shown to be able to modify sonic hedgehog (SHH) signaling. In the absence of SHH signal the Gli3 transcription factor normally activates downstream targets. SHH signal, however, results in cleavage of Gli3 into a form that binds to the same targets but now represses transcription. Surprisingly, it was shown that the proteins encoded by the 5′ HoxD genes were capable of directly binding to the truncated Gli3, with no DNA interaction required, and converting the repressor into an activator (Chen *et al.*, 2004). It was also shown that multiple Hox proteins can interact with the protein Geminin (Luo *et al.*, 2004). This interaction occurred through the homeodomain, and resulted in blocking of the ability of the Hox protein to bind DNA.

It is likely that further studies will show that Hox proteins are able to interact with many more proteins, not just Pbx, Exd, En, Gli3 and Geminin. Different specificities for these multiple interactions will contribute greatly to Hox functional precision.

It is also important to note that at least in some cases Hox proteins bind to target genes as monomers, and not as a complex with cofactors. For example, in *Drosophila* it has been shown that the Ubx Hox protein regulates the downstream target Spalt gene by binding as a monomer, with no requirement for Exd or Hth (Galant *et al.*, 2002). It was proposed that this mechanism provides some evolutionary advantages, as a simple accumulation of multiple copies of the monomer binding sequence, with a TAAT core, will result in the regulation of Spalt by Ubx. It is easier to imagine the formation of multiple copies of this simple sequence, than the formation by random mutation of a much more complex multimer binding sequence.

Downstream of Hox Genes

The Hox genes encode transcription factors that function by regulating expression levels of downstream target genes. A direct route to understanding how Hox genes work is to identify the genes they regulate. This has been a difficult area of research, however, with slow progress. Nevertheless, some interesting discoveries have been made that shed light on the functional pathways regulated by Hox genes.

It is clear that one mechanism sometimes used by Hox genes to sculpt tissue shape is the regulation of cell death, or apoptosis. In *Drosophila*, it has been shown that the Hox gene Dfd directly regulates the apoptosis gene reaper, thereby maintaining boundaries between developing segments, and that another Hox gene Abd-B, similarly regulates segment boundaries through the activation of apoptosis (Lohmann *et al.*, 2002). Therefore Hox genes are not only important in defining the identities of segments once established, but are also important in establishing and maintaining segments. In another study Abd-B was also shown to regulate apoptosis in the developing *Drosophila* nervous system (Miguel-Aliaga and Thor, 2004). Of interest, a connection between Abd-B type Hox genes and apoptosis has also been made in vertebrates, with Hoxa 13 homozygous mutants, for example, showing loss of the apoptosis that normally separates the digits of the developing limbs (Stadler *et al.*, 2001).

Hox genes have also been implicated in the control of cell proliferation. One phenotype often observed in Hox mutants is a reduction in structure size. For example, the Hoxa 11/Hoxd 11 double homozygous mutant mice show a severe reduction in the size of the forearm, or zeugopod, of the limb (Davis *et al.*, 1995). As mentioned previously, multiple observations of this sort have suggested to some that a common function of most Hox genes is the regulation of cell proliferation rates (Duboule, 1995). This combined function in the regulation of cell proliferation and cell death can drive the shaping of segments and tissues.

Another important group of Hox targets are the Hox genes themselves. In some cases this represents autoregulation, with the Hox gene activating itself, as is seen for the *Drosophila* Dfd and Lab genes. This is thought to represent a mechanism for maintaining the determined state. Once the Hox gene is activated it keeps itself in the state. There is also cross-regulation among Hox genes, with the Proboscipedia gene regulated by Dfd, for example (Rusch and Kaufman, 2000).

It is clear that Hox genes can regulate genes that have important roles in developmental patterning as well as genes that encode terminal differentiation products. Ubx, for example, appears to directly regulate connectin, a homophilic cell adhesion molecule in muscle development (Gould and White, 1992). On the other hand, in mice, Hoxc13 has been shown to regulate the terminal differentiation hair keratin genes by binding to multiple copies of TAAT and TTAT recognition motifs present in their promoters (Jave-Suarez *et al.*, 2002). Other interesting downstream targets of Hox genes include the Hoxa 2 regulation of the transcription factor Six2 (Kutejova *et al.*, 2005), and Hoxa 13 regulation of BMP2 and BMP7 (Knosp *et al.*, 2004).

It is generally thought that each Hox gene regulates a large number of downstream targets. Microarrays offer a technology that allows universal screens, potentially giving the rapid identification of large numbers of genes with transcription levels responsive to the expression of a Hox gene. Microarray approaches have been used to look for targets of Hoxa 11 (Valerius *et al.*, 2002) and Hoxc 8 (Lei *et al.*, 2005), and Hoxd 10 (Hedlund *et al.*,

2004), finding integrin genes, osteopontin and frizzled as target genes, among others.

Regulation of Hox Genes

What is upstream of the Hox genes? What regulates these regulators? In this section we will present some of the general concepts that have emerged from many studies in this area.

First, it is often proposed that the clustered organization of the Hox genes reflects a required sharing of enhancers. That is, a single enhancer in a Hox cluster could influence the expression of many Hox genes, and the physical break up of the cluster would separate some Hox genes from their required enhancers, resulting in a dramatic and likely lethal change in expression. This sharing of regulatory elements would drive preservation of the clustered arrangement of Hox genes, which is seen in most but not all organisms.

It is also interesting to note, once again, that the mammalian Hox clusters are almost entirely devoid of interspersed repeat sequences. These repetitive DNAs, such as the B1, B2 and L1 sequences in the mouse genome, are present in very high copy number, and are usually found every couple of thousand base pairs on average. The relative absence of these mobile sequences in the Hox clusters argues that their insertion in these regions has harmful consequences that are strongly selected against. It is reasonable to suppose that the function of this noncoding DNA in the Hox clusters is regulatory in nature.

The Hox genes show an interesting pattern of response to retinoic acid, with the more 3' Hox genes activated by lower levels of retinoic acid, and the more 5' genes more refractory to retinoic acid activation (Simeone *et al.*, 1990). These responses are partly mediated through shared retinoic acid response elements located in the 3' regions of the Hox clusters.

Many upstream transcription factors interact with the Hox promoters/enhancers. The Cdx homeodomain proteins, however, are among the most interesting. Cdx binding sites are found in the promoters of a number of Hox genes and many experiments have indicated that Cdx proteins are direct regulators of Hox transcription (Charite *et al.*, 1998). Mice with mutations in Cdx genes can display homeotic transformations from the resulting mis-expression of Hox genes. It is particularly interesting to note that the Cdx1 gene is itself retinoic acid responsive, with a retinoic acid response element (RARE). Further, targeted mutation of its RARE alters not only Cdx1 expression but results in vertebral homeotic transformations caused by alterations in Hox expression patterns (Houle *et al.*, 2003). This illustrates how retinoic acid can influence Hox expression both directly, through Hox cluster RAREs, and indirectly, through RAREs associated with genes upstream of Hox genes. While the more 3' paralogous groups of Hox genes are more sensitive to retinoic acid, the more 5' Hox genes respond to FGF signaling, and once again there is evidence that this could be mediated at least in part through the Cdx genes (Bel-Vialar *et al.*, 2002; Isaacs *et al.*, 1998).

Once a Hox gene expression pattern is established it must be maintained through an epigenetic memory system mediated through the Polycombs and Trithorax groups of genes. These protein complexes maintain Hox gene expression states through multiple cell divisions. The Polycombs and Trithorax proteins control repressed and active transcriptional states, respectively. Both groups of proteins regulate chromatin accessibility, but additional proteins, including general transcription factors, the RNA polymerase II complex, and perhaps noncoding RNAs contribute to this memory process. Mutations in genes of the Polycombs and Trithorax systems can cause extensive changes in Hox expression, resulting in compound sets of homeotic transformations.

Hox Genes and Disease

Because Hox genes are known to play an important role in the regulation of cell proliferation it is not surprising to find that they can also play a role in cancer. For most cancers the evidence for a causative role is still relatively weak, with many studies simply showing increased expression of specific Hox genes in cancer tissue. But for the leukemias the evidence is quite convincing that several Hox genes can play an important role. Overexpression of Hoxb8, for example, resulting from insertion of an intracisternal A-particle (IAP) proviral element, can contribute to myeloid leukemia (Perkins *et al.*, 1990). In another set of studies using the BXH-2 mice, which have a naturally high incidence of myeloid leukemia due to an active endogenous leukemia virus, it was found that three virus integration hotspots in the genome were Hoxa 7, Hoxa 9 and a gene encoding the Hox interacting protein Meis1 (Nakamura *et al.*, 1996). The co-expression of these three genes has also been seen in human leukemias (Lawrence *et al.*, 1999). Genes encoding other proteins that interact with or regulate Hox genes are also found to be frequently altered in human leukemias, including Pbx and Trithorax genes.

Conclusion

The Hox genes encode transcription factors that can sometimes initiate genetic programs that drive segment identity determination. The 39 mammalian Hox genes are located on four clusters and can be divided into 13 paralogous groups. There is considerable functional redundancy among the mammalian Hox genes, making it difficult to dissect out their functions. It is striking, however, that misexpression of the mammalian ortholog of the *Drosophila* Antennapedia gene can also cause homeotic transformation of antennae to legs, strongly arguing that the functions of these genes are highly conserved during evolution.

The sources of the functional specificity of the Hox genes are being discovered. The Hox homeodomains bind to similar sequences in test tube DNA binding assays, suggesting that they might be functionally equivalent. Nevertheless, homeobox swap experiments show that there is indeed functional specificity residing in the homeobox, suggesting the presence of *in vivo* sequence preferences not detected in the *in vitro* assays. Additional specificity is achieved through distinct interactions with Pbx, Meis and other cofactors.

Ultimately we would like to understand the genetic regulatory networks both upstream and downstream of Hox genes. High throughput strategies would speed this effort. Spotted promoter microarrays can allow the rapid analysis of transcription factor interactions with multiple promoters, and standard oligonucleotide microarrays are being harnessed to identify downstream targets of Hox genes. Accelerating progress will soon define the regulation of the Hox genes and in turn the downstream pathways that they control during development.

References

Bel-Vialar, S., Itasaki, N., and Krumlauf, R. (2002). Initiating Hox gene expression: in the early chick neural tube differential sensitivity to FGF and RA signaling subdivides the HoxB genes in two distinct groups. Development *129*, 5103-5115.

Branford, W. W., Benson, G. V., Ma, L., Maas, R. L., and Potter, S. S. (2000). Characterization of Hoxa-10/Hoxa-11 transheterozygotes reveals functional redundancy and regulatory interactions. Dev Biol *224*, 373-387.

Casares, F., Calleja, M., and Sanchez-Herrero, E. (1996). Functional similarity in appendage specification by the Ultrabithorax and abdominal-A *Drosophila* HOX genes. Embo J *15*, 3934-3942.

Chan, S. K., and Mann, R. S. (1996). A structural model for a homeotic protein-extradenticle-DNA complex accounts for the choice of HOX protein in the heterodimer. Proc Natl Acad Sci USA *93*, 5223-5228.

Chang, C. P., Brocchieri, L., Shen, W. F., Largman, C., and Cleary, M. L. (1996). Pbx modulation of Hox homeodomain amino-terminal arms establishes different DNA-binding specificities across the Hox locus. Mol Cell Biol *16*, 1734-1745.

Charite, J., de Graaff, W., Consten, D., Reijnen, M. J., Korving, J., and Deschamps, J. (1998). Transducing positional information to the Hox genes: critical interaction of cdx gene products with position-sensitive regulatory elements. Development *125*, 4349-4358.

Chen, Y., Knezevic, V., Ervin, V., Hutson, R., Ward, Y., and Mackem, S. (2004). Direct interaction with Hoxd proteins reverses Gli3-repressor function to promote digit formation downstream of Shh. Development *131*, 2339-2347.

Davis, A. P., Witte, D. P., Hsieh-Li, H. M., Potter, S. S., and Capecchi, M. R. (1995). Absence of radius and ulna in mice lacking hoxa-11 and hoxd-11. Nature *375*, 791-795.

Duboule, D. (1995). Vertebrate Hox genes and proliferation: an alternative pathway to homeosis? Curr Opin Genet Dev *5*, 525-528.

Favier, B., Rijli, F. M., Fromental-Ramain, C., Fraulob, V., Chambon, P., and Dolle, P. (1996). Functional cooperation between the non-paralogous genes Hoxa-10 and Hoxd-11 in the developing forelimb and axial skeleton. Development *122*, 449-460.

Fromental-Ramain, C., Warot, X., Messadecq, N., LeMeur, M., Dolle, P., and Chambon, P. (1996). Hoxa-13 and Hoxd-13 play a crucial role in the patterning of the limb autopod. Development *122*, 2997-3011.

Galant, R., Walsh, C. M., and Carroll, S. B. (2002). Hox repression of a target gene: extradenticle-independent, additive action through multiple monomer binding sites. Development *129*, 3115-3126.

Gebelein, B., Culi, J., Ryoo, H. D., Zhang, W., and Mann, R. S. (2002). Specificity of Distalless repression and limb primordia development by abdominal Hox proteins. Dev Cell *3*, 487-498.

Gebelein, B., McKay, D. J., and Mann, R. S. (2004). Direct integration of Hox and segmentation gene inputs during *Drosophila* development. Nature *431*, 653-659.

Gellon, G., and McGinnis, W. (1998). Shaping animal body plans in development and evolution by modulation of Hox expression patterns. Bioessays *20*, 116-125.

Gould, A. P., and White, R. A. (1992). Connectin, a target of homeotic gene control in *Drosophila*. Development *116*, 1163-1174.

Greer, J. M., Puetz, J., Thomas, K. R., and Capecchi, M. R. (2000). Maintenance of functional equivalence during paralogous Hox gene evolution. Nature *403*, 661-665.

Greig, S., and Akam, M. (1993). Homeotic genes autonomously specify one aspect of pattern in the *Drosophila* mesoderm. Nature *362*, 630-632.

Hedlund, E., Karsten, S. L., Kudo, L., Geschwind, D. H., and Carpenter, E. M. (2004). Identification of a Hoxd10-regulated transcriptional network and combinatorial interactions with Hoxa10 during spinal cord development. J Neurosci Res *75*, 307-319.

Hirth, F., Loop, T., Egger, B., Miller, D. F., Kaufman, T. C., and Reichert, H. (2001). Functional equivalence of Hox gene products in the specification of the tritocerebrum during embryonic brain development of *Drosophila*. Development *128*, 4781-4788.

Houle, M., Sylvestre, J. R., and Lohnes, D. (2003). Retinoic acid regulates a subset of Cdx1 function *in vivo*. Development *130*, 6555-6567.

Isaacs, H. V., Pownall, M. E., and Slack, J. M. (1998). Regulation of Hox gene expression and posterior development by the *Xenopus* caudal homologue Xcad3. Embo J *17*, 3413-3427.

Jave-Suarez, L. F., Winter, H., Langbein, L., Rogers, M. A., and Schweizer, J. (2002). HOXC13 is involved in the regulation of human hair keratin gene expression. J Biol Chem *277*, 3718-3726.

Knosp, W. M., Scott, V., Bachinger, H. P., and Stadler, H. S. (2004). HOXA13 regulates the expression of bone morphogenetic proteins 2 and 7 to control distal limb morphogenesis. Development *131*, 4581-4592.

Kobayashi, M., Fujioka, M., Tolkunova, E. N., Deka, D., Abu-Shaar, M., Mann, R. S., and Jaynes, J. B. (2003). Engrailed cooperates with extradenticle and homothorax to repress target genes in *Drosophila*. Development *130*, 741-751.

Kutejova, E., Engist, B., Mallo, M., Kanzler, B., and Bobola, N. (2005). Hoxa2 downregulates Six2 in the neural crest-derived mesenchyme. Development *132*, 469-478.

Lamka, M. L., Boulet, A. M., and Sakonju, S. (1992). Ectopic expression of UBX and ABD-B proteins during *Drosophila* embryogenesis: competition, not a functional hierarchy, explains phenotypic suppression. Development *116*, 841-854.

Lawrence, H. J., Rozenfeld, S., Cruz, C., Matsukuma, K., Kwong, A., Komuves, L., Buchberg, A. M., and Largman, C. (1999). Frequent co-expression of the HOXA9 and MEIS1 homeobox genes in human myeloid leukemias. Leukemia *13*, 1993-1999.

Lawrence, P. A., and Morata, G. (1994). Homeobox genes: their function in *Drosophila* segmentation and pattern formation. Cell *78*, 181-189.

Lei, H., Wang, H., Juan, A. H., and Ruddle, F. H. (2005). The identification of Hoxc8 target genes. Proc Natl Acad Sci USA *102*, 2420-2424.

Lewis, E. B. (1978). A gene complex controlling segmentation in *Drosophila*. Nature *276*, 565-570.

Lohmann, I., McGinnis, N., Bodmer, M., and McGinnis, W. (2002). The *Drosophila* Hox gene deformed sculpts head morphology via direct regulation of the apoptosis activator reaper. Cell *110*, 457-466.

Luo, L., Yang, X., Takihara, Y., Knoetgen, H., and Kessel, M. (2004). The cell-cycle regulator geminin inhibits Hox function through direct and polycomb-mediated interactions. Nature *427*, 749-753.

Maclean, J. A., 2nd, Chen, M. A., Wayne, C. M., Bruce, S. R., Rao, M., Meistrich, M. L., Macleod, C., and Wilkinson, M. F. (2005). Rhox: a new homeobox gene cluster. Cell *120*, 369-382.

Malicki, J., Schughart, K., and McGinnis, W. (1990). Mouse Hox-2.2 specifies thoracic segmental identity in *Drosophila* embryos and larvae. Cell *63*, 961-967.

Mann, R. S., and Chan, S. K. (1996). Extra specificity from extradenticle: the partnership between HOX and PBX/EXD homeodomain proteins. Trends Genet *12*, 258-262.

Mann, R. S., and Morata, G. (2000). The developmental and molecular biology of genes that subdivide the body of *Drosophila*. Annu Rev Cell Dev Biol *16*, 243-271.

McGinnis, W., Garber, R. L., Wirz, J., Kuroiwa, A., and Gehring, W. J. (1984a). A homologous protein-coding sequence in *Drosophila* homeotic genes and its conservation in other metazoans. Cell *37*, 403-408.

McGinnis, W., Hart, C. P., Gehring, W. J., and Ruddle, F. H. (1984b). Molecular cloning and chromosome mapping of a mouse DNA sequence homologous to homeotic genes of *Drosophila*. Cell *38*, 675-680.

Miguel-Aliaga, I., and Thor, S. (2004). Segment-specific prevention of pioneer neuron apoptosis by cell-autonomous, postmitotic Hox gene activity. Development *131*, 6093-6105.

Moskow, J. J., Bullrich, F., Huebner, K., Daar, I. O., and Buchberg, A. M. (1995). Meis1, a PBX1-related homeobox gene involved in myeloid leukemia in BXH-2 mice. Mol Cell Biol *15*, 5434-5443.

Nakamura, T., Largaespada, D. A., Shaughnessy, J. D., Jr., Jenkins, N. A., and Copeland, N. G. (1996). Cooperative activation of Hoxa and Pbx1-related genes in murine myeloid leukaemias. Nat Genet *12*, 149-153.

Peifer, M., and Wieschaus, E. (1990). Mutations in the *Drosophila* gene extradenticle affect the way specific homeo domain proteins regulate segmental identity. Genes Dev *4*, 1209-1223.

Perkins, A., Kongsuwan, K., Visvader, J., Adams, J. M., and Cory, S. (1990). Homeobox gene expression plus autocrine growth factor production elicits myeloid leukemia. Proc Natl Acad Sci USA *87*, 8398-8402.

Phelan, M. L., Rambaldi, I., and Featherstone, M. S. (1995). Cooperative interactions between HOX and PBX proteins mediated by a conserved peptide motif. Mol Cell Biol *15*, 3989-3997.

Pollard, S. L., and Holland, P. W. (2000). Evidence for 14 homeobox gene clusters in human genome ancestry. Curr Biol *10*, 1059-1062.

Rusch, D. B., and Kaufman, T. C. (2000). Regulation of proboscipedia in *Drosophila* by homeotic selector genes. Genetics *156*, 183-194.

Ryoo, H. D., Marty, T., Casares, F., Affolter, M., and Mann, R. S. (1999). Regulation of Hox target genes by a DNA bound Homothorax/Hox/Extradenticle complex. Development *126*, 5137-5148.

Simeone, A., Acampora, D., Arcioni, L., Andrews, P. W., Boncinelli, E., and Mavilio, F. (1990). Sequential activation of HOX2 homeobox genes by retinoic acid in human embryonal carcinoma cells. Nature *346*, 763-766.

Stadler, H. S., Higgins, K. M., and Capecchi, M. R. (2001). Loss of Eph-receptor expression correlates with loss of cell adhesion and chondrogenic capacity in Hoxa13 mutant limbs. Development *128*, 4177-4188.

Valerius, M. T., Patterson, L. T., Feng, Y., and Potter, S. S. (2002). Hoxa 11 is upstream of Integrin alpha8 expression in the developing kidney. Proc Natl Acad Sci USA *99*, 8090-8095.

Wellik, D. M., Hawkes, P. J., and Capecchi, M. R. (2002). Hox11 paralogous genes are essential for metanephric kidney induction. Genes Dev *16*, 1423-1432.

Zhao, Y., and Potter, S. S. (2001). Functional specificity of the Hoxa13 homeobox. Development *128*, 3197-3207.

Zhao, Y., and Potter, S. S. (2002). Functional comparison of the Hoxa 4, Hoxa 10, and Hoxa 11 homeoboxes. Dev Biol *244*, 21-36.

Chapter 19
Nuclear Factor-kappa B

Keith W. Clem and Y. Tony Ip

Program in Molecular Medicine, University of Massachusetts Medical School, 373 Plantation Street, Worcester, MA 01605, USA

Key Words: *Drosophila*, inflammation, immunity, NF-kappa B, signaling, Toll, transcription

Summary

The Nuclear Factor-kappa B (NF-κB) represents a family of transcription factors that controls a variety of processes related to immunity and development. This family of transcription factors is conserved from insects to humans. Five different proteins, encoded by five different genes, of this family are present in mammals and three related proteins are present in *Drosophila*. These factors act downstream of many receptors that receive signals from cytokines, microorganisms, and developmental cues. Receptor activation leads to modification of downstream signaling components, and degradation of the cytoplasmic inhibitors that normally act to retain NF-κB proteins in the cytoplasm. Once the NF-κB proteins are freed from the inhibitor complex, they enter the nucleus to regulate gene expression by binding to target κB sites present on many promoters. Misregulation of the NF-κB family of proteins can cause severe developmental defects, inflammatory diseases, and cancers.

Introduction

Nuclear Factor-kappa B was first identified as a DNA binding activity present in nuclear extracts of B cell lines (Sen and Baltimore, 1986). The binding assay used a small piece of genomic DNA fragment that resembled the enhancer sequence present in the kappa light chain gene, which encodes the kappa light chain of the antibody complex produced by B cells. The enhancer sequence had been shown to regulate the transcription of the kappa light chain gene in pre-B cells, thus the identification of the protein factor that interacted with this enhancer was presumed to provide important insights into B cell function. The protein factor that bound to this kappa light chain enhancer was thus named Nuclear Factor-kappa B (NF-κB) (Bonizzi and Karin, 2004; Chen and Greene, 2004).

Characterization of the NF-κB activity in B cell extracts demonstrated that the factor contained two subunits. The most common combination was a 50 kD subunit and a 65 kD subunit (Baeuerle and Baltimore, 1989). Purification of the binding activity and protein sequencing revealed that both p50 and p65 (now called RelA) subunits have sequence homology to the previously identified avian oncoprotein Rel and the *Drosophila* developmental regulator Dorsal (Bours *et al.*, 1990; Ghosh *et al.*, 1990; Gilmore, 1990; Kieran *et al.*, 1990; Meyer *et al.*, 1991; Nolan *et al.*, 1991; Steward, 1987). These factors constitute what we now know as the NF-κB/Rel family of transcription factors. The nomenclature of these factors in earlier publications was, understandably, not consistent. Today, most scientists have agreed to call these proteins NF-κB or Rel family. In this review, we will use the term NF-κB to represent this family of transcription factors.

NF-κB Family of Proteins in Mammals and Insects

Mouse is the most common mammalian model and fruit fly (*Drosophila melanogaster*) is the most common insect model for experimentations, thus we will focus the discussion of the NF-κB factors in these two organisms. Three NF-κB-related proteins are present in *Drosophila*, named Dorsal, Dif, and Relish (Fig. 19.1).

Corresponding Author: Y. Tony Ip, Tel: +1-(508) 856-5136, Fax: +1-(508) 856-4289, E-mail: Tony.Ip@umassmed.edu

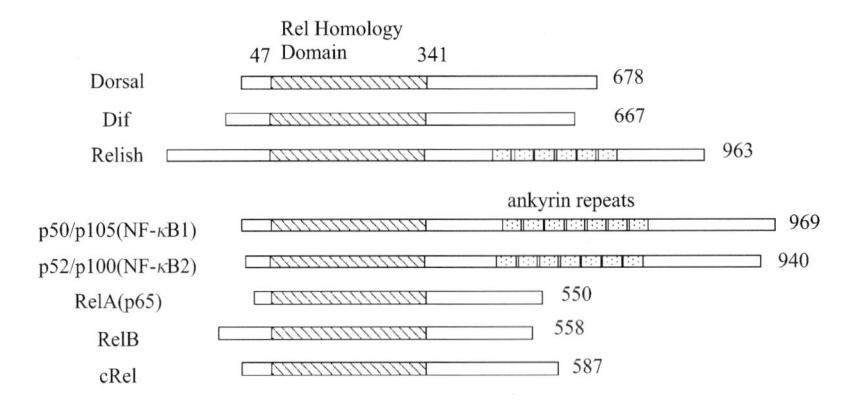

Fig.19.1 The NF-κB proteins in mammals and *Drosophila*. Three NF-κB-related proteins, Dorsal, Dif, and Relish, are present in *Drosophila*. Five NF-κB proteins are present in mammals, and the nomenclature for these proteins in this review is p50/p105, p52/p100, RelA, RelB, and cRel. The names in parenthesis are also used in many publications. All these proteins contain the Rel Homology domain, which is approximately 300 amino acids long and the degree of identity varies from 30% to 50%. The ankyrin repeats in p105, p100, and Relish act as intramolecular inhibitory domains. Other NF-κB proteins are inhibited by IκB proteins, which also contain ankyrin repeats.

In mouse, the NF-κB family is comprised of five proteins: RelA, RelB, c-Rel, p50/p105, and p52/p100 (Bonizzi and Karin, 2004; Chen and Greene, 2004; Gilmore and Ip, 2003).

The full-length proteins of the members of the NF-κB family varies from approximately 500 amino acids to approximately 900 amino acids. Most importantly, all NF-κB proteins contain the conserved domain called the Rel Homology (RH) domain (Steward, 1987). The RH domain is located at the N terminus and is approximately 300 amino acids long. Among the known RH domains the degree of homology is approximately 50% identical. The RH domain is required for the formation of dimers, DNA binding, nuclear translocation, and inhibitor binding (Ganchi *et al*., 1992; Moore *et al*., 1993; Ruben *et al*., 1992).

The C-terminal halves of NF-κB proteins are divergent, but the nature of the sequences can be used to divide NF-κB proteins into two groups. The first group includes p50/p105, p52/p100, and Relish. The C-termini of these proteins have 7-8 copies of an approximately 30 amino acid sequence, called the ankyrin repeat. Ankyrin repeats serve as protein-protein interaction domain, and in NF-κB proteins these repeats fold back and block DNA binding by the RH domain (Capobianco *et al*., 1992; Dushay *et al*., 1996; Hatada *et al*., 1992; Rice *et al*., 1992; Stoven *et al*., 2003). Therefore, the ankyrin repeat domains must be removed during normal activation to enable the NF-κB proteins to bind to DNA (Betts and Nabel, 1996; Stoven *et al*., 2000). Thus, this group of NF-κB proteins exist either in long forms (p105, p100, or full length Relish) that cannot bind to DNA, or as short forms (p50, p52, or Relish-N terminus) that can bind to DNA. The short forms are generated by proteolytic processing of the C-termini of the long forms.

The second group of NF-κB proteins includes RelA, RelB, and c-Rel in mammals, and Dif and Dorsal in *Drosophila*. The sequences of the C-terminal halves of these proteins are quite divergent, but they all function as transcriptional regulatory domains (Schmitz *et al*., 1994). These proteins do not require proteolysis to gain the active conformation, but instead must be released from the complex that contains the inhibitor (see below) (Henkel *et al*., 1992).

All NF-κB proteins are DNA-binding transcriptional regulators (Anrather *et al*., 2005; Fan *et al*., 2004; Hou *et al*., 2003; Xia *et al*., 2004). They bind DNA as dimers, which can be homodimers or heterodimers (Berkowitz *et al*., 2002; Chen *et al*., 1998; Chen *et al*., 2000; Huang *et al*., 2001; Liu *et al*., 1994; Yang and Steward, 1997). In vertebrates, the most common dimer is a p50-RelA (p65) heterodimer, which constitutes the originally identified nuclear binding activity that named NF-κB. All NF-κB dimers can bind to DNA target sites that have similarity to the consensus sequence 5′-GGGRNYYYCC-3′ (R is a purine, N is any nucleotide, Y is pyrimidine). These DNA-binding sequences are generically called κB sites. κB sites are usually found upstream of target genes, and the binding of a NF-κB dimer to a κB site usually results in increased transcription of the gene. However, κB sites have been found in introns, and repression can occur if certain combination of dimer is used or a co-repressor is present.

Regulation of NF-κB Activity

An important characteristic of the NF-κB proteins

is that they are regulated at the level of nuclear translocation (Bonizzi and Karin, 2004; Chen and Greene, 2004). Under most circumstances, NF-κB proteins are located in the cytoplasm, bound to an inhibitor called IκB (Inhibitor of κB binding) (Ganchi et al., 1992; Hatada et al., 1992; Latimer et al., 1998). Because NF-κB proteins are transcription factors, the cytoplasmic localization renders them inactive. In vertebrates, several IκB proteins (IκBα, IκBβ, IκB ε, and Bcl-3) have been identified, and in Drosophila there appears to be a single IκB protein, called Cactus (Bonizzi and Karin, 2004; Chen and Greene, 2004; Gilmore and Ip, 2003). IκB proteins contain a central domain with approximately 6-8 ankyrin repeats, which mediate interaction with the RH domain. As discussed above, the C-terminal ankyrin repeat domains of p100, p105 and Relish can also act as intramolecular IκB sequences. Interaction of an NF-κB protein with the ankyrin repeat domain of IκB or of its own C-terminus causes the complex to reside in the cytoplasm, therefore preventing the NF-κB protein from moving into the nucleus where DNA binding can occur. The mechanism by which IκB inhibits NF-κB is masking of the nuclear localization signal on NF-κB, so that the nuclear import machinery cannot interact with the transcription factor (Ganchi et al., 1992; Hatada et al., 1992; Latimer et al., 1998).

The cytoplasmic NF-κB/IκB complex serves as a fast responding sensor to many extracellular stimuli, thus translating the stimulating signal into genetic activities within the cell (Huguet et al., 1997; Jacobs and Harrison, 1998; Malek et al., 1998; Ruben et al., 1992; Sun et al., 1994). Activation of the signal transduction pathway by binding of an extracellular ligand to its cell surface receptor leads to activation of an IκB kinase (IKK) complex (Bonizzi and Karin, 2004; Chen and Greene, 2004; Gilmore and Ip, 2003). The activated IKK complex phosphorylates two closely-spaced serine residues on IκB (Traenckner et al., 1995). The serine phosphorylation serves as a signal for IκB ubiquitination at nearby lysine residues, which then leads to IκB degradation by the 26S proteasome. The NF-κB transcription factor is therefore released from the inhibitor, and allowed to enter the nucleus to bind to the appropriate κB target sites and modulate gene expression. It should be noted that in B cells and several types of cancer cells, NF-κB proteins are constitutively located in the nucleus, due to a high rate of proteolysis of IκB.

The IKK complex is composed of at least three polypeptides: IKKα, IKKβ, and IKKγ (Bonizzi and Karin, 2004; Chen and Greene, 2004; Gilmore and Ip, 2003). IKKα and IKKβ are catalytically active kinases, while IKKγ does not contain kinase activity but serves as a scaffolding protein for the complex. Activation of the IKK kinases requires phosphorylation of two serine residues within the activation loop of IKKα or IKKβ. IKKβ primarily mediates the phosphorylation and degradation of IκBα complexes, while IKKκ controls the phosphorylation and processing of p100 to p52. In Drosophila, the IκB homologue Cactus is also phosphorylated and degraded upon signal stimulation (Belvin and Anderson, 1996). However, whether an IKK complex is involved in Cactus regulation is still unknown. There is an IKK complex identified in Drosophila. This complex contains at least the homologues of IKKβ (ird5 gene product) and IKKγ (kenny gene product) (Lu et al., 2001; Rutschmann et al., 2000; Silverman et al., 2000) The Drosophila IKK complex is required for the proteolytic cleavage and activation of Relish.

Many receptors of the innate and acquired immune systems in mammals act upstream in the NF-κB signaling pathways (Fig. 19.2). Receptors of the innate immune system such as the Toll-like receptors (TLR) and receptors for the inflammatory cytokines such as TNF and Interleukin-1 (IL-1) all can activate NF-κB (Beutler, 2004; Gilmore and Ip, 2003). After ligand stimulation, these receptors either multimerize or go through conformational change. The recruitment of adaptor proteins and kinases then stimulate the downstream events, the IKK and IκB as described above. In Drosophila, there are no B cells or T cells, and their antimicrobial response relies solely on the innate immune system (Hoffmann, 2003). The recognition and response to microbial infection in Drosophila depends on the Toll and the Immune deficiency (Imd) signaling pathways (Fig. 19.3). The Toll and Imd pathways in Drosophila are homologous to the TLR and TNFR pathways in mammals, respectively. For example, the mammalian TLR recruits the adaptor protein MyD88 and the kinase IRAK4 to stimulate the IKK complex, while the Drosophila Toll recruits a MyD88 homologue and the kinase Pelle, which is a homologue of IRAK4. In the TNFR pathway, the adaptor protein RIP is recruited to the receptor during activation. In the Imd pathway, the peptidoglycan recognition protein (PGRP)-LC serves as the receptor to recruit Imd, which is a homologue of RIP in Drosophila. Thus, the mechanism of signaling and the molecules involved in the mammalian and Drosophila NF-κB pathways are conserved (Brennan and Anderson, 2004; Gilmore and Ip, 2003).

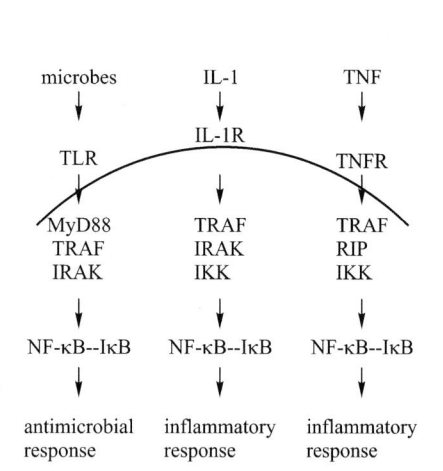

Fig.19.2 The mammalian NF-κB pathways. Cytokines such as TNF and Interleukin-1 are potent stimulators of signaling pathways that lead to NF-κB activation. The recognition of microbial compounds by Toll-like receptors (TLR) also leads to activation of NF-κB. These signaling pathways use many common components, and most combinations of the five NF-κB proteins can act downstream of these pathways. Furthermore, stimulation by different pathways leads to induction of distinct and overlapping target genes. The detailed mechanism that governs the signaling specificity and appropriate response through the five mammalian NF-κB proteins is still being investigated.

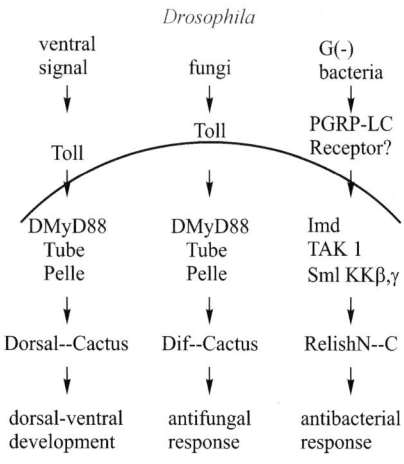

Fig.19.3 The signaling pathways that regulate NF-κB proteins in *Drosophila*. Dorsal is the key factor in dorsal-ventral embryonic development. Stimulation of the receptor Toll activates the cytoplasmic components and leads to the degradation of Cactus, the IκB homologue in *Drosophila*. This results in the release of Dorsal to enter the nucleus to regulate genes important for the establishment of ventral cell fate. During innate immune response in adults, essentially the same Toll pathway is used again, but with Dif acting as the key transcription factor. This Toll-Dif pathway responds to Gram-positive bacterial and fungal infections. Relish acts in the separate Imd signaling pathway, which regulates other aspects of innate immune response, primarily Gram-negative bacterial infection.

Mammalian NF-κB in Inflammatory and Immune Response

The NF-κB proteins in mammals play crucial roles in many processes related to inflammation, infection, and immune system activation. While numerous experiments have demonstrated the involvement of NF-κB in these immune-related processes, gene knockout experiments have provided invaluable insights into the *in vivo* functions of this family of transcription factors (Bonizzi and Karin, 2004; Chen and Greene, 2004; Gilmore and Ip, 2003). Targeted disruption of the genes encoding p50/p105, p52/p100, RelA, RelB, or c-Rel all lead to immune system defects. The knockout of *c-rel* in mice have a reduced immune response primarily because their B cells fail to proliferate in response to antigen (Kontgen *et al.*, 1995). The knockout of *nfkb1* (p50/p105) in mice caused reduced B cell proliferation and increased apoptosis in response to certain antigens (Sha *et al.*, 1995), *rela* (encoding RelA/p65) knockout mice die during embryonic development, due to massive apoptosis in the developing liver (Beg *et al.*, 1995). The RelA protein is required to protect the embryonic liver from tumor necrosis factor (TNF)-induced apoptosis. Genetic suppression experiments performed by crossing *TNF* mutant together with *rela* knockout showed that such combination prevented the massive apoptosis in the embryonic liver (Doi *et al.*, 1999). Many NF-κB target genes encode anti-apoptotic proteins, such as Bcl2 family proteins and IAP proteins. Furthermore, some tumor cells have high levels of nuclear NF-κB activity, probably providing resistance of these tumor cells to apoptosis. Overall, the individual knockout experiments demonstrate that each family member of NF-κB has a unique role in immune cell function.

To gain further insights into the function of NF-κB proteins, some laboratories combined the individual knockout mice strains and analyzed their phenotypes further. Mice with both the *nfkb1* (p50/p105) and *nfkb2* (p52/p100) genes disrupted have defects in B cell, macrophage, spleen, and thymus functions. The double mutants also develop bone defects, namely osteopetrosis, due to a failure of osteoclasts to mature properly (Iotsova *et al.*, 1997). Thus, the primary function of NF-κB is in the immune/inflammatory system, while the development of liver, skin, bone, and skeleton also involves these transcription factors.

Contrary to the loss-of-function phenotypes, constitutive activation of NF-κB causes or has been associated with certain human disease states, such as inflammatory bowel disease and arthritis. Moreover, some common anti-inflammatory agents, including aspirin and glucocorticoids, act partly by blocking the activation of NF-κB. Aspirin appears to block the induction of NF-κB by directly inhibiting the IκB kinase, and glucocorticoids acts by a variety of mechanisms, such as blocking nuclear translocation and antagonizing target gene activation by NF-κB. Many folk medicines that have anti-inflammatory or anti-cancer properties may act by inhibiting NF-κB.

Biological Functions of *Drosophila* NF-κB Proteins

The three NF-κB proteins in *Drosophila* have largely independent, though occasionally redundant, functions (Gilmore and Ip, 2003). Dorsal was originally identified as an important gene required for dorsal-ventral patterning in the early embryo (Steward, 1987). The maternally deposited Dorsal proteins forms a nuclear gradient in the ventral cells of the early embryo and this gradient is required to activate and repress many zygotic genes needed for proper embryonic development. This Dorsal gradient is set up by the Toll signaling pathway, which is constituted by components including MyD88, Tube (another adaptor protein), Pelle, and Cactus (Belvin and Anderson, 1996) (Fig. 19.3).

Dif was identified as the second NF-κB protein in *Drosophila* and was named Dorsal-related immunity factor (Ip *et al.*, 1993). During larval and adult stages, Dif act as the transcription factor downstream of the Toll pathway to stimulate the innate immune response. All the components mentioned above in the embryonic Toll pathway are utilized again in larvae and adults to regulate Dif during immune response (Brennan and Anderson, 2004). The Toll-Dif pathway in adults is stimulated by Gram-positive bacteria and fungi. Such infections activate a series of proteases, which in turn cleave the ligand called Spaetzle. It is noteworthy that the proteases used by the embryo to cleave Spaetzle are different from those used in adult innate immune response. Dif does not have substantial expression in the early embryo. However, if artificially expressed in the early embryo by genetic manipulation, Dif can act similarly to Dorsal by activating many zygotic genes required for dorsal-ventral embryonic development. During innate immune response in larvae, Dorsal and Dif probably have redundant functions. However, in adult flies, Dif is the primary factor in the Toll pathway while Dorsal is not essential. Why the requirements of these two NF-κB proteins are different in different developmental stages is still not understood.

The third *Drosophila* NF-κB protein, Relish, acts as the downstream transcription factor for the Imd pathway (Hedengren *et al.*, 1999). As describe above, Imd is an adaptor protein homologous to the mammalian TNFR-interacting protein RIP. The Imd pathway governs the response to Gram-negative bacterial infection. Gram-negative peptidoglycan binds to the receptor PGRP-LC, which then stimulates Imd. Imd relays the signal to downstream components such as TAK-1 and IKK, which probably activate a caspase and lead to the proteolytic processing of Relish.

The genetic components involved in the Toll pathway are largely different from those in the Imd pathway. The utilization of the three *Drosophila* NF-κB proteins is also rather distinct. However, some evidence suggests that there is interaction of the three NF-κB proteins in these two pathways. Double mutants of Dif and Relish have more severe defects in the activation of some downstream target genes. Moreover, some target genes are defective in their expression only when both pathways are blocked at the same time, while other genes are defective when either pathway is blocked (De Gregorio *et al.*, 2002). It has also been demonstrated that heterodimers are formed among the three NF-κB proteins (Han and Ip, 1999). These heterodimers may increase the complexity of the immune response in *Drosophila*.

Conclusion

The NF-κB proteins in mammals and insects have evolutionarily conserved structures and functions. These proteins play essential roles in development, immune response, and cancer. An important feature of this family of transcription factors is the cytoplasmic-nuclear translocation in response to stimulation. This mechanism provides a sensitive regulation when the animals face environmental challenges, such as microbial infection. There are multiple signaling pathways that feed into these transcription factors, and the many components involved are also subject to multifaceted modulation. These elaborate mechanisms control not only the fine-tuning of the system but also the complexity versus the specificity required for the response. How the cells and the whole animals respond to various environmental challenges to generate the appropriate genetic activity is what we are trying to understand.

References

Anrather, J., Racchumi, G., and Iadecola, C. (2005). cis-acting, element-specific transcriptional activity of differentially phosphorylated nuclear factor-kappa B. J Biol Chem *280*, 244-52.

Baeuerle, P.A., and Baltimore, D. (1989). A 65-kappa D subunit of active NF-kappa B is required for inhibition of NF-kappa B by I kappa B. Genes Dev. *3*, 1689-98.

Beg, A.A., Sha, W.C., Bronson, R.T., Ghosh, S., and Baltimore, D. (1995). Embryonic lethality and liver degeneration in mice lacking the RelA component of NF-kappa B. Nature *376*, 167-70.

Belvin, M.P., and Anderson, K.V. (1996). A conserved signaling pathway: the *Drosophila* Toll-dorsal pathway. Ann. Rev. Cell Dev. Biol. *12*, 393-416.

Berkowitz, B., Huang, D.B., Chen-Park, F.E., Sigler, P.B., and Ghosh, G. (2002). The x-ray crystal structure of the NF-kappa B p50.p65 heterodimer bound to the interferon beta -kappa B site. J Biol Chem *277*, 24694-700.

Betts, J.C., and Nabel, G.J. (1996). Differential regulation of NF-kappa B2(p100) processing and control by amino-terminal sequences. Mol Cell Biol *16*, 6363-71.

Beutler, B. (2004). Inferences, questions and possibilities in Toll-like receptor signalling. Nature *430*, 257-63.

Bonizzi, G., and Karin, M. (2004). The two NF-kappa B activation pathways and their role in innate and adaptive immunity. Trends Immunol *25*, 280-8.

Bours, V., Villalobos, J., Burd, P.R., Kelly, K., and Siebenlist, U. (1990). Cloning of a mitogen-inducible gene encoding a kappa B DNA-binding protein with homology to the rel oncogene and to cell-cycle motifs. Nature *348*, 76-80.

Brennan, C.A., and Anderson, K.V. (2004). *Drosophila*: the genetics of innate immune recognition and response. Annu Rev Immunol. *22*, 457-483.

Capobianco, A.J., Chang, D., Mosialos, G., and Gilmore, T.D. (1992). p105, the NF-kappa B P50 precursor protein, is one of the cellular proteins complexed with the v-Rel oncoprotein in transformed chicken spleen cells. J Virol *66*, 3758-67.

Chen, F.E., Huang, D.B., Chen, Y.Q., and Ghosh, G. (1998). Crystal structure of p50/p65 heterodimer of transcription factor NF-kappa B bound to DNA. Nature *391*, 410-3.

Chen, L.-F., and Greene, W.C. (2004). Shaping the nuclear action of NF-kappa B. Nat Rev Mol Cell Biol. *5*, 392-401.

Chen, Y. Q., Sengchanthalangsy, L. L., Hackett, A., and Ghosh, G. (2000). NF-kappa B p65 (RelA) homodimer uses distinct mechanisms to recognize DNA targets. Structure Fold Des *8*, 419-28.

De Gregorio, E., Spellman, P.T., Tzou, P., Rubin, G.M., and Lemaitre, B. (2002). The Toll and Imd pathways are the major regulators of the immune response in *Drosophila*. EMBO J. *21*, 2568-2579.

Doi, T.S., Marino, M.W., Takahashi, T., Yoshida, T., Sakakura, T., Old, L.J., and Obata, Y. (1999). Absence of tumor necrosis factor rescues RelA-deficient mice from embryonic lethality. Proc Natl Acad Sci USA *96*, 2994-9.

Dushay, M.S., Asling, B., and Hultmark, D. (1996). Origins of immunity: Relish, a compound Rel-like gene in the antibacterial defense of *Drosophila*. Proc. Natl. Acad. Sci. USA *93*, 10343-10347.

Fan, Y., Rayet, B., and Gelinas, C. (2004). Divergent C-terminal transactivation domains of Rel/NF-kappa B proteins are critical determinants of their oncogenic potential in lymphocytes. Oncogene *23*, 1030-42.

Ganchi, P.A., Sun, S.C., Greene, W.C., and Ballard, D.W. (1992). I kappa B/MAD-3 masks the nuclear localization signal of NF-kappa B p65 and requires the transactivation domain to inhibit NF-kappa B p65 DNA binding. Mol Biol Cell *3*, 1339-52.

Ghosh, S., Gifford, A.M., Riviere, L.R., Tempst, P., Nolan, G.P., and Baltimore, D. (1990). Cloning of the p50 DNA binding subunit of NF-kappa B: homology to rel and dorsal. Cell *62*, 1019-29.

Gilmore, T.D. (1990). NF-kappa B, KBF1, dorsal, and related matters. Cell *62*, 841-3.

Gilmore, T.D., and Ip, Y.T. (2003). Signal Transduction Pathways in Development and Immunity: Rel Pathways. In: **Nature Encyclopedia of Life Sciences**. *London: Nature Publishing Group, http://www.els.net/ doi:10.1038/npg.els.0002332.*

Han, Z.S., and Ip, Y.T. (1999). Interaction and specificity of Rel-related proteins in regulating *Drosophila* immunity gene expression. J. Biol. Chem. *274*, 21355-21361.

Hatada, E.N., Nieters, A., Wulczyn, F.G., Naumann, M., Meyer, R., Nucifora, G., McKeithan, T.W., and Scheidereit, C. (1992). The ankyrin repeat domains of the NF-kappa B precursor p105 and the protooncogene bcl-3 act as specific inhibitors of NF-kappa B DNA binding. Proc Natl Acad Sci USA *89*, 2489-93.

Hedengren, M., Asling, B., Dushay, M.S., Ando, I., Ekengren, S., Wihlborg, M., and Hultmark, D. (1999). Relish, a central factor in the control of humoral but not cellular immunity in *Drosophila*. Mol. Cell *4*, 827-837.

Henkel, T., Zabel, U., van Zee, K., Muller, J.M., Fanning, E., and Baeuerle, P.A. (1992). Intramolecular masking of the nuclear location signal and dimerization domain in the precursor for the p50 NF-kappa B subunit. Cell *68*, 1121-33.

Hoffmann, J.A. (2003). The immune response of *Drosophila*. Nature *426*, 33-38.

Hou, S., Guan, H., and Ricciardi, R.P. (2003). Phosphorylation of serine 337 of NF-kappa B p50 is critical for DNA binding. J Biol Chem *278*, 45994-8.

Huang, D.B., Chen, Y.Q., Ruetsche, M., Phelps, C.B., and Ghosh, G. (2001). X-ray crystal structure of proto-oncogene product c-Rel bound to the CD28 response element of IL-2. Structure (Camb) *9*, 669-78.

Huguet, C., Crepieux, P., and Laudet, V. (1997). Rel/NF-kappa B transcription factors and I kappa B inhibitors: evolution from a unique common ancestor. Oncogene *15*, 2965-74.

Iotsova, V., Caamano, J., Loy, J., Yang, Y., Lewin, A., and Bravo, R. (1997). Osteopetrosis in mice lacking NF-kappa B1 and NF-kappa B2. Nat Med *3*, 1285-9.

Ip, Y.T., Reach, M., Engstrom, Y., Kadalayil, L., Cai, H., Gonzalez-Crespo, S., Tatei, K., and Levine, M. (1993). *Dif*, a *dorsal*-related gene that mediates an immune response in *Drosophila*. Cell *75*, 753-763.

Jacobs, M.D., and Harrison, S.C. (1998). Structure of an IkappaBalpha/NF-kappa B complex. Cell *95*, 749-58.

Kieran, M., Blank, V., Logeat, F., Vandekerckhove, J., Lottspeich, F., Le Bail, O., Urban, M.B., Kourilsky, P., Baeuerle, P.A., and Israel, A. (1990). The DNA binding subunit of NF-kappa B is identical to factor KBF1 and homologous to the rel oncogene product. Cell 62, 1007-18.

Kontgen, F., Grumont, R.J., Strasser, A., Metcalf, D., Li, R., Tarlinton, D., and Gerondakis, S. (1995). Mice lacking the c-rel proto-oncogene exhibit defects in lymphocyte proliferation, humoral immunity, and interleukin-2 expression. Genes Dev 9, 1965-77.

Latimer, M., Ernst, M.K., Dunn, L.L., Drutskaya, M., and Rice, N.R. (1998). The N-terminal domain of IkappaB alpha masks the nuclear localization signal(s) of p50 and c-Rel homodimers. Mol Cell Biol 18, 2640-9.

Liu, J., Fan, Q.R., Sodeoka, M., Lane, W.S., and Verdine, G.L. (1994). DNA binding by an amino acid residue in the C-terminal half of the Rel homology region. Chem Biol 1, 47-55.

Lu, Y., Wu, L.P., and Anderson, K.V. (2001). The antibacterial arm of the *Drosophila* innate immune response requires an IkappaB kinase. Genes Dev 15, 104-10.

Malek, S., Huxford, T., and Ghosh, G. (1998). Ikappa Balpha functions through direct contacts with the nuclear localization signals and the DNA binding sequences of NF-kappa B. J Biol Chem 273, 25427-35.

Meyer, R., Hatada, E.N., Hohmann, H.P., Haiker, M., Bartsch, C., Rothlisberger, U., Lahm, H.W., Schlaeger, E.J., van Loon, A.P., and Scheidereit, C. (1991). Cloning of the DNA-binding subunit of human nuclear factor kappa B: the level of its mRNA is strongly regulated by phorbol ester or tumor necrosis factor alpha. Proc Natl Acad Sci USA 88, 966-70.

Moore, P.A., Ruben, S.M., and Rosen, C.A. (1993). Conservation of transcriptional activation functions of the NF-kappa B p50 and p65 subunits in mammalian cells and *Saccharomyces cerevisiae*. Mol Cell Biol 13, 1666-74.

Nolan, G.P., Ghosh, S., Liou, H.C., Tempst, P., and Baltimore, D. (1991). DNA binding and I kappa B inhibition of the cloned p65 subunit of NF-kappa B, a rel-related polypeptide. Cell 64, 961-9.

Rice, N.R., MacKichan, M.L., and Israel, A. (1992). The precursor of NF-kappa B p50 has I kappa B-like functions. Cell 71, 243-53.

Ruben, S.M., Klement, J.F., Coleman, T.A., Maher, M., Chen, C.H., and Rosen, C.A. (1992). I-Rel: a novel rel-related protein that inhibits NF-kappa B transcriptional activity. Genes Dev 6, 745-60.

Ruben, S.M., Narayanan, R., Klement, J.F., Chen, C.H., and Rosen, C.A. (1992). Functional characterization of the NF-kappa B p65 transcriptional activator and an alternatively spliced derivative. Mol Cell Biol 12, 444-54.

Rutschmann, S., Jung, A.C., Zhou, R., Silverman, N., Hoffmann, J.A., and Ferrandon, D. (2000). Role of *Drosophila* IKK gamma in a toll-independent antibacterial immune response. Nat Immunol 1, 342-7.

Schmitz, M.L., dos Santos Silva, M.A., Altmann, H., Czisch, M., Holak, T.A., and Baeuerle, P.A. (1994). Structural and functional analysis of the NF-kappa B p65 C terminus. An acidic and modular transactivation domain with the potential to adopt an alpha-helical conformation. J Biol Chem 269, 25613-20.

Sen, R., and Baltimore, D. (1986). Multiple nuclear factors interact with the immunoglobulin enhancer sequences. Cell 46, 705-16.

Sha, W.C., Liou, H.C., Tuomanen, E.I., and Baltimore, D. (1995). Targeted disruption of the p50 subunit of NF-kappa B leads to multifocal defects in immune responses. Cell 80, 321-30.

Silverman, N., Zhou, R., Stoven, S., Pandey, N., Hultmark, D., and Maniatis, T. (2000). A *Drosophila* IkappaB kinase complex required for Relish cleavage and antibacterial immunity. Genes Dev 14, 2461-71.

Steward, R. (1987). Dorsal, an embryonic polarity gene in *Drosophila* is homologous to the vertebrate proto-oncogen, c-rel. Science 238, 692-694.

Stoven, S., Ando, I., Kadalayil, L., Engstrom, Y., and Hultmark, D. (2000). Activation of the *Drosophila* NF-kappa B factor Relish by rapid endoproteolytic cleavage. EMBO Rep 1, 347-52.

Stoven, S., Silverman, N., Junell, A., Hedengren-Olcott, M., Erturk, D., Engstrom, Y., Maniatis, T., and Hultmark, D. (2003). Caspase-mediated processing of the *Drosophila* NF-kappa B factor Relish. Proc. Natl. Acad. Sci. USA. 100, 5991-5996.

Sun, S.C., Ganchi, P.A., Beraud, C., Ballard, D.W., and Greene, W.C. (1994). Autoregulation of the NF-kappa B transactivator RelA (p65) by multiple cytoplasmic inhibitors containing ankyrin motifs. Proc Natl Acad Sci USA 91, 1346-50.

Traenckner, E. B., Pahl, H. L., Henkel, T., Schmidt, K. N., Wilk, S., and Baeuerle, P. A. (1995). Phosphorylation of human I kappa B-alpha on serines 32 and 36 controls I kappa B-alpha proteolysis and NF-kappa B activation in response to diverse stimuli. EMBO J. 14, 2876-83.

Xia, C., Watton, S., Nagl, S., Samuel, J., Lovegrove, J., Cheshire, J., and Woo, P. (2004). Novel sites in the p65 subunit of NF-kappa B interact with TFIIB to facilitate NF-kappa B induced transcription. FEBS Lett 561, 217-22.

Yang, J., and Steward, R. (1997). A multimeric complex and the nuclear targeting of the *Drosophila* Rel protein Dorsal. Proc Natl Acad Sci USA 94, 14524-9.

Chapter 20
The ATF Transcription Factors in Cellular Adaptive Responses

Tsonwin Hai

Department of Molecular and Cellular Biochemistry, Center for Molecular Neurobiology, Ohio State University, Columbus OH, USA

Key Words: bZip proteins, CREB, ATF1, ATF2, ATF3, ATF4, ATF6, stress response, homeostasis

Summary

The mammalian ATF/CREB family of transcription factors represents a group of basic region-leucine zipper (bZip) proteins consisted of almost 20 members. The name ATF or CREB was originally defined in the late 1980's by their ability to bind to the consensus ATF/CRE site "TGACGTCA." Over the years, cDNA clones encoding identical or homologous proteins have been isolated. Dendrogram analysis of these proteins on the basis of their amino acid sequences in the bZip region indicates that some of them are more similar to the other bZip proteins-AP-1 (Fos/Jun) and C/EBP – than to the other ATF/CREB proteins. Furthermore, members of the ATF/CREB proteins form heterodimers with the AP-1 or C/EBP proteins and the resulting dimers have altered DNA binding specificity. Therefore, the ATF prefix of these bZip proteins reflects the history of discovery, rather than the real similarity between them. In this chapter, I will briefly describe the classification of the ATF/CREB proteins with a historical perspective of their nomenclature, and then briefly review three ATF proteins – ATF3, ATF4 and ATF6. One common feature of these proteins is their involvement in cellular responses to extracellular signals, suggesting a role for these ATF proteins in adaptation and homeostasis.

Introduction

A: ATF/CREB Proteins- a Historical Perspective and the Subgroups

Activating Transcription Factor (ATF) was first named in 1987 to refer to a putative protein with the activity to bind to the adenovirus early promoters E2, E3 and E4 at sites with a common core sequence "CGTCA" (Lee *et al.*, 1987). cAMP responsive element binding protein (CREB) was named in 1987 to refer to a putative protein with the activity to bind to the cAMP responsive element (CRE) on the somatostatin promoter (Montminy and Bilezsikjian, 1987). The consensus binding site for ATF was later defined as TGACGT(C/A)(G/A) (Lin and Green, 1988), a sequence identical to the CRE consensus (TGACGTCA) (Deutsch *et al.*, 1988 and references therein, for a review see Roesler *et al.*, 1988). The identification of identical consensus sequences on two seemingly different sets of promoters - one on viral promoters and the other one on cellular promoters - generated much confusion in the early days and prompted many groups to purify the corresponding binding proteins. After the dust had settled, the names ATF and CREB were used to refer to a group of bZip proteins that conform to the following criteria: (1) they bind to the consensus ATF/CRE sequence "TGACGTCA" *in vitro*, and (2) they form homo- or heterodimers. Around 20 different mammalian proteins with the prefix ATF or CREB have been described, and can be grouped into subgroups on the basis of their amino acid similarity: the CREB/CREM, CRE-BP1 (commonly known as ATF2), ATF3, ATF4, ATF6 and B-ATF subgroups. Proteins within each subgroup share significant similarity both inside and outside the bZip domain. Proteins between

---

Corresponding Author: Room 148, Rightmire Hall, 1060 Carmack Road, Center for Molecular Neurobiology, Ohio State University, Columbus OH 43210, Tel: (614) 292-2910, Fax: (614) 292-5379, E-mail: hai.2@osu.edu

the subgroups, however, do not share much similarity other than the bZip "motif." Therefore, they should be considered as distinct proteins despite their common prefix (ATF or CREB). Table 20.1 lists the subgroups and their corresponding members (some original references are cited in a previous review, Hai et al., 1999).

Table 20.1 The mammalian ATF/CREB family of transcription factors.

Subgroup	Members	Alternative Names
CREB	CREB	CREB-1, CREB-327, ATF-47
	CREM*	
	ATF1	ATF-43, TREB, TREB36, TCRATF1
CRE-BP1	CRE-BP1	ATF2, CREB-2, TCR-ATF2, mXBP, TREB, HB16
	ATFa	ATF7
	CREBPA	
ATF3	ATF3	LRF-1, LRG-21, CRG-5, TI-241
	JDP2	
ATF4	ATF4	CREB2, TAXREB67, mATF4, C/ATF, mTR67
	ATF4L1	
	ATFx	hATF5, ATF7
ATF6	ATF6	ATF6α
	CREB-RP	ATF6β, CREBL1, G13
B-ATF	B-ATF	
	JDP1	p21SNFT, DNAJC12

* The CREM gene also encodes a protein product ICER from an alternative intronic promoter.

B: The Nomenclature of the ATF/CREB Proteins

The nomenclature in the literature for this family of proteins has been confusing. Not only alternative names are used to refer to the same proteins (see Table 20.1), in some cases, the same name is used to refer to different proteins. As an example, the term CREB2 has been used to refer to three different proteins: an alternatively spliced CREB , CRE-BP1 (ATF2), and ATF4. Another example is ATF5, it could refer to Fos or ATFx, an ATF4-homologous protein. The third example is ATF7, the clone isolated and named as ATF7 in 2001 (Peters et al., 2001) is the same as ATFx (also called hATF5). Furthermore, ATFa was renamed as ATF7 (Hamard et al., 2005). It is not clear whether either of these two clones has any resemblance to the original ATF7 cDNA isolated in 1989 (Hai et al., 1989), since the sequence of that clone is not available. For more details on the nomenclature, see (Hai et al., 1999). Therefore, to ensure the identity of a given protein, the best way is to inspect the amino acid sequence.

C: The ATF/CREB Proteins versus other bZip Proteins

Thus far, more than 50 bZip proteins have been identified. Despite the different names for these bZip proteins - ATF/CREB, AP-1 (Fos and Jun proteins), Maf, and C/EBP, several lines of evidence indicate that the distinction between them is blurred. First, dendrogram analysis (Fig. 20.1) of the bZip region indicates that some ATF proteins are more closely related to bZip proteins without the ATF prefix than those with the prefix. As an example, ATF3 is more homologous to the Fos proteins (members of the AP-1 family) than to other subgroups of the ATF/CREB proteins as pointed out previously (Meyer and Habener, 1993). Second, the ATF/CREB consensus binding site (TGAC$\underline{G}$TCA) has only one nucleotide difference from the AP-1 and Maf consensus binding sites (TGACTCA), and some naturally occurring binding sites are composite sites, such as a composite of half ATF/CRE site and half C/EBP site. Thus, the names of the DNA binding sites (the ATF/CRE, AP-1 and C/EBP sites) do not reflect the intrinsic biological distinctions; rather, they reflect the artificial nature imposed by the researchers who studied them. Third, these bZip proteins can bind to each other's consensus sites and regulate transcription in a manner characteristic of the other family (for specific examples see Hai et al., 1999). Fourth, overwhelming evidence indicates that the ATF/CREB proteins form selective heterodimers with each other, and with other bZip proteins such as the AP-1 and C/EBP families of proteins. For a list of the heterodimeric partners, see an analysis of bZip association by coil-coil array (Newman and Keating, 2003), and previous reviews (Hai et al., 1999; Hai and Hartman, 2001). Although the physiological significance of these heterodimers is not clear, the general view is that heterodimer formation can alter DNA binding specificity and transcriptional activities, thus expanding the ability of these bZip proteins to regulate gene expression.

Taken together, various names have been used to refer to the mammalian bZip proteins. However, from their amino acids sequences, DNA binding sites, and heterodimer formation preferences, the distinction in their names is arbitrary; it reflects more the history of discovery than the fundamental biological differences between them.

The ATF Proteins

Although much remains to be elucidated, one common theme of the ATF proteins is their involvement in responding to extracellular signals. Due to the vast number of papers in the literature, a comprehensive review

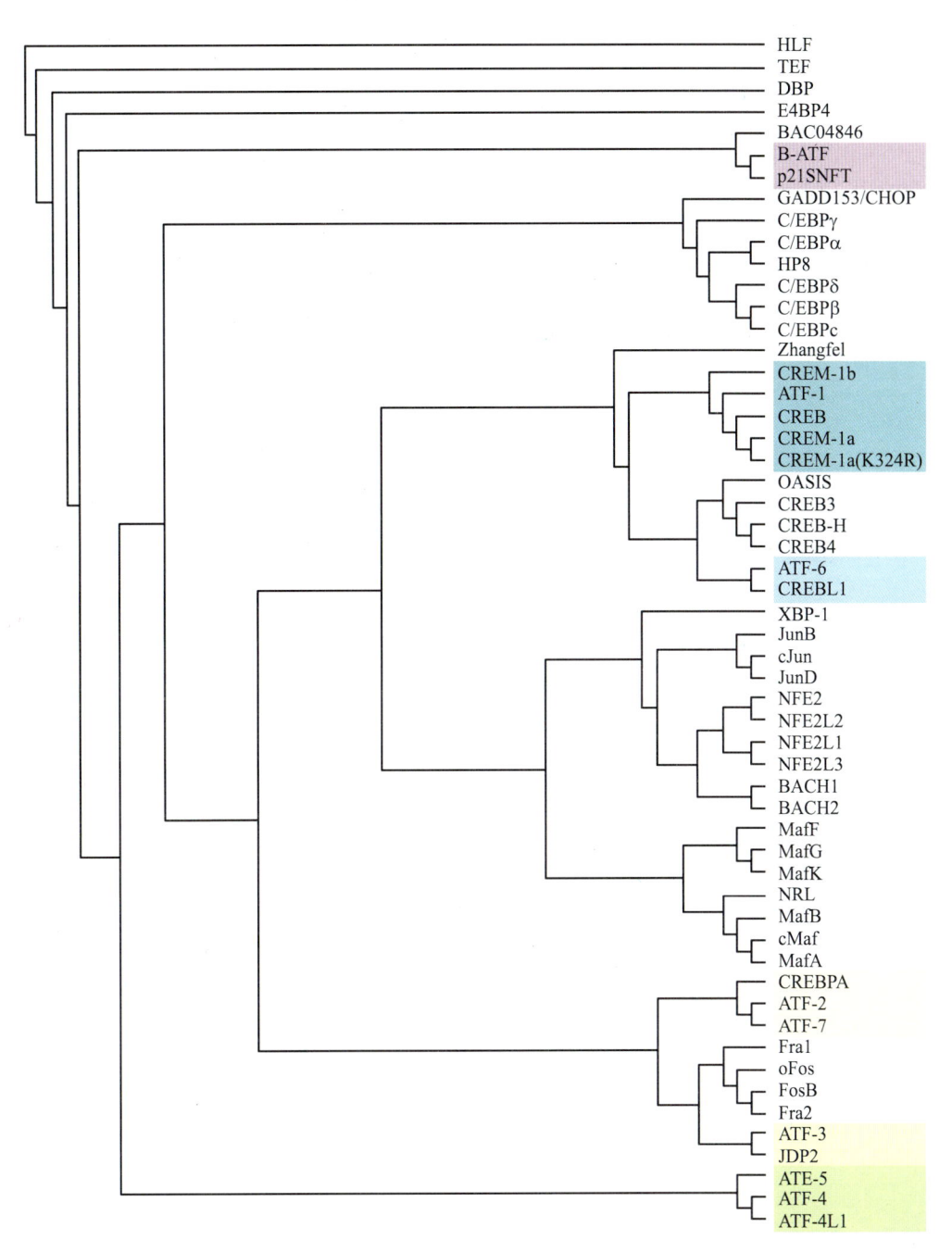

```
                            ┌── HLF
                        ┌───┤
                        │   └── TEF
                    ┌───┤
                    │   ├── DBP
                    │   └── E4BP4
                    │       ┌── BAC04846
                    │   ┌───┤
                    │   │   ├── B-ATF
                    │   │   └── p21SNFT
                    │   │       ┌── GADD153/CHOP
                    │   │   ┌───┤
                    │   │   │   ├── C/EBPγ
                    │   │   │   ├── C/EBPα
                    │   │   │   ├── HP8
                    │   │   │   ├── C/EBPδ
                    │   │   │   ├── C/EBPβ
                    │   │   │   └── C/EBPε
                    │   │   │       ├── Zhangfel
                    │   │   │       ├── CREM-1b
                    │   │   │       ├── ATF-1
                    │   │   │       ├── CREB
                    │   │   │       ├── CREM-1a
                    │   │   │       └── CREM-1a(K324R)
                    │   │   │           ├── OASIS
                    │   │   │           ├── CREB3
                    │   │   │           ├── CREB-H
                    │   │   │           ├── CREB4
                    │   │   │           ├── ATF-6
                    │   │   │           └── CREBL1
                    │   │   │               ├── XBP-1
                    │   │   │               ├── JunB
                    │   │   │               ├── cJun
                    │   │   │               ├── JunD
                    │   │   │               ├── NFE2
                    │   │   │               ├── NFE2L2
                    │   │   │               ├── NFE2L1
                    │   │   │               ├── NFE2L3
                    │   │   │               ├── BACH1
                    │   │   │               └── BACH2
                    │   │   │                   ├── MafF
                    │   │   │                   ├── MafG
                    │   │   │                   ├── MafK
                    │   │   │                   ├── NRL
                    │   │   │                   ├── MafB
                    │   │   │                   ├── cMaf
                    │   │   │                   └── MafA
                    │   │   │                       ├── CREBPA
                    │   │   │                       ├── ATF-2
                    │   │   │                       └── ATF-7
                    │   │   │                           ├── Fra1
                    │   │   │                           ├── oFos
                    │   │   │                           ├── FosB
                    │   │   │                           └── Fra2
                    │   │   │                               ├── ATF-3
                    │   │   │                               └── JDP2
                    │   │   │                                   ├── ATF-5
                    │   │   │                                   ├── ATF-4
                    │   │   │                                   └── ATF-4L1
```

Fig. 20.1 A dendrogram of bZip proteins based on their amino acid sequences in the bZip region. The ATF proteins listed in Table 20.1 are highlighted. Some ATF proteins are more closely related to bZip proteins without the ATF prefix than those with the prefix. As an example, ATF3 is more closely related to the Fos proteins than to ATF1.

of the ATF/CREB family of transcription factors is not possible. Below, I will only review ATF3, ATF4 and ATF6 to illustrate the above point of their commonality. For reviews on the CREB/ATF1 proteins, see (Andrisani, 1999; Daniel et al., 1998; De Cesare et al., 1999; Montminy, 1997; Sassone-Corsi, 1998); for reviews on the ATF2 proteins, see (van Dam and Castellazzi, 2001; Gupta et al., 1995; Raingeaud et al., 1995; Firestein and Feuerstein, 1998; Fuchs et al., 1997). For the transcriptional activity, interacting proteins and potential target genes of ATF2, ATF4 and ATF6, see a previous review (Hai and Hartman, 2001). In this chapter, I will focus on their biological functions with an emphasis on stress response.

A: ATF3
A1: A Stress-inducible Gene

ATF3 was suggested to be a stress-inducible gene based on the observation that its mRNA level greatly increases in various stress models (Chen et al., 1996).

Since then, overwhelming evidence from many laboratories supports the notion that ATF3 is induced by stress signals (a review, Hai et al., 1999). The wide use of the DNA microarray technique added to the long list of signals that can increase the steady-state mRNA levels of ATF3. Table 20.2 summarizes the signals that have been demonstrated to induce ATF3.

Table 20.2 A partial list of the treatments that induce the expression of ATF3 In whole organisms.

In whole organisms

Tissues	Treatments
Liver	Partial hepatectomy, Alcohol, Carbon tetrachloride, Acetaminophen, Cycloheximide, Hepatic ischemia, Lipopolysaccharide (LPS)
Lung	Ventilation induced acute injury
Heart	Ischemia, Ischemia-reperfusion
Kidney	Ischemia-reperfusion
Skin	Wounding
Muscle	Eccentric contraction
Thymus	Anti-CD3 ε
Pancreas	Ischemia-reperfusion, Partial pancreatectomy, Streptozotocin, Caerulein, Taurocholate
Prostate	Ischemia
Spinal cord neurons	Axotomy
Hypoglossal motor neurons	Nerve transection
Corticospinal neurons	Intracortical axotomy
Dentate gyrus	Seizure
Geniculate ganglion	Chorda tympani injury
Dorsal root ganglion	Chronic constriction injury

In cultured cells

Cells	Treatments
Hepatocytes	Cycloheximide, EGF, HGF
Leukemia cells	Doxorubicin
Macrophages	Cytokines, LPS, BCG, PMA, A23187
Myeloid cells	Fas antibody
Neuroblastoma	Forskolin, FGF, A23187
PC12 cells	Arsenite
PC6-3 cells (a subline of PC12)	Depletion of NGF
MCF7 cells	Proteasome inhibitors, Adipokines
HUVECs	PPAR activators, Homocysteine, TNF-α, LDL, oxLDL, LPC, A23187, Thapsigargin, Tunicamycin, H_2O_2, Nitric oxide
Pancreatic β cells	H_2O_2, Elevated palmitate, Elevated glucose, Cytokines, Nitric oxide
HeLa cells	Camptothecin, Adenovirus infection
Cardiomyocytes	NaCN, Deoxyglucose, Doxorubicin, Adenylylcyclase type VI
LβT2 gonadotrope cell line	Gonadotropin-releasing hormone (GnRH)
Gastric cancer AGS cells	*H. pylori* infection
Keratinocytes	Diindolylmethane (DIM)
Hepatoma FaO cells	TGF-β
HepG2 cells	Amino acid or glucose deprivation, ER stress
Mouse embryonic fibroblasts	Amino acid depletion, ER stress
HME87	benzo[a]pyrene diol epoxide (BPDE)
αTC1.6 cells	Forskolin/IBMX
Epithelial cells	TGF-β
HCT-116 cells	Cox-1 and Cox-2 inhibitors
Peripheral blood monocytes	PEG-IFN-α
Various cell types	Serum, Anisomycin, E1A, Genotoxic agents (ionizing radiation, UV, MMS)

One dramatic feature is that the induction of ATF3 is neither tissue-specific nor stimulus-specific. Furthermore, its induction is not limited to a specific species: when data are available, the same stimulus can induce ATF3 in rat, mouse and human. This "non-specificity" is not unique to ATF3. Other genes such as Fos, Jun and Erg-1 are also induced by a variety signals in many different tissues. Interestingly, these inducible genes (known as the immediate early genes) are usually induced in the same cluster as shown by DNA microarray analysis. Therefore, the initial genome response to extracellular signals appears to turn on a set of common genes, irrespective of the nature of the signals or the cell type exposed to the signals. The diversity in the final readouts is most likely determined by the context of the cells.

A2: An Adaptive-response Gene

One common theme of the signals that can induce ATF3 is that, to the first approximation, they all induce cellular damages and can be viewed as stress signals. However, some ATF3-inducing signals do not obviously fit into the category of stress signals. As an example, ATF3 is induced in the MCF-7 breast cancer cells by adipokines (Iyengar et al., 2003), secreted factors from adipocytes that increase cell growth and migration rather than cellular damage. Furthermore, ATF3 expression is induced in S-phase: microarray analyses of genes in different phases of the cell cycle from HeLa cells or primary human fibroblasts indicated that ATF3 expression is induced in the S-phase (Cho et al., 2001; van der Meijden et al., 2002). Since adipokines and S phase transition do not fit the conventional definition of stress signals, their ability to induce ATF3 indicates that the characterization of ATF3 as a stress-inducible gene is overly simplistic. A more complete view is that ATF3 is an "adaptive-response gene," a gene that participates in the processes for the cells to adapt to changes-extracellular and/or intracellular changes. In summary, the lesson we learned from the expression pattern of ATF3 is that it is one of the common subset of "adaptive response" genes that plays a role in a variety of cellular processes. Clearly, a key question then arises: what is the functional significance of ATF3 expression?

A3: Biological Consequences of ATF3 Expression – the Dichotomy of ATF3

Although the functions of ATF3 are not well-understood, one emerging theme is that it has opposite effects on various biological processes, including cell death and cell cycle machineries. As an example, both pro- and anti-apoptotic effects of ATF3 have been reported. Ectopic expression of ATF3 enhanced the ability of etoposide or camptothecin to induce apoptosis in HeLa cells (Mashima et al., 2001) and the ability of curcumin, an anti-cancer compound, to induce apoptosis (Yan et al., 2005), suggesting a pro-apoptotic role of ATF3. Consistently, transgenic mice expressing ATF3 have functional defects in the corresponding tissues: mice expressing ATF3 in the heart have conduction abnormalities and contractile dysfunction (Okamoto et al., 2001); Mice expressing ATF3 in the liver and pancreatic ductal epithelium have liver dysfunction and defects in endocrine pancreas development (Allen-Jennings et al., 2001; Allen-Jennings et al., 2002). Furthermore, primary islets derived from ATF3 knockout mice are partially protected from stress-induced apoptosis (Hartman et al., 2004), and consistently, antisense approach to inhibit ATF3 suppressed stress-induced apoptosis in endothelial cells (Nawa et al., 2002). Therefore, both gain-of-function (ectopic expression) and loss-of-function (knockout or anti-sense) approaches support a pro-apoptotic role of ATF3.

However, several reports suggest a protective effect of ATF3. ATF3 was demonstrated to protect neuronal cells from NGF withdraw-induced death (Nakagomi et al., 2003) and kainic acid-induced death. Furthermore, ATF3 enhanced c-Jun induced-neurite sprouting in culture (Pearson et al., 2003). Even in cardiac myocytes and endothelial cells, two cell types that ATF3 had been demonstrated to have detrimental effects (see above), ATF3 was demonstrated to have protective roles (Nobori et al., 2002; Kawauchi et al., 2002). Therefore, ATF3 has a dichotomous role in apoptosis, it can be either pro- or anti-apoptotic, presumably in a context-dependent manner.

TF3 has also been implicated to regulate the cell cycle machinery. Similar to the situation in apoptosis, ATF3 has been demonstrated to have opposite effects: it can either promote or suppress cell cycle progression. Ectopic expression of ATF3 moderately induced DNA synthesis and cyclin D1 gene expression in hepatoma cells (Allan et al., 2001). Furthermore, ectopic expression of ATF3 partially transformed chick embryo fibroblasts by promoting proliferation under low serum concentration (Perez et al., 2001). These, in combination with its induction by viral transforming proteins adenoviral E1A (Hagmeyer et al., 1996) and hepatitis B virus X protein (Tarn et al., 1999), suggest that ATF3 promotes cell proliferation. However, Fan et al. reported that ectopic expression of ATF3 suppresses cell cycle progression in HeLa cells (Fan et al., 2002). Therefore, conflicting data exist for the roles of ATF3 in cell cycle regulation. Again, one explanation for the apparent discrepancy is the difference in cellular context.

A4: Does ATF3 Play a Role in Human Diseases

Taken together, several lines of evidence described above support the notion that ATF3 contributes to the pathogenesis of some stress-associated diseases. First, it is induced by stress signals and is expressed in the diseases tissues, such as the diabetic islets (Hartman *et al.*, 2004) and atherosclerotic vessels (Nawa *et al.*, 2002). Second, it can be pro-apoptotic by both gain- and loss-of function approaches in cultured cells. Third, transgenic mice expressing ATF3 have dysfunction of the corresponding tissues expressing it. Although further studies are required to prove the causal relationship, these lines of evidence support a role of ATF3 in the pathogenesis of stress-associated diseases. In this context, it is interesting to note that ATF3 was demonstrated to promote cell migration of advanced tumor cells (Ishiguro *et al.*, 2000). Therefore, even in the context that ATF3 is not pro-apoptotic, its expression facilitates the progression of the disease. Does this mean that ATF3 always promotes disease progression? The answer is probably no. The consistent protective effects of ATF3 in the neuronal injury models suggest that ATF3 is likely a pro-regeneration and pro-neurite growth factor in the context of neuronal injuries (in peripheral neurons).

A5: A Proposed Model for ATF3 (Fig.20. 2)

In summary, ATF3 is an adaptive response gene. Its expression - induced by different stimuli in different cell types - contributes to the modulation of the cell death and/or cell cycle machineries in a context-dependent manner. Thus far, the data support the following speculation: in many cell types, ATF3 contributes to tissue dysfunction and disease progression; however, in the peripheral neurons, ATF3 appears to be beneficial by facilitating neurite outgrowth and regeneration.

B: ATF4

B1: ATF4 Protein Stability

ATF4 mRNA is present in all tissues examined thus far. Although its level can be up-regulated by a variety of extracellular signals in different cell types (review, Hai and Hartman, 2001), the mode of regulation for ATF4 is primarily at the translational initiation (see below, ATF4 in stress pathways) and post-translational levels. ATF4 subgroup of proteins (ATF4 and ATFx) interact with components of the proteasome system: ATF4 interacts the βTrCP, an F-box protein of the E3 ubiquitin ligase complex (Lassot *et al.*, 2001), and ATFx interacts with Cdc34, an E2 ubiquitin conjugating enzyme (Pati *et al.*, 1999). A current model is that phosphorylation of ATF4 targets ATF4 to the proteasome system via its interaction with βTrCP, this in turn results in the ubiquitination and subsequent degradation of ATF4 (Lassot *et al.*, 2001). Consistent with this model, the half-life of ATF4 is short (between 30 to 60 minutes), and inhibitors of the proteasome system increase its stability (Lassot *et al.*, 2001). ATF4 contains the sequence DSGXXXS, which is similar to the DSGXXS motif found in other βTrCP substrates, β-catenin, IκBα, and HIV-1 Vpu (see Lassot *et al.*, 2001 for references). Significantly, phosphorylation of IκBα at this motif by IKKα and IKKβ has been demonstrated to target IκBα for degradation by the proteasome system (for a review, see Karin, 1999). Therefore, the stability of ATF4 is regulated in a manner similar to that for the regulation of IκB. Clearly, identification of the kinase(s) and the signals that regulate this process will shed light on the functions of ATF4.

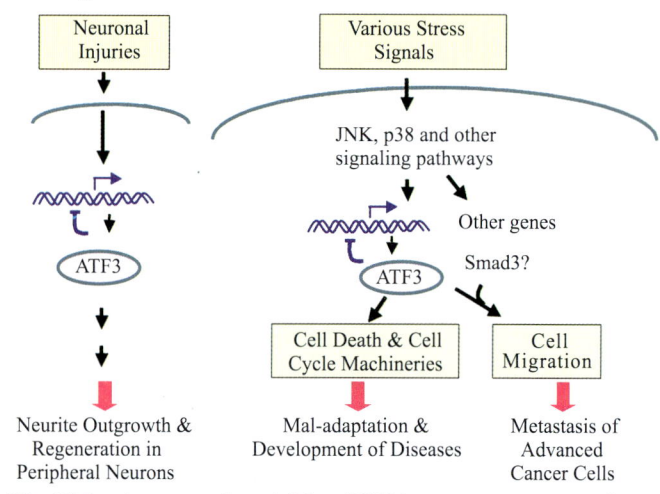

Fig. 20.2 A proposed model for ATF3 in stress response and adaptation. Two features in the figure are not described in the text: the involvement of the JNK and p38 stress pathways in the induction of ATF3 by stress signals, and the auto-repression of ATF3 gene expression by itself.

B2: ATF4 as an Integrator for Multiple Stress Pathways

ATF4 mRNA contains short upstream open reading frames (uORFs) in its 5′ untranslated region (5′UTR). These uORFs confer the regulation of ATF4 protein synthesis in a way much like the one observed for the yeast protein GCN4: translational initiation from the coding AUG is repressed under unstressed condition, but is facilitated under stressed conditions that result in eIF2-α phosphorylation (Harding *et al.*, 2000). Since various stress conditions including endoplasmic reticulum (ER) stress, nutrient starvation, viral infection, and oxidative stress, can result in eIF2-α phosphorylation (a review, Clemens, 2001), ATF4 serves as an integration point for various stress responses (reviews, Rutkowski and Kaufman, 2003; Ron, 2002). Because it is a transcription factor, ATF4 functions as a master switch to regulate downstream gene expression, presumably to help the cells to cope with the stress.

Identification of ATF4 target genes shed some light on the potential roles of ATF4 in these stress responses. Using cDNA microarray analysis to compare wild type and ATF4 knockout fibroblasts exposed to ER stress, Harding et al. identified potential ATF4 target genes (Harding et al., 2003). Among these are the genes involved in amino acid metabolism, amino acid transporter, and redox chemistry. They proposed the following logic for the activation of these genes during ER stress. The main functions of ER are to fold proteins and to initiate protein secretion. These processes result in the loss of reducing equivalents (due to oxidative protein folding) thus an increase in oxidants, and the loss of amino acids (due to secretion of proteins to the extracellular milieu). They proposed that ER stress represents an extra burden of losing reducing equivalents and amino acids. By activating genes for antioxidant and amino acid metabolism, ATF4 helps to cope with ER stress.

Other ATF4 target genes include transcription factors, signaling molecules and membrane proteins (Harding et al., 2003). Of particular interest to this chapter is the identification of ATF3 as a target gene of ATF4 (Jiang et al., 2004); ATF3 in turn regulates Gadd153/CHOP10 (Chen et al., 1996), a bZip protein in the C/EBP family of transcription factors, indicating a cascade of transcriptional program involving various bZip proteins in response to stress signals.

B3: ATF4 in Redox Homeostasis

The up-regulation of genes involved in redox chemistry by ATF4 deserves a special discussion. In addition to the report by Harding et al., ATF4 has been demonstrated to dimerize with the bZip proteins NF-E2 related factor 1 (Nrf1, also called NF-E2L1) and Nrf2 (NF-E2L2), and the resulting dimers can bind to the antioxidant responsive element (ARE), TGA(C/T)NNNGC (reviewed in Hayes and McMahon, 2001), which contains a half ATF/CRE site (TGAC). Although the physiological significance of the dimerization between ATF4 and Nrf proteins is not clear, the importance of ARE in regulating antioxidant genes is well accepted. The AREs are present in the promoters of many antioxidant genes, such as NAD(P) H:quinone oxidoreductase, glutathione S-transferase, heme oxygenase, and glutathione synthetase (a review, Hayes and McMahon, 2001). Thus, various clues point to a role of ATF4 in regulating the redox state of the cells. In this context, it is interesting to note that ATF4 protein is stabilized by anoxia in human cancer cells (Ameri et al., 2004), further supporting a role of ATF4 in redox

homeostasis. Functionally, ectopic expression of ATF4 was demonstrated to accelerate apoptosis in mammary epithelium (Bagheri-Yarmand et al., 2003). However, ectopic expression of ATFx–an ATF4 homologous protein with ~55% identity–was found to repress apoptosis in hematopoietic cells (Persengiev et al., 2002; a review Persengiev and Green, 2003). The opposite effect may be due to the differences in their amino acid sequences, or the differences in the experimental paradigms, or both. Clearly, much remains to be elucidated for the functionality of ATF4 and its homologous protein.

B4: ATF4 in Neuronal Plasticity

Some unexpected results from the studies of the γ-aminobutyric acid (GABA) receptors suggested that ATF4 may play a role in modulating neuronal plasticity. Using the R1 subunit of the $GABA_B$ receptor as a bait, three laboratories independently isolated cDNA encoding ATF4 by the yeast two hybrid screen (Nehring et al., 2000; Vernon et al., 2001; White et al., 2000). The interaction between $GABA_B$ R1 and ATF4 was confirmed by co-immunoprecipitation, *in situ* co-distribution and *in vitro* pull down assay; however, the consequence of this interaction is controversial and not well understood. On the one hand, ATF4 was demonstrated to translocate from the cytoplasmic membrane to the nucleus upon the activation of the receptors (White et al., 2000); on the other hand, it was demonstrated to translocate from the nucleus to the cytoplasm (Vernon et al., 2001). Therefore, much work is required to clarify the cellular consequence of this $GABA_B$ RI-ATF4 interaction. However, this interaction is intriguing in light of the following work. Using gain-of-function (via ectopic expression) and loss-of-function (using dominant negative molecule or RANi knockdown) approaches in mouse or Aplysia model, researchers suggested that ATF4 suppresses long-term facilitation, synaptic plasticity and hippocampal-based spatial memory (Bartsch et al., 1995; Chen et al., 2003; Lee et al., 2003). Since $GABA_B$ receptors are present in both pre- and post-synaptic neurons and are widely recognized to play a role in neurotransmission and synaptic plasticity (Bowery and Enna, 2000), the $GABA_B$ RI-ATF4 interaction may provide a mechanistic basis for the above learning and memory studies. By directly binding to the receptor, ATF4 may function as a switch to directly couple (versus indirectly via various signaling cascades) neuronal activity to gene expression, a key step for long-term changes, thus affecting learning and memory.

B5: A Proposed Model for ATF4 Function (Fig.20.3)

Taken together, ATF4 is an integrator for multiple

stress responses. Upon stimulation, its expression is up-regulated primarily by increased translation or increased protein stability. By up-regulating genes involved in redox chemistry and amino acid metabolism, ATF4 helps the cells to cope with stress. In neuronal cells, ATF4 is most likely involved in modulating neuronal plasticity by coupling receptor activity to gene expression.

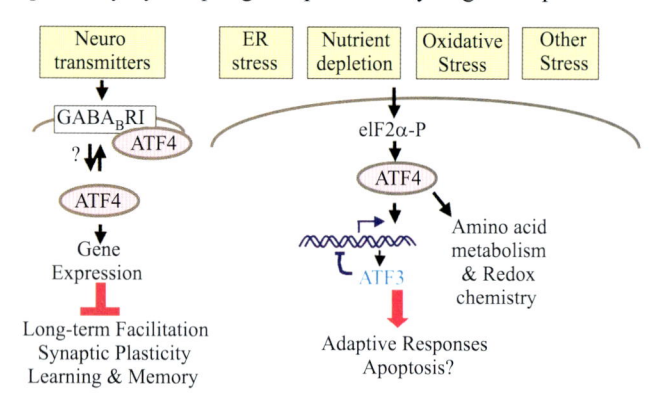

Fig.20.3 **A proposed model for ATF4 in various stress responses and neuronal plasticity.**

C: ATF6

cDNA encoding ATF6 was isolated independently by three experimental approaches addressing different questions. It was isolated as a protein that weakly binds to the ATF/CRE consensus sequence (Hai et al., 1989), a protein that interacts with the serum response factor (SRF, Zhu et al., 1997), or a protein that binds to the ER stress response element (ERSE, Yoshida et al., 1998). Although ATF6 was isolated based on its ability to bind to the palindromic ATF/CRE consensus (Hai et al., 1989) and was demonstrated to bind to the ATF/CRE site (Wang et al., 2000), a report by Mori and colleagues demonstrated that ATF6 prefers to bind to the ERSE which contains a half ATF/CRE site (in a NF-Y-dependent manner) than to the consensus ATF/CRE site (Yoshida et al., 2001). This again illustrates the arbitrary nature of the nomenclature of these bZip proteins as discussed in the beginning of this chapter. Below, I will review the roles of ATF6 in cellular responses to ER stress, the best documented function of ATF6 thus far.

C1: ATF6 in Unfolded Protein Response (UPR)

In response to ER stress, the cells respond by up-regulating genes encoding the chaperone proteins to facilitate protein folding and by down-regulating translational initiation of most mRNAs (except a few mRNAs, such as ATF4 mRNA discussed above) to reduce the ER load. These responses are collectively referred to as the unfolded protein response (UPR) (reviews, Kaufman, 1999; Mori, 2000; Urano et al., 2000). The first clue for the role of ATF6 in UPR came from the cloning of ATF6 cDNA based on its ability to bind to the ERSE (Yoshida et al., 1998), a site found in many ER-induced chaperone genes. Although the cDNA encodes a 90-kDa protein (Zhu et al., 1997; Yoshida et al., 1998), upon ER stress the protein is converted to around 50 kDa (Yoshida et al., 1998). Immunofluorescent and cell fractionation experiments indicated that ATF6 locates in the ER and translocates to the nucleus upon ER stress (Haze et al., 1999). This translocation from ER to nucleus coupled with protein cleavage is reminiscent of the regulation of sterol regulatory element binding proteins (SREBPs) which are cleaved by two proteases: Site-1 protease (S1P) and Site-2 protease (S2P) (a review, Brown et al., 2000). This, in combination with the presence of the S1P-like and S2P-like sites in ATF6, led to the discovery that ATF6 is regulated by the same proteases that process SREBPs (Ye et al., 2000; Chen et al., 2002; Shen and Prywes, 2004).

Three key observations added important pieces of the puzzle and revealed an intricate role of ATF6 in UPR. First, in an attempt to address how ATF6 is anchored at the ER before stress induction, Prywes and colleagues (Shen et al., 2002) investigated ATF6-interacting proteins by co-immunoprecipitation and found that ATF6 interacts with the ER chaperone Bip/grp78. Importantly, upon ER stress, ATF6 dissociates from Bip/grp78 and translocates to Golgi where the proteases S1P and S2P reside. Thus, this observation provided a mechanistic understanding of ATF6 retention in ER. Second, in their investigation of XBP-1, another ERSE binding protein (Yoshida et al., 1998), Mori and colleagues discovered that the XBP-1 mRNA is spliced by IRE1 (Yoshida et al., 2001), a protein well-known for its important role in UPR (reviews, Kaufman, 1999; Mori, 2000; Urano et al., 2000). Importantly, the amount of the XBP1 mRNA appears to be low and must be elevated first via the activation of its corresponding promoter by ATF6 (Yoshida et al., 2001). Therefore, to produce sufficient amount of XBP1 protein, two pathways are required: up-regulation of the gene by ATF6 and splicing of the mRNA by IRE1. Third, although ATF6 plays an important role in up-regulating many ER chaperone genes, it can also down-regulate gene expression: it binds to the sterol regulatory element-binding protein 2 (SREBP2) and represses the target genes of SREBP2, resulting in the attenuation of the lipogenic effects of SREBP2 (Zeng et al., 2004). This action has significant physiological implications, since prolonged nutrient deprivation, such as amino acid or glucose deficiency, induces ER stress. By activating

ATF6, the cells reduce the lipogenic effect of SREBP2, and thus save energy sources to withstand the stress. Therefore, ATF6 coordinates stress response with energy homeostasis.

C2: A Current Model for ATF6 (Fig.20.4)

In summary, ATF6 is synthesized as a precursor protein, which binds to the ER chaperone Bip/grp78 and localizes on the ER membrane. During ER stress, ATF6 dissociates from Bip/grp78 and translocates to the Golgi where it is cleaved by proteases to liberate its active N' terminal domain (ATF6N or p50). ATF6N is then translocated to the nucleus and up-regulates its target genes, including those encoding ER chaperone proteins and XBP1 which also activates various ER chaperone genes. However, ATF6 can also suppress SREBP2-mediated gene expression by directly interacting with it.

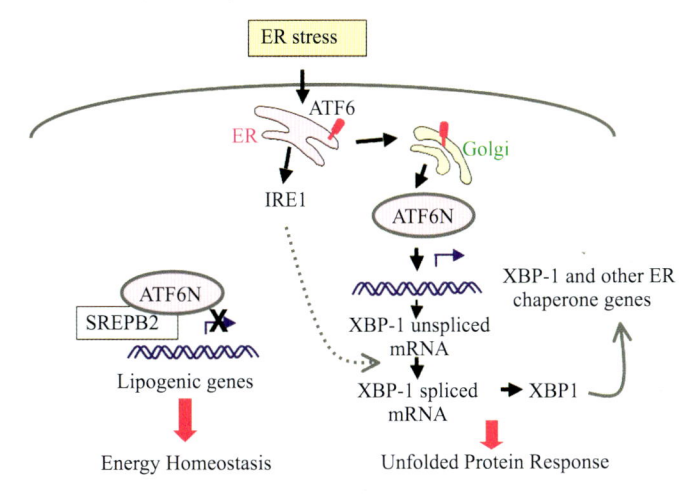

Fig. 20.4 A proposed model for ATF6 in cellular responses to ER stress.

Concluding Remarks

In conclusion, all three ATF proteins reviewed above (ATF3, ATF4 and ATF6) are modulated by extracellular signals: ATF3 is induced by transcriptional activation, ATF4 by translational initiation and protein stabilization, and ATF6 by proteolytic cleavage. Although not reviewed in this chapter, ATF1, CREB and ATF2 are regulated primarily by post-translational modifications: phosphorylation by various kinases. Thus, one common theme of the ATF proteins is their involvement in responding to extracellular signals. Interestingly, AP-1 (Fos/Jun) and C/EBP bZip proteins are also regulated by intra-/extra-cellular signals (see the chapter on AP1 in this book and some previous reviews: Darlington *et al.*, 1995; Yeh and McKnight, 1995; Karin *et al.*, 1997; Xanthoudakis and Curran, 1996), suggesting a common function for these bZip proteins.

Although sharing the bZip motif, these proteins are vastly divergent in their amino acid sequences. Intriguingly, their binding sites are relatively similar, suggesting that the DNA binding sites are conserved through the evolution to impart genomic responses to extracellular signals. The diversity of the proteins that can bind to these sites allows multifaceted and divergent biological responses.

Acknowledgement

I thank Chris Wolford for making the dendrogram and Dan Lu for making Table 20.2. Supports from the National Institute of Health (DK59605) and the American Diabetes Association (to T.H.).

Abbreviation

ATF, activating transcription factor; CREB, cAMP responsive element binding; bZip, basic region leucine zipper

References

Allan, A. L., Albanese, C., Pestell, R. G., and LaMarre, J. (2001). Activating transcription factor 3 induces DNA synthesis and expression of cyclin D1 in hepatocytes. J Biol Chem *276*, 27272-27280.

Allen-Jennings, A. E., Hartman, M. G., Kociba, G. J., and Hai, T. (2001). The roles of ATF3 in glucose homeostasis: A transgenic mouse model with liver dysfunction and defects in endocrine pancreas. J Biol Chem *276*, 29507-29514.

Allen-Jennings, A. E., Hartman, M. G., Kociba, G. J., and Hai, T. (2002). The roles of ATF3 in liver dysfunction and the regulation of phosphoenolpyruvate carboxykinase gene expression. J Biol Chem *277*, 20020-20025.

Ameri, K., Lewis, C. E., Raida, M., Sowter, H., Hai, T., and Harris, A. L. (2004). Anoxic induction of ATF-4 through HIF-1-independent pathways of protein stabilization in human cancer cells. Blood *103*, 1876-1882.

Andrisani, O. M. (1999). CREB-mediated transcriptional control. Crit Rev Euk Gene Exp *9*, 19-32.

Bagheri-Yarmand, R., Vadlamudi, R. K., and Kumar, R. (2003). Activating transcription factor 4 overexpression inhibits proliferation and differentiation of mammary epithelium resulting in impaired lactation and accelerated involution. J Biol Chem *278*, 17421-17429.

Bartsch, D., Ghirardi, M., Skehel, P. A., Karl, K. A., Herder, S. P., Chen, M., Bailey, C. H., and Kandel, E. R. (1995). Aplysia CREB2 represses long-term facilitation: Relief of repression converts transient facilitation into long-term functional and structural change. Cell *83*, 979-992.

Bowery, N. G., and Enna, S. J. (2000). Gamma-aminobutyric acid(B) receptors: first of the functional metabotropic heterodimers. J Pharmacol Exp Ther *292*, 2-7.

Brown, M. S., Ye, J., Rawson, R. B., and Goldstein, J. L. (2000). Regulated intramembrane proteolysis: a control mechanism conserved from bacteria to humans. Cell *100*, 391-398.

Chen, A., Muzzio, I. A., Malleret, G., Bartsch, D., Verbitsky, M., Pavlidis, P., Yonan, A. L., Vronskaya, S., Grody, M. B., Cepeda, I., *et al.* (2003). Inducible enhancement of memory storage and synaptic plasticity in transgenic mice expressing an inhibitor of ATF4 (CREB-2) and C/EBP proteins. Neuron *39*, 655-669.

Chen, B. P. C., Wolfgang, C. D., and Hai, T. (1996). Analysis of ATF3: a transcription factor induced by physiological stresses and modulated by gadd153/Chop10. Mol Cell Biol *16*, 1157-1168.

Chen, X., Shen, J., and Prywes, R. (2002). The luminal domain of ATF6 senses endoplasmic reticulum (ER) stress and causes translocation of ATF6 from the ER to the Golgi. J Biol Chem *277*, 13045-13052.

Cho, R. J., Huang, M., Campbell, M. J., Dong, H., Steinmetz, L., Sapinoso, L., Hampton, G., Elledge, S. J., Davis, R. W., and Lockhart, D. J. (2001). Transcriptional regulation and function during the human cell cycle. Nat Genet *27*, 48-54.

Clemens, M. J. (2001). Initiation factor eIF2 alpha phosphorylation in stress responses and apoptosis. Prog Mol Subcell Biol *27*, 57-89.

Daniel, P. B., Walker, W. H., and Habener, J. F. (1998). Cyclic AMP signaling and gene regulation. Annu Rev Nutr *18*, 353-383.

Darlington, G. J., Wang, N., and Hanson, R. W. (1995). C/EBPa: a critical regulator of genes governing integrative metabolic processes. Curr Opin Genet Dev *5*, 565-570.

De Cesare, D., Fimia, G. M., and Sassone-Corsi, P. (1999). Signaling routes to CREM and CREB: plasticity in transcriptional activation. Trends Biochem Sci *24*, 281-285.

Deutsch, P. J., Hoeffler, J. P., Jameson, J. L., Lin, J. C., and Habener, J. F. (1988). Structural determinants for transcriptional activation by cAMP-responsive DNA elements. J Biol Chem *263*, 18466-18472.

Fan, F., Jin, S., Amundson, S. A., Tong, T., Fan, W., Zhao, H., Zhu, X., Mazzacurati, L., Li, X., Petrik, K. L., *et al.* (2002). ATF3 induction following DNA damage is regulated by distinct signaling pathways and over-expression of ATF3 protein suppresses cells growth. Oncogene *21*, 7488-7496.

Firestein, R., and Feuerstein, N. (1998). Association of activating transcription factor (ATF2) with the ubiquitin-conjugating enzyme hUBC9. J Biol Chem *273*, 5892-5902.

Fuchs, S. Y., Xie, B., Adler, V., Fried, V. A., Davis, R. J., and Ronai, Z. (1997). c-Jun NH_2-terminal kinases target the ubiquitination of their associated transcription factors. J Biol Chem *272*, 32163-32168.

Gupta, S., Campbell, D., Derijard, B., and Davis, R. J. (1995). Transcription factor ATF2 regulation by the JNK signal transduction pathway. Science *267*, 389-393.

Hagmeyer, B. M., Duyndam, M. C., Angel, P., de Groot, R. P., Verlaan, M., Elfferich, P., van der Eb, A., and Zantema, A. (1996). Altered AP-1/ATF complexes in adenovirus-E1-transformed cells due to E1A-dependent induction of ATF3. Oncogene *12*, 1025-1032.

Hai, T., and Hartman, M. G. (2001). The molecular biology and nomenclature of the ATF/CREB family of transcription factors: ATF proteins and homeostasis. Gene *273*, 1-11.

Hai, T., Liu, F., Coukos, W. J., and Green, M. R. (1989). Transcription factor ATF cDNA clones: an extensive family of leucine zipper proteins able to selectively form DNA-binding heterodimers. Genes Dev *3*, 2083-2090.

Hai, T., Wolfgang, C. D., Marsee, D. K., Allen, A. E., and Sivaprasad, U. (1999). ATF3 and stress responses. Gene Expression *7*, 321-335.

Hamard, P. J., Dalbies-Tran, R., Hauss, C., Davidson, I., Kedinger, C., and Chatton, B. (2005). A functional interaction between ATF7 and TAF12 that is modulated by TAF4. Oncogene.

Harding, H. P., Novoa, I., Zhang, Y., Zeng, H., Wek, R., Schapira, M., and Ron, D. (2000). Regulated translation initiation controls stress-induced gene expression in mammalian cells. Mol Cell *6*, 1099-1108.

Harding, H. P., Zhang, Y., Zeng, H., Novoa, I., Lu, P. D., Calfon, M., Sadri, N., Yun, C., Popko, B., Paules, R., *et al.* (2003). An integrated stress response regulates amino acid metabolism and resistance to oxidative stress. Mol Cell *11*, 619-633.

Hartman, M. G., Lu, D., Kim, M. L., Kociba, G. J., Shukri, T., Buteau, J., Wang, X., Frankel, W. L., Guttridge, D., Prentki, M., *et al.* (2004). Role for activating transcription factor 3 in stress-induced beta-cell apoptosis. Mol Cell Biol *24*, 5721-5732.

Hayes, J. D., and McMahon, M. (2001). Molecular basis for the contribution of the antioxidant responsive element to cancer chemoprevention. Cancer Lett *174*, 103-113.

Haze, K., Yoshida, H., Yanagi, H., Yura, T., and Mori, K. (1999). Mammalian transcription factor ATF6 is synthesized as a transmembrane protein and activated by proteolysis in response to endoplasmic reticulum stress. Mol Cell Biol *10*, 3787-3399.

Ishiguro, T., Nagawa, H., Naito, M., and Tsuruo, T. (2000). Inhibitory effect of ATF3 antisense oligonucleotide on ectopic growth of HT29 human colon cancer cells. Jpn J Cancer Res *90*, 833-836.

Iyengar, P., Combs, T. P., Shah, S. J., Gouon-Evans, V., Pollard, J. W., Albanese, C., Flanagan, L., Tenniswood, M. P., Guha, C., Lisanti, M. P., *et al.* (2003). Adipocyte-secreted factors synergistically promote mammary tumorigenesis through induction of anti-apoptotic transcriptional programs and proto-oncogene stabilization. Oncogene *22*, 6408-6423.

Jiang, H. Y., Wek, S. A., McGrath, B. C., Lu, D., Hai, T., Harding, H. P., Wang, X., Ron, D., Cavener, D. R., and Wek, R. C. (2004). Activating transcription factor 3 is integral to the eukaryotic initiation factor 2 kinase stress response. Mol Cell Biol *24*,

1365-1377.

Karin, M. (1999). The beginning of the end: IkB kinase (IKK) and NFkB activation. J Biol Chem 274, 27339-27342.

Karin, M., Liu, Z.-G., and Zandi, E. (1997). AP-1 function and regulation. Curr Opin Cell Biol 9, 240-246.

Kaufman, R. J. (1999). Stress signaling from the lumen of the endoplasmic reticulum: coordination of gene transcriptional and translational controls. Genes Dev 13, 1211-1233.

Kawauchi, J., Zhang, C., Nobori, K., Hashimoto, Y., Adachi, M. T., Noda, A., Sunamori, M., and Kitajima, S. (2002). Transcriptional repressor activating transcription factor 3 protects human umbilical vein endothelial cells from tumor necrosis factor-alpha-induced apoptosis through down-regulation of p53 transcription. J Biol Chem 277, 39025-39034.

Lassot, I., Ségéral, E., Berlioz-Torrent, C., Durand, H., Groussin, L., Hai, T., Benarous, R., and Margottin-Goguet, F. (2001). ATF4 degradation relies on a phosphorylation-dependent interaction with the $SCF\beta^{TrCP}$ ubiquitin ligase. Mol Cell Biol 21, 2192-2202.

Lee, J. A., Kim, H., Lee, Y. S., and Kaang, B. K. (2003). Overexpression and RNA interference of Ap-cyclic AMP-response element binding protein-2, a repressor of long-term facilitation, in Aplysia kurodai sensory-to-motor synapses. Neurosci Lett 337, 9-12.

Lee, K. A. W., Hai, T. Y., SivaRaman, L., Thimmappaya, B., Hurst, H. C., Jones, N. C., and Green, M. R. (1987). A cellular protein, activating transcription factor, activates transcription of multiple E1a-inducible adenovirus early promoters. Proc Natl Acad Sci USA 84, 8355-8359.

Lin, Y. S., and Green, M. R. (1988). Interaction of a common transcription factor, ATF, with regulatory elements in both E1a- and cyclic AMP-inducible promoters. Proc Natl Acad Sci USA 85, 3396-3400.

Mashima, T., Udagawa, S., and Tsuruo, T. (2001). Involvement of transcriptional repressor ATF3 in acceleration of caspase protease activation during DNA damaging agent-induced apoptosis. J Cell Physiol 188, 352-358.

Meyer, T. E., and Habener, J. F. (1993). Cyclic adenosine 3′,5′-monophosphate response element binding protein (CREB) and related transcription-activating deoxyribonucleic acid-binding proteins. Endocrine Reviews 14, 269-290.

Montminy, M. (1997). Transcriptional regulation by cyclic AMP. Annu Rev Biochem 66, 807-822.

Montminy, M. R., and Bilezsikjian, L. M. (1987). Binding of a nuclear protein to the cyclic AMP response element of the somatostatin gene. Nature 328, 175-178.

Mori, K. (2000). Tripartite management of unfolded proteins in the endoplasmic reticulum. Cell 101, 451-454.

Nakagomi, S., Suzuki, Y., Namikawa, K., Kiryu-Seo, S., and Kiyama, H. (2003). Expression of the Activating Transcription Factor 3 Prevents c-Jun N-Terminal Kinase-Induced Neuronal Death by Promoting Heat Shock Protein 27 Expression and Akt Activation. J Neurosci 23, 5187-5196.

Nawa, T., Nawa, M. T., Adachi, M. T., Uchimura, I., Shimokawa, R., Fujisawa, K., Tanaka, A., Numano, F., and Kitajima, S. (2002). Expression of transcriptional repressor ATF3/LRF1 in human atherosclerosis: colocalization and possible involvement in cell death of vascular endothelial cells. Atherosclerosis 161, 281-291.

Nehring, R. B., Horikawa, H. P., El Far, O., Kneussel, M., Brandstatter, J. H., Stamm, S., Wischmeyer, E., Betz, H., and Karschin, A. (2000). The metabotropic GABAB receptor directly interacts with the activating transcription factor 4. J Biol Chem 275, 35185-35191.

Newman, J. R. S., and Keating, A. E. (2003). Comprehensive Identification of Human bZIP Interactions with Coiled-Coil Arrays. Science 300, 2097-2101.

Nobori, K., Ito, H., Tamamori-Adachi, M., Adachi, S., Ono, Y., Kawauchi, J., Kitajima, S., Marumo, F., and Isobe, M. (2002). ATF3 inhibits doxorubicin-induced apoptosis in cardiac myocytes: a novel cardioprotective role of ATF3. J Mol Cell Cardiol 34, 1387-1397.

Okamoto, Y., Chaves, A., Chen, J., Kelley, R., Jones, K., Weed, H. G., Gardner, K. L., Gangi, L., Yamaguchi, M., Klomkleaw, W., et al. (2001). Transgenic mice expressing ATF3 in the heart have conduction abnormalities and contractile dysfunction. Am J Pathol 159, 639-650.

Pati, D., Meistrich, M. L., and Plon, S. E. (1999). Human Cdc34 and Rad6B ubiquitin-conjugating enzymes target repressors of cyclic AMP-induced transcription for proteolysis. Mol Cell Biol 19, 5001-5013.

Pearson, A. G., Gray, C. W., Pearson, J. F., Greenwood, J. M., During, M. J., and Dragunow, M. (2003). ATF3 enhances c-Jun-mediated neurite sprouting. Brain Res Mol Brain Res 120, 38-45.

Perez, S., Vial, E., van Dam, H., and Castellazzi, M. (2001). Transcription factor ATF3 partially transforms chick embryo fibroblasts by promoting growth factor-independent proliferation. Oncogene 20, 1135-1141.

Persengiev, S. P., Devireddy, L. R., and Green, M. R. (2002). Inhibition of apoptosis by ATFx: a novel role for a member of the ATF/CREB family of mammalian bZIP transcription factors. Genes Dev 16, 1806-1814.

Persengiev, S. P., and Green, M. R. (2003). The role of ATF/CREB family members in cell growth, survival and apoptosis. Apoptosis 8, 225-228.

Peters, C. S., Liang, X., Li, S., Kannan, S., Peng, Y., Taub, R., and Diamond, R. H. (2001). ATF-7, a novel bZIP protein, interacts with the PRL-1 protein-tyrosine phosphatase. J Biol Chem 276, 13718-13726.

Raingeaud, J., Gupta, S., Rogers, J. S., Dickens, M., Han, J., Ulevitch, R. J., and Davis, R. J. (1995). Pro-inflammatory cytokines and environmental stress cause p38 mitogen-activated protein kinase activation by dual phosphorylation on tyrosine and threonine. J Biol Chem 270, 7420-7426.

Roesler, W. J., Vandenbark, G. R., and Hanson, R. W. (1988).

Cyclic AMP and the induction of eukaryotic gene transcription. J Biol Chem 263, 9063-9066.

Ron, D. (2002). Translational control in the endoplasmic reticulum stress response. J Clin Invest 110, 1383-1388.

Rutkowski, D. T., and Kaufman, R. J. (2003). All roads lead to ATF4. Dev Cell 4, 442-444.

Sassone-Corsi, P. (1998). Coupling gene expression to cAMP signalling: role of CREB and CREM. Int J Biochem Cell Biol 30, 27-38.

Shen, J., Chen, X., Hendershot, L., and Prywes, R. (2002). ER stress regulation of ATF6 localization by dissociation of BiP/GRP78 binding and unmasking of Golgi localization signals. Dev Cell 3, 99-111.

Shen, J., and Prywes, R. (2004). Dependence of site-2 protease cleavage of ATF6 on prior site-1 protease digestion is determined by the size of the luminal domain of ATF6. J Biol Chem 279, 43046-43051.

Tarn, C., Bilodeau, M. L., Hullinger, R. L., and Andrisani, O. M. (1999). Differential immediate early gene expression in conditional hepatitis B virus pX-transforming versus nontransforming hepatocyte cell lines. J Biol Chem 274, 2327-2336.

Urano, F., Bertolotti, A., and Ron, D. (2000). IRE1 and efferent signaling from the endoplasmic reticulum. J Cell Sci 113 Pt 21, 3697-3702.

van Dam, H., and Castellazzi, M. (2001). Distinct roles of Jun : Fos and Jun: ATF dimers in oncogenesis. Oncogene 20, 2453-2464.

van der Meijden, C. M., Lapointe, D. S., Luong, M. X., Peric-Hupkes, D., Cho, B., Stein, J. L., van Wijnen, A. J., and Stein, G. S. (2002). Gene profiling of cell cycle progression through S-phase reveals sequential expression of genes required for DNA replication and nucleosome assembly. Cancer Res 62, 3233-3243.

Vernon, E., Meyer, G., Pickard, L., Dev, K., Molnar, E., Collingridge, G. L., and Henley, J. M. (2001). GABA(B) receptors couple directly to the transcription factor ATF4. Mol Cell Neurosci 17, 637-645.

Wang, Y., Shen, J., Arenzana, N., Tirasophon, W., Kaufman, R., and Prywes, R. (2000). Activation of ATF6 and an ATF6 DNA binding site by the endoplasmic reticulum stress response. J Biol Chem 275, 27013-27020.

White, J. H., McIllhinney, R. A., Wise, A., Ciruela, F., Chan, W. Y., Emson, P. C., Billinton, A., and Marshall, F. H. (2000). The GABAB receptor interacts directly with the related transcription factors CREB2 and ATFx. Proc Natl Acad Sci USA 97, 13967-13972.

Xanthoudakis, S., and Curran, T. (1996). Redox regulation of AP-1: a link between transcription factor signaling and DNA repair. Adv Exp Med Biol 387, 69-75.

Yan, C., Jamaluddin, M. S., Aggarwal, B., Myers, J., and Boyd, D. D. (2005). Gene expression profiling identifies activating transcription factor 3 as a novel contributor to the proapoptotic effect of curcumin. Mol Cancer Ther 4, 233-241.

Ye, J., Rawson, R. B., Komuro, R., Chen, X., Dave, U. P., Prywes, R., Brown, M. S., and Goldstein, J. L. (2000). ER stress induces cleavage of membrane-bound ATF6 by the same proteases that process SREBPs. Mol Cell 6, 1355-1364.

Yeh, W. C., and McKnight, S. L. (1995). Regulation of adipose maturation and energy homeostasis. Curr Opin Cell Biol 7, 885-890.

Yoshida, H., Haze, K., Yanagi, H., Yura, T., and Mori, K. (1998). Identification of the cis-acting endoplasmic reticulum stress response element responsible for transcriptional induction of mammalian glucose-regulated proteins. Involvement of basic leucine zipper transcription factors. J Biol Chem 273, 33741-33749.

Yoshida, H., Matsui, T., Yamamoto, A., Okada, T., and Mori, K. (2001). XBP1 mRNA is induced by ATF6 and spliced by IRE1 in response to ER stress to produce a highly active transcription factor. Cell 107, 881-891.

Zeng, L., Lu, M., Mori, K., Luo, S., Lee, A. S., Zhu, Y., and Shyy, J. Y. (2004). ATF6 modulates SREBP2-mediated lipogenesis. EMBO J 23, 950-958.

Zhu, C., Johansen, F.-E., and Prywes, R. (1997). Interaction of ATF6 and serum response factor. Mol Cell Biol 17, 4957-4966.

Section IV

The Genome

Chapter 21

Function and Mechanism of Chromatin Boundaries

Haini N. Cai

Department of Cellular Biology, University of Georgia, 724 Biological Sciences Building, 1000 Cedar Street, Athens, GA30602, USA

Key Words: promoter, enhancer-blocking, chromatin domain, barrier, nuclear organization, suHw, scs, cHS4, SF1, CTCF

Summary

In recent years, a novel class of DNA elements called chromatin boundaries or insulators has drawn special attention in the field of gene regulation and chromatin organization. Chromatin boundaries were first identified in *Drosophila* as genomic regions that separate neighboring chromatin domains of distinctive structure and function. Boundary elements are experimentally defined by two activities: they block the regulatory effects of enhancers from promoters when positioned interveningly, and when placed on both sides of integrated transgenes, they insulate these transgenes from the transcriptional influences of the surrounding genome. Elements with similar properties have been reported in a variety of species from yeast to humans. Molecular genetic and cell biological studies suggest that boundary activity may represent a general mechanism of gene regulation, genome organization, as well as chromatin and nuclear organization. However, the biological functions and the protein components of many boundary elements remain to be characterized. More importantly, the mechanism through which boundaries affect various forms of transcription regulation is still poorly understood.

Transcription Activation by Regulatory Enhancers

Transcriptional regulation represents the most important step in gene regulation. Studies in the last twenty years have established that tissue-specific transcriptional activation involves two different classes of *cis* regulatory elements: a basal/core promoter element near the transcription start site; and more distally linked *cis*-regulatory enhancers (for review see Arnosti, 2003; Levine and Tjian, 2003). These enhancers, typically a few hundred base pairs (bp) in length and contain clusters of binding sites for both positive and negative regulatory factors, can each specify a subset of the expression pattern. They integrate regulatory information at the endpoint of a network of signaling and regulatory cascades to direct localized patterns of gene expression in the animal body (Levine and Davidson, 2005).

A: Mechanism of Enhancer Function

In spite of the important role that enhancers play in regulating gene expression, the mechanisms by which they control transcription from a promoter is not well understood. Evidence suggests that activator proteins bound at a distal enhancer could influence the rate of transcription initiation through multiple mechanisms, and at different stages of transcription initiation. They may change the local chromatin structure to increase the accessibility of enhancer or promoter DNA to transcription factors or the basal machinery, by recruiting enzymes that modify the histone tails or remodel nucleosomes. They may directly recruit or stabilize the association of RNA polymerase II preinitiation complex to promoter DNA. Regulatory factors can even modulate the structure of the RNA polymerase II initiation complex, causing increased mRNA synthesis.

A general characteristic of the enhancer function is their relative indifference to the distance, location and orientation of the promoter. Many enhancers act from kilobases or even tens of kilobases away from the target gene promoter. Many also activate transcription from either upstream or downstream positions. One hypothesis

---

Corresponding Author: Tel: (706) 542-3329, Fax: (706) 542-4271, E-mail: hcai@uga.edu

suggests that enhancers serve as "funnels" to concentrate regulatory proteins, which then traverse the intervening DNA to affect transcription at the promoter (processive/ tracking/ facilitated tracking models, for reviews see Blackwood and Kadonaga, 1998). Indeed, it has been reported that RNA Polymerase II or enhancer-binding proteins can be found accumulating in the intervening regions between the enhancer and the promoter (Martens *et al.*, 2004; Schmitt and Paro, 2004). In addition, changes in chromatin structure or accumulation of non-coding transcripts corresponding to enhancers or intervening regions between the enhancer and the promoter have also been reported (Ashe *et al.*, 1997; Bae *et al.*, 2002).

However, not all evidence supports the processive model, other observations suggest that physical linkage between enhancers and the promoter may not be necessary for gene activation (Dunaway and Droge, 1989). Moreover, recent reports suggest that enhancer DNA may be in proximity of the promoter while the intervening DNA is "looped" out (Blackwood and Kadonaga, 1998; Ronshaugen and Levine, 2004; Sayegh *et al.*, 2005; Tolhuis *et al.*, 2002). The ultimate evidence for the linkage-independent gene regulation is the phenomenon of transvection where enhancer residing on one chromosome can interact with a gene promoter on a different chromosome (Duncan, 2002; Hopmann *et al.*, 1995; Pirrotta, 1999; Wu and Morris, 1999). However, special DNA sequences such as the "promoter targeting sequences" (PTS) or "promoter tethering element" (PTE) are necessary to facilitate such long distance interaction (Calhoun *et al.*, 2002; Zhou and Levine, 1999). Other models combine the features from both hypotheses and suggest that enhancers could regulate transcription over long distance through certain relay mechanisms, either by facilitator proteins (Dorsett, 1999), or intermediate DNA regions (Bulger and Groudine, 1999; Pirrotta, 1998). It is possible that depending on the distance, chromatin landscape, and the type of activation, different mechanism could be involved to mediate the regulatory effect of an enhancer to a promoter.

B: Regulation of Enhancer-promoter Specificity by Promoter Competition

A second consideration about enhancer-promoter interaction is their specificity. From the above discussion, it is clear that many enhancers can activate transcription from promoters in a long range, and from both upstream and downstream directions. This poses a serious problem for the independent regulation of neighboring genes, especially in compact genomes or complex genetic loci where genes and their closely regulatory enhancers are positioned. What controls the specificity of enhancer-promoter interactions? Two complementary mechanisms have been proposed. Enhancers may selectively interact with their cognate promoters among other equally accessible promoters due to their intrinsic properties. Alternatively, specialized DNA elements such as chromatin boundaries or insulators can restrict the access of a given enhancer to inappropriate promoters.

Several studies provided evidence that preferential interaction between an enhancer and one neighboring promoter could reduce or preclude its interaction with other available promoters. This is well illustrated by the developmental switch of the vertebrate β-globin genes in which two *cis*-linked ε and β globin promoters compete for a shared β/ε enhancer. During the embryonic stage, the ε gene promoter out-competes the β gene promoter and is preferentially activated by the β/ε enhancer. In adults, binding of adult specific transcription factors near the basal promoter of the β gene augments the enhancer-β promoter interaction, resulting in activation of the β gene and concomitant shut-off of the ε gene (Choi and Engel, 1988; Foley and Engel, 1992). Another example is from the observation that the *Drosophila* genes *gooseberry* (*gsb*) and *gooseberry neuro* (*gsbn*) are closely linked and divergently transcribed (Li and Noll, 1994). The two genes are expressed in completely distinct patterns, each regulated by its own tissue-specific enhancers located in a 10-kb intergenic region. In transgenic assays, these enhancers can activate their cognate promoters but not the heterologous promoter from the other gene. Similarly, promoter specificity is observed for *Drosophila* decapentaplegic *(dpp)* enhancers, which are positioned close to neighboring *SLY1 homologous* (*Slh*) and *out at first (oaf)* genes (Merli *et al.*, 1996). Normally, *Slh* and *oaf* are not affected by the *dpp* enhancers. However, when a limited region of the *oaf* promoter was replaced by the *hsp70* promoter through *in vivo* gene targeting, the chimeric *oaf* gene became activated by the *dpp* enhancers.

Study of regulatory autonomy of the *Drosophila fushi tarazu (ftz)* gene provided another example that promoter competition could determine enhancer specificity (Ohtsuki *et al.*, 1998). One of the *ftz* enhancer is Auto-regulatory Enhancer 1 (AE1), positioned in the intergenic region between *ftz* and the divergently transcribed neighboring gene *Scr comb reduced* (*Scr*). In a transgenic reporter assay, AE1 placed between two divergently directed promoters, both of which contain the same key promoter motifs such as TATA or Inr, could simultaneously activate both promoters. However, when a TATA-containing promoter is paired with a TATA-less/Inr-containing promoter, AE1 activates only the TATA promoter but not the TATA-less/Inr promoter, suggesting that the specificity between AE1 and the TATA-containing promoter is based on promoter preference, or promoter competition. Indeed, when the TATA promoter is replaced by a third type, which contains neither TATA nor Inr motifs, the AE1 enhancer preferentially activates the TATA-less/Inr promoter. These results demonstrate that both enhancers and core promoter sequences play critical roles in specifying enhancer-promoter interactions.

Chromatin Boundaries Influence Transcription

An increasing body of evidence suggests that another important mechanism for selective enhancer-promoter interaction involves the function of chromatin boundaries. These DNA elements were first identified in *Drosophila* as sequences that functionally separate the neighboring genes. Elements of similar function have also been identified in diverse organisms from yeast to humans.

A: Discovery of Chromatin Boundary

The first members of this novel group of regulatory DNAs are the *Drosophila* scs, scs' and suHw elements, all discovered more than a decade ago. The scs and scs' DNA elements flank the *Drosophila hsp70* locus and demarcate the limit of the heatshock puff on the polytene chromosome during the transcription induction (Kellum and Schedl, 1992; Udvardy *et al.*, 1985). The 900 bp scs and 500 bp scs' each contains a GC-rich and nuclease hypersensitive core surrounded by AT-rich nuclease resistant sites. These and other structural features led to the initial hypothesis that insulators may organize "specialized chromatin structure" that form the boundary of higher order chromatin domains (Jupe *et al.*, 1995; Kellum and Schedl, 1991). Later studies showed that insulators from many different species often appear as nuclease-hypersensitive sites near the border of neighboring genetic loci or chromatin domains.

The biological activity of these boundary elements, however, was first discovered using a gene protection assay. In this assay, scs and scs', placed on both sides of a *Drosophila miniwhite* reporter, were shown to reduce the variation in the eye color of the transgenic flies. Such variation, normally seen in flies with *miniwhite*, is believed to be due to the transcriptional influences from the genome surrounding the transgene insertion site, is thus called chromatin position effect (CPE, see Fig. 21.1A). It was proposed that the units of gene regulation as well as genome organization may correspond to individual domains of chromatin. DNA elements like scs and scs' elements could based on the above evidence, function as the boundaries between such distinct chromatin domains. They are proposed to insulate genes in one domain from the regulatory influences of the neighboring domains (Kellum and Schedl, 1991).

The studies of another *Drosophila* element called suHw have revealed a second important property of chromatin boundaries. The gypsy retrotransposon, which contains the 340 bp DNA suHw element, was found to disrupt enhancer function when it inserts into the regulatory region of genes such as *yellow* and *forked*. Analysis of these mutations revealed a marked directionality in the gypsy mutagenic effect: it only disrupt the function of distal enhancers relative to the gene promoter, without affecting the proximal enhancers (Cai and Levine, 1995; Dorsett, 1993; Geyer and Corces, 1992; Harrison *et al.*, 1989; Holdridge and Dorsett, 1991; Modolell *et al.*, 1983). For example, the *yellow^2* (*y^2*) phenotype is caused by *gypsy* insertion in the regulatory region of *yellow*. The function of the body and wing enhancers located upstream of the insertion site are disrupted, while other downstream tissue-specific enhancers are unaffected. The positional dependence of the interference is a key characteristic of boundary elements. Further studies indicate that the enhancer-blocking effect of the many chromatin boundaries is independent of the distance between the enhancer and promoter, the position or the orientation of the boundary itself so long as it is between the enhancer and the promoter (Dorsett, 1993).

The scs and suHw elements were initially defined by only one of the two key boundary properties, scs by CPE-blocking and suHw by enhancer-blocking, they were both shown in subsequent studies to possess activities to protect *miniwhite* from *Drosophila* position effect, and to block various enhancers as well as silencers (Kellum and Schedl, 1992; Mallin *et al.*, 1998; Roseman *et al.*, 1993; Sigrist and Pirrotta, 1997). These properties are consistent with the notion that boundary elements insulate local genes from various transcriptional influences, hence the name insulator. In this review, the terms of insulator, chromatin boundary or boundary are used interchangeably. The two activities defined in these initial studies, namely, the ability to block CPE and other position effect (PE), and the ability to block enhancers and silencers, are now considered the defining properties of boundary/insulator elements.

Some protein factors mediating the function of these boundaries have been identified. Proteins that interact with the scs and scs' elements and mediate their activities include zw5, a zinc finger BTB domain protein, and BEAF32, respectively (Blanton *et al.*, 2003; Gaszner *et al.*, 1999; Hart *et al.*, 1997; Zhao *et al.*, 1995). Proteins mediating the enhancer-blocking activity

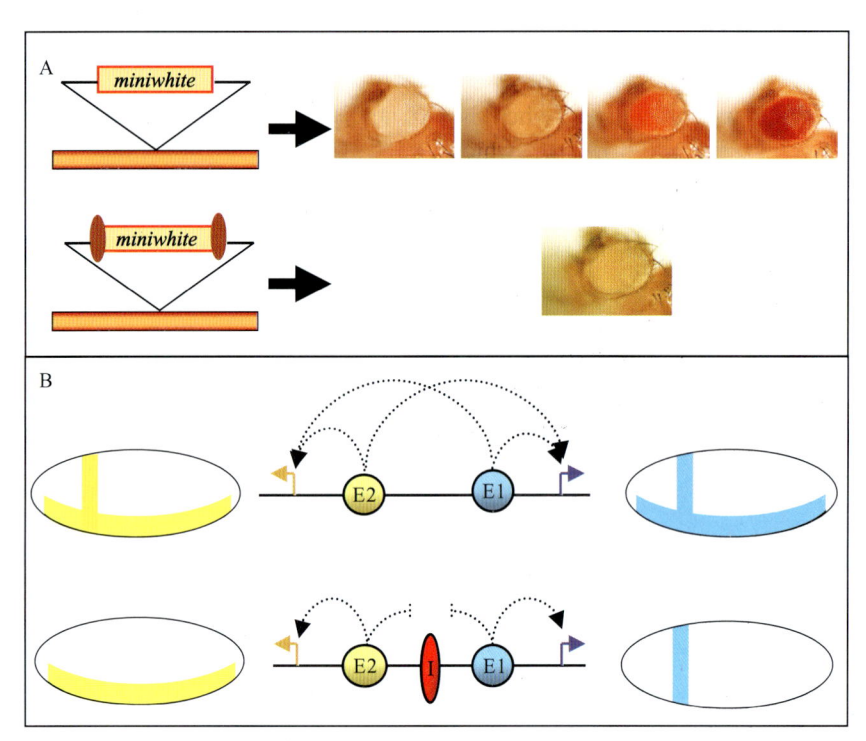

Fig.21.1A Boundaries can protect transgenes from chromosomal positional effects (CPE) in *Drosophila*. Upper panel: independent *Drosophila* strains containing a transgenic *miniwhite* reporter gene (diagramed on the left, brown bars represent a chromosome) exhibit diverse eye color phenotypes in an insertion site independent fashion (right). Lower panel: independent *Drosophila* strains containing transgenic *miniwhite* reporter gene flanked on both sides by boundary elements (dark brown swirls) exhibit reduced variation in the eye color phenotype.

Fig.21.1B Boundaries can block enhancer-promoter interactions. Upper panel: a schematic of two *Drosophila* embryos with enhancer-driven expression of the two transgenic reporters (diagramed in the center). The two reporter genes are represented by the yellow and blue arrows, their activity corresponding the yellow and blue domains in the embryos. E1 and E2 are two tissue-specific enhancers, directing transcription in the vertical and horizontal domains in the embryos, respectively. Arched arrows present enhancer-promoter interactions during gene activation. Arched lines with a short dash represent blocked enhancer-promoter interactions. Both enhancers can activate both reporter genes as shown in the expression pattern. Lower panel: insertion of a single boundary element blocks the E1-yellow promoter and E2-blue promoter interactions, resulting in the loss these tissue-specific expression driven by these enhancers.

of the suHw insulator include SUHW, a zinc-finger protein that binds to the multiple AT-rich repeats within the suHw element, and Mod(mdg4), a BTB/POZ chromo-domain protein (Gdula and Corces, 1997; Georgiev and Kozycina, 1996; Gerasimova *et al.*, 1995; Harrison *et al.*, 1993; Parkhurst *et al.*, 1988; Spana and Corces, 1990; Spana *et al.*, 1988). Mod(mdg4) is also involved in other chromatin modification processes as a Trx-G and E(var) protein (Dorn *et al.*, 1993; Gerasimova and Corces, 1998). Structural and functional analyses indicate that the leucine zipper domain of SuHw interacts with Mod(mdg4) via its C-terminal acidic domain and recruits it to the chromosome (Gdula and Corces, 1997; Ghosh *et al.*, 2001). Certain mutations in the Mod(mdg4) protein result in promoter-specific silencing of transcription in a suHw-dependent fashion (Cai and Levine, 1997), reinforcing the notion that insulators may function at the level of chromatin structure and organization (Gerasimova *et al.*, 1995). It was recently reported that a third protein, CP190, which binds to DNA near some but not all boundary sites that SuHw and Mod(mdg4) bind to, and appears to strengthen the boundary activity by interacting with both SuHw and Mod(mdg4) proteins (Pai *et al.*, 2004). In addition to the sites within the gypsy retrotransposon, the SuHw protein also binds to several hundred sites within the *Drosophila* genome, as revealed by immunostaining on polytene chromosome (Gerasimova *et al.*, 1995). These sites, some of which have been shown to contain insulator activity in transgenic assays, are believed to correspond to endogenous boundary elements (Golovnin *et al.*, 2003; Pai *et al.*, 2004; Parnell *et al.*, 2003).

B: Chromatin Boundaries Regulate Endogenous Gene Expression

The first chromatin boundary implicated in endogenous genes regulation is the Fab7 element from the *Abdominal B (Abd B)* regulatory region of the *Drosophila* bithorax gene complex (Bx-C, see Fig.21.2). *Abd B*, a member of the *Drosophila* homeotic gene family, is controlled by several tissue-specific enhancers (iabs). Its expression, as other homeotic genes, is established during early development by activator and repressor proteins recruited to these enhancers, and maintained during late development by the active or repressive chromatin organized by the Polycomb group (Pc-G) and trithorax group (Trx-G) proteins, respectively. The active or silent chromatin, unlike the early transcription factors recruited to specific DNA sequences, is believed to spread along chromosome and could affect neighboring enhancers or genes.

Genetic and molecular analysis of the *Abd B* regulatory region revealed multiple chromatin boundaries including Fab-7 (Galloni *et al.*, 1993; Gyurkovics *et al.*, 1990), Mcp (Karch *et al.*, 1994), and Fab 8 (Barges *et al.*, 2000; Zhou *et al.*, 1999). A unique feature of the Bx-C boundaries is that they are positioned intragenically, flanking the tissue-specific iab- enhancers (see Fig. 21.2A). These boundaries block enhancer-promoter interactions when tested in transgenic flies, and can protect transgenes from chromatin-mediated transcriptional influences (Hagstrom *et al.*, 1996; Zhou *et al.*, 1999; Zhou *et al.*, 1996). It has been proposed that Mcp and Fab-7 may specify interactions between the parasegment-specific iab enhancers and the *Abd B* promoter by organizing *Abd-B* regulatory region into independent chromatin domains, while insulating each enhancer from the regulatory influences of adjacent enhancers (Galloni *et al.*, 1993; Gyurkovics *et al.*, 1990; Mihaly *et al.*, 1998). Indeed, deletions of Fab7 boundary resulted in loss of distinctive *Abd B* expression pattern directed by the two iab enhancers on either side of the boundary. Instead, an intermediate *Abd B* expression pattern appeared as if the activities of the two enhancers were mixed (Galloni *et al.*, 1993; Karch *et al.*, 1994; Mihaly *et al.*, 1998).

However, the intragenic position of these boundaries presents a problem for the iab enhancers, which would have to overcome these boundaries in order to activate the *Abd B* promoter. The solution to this problem may lie in the recent discovery of another type of specialized DNA near certain homeotic gene promoters. These elements, which include the "transvection mediating region" (TMR), the "promoter targeting sequences" (PTS) near the *Abd B* promoter, the "pairing sensitive sites" in the *engrailed* promoter, the "promoter tethering element" (PTE) near the *Scr* promoter, can facilitate long range enhancer-promoter interactions, even across different chromosomes, or over chromatin boundaries (Calhoun and Levine, 2003; Calhoun *et al.*, 2002; Kassis, 1994; Kassis, 2002; Lin *et al.*, 2004; Zhou *et al.*, 1999; Zhou and Levine, 1999). It was proposed that such pairing elements may be present near each iab enhancer to interact with boundaries and to facilitate specific long-range enhancer-promoter interactions (see Fig. 21.2B).

Another boundary element directly involved in endogenous gene regulation is the vertebrate CTCF binding sites. CTCF, a zinc finger DNA-binding protein, was originally discovered as a transcription regulator. It interacts with multiple boundary elements, including the 5' hypersensitive sites (cHS4) within the locus control region of the chicken beta-globin loci and is critical for its enhancer-blocking activity in cell culture assays (Bell *et al.*, 1999; Chung *et al.*, 1993; Reitman *et al.*, 1990). The role of CTCF in endogenous gene regulation was discovered in the imprinted regulation of the mouse *H19/Igf2* locus (Bell and Felsenfeld, 2000; Hark *et al.*, 2000). The *H19* and *Igf2* gene expression are differentially regulated according to the parental origin of the chromosome: the *H19* gene is activated only from the maternal allele, whereas the *Igf2* gene is transcribed only from the paternal allele (see Fig. 21.3). The activation of both genes depends on a shared global enhancer 3' to the *H19* gene. An imprint control region (ICR) containing four CTCF binding sites, positioned between the two genes, is responsible for the allelic-specific gene activation by the global enhancer. On the maternal allele, interaction between the distal *Igf2* promoter and the enhancer is blocked by a CTCF-mediated chromatin boundary assembled in the intervening ICR. On the paternal allele, methylation of the CpG islands within the ICR during gametogenesis prevents the binding of the CTCF and thus the boundary assembly. As a result, the distal *Igf2* gene is accessible and preferentially activated by the shared enhancer at the expense of the *H19* gene. Interestingly, replacing ICR with chicken globin LCR (see below), both of which contain CTCF binding sites, conferred boundary activity (and therefore maternal expression pattern) on alleles from both parental original. This result indicates that although the parental-specific gene activation in the region in the zygote requires the CTCF-mediated boundary activity, the parental specific methylation in the germline is

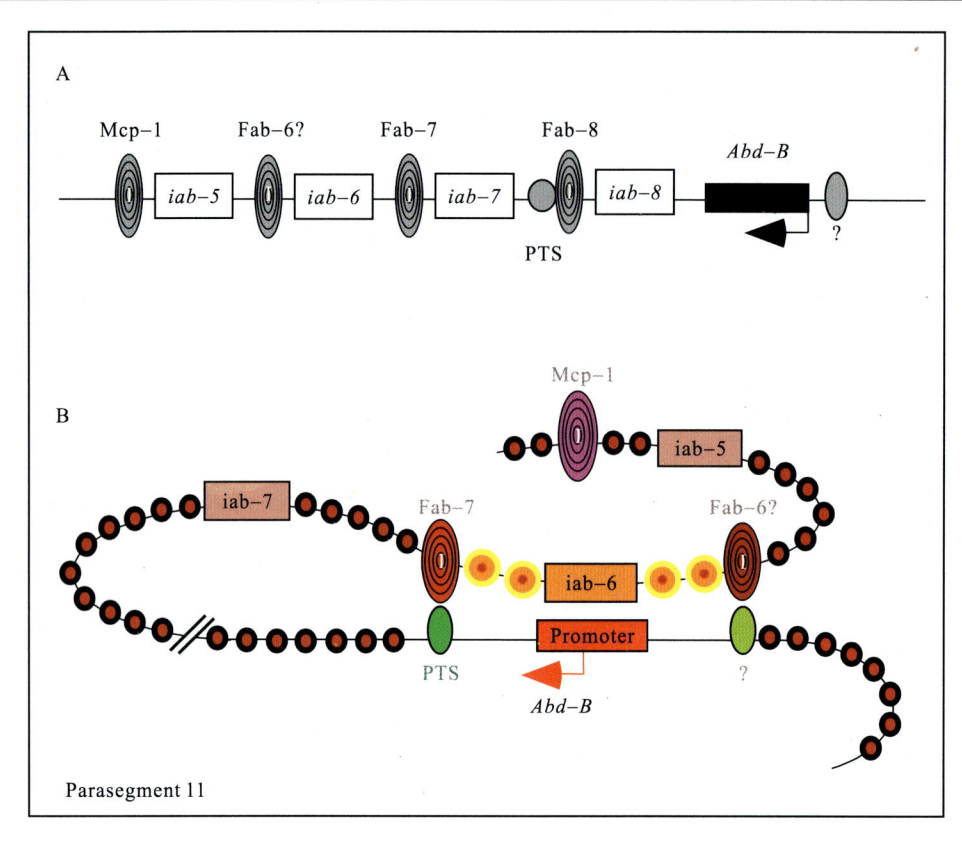

Fig.21.2 Boundaries in the *Drosophila* bithorax complex (Bx-C). Upper panel: a schematic of the *Drosophila Abd-B* gene (filled rectangle on right, black arrow indicates *Abd-B* transcription start site). Regulatory elements are as marked (see text). Swirls represent boundary elements and open rectangles represent enhancers. PTS and PTE are represented by small ovals near the *Abd-B* gene. Lower panel: interactions of Fab-7 and a postulated Fab-6 boundaries with the PTS/PTE elements near the *Abd-B* promoter mediate the iab-6 enhancer activation of *Abd-B* in parasegment 11 (light colored circles represent open and active chromatin), while insulating *Abd-B* against the repressive chromatin (dark brown circles) from the neighboring chromatin.

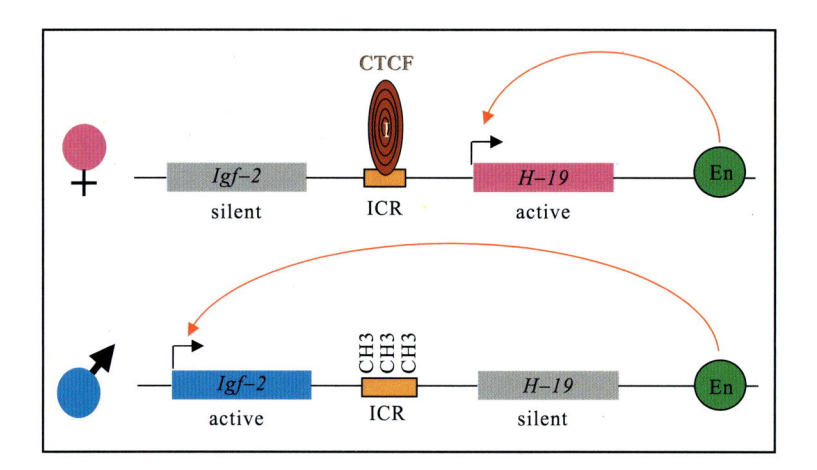

Fig.21.3 A CTCF-mediated boundary regulates parental-specific activation of the imprinted *H19/Igf2* loci in human and mouse. The transcriptional states of the *H19/Igf2* loci are diagram schematically. The green circle on the far right represents a global enhancer (see text) responsible for the activation of both *H19* and *Igf2* genes. Arched arrows represent enhancer-promoter interactions during gene activation. Arched lines with a short dash represent blocked enhancer-promoter interactions. Upper panel: on the maternal allele, a CTCF-meditated boundary assembled at the intergenic ICR region blocks the access of the more preferred *Igf2* promoter, allowing the activation of the *H19* gene by the shared enhancer. Lower panel: on the paternal allele, germline specific methylation (CH$_3$) of the CpG island within the ICR region abolishes the boundary activity, allowing the *Igf-2*-enhancer interaction and the activation of *Igf-2*.

controlled by factors other than CTCF (Szabo et al., 2002).

C: Compound Chromatin Boundary

In addition to the CTCF-mediated enhancer-blocking activity, the chicken beta-globin boundary also contains activity that shield transgenes from transcriptional influences from the surrounding genome. In fact, the cHS4 boundary was initially identified by its ability to protect stably integrated transgenes in cultured cells from the genomic silencing effect. Such silencing effects, widely seen with transgenes inserted into the mammalian genome, are characterized by progressive repression of transgene expression in a copy number and insert site dependent fashion, hence called position effect (PE, Chung et al., 1993; Pikaart et al., 1998; Reitman et al., 1990). Although often compared to *Drosophila* CPE, PE in mammalian cells is characteristically repressive in nature, whereas observed CPE in *Drosophila* is often positive. PE has been observed with many genes and genomic locations (for review see Hawley, 2001; Kioussis and Festenstein, 1997; Martin and Whitelaw, 1996; Yee and Zaia, 2001), whereas CPE is defined predominantly by the behavior of *miniwhite* (Hazelrigg et al., 1984; Pirrotta et al., 1985; Spradling and Rubin, 1983). The globin boundary can block CPE in *Drosophila*, as assayed by the protection of the *miniwhite* expression (Chung et al., 1993; Reitman et al., 1990). The converse, namely, blocking of the repressive PE by *Drosophila* boundary in vertebrate cells, has also been reported (van der Vlag et al., 2000). Nevertheless, it should also be noted that, although both called position effect, the nature and extent of transcriptional influences from the neighboring genome in these two cases are likely to be distinct, so could the ability and mechanisms of boundaries to block them.

Chromatin boundaries were found to flank the globin loci in both mouse and human (Bulger et al., 1999; Farrell et al., 2002, see Fig.21.4). The homology in DNA sequence, as well as similarity in genome and chromatin organization, extends well beyond the beta-globin loci itself. In fact, the globin loci in both these species are embedded in a conserved array of odorant receptor genes, which are active in different tissues different from those for globin genes (Bulger et al., 2000). The more distantly related chicken globin loci are also bound at the 3' end by an olfactory receptor gene (Bulger et al., 2000), and at the 5' end by a 16kb constitutively condensed chromatin containing the folate receptor gene. Examination of histone modification pattern of a 54 kb region surrounding the globin loci revealed a positive correlation between the level of hisone acetylation with the developmental- and tissue-specific gene activation, with the lowest level of acetylation near the condensed folate gene, and highest level of acetylation near the 5′HS4 boundary (Hebbes et al., 1994; Litt et al., 2001, see Fig.21.4). These results suggest that the chromatin organization at the boundary may counter the effect or the spread of the silent chromatin from the neighboring regions and preserve the independent regulation of the globin genes.

The chicken globin 5′HS4 boundary was also the first reported example of a compound chromatin boundary, in which the enhancer-blocking activity and the CPE-blocking or barrier activity are physically separable. The two activities are located in distinct regions of the chicken 5' HS4 element: whereas the FtII region with a CTCF site is necessary and sufficient for enhancer-blocking activity, it cannot insulate transgenes from progressive silencing by the surrounding genome. In

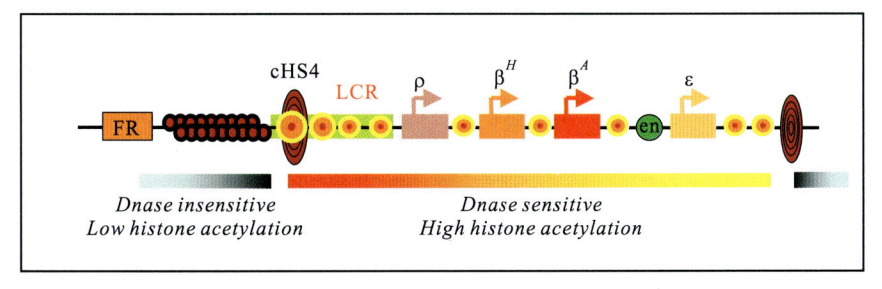

Fig.21.4 Boundaries in the chicken globin loci. The cHS4 compound boundary near the 5' LCR of the beta globin loci contains both enhancer-blocking and barrier activities and is responsible for shielding the loci from the repressive effects of the FR heterochromatin. Filled rectangles with arrows represent developmentally regulated beta globin genes as marked. Brown swirls represent the 5' and 3' boundaries. Chains of small brown circles represent the 16kb silent chromatin near the *folate receptor* gene (FR), which exhibits DNase resistance and low histone acetylation (black-grey graded bar underneath). Colored circles throughout the globin loci represent open and active chromatin with high sensitivity to DNase and high histone acetylation (red-yellow graded bar underneath). The 3' boundary is postulated to protect the globin loci from the 3' repressive chromatin (black-grey graded bar underneath).

contrast, the FtIV region in the element is responsible for the barrier activity (Recillas-Targa et al., 2002), and for blocking transcriptional silencing by a MoMLV LTR silencer (Yao et al., 2003). Furthermore, the enhancer-blocking and barrier activities depend on distinct protein factors: CTCF is required for the enhancer-blocking but not barrier function. Recent work from the Felsenfeld group showed that CACGGG sequences in FtIV interact with the USF protein: binding of USF is necessary for the formation of a protein complex including PCAF (H3 acetylation) and SET7/9 (H3 K4 methylation (West et al., 2004)). USF binding mediates the characteristic histone modification pattern in the cHS4 boundary and is responsible for assembly of the barrier against heterochromatin (West et al., 2004; Zhou and Berger, 2004).

Although both insulator and barrier activities are known to coexist in previously characterized boundaries like suHw, scs, and Fab7, the two activities are not separate in most of these cases. In particular, both activities mediated by the suHw boundary depend on simple DNA sequence motifs and well-characterized factors. Antibody staining of the polytene chromosomes showed that CP190 and Mod(mdg4) colocalize with most if not all SuHw protein staining sites, suggesting that a single protein complex and possibly a shared mechanism that mediating both insulator and barrier activities (Gerasimova and Corces, 1998; Gerasimova et al., 1995; Pai et al., 2004; Spana et al., 1988).

However, compound boundaries may represent a mechanism specialized for genomic regions where multiple modes of gene regulation are at play. Recent studies of the novel SF1 boundary, in the *Drosophila* Antennapedia homeotic gene complex (ANT-C, see Fig.21.5), provided additional support for this notion. As discussed in the previous section, expression of homeotic genes during late development is maintained by the active or silent chromatin domains organized by Trx-G or Pc-G proteins, respectively (Pirrotta, 1998; Ringrose and Paro, 2004). Limiting the extent of such organized chromatin domains is critical to homeotic gene regulation especially in ANT-C, unique among other Hox clusters in its inclusion of several non-homeotic genes, such as the aforementioned *ftz*, among homeotic genes.

To further compound the regulatory complexity, the *ftz* transcription unit is in fact embedded in the regulatory region of the neighboring homeotic gene *Scr* (see Fig.21.5, also see Maeda and Karch, 2003), which contains a series of long range enhancers both upstream and downstream of *ftz*. Although promoter competition can account for the exclusive interaction between AE1 and the *ftz* promoter, the same may not apply to other enhancers in the region. For instance, ftz distal, a *ftz* enhancer positioned equidistance from both *ftz* and *Scr* promoters, exhibited little promoter preference between the TATA and TATA-less promoters in transgenic assays. Thus the challenge to independent gene regulation in this genomic interval is at least two fold (see Fig. 21.5): what prevents the encroachment of active *ftz* transcription domain by the repressive chromatin in body regions where neighboring homeotic genes are silent? What controls the enhancer trafficking in the region so that these two neighboring genes are specifically activated by their respective enhancers from such juxtaposed positions?

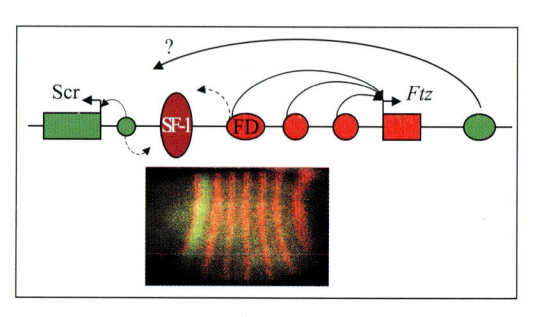

Fig.21.5 SF1 boundary regulates enhancer-traffic in the *Scr-ftz* region in *Drosophila*. Upper panel: schematic diagram of the *Scr-ftz* region (see text). Filled red rectangle represents the *ftz* transcription unit with the arrow indicating the transcription start site. Filled red circles represent *ftz* enhancers. FD is the ftz distal enhancer. Filled green rectangle represents the *Scr* transcription unit with the arrow indicating the transcription start site. Filled green circles represent *Scr* enhancers. Arched arrows represent enhancer-promoter interactions during gene activation. The large brown oval represents the intergenic SF1 boundary. Dashed arches represent potential interactions blocked by the SF1 boundary element. Lower panel: expression pattern of the *ftz* and *Scr* genes in an early *Drosophila* embryo as visualized by double in situ hybridizations (red: *ftz* mRNA, green: *Scr* mRNA, HN. Cai, unpublished data), indicating that *ftz* gene is expressed in regions anterior of *Scr* expression domain.

SF1, a chromatin boundary between the *Scr* and *ftz* promoters, may provide part of the answer to the above questions. In transgenic assays, SF1 can block the promiscuous ftz distal enhancer from interacting with an *Scr*-like promoter, and restrict it to a TATA-containing *ftz*-like promoter. The enhancer-blocking activity of SF1 persists through later stages of development, consistent with its potential role in regulating homeotic genes (Belozerov et al., 2003). SF1 contains a potent barrier, shown by its ability to protect *miniwhite* from CPE, as well as from the Pc-G-mediated silencing (Majumder and Cai). Interestingly, the enhancer-blocking and the barrier activities of SF1 appear to be located in distinct

DNA regions of the full-length boundary, and depend on different protein factors, suggesting that SF1 is a compound boundary, the first such example in *Drosophila*. The SF1b fragment exhibits an enhancer-blocking activity approximately 80% that of the full-length boundary, but no detectable barrier activity. In contrast, the SF1c region contains the highest level of barrier activity but the lowest enhancer-blocking activity among three SF1 subfragments. Although binding sites for the GAF protein were present in both regions, they are critical for the enhancer-blocking activity of SF1b but not required for the barrier activity of SF1c (Belozerov et al., 2003; Farkas et al., 1994; Majumder and Cai).

It might be more than a mere coincidence that a compound chromatin boundary is found in the *Scr-ftz* genomic interval, where, as in the vertebrate beta-globin region, proper gene regulation requires modulation of both enhancer-promoter interaction as well as transcriptional influences by organized chromatin. Compound boundaries with separate enhancer-blocking and barrier activities can block different transcription influences and allow these activities to be independently metered to meet the regulatory demands of local gene regulation. Indeed, separation and selective association of barrier and insulator activities in compound boundaries provide flexibility to boundary function, and may reflect conserved strategy in gene regulation and genome organization in both vertebrates and invertebrates.

However, questions remain as to how the *Scr* enhancers located downstream of *ftz* transcription unit overcome both the interference of a highly competitive *ftz* promoter, as well as the block of the SF1 insulator. Answers to this dilemma may lie in the precise mechanism through which chromatin boundary function (see Section III).

D: Boundary Elements in Yeast

In nowhere is the organization and propagation of silent chromatin domains better understood than in the telomeric region and silent mating-type (HM) loci in yeast *Saccharomyces cerevisiae*. The process begins with the binding of multiple Rap1p to the telomeric TG_{1-3} repeats or the HM silencer region (Loo and Rine, 1995). The subsequent recruitment of Sir3p, Sir4p and other components of the silent complex including ORC and Sir2p, a NAD-dependent histone deacetylase, allows Sir proteins to interact with H3 and H4 core histones via their hypoacetylated N termini (Braunstein et al., 1993). The chain reaction results in chromatin containing highly compact nucleosome array, which is inaccessible to external proteins and transcriptionally silent. The Sir2p-Sir2p interaction further stabilizes the heterochromatin structure and facilitates its spread along the DNA fiber for as far as 3kb beyond the initiation site (for reviews, see Kurdistani and Grunstein, 2003; Shore, 2001).

It is not surprising then that the first examples of protozoan boundary activities were also identified from the yeast telomere and silent HM loci. A pair of boundary elements was found to bracket the silencers that flank the HMR silent mating locus in *Saccharomyces cerevisiae* (Donze et al., 1999). Deletion of these elements resulted in the expansion of the heterochromatin domain. Insertion of these elements between a reporter gene and the HMR silencer or the telomere region can block the repressive effects of heterochromatin. Further analysis of the HMR right boundary revealed a TY1 LTR and a tRNA gene, $tRNA^{Thr}$, whose transcription potential was later shown to be critical for the barrier activity. Mutation in the internal promoter elements of $tRNA^{Thr}$, a Pol III gene, as well as mutations in two Pol III basal factors that bind to the promoter, abolishes the barrier activity (Donze and Kamakaka, 2001). However, similar activity was not found in other Pol III genes, nor was RNA Pol III required for the $tRNA^{Thr}$ barrier activity, suggesting that barrier function may be the result of the specific protein complex at this promoter, rather than the transcription of a Pol III gene *per se* (Donze and Kamakaka, 2001).

Silencer-blocking activity has also been associated with promoter regions of yeast ribosomal protein genes (UAS_{rpg}/TEF) from both *A. gossypli* and *S. cerevisiae* (Bi and Broach, 1999). These elements can block the spread of heterochromatin emanated from the HML and HMR silencers, shown by the recovery of reporter genes expression and in reduced negative supercoiling in the local chromatin topology. The block is dependent on the position of the UAS_{rpg} elements, as expected from a boundary element, and also dependent on the orientation of the elements. The minimum UAS_{rpg} boundary consists of three binding sites for, paradoxically, Rap1p, the protein responsible for initiating the propagation of the silent chromatin at the telomeric and HML/HMR loci. However, the authors proposed that the binding of Rap1p, which is also known to displace nucleosomes, could create a gap in the nucleosome array along the chromosome, interrupting the relay of Sir protein-hypoacetylated H3/H4 chain reaction and stop the spread of heterochromatin.

Additional boundary-like activities in yeast also include the subtelomeric anti-silencing regions (STAR), which were shown to block the telomeric-mediated position effect (TPE) and protect reporter gene expression

(Fourel *et al.*, 2001). Two transcription activators, Reb1 and Tbf1, interact with STAR. Although the silencer-blocking effect is recapitulated by tethering several activator proteins, including Reb1 and Tbf1, though Gal4 fusion proteins, the author showed that the effect of these barrier proteins on reporter gene expression is observed only in genetic background where silencing is intact, suggesting that STARs block the spread of heterochromatin, rather than activating transcription of the reporter genes.

Mechanism of Action of Chromatin Boundaries

The mechanisms through which chromatin boundaries function remain largely unknown. In particular, our ignorance about how boundaries block enhancer-promoter interactions is due to the lack of a clear understanding how enhancers interact with promoters in the first place. The complexity or even confusion about how boundaries work also arises from the broad definition of the group. DNA elements could be characterized as boundaries based on their chromosomal locations such as positioned between open and closed chromatin domains; on biochemical or cell biological properties such as nuclease sensitivity, nuclear localization or partition after nuclear extraction; or on functional criteria such as ability to block various transcriptional influences. The latter criterion is by no means unifying, for transcriptional influences from various enhancers, silencer, and active or repressive chromatin are oftentimes defined by different assays, in different systems, and are mostly poorly understood themselves. The diversity in the DNA sequences and protein factors involved in boundary function provided little clues about their mechanism of action. The diverse genomic and regulatory contexts in which boundary elements function suggest that a multitude of mechanisms may be at work. We will discuss several hypotheses and related experimental observations that prompted them.

A: Barrier Function Against Organized Chromatin

Of all transcriptional influences affected by chromatin boundaries, those mediated by organized chromatin are the best understood mechanistically. Chromatin structure and organization has long been known to influence gene activity. Regional and regulated differences in histone composition or modification affect nucleosomal organization and gene accessibility. In many cases the *cis* and *trans*-components involved in the establishment, stabilization, and spread or active or silent chromatin structure are well characterized (for recent reviews see Elgin and Grewal, 2003; Felsenfeld, 2003; Grewal and Elgin, 2002; Lusser and Kadonaga, 2003; Perrod and Gasser, 2003; Peterson and Laniel, 2004; Richards and Elgin, 2002).

A1: Unidirectional Propagation of Silent Chromatin

DNA sequences could end up at the border of distinct chromatin domains for different reasons: if the neighboring domains are generated from nucleation of two opposing chromatin states, both in activity and in direction, the border of the two domains, where the influences from either sides are neutralize by the other, could be occupied by nothing more than random DNA (Kimura and Horikoshi, 2004). Other DNA elements play a more active role in creating the border: they could initiate directional propagation of one chromatin amidst a different chromatin, or they could impede the spread of one chromatin into a different chromatin. Either event would lead to discontinuity in the state of chromatin structure, and along with it, chromatin boundaries.

The best example for the former is the yeast HML domain boundary, which seems to operate via a different mechanism from that of the HMR boundary. Rather than confining the repressive chromatin within the silent mating-type locus, the HML boundary correspond to the initiation site of unidirectional propagation of silencing (Bi *et al.*, 1999, see Fig.21.6). HML silent domain is defined at both ends by a sharp transition in reporter gene activity and in histone acetylation profile, as is the HMR domain. However, DNA sequences flanking the silent domain exhibit no barrier activity. Deletion of these flanking sequences did not result in expansion of the silent domain, as determined by reporter genes expression. Furthermore, inverting the *I* silencer at the right end of the HML domain increases silencing outside of the domain while decreasing it within the domain. These observations suggest the *I* silencer establish the domain by unidirectional nucleation of repressive chromatin towards the MAT genes (Bi *et al.*, 1999; Shei and Broach, 1995). It's interesting that only initiators of silent domain, not active domain, were associated with boundary function in yeast. As most of the yeast genome resembles active chromatin, boundaries of silent domains are functionally relevant and readily discernable. By the same notion, sequences responsible for initiating active chromatin, which could escape detection if existed in yeast, are more relevant in a mostly silent mammalian genome.

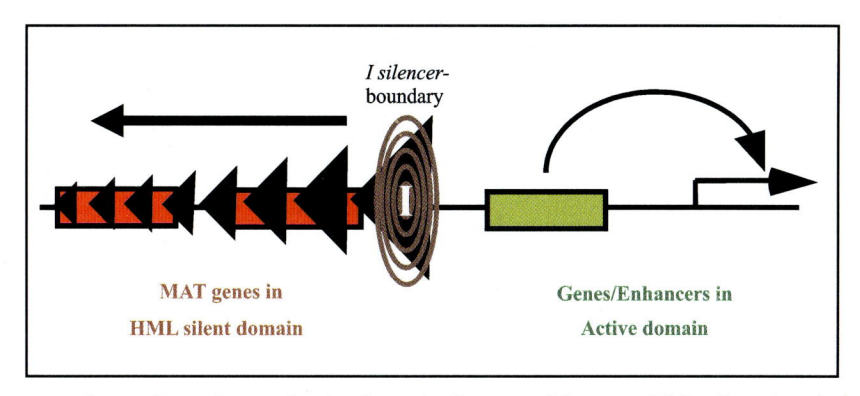

Fig. 21.6 Boundary can nucleate silent chromatin. A schematic diagram of the yeast HML silent domain is shown. Fill rectangles represent reporter genes inserted in the vicinity of the silent domain. Red rectangle, a repressed reporter gene within the silent domain, green rectangle, an active reporter gene outside of the silent domain. Brown swirl represent the *I* silencer / boundary element. Black triangles of decreasing size represent heterochromatin nucleating from the boundary element. The black arrow represents the direction of silence.

A2: Counteraction or Impediment to the Spread of Heterochromatin

In contrast to the directional nucleation of organized chromatin, majority of the barrier elements discussed so far appear to delimit the silent chromatin domains by preventing its linear propagation along the chromosome. Some barriers appear to actively counter the spread of heterochromatin by recruiting histone modification enzymes that establishing active open chromatin. It was shown in yeast that tethering of SAS2 or GCN5 histone acetyl transferases (HAT) to chromatin with Gal4 fusion proteins could recapitulate barrier activity against heterochromatin (Donze and Kamakaka, 2001). In this case, mutation in the histone acetyl transferase activity of SAS2 and GCN5 reduced the barrier activity. Counteraction to heterochromatin could also underlie the barrier activity in the chicken beta-globin HS4 boundary: USF bound at the barrier appear to recruit PCAF (H3 acetylation) and HSET7/9 (H3K4 methylation), which may organize active chromatin to prevent the spread of silencing from the neighboring folate receptor heterochromatin (West *et al*., 2004). Active transcription could be part of barrier mechanism in disrupting the progression of silent chromatin. Cryptic non-coding transcripts corresponding to intergenic enhancers and boundaries have been associated with anti-silencing effects (Fourel *et al*., 2002; Hogga and Karch, 2002; Schmitt *et al*., 2005; Sutter *et al*., 2003). Chromatin remodeling associated with transcription initiation could disrupt nucleosomal arrangement necessary for the formation of heterochromatin (Ashe *et al*., 1997; Cho *et al*., 1998; Gribnau *et al*., 2000). In cases where promoter or transcribed genes are part of boundary there is evidence that a complete transcription unit function as a stronger insulator than simply the promoter region (Bi and Broach, 2001).

Other barriers could act through a more passive mechanism of merely stopping the progression of heterochromatin without dismantling the silent domain. As mentioned above, the interruption of the regular nucleosomal organization, even by the binding of a transcription repressor Rap1p in the case of yeast UAS$_{rpg}$ boundary could create a gap in the nucleosomal array and be responsible for the passive boundary activity. The differences between the active and passive barriers were explored in a study using two reporter genes inserted near the yeast HML silent domain (Ishii and Laemmli, 2003). Transcription activator or boundary proteins are fused to Gal4 DNA-binding domain and tethered to the UAS sites flanking only one of the two reporter genes between the E and I silencers. While transcription activators (TA) are expected to dismantle the silent chromatin domain on both side of the UAS sites and activate both reporters, the passive boundaries (BA) are expected to protect the reporter between the two UAS sites without affecting the silencing of the other reporter. In comparison, short-ranged desilencing activities (DA) can establish a local island of euchromatin and also relieve repression of both reporters. By these criteria, the *Drosophila* BEAF protein, which interact with the scs' boundary behaved as a boundary activity, whereas GAGA and CTCF acted as desilencing activities (Ishii and Laemmli, 2003).

A3: Barriers Against Active Chromatin

As mentioned above, expression of integrated transgenes are often influenced by the host genome near the insertion site. While transgenes inserted near the yeast telomeres or in vertebrate genomes are frequently repressed by the surrounding chromatin, the *miniwhite* transgene integrated into the *Drosophila* genome are often activated. This phenomenon has been attributed to the generally more active genomic environment in *Drosophila*. However, the extent of eye color variation in *miniwhite* flies is rarely duplicated in the expression of other fly transgenes. The frequency with which *miniwhite* is activated is also much higher than expected for eye-specific "enhancer trapping", and is not observed with the full-length *white* promoter, which should respond comparably to eye-specific enhancers. The *miniwhite* promoter may be in a potentiated state that renders it responsive to subtle differences in chromatin environment that escape the detection of other transgene promoters. Therefore the chromosomal position effect exhibited by *miniwhite* may not be equivalent to the repressive position effect observed in mammalian genome. Nevertheless, the lack of consistent silencing of transgenes still could be taken as an indication that the genomic environment in *Drosophila* is more permissive than that in vertebrates.

Boundary elements from *Drosophila* and the chicken beta-globin loci can shield the *miniwhite* gene from CPE. Although barriers against active chromatin are not as well characterized as those against heterochromatin, mechanisms that disrupt open chromatin or prevent its propagation could be responsible for such activity. The non-overlapping insulator and barrier activities within the chicken globin boundary suggest that transcriptional influences from enhancers could be mechanistically distinct from that of silent chromatin. Presence of separate enhancer-blocking and CPE-blocking activities within the *Drosophila* SF1 compound boundary indicates that transcription activation could also be mediated through different mechanisms that are blockable by different boundary components. Further, the SF1 barrier appears to be effective blocking both positive (CPE) and negative (PRE repression) regulation mediated by chromatin.

B: Enhancer-blocking Mechanisms by Insulators

How insulators block enhancers is not clear, partly because how enhancers affect transcription at distant promoters is currently not fully understood. The fact that insulators can only disrupt the function of distal enhancers resembles position effect variegation (PEV) in that the repressive state appears to emanate in a directional fashion (Pirrotta, 1997; Pirrotta and Rastelli, 1994). This has prompted hypotheses that insulators disrupt enhancer function by unidirectional or bidirectional repression (Chen and Corces, 2001; Gerasimova et al., 1995). However, most insulators do not exhibit orientation dependence in their enhancer-blocking activity. In addition, studies in transgenic *Drosophila* also indicated that the blocked enhancers remain accessible to transcription factors and can activate transcription from a distal promoter with respect to the insulator (Cai and Levine, 1995; Scott and Geyer, 1995). These results suggest that enhancer blocking does not involve inactivation of the enhancer or repression of the enhancer function *per se*, but rather restriction of its effect across the insulator, through a mechanism that is more interceptive rather than nucleating in nature. Although genes are known under certain circumstances to be regulated by unlinked DNA elements through transvection-like mechanisms, vast majority of the promoters are controlled by *cis*-linked, and in most cases closely linked, regulatory elements. Enhancer-promoter communication thus may involve processive mechanisms that are susceptible to interception by insulators. Several models that resemble "traps" or "road-blocks" have been proposed to account for enhancer-blocking activity. They are grouped into the following two general categories according to the nature of the interference.

B1: Regulatory Interference: Promoter Decoy or Facilitator Trap

Although little sequence homology is seen among different insulator DNA elements, several insulators do contain AT-rich regions that resemble the TATA motif of a promoter (Kellum and Schedl, 1991; Spana et al., 1988). Some boundary elements indeed correspond to gene promoters or transcription origin of nongenic RNAs (Avramova and Tikhonov, 1999; Bi and Broach, 1999; Hogga and Karch, 2002; Ohtsuki and Levine, 1998). The promoter-like properties of insulators have prompted the model that they "mimic" promoters and "trap" either enhancers or transcription factors, preventing them or their effects from reaching the downstream target promoter (Geyer and Clark, 2002, see Fig.21.7B). The "promoter-decoy" model is attractive because it is mechanistically similar to the experimentally documented transcription interference by promoter competition, where enhancer-activated transcription from one gene/promoter can be reduced or abolished by the placement or activation of a competing neighboring gene/promoter (Cai et al., 2001; Choi and Engel, 1988; Foley and Engel, 1992). Although competition was seen only among activated promoters, it is possible that transcription *pe se* is not a requirement for a boundary DNA to act as promoter decoy.

A different but related model suggests that insulators block enhancers by interacting and intercepting proteins that normally facilitate long-range enhancer-promoter interactions (Dorsett, 1999, see Fig.21.7C). This model is supported by the finding that SuHw interacts with several proteins, including *Chip* and *Nipped-B*, which are involved in mediating general long-range enhancer-promoter interactions (Morcillo *et al.*, 1997; Rollins *et al.*, 1999). Mutations in these proteins disrupt the function of remote enhancers in loci such as *cut* and *Ubx*. Gypsy insertion between these distal enhancers and the gene promoters allows SuHw to interact with the facilitator proteins and interfere with their normal function. In addition to playing an antagonistic role to the SuHw protein, Chip is also involved in cross-linking and promoting homeodomain proteins binding to DNA, and Nipped-B in mediating chromatid cohesion, which could also facilitate enhancer looping to promoters (Rollins *et al.*, 2004; Torigoi *et al.*, 2000). The only caveat against this model is that SuHw can block both nearby and remote enhancers, and if the facilitator proteins are involved only in the function of remote enhancers, a different mechanism or an additional set of proteins would have to mediate nearby enhancers for SuHw to affect enhancer function in its general fashion.

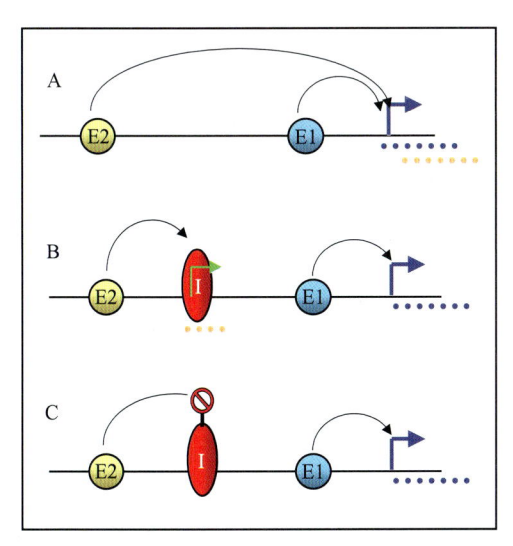

Fig.21.7 Regulatory interference by promoter decoy or facilitator captures. Upper panel: a hypothetical gene (blue arrow) and its two regulatory enhancers, E1 and E2, are diagramed. Arched arrows represent enhancer-promoter interactions during gene activation. Dashed lines near the promoter indicate transcription directed by each enhancer. Middle: a boundary element (brown oval) containing promoter-like motifs or a cryptic promoter (green arrow within the brown oval) can trap the distal E2 enhancer, disrupting or abolishing its activation of the downstream gene (loss of yellow dash). The cryptic promoter may produce non-genic RNA (yellow dash). Lower panel: boundary-binding proteins may intercept (stop sign) the upstream enhancer by interacting with the facilitator proteins, blocking its activation of the downstream gene (loss of yellow dash).

B2.1: Structural Interference: Chromatin Looping through Insulator Interactions

A second type of models suggests that insulators restrict enhancer function or enhancer-promoter interactions through physical organization of DNA/chromatin fibers. This model was prompted by the observation from transgenic *Drosophila* that two tandem copies of suHw allow the distal enhancer to bypass the insulator block and activate a downstream reporter gene (X, see Fig. 21.8A, Cai and Shen, 2001). Further, if a second reporter gene (Y) is inserted between the two suHw elements, the distal enhancer would be blocked from Y but can still activate X (see Fig. 21.8A, Muravyova *et al.*, 2001). Based on these observations it was proposed that suHw element might interact with each other to form chromatin loop domains, thereby restricting interactions between neighboring regulatory elements. As seen in Fig.21.8B, a single insulator inserted between a gene and its enhancer may interact with another insulator elsewhere in the genome and sequester the enhancer within a topologically distinct chromatin loop, restricting its ability to interact with promoters located outside of the loop. Interactions between suHw elements placed in tandem could impede interactions necessary to form such loop domains, leading to the loss of enhancer-blocking activity.

Consistent with the insulator looping model, several protein components of chromatin boundaries, such as GAF and Mod(mdg4), the latter being required for the suHw insulator activity, contain BTB domain, which is known to mediate protein-protein interaction and oligomerization (Farkas *et al.*, 1994; Gerasimova *et al.*, 1995; Yoshida *et al.*, 1999). In fact, the GAF protein, which is required for the insulator activity of SF1, Fab-7 and the *evenskipped* promoter, was shown to tether together distant DNA elements through its BTB-mediated interaction (Belozerov *et al.*, 2003; Mahmoudi *et al.*, 2002; Ohtsuki and Levine, 1998; Schweinsberg *et al.*, 2004). BTB-mediated insulator interactions are also likely to be responsible for insulator-bypass of paired Mcp-1 elements, which contains GAF binding sites, and the heterologous boundary combination of suHw and Mcp-1 (Gruzdeva *et al.*, 2005; Melnikova *et al.*, 2004). However, other protein-protein dimerization or oligomerization could also mediate insulator interactions. It was recently shown by chromatin conformation capture technique (3C, Dekker *et al.*, 2002), or ligation-mediated PCR, that scs and scs' boundaries, which flank the *Drosophila* heatshock loci, are present in close physical proximity in the nuclear space, possibly through BEAF-zw5 interaction (Blanton *et al.*, 2003).

It is not clear how chromatin looping restricts

enhancer-promoter interaction, given the fact that the four strands of DNA fiber emerging from the insulator complex, or the "knot", at the base of the chromatin loop are considered to be more or less equivalent. It is possible that the protein complex responsible for "tying up" the "knot" renders asymmetry in the local conformation or topology of the emerging DNA strands, so that cross-strand transfer of protein factors or other forms of regulatory effects is energetically more favored. It is interesting that such tether-mediated enhancer-block can be replicated with non-chromosomal DNA and small simple proteins in bacteria or cultured human cells (Ameres et al., 2005; Bondarenko et al., 2003).

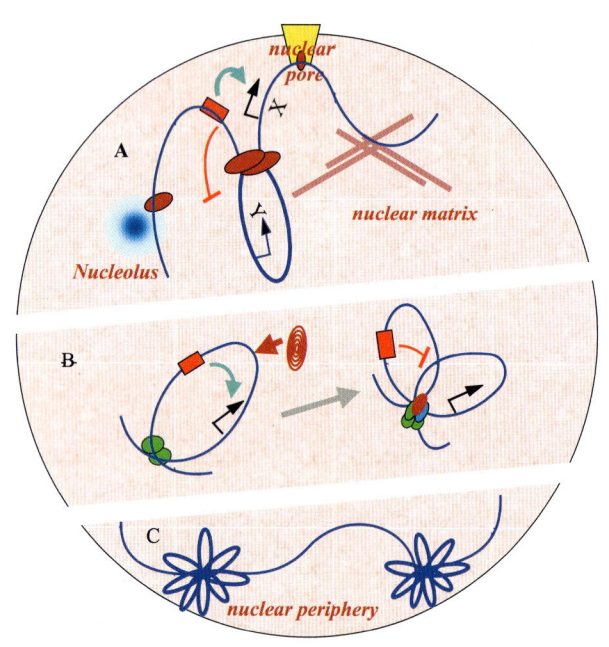

Fig.21.8 Structural interference by boundary-mediated chromatin loop domain formation. A hypothetical nucleus is diagramed with blue lines and ovals representing chromatin fibers and chromatin loops. Brown swirls and ovals represent chromatin boundaries. Arched arrows represent enhancer-promoter interactions during gene activation. Arched lines with a short dash represent blocked enhancer-promoter interactions. A: boundary bypass mediated by boundary-boundary interaction (see text). Two boundaries represented by the brown ovals interact with each other, looping out the intervening DNA region and block the red enhancer from activating gene Y within the loop, but allowing it to activate the distal gene X (insulator bypass). In addition, boundaries can interact with other nuclear structures such as the nuclear pores (yellow funnel), the nucleolus (light blue circle) or nuclear matrix or nuclear scaffold proteins (pink rods). B: a boundary inserted between an enhancer and a promoter interacts with other boundaries or regions of chromatin, bisecting the local chromatin loop into two separate loops, and thereby restricting enhancer-promoter interactions across the loop (see text). Green ovals represent boundary-binding proteins. C: insulators bodies formed near the nuclear periphery (see text).

B2.2: Structural Interference: Chromatin Looping through Insulator-nuclear Structure Interactions

Tethering of distant DNA elements is not the only means to form chromatin loops. The suHw-mediated chromatin loop formation was also suggested by cell biological studies. The Corces group reported that gypsy retrotransposon inserted in different regions of the X chromosome colocalized to the nuclear periphery in a SuHw and Mod(mdg4) dependent fashion (Gerasimova et al., 2000). It was proposed that the interaction with structures at the nuclear periphery allows multiple insulators to coalesce and form "insulator bodies", a rosette-like structure with insulators in the center and the intervening DNAs looped out (Gerasimova et al., 2000). It does not appear, however, that the peripheral association is essential for the insulator function since the 340-bp suHw element, which is sufficient for enhancer-blocking activity of gypsy, is found both at the periphery as well as interior of the nucleus (Xu et al., 2004). Other boundary elements are also found to localize to certain subnuclear compartments (see Fig. 21.8A). For example, a nuclear pore protein was found to be an essential component of a yeast boundary (Ishii et al., 2002). The vertebrate insulator protein CTCF was found to interact and colocalize with nucleophosmin, a nucleolar protein, at insulator regions on the chromosome. The CTCF binding sites also localized to nucleolus (Yusufzai et al., 2004). These results suggest that attachment and clustering of boundary elements to each other or to various nuclear structures may represent a general mechanism through which boundaries partition the genomes into independent loop domains.

The idea of topologically independent chromatin loops had much support from biochemical and cell biological study of nuclear structure over the last fifty years. It was envisioned that higher eukaryotic genomes are organized through multiple levels of compaction. At each level, units of DNA are folded up with special protein structures, be it hundreds of base pairs around nucleosomes, or tens of kilobases anchored to nuclear scaffolds. These units of genome can be accessed in an orderly and regulated fashion during nuclear and chromosomal functions such as transcription and replication. Glimpses of these units afforded by various nuclear extraction procedures revealed loops of chromatin attached at their base to the protein skeletons through DNAs called matrix/scaffold attachment regions (MAR/SAR, Mirkovitch et al., 1987; Mirkovitch et al., 1984). The best examples of these loops are seen in the "lampbrush" chromosomes from the amphibian oocyte and nuclear halo from mammalian cells (Dijkwel and Hamlin, 1995; Srivastava, 1951). Studies suggest that these "loops" between attachment points may form topologically separate chromatin domains (Attal et al., 1995; Bode et al., 2003; Greally et al., 1997; Heng et al., 2001). In addition to the organizational roles, these loop domains also correspond to functional units of the genome, which are regulated during cellular processes and development (Chambeyron et al., 2005). It is not surprising in this light that many insulator elements are also experimentally defined as MAR/SAR elements (Forrester et al., 1999; Nabirochkin et al., 1998; Namciu et al., 1998; Yusufzai et al., 2004).

Evolutionary Conservation of Boundary Function

Boundary elements appear to serve essential function in diverse organisms from yeast to human. However, in spite of the similar activities they exhibit in blocking transcriptional influences from regulatory enhancers or organized chromatin, surprisingly little conservation has been documented in their DNA sequences and protein components. This could be due to the still limited number of cases of boundary discovered, especially where both the cis- and trans- components are well characterized. Although zinc-finger-domain and BTB/POZ domain proteins are found to associate with several boundary elements, it could be due to functional convergence rather than evolutionary conservation. However, cases of cross-species boundary function have been reported. For example, the chicken globin HS4 element was shown to function in the Drosophila CPE-blocking assay, likely due to its barrier activity (Chung et al., 1993). CTCF was also reported to block telomere silencing in yeast in a boundary-like fashion (BA) (Defossez and Gilson, 2002), although a similar test in the HML silent loci suggests that CTCF and GAGA act as desilencing activity (DA, Ishii and Laemmli, 2003). The best case of boundary conservation comes from the recent report of dCTCF, a Drosophila ortholog of the vertebrate CTCF. It could repress transcription in monkey COS cells when tethered through Gal4 fusion, as could the vertebrate CTCF protein. In addition, the Drosophila Fab-8 boundary, which contains the dCTCF binding sites, can block enhancers in a dCTCF-dependent fashion both in transgenic Drosophila as well as in human K562 cells (Moon et al., 2005). SuHw, a well-characterized Drosophila insulator protein, was found to function as a barrier against heterochromatin and appears to do so in the absence of Mod(mdg4), its obligatory partner in Drosophila (Donze and Kamakaka, 2001). Although an early study indicated that SuHw exerted no insulator effect on enhancer-mediate activation in yeast, the difference could be due to the lack of Mod(mdg4), or insulator bypass since multiple copies of the suHw element was used in the earlier study (Kim et al., 1993). The Drosophila scs element was also reported to function as an enhancer blocker in Xenopus, and BEAF32, which interacts with the Drosophila scs' element, was shown to function as a boundary activity when tethered to yeast silence domains (Dunaway et al., 1997; Ishii and Laemmli, 2003; Krebs and Dunaway, 1998).

Conclusion

Chromatin boundaries or insulators play important roles in gene regulation. Although we have come a long way in understanding their functions and mechanisms, many significant questions remain unanswered, especially in regards to how they participate in genome, chromatin and nuclear organization. For example, how are boundary-like elements distributed in the genome? What proportions of boundary phenomena play causal roles in effecting changes in gene activity and nuclear property, and what proportions merely reflect the consequences of these changes? Is there tissue-specific or developmental stage specific boundary function? Are different boundary mechanisms (such as directional nucleation, chain-breaking, or looping) selectively used for distinct regulatory tasks and in different types of genetic loci? What is the significance of compound boundaries in coordinating gene regulation in complex genetic loci? Although models have been proposed for the function of intragenic boundaries, how they govern enhancer accessibility by interacting with PTS/PTE elements is not clear. As more boundaries are identified and characterized, we hope to gain better understanding of their functions and mechanisms in the future.

References

Ameres, S. L., Drueppel, L., Pfleiderer, K., Schmidt, A., Hillen, W., and Berens, C. (2005). Inducible DNA-loop formation blocks transcriptional activation by an SV40 enhancer. Embo J 24, 358-367.

Arnosti, D. N. (2003). Analysis and function of transcriptional regulatory elements: insights from Drosophila. Annu Rev Entomol 48, 579-602.

Ashe, H. L., Monks, J., Wijgerde, M., Fraser, P., and Proudfoot, N. J. (1997). Intergenic transcription and transinduction of the human beta-globin locus. Genes Dev 11, 2494-2509.

Attal, J., Cajero-Juarez, M., Petitclerc, D., Theron, M. C., Stinnakre, M. G., Bearzotti, M., Kann, G., and Houdebine, L. M. (1995). The effect of matrix attached regions (MAR) and specialized chromatin structure (SCS) on the expression of gene constructs in cultured cells and in transgenic mice. Mol Biol Rep *22*, 37-46.

Avramova, Z., and Tikhonov, A. (1999). Are scs and scs' 'neutral' chromatin domain boundaries of the locus? Trends Genet *15*, 138-139.

Bae, E., Calhoun, V. C., Levine, M., Lewis, E. B., and Drewell, R. A. (2002). Characterization of the intergenic RNA profile at abdominal-A and Abdominal-B in the *Drosophila* bithorax complex. Proc Natl Acad Sci USA *99*, 16847-16852.

Barges, S., Mihaly, J., Galloni, M., Hagstrom, K., Muller, M., Shanower, G., Schedl, P., Gyurkovics, H., and Karch, F. (2000). The Fab-8 boundary defines the distal limit of the bithorax complex iab-7 domain and insulates iab-7 from initiation elements and a PRE in the adjacent iab-8 domain. Development *127*, 779-790.

Bell, A. C., and Felsenfeld, G. (2000). Methylation of a CTCF-dependent boundary controls imprinted expression of the Igf2 gene. Nature *405*, 482-485.

Bell, A. C., West, A. G., and Felsenfeld, G. (1999). The protein CTCF is required for the enhancer blocking activity of vertebrate insulators. Cell *98*, 387-396.

Belozerov, V. E., Majumder, P., Shen, P., and Cai, H. N. (2003). A novel boundary element may facilitate independent gene regulation in the Antennapedia Complex of *Drosophila*. EMBO J *22*.

Bi, X., Braunstein, M., Shei, G. J., and Broach, J. R. (1999). The yeast HML I silencer defines a heterochromatin domain boundary by directional establishment of silencing. Proc Natl Acad Sci USA *96*, 11934-11939.

Bi, X., and Broach, J. R. (1999). UASrpg can function as a heterochromatin boundary element in yeast. Genes Dev *13*, 1089-1101.

Bi, X., and Broach, J. R. (2001). Chromosomal boundaries in *S. cerevisiae*. Curr Opin Genet Dev *11*, 199-204.

Blackwood, E. M., and Kadonaga, J. T. (1998). Going the distance: a current view of enhancer action. Science *281*, 61-63.

Blanton, J., Gaszner, M., and Schedl, P. (2003). Protein:protein interactions and the pairing of boundary elements *in vivo*. Genes Dev *17*, 664-675.

Bode, J., Goetze, S., Heng, H., Krawetz, S. A., and Benham, C. (2003). From DNA structure to gene expression: mediators of nuclear compartmentalization and dynamics. Chromosome Res *11*, 435-445.

Bondarenko, V. A., Jiang, Y. I., and Studitsky, V. M. (2003). Rationally designed insulator-like elements can block enhancer action *in vitro*. Embo J *22*, 4728-4737.

Braunstein, M., Rose, A. B., Holmes, S. G., Allis, C. D., and Broach, J. R. (1993). Transcriptional silencing in yeast is associated with reduced nucleosome acetylation. Genes Dev *7*, 592-604.

Bulger, M., Bender, M. A., van Doorninck, J. H., Wertman, B., Farrell, C. M., Felsenfeld, G., Groudine, M., and Hardison, R. (2000). Comparative structural and functional analysis of the olfactory receptor genes flanking the human and mouse beta-globin gene clusters. Proc Natl Acad Sci USA *97*, 14560-14565.

Bulger, M., and Groudine, M. (1999). Looping versus linking: toward a model for long-distance gene activation. Genes Dev *13*, 2465-2477.

Bulger, M., van Doorninck, J. H., Saitoh, N., Telling, A., Farrell, C., Bender, M. A., Felsenfeld, G., Axel, R., and Groudine, M. (1999). Conservation of sequence and structure flanking the mouse and human beta-globin loci: the beta-globin genes are embedded within an array of odorant receptor genes. Proc Natl Acad Sci USA *96*, 5129-5134.

Cai, H., and Levine, M. (1995). Modulation of enhancer-promoter interactions by insulators in the *Drosophila* embryo. Nature *376*, 533-536.

Cai, H. N., and Levine, M. (1997). The gypsy insulator can function as a promoter-specific silencer in the *Drosophila* embryo. EMBO J *16*, 1732-1741.

Cai, H. N., and Shen, P. (2001). Effects of cis arrangement of chromatin insulators on enhancer-blocking activity. Science *291*, 493-495.

Cai, H. N., Zhang, Z., Adams, J. R., and Shen, P. (2001). Genomic context modulates insulator activity through promoter competition. Development *128*, 4339-4347.

Calhoun, V. C., and Levine, M. (2003). Long-range enhancer-promoter interactions in the Scr-Antp interval of the *Drosophila* Antennapedia complex. Proc Natl Acad Sci USA *100*, 9878-9883.

Calhoun, V. C., Stathopoulos, A., and Levine, M. (2002). Promoter-proximal tethering elements regulate enhancer-promoter specificity in the *Drosophila* Antennapedia complex. Proc Natl Acad Sci USA *99*, 9243-9247.

Chambeyron, S., Da Silva, N. R., Lawson, K. A., and Bickmore, W. A. (2005). Nuclear re-organisation of the Hoxb complex during mouse embryonic development. Development *132*, 2215-2223.

Chen, S., and Corces, V. G. (2001). The gypsy insulator of *Drosophila* affects chromatin structure in a directional manner. Genetics *159*, 1649-1658.

Cho, H., Orphanides, G., Sun, X., Yang, X. J., Ogryzko, V., Lees, E., Nakatani, Y., and Reinberg, D. (1998). A human RNA polymerase II complex containing factors that modify chromatin structure. Mol Cell Biol *18*, 5355-5363.

Choi, O. R., and Engel, J. D. (1988). Developmental regulation of beta-globin gene switching. Cell *55*, 17-26.

Chung, J. H., Whiteley, M., and Felsenfeld, G. (1993). A 5′ element of the chicken beta-globin domain serves as an insulator in human erythroid cells and protects against position effect in *Drosophila*. Cell *74*, 505-514.

Defossez, P. A., and Gilson, E. (2002). The vertebrate protein CTCF functions as an insulator in *Saccharomyces cerevisiae*. Nucleic Acids Res *30*, 5136-5141.

Dekker, J., Rippe, K., Dekker, M., and Kleckner, N. (2002). Capturing chromosome conformation. Science 295, 1306-1311.

Dijkwel, P. A., and Hamlin, J. L. (1995). Origins of replication and the nuclear matrix: the DHFR domain as a paradigm. Int Rev Cytol 162A, 455-484.

Donze, D., Adams, C. R., Rine, J., and Kamakaka, R. T. (1999). The boundaries of the silenced HMR domain in *Saccharomyces cerevisiae*. Genes Dev 13, 698-708.

Donze, D., and Kamakaka, R. T. (2001). RNA polymerase III and RNA polymerase II promoter complexes are heterochromatin barriers in *Saccharomyces cerevisiae*. Embo J 20, 520-531.

Dorn, R., Krauss, V., Reuter, G., and Saumweber, H. (1993). The enhancer of position-effect variegation of *Drosophila*, E(var)3-93D, codes for a chromatin protein containing a conserved domain common to several transcriptional regulators. Proc Natl Acad Sci USA 90, 11376-11380.

Dorsett, D. (1993). Distance-independent inactivation of an enhancer by the suppressor of Hairy-wing DNA-binding protein of *Drosophila*. Genetics 134, 1135-1144.

Dorsett, D. (1999). Distant liaisons: long-range enhancer-promoter interactions in *Drosophila*. Curr Opin Genet Dev 9, 505-514.

Dunaway, M., and Droge, P. (1989). Transactivation of the *Xenopus* rRNA gene promoter by its enhancer. Nature 341, 657-659.

Dunaway, M., Hwang, J. Y., Xiong, M., and Yuen, H. L. (1997). The activity of the scs and scs' insulator elements is not dependent on chromosomal context. Mol Cell Biol 17, 182-189.

Duncan, I. W. (2002). Transvection effects in *Drosophila*. Annu Rev Genet 36, 521-556.

Elgin, S. C., and Grewal, S. I. (2003). Heterochromatin: silence is golden. Curr Biol 13, R895-898.

Farkas, G., Gausz, J., Galloni, M., Reuter, G., Gyurkovics, H., and Karch, F. (1994). The Trithorax-like gene encodes the *Drosophila* GAGA factor. Nature 371, 806-808.

Farrell, C. M., West, A. G., and Felsenfeld, G. (2002). Conserved CTCF insulator elements flank the mouse and human beta-globin loci. Mol Cell Biol 22, 3820-3831.

Felsenfeld, G. (2003). Quantitative approaches to problems of eukaryotic gene expression. Biophys Chem 100, 607-613.

Foley, K. P., and Engel, J. D. (1992). Individual stage selector element mutations lead to reciprocal changes in beta- vs. epsilon-globin gene transcription: genetic confirmation of promoter competition during globin gene switching. Genes Dev 6, 730-744.

Forrester, W. C., Fern#ndez, L. A., and Grosschedl, R. (1999). Nuclear matrix attachment regions antagonize methylation-dependent repression of long-range enhancer-promoter interactions. Genes Dev 13, 3003-3014.

Fourel, G., Boscheron, C., Revardel, E., Lebrun, E., Hu, Y. F., Simmen, K. C., Muller, K., Li, R., Mermod, N., and Gilson, E. (2001). An activation-independent role of transcription factors in insulator function. EMBO Rep 2, 124-132.

Fourel, G., Miyake, T., Defossez, P. A., Li, R., and Gilson, E. (2002). General regulatory factors (GRFs) as genome partitioners. J Biol Chem 277, 41736-41743.

Galloni, M., Gyurkovics, H., Schedl, P., and Karch, F. (1993). The bluetail transposon: evidence for independent cis-regulatory domains and domain boundaries in the bithorax complex. Embo J 12, 1087-1097.

Gaszner, M., Vazquez, J., and Schedl, P. (1999). The Zw5 protein, a component of the scs chromatin domain boundary, is able to block enhancer-promoter interaction. Genes Dev 13, 2098-2107.

Gdula, D. A., and Corces, V. G. (1997). Characterization of functional domains of the su(Hw) protein that mediate the silencing effect of mod(mdg4) mutations. Genetics 145, 153-161.

Georgiev, P., and Kozycina, M. (1996). Interaction between mutations in the suppressor of Hairy wing and modifier of mdg4 genes of *Drosophila* melanogaster affecting the phenotype of gypsy-induced mutations. Genetics 142, 425-436.

Gerasimova, T. I., Byrd, K., and Corces, V. G. (2000). A chromatin insulator determines the nuclear localization of DNA. Mol Cell 6, 1025-1035.

Gerasimova, T. I., and Corces, V. G. (1998). Polycomb and trithorax group proteins mediate the function of a chromatin insulator. Cell 92, 511-521.

Gerasimova, T. I., Gdula, D. A., Gerasimov, D. V., Simonova, O., and Corces, V. G. (1995). A *Drosophila* protein that imparts directionality on a chromatin insulator is an enhancer of position-effect variegation. Cell 82, 587-597.

Geyer, P. K., and Clark, I. (2002). Protecting against promiscuity: the regulatory role of insulators. Cell Mol Life Sci 59, 2112-2127.

Geyer, P. K., and Corces, V. G. (1992). DNA position-specific repression of transcription by a *Drosophila* zinc finger protein. Genes Dev 6, 1865-1873.

Ghosh, D., Gerasimova, T. I., and Corces, V. G. (2001). Interactions between the Su(Hw) and Mod(mdg4) proteins required for gypsy insulator function. Embo J 20, 2518-2527.

Golovnin, A., Birukova, I., Romanova, O., Silicheva, M., Parshikov, A., Savitskaya, E., Pirrotta, V., and Georgiev, P. (2003). An endogenous Su(Hw) insulator separates the yellow gene from the Achaete-scute gene complex in *Drosophila*. Development 130, 3249-3258.

Greally, J. M., Guinness, M. E., McGrath, J., and Zemel, S. (1997). Matrix-attachment regions in the mouse Chromosome 7F imprinted domain. Mamm Genome 8, 805-810.

Grewal, S. I., and Elgin, S. C. (2002). Heterochromatin: new possibilities for the inheritance of structure. Curr Opin Genet Dev 12, 178-187.

Gribnau, J., Diderich, K., Pruzina, S., Calzolari, R., and Fraser, P. (2000). Intergenic transcription and developmental remodeling of chromatin subdomains in the human beta-globin locus. Mol Cell 5, 377-386.

Gruzdeva, N., Kyrchanova, O., Parshikov, A., Kullyev, A., and Georgiev, P. (2005). The Mcp Element from the bithorax Complex Contains an Insulator That Is Capable of Pairwise

Interactions and Can Facilitate Enhancer-Promoter Communication. Mol Cell Biol *25*, 3682-3689.

Gyurkovics, H., Gausz, J., Kummer, J., and Karch, F. (1990). A new homeotic mutation in the *Drosophila* bithorax complex removes a boundary separating two domains of regulation. Embo J *9*, 2579-2585.

Hagstrom, K., Muller, M., and Schedl, P. (1996). Fab-7 functions as a chromatin domain boundary to ensure proper segment specification by the *Drosophila* bithorax complex. Genes Dev *10*, 3202-3215.

Hark, A. T., Schoenherr, C. J., Katz, D. J., Ingram, R. S., Levorse, J. M., and Tilghman, S. M. (2000). CTCF mediates methylation-sensitive enhancer-blocking activity at the H19/Igf2 locus. Nature *405*, 486-489.

Harrison, D. A., Gdula, D. A., Coyne, R. S., and Corces, V. G. (1993). A leucine zipper domain of the suppressor of Hairy-wing protein mediates its repressive effect on enhancer function. Genes Dev *7*, 1966-1978.

Harrison, D. A., Geyer, P. K., Spana, C., and Corces, V. G. (1989). The gypsy retrotransposon of *Drosophila melanogaster*: mechanisms of mutagenesis and interaction with the suppressor of Hairy-wing locus. Dev Genet *10*, 239-248.

Hart, C. M., Zhao, K., and Laemmli, U. K. (1997). The scs' boundary element: characterization of boundary element-associated factors. Mol Cell Biol *17*, 999-1009.

Hawley, R. G. (2001). Progress toward vector design for hematopoietic stem cell gene therapy. Curr Gene Ther *1*, 1-17.

Hazelrigg, T., Levis, R., and Rubin, G. M. (1984). Transformation of white locus DNA in *Drosophila*: dosage compensation, zeste interaction, and position effects. Cell *36*, 469-481.

Hebbes, T. R., Clayton, A. L., Thorne, A. W., and Crane-Robinson, C. (1994). Core histone hyperacetylation co-maps with generalized DNase I sensitivity in the chicken beta-globin chromosomal domain. Embo J *13*, 1823-1830.

Heng, H. H., Krawetz, S. A., Lu, W., Bremer, S., Liu, G., and Ye, C. J. (2001). Re-defining the chromatin loop domain. Cytogenet Cell Genet *93*, 155-161.

Hogga, I., and Karch, F. (2002). Transcription through the iab-7 cis-regulatory domain of the bithorax complex interferes with maintenance of Polycomb-mediated silencing. Development *129*, 4915-4922.

Holdridge, C., and Dorsett, D. (1991). Repression of hsp70 heat shock gene transcription by the suppressor of hairy-wing protein of *Drosophila melanogaster*. Mol Cell Biol *11*, 1894-1900.

Hopmann, R., Duncan, D., and Duncan, I. (1995). Transvection in the iab-5,6,7 region of the bithorax complex of *Drosophila*: homology independent interactions in trans. Genetics *139*, 815-833.

Ishii, K., Arib, G., Lin, C., Van Houwe, G., and Laemmli, U. K. (2002). Chromatin boundaries in budding yeast: the nuclear pore connection. Cell *109*, 551-562.

Ishii, K., and Laemmli, U. K. (2003). Structural and dynamic functions establish chromatin domains. Mol Cell *11*, 237-248.

Jupe, E. R., Sinden, R. R., and Cartwright, I. L. (1995). Specialized chromatin structure domain boundary elements flanking a *Drosophila* heat shock gene locus are under torsional strain *in vivo*. Biochemistry *34*, 2628-2633.

Karch, F., Galloni, M., Sipos, L., Gausz, J., Gyurkovics, H., and Schedl, P. (1994). Mcp and Fab-7: molecular analysis of putative boundaries of cis-regulatory domains in the bithorax complex of *Drosophila melanogaster*. Nucleic Acids Res *22*, 3138-3146.

Kassis, J. A. (1994). Unusual properties of regulatory DNA from the *Drosophila* engrailed gene: three "pairing-sensitive" sites within a 1.6-kb region. Genetics *136*, 1025-1038.

Kassis, J. A. (2002). Pairing-sensitive silencing, polycomb group response elements, and transposon homing in *Drosophila*. Adv Genet *46*, 421-438.

Kellum, R., and Schedl, P. (1991). A position-effect assay for boundaries of higher order chromosomal domains. Cell *64*, 941-950.

Kellum, R., and Schedl, P. (1992). A group of scs elements function as domain boundaries in an enhancer-blocking assay. Mol Cell Biol *12*, 2424-2431.

Kim, J., Shen, B., and Dorsett, D. (1993). The *Drosophila melanogaster* suppressor of Hairy-wing zinc finger protein has minimal effects on gene expression in *Saccharomyces cerevisiae*. Genetics *135*, 343-355.

Kimura, A., and Horikoshi, M. (2004). Partition of distinct chromosomal regions: negotiable border and fixed border. Genes Cells *9*, 499-508.

Kioussis, D., and Festenstein, R. (1997). Locus control regions: overcoming heterochromatin-induced gene inactivation in mammals. Curr Opin Genet Dev *7*, 614-619.

Krebs, J. E., and Dunaway, M. (1998). The scs and scs' insulator elements impart a cis requirement on enhancer-promoter interactions. Mol Cell *1*, 301-308.

Kurdistani, S. K., and Grunstein, M. (2003). Histone acetylation and deacetylation in yeast. Nat Rev Mol Cell Biol *4*, 276-284.

Levine, M., and Davidson, E. H. (2005). From the Cover. Gene Regulatory Networks Special Feature: Gene regulatory networks for development. Proc Natl Acad Sci USA *102*, 4936-4942.

Levine, M., and Tjian, R. (2003). Transcription regulation and animal diversity. Nature *424*, 147-151.

Li, X., and Noll, M. (1994). Compatibility between enhancers and promoters determines the transcriptional specificity of gooseberry and gooseberry neuro in the *Drosophila* embryo. Embo J *13*, 400-406.

Lin, Q., Chen, Q., Lin, L., and Zhou, J. (2004). The Promoter Targeting Sequence mediates epigenetically heritable transcription memory. Genes Dev *18*, 2639-2651.

Litt, M. D., Simpson, M., Recillas-Targa, F., Prioleau, M. N., and Felsenfeld, G. (2001). Transitions in histone acetylation reveal boundaries of three separately regulated neighboring loci. Embo J *20*, 2224-2235.

Loo, S., and Rine, J. (1995). Silencing and heritable domains of gene expression. Annu Rev Cell Dev Biol *11*, 519-548.

Lusser, A., and Kadonaga, J. T. (2003). Chromatin remodeling by ATP-dependent molecular machines. Bioessays *25*, 1192-1200.

Maeda, R. K., and Karch, F. (2003). Ensuring enhancer fidelity. Nat Genet *34*, 360-361.

Mahmoudi, T., Katsani, K. R., and Verrijzer, C. P. (2002). GAGA can mediate enhancer function in trans by linking two separate DNA molecules. Embo J *21*, 1775-1781.

Majumder, P., and Cai, H. N. unpublished data.

Mallin, D. R., Myung, J. S., Patton, J. S., and Geyer, P. K. (1998). Polycomb group repression is blocked by the *Drosophila* suppressor of Hairy-wing [su(Hw)] insulator. Genetics *148*, 331-339.

Martens, J. A., Laprade, L., and Winston, F. (2004). Intergenic transcription is required to repress the *Saccharomyces cerevisiae* SER3 gene. Nature *429*, 571-574.

Martin, D. I., and Whitelaw, E. (1996). The vagaries of variegating transgenes. Bioessays *18*, 919-923.

Melnikova, L., Juge, F., Gruzdeva, N., Mazur, A., Cavalli, G., and Georgiev, P. (2004). Interaction between the GAGA factor and Mod(mdg4) proteins promotes insulator bypass in *Drosophila*. Proc Natl Acad Sci USA *101*, 14806-14811.

Merli, C., Bergstrom, D. E., Cygan, J. A., and Blackman, R. K. (1996). Promoter specificity mediates the independent regulation of neighboring genes. Genes Dev *10*, 1260-1270.

Mihaly, J., Hogga, I., Barges, S., Galloni, M., Mishra, R. K., Hagstrom, K., Muller, M., Schedl, P., Sipos, L., Gausz, J., *et al.* (1998). Chromatin domain boundaries in the Bithorax complex. Cell Mol Life Sci *54*, 60-70.

Mirkovitch, J., Gasser, S. M., and Laemmli, U. K. (1987). Relation of chromosome structure and gene expression. Philos Trans R Soc Lond B Biol Sci *317*, 563-574.

Mirkovitch, J., Mirault, M. E., and Laemmli, U. K. (1984). Organization of the higher-order chromatin loop: specific DNA attachment sites on nuclear scaffold. Cell *39*, 223-232.

Modolell, J., Bender, W., and Meselson, M. (1983). *Drosophila melanogaster* mutations suppressible by the suppressor of Hairy-wing are insertions of a 7.3-kilobase mobile element. Proc Natl Acad Sci USA *80*, 1678-1682.

Moon, H., Filippova, G., Loukinov, D., Pugacheva, E., Chen, Q., Smith, S. T., Munhall, A., Grewe, B., Bartkuhn, M., Arnold, R., *et al.* (2005). CTCF is conserved from *Drosophila* to humans and confers enhancer blocking of the Fab-8 insulator. EMBO Rep *6*, 165-170.

Morcillo, P., Rosen, C., Baylies, M. K., and Dorsett, D. (1997). Chip, a widely expressed chromosomal protein required for segmentation and activity of a remote wing margin enhancer in *Drosophila*. Genes Dev *11*, 2729-2740.

Muravyova, E., Golovvin, A., Gracheva, E., Parshikov, T., Belenkaya, T., Pirrotta, V., and Georgiev, P. (2001). Paired su(Hw) insulators lose enhancer blocking activity and may instead facilitate enhancer-promoter interaction. Science *291*.

Nabirochkin, S., Ossokina, M., and Heidmann, T. (1998). A nuclear matrix/scaffold attachment region co-localizes with the gypsy retrotransposon insulator sequence. J Biol Chem *273*, 2473-2479.

Namciu, S. J., Blochlinger, K. B., and Fournier, R. E. (1998). Human matrix attachment regions insulate transgene expression from chromosomal position effects in *Drosophila melanogaster*. Mol Cell Biol *18*, 2382-2391.

Ohtsuki, S., and Levine, M. (1998). GAGA mediates the enhancer blocking activity of the eve promoter in the *Drosophila* embryo. Genes Dev *12*, 3325-3330.

Ohtsuki, S., Levine, M., and Cai, H. N. (1998). Different core promoters possess distinct regulatory activities in the *Drosophila* embryo. Genes Dev *12*, 547-556.

Pai, C. Y., Lei, E. P., Ghosh, D., and Corces, V. G. (2004). The centrosomal protein CP190 is a component of the gypsy chromatin insulator. Mol Cell *16*, 737-748.

Parkhurst, S. M., Harrison, D. A., Remington, M. P., Spana, C., Kelley, R. L., Coyne, R. S., and Corces, V. G. (1988). The *Drosophila* su(Hw) gene, which controls the phenotypic effect of the gypsy transposable element, encodes a putative DNA-binding protein. Genes Dev *2*, 1205-1215.

Parnell, T. J., Viering, M. M., Skjesol, A., Helou, C., Kuhn, E. J., and Geyer, P. K. (2003). An endogenous suppressor of hairy-wing insulator separates regulatory domains in *Drosophila*. Proc Natl Acad Sci USA *100*, 13436-13441.

Perrod, S., and Gasser, S. M. (2003). Long-range silencing and position effects at telomeres and centromeres: parallels and differences. Cell Mol Life Sci *60*, 2303-2318.

Peterson, C. L., and Laniel, M. A. (2004). Histones and histone modifications. Curr Biol *14*, R546-551.

Pikaart, M. J., Recillas-Targa, F., and Felsenfeld, G. (1998). Loss of transcriptional activity of a transgene is accompanied by DNA methylation and histone deacetylation and is prevented by insulators. Genes Dev *12*, 2852-2862.

Pirrotta, V. (1997). Chromatin-silencing mechanisms in *Drosophila* maintain patterns of gene expression. Trends Genet *13*, 314-318.

Pirrotta, V. (1998). Polycombing the genome: PcG, trxG, and chromatin silencing. Cell *93*, 333-336.

Pirrotta, V. (1999). Transvection and chromosomal trans-interaction effects. Biochim Biophys Acta *1424*, M1-8.

Pirrotta, V., and Rastelli, L. (1994). White gene expression, repressive chromatin domains and homeotic gene regulation in *Drosophila*. Bioessays *16*, 549-556.

Pirrotta, V., Steller, H., and Bozzetti, M. P. (1985). Multiple upstream regulatory elements control the expression of the *Drosophila* white gene. Embo J *4*, 3501-3508.

Recillas-Targa, F., Pikaart, M. J., Burgess-Beusse, B., Bell, A. C., Litt, M. D., West, A. G., Gaszner, M., and Felsenfeld, G. (2002). Position-effect protection and enhancer blocking by the chicken beta-globin insulator are separable activities. Proc Natl Acad Sci USA *99*, 6883-6888.

Reitman, M., Lee, E., Westphal, H., and Felsenfeld, G. (1990). Site-independent expression of the chicken beta A-globin gene in transgenic mice. Nature *348*, 749-752.

Richards, E. J., and Elgin, S. C. (2002). Epigenetic codes for heterochromatin formation and silencing: rounding up the usual suspects. Cell *108*, 489-500.

Ringrose, L., and Paro, R. (2004). Epigenetic regulation of cellular memory by the Polycomb and Trithorax group proteins. Annu Rev Genet *38*, 413-443.

Rollins, R. A., Korom, M., Aulner, N., Martens, A., and Dorsett, D. (2004). *Drosophila* nipped-B protein supports sister chromatid cohesion and opposes the stromalin/Scc3 cohesion factor to facilitate long-range activation of the cut gene. Mol Cell Biol *24*, 3100-3111.

Rollins, R. A., Morcillo, P., and Dorsett, D. (1999). Nipped-B, a *Drosophila* homologue of chromosomal adherins, participates in activation by remote enhancers in the cut and Ultrabithorax genes. Genetics *152*, 577-593.

Ronshaugen, M., and Levine, M. (2004). Visualization of trans-homolog enhancer-promoter interactions at the Abd-B Hox locus in the *Drosophila* embryo. Dev Cell *7*, 925-932.

Roseman, R. R., Pirrotta, V., and Geyer, P. K. (1993). The su(Hw) protein insulates expression of the *Drosophila melanogaster* white gene from chromosomal position-effects. Embo J *12*, 435-442.

Sayegh, C., Jhunjhunwala, S., Riblet, R., and Murre, C. (2005). Visualization of looping involving the immunoglobulin heavy-chain locus in developing B cells. Genes Dev *19*, 322-327.

Schmitt, S., and Paro, R. (2004). Gene regulation: a reason for reading nonsense. Nature *429*, 510-511.

Schmitt, S., Prestel, M., and Paro, R. (2005). Intergenic transcription through a polycomb group response element counteracts silencing. Genes Dev *19*, 697-708.

Schweinsberg, S., Hagstrom, K., Gohl, D., Schedl, P., Kumar, R. P., Mishra, R., and Karch, F. (2004). The enhancer-blocking activity of the Fab-7 boundary from the *Drosophila* bithorax complex requires GAGA-factor-binding sites. Genetics *168*, 1371-1384.

Scott, K. S., and Geyer, P. K. (1995). Effects of the su(Hw) insulator protein on the expression of the divergently transcribed *Drosophila* yolk protein genes. Embo J *14*, 6258-6267.

Shei, G. J., and Broach, J. R. (1995). Yeast silencers can act as orientation-dependent gene inactivation centers that respond to environmental signals. Mol Cell Biol *15*, 3496-3506.

Shore, D. (2001). Telomeric chromatin: replicating and wrapping up chromosome ends. Curr Opin Genet Dev *11*, 189-198.

Sigrist, C. J., and Pirrotta, V. (1997). Chromatin insulator elements block the silencing of a target gene by the *Drosophila* polycomb response element (PRE) but allow trans interactions between PREs on different chromosomes. Genetics *147*, 209-221.

Spana, C., and Corces, V. G. (1990). DNA bending is a determinant of binding specificity for a *Drosophila* zinc finger protein. Genes Dev *4*, 1505-1515.

Spana, C., Harrison, D. A., and Corces, V. G. (1988). The *Drosophila melanogaster* suppressor of Hairy-wing protein binds to specific sequences of the gypsy retrotransposon. Genes Dev *2*, 1414-1423.

Spradling, A. C., and Rubin, G. M. (1983). The effect of chromosomal position on the expression of the *Drosophila* xanthine dehydrogenase gene. Cell *34*, 47-57.

Srivastava, M. D. (1951). 'Lampbrush' fibers in the chromosomes of Chrotogonus incertus Bolivar. Nature *167*, 775-776.

Sutter, N. B., Scalzo, D., Fiering, S., Groudine, M., and Martin, D. I. (2003). Chromatin insulation by a transcriptional activator. Proc Natl Acad Sci USA *100*, 1105-1110.

Szabo, P. E., Tang, S. H., Reed, M. R., Silva, F. J., Tsark, W. M., and Mann, J. R. (2002). The chicken beta-globin insulator element conveys chromatin boundary activity but not imprinting at the mouse Igf2/H19 domain. Development *129*, 897-904.

Tolhuis, B., Palstra, R. J., Splinter, E., Grosveld, F., and de Laat, W. (2002). Looping and interaction between hypersensitive sites in the active beta-globin locus. Mol Cell *10*, 1453-1465.

Torigoi, E., Bennani-Baiti, I. M., Rosen, C., Gonzalez, K., Morcillo, P., Ptashne, M., and Dorsett, D. (2000). Chip interacts with diverse homeodomain proteins and potentiates bicoid activity in vcivo. Proc Natl Acad Sci USA *97*, 2686-2691.

Udvardy, A., Maine, E., and Schedl, P. (1985). The 87A7 chromomere. Identification of novel chromatin structures flanking the heat shock locus that may define the boundaries of higher order domains. J Mol Biol *185*, 341-358.

van der Vlag, J., den Blaauwen, J. L., Sewalt, R. G., van Driel, R., and Otte, A. P. (2000). Transcriptional repression mediated by polycomb group proteins and other chromatin-associated repressors is selectively blocked by insulators. J Biol Chem *275*, 697-704.

West, A. G., Huang, S., Gaszner, M., Litt, M. D., and Felsenfeld, G. (2004). Recruitment of histone modifications by USF proteins at a vertebrate barrier element. Mol Cell *16*, 453-463.

Wu, C. T., and Morris, J. R. (1999). Transvection and other homology effects. Curr Opin Genet Dev *9*, 237-246.

Xu, Q., Li, M., Adams, J., and Cai, H. N. (2004). Nuclear location of a chromatin insulator in *Drosophila melanogaster*. J Cell Sci *117*, 1025-1032.

Yao, S., Osborne, C. S., Bharadwaj, R. R., Pasceri, P., Sukonnik, T., Pannell, D., Recillas-Targa, F., West, A. G., and Ellis, J. (2003). Retrovirus silencer blocking by the cHS4 insulator is CTCF independent. Nucleic Acids Res *31*, 5317-5323.

Yee, J. K., and Zaia, J. A. (2001). Prospects for gene therapy using HIV-based vectors. Somat Cell Mol Genet *26*, 159-174.

Yoshida, C., Tokumasu, F., Hohmura, K. I., Bungert, J., Hayashi, N., Nagasawa, T., Engel, J. D., Yamamoto, M., Takeyasu, K., and Igarashi, K. (1999). Long range interaction of cis-DNA elements mediated by architectural transcription factor Bach1. Genes Cells *4*, 643-655.

Yusufzai, T. M., Tagami, H., Nakatani, Y., and Felsenfeld, G.

(2004). CTCF tethers an insulator to subnuclear sites, suggesting shared insulator mechanisms across species. Mol Cell *13*, 291-298.

Zhao, K., Hart, C. M., and Laemmli, U. K. (1995). Visualization of chromosomal domains with boundary element-associated factor BEAF-32. Cell *81*, 879-889.

Zhou, J., Ashe, H., Burks, C., and Levine, M. (1999). Characterization of the transvection mediating region of the abdominal-B locus in *Drosophila*. Development *126*, 3057-3065.

Zhou, J., Barolo, S., Szymanski, P., and Levine, M. (1996). The Fab-7 element of the bithorax complex attenuates enhancer-promoter interactions in the *Drosophila* embryo. Genes Dev *10*, 3195-3201.

Zhou, J., and Berger, S. L. (2004). Good fences make good neighbors: barrier elements and genomic regulation. Mol Cell *16*, 500-502.

Zhou, J., and Levine, M. (1999). A novel cis-regulatory element, the PTS, mediates an anti-insulator activity in the *Drosophila* embryo. Cell *99*, 567-575.

Chapter 22

Heterochromatin and X Inactivation

Rebecca Kellum

Department of Biology, 101 T.H. Morgan Building, University of Kentucky, Lexington, KY 40506-0225

Key Words: heterochromatin, X inactivation, histone H3 methyltransferase, HP1, Polycomb Group proteins, dsRNA interference, centromere telomere

Summary

Cytological examinations of nuclear structure in the early 1900s led to the discovery of heterochromatic regions that retain the deeply staining properties of metaphase chromatin throughout the cell cycle. The molecular determinants of heterochromatin have largely remained a mystery. Genetic experiments conducted throughout the twentieth century have led the way in identifying DNA and protein components of both the constitutive heterochromatin of centromeres and telomeres and facultative heterochromatin of the inactive X chromosome of mammals and distributed throughout the chromosomes of many species. Several features of heterochromatin structures have emerged from these studies. Formation of each structure involves targeting of specific histone modifications that provide binding sites for a heterochromatin protein. RNA molecules have also been found to play critical roles in forming each structure. These features will be discussed in this chapter in the context of each category of heterochromatin.

In the late 1920s, Heitz (1928) recognized regions of chromosomes that differ from the bulk of the genome in failing to undergo regular cycles of condensation and de-condensation during cell cycle progression. Rather, these regions, which he was able to identify as centromeres, remain tightly compacted along the nuclear envelope throughout the cell cycle. He coined the term heterochromatin to differentiate them from the more typical chromatin that he designated as euchromatin. The heterochromatic nature of both centromeres and telomeres is now recognized as an almost universal feature of these regions. In the 1960s, other regions of the genome were found to be capable of assuming a heterochromatic state under specific developmental conditions. The most notable example of this is the process observed by Mary Lyons (1961) in which X chromosomes of female mammals adopt a heterochromatic state in order to equalize expression of X-linked genes relative to males carrying a single X chromosome. Brown (1966) applied the term facultative heterochromatin to describe this type of heterochromatin, in contrast to the constitutive heterochromatin of centromeric and telomeric regions. Molecular characterizations of each type of heterochromatin over the past several decades have revealed both similarities and differences between them.

Some of the earliest genetic experiments demonstrated the gene silencing properties of heterochromatin (Muller, 1930); euchromatic genes are subjected to mosaic repression when translocated next to a block of pericentric heterochromatin. The mosaic character of heterochromatin-induced silencing is thought to reflect a cell memory mechanism that maintains the heterochromatic state of the gene during cell division. Radiolabeled-ribonucleotide incorporation experiments in the 1970s further established a correlation between heterochromatin and transcriptional silence (Fakan and Bernard, 1973; Lakhotia and Jacob, 1974). Similar experiments with radiolabeled deoxyribonucleotides showed heterochromatic regions to also be replicated later in S phase than euchromatic regions (Lima de Faria and Jawarski, 1968). Although heterochromatic regions are, in general, transcriptionally inert and later replicating than euchromatin, it is important to note there is evidence from a number of biological systems that heterochromatin is not entirely incompatible with transcription (Weiler and Wakimoto, 1995) and that late

---

Corresponding Author: Tel: (859) 257-9741, E-mail: rkellum@uky.edu

replication is not a universal feature of heterochromatin (Hsu et al., 1964). Nevertheless, these general properties apply to both constitutive and facultative forms of heterochromatin. In addition to the effects of heterochromatin on transcription and replication, the constitutive heterochromatin of centromeres and telomeres also has vital roles in maintaining chromosome stability. These roles were uncovered through mutant phenotypes resulting from loss of either heterochromatin sequences or heterochromatin-associated proteins and will be discussed in more detail in the following sections in the context of the particular heterochromatin component.

Many proteins known to be associated with either form of heterochromatin were initially discovered in genetic screens with *Drosophila melanogaster*. Protein components of constitutive heterochromatin were identified through screens for suppressors of the heterochromatin-induced silencing of euchromatic genes, also known as position effect variegation (Wustmann et al., 1989; Sinclair et al., 1992). Many proteins identified in these screens are conserved components of constitutive heterochromatin from fission yeast to humans, the best known of example being the heterochromatin protein 1 (HP1). HP1 shares a region of homology with the conserved Polycomb (PC) protein that functions in facultative heterochromatin (James and Elgin, 1986; Sinclair et al., 1992; Eissenberg et al., 1992; Paro and Hogness, 1991). PC was also identified through *Drosophila* genetics, in a screen aimed at identifying genes functioning in body plan specification (Lewis, 1978; reviewed in Ringrose and Paro, 2004). The domain of homology between HP1 and PC, known as the chromodomain, is found in a range of chromatin modulator proteins (Eissenberg, 2001) and specifies binding to specific covalent modifications in the amino-terminal tail of histone H3 (Bannister et al., 2001; Lachner et al., 2001). A vast body of literature now links a variety of specific histone modifications (acetylation, methylation, phosphorylation, and ubiquitylation) to specific modes of transcription, replication, and other chromosomal behaviors (reviewed in Strahl and Allis, 2000; Jenuwein and Allis, 2001). The genetic screens in which HP1 and Polycomb were identified also recovered mutations in the enzymes responsible for generating the specific histone binding site for each protein in constitutive and facultative heterochromatin, respectively. The SU(VAR)3-9 protein is responsible for providing a binding site of lysine 9 methylated histone H3 for HP1 in constitutive heterochromatin (Bannister et al., 2001; Lachner et al., 2001, Jacobs et al., 2001; Schotta et al., 2002). The Enhancer of zeste/extra sexcombs (E(z)/Esc) Polycomb Group protein complex is thought to provide a binding site of lysine 27 methylated histone H3 for Polycomb in facultative heterochromatin (Czermin et al., 2002; Muller et al., 2002; reviewed by Ringrose and Paro, 2004). Each of these histone modifications was recently shown to require prior recruitment of a histone H2A.Z variant (H2Av) and acetylation of histone H4 on lysine-12 in *Drosophila* (Swaminathan et al., 2005).

Centromeric Constitutive Heterochromatin

The earliest chromosome *in situ* hybridization experiments showed constitutive heterochromatin of mouse chromosomes to be composed largely of repetitive DNA (α-satellite DNA) (Pardue and Gall, 1970). This repetitive sequence composition is now recognized as an almost universal feature of constitutive heterochromatin, although little conservation is observed for these repeats at the primary sequence level (reviewed in John, 1988). It is also clear that repetitive sequence is not sufficient in determining the heterochromatin state.

Human centromeric proteins (CENP-A,-B,-C,-H,-I) were identified in the 1980s through autoimmune sera from scleroderma patients (Brenner et al., 1981; Earnshaw et al., 1985). Studies in a variety of systems, including genetically tractable ones, have now shown conserved roles for many of these proteins in centromere function (reviewed in Amor et al., 2004). These studies have allowed a distinction to be drawn between the central core of the centromere that is required for assembly of a functional kintochore in sister chromatid separation and the flanking or underlying pericentric heterochromatin. It has been possible to identify specific sequences defining the central core in unicellular yeasts, but studies in multicellular eukaryotes point to specific chromatin structure rather than a specific sequence in defining this core (Kniola et al., 2001; Lo et al., 2001; Blower et al., 2002; Craig et al., 2003; Sun et al., 2003; Pidoux and Allshire, 2004; Sullivan and Karpen, 2004). An almost universal marker of central core chromatin from humans to budding yeast is the CENP-A protein (Palmer et al., 1991; Meluh et al., 1998; Takahashi et al., 2000; Blower and Karpen, 2001; Oegema et al., 2001; Sullivan, 2001). This protein contains a C-terminal histone-fold domain that allows it to replace histone H3 and confer a uniquely rigid structure to centromeric nucleosomes (Malik and Henikoff, 2003; Black et al., 2004). The mere presence of CENP-A, however, is not sufficient to trigger kinetochore assembly, as shown through CENP-A over-expression studies that result in its ectopic localization (Van Hooser et al., 2001). In many species (except Drosophila, C. elegans, and Maize) the Mis12 protein also localizes to this region, independently of CENP-A (Goshima et al., 2003).

The remaining CENP proteins are found in pericentric heterochromatin flanking the centromere central core. Heitz (1934) was able to distinguish two categories of pericentric heterochromatin (α and β) in examinations of polytene nuclei from Drosophila larval salivary glands. The α- class lies closest to, and may even overlap, the central core of the centromere and is composed of linear arrays of short (7-10 bp) A/T-rich repeats (Peacock and Lohe, 1980; Lohe et al., 1993). The α-satellite sequence of mammalian chromosomes might be considered closest to this category of Drosophila heterochromatin and is the binding site for the mammalian CENP-B protein (Masumoto et al., 1989). Homologues of this protein are also found in pericentric heterochromatin of Schizosaccharomyces pombe chromosomes (Murakami et al., 1996; Halverson et al., 1997; Lee et al., 1997; Baum and Clarke, 2000; Irelan et al., 2001), but no clear homologue of this protein has been observed in many species (Amor et al., 2004). CENP-B also does not appear to be absolutely required for centromere function in mice (Hudson et al., 1998; Kapoor et al., 1998; Perez-Castro et al., 1998).

The second category of Drosophila heterochromatin (α-heterochromatin) lies distal to α-heterochromatin on the chromosome and is composed of a mixture of moderately abundant transposable element and unique gene coding sequences (Miklos et al., 1988; Pimpinelli et al., 1995). The pericentric regions of S. pombes chromosomes that are associated with CENP-B more closely resemble this category of Drosophila heterochromatin (Baum and Clarke, 2000), as it is composed of outer centromere repeat transposons. Like Drosophila β-heterochromatin, it is also associated with the S. pombe HP1 homologue (Swi6). The association of Swi6 with these repeats is disrupted in a double mutant for two of three CENP-B homologues (Nakagawa et al., 2002). Genetic studies have also shown Swi6 association with these repeats to be required for recruitment of cohesin subunits to these regions for sister chromatid cohesion during mitosis (Bernard et al., 2001; Nonaka et al., 2001). In addition to this role for HP1 in mitotic chromosome segregation, human HP1 is also required for association of human Mis 12 proteins with the central centromeric core (Obuse et al., 2004).

Association of HP1 with pericentric heterochromatin also requires lysine 9-methylated histone H3 modifications that are generated by SU(VAR)3-9 proteins in species ranging from humans to fission yeast (Bannister et al., 2001; Lachner et al., 2001; Jacobs et al., 2001; Schotta et al., 2002; Cowell et al., 2002). Studies in fission yeast showed a role for transcription of sense and anti-sense transcripts from outer centromeric repeat transposons and the double-stranded RNA interference (dsRNAi) machinery in recruiting these modifications and Swi6 to these regions (Volpe et al., 2002; Reinhart, 2002). Cytogenetic studies in Drosophila also point to a role for this machinery in HP1 association with pericentric heterochromatin (Pal-Bhadra et al., 2003). Moreover, the Dicer component of the dsRNAi machinery is also required to localize the Rad21 cohesin subunit, but not CENP-A or –C, to centromeres of vertebrate DT40 cell lines (Fukagawa et al., 2004). An RNA-induced transcriptional silencing (RITS) complex that is responsible for linking the dsRNAi machinery to SU(VAR)3-9 has been purified from S. pombe (Verdel et al., 2004; Noma et al., 2004). This complex contains small-interfering RNAs (siRNAs), the conserved Ago1 dsRNAi protein, the Chp1 chromodomain containing protein, and a novel Tas3 protein. An RNA-directed RNA polymerase complex (RDRC) and Dicer ribonuclease activity of RITS are required to produce siRNAs. Targeting of RITS to outer centromeric repeats occurs through association of the siRNAs with a platform of non-coding RNA at the chromosomal site of the repeat sequence and Chp1-binding to lysine 9-methylated histone H3 (Motamedi et al., 2004; Sugiyama et al., 2005). RNAase treatments have shown a role for RNA in the association of HP1 with heterochromatin of mammalian (Maison et al., 2002) and Drosophila cells (Badugu et al., 2005), perhaps through a similar mechanism.

Telomeric Constitutive Heterochromatin

Shortly after the discovery of heterochromatin by Heitz, Muller (1938) recognized that telomeres share morphological features with pericentric β-heterochromatin. It is now appreciated that the telomeres provide a mechanism for compensating with the inability of DNA polymerase to complete replication at the ends of chromosomes (Greider and Blackburn, 1987). Correlations between decreased telomere lengths and cellular senescence and organismal aging have been made, and there is evidence that the heterochromatic structure of telomeres reflects a mechanism for regulating telomere length (Smogorzewska and de Lange, 2004; Karlseder et al., 2002).

Most eukaryotes use an RNA-dependent DNA polymerase activity of the telomerase enzyme to replicate chromosome ends. The result is the addition of TTAGGG repeats, as specified by the RNA template for the telomerase enzyme, to chromosome ends (revewied in Smorgorzewska and de Lange, 2004). Some of the earliest studies of telomere length regulation focused on the *Sacharomyces cerevisiase* (budding yeast) Repressor/Activator protein 1 (RAP1) (Lustig *et al.*, 1990). This protein functions at a variety of loci in budding yeast, most notably, in nucleating silenced chromatin at mating type genes and telomeres (Hardy *et al.*, 1992; reviewed in Moazed, 2001). RAP1 is a conserved component of telomeres from budding yeast to vertebrates that regulates telomere length through directly or indirectly binding TTAGGG repeats and forming a telomere capping complex that limits access of telomerase to the chromosome end (Cooper *et al.*, 1997; Smogorzewska *et al.*, 2000; Li and de Lange, 2003). *Drosophila* uses a different mechanism for maintaining and protecting telomere ends, as it lacks a telomerase-encoding gene. It, instead, uses the RNA-dependent DNA polymerase activity of a telomeric retrotransposon toward this end (reviewed in Pardue and DeBaryshe, 2003). HP1 and the HP1-associated HOAP protein are responsible for capping *Drosophila* telomeres. Mutants for either protein display aberrant telomere-telomere fusions and up-regulated transcription of telomeric retrotransposons (Fanti *et al.*, 1998; Cenci *et al.*, 2003). SU(VAR)3-9 histone H3 modifications are also present at telomeres and association of HP1 with these modifications is required for proper silencing of these transposons (Perrini *et al.*, 2004). HP1 and SU(VAR)3-9 modifications are also found at mouse telomeres. Moreover, mice that are double null for both SU(VAR)3-9 homologues have abnormally long telomeres (Garcia-Cao *et al.*, 2004).

Facultative Heterochromatin

Genetic screens initiated in the 1920s to understand the process of *Drosophila* body plan specification led to the identification of proteins that regulate homeotic gene expression through formation of a facultative heterochromatin-like state (Lewis, 1978). Mutations in Polycomb Group (PcG) genes cause transformations of body segment into more posterior ones as a result of de-repressing homeotic genes, whereas mutations in a second group of trithorax group (trxG) genes have the reciprocal phenotype of reducing expression of homeotic genes and transforming body segments into more anterior fates. Members of each group of regulatory protein have conserved roles in homeotic gene (Hox clusters) regulation of vertebrates (Wilkinson *et al.*, 1989). PcG and TrxG proteins provide a mechanism for maintaining a transcriptional regulatory program established in the early embryo through their activities on Polycomb Responsive Elements (PRE) located in the intergenic regions of homeotic genes (reviewed in Ringrose and Paro, 2004). Two categories of PcG protein complexes have been purified from both *Drosophila* and vertebrates. Polycomb Repressive Complex 1 (PRC1) contains the conserved PcG proteins, Polycomb, Polyhomeotic, Posterior sex combs, and sex combs extra/dRING, and confers resistance of *in vitro*-assembled chromatin templates to nucleosome remodeling by SWI/SNF complexes (Francis *et al.*, 2001; Lavigne *et al.*, 2004). A second PC complex (PRC2) contains the conserved ESC, E(Z), and SU(Z)12 proteins and is capable of catalyzing lysine 9- and/or 27-specific histone H3 methylation that presumably provides a chromatin binding site for the Polycomb protein of the PRC1 complex (Czermin *et al.*, 2002; Muller *et al.*, 2002). TrxG complexes counteract the effect or assembly of PcG complexes at PREs and are even more broadly conserved than PcG complexes (reviewed in Simon and Tamkun, 2002). Four different TrxG complexes have been purified from *Drosophila* (reviewed by Ringrose *et al.*, 2005 and Simon and Tamkun, 2002). One of these (BRM complex) is highly related to the budding yeast SWI/SNF complex that functions in ATP-dependent nucleosome remodeling and has roles in transcriptional activation throughout the genome. Three other TrxG complexes contain histone modifying activities of ASH and TRX proteins that function in providing lysine 4-specific histone H3 and acetylated histone modifications associated with active chromatin more specifically at homeotic genes.

A remarkable feature of homeotic gene clusters is co-linearity between the order of homeotic genes along the chromosome axis and the anterioposterior pattern of expression of the genes along the animal body axis (Karch *et al.*, 1985; reviewed in Mihaly *et al.*, 1998). Even more remarkably, this co-linearity is maintained in the Hox clusters of vertebrates (Wilkinson *et al.*, 1989). In the bithorax complex, this colinearity is known to reflect the order of *cis*-regulatory elements (enhancers, insulators, and PREs) throughout an extensive (~100 kb) intergenic region that directs segment-specific expression of only two gene-coding sequences for transcription factors abd-A and Abd-B. It has been known since the original molecular characterization of this locus that this intergenic region harbors a complex pattern of transcription during early embryogenesis, although the exact function of these RNAs was not completely understood (Sanchez-Herrero and Akam, 1989). High resolution mapping of these RNAs in recent years has shown many of these RNAs to be non-coding RNAs and to originate from the gene regulatory regions in a manner that precedes transcription initiated from the abd-A and Abd-B promoters (Bae *et al.*, 2002; Drewell *et al.*, 2002). This pattern of non-coding RNA transcription is also co-linear with the order of regulatory domains across this region. Transcripts initiated within one domain were also found not to spread across insulator/boundary sequences between regulatory domains that have been identified through genetic and molecular experiments. Anti-sense transcripts originate from the vicinity of some insulator sequences that are closely associated with PREs, suggesting a role for them in either preventing the use of the opposing strand as a template by RNA polymerase or in specifying PcG assembly through a dsRNAi-dependent mechanism. The complex pattern of non-coding transcription in the early embryo is thought to prime the segment-specific activity of PREs in later development. Transcription through a PRE has, indeed, been shown to counteract the silencing activity of a PRE (Hogga and Karch, 2002; Schmitt *et al.*, 2005). These new data point to exciting new directions in the studies of mechanisms for forming PcG heterochromatin. New perspectives on the mechanism of silencing by PcG proteins have also emerged in recent years. New studies point to an effect of PcG complexes on modulating the activity of the transcriptional machinery at promoters rather than altering the accessibility of that machinery to promoters (Dellino *et al.*, 2004).

It should also be noted that it is becoming increasingly clear that HP1 proteins have roles in regulating expression of euchromatic genes and, perhaps, in forming facultative heterochromatin. One of the best-characterized examples of this is Swi6-dependent silencing of mating type gene silencing in *S. pombe* (Hall *et al.*, 2002). Mammalian cells contain three isoforms of HP1 (α, β, and γ), and there are immunological data suggesting more specific roles for HP1-β, and HP1-γ in facultative heterochromatin (reviewed in Kellum, 2003). However, each of these isoforms is capable of associating with the transcriptional intermediary factors TIF1α and TIF1β that could direct HP1 to sites of KRAB Zinc-finger regulated genes (Le Douarin *et al.*, 1996; Lechner *et al.*, 2000; Nielsen *et al.*, 1999; reviewed in Kellum, 2003). Chromatin immunoprecipitation experiments also show association of HP1 with the Retinoblastoma-regulated promoter of the human Cyclin A gene (Nielsen *et al.*, 2001b). Ample documentation has also been given for roles of HP1 in regulating euchromatic genes in *Drosophila* (Hwang *et al.*, 2001; Cryderman *et al.*, 2005).

X-Chromosome Inactivation

There is evidence that PcG complexes also function in facultative heterochromatin associated with X inactivation. The heterochromatinization of one of the pair of X chromosomes in female mammals achieves dosage equivalence between XX females and XY males. X inactivation is mediated through coating of the X chromosome with a non-coding *Xist* RNA transcribed from the Xic center located on the transcribed X chromosome (Penn *et al.*, 1996). X inactivation can be inhibited by expression of a *Tsix* RNA that is anti-sense to *Xist* that initiates transcription downstream of *Xist* in *cis* (Lee *et al.*, 1999). Thus, unlike in constitutive heterochromatin assembly where synthesis of both sense and anti-sense transcripts favors assembly, synthesis of an RNA that is anti-sense to *Xist* interferes with *Xist* function in X-inactivation. Multiple models have been put forth for how *Tsix* interferes with *Xist* function. Recent data support a mechanism that is not simply RNA-based. Transcription of *Tsix* in *trans* does not interfere with *Xist* function, and termination of *Tsix* transcription short of crossing *Xist* sequence results in loss of *Tsix* function. Thus, *Tsix* function requires antiparallel transcription through *Xist* (Shibata and Lee, 2004).

A combination of allele-specific gene expression assays and histone modification immunostaining experiments have been used to monitor the process of X inactivation during early embryogenesis (Huynh and Lee, 2003; Okamoto *et al.*, 2004; Mak *et al.*, 2004; reviewed in Cheng and Disteche, 2004). These studies have been informative, not only with regards to the sequence of histone modifications accompanying the process, but also regarding the state of activation of maternally (X^m) and paternally (X^p)-derived X chromosomes during these early stages. The X^p of the paternal germ cell is known to be inactive at the time of fertilization, and therefore, must undergo a reactivation after fertilization of the maternal germ cell with an active X^m. In the 1970s, it was learned that the X^p is imprinted such that it preferentially undergoes inactivation in the trophoectoderm cells of the 32-cell blastocystth, give rise to extra-embryonic tissues (West *et al.*, 1977; Takagi and Sasaki, 1975). The inner cell mass of the blastocyst that will give rise to the embryo proper will undergo random X-inactivation. These more recent studies point to a re-activation, followed by an imprinted re-inactivation of X^p, much earlier in embryonic development than previously thought. This inactivation is accompanied by a sequence of histone modifications appearing on X^p beginning in the 8-cell stage. These modifications remain in the trophoectroderm cells during the 32-cell stage but disappear in the inner cell mass when the X^p becomes re-activated. During the 8-cell stage the coating of X^p with *Xist* RNA is followed by histone H3K4 hypomethylation and H3K9 hypoacetylation. This is followed by recruitment of PcG protein complex PRC2 (Eed-Ezh2), histone H3K27 methylation, and macro histone H2A during the 16-cell stage, and finally by histone H3K9 methylation during the 32-cell blastocyst stage. In the inner cell mass, the reactivation of X^p prior to random inactivation is accompanied by loss of Eed-Ezh2 association and histone H3K27 and H3K9 methylation. The accumulation of PRC2 proteins on the Xi is transient, whereas histone H3K27 modifications are more stable (Plath *et al.*, 2003; Silva *et al.*, 2003). Transient accumulation of mPRC1 proteins is also observed on the inactive X in a manner that is *Xist* RNA-dependent but not completely H3K27 methylation-dependent (Plath *et al.*, 2004). There are also recent reports of H2A ubiquitination on the inactive X, catalyzed by Ring 1 A/B proteins of mPRC1 (de Napoles *et al.*, 2004).

The process of X inactivation has also been recapitulated on a foreign chromosome by expression of *Xist* RNA from a transgene on that chromosome in embryonic stem cells (Wutz and Jaenisch, 2000). Through this approach, a repeat sequence at the 5′ end of *Xist* was identified that is required for silencing but not for association of Xist RNA or H3K27m3 and H4K20m1 with the chromosome (Wutz *et al.*, 2002). This approach has also shown that *Xist* expression early in embryonic development establishes a chromosomal memory that is maintained in the absence of this repeat sequence or silencing (Kohlamaier *et al.*, 2004). This memory allows more efficient introduction of H3K27m3 and H4K20m1 modifications after a developmental time when *Xist* expression is no longer capable of initiating X inactivation. The nature of this chromosomal memory is currently not known. The nature of the imprint on X^p that causes it to preferentially undergo inactivation in the trophoectoderm is also not understood, although a DNA methylation-sensitive CTCF chromatin insulator located in an imprinting/choice center at the 5′ end of *Tsix* has been proposed to play a role (Chao *et al.*, 2002).

Future Directions

A remarkable feature of each type of heterochromatin discussed in this chapter is the shared involvement of, not only histone modifications, but non-coding RNAs in their formation. The future awaits further elucidation of these mechanisms. The factors responsible for imprinting a heterochromatic state, and establishing both transient and long-term memories of that state also remain unresolved. DNA methylation appears to be one factor in long-term memory of X-inactivation, as discussed in more detail in another chapter of this volume (Csankovszki *et al.*, 2001). The association of HP1 with replication-dependent chromatin assembly factors (CAF-1) (Verreault *et al.*, 1996; Murzina *et al.*, 1999) and origin replication complex (ORC) (Pak *et al.*, 1997; Huang *et al.*, 1998) subunits suggest possible mechanisms for maintaining constitutive forms of heterochromatin.

References

Amor, D.J., Kalitsis, P., Sumer, H., and Choo, K.H.A. (2004). Building the centromere from foundation proteins to 3D organization. Trends Cell Biol. *14*, 359-368.

Badugu, R., Yoo, Y., Singh, P.B., and Kellum, R. (2005). Mutations in the heterochromatin protein 1 (HP1) hinge domain affect HP1 protein interactions and chromosomal distribution. Chromosoma *113*, 370-384.

Bae, E., Calhoun, V.C., Levine, M., Lewis, E.B., and Drewell, R.A. (2002). Characterization of the intergenic RNA profile at *abdominal-A* and *Abdominal-B* in the *Drosophila* bithorax complex. Proc. Natl. Acad. Sci. USA *99*, 16847-16852.

Bannister, A.J., Zegerman, P., Partridge, J.F., Miska, E.A., Thomas, J.O., Allshire, R.C., and Kouzarides, T. (2001). Selective recognition of methylated lysine 9 on histone H3 by the HP1 chromodomain. Nature *410*, 120-124.

Baum, M, and Clarke, L. (2000). Fission yeast homologs of human CENP-B have redundant functions affecting cell growth and chromosome segregation. Mol. Cell. Biol. *20*, 2852-2864.

Bernard, P., Maure, J.F., Partridge, J.F., Genier, S., Javerzat, J.P., and Allshire, R.C. (2001). Requirement of heterochromatin for cohesion at centromeres. Science *294*, 2539-2542.

Black, B.E., Foltz, D.R., Chakravarthy, S., Luger, K., Woods, V.L., and Cleveland, D.W. (2004). Structural determinants for generating centromeric chromatin. Nature *430*, 578-582.

Blower, M.D., and Karpen, G.H. (2001). The role of *Drosophila* CID in kinetochore formation, cell-cycle progression and heterochromatin interactions. Nat. Cell Biol. *3*, 730-739.

Blower, M.D., Sullivan, B.A., and Karpen, G.H. (2002). Conserved organization of centromeric chromatin in flies and humans. Dev Cell. *2*, 319-330.

Brenner, S. Pepper, D., Berns, M.W., Tan, E., and Brinkley, B.R. (1981). Kinetochore structure, duplication, and distribution in mammalian cells: analylsis by human autoantibodies from scleroderma patients. J. Cell Biol. *91*, 95-102.

Brown, S.W. (1966). Heterochromatin. Science *151*, 417-425.

Csankovszki, G., Nagy, A., Jaenisch, R. (2001). Synergism of Xist RNA, DNA, methylation, and histone hypoacetylation in maintaining X chromosome inactivation. J. Cell Biol. *153*, 773-784.

Cenci, G., Siriaco, G., Raffa, G.D., Kellum, R., and Gatti, M. (2003). The *Drosophila* HOAP protein is required for telomere capping. Nat Cell Biol. *5*, 82-84.

Chao, W., Huynh, K.D., Spencer, R.J., Davidow, L.S., and Lee, J.T. (2002). CTCF, a candidate trans-acting factor for X-inactivation choice. Science *295*, 345-347.

Cheng, M.K., and Disteche, C.M. (2004). Silence of the fathers: early X inactivation. BioEssays *26*, 821-824.

Chinwalla V., Jane, E.P., and Harte, P.J. (1995). The *Drosophila* Trithorax protein binds to specific chromosomal sites and is co-localized with Polycomb at many sites. EMBO J. *14*, 2056-2065.

Cooper, J.P., Nimmo, E.R., Allshire, R.C., and Cech, T.R. (1997). Regulation of telomere length and function by a Myb-domain protein in fission yeast. Nature *385*, 744-747.

Cowell, I.G., Aucott, R., Mahadevaiah, S.K., Burgoyne, P.S., Huskisson, N., Bongiorni, S., Prantera, G., Fanti, L., Pimpinelli, S., Wu, R., Gilbert, D.M., Shi, W., Fundele, R., Morrison, H., Jeppesen, P., and Singh, P.B. (2002). Heterochromatin, HP1, and methylation at lysine 9 of histone H3 in animals. Chromosoma *111*, 22-36.

Craig, J.M., Wong, L.H., Lo, A.W., Earle, E., and Choo, K.H. (2003). Centromeric chromatin pliability and memory at a human neocentromere. EMBO J. *22*, 2495-504.

Cryderman, D.E., Grade, S.K., Li, Y., Fanti, L., Pimpinelli, S., and Wallrath, L.L. (2005). Role of *Drosophila* HP1 in euchromatic gene expression. Dev Dyn. *232*, 767-774.

Czermin, B., Melfi, R., McCabe, D., Steitz, V., Imhof, A., Pirrotta, V. (2002). *Drosophila* Enhancer of Zeste/ESC complexes have a histone H3 methyltransferase activity that marks chromosomal Polycomb sites. Cell *111*, 185-196.

Dellino, G.I., Schwartz, Y.B., Farkas, G., McCabe, D., Elgin, S.C.R., and Pirrotta, V. (2004). Polycomb silencing blocks transcription initiation. Mol. Cell *13*, 887-893.

de Napoles, M., Mermoud, J.E., Wakao, R., Tang, A., Endoh, M., Appanah, R., Nesterova, T.B., Silva, J., Otte, A.P., Vidal, M., Koseki, H., and Brockdorff, N. (2004). Polycomb Group protein Ring1A/B link ubiquitylation of histone H2A to heritable gene silencing and X inactivation. Dev. Cell *7*, 663-676.

Drewell, R.A., Bae, E., Burr, J., Lewis, E.B. (2002). Transcription defines the embryonic domains of *cis*-regulatory activity at the *Drosophila* bithorax complex. Proc. Natl. Acad. Sci. USA *99*, 16853-16858.

Earnshaw, W.C., and Rothfield, N. (1985). Identification of a family of human centromere proteins using autoimmune sera from patients with scleroderma. Chromosoma *91*, 313-321.

Eissenberg, J.C., Morris, G.D., Reuter, G., and Hartnett, T. (1992) The heterochromatin-associated protein HP-1 is an essential protein in *Drosophila* with dosage-dependent effects on position-effect variegation. Genetics *131*, 345-352.

Eissenberg, J.C. (2001). Molecular biology of the chromo domain: an ancient chromatin module comes of age. Gene. *275*, 19-29.

Fakan, S., and Bernhard, W. (1973). Nuclear labeling after prolonged ^{3}H-uridine incorporation as visualized by high resolution autoradiography. Exper. Cell Res. *79*, 431-444.

Fanti, L., Giovinazzo, G., Berloco, M., and Pimpinelli, A. (1998). The heterochromatin protein 1 prevents telomere fusions in *Drosophila*. Mol Cell *2*, 527-538.

Francis, N.J., Saurin, A.J., Shao, Z., Kingston, R.E. (2001). Reconstitution of a functional core Polycomb repressive complex. Mol. Cell *8*, 545-556.

Garcia-Cao, M., O'Sullivan, R., Peters, A.H.F.M., Jenuwein, T., and Blasco, M.A. (2004). Epigenetic regulation of telomere length in mammalian cells by the Suv39h1 and Suv39h2 histone methyltransferases. Nature Gen. *36*, 94-99.

Goshima, G., Kiyomitsu, T., Yoda, K., and Yanagida, M. (2003). Human centromere chromatin protein hMis12, essential for equal segregation, is independent of CENP-A loading pathway. J Cell Biol. *160*, 25-39.

Greider, C.W., and Blackburn, E.H. (1987). The telomere terminal transferase of *Tetrahymena* is a ribonucleoprotein enzyme with two kinds of primer specificity. Cell *51*, 887-898.

Hall, I.M., Shankaranarayana, G.D., Noma, K., Ayoub, N., Cohen, A., and Grewal, S.I. (2002). Establishment and maintenance of a heterochromatin domain. Science *297*, 2232-2237.

Halverson, D., Baum, M., Stryker, J., Carbon, J., and Clarke, L. (1997). A centromere DNA-binding protein from fission yeast affects chromosome segregation and has homology to human CENP-B. J. Cell Biol. *136*, 487-500.

Hardy, C.F., Sussel, L., and Shore, D. (1992). A RAP1-interacting protein involved in transcriptional silencing and telomere length regulation. Genes Dev. *6*, 801-814.

Heitz, E. (1928). Das Heterochromatin der Moose. Jahrb. Wiss. Bot. *69*, 762-818.

Heitz, E. (1934). über amd Heterochromatinization sowie Konstahz and Bau der Chromomeren bei *Drosophila*. Biol. Zentralbl. *54*, 588-609.

Hogga, I, and Karch, F. (2002). Transcription through the iab-7 cis-regulatory domain of the bithorax complex interferes with maintenance of Polycomb-mediated silencing. Devel. *21*, 4915-4922.

Huang, D.W., Fanti, L., Pak, D.T.S., Botchan, M.R., Pimpinelli, S., and Kellum, R. (1998) Distinct cytoplasmic and nuclear fractions of *Drosophila* heterochromatin protein 1: their phosphorylation levels and associations with origin recognition complex proteins. J. Cell Biol. *142*, 307-318.

Hudson, D.F., Fowler, K.J., Earle, E., Saffery, R., kalitsis, P., Trowell, H., Hill, J., Wreford, N.G., de Kretser, D.M., Cancilla, M.R., Howman, E., Hii, L., Cutts, S.M., Irvine, D.V., and Choo,

K.H. (1998). Centromere protein B null mice are mitotically and meiotically normal but have lower body and testis weights. J Cell Biol. *141*, 309-319.

Hsu, T.C., Schmidt, W. and Stubblefield, E. (1964). DNA replication sequences in higher animals. In The Role of Chromosomes in Development, ed. M. Locke, pp. 83-112. New York: Academic Press.

Hwang, K.-K., Eissenberg, J.C., and Worman, H.J. (2001). Transcriptional repression of euchromatic genes by *Drosophila* heterochromatin protein 1 and histone modifiers Proc Natl Acad Sci USA *98*, 11423-11427.

Irelan, J.T., Gutkin, G.I., and Clarke, L. (2001). Functional redundancies, distinct localizations and interactions among three fission yeast homologs of centromere protein-B. Genetics *157*, 1191-1203.

Jacobs, S.A., Taverna, S.D., Zhang, Y., Briggs, S.D., Li, J., Eissenberg, J.C., Allis, C.D., and Khorasanizadeh, S. (2001). Specificity of the HP1 chromo domain for the methylated N-terminus of histone H3. EMBO J *20*, 5232-5241.

James, T.C., and Elgin, S.C.R. (1986). Identification of a nonhistone chromosomal protein associated with heterochromatin in *Drosophila melanogaster* and its gene. Mol. Cell. Biol. *6*, 3862-3872.

Jenuwein, T., and Allis, C.D. (2001). Translating the histone code. Science *293*, 1074-1080.

John, B. (1988). The biology of heterochromatin. *In* Heterochromatin: Molecular and Structural Aspects (ed. R.S. Verma), pp. 1-128. Cambridge University Press, Cambridge.

Kapoor, M., Montes de Oca, L.R., Liu, G., Lozano, G., Cummings, C., Mancini, M., Ouspenski, I., Brinkley, B.R., and May, G.S. (1998). The cenpB gene is not essential in mice. Chromosoma *107*, 570-576.

Karlseder, J., Smogorzewska, A., and de Lange, T. (2002). Senescence induced by altered telomeric state, not telomere loss. Science *295*, 2446-2449.

Kniola, B., O'Toole, E., McIntosh, R., Mellone, B., Allshire, R., Mengarelli, S., Hultenby, K, and Ekwall, K. (2001). The domain structure of centromeres is conserved from fission yeast to humans. Mol. Biol. Cell *12*, 2767-2775.

Kohlmaier, A., Savarese, F., Lachner, M., Martens, J., Jenuwein, T., and Wutz, A. (2004). A chromosomal memory triggered by *Xist* regulates histone methylation in X inactivation. PloS Biol. *2*, 991-1003.

Karch, F., Weiffenbach, B., Peifer, M., Bender, W., Duncan, I., Celniker, S., Crosby, M., and Lewis, E.B. (1985). The abdominal region of the Bithorax Complex. Cell *43*, 81-95.

Lachner, M., O'Carroll, D., Rea, S., Mechtler, K., and Jenuwein, T. (2001). Methylation of histone H3 lysine 9 creates a binding site for HP1 proteins. Nature *410*, 116-120.

Lavigne, M., Francis, N.J., King, I.F., and Kingston, R.E. (2004). Propagation of silencing: recruitment and repression of naïve chromatin in *trans* by Polycomb repressed chromatin. Mol. Cell *13*, 415-425.

Lakhotia, S.C., and Jacob, J. (1974). EM autoradiographic studies on polytene nuclei of *Drosophila melanogaster*. II. Organization and transcriptive activity of the chromocentre. Exp. Cell. Res. *86*, 253-263.

Lechner, M.S., Begg, G.E., Speicher, D.W., and Rauscher, F.J. (2000). Molecular determinants for targeting heterochromatin protein 1-mediated gene silencing: Direct chromoshadow domain-KAP-1 corepressor interaction is essential. Mol Cell Biol *20*, 6449-6465

Le Douarin, B., Nielsen, A., Garnier, J.-M., Ichinose, H., Jeanmougin, F., Losson, R., and Chambon, P. (1996). A possible involvement of TIF1α and TIF1β in the epigenetic control of transcription by nuclear receptors. EMBO J. *15*, 6701-6715.

Lee, J.T., Davidow, L.S., and Warshawsky, D. (1999). *Tsix*, a gene antisense to *Xist* at the X-inactivation center. Nat. Genet. *21*, 400-404.

Lee, J.K., Huberman, J.A., and Hurwitz, J. (1997). Purification and characterization of a CENP-B homologue protein that binds to the centromeric K-type repeat DNA of *Schizosaccharomyces pombe*. Proc. Natl. Acad. Sci. USA *94*, 8427-8432.

Lewis, E.B. (1978). A gene complex controlling segmentation in *Drosophila*. Nature *276*, 565-570.

Li, B., and de Lange, T. (2003). Rap1 affects the length and heterogeneity of human telomeres. Mol. Biol. Cell *12*, 5060-5068.

Lima de Faria, A., and Jaworska, H. (1968). Late DNA synthesis in heterochromatin. Nature *217*, 138-142.

Lo, A.W., magliano, D.J., Sibson, M.C., Kalitsis, P., Craig, J.M., and Choo, K.H. (2001). A novel chromatin immunoprecipitation and array (CIA) analysis identifies a 460-kb CENP-A-binding neocentromere DNA.Genome Res. *11*, 448-457.

Lohe, A.R., Hilliker, A.J., and Roberts, P.A. (1993). Mapping simple repeated DNA sequences in heterochromatin of *Drosophila melanogaster*. Genetics *134*, 1149-1174.

Lustig, A.J., Kurtz, S., and Shore, D. (1990). Involvement of the silencer and UAS binding protein RAP1 in regulation of telomere length. Science *250*, 549-553.

Lyon, M.F. (1961). Gene action in the X-chromosome of the mouse (*Mus musculus L.)* Nature *190*, 372-373.

Maison, C., Bailly, D., Peters, A.H.F.M., Ouivv, J.P., Roche, D., Taddei, A., Lachner, M., Jenuwein, T., and Almouzni, G. (2002). Higher-order structure in pericentric heterochromatin involves a distinct pattern of histone modification and an RNA component. Nature Genet. *30*, 329-334.

Malik, H.S., and Henikoff, S. (2003). Phylogenomics of the nucleosome. Nat. Struct. Biol. 10, 882-891.

Masumoto, H., Masukata, H., Muro, Y., Nozaki, N., and Okazaki, T. (1989). A human centromere antigen (CENP-B) interacts with a short specific sequence in alphoid DNA, a human centromeric satellite. J. Cell Biol. *109*, 1963-1973.

Meluh, P.B., Yang, P., Glowczewski, L., Koshland, D., and Smith,

M.M. (1998). Cse4p is a component of the core centromere of *Saccharomyces cerevisiae.* Cell. 94, 607-513.

Mihaly, J., Hogga, I., Barges, S., Galloni, M., Mishra, R.K., Hagstrom, K., Muller, M., Schedl, P., Sipos, L., Gausz, J., Gyurkovics, H., and Karch, F. (1998). Chromatin domain boundaries in the Bithorax complex. Cell Mol Life Sci. 54, 60-70.

Miklos, G.L.G., Yamamoto, M.-T., Davies, J., and Pirrotta, V. (1988) Microcloning reveals a high frequency of repetitive sequences characteristic of chromosome 4 and the β-heterochromatin of *Drosophila melanogaster.* Proc. Natl. Acad. Sci. USA *85,* 2051-2055.

Moazed, D. (2001). Common themes in mechanisms of gene silencing. Mol Cell *8,* 489-498.

Motamedi, M.R., Verdel, A., Colmenares, S.U., Gerber, S.A., Gygi, S.P., and Moazed, D. (2004). Two RNAi complexes, RITS and RDRC, physically interact and localize to noncoding centromeric RNAs. Cell *119,* 789-802.

Muller, H.J. (1930). Types of visible variations induced by X-rays in *Drosophila.* J. Genet. *22,* 299-334.

Muller, H.J. (1938). The remaking of chromosomes. Collecting Net. XIII *198,* 181-195.

Muller, J., Hart, C., Francis, N., Vargas, M., Sengupta, A., Wild, B., Miller, E.L., O'Connor, M.B., Kingston R.E., Simon, J.A. (2002). Histone methyltransferase activity of a *Drosophila* Polycomb group repressor complex. Cell 111, 197-208.

Murakami, Y., Huberman, J.A., and Hurwitz, J. (1996). Identification, purification, and molecular cloning of autonomously replicating sequence-binding protein 1 from fission yeast *Schizosaccharomyces pombe.* Proc. Natl. Acad. Sci. USA *93,* 502-507.

Murzina, N., Verreault, A., Laue, E., and Stillman, B. (1999). Heterochromatin dynamics in mouse cells: interaction between chromatin assembly factor 1 and HP1 proteins. Mol. Cell *4,* 529-540.

Nakagawa, H., Lee, J.K., Hurwitz, J., Allshire, R.C., Nakayama, J., Grewal, S.I., Tanaka, K., and Murakami, Y. (2002). Fission yeast CENP-B homologs nucleate centromeric heterochromatin by promoting heterochromatin-specific histone tail modifications. Genes Dev. *16,* 1766-1778.

Nielsen, A.L., Ortiz, J.A., You, J., Oulad-Abdelghani, M., Khechumian, R., Gansmuller, A., Chambon, P., and Losson, R. (1999). Interaction with members of the heterochromatin protein 1 (HP1) family and histone deacetylation are differentially involved in transcriptional silencing by members of the TIF1 family. EMBO J. *18,* 6385-6395

Nielsen, S.J., Schneider, R., Bauer, U.M., Bannister, A.J., Morrison, A., O'Carroll, D., Firestein, R., Cleary, M., Jenuwein, T., Herrera, R.E., and Kouzarides, T. (2001). Rb targets histone H3 methylation and HP1 to promoters. Nature 412, 561-565

Noma, K., Sugiyama, T., Cam, H., Verdel, A., Zofall, M., Jia, S., Moazed, D., and Grewal, S.I. (2004). RITS acts in cis to promote RNA interference-mediated transcriptional and post-transcriptional silencing. Nat Genet. *36,* 1174-1180.

Nonaka, N., Kitajima, T., Yokobayashi, S., Xiao, G., Yamamoto, M., Grewal, S.I.S., and Watanabe, Y. (2001). Recruitment of cohesin to heterochromatic regions by Swi6/HP1 in fission yeast. Nat. Cell Biol. *4,* 89-93.

Obuse, C., Iwasaki, O., Kiyomitsu, T., Goshima, G., Toyoda, Y., and Yanagida, M. (2004). A conserved Mis12 centromere complex is linked to heterochromatic HP1 and outer kinetochore protein Zwint-1. Nat Cell Biol. *6,* 1135-1141.

Oegema, K., Desai, A., Rybina, S., Kirkham, M., and Hyman, A.A. (2001). Functional analysis of kinetochore assembly in *Caenorhabditis elegans.* J Cell Biol. 153, 1209-1226.

Pak, D.T.S., Pflumm, M., Chesnokov, I., Huang, D.W., Kellum, R., Marr, J., Romanowski, P., and Botchan, M. (1997). Association of the origin recognition complex with heterochromatin and HP1 in higher eukaryotes. Cell *91,* 311-323.

Pal-Bhadra, M., Leibovitch, B.A., Gandhi, S.G., Rao, M., Bhadra, U., Birchler, J.A., and Elgin, S.C. (2004). Heterochromatic silencing and HP1 localization in *Drosophila* are dependent on the RNAi machinery. Science *303,* 669-672.

Palmer, D.K., O'Day, K., Trong, H.L., Charbonneau, H., and Margolis, R.L. (1991). Purification of the centromere-specific protein CENP-A and demonstration that it is a distinctive histone. Proc.Natl.Acad.Sci. USA *88,* 3734-3738.

Pardue, M.L., and DeBaryshe, P.G. (2003). Retrotransposons provide an evolutionarily robust non-telomerase mechanism to maintain telomeres. Annu. Rev. Genet.*37,* 485-511.

Pardue, M.L., and Gall, J.G. (1970). Chromosomal localization of mouse satellite DNA. Science 168: 1356-1358.

Paro, R, and Hogness, D.S. (1991). The Polycomb protein shares a homologous domain with a heterochromatin-associated protein of *Drosophila.*Proc Natl Acad Sci USA. 88, 263-267.

Partridge, J.F., Scott, K.S.C., Bannister, A.J., Kouzarides, T., and Allshire, R.C. (2002). *Cis*-Acting DNA from fission yeast centromeres mediates histone H3 methylation and recruitment of silencing factors and cohesin to an ectopic site. Curr. Biol. *12,* 1652-1660.

Peacock, W.J., and Lohe, A.R. (1980). Repeated sequences in the chromosomes of *Drosophila.* Symp. R. Entomol. Soc. Lond. *10,* 1-12.

Penny, G.D., Kay, G.F., Sheardown, S.A., Rastan, S., Brockdorff, N. (1996). Requirement for *Xist* in X chromosome inactivation. Nature *379,* 131-137.

Perez-Castro, A.V., Shamanski, F.L., Meneses, J.J., Lovato, T.L., Vogel, K.G., Moyzis, R.K., and Pedersen, R. (1998). Centromeric protein B null mice are viable with no apparent abnormalities.Dev Biol. *201,* 135-143.

Perrini, B., Piacentini, L., Fanti, L., Altieri, F., Chichiarelli, S., Berloco, M., Turano, C., Ferraro, A., and Pimpinelli, S. (2004). HP1 controls telomere capping, telomere elongation, and telomere silencing by two different mechanisms in *Drosophila.*

Mol. Cell *15,* 467-476.

Petrie, V.J., Wuitschick, J.D., Givens, C.D., Kosinski. A.M., and Partridge, J.F. (2005). RNA interference (RNAi)-dependent and RNAi-independent association of the Chp1 chromodomain protein with distinct heterochromatic loci in fission yeast. Mol. Cell. Biol. *25,* 2331-2346.

Pidoux, A.L., and Allshire, R.C. (2004). Kinetochore and heterochromatin domains of the fission yeast centromere. Chrom. Res. *12,* 521-534.

Pimpinelli, S., Berloco, M., Fanti, L., Dimitri, P., Bonaccorsi, S., Marchetti, E., Caizzi, R., Caggese, C., and Gatti, M. (1995). Transposable elements are stable structural components of *Drosophila melanogaster* heterochromatin. Proc. Natl. Acad. Sci. USA *92,* 3804-3808.

Plath, K., Talbot, D., Hamer, K.M., Otte, A.P., Yang, T.P., Jaenisch, R., and Panning, B. (2004). Developmentally regulated alterations in Polycomb repressive complex 1 proteins on the inactive X chromosome. J. Cell Biol. *167,* 1025-1035.

Plath, K.J., Fang, S.K., Mlynarczyk-Evans, R., Cao, K.A., Worringer, H., Wang, de la, Cruz, C.C., Otte, A.P., Panning, B., and Zhang, Y. (2003). Role of histone H3 lysine 27 methylation in X inactivation. Science *300,* 131-137.

Reinhart, B.J., Bartel, D.P. (2002). Small RNAs correspond to centromere heterochromatic repeats. Science *297,* 1831.

Ringrose, L., and Paro, R. (2004). Epigenetic regulation of cellular memory by the Polycomb and Trithorax Group proteins. Annu. Rev. Genet. *38,* 413-443.

Rozovskaia, T., Tillib, S., Smith, S., Sedkov, Y., Rozenblatt-Rosen, O. (1999). Trithorax and ASH1 interact directly and associate with the trithorax group-responsive bxd region of the Ultrabithorax promoter. Mol. Cell. Biol. 19, 6441-6447.

Sanchez-Herrero, E. and Akam, M. (1989). Spatially ordered transcription of regulatory DNA in the bithorax complex of *Drosophila*. Develop. *107,* 321-329.

Schmitt, S., Prestel, M., and Paro, R. (2005). Intergenic transcription through a Polylcomb group response element counteracts silencing. Genes Dev. *19,* 1-12.

Schotta, G., Ebert, A., Krauss, V., Fischer, A., Hoffmann, J., Rea, S., Jenuwein, T., Dorn, R., and Reuter, G. (2002). Central role *of Drosophila* SU(VAR)3-9 in histone H3-K9 methylation and heterochromatic gene silencing. EMBO J. *21,* 1121-1131.

Shibata, S., and Lee, J.T. (2004). *Tsix* transcription-versus RNA-based mechanisms in *Xist* repression and epigenetic choice. Curr. Biol. *14,* 1747-1754.

Silva, J., Mak, W., Zvetkova, I., Appanah, R., and Nesterova, T.B. (2003). Establishment of histone H3 methylation on the inactive X chromosome requires transient recruitment of Eed-Enx1 polycomb group complex. Dev. Cell *4,* 481-495.

Sinclair, D.A., Ruddell, A.A., Brock, J.K., Clegg, N.J., Lloyd, V.K., and Grigliatti, T.A. (1992). A cytogenetic and genetic position-effect variegation in *Drosophila melanogaster.* Genetics *130,* 333-344.

Smogorzewska, A., van Steensel, B., Bianchi, A., Oelmann, S., Schaefer, M.R., Schnapp, G., and de Lange, T. (2000). Control of human telomere length by TRF1 and TRF2. Mol Cell Biol *20,* 1659-1668

Smogorzewska, A., and de Lange, T. (2004). Regulation of telomerase by telomeric proteins. Annu. Rev. Biochem. *73,* 177-208.

Strahl, B.D., and Allis, C.D. (2000). The language of covalent histone modifications. Nature *403,* 41-45.

Sugiyama, T., Cam, H., Verdel, A., Moazed, D., Grewal, S.I.S. (2005). RNA-dependent RNA polymerase is an essential component of a self-enforcing loop coupling heterochromatin assembly to siRNA production. Proc. Natl. Acad. Sci. USA *102,* 152-157.

Sullivan, B.A., and Karpen, G.H. (2004). Centromeric chromatin exhibits a histone modification pattern that is distinct from both euchromatin and heterochromatin. Nat. Struct. Mol. Biol. *11,* 1076-1083.

Sullivan, K.F. (2001). A solid foundation: functional specialization of centromeric chromatin. Curr. Opin. Genet. Dev. *11,* 182-188.

Sun, X., Le, H.D., Wahlstrom, J.M., and Karpen, G.H. (2003). Sequence analysis of a functional *Drosophila* centromere. Genome Res. *13,* 182-194.

Swaminathan, J., Baxter E.M., and Corces, V.G. (2005). The role of histone H2Av variant replacement and histone H4 acetylation in the establishment of *Drosophila* heterochromatin. Genes Devel. *19,* 65-76.

Takagi, N., and Sasaki, M. (1975). Preferential inactivation of the paternally derived X chromosome in the extraembryonic membranes of the mouse. Nature *256,* 640-642.

Takahashi, K., Chen, E.S., and Yanagida, M. (2000). Requirement of Mis6 centromere connector for localizing a CENP-A-like protein in fission yeast. Science *288,* 2215-2219.

Van Hooser, A.A., Ouspenski, I.I., Gregson, H.C., Starr, D.A., Yen, T.J., Goldberg, M.L., Yokomori, K., Earnshaw, W.C., Sillivan, K.F., and Brinkley, B.R. (2001). Specification of kinetochore-forming chromatin by the histone H3 variant CENP-A. J Cell Sci. *114,* 3529-3542.

Verdel, A., Jia, S., Gerber, S., Sugiyama, T., Gygi, S., Grewal, S.I., and Moazed, D. (2004). RNAi-mediated targeting of heterochromatin by the RITS complex. Science. *303,* 672-676.

Verreault, A., Kaufman, P.D., Kobayashi, R., and Stillman, B. (1996). Nucleosome assembly by a complex of CAF-1 and acetylated histones H3/H4. Cell *87,* 95-104.

Volpe, T.A., Kidner, C., Hall, I.M., Ten,g G., Grewal, S.I., Martienssen, R.A. (2002). Regulation of heterochromatic silencing and histone H3 lysine-9 methylation by RNAi. Science *297,* 1833-1837.

Weiler, K.S., and Wakimoto, B.T. (1995). Heterochromatin and gene expression in *Drosophila.* Annu. Rev. Genet. *29,* 577-605.

West, J.D., Frels, W.I., Chapman, V.M., and Papaioannou, V.E.

(1977). Preferential expression of the maternallyl derived X chromosome in the mouse yolk sac. Cell *12,* 873-882.

Wilkinson, D.G., Bhatt, S., Cook, S., Bonicelli, E., and Krumlauf, R. (1989). Segmental expression of Hox-2 homoeobox-containing genes in the developing mouse hindbrain. Nature *341,* 405-409.

Wustmann, G., Szidonya, J., Taubert, H., and Reuter, G. (1989). The genetics of position-effect variegation modifying loci in *Drosophila melanogaster.* Mol. Gen. Genet. *217,* 520-527.

Yang, C.H., Tomkiel, J., Saitoh, H., Johnson, D.H., and Earnshaw, W.C. (1996). Identification of overlapping DNA-binding and centromere-targeting domains in the human kinetochore protein CENP-C. Mol. Cell. Biol. *16,* 3576-3586.

Chapter 23

DNA Methylation Regulates Genomic Imprinting, X Inactivation, and Gene Expression during Mammalian Development

Taiping Chen and En Li

Epigenetics Program, Novartis Institutes for Biomedical Research, 250 Massachusetts Avenue, Cambridge, MA 02139, USA

Key Words: DNA methylation, DNA methyltransferases, Dnmt1, Dnmt3a, Dnmt3b, Dnmt3L, genomic imprinting, X chromosome inactivation, epigenetics

Summary

DNA methylation is a major form of epigenetic modification found in the genome of many eukaryotic organisms. In mammals, DNA methylation patterns are established and maintained during development by three distinct DNA cytosine methyltransferases: Dnmt1, Dnmt3a, and Dnmt3b. Genetic manipulation of these enzymes in mice have revealed that DNA methylation is essential for mammalian development and play crucial roles in a variety of biological processes, such as gene regulation, genomic imprinting, and X chromosome inactivation. An emerging theme from recent studies is that DNA methylation, in collaboration with other epigenetic mechanisms, regulates higher-order chromatin structure, resulting in individual and global gene silencing. Understanding the functions of DNA methylation in mammalian development will help to elucidate the role of epigenetic mechanisms in human diseases, such as neurobehavioral disorders and cancer.

Introduction

In mammals, almost all types of cells in an individual have an identical genome, yet their morphology and behavior differ greatly due to specific patterns of gene expression. For a long time, the transcriptional status and therefore the character of a cell were thought to be determined only by "transcription factors" that bind to specific DNA sequences. Such a model, however, could not explain the fact that most types of cells maintain their differentiated states quite stably. Over the last decade, ample evidence has gradually moved chromatin structure to a prominent position in the field of transcription regulation. Epigenetic modifications of genomic DNA and its associated proteins (e.g. histones) play crucial roles in the regulation of chromatin structure.

Cytosine methylation is a common modification found in the genomic DNA of many eukaryotic organisms, from fungi to plants, from invertebrates to vertebrates. However, the levels and patterns of DNA methylation vary significantly among these organisms, ranging from lack of methylation in the nematode worm *Caenorhabditis elegans* to heavy methylation in vertebrates. Based on genetic studies of various model organisms, it appears that DNA methylation has become increasingly integrated into the eukaryotic developmental programs during evolution (Chen and Li, 2004). In this chapter, we focus on advances in understanding the function of DNA methylation in mammalian development. Of particular interest are recent studies addressing the roles of DNA methylation in genomic imprinting, X chromosome inactivation, and regulation of tissue-specific gene expression.

DNA Methylation Patterns of the Mammalian Genome

In mammalian somatic cells, cytosine methylation occurs in 60%-80% of all CpG dinucleotides. However, methylated cytosines are not randomly distributed in the

---

Corresponding Author: En Li, Tel: (617) 871-7072, Fax: (617) 871-7263, E-mail: en.li@pharma.novartis.com

genome, but rather are compartmentalized within specific regions. Heterochromatin, including centromeric and pericentric regions, transposable elements, and repetitive sequences are heavily methylated, which contributes to the transcriptionally repressed, highly condensed chromatin structure characteristic of these regions. In contrast, the majority of CpG islands, which are GC-rich regions that contain high densities of CpG dinucleotides and are located at the promoter regions of many genes, are methylation-free, regardless of the expression status of the associated genes. However, in genomic regions where transcription is stably silenced, such as the inactive X chromosome in females and the silenced allele of imprinted genes, promoter-associated CpG islands are generally methylated, and this methylation is essential for maintaining the silenced state (Beard et al., 1995; Li et al., 1993; Panning and Jaenisch, 1996). CpG island methylation is also observed in certain human tissues during aging, or in abnormal cells such as cancer cells and immortalized cell lines (Baylin and Bestor, 2002; Issa, 2000; Jones et al., 1990).

Aberrant changes of DNA methylation patterns are associated with a number of human diseases, most notably the immunodeficiency, centromeric instability, and facial anomalies (ICF) syndrome and cancer (Robertson and Wolffe, 2000). ICF syndrome is a rare autosomal recessive disorder caused by mutations in *DNMT3B*, one of the DNA methyltransferase genes (Hansen et al., 1999; Okano et al., 1999; Shirohzu et al., 2002; Wijmenga et al., 2000; Xu et al., 1999). Lymphocytes from patients with ICF syndrome show demethylation of classical satellites 2 and 3 and chromosomal abnormalities (Ehrlich, 2003; Jeanpierre et al., 1993). In cancer cells, global loss of methylation and regional gain of methylation are frequently observed. Hypomethylation of repetitive sequences may predispose cells to chromosomal defects and rearrangements that result in genomic instability (Chen et al., 1998; Eden et al., 2003; Gaudet et al., 2003). Hypermethylation may promote tumorigenesis through transcriptional silencing of key genes. Indeed, many tumor suppressor genes are abnormally methylated and silenced in various types of cancer (Baylin and Bestor, 2002; Jones and Baylin, 2002).

DNA Methyltransferases

DNA (cytosine-5) methyltransferases (C5-MTases) catalyze the transfer of a methyl group ($-CH_3$) from *S*-adenosyl-L-methionine (AdoMet) to the C-5 position of cytosine residues in DNA. Four C5-MTases, namely Dnmt1, Dnmt2, Dnmt3a, and Dnmt3b, have been identified in humans and mice (Bestor et al., 1988; Okano et al., 1998a; Okano et al., 1998b; Van den Wyngaert et al., 1998; Xie et al., 1999; Yoder and Bestor, 1998) (Fig. 23.1). Studies have shown that these enzymes have distinct expression patterns, biochemical properties, and biological functions (Chen and Li, 2004).

A: Dnmt1

Dnmt1 is expressed constitutively in proliferating cells and ubiquitously in somatic tissues throughout mammalian development. Purified Dnmt1 protein methylates DNA containing hemimethylated CpG dinucleotides more efficiently than unmethylated DNA *in vitro* (Pradhan et al., 1999; Yoder et al., 1997a). The

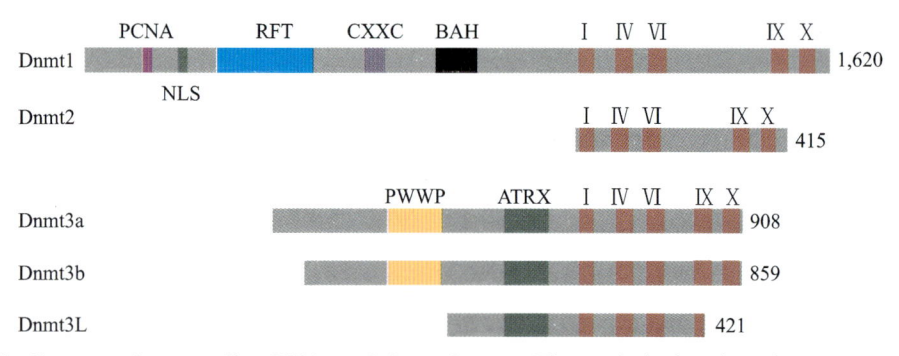

Fig.23.1 Schematic diagram of mammalian DNA methyltransferases. The catalytic domains of Dnmt1, Dnmt2, and the Dnmt3 family members are conserved (the most conserved signature motifs, I, IV, VI, IX, and X are shown), but there is little similarity among their N-terminal regulatory domains. PCNA, PCNA-interacting domain; NLS, nuclear localization signal; RFT, replication foci-targeting domain; CXXC, a cysteine-rich domain implicated in binding DNA sequences containing CpG dinucleotides; BAH, bromo-adjacent homology domain implicated in protein-protein interactions; PWWP, a domain containing a highly conserved "proline-tryptophan-tryptophan-proline" motif involved in heterochromatin association; ATRX, an ATRX-related cysteine-rich region containing a C2-C2 zinc finger and an atypical PHD domain implicated in protein-protein interactions.

Dnmt1 protein has been shown to localize to DNA replication foci during S phase (Leonhardt *et al.*, 1992), indicating that its function is coupled to DNA replication. Inactivation of the mouse *Dnmt1* gene by gene targeting results in extensive demethylation of all sequences examined, but has little effect on de novo methylation of newly integrated retrovirus DNA (Lei *et al.*, 1996; Li *et al.*, 1992). Furthermore, overexpression of Dnmt1 alone fails to induce de novo methylation in mouse ES cells or in *Drosophila* (Chen *et al.*, 2003a; Lyko *et al.*, 1999). These findings suggest that Dnmt1 functions primarily as a maintenance methyltransferase, which copies the parental-strand methylation pattern onto the daughter strand after each round of DNA replication, and that it alone has little or no de novo methyltransferase activity (Fig.23.2).

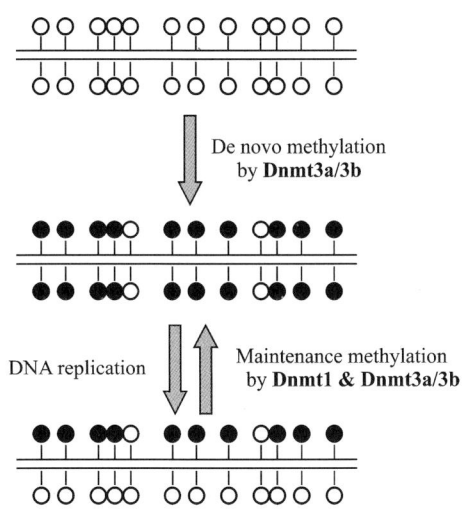

Fig.23.2 De novo and maintenance methylation. De novo methyltransferases, Dnmt3a and Dnmt3b, methylate unmodified DNA and establish methylation patterns. After each round of DNA replication, maintenance methyltransferases copy the parental-strand methylation pattern onto the daughter stand. Dnmt1, which recognizes hemimethylated sites, is the major maintenance enzyme, but Dnmt3a and Dnmt3b are also required for stable and faithful inheritance of methylation patterns. Open circles represent unmethylated CpG sites and filled circles represent methylated CpG sites.

Because most bacterial C5-MTases, which do not display maintenance activity, lack an N-terminal domain, the unique maintenance methyltransferase activity of Dnmt1 is thought to be determined by its large N-terminal region. Indeed, a number of functional domains have been defined in this region (Fig. 23.1). These include a proliferating cell nuclear antigen (PCNA)-interacting domain, a nuclear localization signal (NLS), a replication foci-targeting (RFT) domain, a CXXC domain, and a bromo-adjacent homology (BAH) domain. The RFT, BAH, and PCNA-interacting domains are involved in targeting Dnmt1 to the replication foci during the S phase (Leonhardt *et al.*, 1992; Chuang *et al.*, 1997; Callebaut *et al.*, 1999; Liu *et al.*, 1998). The location of the target recognition domain, the domain responsible for recognizing hemimethylated CpG sites, is still controversial. Fatemi *et al.* showed that the catalytic domain has a preference for binding hemimethylated CpG sites (Fatemi *et al.*, 2001), whereas Araujo *et al.* mapped the target recognition domain to a region in the N terminus (amino acids 122-417, in proximity to the PCNA-interacting domain) (Araujo *et al.*, 2001). The CXXC domain, a cysteine-rich Zn^{2+}-binding motif, has been shown to bind DNA sequences containing CpG dinucleotides and, thus, may also be involved in target recognition (Birke *et al.*, 2002; Fujita *et al.*, 1999; Lee *et al.*, 2001a; Voo *et al.*, 2000).

B:Dnmt2

Dnmt2 is a member of a protein family conserved from yeast (*Schizosaccharomyces pombe*) and fruit fly (*Drosophila melanogaster*) to mammals. The yeast pmt1 has been demonstrated to be enzymatically inactive due to an amino acid change at the catalytic site (Pinarbasi *et al.*, 1996; Wilkinson *et al.*, 1995). The Dnmt2 homologues from *Drosophila* (dDnmt2), mouse (mDnmt2), and human (hDNMT2) all contain the conserved DNA MTase motifs, but unlike the Dnmt1 and Dnmt3 families of MTases, lack an N-terminal regulatory domain (Hung *et al.*, 1999; Lyko *et al.*, 2000; Okano *et al.*, 1998b; Tweedie *et al.*, 1999; Van den Wyngaert *et al.*, 1998; Yoder and Bestor, 1998) (Fig. 23.1), hDNMT2 has been shown to form a two-domain structure that is similar to that of M.HhaI, an active prokaryotic C5 MTase (Dong *et al.*, 2001). Recent studies have shown that dDnmt2, mDnmt2, and hDNMT2 are all genuine DNA methyltransferases, and their primary target appears to be non-CpG sites (Hermann *et al.*, 2003; Kunert *et al.*, 2003; Liu *et al.*, 2003; Tang *et al.*, 2003). While the biological function of Dnmt2 remains to be determined, genetic studies have demonstrated that it does not play a major role in de novo or maintenance methylation of CpG sites in mammals. Targeted disruption of the *Dnmt2* gene in mouse embryonic stem (ES) cells has no effect on preexisting genomic methylation patterns or on the ability to methylate newly integrated retrovirus DNA de novo (Okano *et al.*, 1998b). Dnmt2 is also not essential for mammalian and *Drosophila* development (Kunert *et al.*, 2003)(M. Okano and E. Li, unpublished results).

C:Dnmt3

The Dnmt3 family has three members: Dnmt3a, Dnmt3b, and Dnmt3L (Dnmt3-like). Dnmt3L, which shows sequence similarity to Dnmt3a and Dnmt3b, does not have C5-MTase activity due to the lack of some critical catalytic motifs, but may function as a regulator of DNA methylation (see below) (Aapola et al., 2001; Hata et al., 2002) (Fig.23.1). Unlike Dnmt1, Dnmt3a and Dnmt3b are highly expressed in ES cells, early embryos, and developing germ cells, where de novo methylation is known to take place, but are downregulated in somatic tissues of postnatal animals (Okano et al., 1998a). Recombinant Dnmt3a and Dnmt3b proteins methylate both unmethylated and hemimethylated DNA substrates with similar efficiencies in vitro (Aoki et al., 2001; Okano et al., 1998a). Inactivation of both *Dnmt3a* and *Dnmt3b* by gene targeting blocks de novo methylation in ES cells and early embryos (Okano et al., 1999). Recently, Dnmt3a and Dnmt3L, but not Dnmt3b, have been shown to be essential for the establishment of methylation imprints during gametogenesis (see below) (Bourc'his et al., 2001; Hata et al., 2002; Kaneda et al., 2004). Moreover, Dnmt3a and Dnmt3b cause de novo methylation when overexpressed in mammalian cells or transgenic flies (Chen et al., 2003a; Hsieh, 1999; Lyko et al., 1999). These findings strongly support the notion that Dnmt3a and Dnmt3b function primarily as de novo methyltransferases, which are responsible for the establishment of DNA methylation patterns during embryogenesis and gametogenesis (Fig.23.2 and 23.3).

In addition to establishing DNA methylation patterns, Dnmt3a and Dnmt3b play a role in maintaining global DNA methylation levels as well. Based on genetic studies in mice, the Dnmt1 and Dnmt3 families of methyltransferases have distinct and non-redundant functions but they act cooperatively to maintain hypermethylation of the genome. While Dnmt1 is the major maintenance enzyme, the relative contributions of Dnmt1, Dnmt3a, and Dnmt3b to the maintenance of global methylation appear to differ in a cell-type-specific manner. Targeted disruption of *Dnmt3a* or *Dnmt3b* in mouse ES cells has only minor effect on global methylation (despite demethylation of specific sequences), whereas disruption of both *Dnmt3a* and *Dnmt3b* results in progressive loss of methylation throughout the genome, indicating that the two enzymes have largely redundant functions in ES cells (Chen et al., 2003a; Liang et al., 2002). In mouse embryonic fibroblasts (MEFs), however, Dnmt3b, but not Dnmt3a, is required for the maintenance of DNA methylation levels (Dodge et al., 2005). Unlike *Dnmt3a-/-Dnmt3b-/-* ES cells, *Dnmt3a-/-Dnmt3b-/-* MEFs do not show progressive loss of methylation in culture, suggesting that in the absence of Dnmt3a and Dnmt3b, Dnmt1 is capable of maintaining a higher level of DNA methylation in MEFs than in ES cells (Dodge et al., 2005). Genetic studies of human DNMTs have also been carried out in various cancer cell lines. It has been reported that the HCT116 colon cancer cells lacking DNMT1 or DNMT3B retain significant genomic

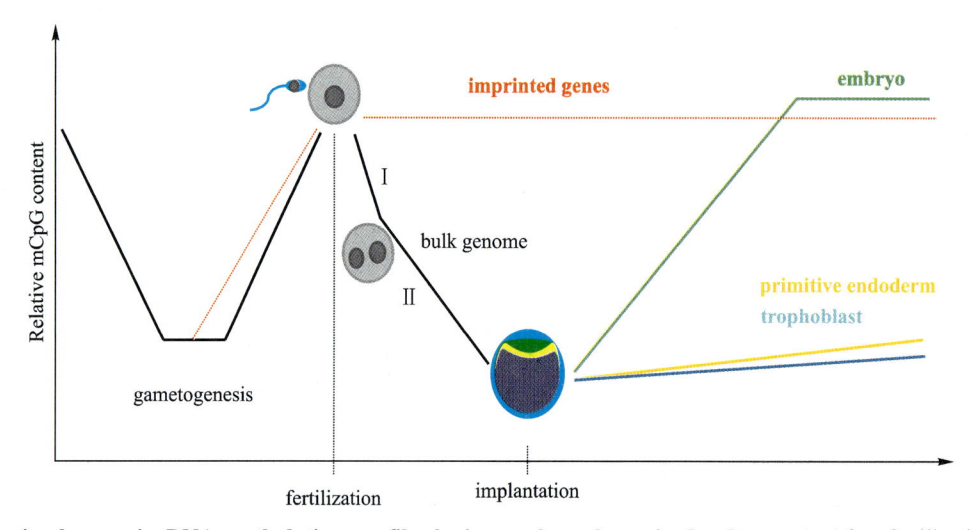

Fig.23.3 Dynamic changes in DNA methylation profile during early embryonic development. After fertilization, the paternal genome is demethylated by an active process within hours (I) and the maternal genome is passively demethylated during subsequent cleavage divisions (II). This wave of demethylation erases all inherited parental methylation marks except for those at imprinted loci. After implantation, the embryo undergoes a wave of de novo methylation that establishes a new embryonic methylation pattern, whereas the primitive endoderm and trophoblast remain hypomethylated. Demethylation and de novo methylation also occur during gametogenesis and play critical roles in the establishment of genomic imprinting.

methylation and associated gene silencing, whereas cells with both DNMT1 and DNMT3B inactivated show much lower levels of DNA methylation, suggesting that the two enzymes function redundantly to maintain CpG methylation (Rhee et al., 2002; Rhee et al., 2000). However, recent studies have shown that depletion of DNMT1 alone by either antisense or siRNA in HCT116 cells and other human cancer cells results in global and gene-specific demethylation and re-expression of tumor suppressor genes (Robert et al., 2003). Further studies are necessary to determine whether normal and cancer cells use different mechanisms to maintain DNA methylation patterns.

Dnmt3a and Dnmt3b are very similar in structural organizations. Their N-terminal regulatory domains contain a variable region (~280 amino acids in Dnmt3a and ~220 amino acids in Dnmt3b), a PWWP domain, and a cysteine-rich region that shares homology with a region in the SNF2/SWI family member ATRX (Okano et al., 1998a) (Fig. 23.1). The PWWP domain is a protein module of 100-150 amino acids containing a highly conserved "proline-tryptophan-tryptophan-proline" motif (Qiu et al., 2002; Stec et al., 2000). Its functional significance is highlighted by the finding that a missense mutation (S270P) in the human DNMT3B PWWP domain causes ICF syndrome (Shirohzu et al., 2002). Recent studies have demonstrated that the PWWP domains of Dnmt3a and Dnmt3b are involved in targeting these enzymes to pericentric heterochromatin (Chen et al., 2004; Ge et al., 2004). While the molecular mechanism remains to be determined, one possibility is that the PWWP domain interacts with one or more components of pericentric heterochromatin. The ATRX-homology domain consists of a C2-C2 zinc finger and a plant homeodomain (PHD)-like sequence. This domain has been shown to interact with the transcriptional repressor RP58, the heterochromatin protein HP-1, histone deacetylases (HDACs), and the histone methyltransferase Suv39h1 (Bachman et al., 2001; Fuks et al., 2001). Dnmt3a and Dnmt3b show high sequence homology except for their variable regions. Interestingly, Dnmt3a2, a Dnmt3a isoform that lacks the variable region, displays a diffuse nuclear localization pattern, in contrast to Dnmt3a, which is concentrated in heterochromatin regions (Chen et al., 2002). This indicates that the variable region of Dnmt3a is also involved in targeting the protein to heterochromatin.

Role of DNA Methylation in Mammalian Development

In mammals, complex changes in DNA methylation levels occur during embryonic development (Fig. 23.3). Shortly after fertilization, demethylation occurs in the male pronucleus, which seems to be independent of DNA replication (Mayer et al., 2000; Oswald et al., 2000). After formation of the zygote, both maternal and paternal chromosomes undergo progressive demethylation by a passive mechanism, which erases the methylation marks (except those at imprinted loci) inherited from the gametes (Howlett and Reik, 1991; Kafri et al., 1992; Rougier et al., 1998). Whether extensive demethylation of the genome during pre-implantation development is essential for normal development is unknown. Embryonic DNA methylation patterns are established after implantation through lineage-specific de novo methylation that begins in the inner cell mass of a blastocyst (Howlett and Reik, 1991; Kafri et al., 1992; Santos et al., 2002). DNA methylation levels increase rapidly in the primitive ectoderm, which gives rise to the entire embryo, whereas methylation is either inhibited or not maintained in the trophoblast and the primitive endoderm lineage, which give rise to the placenta and yolk sac membrane, respectively (Chapman et al., 1984; Rossant et al., 1986). During differentiation of somatic tissues, a subset of tissue-specific genes become demethylated. Demethylation and de novo methylation also occur during gametogenesis and have been shown to play a critical role in the establishment of parental-specific methylation marks in imprinted loci (Reik et al., 2001) (Fig. 23.3).

Our knowledge about the significance of DNA methylation in mammalian development comes mainly from genetic manipulations of DNA methyltransferase genes in mice. Studies of the zygotic functions of DNA methyltransferases have shown that the establishment of embryonic methylation patterns requires both de novo and maintenance methyltransferase activities, and that the maintenance of genomic methylation above a threshold level is essential for embryonic development (Li et al., 1992; Okano et al., 1999). A hypomorphic Dnmt1 mutant shows embryonic lethality at embryonic day (E)10.5 (Li et al., 1992). Complete elimination of the Dnmt1 function by disrupting the catalytic domain results in more severe phenotypes. Development of the homozygous embryos is arrested between presomite stage and 8-somite stage around E9.5 (Lei et al., 1996). Dnmt1 mutant embryos show extensive demethylation of various sequences, including repetitive elements and individual genes (Beard et al., 1995; Chen et al., 2003a; Li et al., 1993; Li et al., 1992; Walsh et al., 1998). Disruption of Dnmt3b also results in embryonic lethality with multiple developmental defects and growth impairment after E9.5 and lethality after E12.5. DNA methylation analysis has shown demethylation of the minor satellite repeats in *Dnmt3b* mutant embryos (Okano *et al.*, 1999). In agreement with this observation, patients with ICF syndrome, which is caused by partial loss-of-function mutations of *DNMT3B*, show hypomethylation of classical satellite DNA (Jeanpierre *et al.*, 1993). *Dnmt3a* mutant mice die around 4 weeks of age (Okano *et al.*, 1999). *Dnmt3a* and *Dnmt3b* double mutant embryos die around E9.5 (Okano *et al.*, 1999), similar to *Dnmt1*-null mutants.

The underlying mechanisms for the developmental defects observed in the *Dnmt* mutants remain largely unknown. Loss of DNA methylation does not affect embryonic stem (ES) cell proliferation and viability, and the effect of disrupting methylation patterns only becomes apparent during or after gastrulation when the pluripotent embryonic cells begin to differentiate (Lei *et al.*, 1996; Li *et al.*, 1992; Okano *et al.*, 1999). Consistent with these results, *Dnmt1*-deficient ES cells die upon induction of differentiation and *Dnmt3a* and *Dnmt3b* double mutant ES cells fail to differentiate (Chen *et al.*, 2003a; Jackson *et al.*, 2004; Lei *et al.*, 1996; Tucker *et al.*, 1996). Studies using conditional *Dnmt* mutants have shown that inactivation of *Dnmt1* in mouse embryonic fibroblasts (MEFs) results in severe demethylation and cell death, whereas inactivation of *Dnmt3b* in MEFs results in moderate demethylation, chromosomal instability, and abnormal cell proliferation (Dodge *et al.*, 2005; Jackson-Grusby *et al.*, 2001). These results suggest that DNA methylation is critical for cellular differentiation and the normal functioning of differentiated cells.

DNA Methylation in Genomic Imprinting

Nuclear transplantation experiments have demonstrated that the maternal and paternal genomes are nonequivalent and both are required for normal development of the mouse embryo (Barton *et al.*, 1984; McGrath and Solter, 1984; Surani *et al.*, 1984). The functional differences between the paternal and maternal genomes are attributed to genomic imprinting, an epigenetic process in the germ line that leads to differential modification of the genome in the male and female gametes (Reik and Walter, 2001b). Genomic imprinting results in differential expression of the paternal and maternal alleles of a small set of genes known as imprinted genes. Among the 70 or so known imprinted genes, some, such as *H19*, *Igf2r*, and $p57^{kip2}$, are expressed when inherited from the mother, while others, such as *Snrpn*, *Peg1*, and *Peg3*, are expressed when inherited from the father. Imprinted genes are involved in a variety of developmental processes such as embryonic development, placenta function, fetal growth, and maternal behaviors (Jaenisch, 1997; Tilghman, 1999). Defects in genomic imprinting are associated with a number of human disorders, including Beckwith-Wiedemann syndrome, Prader-Willi/Angelman syndromes, Russell-Silver syndrome, and numerous types of cancer.

Almost all imprinted genes show allele-specific DNA methylation within specific domains. Most imprinted genes acquire their methylation imprints in the female germ cells, whereas a few imprinted genes, such as *H19* and *Rasgrf1*, acquire their methylation imprints in the male germ cells (Reik and Walter, 2001a). Deletion of such differentially methylated regions (DMRs) has been shown to cause loss of imprinting (Tremblay *et al.*, 1995; Thorvaldsen *et al.*, 1998; Yoon *et al.*, 2002; Stoger *et al.*, 1993; Wutz *et al.*, 1997; Shemer *et al.*, 1997). Genetic studies of the Dnmt3 family members have provided compelling evidence that DNA methylation is an essential epigenetic mark for the establishment of genomic imprinting. Kaneda *et al.* have recently shown that disruption of *Dnmt3a*, but not *Dnmt3b*, in primordial germ cells (PGCs) (by conditional knockout technology) results in loss of paternal and maternal imprinting. Offspring from *Dnmt3a* conditional mutant female mice die *in utero* and lack methylation and allele-specific expression at all maternally imprinted loci examined. *Dnmt3a* conditional mutant male mice show impaired spermatogenesis and lack methylation at two paternally imprinted loci examined in spermatogonia (Kaneda *et al.*, 2004). Similar phenotype is observed in *Dnmt3L* knockout mice. Although the zygotic function of *Dnmt3L* is not essential for embryonic development, *Dnmt3L*-/- females fail to establish maternal methylation imprints in the oocytes, which leads to loss of monoallelic expression of maternally imprinted genes and developmental defects in the offspring, and *Dnmt3L*-/- males show defects in spermatogenesis (Bourc'his *et al.*, 2001; Hata *et al.*, 2002; Bourc'his and Bestor, 2004). As discussed above, *Dnmt3L* encodes a protein that shares homology with Dnmt3a and Dnmt3b, but lacks methyltransferase activity. Dnmt3L has been shown to stimulate the activity of Dnmt3a through a direct interaction (Hata *et al.*, 2002; Chedin *et al.*, 2002; Suetake *et al.*, 2004; Margot *et al.*, 2003; Gowher *et al.*, 2005). Taken together, these findings suggest that Dnmt3a is the major DNA methyltransferase responsible for the establishment of methylation imprints in germ cells and Dnmt3L is an essential co-factor.

Prior to the establishment of methylation imprints in the germ cells, the parental imprinting memory must be erased. This occurs in PGCs of the developing embryo (Reik *et al*., 2001) (Fig. 23.3). A number of studies indicate that the erasure of DNA methylation of regions within imprinted loci starts around day 10.5 post coitum, following the entry of PGCs into the genital ridge, and the process is completed within one day of development (Hajkova *et al*., 2002; Lee *et al*., 2002). The kinetics of the erasing process suggests that active demethylation probably occurs in PGCs. However, little is known about the biochemical process underlying DNA demethylation.

Unlike most methylation marks inherited from the parents, which are erased during pre-implantation development, the gamete-derived methylation patterns of imprinted genes are maintained in the somatic tissues throughout embryonic development (Fig.23.3). Although the mechanism by which imprinted genes escape the pre-implantation wave of demethylation is not completely understood, maternally derived Dnmt1o, a Dnmt1 isoform expressed in the oocytes, has been shown to be essential for maintaining the allele-specific methylation at imprinted loci (Howell *et al*., 2001). After implantation of the blastocyst, methylation of imprinted genes is probably maintained mainly by the somatic form of Dnmt1 (Li *et al*., 1993). Genetic studies have demonstrated that maintaining the differential methylation patterns in somatic cells is crucial for monoallelic expression of imprinted genes. In Dnmt1-deficient embryos, alleles of both *Igf2* and *Igf2r*, which are normally expressed from the paternal and maternal alleles, respectively, are silenced, whereas the *H19* gene, which is normally maternally expressed, is biallelically transcribed (Li *et al*., 1993). The mechanisms by which DNA methylation regulates imprinted gene expression have been studied extensively. Methylation of DMRs in imprinted loci can directly silence some imprinted genes (e.g. *H19*) and indirectly activate others by secondary mechanisms, such as blocking the binding of CTCF (CCCTC-binding factor) to insulators (e.g. *Igf2*) and repressing the production of antisense transcripts (e.g. *Igf2r*) (Bell and Felsenfeld, 2000; Hark *et al*., 2000). In addition to DNA methylation, other epigenetic mechanisms are involved in the regulation of imprinted gene expression. Studies over the last several years have shown that histone acetylation, histone methylation, and the Polycomb group family of chromatin regulators play important roles in the maintenance of imprinting in somatic cells (Li, 2002; Xin *et al*., 2003; Mager *et al*., 2003).

DNA Methylation in X Chromosome Inactivation.

X chromosome inactivation is a dosage-compensation mechanism in mammals that results in the transcriptional silencing of one of the two X chromosomes in the female during early embryogenesis. While X inactivation is imprinted with the paternal X chromosome being inactivated in the extraembryonic tissues, it is random in the embryonic lineages. The X-inactivation process, which converts an X chromosome from active euchromatin into highly condensed and silent heterochromatin, can be divided into three steps: 1) initiation of X inactivation, 2) spreading of heterochromatin along the length of entire X chromosome, and 3) maintenance of the inactive state in somatic cells. Studies of *Dnmt* knockout mice have revealed that DNA methylation is not essential for the initiation and propagation of X inactivation, but is required for the stable maintenance of the silent state of X-linked genes (Beard *et al*., 1995; Panning and Jaenisch, 1996; Sado *et al*., 2004).

X inactivation occurs shortly after the implantation of female embryos, or upon the induction of differentiation of female ES cells. This process is regulated by a region on the X chromosome called X-inactivation center (*Xic*), a complex locus which determines how many (counting step) and which (choice step) X chromosomes will be silenced. Non-coding RNAs, *Xist* and its anti-sense *Tsix*, which are mapped in *Xic*, play a crucial role in this process (Avner and Heard, 2001). At the onset of X inactivation, *Xist* is expressed from the presumptive inactive X chromosome and then coats the same chromosome in *cis*. This step is thought to be necessary and sufficient for the initiation of X inactivation. Targeted disruption of the *Xist* gene abrogates X inactivation (Penny *et al*., 1996; Marahrens *et al*., 1997), whereas expression of *Xist* RNA results in ectopic silencing of autosomes in *cis* in transgenic cell lines (Heard *et al*., 1999; Lee and Jaenisch, 1997; Lee *et al*., 1996; Wutz and Jaenisch, 2000). *Tsix* negatively regulates *Xist* expression in cis, possibly by an RNA interference mechanism or by silencing the *Xist* promoter (Luikenhuis *et al*., 2001; Sado *et al*., 2001; Stavropoulos *et al*., 2001). Genetic studies have shown that *Tsix* plays important roles in both imprinted X inactivation in extraembryonic tissues and random X inactivation in embryonic tissues. In mice, the *Tsix* transcript first appears in the blastocyst, and only the maternal allele of *Tsix* is expressed at this stage. The imprinted expression of *Tsix* persists in the extra-embryonic tissues after implantation, but is erased in embryonic tissues (Sado *et al*., 2001). Deletion of the main *Tsix* promoter or disruption of *Tsix* transcripts from the maternal allele of *Tsix* results in the expression of maternal *Xist*. It also leads to the inactivation of both paternal and maternal X chromosomes in the trophoblast cells, which results in embryonic lethality (Sado *et al.*, 2001; Lee, 2000). These results indicate that the expression of maternal *Tsix* at the blastocyst stage contributes to the exclusive expression of the paternal allele of *Xist* and, thus, the inactivation of the paternal X chromosome in the trophoblast cells of the placenta. Deletion of *Tsix* results in non-random X inactivation in embryonic tissues (Lee and Lu, 1999; Sado *et al.*, 2001), suggesting that random *Tsix* expression in embryonic cells contributes to the random expression of the two *Xist* alleles.

Xist expression is required not only for the initiation of X inactivation, but also for the spreading of X inactivation from the *Xic* to the rest of the chromosome. Deletion analysis indicates that a 5' repeat element of the *Xist* RNA is crucial for long-range silencing on the inactive X, whereas several spatially separated sequences throughout *Xist* are required for coating the X chromosome (Wutz *et al.*, 2002). While the molecular mechanisms whereby *Xist* directs spreading of X inactivation are largely unknown, *Xist* RNA might recruit factors that are involved in heterochromatin formation. The inactive X is characterized by a number of epigenetic chromatin modifications, including histone H3-K9 and H3-K27 methylations, hypoacetylation of core histones, enrichment of the histone variant macroH2A, and DNA methylation (Park and Kuroda, 2001; Brockdorff, 2002). These epigenetic mechanisms may function sequentially and synergistically to cause heterochromatinization of the entire X chromosome.

In mice, the 5' region of the *Xist* gene harbors a CpG island which is unmethylated on the inactive X chromosome and highly methylated on the active X chromosome, raising the possibility that DNA methylation may control *Xist* expression (Norris *et al.*, 1994). Studies of *Dnmt1*-deficient ES cells and embryos have shown that although X inactivation can occur in the absence of DNA methylation, maintenance of the methylation at the *Xist* promoter seems to be necessary for its stable silencing at the active X chromosome (Beard *et al.*, 1995; Panning and Jaenisch, 1996). Because some residual methylation is still detected at the *Xist* promoter in *Dnmt1*-/- embryos, the role of de novo methylation in the establishment of this differential pattern and its correlation with the monoallelic expression of *Xist* have been studied. The promoter of *Xist* is extensively demethylated in (*Dnmt3a*-/-, *Dnmt3b*-/-) ES cells and embryos (Okano *et al.*, 1999; Sado *et al.*, 2004). However, *Xist* expression remains monoallelic and consequently, X inactivation can initiate and propagate properly along the chromosome in the female mutants (Sado *et al.*, 2004). Taken together, these findings suggest that a mechanism(s) other than DNA methylation plays a principle role in the initiation and propagation of X inactivation.

Once established, the globally silent state and heterochromatin architecture of the inactive X are clonally inherited in somatic cells. In contrast to its essential role in the establishment of X inactivation, *Xist* is not required for maintenance of X inactivation, as the inactive state is maintained in human and mouse somatic cells lacking the *Xist* locus (Brown and Willard, 1994; Csankovszki *et al.*, 1999). However, DNA methylation is crucial for the stable silencing of the inactive X chromosome in somatic cells. In *Dnmt1* mutant embryos, random X inactivation in the embryonic lineage is unstable as a result of hypomethylation, as indicated by the reactivation of an X-linked transgene that was initially repressed (Sado *et al.*, 2000). Hypomethylation and reactivation of some genes on the inactive X are also observed in patients with ICF syndrome (Hansen *et al.*, 1999). Maintenance of imprinted X inactivation in the extraembryonic lineage, however, appear to be immune to hypomethylation, as X inactivation in the visceral endoderm in *Dnmt1* mutant embryos takes place properly on the paternal X (Sado *et al.*, 2000).

DNA Methylation in Regulation of Tissue-specific Gene Expression

While the primary function of DNA methylation has been a subject for debate (Bird, 2002; Yoder *et al.*, 1997b), there is accumulating evidence that DNA methylation is involved in the regulation of gene expression during development. In *Xenopus*, gene expression is suppressed during the first 12 cleavages of the zygote, and transcription normally initiates at the midblastula transition. In embryos depleted of xDnmt1, many developmentally regulated genes are activated two cell cycles earlier due to loss of methylation at promoter regions, suggesting a role for DNA methylation in controlling the timing of gene expression in *Xenopus* development (Stancheva *et al.*, 2002; Stancheva and Meehan, 2000). As discussed above, genetic studies in mice have shown that DNA methylation is essential for cellular differentiation and for the survival of differentiated cells. Deletion of *Dnmt1* in cultured fibroblasts results in widespread gene activation, with ~10% of all genes being activated aberrantly (Jackson-Grusby et al., 2001). These results suggest that DNA methylation may play an important role in the regulation of tissue-specific and developmental stage-specific gene expression in mammalian development. Indeed, numerous examples of tissue-specific genes regulated by promoter methylation have been reported. One study has shown that the gene encoding the glial fibrillary acidic protein (GFAP) is regulated by methylation of a CpG dinucleotide within the STAT3-binding element in the GFAP promoter region during astrocyte differentiation. The CpG site is methylated in neuronal epithelial cells prior to differentiation and becomes demethylated upon astrocyte differentiation and accessible to STAT3, which activates transcription of the GFAP gene (Takizawa et al., 2001). Another study has provided evidence that cytosine methylation is important in the establishment and maintenance of cell type-specific expression of the maspin gene (SERPINB5) in human cells (Futscher et al., 2002). DNA methylation has also been shown to regulate interleukin-4 expression during T-cell differentiation (Lee et al., 2001b) and Oct-4 expression during mouse development as well as ES and embryonal carcinoma (EC) cell differentiation (Gidekel and Bergman, 2002; Hattori et al., 2004; Deb-Rinker et al., 2005).

DNA methylation regulates gene expression through two major mechanisms (Fig.23.4). The first mechanism involves direct interference of the methyl group in binding of transcription regulatory factors to their target sequences. For example, as discussed above, methylation of a CpG site in the GFAP promoter blocks STAT3 binding and thus, prevents GFAP expression (Takizawa et al., 2001), and methylation of DMRs of some imprinted genes blocks CTCF binding and controls monoallelic expression of these genes (Bell and Felsenfeld, 2000; Hark et al., 2000). The second mechanism, which is probably more prevalent, involves changes of chromatin structure mediated by methyl-CpG-binding proteins. Five methyl-CpG-binding proteins, MeCP2, MBD1, MBD2, MBD3, and MBD4, have been characterized, and all but MBD4 have been implicated in methylation-dependent repression of transcription (Bird and Wolffe, 1999). An unrelated protein Kaiso, a member of the BTB/POZ family of transcription factors, has also been shown to bind methylated DNA and bring about methylation-dependent gene silencing (Prokhortchouk et al., 2001). Methyl-CpG-binding proteins repress transcription through recruitment of repressive complexes to methylated DNA (Fig.23.4). For instance, MeCP2 interacts with the Sin3a co-repressor complex containing HDACs (Jones et al., 1998; Nan et al., 1998), MBD2 and MBD3 associate with the multiunit NuRD complex, which contains an ATP-dependent chromatin-remodeling protein, Mi-2, and HDACs (Feng and Zhang, 2001; Wade et al., 1999), and Kaiso is a component of the N-CoR co-repressor complex containing HDAC3 (Yoon et al., 2003).

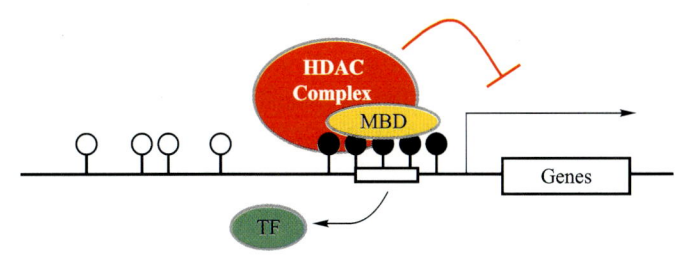

Fig.23.4 Transcriptional silencing by DNA methylation. Methylated CpGs at the 5′ region of genes inhibit transcription either by blocking the binding of transcription factors (TFs) to their target sequences or by interacting with methyl-CpG-binding proteins (MBDs), which in turn recruit co-repressor complexes containing histone deacetylases (HDACs).

Mutations in MeCP2, an X-linked gene, have been shown to cause Rett syndrome (RTT), a neurodevelopmental disorder that affects girls almost exclusively (Amir et al., 1999). Transcriptional profiling analyses of brains from RTT patients and MeCP2-deficient mice have revealed that MeCP2 probably regulates only a limited number of genes, in contrast to its predicted role as a global transcriptional regulator (Colantuoni et al., 2001; Johnston et al., 2001; Traynor et al., 2002; Tudor et al., 2002). Recent studies have shown that MeCP2 regulates the expression of brain-derived neurotrophic factor (BDNF) and the imprinting of DLX5, a maternally expressed gene involved in GABAergic neuron activity (Martinowich et al., 2003; Chen et al., 2003b; Horike et al., 2005).

Concluding Remarks

Over the last decade, great progress has been made in the understanding of the role of DNA methylation in mammalian development. Studies using mice and murine cells lacking the DNA methyltransferase genes have demonstrated the involvement of DNA methylation in various biological processes. The identification and characterization of methyl-CpG-binding proteins have shed light on how the methylation signal is interpreted. However, much less is understood about the molecular mechanisms by which DNA methylation patterns are generated during development. For instance, despite our knowledge of the involvement of Dnmt3 family members in the establishment of methylation imprints, we are almost entirely ignorant about what determines the differential methylation of DMRs of imprinted genes in the male and female germ cells. Studies in recent years have established that DNA methylation is regulated by other epigenetic mechanisms such as the histone modification and chromatin remodeling systems (Dennis et al., 2001; Gibbons et al., 2000; Jackson et al., 2002; Jeddeloh et al., 1999; Lehnertz et al., 2003; Tamaru and Selker, 2001). In the future, we expect to see intense study of the mechanistic links between DNA methyltransferases and other epigenetic regulators and, in particular, the factors that target the DNA methylation machinery to specific genes and chromosomal domains. Another area of great interest is the elucidation of the signaling pathways that regulate the expression of DNA methyltransferases during development. Alterations to DNA methylation have been linked to various human diseases, such as mental retardation syndromes and cancer (Jones and Baylin, 2002; Robertson and Wolffe, 2000). A better understanding of the upstream and downstream events of DNA methylation will help to elucidate, and perhaps manipulate, the underlying causes of these diseases.

References

Aapola, U., Lyle, R., Krohn, K., Antonarakis, S. E., and Peterson, P. (2001). Isolation and initial characterization of the mouse Dnmt3l gene. Cytogenet Cell Genet *92*, 122-126.

Amir, R. E., Van den Veyver, I. B., Wan, M., Tran, C. Q., Francke, U., and Zoghbi, H. Y. (1999). Rett syndrome is caused by mutations in X-linked MECP2, encoding methyl-CpG-binding protein 2. Nat Genet *23*, 185-188.

Aoki, A., Suetake, I., Miyagawa, J., Fujio, T., Chijiwa, T., Sasaki, H., and Tajima, S. (2001). Enzymatic properties of de novo-type mouse DNA (cytosine-5) methyltransferases. Nucleic Acids Res *29*, 3506-3512.

Araujo, F. D., Croteau, S., Slack, A. D., Milutinovic, S., Bigey, P., Price, G. B., Zannis-Hajopoulos, M., and Szyf, M. (2001). The DNMT1 target recognition domain resides in the N terminus. J Biol Chem *276*, 6930-6936.

Avner, P., and Heard, E. (2001). X-chromosome inactivation: counting, choice and initiation. Nat Rev Genet *2*, 59-67.

Bachman, K. E., Rountree, M. R., and Baylin, S. B. (2001). Dnmt3a and Dnmt3b are transcriptional repressors that exhibit unique localization properties to heterochromatin. J Biol Chem *276*, 32282-32287.

Barton, S. C., Surani, M. A., and Norris, M. L. (1984). Role of paternal and maternal genomes in mouse development. Nature *311*, 374-376.

Baylin, S. B., and Bestor, T. H. (2002). Altered methylation patterns in cancer cell genomes: cause or consequence? Cancer Cell *1*, 299-305.

Beard, C., Li, E., and Jaenisch, R. (1995). Loss of methylation activates Xist in somatic but not in embryonic cells. Genes Dev *9*, 2325-2334.

Bell, A. C., and Felsenfeld, G. (2000). Methylation of a CTCF-dependent boundary controls imprinted expression of the Igf2 gene. Nature *405*, 482-485.

Bestor, T., Laudano, A., Mattaliano, R., and Ingram, V. (1988). Cloning and sequencing of a cDNA encoding DNA methyltransferase of mouse cells. The carboxyl-terminal domain of the mammalian enzymes is related to bacterial restriction methyltransferases. J Mol Biol *203*, 971-983.

Bird, A. (2002). DNA methylation patterns and epigenetic memory. Genes Dev *16*, 6-21.

Bird, A. P., and Wolffe, A. P. (1999). Methylation-induced repression--belts, braces, and chromatin. Cell *99*, 451-454.

Birke, M., Schreiner, S., Garcia-Cuellar, M. P., Mahr, K., Titgemeyer, F., and Slany, R. K. (2002). The MT domain of the proto-oncoprotein MLL binds to CpG-containing DNA and discriminates against methylation. Nucleic Acids Res *30*, 958-965.

Bourc'his, D., and Bestor, T. H. (2004). Meiotic catastrophe and retrotransposon reactivation in male germ cells lacking Dnmt3L. Nature *431*, 96-99.

Bourc'his, D., Xu, G. L., Lin, C. S., Bollman, B., and Bestor, T. H. (2001). Dnmt3L and the establishment of maternal genomic imprints. Science *294*, 2536-2539.

Brockdorff, N. (2002). X-chromosome inactivation: closing in on proteins that bind Xist RNA. Trends Genet *18*, 352-358.

Brown, C. J., and Willard, H. F. (1994). The human X-inactivation centre is not required for maintenance of X-chromosome inactivation. Nature *368*, 154-156.

Callebaut, I., Courvalin, J. C., and Mornon, J. P. (1999). The BAH (bromo-adjacent homology) domain: a link between DNA methylation, replication and transcriptional regulation. FEBS Lett *446*, 189-193.

Chapman, V., Forrester, L., Sanford, J., Hastie, N., and Rossant, J. (1984). Cell lineage-specific undermethylation of mouse repetitive DNA. Nature *307*, 284-286.

Chedin, F., Lieber, M. R., and Hsieh, C. L. (2002). The DNA methyltransferase-like protein DNMT3L stimulates de novo methylation by Dnmt3a. Proc Natl Acad Sci USA *99*, 16916-16921.

Chen, R. Z., Pettersson, U., Beard, C., Jackson-Grusby, L., and Jaenisch, R. (1998). DNA hypomethylation leads to elevated mutation rates. Nature *395*, 89-93.

Chen, T., and Li, E. (2004). Structure and function of eukaryotic DNA methyltransferases. Curr Topics Dev Biol *60*, 55-89.

Chen, T., Tsujimoto, N., and Li, E. (2004). The PWWP domain of Dnmt3a and Dnmt3b is required for directing DNA methylation to the major satellite repeats at pericentric heterochromatin. Mol Cell Biol 24, 9048-9058.

Chen, T., Ueda, Y., Dodge, J. E., Wang, Z., and Li, E. (2003a). Establishment and maintenance of genomic methylation patterns in mouse embryonic stem cells by Dnmt3a and Dnmt3b. Mol Cell Biol 23, 5594-5605.

Chen, T., Ueda, Y., Xie, S., and Li, E. (2002). A novel Dnmt3a isoform produced from an alternative promoter localizes to euchromatin and its expression correlates with active *de novo* methylation. J Biol Chem 277, 38746-38754.

Chen, W. G., Chang, Q., Lin, Y., Meissner, A., West, A. E., Griffith, E. C., Jaenisch, R., and Greenberg, M. E. (2003b). Derepression of BDNF transcription involves calcium-dependent phosphorylation of MeCP2. Science 302, 885-889.

Chuang, L. S., Ian, H. I., Koh, T. W., Ng, H. H., Xu, G., and Li, B. F. (1997). Human DNA-(cytosine-5) methyltransferase-PCNA complex as a target for p21WAF1. Science 277, 1996-2000.

Colantuoni, C., Jeon, O. H., Hyder, K., Chenchik, A., Khimani, A. H., Narayanan, V., Hoffman, E. P., Kaufmann, W. E., Naidu, S., and Pevsner, J. (2001). Gene expression profiling in postmortem Rett Syndrome brain: differential gene expression and patient classification. Neurobiol Dis 8, 847-865.

Csankovszki, G., Panning, B., Bates, B., Pehrson, J. R., and Jaenisch, R. (1999). Conditional deletion of Xist disrupts histone macroH2A localization but not maintenance of X inactivation. Nat Genet 22, 323-324.

Deb-Rinker, P., Ly, D., Jezierski, A., Sikorska, M., and Walker, P. R. (2005). Sequential DNA methylation of the Nanog and Oct-4 upstream regions in human NT2 cells during neuronal differentiation. J Biol Chem 280, 6257-6260.

Dennis, K., Fan, T., Geiman, T., Yan, Q., and Muegge, K. (2001). Lsh, a member of the SNF2 family, is required for genome-wide methylation. Genes Dev 15, 2940-2944.

Dodge, J. E., Okano, M., Dick, F., Tsujimoto, N., Chen, T., Wang, S., Ueda, Y., Dyson, N., and Li, E. (2005). Inactivation of Dnmt3b in mouse embryonic fibroblasts results in DNA hypomethylation, chromosomal instability, and spontaneous immortalization. J Biol Chem 280, 17986-17991.

Dong, A., Yoder, J. A., Zhang, X., Zhou, L., Bestor, T. H., and Cheng, X. (2001). Structure of human DNMT2, an enigmatic DNA methyltransferase homolog that displays denaturant-resistant binding to DNA. Nucleic Acids Res 29, 439-448.

Eden, A., Gaudet, F., Waghmare, A., and Jaenisch, R. (2003). Chromosomal instability and tumors promoted by DNA hypomethylation. Science 300, 455.

Ehrlich, M. (2003). The ICF syndrome, a DNA methyltransferase 3B deficiency and immunodeficiency disease. Clin Immunol 109, 17-28.

Fatemi, M., Hermann, A., Pradhan, S., and Jeltsch, A. (2001). The activity of the murine DNA methyltransferase Dnmt1 is controlled by interaction of the catalytic domain with the N-terminal part of the enzyme leading to an allosteric activation of the enzyme after binding to methylated DNA. J Mol Biol 309, 1189-1199.

Feng, Q., and Zhang, Y. (2001). The MeCP1 complex represses transcription through preferential binding, remodeling, and deacetylating methylated nucleosomes. Genes Dev 15, 827-832.

Fujita, N., Takebayashi, S., Okumura, K., Kudo, S., Chiba, T., Saya, H., and Nakao, M. (1999). Methylation-mediated transcriptional silencing in euchromatin by methyl-CpG binding protein MBD1 isoforms. Mol Cell Biol 19, 6415-6426.

Fuks, F., Burgers, W. A., Godin, N., Kasai, M., and Kouzarides, T. (2001). Dnmt3a binds deacetylases and is recruited by a sequence-specific repressor to silence transcription. EMBO J 20, 2536-2544.

Futscher, B. W., Oshiro, M. M., Wozniak, R. J., Holtan, N., Hanigan, C. L., Duan, H., and Domann, F. E. (2002). Role for DNA methylation in the control of cell type specific maspin expression. Nat Genet 31, 175-179.

Gaudet, F., Hodgson, J. G., Eden, A., Jackson-Grusby, L., Dausman, J., Gray, J. W., Leonhardt, H., and Jaenisch, R. (2003). Induction of tumors in mice by genomic hypomethylation. Science 300, 489-492.

Ge, Y. Z., Pu, M. T., Gowher, H., Wu, H. P., Ding, J. P., Jeltsch, A., and Xu, G. L. (2004). Chromatin targeting of de novo DNA methyltransferases by the PWWP domain. J Biol Chem 279, 25447-25454.

Gibbons, R. J., McDowell, T. L., Raman, S., O'Rourke, D. M., Garrick, D., Ayyub, H., and Higgs, D. R. (2000). Mutations in ATRX, encoding a SWI/SNF-like protein, cause diverse changes in the pattern of DNA methylation. Nat Genet 24, 368-371.

Gidekel, S., and Bergman, Y. (2002). A unique developmental pattern of Oct-3/4 DNA methylation is controlled by a cis-demodification element. J Biol Chem 277, 34521-34530.

Gowher, H., Liebert, K., Hermann, A., Xu, G., and Jeltsch, A. (2005). Mechanism of Stimulation of Catalytic Activity of Dnmt3A and Dnmt3B DNA-(cytosine-C5)-methyltransferases by Dnmt3L. J Biol Chem 280, 13341-13348.

Hajkova, P., Erhardt, S., Lane, N., Haaf, T., El-Maarri, O., Reik, W., Walter, J., and Surani, M. A. (2002). Epigenetic reprogramming in mouse primordial germ cells. Mech Dev 117, 15-23.

Hansen, R. S., Wijmenga, C., Luo, P., Stanek, A. M., Canfield, T. K., Weemaes, C. M., and Gartler, S. M. (1999). The DNMT3B DNA methyltransferase gene is mutated in the ICF immunodeficiency syndrome. Proc Natl Acad Sci USA 96, 14412-14417.

Hark, A. T., Schoenherr, C. J., Katz, D. J., Ingram, R. S., Levorse, J. M., and Tilghman, S. M. (2000). CTCF mediates methylation-sensitive enhancer-blocking activity at the H19/Igf2 locus. Nature 405, 486-489.

Hata, K., Okano, M., Lei, H., and Li, E. (2002). Dnmt3L cooperates with the Dnmt3 family of de novo DNA methyltransferases to establish maternal imprints in mice.

Development *129*, 1983-1993.

Hattori, N., Nishino, K., Ko, Y. G., Ohgane, J., Tanaka, S., and Shiota, K. (2004). Epigenetic control of mouse Oct-4 gene expression in embryonic stem cells and trophoblast stem cells. J Biol Chem *279*, 17063-17069.

Heard, E., Mongelard, F., Arnaud, D., and Avner, P. (1999). Xist yeast artificial chromosome transgenes function as X-inactivation centers only in multicopy arrays and not as single copies. Mol Cell Biol *19*, 3156-3166.

Hermann, A., Schmitt, S., and Jeltsch, A. (2003). The human Dnmt2 has residual DNA-(cytosine-C5) methyltransferase activity. J Biol Chem *278*, 31717-31721.

Horike, S., Cai, S., Miyano, M., Cheng, J. F., and Kohwi-Shigematsu, T. (2005). Loss of silent-chromatin looping and impaired imprinting of DLX5 in Rett syndrome. Nat Genet *37*, 31-40.

Howell, C. Y., Bestor, T. H., Ding, F., Latham, K. E., Mertineit, C., Trasler, J. M., and Chaillet, J. R. (2001). Genomic imprinting disrupted by a maternal effect mutation in the Dnmt1 gene. Cell *104*, 829-838.

Howlett, S. K., and Reik, W. (1991). Methylation levels of maternal and paternal genomes during preimplantation development. Development *113*, 119-127.

Hsieh, C. L. (1999). *In vivo* activity of murine de novo methyltransferases, Dnmt3a and Dnmt3b. Mol Cell Biol *19*, 8211-8218.

Hung, M. S., Karthikeyan, N., Huang, B., Koo, H. C., Kiger, J., and Shen, C. J. (1999). *Drosophila* proteins related to vertebrate DNA (5-cytosine) methyltransferases. Proc Natl Acad Sci USA *96*, 11940-11945.

Issa, J. P. (2000). CpG-island methylation in aging and cancer. Curr Top Microbiol Immunol *249*, 101-118.

Jackson, J. P., Lindroth, A. M., Cao, X., and Jacobsen, S. E. (2002). Control of CpNpG DNA methylation by the KRYPTONITE histone H3 methyltransferase. Nature *416*, 556-560.

Jackson, M., Krassowska, A., Gilbert, N., Chevassut, T., Forrester, L., Ansell, J., and Ramsahoye, B. (2004). Severe global DNA hypomethylation blocks differentiation and induces histone hyperacetylation in embryonic stem cells. Mol Cell Biol *24*, 8862-8871.

Jackson-Grusby, L., Beard, C., Possemato, R., Tudor, M., Fambrough, D., Csankovszki, G., Dausman, J., Lee, P., Wilson, C., Lander, E., and Jaenisch, R. (2001). Loss of genomic methylation causes p53-dependent apoptosis and epigenetic deregulation. Nat Genet *27*, 31-39.

Jaenisch, R. (1997). DNA methylation and imprinting: why bother? Trends Genet *13*, 323-329.

Jeanpierre, M., Turleau, C., Aurias, A., Prieur, M., Ledeist, F., Fischer, A., and Viegas-Pequignot, E. (1993). An embryonic-like methylation pattern of classical satellite DNA is observed in ICF syndrome. Hum Mol Genet *2*, 731-735.

Jeddeloh, J. A., Stokes, T. L., and Richards, E. J. (1999). Maintenance of genomic methylation requires a SWI2/SNF2-like protein. Nat Genet *22*, 94-97.

Johnston, M. V., Jeon, O. H., Pevsner, J., Blue, M. E., and Naidu, S. (2001). Neurobiology of Rett syndrome: a genetic disorder of synapse development. Brain Dev *23 Suppl 1*, S206-213.

Jones, P. A., and Baylin, S. B. (2002). The fundamental role of epigentic events in cancer. Nat Rev Genet *3*, 415-428.

Jones, P. A., Wolkowicz, M. J., Rideout, W. M., 3rd, Gonzales, F. A., Marziasz, C. M., Coetzee, G. A., and Tapscott, S. J. (1990). De novo methylation of the MyoD1 CpG island during the establishment of immortal cell lines. Proc Natl Acad Sci USA *87*, 6117-6121.

Jones, P. L., Veenstra, G. J., Wade, P. A., Vermaak, D., Kass, S. U., Landsberger, N., Strouboulis, J., and Wolffe, A. P. (1998). Methylated DNA and MeCP2 recruit histone deacetylase to repress transcription. Nat Genet *19*, 187-191.

Kafri, T., Ariel, M., Brandeis, M., Shemer, R., Urven, L., McCarrey, J., Cedar, H., and Razin, A. (1992). Developmental pattern of gene-specific DNA methylation in the mouse embryo and germ line. Genes Dev *6*, 705-714.

Kaneda, M., Okano, M., Hata, K., Sado, T., Tsujimoto, N., Li, E., and Sasaki, H. (2004). Essential role for de novo DNA methyltransferases Dnmt3a in paternal and maternal imprinting. Nature *429*, 900-903.

Kunert, N., Marhold, J., Stanke, J., Stach, D., and Lyko, F. (2003). A Dnmt2-like protein mediates DNA methylation in *Drosophila*. Development *130*, 5083-5090.

Lee, J., Inoue, K., Ono, R., Ogonuki, N., Kohda, T., Kaneko-Ishino, T., Ogura, A., and Ishino, F. (2002). Erasing genomic imprinting memory in mouse clone embryos produced from day 11.5 primordial germ cells. Development *129*, 1807-1817.

Lee, J. H., Voo, K. S., and Skalnik, D. G. (2001a). Identification and characterization of the DNA binding domain of CpG-binding protein. J Biol Chem *276*, 44669-44676.

Lee, J. T. (2000). Disruption of imprinted X inactivation by parent-of-origin effects at Tsix. Cell *103*, 17-27.

Lee, J. T., and Jaenisch, R. (1997). Long-range cis effects of ectopic X-inactivation centres on a mouse autosome. Nature *386*, 275-279.

Lee, J. T., and Lu, N. (1999). Targeted mutagenesis of Tsix leads to nonrandom X inactivation. Cell *99*, 47-57.

Lee, J. T., Strauss, W. M., Dausman, J. A., and Jaenisch, R. (1996). A 450 kb transgene displays properties of the mammalian X-inactivation center. Cell *86*, 83-94.

Lee, P. P., Fitzpatrick, D. R., Beard, C., Jessup, H. K., Lehar, S., Makar, K. W., Perez-Melgosa, M., Sweetser, M. T., Schlissel, M. S., Nguyen, S., *et al.* (2001b). A critical role for Dnmt1 and DNA methylation in T cell development, function, and survival. Immunity *15*, 763-774.

Lehnertz, B., Ueda, Y., Derijck, A. A., Braunschweig, U., Perez-Burgos, L., Kubicek, S., Chen, T., Li, E., Jenuwein, T., and

Peters, A. H. (2003). Suv39h-mediated histone H3 lysine 9 methylation directs DNA methylation to major satellite repeats at pericentric heterochromatin. Curr Biol *13*, 1192-1200.

Lei, H., Oh, S. P., Okano, M., Juttermann, R., Goss, K. A., Jaenisch, R., and Li, E. (1996). De novo DNA cytosine methyltransferase activities in mouse embryonic stem cells. Development *122*, 3195-3205.

Leonhardt, H., Page, A. W., Weier, H. U., and Bestor, T. H. (1992). A targeting sequence directs DNA methyltransferase to sites of DNA replication in mammalian nuclei. Cell *71*, 865-873.

Li, E. (2002). Chromatin modification and epigenetic reprogramming in mammalian development. Nat Rev Genet *3*, 662-673.

Li, E., Beard, C., and Jaenisch, R. (1993). Role for DNA methylation in genomic imprinting. Nature *366*, 362-365.

Li, E., Bestor, T. H., and Jaenisch, R. (1992). Targeted mutation of the DNA methyltransferase gene results in embryonic lethality. Cell *69*, 915-926.

Liang, G., Chan, M. F., Tomigahara, Y., Tsai, Y. C., Gonzales, F. A., Li, E., Laird, P. W., and Jones, P. A. (2002). Cooperativity between DNA methyltransferases in the maintenance methylation of repetitive elements. Mol Cell Biol *22*, 480-491.

Liu, K., Wang, Y. F., Cantemir, C., and Muller, M. T. (2003). Endogenous assays of DNA methyltransferases: Evidence for differential activities of DNMT1, DNMT2, and DNMT3 in mammalian cells *in vivo*. Mol Cell Biol *23*, 2709-2719.

Liu, Y., Oakeley, E. J., Sun, L., and Jost, J. P. (1998). Multiple domains are involved in the targeting of the mouse DNA methyltransferase to the DNA replication foci. Nucleic Acids Res *26*, 1038-1045.

Luikenhuis, S., Wutz, A., and Jaenisch, R. (2001). Antisense transcription through the Xist locus mediates Tsix function in embryonic stem cells. Mol Cell Biol *21*, 8512-8520.

Lyko, F., Ramsahoye, B. H., Kashevsky, H., Tudor, M., Mastrangelo, M. A., Orr-Weaver, T. L., and Jaenisch, R. (1999). Mammalian (cytosine-5) methyltransferases cause genomic DNA methylation and lethality in *Drosophila*. Nat Genet *23*, 363-366.

Lyko, F., Whittaker, A. J., Orr-Weaver, T. L., and Jaenisch, R. (2000). The putative *Drosophila* methyltransferase gene dDnmt2 is contained in a transposon-like element and is expressed specifically in ovaries. Mech Dev *95*, 215-217.

Mager, J., Montgomery, N. D., de Villena, F. P., and Magnuson, T. (2003). Genome imprinting regulated by the mouse Polycomb group protein Eed. Nat Genet *33*, 502-507.

Marahrens, Y., Panning, B., Dausman, J., Strauss, W., and Jaenisch, R. (1997). Xist-deficient mice are defective in dosage compensation but not spermatogenesis. Genes Dev *11*, 156-166.

Margot, J. B., Ehrenhofer-Murray, A. E., and Leonhardt, H. (2003). Interactions within the mammalian DNA methyltransferase family. BMC Mol Biol *4*, 7.

Martinowich, K., Hattori, D., Wu, H., Fouse, S., He, F., Hu, Y., Fan, G., and Sun, Y. E. (2003). DNA methylation-related chromatin remodeling in activity-dependent BDNF gene regulation. Science *302*, 890-893.

Mayer, W., Niveleau, A., Walter, J., Fundele, R., and Haaf, T. (2000). Demethylation of the zygotic paternal genome. Nature *403*, 501-502.

McGrath, J., and Solter, D. (1984). Completion of mouse embryogenesis requires both the maternal and paternal genomes. Cell *37*, 179-183.

Nan, X., Ng, H. H., Johnson, C. A., Laherty, C. D., Turner, B. M., Eisenman, R. N., and Bird, A. (1998). Transcriptional repression by the methyl-CpG-binding protein MeCP2 involves a histone deacetylase complex. Nature *393*, 386-389.

Norris, D. P., Patel, D., Kay, G. F., Penny, G. D., Brockdorff, N., Sheardown, S. A., and Rastan, S. (1994). Evidence that random and imprinted Xist expression is controlled by preemptive methylation. Cell *77*, 41-51.

Okano, M., Bell, D. W., Haber, D. A., and Li, E. (1999). DNA methyltransferases Dnmt3a and Dnmt3b are essential for de novo methylation and mammalian development. Cell *99*, 247-257.

Okano, M., Xie, S., and Li, E. (1998a). Cloning and characterization of a family of novel mammalian DNA (cytosine-5) methyltransferases. Nat Genet *19*, 219-220.

Okano, M., Xie, S., and Li, E. (1998b). Dnmt2 is not required for de novo and maintenance methylation of viral DNA in embryonic stem cells. Nucleic Acids Res *26*, 2536-2540.

Oswald, J., Engemann, S., Lane, N., Mayer, W., Olek, A., Fundele, R., Dean, W., Reik, W., and Walter, J. (2000). Active demethylation of the paternal genome in the mouse zygote. Curr Biol *10*, 475-478.

Panning, B., and Jaenisch, R. (1996). DNA hypomethylation can activate Xist expression and silence X-linked genes. Genes Dev *10*, 1991-2002.

Park, Y., and Kuroda, M. I. (2001). Epigenetic aspects of X-chromosome dosage compensation. Science *293*, 1083-1085.

Penny, G. D., Kay, G. F., Sheardown, S. A., Rastan, S., and Brockdorff, N. (1996). Requirement for Xist in X chromosome inactivation. Nature *379*, 131-137.

Pinarbasi, E., Elliott, J., and Hornby, D. P. (1996). Activation of a yeast pseudo DNA methyltransferase by deletion of a single amino acid. J Mol Biol *257*, 804-813.

Pradhan, S., Bacolla, A., Wells, R. D., and Roberts, R. J. (1999). Recombinant human DNA (cytosine-5) methyltransferase. I. Expression, purification, and comparison of de novo and maintenance methylation. J Biol Chem *274*, 33002-33010.

Prokhortchouk, A., Hendrich, B., Jorgensen, H., Ruzov, A., Wilm, M., Georgiev, G., Bird, A., and Prokhortchouk, E. (2001). The p120 catenin partner Kaiso is a DNA methylation-dependent transcriptional repressor. Genes Dev *15*, 1613-1618.

Qiu, C., Sawada, K., Zhang, X., and Cheng, X. (2002). The PWWP domain of mammalian DNA methyltransferase Dnmt3b defines a new family of DNA-binding folds. Nat Struct Biol *9*, 217-224.

Reik, W., Dean, W., and Walter, J. (2001). Epigenetic reprogramming in mammalian development. Science *293*, 1089-1093.

Reik, W., and Walter, J. (2001a). Evolution of imprinting mechanisms: the battle of the sexes begins in the zygote. Nat Genet *27*, 255-256.

Reik, W., and Walter, J. (2001b). Genomic imprinting: parental influence on the genome. Nat Rev Genet *2*, 21-32.

Rhee, I., Bachman, K. E., Park, B. H., Jair, K. W., Yen, R. W., Schuebel, K. E., Cui, H., Feinberg, A. P., Lengauer, C., Kinzler, K. W.*, et al.* (2002). DNMT1 and DNMT3b cooperate to silence genes in human cancer cells. Nature *416*, 552-556.

Rhee, I., Jair, K. W., Yen, R. W., Lengauer, C., Herman, J. G., Kinzler, K. W., Vogelstein, B., Baylin, S. B., and Schuebel, K. E. (2000). CpG methylation is maintained in human cancer cells lacking DNMT1. Nature *404*, 1003-1007.

Robert, M. F., Morin, S., Beaulieu, N., Gauthier, F., Chute, I. C., Barsalou, A., and MacLeod, A. R. (2003). DNMT1 is required to maintain CpG methylation and aberrant gene silencing in human cancer cells. Nat Genet *33*, 61-65.

Robertson, K. D., and Wolffe, A. P. (2000). DNA methylation in health and disease. Nat Rev Genet *1*, 11-19.

Rossant, J., Sanford, J. P., Chapman, V. M., and Andrews, G. K. (1986). Undermethylation of structural gene sequences in extraembryonic lineages of the mouse. Dev Biol *117*, 567-573.

Rougier, N., Bourc'his, D., Gomes, D. M., Niveleau, A., Plachot, M., Paldi, A., and Viegas-Pequignot, E. (1998). Chromosome methylation patterns during mammalian preimplantation development. Genes Dev *12*, 2108-2113.

Sado, T., Fenner, M. H., Tan, S. S., Tam, P., Shioda, T., and Li, E. (2000). X inactivation in the mouse embryo deficient for Dnmt1: distinct effect of hypomethylation on imprinted and random X inactivation. Dev Biol *225*, 294-303.

Sado, T., Okano, M., Li, E., and Sasaki, H. (2004). De novo DNA methylation is dispensable for the initiation and propagation of X chromosome inactivation. Development *131*, 975-982.

Sado, T., Wang, Z., Sasaki, H., and Li, E. (2001). Regulation of imprinted X-chromosome inactivation in mice by Tsix. Development *128*, 1275-1286.

Santos, F., Hendrich, B., Reik, W., and Dean, W. (2002). Dynamic reprogramming of DNA methylation in the early mouse embryo. Dev Biol *241*, 172-182.

Shemer, R., Birger, Y., Riggs, A. D., and Razin, A. (1997). Structure of the imprinted mouse Snrpn gene and establishment of its parental-specific methylation pattern. Proc Natl Acad Sci USA *94*, 10267-10272.

Shirohzu, H., Kubota, T., Kumazawa, A., Sado, T., Chijiwa, T., Inagaki, K., Suetake, I., Tajima, S., Wakui, K., Miki, Y.*, et al.* (2002). Three novel DNMT3B mutations in Japanese patients with ICF syndrome. Am J Med Genet *112*, 31-37.

Stancheva, I., El-Maarri, O., Walter, J., Niveleau, A., and Meehan, R. R. (2002). DNA methylation at promoter regions regulates the timing of gene activation in *Xenopus* laevis embryos. Dev Biol *243*, 155-165.

Stancheva, I., and Meehan, R. R. (2000). Transient depletion of xDnmt1 leads to premature gene activation in *Xenopus* embryos. Genes Dev *14*, 313-327.

Stavropoulos, N., Lu, N., and Lee, J. T. (2001). A functional role for Tsix transcription in blocking Xist RNA accumulation but not in X-chromosome choice. Proc Natl Acad Sci USA *98*, 10232-10237.

Stec, I., Nagl, S. B., van Ommen, G. J., and den Dunnen, J. T. (2000). The PWWP domain: a potential protein-protein interaction domain in nuclear proteins influencing differentiation? FEBS Lett *473*, 1-5.

Stoger, R., Kubicka, P., Liu, C. G., Kafri, T., Razin, A., Cedar, H., and Barlow, D. P. (1993). Maternal-specific methylation of the imprinted mouse Igf2r locus identifies the expressed locus as carrying the imprinting signal. Cell *73*, 61-71.

Suetake, I., Shinozaki, F., Miyagawa, J., Takeshima, H., and Tajima, S. (2004). DNMT3L stimulates the DNA methylation activity of Dnmt3a and Dnmt3b through a direct interaction. J Biol Chem *279*, 27816-27823.

Surani, M. A., Barton, S. C., and Norris, M. L. (1984). Development of reconstituted mouse eggs suggests imprinting of the genome during gametogenesis. Nature *308*, 548-550.

Takizawa, T., Nakashima, K., Namihira, M., Ochiai, W., Uemura, A., Yanagisawa, M., Fujita, N., Nakao, M., and Taga, T. (2001). DNA methylation is a critical cell-intrinsic determinant of astrocyte differentiation in the fetal brain. Dev Cell *1*, 749-758.

Tamaru, H., and Selker, E. U. (2001). A histone H3 methyltransferase controls DNA methylation in Neurospora crassa. Nature *414*, 277-283.

Tang, L. Y., Reddy, M. N., Rasheva, V., Lee, T. L., Lin, M. J., Hung, M. S., and Shen, C. K. (2003). The eukaryotic DNMT2 genes encode a new class of cytosine-5 DNA methyltransferases. J Biol Chem *278*, 33613-33616.

Thorvaldsen, J. L., Duran, K. L., and Bartolomei, M. S. (1998). Deletion of the H19 differentially methylated domain results in loss of imprinted expression of H19 and Igf2. Genes Dev *12*, 3693-3702.

Tilghman, S. M. (1999). The sins of the fathers and mothers: genomic imprinting in mammalian development. Cell *96*, 185-193.

Traynor, J., Agarwal, P., Lazzeroni, L., and Francke, U. (2002). Gene expression patterns vary in clonal cell cultures from Rett syndrome females with eight different MECP2 mutations. BMC Med Genet *3*, 12.

Tremblay, K. D., Saam, J. R., Ingram, R. S., Tilghman, S. M., and Bartolomei, M. S. (1995). A paternal-specific methylation imprint marks the alleles of the mouse H19 gene. Nat Genet *9*, 407-413.

Tucker, K. L., Talbot, D., Lee, M. A., Leonhardt, H., and Jaenisch, R. (1996). Complementation of methylation deficiency in embryonic stem cells by DNA methyltransferase minigene. Proc Natl Acad Sci USA *93*, 12920-12925.

Tudor, M., Akbarian, S., Chen, R. Z., and Jaenisch, R. (2002). Transcriptional profiling of a mouse model for Rett syndrome reveals subtle transcriptional changes in the brain. Proc Natl Acad Sci USA *99*, 15536-15541.

Tweedie, S., Ng, H. H., Barlow, A. L., Turner, B. M., Hendrich, B., and Bird, A. (1999). Vestiges of a DNA methylation system in *Drosophila melanogaster*? Nat Genet *23*, 389-390.

Van den Wyngaert, I., Sprengel, J., Kass, S. U., and Luyten, W. H. (1998). Cloning and analysis of a novel human putative DNA methyltransferase. FEBS Lett *426*, 283-289.

Voo, K. S., Carlone, D. L., Jacobsen, B. M., Flodin, A., and Skalnik, D. G. (2000). Cloning of a mammalian transcriptional activator that binds unmethylated CpG motifs and shares a CXXC domain with DNA methyltransferase, human trithorax, and methyl-CpG binding domain protein 1. Mol Cell Biol *20*, 2108-2121.

Wade, P. A., Gegonne, A., Jones, P. L., Ballestar, E., Aubry, F., and Wolffe, A. P. (1999). Mi-2 complex couples DNA methylation to chromatin remodelling and histone deacetylation. Nat Genet *23*, 62-66.

Walsh, C. P., Chaillet, J. R., and Bestor, T. H. (1998). Transcription of IAP endogenous retroviruses is constrained by cytosine methylation. Nat Genet *20*, 116-117.

Wijmenga, C., Hansen, R. S., Gimelli, G., Bjorck, E. J., Davies, E. G., Valentine, D., Belohradsky, B. H., van Dongen, J. J., Smeets, D. F., van den Heuvel, L. P., et al. (2000). Genetic variation in ICF syndrome: evidence for genetic heterogeneity. Hum Mutat *16*, 509-517.

Wilkinson, C. R., Bartlett, R., Nurse, P., and Bird, A. P. (1995). The fission yeast gene pmt1+ encodes a DNA methyltransferase homologue. Nucleic Acids Res *23*, 203-210.

Wutz, A., and Jaenisch, R. (2000). A shift from reversible to irreversible X inactivation is triggered during ES cell differentiation. Mol Cell *5*, 695-705.

Wutz, A., Rasmussen, T. P., and Jaenisch, R. (2002). Chromosomal silencing and localization are mediated by different domains of Xist RNA. Nat Genet *30*, 167-174.

Wutz, A., Smrzka, O. W., Schweifer, N., Schellander, K., Wagner, E. F., and Barlow, D. P. (1997). Imprinted expression of the Igf2r gene depends on an intronic CpG island. Nature *389*, 745-749.

Xie, S., Wang, Z., Okano, M., Nogami, M., Li, Y., He, W. W., Okumura, K., and Li, E. (1999). Cloning, expression and chromosome locations of the human DNMT3 gene family. Gene *236*, 87-95.

Xin, Z., Tachibana, M., Guggiari, M., Heard, E., Shinkai, Y., and Wagstaff, J. (2003). Role of histone methyltransferase G9a in CpG methylation of the Prader-Willi syndrome imprinting center. J Biol Chem *278*, 14996-15000.

Xu, G. L., Bestor, T. H., Bourc'his, D., Hsieh, C. L., Tommerup, N., Bugge, M., Hulten, M., Qu, X., Russo, J. J., and Viegas-Pequignot, E. (1999). Chromosome instability and immunodeficiency syndrome caused by mutations in a DNA methyltransferase gene. Nature *402*, 187-191.

Yoder, J. A., and Bestor, T. H. (1998). A candidate mammalian DNA methyltransferase related to pmt1p of fission yeast. Hum Mol Genet *7*, 279-284.

Yoder, J. A., Soman, N. S., Verdine, G. L., and Bestor, T. H. (1997a). DNA (cytosine-5)-methyltransferases in mouse cells and tissues. Studies with a mechanism-based probe. J Mol Biol *270*, 385-395.

Yoder, J. A., Walsh, C. P., and Bestor, T. H. (1997b). Cytosine methylation and the ecology of intragenomic parasites. Trends Genet *13*, 335-340.

Yoon, B. J., Herman, H., Sikora, A., Smith, L. T., Plass, C., and Soloway, P. D. (2002). Regulation of DNA methylation of Rasgrf1. Nat Genet *30*, 92-96.

Yoon, H. G., Chan, D. W., Reynolds, A. B., Qin, J., and Wong, J. (2003). N-CoR mediates DNA methylation-dependent repression through a methyl CpG binding protein Kaiso. Mol Cell *12*, 723-734.

Chapter 24

Comparative Genomics of Tissue Specific Gene Expression

Anil G. Jegga[1], Sue Kong[1], Jianhua Zhang[2], Amy Moseley[3], Ashima Gupta[1], Sarah S. Williams[1], Mary Beth Genter[4] and Bruce J. Aronow[1,5]

[1]*Division of Biomedical Informatics*
[2]*Department of Cell Biology, Neurobiology and Anatomy*
[3]*Department of Molecular Genetics, Microbiology and Biochemistry*
[4]*Department of Environmental Health, University of Cincinnati College of Medicine, Cincinnati, OH 45267*
[5]*Division of Molecular and Developmental Biology, Children's Hospital Research Foundation, Cincinnati, OH 45229*

Key Words: microarray, orthologenomics, functional genomics, gene regulation, cis regulatory modules, comparative genomics, CNS evolution, gene expression

Abstract

Specification and specialization of cell and tissue structure and function is the result of the informational content and transcriptional programming of the entire genome. Because of this, whole genome/whole organism expression profiling has the potential to reveal tremendous amounts of information about the genome, specific cell types and tissues, and the evolution biological systems. Systems evolution can be studied by identifying and correlating conserved and diverged gene expression patterns of homologous and non-homologous cells and tissues, the functions of these implicated genes, and the corresponding genomic sequences and gene structural features that are conserved or diverged between multiple species' genomes. The combination of these approaches represents comparative transcriptomics or "orthologenomics", which has as its general hypothesis that improved understanding of any particular biological system can be derived through multispecies comparative analyses of genomes, transcriptionally active genes, and corresponding gene features. In this chapter we have sought to illustrate a test of this hypothesis using genes over-expressed in the central nervous system (CNS) tissues of humans and mice. The exercise lends strong support to the conjecture that these types of approaches will provide powerful new insights into specific systems that determine the health and disease of complex organisms.

Introduction

The availability of deeply annotated genome sequences collected across multiple organisms along with microarray-based gene expression profiles from corresponding cells and tissue types from each organism has enabled researchers to begin to evaluate conserved and divergent structures and functions of individual genes and entire chromosomal regions. Using gene expression profiling technology, large-scale views can be assembled of the fractions of genes that exhibit tissue and cell type specific regulation with respect to constitutive or inducible gene expression patterns. As more and more distinct biological samples are profiled and compared to each other using gene expression measurements, a variety of analyses can provide valuable clues into the biological roles and biological contexts within which the more than 25,000 human genes operate. Not only can the tissue specific and biological state-specific patterns of individual genes provide valuable insights, but the groups of genes that share coordinate regulation across a series of related biological conditions can provide even more powerful clues. The value of these types of gene clusters increases as these groupings are subjected to biological

---

Corresponding Author: Bruce Aronow, Tel: (513)636-4865, Fax: (513) 636-2056, E-mail: bruce.aronow@cchmc.org

validation, ranging from numerical verification of expression levels, cell type specific localization, the ascertainment of shared regulatory mechanisms, as well as group-correlated functional roles in biological processes. In this chapter, we discuss the various issues involved in assessing independent gene expression experiments across species, and then highlight early insights into the question of how frequently and under what circumstances functionally related genes occur in physical proximity on chromosomes and to what extent these orthologously regulated genes share cis-regulatory element structures.

Gene expression profiling, an experimental method whereby RNA accumulation in cells and tissues is assayed for several thousands of genes simultaneously in a single experiment, is well suited to unravel the complex regulation and/or interaction of both genes and proteins likely involved in most physiological processes. The advent of DNA microarrays has revolutionized gene expression studies and made gene expression profiling feasible, at least for those species whose genomes are well characterized. The number of publications reporting microarray data according to PubMed search was about 410 from 1990 to 2000, compared to 11800 reports from 2001 to date. The identification and characterization of genes underlying a range of physiological responses will accelerate the elucidation of the pathways and processes regulating the physiological processes, and will give us the functional knowledge of the gene in a particular tissue under particular conditions. Oligonucleotide arrays (e.g. Affymetrix GeneChips) (Lipshutz et al., 1999), and the spotted cDNA microarrays (Duggan et al., 1999; Brown et al., 1999; Cheung et al., 1999) are the two most commonly used experimental platforms for expression profiling . Since the uniform nature of the arrays permits data-banking of individual profiles, comparison of expression data generated at different laboratories can be explored. At the same time, systematic probing of multiple gene ontologies - the functional categorization of genes and proteins - and other gene groupings based on shared biologic relationships allows identification of expression-pattern-correlated ontology sub-clusters. These ontology-driven sub-clusters can serve as "pattern-probes" to implicate additional genes outside of the known ontology members.

Variation and Intercomparability of Gene Expression Experiments

For analyzing gene expression, the simplicity of microarray approaches makes it an attractive and viable option for the biomedical research community compared to the other non-array methods like SAGE (Serial analysis of gene expression) (Liang, 2002; Evans et al., 2002; Soulet and Rivest, 2002). The conceptual availability of nearly all predicted proteins based on genome sequence analysis, for instance, means that every possible drug target encoded in the genome is available for testing. Ironically, drug discovery is no longer hindered by a shortage of targets but, rather, mired by an excess of targets. In this changing scenario, microarray-based profiling of biological processes can play an essential role in surmounting target identification hurdles and can also aid in the process of drug discovery and development through structure activity refinement based on evaluating the effects of variant chemical structures on biological response as measured in part through microarray profiling (Somogyi and Greller, 2001; Clarke et al., 2001). With a drop in the cost of the microarrays, their use in functional genomics basic research is becoming increasingly popular in addition to their widespread use in drug discovery. However, microarray data analyses have often been subjected to criticism that the results are "quite elusive about measurement reproducibility" (Claverie, 1999), a direct consequence of the large number of uncontrolled or unknown variables that limit the inter-comparability of gene expression measurements (Box 1) along with the high cost of isolating and investigating each of the variables. Estimating the relative contribution, and then the subtraction or correction of each source of experimental variability plays a critical role for proper experimental design, and then subsequently, the ability to combine separate experiments and studies, and then in performing integrative analyses. Thus, to maximize our ability to combine multiple microarray experiments from diverse conditions and organisms into a single set of statistical analyses, critical tasks are to establish cross-species gene orthologs, the relatedness within organism and between organism of the biological samples, and then the corrections of the microarray platform with respect to systematic variation in gene expression measurements per ortholog, per RNA labeled, and per individual microarray. All of these effects need to be eliminated through normalization strategies such that expression levels can be compared across the breadth of biology. The requirements poised by this level of data processing underscore the urgent need to standardize both measurements and the description of experimental approaches that can allow for maximal synergisms of these types of whole genome-based analyses of biological systems and disease processes.

> **Box1: Critical Steps and Challenges for the Genome-wide based Discovery of Phylogenetically Conserved Transcriptional Programs**
>
> Biological Issues
> - Between-species problems
>   - Whole Genome Identification of True Gene Orthologs
>   - Identifying Divergent Gene Paralogs
>   - Orthology of Biological States: cell, tissue, and stimulated states
>   - Orthologous Regulatory Mechanism; transcriptional, post-transcriptional
> - Within Species Problems
>   - Phenotype (underlying pathological variable)
>   - Genotype (including polymorphisms)
>   - Environmental Variation
>   - Tissue heterogeneity
>   - Age
>   - Ploidy
>   - Gender
>
> Overcoming Technical Issues to Detect Biologically Significant Variables and Patterns
> - Variation due to independent array platforms
> - Variation correctable through universal reference normalization
> - Optimizing analytical methods to discern phylogenetically conserved expression patterns
>   - Hierarchical clustering
>   - Relevance networks
>   - Principal component analysis
>   - Nearest neighbors
>   - Support vector machines
>   - Pool Enrichment
>   - Shared cis-element module detection
>   - Detection of phylogenetically conserved differentiation, pathway activation, and biological processes

Of the several sources of variation in expression experiments, the type of platform used is a major one. Possible causes for platform-dependent differential gene expression results may include: probe sequence differences, variations in labeling and hybridization conditions and ultimately factors that derive from an overall lack of calibration and validation standards across multiple technologies. Apparently reduced agreement between separate measurement platforms can also be the result of biological variations when the separate platforms have probes that map to the same gene but are probing alternately spliced exons. For example, Kothapalli et al (2002), in their comparative study of microarray data obtained from cDNA and oliogonucleotide arrays, found differential expression of PAC-1, which was subsequently confirmed by northern blot analysis. However, they failed to see any expression of protein corresponding to PAC-1 by Western blot analysis. They were also unable to amplify this gene using gene-specific primers by RT-PCR. These results indicate that the quality of probe sequences and the location of the probes within the gene that are selected for incorporation into the array are also important. If the probes are selected only from the 3' end of a given gene, then different splice variants of that gene may not be identified when alternative splicing occurs at a more 5' region of the gene.

Maximal biological information derivation from a microarray experimental dataset makes optimal experimental design a fundamental pre-requisite (reviewed by Yang and Speed, 2002; Simon et al., 2002; Churchill, 2002). A general lack of standardized experimental design or controls can also create stumbling blocks that make it difficult to compare results across platforms or even within platforms. The Microarray Gene Expression Data Group (MGED group) has addressed these issues resulting in the guidelines called MIAME (Minimum Information About a Microarray Experiment) guidelines (Brazma et al., 2001). The MIAME guidelines include details of experimental design, array design or the name and location of spots on arrays, sample name, extraction and labeling, hybridization protocols, methods for image measurements, and the controls used. These and similar guidelines are critical to permit the most sensitive comparisons of multiple microarray expression data-based experiments performed in a single organism and as well, between independent organisms. However, MIAME standards do not really provide guidance in the areas of optimal experimental design, reference standard calibrations, or other additional experimental or procedural modifications that could enable one data set to be optimally combined with others.

A: Maximizing the Comparability of Independent Microarray Platforms

Adjusting gene expression measurements derived from separate platforms or across species in order to achieve a high level of inter-platform agreement is a sizeable challenge. Following a comparative expression study of Stanford-type cDNA arrays and Affymetrix chips, Kuo et al (2002) concluded that there was poor correlation between the two platforms in all measurements of similarity, and suggested that cDNA array data cannot be combined with the short oligonucleotide array results. Even though the data provide an overall idea of gene expression and contribute to understanding the biological and molecular mechanisms involved in various physiological processes, as far as the characterization of tissue compartmentalization goes, the data were considered almost useless. Similar conclusions were made by Kothapalli et al (2002) in their study of the differences between normal peripheral blood mononuclear cells and large granular lymphocytic leukemia using cDNA arrays from Incyte Genomics versus the Affymetrix platform. For example, in the Affymetrix array, the gene encoding perforin showed a 103.0-fold increase while the increase was only 3.8-fold in the cDNA array. When a Northern blot analysis was performed using a probe identical to the one spotted on the cDNA microarrays, the fold increase was somewhere between the two extreme values. While the direction of change could be argued to be the most important aspect, using these types of data, it is also possible to completely misinterpret one sample or set of samples from another, falsely concluding that an entire cluster of genes corresponds to an activation state or cellular component, when in fact, these are simply much more strongly represented in a particular sample series. Another comparative study measuring the gene expression in a human neuroblastoma cell line (Li et al, 2002) using Affymetrix chips and Incyte cDNA arrays, revealed an increase in the mRNAs of 218 genes and 4 genes (subset of 218 RNAs), respectively. Nine of the 218 genes were confirmed as upregulated by RT-PCR. It was therefore concluded that short oligonucleotide arrays were more reliable than cDNA microarrays for measuring gene expression changes. However, as other studies show, this is not always the case. Yuen et al (2002) and Rhodes et al (2002) reported that the cDNA platforms perform as well or better than short oligonucleotide arrays. There are also reports, though few, comparing the expression data from long oligonucleotide arrays with other formats (Kane et al., 2000). The Rosetta Inpharmatics group, in a comparative study of short and long oligonucleotide arrays, demonstrated that the 60mer length gave the best combination of sensitivity and specificity. Assuming 100,000 transcripts per cell, the reported sensitivity was as close to one in one-million, or ~0.1 mRNA copies per cell (Hughes et al., 2001). Comparing the 60mer arrays with cDNA arrays also gave good results. Using an array of one oligonucleotide per gene, they found a close correlation with results from the cDNA array (r=0.97). The observation that one long oligonucleotide per gene was sufficient for expression studies was an important finding as it significantly simplifies both the in situ synthesis and robotic printing approaches to making arrays. The long oligonucleotide platform was also used to show that a particular gene expression profile can be used to predict the clinical outcome of breast cancer (van Veer et al., 2002). The group studied samples from primary breast tumors and determined a gene expression signature strongly predictive of poor prognosis, in addition to establishing an expression signature of BRCA1 carriers. These data, along with other publications (Relogio et al., 2002; Wagner et al., 2002) indicate that the use of long oligonucleotides is an excellent approach for the measurement of gene expression levels (Barrett and Kawasaki, 2003). In performing a meta-analysis of a large series of breast cancer cases using Affymetrix, long oligo, and Stanford cDNA array technology, Sorlie et al. (2003) showed that all platforms agreed with respect to the identification of patient subtypes.

An inherent problem with short oligonucleotide approach, when used for complex eukaryotic genomes, is cross-hybridization of the oligonucleotides to unrelated probes because of the short length of the target. This problem has been alleviated to a large extent by using several oligonucleotides per gene. The principal disadvantage with this approach is that changes are not readily accommodated, for instance, when new sequence information becomes available or if different arrays are preferred.

The cDNA approach has the advantage that no sequence information is necessary before setting up the arrays. The PCR products from cDNA banks can be synthesized using universal primers, and interesting genes can be sequenced after array analysis. The large size of the PCR product is also helpful in enabling stringent hybridization conditions and lowering cross-hybridization of unrelated genes, although closely related gene families will still be able to anneal to some extent. Although PCR production of DNAs for microarrays is not difficult *per se*, the large number of DNAs required for complete coverage of a complex genome is taxing for most laboratories. To produce, analyze, purify, aliquot and keep track of 20,000–40,000 different PCR products is not easy, even for commercial sources dedicated to such a task (Barrett and Kawasaki, 2003).

Sharing of Microarray Gene Expression Data

Sharing of microarray data has manifold advantages, including the potential for improving analysis and confidence in results, and facilitating global comparisons between experiments and between laboratories (Geschwind, 2001). Several centralized non-commercial databases —platform-specific (Stanford microarray database), organism-specific (Yeast microarray global viewer), and project-specific (*Drosophila* development, HugeIndex) are also available and being developed to facilitate this process. The sharing poses unique challenges albeit surmountable, and various sharing formats have been proposed and explored. The multiple steps associated with an array experiment - from the initial sample characteristics and RNA preparation, to the array fabrication, array platform, scanner characteristics and image analysis - contribute significantly to the experimental outcome (Luo and Geschwind, 2001). With all of the steps involved in obtaining interpretable array data, it is no surprise that differences in methodology at any step can introduce variability in experimental results. So, common protocols and language, as well as the minimal information about a microarray experiment (MIAME), are absolutely necessary to facilitate widespread data sharing (Brazma *et al.*, 2001).

A: Motivations for Sharing Microarray Gene Expression Data

There are a variety of motivations for complete sharing of gene expression data. First, different laboratories focus on the functional relevance of a few identified signature genes, on the clusters of genes in a particular tissue, and in their specific experimental model organism. The analysis of microarray data in the context of new knowledge has the potential to provide critical new insights. It is thus wasteful or even unethical for investigators not to share this information.

Second, any single experiment is often composed of a small number of replications. One could obtain an increased confidence in results (Lee *et al.*, 2000) and an increase in analytical flexibility by combining data from different laboratories that are studying related phenomena. Third, making data available to mathematicians may aid the development of improved analytical algorithms and methods, as well as aid the comparison of different methodologies. So far very little has been published comparing various analytical platforms to demonstrate their different strengths and weaknesses. Making data available to mathematicians and others will promote the development of new tools that may be much more powerful than existing approaches. Together, these rationales strongly support that microarray data should be shared, ideally in centralized databases (Brazma *et al.*, 2000).

B: Common Standards for Expression Data Sharing

Geschwind (2001) proposed four basic paradigms for sharing microarray data. These four paradigms differ in terms of how much processing of the data is needed before sharing. First, the raw TIFF (tag information file format) images can be shared. Second, the extracted raw spot intensity values with background measurements can be shared. Third, the processed data such as averaged intensity ratios can be shared. Fourth, a list of genes that show clear differential expression can be shared (see Geschwind, 2001 for a detailed review). Each of these four paradigms has its advantages and disadvantages. Depending on the data processing method, the final gene list indicative of a biological function can be biased. However, the raw data is often overwhelming and hard to understand. It is also essential that the process does not impose practical limitations on those generating the data. A well accepted approach is thus to not only include the gene lists in publications for biologists, but to also deposit the raw, the extracted raw spot intensity values with background measurements, as well as the averaged intensity ratios into public databases for re-analysis or methodological development.

C: Challenges in Expression Data Sharing and Cross-analyzing.

Potential difficulties in effective data sharing include improper use or understanding of the data, and problems with interpretation of ambiguities in the experimental methods or sample characteristics. Geschwind (2001) discussed several issues that are especially important when expression data are presented as a simple list of genes. First, different microarray platforms generate data differently. Second, different methods of image analysis and processing, including different methods of normalization, may produce discrepancies in results. Last, the differences in dye intensity, labeling efficiency, RNA quality, or signal-amplification methods may also alter the data outputs. These issues emphasize the importance of establishing a common descriptive language and protocol so that the methodology, equipment, and samples used to generate the microarray data are properly described when the data are put into a public repository. Thus, data sharing must include a detailed description of the experimental protocols, including those on array fabrication, hybridization, scanning, statistical analysis, as well as the final list of genes said to be significantly regulated in the study (Geschwind, 2001).

Inter and Intra-species Variability in Gene Expression

Existing microarray data from mouse and human tissues exhibit a significant similarity of gene expression profiles in organ development and in various physiological processes. However, in some cases, human tissues have genes that are not evident in mouse tissues, perhaps indicating a more complex or precise regulation of physiological functions in humans. Although the mouse genome sequencing is yet to be completed, the existing information is consistent with the notion that the absence of certain genes in mouse may be due to a divergence of mouse and human ancestors. Conversely, mouse tissues on occasion express genes not found in human tissues, suggesting a redundancy in some aspects of physiological regulatory processes in mice. Comparison of the repertoire of expressed genes involved in development and adult biological processes in humans and mice lays a foundation for understanding the benefits and limitations of using mouse models for understanding the pathogenesis of various human diseases, as well as for designing and testing new drugs and treatment strategies against these diseases prior to human clinical trials.

A: Gene Expression in Relation to their Chromosomal Locations

Adjacent genes on the same chromosome often exhibit similar expression and regulatory patterns in different tissues. The organization of genes in the genome into definite loci or clusters or pathogenicity islands suggests that many functionally related genes exhibit physical proximity. However, the mechanism underlying how the observed similar expression pattern of adjacent genes is established is unknown. Furthermore, many co-expressed genes are syntenic, i.e. the same genes cluster on the same chromosome in more than one species. What makes a cluster of genes resist translocations, inversions, transpositions and other chromosome rearrangements to conserve ancestral linkages and syntenies is an interesting biological question to investigate.

Nadeau and Taylor were the first to demonstrate that about 180 conserved segments of preserved gene orders exist in human and mouse (Nadeau and Taylor, 1984). Later, several reports confirmed this finding (Copeland *et al.*, 1993; DeBry and Seldin, 1996; Waterston *et al.*, 2002; and Gregory *et al.*, 2002). Pevzner and Tesler (2003) demonstrated that human and mouse genomes share 281 synteny blocks of size at least 1 megabase and that at least 245 rearrangements of these blocks occurred since the divergence of human and mouse.

Conserved genes and gene homologs frequently cluster together on the same chromosome. For example, the gamma-crystallin (*CRYGA, CRYGB, CRYGC, and CRYGD*) genes cluster on human chromosome 2q33-q35. Genes encoding blood coagulation factors VII (*F7*) and X (*F10*) cluster on human chromosome 13q34. The seven alcohol dehydrogenase (*ADH1A, ADH1B, ADH1C, ADH4, ADH5, ADH6,* and *ADH7*) genes cluster on human chromosome 4q22. Interestingly, *F7* and *F10* are syntenic on mouse chromosome 8. Similarly, the *ADH* genes are syntenic on mouse chromosome 3, although with *ADH1B, ADH1C* and *ADH6* absent in mouse genome. The gamma-crystallin genes are syntenic on mouse chromosome 1.

The linear order of members of a family of related genes can reflect the order in which they become activated during development (Cooper, 1999). For example, the expression of *HBE1* (embryonic), *HBG2, HBG1* (fetal), *HBD, HBB* (postnatal) genes is controlled by an upstream locus control region. Correct gene order for this human beta-globin cluster is required for the normal temporal order of gene expression during development (Hanscombe *et al.*, 1991). Likewise, the chromosomal organization of HOX genes is in tandem with their order of expression (van der Hoeven *et al.*, 1996). Interestingly, the human *ADH* genes' organization on chromosome 4q21, 5'-*ADH1C-ADH1B- ADH1A*-3', is in reverse order of their transcriptional activation during liver development (Yasunami, *et al.*, 1990).

Clustering of certain genes may also be essential for their coordinate regulation by a common set of control elements. Such gene clusters include the albumin gene family (*ALB, AFP, AFM, GC*) on human chromosome 4q11-q13 and mouse chromosome 5; the eleven pregnancy-specific glycoproteins (*PSG1, PSG2, PSG3, PSG4, PSG5, PSG6, PSG7, PSG8, PSG9, PSG10, PSG11*) on human chromosome 19q13.2, and the fibrinogens alpha, beta and gamma (*FGA, FGB AND FGG*) on human chromosome 4q31 (Cooper, 1999). The correlation of temporal order of gene expression with the physical order of the same genes on the chromosome is not a universal occurrence. For example, the myosin heavy chain (*MYH1, MYH2, MYH3, MYH4, MYH8, and MYH13*) gene clusters in human and mouse on chromosomes 17 and 11 respectively are not expressed in the same temporal order as these genes are in physical order (Weiss *et al.*, 1999). Overall, definitive knowledge of temporal regulation of gene expression needs to be based on experimental data, although the sequence data may provide interesting testable hypothesis in terms of gene expression regulation.

B: Genome-wide Organization of Genes in Relation to their Expression Profiles

Many genes seem to benefit by virtue of their chromosomal location with respect to the pattern and level of expression that they are able to achieve. To provide an example of addressing this at a whole genome level, we have focused a microarray data analysis on the chromosomal location analysis of genes that tend to share similar expression profiles. For this, significant power is provided by full genome and diverse tissue coverage. We combined data sets from human and mouse microarray platforms to make a single expression meta-database (Fig.24.1, 24.2 and 24.3). We found that a significant percentage of genes over-expressed in the nervous system occur in small clusters scattered throughout the genome. Many of the co-expressed small clusters of genes are associated with highly specialized functions of the nervous system and may also share genomic regulatory mechanisms.

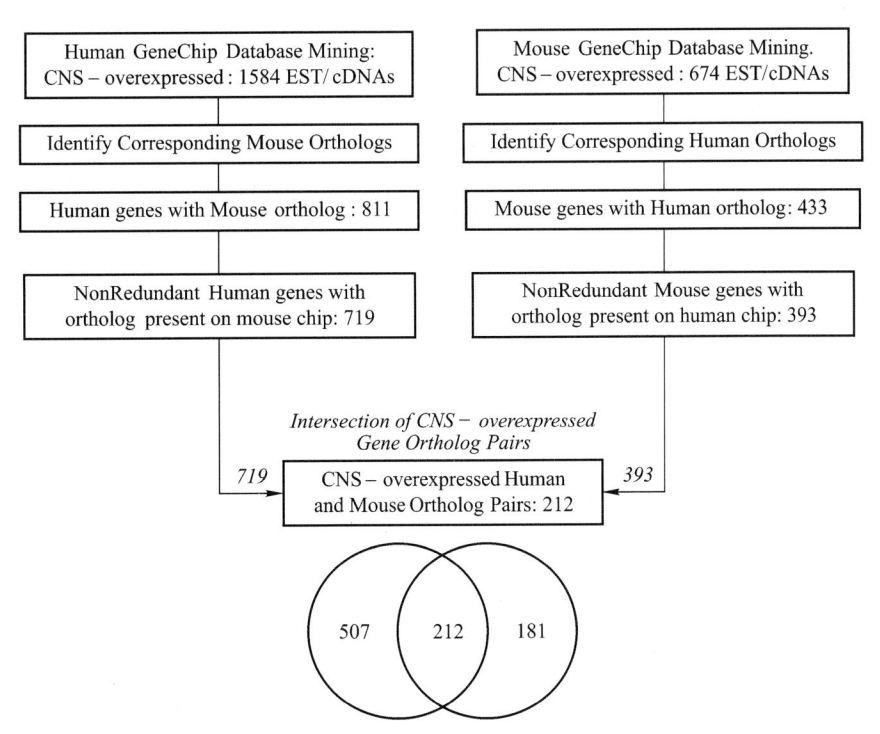

Fig.24.1 Identification of CNS-overexpressed human and mouse ortholog pairs. Human and mouse CNS-related genes were identified from 101 human and 89 mouse tissue samples and cell lines obtained from previously published data set at the Genomics Institute of Novartis Research Foundation using Affymetrix Human U95A and Mouse U74A oligonucleotide arrays. 1584 human genes and 674 mouse genes highly expressed in CNS relative to their expression in most other tissues were identified using ANOVA and fold-change criteria. Tissues used in human CNS gene selection include amygdala, caudate nucleus, cerebellum, corpus callosum, cortex, DRG, fetal brain, spinal cord, thalamus and whole brain. Amygdala, cerebellum, cortex, DRG, eye, frontal cortex, hippocampus, hypothalamus, olfactory bulb, spinal cord lower, spinal cord upper, striatum and trigeminal tissues were used from mouse for CNS gene identification. Among the 1584 human CNS genes, 811 genes were identified with corresponding mouse ortholog which results in 719 nonredundant genes on MG-U74A chip. For the 674 mouse CNS genes, 433 genes have corresponding human orthologs on HG-U95Av2 chip and 393 are nonredundant mouse genes. Comparing the 719 Human CNS and 393 mouse CNS unique genes gave 212 CNS-overexpressed human and mouse ortholog pairs.

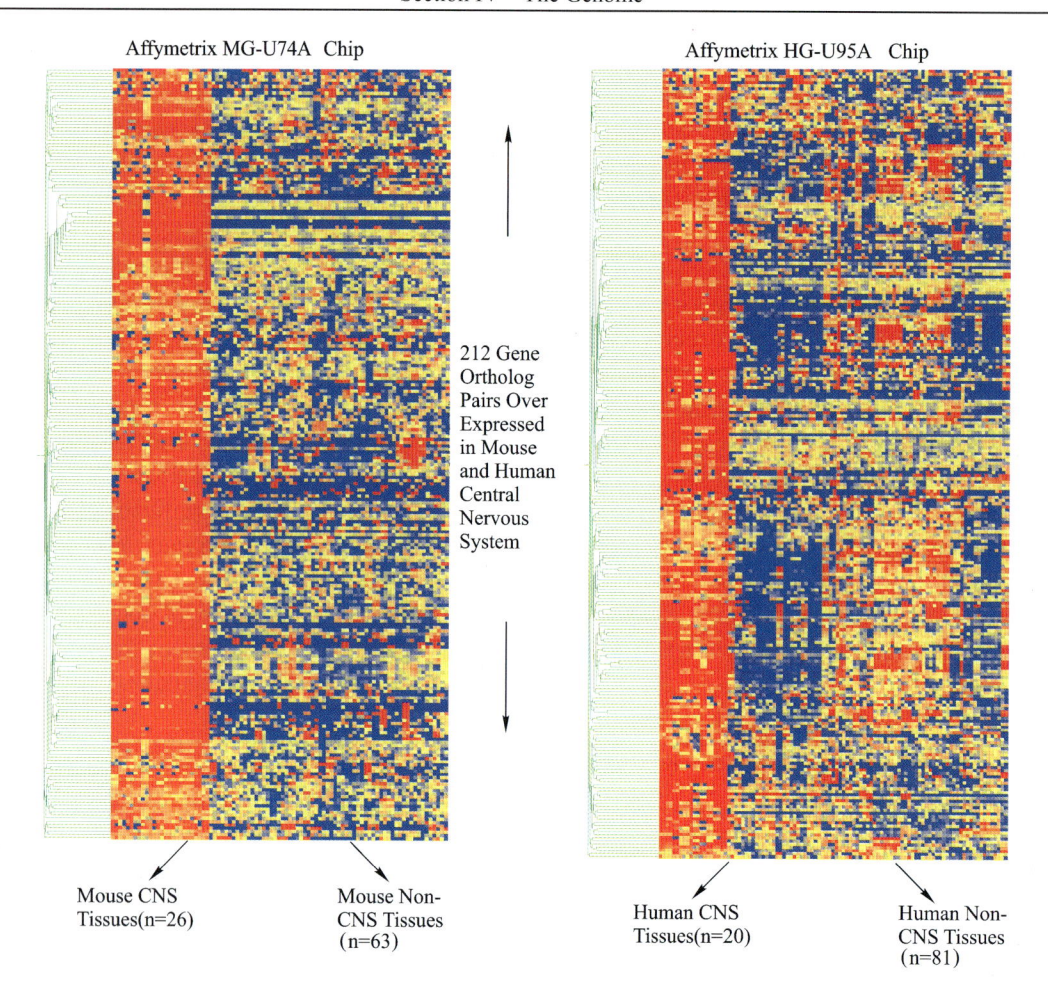

Fig.24.2 Expression profile of 212 unique CNS genes in human and mouse tissues. Hierarchical tree clustering of the 212 unique human genes in HG-U95A (259 probesets) and the corresponding 212 unique mouse genes in MG-U74A (246 probesets) was carried out using Pearson correlation using the log of the average of the relative expression ratio of each gene as measured in replicate arrays. Sequences with similar expression patterns across all tissues are clustered together in the resulting trees, the closeness of the sequences in sub trees is a measure of how closely correlated their expression is.

Tight clustering of co-expressed genes, for example, operons, is common in prokaryotes. Genes that encode interactive proteins tend to be linked and to stay linked through evolution (Dandekar et al., 1998). Although eukaryotic genes are typically assumed to be randomly distributed, the occurrence of operons has been described in *Caenorhabditis elegans* (Blumenthal et al., 1998). Further reports have suggested that eukaryotic genes are not randomly located (Ko et al., 1998; Bortoluzzi et al., 1998; Caron et al., 2001; Spellman and Rubin, 2002), although few reports provide systematic analyses of the chromosomal clustering of human or mouse genes that are similarly expressed in multiple normal tissues.

Caron et al. (2001) analyzed the gene expression profiles for many chromosomal regions in various tissue types (Human Transcriptome Map). The genes studied corresponded to about 24,000 UniGene clusters. The expression levels of these genes were estimated from 12 SAGE libraries made in different conditions. About 50 large regions, called RIDGEs (*R*egion of *I*ncrease*D* *G*ene *E*xpression) were found to be clusters of highly expressed genes. However, this report incorporated data from several cancerous tissues, and did not control for the correlation between tandem duplicates. A similar study by Lercher et al. (2002) was based on 11,000 UniGene clusters and 14 SAGE libraries. This study suggested that RIDGEs might mostly consist of housekeeping genes and no clusters of genes with similar tissue expression profiles were identified.

In order to concentrate a genomic analysis on tissue-specific expression, others have used sets of genes highly expressed in a given tissue. Gabrielsson et al. (2000) performed a microarray analysis of genes expressed in the adipose tissue. When these genes were mapped back to the human genome, clusters of adipose

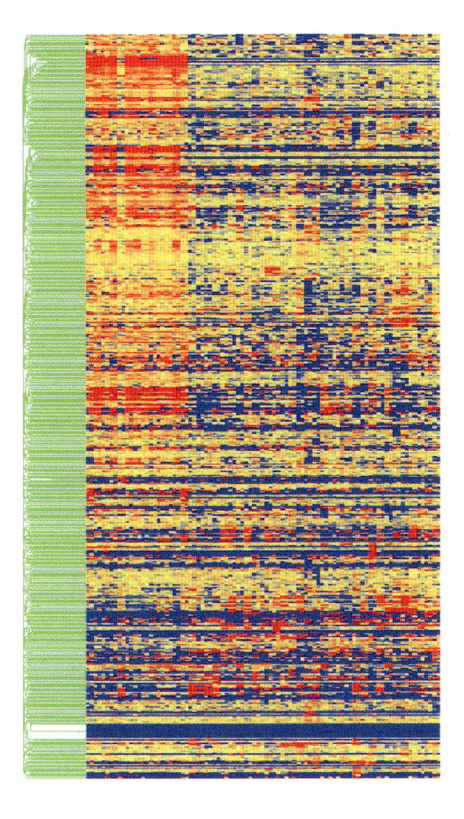

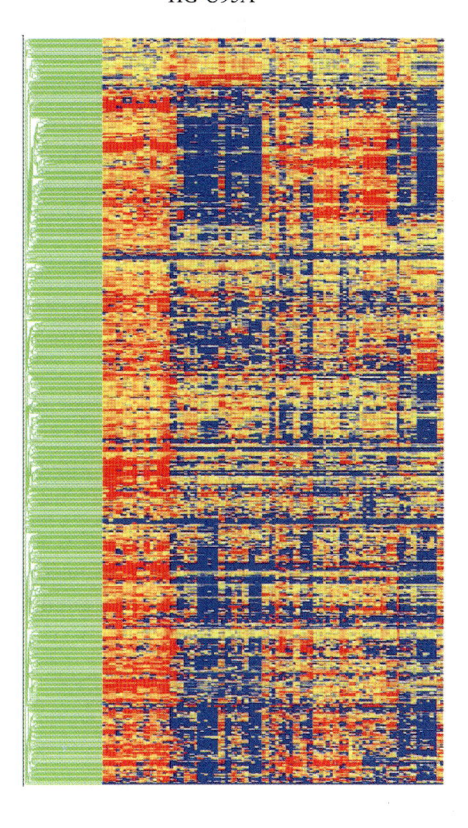

Fig.24.3 Expression profile of the pool of mouse CNS only and human CNS only unique genes in human and mouse tissues. The human CNS only and mouse CNS only genes were combined in both HG-U95A and MG-U74A genomes. Hierarchical tree clustering of the 693 unique human genes in HG-U95A and the 628 unique mouse genes in MG-U74A was carried out using Pearson correlation using the log of the average of the relative expression ratio of each gene as measured in replicate arrays. Sequences with similar expression patterns across all tissues are clustered together in the resulting trees, the closeness of the sequences in sub trees is a measure of how closely correlated their expression is.

tissue specific genes were revealed on chromosomes 11, 19 and 22. Using ESTs, Dempsey et al. (2001) focused on genes from the q arm of chromosomes 21 and 22 since those chromosomes have been completely sequenced. They found that of all of the sequences encoding previously identified genes, an average of 59% of the genes on 21q and 22q are expressed in the cardio-vascular system (CVS). Bortoluzi et al. (1998) performed a similar study on genes expressed in the skeletal muscle. They identified positional clusters of skeletal muscle genes on chromosomes 17, 19 and X. An EST analysis of the murine placenta by Ko et al. (1998) identified clusters of placenta specific genes on chromosomes 2, 7, 9 and 17. Overall, these studies suggest that clusters of tissue specific genes do exist, and might be more frequent than initially thought. But are syntenic blocks of a multitude of genes co-expressed? Glusman et al. (2001) reported that 80% of over 900 olfactory receptor genes are found in clusters of 6 to 138 genes. The idea that genomes may be divided into domains important for controlling the expression of groups of adjacent genes received further weight by two other observations in yeast and human. Pairs or triplets of co-occurring genes in the budding yeast genome displayed similar expression pattern (Cohen et al., 2000). Second, Caron et al., (2001) reported at least 50 regions on the human genome displaying strong clustering of genes that are ubiquitously and strongly expressed in nearly all tissues (Lercher et al., 2002).

Oliver et al (2002) proposed three models to account for coordinate regulation of expression of neighboring genes. (a) Incidental regulation. In this model, a transcription factor binds at a target gene and incidentally up-regulates its neighboring genes. Furthermore, the level of expression of neighboring genes is determined by proximity to the target gene and expression is expected to decrease as the distance of the neighboring gene increases from the target gene. (b)

Structural domain model. This model describes a discrete 'open' chromatin domain that is created as a result of activation of a target gene within the domain. Flanking boundary or insulator elements define the neighborhood and the limits of the open chromatin domain. (c) Expression neighborhoods in three-dimensional space. In this model, activation of a target gene results in its recruitment to a specific nuclear location. This would necessarily involve the co-recruitment of neighboring genes. The particular sub-nuclear location exposes the neighborhood to increased concentrations of components of the transcriptional machinery.

C: Genomic Approaches to the Detection of Conserved Tissue-and Cell-type Specific Gene Expression Patterns

The functional potential of a specific cell is determined to a large extent - though certainly not exclusive - by its selective expression, and sometimes repression, of an ensemble of genes whose functions encompass many different categories. In order to identify genes that exhibit cell-type, or tissue-specific, or activation state-specific expression on a genomic scale, the construction of large-scale gene expression databases is particularly valuable. An ideal gene expression database would sample all possible biological states and could reveal a wealth of coordinately regulated gene groups that reflect the interplay of a myriad of gene functional relationships and transcriptional control mechanisms. To test this hypothesis, we built a database comprised of the mRNA expression profiles of 100 normal, developing and disease C57Bl/6 mouse tissues using the Incyte Mouse GEM1 8734 element clone set. Tissues from the central nervous system, immuno-hematologic, renal, and GI systems exhibited impressive diversity and specificity of their expression repertoire, with each organ system exhibiting strong compartment-specific expression of up to several hundred of the cDNA elements represented on the microarray. Systematic probing of multiple gene ontologies and other gene groupings based on shared biologic relationships allowed for identification of expression-pattern-correlated ontology sub-clusters, which could then serve as "pattern-probes" to implicate additional genes outside of the known ontology members. For example, more than 100 genes have been identified that correlated with collagen-related sub-clusters that were strongly expressed in skin and a variety of other epithelial tissues. Within these were genes highly associated with FGF, EGF, WNT, and TGF regulated pathways. This ontology-driven analytic approach applied to large-scale expression databases has the potential to implicate pathway relationships and functional associations.

C1: Identification of Conserved Central Nervous System Specific Genes

To identify genes whose expression is much higher in the central nervous system (CNS) versus non-CNS tissues, a variety of approaches can be applied, for example, the use of statistical Analysis of Variance (ANOVA) combined with maximal fold-difference approaches allows for the identification of top ranked general CNS genes (Fig.24.1, 24.2 and 24.3). These can be further filtered for genes that are clustered in chromosomal regions, uninterrupted by non-CNS genes, as well as those that are phylogenetically syntenic. We have followed this approach to examine CNS-specific genes using the compendium of gene expression profiles generated by the Novartis Research Institute using human and mouse tissue and cell line samples applied to Affymetrix GeneChips (Su *et al.* 2002). Organizing CNS-high versus CNS-low expressed genes, according to their positions along chromosomes, revealed numerous groups of physically adjacent CNS-high genes, that shared markedly similar expression patterns within the CNS tissues. We further analyzed these clusters to see whether they were functionally related or disease-associated. The numbers of genes that comprised the chromosomal position clusters varied, but appeared to average about four genes. The length of genomic sequences occupied by similarly expressed gene groups was highly variable, but varied from 1-5 million base pairs.

The fact that specific co-expressed gene clusters are syntenic in human and mouse supports the idea, that coordinated expression may be advantageous for the organism. There are several brain-enriched syntenic clusters across various human chromosomes in our dataset. However, surprisingly we have not found a cluster group of more than four genes in human and mouse suggesting that the average conserved segment is small, and perhaps regional differences in the rate of chromosomal re-arrangements exist. Though simple structural association in a chromosomal context can not reveal the mechanism(s) responsible for the observed similarities in expression of adjacent genes, the findings are most consistent with either regulation at the level of chromatin structure or the presence of a common control region like a locus control region (LCR). Examination of conserved clusters of *cis*-element motifs associated with promoters or other conserved non-coding sequences may provide significant clues to gene regulatory mechanisms, particularly when this can be combined with other knowledge of the structure and composition of known functional regulatory regions active in the same cell types. In addition, examination of the biochemical function or biological process-association of each of the constituent genes within a cluster may suggest previously unappreciated functional collaborations.

The mouse-CNS-only genes and human-CNS-only genes were identified and pooled into one gene list in both HG-U95A and MG-U74A genomes. As a result, 693 unique genes were identified in human U95A and 628 unique genes were identified in mouse U74A (Fig. 24.3). There was a general agreement with the CNS genes identified in our microarray experiments with the literature citations. However, there are instances wherein genes with a high expression in CNS didn't behave the same way in the ortholog. In most such cases, genes with diagonally opposite expression patterns occurred either because the criteria used in the identification of brain specific genes on either human or mouse chip were low or the variances were large on one of the array platforms. In the case of *TDE2*, identified as mouse CNS gene but not human CNS gene in our data, there was indeed a reference mentioning expression of *TDE2* restricted to the central nervous system in mouse brain (Grossman *et al.*, 2000). Additionally, the expression of its human ortholog *TDE2* was reported to be undetectable in brain according to serial analysis of gene expression (SAGE) (Player *et al.*, 2003). Based on the data we used, the average expression of *TDE2* in human brain tissues is only slightly higher than the average expression in non-brain tissues. However, another gene (Amyloid precursor-like protein 1, *APLP1*), reportedly involved in the pathogenesis of Alzheimer's disease with a predominant expression in brain in humans (Lenkkeri *et al.*, 1998) and mouse cerebral cortex (Zhong *et al.*, 1996) showed up as a CNS-high in humans only in our microarray data. The failure of this gene to be classified as CNS-high in mouse is probably because of very low signal (or negative values in MAS 4.0) and the variances in both brain tissue and non-brain tissue groups are relatively large.

C2: Immunohematopoietic Genes

As another example of mining expression databases for genes active in specific biological systems, we identified 360 genes active in immune system tissues and cell types (Hutton *et al.* 2004). These were further divided as a function of their expression pattern across the more focused immunohematopoietic tissue series. Clusters that resulted from this approach were strongly enriched for genes known to functionally collaborate in biological processes that include cell cycle, immunologic specific signal transduction pathways, and transcriptionally regulated pathways related to chemokine-mediated regulation, histocompatibility, and immuno-effector functions.

D: Finding Conserved Cis-element Clusters in Co-expressed Genes

A variety of computational approaches can be followed in order to identify transcriptional mechanisms that may be shared by co-expressed genes. Until recently, this approach has largely been focused on identification of potential regulatory factor-binding sites in the DNA sequences upstream of genes. However, the expanding number of completely sequenced eukaryotic genomes combined with a wealth expression microarray data has provided new opportunities for computationally-based strategies for deciphering genetic regulatory networks. Using the strong filter of identifying evolutionarily conserved non-coding genomic sequences and then further identifying those cis-elements that are also conserved within conserved sequences is feasible because functional sequences tend to evolve at a slower rate than non-functional sequences. In fact, searches for phylogenetic footprints, the clusters of invariant or slowly changing positions in the aligned sequences of related but divergent organisms, has essentially become a standard approach to examine chromosomal coding regions that surround coding regions (Gumucio *et al.*, 1996; Hardison *et al.* 1997; Oeltjen *et al.*, 1997; Brickner *et al.*, 1999; Loots *et al.*, 2002; Jegga *et al.*, 2002). However, a singular efficient method to identify regulatory regions, predict their underlying transcriptional machinery, and decipher their cell-type specific activities is still elusive. We have explored the extension of comparative genomics approach to tackle this problem. Starting with an earlier developed method for identification of cis-clusters in phylogenetic footprints (http://trafac.chmcc.org), we extended the binary ortholog query to identify compositionally similar cis regulatory element clusters that occur in multiple co-regulated genes within each of their ortholog-pair evolutionarily conserved cis- regulatory regions (http://cismols.cchmc.org). To do this, we first identify the conserved cis clusters (Jegga *et al.*, 2005) in each ortholog, and then compare these hits against different genes that have similar expression patterns. We successfully validated our algorithm on several data sets comprising 1) skeletal muscle specific genes (Fig.24.4), 2) liver specific genes and, 3) T-cell specific genes. The computationally predicted cis-clusters, which we call as

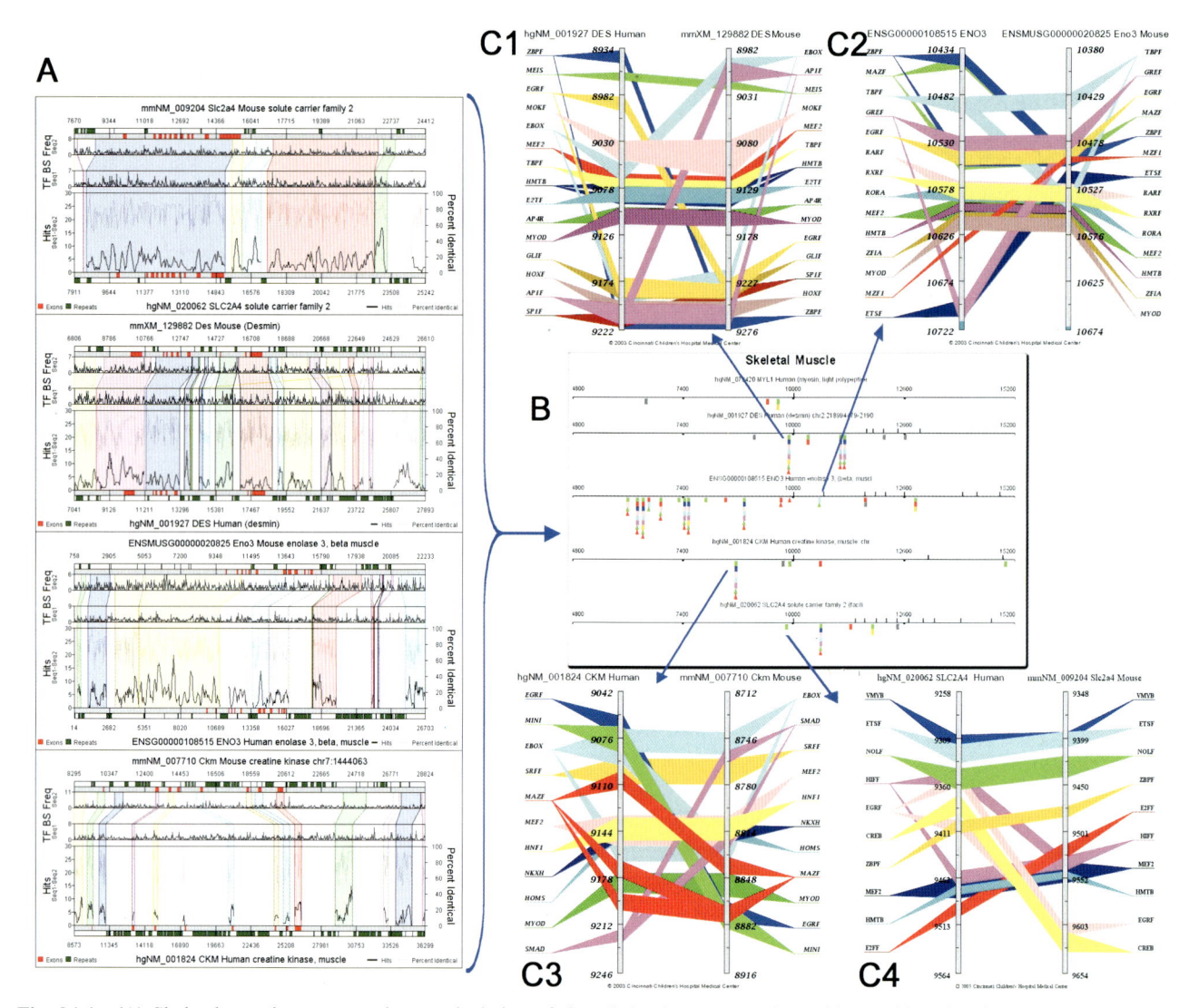

Fig. 24.4 (A) Skeletal muscle genes-regulogram depiction of shared cis-elements. Horizontal bars with colored segments (exons) are human and mouse genomic sequences. The different colored quadrilaterals are regions of alignment. Within each of these blocks, the percent sequence similarity and the number of TF-binding sites are represented as two separate line graphs.
(B): CisMols (Cis Regulatory Modules). The ortholog "peaks" or "hits" are compared to identify shared cis-regulatory modules. The horizontal lines are the genomic sequences. Yellow vertical bars are the exons. The different colored boxes represent the different cis-clusters.
(C) (1 to 4): TraFaC images of the experimentally validated regulatory regions of Skeletal Muscle genes (represented as blue circle on Regulograms). The two gray vertical bars are the two genes compared. The TF-binding sites occurring in both the genes are highlighted as various colored bars drawn across the two genes.

cismols, are rich in factor binding sites known to be active in the respective compartments. These cis-element clusters could serve as valuable probes for genome wide identification of regulatory regions and new gene targets too. These types of results - which really must be considered as early clues to the potential value of the approach - suggest that the combinatorial approach is broadly capable of identifying known regulatory regions and potential regulatory regions that contain cis-elements for known compartment-specific trans-acting factors. Improved integration and refinement of these algorithms should allow for the development of more efficient methods to identify the regulatory switches and the development of hypotheses amenable for experimental validation and refinement.

Conclusion

The availability of complete genomic sequences from many different organisms along with rapid advances in microarray technology has provided a wide array of platforms to select for gene expression profiling. However, to overcome growing pains of the field (Barrett and Kawasaki, 2003), it is critical to adopt standards that can enhance data set intercomparability.

Commonly experienced failures to reproduce or replicate the studies in different laboratories or within the same laboratory is a roadblock to success in genomic research. Guidelines for data standardization such as those specified in MIAME guidelines (Brazma et al., 2001; Stoeckert et al., 2002; Ball et al., 2002, Spellman et al. 2002) include specific details the researcher should provide about the experimental design, arrays' design or the name and location of spots on arrays, sample name, RNA extraction and labeling, hybridization protocols, methods for image measurements, and the controls used. Additional challenges are in the large number of microarray platforms that use different probes and exhibit different response characteristics. Further, it is also relatively clear that sensitive implementations of oligonucleotide array approaches appear to offer higher resolution views of the transcriptome and may rather efficiently replace cDNA arrays. In order to achieve a high degree of intercomparability of expression profile data between separate experiment series, the use of universal reference RNAs that can provide some degree of inter-experimental calibration. A large scale test of this hypothesis is currently not reported, but pilot experiments to this issue are quite promising (http://emice.nci.nih.gov/ Freudenberg, et al., 2005, submitted).

Comparisons of cDNA or EST profiles between mice and humans following the microarray studies reveal remarkable conservation of expressed genes involved in the development of various organs and physiological processes. In some cases, humans tend to have genes that are not evident in mouse, indicating probably a more complex or more precise regulation of physiological functions. Despite the incompleteness of the mouse genome with respect to its ability to reflect the complete wealth of human genetic programs, the absence of certain genes in mouse can be correlated with duplications in the human genome, which may have happened after divergence of mouse and human ancestors. In some cases, the mouse genome exhibits additional genes relative to that of humans. Analyzing the similarities and differences in the repertoire of expressed genes involved in development and biological processes in humans and mice lays a foundation for understanding not just the utility but the limitations of using mouse models for various human diseases, drug discovery and for testing of novel therapeutic agents prior to clinical trials in humans.

Finally, expanding the use of DNA microarrays to non-model species could be critical in elucidating certain physiological pathways and will be valuable in determining the genes associated with these processes. Approaches that do not require complete genome information have recently been applied to "non-model" organisms. As whole genomes are sequenced for non-model organisms, the application of DNA microarrays to comparative physiology will expand even further. The recent development of protein microarrays (surveyed by Kricka et al., 2003), and tissue arrays (Nishizuka et al., 2003) will be critical in understanding the regulation of physiological processes not accounted for at the genomic level. Together, DNA, protein, and tissue microarrays, with results filtered through the comparative genomics multi-organism sieve, may provide the most thorough and efficient method of understanding the molecular basis of physiological processes to date. In turn, classical biochemical, histological and physiological approaches will be vital in characterizing and verifying the function of the novel genes and pathways implicated by functional genomics analyses. Ultimately, a compendium of high throughput methods may be used to predict physiological responses and novel therapeutic approaches to biological process profiles discernible in clinical samples.

Acknowledgement

Supported in part by the NIEHS UO1 Mouse Comparative Genomics Centers Consortium grant, NCI UO1 Mouse Models of Human Cancer Consortium grant and the Howard Hughes Medical Institute.

References

Ball, C.A., Sherlock, G., Parkinson, H., Rocca-Sera, P., Brooksbank, C., Causton, H.C., Cavalieri, D., Gaasterland, T., Hingamp, P., Holstege, F., Ringwald, M., Spellman, P., Stoeckert, C.J., Jr., Stewart, J.E., Taylor, R., Brazma, A., and Quackenbush, J. (2002). Microarray Gene Expression Data (MGED) Society. Standards for microarray data. Science *298*, 539.

Barrett, J.C., and Kawasaki, E.S. (2003). Microarrays: the use of oligonucleotides and cDNA for the analysis of gene expression. Drug Discov Today *8*, 134-141.

Blumenthal, T. (1998). Gene clusters and polycistronic transcription in eukaryotes. BioEssays *20*, 480-487.

Bortoluzzi, S., Rampoldi, L., Simionati, B., Zimbello, R., Barbon, A., d'Alessi, F., Tiso, N., Pallavicini, A., Toppo, S., Cannata, N., Valle, G., Lanfranchi, G., and Danieli, G.A. (1998). A comprehensive, high-resolution genomic transcript map of human skeletal muscle. Genome Research *8*, 817-825.

Brazma, A., Hingamp, P., Quackenbush, J., Sherlock, G., Spellman, P., Stoeckert, C., Aach, J., Ansorge, W., Ball, C.A., Causton, H.C., Gaasterland, T., Glenisson, P., Holstege, F.C.,

Kim, I.F., Markowitz, V., Matese, J.C., Parkinson, H., Robinson, A., Sarkans, U., Schulze-Kremer, S., Stewart, J., Taylor, R., Vilo, J., and Vingron, M. (2001). Minimum information about a microarray experiment (MIAME)-toward standards for microarray data. Nature Genetics 29, 365-371.

Brazma, A., Robinson, A., Cameron, G., and Ashburner, M. (2000). One-stop shop for microarray data. Nature 403, 699-700.

Brickner, A.G., Koop, B.F., Aronow, B.J., and Wiginton, D.A. (1999). Genomic sequence comparison of the human and mouse adenosine deaminase gene regions. Mamm. Genome 10, 95-101.

Brown, P.O., and Botstein, D. (1999). Exploring the new world of the genome with DNA microarrays. Nature Genetics 21, 33-37.

Caron, H., van Schaik, B., van der Mee, M., Baas, F., Riggins, G., van Sluis, P., Hermus, M.C., van Asperen, R., Boon, K., Voute, P.A., Heisterkamp, S., van Kampen, A., and Versteeg, R. (2001). The human transcriptome map: clustering of highly expressed genes in chromosomal domains. Science 291, 1289-1292.

Cheung, V.G., Morley, M., Aguilar, F., Massimi, A., Kucherlapati, R., and Childs, G. (1999). Making and reading microarrays. Nature Genetics 21, 15-19.

Churchill, G.A. (2002). Fundamentals of experimental design for cDNA microarrays. Nature Genetics 32 Suppl, 490-495.

Clarke, P.A., te Poele, R., Wooster, R., and Workman, P. (2001). Gene expression microarray analysis in cancer biology, pharmacology, and drug development: progress and potential. Biochem Pharmacol. 62, 1311-1336.

Claverie, J.M. (1999). Computational methods for the identification of differential and coordinated gene expression. Hum Mol Genet 8, 1821-1832.

Cohen, B.A., Mitra, R.D., Hughes, J.D., and Church, G.M. (2000). A computational analysis of whole-genome expression data reveals chromosomal domains of gene expression. Nature Genetics 26, 183-186.

Cooper, D.N. (1999). Structure and function in the human genome. In: Cooper DN (ed) Human Gene Evolution, 3-53. Oxford, UK: Bios Scientific.

Copeland, N., Jenkins, N.A., Gilbert, D.J., Eppig, J.T., Maltais, L.J., Miller, J.C., Dietrich, W.F., Weaver, A., Lincoln, S.E., and Steen, R.G. (1993). A genetic linkage map of the mouse: Current applications and future prospects. Science 262, 57-66.

Dandekar, T., Snel, B., Huynen, M., and Bork, P. (1998). Conservation of gene order: a fingerprint of proteins that physically interact. Trends Biochem. Sci. 23, 324-328.

DeBry, R.W., and Seldin, M.F., (1996). Human/mouse homology relationships. Genomics 33, 337-351.

Dempsey, A.A., Pabalan, N., Tang, H.C., and Liew, C.C., (2001). Organization of human cardiovascular-expressed genes on chromosomes 21 and 22. J Mol Cell Cardiol 33, 587-591.

Duggan, D.J., Bittner, M., Chen, Y., Meltzer, P., and Trent, J.M. (1999). Expression profiling using cDNA microarrays. Nature Genetics 21 (Suppl), 10-14.

Evans, S.J., Datson, N.A., Kabbaj, M., Thompson, R.C., Vreugdenhil, E., De Kloet, E.R., Watson, S.J., and Akil, H. (2002). Evaluation of Affymetrix Gene Chip sensitivity in rat hippocampal tissue using SAGE analysis. Serial Analysis of Gene Expression. Eur J Neurosci 16, 409-413.

Gabrielsson, B.L., Carlsson, B., and Carlsson, L.M. (2000). Partial genome scale analysis of gene expression in human adipose tissue using DNA array. Obesity Research 8, 374-384.

Geschwind, D.H. (2000). Mice, microarrays, and the genetic diversity of the brain. Proc. Natl Acad. Sci. USA 97, 10676-10678.

Glusman, G., Yanai, I., Rubin, I., and Lancet, D. (2001). The complete human olfactory subgenome. Genome Res 11, 685-702.

Gregory, S.G., Sekhon, M., Schein, J., Zhao, S., Osoegawa, K., Scott, C.E., Evans, R.S., Burridge, P.W., Cox, T.V., Fox, C.A. et al. (2002). A physical map of the mouse genome. Nature 418, 743-750.

Grossman, T.R., Luque, J.M., Nelson, N. (2000). Identification of a ubiquitous family of membrane proteins and their expression in mouse brain. J Exp Biol. 203 Pt 3, 447-457.

Gumucio, D., Shelton, D., Zhu, W., Millinoff, D., Gray, T., Bock, J., Slightom, J., and Goodman, M. (1996) Evolutionary strategies for the elucidation of cis and trans factors that regulate the developmental switching programs of the beta-like globin genes. Mol. Phylogenet. and Evol 5, 18–32.

Hanscombe, O., Whyatt, D., Fraser, P., Yannoutsos, N., Greaves, D., Dillon. N., and Grosveld, F. (1991). Importance of globin gene order for correct developmental expression. Genes and Development 5, 1387-1394.

Hardison, R. (2000) Conserved noncoding sequences are reliable guides to regulatory elements. Trends Genet 16, 369–372.

Hughes, T.R., Mao, M., Jones, A.R., Burchard, J., Marton, M.J., Shannon, K.W., Lefkowitz, S.M., Ziman, M., Schelter, J.M., Meyer, M.R., Kobayashi, S., Davis, C., Dai, H., He, Y.D., Stephaniants, S.B., Cavet, G., Walker, W.L., West, A., Coffey, E., Shoemaker, D.D., Stoughton, R., Blanchard, A.P., Friend, S.H., and Linsley, P.S. (2001). Expression profiling using microarrays fabricated by an ink-jet oligonucleotide synthesizer. Nat Biotechnol 19, 342-347.

Jegga, A.G., Sherwood, S.P., Carman, J.W., Pinski, A.T., Phillips, J.L., Pestian, J.P., and Aronow, B.J. (2002). Detection and visualization of compositionally similar cis-regulatory element clusters in orthologous and coordinately controlled genes. Genome Res 12, 1408-1417.

Jegga, A.G., Gupta, A., Gowrisankar, S., Deshmukh, M.A., Gonnolly, S., Finley, K., Aronow, B.J. (2005). CisMols Analyzer: identification of compositionally similar cis-element clusters in ortholog conserved regions of coordinately expressed genes. Nucleic Acids Res. 33 (Web Server issue), W408-W411.

Kane, M.D., Jatkoe, T.A., Stumpf, C.R., Lu, J., Thomas, J.D., and Madore, S.J. (2000). Assessment of the sensitivity and specificity of oligonucleotide (50mer) microarrays. Nucleic Acids Res 28, 4552-4557.

Ko, M.S., Threat, T.A., Wang, X., Horton, J.H., Cui, Y., Wang, X., Pryor, E., Paris, J., Wells-Smith, J., Kitchen, J.R., Rowe, L.B., Eppig, J., Satoh, T., Brant, L., Fujiwara, H., Yotsumoto, S., and Nakashima, H. (1998). Genome-wide mapping of unselected transcripts from extraembryonic tissue of 7.5-day mouse embryos reveals enrichment in the t-complex and under-representation on the X chromosome. Hum. Mol. Genet 7, 1967-1978.

Kothapalli, R., Yoder, S.J., Mane, S., and Loughran, T.P., Jr. (2002). Microarray results: how accurate are they? BMC Bioinformatics 3, 22.

Kricka, L.J., Joos, T., and Fortina, P. (2003). Protein microarrays: a literature survey. Clin Chem 49, 2109.

Lee, M.L., Kuo, F.C., Whitmore, G.A., and Sklar, J. (2000). Importance of replication in microarray gene expression studies: statistical methods and evidence from repetitive cDNA hybridizations. Proc. Natl Acad. Sci. USA 97, 9834-9839.

Lenkkeri, U., Kestila, M., Lamerdin, J., McCready, P., Adamson, A., Olsen, A., Tryggvason, K.(1998). Structure of the human amyloid-precursor-like protein gene APLP1 at 19q13.1. Hum Genet 102, 192-196.

Lercher, M.J., Urrutia, A.O., and Hurst, L.D. (2002). Clustering of housekeeping genes provides a unified model of gene order in the human genome. Nat Genet 31,180-183.

Li, J., Pankratz, M., and Johnson, J.A. (2002). Differential gene expression patterns revealed by oligonucleotide versus long cDNA arrays. Toxicol Sci 69, 383-390.

Liang, P. (2002). SAGE Genie: a suite with panoramic view of gene expression. Proc Natl Acad Sci USA 99, 11547-11548.

Lipshutz, R.J., Fodor, S.P., Gingeras, T.R., and Lockhart, D.J. (1999). High density synthetic oligonucleotide arrays. Nature Genetics 21, 20-24.

Loots,G., Ovcharenko,I., Pachter,L., Dubchak,I., and Rubin,E. (2002). RVISTA for comparative sequence-based discovery of functional transcription factor binding sites. Genome Res. 12, 832–839.

Luo, Z., and Geschwind, D.H. (2001). Microarray applications in neuroscience. Neurobiol. Dis. 8, 183-193.

Nadeau, J.H., and Taylor, B.A. (1984). Lengths of chromosoamal segments conserved since divergence of man and mouse. Proc Natl Acad Sci USA 81, 814-818.

Nishizuka, S., Chen, S.T., Gwadry, F.G., Alexander, J., Major, S.M., Scherf, U., Reinhold, W.C.,Waltham, M., Charboneau, L., Young, L., Bussey, K.J., Kim, S., Lababidi, S., Lee, J.K. ,Pittaluga, S., Scudiero, D.A., Sausville, E.A., Munson, P.J., Petricoin, E.F., 3rd, Liotta, L.A.,Hewitt, S.M., Raffeld, M., and Weinstein, J.N. (2003). Diagnostic markers that distinguish colon and ovarian adenocarcinomas:identification by genomic, proteomic, and tissue array profiling. Cancer Res. 63, 5243-5250.

Oeltjen, J.C., Malley, T.M., Muzny, D.M., Miller, W., Gibbs, R.A., and Belmont, J.W. (1997) Large-scale comparative sequence analysis of the human and murine Bruton's tyrosine kinase loci reveals conserved regulatory domains. Genome Res. 7, 315-329.

Oliver, B., Parisi, M., and Clark, D. (2002). Gene expression neighborhoods. J Biol. 1, 4.1-4.3.

Pevzner, P., and Tesler, G. (2003). Genome rearrangements in mammalian evolution: lessons from human and mouse genomes. Genome Res. 13, 37-45.

Player, A., Gillespie, J., Fujii, T., Fukuoka, J., Dracheva, T., Meerzaman, D., Hong, K.M., Curran, J., Attoh, G., Travis, W., Jen, J. (2003). Identification of TDE2 gene and its expression in non-small cell lung cancer. Int J Cancer 107, 238-243.

Relogio, A., Schwager, C., Richter, A., Ansorge, W., and Valcarcel, J. (2002). Optimization of oligonucleotide-based DNA microarrays. Nucleic Acids Res. 30, e51.

Rhodes, D.R., Barrette, T.R., Rubin, M.A., Ghosh, D., and Chinnaiyan, A.M. (2002). Meta-analysis of microarrays: interstudy validation of gene expression profiles reveals pathway dysregulation in prostate cancer. Cancer Res. 62, 4427-4433.

Simon, R., Radmacher, M.D., and Dobbin, K. (2002). Design of studies using DNA microarrays. Genet Epidemiol 23, 21-36.

Somogyi, R., and Greller, L.D. (2001). The dynamics of molecular networks: applications to therapeutic discovery. Drug Discov. Today 6, 1267-1277.

Sorlie, T., Tibshirani, R., Parker, J., Hastie, T., Marron, J.S., Nobel, A., Deng, S., Johnsen, H., Pesich, R., Geisler, S., Demeter, J., Perou, C.M., Lonning, P.E., Brown, P.O., Borresen-Dale, A.L., and Botstein, D. (2003). Repeated observation of breast tumor subtypes in independent gene expression data sets. Proc Natl Acad Sci USA 100, 8418-8423.

Spellman, P.T., Rubin, G.M. (2002). Evidence, for large domains of similarly expressed genes in the *Drosophila* genome. J Biol. Epub 2002 Jun 18. *1(1)*, 5.

Spellman, P.T., Miller, M., Stewart, J., Troup, C., Sarkans, U., Chervitz, S., Bernhart, D., Sherlock, G., Ball, C., Lepage, M., Swiatek, M., Marks, WL., Goncalves, J., Markel, S., Iordan, D., Shojatalab, M., Pizarro, A., White, J., Hubley, R., Deutsch, E., Senger, M., Aronow, B.J.,Robinson, A., Bassett, D., Stoeckert, C.J., Jr., and Brazma, A. (2002). Design and implementation of microarray gene expression markup language (MAGE-ML). Genome Biol. 3, RESEARCH0046.

Stoeckert, C.J., Jr., Causton, H.C., and Ball, C.A. (2002). Microarray databases: standards and ontologies. Nat Genet 32 Suppl, 469-473.

Su, A.I., Cooke, M.P., Ching, K.A., Hakak, Y., Walker, J.R., Wiltshire, T., Orth, A.P., Vega, R.G., Sapinoso, L.M., Moqrich, A., Patapoutian, A., Hampton, G.M., Schultz, P.G., and Hogenesch, J.B. (2002). Large-scale analysis of the human and mouse transcriptomes. Proc Natl Acad Sci USA 99, 4465-4470.

Van der Hoeven, F., Zakany, J., and Duboule, D. (1996). Gene transpositions in the *HoxD* complex reveal a hierarchy of regulatory controls. Cell 85, 1025-1035.

van't Veer, L.J., Dai, H., van de Vijver, M.J., He, Y.D., Hart, A.A., Mao, M., Peterse, H.L., van der Kooy, K., Marton, M.J., Witteveen, A.T., Schreiber, G.J., Kerkhoven, R.M., Roberts, C.,

Linsley, P.S., Bernards, R., and Friend, S.H. (2002). Gene expression profiling predicts clinical outcome of breast cancer. Nature *415*, 530-536.

Wagner, E.K., Ramirez, J.J., Stingley, S.W., Aguilar, S.A., Buehler, L., Devi-Rao, G.B., and Ghazal, P. (2002). Practical approaches to long oligonucleotide-based DNA microarray: lessons from herpesviruses. Prog Nucleic Acid Res Mol Biol. *71*, 445-491.

Waterston, R.H., Lindblad-Toh, K., Birney, E., Rogers, J., Abril, J.F., Agarwal, P., Agarwala, R., Ainscough, R., Alexandersson, M., and An, P. (2002). Initial sequencing and comparative analysis of the mouse genome. Nature *420*, 520-562

Weiss, A., McDonough, D., Wertman, B., Acakpo-Satchivi, L., Montgomery, K., Kucherlapati, R., Leinwand, L. and Krauter, K. (1999). Organization of human and mouse skeletal myosin heavy chain gene clusters is highly conserved. Proc Natl Acad Sci USA *96*, 2958-2963.

Yang, Y.H., and Speed, T. (2002). Design issues for cDNA microarray experiments. Nat. Rev. Genet. *3*, 579-588.

Yasunami, M., Kikuchi, I., Sarapata, D., and Yoshida, A. (1990). The human class I alcohol dehydrogenase gene cluster: three genes are tandemly organized in an 80-kb-long segment of the genome. Genomics *7*, 152-158.

Yuen, T., Wurmbach, E., Pfeffer, R.L., Ebersole, B.J., and Sealfon, S.C. (2002). Accuracy and calibration of commercial oligonucleotide and custom cDNA microarrays. Nucleic Acids Res. *30*, e48.

Zhong, S., Wu, K., Black, I.B., Schaar, D.G. (1996). Characterization of the genomic structure of the mouse APLP1 gene. Genomics *32*, 159-162.

Chapter 25
Transcription and Genomic Integrity

Julie M. Poisson, Yinhuai Chen and Yolanda Sanchez

Department of Molecular Genetics, Biochemistry and Microbiology, University of Cincinnati College of Medicine, Cincinnati, OH 45267-0524

Key Words: stalled RNA polymerases, transcription-coupled repair, DNA damage-regulated transcription, checkpoints, DNA and RNA polymeras collision, coodination of transcription and DNA replication

Summary

The numerous discoveries that have elucidated the basic mechanisms of the assembly of the transcription machinery and the factors that regulate gene expression have been discussed elsewhere in this book. Here we address another layer of complexity intrinsic to the transcription machinery, namely, its role in maintaining genomic integrity. First, the transcription machinery specifically facilitates the repair of damaged template DNA in the transcribed strand of active genes. In addition to damaged DNA, collisions between the transcription and replication machineries represent a source of genomic instability and are thus regulated by the coordination of these two processes. Finally, the transcription machinery plays an active role during the cell's response to DNA damage by up-regulating certain genes to cope with the damage and down-regulating other genes. We will discuss the detailed mechanisms of these processes to highlight the contribution of the transcription machinery to the maintenance of the genomic integrity of the cell.

Part I: Regulation of Transcription and Repair of Templates when RNA Polymerases Stall

A mechanism to resolve damaged DNA that is being transcribed is important because lesions in DNA can lead to the stalling of RNA polymerases (Tornaletti and Hanawalt, 1999). If not resolved, the prolonged stalling of an RNA polymerase represents a potentially harmful event to the cell, especially if the block occurs in an essential gene. In this case, the stalled RNA polymerase fails to produce complete transcript from the essential gene, leading to loss of viability (Hanawalt, 1994; Svejstrup, 2002). To prevent such a situation, cells must have mechanisms to repair the DNA lesions that block the RNA polymerases and allow the resumption of transcription.

In the 1980s, much evidence was gathered to substantiate the existence of a connection between the processes of transcription and DNA repair. Hanawalt and colleagues observed that damage in an actively transcribed gene was repaired with greater efficiency than in the genome overall (Bohr *et al.*, 1985; Mellon *et al.*, 1986). Specifically, they reported that the removal of UV-induced cyclobutane pyrimidine dimers (CPDs) from the dihydrofolate reductase (DHFR) gene in Chinese hamster ovary (CHO) cells was more efficient than the repair of the same type of lesion in the genome as a whole or in non-transcribed sequences (Bohr *et al.*, 1985; Mellon *et al.*, 1986). Furthermore, the Hanawalt group showed that within the DHFR gene in human and CHO cells, only damage within the transcribed strand but not in the non-transcribed strand was repaired with faster kinetics (Mellon *et al.*, 1987). These observations led to the concept of transcription-coupled repair or TCR. In contrast to global genomic repair (GGR), the process that facilitates repair of lesions throughout the genome (de Laat *et al.*, 1999), TCR is a process that preferentially resolves DNA damage lesions in the transcribed strand of actively transcribed genes (Mellon *et al.*, 1987). TCR was later shown to occur in mammalian cells (Mellon *et al.*, 1986; Mellon *et al.*,

Corresponding Author: Yolanda Sanchez, Tel: (513) 558-3275, Fax: (513) 558-8474, E-mail: Yolanda.Sanchez@uc.edu

1987), *Saccharomyces cerevisiae* (Leadon and Lawrence, 1992; Smerdon and Thoma, 1990; Sweder and Hanawalt, 1992), and *Escherichia coli* (Mellon and Hanawalt, 1989), and its importance is illustrated in patients carrying mutations that render TCR defective. These patients develop Cockayne syndrome, an autosomal recessive disorder characterized by symptoms of sun- sensitivity and a physical appearance of premature aging without a predisposition to cancer (Spivak, 2004).

With the subsequent knowledge gained regarding the mechanism of the TCR process, a general model has been developed and is illustrated in Fig.25.1. This model proposes that the RNA polymerase, stalled by a damage in the transcribed strand to be resolved rapidly

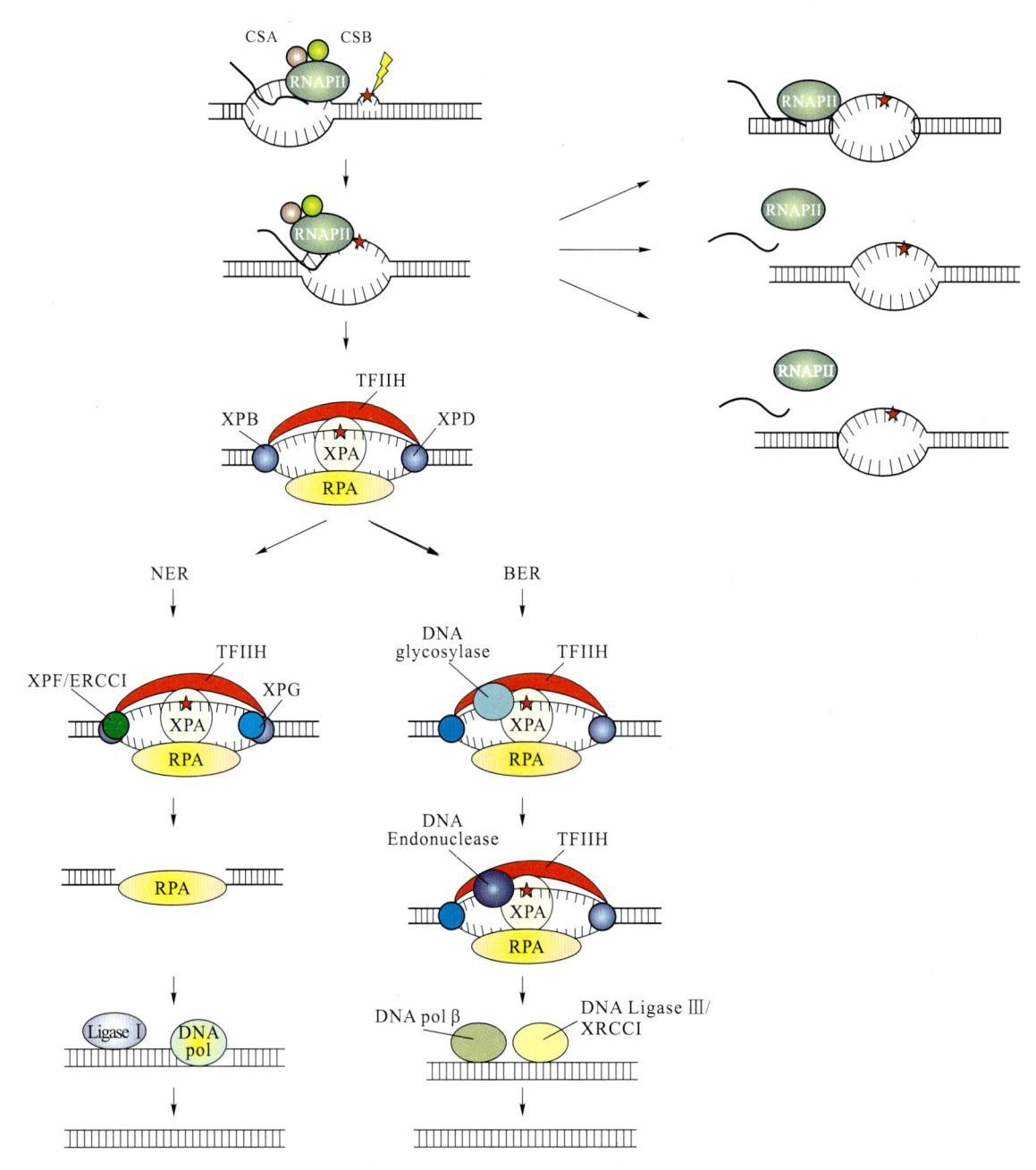

Fig.25.1 Model for transcription-coupled repair. The proposed steps and proteins involved in each step of TCR are shown. Details are described in the text of Part I. Briefly, damage occurs in the transcribed DNA strand and the RNA polymerase stalls. Three possible fates of the ternary complex composed of the DNA template, RNA polymerase, and nascent transcript are illustrated: regression of the polymerase and maintenance of a stable complex (right, top); dissociation of the polymerase and nascent transcript from the DNA template (right, middle); ubiquitin-mediated degradation of the polymerase and dissociation of the nascent transcript (right, bottom). Factors are then recruited and repair ensues by nucleotide excision repair (NER) or base excision repair (BER). Modified from (de Boer and Hoeijmakers, 2000; de Laat *et al.*, 1999; Svejstrup, 2002).

lesion in the DNA of the gene that is being transcribed, acts as a signal for the recruitment of factors that allow (Donahue et al., 1996; Hanawalt, 1994). The model suggests that the RNA polymerase must then be removed from the site of damage in order to allow access of repair factors (Hanawalt, 1994). The lesion is then repaired by the recruited factors. Finally, the RNA polymerase can resume transcription of the repaired template. Evidence that supports each of these proposed steps in the TCR process will be discussed below.

Stalling of RNA Polymerases

A: Active RNA Polymerase is Required for TCR

Several lines of evidence support the first step in the TCR model that a stalled RNA polymerase is necessary to initiate repair. In order for the RNA polymerase to stall, it must first initiate transcription before encountering a lesion in the DNA template. In mammalian, yeast, and E. coli cells, several methods showed that an active RNA polymerase elongation complex is required for the preferential repair of damage in the transcribed DNA strand (Christians and Hanawalt, 1992; Leadon and Lawrence, 1992; Mellon and Hanawalt, 1989; Sweder and Hanawalt, 1992; Tornaletti and Hanawalt, 1999). In E. coli, it was demonstrated that induction of the lac operon was necessary for the preferential repair of cyclobutane pyrimidine dimers (CPDs) in the transcribed strand of the genes within the operon (Mellon and Hanawalt, 1989). Using the yeast model system, it was observed that cells containing a temperature-sensitive mutation in the largest subunit of RNA polymerase II (RNAPII), Rpb1p, experienced loss of TCR when Rpb1p was inactivated by growth at the non-permissive temperature (Leadon and Lawrence, 1992; Sweder and Hanawalt, 1992). CHO cells treated with α-amanitin, an inhibitor of RNAPII elongation (Nguyen et al., 1996), exhibited a loss of strand-specific repair of the DFHR gene (Christians and Hanawalt, 1992). Finally, when human cells defective in GGR were treated with α-amanitin, the preferential repair of CPDs in expressed genes was abolished (Carreau and Hunting, 1992). These observations imply that the RNA polymerase must initiate elongation of a nascent RNA transcript in order to stall at a lesion in the transcribed DNA strand and coordinate repair of that damage.

B: Lesions in the Transcribed Strand of the DNA Template Stall RNA Polymerase II

Once transcription has entered the elongation phase, initiation of the repair process requires the stalling of RNA polymerase, which acts as a signal for recruitment of factors involved in subsequent repair steps (Hanawalt, 1994). Extensive studies have been undertaken to define the types of DNA lesions that cause specific RNA polymerases to stall. Most of these studies have been carried out using mammalian RNAPII, E. coli RNA polymerase, or T7 RNA polymerase (RNAP). However, several studies have investigated the impact of DNA lesions on the progression of RNA polymerases I and III and the subsequent initiation of TCR. We will first focus on the consequences of the stalling of T7 RNAP, E. coli RNAP, and eukaryotic RNAPII and then briefly discuss the outcome of the stalling of RNA polymerases I and III.

Studies that have investigated the ability of DNA lesions to impede the progression of RNA polymerase utilize run-off assays, in which the length of the transcript produced from a single initiation event by an RNA polymerase is monitored. Production of a full-length transcript from template containing a lesion is indicative of a lack of polymerase blockage, while production of a truncated transcript is indicative of polymerase blockage. The run-off experiments can be carried out in vitro with purified proteins or in cells. The stability of the ternary complex composed of the RNA polymerase, lesion-containing DNA template, and nascent RNA transcript is also monitored as evidence that the RNA polymerase in fact stalls at the site of damage.

The impact of DNA lesions on transcription progression shown by these studies has been critical to understand the TCR process. Most importantly, the studies indicated that the lesion, which may be caused by various insults such as exposure to endogenous stress, sunlight, or environmental toxins, must be located on the transcribed strand in order for the RNA polymerase to be blocked. It was shown that transcription elongation by T7 RNAP, E. coli RNAP, mammalian RNAPII, and human RNAPII was blocked, in vitro, to varying degrees by lesions caused by chemicals that induce DNA adducts when the damage was located in the transcribed strand. However, transcription was not blocked by the same lesions in the non-transcribed strand (Chen and Bogenhagen, 1993; Choi et al., 1994; Cline et al., 2004; Corda et al., 1993; Cullinane et al., 1999; Donahue et al., 1996; Donahue et al., 1994; Hatahet et al., 1994; Kalogeraki et al., 2003; Mei Kwei et al., 2004; Nath and Romano, 1991; Perlow et al., 2002; Roth et al., 2001; Schinecker et al., 2003; Shi et al., 1988; Smith et al., 1998; Tornaletti et al., 1997; Tornaletti and Hanawalt, 1999; Tornaletti et al., 2001; Tornaletti et al., 2003; Tornaletti et al., 1999). The blockage of RNA polymerase by such lesions in these *in vitro* experiments is consistent with observations that a blockage of RNA polymerase occurred when templates with damage were introduced into cells. For example, when plasmids containing cisplatin-induced lesions were transfected into human or CHO cells, RNAPII transcription from a reporter gene was inhibited (Mello *et al.*, 1995). Additionally, RNAPII transcription of a reporter gene in human cells was blocked by the presence of a single 8-oxoguanine lesion in the vector (Le Page *et al.*, 2000). When CHO or human cells were transfected with DNA constructs that contained a CPD or a cyclo-deoxyadenosine lesion, transcription of these templates was blocked at the site of the lesion (Brooks *et al.*, 2000). These data support the model that DNA damage in the transcribed strand acts as a block to RNA polymerase transcription in cells.

Not only must RNA polymerase stall for initiation of TCR, but it is also important for the polymerase to remain at the lesion to facilitate the recruitment of other factors necessary to perform TCR. Several groups have provided evidence that the nascent RNA transcript, lesion-containing DNA template, and RNA polymerase exist as a stable ternary complex. The stability of RNAPII complexes stalled at a DNA lesion has been assessed using several different methods, including footprinting assays, immunoprecipitation/PCR, immunoprecipitation/RT-PCR assays, and atomic force microscopy (AFM) imaging of stalled complexes (Mei Kwei *et al.*, 2004; Tornaletti *et al.*, 1999). Footprinting analysis of mammalian and human RNAPII stalled at a CPD showed that the RNA polymerase covered approximately 35 or 40 nucleotides respectively, surrounding the lesion (Tornaletti *et al.*, 1999). Immunoprecipitation/PCR and immunoprecipitation/RT-PCR assays demonstrated that human RNAPII was located stably at the lesion site caused by a CPD or 6-4 photoproduct (6-4 PP) (Mei Kwei *et al.*, 2004). Finally, AFM images of the ternary complex showed that human RNAPII bound templates containing a UV-induced CPD or 6-4 PP in the vicinity of the lesion (Mei Kwei *et al.*, 2004). These studies confirm that the RNA polymerase stalled at a lesion in the transcribed strand is stable and therefore likely competent to initiate and coordinate the subsequent repair process, as proposed in the TCR model.

If TCR is a process that is elicited by RNA polymerase stalling, RNA polymerases other than RNAPII may serve as facilitators of the repair of the template DNA they transcribe. It has been documented that polymerases other than RNAPII do in fact stall in response to DNA damage. For example, *in vitro* run-off experiments confirmed that human RNA polymerase I (RNAPI) stalled when transcribing a template that contained a CPD (Hara *et al.*, 1999). Furthermore, DNaseI and exonuclease footprinting showed that 29-43 nucleotides around a CPD were protected, indicating the presence of a stable ternary complex containing RNAPI (Hara *et al.*, 1999). Despite the stalling and formation of a stable complex at the lesion, preferential repair of the transcribed strand did not occur in ribosomal DNA genes transcribed by RNAPI in CHO or human cells (Christians and Hanawalt, 1993; Christians and Hanawalt, 1994; Conconi *et al.*, 2002; Fritz and Smerdon, 1995; Vos and Wauthier, 1991; Yu *et al.*, 2000), but has been reported to occur in rRNA genes in yeast (Conconi *et al.*, 2002; Verhage *et al.*, 1996). Similarly, preferential repair of the transcribed strand of human tRNA genes did not occur (Dammann and Pfeifer, 1997), implying that RNAPIII stalling did not elicit TCR. These data suggest that the process of TCR is coordinated by specific polymerases and is thereby directed toward the repair of only the genes transcribed by those polymerases.

Removal of the RNA Polymerase from the Site of Damage

Once an RNA polymerase stalls at a DNA lesion, it must eventually be displaced or removed, at least temporarily, from the site of damage so that repair factors can gain access to the damage (Hanawalt and Mellon, 1993). This removal must be regulated to coordinate repair and resumption of transcription. Furthermore, in multicellular organisms, proper coordination of this process is essential because stalling of RNA polymerase may induce apoptosis (Ljungman and Zhang, 1996; McKay *et al.*, 1998). The process of RNA polymerase removal is well characterized in *E. coli*, and data regarding the process of RNA polymerase displacement from the lesion site also exist for eukaryotic organisms.

A: Mechanism of RNA Polymerase Removal

When *E. coli* RNAP stalls due to the presence of a lesion in the transcribed strand of an active gene, the product of the *mfd* gene, Mfd or transcription-repair coupling factor (TRCF), was shown to displace RNA polymerase (Mellon *et al.*, 1986; Mellon *et al.*, 1996) and to promote the release of the nascent RNA transcript from the ternary complex (Mellon *et al.*, 1996; Park *et al.*, 2002). It was observed that Mfd then targeted repair proteins to the site of DNA damage (Mellon *et al.*, 1996; Park *et al.*, 2002), including UvrA, a damage recognition protein (Mellon et al., 1996). This process was shown to occur only when transcription was actively occurring. Mfd therefore plays an important role by connecting recognition of the stalled RNA polymerase to the recruitment of repair proteins to the site of the stalled RNA polymerase and thus the site of DNA damage.

A similar model has been proposed for eukaryotic repair of DNA lesions in the transcribed strand. It has been suggested that the RNA polymerase may be translocated away from the lesion to allow repair to take place and transcription to resume (Hanawalt, 1994; Schinecker et al., 2003; Woudstra et al., 2002). However, for human RNAPII, it is unclear whether the polymerase is actually released or merely translocated away from the damage site without dissociating (Svejstrup, 2002). The characteristics of selected proteins with known roles in the control of RNA polymerase translocation and dissociation are described below.

A1: Transcription Elongation Factor SII (TFIIS)

One mechanism by which RNAPII could be removed from the damage site to allow access of repair factors involves translocation of the polymerase promoted by transcription elongation factor SII (TFIIS). This protein facilitates the ability of RNAPII to read through various impediments encountered during normal transcription, such as natural pause sites (Wind and Reines, 2000). The mechanism by which SII allows bypass of these sites was shown to occur through activation of the cryptic 3′ to 5′ endonuclease function of RNAPII arrested in a ternary complex (Gu et al., 1993; Izban and Luse, 1992; Nudler et al., 1994; Samkurashvili and Luse, 1996; Wang et al., 1995). This resulted in the hydrolysis of the nascent RNA transcript at internal phosphodiester bonds (Izban and Luse, 1992). Subsequently, the 3′ hydroxyl group of the processed RNA transcript was placed correctly in the catalytic site of the enzyme, allowing nucleotide polymerization to resume (Izban and Luse, 1992). In addition to its role in bypassing pause sites, a role for SII in the situation of RNAPII stalled at DNA lesions has been documented. For example, it was shown by nuclear run-off assay that SII-induced shortening of the nascent RNA transcript occurred when SII was added to mammalian RNAPII complexes arrested in a ternary complex at a CPD (Donahue et al., 1994; Tornaletti et al., 1999) or at an intrastrand crosslink caused by cisplatin in the DNA template (Donahue et al., 1994; Tornaletti et al., 2003). During this process, the transcript was not released from the arrested complex (Donahue et al., 1994). This indicated that the ternary complex was stable to allow for the eventual resumption of transcription by the RNA polymerase. The shortening of the nascent RNA template supports the TCR model by providing evidence that the RNA polymerase can be translocated from the site of the DNA damage lesion. Thus, the steric hindrance caused by the RNA polymerase that may prevent the access of repair factors to the damage is removed.

A2: Transcription Release Factor 2

Factor 2 is a negative transcription elongation factor (N-TEF) (Cullinane et al., 1999; Xie and Price, 1996). More specifically, it is an ATP-dependent RNA polymerase II termination factor (Xie and Price, 1996). During the normal process of transcription, the N-TEF2 complex formed after transcription initiation results in the formation of short transcripts due to abortive elongation of the transcripts (Marshall et al., 1996). The factor 2 component of N-TEF2 promotes the premature dissociation of the RNA polymerase from the transcripts (Liu et al., 1998; Xie and Price, 1996). In addition to its role in the production of these short transcripts in the absence of damage, it was shown in *vitro* that human factor 2 (HuF2) promoted the release of both RNA polymerase I and II from the ternary complex in which the polymerase was stalled at a CPD (Hara et al., 1999). Furthermore, footprinting analysis showed that HuF2 promoted the dissociation of the entire ternary complex containing RNAPI stalled at a CPD (Hara et al., 1999). These data suggest that HuF2 may play a role in removal of a stalled RNA polymerase from a site of damage. Since in this case the entire ternary complex is dissociated, transcription initiation must start entirely anew rather than resuming with the nascent RNA transcript that was already produced.

A3: Degradation of RNA Polymerase

One alternative to translocation or dissociation of the RNA polymerase from the DNA lesion site involves the destruction of RNA polymerase. This can be achieved by ubiquitin-mediated proteolysis of the RNA polymerase, which has been shown to occur in response to DNA damage (Beaudenon et al., 1999; Bregman et al., 1996; Luo et al., 2001; Ratner et al., 1998). Additionally, data suggest that RNAPII ubiquitination is induced when the polymerase is inhibited by α-amanitin and when arrested at a cisplatin lesion in the DNA template (Lee et al., 2002). Degradation of the polymerase likely results in the dissociation of the entire ternary complex, resulting in the loss of the nascent RNA transcript. Accordingly, RNAPII degradation is probably a "last resort" if the stalled RNA polymerase cannot be removed from the damage site by another mechanism (Woudstra et al., 2002). However, if degradation of the polymerase allows repair of the template, subsequent RNAPII molecules can continue transcription of the gene.

Recruitment of Factors Necessary for Repair in Eukaryotes

The recruitment of factors necessary for repair and the actual repair event itself comprise the next step in TCR. Studies in both mammalian and yeast cells have been essential to the understanding of the repair process. Although repair of lesions by TCR was initially thought to occur exclusively by nucleotide excision repair (NER), it has been shown that certain lesions are repaired by base excision repair (BER). The general model for recruitment of repair proteins proposes a common mechanism for the recognition of damage and preparation for recruitment of DNA repair proteins, regardless of whether the lesion will eventually be repaired by the NER or BER machinery. This model proposes that once DNA damage is encountered by the RNA polymerase in association with CSA and CSB proteins, the DNA surrounding the lesion is opened by the action of RPA, XPA, and the bi-directional XPB/XPD helicase subunits of the TFIIH complex (de Boer and Hoeijmakers, 2000). Once the DNA is opened, producing two single-stranded regions of DNA, the BER or NER machinery is recruited to repair the damage in the transcribed strand by distinct mechanisms that are discussed below and are illustrated in Fig. 25.1.

A: CSA and CSB Proteins

Two factors required for TCR include the Cockayne syndrome (CS) factors CSA and CSB (Leadon and Cooper, 1993; van Hoffen et al., 1993; Venema et al., 1990a), regardless of whether the repair is coordinated by NER or BER. These proteins were initially identified to have a specific role in the TCR process because CS cell lines that express defective forms of these proteins repaired UV- and IR-induced damage inefficiently by TCR but were still competent to repair damage through GGR (Leadon and Cooper, 1993; van Hoffen et al., 1993; Venema et al., 1990a; Venema et al., 1990b). The yeast homologs of CSA and CSB are Rad28 and Rad26, respectively (Bhatia et al., 1996; van Gool et al., 1994). While *rad26* mutant cells are defective in TCR (van Gool et al., 1994), *rad28* mutants are not (Bhatia et al., 1996). However, mutations in *RAD28* do increase the UV sensitivity of the cells in certain genetic backgrounds (Bhatia et al., 1996), consistent with a role for Rad28p in DNA repair.

Although the roles of CSA and CSB have not been precisely defined, much evidence exists to support their involvement in transcription and transcription-coupled repair. CSA was shown to interact in co-immunoprecipitation experiments with CSB and the p44 subunit of TFIIH, consistent with a role in transcription and/or repair (Henning et al., 1995). It was also demonstrated that CSA has ATPase activity and contains WD-repeats (Henning et al., 1995), which are implicated in the facilitation of protein-protein interactions (Donahue et al., 1994). CSB and its yeast homolog *RAD26* are SWI2/SNF2 protein family members and were shown to be DNA-dependent ATPases and to contain helicase motifs (Eisen et al., 1995; Iyer et al., 1996; Selby and Sancar, 1997a; Troelstra et al., 1992). Recombinant CSB protein was observed to have chromatin remodeling activity (Citterio et al., 2000), which has been proposed to alter the interface between RNAPII and DNA, thus allowing the recruitment of repair proteins (Svejstrup, 2003). Consistent with a role in transcription and TCR, CSB was shown to co-fractionate with RNAPII (van Gool et al., 1997) and to stimulate transcript elongation by RNAPII (Selby and Sancar, 1997a; Selby and Sancar, 1997b). A key finding that further implicated CSB in TCR was the identification of its interaction with XPA and XPG. XPA is thought to be involved in the initial recognition of DNA damage and organization of repair factors around the lesion (de Boer and Hoeijmakers, 2000), and XPG, a structure-specific endonuclease, is involved in a later step of NER (Iyer et al., 1996; Troelstra et al., 1992). Finally, CSB has been implicated in the recruitment of TFIIH for TCR from its role in basal transcription initiation (Tantin, 1998; Tijsterman et al., 1997; Tu et al., 1997; You et al., 1998). Collectively, these data provide support that the CSA and CSB proteins play an important role in transcription-coupled repair.

B: Other Factors

Another protein required for TCR in all cases is TFIIH, which is a component of the basic RNAPII transcription machinery (Frit et al., 1999; Svejstrup et al., 1996) and is required for all repair that occurs through the NER pathway (Drapkin et al., 1994; Feaver et al., 1993; Schaeffer et al., 1994; Schaeffer et al., 1993; Wang et al., 1994). TFIIH has been shown to be required for TCR of oxidative lesions by BER as well (Le Page et al., 2000). TFIIH is composed of 9 subunits, including two helicases, a protein kinase, and a RING finger motif (Frit et al., 1999). The subunits XPB and XPD are the helicases (Ma et al., 1994; Schaeffer et al., 1994; Schaeffer et al., 1993; Sung et al., 1993), and this helicase activity is required for the unwinding of the DNA helix around the lesion in NER (Evans et al., 1997; Frit et al., 1999; Svejstrup et al., 1996).

The next steps in TCR depend on the type of lesion, which determines whether the damage is repaired by NER or by BER. Studies showed that bulky lesions that distort the DNA helix were generally repaired by NER (Buschta-Hedayat et al., 1999; Wood, 1999), while oxidized and alkylated bases were generally repaired by BER (Memisoglu and Samson, 2000; Plosky et al., 2002). It was observed that NER in the process of GGR began when the lesion was recognized by binding of XPC/HHR23B complex (Rad4/Rad23 in yeast) (Batty et al., 2000; Sugasawa et al., 1998; Volker et al., 2001). XPC, and hence this initial step, was not required for TCR because of the presumption that RNA polymerase acts as the lesion sensor in TCR. Consistent with this, it was shown that if DNA was already unwound, XPC/hHR23B was not required to initiate TCR (Mu and Sancar, 1997). Subsequently, the formation of an open complex ensued, which required unwinding of the DNA helix (de Boer and Hoeijmakers, 2000). This unwinding of the DNA in the vicinity of the lesion requires the basal transcription/repair factor TFIIH (Evans et al., 1997; Frit et al., 1999; Svejstrup et al., 1996). Both ATP-dependent helicase components of TFIIH, namely XPB (yeast Rad25) and XPD (yeast Rad3), were necessary for nucleotide excision repair (Evans et al., 1997; Frit et al., 1999; Svejstrup et al., 1996).

The remainder of the process of nucleotide excision repair is indistinguishable between GGR and TCR once the open complex is formed. Other factors that were shown to be necessary for the subsequent steps of NER included the damage-binding factor XPA (yeast Rad14) (Evans et al., 1997), the XPA binding protein XAB2, which interacted with XPA, CSA, CSB, and RNAPII (164), and the single-strand DNA-binding protein RPA (de Laat et al., 1998; Kim et al., 1992). This allowed the recruitment of the structure-specific DNA endonucleases XPG (yeast Rad2) and XPF/ERCC1 (yeast Rad1/Rad10) (O'Donovan et al., 1994; Sijbers et al., 1996). These enzymes catalyzed the formation of single-strand DNA breaks at the 3' and 5' sides of the open complex, respectively (O'Donovan et al., 1994; Sijbers et al., 1996). The 24-32 nucleotide oligonucleotide that contained the DNA lesion was subsequently removed (Mu and Sancar, 1997). Finally, DNA polymerases delta and epsilon were shown to fill in the gap and nicks were sealed with DNA ligase I (Barnes et al., 1992; Shivji et al., 1995).

TCR-NER has been shown to have a connection to the mismatch repair (MMR) proteins MLH1 and MSH2 (Mellon et al., 1996). Mutation of the E. coli MLH1 and MSH2 homologs mutS and mutL caused a defect in TCR of CPDs in the lac operon (Mellon and Champe, 1996). This connection was also shown in human cells with mutations in the MLH1, MSH2, or PMS2 genes. Such cells failed to exhibit preferential repair of UV-induced DNA lesions in the active strand of a transcribed gene (Mellon et al., 1996).

The field of research investigating TCR expanded when it was learned that lesions located in the transcribed strand of an active gene were repaired by BER. Such lesions normally repaired by BER include spontaneously formed, oxidized, alkylated, or mismatched bases (Lindahl et al., 1997; Memisoglu and Samson, 2000). The first evidence coupling of BER to transcription was gathered from yeast studies, in which thymine glycols that are normally repaired by BER were shown to be repaired by TCR (Leadon et al., 1995). Subsequently, it was shown that thymine glycols were substrates for TCR in human cells (Cooper et al., 1997; Le Page et al., 2000). Further links of the repair of oxidative damage to the process of TCR included the observation that mutations impairing the function of CSA or CSB resulted in the inability to repair lesions caused by thymine glycol (Cooper et al., 1997; Le Page et al., 2000), 8-oxoguanine (Le Page et al., 2000), and IR (Leadon and Cooper, 1993). Additionally, TCR of thymine glycols occurred in XP-A, XPF, and XPG patients, who had NER defects but were still competent for TCR by BER (Cooper et al., 1997).

As mentioned previously, TCR coordinated through BER initiates through the same steps as that coordinated through NER; however, rather than NER repair proteins being recruited, BER proteins are recruited. The model for BER proposes that the process initiates when the damaged base is removed by a specific DNA glycosylase to form an abasic site (Krokan et al., 1997; Scharer and Jiricny, 2001). This abasic site is recognized by an apurinic/apyrimidinic endonuclease, resulting in the cleavage of the phosphodiester bond 5' to the lesion (Boiteux and Guillet, 2004). Consequently, a strand break is formed that has a 3'-hydroxyl group and a 5'-abasic terminus (Boiteux and Guillet, 2004). Repair ensues as DNA polymerase β replaces the abasic residue with the appropriate nucleotide, which requires both endonuclease and polymerization activity of the polymerase (Matsumoto and Kim, 1995). Alternatively, Fen1 may excise the 5′-abasic terminus, producing a gap that is then filled by a DNA polymerase (Boiteux and Guillet, 2004). Finally, ligase I or XRCC1/DNA ligase III ligates nicks and thus completes the repair (Kubota et al., 1996; Nicholl et al., 1997; Srivastava et al., 1998).

Resumption of Transcription after Repair

After repair of the DNA lesion is accomplished, transcription of the previously damaged template DNA can resume. Evidence from nuclear run-off assays suggested that the RNA polymerase could resume transcription after it arrested at a DNA lesion and then regressed due to the presence of SII. In these assays, RNA polymerase was able to transcribe the recessed nascent transcript to the site of the damage, which was still present because no repair factors were present in the *in vitro* assay (Tornaletti et al., 2003). Furthermore, RNAPII catalyzed elongation of the nascent transcript past the site of damage after the lesion was repaired. To show this, RNAPII complexes were arrested at a CPD and treated with photolyase and light after transcript cleavage with SII. Transcription was then allowed to proceed. Some transcripts were elongated past the CPD site (Tornaletti et al., 2003). This suggested that repair of the lesion had allowed the RNA polymerase to progress past that site. These data support the model that RNA polymerase can resume transcription after DNA repair has taken place (Tornaletti et al., 2003).

Conclusion

The evidence gathered thus far regarding the process of TCR has allowed the development of a testable model. Because many of the factors involved in TCR have now been identified, a more detailed understanding of the molecular mechanism by which these proteins facilitate the detection and repair of distinct lesions can be acquired. Gaining a more precise understanding of the mechanism of TCR will lead to the identification of targets and therapeutic strategies for Cockayne syndrome patients.

Part II: Transcription and Replication Collisions

It has been long recognized that DNA replication and transcription are coordinated. For example, histone gene expression is tightly coupled with DNA synthesis under normal growth conditions or when there is DNA damage (Zhao, 2004). Coordination of transcription and replication also serves to control collisions between replication and transcription machineries. In cases where such encounters are unavoidable, cells have developed mechanisms either to halt replication fork progression or to allow replication forks to bypass the RNA polymerase complex. In this section, we will discuss the collisions of transcription and replication occurring in bacteria and the mechanisms to avoid them in yeast and higher organisms.

Head-on and Co-directional Collisions

In bacteria, DNA replication and transcription take place continuously and simultaneously through the life-cycle. Head-on collisions occur when the DNA replication machinery and the RNA polymerase progress in opposite directions. In this case, the front edge of RNA polymerase meets the DNA helicase DnaB. Also, since DNA polymerase moves along the DNA template at least 10 times faster than RNA polymerase, collisions can happen when both move in the same direction. In the co-directional case, the front edge of DNA polymerase of the leading strand collides with the rear edge of RNA polymerase (Mirkin and Mirkin, 2005).

How collisions are resolved is not completely clear. Studies *in vitro* using bacteriophage replication apparatus and *E. coli* RNA polymerase showed that in both types of collisions, replication forks could pass the stalled transcription complex. However, the fork experienced a longer pause in head-on collisions. Furthermore, it was found that the RNA polymerase switched its template to the newly synthesized DNA strand after being passed by the replication complex in a head-on collision, while it remained bound to its original DNA template in the co-directional collision case (Liu and Alberts, 1995; Liu et al., 1993). Different results were obtained from *in vitro* studies with bacteriophage *B. subtilis* phi29. In these studies, it was observed that the replication forks were blocked by the stalled transcription complex oriented either co-directionally or against the direction of replication. When the stalled transcription was allowed to proceed, however, the DNA polymerase was able to pass the transcription complex oriented in the opposite direction and resume its normal speed. In the co-directional case replication also resumed but to a much lower rate probably due to lack of a mechanism that allows DNA polymerase to bypass RNA polymerase when both are traveling in the same direction (Elias-Arnanz and Salas, 1997; Elias-Arnanz and Salas, 1999).

In vivo evidence supports the idea that head-on collisions are more difficult to resolve than co-directional ones. When an inducible replication origin was inserted either upstream or downstream of a ribosomal RNA operon in *E. coli* to allow replication to proceed either co-directionally or against transcription, it was observed that the fork movement was not affected when transcription and replication were co-directional, but was slowed down when transcription and replication occurred in opposite directions (French, 1992). A more recent study with *E. coli* plasmids also showed that co-directional transcription had no effect on replication progression while head-on transcription severely compromised replication fork movement (Mirkin and Mirkin, 2005).

The slower replication rate caused by head-on collision with the transcription machinery could be due to DNA topological constraints or knotting produced by highly supercoiled DNA generated by the inward movements of replication and transcription machineries. Indeed, there was an accumulation of positively supercoiled DNA detected between the replisome and the RNA polymerase complex in artificial plasmids derived from pBR322 (Olavarrieta *et al.*, 2002). However, there is evidence demonstrating that the slow movements of replication forks in head-on collisions are more likely due to direct physical contacts between the replication and transcription machineries, rather than to excessive topological stress (Mirkin and Mirkin, 2005). Nevertheless, the slowed replication caused by transcription-replication head-on collision imposes a disadvantage upon the organism given the rapid growth of bacteria. In fact, it has been suggested that the organization of bacterial genomes reflects the natural selection against head-on collisions, especially for genes that are frequently transcribed (Brewer, 1988; French, 1992; Mirkin and Mirkin, 2005). In *E. coli*, all seven ribosomal operons are transcribed in the same direction of replication, so are 62% of tRNA genes and 55% of protein-coding genes (Blattner *et al.*, 1997). More importantly, 70% of essential genes in *E. coli* and 90% in *B. subtilis* are transcribed co-directionally with replication (Rocha and Danchin, 2003).

Replication Fork Pause

In contrast to bacteria, transcribed genes in the eukaryotic genomes are much less densely distributed. Also, DNA synthesis in eukaryotic cells takes place only during a portion of the life-cycle. It therefore raises a question whether transcription-replication collisions actually happen in eukaryotic cells. However, recent studies in *Drosophila* using genomic microarray showed that the timing of replication correlates with active transcription over chromosomal domains (MacAlpine *et al.*, 2004; Schubeler *et al.*, 2002). This indicates the possibility that the replication machinery may collide with transcription.

The discovery of a replication fork pause (RFP) site at tRNA genes in the budding yeast (*S. cerevisiae*) provided evidence for replication-transcription collisions in eukaryotic cells (Deshpande and Newlon, 1996). In the *S. cerevisiae* genome, an active replication origin, *ARS307*, is located downstream of a tRNA gene, *SUP53*. It was observed that replication forks moving upward from *ARS307*, against the direction of the transcription of *SUP53*, stalled at *SUP53*. Further analysis using plasmids containing *SUP53* and its flanking sequence and an *ARS* revealed that the pause of replication forks depended on the promoter of *SUP53* and RNA polymerase III, indicating transcription initiation and active transcription are required. Interestingly, the RFP site is polar—stalling of replication forks did not happen when *SUP53* was placed on the plasmid co-directionally with replication, suggesting that, similar to bacteria, only head-on collisions may result in RFP.

The stalling of replication forks at the RFP site is transient, its resolution may depend on Rrm3, a DNA helicase that facilitates replication through chromosomal sites with large non-nucleosomal protein-DNA complexes. In cells that lack of Rrm3, much stronger pauses of replication forks at tRNA genes were detected than in wild-type cells when the transcription and replication proceeded in opposite directions. In addition, the absence of Rrm3 resulted in fork pauses even when the transcription and replication of tRNA genes proceeded co-directionally (Ivessa *et al.*, 2003). It is not clear, though, how Rrm3 helps resolve RFPs caused by replication-transcription collisions, hence allowing both processes to take place simultaneously.

Replication Fork Barrier

In addition to factors such as Rrm3 that may help resolve transcription-replication collisions, cells appear to have developed a strategy, replication fork barrier or RFB, to prevent the occurrence of such events. RFB sites are found in the ribosomal RNA gene repeats. They block the movement of replication forks against rRNA transcription but allow co-directional replication with transcription of rRNA genes (Brewer and Fangman, 1988; Kobayashi *et al.*, 1992).

RFB sites have been identified in diverse eukaryotes from yeasts to mammals and plants. Among them, the RFB in *S. cerevisiae* rDNA is the most well-characterized. The *S. cerevisiae* genome carries about 150 copies of tandemly repeated rDNA units, with each unit encoding a 35S and a 5S rRNA gene in addition to two nontranscribed spacers (NTS) (Fig. 25.2). The RFB site contains about 100 bp of specific nucleotide sequence located in NTS1, at the 3′ end of both rRNA genes. The replication fork originating from a replication origin (ARS) in NTS2 moves through 5S rRNA gene co-directionally with its transcription and then stalls at the RFB site, thereby avoiding head-on collisions with transcription of the downstream 35S rDNA. Interestingly, the fork-blocking activity of RFBs does not require active transcription (Brewer et al., 1992), but does require the RFB sequence and a *trans*-acting factor, Fob1 that binds to the RFB site (Kobayashi, 2003; Kobayashi et al., 1998).

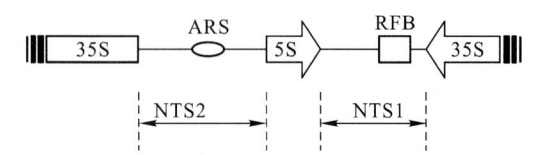

Fig.25.2 Organization of rRNA genes (5S and 35S), nontranscribed spacers (NTS), replication origin or autonomous replication sequence (ARS) and replication fork barrier (RFB) site of rDNA repeats in *S. cerevisiae*. The arrows of the boxes for rRNA genes indicate the directions of transcription.

Fob1 represents a class of replication/transcription termination proteins called contrahelicases that also include Tus in *E. coli* (Mulugu et al., 2001) and TTF-1 in mammals (Evers and Grummt, 1995; Putter and Grummt, 2002). TTF-1 was originally identified for termination of transcription by RNA polymerase I. Recently it has been shown TTF-1 can also block replication (Gerber et al., 1997). In mice, TTF-1 binds to the Sal box found downstream of the 3′ end of the rRNA coding region, arresting replication forks moving in the direction opposite to transcription (Lopez-estrano et al., 1998). Similar results were obtained using an SV40-based cell free replication system which showed that binding of TTF-1 to Sal box 2 is required for replication fork arrest and, like yeast, RFB activity occurs independent of transcription (Gerber et al., 1997; Putter and Grummt, 2002). Taken together, the existence of RFB sites and contrahelicases across species demonstrate the importance of protecting actively transcribed genes from disruption by replication.

Biological Consequences of Replication-transcription Collisions

It appears that whereas co-directional transcription and replication is permitted by cells from bacteria to mammals, the two processes occurring head to head should be prevented or modulated. The latter results in head-on collisions between the replication apparatus and the RNA polymerase complex. Such collisions, if not timely resolved, slow down replication and/or cause abortion of transcription. Also, when a collision takes place, it provides a hot spot for recombination. In yeast, collisions were detected when cells with reduced copy number of rDNA lacked Fob1. Notably, these cells had increased production of extrachromosomal rDNA circles (ERCs) and variation of rDNA copy numbers, both indicating chromosomal re-arrangement or recombination (Takeuchi et al., 2003). A more recent study using yeast plasmids showed that transcription by RNA polymerase II heading against the direction of replication induced a significant increase in recombination while transcription co-directional with replication had little effect on recombination (Prado and Aguilera, 2005). These results further demonstrate that replication-transcription collisions not only affect the genes being transcribed but also the stability of genome as a whole.

Part III: DNA Damage Modulates Transcription: (Ying and Yang)

The genetic material of all organisms is constantly challenged by DNA damaging agents from endogenous sources, such as the byproducts of respiration and from exogenous sources, such as UV radiation and environmental toxicants. In addition, during every S phase a number of the replication forks stall or collapse invoking mechanisms under study today to reactivate the fork or fix the damage resulting from the collapsed fork. Accurate transmission of chromosomes to each daughter cell requires that cells do not begin anaphase until all chromosomes have been completely replicated and correctly aligned on the spindle. Similarly, cells that have incurred DNA damage in G1 delay DNA replication until the damage has been repaired. Cells that have incurred damage in S phase slow down DNA replication and repair damage before they progress through mitosis (Clarke and Gimenez-Abian, 2000; Hartwell and Weinert, 1989; Lowndes and Murguia, 2000). In addition to cell cycle delay the response to DNA damage also has a key role in the regulation of transcription, which involves both down-regulation of large numbers of transcripts by control of large transcriptional networks as well as up-regulation of key genes that are important for both cell cycle arrest, DNA metabolism and DNA repair. In metazoans the latter category also includes genes that regulate cell death or apoptosis. Genetic analyses in yeast have shown that checkpoint pathways regulate cell cycle transitions in response to perturbations and also mount a transcriptional response. One of the key players in this response in metazoans, the tumor suppressor p53, has been covered in a separate chapter, so here we will focus on the lessons learned from the studies in yeast.

The S-phase and DNA Damage Checkpoints. The Genetic Screens that Uncovered the Signal Transduction Pathways that up Regulate Transcription Following Gentoxic Stress

A: The Rad Screen

By the late 1960s, many groups had carried out screens to identify yeast mutants that were sensitive to ionizing radiation, ultraviolet radiation, or both, these mutants were named *rad* mutants. Brian Cox, John Game and Robert Mortimer, among others, identified a collection of radiation sensitive mutants (Cox and Parry, 1968). In 1970 scientists came to an agreement that mutations that conferred sensitivity to ionizing radiation would continue to be named *rad* mutants, numbered from *RAD50* upward, and they would be loci separate from those identified in screens for UV sensitive mutants.

There are at least three explanations as to why mutations would lead to a radiation-sensitive phenotype. The mutated gene could encode a protein that is required in order for the cell to 1) repair DNA damage; 2) stop the cell cycle before a critical transition (i.e. mitosis) in order to allow DNA repair or 3) regulate the increased expression of proteins involved in DNA repair or cell cycle arrest. Mutation in the second class of genes usually confers sensitivity to both types of radiation, although this cannot be used as the only criterion for assigning them to the cell cycle arrest category.

B: The Mec Screen

Although many groups embarked on studies of the *rad* mutants that affected repair of the UV- and IR-induced lesions, Lee Hartwell and Ted Weinert analyzed the *rad* mutants for their ability to stop the cell cycle in G2 in response to DNA damage and called the surveillance mechanism responsible for monitoring the successful completion of cell cycle events checkpoints. Investigators had long noted that yeast cells delayed cell cycle progression in response to DNA damage (refs in (Weinert *et al.*, 1994)), but the genes required for the G2/M arrest had not been uncovered. Ted Weinert, while working with Lee Hartwell, characterized the *rad* mutations that were required for cell cycle arrest following a DNA damage signal. He also carried out a screen that would identify mutations in additional genes involved in this checkpoint response and named them *mec* mutants (mitotic entry checkpoint). Using *cdc* mutants that accumulate damage lesions in G2 at the restrictive temperature, Weinert showed that a *rad9* mutation (originally identified as a UV-sensitive mutant (Cox and Parry, 1968)) and subsequently mutations in *RAD17* and *RAD24* allowed cells to go through mitosis in the presence of DNA damage. Weinert then used a genetic screen with the damage-inducing *cdc* mutants and identified the *mec1*, *mec2* and *mec3* mutants as defective for the G2/M checkpoint (Weinert, 1992).

C: The Sad, Dun and Crt Screens

Meanwhile Stephen Elledge's group carried out a screen to identify mutants that were defective for the S-phase checkpoint-induced arrest (Sad= S phase arrest defective) that would lead to catastrophic mitosis of unreplicated chromosomes. That screen identified *SAD1* and *SAD3* as essential components for this response (Allen *et al.*, 1994). The Elledge lab also carried out a screen for mutants that failed to turn on the transcriptional response as measured by the upregulation of ribonucleotide reductase 3 mRNA (*RNR3*). These mutants were called dun (DNA damage uninducible) and led to the identification of *DUN1* and *DUN2* as genes required for the transcriptional response following replication blocks and DNA damage (Navas *et al.*, 1995; Zhou and Elledge, 1993). *RNR3* transcription is induced by DNA damage and replication blocks, therefore, Zheng Zhou and Stephen Elledge also carried out a screen for mutants that had constitutive *RNR* expression (Constitutive *RNR* transcription, CRT mutants) in order to identify the factors that regulated *RNR* transcription.

At around the same time, Kato and Ogawa characterized a mutant named *esr1*, which showed sensitivity to the alkylating agent methyl methanesulfonate (MMS) and to UV radiation (Kato and Ogawa, 1994). In addition, the *esr1* mutants displayed meiotic defects. David Stern's group took a biochemical approach and isolated Spk1p in a screen that used a bacterial system to identify dual specificity kinases (kinases that can phosphorylate proteins on serines/threonine and tyrosine residues) from yeast (Zheng *et al.*, 1993).

When all of the genes mutated in these strains were cloned it turned out that many alleles of the same genes had been identified in the different screens and this pointed not only to their important role in various responses but also provided hints as to their biochemical functions. *RAD53* encoded an essential kinase that is required for the cell to arrest not only in response to DNA damage but also when DNA replication is slowed down or blocked (Allen *et al.*, 1994; Weinert *et al.*, 1994; Zheng *et al.*, 1993). *RAD53* alleles were identified not only in the rad screens but also in the sad (*sad1*) and mec (*mec2*) screens, and Rad53p was the kinase identified in David Stern's screen for dual specificity kinases (Spk1p). Another essential kinase that regulates this response is Mec1p, and it was identified in the mec screen, sad screen (sad3) and in Ogawa's screen as the gene mutated in *esr1*.

Dun1p turned out to be another kinase (Zhou and Elledge, 1993) that shared phospho-peptide recognition domains (FHA forkhead-associated domains (Hofmann and Bucher, 1995)) with Rad53p outside of the kinase domain. Dun2p is the catalytic subunit of DNA polymerase epsilon (Navas *et al.*, 1995). The fact that a component of DNA polymerase showed defects in a checkpoint response sparked enthusiasm and speculation that the DNA polymerase complexes, by the nature of their function, made attractive candidates for sensors and scanning machines (Navas *et al.*, 1996). Three kinases had been identified that are required in order to signal DNA damage and replication blocks. Many groups began to organize the *rad*, *mec*, and *sad* mutants (and other mutants that displayed sensitivity to agents that damage DNA or cause replication blocks) into the DNA Repair category or as components of signal transduction pathways that signal the presence of these lesions. As we mentioned in the first section of this chapter, two such proteins, Rad26p and Rad28p, have roles in TCR.

There were several criteria and several readouts that were used by many laboratories in order to piece together the checkpoint signal transduction pathways. The readouts included the effect of mutations on 1) the cell's ability to delay cell cycle progression when encountering damage at different stages of the cell cycle; 2) the activation of the kinases by phosphorylation, and 3) the up-regulation of *RNR3* mRNA and other damage-inducible gene transcripts. Using these readouts, proteins required for the checkpoint response with similarity to proteins involved in DNA replication (DNA polymerase or polymerase associated proteins) or lesion processing (proteins with homology to nucleases or other DNA associated complexes) were considered to act as sensors that would trigger the response. Proteins that acted by regulating phosphorylation status of targets, such as protein kinases, were assigned the role of transducers due to their biochemical function. The proteins phosphorylated by the kinases (and presumably also regulated by phosphatases) were considered as effectors. These turned out to be crude assignments since most signal transduction pathways involve feedback loops that cause the kinases to phosphorylate upstream components. Nevertheless, at the time, it was a good plan of attack.

One role for the checkpoint pathways is the transcriptional induction of genes encoding products involved in DNA metabolism and DNA repair. The ribonucleotide reductase (*RNR*) genes are transcriptional targets of the S-phase and DNA damage checkpoints that have been studied extensively in yeast and are conserved in mammals. The Rnr proteins catalyze the rate-limiting step of DNA synthesis, which is the generation of the deoxyribonucleotide pools. The S-phase and DNA damage checkpoint mediated by Mec1p, Rad53p and Dun1p controls the transcriptional upregulation of the *RNR* genes in yeast (Huang *et al.*, 1998; Zhao *et al.*, 1998; Zhou and Elledge, 1993). Two of the effectors for the checkpoint-induced upregulation of Rnr activity are the transcriptional repressor Crt1p and the Rnr regulatory subunit Sml1p (Zhao *et al.*, 1998; Zhou and Elledge, 1992). In response to DNA damage or stalled replication forks, Crt1p is hyperphosphorylated in a Dun1p-dependent manner and released from the DNA allowing transcription of the *RNR* genes (Huang *et al.*, 1998). In addition, phosphorylation of Sml1p in a Dun1p-dependent fashion allows the activation of ribonucleotide reductase, presumably by mediating the degradation of Sml1p (Zhao and Rothstein, 2002).

The genes encoding the RNRps are regulated by upstream repressor sequences and damage-responsive elements. These sequences are bound by the damage-responsive transcription factor Crt1p, which recruits the co-repressor proteins Tup1p-Ssn6p to the promoter of the *RNR* genes (Huang *et al.*, 1998). Following checkpoint activation by DNA damage or replication blocks, Dun1p-dependent phosphorylation of Crt1p results in reduced binding of Crt1p to the *RNR* promoters assayed by chromatin immunoprecipitation (Chip) (Huang *et al.*, 1998). The *CRT1* promoter also has a binding site for Crt1p, therefore *CRT1* expression is also increased after DNA damage. This allows re-establishment of repression once the DNA damage has been repaired or the replication blocks have been resolved (Fig.25.3) (Huang *et al.*, 1998).

Work from Joseph Reese's laboratory mapped out the events that occurred at the chromatin level to elucidate the molecular mechanism by which Crt1 along with the transcriptional machinery regulated de-repression of the *RNR3* gene in budding yeast (Li and Reese, 2000; Li and Reese, 2001; Sharma *et al.*, 2003; Zhang and

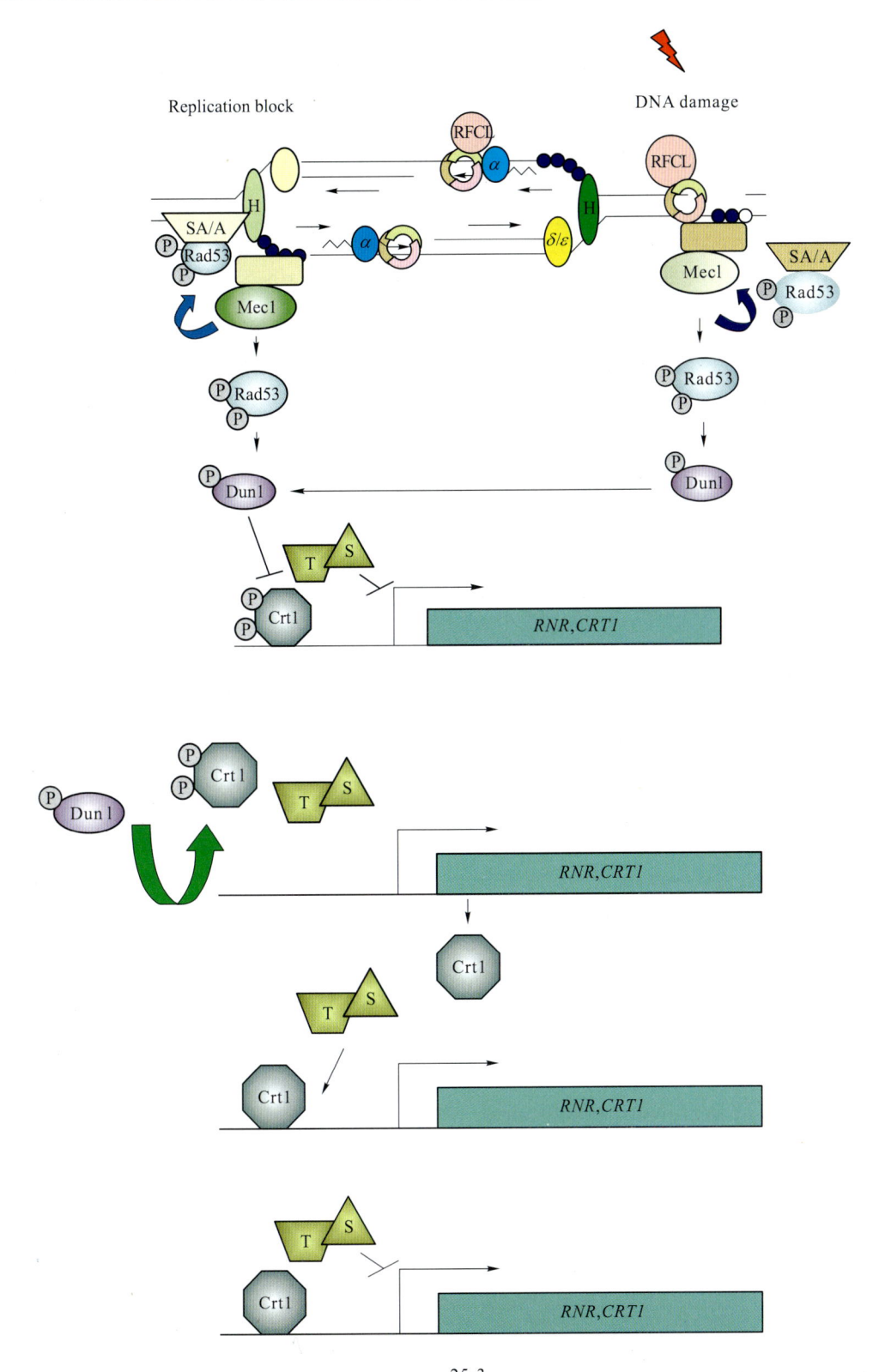

Fig.25.3 Regulation of DNA damage-inducible genes by the 25.3 checkpoint pathway involves phosphorylation of the transcriptional repressor Crt1, which has been proposed to cause loss of interaction with the promoter element. BOTTOM: Crt1 is thought to repress transcription of these genes by recruiting the co-repressor complex containing Tup1 (T) and Ssn6 (S) to the promoter of DNA damage-inducible genes.

Reese, 2004a; Zhang and Reese, 2004b). Reese and co-workers showed that Tup1p-Ssn6p recruitment to chromatin established a nucleosomal array at the *RNR3* promoter, which was disrupted when *RNR3* was de-repressed by a checkpoint signal (Li and Reese, 2001). Furthermore, Reese's lab used the *RNR3* model to examine the relative contributions of general transcription factors, RNA polymerase II (RNAPII), and acetylation to the nucleosome remodeling and recruitment of the SWI/SNF complex. Their studies suggested that remodeling of the *RNR3* promoter *in vivo* in response to DNA damage required both the general transcription factors TAF1, TAF12 (TFIID) and the large subunit of RNAPII. However, acetylation of histone H3 did not require TFIID or RNAPII (Sharma *et al.*, 2003). These studies underscore the power of the yeast genetic model system for dissecting the events that regulate gene expression.

Global Downregulation of Transcription of Ribosomal Genes Following Different Forms of Stress including DNA Damage

Another level of transcriptional regulation following genotoxic stress is the downregulation of genes involved in ribosome synthesis and biogenesis. By the year 2001, not only had the yeast genome been sequenced, but there was a collection of deletion mutants in every non-essential yeast gene which has been used in screens to identify genes involved in several physiological processes, including DNA repair. Michael Resnick's group, who had identified some of the original *RAD* genes in standard genetic screens, used this collection of deletion mutants to identify additional genes involved in the response to ionizing radiation (Bennett *et al.*, 2001). This screen uncovered old favorites and new mutants that had varying degrees of sensitivity to ionizing radiation and to other agents that damage DNA or block DNA replication. The sensitive strains had mutations in genes that encode proteins involved in DNA repair, cell cycle arrest, transcription, chromatin remodeling, nuclear architecture and endocytosis. In these studies the budding yeast system proved once again to be a powerful genetic tool for the identification of proteins that function in pathways that regulate cell cycle progression and genomic stability by regulating gene expression. One such mutant identified in the genome-wide screen was in the gene that encodes for the transcription factor Sfp1p (Bennett *et al.*, 2001; Xu and Norris, 1998). Curiously *SFP1* had been identified by Mike Tyer's group as a whi mutant that was defective in regulating cell size and further showed to be involved in the regulation of ribosomal gene expression (Jorgensen *et al.*, 2002). Recently, the Tyers, O'Shea, Hall, Warner, Shore and Struhl among other groups carried out beautiful studies to address the mechanism by which so many signals including nutrient availability and genotoxic stress converged on Sfp1p-regulated genes. The best understood role for Sfp1p is coordination of cell size with nutrient availability by controls of translation and ribosome biogenesis. Three pathways that respond to nutrients, TOR, AKT and PKA, converged on Sfp1p and the affiliated transcription factors Fhl1p and Ifh1p by regulating their protein-protein interactions and cellular distribution (Jorgensen *et al.*, 2004; Marion *et al.*, 2004; Martin *et al.*, 2004; Schawalder *et al.*, 2004; Wade *et al.*, 2004). While nutrient conditions are favorable Sfp1p is bound to chromatin directing expression of Ribosomal protein genes (RP); however, when the nutrients are removed or unavailable, Sfp1p no longer associates with chromatin and is excluded from the nucleus. Sfp1p in turn regulates the nuclear localization of Fhl1p and Ifh1p, both of which bind to RP gene promoters (Jorgensen *et al.*, 2004; Marion *et al.*, 2004; Martin *et al.*, 2004; Schawalder *et al.*, 2004; Wade *et al.*, 2004). An interesting find from these studies which could explain why *SFP1* was identified in the Resnick rad screen was that Sfp1p nuclear localization was also regulated by other stress signals including oxidative stress and treatment with the alkylating agent MMS, both of which cause DNA damage (Jorgensen *et al.*, 2004; Marion *et al.*, 2004). The localization of Sfp1p in response to DNA damage did not require the known pathways that regulated its localization in response to nutrients.

Important questions that remain to be addressed are the implications of regulating RP gene transcription and/or cell size in response to genotoxic stress. It could be that transient down-regulation of transcription would prevent activation of TCR by preventing the formation of blocked RNA polymerase complexes (see section I). Furthermore, future studies will determine whether the checkpoint pathways also regulate the localization of Sfp1p (and other related factors) in order to downregulate expression of RP genes (Wade *et al.*, 2004).

Conclusion

In this chapter we have highlighted several mechanisms that impinge on the transcriptional machinery in order to maintain genomic integrity. First, we discussed mechanisms that the cell utilizes to repair damaged templates that are being actively transcribed;

second, we discussed the coordination of DNA replication and transcription to control collisions between the DNA and RNA polymerase complexes; and third, we discussed how the transcriptional machinery itself is regulated to modulate expression of genes that coordinate the response to DNA damage and similar stressors. Finally, we would like to end by indicating that these studies underscore the critical role that model organisms such as yeast have played and continue to play in our understanding of gene expression and homeostasis.

Acknowledgement

We are grateful to members of the Sanchez lab for helpful comments and discussions. Work in our laboratory is funded in part by NIH/NCI RO1 CA84463, the Department of Defense DAMD 17-01-1-020, and the Pew Scholars Program in the Biomedical Sciences to YS, and by grants P30 ES06096 and U01 ES11038 Comparative Mouse Genomics Centers Consortium from the National Institute of Environmental Health Sciences, NIH. J.P. is a recipient of the University Distinguished Graduate Fellowship.

Key abbreviations used in this chapter

6-4 PP	6-4 photoproduct
BER	base excision repair
cdc	cell division cycle
CHO	Chinese hamster ovary
CPD	cyclobutane pyrimidine dimer
CRT	constitutive RNR transcription
CS	Cockayne syndrome
DHFR	dihydrofolate reductase
ERC	extrachromosomal rDNA circle
FHA	forkhead-associated
G2/M	G2/mitosis
GGR	global genomic repair
HuF2	human factor 2
MEC	mitotic entry checkpoint
MMS	methyl methanesulfonate
NER	nucleotide excision repair
N-TEF	negative transcription elongation factor
NTS	nontranscribed spacer
RFB	replication fork barrier
RFP	replication fork pause
RNAP	RNA polymerase
RP	ribosomal protein
RNR	ribonucleotide reductase
SAD	S phase arrest defective
TCR	transcription-coupled repair
TFIIS or SII	transcription elongation factor SII
TRCF	transcription-repair coupling factor

References

Allen, J. B., Zhou, Z., Siede, W., Friedberg, E. C., and Elledge, S. J. (1994). The SAD1/RAD53 protein kinase controls multiple checkpoints and DNA damage-induced transcription in yeast. Genes Dev *8*, 2401-2415.

Barnes, D. E., Tomkinson, A. E., Lehmann, A. R., Webster, A. D., and Lindahl, T. (1992). Mutations in the DNA ligase I gene of an individual with immunodeficiencies and cellular hypersensitivity to DNA-damaging agents. Cell *69*, 495-503.

Batty, D., Rapic'-Otrin, V., Levine, A. S., and Wood, R. D. (2000). Stable binding of human XPC complex to irradiated DNA confers strong discrimination for damaged sites. J Mol Biol *300*, 275-290.

Beaudenon, S. L., Huacani, M. R., Wang, G., McDonnell, D. P., and Huibregtse, J. M. (1999). Rsp5 ubiquitin-protein ligase mediates DNA damage-induced degradation of the large subunit of RNA polymerase II in *Saccharomyces cerevisiae*. Mol Cell Biol *19*, 6972-6979.

Bennett, C. B., Lewis, L. K., Karthikeyan, G., Lobachev, K. S., Jin, Y. H., Sterling, J. F., Snipe, J. R., and Resnick, M. A. (2001). Genes required for ionizing radiation resistance in yeast. Nat Genet *29*, 426-434.

Bhatia, P. K., Verhage, R. A., Brouwer, J., and Friedberg, E. C. (1996). Molecular cloning and characterization of *Saccharomyces cerevisiae* RAD28, the yeast homolog of the human Cockayne syndrome A (CSA) gene. J Bacteriol *178*, 5977-5988.

Blattner, F. R., Plunkett, G., 3rd, Bloch, C. A., Perna, N. T., Burland, V., Riley, M., Collado-Vides, J., Glasner, J. D., Rode, C. K., Mayhew, G. F., *et al*. (1997). The complete genome sequence of *Escherichia coli* K-12. Science *277*, 1453-1474.

Bohr, V. A., Smith, C. A., Okumoto, D. S., and Hanawalt, P. C. (1985). DNA repair in an active gene: removal of pyrimidine dimers from the DHFR gene of CHO cells is much more efficient than in the genome overall. Cell *40*, 359-369.

Boiteux, S., and Guillet, M. (2004). Abasic sites in DNA: repair and biological consequences in *Saccharomyces cerevisiae*. DNA Repair (Amst) *3*, 1-12.

Bregman, D. B., Halaban, R., van Gool, A. J., Henning, K. A., Friedberg, E. C., and Warren, S. L. (1996). UV-induced ubiquitination of RNA polymerase II: a novel modification deficient in Cockayne syndrome cells. Proc Natl Acad Sci USA *93*, 11586-11590.

Brewer, B. J. (1988). When polymerases collide: replication and the transcriptional organization of the *E. coli* chromosome. Cell *53*, 679-686.

Brewer, B. J., and Fangman, W. L. (1988). A replication fork barrier at the 3′ end of yeast ribosomal RNA genes. Cell *55*,

637-643.

Brewer, B. J., Lockshon, D., and Fangman, W. L. (1992). The arrest of replication forks in the rDNA of yeast occurs independently of transcription. Cell 71, 267-276.

Brooks, P. J., Wise, D. S., Berry, D. A., Kosmoski, J. V., Smerdon, M. J., Somers, R. L., Mackie, H., Spoonde, A. Y., Ackerman, E. J., Coleman, K., et al. (2000). The oxidative DNA lesion 8,5'-(S)-cyclo-2'-deoxyadenosine is repaired by the nucleotide excision repair pathway and blocks gene expression in mammalian cells. J Biol Chem 275, 22355-22362.

Buschta-Hedayat, N., Buterin, T., Hess, M. T., Missura, M., and Naegeli, H. (1999). Recognition of nonhybridizing base pairs during nucleotide excision repair of DNA. Proc Natl Acad Sci USA 96, 6090-6095.

Carreau, M., and Hunting, D. (1992). Transcription-dependent and independent DNA excision repair pathways in human cells. Mutat Res 274, 57-64.

Chen, Y. H., and Bogenhagen, D. F. (1993). Effects of DNA lesions on transcription elongation by T7 RNA polymerase. J Biol Chem 268, 5849-5855.

Choi, D. J., Marino-Alessandri, D. J., Geacintov, N. E., and Scicchitano, D. A. (1994). Site-specific benzo[a]pyrene diol epoxide-DNA adducts inhibit transcription elongation by bacteriophage T7 RNA polymerase. Biochemistry 33, 780-787.

Christians, F. C., and Hanawalt, P. C. (1992). Inhibition of transcription and strand-specific DNA repair by alpha-amanitin in Chinese hamster ovary cells. Mutat Res 274, 93-101.

Christians, F. C., and Hanawalt, P. C. (1993). Lack of transcription-coupled repair in mammalian ribosomal RNA genes. Biochemistry 32, 10512-10518.

Christians, F. C., and Hanawalt, P. C. (1994). Repair in ribosomal RNA genes is deficient in xeroderma pigmentosum group C and in Cockayne's syndrome cells. Mutat Res 323, 179-187.

Citterio, E., Van Den Boom, V., Schnitzler, G., Kanaar, R., Bonte, E., Kingston, R. E., Hoeijmakers, J. H., and Vermeulen, W. (2000). ATP-dependent chromatin remodeling by the Cockayne syndrome B DNA repair-transcription-coupling factor. Mol Cell Biol 20, 7643-7653.

Clarke, D. J., and Gimenez-Abian, J. F. (2000). Checkpoints controlling mitosis. Bioessays 22, 351-363.

Cline, S. D., Riggins, J. N., Tornaletti, S., Marnett, L. J., and Hanawalt, P. C. (2004). Malondialdehyde adducts in DNA arrest transcription by T7 RNA polymerase and mammalian RNA polymerase II. Proc Natl Acad Sci USA 101, 7275-7280.

Conconi, A., Bespalov, V. A., and Smerdon, M. J. (2002). Transcription-coupled repair in RNA polymerase I-transcribed genes of yeast. Proc Natl Acad Sci USA 99, 649-654.

Cooper, P. K., Nouspikel, T., Clarkson, S. G., and Leadon, S. A. (1997). Defective transcription-coupled repair of oxidative base damage in Cockayne syndrome patients from XP group G. Science 275, 990-993.

Corda, Y., Job, C., Anin, M. F., Leng, M., and Job, D. (1993). Spectrum of DNA--platinum adduct recognition by prokaryotic and eukaryotic DNA-dependent RNA polymerases. Biochemistry 32, 8582-8588.

Cox, B. S., and Parry, J. M. (1968). The isolation, genetics and survival characteristics of ultraviolet light-sensitive mutants in yeast. Mutat Res 6, 37-55.

Cullinane, C., Mazur, S. J., Essigmann, J. M., Phillips, D. R., and Bohr, V. A. (1999). Inhibition of RNA polymerase II transcription in human cell extracts by cisplatin DNA damage. Biochemistry 38, 6204-6212.

Dammann, R., and Pfeifer, G. P. (1997). Lack of gene- and strand-specific DNA repair in RNA polymerase III-transcribed human tRNA genes. Mol Cell Biol 17, 219-229.

de Boer, J., and Hoeijmakers, J. H. (2000). Nucleotide excision repair and human syndromes. Carcinogenesis 21, 453-460.

de Laat, W. L., Appeldoorn, E., Sugasawa, K., Weterings, E., Jaspers, N. G., and Hoeijmakers, J. H. (1998). DNA-binding polarity of human replication protein A positions nucleases in nucleotide excision repair. Genes Dev 12, 2598-2609.

de Laat, W. L., Jaspers, N. G., and Hoeijmakers, J. H. (1999). Molecular mechanism of nucleotide excision repair. Genes Dev 13, 768-785.

Deshpande, A. M., and Newlon, C. S. (1996). DNA replication fork pause sites dependent on transcription. Science 272, 1030-1033.

Donahue, B. A., Fuchs, R. P., Reines, D., and Hanawalt, P. C. (1996). Effects of aminofluorene and acetylaminofluorene DNA adducts on transcriptional elongation by RNA polymerase II. J Biol Chem 271, 10588-10594.

Donahue, B. A., Yin, S., Taylor, J. S., Reines, D., and Hanawalt, P. C. (1994). Transcript cleavage by RNA polymerase II arrested by a cyclobutane pyrimidine dimer in the DNA template. Proc Natl Acad Sci USA 91, 8502-8506.

Drapkin, R., Reardon, J. T., Ansari, A., Huang, J. C., Zawel, L., Ahn, K., Sancar, A., and Reinberg, D. (1994). Dual role of TFIIH in DNA excision repair and in transcription by RNA polymerase II. Nature 368, 769-772.

Eisen, J. A., Sweder, K. S., and Hanawalt, P. C. (1995). Evolution of the SNF2 family of proteins: subfamilies with distinct sequences and functions. Nucleic Acids Res 23, 2715-2723.

Elias-Arnanz, M., and Salas, M. (1997). Bacteriophage phi29 DNA replication arrest caused by codirectional collisions with the transcription machinery. Embo J 16, 5775-5783.

Elias-Arnanz, M., and Salas, M. (1999). Resolution of head-on collisions between the transcription machinery and bacteriophage phi29 DNA polymerase is dependent on RNA polymerase translocation. Embo J 18, 5675-5682.

Evans, E., Moggs, J. G., Hwang, J. R., Egly, J. M., and Wood, R. D. (1997). Mechanism of open complex and dual incision formation by human nucleotide excision repair factors. Embo J 16, 6559-6573.

Evers, R., and Grummt, I. (1995). Molecular coevolution of

mammalian ribosomal gene terminator sequences and the transcription termination factor TTF-I. Proc Natl Acad Sci USA 92, 5827-5831.

Feaver, W. J., Svejstrup, J. Q., Bardwell, L., Bardwell, A. J., Buratowski, S., Gulyas, K. D., Donahue, T. F., Friedberg, E. C., and Kornberg, R. D. (1993). Dual roles of a multiprotein complex from S. cerevisiae in transcription and DNA repair. Cell 75, 1379-1387.

French, S. (1992). Consequences of replication fork movement through transcription units in vivo. Science 258, 1362-1365.

Frit, P., Bergmann, E., and Egly, J. M. (1999). Transcription factor IIH: a key player in the cellular response to DNA damage. Biochimie 81, 27-38.

Fritz, L. K., and Smerdon, M. J. (1995). Repair of UV damage in actively transcribed ribosomal genes. Biochemistry 34, 13117-13124.

Gerber, J. K., Gogel, E., Berger, C., Wallisch, M., Muller, F., Grummt, I., and Grummt, F. (1997). Termination of mammalian rDNA replication: polar arrest of replication fork movement by transcription termination factor TTF-I. Cell 90, 559-567.

Gu, W., Powell, W., Mote, J., Jr., and Reines, D. (1993). Nascent RNA cleavage by arrested RNA polymerase II does not require upstream translocation of the elongation complex on DNA. J Biol Chem 268, 25604-25616.

Hanawalt, P., and Mellon, I. (1993). Stranded in an active gene. Curr Biol 3, 67-69.

Hanawalt, P. C. (1994). Transcription-coupled repair and human disease. Science 266, 1957-1958.

Hara, R., Selby, C. P., Liu, M., Price, D. H., and Sancar, A. (1999). Human transcription release factor 2 dissociates RNA polymerases I and II stalled at a cyclobutane thymine dimer. J Biol Chem 274, 24779-24786.

Hartwell, L. H., and Weinert, T. A. (1989). Checkpoints: controls that ensure the order of cell cycle events. Science 246, 629-634.

Hatahet, Z., Purmal, A. A., and Wallace, S. S. (1994). Oxidative DNA lesions as blocks to in vitro transcription by phage T7 RNA polymerase. Ann N Y Acad Sci 726, 346-348.

Henning, K. A., Li, L., Iyer, N., McDaniel, L. D., Reagan, M. S., Legerski, R., Schultz, R. A., Stefanini, M., Lehmann, A. R., Mayne, L. V., and Friedberg, E. C. (1995). The Cockayne syndrome group A gene encodes a WD repeat protein that interacts with CSB protein and a subunit of RNA polymerase II TFIIH. Cell 82, 555-564.

Hofmann, K., and Bucher, P. (1995). The FHA domain: a putative nuclear signalling domain found in protein kinases and transcription factors. Trends Biochem Sci 20, 347-349.

Huang, M., Zhou, Z., and Elledge, S. J. (1998). The DNA replication and damage checkpoint pathways induce transcription by inhibition of the Crt1 repressor. Cell 94, 595-605.

Ivessa, A. S., Lenzmeier, B. A., Bessler, J. B., Goudsouzian, L. K., Schnakenberg, S. L., and Zakian, V. A. (2003). The Saccharomyces cerevisiae helicase Rrm3p facilitates replication past nonhistone protein-DNA complexes. Mol Cell 12, 1525-1536.

Iyer, N., Reagan, M. S., Wu, K. J., Canagarajah, B., and Friedberg, E. C. (1996). Interactions involving the human RNA polymerase II transcription/nucleotide excision repair complex TFIIH, the nucleotide excision repair protein XPG, and Cockayne syndrome group B (CSB) protein. Biochemistry 35, 2157-2167.

Izban, M. G., and Luse, D. S. (1992). The RNA polymerase II ternary complex cleaves the nascent transcript in a 3′----5′ direction in the presence of elongation factor SII. Genes Dev 6, 1342-1356.

Jorgensen, P., Nishikawa, J. L., Breitkreutz, B. J., and Tyers, M. (2002). Systematic identification of pathways that couple cell growth and division in yeast. Science 297, 395-400.

Jorgensen, P., Rupes, I., Sharom, J. R., Schneper, L., Broach, J. R., and Tyers, M. (2004). A dynamic transcriptional network communicates growth potential to ribosome synthesis and critical cell size. Genes Dev 18, 2491-2505.

Kalogeraki, V. S., Tornaletti, S., and Hanawalt, P. C. (2003). Transcription arrest at a lesion in the transcribed DNA strand in vitro is not affected by a nearby lesion in the opposite strand. J Biol Chem 278, 19558-19564.

Kato, R., and Ogawa, H. (1994). An essential gene, ESR1, is required for mitotic cell growth, DNA repair and meiotic recombination in Saccharomyces cerevisiae. Nucleic Acids Res 22, 3104-3112.

Kim, C., Snyder, R. O., and Wold, M. S. (1992). Binding properties of replication protein A from human and yeast cells. Mol Cell Biol 12, 3050-3059.

Kobayashi, T. (2003). The replication fork barrier site forms a unique structure with Fob1p and inhibits the replication fork. Mol Cell Biol 23, 9178-9188.

Kobayashi, T., Heck, D. J., Nomura, M., and Horiuchi, T. (1998). Expansion and contraction of ribosomal DNA repeats in Saccharomyces cerevisiae: requirement of replication fork blocking (Fob1) protein and the role of RNA polymerase I. Genes Dev 12, 3821-3830.

Kobayashi, T., Hidaka, M., Nishizawa, M., and Horiuchi, T. (1992). Identification of a site required for DNA replication fork blocking activity in the rRNA gene cluster in Saccharomyces cerevisiae. Mol Gen Genet 233, 355-362.

Krokan, H. E., Standal, R., and Slupphaug, G. (1997). DNA glycosylases in the base excision repair of DNA. Biochem J 325 (Pt 1), 1-16.

Kubota, Y., Nash, R. A., Klungland, A., Schar, P., Barnes, D. E., and Lindahl, T. (1996). Reconstitution of DNA base excision-repair with purified human proteins: interaction between DNA polymerase beta and the XRCC1 protein. Embo J 15, 6662-6670.

Le Page, F., Kwoh, E. E., Avrutskaya, A., Gentil, A., Leadon, S. A., Sarasin, A., and Cooper, P. K. (2000). Transcription-coupled repair of 8-oxoguanine: requirement for XPG, TFIIH, and CSB and implications for Cockayne syndrome. Cell *101*, 159-171.

Leadon, S. A., Barbee, S. L., and Dunn, A. B. (1995). The yeast RAD2, but not RAD1, gene is involved in the transcription-coupled repair of thymine glycols. Mutat Res *337*, 169-178.

Leadon, S. A., and Cooper, P. K. (1993). Preferential repair of ionizing radiation-induced damage in the transcribed strand of an active human gene is defective in Cockayne syndrome. Proc Natl Acad Sci USA *90*, 10499-10503.

Leadon, S. A., and Lawrence, D. A. (1992). Strand-selective repair of DNA damage in the yeast GAL7 gene requires RNA polymerase II. J Biol Chem *267*, 23175-23182.

Lee, K. B., Wang, D., Lippard, S. J., and Sharp, P. A. (2002). Transcription-coupled and DNA damage-dependent ubiquitination of RNA polymerase II *in vitro*. Proc Natl Acad Sci USA *99*, 4239-4244.

Li, B., and Reese, J. C. (2000). Derepression of DNA damage-regulated genes requires yeast TAF(II)s. Embo J *19*, 4091-4100.

Li, B., and Reese, J. C. (2001). Ssn6-Tup1 regulates RNR3 by positioning nucleosomes and affecting the chromatin structure at the upstream repression sequence. J Biol Chem *276*, 33788-33797.

Lindahl, T., Karran, P., and Wood, R. D. (1997). DNA excision repair pathways. Curr Opin Genet Dev *7*, 158-169.

Liu, B., and Alberts, B. M. (1995). Head-on collision between a DNA replication apparatus and RNA polymerase transcription complex. Science *267*, 1131-1137.

Liu, B., Wong, M. L., Tinker, R. L., Geiduschek, E. P., and Alberts, B. M. (1993). The DNA replication fork can pass RNA polymerase without displacing the nascent transcript. Nature *366*, 33-39.

Liu, M., Xie, Z., and Price, D. H. (1998). A human RNA polymerase II transcription termination factor is a SWI2/SNF2 family member. J Biol Chem *273*, 25541-25544.

Ljungman, M., and Zhang, F. (1996). Blockage of RNA polymerase as a possible trigger for u.v. light-induced apoptosis. Oncogene *13*, 823-831.

Lopez-estrano, C., Schvartzman, J. B., Krimer, D. B., and Hernandez, P. (1998). Co-localization of polar replication fork barriers and rRNA transcription terminators in mouse rDNA. J Mol Biol *277*, 249-256.

Lowndes, N. F., and Murguia, J. R. (2000). Sensing and responding to DNA damage. Curr Opin Genet Dev *10*, 17-25.

Luo, Z., Zheng, J., Lu, Y., and Bregman, D. B. (2001). Ultraviolet radiation alters the phosphorylation of RNA polymerase II large subunit and accelerates its proteasome-dependent degradation. Mutat Res *486*, 259-274.

Ma, L., Siemssen, E. D., Noteborn, H. M., and van der Eb, A. J. (1994). The xeroderma pigmentosum group B protein ERCC3 produced in the baculovirus system exhibits DNA helicase activity. Nucleic Acids Res *22*, 4095-4102.

MacAlpine, D. M., Rodriguez, H. K., and Bell, S. P. (2004). Coordination of replication and transcription along a *Drosophila* chromosome. Genes Dev *18*, 3094-3105.

Marion, R. M., Regev, A., Segal, E., Barash, Y., Koller, D., Friedman, N., and O'Shea, E. K. (2004). Sfp1 is a stress- and nutrient-sensitive regulator of ribosomal protein gene expression. Proc Natl Acad Sci USA *101*, 14315-14322.

Marshall, N. F., Peng, J., Xie, Z., and Price, D. H. (1996). Control of RNA polymerase II elongation potential by a novel carboxyl-terminal domain kinase. J Biol Chem *271*, 27176-27183.

Martin, D. E., Soulard, A., and Hall, M. N. (2004). TOR regulates ribosomal protein gene expression via PKA and the Forkhead transcription factor FHL1. Cell *119*, 969-979.

Matsumoto, Y., and Kim, K. (1995). Excision of deoxyribose phosphate residues by DNA polymerase beta during DNA repair. Science *269*, 699-702.

McKay, B. C., Ljungman, M., and Rainbow, A. J. (1998). Persistent DNA damage induced by ultraviolet light inhibits p21waf1 and bax expression: implications for DNA repair, UV sensitivity and the induction of apoptosis. Oncogene *17*, 545-555.

Mei Kwei, J. S., Kuraoka, I., Horibata, K., Ubukata, M., Kobatake, E., Iwai, S., Handa, H., and Tanaka, K. (2004). Blockage of RNA polymerase II at a cyclobutane pyrimidine dimer and 6-4 photoproduct. Biochem Biophys Res Commun *320*, 1133-1138.

Mello, J. A., Lippard, S. J., and Essigmann, J. M. (1995). DNA adducts of cis-diamminedichloroplatinum(II) and its trans isomer inhibit RNA polymerase II differentially *in vivo*. Biochemistry *34*, 14783-14791.

Mellon, I., Bohr, V. A., Smith, C. A., and Hanawalt, P. C. (1986). Preferential DNA repair of an active gene in human cells. Proc Natl Acad Sci USA *83*, 8878-8882.

Mellon, I., and Champe, G. N. (1996). Products of DNA mismatch repair genes mutS and mutL are required for transcription-coupled nucleotide-excision repair of the lactose operon in *Escherichia coli*. Proc Natl Acad Sci USA *93*, 1292-1297.

Mellon, I., and Hanawalt, P. C. (1989). Induction of the *Escherichia coli* lactose operon selectively increases repair of its transcribed DNA strand. Nature *342*, 95-98.

Mellon, I., Rajpal, D. K., Koi, M., Boland, C. R., and Champe, G. N. (1996). Transcription-coupled repair deficiency and mutations in human mismatch repair genes. Science *272*, 557-560.

Mellon, I., Spivak, G., and Hanawalt, P. C. (1987). Selective removal of transcription-blocking DNA damage from the transcribed strand of the mammalian DHFR gene. Cell *51*, 241-249.

Memisoglu, A., and Samson, L. (2000). Base excision repair in yeast and mammals. Mutat Res *451*, 39-51.

Mirkin, E. V., and Mirkin, S. M. (2005). Mechanisms of transcription-replication collisions in bacteria. Mol Cell Biol *25*, 888-895.

Mu, D., and Sancar, A. (1997). Model for XPC-independent transcription-coupled repair of pyrimidine dimers in humans. J Biol Chem *272*, 7570-7573.

Mulugu, S., Potnis, A., Shamsuzzaman, Taylor, J., Alexander, K., and Bastia, D. (2001). Mechanism of termination of DNA replication of *Escherichia coli* involves helicase-contrahelicase interaction. Proc Natl Acad Sci USA *98*, 9569-9574.

Nath, S. T., and Romano, L. J. (1991). Transcription by T7 RNA polymerase using benzo[a]pyrene-modified templates. Carcinogenesis *12*, 973-976.

Navas, T. A., Sanchez, Y., and Elledge, S. J. (1996). RAD9 and DNA polymerase epsilon form parallel sensory branches for transducing the DNA damage checkpoint signal in *Saccharomyces cerevisiae*. Genes Dev *10*, 2632-2643.

Navas, T. A., Zhou, Z., and Elledge, S. J. (1995). DNA polymerase epsilon links the DNA replication machinery to the S phase checkpoint. Cell *80*, 29-39.

Nguyen, V. T., Giannoni, F., Dubois, M. F., Seo, S. J., Vigneron, M., Kedinger, C., and Bensaude, O. (1996). In vivo degradation of RNA polymerase II largest subunit triggered by alpha-amanitin. Nucleic Acids Res *24*, 2924-2929.

Nicholl, I. D., Nealon, K., and Kenny, M. K. (1997). Reconstitution of human base excision repair with purified proteins. Biochemistry *36*, 7557-7566.

Nudler, E., Goldfarb, A., and Kashlev, M. (1994). Discontinuous mechanism of transcription elongation. Science *265*, 793-796.

O'Donovan, A., Davies, A. A., Moggs, J. G., West, S. C., and Wood, R. D. (1994). XPG endonuclease makes the 3' incision in human DNA nucleotide excision repair. Nature *371*, 432-435.

Olavarrieta, L., Hernandez, P., Krimer, D. B., and Schvartzman, J. B. (2002). DNA knotting caused by head-on collision of transcription and replication. J Mol Biol *322*, 1-6.

Park, J. S., Marr, M. T., and Roberts, J. W. (2002). E. coli Transcription repair coupling factor (Mfd protein) rescues arrested complexes by promoting forward translocation. Cell *109*, 757-767.

Perlow, R. A., Kolbanovskii, A., Hingerty, B. E., Geacintov, N. E., Broyde, S., and Scicchitano, D. A. (2002). DNA adducts from a tumorigenic metabolite of benzo[a]pyrene block human RNA polymerase II elongation in a sequence- and stereochemistry-dependent manner. J Mol Biol *321*, 29-47.

Plosky, B., Samson, L., Engelward, B. P., Gold, B., Schlaen, B., Millas, T., Magnotti, M., Schor, J., and Scicchitano, D. A. (2002). Base excision repair and nucleotide excision repair contribute to the removal of N-methylpurines from active genes. DNA Repair (Amst) *1*, 683-696.

Prado, F., and Aguilera, A. (2005). Impairment of replication fork progression mediates RNA polII transcription-associated recombination. Embo J *24*, 1267-1276.

Putter, V., and Grummt, F. (2002). Transcription termination factor TTF-I exhibits contrahelicase activity during DNA replication. EMBO Rep *3*, 147-152.

Ratner, J. N., Balasubramanian, B., Corden, J., Warren, S. L., and Bregman, D. B. (1998). Ultraviolet radiation-induced ubiquitination and proteasomal degradation of the large subunit of RNA polymerase II. Implications for transcription-coupled DNA repair. J Biol Chem *273*, 5184-5189.

Rocha, E. P., and Danchin, A. (2003). Essentiality, not expressiveness, drives gene-strand bias in bacteria. Nat Genet *34*, 377-378.

Roth, R. B., Amin, S., Geacintov, N. E., and Scicchitano, D. A. (2001). Bacteriophage T7 RNA polymerase transcription elongation is inhibited by site-specific, stereospecific benzo[c]phenanthrene diol epoxide DNA lesions. Biochemistry *40*, 5200-5207.

Samkurashvili, I., and Luse, D. S. (1996). Translocation and transcriptional arrest during transcript elongation by RNA polymerase II. J Biol Chem *271*, 23495-23505.

Schaeffer, L., Moncollin, V., Roy, R., Staub, A., Mezzina, M., Sarasin, A., Weeda, G., Hoeijmakers, J. H., and Egly, J. M. (1994). The ERCC2/DNA repair protein is associated with the class II BTF2/TFIIH transcription factor. Embo J *13*, 2388-2392.

Schaeffer, L., Roy, R., Humbert, S., Moncollin, V., Vermeulen, W., Hoeijmakers, J. H., Chambon, P., and Egly, J. M. (1993). DNA repair helicase: a component of BTF2 (TFIIH) basic transcription factor. Science *260*, 58-63.

Scharer, O. D., and Jiricny, J. (2001). Recent progress in the biology, chemistry and structural biology of DNA glycosylases. Bioessays *23*, 270-281.

Schawalder, S. B., Kabani, M., Howald, I., Choudhury, U., Werner, M., and Shore, D. (2004). Growth-regulated recruitment of the essential yeast ribosomal protein gene activator Ifh1. Nature *432*, 1058-1061.

Schinecker, T. M., Perlow, R. A., Broyde, S., Geacintov, N. E., and Scicchitano, D. A. (2003). Human RNA polymerase II is partially blocked by DNA adducts derived from tumorigenic benzo[c]phenanthrene diol epoxides: relating biological consequences to conformational preferences. Nucleic Acids Res *31*, 6004-6015.

Schubeler, D., Scalzo, D., Kooperberg, C., van Steensel, B., Delrow, J., and Groudine, M. (2002). Genome-wide DNA replication profile for *Drosophila melanogaster*: a link between transcription and replication timing. Nat Genet *32*, 438-442.

Selby, C. P., and Sancar, A. (1997a). Cockayne syndrome group B protein enhances elongation by RNA polymerase II. Proc Natl Acad Sci USA *94*, 11205-11209.

Selby, C. P., and Sancar, A. (1997b). Human transcription-repair coupling factor CSB/ERCC6 is a DNA-stimulated ATPase but is not a helicase and does not disrupt the ternary transcription complex of stalled RNA polymerase II. J Biol Chem *272*, 1885-1890.

Sharma, V. M., Li, B., and Reese, J. C. (2003). SWI/SNF-dependent chromatin remodeling of RNR3 requires TAF(II)s and the general transcription machinery. Genes Dev *17*, 502-515.

Shi, Y. B., Gamper, H., and Hearst, J. E. (1988). Interaction of T7 RNA polymerase with DNA in an elongation complex arrested at a specific psoralen adduct site. J Biol Chem *263*, 527-534.

Shivji, M. K., Podust, V. N., Hubscher, U., and Wood, R. D. (1995). Nucleotide excision repair DNA synthesis by DNA polymerase epsilon in the presence of PCNA, RFC, and RPA. Biochemistry *34*, 5011-5017.

Sijbers, A. M., de Laat, W. L., Ariza, R. R., Biggerstaff, M., Wei, Y. F., Moggs, J. G., Carter, K. C., Shell, B. K., Evans, E., de Jong, M. C., *et al.* (1996). Xeroderma pigmentosum group F caused by a defect in a structure-specific DNA repair endonuclease. Cell *86*, 811-822.

Smerdon, M. J., and Thoma, F. (1990). Site-specific DNA repair at the nucleosome level in a yeast minichromosome. Cell *61*, 675-684.

Smith, C. A., Baeten, J., and Taylor, J. S. (1998). The ability of a variety of polymerases to synthesize past site-specific cis-syn, trans-syn-II, (6-4), and Dewar photoproducts of thymidylyl-(3'-->5')-thymidine. J Biol Chem *273*, 21933-21940.

Spivak, G. (2004). The many faces of Cockayne syndrome. Proc Natl Acad Sci USA *101*, 15273-15274.

Srivastava, D. K., Berg, B. J., Prasad, R., Molina, J. T., Beard, W. A., Tomkinson, A. E., and Wilson, S. H. (1998). Mammalian abasic site base excision repair. Identification of the reaction sequence and rate-determining steps. J Biol Chem *273*, 21203-21209.

Sugasawa, K., Ng, J. M., Masutani, C., Iwai, S., van der Spek, P. J., Eker, A. P., Hanaoka, F., Bootsma, D., and Hoeijmakers, J. H. (1998). Xeroderma pigmentosum group C protein complex is the initiator of global genome nucleotide excision repair. Mol Cell *2*, 223-232.

Sung, P., Bailly, V., Weber, C., Thompson, L. H., Prakash, L., and Prakash, S. (1993). Human xeroderma pigmentosum group D gene encodes a DNA helicase. Nature *365*, 852-855.

Svejstrup, J. Q. (2002). Mechanisms of transcription-coupled DNA repair. Nat Rev Mol Cell Biol *3*, 21-29.

Svejstrup, J. Q. (2003). Rescue of arrested RNA polymerase II complexes. J Cell Sci *116*, 447-451.

Svejstrup, J. Q., Vichi, P., and Egly, J. M. (1996). The multiple roles of transcription/repair factor TFIIH. Trends Biochem Sci *21*, 346-350.

Sweder, K. S., and Hanawalt, P. C. (1992). Preferential repair of cyclobutane pyrimidine dimers in the transcribed strand of a gene in yeast chromosomes and plasmids is dependent on transcription. Proc Natl Acad Sci USA *89*, 10696-10700.

Takeuchi, Y., Horiuchi, T., and Kobayashi, T. (2003). Transcription-dependent recombination and the role of fork collision in yeast rDNA. Genes Dev *17*, 1497-1506.

Tantin, D. (1998). RNA polymerase II elongation complexes containing the Cockayne syndrome group B protein interact with a molecular complex containing the transcription factor IIH components xeroderma pigmentosum B and p62. J Biol Chem *273*, 27794-27799.

Tijsterman, M., Verhage, R. A., van de Putte, P., Tasseron-de Jong, J. G., and Brouwer, J. (1997). Transitions in the coupling of transcription and nucleotide excision repair within RNA polymerase II-transcribed genes of *Saccharomyces cerevisiae*. Proc Natl Acad Sci USA *94*, 8027-8032.

Tornaletti, S., Donahue, B. A., Reines, D., and Hanawalt, P. C. (1997). Nucleotide sequence context effect of a cyclobutane pyrimidine dimer upon RNA polymerase II transcription. J Biol Chem *272*, 31719-31724.

Tornaletti, S., and Hanawalt, P. C. (1999). Effect of DNA lesions on transcription elongation. Biochimie *81*, 139-146.

Tornaletti, S., Maeda, L. S., Lloyd, D. R., Reines, D., and Hanawalt, P. C. (2001). Effect of thymine glycol on transcription elongation by T7 RNA polymerase and mammalian RNA polymerase II. J Biol Chem *276*, 45367-45371.

Tornaletti, S., Patrick, S. M., Turchi, J. J., and Hanawalt, P. C. (2003). Behavior of T7 RNA polymerase and mammalian RNA polymerase II at site-specific cisplatin adducts in the template DNA. J Biol Chem *278*, 35791-35797.

Tornaletti, S., Reines, D., and Hanawalt, P. C. (1999). Structural characterization of RNA polymerase II complexes arrested by a cyclobutane pyrimidine dimer in the transcribed strand of template DNA. J Biol Chem *274*, 24124-24130.

Troelstra, C., van Gool, A., de Wit, J., Vermeulen, W., Bootsma, D., and Hoeijmakers, J. H. (1992). ERCC6, a member of a subfamily of putative helicases, is involved in Cockayne's syndrome and preferential repair of active genes. Cell *71*, 939-953.

Tu, Y., Bates, S., and Pfeifer, G. P. (1997). Sequence-specific and domain-specific DNA repair in xeroderma pigmentosum and Cockayne syndrome cells. J Biol Chem *272*, 20747-20755.

van Gool, A. J., Citterio, E., Rademakers, S., van Os, R., Vermeulen, W., Constantinou, A., Egly, J. M., Bootsma, D., and Hoeijmakers, J. H. (1997). The Cockayne syndrome B protein, involved in transcription-coupled DNA repair, resides in an RNA polymerase II-containing complex. Embo J *16*, 5955-5965.

van Gool, A. J., Verhage, R., Swagemakers, S. M., van de Putte, P., Brouwer, J., Troelstra, C., Bootsma, D., and Hoeijmakers, J. H. (1994). RAD26, the functional *S. cerevisiae* homolog of the Cockayne syndrome B gene ERCC6. Embo J *13*, 5361-5369.

Van Hoffen, A., Natarajan, A. T., Mayne, L. V., van Zeeland, A. A., Mullenders, L. H., and Venema, J. (1993). Deficient repair of the transcribed strand of active genes in Cockayne's syndrome cells. Nucleic Acids Res *21*, 5890-5895.

Venema, J., Mullenders, L. H., Natarajan, A. T., van Zeeland, A. A., and Mayne, L. V. (1990a). The genetic defect in Cockayne syndrome is associated with a defect in repair of UV-induced DNA damage in transcriptionally active DNA. Proc Natl Acad Sci USA *87*, 4707-4711.

Venema, J., van Hoffen, A., Natarajan, A. T., van Zeeland, A. A., and Mullenders, L. H. (1990b). The residual repair capacity of xeroderma pigmentosum complementation group C fibroblasts is highly specific for transcriptionally active DNA. Nucleic Acids Res *18*, 443-448.

Verhage, R. A., Van de Putte, P., and Brouwer, J. (1996). Repair of rDNA in *Saccharomyces cerevisiae*: RAD4-independent strand-specific nucleotide excision repair of RNA polymerase I transcribed genes. Nucleic Acids Res *24*, 1020-1025.

Volker, M., Mone, M. J., Karmakar, P., van Hoffen, A., Schul, W., Vermeulen, W., Hoeijmakers, J. H., van Driel, R., van Zeeland, A. A., and Mullenders, L. H. (2001). Sequential assembly of the nucleotide excision repair factors *in vivo*. Mol Cell *8*, 213-224.

Vos, J. M., and Wauthier, E. L. (1991). Differential introduction of DNA damage and repair in mammalian genes transcribed by RNA polymerases I and II. Mol Cell Biol *11*, 2245-2252.

Wade, J. T., Hall, D. B., and Struhl, K. (2004). The transcription factor Ifh1 is a key regulator of yeast ribosomal protein genes. Nature *432*, 1054-1058.

Wang, D., Meier, T. I., Chan, C. L., Feng, G., Lee, D. N., and Landick, R. (1995). Discontinuous movements of DNA and RNA in RNA polymerase accompany formation of a paused transcription complex. Cell *81*, 341-350.

Wang, Z., Svejstrup, J. Q., Feaver, W. J., Wu, X., Kornberg, R. D., and Friedberg, E. C. (1994). Transcription factor b (TFIIH) is required during nucleotide-excision repair in yeast. Nature *368*, 74-76.

Weinert, T. A. (1992). Dual cell cycle checkpoints sensitive to chromosome replication and DNA damage in the budding yeast *Saccharomyces cerevisiae*. Radiat Res *132*, 141-143.

Weinert, T. A., Kiser, G. L., and Hartwell, L. H. (1994). Mitotic checkpoint genes in budding yeast and the dependence of mitosis on DNA replication and repair. Genes Dev *8*, 652-665.

Wind, M., and Reines, D. (2000). Transcription elongation factor SII. Bioessays *22*, 327-336.

Wood, R. D. (1999). DNA damage recognition during nucleotide excision repair in mammalian cells. Biochimie *81*, 39-44.

Woudstra, E. C., Gilbert, C., Fellows, J., Jansen, L., Brouwer, J., Erdjument-Bromage, H., Tempst, P., and Svejstrup, J. Q. (2002). A Rad26-Def1 complex coordinates repair and RNA pol II proteolysis in response to DNA damage. Nature *415*, 929-933.

Xie, Z., and Price, D. H. (1996). Purification of an RNA polymerase II transcript release factor from *Drosophila*. J Biol Chem *271*, 11043-11046.

Xu, Z., and Norris, D. (1998). The SFP1 gene product of *Saccharomyces cerevisiae* regulates G2/M transitions during the mitotic cell cycle and DNA-damage response. Genetics *150*, 1419-1428.

You, Z., Feaver, W. J., and Friedberg, E. C. (1998). Yeast RNA polymerase II transcription *in vitro* is inhibited in the presence of nucleotide excision repair: complementation of inhibition by Holo-TFIIH and requirement for RAD26. Mol Cell Biol *18*, 2668-2676.

Yu, A., Fan, H. Y., Liao, D., Bailey, A. D., and Weiner, A. M. (2000). Activation of p53 or loss of the Cockayne syndrome group B repair protein causes metaphase fragility of human U1, U2, and 5S genes. Mol Cell *5*, 801-810.

Zhang, Z., and Reese, J. C. (2004a). Redundant mechanisms are used by Ssn6-Tup1 in repressing chromosomal gene transcription in *Saccharomyces cerevisiae*. J Biol Chem *279*, 39240-39250.

Zhang, Z., and Reese, J. C. (2004b). Ssn6-Tup1 requires the ISW2 complex to position nucleosomes in *Saccharomyces cerevisiae*. Embo J *23*, 2246-2257.

Zhao, J. (2004). Coordination of DNA synthesis and histone gene expression during normal cell cycle progression and after DNA damage. Cell Cycle *3*, 695-697.

Zhao, X., Muller, E. G., and Rothstein, R. (1998). A suppressor of two essential checkpoint genes identifies a novel protein that negatively affects dNTP pools [In Process Citation]. Mol Cell *2*, 329-340.

Zhao, X., and Rothstein, R. (2002). The Dun1 checkpoint kinase phosphorylates and regulates the ribonucleotide reductase inhibitor Sml1. Proc Natl Acad Sci USA *99*, 3746-3751.

Zheng, P., Fay, D. S., Burton, J., Xiao, H., Pinkham, J. L., and Stern, D. F. (1993). SPK1 is an essential S-phase-specific gene of *Saccharomyces cerevisiae* that encodes a nuclear serine/threonine/tyrosine kinase. Mol Cell Biol *13*, 5829-5842.

Zhou, Z., and Elledge, S. J. (1992). Isolation of crt mutants constitutive for transcription of the DNA damage inducible gene RNR3 in *Saccharomyces cerevisiae*. Genetics *131*, 851-866.

Zhou, Z., and Elledge, S. J. (1993). DUN1 encodes a protein kinase that controls the DNA damage response in yeast. Cell *75*, 1119-1127.

Chapter 26
Cell Death and Transcription

Jianhua Zhang[1] and Wei-Xing Zong[2]

[1]Department of Pathology, 961 Sparks Center, 1530 3rd Ave S, University of Alabama at Birmingham, Birmingham, AL 35294-0017
[2]Abramson Cancer Research Institute, University of Pennsylvania, BRB II/III, Room 445, 421 Curie Blvd, Philadelphia, PA 19104-6160

Key Words: transcription, cell death, Bcl-2, death receptor, caspase, p53, E2F, NF-κB, chromatin remodeling

Summary

Cell death is required for development and tissue homeostasis of all multicellular organisms. Cell death regulation is highly dependent on cell types and the physiological, pharmacological and pathological stimuli and can take distinctive forms. One mechanism that is conserved from *C. elegans* to humans to ensure desired cell death and to avoid unwanted cell death is through transcriptional regulation of cell death genes. Transcription regulation may be conferred by specific transcription factors or at a global level by chromatin remodeling activities.

Introduction

All living organisms experience growth by multiplication of cell numbers and undergo renewal by replacing mutated, infected, damaged, excessive, old, or outdated cells. Normal development of multicellular organisms needs appropriate cell death to carve out hollow structures, to eliminate supernumerary cells, and to ensure useful cells to survive the growth and differentiation environment (Kerr *et al.*, 1972; Vaux and Korsmeyer, 1999; Kuan *et al.*, 2000). Excessive cell death can lead to human diseases including neurodegeneration and immunodeficiency, whereas insufficient cell death can contribute to autoimmunity and cancer (Thompson, 1995; Opferman and Korsmeyer, 2003; Okada and Mak, 2004). To ensure appropriate timing and extent of cell death in developing and mature organisms, the majority of cells die through intrinsically controlled programs that are largely conserved during evolution (Horvitz, 1999; Danial and Korsmeyer, 2004). Understanding regulation of programmed cell death in different developmental and cell type contexts, in response to different physiological, pharmacological and pathological stimuli is crucial for prevention, and management of human diseases.

In *C. elegans* and in *Drosophila*, a prominent mechanism of cell death regulation is at the level of transcription. In higher organisms, significant pre- and post-transcriptional mechanisms add to the complexity of regulation of cell death. These include epigenetics, translation, protein modification, sequestration, and degradation. Extensive reviews exist that cover many of these aspects. This chapter examines how transcription and programmed cell death are coupled and discusses implications of current observations. Other related recent comments and reviews can be found in Tran *et al.* (2004), and Kumar and Cakouros (2004).

Cell Death Pathways

Cell death takes distinctive forms in response to different intracellular and extracellular signals. The most extensively studied form of programmed cell death is termed apoptosis, which typically refers to cell death with characteristic morphological changes including cell and nuclear shrinkage, chromatin condensation, and the formation of apoptotic bodies (Kerr, 1972; Vaux and

---

Corresponding Author: Jianhua Zhang, Tel: (205) 996-5153, Fax: (205) 934-6700, E-mail: jzhang@path.uab.edu

Korsmeyer, 1999). One of the biochemical hallmarks of apoptosis is the cleavage of chromosomal DNA into oligonucleosomal-sized fragments (Wyllie, 1980). The dying cells are phagocytosed by scavenger cells (Henson et al., 2001). The genetic program of apoptosis is highly conserved from C. elegans to humans (Fig. 26.1) (Horvitz, 1999; Wang, 2001; Hay et al., 2004; Danial and Korsmeyer, 2004).

In mammals, apoptosis is executed through two general molecular pathways. The intrinsic pathway involves the activation of multi-BH (Bcl-2 homology) domain proteins Bax and Bak, which results in the release of cytochrome c from the mitochondrial intermembrane space into the cytosol in response to a wide variety of death signals including growth factor deprivation and DNA damage (Lindsten et al., 2000; Wei et al., 2001; Adams and Cory, 2001; Wang, 2001; Danial and Korsmeyer, 2004). The translocation of cytochrome c triggers a cascade of reaction, beginning with the activation of the apoptosome that contains Apaf-1 and caspase-9, proceeding to the activation of downstream caspases that leads to the cleavage of proteins essential for normal cell function and survival (Li et al., 1997; Zou et al., 1997; Wang, 2001). One of these proteins is DNA fragmentation factor 45 (DFF45, also called the inhibitor of caspase-activated DNase, or ICAD). Cleavage of DFF45 results in the activation of DFF40 (also called caspase-activated DNase or CAD) that degrades chromosomal DNA in a massive scale (Nagata, 2000; Zhang and Xu, 2002). In addition, several other apoptogenic factors such as Smac, apoptosis-inducing factor (AIF), and endonuclease G (endoG) also translocate from mitochondria into different subcellular compartments to facilitate apoptosis. The latter two are of particular interest as they may induce DNA fragmentation and cell death in a caspase-independent manner (Wang, 2001). Inhibitor of apoptosis (IAP) family proteins keep caspases in check by inhibition of their activities (Salvesen and Duckett, 2002).

A second apoptosis pathway, the extrinsic pathway, is important for cell death regulation under many physiological conditions, especially during the development, homeostasis, and for the proper function of the immune system. The extrinsic pathway involves interaction of death ligand molecules with their respective receptors on the cell surface, such as the binding of FasL to Fas or TNF to the TNF receptor. Ligand binding leads to the oligomerization of the death receptors, the formation of the Death-Inducing-Signaling Complex (DISC), and subsequently the activation of caspase-8 (Muzio et al., 1996). Intracellular signal transduction such as the MAP kinase- and the NF-κB-mediated mechanisms play an important role in modulating the extrinsic cell death pathway (Baud and Karin, 2001; Deng et al., 2003; Varfolomeev and Ashkenazi, 2004). The extrinsic apoptosis pathways are less well defined in C. elegans and Drosophila compared to mammals.

The Bcl-2 family proteins play pivotal roles in regulating apoptosis. As Bcl-2 family proteins function primarily at mitochondria, they also coordinate the cross-talk of death signaling among intracellular compartments including the mitochondrion, the nucleus and the endoplasmic reticulum (Gross et al., 1999; Adams and Cory, 2001; Zong et al., 2003; Scorrano et al., 2003). In addition to its central role in mediating intrinsic cell death, the mitochondria-mediated pathway can also amplify death signals from the extrinsic pathway through a Bid-mediated mechanism (Li et al., 1998; Luo et al., 1998; Yin et al., 1999). Apoptosis in response to DNA damage is accompanied by an up-regulation of p53, followed by p53-mediated transcription of Puma, Noxa and Bax (Haupt et al., 2003).

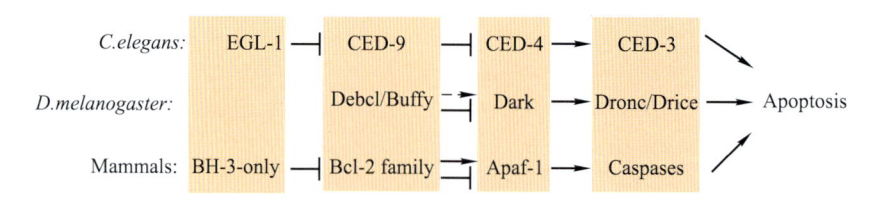

```
C.elegans:      EGL-1 ──┤ CED-9 ──┤ CED-4 ──→ CED-3        ╲
D.melanogaster:         Debcl/Buffy ⇉ Dark ──→ Dronc/Drice ──→ Apoptosis
Mammals:     BH-3-only ──┤ Bcl-2 family ⇉ Apaf-1 ──→ Caspases  ╱
```

Fig.26.1 The conserved apoptosis machinery. Caspases are crucial mediators of apoptosis in C. elegans, D. melanogaster as well as in mammals. Caspase (CED-3, Dronc/Drice) activation is promoted by adaptor molecules CED-4, Dark, or Apaf-1. Upstream Bcl-2 homologs (Debcl/Buffy, CED-9) regulate functions of Apaf-1. The Bcl-2 homologous proteins can be further divided into BH3-only proteins (EGL-1 in C. elegans), as well as multi-domain pro- and anti-apoptotic proteins. The function of Drosophila Bcl-2 family members, Debcl and Buffy, has not been firmly established in apoptosis regulation.

Protein cleavage and degradation are an imperative aspect of apoptosis. More than a dozen caspases mediate apoptosis by cleaving downstream molecules (Fischer et al., 2003). The activities of the lysosomal cysteine protease, cathepsins B, D, and L also contribute to the intrinsic apoptotic pathway (Guicciardi et al., 2000; Ferri and Kroemer, 2001; Jaattela and Tschopp, 2003; Guicciardi et al., 2004). Deficiencies and overabundances of these regulators and executioners of apoptosis can lead to various developmental abnormalities and tissue homeostasis phenotypes in mice and humans (Ranger et al., 2001).

Cells may die by alternative paths distinctive from the typical apoptotic route, depending on death stimuli, nutrient availability and whether the canonical apoptosis pathway are blocked at the levels of activation of Bax and Bak or caspases (Ferrari et al., 1998; Los et al., 2002; Schwab et al., 2002; Zong et al., 2004). In the shortage of ATP, or in response to profound pathological damage or physical insults, cells may commit to necrotic cell death characteristic of cell body swelling and cellular membrane ruptures (Zeiss, 2003; Proskuryakov et al., 2003; Nelson and White, 2004). Necrosis was thought to be solely a passive response to extracellular physical and chemical damage. Contrary to the conventional wisdom, recent work has found that necrosis can be regulated by not only extracellular signals but also intracellular signals. Understanding the regulation of necrosis is particularly relevant to cancer therapy, because most cancers have acquired mutations that confer resistance to apoptosis. Furthermore, the pro-inflammatory response triggered by necrosis may also elicit systemic reactions to cancer therapy and thereby contribute to cancer regression. Notably, because cancer cells are proliferating cells highly dependent on glycolysis to generate cellular energy, they are more susceptible to poly(ADP-ribose) polymerase (PARP)-mediated NAD depletion, which leads to necrosis, in response to DNA-alkylating damage (Zong et al., 2004).

When deprived of growth factors or nutrient source, or in response to certain cellular stress, eukaryotic cells may also start to form double membrane vesicles termed autophagosomes that enclose the excessive or damaged organelles and macromolecules. The autophagosomes fuse with lysosomes, and the enclosed intracellular materials are digested to provide the cell with energy and molecular components to sustain the minimal cellular function and survival. This "self-eating" process is termed autophagy. Although considered a cell survival mechanism under nutrient starvation and other stress, autophagy results in cell death under extensive self-digestion (Nelson and White, 2004; Levine and Klionsky, 2004; Levine, 2005; Klionsky, 2005). Autophagic cell death appears to be coordinated with apoptotic cell death by the PI3K and mTOR pathways that sense nutrient availability (Shintani and Klionsky, 2004; Asnaghi et al., 2004).

Transcription Factors in Regulating Invertebrate Cell Death

In *C. elegans*, developmental cell death involves an invariant number and preset lineage of cells. Of 1090 somatic cells generated through cell division in the entire lifespan of the hermaphrodites, 131 die through programmed cell death (Horvitz, 1999). The neurosecretory motor neuron (NSM) sister cells undergo lineage-invariant developmental cell death. In NSM sister cell death, the expression of EGL-1, the homolog of the mammalian BH3-only proteins, is regulated at the level of transcription by coordinated functions of CES-1, CES-2, HLH-2 and HLH3 (Thellmann et al., 2003). EGL-1 can also be regulated by transcription factor TRA-1 to mediate death of hermaphrodite-specific neurons (HSNs) in the males (Conradt and Horvitz, 1999). Genotoxic stress induces EGL-1 expression to regulate germline cell death in a manner that is dependent on CEP-1, a p53 homolog (Hofmann et al., 2002). Other transcription factors may also regulate programmed cell death in *C. elegans*. For example, PAG-3 mutation results in extra cell corpses due to reiterated neuroblast cell death. However, the target genes and the mechanisms of how PAG-3 influences cell death are unknown (Cameron et al., 2002). Additionally, the EOR-1 putative transcription factor may be involved in chromatin remodeling and influence cell death (Hoeppner et al., 2004). Over all, many of the decisions of programmed cell death can be regulated at the level of transcription in *C. elegans* (Fig.26.2).

In *Drosophila*, one key point of apoptosis regulation is at the transcription level for Reaper, Hid, Grim (RHG) family of pro-apoptotic proteins that also include Sickle and JAFRAC2. Up-regulation of RHG proteins can initiate apoptosis by disrupting the inhibition of caspase activities (Dronc/Drice) by the inhibitors of apoptosis (DIAP) (Fig. 26.3) (Bergmann et al., 1998; Hay et al., 2004). During development, the steroid ecdysone induces a two-step cell death program in the larval midgut and the salivary gland (Thummel, 1996; Baehrecke, 2000). Ecdysone can bind to the EcR and Usp (*ultraspiracle*, a RXR homolog) heterodimeric receptor and transcriptionally regulate the expression of Broad-Complex (BR-C), E74 and D75. BR-C (a zinc finger transcription factor), E74 (an ETS-like transcription factor) and E75 (an orphan nuclear receptor transcription factor) can then regulate Reaper, Hid, Dark, and Dronc expression to modulate the hormone-induced developmental cell death (Fig. 26.3) (Hall and Thummel, 1998; Jiang et al., 2000; Lee et al., 2000; Lee et al., 2002; Cakouros et al., 2002; Kilpatrick et al., 2005). Other transcription factors, such as dfos, can modulate the induction of Reaper and Hid to ensure the proper timing of larval salivary gland cell death (Lehmann et al., 2002). To maintain maxillary and mandibular segment boundaries, transcriptional activation of Reaper by the Hox gene, Deformed (Dfd), occurs to ensure the location-specific cell death (Lohmann et al., 2002).

Abdominal segment boundary cell death is stimulated by transcriptional activation of Reaper by the Hox gene, Abdominal B (Abd-B) (Lohmann et al., 2002). Recent findings suggest that Abd-B also has an anti-apoptotic role in preventing pioneer neuronal cell death in posterior segments by repressing Reaper and Grim expression (Miguel-Aliaga and Thor, 2004). Abdominal A (Abd-A) is important for neuroblast cell death, although which of the RHG genes is regulated by

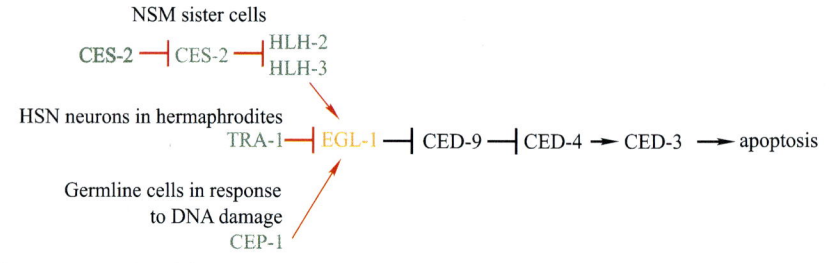

Fig.26.2 Decisions of programmed cell death can be regulated at the level of transcription in *C. elegans*. In neurosecretory motor neuron (NSM) sister cells, EGL-1 expression is regulated at the level of transcription by coordinate functions of transcription factor CES-1, CES-2, HLH-2 and HLH3. In hermaphrodite-specific neurons (HSNs) in the males, EGL-1 can be regulated by transcription factor TRA-1 to mediate HSN cell death. During germline cell death in response to DNA damage, EGL-1 expression is regulated by a p53 homolog, CEP-1. Transcription regulation of apoptotic genes is indicated by red lines. Transcription factors are in green. Cell death genes that are regulated at the level of transcription are in orange.

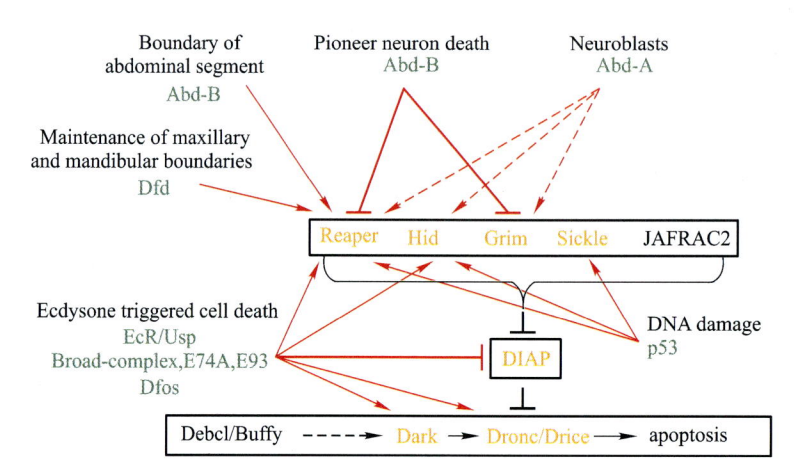

Fig.26.3 Transcriptional regulation of apoptosis in *D. melanogaster*. In *Drosophila*, a key point of apoptosis regulation is mediated at the transcriptional level for RHG family of pro-apoptotic proteins: Reaper, Hid, Grim, Sickle and JAFRAC2. Up-regulation of RHG proteins disrupts the inhibition of caspases (Dronc/Drice) by DIAP. The ecdysone-induced larval midgut and salivary gland cell death is mediated by EcR/Usp heterodimeric transcription factor and downstream transcription factors, BR-C, E74 and D75. dfos modulates the proper timing of larval salivary gland cell death via regulation of Reaper and Hid expression. Cell death in maxillary and mandibular segment boundaries is controlled by transcriptional activation of Reaper by Dfd. Cell death in abdominal segment boundaries is controlled by transcriptional activation of Reaper by Abd-B. Abd-B also has an anti-apoptotic role in preventing pioneer neuron cell death in posterior segments by repressing Reaper and Grim expression. Abd-A is important for neuroblast cell death, although which of the RHG gene is regulated by Abd-A is currently unclear. Midline glia cell death is controlled by Hid through a RAS-MAPK-dependent mechanism. DNA damage caused by γ-irradiation is mediated by up-regulation of Reaper, Hid, Sickle via a p53-dependent pathway, whereas UV induced Dark expression is dependent on E2F. Transcription regulation of apoptotic genes is indicated by red lines. Transcription factors are in green. Cell death genes that are regulated at the level of transcription are in orange.

Abd-A is currently unclear (Bello *et al.*, 2003). Hid regulation, either transcriptionally or post-translationally by the RAS-MAPK pathway, is important in initiating midline glia cell death (Bergmann *et al.*, 1998; Kurada and White, 1998; Bergmann *et al.*, 2002). Furthermore, dMyc-induced cell competition and neighboring cell death is associated with Hid mRNA up-regulation (de la Cova *et al.*, 2004; Moreno and Basler, 2004). Interestingly, DNA damage caused by γ-irradiation induces cell death by up-regulating Reaper, Hid, Sickle via a p53-dependent pathway (Sogame *et al.*, 2003; Brodsky *et al.*, 2004), whereas UV-induced Dark expression is dependent on E2F (Zhou and Steller, 2003).

Transcriptional Regulation of Cell Death in Mammals

A: Transcriptional Regulation of the Core Components of the Apoptosis Machinery

Transcriptional regulation of the core components of the apoptosis machinery also provides an important mechanism for controlling cell death in mammals. Transcriptional regulation of Bcl-2 family proteins is a conserved mechanism from *C. elegans* to humans. As discussed above, Bcl-2 family proteins play an important role in regulating mitochondrial membrane permeabilization, the initiation of apoptosis from multiple intracellular compartments, as well as the cross-talk from the extrinsic to the intrinsic apoptotic pathways. Three classes of Bcl-2 family proteins exist. One class is the pro-apoptotic BH3-only proteins that are homologs of *C. elegans* protein EGL-1. This class includes Bad, Bik/Nbk, Bim/Bok, Bmf, Bnip3/Nix, Noxa, Puma, and Bid that act upstream of the multi-BH domain proteins. The second class includes pro-apoptotic "BH1-3-multi-domain" proteins, such as Bax, Bak, and Bok. The third class includes anti-apoptotic BH1-4-multi-domain proteins such as Bcl-xL, Bcl-w, Bcl-2, Mcl-1, A1/Bfl-1, and Boo/Diva (Adams and Cory, 2001; Danial and Korsmeyer, 2004).

Correlating with their crucial roles in regulating cell death and survival, the expression of the Bcl-2 family proteins is controlled by multiple factors at multiple levels and fluctuates under different developmental, differentiation and environmental contexts (Fig.26.4) (Mayo *et al.*, 1999; Margue *et al.*, 2000; Ha *et al.*, 2001; Sevilla *et al.*, 2001; Russell *et al.*, 2002; Heckman *et al.*, 2003; Vickers *et al.*, 2004; Soleymanlou *et al.*, 2005; Meller *et al.*, 2005). Dysregulated or unbalanced expression of pro-and anti-apoptotic Bcl-2 family proteins has been noted in many human diseases including cancers (Coultas and Strasser, 2003; Shacka and Roth, 2005).

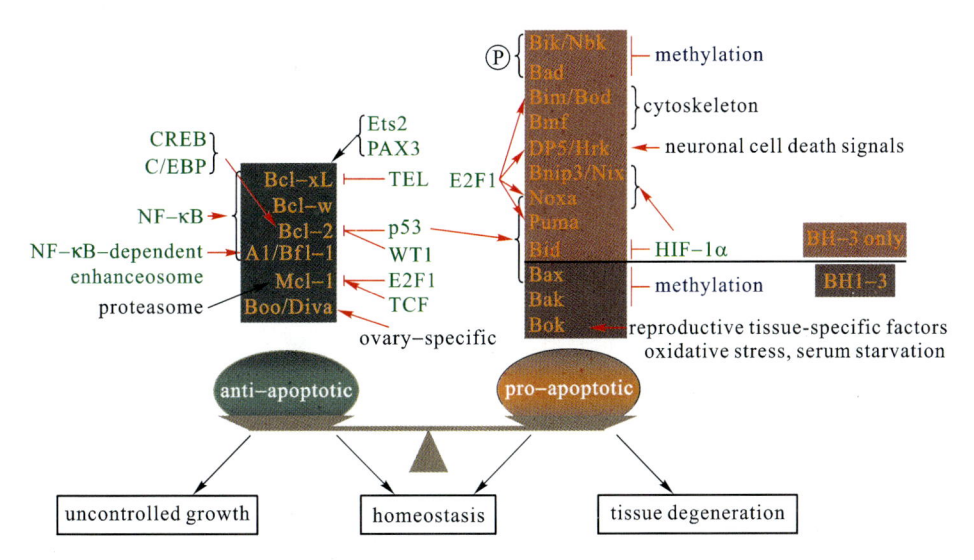

Fig.26.4 Regulation of Bcl-2 family proteins in mammals. The relative expression levels of the Bcl-2 family proteins determine the balance between cell death and survival. Three classes of Bcl-2 family proteins exist. One class is the pro-apoptotic BH3-only proteins that are homologs of *C. elegans* protein EGL-1. This class includes: Bim/Bok, Bmf, Bad, Bik/Nbk, Bid, Noxa, Puma, and Bnip3/Nix. The second class includes pro-apoptotic BH1-BH3-domain-containing proteins, such as Bax, Bak and Bok. The third class includes anti-apoptotic multi-BH domain proteins such as Bcl-xL, Bcl-w, Bcl-2, Mcl-1, A1/Bfl-1 and Boo/Diva. Bid, Puma, Noxa, Bax are direct p53 targets, whereas the expression of anti-apoptotic Bcl-2 gene is inhibited by p53. Bcl-xl, Bcl-w, Bcl-2 and A1/Bfl-1 are regulated by NF-κB. Mcl-1, pro-apoptotic Bim, DP5, Noxa and Puma are regulated by E2F-1. HIF-1α regulates the expression of Noxa, Bid, Bnip3/Nix and NF-κB. Methylation and histone acetylation also regulate the expression of some of the Bcl-2 family genes. Additional levels of regulation are provided by phosphorylation for Bad and Bik, and by sequestration to cytoskeletal structures for Bim and Bmf. Mcl-1 is regulated by proteasome-mediated degradation in response to UV.

Of particular interest, transcription factors p53 plays an important role in regulating the expression of a number of Bcl-2 family proteins. The pro-apoptotic BH3-only members Bid, Puma and Noxa, and the pro-apoptotic BH1-3 protein Bax, are direct p53 transactivation targets, whereas the expression of the anti-apoptotic Bcl-2 protein is inhibited by p53 (Fig. 26.4) (Puthalakath and Strasser, 2002; Koutsodontis and Kardassis, 2004; Gu et al., 2004; Schuler and Green, 2005). The altered expression of these proteins correlates with p53-mediated cell death in response to DNA damage. As transcriptional control of Bcl-2 family genes by p53 is of requisite importance in determining cell fate, caution needs to be made in interpreting experimental results demonstrating cell death regulation through p53 activation, because an alternative route of p53 function may be through transcription-independent, mitochondrial translocation-dependent mechanisms (Marchenko et al., 2002; Mihara et al., 2003; Erster et al., 2004; Leu et al., 2004; Chipuk et al., 2004).

Transcriptional regulation of Bcl-2 family proteins can couple with several other signaling pathways. The expression of several anti-apoptotic Bcl-2 family members, such as Bcl-xL, Bcl-w, Bcl-2 and A1/Bfl-1, is regulated by NF-κB transcription factors, correlating with the cell survival-promoting activity of NF-κB (Zong et al., 1999; Lee et al., 1999; Grumont et al., 1999; Chen et al., 2000; Kurland et al., 2001; Varfolomeev and Ashkenazi, 2004; Tran et al., 2005). In response to hypoxia conditions, such as neurons and cardiomyocytes in cerebral and myocardial ischemia, cells in the center of solid tumors, and neutrophils exiting from the circulatory system to combat inflammation, hypoxia-inducible factor α (HIF-1α) regulates the expression of Bnip3/Nix (Bruick, 2000; Sowter et al., 2001), Noxa (Kim et al., 2004), Bid (Erler et al., 2004) and NF-κB (Walmsley et al., 2005), thereby modulating cell survival and cell death.

Cancer cells acquire dysregulated cell cycle regulation to sustain their abnormal proliferation. The transcription factors controlling cell cycle may be involved in the regulation of apoptosis. The expression of anti-apoptotic Bcl-2 is positively regulated by the retinoblastoma protein (Rb) (Decary et al., 2002). The expression of the anti-apoptotic Mcl-1 is repressed, and that of the pro-apoptotic Bim, DP5, Noxa, and Puma is up-regulated by E2F-1 respectively (Croxton et al., 2002; Hershko and Ginsberg, 2004). Because the Rb-E2F pathway monitors normal cell cycle progression, as well as cell proliferation in response to DNA damage by regulating genes involved in cell cycle and DNA replication, the ability for Rb and E2F to regulate apoptotic gene expression enables a coordination of cell cycle and apoptosis control. The up-regulation of pro-apoptotic proteins by E2F may render the fast cycling cancer cells lower threshold in tolerating apoptosis and therefore the susceptibility to chemotherapeutic treatments.

Transcriptional regulation also affects the expression of apoptotic factors in the extrinsic death pathway in mammals. The expression of the Fas ligand in the immune system is regulated by a large number of transcription factors, including AP-1, NF-κB, Egr, and NF-AT (Li-Weber and Krammer, 2003). The expression of Fas is regulated by p53, c-Jun and Stat3 (Owen-Schaub et al., 1995; Ivanov et al., 2001). The expression of the TNF receptor associated factor (TRAF), is regulated by NF-κB (Poppelmann et al., 2005). The expression of caspase-8 inhibitor FLIP is regulated by FOXO3a, NF-κB, and c-Myc (Micheau et al., 2001; Gerondakis S, Strasser A, 2003; Skurt et al., 2004; Ricci et al., 2004). TRAIL receptor DR5, as well as caspase-8 are also regulated by p53 (Owen-Schaub et al., 1995; Wu et al., 1997; Haupt et al., 2003; Schuler and Green, 2005).

Other key components of the apoptosis pathway subject to transcriptional control include Apaf-1, caspases and inhibitors of apoptosis (IAPs). Apaf-1 and caspase-6 are activated, whereas an IAP family member, survivin, is repressed by p53 and may contribute to p53-mediated apoptosis in response to DNA damage (Moroni et al., 2001; MacLachlan and El-Deiry, 2002; Hoffman et al., 2002; Shishodia and Aggarwal, 2004). Apaf-1 and caspases 3,7,8,9 are regulated by E2F, and thereby providing a mechanism for coupling cell proliferation to sensitization to apoptosis (Moroni et al., 2001; Nahle et al. 2002). Appropriate coupling of S phase entry and cell death sensitization may be crucial for limiting the transforming potential of oncogene activation or tumor suppressor gene inactivation, as well as sensing and responding to DNA damage (Evan and Vousden, 2001; Lin and Lowe, 2001; Nahle et al. 2002; Stevens et al., 2003). IAP family proteins, c-IAP1 and c-IAP2, are activated by NF-κB and may facilitate the protective role of NF-κB against TNF-α-induced apoptosis (Chu et al., 1997; Wang et al., 1998).

B: Transcriptional Regulation of Cell Death by Chromatin Remodeling

From the fruitful investigation of the transcription mechanisms in eukaryotes, it became clear that transcription is regulated by co-activators, co-repressors, histone modification enzymes, as well as by sequence-specific transcription factors. These mechanisms turn out to also play important roles in regulating cell death (Fig.26.5). First, histone phosphorylation and histone export from the nucleus may have significant impact on transcription as well as serve as signals for cell death induction (Konishi et al., 2003; Fernandez-Capetillo et al., 2004; Ahn et al., 2005). Second, histone acetylation and deacetylation affect condensation of nucleosome structure, and histone acetylase and histone deacetylases can act as transcriptional coactivators and corepressors (Grunstein, 1997; Gregory et al., 2001). Inhibition of histone deacetylase induces cell death by regulating gene expression in multiple cell death pathways. To take advantage of these findings, inhibitors of histone deacetylase are being developed as potential cancer therapeutic agents (Peart et al., 2005; Duan et al., 2005). Third, transcription of a number of genes involved in apoptosis, such as Bad, Bak, Bik, and Bax are suppressed by DNA methylation (Pompeia et al., 2004). Epigenetic silencing of apoptotic gene transcription may contribute to chemoresistance of cancer cells.

Chromatin remodeling and transcriptional regulation by co-activators or co-repressors play a crucial role in cell death regulated by p53 and NF-κB (Lill et al., 1997; Thomas and White, 1998; Yamit-Hezi and Dikstein, 1998; Shikama et al., 1999; Yamit-Hezi et al., 2000; Mujtaba et al., 2004; Banerjee et al., 2004; Hoberg et al., 2004). DNA damage-induced transcription blockade and subsequent pausing of RNA polymerase II may play a causative role to p53 modification, loss of nuclear export of RNA, and preferential inhibition of transcription of survival genes that coincidently have large gene sizes (Ljungman and Lane, 2004).

Global transcription can be modulated by chromatin condensation or histone polyADP- ribosylation (Slattery et al., 1983; Chiarugi, 2002). PolyADP-ribose polymerase-1 (PARP-1), the enzyme that catalyzes polyADP-ribosylation of histones and other chromatin binding proteins involved in DNA repair and transcriptional regulation, is activated in response to DNA damage and has been demonstrated to be involved in the regulation of both apoptosis and necrosis (Berger, 1985; Wang et al. 1997; Ha and Snyder 1999; Yu et al. 2002; Chiaguri, 2002; Zong et al., 2004). PARP-1 is also a critical regulator of NF-κB activity, and is involved in cell death mediated by AIF (D'Amours et al., 1999, Oliver et al., 1999; Yu et al., 2002).

C: Are Necrosis and Autophagy Regulated at the Transcription Level?

Transcription regulation of necrotic and autophagic cell death is not well studied. Nonetheless, many of the players involved in apoptotic regulation also participate in necrotic cell death, including TNF receptor family, Bcl-2 family, caspases, MAPK and JNK pathways, and PARP (Proskuryakov et al., 2003). The transcription regulation of these molecules will predictably influence the susceptibility of necrotic cell death. In addition, factors that are specifically involved in necrotic cell death, such as glutamate receptors in neurons, calcium binding proteins in a variety of cell types, components of the mitochondrial respiratory chain, and necrosis-specific DNases, may be regulated at the transcription level (Proskuryakov et al., 2003). Future work is necessary to establish the role of transcription in regulating necrosis.

Autophagy is concurrently regulated with apoptosis, transcription, translation and posttranslational protein modification by Akt- and mTOR-mediated signaling

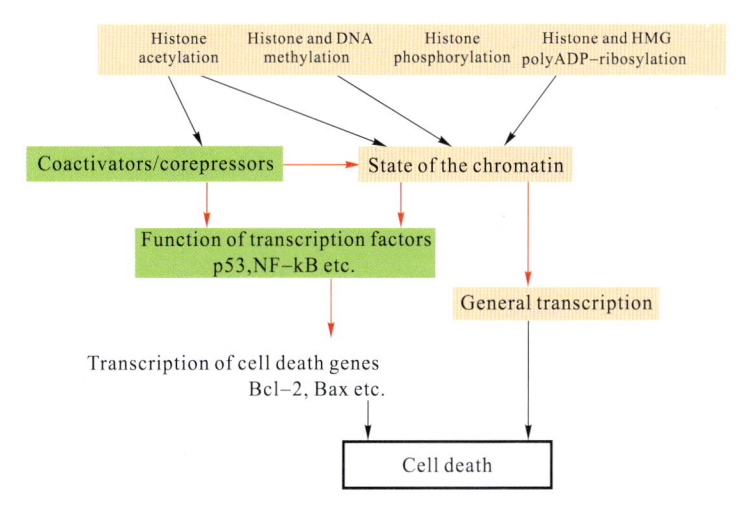

Fig.26.5 Chromatin remodeling influences global and cell death-specific gene transcription. Chromatin structure can be affected by histone acetylation, histone phosphorylation, DNA methylation as well as histone polyADP-ribosylation. Chromatin structure and other RNA polymerase II coactivators and corepressors can influence transcription of genes that play important roles in cell death.

pathways in response to nutrient deprivation (Datta et al., 1999; Cardenas et al., 1999; Chan et al., 2001; Vivanco and Sawyers, 2002; Baehrecke, 2003; Rohde and Cardenas, 2003; Rohde et al., 2004; Hay and Sonenberg, 2004). Akt regulates apoptosis through phosphorylation of factors involved in transcriptional regulation such as Foxo, IKKs, and p53 (Vivanco and Sawyers, 2002). mTOR modulates transcription of rRNA by RNA Pol I, transcription of ribosomal protein by RNA Pol II, and transcription of tRNA and 5S RNA by RNA Pol III (Tsang et al., 2003; Hay and Sonenberg, 2004). How these transcription events play a role in regulating autophagic cell death needs further investigation.

Unresolved Issues

Understanding of cell type-specific and stimulus-specific regulation of transcription of various apoptotic factors is still incomplete. Even more so are the epigenetic and biochemical mechanisms of transcription activities on promoters of the apoptotic factors. Although many studies demonstrate a concerted activation of pro- apoptotic genes with inhibition of anti-apoptotic genes, examples do exist when pro- and anti-apoptotic genes are activated at the same time. Conceivably, at the apical point of any death initiation process, two dividing forces are provoked into action, to fight for life or to give it up for good. Complete transcription shutdown may speed up the death process. Prolonged caspase activation results in degradation of many factors influencing transcription, thereby playing a role in shutting down transcription in anticipation of death (Fischer et al., 2003). However, transient activation of caspase-3 without DNA fragmentation results in cell survival as evidenced in ischemic pre-conditioning in neurons, perhaps by preserving DNA integrity and transcription (Tanaka et al., 2004). The ultimate cell fate obviously depends on the tug-of-war between the two competing forces. Although at the completion of cell death, DNA is destroyed to eliminate any new RNA synthesis, it remains unclear whether any concurrent transcription activity continues during DNA fragmentation.

Gene expression by *de novo* mRNA synthesis can serve as the first line of defense in determining cell death and survival. Still, post-transcriptional regulatory strategies supply speedy response to death stimuli in many occasions (Clemens et al., 2000; Salvesen and Duckett, 2002; Fischer et al., 2003; Tran et al., 2004; Holcik and Sonenberg, 2005; Vaux and Silke, 2005). Whether and how transcriptional regulation of cell death coordinates with post-transcriptional mechanisms is still unclear.

Concluding Remarks

Cell death is a highly regulated biological process. Transcriptional control of specific cell death factors is an evolutionarily conserved mechanism observed in different cell types in distant species. Transcriptional regulation of certain cell death/survival factors is a prerequisite in many death-associated situations. In addition to transcriptional control of the core cell death factors, global modulation of transcription also contributes to whether, when, and how cells die. Global transcription levels seem to be reduced by chromatin modification such as histone acetylation, methylation, phosphorylation, ubiquitination, or polyADP-ribosylation, while at the same time specific cell death factors are synthesized.

Disjointed cell death and transcription regulation may contribute to developmental, proliferative, and degenerative diseases, such as tumorigenesis, acute, and chronic neurodegeneration, and immunological diseases (Fridman and Lowe, 2003; Kucharczak et al., 2003; Shacka and Roth, 2005). Delivery of transcription factors such as p53, and use of inhibitors of histone deacetylation, targeting NF-κB, E2F-1 or PARP to treat cancer or neurodegeneration, are being investigated (Virag and Szabo, 2002; Fang and Roth, 2003; Lin and Karin, 2003; Wang et al., 2003; Marks et al., 2004; Bell and Ryan, 2005; Fischer and Schulze-Osthoff, 2005). These strategies target multiple downstream cell death pathways to switch on or off cell death, therefore may help overcome the limitation of single target therapeutic strategies that may not be effective in cells defective in certain cell death pathways. Challenges remain to understand the sophistication and impact of coupling regulation of transcription and cell death, and to devise effective strategies to treat diseases.

Acknowledgment

We thank Drs. Jun Ma, Johanna Meij, and Xiao-Ming Yin for discussions and critical reading of this chapter. J.Z. is supported by NIH, DOD, Lupus Research Institute, Epilepsy Foundation, Ohio ACS, University of Cincinnati Dean's Discovery Fund, and University of Cincinnati Center for Environmental Genetics. W.X.Z. is supported by the Leukemia and Lymphoma Society.

References

Adams, J.M., and Cory, S. (2001). Life-or-death decisions by the Bcl-2 protein family. Trends Biochem Sci 26, 61-66.

Ahn, S.H., Cheung, W.L., Hsu, J.Y., Diaz, R.L., Smith, M.M., and Allis, C.D. (2005). Sterile 20 kinase phosphorylates histone H2B at serine 10 during hydrogen peroxide-induced apoptosis in S. cerevisiae. Cell 120, 25-36.

Baehrecke, E. H. (2000). Steroid regulation of programmed cell death during Drosophila development. Cell Death Differ 7, 1057-1062.

Baehrecke, E.H. (2003). Autophagic programmed cell death in Drosophila. Cell Death Differ 10, 940-945.

Banerjee, S., Kumar, B.R., and Kundu, T.K. (2004). General transcriptional coactivator PC4 activates p53 function. Mol Cell Biol 24, 2052-2062.

Baud, V., and Karin, M. (2001). Signal transduction by tumor necrosis factor and its relatives. Trends Cell Biol 11, 372-327.

Bell, H.S., and Ryan, K.M. (2005). Intracellular signalling and cancer: complex pathways lead to multiple targets. Eur J Cancer 41, 206-215.

Bello, B.C., Hirth, F., and Gould, A.P. (2003). A pulse of the Drosophila Hox protein Abdominal-A schedules the end of neural proliferation via neuroblast apoptosis. Neuron 37, 209–219.

Berger, N.A. (1985). Poly(ADP-ribose) in the cellular response to DNA damage. Radiat Res 101, 4-15.

Bergmann, A., Agapite, J., McCall, K.A., and Steller, H. (1998). The Drosophila gene hid is a direct molecular target of Ras-dependent survival signaling. Cell 95, 331–341.

Bergmann, A., Tugentman, M., Shilo, B.Z., and Steller, H. (2002). Regulation of cell number by MAPK-dependent control of apoptosis: a mechanism for trophic survival signaling. Dev Cell 2, 159-170.

Brodsky, M.H., Weinert, B.T., Tsang, G., Rong, Y.S., McGinnis, N.M., Golic, K.G., Rio, D.C., and Rubin, G.M. (2004) .Drosophila melanogaster MNK/Chk2 and p53 regulate multiple DNA repair and apoptotic pathways following DNA damage. Mol Cell Biol 24, 1219-1231.

Bruick, R.K. (2000). Expression of the gene encoding the proapoptotic Nip3 protein is induced by hypoxia. PNAS 97, 9082-9087.

Cakouros, D., Daish, T., Martin, D., Baehrecke, E.H., and Kumar, S. (2002). Ecdysone-induced expression of the caspase DRONC during hormone-dependent programmed cell death in Drosophila is regulated by Broad-Complex. J Cell Biol 157, 985-995.

Cameron, S., Clark, S.G., McDermott, J.B., Aamodt, E., and Horvitz, H.R. (2002). PAG-3, a Zn-finger transcription factor, determines neuroblast fate in C. elegans. Development 129, 1763-1774.

Cardenas, M.E., Cutler, N.S., Lorenz, M.C., Di Como, C.J., and Heitman, J. (1999). The TOR signaling cascade regulates gene expression in response to nutrients. Genes Dev 13, 3271-3279.

Chan, T.F., Bertram, P.G., Ai, W., and Zheng, X.F. (2001). Regulation of APG14 expression by the GATA-type transcription factor Gln3p. J Biol Chem 276, 6463-6467.

Chen, C., Edelstein, L.C., and Gelinas, C. (2000). The Rel/NF-kappa B family directly activates expression of the apoptosis inhibitor Bcl-xL. Mol Cell Biol 20, 2687-2695.

Chiarugi, A. (2002). Poly(ADP-ribose) polymerase: killer or conspirator? The 'suicide hypothesis' revisited. Trends Pharmacol Sci 23, 122-129.

Chipuk, J.E., Kuwana, T., Bouchier-Hayes, L., Droin, N.M., Newmeyer, D.D., Schuler, M., and Green, D.R. (2004). Direct activation of Bax by p53 mediates mitochondrial membrane permeabilization and apoptosis. Science 303, 1010-1014.

Chu, Z.L., McKinsey, T.A., Liu, L., Gentry, J.J., Malim, M.H., and Ballard, D.W. (1997). Suppression of tumor necrosis factor-induced cell death by inhibitor of apoptosis c-IAP2 is under NF-kappa B control. PNAS 94, 10057-10062.

Clemens, M.J., Bushell, M., Jeffrey, I.W., Pain, V.M., and Morley, S.J. (2000). Translation initiation factor modifications and the regulation of protein synthesis in apoptotic cells. Cell Death Differ 7, 603-615.

Conradt, B., and Horvitz, H.R. (1999). The TRA-1A sex determination protein of C. elegans regulates sexually dimorphic cell deaths by repressing the egl-1 cell death activator gene. Cell 98, 317-327.

Coultas, L., and Strasser, A. (2003). The role of the Bcl-2 protein family in cancer. Semin Cancer Biol 13, 115-123.

Croxton, R., Ma, Y., Song, L., Haura, E.B., and Cress, W.D. (2002). Direct repression of the Mcl-1 promoter by E2F1. Oncogene 21, 1359-1369.

D'Amours, D., Desnoyers, S., D'Silva, I., and Poirier, G.G. (1999). Poly(ADP-ribosyl)ation reactions in the regulation of nuclear functions. Biochem J 342, 249-268.

Danial, N.N., and Korsmeyer, S.J. (2004). Cell Death: Critical Control Points. Cell 116, 205-219.

Datta, S.R., Brunet, A., and Greenberg, M.E. (1999). Cellular survival: a play in three Akts. Genes Dev 13, 2905-2927.

de la Cova, C., Abril, M., Bellosta, P., Gallant, P., and Johnston, L. (2004). Drosophila myc regulates organ size by inducing cell competition. Cell 117, 107-116.

Decary, S., Decesse, J.T., Ogryzko, V., Reed, J.C., Naguibneva, I., Harel-Bellan, A., and Cremisi, C.E. (2002). The retinoblastoma protein binds the promoter of the survival gene bcl-2 and regulates its transcription in epithelial cells through transcription factor AP-2. Mol Cell Biol 22, 7877-7888.

Deng, Y., Ren, X., Yang, L., Lin, Y., and Wu, X. (2003). A JNK-dependent pathway is required for TNFα-induced apoptosis. Cell 115, 61-70.

Duan, H., Heckman, C.A., and Boxer, L.M. (2005). Histone deacetylase inhibitors down-regulate bcl-2 expression and induce apoptosis in t(14;18) lymphomas. Mol Cell Biol 25, 1608-1619.

Erler, J.T., Cawthorne, C.J., Williams, K.J., Koritzinsky, M., Wouters, B.G., Wilson, C., Miller, C., Demonacos, C., Stratford, I.J., and Dive, C. (2004). Hypoxia-mediated down-regulation of Bid and Bax in tumors occurs via Hypoxia-Inducible Factor 1-dependent and -independent mechanisms and contributes to drug resistance. Mol Cell Biol 24, 2875-2889.

Erster, S., Mihara, M., Kim, R.H., Petrenko, O., and Moll, U.M. (2004). *In vivo* mitochondrial p53 translocation triggers a rapid first wave of cell death in response to DNA damage that can precede p53 target gene activation. Mol Cell Biol 24, 6728-6741.

Evan, G.I., and Vousden, K.H. (2001). Proliferation, cell cycle and apoptosis in cancer. Nature 411, 342-348.

Fang, B., and Roth, J.A. (2003). Tumor-suppressing gene therapy. Cancer Biol Ther (4 Suppl 1), S115-121.

Fernandez-Capetillo, O., Allis, C.D., and Nussenzweig, A. (2004). Phosphorylation of histone H2B at DNA double-strand breaks. J Exp Med 199, 1671-1677.

Ferrari, D., Stepczynska, A., Los, M., Wesselborg, S., and Schulze-Osthoff, K. (1998). Differential regulation and ATP requirement for caspase-8 and caspase-3 activation during CD95- and anticancer drug-induced apoptosis. J Exp Med 188, 979-984.

Ferri, K.F., and Kroemer, G. (2001). Organelle-specific initiation of cell death pathways. Nat Cell Biol 3, E255-263.

Fischer, U., Janicke, R.U., and Schulze-Osthoff, K. (2003). Many cuts to ruin: a comprehensive update of caspase substrates. Cell Death Differ 10, 76-100.

Fischer, U., and Schulze-Osthoff, K. (2005). Apoptosis-based therapies and drug targets. In press.

Fridman, J.S., and Lowe, S.W. (2003). Control of apoptosis by p53. Oncogene 22, 9030-9040.

Gerondakis, S., and Strasser, A. (2003). The role of Rel/NF-kappaB transcription factors in B lymphocyte survival. Semin Immunol 15, 159-166.

Gregory, P.D., Wagner, K., and W. Horz. (2001). Histone acetylation and chromatin remodeling. Exp Cell Res 265, 195-202.

Gross, A., McDonnell, J.M., and Korsmeyer, S.J. (1999). BCL-2 family members and the mitochondria in apoptosis. Genes Dev 13, 1899-1911.

Grumont, R.J., Rourke, I.J., and Gerondakis, S. (1999). Rel-dependent induction of A1 transcription is required to protect B cells from antigen receptor ligation-induced apoptosis. Genes Dev 13, 400-411.

Grunstein, M. (1997). Histone acetylation in chromatin structure and transcription. Nature 389, 349-352.

Gu, J., Zhang, L., Swisher, S.G., Liu, J., Roth, J.A., and Fang, B. (2004). Induction of p53-regulated genes in lung cancer cells: implications of the mechanism for adenoviral p53-mediated apoptosis. Oncogene 23, 1300-1307.

Guicciardi, M.E., Deussing, J., Miyoshi, H., Bronk, S.F., Svingen, P.A., Peters, C., Kaufmann, S.H., and Gores, G.J. (2000). Cathepsin B contributes to TNF-alpha-mediated hepatocyte apoptosis by promoting mitochondrial release of cytochrome c. J Clin Invest 106, 1127-3117.

Guicciardi, M.E., Leist, M., and Gores, G.J. (2004). Lysosomes in cell death. Oncogene 23, 2881-2890.

Ha, H.C., and Snyder, S.H. (1999). Poly(ADP-ribose) polymerase is a mediator of necrotic cell death by ATP depletion. Proc Natl Acad Sci 96, 13978-13982.

Ha, S.H., Lee, S.R., Lee, T.H., Kim, Y.M., Baik, M.G., and Choi, Y.J. (2001). The expression of Bok is regulated by serum in HC11 mammary epithelial cells. Mol Cells 12, 368-371.

Hall, B.L., and Thummel, C.S. (1998). The RXR homolog ultraspiracle is an essential component of the *Drosophila* ecdysone receptor. Development 125, 4709-4717.

Haupt, S., Berger, M., Goldberg, Z., and Haupt, Y. (2003). Apoptosis - the p53 network. J Cell Sci 116, 4077-4085.

Hay, B.A., Huh, J.R., and Guo, M. (2004). The genetics of cell death: approaches, insights and opportunities in *Drosophila*. Nat Rev Genet 5, 911-922.

Hay, N., and Sonenberg, N. (2004). Upstream and downstream of mTOR. Genes Dev 18, 1926-1945.

Heckman, C.A., Wheeler, M.A., and Boxer, L.M. (2003). Regulation of Bcl-2 expression by C/EBP in t(14;18) lymphoma cells. Oncogene 22, 7891-7899.

Henson, P.M., Bratton, D.L., and Fadok, V.A. (2001). Apoptotic cell removal. Curr Biol 11, R795-805.

Hershko, T., and Ginsberg, D. (2004). Up-regulation of Bcl-2 homology 3 (BH3)-only proteins by E2F1 mediates apoptosis. J Biol Chem 279, 8627-8634.

Hoberg, J.E., Yeung, F., and Mayo, M.W. (2004). SMRT derepression by the IkappaB kinase alpha: a prerequisite to NF-kappaB transcription and survival. Mol Cell 16, 245-255.

Hoeppner, D.J., Spector, M.S., Ratliff, T.M., Kinchen, J.M., Granat, S., Lin, S.C., Bhusri, S.S., Conrad,t B., Herman, M.A., and Hengartner, M.O. (2004). eor-1 and eor-2 are required for cell-specific apoptotic death in *C. elegans*. Dev Biol 274, 125-138.

Hoffman, W.H., Biade, S., Zilfou, J.T., Chen, J., and Murphy, M. (2002). Transcriptional repression of the anti-apoptotic survivin gene by wild type p53. J Biol Chem 277, 3247-3257.

Hofmann, E.R., Milstein, S., Boulton, S.J., Ye, M., Hofmann, J.J., Stergiou, L., Gartner, A., Vidal, M., and Hengartner, M.O. (2002). *Caenorhabditis elegans* HUS-1 is a DNA damage checkpoint protein required for genome stability and EGL-1-mediated apoptosis. Curr Biol 19, 1908–1918.

Holcik, M., and Sonenberg, N. (2005). Translational control in stress and apoptosis. Nat Rev Mol Cell Biol 6, 318-327.

Horvitz, H.R. (1999). Genetic control of programmed cell death in the nematode *Caenorhabditis elegans*. Cancer Res 59, 1701s-1706s.

Ivanov, V.N., Bhoumik, A., Krasilnikov, M., Raz, R., Owen-Schaub, L.B., Levy, D., Horvath, C.M., and Ronai, Z. (2001).

Cooperation between STAT3 and c-Jun suppresses Fas transcription. Molecular Cell 7, 517-528.

Jaattela, M., and Tschopp, J. (2003). Caspase-independent cell death in T lymphocytes. Nat Immunol 4, 416-423.

Jiang, C., Lamblin, A.-F.J., Steller, H., and Thummel, C.S. (2000). A steroid-triggered transcriptional hierarchy controls salivary gland cell death during Drosophila metamorphosis. Mol Cell 5, 445-455.

Kerr, J.F.R., Wyllie, A.H., and Currie, A.R. (1972). Apoptosis: a basic biological phenomenon with wide-ranging implication in tissue kinetics. Br J Cancer 26, 239-257.

Kilpatrick, Z.E., Cakouros, D., and Kumar, S. (2005). Ecdysone-mediated up-regulation of the effector caspase DRICE is required for hormone-dependent apoptosis in Drosophila cells. J Biol Chem 280, 11981-11986.

Kim, J.Y., Ahn, H.J., Ryu, J.H., Suk, K., and Park, J.H. (2004). BH3-only protein Noxa is a mediator of hypoxic cell death induced by hypoxia-inducible factor 1α. J Exp Med 199, 113-124.

Kim, M.Y., Mauro, S., Gevry, N., Lis, J.T., and Kraus, W.L. (2004). NAD+-dependent modulation of chromatin structure and transcription by nucleosome binding properties of PARP-1. Cell 119, 803-814.

Klionsky, D.J. (2005). The molecular machinery of autophagy: unanswered questions. J Cell Sci 118, 7-18.

Konishi, A., Shimizu, S., Hirota, J., Takao, T., Fan, Y., Matsuoka, Y., Zhang, L., Yoneda, Y., Fujii, Y., Skoultchi, A.I., and Tsujimoto, Y. (2003). Involvement of histone H1.2 in apoptosis induced by DNA double-strand breaks. Cell 114, 673-688.

Koutsodontis, G., and Kardassis, D. (2004). Inhibition of p53-mediated transcriptional responses by mithramycin A. Oncogene 23, 9190-9200.

Kuan, C.Y., Roth, K.A., Flavell, R.A., and Rakic, P. (2000). Mechanisms of programmed cell death in the developing brain. Trends Neurosci 23, 291-297.

Kucharczak, J., Simmons, M.J., Fan, Y., and Gelinas, C. (2003). To be, or not to be: NF-kappaB is the answer-role of Rel/NF-kappaB in the regulation of apoptosis. Oncogene 22, 8961-8982.

Kumar, S., and Cakouros, D. (2004). Transcriptional control of the core cell-death machinery. Trends Biochem Sci 29, 193-199.

Kurada, P., and White, K. (1998). Ras promotes cell survival in Drosophila by downregulating hid expression. Cell 95, 319–329.

Kurland, J.F., Kodym, R., Story, M.D., Spurgers, K.B., McDonnell, T.J., and Meyn, R.E. (2001). NF-kappaB1 (p50) homodimers contribute to transcription of the bcl-2 oncogene. J Biol Chem 276, 45380-45386.

Lee, C.Y., Wendel, D.P., Reid, P., Lam, G., Thummel, C.S., and Baehrecke, E.H. (2000). E93 directs steroid-triggered programmed cell death in Drosophila. Mol Cell 6, 433-443.

Lee, C.Y., Simon, C.R., Woodard, C.T., and Baehrecke, E.H. (2002). Genetic mechanism for the stage- and tissue-specific regulation of steroid-triggered programmed cell death in Drosophila. Dev Biol 252, 138-148.

Lee, H.H., Dadgostar, H., Cheng, Q., Shu, J. and Cheng, G. (1999). NF- B-mediated up-regulation of Bcl-x and Bfl-1/A1 is required for CD40 survival signaling in B lymphocytes. Proc Natl Acad Sci 96, 9136-9141.

Lehmann, M., Jiang, C., Ip, Y.T., and Thummel, C.S. (2002). AP-1, but not NF-kappa B, is required for efficient steroid-triggered cell death in Drosophila. Cell Death Differ 9, 581-590.

Leu, J.I.J., Dumont, P., Hafey, M., Murphy, M.E., and George, D.L. (2004). Mitochondrial p53 activates Bak and causes disruption of a Bak-Mcl1 complex. Nat Cell Biol 6, 443-450.

Levine, B. (2005). Eating oneself and uninvited guests: autophagy-related pathways in cellular defense. Cell 120, 159-162.

Levine, B., and Klionsky, D.J. (2004). Development by self-digestion: molecular mechanisms and biological functions of autophagy. Dev Cell 6, 463-477.

Li, P., Nijhawan, D., Budihardjo, I., Srinivasula, S.M., Ahmad, M., Alnemri, E.S., and Wang, X. (1997). Cytochrome c and dATP-dependent formation of Apaf-1/caspase-9 complex initiates an apoptotic protease cascade. Cell 91, 479–489.

Li, H., Zhu, H., Xu, C.J. and Yuan, J. (1998). Cleavage of BID by caspase 8 mediates the mitochondrial damage in the Fas pathway of apoptosis. Cell 94, 491–501.

Lill, N.L., Grossman, S.R., Ginsberg, D., DeCaprio, J., and Livingston, D.M. (1997). Binding and modulation of p53 by p300/CBP coactivators. Nature 387, 823-827.

Lin, A., and Karin, M. (2003). NF-κB in cancer: a marked target. Semin Cancer Biol 13, 107-114.

Lin, A.W., and Lowe, S.W. (2001). Oncogenic ras activates the ARF-p53 pathway to suppress epithelial cell transformation. Proc Natl Acad Sci USA 98, 5025-5030.

Lindsten, T., Ross, A.J., King, A., Zong, W.X., Rathmell, J.C., Shiels, H.A., Ulrich, E., Waymire, K.G., Mahar, P., Frauwirth, K. et al., (2000). The combined functions of pro-apoptotic Bcl-2 family members bak and bax are essential for normal development of multiple tissues. Mol Cell 6, 1389–1399.

Liu, X., Kim, C.N., Yang, J., Jemmerson, R., and Wang, X. (1996). Induction of apoptotic program in cell-free extracts: requirement for dATP and cytochrome c. Cell 86, 147–157.

Li-Weber, M., and Krammer, P.H. (2003). Function and regulation of the CD95 (APO-1/Fas) ligand in the immune system. Semin Immunol 15, 145-157.

Ljungman, M., and Lane, D.P. (2004). Transcription - guarding the genome by sensing DNA damage. Nat Rev Cancer 4, 727-737.

Lohmann, I., McGinnis, N., Bodmer, M., and McGinnis, W. (2002). The Drosophila Hox gene deformed sculpts head morphology via direct regulation of the apoptosis activator reaper. Cell 110, 457-466.

Los, M., Mozoluk, M., Ferrari, D., Stepczynska, A., Stroh, C., Renz, A., Herceg, Z., Wang, Z.Q., and Schulze-Osthoff, K. (2002). Activation and caspase-mediated inhibition of PARP: a molecular switch between fibroblast necrosis and apoptosis in death receptor signaling. Mol Biol Cell 13, 978-988

Luo, X., Budihardjo, I., Zou, H., Slaughter, C. and Wang, X. (1998). Bid, a Bcl2 interacting protein, mediates cytochrome c release from mitochondria in response to activation of cell surface death receptors. Cell 94, 481-490.

MacLachlan, T.K., and El Deiry, W.S. (2002). Apoptotic threshold is lowered by p53 transactivation of caspase-6. PNAS 99, 9492-9497.

Marchenko, N.D., Zaika, A., and Moll, U.M. (2000). Death signal-induced localization of p53 protein to mitochondria. a potential role in apoptotic signaling. J Biol Chem 275, 16202-16212.

Margue, C.M., Bernasconi, M., Barr, F.G., and Schafer, B.W. (2000). Transcriptional modulation of the anti-apoptotic protein BCL-XL by the paired box transcription factors PAX3 and PAX3/FKHR. Oncogene 19, 2921-2929.

Marks, P.A., Richon, V.M., Miller, T., Kelly, W.K. (2004). Histone deacetylase inhibitors. Adv Cancer Res 91, 137-168.

Mayo, M.W., Wang, C.Y., Drouin, S.S., Madrid, L.V., Marshall, A.F., Reed, J.C., Weissman, B.E., and Baldwin, A.S. (1999). WT1 modulates apoptosis by transcriptionally upregulating the bcl-2 proto-oncogene. EMBO J 18, 3990-4003.

Meller, R., Minami, M., Cameron, J.A., Impey, S., Chen, D., Lan, J.Q., Henshall, D.C., and Simon, R.P. (2005). CREB-mediated Bcl-2 protein expression after ischemic preconditioning. J Cereb Blood Flow Metab 25, 234-246.

Micheau, O., Lens, S., Gaide, O., Alevizopoulos, K., and Tschopp, J. (2001). NF-κB signals induce the expression of c-FLIP. Mol Cell Biol 21, 5299-5305.

Mihara, M., Erster, S., Zaika, A., Petrenko, O., Chittenden, T., Pancoska, P., and Moll, U.M. (2003). p53 has a direct apoptogenic role at the mitochondria. Molecular Cell 11, 577-590.

Moreno, E., and Basler, K. (2004). dMyc transforms cells into super-competitors. Cell 117, 117–129.

Moroni, M.C., Hickman, E.S., Denchi, E.L., Caprara, G., Colli, E., Cecconi, F., Muller, H., and Helin, K. (2001). Apaf-1 is a transcriptional target for E2F and p53. Nat Cell Biol 3, 552-558.

Miguel-Aliaga, I., and Thor, S. (2004). Segment-specific prevention of pioneer neuron apoptosis by cell-autonomous, postmitotic Hox gene activity. Development 131, 6093-6105.

Mujtaba, S., He, Y., Zeng, L., Yan, S., Plotnikova, O., Sachchidanand, Sanchez, R., Zeleznik-Le, N.J., Ronai, Z., and Zhou, M.M. (2004). Structural mechanism of the bromodomain of the coactivator CBP in p53 transcriptional activation. Mol Cell 13, 251-263.

Muzio, M., Chinnaiyan, A.M., Kischkel, F.C., O'Rourke, K., Shevchenko, A., Ni, J., Scaffidi, C., Bretz, J.D., Zhang, M., Gentz, R. et al. (1996). FLICE, a novel FADD-homologous ICE/CED-3-like protease, is recruited to the CD95 (Fas/APO-1) death–inducing signaling complex. Cell 85, 817–827.

Nagata, S. (2000). Apoptotic DNA fragmentation. Exp. Cell Res. 256, 12-18.

Nahle, Z., Polakoff, J., Davuluri, R.V., McCurrach, M.E., Jacobson, M.D., Narita, M., Zhang, M.Q., Lazebnik, Y., Bar-Sagi, D., and Lowe, S.W. (2002). Direct coupling of the cell cycle and cell death machinery by E2F. Nat Cell Biol 4, 859-864.

Nelson, D.A., and White, E. (2004). Exploiting different ways to die. Genes Dev 18, 1223-1226.

Nijhawan, D., Fang, M., Traer, E., Zhong, Q., Gao, W., Du, F., and Wang, X. (2003). Elimination of Mcl-1 is required for the initiation of apoptosis following ultraviolet irradiation. Genes Dev 17, 1475-1486.

Okada, H., and Mak, T.W. (2004). Pathways of apoptotic and non-apoptotic death in tumour cells. Nat Rev Cancer 4, 592-603.

Oliver, F.J., Menissier-de Murcia, J., Nacci, C., Decker, P., Andriantsitohaina, R., Muller, S., de la Rubia, G., Stoclet, J.C., and de Murcia, G. (1999). Resistance to endotoxic shock as a consequence of defective NF-kappa B activation in poly (ADP-ribose) polymerase-1 deficient mice. EMBO J. 18, 4446-4454.

Opferman, J.T., and Korsmeyer, S.J. (2003). Apoptosis in the development and maintenance of the immune system. Nat Immunol 4, 410-415.

Owen-Schaub, L.B., Zhang, W., Cusack, J.C., Angelo, L.S., Santee, S.M., Fujiwara, T., Roth, J.A., Deisseroth, A.B., Zhang, W.W., and Kruzel, E. (1995). Wild-type human p53 and a temperature-sensitive mutant induce Fas/APO-1 expression. Mol Cell Biol 15, 3032-3040.

Peart, M.J., Smyth, G.K., van Laar, R.K., Bowtell, D.D., Richon, V.M., Marks, P.A., Holloway, A.J., and Johnstone, R.W. (2005) .Identification and functional significance of genes regulated by structurally different histone deacetylase inhibitors. Proc Natl Acad Sci 102, 3697-3702.

Pompeia, C., Hodge, D.R., Plass, C., Wu, Y.Z., Marquez, V.E., Kelley, J.A., and Farrar, W.L. (2004). Microarray analysis of epigenetic silencing of gene expression in the KAS-6/1 multiple myeloma cell line. Cancer Res 64, 3465-3473.

Poppelmann, B., Klimmek, K., Strozyk, E., Voss, R., Schwarz, T., and Kulms, D. (2005). NFκB-dependent down-regulation of Tumor Necrosis Factor Receptor-associated proteins contributes to Interleukin-1-mediated enhancement of ultraviolet B-induced apoptosis. J Biol Chem 280, 15635-15643.

Proskuryakov, S.Y., Konoplyannikov, A.G., and Gabai, V.L. (2003). Necrosis: a specific form of programmed cell death? Exp Cell Res 283, 1-16.

Puthalakath, H., and Strasser, A. (2002). Keeping killers on a tight leash: transcriptional and post-translational control of the pro-apoptotic activity of BH3-only proteins. Cell Death Differ 9, 505-512.

Ranger, A.M., Malynn, B.A., and Korsmeyer, S.J. (2001). Mouse models of cell death. Nat Genet 28, 113-118.

Ricci, M.S., Jin, Z., Dews, M., Yu, D., Thomas-Tikhonenko, A., Dicker, D.T., and El Deiry, W.S. (2004). Direct repression of FLIP expression by c-myc is a major determinant of TRAIL sensitivity. Mol Cell Biol 24, 8541-8555.

Rohde, J.R., Campbell, S., Zurita-Martinez, S.A., Cutler, N.S., Ashe, M., and Cardenas, M.E. (2004). TOR controls transcriptional and translational programs via Sap-Sit4 protein phosphatase signaling effectors. Mol Cell Biol 24, 8332-8341.

Rohde, J.R., and Cardenas, M.E. (2003). The tor pathway regulates gene expression by linking nutrient sensing to histone acetylation. Mol Cell Biol 23, 629-635.

Russell, H.R., Lee, Y., Miller, H.L., Zhao, J., and McKinnon, P.J. (2002). Murine ovarian development is not affected by inactivation of the bcl-2 family member diva. Mol Cell Biol 22, 6866-6870.

Salvesen, G.S., and Duckett, C.S. (2002). IAP proteins: blocking the road to death's door. Nat Rev Mol Cell Biol 3, 401-410.

Schuler, M., and Green, D.R. (2005). Transcription, apoptosis and p53: catch-22. Trends Genet 21, 182-187.

Schwab, B.L., Guerini, D., Didszun, C., Bano, D., Ferrando-May, E., Fava, E., Tam, J., Xu, D., Xanthoudakis, S., Nicholson, D.W., et al. (2002). Cleavage of plasma membrane calcium pumps by caspases: a link between apoptosis and necrosis. Cell Death Differ 9, 818-831

Scorrano, L., Oakes, S.A., Opferman, J.T., Cheng, E.H., Sorcinelli, M.D., Pozzan, T., and Korsmeyer, S.J. (2003). BAX and BAK regulation of endoplasmic reticulum Ca2+: a control point for apoptosis. Science 300, 135-139.

Sevilla, L., Zaldumbide, A., Pognonec, P., and Boulukos, K.E. (2001). Transcriptional regulation of the bcl-x gene encoding the anti-apoptotic Bcl-xL protein by Ets, Rel/NFkappaB, STAT and AP1 transcription factor families. Histol Histopathol 16, 595-601.

Shacka, J.J., and Roth, K.A. (2005). Regulation of neuronal cell death and neurodegeneration by members of the Bcl-2 family: therapeutic implications. Curr Drug Targets CNS Neurol Disord 4, 25-39.

Shikama, N., Lee, C.W., France, S., Delavaine, L., Lyon, J., Krstic-Demonacos, M., and La Thangue, N.B. (1999). A novel cofactor for p300 that regulates the p53 response. Mol Cell 4, 365-76.

Shishodia, S., and Aggarwal, B.B. (2004). Guggulsterone inhibits NF-κB and IκBα kinase activation, suppresses expression of anti-apoptotic gene products, and enhances apoptosis. J Biol Chem 279, 47148-47158.

Shintani, T., and Klionsky, D.J. (2004). Autophagy in health and disease: a double-edged sword. Science 306, 990-995.

Skurk, C., Maatz, H., Kim, H.S., Yang, J., Abid, M.R., Aird, W.C., and Walsh, K. (2004). The Akt-regulated forkhead transcription factor FOXO3a controls endothelial cell viability through modulation of the caspase-8 inhibitor FLIP. J Biol Chem 279, 1513-1525.

Slattery, E., Dignam, J.D., Matsui, T., and Roeder, R.G. (1983) .Purification and analysis of a factor which suppresses nick-induced transcription by RNA polymerase II and its identity with poly(ADP-ribose) polymerase. J Biol Chem 258, 5955-5959.

Sogame, N., Kim, M., and Abrams, J.M. (2003). Drosophila p53 preserves genomic stability by regulating cell death. Proc Natl Acad Sci 100, 4696-4701.

Soleymanlou, N., Wu, Y., Wang, J.X., Todros, T., Ietta, F., Jurisicova, A., Post, M., and Caniggia, I. (2005). A novel Mtd splice isoform is responsible for trophoblast cell death in pre-eclampsia. Cell Death Differ 1-12.

Sowter, H.M., Ratcliffe, P.J., Watson, P., Greenberg, A.H., and Harris, A.L. (2001). HIF-1-dependent regulation of hypoxic induction of the cell death factors BNIP3 and NIX in human tumors. Cancer Res 61, 6669-6673.

Stevens, C., Smith, L., and La Thangue, N.B. (2003). Chk2 activates E2F-1 in response to DNA damage. Nat Cell Biol 5, 401-409.

Tanaka, H., Yokota, H., Jover, T., Cappuccio, I., Calderone, A., Simionescu, M., Bennett, M.V., and Zukin, R.S. (2004). Ischemic preconditioning: neuronal survival in the face of caspase-3 activation. J Neurosci 24, 2750-2759.

Thellmann, M., Hatholz, J., and Conradt, B. (2003). The Snail-like Ces-1 protein of C. elegans can block the expression of the BH3-only cell-death activator gene egl-1 by antagonizing the function of basic HLH proteins. Development 130, 4057-4071.

Thomas, A., and White, E. (1998). Suppression of the p300-dependent mdm2 negative-feedback loop induces the p53 apoptotic function. Genes Dev 12, 1975-1985.

Thompson, C.B. (1995). Apoptosis in the pathogenesis and treatment of disease. Science 267, 1456-1462

Thummel, C.S. (1996). Flies on steroids-Drosophila metamorphosis and the mechanisms of steroid hormone action. Trends Genet 12, 306–310.

Tran, N.L., McDonough, W.S., Savitch, B.A., Sawyer, T.F., Winkles, J.A., and Berens, M.E. (2005). The tumor necrosis factor-like weak inducer of apoptosis (TWEAK)-fibroblast growth factor-inducible 14 (Fn14) signaling system regulates glioma cell survival via NFkappaB pathway activation and BCL-XL/BCL-W expression. J Biol Chem 280, 3483-3492.

Tran, S.E., Meinander, A., and Eriksson, J.E. (2004). Instant decisions: transcription-independent control of death-receptor-mediated apoptosis. Trends Biochem Sci 29, 601-608.

Tsang, C.K., Bertram, P.G., Ai, W., Drenan, R., and Zheng, X.F. (2003). Chromatin-mediated regulation of nucleolar structure and RNA Pol I localization by TOR. EMBO J 22, 6045-6056.

Varfolomeev, E.E., and Ashkenazi, A. (2004). Tumor necrosis factor: an apoptosis JuNKie? Cell 116, 491-497.

Vaux, D.L., and Korsmeyer, S.J. (1999). Cell death in development. Cell 96, 245-254.

Vaux, D.L., and Silke, J. (2005). IAPs, RINGs and ubiquitylation. Nat Rev Mol Cell Biol 6, 287-297.

Vickers, E.R., Kasza, A., Aksan-Kurnaz, I., Seifert, A., Zeef, L., O'Donnell, A., Hayes, A., and Sharrocks, A.D. (2004). Ternary complex factor-serum response factor complex-regulated gene activity is required for cellular proliferation and inhibition of apoptotic cell death. Mol Cell Biol 24, 10340-10351.

Virag, L., and Szabo, C. (2002). The Therapeutic Potential of Poly(ADP-Ribose) Polymerase Inhibitors. Pharmacol Rev 54, 375-429.

Vivanco, I., and Sawyers, C.L. (2002). The phosphatidylinositol 3-Kinase AKT pathway in human cancer. Nat Rev Cancer 2, 489-501.

Walisser, J.A., and Thies, R.L. (1999). Poly(ADP-ribose) polymerase inhibition in oxidant-stressed endothelial cells prevents oncosis and permits caspase activation and apoptosis. Exp Cell Res 251, 401-413.

Walmsley, S.R., Print, C., Farahi, N., Peyssonnaux, C., Johnson, R.S., Cramer, T., Sobolewski, A., Condliffe, A.M., Cowburn, A.S., Johnson, N. et al., (2005). Hypoxia-induced neutrophil survival is mediated by HIF-1α-dependent NF-κB activity. J Exp Med 201, 105-115.

Wang, C.Y., Mayo, M.W., Korneluk, R.G., Goeddel, D.V., and Baldwin, A.S.Jr. (1998). NF-κB antiapoptosis: induction of TRAF1 and TRAF2 and c-IAP1 and c-IAP2 to suppress caspase-8 activation. Science 281, 1680-1683.

Wang, W., Rastinejad, F., and El-Deiry, W.S. (2003). Restoring p53-dependent tumor suppression. Cancer Biol Ther 2(4 Suppl 1), S55-63.

Wang, X. (2001). The expanding role of mitochondria in apoptosis. Genes Dev 15, 2922-2933.

Wang, Z.Q., Stingl, L., Morrison, C., Jantsch, M., Los, M., Schulze-Osthoff, K., and Wagner, E.F. (1997). PARP is important for genomic stability but dispensable in apoptosis. Genes Dev 11, 2347-2358.

Wei, M.C., Zong, W.X., Cheng, E.H., Lindsten, T., Panoutsakopoulou, V., Ross, A.J., Roth, K.A., MacGregor, G.R., Thompson, C.B., and Korsmeyer, S.J. (2001). Pro-apoptotic BAX and BAK: a requisite gateway to mitochondrial dysfunction and death. Science 292, 727-730.

Wu, G.S., Burns, T.F., McDonald, E.R., Jiang, W., Meng, R., Krantz, I.D., Kao, G., Gan, D.D., Zhou, J.Y., Muschel, R. et al. (1997). KILLER/DR5 is a DNA damage-inducible p53-regulated death receptor gene. Nat Genet 17, 141-143.

Wyllie, A.H. (1980). Glucocorticoid induced thymocyte apoptosis is associated with endogenous endonuclease activation. Nature 284, 555-556.

Yamit-Hezi, A., and Dikstein, R. (1998). TAFII105 mediates activation of anti-apoptotic genes by NF-kappa B. EMBO J 17, 5161-5169.

Yamit-Hezi, A., Nir, S., Wolstein, O., and Dikstein, R. (2000). Interaction of TAFII105 with selected p65/RelA dimers is associated with activation of subset of NF-kappa B genes. J Biol Chem 275, 18180-18187.

Yin, X.M., Wang, K., Gross, A., Zhao, Y., Zinkel, S., Klocke, B., Roth, K.A., and Korsmeyer, S.J. (1999). Bid-deficient mice are resistant to Fas-induced hepatocellular apoptosis. Nature 400, 886–891.

Yu, S.W., Wang, H., Poitras, M.F., Coombs, C., Bowers, W.J., Federoff, H.J., Poirier, G.G., Dawson, T.M., and Dawson, V.L. (2002). Mediation of poly(ADP-ribose) polymerase-1-dependent cell death by apoptosis-inducing factor. Science 297, 259-263.

Zeiss, C.J. (2003). The apoptosis-necrosis continuum: insights from genetically altered mice. Vet Pathol 40, 481-495.

Zhang, J, and Xu, M. (2002). Apoptotic DNA degradation and tissue homeostasis. Trends Cell Biol 12, 84-89.

Zhou, L., and Steller, H. (2003). Distinct pathways mediate UV-induced apoptosis in Drosophila embryos. Dev Cell 4, 599-605.

Zong, W.X., Ditswo doplasmic reticulum to initiate apoptosis. J Cell Biol 162, 59-69. rth, D., Bauer, D.E., Wang, Z.Q., and Thompson, C.B. (2004). Alkylating DNA damage stimulates a regulated form of necrotic cell death. Genes Dev 18, 1272-1282.

Zong, W. X., Edelstein, L.C., Chen, C., Bash, J., and Gélinas, C. (1999). The prosurvival Bcl-2 homolog Bfl-1/A1 is a direct transcriptional target of NF-κB that blocks TNFα-induced apoptosis. Genes Dev 13, 382-387.

Zong, W. X., Li, C., Hatzivassiliou, G., Lindsten, T., Yu, Q.C., Yuan, J., and Thompson, C.B. (2003). Bax and Bak can localize to the endoplasmic reticulum to initiate apoptosis. J Cell Biol 162, 59-69.

Zou, H., Henzel, W.J., Liu, X., Lutschg, A., and Wang, X. (1997). Apaf-1, a human protein homologous to C. elegans CED-4, participates in cytochrome c-dependent activation of caspase-3. Cell 90, 405-413.

Section V

Special Topics

Chapter 27
Pre-mRNA Splicing in Eukaryotic Cells

Xiang-Dong Fu

Department of Cellular and Molecular Medicine, University of California, San Diego, 9500 Gilman Drive, La Jolla, CA 92093-0651

Key Words: the RNA world, coding and non-coding RNA, mRNA processing, alternative splicing, RNA stability and transport, transcription-RNA processing coupling in the nucleus

Summary

Gene expression in eukaryotic cells is a collective outcome of transcription, RNA processing, and protein translation. In this chapter, I focus on the mechanism and regulation of gene expression at the level of RNA metabolism. I begin with the introduction of different kinds of RNA expressed in mammalian cells, including mRNAs, rRNAs, tRNAs, small RNAs, miRNAs, and RNAs of unknown function. Realizing that the RNA world is a big topic, which is beyond the scope of a single chapter, I am forced to concentrate on the pathway and regulation of pre-mRNA processing, which is arguably the most important step in gene expression and regulation. Readers are referred to outstanding reviews on other RNA categories for detailed information. Key features in this chapter also include the discussion on integration of RNA processing with other critical nuclear events and on genomics of splicing in the current post-genome era. Most of the information covered here is a condensed version of a number of lectures given annually to graduate students in Beijing and Shanghai as part of the Molecular and Cell Biology course contributed by members of the Ray Wu society.

Overview of the RNA World

The central dogma in gene expression is the flow of genetic information from DNA to RNA to protein. DNA serves to store and pass on genetic information, but cannot function as an enzyme during the expression and replication of genetic information. Proteins, on the other hand, are enzymes and structural components of cells and organisms, but do not have the capacity to store genetic information. RNA is the bridge between DNA and protein. Remarkably, RNA is capable of carrying genetic information and processing catalytic functions much like protein enzymes, a property suspected to represent the earliest form of life (Cech, 1985).

A: Synthesis and Processing of the Major Three RNA Classes

Traditionally, RNAs are roughly divided into three major classes, which are transcribed by one of the three RNA polymerases in the cell. rRNAs are transcribed by Pol I, mRNAs by Pol II, and tRNAs by Pol III respectively. rRNAs are transcribed as a long 45s precursor transcript, which is then processed into 28s, 18s, and 5.8s rRNAs by a large number of proteins, including endo- and exo-nucleases and specificity factors (Lafontaine and Tollervey, 1995). A separate 5s rRNA gene, however, is transcribed by Pol III. Many sites within rRNAs are modified (i.e. methylation, pesudouridination, etc.), which is essential for their functions in translation.

tRNAs are individually transcribed as pre-tRNA precursors. The 5′-end sequence is removed by RNase P, which is a ribozyme consisting of a catalytic RNA component and a structural protein component (Kirsebom, 2002). The 3′-end of pre-tRNA is first processed by RNase-mediated cleavage followed by CCA tri-nucleotide addition, which is catalyzed in the absence of a template (Schurer et al., 2001). A subclass of tRNAs also contains a short intron, which is removed by concerted actions of endonucleases and ligases

Corresponding Author: Tel: (858) 534-4937, Fax: (858) 534-8549, E-mail: xdfu@ucsd.edu

(Deutscher, 1984). tRNAs are extensively modified to become competent for amino acid charging by aminoacyl-tRNA synthetases (Martinis et al., 1999).

mRNAs are first transcribed as intron-containing pre-mRNAs. A pre-mRNA is processed into a mature and functional mRNA in three steps: (1) capping, (2) splicing, and (3) polyadenylation. Capping takes place co-transcriptionally during which a monomethylated guanylate (G) is linked to the first nucleotide at the 5' end via a 5'-5' phosphodiester bond (Shatkin and Manley, 2000). Splicing occurs in the spliceosome, a multi-component complex, to remove intervening sequences (or introns), which is the main topic of this chapter. Polyadenylation initiates with (1) the recognition of the polyadenylation signal (AAUAA), which is approximately 30 nt upstream of the ploy(A) tail, by cleavage and polyadenylation specificity factors (CPSFs), (2) and binding of the GU-rich sequences downstream of the poly(A) site by cleavage stimulation factors (CstFs) (Keller and Minvielle-Sebastia, 1997). The poly(A) polymerase, with the aid of poly(A) binding protein II, which keeps the length of the poly(A) tail relatively constant, catalyzes the addition of approximately 200 adenosines. Splicing and polyadenylation can take place co-transcriptionally or after the precursor transcript is released from the chromatin to the nucleoplasm (Minvielle-Sebastia and Keller, 1999).

B: Non-coding RNAs

Besides the three major RNA classes, many small non-coding RNAs are expressed in eukaryotic cells, and their diverse biological roles are being increasingly recognized and appreciated. Small nuclear RNAs (or snRNAs) have been extensively studied as part of the RNA processing machinery (see further details below), snoRNAs are a special class of small RNAs localized in the nucleolus. These snoRNAs, which may be transcribed from individual genes or part of introns in pre-mRNAs, play important roles in guiding site-specific modifications of rRNAs and snRNAs (Decatur and Fournier, 2003).

More recently, the world of microRNAs (miRNAs) was brought forward as a class of ~21 nt small RNAs involved in a variety of gene expression paradigms (Novina and Sharp, 2004). miRNAs are found in intragenic regions or within introns of other genes. The organization and transcription of miRNAs are poorly understood, although recent evidence suggests that they are transcribed by Pol II (Lee et al., 2004). miRNAs are the final products processed from transcribed precursors in two key steps: (1) The initial transcripts, known as primary miRNAs (pri-miRNAs), are processed by the Drosha-containing complex (Drosha is a RNase) into pre-miRNAs, which all have a similar stem-loop structure, and (2) pre-miRNAs are processed by Dicer (another RNase) into the final single-stranded miRNAs (Cullen, 2004). Mature miRNAs are incorporated into the RISC complex (RNA induced silencing complex) to mediate target degradation (if a miRNA base-pairs perfectly with a target, similar to the action of RNAi) or in most cases translational repression (if a miRNA forms partial base-pairing with its target). The function of miRNAs has been implicated in the regulation of development, cell proliferation, and apoptosis (Hartmann et al., 2004).

Another class of small RNAs is referred to as small heterogeneous RNAs (or shRNAs) because their lengths are not as confined as miRNAs. This class of non-coding RNAs is mostly transcribed from repeat-containing intragenic regions. shRNAs help eliminate transposable elements (an innate cellular defense mechanism against genomic instability) and induce the formation of heterochromatin, a mechanism thought to be critical in programming cell differentiation (Mochizuki et al., 2002; Taverna et al., 2002; Volpe et al., 2002, 2003;Volpe et al., 2003; Volpe et al., 2002).

C: TUFs: A Large Number of RNAs to be Understood

Sequencing the genome of many model organisms, has brought attention to the observation that the number of genes in individual genomes cannot explain the complexity and functional diversity of these organisms. In fact, it is difficult to determine the total number of genes encoded in a given genome. For example, a tiling array analysis of total RNA from human cells detected a far higher gene count than previously reported (Kapranov et al., 2002). Some of these "extra" counts may be pseudogenes. However, many clearly correspond to previously unrecognized genes, some of which show the multi-exon arrangement typical of a eukaryotic gene. In a recent genome-decoding consortium meeting, these unknown transcripts were named transcripts of unknown functions (TUFs) (The ENCODE Project Consortium). It will be interesting to learn how many of these TUFs actually correspond to real genes and whether some of them represent new classes of genes that have escaped recognition by conventional molecular biology.

Pre-mRNA Splicing: Pathway and Factors

Pre-mRNA splicing has been extensively studied in the past three decades, beginning with the discovery of "split genes" in the 70s (Sharp, 1994). The development of the *in vitro* splicing system and the power of yeast genetics allowed for the dissection of the splicing mechanism.

A: Consensus Splicing Signals and the Splicing Pathway

In the majority of eukaryotic genes, exons are relatively short in length, typically ranging from 100 to 300 nts, whereas introns are relatively long and variable in length, with lengths up to 100 kb. Key splicing signals mostly reside in the intron side of the exon/intron junction. As shown in Fig.27.1, the 5′ splice site is composed of an invariant GT dinucleotide flanked by conserved nucleotides, with the most important one being a G in the fifth position in the intron side. The 3′ splice site is more complicated and can be divided into three important regions: the branchpoint sequence, the polypyrimidine tract, and the 3′ invariant AG dinucleotide (Sharp and Burge, 1997). These splicing signals are loosely conserved in mammalian genes. In yeast, however, the splicing signals are much more conserved and the branchpoint sequence shows no variation. The consensus sequence shown in Fig.27.1 represents the majority of introns, conveniently referred to as major introns. In addition to major introns, there exists a minor class of introns, which is characterized by the conserved AT and AC dinucleotide at the 5′ and 3′ splice site, respectively (Burge et al., 1998). The major class is thus referred to as GT-AG introns and the minor class to AT-AC introns. The major and minor classes utilize both overlapping and distinct factors to build the spliceosome (see below).

The major and minor classes of introns follow the same chemical pathway for intron removal. As shown in Fig.27.2, the splicing reaction proceeds in two steps. In the first step, the 5′ exon is cleaved in a nucleophilic attack by the 2′-OH group of the branchpoint nucleotide (adenosine being the most common one), resulting in the release of the first exon and the formation of a lariat intermediate. In the second step, the 3′ exon is cleaved during the second nucleophilic attack by the 2′-OH group of the last nucleotide of the released 5′ exon, resulting in a lariat intron and ligated exons. The lariat intron is quickly degraded (with the aid of a debranching

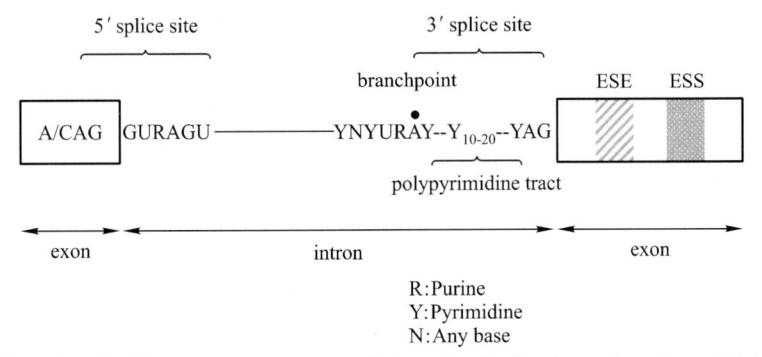

Fig.27.1 Consensus splicing signals. Shown are consensus splicing signals for the major class of introns. ESE: exonic splicing enhancer; ESS: exonic splicing silencer.

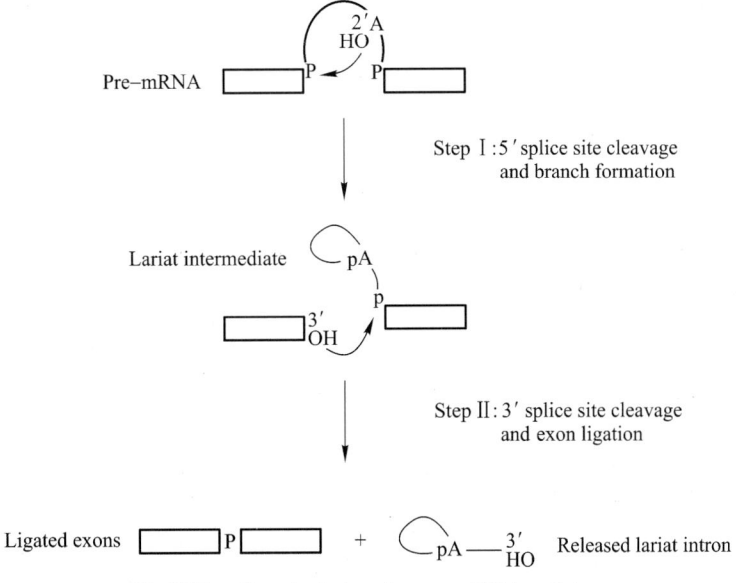

Fig.27.2 Chemical steps in pre-mRNA splicing.

enzyme) in the nucleus and the ligated exons are exported from the nucleus after the removal of all introns from the pre-mRNA.

B: Exon Definition

In mammalian systems, exons are short whereas introns are long in length, and splicing signals flanking each exon are recognized first by the exon definition mechanism (Berget, 1995). Interactions between specific factors binding independently at the 3′ and the downstream 5′ splice sites result in each exon being recognized as a unit. As predicted from the exon definition model, a functional downstream 5′ splice site was found to stimulate the upstream splicing event (Hoffman and Grabowski, 1992). Interactions between specific factors at the two splice sites are limited by the physical distance between the factors; therefore the length of the exon has a major impact. In addition to the mechanism of exon definition, splicing factors can also interact across an intron if the intron length is short enough (intron definition). This two mechanisms work in parallel, where small exons are recognized by exon definition whereas small introns are recognized by intron definition.

C: SnRNP and Non-snRNP Splicing Factors

Besides consensus splicing signals at the 5′ and the 3′ splice sites, other sequence elements within exons and introns can positively and negatively influence the splicing efficiency. Exonic sequences that can stimulate or inhibit splicing are referred to as exonic splicing enhancers (ESEs) or exonic splicing silencers (ESSs) respectively. Likewise, intronic sequences that have positive or negative affects on splicing are called intronic splicing enhancers (ISEs) or intronic splicing silencers (ISSs), respectively.

Small nuclear ribonucleoprotein particles (snRNPs) and non-snRNP protein factors mediate the recognition of consensus splicing signals and regulatory sequences (Kramer, 1996). Many non-snRNP protein factors play essential roles during splicing while others are only involved in alternative splicing. Splicing of major introns is mediated by U1, U2, U4/6, and U5 snRNPs, while splicing of minor introns is mediated by U11, U12, U4atac/6atac, and U5 (Tarn and Steitz, 1997; Patel and Steitz, 2003). Thus, only U5 snRNP is common to both classes of introns. Individual snRNPs consist of one uridine-rich small nuclear RNA (snRNA) and a group of associated proteins with the exception of U4/6 and U4atac/6atac di-snRNPs. They consist of two snRNAs packed in one snRNP particle because of extensive base-pairing between the two snRNAs, which is disrupted during splicing to establish a RNA-based catalytic core (see below).

Non-snRNP splicing factors are individual protein factors that are not part of snRNP complexes. Many essential non-snRNP splicing factors are RNA binding proteins while others are RNA helicases. The family of SR proteins, which is characterized by one or two RNA Recognition Motifs (RRM) at the N-terminus and an arginine/serine dipeptide repeat (RS domain) at the C-terminus (Fu, 1995; Graveley, 2000), are well-characterized non-snRNPs. RNA helicases play a central role in RNA rearrangement along the splicing pathway (Staley and Guthrie, 1998). Mass spectrometry has successfully identified proteins associated with purified spliceosomes and subspliceosomal complexes (Gottschalk et al., 1999; Stevens, 2000; Zhou et al., 2002). Many newly identified spliceosome associated proteins, however, remain to be functionally characterized.

D: Spliceosome Assembly

Splicing takes place in the spliceosome, which is assembled in a step-wise manner as illustrated in Fig.27.3. RNA binding proteins rapidly bind to RNA when an RNA is mixed with nuclear extracts, forming heterogeneous ribonucleoprotein particles known as hnRNPs (or H complex). U1 snRNP base-pairs with the 5′ splice site, forming the E complex (E for early) (Michaud and Reed, 1993). The E complex is a commitment complex, meaning that the pre-mRNA in the complex is committed to the splicing pathway in an irreversible manner.

In the next step, U2 snRNP joins the E complex by base-pairing with the branchpoint sequence, resulting in the formation of the A complex. The mechanism for A complex assembly is more complicated than the formation of the E complex and requires several key factors. First, U2AF, a heterodimer, binds to the 3′ splice site (Zamore and Green, 1991). The large subunit U2AF65 binds to the polypyrimdine tract while the small subunit U2AF35 touches the conserved AG dinucleotide (Wu et al., 1999). A number of other protein factors (SF1, SF3, and a number of spliceosome associated proteins or SAPs) are also important for 3′ splice site specification. SF1, which was later renamed as the branchpoint binding protein BBP, directly binds to the branchpoint sequence (Abovich and Rosbash, 1997; Berglund et al., 1998). In contrast to U1 binding to the 5′ splice site, U2 base-pairing with the branchpoint is an ATP-dependent process. The requirement for ATP likely reflects the essential role of the RNA helicase UAP56 in facilitating U2 binding to the branchpoint sequence (Fleckner et al., 1997; Kistler and Guthrie, 2001).

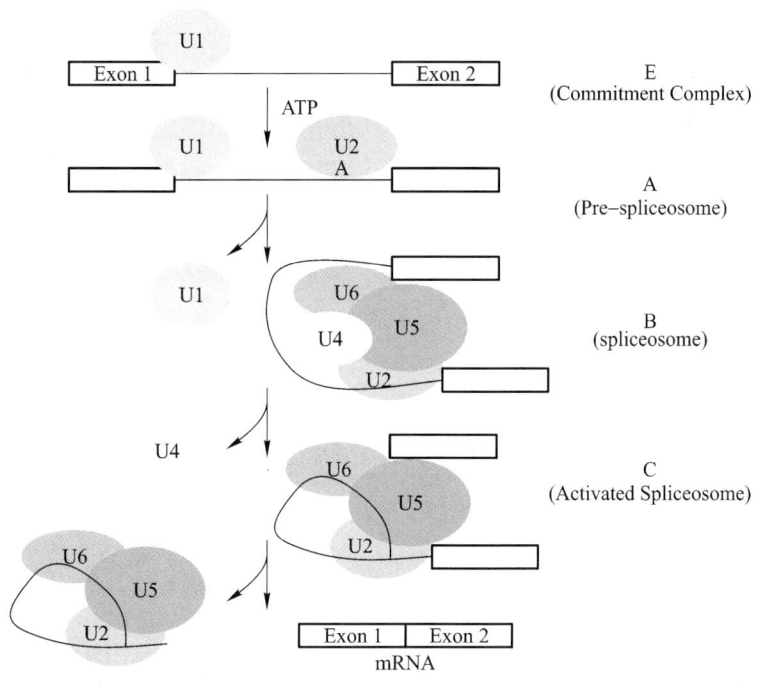

Fig.27.3 The spliceosome assembly pathway for the removal of the major class of introns. The spliceosome assembly for minor introns are slightly different: U1 is replaced by U11; U2 is replaced by U12; U4/6 is replaced by U4atac/6atac. Furthermore, U11 and U12 may jointly recognize the 5′ and 3′ splice site during spliceosome assembly.

The A complex, containing U1 and U2, is conditioned for further spliceosome assembly via the addition of the U4/6.U5 tri-snRNPs, resulting in the formation of the B complex known as the mature spliceosome. U4 and U6 are extensively base-paired and jointly packaged into a di-snRNP particle. U5, however, can exist as a single snRNP particle, and before joining the spliceosome, it forms a complex with U4/6 to generate a tri-snRNP particle. U1 is released from the A complex during the joining of the tri-snRNP particle to the spliceosome in which the tri-snRNP interacts with the pre-mRNA and U2, thereby forming an RNA-based catalytic core. A number of RNA helicases are involved in this process. As a result, U4 is released from the spliceosome, giving rise to the formation of the active spliceosome or the C complex (Ares and Weiser, 1995; Staley and Guthrie, 1998). The two-step splicing reaction than takes place in the C complex. In the end, U2/5/6 snRNPs are released with the lariat intron, thereby allowing ligated exons to dissociate from snRNPs in preparation for nuclear export.

E: Role of SR Proteins in Constitutive Splicing

Highlight of protein factors involved in spliceosome assembly has been briefly described above. The family of SR proteins deserves further attention. In baking yeast, intron-containing genes account for approximately 5% of the total number of genes encoded in the genome with the vast majority containing a single intron. Fewer protein factors are needed for the recognition of yeast splice sites because they are strongly conserved. As a result, SR proteins are not present in yeast, except for a few SR-like RNA binding proteins. Thus, SR proteins are specialized factors for splicing in higher eukaryotic cells.

The participation of SR proteins in a number of critical steps during constitutive splicing leads to the conclusion that they are essential splicing factors (Fu, 1995). It is interesting that, although SR proteins are collectively essential, the majority of splicing events can take place in the presence of just one SR protein. This built-in functional redundancy may allow vast constitutive splicing events to proceed in different cell types and under a variety of growth conditions where SR proteins are differentially expressed. SR proteins become essential at the very earliest step of spliceosome assembly. In fact, SR proteins are sufficient to commit pre-mRNAs to the splicing pathway (Fu, 1993). Although SR proteins are not essential for U1 binding to the 5′ splice site, they were found to facilitate efficient and accurate selection of functional 5′ splice sites and avoid cryptic 5′ splice sites frequently found in intronic sequences. During the formation of the A complex, SR proteins play an important role in the recruitment of U2 to the 3′ splice site (Fu and Maniatis, 1992). This is mediated by the interaction between SR proteins and the U2AF heterodimer (Wu and Maniatis, 1993). The joining of U4/6.5 tri-snRNPs to the A complex also depends on SR proteins (Roscigno and Garcia-Blanco, 1995). SR protein-dependent snRNP recruitment is likely mediated by the RS domain-mediated protein-protein interactions (Yeakley *et al.*, 1999). Interestingly, the mammalian orthologs of RNA helicases involved in splicing all carry an RS domain. More recent studies suggest that the RS domain of SR proteins may also interact with RNA in assembled spliceosomes (Shen *et al.*, 2004).

Phosphorylation is essential for SR proteins to function during spliceosome assembly (Mermoud *et al.*, 1992; Mermoud *et al.*, 1994). SR proteins are phosphorylated by two families of SR protein specific kinases known as SRPKs and Clks (Gui *et al.*, 1994b; Colwill *et al.*, 1996a; Colwill *et al.*, 1996b). During splicing in the assembed spliceosome, however, SR proteins are dephosphorylated. Prevention of SR protein dephosphorylation blocks splicing, suggesting that dephosphorylation of SR proteins is essential for spliceosome resolution (Tazi *et al.*, 1993; Xiao and Manley, 1998; Prasad *et al.*, 1999). It is important to point out here that a full phosphorylation/dephosphorylation cycle in a single SR protein may not be essential to carry out the splicing reaction: a particular phosphorylated SR protein may be used to initiate spliceosome assembly while another different SR protein, upon dephosphorylation, may drive the splicing reaction to completion (Xiao and Manley, 1998). At the cellular level, phosphorylation appears to mediate SR protein trafficking from "storage" sites within the nucleus to nascent transcripts whereas dephosphorylation seems to be responsible for the reversal of this process (Gui *et al.*, 1994a; Misteli *et al.*, 1998).

F: RNA Catalysis

During spliceosome assembly and within fully assembled spliceosome, RNA elements in pre-mRNA and in snRNPs engage in base-pairing and non-Watson/Crick interactions to establish a core for RNA-based catalysis. This concept has yet to receive full experimental proof. However mounting evidence points in this direction. One piece of such evidence comes from the similar splicing pathways for group II introns and pre-mRNAs (Sharp, 1994). Group II introns are self-splicing introns found mostly in low eukaryotic cells, such as nematodes and trypanosomes, and in mitochondrial and chloroplasts of higher eukaryotic cells. Splicing of group II introns, which can take place in the absence of protein factors, follows a two-step chemical reaction identical to that of pre-mRNA splicing, lending a strong support for an RNA-based catalytic core in both the splicing pathways.

Further evidence for RNA-based catalysis comes from elegant yeast genetics combined with biochemical approaches, such as induced RNA-RNA cross-linking and mapping of cross-linked sites. These approaches led to the elucidation of conserved and functionally important RNA-RNA interactions in the spliceosome (Guthrie, 1991; Guthrie, 1994). For example, as illustrated in Fig.27.4, the U1-5′ splice site base pairing established initially in early splicing complexes are later disrupted, giving ways for base- pairing between U6 and the 5′ splice site. U2 remains bound at the branchpoint sequence, and after U4 is released from the spliceosome, part of U2 RNA is then engaged in base-pairing with the U6 RNA. Thus, the 5′ splice site and the 3′ splice site are linked through a network of RNA-RNA interactions. These interactions are further strengthened by a conserved loop in the U5 RNA via its base-pairing with the 5′ splice site in one side and with the 3′ splice site in the other. This final core structure closely resembles the minimal catalytic core in group II introns, strongly suggesting that both types of the splicing reaction use the same catalytic mechanism. Finally, attempts have been made to reconstitute the entire splicing or at least part of the process *in vitro* using synthetic RNAs resembling different regions of pre-mRNA and snRNAs (Valadkhan and Manley, 2001; Valadkhan and Manley, 2003). This would provide the ultimate proof for the RNA-based catalytic mechanism.

Alternative Splicing: Mechanisms and Regulation

Unlike yeast, splice sites in higher eukaryotic cells are considerably more variable. As a result, a pre-mRNA may give rise to more than one mRNA product via alternative splicing. In some dramatic cases, a pre-mRNA can, in theory, produce thousands of mRNAs, indicating that alternative splicing has the potential to significantly enlarge the proteome (Black, 2000; Black, 2003). Alternative splicing is a widespread phenomenon in eukaryotic genomes such that more than half of the genes in humans are alternatively spliced (Modrek and Lee, 2002; Johnson *et al.*, 2003). Thus, alternative splicing has become more a rule rather than an exception. This provides additional dimensions to the regulation of gene expression as different mRNA isoforms may have distinct half-lives, follow different export pathways, and interact with different factors for intracellular targeting and regulated translation (Black, 2003). Due to its scale and complexity, alternative splicing has become a major challenge in post-genome research.

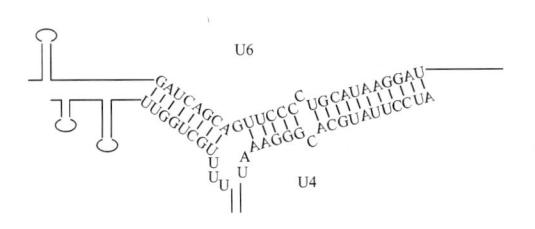

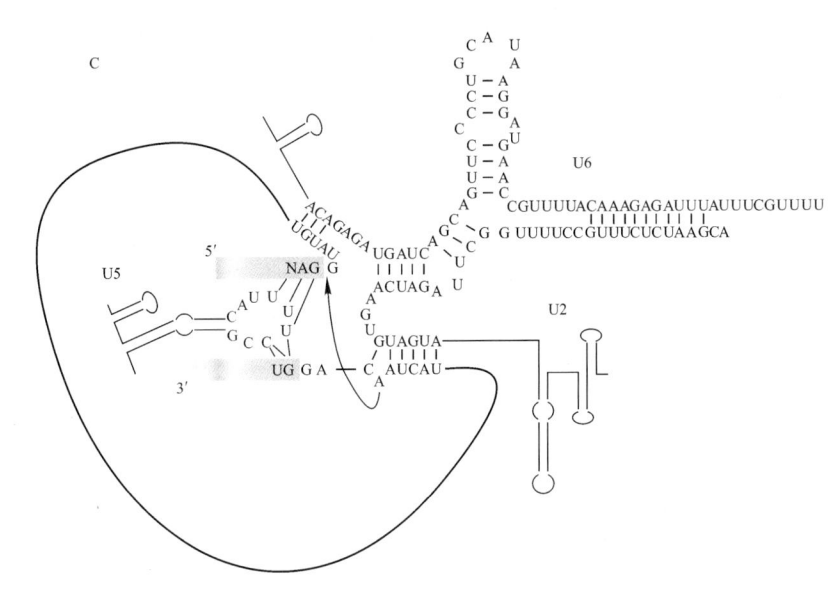

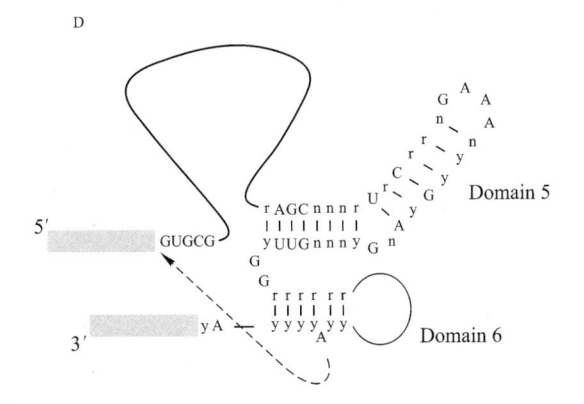

A: Multiple Ways to Splice Alternatively

A pre-mRNA may be alternatively spliced in many different ways. As illustrated in Fig.27.5, commonly found alternative splicing modes include (1) alternative 5′ choices, (2) alternative 3′ choices, (3) intron retention, (4) exon skipping, (5) mutually selected exons, and (6) combinatory exons. In addition, alternative splicing may also be paired with alternative promoters or the use of alternative polyadenylation sites. Examples of these modes can be found in the UCSC Genome Browser (http://genome.ucsc.edu), which is a compilation of cloned cDNAs aligned with their corresponding genomic loci.

Fig.27.4 Catalytic core for RNA splicing. A. Binding of U1 to the 5′ splice site and U2 to the branchpoint via base-pairing with the conserved splicing signals. B. Intra-snRNP base-pairing between U4 and U6. C. Deduced RNA-RNA interactions in the spliceosome. D. The catalytic core of a group II self-splicing intron. The core is structurally related to the RNA network within the spliceosome, which provides a strong support for a related, RNA-based chemical mechanism in pre-mRNA splicing.

Key questions in the study of alternative splicing include (1) how alternative splicing is regulated and (2) what is the function of individual mRNA isoforms. The understanding of splicing regulation requires the identification of cis-acting regulatory elements and trans-acting factors. Table 27.1 lists some well-studied splicing regulators, most of which were identified by biochemical approaches on model systems. More recently, a large-scale RNAi screening against RNA binding proteins and splicing factors revealed the involvement of many previously conceived constitutive splicing factors in alternative splicing (Park *et al.*, 2004). Thus, our knowledge on regulated splicing is quite limited, despite intensive research in the last two decades. Given the prevalence of mRNA isoforms in mammalian cells, it is a great challenge to determine which isoforms are functionally important and which ones are just noise in gene expression as a reflection of defects in the RNA processing machinery.

B: Sex Determination in the Fly: the Best Understood Case

The most understood alternative splicing pathway, in terms of functionality and regulatory mechanisms, is the sex determination pathway in *Drosophila* (Fig. 27.6). A series of powerful genetic studies has helped in the dissection of the pathway (Baker, 1989), while elegant biochemical experiments have contribute to the elucidation of the molecular mechanisms involved (Maniatis and Tasic, 2002). In this pathway, the sex lethal gene (Sxl) is expressed only in females. Sxl, a RNA binding protein, regulates alternative splicing of its own RNA by binding to intronic sequences upstream the alterative exon, resulting in skipping of the exon. The skipped exon contains a stop coden, therefore, skipping of this exon in females leads to the production of a functional Sx1 protein, whereas in males to the expression of a non-functional truncated Sx1 protein. This pathway helps decrease the chance of accidental expression of Sxl in males. The Sxl protein also acts on a key downstream

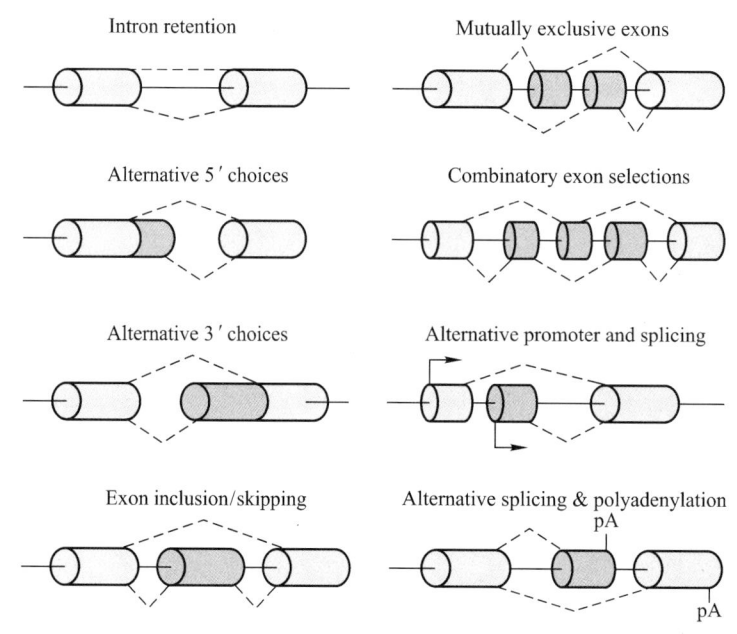

Fig. 27.5 Modes of alternative splicing.

Table 27.1 RNA binding proteins involved in the regulation of alternative splicing.

Family Name	Examples	Key Domain	Essential Splicing Factors?
SR Proteins	SC35,ASF/SF2,9G8,hTar2-β,hTra2-β, SRp20,30c,40,46,54,SRp55,75,86	RRM and RS	Yes
HnRNPs	HnRNP A/B,hnRNP F,hnRNP H,hnRNP I/PTB,Nptb,TIA-1,Eiva,Fox-1,2,3	RRM,some with RGG boxes	No
KH-type	KSRP,Nova-1,2,PSI	KH	No
CELF Factors	CUG-BP1,CUG-BP2/ETR-3,NAPOR	RRM	No
MBNL	MBNL-1,2,3	C_3H zinc finger	No

target, the transformer gene (Tra). Sxl binds to an intronic regulatory element within Tra, thereby blocking the usage of a nearby 3' splice site and shifting the splicing donor to the next 3' splice site. This leads to the expression of a full-length Tra protein (instead of a truncated one) in females. Tra then forms a heterdimer with Tra-2 protein expressed in a non-sex specific manner, which together bind to an exonic regulatory element in the downstream doublesex (Dsx) pre-mRNA. This binding event, together with other cellular splicing factors, is responsible for the activation of an upstream 3' splice site, thereby giving rise to a female-specific Dsx protein. In the absence of Tra/Tra2 binding, a downstream 3' splice site is used, thereby giving rise to a male-specific Dsx protein. Thus, both alternative products of Dsx result in functional proteins. The female Dsx protein negatively regulates genes involved in male differentiation whereas the male Dsx protein negatively regulates genes involved in female differentiation. This pathway illustrates the importance of regulated splicing in a crucial biological process.

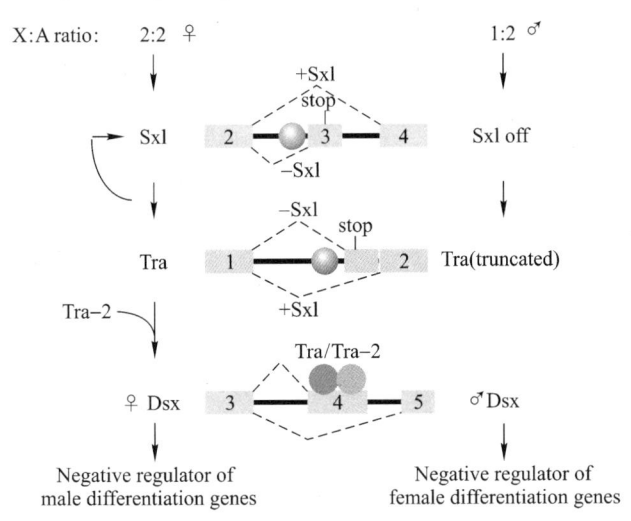

Fig.27.6 The sex determination pathway in *Drosophila*. Fly sex phenotype is determined by the X chromosome to Autosome ratio (2:2 for female and 1:2 for male). This ratio dictates Sxl expression in early development. The female phenotype is maintained by a cascade of regulated splicing. Intronic regulatory elements regulate Sxl and Tra splicing whereas exonic regulatory elements regulate the splicing of Dsx.

C: ESEs and ESSs and Their Effectors

The *Drosophila* sex determination pathway provides a road map to the understanding of regulated splicing in mammalian systems. Many cis-acting elements involved in splicing regulation have been identified through mutagenesis and functional studies in *in vitro* splicing or in transfected cells. Exonic sequences positively regulating splicing are referred to as exonic splicing enhancers (ESEs) whereas those negatively regulating splicing are called exonic splicing silencers (ESSs). Interestingly, ESEs and ESSs are not unique to alternative exons; these regulatory elements are also widely found within constitutive exons (Schaal and Maniatis, 1999). The ESE/ESS ratio, in addition to other intronic regulatory elements, may be the distinguishing characteristics between alternative and constitutive exons (Fu, 2004).

In general, ESEs are recognized by SR proteins whereas ESSs are recognized by hnRNP proteins (Fig. 27.7). In this sense, Tra-2 is a SR protein. In fact, Tra-2 contains two RNA recognition motifs and two RS domains (one at the C-terminus, like a typical SR protein, and the other at the N-terminus). Tra is also a SR-like protein because it contains a RS domain, but not RRM. The binding of SR proteins to ESEs results in the recruitment of U1 to a nearby downstream 5' splice site and U2 binding to a nearby upstream 3' splice site during exon definition. The recruitment of U2 is accomplished by U2AF binding at the 3' splice site. Because both U2AF subunits also have a RS domain, it is believed that the U1 and U2 recruitment is facilitated by RS domain-mediated protein-protein interactions (Wu and Maniatis, 1993). In addition to the recruitment of spliceosome components, SR proteins are also able to antagonize the interaction of hnRNP proteins with ESSs, which prevents the recruitment of U1 and U2 to their target splice sites during spliceosome assembly (Zhu *et al.*, 2001).

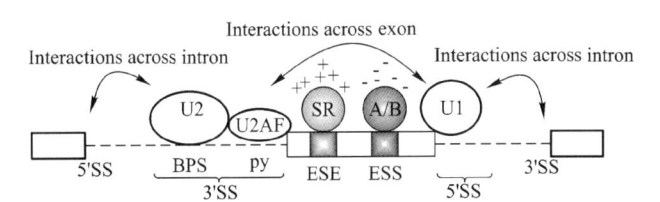

Fig.27.7 Positive and negative regulation of splicing via exonic regulatory elements. SR proteins bind to ESEs, thereby promoting the binding of U1 to the downstream 5' splice site and recruiting U2 to the upstream 3' splice site. HnRNP proteins antagonize the action of SR proteins by binding to ESSs in this process. Protein interactions across the exon are part of the early exon definition process. After that, protein interactions are established across the intron to initiate spliceosome assembly.

D: ISEs and ISSs and Their Effectors

Intron sequences also harbor splicing regulatory elements. Intronic splicing enhancers (ISEs) and silencers (ISSs) are not as well understood as ESEs and ESSs. The c-src model, where a small exon (N) is only included in neurons and is skipped in most other cell types, represents the most understood case. Extensive mutagenesis studies have revealed intronic control elements in both sides of the alternative exon (Fig.27.8). Here, the definition of ISEs and ISSs becomes a little fuzzy. An ISE in one cell type may be an ISS in another. Intronic sequences downstream of the N exon are important for the inclusion of the exon in neuronal cells, therefore these would be considered neuron specific ISEs. However, in non-neuronal cells, the same sequences serve to prevent the selection of the splice sites flanking the N exon. Thus, those sequences are ISSs in non-neuronal cells.

Many protein factors have been shown to interact ISEs and ISSs. PTB was originally characterized as a polypyrimidine tract binding protein with strong preference for U-rich sequences, which shares some sequence binding specificity with U2AF65 (Garcia-Blanco *et al.*, 1989). PTB is not required for splicing, and thus, it is not an essential splicing factor. Because of its competitive binding with U2AF, PTB is regarded as a negative regulator for splicing. In non-neuronal cells, PTB binds to both sides of the N exon, resulting in the formation of a multi-component complex, which shields the N-exon from the splicing machinery. In neuronal cells, on the other hand, the multi-component complex has a different composition. One of the key molecules is a neuron-specific PTB homologue known as nPTB. nPTB appears sensitive to an ATP-dependent process, which opens up the inhibitory complex and converts the ISS into ab ISE resulting in the inclusion of the N exon (Chan and Black, 1997; Modafferi and Black, 1999).

The identification of tissue specific splicing regulators has been a major goal in the splicing field. The first successful example is from the study of a disease gene known as Nova (Jensen *et al.*, 2000a). The Nova family of RNA binding proteins is characterized by a signature KH domain, which is also present in many other RNA binding proteins (Table 27.1). Via *in vitro* selection for high affinity RNA elements, Nova was found to interact with YCAY (Y being U or C) elements in pre-mRNA (Jensen *et al.*, 2000b). Multiple copies of this consensus sequence are frequently found in introns of many neuron-specific genes (Dredge and Darnell, 2003). Biochemical and mutagenesis studies demonstrate that the Nova family of RNA binding proteins act through these RNA motifs to regulate neuron-specific alternative splicing events (Ule *et al.*, 2003). However, it is important to note that Nova binding can activate splicing in some cases and inhibit splicing in others. The mechanism of either regulatory pathway remains elusive. Nova RNA binding protein knock-out mice have provided the strongest evidence for its role in neuron-specific splicing regulation (Jensen *et al.*, 2000a). The phenotype of the knock-out mouse mimics the human disease and shows the mis-regulation of many neuron-specific genes. Because many alternative splicing events are altered in the knockout, it is presently unclear which genes directly contribute to the neuronal disorders in mice and humans.

E: Recursive Splicing

As mentioned earlier, introns vary dramatically in length, some being 100 kb or longer. The ability of the splicing machinery to correctly identify functional splice sites within a boundless sea of highly related intronic sequences remains unsolved. The ratio of ESE/ESS has been suggested to aid this process. In addition, recent evidence suggests that splice site selection may be a co-transcriptional event *in vivo*, such that a functional splice site may be recognized right after its emergence from the Pol II complex, which then waits for the appearance of its downstream pair.

In the past several years, a new cellular mechanism, recursive splicing, has been shown to deal with splice site selection within long introns in *Drosophila* (Lopez, 1998). Recursive splicing may simply be viewed as multiple splicing events within the same intron, which eventually results in the removal of the long intron (Fig. 27.9). In this process, the 5′ splice site finds a downstream 3′ splice site to begin the initial splicing reaction. After the removal of the first piece of the intron, the resultant spliced product has a reconstituted 5′ splice site. This allows the next splicing event to take place. After all splicing reactions are finished in the long intron, the end product appears to have been a result of direct splicing of the upstream and downstream exons. This mechanism not only illustrates a solution to the removal of long introns, but may also explain some puzzling splicing products carrying one or a few nucleotide insertions between two exon sequences. It is suspected that these short nucleotide insertions may result from a recursive splicing situation where the 3′ splice site is followed by one or a few nucleotides before connecting to the downstream consensus 5′ splice site. This would create so-called "mini exons" in the length range of zero (a case where there are no nucleotides separating the upstream 3′ splice site and the downstream 5′ splice site) to a few nucleotides.

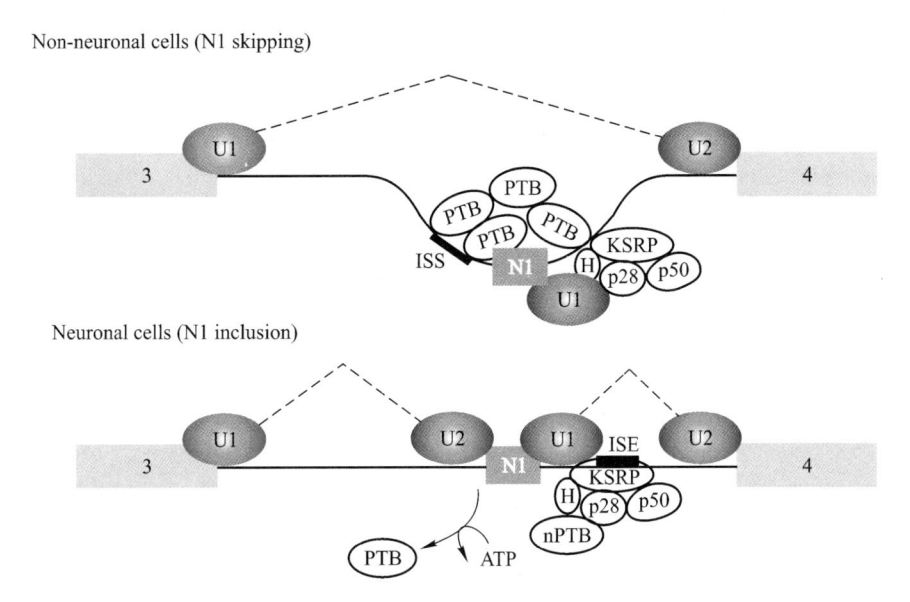

Fig.27.8 Positive and negative regulation of splicing via intronic regulatory elements. Example of PTB-regulated alternative splicing of c-src. PTB interacts with both up- and downstream intronic elements (ISS) in non-neuronal cells. In neurons, nPTB may be responsible for the reorganization of the suppression complex (an ATP-dependent process) therefore allowing the inclusion of the N exon. Other factors binding to a downstream intronic splicing enhancer element (ISE) also play a crucial role in the inclusion of the N exon in neuronal cells.

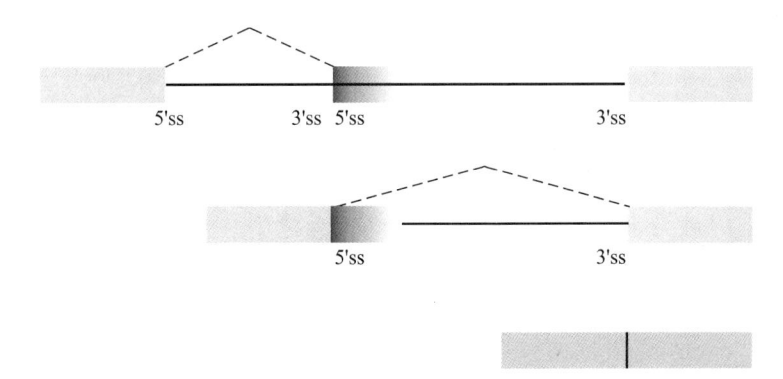

Fig.27.9 Removal of long introns by recursive splicing. The 5' splice site is spliced to a downstream 3' splice site within the long intron. After the splicing reaction, a 5' splice site is reconstituted and ready to pair with a subsequent downstream 3' splice site. Recursive splicing has been reported in the Ubx gene in *Drosophila*. Recursive splicing has not yet been reported in mammalian systems.

F: Regulated Splicing in Development and Disease

This is a big topic, which has been recently reviewed (Cartegni et al., 2002; Faustino and Cooper, 2003). In *Drosophila*, the role of both isoform expression and trans-acting splicing regulators in development has been well documented, with the sex determination pathway being best understood. Relatively little is known about the role of alternative splicing in development in mammalian systems. This may be largely due to the difficulty of conducting large-scale forward genetic studies in mammals. In a few reported cases, reverse genetic approaches (knock-out or knock-in) have been used to determine the function of specific isoforms or specific splicing regulators. For example, inactivation of WT1 isoform expression led to kidney failure (Hammes et al., 2001) and inactivation of FGFR2 isoforms caused abnormal limb development (De Moerlooze et al., 2000; Hajihosseini et al., 2001). Furthermore, knock-out of a number of SR proteins resulted in early embryonic lethality (Jumaa et al., 1999; Wang et al., 2001; Ding et al., 2004). More recently, it was shown that conditional ablation of prototypical SR proteins SC35 and ASF/SF2 in the heart had no effect on heart development, but caused a typical heart disease known as dilated cardiomyopathy (Ding et al., 2004; Xu et al., 2005). Interestingly, ASF/SF2 was found to be a critical regulator in a postnatal splicing reprogramming pathway, which is essential for heart remodeling during the juvenile to adult transition (Xu *et al.*, 2005). These studies demonstrate the importance of regulated splicing in animal development. The examples reported thus far only mark the beginning in understanding the biology of alternative splicing in mammalian systems.

The impact of splicing defects on diseases has long been recognized. An early survey indicates that about 15% of disease-causing mutations are because of mutations in exon/intron splicing signals in a wide range of cellular genes (Krawczak *et al.*, 1992; Stenson *et al.*, 2003). More recently, it was found that silent, mis-sense, and even non-sense mutations can contribute to disease phenotype because of induced splicing defects by the mutations (Cartegni *et al.*, 2002). Previously, disease phenotype associated with mis-sense and non-sense mutations was thought to be a result of impaired protein functions. It is now clear that many of these mutations result in mis-splicing and/or accelerated RNA decay, thereby mimicking the effect of null mutations. Two possible mechanisms exist for this effect. First, point mutations within exons may disrupt existing ESEs or create new ESSs, resulting in abnormal splicing. Second, non-sense mutations may trigger non-sense mediated decay (NMD), resulting in dramatic down-regulation of the affected transcript. The first mechanism explains the observation that point mutations causing Duchenne muscular dystrophy are associated with a more severe disease phenotype than deletion mutants (Pillers *et al.*, 1999).

In contrast to mutations in cis-acting regulatory elements, little is known about the effect of trans-acting splicing regulators in disease. The most well-known example is the survival motor neuron (SMN) gene. Mutations in SMN cause spinal muscular atrophy (SMA) (Lefebvre *et al.*, 1997). Biochemical studies revealed that SMN functions in snRNP recycling and therefore plays a role in pre-mRNA splicing in mammalian cells (Pellizzoni *et al.*, 1998). However it is unclear why motor neurons are particularly sensitive to mutations in the SMN gene. Genetic mapping also revealed the role of several essential splicing factors (Prp3, Prp8, and Prp31) in retinitis pigmentosa (Hims *et al.*, 2003). Mutations in these genes selectively cause aberrant splicing of a small number of genes, such as rhodopsin, which is critical for viability of retina cells (Yuan *et al.*, 2005).

Reverse genetics has also contributed to the elucidation of the roles of other trans-acting splicing regulators in disease. As previously mentioned, Nova was shown to induce neurological disorders in knock-out mice similar to those seen in human patients carrying mutations in Nova (Jensen *et al.*, 2000a).

Recently, knock-out mice of a gene named Muscleblind, which encodes for a RNA binding protein, gave rise to the same muscular atrophy phenotype as found in humans (Kanadia *et al.*, 2003). These studies show the relevance of splicing regulators in human disease. Broadly speaking, alternative splicing may be widely associated with human diseases, such as cancer, either directly contributing to a specific disease phenotype or indirectly accompanying the progression of a disease. A recent computational analysis suggests that many mRNA isoforms may be specifically induced during cellular transformation (Modrek and Lee, 2002). Therefore, understanding of the mechanism and regulation of alternative splicing is fundamental to disease research.

Coupling of Splicing with Other Nuclear Events

RNA splicing is not an isolated event in the nucleus, rather it is orchestrated with upstream transcriptional activities and downstream nuclear export steps. Understanding these integrated mechanisms will help to uncover regulated gene expression networks in higher eukaryotic cells. This topic has been recently reviewed (Maniatis and Reed, 2002).

A: Transcription-splicing Integration

EM visualization of co-transcriptional processing of pre-mRNA in spread chromosomes was the first to show the temporal integration of transcription and splicing (Beyer and Osheim, 1988; Beyer and Osheim, 1991). RT-PCR analysis of mRNAs associated with dissected chromosomes or those released into the nucleoplasm was further used to prove the case (Bauren and Wieslander, 1994). Results showed that upstream introns were removed before the completion of transcription (chromosome-associated) and downstream introns could be spliced either before or after transcription (only a fraction of spliced mRNA was chromosome-associated). Furthermore, *in situ* hybridization using specific exon-exon *junction* probes revealed the co-localization of spliced RNA with nascent transcripts at the gene locus in the nucleus (Zhang *et al.*, 1994). Together, these observations strongly support the idea that splicing is temporally paired with transcription in the nucleus, keeping in mind the possibility that some splicing events may be initiated during transcription but not completed until after the release of the transcript into the nucleoplasm.

Promoter use and transcription elongation have been shown to have some impact on splice site selection, further supporting the integration of transcription and splicing in the nucleus. In transfected cells, it was first recognized that alternative splicing of certain genes may be affected by the choice of promoters to drive the expression of a reporter gene (Cramer *et al.*, 1999; Cramer *et al.*, 2001). It was suggested that different promoters may drive gene expression through distinct mechanisms by using different sets of transcription factors. These factors may result in differential recruitment of splicing factors responsible for the recognition of the emerging splice sites during transcription.

One can further image that the transcription elongation rate may play a role in splice site selection. A slow polymerase would give an emerging weak splice site more time to be recognized and paired with the upstream splice site, forming a committed splicing complex before the appearance of a stronger downstream splice site. However, with a fast polymerase, the emergence of a strong downstream splice site may be faster, therefore not giving a weaker upstream splice site a chance to be recognized before the emergence of the stronger downstream splice site. According to this model, alternative splicing events may be modulated by the elongation rate of the Pol II complex. This was recently shown to be the case when a slower Pol II (due to a point mutation in the polymerase) was compared with wt Pol II (Kadener *et al.*, 2002). It is important to note that the models for differential promoter loading and variations in Pol II kinetics may not be mutually exclusive. Transcription elongation rates may directly be dependent on the promoter used and the co-factors recruited.

Factors involved in the integration of transcription and splicing remain to be characterized. SR proteins have been found to able to interact with the C-terminal domain (CTD) of Pol II in a phosphorylation dependent manner (Misteli and Spector, 1999). Interestingly, phosphorylated Pol II was able to stimulate splicing *in vitro*, even though it is not an essential splicing factor (Hirose *et al.*, 1999). A number of other RNA processing factors are also associated with the CTD of Pol II, including enzymes involved in capping and polyadenylation, suggesting that RNA processing events ranging from capping to intron removal to polyadenylation are largely co-transcriptional *in vivo* (Proudfoot, 2000; Proudfoot *et al.*, 2002). CTD may be a crucial docking site for many different processing factors. However, other factors functioning in transcription elongation, such as protein kinases and phosphatases, may dictate the affinity of various processing factors for CTD. At this point, most mechanisms proposed for the integration of transcription and splicing remain largely speculative and wait for further experimental evidence.

B: Coupling of Splicing with Nuclear Export

It has long been observed that unprocessed pre-mRNA cannot be exported out of the nucleus, suggesting the existence of a RNA quality control mechanism. Initially, it was thought that the RNA processing machinery might be localized near the nuclear pore complex to spatially and temporally integrate splicing with RNA export. However this is clearly not the case. Instead, it has been found that transcription factors are able to recruit splicing factors, which then recruit export factors (Reed, 2003). As shown in Fig.27.10, the transcription export complex (TREX), which is part of the transcription elongation complex, appears to directly interact with specific RNA binding proteins, such as UAP65, which plays an important role in mediating U2 binding to the branchpoint sequence during spliceosome assembly. As part of the spliceosome, UAP65 then recruits a small protein called Aly, which is a key component of the complex deposited onto every exon-exon junctions (known as the EJC complex, see below) after splicing. Aly then directly interacts with TAP, a critical mediator involved in RNA export, which in turn interacts with nuclear pore complex. Through this cascade of protein-protein interactions, transcription is apparently integrated with splicing and then with export.

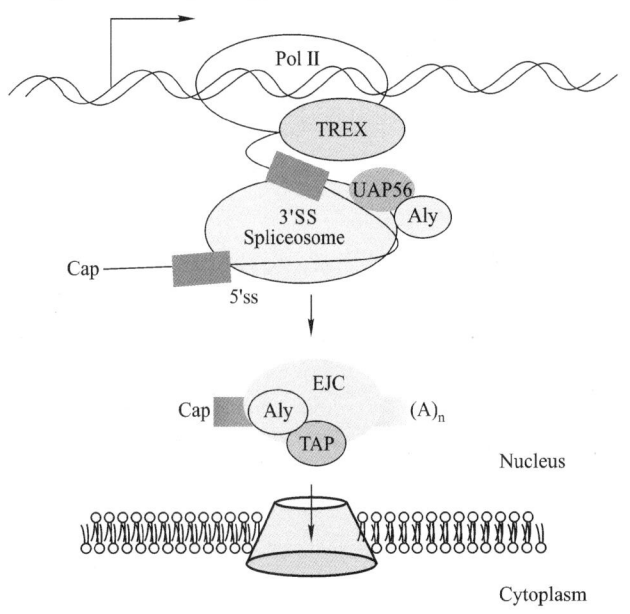

Fig.27.10 Integration of splicing with RNA export. The TREX complex plays a role in the co-transcriptional recruitment of UAP56 to pre-mRNA. During splicing, UAP56 recruits Aly, which then becomes a component of EJC. Aly directly interacts with TAP to initiate RNA export. TAP may also be independently recruited by dephosphorylated SR proteins associated with spliced mRNAs as described in the text.

Other RNA binding proteins, such as the cap binding protein CBP80/20 and SR proteins, also seem to contribute to RNA export (Huang et al., 2003). A subset of SR proteins have been shown to shuttle between the nucleus and the cytoplasm, suggesting a role for shuttling SR proteins during RNA export (Caceres et al., 1998). In addition, shuttling SR proteins may also play a role in translation (Sanford et al., 2004) (see below).

C: Integration of Splicing and Regulation of RNA Stability

A key complex involved in various integration events is the so-called EJC complex, which consist of a group of proteins deposited onto the exon-exon junction after a splicing reaction (Le Hir et al., 2000). The EJC complex connects RNA splicing to RNA export as described above. The EJC complex also plays a role in non-sense mediated RNA decay (NMD) (Le Hir et al., 2001). This is accomplished by the recruitment of Upf proteins, which in turn recruits the decapping enzyme Dcp1. RNA decapping is one of the major pathways for regulated RNA degradation in eukaryotic cells (Hilleren and Parker, 1999).

Two positional rules have come to light in recent years, regarding the influence of upstream RNA processing events on RNA stability. As illustrated in Fig. 27.11, the first positional rule states that activation of the NMD pathway is triggered by a premature stop codon located more than 50 to 55 nt upstream of an exon-exon junction. The second rule states that EJC is deposited on exon sequences 20 to 24 nt upstream of each exon-exon junction as a result of the splicing reaction.

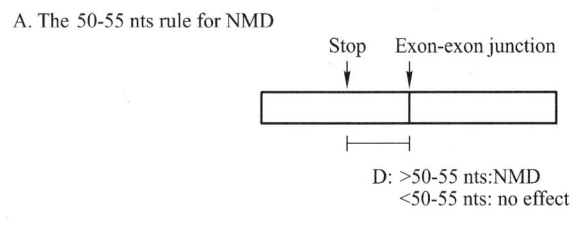

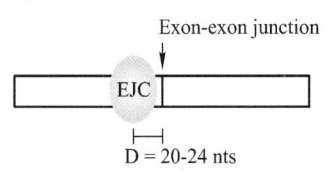

Fig.27.11 The two positional rules for NMD. A. The rule for NMD. A pre-mature stop codon located more than 50 to 55 nt upstream of an exon-exon junction will trigger NMD. B. An EJC complex deposited 20 to 24 nt upstream of an exon-exon junction. The interaction of the EJC with exonic sequences is position-dependent and sequence-independent.

These rules provide critical insights into the impact of RNA splicing on the regulation of RNA stability in the cytoplasm. As illustrated in Fig.27.12, during or right after nuclear export of spliced mRNA to the cytoplasm, the first round of translation (pilot round) rearranges protein complexes deposited onto the mRNA. Most natural stop codons reside in the last exon, therefore during the pilot round, the scanning ribosomes strip off all EJCs before reaching the natural stop codon. However in the presence of a pre-mature stop codon, the scanning ribosomes falls off the mRNA before stripping off all EJC complexes. The remaining EJC complex then recruits Upf proteins, which in turn recruit the decapping enzyme, leading to the down-regulation of the mRNA. It should be noted that additional sequence specific RNA binding proteins in the cytoplasm may add to the regulation of RNA stability (Chen et al., 2001; Gherzi et al., 2004). As a result, individual RNA molecules have distinct half-lives in a given cell under a given experimental condition.

D: Connecting Nuclear Processing to Protein Translation in the Cytoplasm

Proteins associated with mRNA during nuclear export may also play a role in protein translation. This was recently found to be true with shuttling SR proteins such as ASF/SF2 (Sanford et al., 2004). Exported mRNA carrying this splicing factor appears to be more competent in translation, although the mechanism remains to be understood.

Shuttling nuclear RNA processing factors are able to move back and forth to and from the nucleus. This export/import cycle may be regulated by phosphorylation. As previously mentioned, SR proteins need to be phosphorylated to initiate spliceosome assembly and dephosphorylated for spliceosome resolution. For shuttling SR proteins, dephosphorylation is also essential for interaction with the export factor TAP (Huang et al., 2003; Lai and Tarn, 2004). Once shuttling SR proteins reach the cytoplasm, they are phosphorylated by a family of SR protein specific kinase SRPKs (Yun and Fu, 2000). This is an essential step to enable SR proteins to interact with their nuclear import receptor to enter the nucleus (Lai et al., 2001; Yun et al., 2003). Recently, it has been proposed that SRPK-mediated phosphorylation might have a dual effect: (1) phosphorylation may facilitate the release of shuttling SR proteins from exported mRNA, and (2) phosphorylation may then promotes re-entry of the released SR proteins (Gilbert and Guthrie, 2004). Thus, nuclear export of mRNA may be regulated by a nucleus-localized phosphatase and a cytoplasm-localized kinase system. This is in contrast to

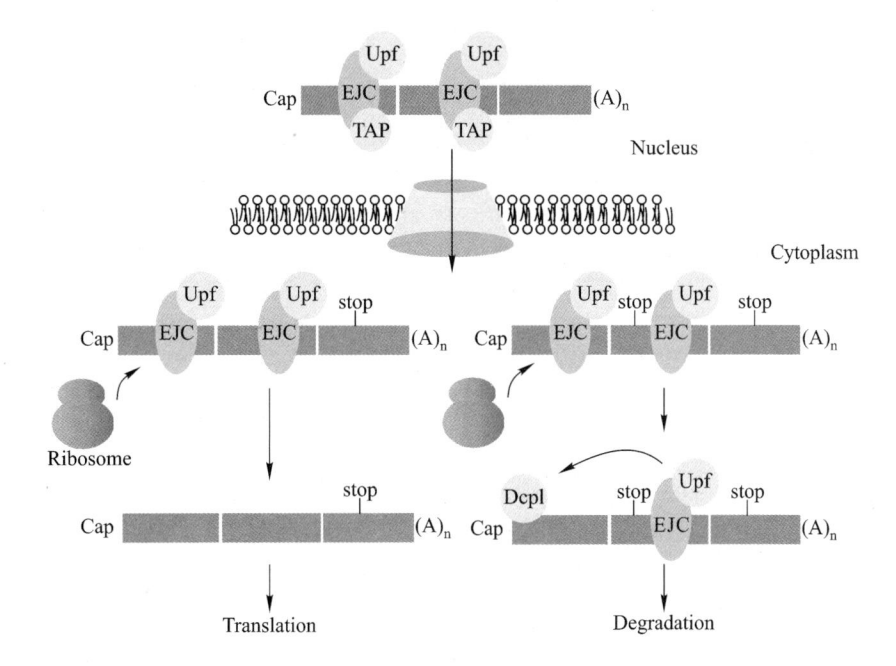

Fig.27.12 Integration of splicing with RNA stability. EJCs, which are deposited onto exon-exon junctions after splicing, play an important role in RNA export. Right after export, ribosomes engage in the first round of translation. During this first round of translation, scanning ribosomes removes EJCs from the mRNA. However, if a pre-mature stop codon is present, it will induce the release of ribosome and all subsequent EJCs will remain on the mRNA. The remaining EJC complexes will recruit Upf proteins to initiate RNA degradation in the cytoplasm. Other sequence specific RNA binding proteins may also interact with mRNA in the cytoplasm to regulate RNA stability, which is independent of EJCs as described in the text.

other export pathways, which requires the RanGTP gradient created by the nucleus-localized RCC1 (a Ran exchange factor to generate RanGTP) and a cytoplasm-localized RanGAP (a Ran activation protein to catalyze GTP hydrolysis to produce RanGDP) (Weis, 2002).

Genomics of RNA Splicing

Like everything else, RNA splicing has become global in the post-genome era. The term "-omics" is used to describe the whole collection of a group, such as proteomics (for proteins), kinomics (for kinases), etc. As discussed earlier, alternatively spliced mRNA isoforms are widespread in higher eukaryotic cells. This presents new opportunities and challenges in understanding the new dimension of gene expression. In the last section of the chapter, features of what one may refer to as RNAomics are briefly depicted.

A: Scale of Alternative Splicing in Eukaryotic Genomes

As described earlier, alternative splicing now appears to be more the rule rather than the exception. About half of the genes in human and mouse express multiple mRNA isoforms. Some of these isoforms may be regulated and thus functionally important whereas others may reflect splicing errors or the products produced in a diseased state. Individual mRNA isoforms from a given gene may encode distinct protein products, be differentially localized in cells or embryos, or have unique half-lives. Therefore, alternative splicing may contribute to the complexity and diversity of the proteome in eukaryotic cells and provide points of differential gene expression regulation. It should be stressed that the functional importance of a given mRNA isoform may not correlate with its abundance. This thus points to a specific problem in the field because most studies focus on the most abundant isoform for functional dissection.

B: Features of Alternatively Spliced Regions

Analysis of alternatively spliced genes reveals both expected and unexpected features. Generally speaking, splice sites involved in alternative exons are relatively weaker (more divergent from the consensus sequence) than constitutive splice sites. Weaker splice sites may allow the splicing machinery to regulate and control their usage. Alternative exons are shorter than constitutive exons, suggesting that short exons may be subject to alternative splicing because of a reduced ESE frequency. Interestingly, skipped exons maintain the reading frame (length is a multiple of 3 nts) of a transcript more frequently than constitutive ones. This property allows preservation of the protein structure encoded by downstream constitutive exons. Surprisingly, sequences surrounding alternative splice sites (both in the exon and the intron sides) seem to be more conserved across different species than constitutive exons (Sorek and Ast, 2003). This finding argues against the assumption that weak splice sites are sites which have not yet fully evolved. Instead this suggests that alternative splicing signals (weaker splice sites) may be preserved on purpose during evolution, and thus are functionally important. These sequence features form a basis for the development of *ab initio* computational tools to predict alternative splicing.

The evolution of alternative splicing is an interesting and open research area. It is postulated that mutually exclusive exons might result from exon duplication events therefore explaining the sequence similarities between pairs of mutually exclusive exons. A fraction of skipped exons, on the other hand, appears to have evolved from intronic sequences carrying Alu elements ("exonization" of Alu elements) (Sorek *et al.*, 2002; Lev-Maor *et al.*, 2003). A more recent analysis indicates that "exonization" of multiple types of transposable elements may also contribute to the evolution of alternative splicing skipped exons (Zheng *et al.*, 2005).

C: Development of Splicing Arrays

Given the abundance of alternative splicing in higher eukaryotic cells, robust isoform-sensitive microarray systems would aid in the understanding of regulated splicing in development and disease. Three basic microarray platforms have been developed. One platform uses short oligonucleotides (40-mer) to detect individual exon-exon junctions (Clark *et al.*, 2002). This strategy has been used to fabricate a high-density array, targeting exon-exon (mostly constitutive exons) junctions in ~10,000 human transcripts (Johnson *et al.*, 2003). More recently, this strategy has been used to interrogate several thousand skipped exons in the mouse (Pan *et al.*, 2004). One major concern with the use of exon-exon probes is the so-called "half-hybridization" phenomenon where half of each exon-exon probe will hybridize to all competing isoforms containing a common donor or acceptor.

The second splicing array platform, the "all-exon" array, is under development at Affymetrix. In constructing this array, all potential exons were identified with the help of several gene prediction programs. This strategy involves design of four oligonucleotides for each exon. Potential alternative portions of an exon are considered as separate exon or exonic regions, therefore four oligonucleotides are also designed for the exon portion.

An advantage of this "all-exon" array approach is the ability to conduct an unbiased search for regulated splicing. However, an obvious disadvantage is the lack of exon-exon linkage information.

The third platform is based on a molecular barcode strategy (Yeakley *et al.*, 2002). In this approach, a total of three oligonucleotides are used to detect two alternative isoforms. One oligonucleotide targets the common exon site (donor or acceptor) of an alternative event and is linked to a universal primer-landing site. Two additional oligonucleotides are separately synthesized to target the alternative exonic regions. These two oligonucleotides are each linked to a unique 20-mer sequence (called zipcodes) followed by another universal primer landing site. These zip codes are printed on a universal array, which are later used to detect different isoforms. The splicing profile experiment begins with a RNA annealing reaction in which total RNA is mixed with pooled oligonucleotides and biotinylated oligo-dT under denaturing and annealing conditions. Annealed oligonucleotides are selected for on streptavidin oligo-dT beads (solid selection phase). During the solid selection phase, the beads capture all polyA mRNAs along with annealed oligonucleotides. Free oligonucleotides are washed away and T4 DNA ligase is then used to ligate adjacent oligonucleotides bridged by the mRNAs. This process converts half amplicons to full amplicons. Only ligated oligonucleotides can be amplified by PCR via the pair of universal primers. One of the primers is end-labeled with a fluorescent dye. Thus the PCR products can be directly applied to the universal zipcode array. Each zipcode reports one mRNA isoform according to the oligonucleotide design. This approach has aided in the discovery of targets for specific splicing factors, alternative splicing events regulated in various signal transduction pathways, and tumor specific mRNA isoform biomarkers (Li *et al.*, 2005).

D: Contributions of Splicing to Biology: Opportunities and Challenges

Given the degree to which we understand the basic mechanisms and various regulatory strategies for alternative splicing, the research in this post-transcriptional step of gene expression is at its infancy. For example, we know little about the biological functions of the majority of isoforms and how alternative splicing may be regulated in development and disease. Understanding the integration of splicing and various upstream and downstream events are still in preliminary stages. As many are chasing fundamental questions on processing and regulation of "regular"

RNAs in this field, the mysterious world of microRNAs has recently surfaced. In light of these recent advances, this chapter may better be viewed as a call for the next generation of scientists to pursue a career in RNA research rather than a summation of what has been accomplished thus far in the world of RNA.

Acknowledgement

Work in the Fu lab is supported by grants from National Institutes of Health, USA.

References

Abovich, N., and Rosbash, M. (1997). Cross-intron bridging interactions in the yeast commitment complex are conserved in mammals. Cell *89*, 403-412.

Ares, M., Jr., and Weiser, B. (1995). Rearrangement of snRNA structure during assembly and function of the spliceosome. Prog Nucleic Acid Res Mol Biol *50*, 131-159.

Baker, B. S. (1989). Sex in flies: the splice of life. Nature *340*, 521-524.

Bauren, G., and Wieslander, L. (1994). Splicing of Balbiani ring 1 gene pre-mRNA occurs simultaneously with transcription. Cell *76*, 183-192.

Berget, S. M. (1995). Exon recognition in vertebrate splicing. J Biol Chem *270*, 2411-2414.

Berglund, J. A., Abovich, N., and Rosbash, M. (1998). A cooperative interaction between U2AF65 and mBBP/SF1 facilitates branchpoint region recognition. Genes Dev *12*, 858-867.

Beyer, A. L., and Osheim, Y. N. (1988). Splice site selection, rate of splicing, and alternative splicing on nascent transcripts. Genes Dev *2*, 754-765.

Beyer, A. L., and Osheim, Y. N. (1991). Visualization of RNA transcription and processing. Semin Cell Biol *2*, 131-140.

Black, D. L. (2000). Protein diversity from alternative splicing: a challenge for bioinformatics and post-genome biology. Cell *103*, 367-370.

Black, D. L. (2003). Mechanisms of alternative pre-messenger RNA splicing. Annu Rev Biochem *72*, 291-336.

Burge, C. B., Padgett, R. A., and Sharp, P. A. (1998). Evolutionary fates and origins of U12-type introns. Mol Cell *2*, 773-785.

Caceres, J. F., Screaton, G. R., and Krainer, A. R. (1998). A specific subset of SR proteins shuttles continuously between the nucleus and the cytoplasm. Genes Dev *12*, 55-66.

Cartegni, L., Chew, S. L., and Krainer, A. R. (2002). Listening to silence and understanding nonsense: exonic mutations that affect splicing. Nat Rev Genet *3*, 285-298.

Cech, T. R. (1985). Self-splicing RNA: implications for evolution. Int Rev Cytol *93*, 3-22.

Chan, R. C., and Black, D. L. (1997). The polypyrimidine tract binding protein binds upstream of neural cell-specific c-src exon N1 to repress the splicing of the intron downstream. Mol Cell Biol *17*, 4667-4676.

Chen, C. Y., Gherzi, R., Ong, S. E., Chan, E. L., Raijmakers, R., Pruijn, G. J., Stoecklin, G., Moroni, C., Mann, M., and Karin, M. (2001). AU binding proteins recruit the exosome to degrade ARE-containing mRNAs. Cell *107*, 451-464.

Clark, T. A., Sugnet, C. W., and Ares, M., Jr. (2002). Genomewide analysis of mRNA processing in yeast using splicing-specific microarrays. Science *296*, 907-910.

Colwill, K., Feng, L. L., Yeakley, J. M., Gish, G. D., Caceres, J. F., Pawson, T., and Fu, X. D. (1996a). SRPK1 and Clk/Sty protein kinases show distinct substrate specificities for serine/arginine-rich splicing factors. J Biol Chem *271*, 24569-24575.

Colwill, K., Pawson, T., Andrews, B., Prasad, J., Manley, J. L., Bell, J. C., and Duncan, P. I. (1996b). The Clk/Sty protein kinase phosphorylates SR splicing factors and regulates their intranuclear distribution. EMBO J *15*, 265-275.

Cramer, P., Caceres, J. F., Cazalla, D., Kadener, S., Muro, A. F., Baralle, F. E., and Kornblihtt, A. R. (1999). Coupling of transcription with alternative splicing: RNA pol II promoters modulate SF2/ASF and 9G8 effects on an exonic splicing enhancer. Mol Cell *4*, 251-258.

Cramer, P., Srebrow, A., Kadener, S., Werbajh, S., de la Mata, M., Melen, G., Nogues, G., and Kornblihtt, A. R. (2001). Coordination between transcription and pre-mRNA processing. FEBS Lett *498*, 179-182.

Cullen, B. R. (2004). Transcription and processing of human microRNA precursors. Mol Cell *16*, 861-865.

De Moerlooze, L., Spencer-Dene, B., Revest, J., Hajihosseini, M., Rosewell, I., and Dickson, C. (2000). An important role for the IIIb isoform of fibroblast growth factor receptor 2 (FGFR2) in mesenchymal-epithelial signalling during mouse organogenesis. Development *127*, 483-492.

Decatur, W. A., and Fournier, M. J. (2003). RNA-guided nucleotide modification of ribosomal and other RNAs. J Biol Chem *278*, 695-698.

Deutscher, M. P. (1984). Processing of tRNA in prokaryotes and eukaryotes. CRC Crit Rev Biochem *17*, 45-71.

Ding, J. H., Xu, X., Yang, D., Chu, P. H., Dalton, N. D., Ye, Z., Yeakley, J. M., Cheng, H., Xiao, R. P., Ross, J., et al. (2004). Dilated cardiomyopathy caused by tissue-specific ablation of SC35 in the heart. EMBO J *23*, 885-896.

Dredge, B. K., and Darnell, R. B. (2003). Nova regulates GABA(A) receptor gamma2 alternative splicing via a distal downstream UCAU-rich intronic splicing enhancer. Mol Cell Biol *23*, 4687-4700.

Faustino, N. A., and Cooper, T. A. (2003). Pre-mRNA splicing and human disease. Genes Dev *17*, 419-437.

Fleckner, J., Zhang, M., Valcarcel, J., and Green, M. R. (1997). U2AF65 recruits a novel human DEAD box protein required for the U2 snRNP-branchpoint interaction. Genes Dev *11*, 1864-1872.

Fu, X. D. (1993). Specific commitment of different pre-mRNAs to splicing by single SR proteins. Nature *365*, 82-85.

Fu, X. D. (1995). The superfamily of arginine/serine-rich splicing factors. RNA *1*, 663-680.

Fu, X. D. (2004). Towards a splicing code. Cell *119*, 736-738.

Fu, X. D., and Maniatis, T. (1992). The 35-kDa mammalian splicing factor SC35 mediates specific interactions between U1 and U2 small nuclear ribonucleoprotein particles at the 3' splice site. Proc Natl Acad Sci USA *89*, 1725-1729.

Garcia-Blanco, M. A., Jamison, S. F., and Sharp, P. A. (1989). Identification and purification of a 62,000-dalton protein that binds specifically to the polypyrimidine tract of introns. Genes Dev *3*, 1874-1886.

Gherzi, R., Lee, K. Y., Briata, P., Wegmuller, D., Moroni, C., Karin, M., and Chen, C. Y. (2004). A KH domain RNA binding protein, KSRP, promotes ARE-directed mRNA turnover by recruiting the degradation machinery. Mol Cell *14*, 571-583.

Gilbert, W., and Guthrie, C. (2004). The Glc7p nuclear phosphatase promotes mRNA export by facilitating association of Mex67p with mRNA. Mol Cell *13*, 201-212.

Gottschalk, A., Neubauer, G., Banroques, J., Mann, M., Luhrmann, R., and Fabrizio, P. (1999). Identification by mass spectrometry and functional analysis of novel proteins of the yeast [U4/U6.U5] tri-snRNP. Embo J *18*, 4535-4548.

Graveley, B. R. (2000). Sorting out the complexity of SR protein functions. RNA *6*, 1197-1211.

Gui, J. F., Lane, W. S., and Fu, X. D. (1994a). A serine kinase regulates intracellular localization of splicing factors in the cell cycle. Nature *369*, 678-682.

Gui, J. F., Tronchere, H., Chandler, S. D., and Fu, X. D. (1994b). Purification and characterization of a kinase specific for the serine- and arginine-rich pre-mRNA splicing factors. Proc Natl Acad Sci USA *91*, 10824-10828.

Guthrie, C. (1991). Messenger RNA splicing in yeast: clues to why the spliceosome is a ribonucleoprotein. Science *253*, 157-163.

Guthrie, C. (1994). The spliceosome is a dynamic ribonucleoprotein machine. Harvey Lect *90*, 59-80.

Hajihosseini, M. K., Wilson, S., De Moerlooze, L., and Dickson, C. (2001). A splicing switch and gain-of-function mutation in FgfR2-IIIc hemizygotes causes Apert/Pfeiffer-syndrome-like phenotypes. Proc Natl Acad Sci USA *98*, 3855-3860.

Hammes, A., Guo, J. K., Lutsch, G., Leheste, J. R., Landrock, D., Ziegler, U., Gubler, M. C., and Schedl, A. (2001). Two splice variants of the Wilms' tumor 1 gene have distinct functions during sex determination and nephron formation. Cell *106*, 319-329.

Hartmann, C., Corre-Menguy, F., Boualem, A., Jovanovic, M., and Lelandais-Briere, C. (2004). (MicroRNAs: a new class of gene expression regulators). Med Sci (Paris) *20*, 894-898.

Hilleren, P., and Parker, R. (1999). Mechanisms of mRNA surveillance in eukaryotes. Annu Rev Genet *33*, 229-260.

Hims, M. M., Diager, S. P., and Inglehearn, C. F. (2003). Retinitis pigmentosa: genes, proteins and prospects. Dev Ophthalmol *37*, 109-125.

Hirose, Y., Tacke, R., and Manley, J. L. (1999). Phosphorylated RNA polymerase II stimulates pre-mRNA splicing. Genes Dev *13*, 1234-1239.

Hoffman, B. E., and Grabowski, P. J. (1992). U1 snRNP targets an essential splicing factor, U2AF65, to the 3' splice site by a network of interactions spanning the exon. Genes Dev *6*, 2554-2568.

Huang, Y., Gattoni, R., Stevenin, J., and Steitz, J. A. (2003). SR splicing factors serve as adapter proteins for TAP-dependent mRNA export. Mol Cell *11*, 837-843.

Jensen, K. B., Dredge, B. K., Stefani, G., Zhong, R., Buckanovich, R. J., Okano, H. J., Yang, Y. Y., and Darnell, R. B. (2000a). Nova-1 regulates neuron-specific alternative splicing and is essential for neuronal viability. Neuron *25*, 359-371.

Jensen, K. B., Musunuru, K., Lewis, H. A., Burley, S. K., and Darnell, R. B. (2000b). The tetranucleotide UCAY directs the specific recognition of RNA by the Nova K-homology 3 domain. Proc Natl Acad Sci USA *97*, 5740-5745.

Johnson, J. M., Castle, J., Garrett-Engele, P., Kan, Z., Loerch, P. M., Armour, C. D., Santos, R., Schadt, E. E., Stoughton, R., and Shoemaker, D. D. (2003). Genome-wide survey of human alternative pre-mRNA splicing with exon junction microarrays. Science *302*, 2141-2144.

Jumaa, H., Wei, G., and Nielsen, P. J. (1999). Blastocyst formation is blocked in mouse embryos lacking the splicing factor SRp20. Curr Biol *9*, 899-902.

Kadener, S., Fededa, J. P., Rosbash, M., and Kornblihtt, A. R. (2002). Regulation of alternative splicing by a transcriptional enhancer through RNA pol II elongation. Proc Natl Acad Sci USA *99*, 8185-8190.

Kanadia, R. N., Johnstone, K. A., Mankodi, A., Lungu, C., Thornton, C. A., Esson, D., Timmers, A. M., Hauswirth, W. W., and Swanson, M. S. (2003). A muscleblind knockout model for myotonic dystrophy. Science *302*, 1978-1980.

Kapranov, P., Cawley, S. E., Drenkow, J., Bekiranov, S., Strausberg, R. L., Fodor, S. P., and Gingeras, T. R. (2002). Large-scale transcriptional activity in chromosomes 21 and 22. Science *296*, 916-919.

Keller, W., and Minvielle-Sebastia, L. (1997). A comparison of mammalian and yeast pre-mRNA 3'-end processing. Curr Opin Cell Biol *9*, 329-336.

Kirsebom, L. A. (2002). RNase P RNA-mediated catalysis. Biochem Soc Trans *30*, 1153-1158.

Kistler, A. L., and Guthrie, C. (2001). Deletion of MUD2, the yeast homolog of U2AF65, can bypass the requirement for sub2, an essential spliceosomal ATPase. Genes Dev *15*, 42-49.

Kramer, A. (1996). The structure and function of proteins involved in mammalian pre-mRNA splicing. Annu Rev Biochem *65*, 367-409.

Krawczak, M., Reiss, J., and Cooper, D. N. (1992). The mutational spectrum of single base-pair substitutions in mRNA splice junctions of human genes: causes and consequences. Hum Genet *90*, 41-54.

Lafontaine, D., and Tollervey, D. (1995). Trans-acting factors in yeast pre-rRNA and pre-snoRNA processing. Biochem Cell Biol *73*, 803-812.

Lai, M. C., Lin, R. I., and Tarn, W. Y. (2001). Transportin-SR2 mediates nuclear import of phosphorylated SR proteins. Proc Natl Acad Sci USA *98*, 10154-10159.

Lai, M. C., and Tarn, W. Y. (2004). Hypophosphorylated ASF/SF2 binds TAP and is present in messenger ribonucleoproteins. J Biol Chem *279*, 31745-31749.

Le Hir, H., Gatfield, D., Izaurralde, E., and Moore, M. J. (2001). The exon-exon junction complex provides a binding platform for factors involved in mRNA export and nonsense-mediated mRNA decay. EMBO J *20*, 4987-4997.

Le Hir, H., Moore, M. J., and Maquat, L. E. (2000). Pre-mRNA splicing alters mRNP composition: evidence for stable association of proteins at exon-exon junctions. Genes Dev *14*, 1098-1108.

Lee, Y., Kim, M., Han, J., Yeom, K. H., Lee, S., Baek, S. H., and Kim, V. N. (2004). MicroRNA genes are transcribed by RNA polymerase II. EMBO J *23*, 4051-4060.

Lefebvre, S., Burlet, P., Liu, Q., Bertrandy, S., Clermont, O., Munnich, A., Dreyfuss, G., and Melki, J. (1997). Correlation between severity and SMN protein level in spinal muscular atrophy. Nat Genet *16*, 265-269.

Lev-Maor, G., Sorek, R., Shomron, N., and Ast, G. (2003). The birth of an alternatively spliced exon: 3' splice-site selection in Alu exons. Science *300*, 1288-1291.

Li, H-R., Yeakley, J.M., Nair, T.M., Kwon, Y.S., Bibikova, M., Zhou, L., Zheng, C., Downs, T., Wang-Rodriguz, J., Fu, X-D., and Fan, J-B. (2005). Two-dimensional transcriptome profiling: Identification of novel prostate cancer biomarkers from archived paraffin-embedded cancer specimens. Submitted.

Lopez, A. J. (1998). Alternative splicing of pre-mRNA: developmental consequences and mechanisms of regulation. Annu Rev Genet *32*, 279-305.

Maniatis, T., and Reed, R. (2002). An extensive network of coupling among gene expression machines. Nature *416*, 499-506.

Maniatis, T., and Tasic, B. (2002). Alternative pre-mRNA splicing and proteome expansion in metazoans. Nature *418*, 236-243.

Martinis, S. A., Plateau, P., Cavarelli, J., and Florentz, C. (1999). Aminoacyl-tRNA synthetases: a new image for a classical family. Biochimie *81*, 683-700.

Mermoud, J. E., Cohen, P., and Lamond, A. I. (1992). Ser/Thr-specific protein phosphatases are required for both catalytic steps of pre-mRNA splicing. Nucleic Acids Res *20*, 5263-5269.

Mermoud, J. E., Cohen, P. T., and Lamond, A. I. (1994). Regulation of mammalian spliceosome assembly by a protein phosphorylation mechanism. EMBO J *13*, 5679-5688.

Michaud, S., and Reed, R. (1993). A functional association between the 5' and 3' splice site is established in the earliest prespliceosome complex (E) in mammals. Genes Dev *7*, 1008-1020.

Minvielle-Sebastia, L., and Keller, W. (1999). mRNA polyadenylation and its coupling to other RNA processing reactions and to transcription. Curr Opin Cell Biol *11*, 352-357.

Misteli, T., Caceres, J. F., Clement, J. Q., Krainer, A. R., Wilkinson, M. F., and Spector, D. L. (1998). Serine phosphorylation of SR proteins is required for their recruitment to sites of transcription *in vivo*. J Cell Biol *143*, 297-307.

Misteli, T., and Spector, D. L. (1999). RNA polymerase II targets pre-mRNA splicing factors to transcription sites *in vivo*. Mol Cell *3*, 697-705.

Mochizuki, K., Fine, N. A., Fujisawa, T., and Gorovsky, M. A. (2002). Analysis of a piwi-related gene implicates small RNAs in genome rearrangement in *tetrahymena*. Cell *110*, 689-699.

Modafferi, E. F., and Black, D. L. (1999). Combinatorial control of a neuron-specific exon. RNA *5*, 687-706.

Modrek, B., and Lee, C. (2002). A genomic view of alternative splicing. Nat Genet *30*, 13-19.

Novina, C. D., and Sharp, P. A. (2004). The RNAi revolution. Nature *430*, 161-164.

Pan, Q., Shai, O., Misquitta, C., Zhang, W., Saltzman, A. L., Mohammad, N., Babak, T., Siu, H., Hughes, T. R., Morris, Q. D., et al. (2004). Revealing global regulatory features of mammalian alternative splicing using a quantitative microarray platform. Mol Cell *16*, 929-941.

Park, J. W., Parisky, K., Celotto, A. M., Reenan, R. A., and Graveley, B. R. (2004). Identification of alternative splicing regulators by RNA interference in *Drosophila*. Proc Natl Acad Sci USA *101*, 15974-15979.

Patel, A. A., and Steitz, J. A. (2003). Splicing double: insights from the second spliceosome. Nat Rev Mol Cell Biol *4*, 960-970.

Pellizzoni, L., Kataoka, N., Charroux, B., and Dreyfuss, G. (1998). A novel function for SMN, the spinal muscular atrophy disease gene product, in pre-mRNA splicing. Cell *95*, 615-624.

Pillers, D. M., Fitzgerald, K. M., Duncan, N. M., Rash, S. M., White, R. A., Dwinnell, S. J., Powell, B. R., Schnur, R. E., Ray, P. N., Cibis, G. W., and Weleber, R. G. (1999). Duchenne/Becker muscular dystrophy: correlation of phenotype by electroretinography with sites of dystrophin mutations. Hum Genet *105*, 2-9.

Prasad, J., Colwill, K., Pawson, T., and Manley, J. L. (1999). The protein kinase Clk/Sty directly modulates SR protein activity: both hyper- and hypophosphorylation inhibit splicing. Mol Cell

Biol *19*, 6991-7000.

Proudfoot, N. (2000). Connecting transcription to messenger RNA processing. Trends Biochem Sci *25*, 290-293.

Proudfoot, N. J., Furger, A., and Dye, M. J. (2002). Integrating mRNA processing with transcription. Cell *108*, 501-512.

Reed, R. (2003). Coupling transcription, splicing and mRNA export. Curr Opin Cell Biol *15*, 326-331.

Roscigno, R. F., and Garcia-Blanco, M. A. (1995). SR proteins escort the U4/U6.U5 tri-snRNP to the spliceosome. RNA *1*, 692-706.

Sanford, J. R., Gray, N. K., Beckmann, K., and Caceres, J. F. (2004). A novel role for shuttling SR proteins in mRNA translation. Genes Dev *18*, 755-768.

Schaal, T. D., and Maniatis, T. (1999). Multiple distinct splicing enhancers in the protein-coding sequences of a constitutively spliced pre-mRNA. Mol Cell Biol *19*, 261-273.

Schurer, H., Schiffer, S., Marchfelder, A., and Morl, M. (2001). This is the end: processing, editing and repair at the tRNA 3'-terminus. Biol Chem *382*, 1147-1156.

Sharp, P. A. (1994). Split genes and RNA splicing. Cell *77*, 805-815.

Sharp, P. A., and Burge, C. B. (1997). Classification of introns: U2-type or U12-type. Cell *91*, 875-879.

Shatkin, A. J., and Manley, J. L. (2000). The ends of the affair: capping and polyadenylation. Nat Struct Biol *7*, 838-842.

Shen, H., Kan, J. L., and Green, M. R. (2004). Arginine-serine-rich domains bound at splicing enhancers contact the branchpoint to promote prespliceosome assembly. Mol Cell *13*, 367-376.

Sorek, R., and Ast, G. (2003). Intronic sequences flanking alternatively spliced exons are conserved between human and mouse. Genome Res *13*, 1631-1637.

Sorek, R., Ast, G., and Graur, D. (2002). Alu-containing exons are alternatively spliced. Genome Res *12*, 1060-1067.

Staley, J. P., and Guthrie, C. (1998). Mechanical devices of the spliceosome: motors, clocks, springs, and things. Cell *92*, 315-326.

Stenson, P. D., Ball, E. V., Mort, M., Phillips, A. D., Shiel, J. A., Thomas, N. S., Abeysinghe, S., Krawczak, M., and Cooper, D. N. (2003). Human Gene Mutation Database (HGMD): 2003 update. Hum Mutat *21*, 577-581.

Stevens, S. W. (2000). Analysis of low-abundance ribonucleoprotein particles from yeast by affinity chromatography and mass spectrometry microsequencing. Methods Enzymol *318*, 385-398.

Tarn, W. Y., and Steitz, J. A. (1997). Pre-mRNA splicing: the discovery of a new spliceosome doubles the challenge. Trends Biochem Sci *22*, 132-137.

Taverna, S. D., Coyne, R. S., and Allis, C. D. (2002). Methylation of histone h3 at lysine 9 targets programmed DNA elimination in *tetrahymena*. Cell *110*, 701-711.

Tazi, J., Kornstadt, U., Rossi, F., Jeanteur, P., Cathala, G., Brunel, C., and Luhrmann, R. (1993). Thiophosphorylation of U1-70K protein inhibits pre-mRNA splicing. Nature *363*, 283-286.

Ule, J., Jensen, K. B., Ruggiu, M., Mele, A., Ule, A., and Darnell, R. B. (2003). CLIP identifies Nova-regulated RNA networks in the brain. Science *302*, 1212-1215.

Valadkhan, S., and Manley, J. L. (2001). Splicing-related catalysis by protein-free snRNAs. Nature *413*, 701-707.

Valadkhan, S., and Manley, J. L. (2003). Characterization of the catalytic activity of U2 and U6 snRNAs. RNA *9*, 892-904.

Volpe, T., Schramke, V., Hamilton, G. L., White, S. A., Teng, G., Martienssen, R. A., and Allshire, R. C. (2003). RNA interference is required for normal centromere function in fission yeast. Chromosome Res *11*, 137-146.

Volpe, T. A., Kidner, C., Hall, I. M., Teng, G., Grewal, S. I., and Martienssen, R. A. (2002). Regulation of heterochromatic silencing and histone H3 lysine-9 methylation by RNAi. Science *297*, 1833-1837.

Wang, H. Y., Xu, X., Ding, J. H., Bermingham, J. R., Jr., and Fu, X. D. (2001). SC35 plays a role in T cell development and alternative splicing of CD45. Mol Cell *7*, 331-342.

Weis, K. (2002). Nucleocytoplasmic transport: cargo trafficking across the border. Curr Opin Cell Biol *14*, 328-335.

Wu, J. Y., and Maniatis, T. (1993). Specific interactions between proteins implicated in splice site selection and regulated alternative splicing. Cell *75*, 1061-1070.

Wu, S., Romfo, C. M., Nilsen, T. W., and Green, M. R. (1999). Functional recognition of the 3' splice site AG by the splicing factor U2AF35. Nature *402*, 832-835.

Xiao, S. H., and Manley, J. L. (1998). Phosphorylation-dephosphorylation differentially affects activities of splicing factor ASF/SF2. EMBO J *17*, 6359-6367.

Xu, X., Yang, D., Ding, J. H., Wang, W., Chu, P. H., Dalton, N. D., Wang, H. Y., Bermingham, J. R., Jr., Ye, Z., Liu, F., *et al.* (2005). ASF/SF2-regulated CaMKIIdelta alternative splicing temporally reprograms excitation-contraction coupling in cardiac muscle. Cell *120*, 59-72.

Yeakley, J. M., Fan, J. B., Doucet, D., Luo, L., Wickham, E., Ye, Z., Chee, M. S., and Fu, X. D. (2002). Profiling alternative splicing on fiber-optic arrays. Nat Biotechnol *20*, 353-358.

Yeakley, J. M., Tronchere, H., Olesen, J., Dyck, J. A., Wang, H. Y., and Fu, X. D. (1999). Phosphorylation regulates *in vivo* interaction and molecular targeting of serine/arginine-rich pre-mRNA splicing factors. J Cell Biol *145*, 447-455.

Yuan, L., Kawada, M., Havlioglu, N., Tang, H., and Wu, J. Y. (2005). Mutations in PRPF31 inhibit pre-mRNA splicing of rhodopsin gene and cause apoptosis of retinal cells. J Neurosci *25*, 748-757.

Yun, C. Y., and Fu, X. D. (2000). Conserved SR protein kinase functions in nuclear import and its action is counteracted by arginine methylation in *Saccharomyces cerevisiae*. J Cell Biol *150*, 707-718.

Yun, C. Y., Velazquez-Dones, A. L., Lyman, S. K., and Fu, X. D. (2003). Phosphorylation-dependent and -independent nuclear import of RS domain-containing splicing factors and regulators. J

Biol Chem *278*, 18050-18055.

Zamore, P. D., and Green, M. R. (1991). Biochemical characterization of U2 snRNP auxiliary factor: an essential pre-mRNA splicing factor with a novel intranuclear distribution. EMBO J *10*, 207-214.

Zhang, G., Taneja, K. L., Singer, R. H., and Green, M. R. (1994). Localization of pre-mRNA splicing in mammalian nuclei. Nature *372*, 809-812.

Zheng, C., Fu, X-D., and Gribskov, M. (2005). Characteristics and regulatory elements defining constitutive splicing and different modes of alternative splicing in mouse and human. RNA, in press.

Zhou, Z., Licklider, L. J., Gygi, S. P., and Reed, R. (2002). Comprehensive proteomic analysis of the human spliceosome. Nature *419*, 182-185.

Zhu, J., Mayeda, A., and Krainer, A. R. (2001). Exon identity established through differential antagonism between exonic splicing silencer-bound hnRNP A1 and enhancer-bound SR proteins. Mol Cell *8*, 1351-1361.

Chapter 28

Genome Organization: The Effects of Transcription-driven DNA Supercoiling on Gene Expression Regulation

Chien-Chung Chen and Hai-Young Wu

Department of Pharmacology, Wayne State University, School of Medicine, Detroit, Michigan, 48201 USA

Key Words: DNA supercoiling, chromosome architecture, heterochromatin, boundary element, Gene silencing, leuO

Summary

The transcriptional outputs of genes are often dependent on their chromosomal localizations. Such positional effects suggest that the architecture of chromatin structures play roles in gene expression regulation. The molecular mechanisms that underlie the position effects of gene expression regulation remain unclear. In particular, the role of DNA supercoiling that plays in this level of transcriptional control is absent. The fact that the transcription process itself generates DNA supercoiling has further complicated the issue and has led to a hypothesis that DNA supercoiling may be involved in coordinating the expression of multiple genes in a region via modulating the chromosomal architecture or directly altering the structures of the DNA elements. This possibility has been built upon the existence of unconstrained DNA supercoiling on the chromosome.

Chromosomal DNAs are under Torsional Stress

Several lines of evidence unambiguously supported that DNA of prokaryotic cells is torsionally constrained. Studies using the DNA photo-crosslinking reagent psoralen demonstrate that prokaryotic DNAs are under negative superhelical tension (Sinden *et al.*, 1980). With the known gyration activity of prokaryotic DNA topoisomerase II, the existence of negative DNA superhelicity in prokaryotes has been commonly accepted. The gyration activity of the bacterial DNA topoisomerase II was evidenced in numerous *in vitro* studies (reviewed in Gellert, 1981; Menzel and Gellert, 1994). The *in vivo* evidences are also compelling. When *Escherichia coli* cells were treated with coumermycin, an inhibitor of DNA gyrase, the superhelical density of DNA was found to be reduced 80%-90% relative to DNA purified from untreated cells. No decrease in the superhelical density of the DNA was observed if DNA was extracted from bacterial strains resistant to coumermycin (Drlica and Snyder, 1978).

In contrast, the existence of unconstrained DNA supercoiling in eukaryotes had been a matter of debate in the past several decades. The overall torsional stress in the chromosome of HeLa and *Drosophila* cells was undetectable in the psoralen-DNA crosslinking experiments (Sinden *et al.*, 1980). However, Weintraub's DNase I hypersensitivity studies suggested the presence of localized superhelical tension in eukaryotic DNA since DNase I-hypersensitive sites were found in the active chromatin domains of chicken nuclei (Groudine and Weintraub, 1982; Larsen and Weintraub, 1982). The conflicting data raised questions about the presence of un-constrained torsional stress on the eukaryotic chromosome. The lack of gyration activity of the eukaryotic DNA topoisomerase has been a strong point of argument for the absence of DNA supercoiling in eukaryotes. The eukaryotic DNA topoisomerases are known to cause only relaxation of DNA supercoiling (Wang, 1996). The debate remained as the twin-domain model of transcription was proposed (Liu and Wang, 1987) and experimentally proven in both prokaryotes and eukaryotes (Brill and Sternglanz, 1988; Dunaway and Ostrander, 1993; Giaever and Wang, 1988; Tsao *et*

Corresponding Author: Hai-Young Wu, Tel: (313) 577-1584, Fax: (313) 577-6739, E-mail: haiwu@med.wayne.edu

al., 1989; Wu *et al.*, 1988). The DNA supercoiling driven by transcription provides a different view of the source of unconstrained torsional stress in cellular DNAs. Based on the transient, local, and directional nature of transcription-driven DNA supercoiling, the superhelical state of chromosomal DNA is expected to be very dynamic. At a given chromosomal location, the superhelical tension of DNA is dependent on the relative directionalities and activities of the neighboring transcriptional units.

Given the abundance of transcriptional activity on chromosomes, transcription-driven DNA supercoiling is likely to be the major source of DNA supercoiling changes on chromosomes. Indeed, using photoactivated 4′-hydroxymethyl-4, 5′, 8-trimethylpsoralen (HMT) as a probe, localized torsional stress was detected upstream of active, transcribing genes while other regions in the human genome were relatively free of torsional stress (Ljungman and Hanawalt, 1992; Ljungman and Hanawalt, 1995). HMT is known to cross-link DNA at a much faster rate compare with that of the earlier psoralen compound. The fast-acting HMT must have detected the transient, local DNA supercoiling dynamic on the chromosome, as compared with the earlier psoralen compound that failed to detect these changes. The new cross-linking data is consistent with the DNase I hypersensitivity found in Weintraub's studies. This is because the detected torsional stress is associated with transcriptional activities in eukaryotic chromosomes (Ljungman and Hanawalt, 1995). The consistent view now is that both prokaryotic and eukaryotic DNAs are under torsional stress and that the local, transient and high degree DNA supercoiling driven by transcription shall be the relevant "DNA supercoiling" for the regulation of DNA-based biological processes.

DNA Supercoiling and Gene Expression Regulation

Most of the DNA-based biological processes are potentially affected by the chromosomal DNA supercoiling dynamics, including replication (Baker and Kornberg, 1988), recombination (Droge, 1993), and transcription (Pruss and Drlica, 1989). In terms of gene expression regulation, the relationship between the expression level of genes and DNA supercoiling has been well established. The herpes simplex virus thymidine kinase gene on circular plasmids appeared to be transcribed at least 500 times more efficiently than the same gene on linear plasmids after both plasmids were injected into *Xenopus* oocytes (Harland *et al.*, 1983). In prokaryotes, perturbation of DNA supercoiling, either caused by mutations of topoisomerase genes or by the inhibition of topoisomerase inhibitors, is capable of altering the expression of many genes (reviewed in Pruss and Drlica, 1989; Wang, 1996). Biochemical studies revealed that negative DNA supercoiling affects the transcriptional initiation by either 1) directly modulating the binding of RNA polymerase itself at the promoter (Amouyal and Buc, 1987), or 2) indirectly modulating the binding of upstream transcriptional regulators (Baliga and Dassarma, 2000; Davis *et al.*, 1999), and/or the abundant nucleoid proteins (Travers *et al.*, 2001). These events eventually affect the open complex formation at the promoter.

While the correlation between the overall changes of the superhelicity of cellular DNA and the expression of genes is generally true for most promoters, there are, however, some contradictory observations *in vitro* and *in vivo*. For example, the expression of *gyrA* is activated upon a decrease of negative superhelicity on the DNA template (Menzel and Gellert, 1987). However, the expression of *gyrA* in the chromosomal DNA context increased whereas the expression of a *gyrA* promoter-fused *galK* gene on a plasmid decreased in the oxolinic acid-treated bacterial cells despite that the negative superhelicity of either chromosomal or plasmid DNA was raised to the same extent (Franco and Drlica, 1989). Contradictory results have also been found in the studies of the DNA supercoiling effect on the expression of bacterial *bgl* operon and the mutant *leu-500* promoter (Higgins *et al.*, 1988; Richardson *et al.*, 1988). The aberrant correlation between overall DNA supercoiling and gene expression is not limited to prokaryotic models since the expression of ribosomal genes from injected ribosomal DNA plasmids and the expression of endogenous ribosomal genes were also observed to respond differently to the same degree of DNA topological constraint in *Xenopus* oocytes (Pruitt and Reeder, 1984). Most of these contradictions were raised when genes were relocated (e.g. the promoter was moved from its chromosomal location to a site on the plasmid DNA context). Hence, we speculated that rather than the correlation with the DNA supercoiling changes, in most of gene expression cases, the variation of DNA supercoiling at a local site may in fact be directly relevant in transcriptional control.

Transcription-driven DNA Supercoiling and Coordinated Gene Expression

Transcriptional elongation is known to generate positive supercoiling ahead of, and negative supercoiling behind, the moving RNA polymerase complex (Liu and Wang, 1987). Transcription-driven DNA supercoiling has been evidenced in both prokaryotes (Tsao et al., 1989; Wu et al., 1988) and eukaryotes (Brill and Sternglanz, 1988; Dunaway and Ostrander, 1993; Giaever et al., 1988; Giaever and Wang, 1988; Krebs and Dunaway, 1996). The effect of local DNA supercoiling changes driven by transcriptional activity has been observed in the regulation of a number of DNA-based cellular processes including recombination, replication and transcription (Droge, 1993; Droge, 1994). The most fascinating effects are those involving transcriptional regulation since the transcription process itself generates DNA supercoiling that may reciprocally affect its own regulation (Chen and Wu, 2003). Hence, the effect of transcription-driven DNA supercoiling on gene expression regulation has been the main focus of our research in the past decade. In particular, we are interested in the possibility that transcription-driven DNA supercoiling may serve as signals in coordinating the expression of multiple genes in a region. The coordination of the expression of a group of functionally related genes (gene cluster) shall be an efficient way of using the limited genomic information.

Based on the fact that transcriptional activity generates torsional stress on the DNA template, the superhelical state of chromosomal DNA is expected to be non-static. Such chromosomal DNA supercoiling dynamics are capable of affecting transcriptional regulation (reviewed in Pruss and Drlica, 1989; Wang, 1996). Because negative DNA supercoiling is generated behind the transcribing RNA polymerase complex during transcription, the DNA located immediately upstream of an active transcriptional unit is under the influence of a high degree of negative DNA supercoiling. A transient increase in negative DNA supercoiling upstream of a promoter would facilitate the DNA flexibility and the melting of the DNA duplex in that region. Subsequently a variety of DNA secondary structures, due to local denaturation, transitions to Z-form and to H-form, and formations of cruciform extrusions, etc may be stabilized (Htun and Dahlberg, 1989; Kowalski et al., 1988 and reviewed in (Dai and Rothman-Denes, 1999)). The transcription-driven DNA supercoiling-dependent DNA structural transitions may then affect transcription by altering the proper binding of RNA polymerase, by allowing or disrupting interaction between RNA polymerase and regulator proteins, by aiding in isomerization from closed to open promoter complexes and by nucleating the formation of higher order structures leading to activation and repression (reviewed in Dai and Rothman-Denes, 1999). In addition, the binding of transcription factors upstream of the promoter may also be modulated by the formation of alternative DNA structures, resulting in changes in gene expression. Alternatively, negative superhelicity was proposed to be able to destabilize the DNA duplex at specific locations distally from the promoter, and the free energy generated from the DNA denaturation will be transmitted to the target promoter, leading to an increase in transcriptional activity by reducing the energy required for open complex formation that has been demonstrated in the regulation of the *E. coli ilvYC* operon and the human *c-myc* gene (Levens et al., 1997; Sheridan et al., 1999).

Such reciprocal DNA supercoiling effects may explain the observation that the relative orientation between a pair of transcribing genes is crucial in the regulation of expression in either one or both of the genes (Opel et al., 2001; Opel and Hatfield, 2001). It appears that the coordinated expression and functional relationship of divergently arrayed genes may be dependent on the effect of transcription-driven DNA supercoiling. Indeed, an increasing number of head-to-head divergently expressed gene pairs have been found (Trinklein et al., 2004; Whitehouse et al., 2004), including α1 and α2 collagen genes, dihydrofolate reductase and human homologue of bacterial *MutS*, murine *surf-1* and *surf-2*, Wilm's tumor *Wt1* and *wit-1* and *brca1* and *nbr1* genes. Understanding the transcriptional regulation of genes by transcription-driven DNA supercoiling would provide a detailed view of complex gene-gene communication that is presently ill-defined.

The gene locations on chromosomes are not random, rather the genome is organized in a manner for efficient use of the limited genomic information. Genome research has revealed that the total number of genes in the human genome is similar with the number of genes in *Caenorhabditis elegans* (International Human Genome Sequencing Consortium, 2004). The sophisticated human physiology must be dependent on the efficiency of using the same number of genes. Noteworthy to mention is that the expression of multiple functionally related genes may be coordinated in as sequential manner so that proper levels of gene expression will be spatially and temporally executed with precision required for optimal physiology.

The Positional Effect of Gene Expression

The coordinated expression of genes from a native chromosomal context indicates the importance of the relative position of functionally related genes. Once a gene is moved to a foreign DNA context, its expression is disturbed, as demonstrated in an early study of position-effect color variegation on mouse coats. This chromosomal position effect in which the rearrangement breakpoint removes one gene from its normal euchromatic location to the proximity of a heterochromatic region can cause transcriptional repression of the normally active gene. Moreover, activation (derepression) of the silent gene could be observed when it is moved away from the heterochromatic domain (Allshire et al., 1994; Butner and Lo, 1986; Gottschling et al., 1990; Hazelrigg et al., 1984). Changes in expression states of genes due to gene re-location strongly suggest that the chromatin structure plays a crucial role in the establishment of complex gene regulation throughout cell differentiation. Indeed, it has been demonstrated that the formation of heterochromatin or a heterochromatin-like structure is responsible for the inactivation of X chromosome in mammals, the position-effect variegation in *Drosophila*, and the repression of the yeast mating-type loci (reviewed in Lewin, 1998). Chromatin architectural changes may determine the position effect of gene expression via modulating the distribution of local DNA supercoiling, e.g. the segregation of supercoiled DNA driven by genes located between different chromosomal domains.

Chromosome Architecture and Gene Expression

It has become clear that modulation of chromatin structures plays an important role in the transcriptional regulation in eukaryotes since transcriptional regulators must overcome the chromatin barrier to gain access to their sites in order to affect transcription. Some transcription factors manage to bind to nucleosomes and to subsequently recruit enzymes (histone acetyltransferase and SWI/SNF-related ATPase), which not only alter chromatin in a way that permits assembly of the basal transcriptional machinery, but also facilitate the binding of other factors (Di Crocc et al., 1999). In some other cases, where there is no obstacle to inhibit access to the critical regulatory elements that are positioned between nucleosomes, chromatin rearrangement triggered by these chromatin-bound factors permits binding of additional regulators to nearby nucleosomes and/or assembly of the basal transcriptional machinery (Agalioti et al., 2000). It is worthy nothing that stable alterations in nucleosome structure are critical ways to create an altered chromatin state by which gene expression is regulated. Recent studies have further identified a large group of proteins whose primary function is to assist active transcription by altering chromatin so that its DNA sequence becomes more accessible to the transcriptional apparatus. Conversely, different proteins help repress transcription by making chromatin less accessible (reviewed in Workman and Kingston, 1998). Taken together, these reports support a model that the expression of genes is mainly determined by the open state of euchromatin structures, which is regulated by numerous functional protein factors, while the remaining heterochromatin usually represents transcriptionally inactive chromosome domains.

Transcriptionally Repressive Heterochromatin and Gene Silencing

Heterochromatin is a specialized chromatin structure that persists throughout the cell cycle to limit access to DNA by machineries that transcribe or recombine the cell's genetic information, which may be inherited by daughter cells following cell division. Thus, some gene loci on the highly compact chromosomal regions that form heterochromatin structures, including centromeres and telomeres, have had periodically inactive expression to ensure that the genetic information is accurate and conserved. Indeed, heterochromatin plays important roles in the maintenance of 1) the differentiated states of cells by consistently repressing gene expression, and 2) chromosome stability by inhibiting hyper-recombination in highly repetitive chromosomal regions (Loo and Rine, 1994; Lustig, 1998).

The formation of heterochromatin is involved in the constitutively transcriptional repression of genes, named gene silencing. Generally, gene silencing can be observed at both a molecular and a chromosomal level. It has been proposed that *cis*-spreading of the heterochromatin structure from the heterochromatin-euchromatin junction is required for the repression of proximal genes. Indeed, a gene rearrangement event that displaces the white eye color gene of *Drosophila* from its normal euchromatic location to the vicinity of heterochromatin causes the heritable position-effect variegation of eye color (Wakimoto, 1998). It has also been postulated that at a chromosomal scale, the heterochromatin structure can block access of the transcriptional machinery, resulting in gene silencing. The effect of heterochromatin structure on gene silencing remains less clear at a molecular level due to the complexity of higher eukaryotic systems, despite the fact that heterochromatin has been cytologically defined.

The yeast model system provides the genetic and molecular approaches for dissecting the effect of heterochromatin structures on gene silencing. The involvement of heterochromatin structures in the transcriptional repression of genes at mating-type loci of *Saccharomyces cerevisiae* and *Schizosaccharomyces pombe* has been extensively studied (Gartenberg, 2000; Laurenson and Rine, 1992). Establishing and maintaining repression at the loci requires a number of trans-acting protein factors and two *cis*-acting regulatory elements as silencers. The extensive interactions among *trans*-acting protein factors support the current model for gene silencing in which the silencer-binding proteins nucleate at the silencers and recruit other silent information repressor proteins to form a protein complex, which propagates along the neighboring nucleosomes, forming a heterochromatin-like structure.

In bacteria, a similar "heterochromatin-like" nucleoprotein structure has also been proposed for gene silencing at sites adjacent to the P1 phage and F plasmid centromeres, which functions similar to centromeric gene silencing in eukaryotes (Kim and Wang, 1999; Rodionov *et al*., 1999). Centromere-bound proteins (ParB and SopB) recognize their cognate binding sites (*parS* and *sopC*) and spread along DNA as far as 10 kb away, resulting in genes at adjacent DNA regions being silenced and inaccessible to *dam* methylase and DNA gyrase (Lynch and Wang, 1995; Rodionov *et al*., 1999). The *E. coli bgl* operon is another classical example used to study the effect of transcriptionally repressive nucleoprotein structures on gene expression. The *bgl* promoter is silent under cell homeostasis so that wild-type *E.coli* cannot metabolize β-glucoside (Schaefler, 1967). Studies have shown that not only is the *bgl* promoter silent in its normal context, but other promoters exchanged for the *bgl* promoter become silenced at the *bgl* locus (Schnetz, 1995). This result is consistent with the fact that both *cis*-acting elements located upstream and downstream of the *bgl* promoter and two *trans*-acting protein factors: the histone-like nucleoid structuring protein (H-NS) and factor for inversion stimulation (FIS) are required for *bgl* silencing (Defez and De Felice, 1981; Ueguchi *et al*., 1996). In this respect, H-NS, an abundant nucleoid-associated protein, was shown to bind to an ~100-bp AT-rich sequence upstream of the promoter, forming a nucleoprotein structure that spreads along the neighboring region via protein-protein interactions, resulting in a heterochromatin-like feature of the promoter that is unfavorable for transcription (Schnetz, 1995; Ueguchi *et al*., 1996). One of the reasons why this region was designated as a "silencer element", and not an operator, was the observation that *bgl* silencing was not significantly relieved either by a point mutation within it or even by small internal insertions (Caramel and Schnetz, 1998). Moreover, there is flexibility as to the orientation of and distance of the silencer element from the promoter, because its silencing activity is not diminished when moved to a further location, 150 bp upstream (Schnetz, 1995; Schnetz and Rak, 1992). The mechanistic similarity of H-NS-mediated *bgl* silencing with eukaryotic silencing mediated by heterochromatin or heterochromatin-like structure suggests that certain microdomains affected by the organization of the chromosome play an important role in gene function.

The Relief of Heterochromatin-mediated Gene Silencing

As described above, the same gene inserted into different sites in the genome can exhibit markedly different levels of expression. This position effect on gene expression may reflect local differences in chromatin structure as well as the particular distribution of regulatory elements through the genome. If heterochromatin repressive domains are not restricted in some fashion, essential genes in neighboring domains may be inappropriately repressed. It raises the question of what prevents the repressed regions from extending indefinitely.

Regional transcriptional silencing in the mating-type loci, centromere, and telomere of yeast provides a model to understand as to how to delimit the silent chromatin. At *HMR* loci and subtelomeric regions, the flanking boundary elements limit the propagation of repressed heterochromatin (Donze *et al*., 1999; Fourel *et al*., 1999). However, the flanking silencers at the *HML* loci impose silencing in a directional manner over a limited distance, thereby repressing only those genes that reside between them (Bi *et al*., 1999). It suggests that boundary elements may recruit a large multiprotein complex to block the spreading of silencer-bound proteins along the chromatin and that the differences of relative availability and binding affinity between the silencer-bound proteins may also determine the directionality of gene silencing. Two different models have been suggested for restricting silenced domains in yeast: the chromatin-modifying model (Donze and Kamakaka, 2002) and topological domain model (Ishii *et al*., 2002). According to the chromatin-modifying model, barrier proteins bound to DNA create regions of open chromatin that prevent the propagation of silenced chromatin. These boundary proteins are believed to function by recruiting chromatin-modifying enzymes that in turn modify nucleosomes and alter the chromatin substrate to a state that is unfavorable for binding of the silencer-bound proteins. The topological domain model, on the other hand, states that boundary elements tether DNA to a nuclear substructure, which then forms a "road block" to the spreading heterochromatin. While these two models are mechanistically distinct, the outcome of both is the creation and maintenance of adjacent chromatin domains with opposing transcriptional activities.

In bacteria, since *cis*-spreading of the H-NS nucleoprotein complex causes transcriptional silencing at some promoters (and thus shares mechanistic similarity with heterochromatin type of gene silencing in eukaryotes), one can expect that a protein-DNA complex located between a H-NS nucleation site and the target promoter may block the propagation of the transcriptionally-repressive nucleoprotein complex. Indeed, the result demonstrated that the integration of insertion elements within AT-rich regions can alleviate *bgl* silencing, but not if the coding sequence is deleted (Reynolds *et al.*, 1986). Furthermore, *lac* and λ operator-repressor complexes situated within the upstream silencer, but far away from the RNA polymerase, can also relieve silencing in a phase-independent manner (Caramel and Schnetz, 1998). It suggested that extensive protein-DNA interaction could abolish silencing by prevention of the H-NS-associated nucleoprotein structure from reaching RNA polymerase, as if it were located at the *cis*-spreading path between the silencer and the target promoter. For example: The overexpressions of the *bglJ* and *leuO* genes are known to cause activation of the *bgl* promoter (Giel *et al.*, 1996; Ueguchi *et al.*, 1998). Similar regulation has also been found in *ompS1* and *ade* genes, but detailed mechanisms underlying the transcriptional repression and derepression of gene expression remain elusive. The proteins that play such roles in a native DNA context need yet to be explored.

The Sequential Activation of Genes in the *S. typhimurium ilvIH-leuO-leuABCD* Gene Cluster

The activation of the mutant *leu-500* promoter of the leucine operon in *Salmonella typhimurium topA* mutants is one of the best examples available for discerning the complexity of DNA supercoiling-dependent transcriptional activation. The *leu-500* mutation is an A to G transition in the -10 region of the promoter of the *S. typhimurium leuABCD* operon, which abolishes the promoter activity and results in leucine auxotrophy (Gemmill *et al.*, 1984; Margolin and Mukai, 1966). Later, a second-site mutation at the DNA topoisomerase I gene, *topA*, was shown to suppress the *leu-500* mutation and restore leucine prototrophy (Margolin and Mukai, 1966; Margolin *et al.*, 1985; Trucksis and Depew, 1981). The A to G transition in the *leu-500* promoter is expected to increase the free energy requirement for the open complex formation of the mutant promoter. Since the hyper-negative DNA supercoiling caused in *topA* mutants is believed to provide extra free energy to overcome such an energy barrier, the transcriptional activity of the *leu-500* promoter was thought to correlate with the level of negative superhelicity on the DNA template. However, suppression of the *leu-500* mutation (*leu-500* activation) only correlated with the absence of *topA* but not the overall negative DNA superhelicity that was measured in various *topA* and *gyr* mutant series (Richardson *et al.*, 1988). Hence, it was suggested that local rather than global DNA supercoiling was important for *leu-500* activation (Lilley and Higgins, 1991; Richardson *et al.*, 1988).

To understand the effect of DNA supercoiling on *leu-500* activation, various laboratories have tested the hypothesis that transcription-mediated negative DNA supercoiling may be the missing regulatory factor responsible for activation of the *leu-500* promoter with a *topA*⁻ background. Our research group has demonstrated that in a *topA*⁻ mutant, the minimal *leu-500* promoter (positions -80 to +87 of *leuABCD* operon) on plasmid DNA can be activated by a *lac* promoter that transcribed divergently, but not convergently towards it (Tan *et al.*, 1994). In addition, with the insertion of random DNA sequences into the region between divergent *leu-500* and *lac* promoters, the negative DNA supercoiling-dependent *leu-500* activation is limited within 250-450 bp. Hence, the effect was named the short-range promoter-promoter interaction (Tan *et al.*, 1994). Both Dr. Lilley's and Dr. Bossi's groups also detected the activation of the plasmid-borne *leu-500* promoter within a similar distance by the transcriptional units transcribing away from the *leu-500* promoter (Chen *et al.*, 1992; Chen *et al.*, 1994; Chen *et al.*, 1993; Spirito and Bossi, 1996).

The negative DNA supercoiling dependence of *leu-500* activation on the plasmid DNA, however, may not reflect the chromosomal situation that was observed in the study of the reversion of leucine phenotype in the *S. typhimurium topA*⁻ strain (Margolin and Mukai, 1966). Based on the discrepancy discussed in the papers (Lilley and Higgins, 1991; Richardson *et al.*, 1988) that some missing DNA elements may account for the *topA*⁻-dependent *leu-500* activation, the hypothesis that the transcriptional activity upstream of the *leuABCD* operon in the chromosomal DNA context may provide the local DNA supercoiling needs for *leu-500* activation, was raised. Intensive studies conducted in this laboratory to search for such potential transcriptional units revealed that activation of the plasmid-borne *leu-500* promoter occurred when the promoter of the *ilvIH* operon from the upstream chromosomal DNA region was also co-inserted (Wu et al., 1995). The involvement of the upstream promoter activity in leu-500 activation strongly suggests that transcription-driven DNA supercoiling is responsible for the unprecedented long-range (1.9 kb) promoter-promoter interaction (Wu et al., 1995).

Since deleting part of the intervening DNA sequence (-554 to -83 of leuABCD operon) abolished the ilvIH promoter activity-dependent leu-500 activation, it was suggested that this region might be involved in communicating the ilvIH-initiated DNA supercoiling effect (Wu et al., 1995). A detailed study in our laboratory showed that the transcriptional activity of an intermediate, putative leuO gene is associated with the ilvIH-dependent leu-500 activation by mediating the DNA supercoiling effect over a long distance (Fang and Wu, 1998a). Loss of the activity of the leuO promoter by the introduction of a 2-bp mutation at the -10 DNA sequence blocked leu-500 activation at the distal end while the ilvIH promoter activity remained intact (Fang and Wu, 1998a). This result confirms that the leuO promoter activity is necessary for relaying the ilvIH-mediated effect on leu-500 activation through a promoter-promoter interaction. Furthermore, the LeuO protein, the product of the leuO gene, is also important for the subsequent activation of the leu-500 promoter at the other end of the 1.9-kb region (Fang and Wu, 1998a). It appears that both a functional leuO promoter and the LeuO protein are required for leu-500 activation. Hence, the promoter relay mechanism was to explain the sequential activation of genes in the ilvIH-leuO-leuABCD gene cluster (Fang and Wu, 1998a). The transcriptional activity of the ilvIH promoter (transcription-driven DNA supercoiling) activates the intermediate leuO gene. Subsequently, the leuO gene product, LeuO, plays a *trans*-acting role that maintains the transcriptional activity of the leuO promoter, which activates the supercoiling-sensitive leu-500 promoter via transcription-driven DNA supercoiling as the final step of this sequential gene activation (Fig. 28.1).

The fact that a mutation in either the ilvIH or leuO promoter knocked out the leu-500 activation in a topA⁻ strain (Wu et al., 1995) clearly indicated that the absence of topA itself is insufficient to activate the leu-500 promoter. By monitoring the growth of a wild-type topA strain harboring the leu-500 mutation in leucine-free medium and the expression condition of originally silent ilvIH and leuO genes during cell growth, the leu-500 activation and the topA⁻ genetic background requirement were decoupled at a chromosomal DNA level. These studies also provided clear evidence that the ilvIH transcriptional activity rather than the topA⁻ genetic background plays an indispensable role in the leu-500 activation by the promoter relay mechanism (Fang and Wu, 1998b).

It is apparent that the promoter relay mechanism involved in the sequential activation of genes in the ilvIH-leuO-leuABCD gene cluster is also a regulatory event of an environmental stress response. Using the expression of the normally silent leuO gene as an indicator for the sequential gene activation in the ilvIH-leuO-leuABCD gene cluster, a study in our laboratory demonstrated that expression of the leuO gene increased transiently when bacterial cells were entering the stationary phase of growth. This increase in leuO expression was presumably due to the promoter relay mechanism in response to the change of the logarithmic to the stationary phase of bacterial cell growth (Fang et al., 2000). In addition, it is clear now that under extreme starvation for branched-chain amino acids during exponential growth, the slow increase of the ppGpp level in the relA1 mutant strain causes the ilvIH activation, which results in leuO expression via the promoter relay mechanism during a two-hour growth arrest (Majumder et al., 2001). Taking these physiological results together, the proposed promoter relay mechanism involved in the sequential gene activation in the ilvIH-leuO-leuABCD gene cluster is a bacterial stress-related response that operates when cells are under specific nutritional limitation. It is consistent

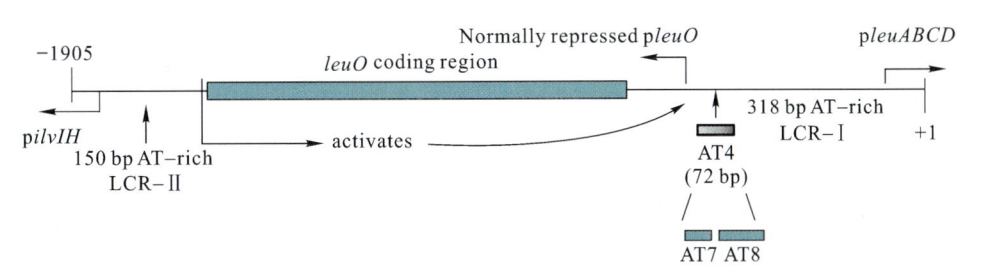

Fig. 28.1 **The ilvIH-leuO-leuABCD gene locus where the promoter relay is at work.** The presently known elements located upstream of the *S. typhimurium* leuABCD (-1905 to +1 position of the leuABCD) are illustrated. The two AT-rich DNA sequences upstream and downstream of the leuO are designated as locus control region-I (LCR-I) and -II (LCR-II), respectively.

with the fact that *ilvIH* and *leuO* expression are undetectable in regular laboratory testing conditions when rich medium is used in most of the laboratory conditions.

The Expression of the *leuO* Gene is Controlled by Two Distinct *Cis*-elements

As part of the regulation of the promoter relay mechanism, maintenance of a physiologically silent state of the *leuO* gene is critical. Furthermore, relief of the repressed *leuO* gene responsive to extracellular stimuli during the growth of bacteria is required for cell viability. While transcription-driven DNA supercoiling mediated from the *ilvIH* promoter was responsible for *leu-500* activation, its roles in controlling inactive and active states of the *leuO* gene was elusive. Transcription-driven DNA supercoiling is indispensable in these processes, but the supercoiling alone is insufficient for triggering coordinated gene expression (Wu *et al.*, 1995). It appears that certain elements in this region must be crucial for relaying the DNA supercoiling effect over a long (1.9 kb) distance. Indeed, our study showed that a 72-bp AT-rich DNA segment, AT4, which is located at the *leuO* promoter end of locus control region-I (LCR-I), is responsible for transcriptional repression of the *leuO* gene (Chen *et al.*, 2001). Based on the characteristics of AT4-mediated transcriptional repression activity, AT4 was designated as a bacterial gene silencer. The AT4-mediated gene silencing activity was optimal when AT4 was situated between a pair of divergently transcribed transcriptional units (Chen *et al.*, 2001). It suggested that transcription-driven DNA supercoiling might regulate the function of the gene silencer by providing extra energy to trigger DNA structural transitions in the AT-rich region. In addition, the fact that the LeuO protein binding to AT4 negated AT4-mediated gene silencing has led to the discovery of two important DNA elements: a 25-bp AT7, the LeuO binding site and a 47-bp AT8, the gene silencer (Chen *et al.*, 2003). While AT8 retains the full gene silencing activity that is no longer relieved by LeuO, AT7 itself has no gene silencing activity. However, LeuO provided *in trans* relieves AT8-mediated gene silencing when AT7 is located within the proximity to AT8 (Chen *et al.*, 2003). Clearly, LeuO binding itself is a transcriptionally neutral and the transcriptional effect of the LeuO protein is determined by the transcriptional repression activity of the gene silencer AT8. This finding has provided clues for studying the molecular details of the LeuO-mediated transcriptional derepression and the gene silencer-mediated transcriptional repression, as well.

To elucidate the molecular mechanism of gene silencer-mediated gene silencing, a genetic screening system was developed to identify genes that encode protein factors responsible for silencing of the *leuO* gene. We found one positive clone, the *hns* gene, which encodes H-NS, is clearly responsible for the gene silencing activity mediated by the gene silencer AT8 (Chen *et al.*, 2005). Biochemical studies revealed that the gene silencer AT8 is a nucleation site that recruits H-NS to form a *cis*-spreading nucleoprotein filament that is responsible for the *leuO* gene silencing. These results are consistent with previous findings in bacterial transcriptional regulation that demonstrated that H-NS has a potential to cooperatively form extended oligomeric binding arrays on DNA, such as the H-NS-regulated loci *bgl*, *hns*, the phage Mu early promoter (P_E) region and the *pap* regulon (Caramel and Schnetz, 1998; Falconi *et al.*, 1993; Spurio *et al.*, 1997; van Ulsen *et al.*, 1996; White-Ziegler *et al.*, 1998). It appears that H-NS is recruited to the nucleation site embedded in a long stretch of AT-rich DNA (the LCR-I DNA in our model system or the AT-rich DNA flanking the *bgl* promoter) as the first step of the transcriptional repression (Chen *et al.*, 2005). The localized H-NS proteins then formed a *cis*-spreading nucleoprotein filament on the AT-rich DNA for repressing the neighboring promoters as suggested in recent reviews (Rimsky, 2004; Yarmolinsky, 2000).

Notably the mechanistic mechanism of transcriptional repression of the *leuO* gene by a *cis*-spreading nucleoprotein filament containing H-NS is reminiscent of heterochromatin-dependent gene silencing in eukaryotes. In the case of heterochromatin like structure-mediated gene silencing found in the yeast mating-type loci, the propagation of repressed heterochromatin established at silencers E and I is blocked by flanking boundary elements. In order to delimit heterochromatin, the boundary elements should be located between the gene silencer and the target gene. Our recent data showed that the relative positions of the gene silencer, the LeuO-binding site and the promoter of the target *leuO* gene were crucial for the efficient LeuO-mediated transcriptional derepression. This prompted us to hypothesized that LeuO might regulate the AT8-mediated *cis*-spreading transcriptionally repressive nucleoprotein filament via a boundary element-like activity. Indeed, the transcriptional derepression activity mediated by the LeuO protein require the LeuO-binding site (AT7) to be positioned between the gene silencer (AT8) and the target promoter. LeuO apparently is blocking the cis-spreading pathway of the transcriptionally repressive nucleoprotein filament from reaching the promoter of the targeted *leuO* gene. In the absence of the gene silencer, LeuO protein itself is a transcriptionally inert element. These results suggested that LeuO functions as a boundary element in bacterial transcriptional regulation (Chen and Wu, 2005). The boundary element activity of LeuO is unprecedented in bacterial gene expression regulation and is basically consistent with the chromosomal barrier functions of the boundary elements/insulators found in the eukaryotic systems. New molecular detail found was that when one LeuO-binding site was positioned between the gene silencer and the target promoter, a second LeuO-binding site in the region resulted in a synergistic effect on LeuO-mediated nucleoprotein blockage (Chen and Wu, 2005). Using a tetrameric Lac repressor for testing, we demonstrated that protein- protein interaction-mediated protein binding cooperativity is responsible for the blocking synergy. This prompted us to speculate that more than one LeuO-binding site in the *ilvIH-leuO-leuABCD* gene cluster may be required for a synergistic effect on LeuO-mediated transcriptional regulation in this region as part of the promoter relay mechanism. Indeed, we found multiple LeuO-binding sites in the LCRs of the *ilvIH-leuO-leuABCD* gene cluster. The distal LeuO-binding site located in the LCR-II is apparently required for a synergistic relief of the gene silencing effect mediated by the gene silencer AT8, which is located 1.5 kb upstream in LCR-I (Chen and Wu, 2005). This is clear evidence that barrier proteins may regulate transcriptional apparatus on the chromosome through barrier-barrier interaction via looping out the intervening DNA between the binding sites of the barrier proteins.

Concluding Remarks and Perspectives

Our research toward understanding the effect of transcription-driven DNA supercoiling on gene expression regulation has led to the finding of the intriguing sequential activation of genes in the bacterial *ilvIH-leuO-leuABCD* gene cluster. Subsequently, the transcriptionally-repressive nucleoprotein filament mediated by H-NS (Chen *et al.*, 2005) and the boundary element activity of the LeuO protein (Chen and Wu, 2005) were found to be important for the repression-derepression processes that are crucial for the promoter relay mechanism in the gene cluster. It seems that gene regulation via modulating chromosomal architecture is a fundamentally important mechanism that highly conserved in all species ranging from bacteria to the more complex eukaryotes including human. The relatively simple bacterial model system will serve an important study system for elucidation of the molecular details of transcriptional regulation at the chromosomal architectural level.

Acknowledgment

We are in debt to Dr. Victoria Kimler for her critical reading of the manuscript. This work is supported by the US National Institutes of Health (grant GM-53617).

Refevences

International Human genome Sequencing Consortium (2004). Finishing the euchromatic sequence of the human genome. Nature *431*, 931-945.

Agalioti, T., Lomvardas, S., Parekh, B., Yie, J., Maniatis, T., and Thanos, D. (2000). Ordered recruitment of chromatin modifying and general transcription factors to the IFN-beta promoter. Cell *103*, 667-678.

Allshire, R. C., Javerzat, J. P., Redhead, N. J., and Cranston, G. (1994). Position effect variegation at fission yeast centromeres. Cell *76*, 157-169.

Amouyal, M., and Buc, H. (1987). Topological unwinding of strong and weak promoters by RNA polymerase. A comparison between the lac wild-type and the UV5 sites of *Escherichia coli*. J Mol Biol *195*, 795-808.

Baker, T. A., and Kornberg, A. (1988). Transcriptional activation of initiation of replication from the *E. coli* chromosomal origin: an RNA-DNA hybrid near oriC. Cell *55*, 113-123.

Baliga, N. S., and Dassarma, S. (2000). Saturation mutagenesis of the haloarchaeal bop gene promoter: identification of DNA supercoiling sensitivity sites and absence of TFB recognition element and UAS enhancer activity. Mol Microbiol *36*, 1175-1183.

Bi, X., Braunstein, M., Shei, G. J., and Broach, J. R. (1999). The yeast HML I silencer defines a heterochromatin domain boundary by directional establishment of silencing. Proc Natl Acad Sci USA *96*, 11934-11939.

Brill, S. J., and Sternglanz, R. (1988). Transcription-dependent DNA supercoiling in yeast DNA topoisomerase mutants. Cell *54*, 403-411.

Butner, K., and Lo, C. W. (1986). Modulation of *tk* expression in mouse pericentromeric heterochromatin. Mol Cell Biol *6*, 4440-4449.

Caramel, A., and Schnetz, K. (1998). Lac and lambda repressors relieve silencing of the *Escherichia coli bgl* promoter. Activation by alteration of a repressing nucleoprotein complex. J Mol Biol *284*, 875-883.

Chen, C. C., Chou, M. Y., Huang, C. H., Majumder, A., and Wu, H. Y. (2005). A *cis*-spreading nucleoprotein filament is responsible for the gene silencing activity found in the promoter relay mechanism. J Biol Chem *280*, 5101-5112.

Chen, C. C., Fang, M., Majumder, A., and Wu, H. Y. (2001). A 72-base pair AT-rich DNA sequence element functions as a bacterial gene silencer. J Biol Chem *276*, 9478-9485.

Chen, C. C., Ghole, M., Majumder, A., Wang, Z., Chandana, S., and Wu, H. Y. (2003). LeuO-mediated transcriptional derepression. J Biol Chem *278*, 38094-38103.

Chen, C. C., and Wu, H. Y. (2003). Transcription-driven DNA supercoiling and gene expression control. Front Biosci *8*, d430-439.

Chen, C. C., and Wu, H. Y. (2005). LeuO protein delimits the transcriptionally-active and -repressive domains on the bacterial chromosome. J Biol Chem. *280*, 15111-15121

Chen, D., Bowater, R., Dorman, C. J., and Lilley, D. M. (1992). Activity of a plasmid-borne *leu-500* promoter depends on the transcription and translation of an adjacent gene. Proc Natl Acad Sci USA *89*, 8784-8788.

Chen, D., Bowater, R., and Lilley, D. M. (1994). Topological promoter coupling in *Escherichia coli*: delta *topA*-dependent activation of the *leu-500* promoter on a plasmid. J Bacteriol *176*, 3757-3764.

Chen, D., Bowater, R. P., and Lilley, D. M. (1993). Activation of the *leu-500* promoter: a topological domain generated by divergent transcription in a plasmid. Biochemistry *32*, 13162-13170.

Dai, X., and Rothman-Denes, L. B. (1999). DNA structure and transcription. Curr Opin Microbiol *2*, 126-130.

Davis, N. A., Majee, S. S., and Kahn, J. D. (1999). TATA box DNA deformation with and without the TATA box-binding protein. J Mol Biol *291*, 249-265.

Defez, R., and De Felice, M. (1981). Cryptic operon for beta-glucoside metabolism in *Escherichia coli* K12: genetic evidence for a regulatory protein. Genetics *97*, 11-25.

Di Croce, L., Koop, R., Venditti, P., Westphal, H. M., Nightingale, K. P., Corona, D. F., Becker, P. B., and Beato, M. (1999). Two-step synergism between the progesterone receptor and the DNA-binding domain of nuclear factor 1 on MMTV minichromosomes. Mol Cell *4*, 45-54.

Donze, D., Adams, C. R., Rine, J., and Kamakaka, R. T. (1999). The boundaries of the silenced HMR domain in *Saccharomyces cerevisiae*. Genes Dev *13*, 698-708.

Donze, D., and Kamakaka, R. T. (2002). Braking the silence: how heterochromatic gene repression is stopped in its tracks. Bioessays *24*, 344-349.

Drlica, K., and Snyder, M. (1978). Superhelical *Escherichia coli* DNA: relaxation by coumermycin. J Mol Biol *120*, 145-154.

Droge, P. (1993). Transcription-driven site-specific DNA recombination *in vitro*. Proc Natl Acad Sci USA *90*, 2759-2763.

Droge, P. (1994). Protein tracking-induced supercoiling of DNA: a tool to regulate DNA transactions *in vivo*? Bioessays *16*, 91-99.

Dunaway, M., and Ostrander, E. A. (1993). Local domains of supercoiling activate a eukaryotic promoter *in vivo*. Nature *361*, 746-748.

Falconi, M., Higgins, N. P., Spurio, R., Pon, C. L., and Gualerzi, C. O. (1993). Expression of the gene encoding the major bacterial nucleotide protein H-NS is subject to transcriptional auto-repression. Mol Microbiol *10*, 273-282.

Fang, M., Majumder, A., Tsai, K. J., and Wu, H. Y. (2000). ppGpp-dependent *leuO* expression in bacteria under stress. Biochem Biophys Res Commun *276*, 64-70.

Fang, M., and Wu, H. Y. (1998a). A promoter relay mechanism for sequential gene activation. J Bacteriol *180*, 626-633.

Fang, M., and Wu, H. Y. (1998b). Suppression of *leu-500* mutation in $topA^+$ *Salmonella typhimurium* strains. The promoter relay at work. J Biol Chem *273*, 29929-29934.

Fourel, G., Revardel, E., Koering, C. E., and Gilson, E. (1999). Cohabitation of insulators and silencing elements in yeast subtelomeric regions. Embo J *18*, 2522-2537.

Franco, R. J., and Drlica, K. (1989). Gyrase inhibitors can increase *gyrA* expression and DNA supercoiling. J Bacteriol *171*, 6573-6579.

Gartenberg, M. R. (2000). The Sir proteins of *Saccharomyces cerevisiae*: mediators of transcriptional silencing and much more. Curr Opin Microbiol *3*, 132-137.

Gellert, M. (1981). DNA topoisomerases. Annu Rev Biochem *50*, 879-910.

Gemmill, R. M., Tripp, M., Friedman, S. B., and Calvo, J. M. (1984). Promoter mutation causing catabolite repression of the *Salmonella typhimurium* leucine operon. J Bacteriol *158*, 948-953.

Giaever, G. N., Snyder, L., and Wang, J. C. (1988). DNA supercoiling *in vivo*. Biophys Chem *29*, 7-15.

Giaever, G. N., and Wang, J. C. (1988). Supercoiling of intracellular DNA can occur in eukaryotic cells. Cell *55*, 849-856.

Giel, M., Desnoyer, M., and Lopilato, J. (1996). A mutation in a new gene, *bglJ*, activates the *bgl* operon in *Escherichia coli* K-12. Genetics *143*, 627-635.

Gottschling, D. E., Aparicio, O. M., Billington, B. L., and Zakian, V. A. (1990). Position effect at *S. cerevisiae* telomeres: reversible repression of Pol II transcription. Cell *63*, 751-762.

Groudine, M., and Weintraub, H. (1982). Propagation of globin DNAase I-hypersensitive sites in absence of factors required for induction: a possible mechanism for determination. Cell *30*, 131-139.

Harland, R. M., Weintraub, H., and McKnight, S. L. (1983). Transcription of DNA injected into *Xenopus* oocytes is influenced by template topology. Nature *302*, 38-43.

Hazelrigg, T., Levis, R., and Rubin, G. M. (1984). Transformation of white locus DNA in *drosophila*: dosage compensation, zeste interaction, and position effects. Cell *36*, 469-481.

Higgins, C. F., Dorman, C. J., Stirling, D. A., Waddell, L., Booth, I. R., May, G., and Bremer, E. (1988). A physiological role for DNA supercoiling in the osmotic regulation of gene expression in

S. typhimurium and *E. coli*. Cell *52*, 569-584.

Ishii, K., Arib, G., Lin, C., Van Houwe, G., and Laemmli, U. K. (2002). Chromatin boundaries in budding yeast: the nuclear pore connection. Cell *109*, 551-562.

Kim, S. K., and Wang, J. C. (1999). Gene silencing via protein-mediated subcellular localization of DNA. Proc Natl Acad Sci USA *96*, 8557-8561.

Krebs, J. E., and Dunaway, M. (1996). DNA length is a critical parameter for eukaryotic transcription *in vivo*. Mol Cell Biol *16*, 5821-5829.

Larsen, A., and Weintraub, H. (1982). An altered DNA conformation detected by S1 nuclease occurs at specific regions in active chick globin chromatin. Cell *29*, 609-622.

Laurenson, P., and Rine, J. (1992). Silencers, silencing, and heritable transcriptional states. Microbiol Rev *56*, 543-560.

Levens, D., Duncan, R. C., Tomonaga, T., Michelotti, G. A., Collins, I., Davis-Smyth, T., Zheng, T., and Michelotti, E. F. (1997). DNA conformation, topology, and the regulation of *c-myc* expression. Curr Top Microbiol Immunol *224*, 33-46.

Lewin, B. (1998). The mystique of epigenetics. Cell *93*, 301-303.

Lilley, D. M., and Higgins, C. F. (1991). Local DNA topology and gene expression: the case of the *leu-500* promoter. Mol Microbiol *5*, 779-783.

Liu, L. F., and Wang, J. C. (1987). Supercoiling of the DNA template during transcription. Proc Natl Acad Sci USA *84*, 7024-7027.

Ljungman, M., and Hanawalt, P. C. (1992). Localized torsional tension in the DNA of human cells. Proc Natl Acad Sci USA *89*, 6055-6059.

Ljungman, M., and Hanawalt, P. C. (1995). Presence of negative torsional tension in the promoter region of the transcriptionally poised dihydrofolate reductase gene *in vivo*. Nucleic Acids Res *23*, 1782-1789.

Loo, S., and Rine, J. (1994). Silencers and domains of generalized repression. Science *264*, 1768-1771.

Lustig, A. J. (1998). Mechanisms of silencing in *Saccharomyces cerevisiae*. Curr Opin Genet Dev *8*, 233-239.

Lynch, A. S., and Wang, J. C. (1995). SopB protein-mediated silencing of genes linked to the *sopC* locus of *Escherichia coli* F plasmid. Proc Natl Acad Sci USA *92*, 1896-1900.

Majumder, A., Fang, M., Tsai, K. J., Ueguchi, C., Mizuno, T., and Wu, H. Y. (2001). LeuO expression in response to starvation for branched-chain amino acids. J Biol Chem *276*, 19046-19051.

Margolin, P., and Mukai, F. H. (1966). A model for mRNA transcription suggested by some characteristics of 2-aminopurine mutagenesis in *Salmonella*. Proc Natl Acad Sci USA *55*, 282-289.

Margolin, P., Zumstein, L., Sternglanz, R., and Wang, J. C. (1985). The *Escherichia coli supX* locus is *topA*, the structural gene for DNA topoisomerase I. Proc Natl Acad Sci USA *82*, 5437-5441.

Menzel, R., and Gellert, M. (1987). Modulation of transcription by DNA supercoiling: a deletion analysis of the *Escherichia coli gyrA* and *gyrB* promoters. Proc Natl Acad Sci USA *84*, 4185-4189.

Menzel, R., and Gellert, M. (1994). The biochemistry and biology of DNA gyrase. Adv Pharmacol *29A*, 39-69.

Opel, M. L., Arfin, S. M., and Hatfield, G. W. (2001). The effects of DNA supercoiling on the expression of operons of the *ilv* regulon of *Escherichia coli* suggest a physiological rationale for divergently transcribed operons. Mol Microbiol *39*, 1109-1115.

Opel, M. L., and Hatfield, G. W. (2001). DNA supercoiling-dependent transcriptional coupling between the divergently transcribed promoters of the *ilvYC* operon of *Escherichia coli* is proportional to promoter strengths and transcript lengths. Mol Microbiol *39*, 191-198.

Pruitt, S. C., and Reeder, R. H. (1984). Effect of topological constraint on transcription of ribosomal DNA in *Xenopus* oocytes. Comparison of plasmid and endogenous genes. J Mol Biol *174*, 121-139.

Pruss, G. J., and Drlica, K. (1989). DNA supercoiling and prokaryotic transcription. Cell *56*, 521-523.

Reynolds, A. E., Mahadevan, S., LeGrice, S. F., and Wright, A. (1986). Enhancement of bacterial gene expression by insertion elements or by mutation in a CAP-cAMP binding site. J Mol Biol *191*, 85-95.

Richardson, S. M., Higgins, C. F., and Lilley, D. M. (1988). DNA supercoiling and the *leu-500* promoter mutation of *Salmonella typhimurium*. Embo J *7*, 1863-1869.

Rimsky, S. (2004). Structure of the histone-like protein H-NS and its role in regulation and genome superstructure. Curr Opin Microbiol *7*, 109-114.

Rodionov, O., Lobocka, M., and Yarmolinsky, M. (1999). Silencing of genes flanking the P1 plasmid centromere. Science *283*, 546-549.

Schaefler, S. (1967). Inducible system for the utilization of beta-glucosides in *Escherichia coli*. I. Active transport and utilization of beta-glucosides. J Bacteriol *93*, 254-263.

Schnetz, K. (1995). Silencing of *Escherichia coli bgl* promoter by flanking sequence elements. Embo J *14*, 2545-2550.

Schnetz, K., and Rak, B. (1992). IS5: a mobile enhancer of transcription in *Escherichia coli*. Proc Natl Acad Sci USA *89*, 1244-1248.

Sheridan, S. D., Benham, C. J., and Hatfield, G. W. (1999). Inhibition of DNA supercoiling-dependent transcriptional activation by a distant B-DNA to Z-DNA transition. J Biol Chem *274*, 8169-8174.

Sinden, R. R., Carlson, J. O., and Pettijohn, D. E. (1980). Torsional tension in the DNA double helix measured with trimethylpsoralen in living *E. coli* cells: analogous measurements in insect and human cells. Cell *21*, 773-783.

Spirito, F., and Bossi, L. (1996). Long-distance effect of downstream transcription on activity of the supercoiling-sensitive *leu-500* promoter in a *topA* mutant of *Salmonella typhimurium*. J

Bacteriol *178*, 7129-7137.

Spurio, R., Falconi, M., Brandi, A., Pon, C. L., and Gualerzi, C. O. (1997). The oligomeric structure of nucleoid protein H-NS is necessary for recognition of intrinsically curved DNA and for DNA bending. Embo J *16*, 1795-1805.

Tan, J., Shu, L., and Wu, H. Y. (1994). Activation of the *leu-500* promoter by adjacent transcription. J Bacteriol *176*, 1077-1086.

Travers, A., Schneider, R., and Muskhelishvili, G. (2001). DNA supercoiling and transcription in *Escherichia coli*: The FIS connection. Biochimie *83*, 213-217.

Trinklein, N. D., Aldred, S. F., Hartman, S. J., Schroeder, D. I., Otillar, R. P., and Myers, R. M. (2004). An abundance of bidirectional promoters in the human genome. Genome Res *14*, 62-66.

Trucksis, M., and Depew, R. E. (1981). Identification and localization of a gene that specifies production of *Escherichia coli* DNA topoisomerase I. Proc Natl Acad Sci USA *78*, 2164-2168.

Tsao, Y. P., Wu, H. Y., and Liu, L. F. (1989). Transcription-driven supercoiling of DNA: direct biochemical evidence from *in vitro* studies. Cell *56*, 111-118.

Ueguchi, C., Ohta, T., Seto, C., Suzuki, T., and Mizuno, T. (1998). The leuO gene product has a latent ability to relieve *bgl* silencing in *Escherichia coli*. J Bacteriol *180*, 190-193.

Ueguchi, C., Suzuki, T., Yoshida, T., Tanaka, K., and Mizuno, T. (1996). Systematic mutational analysis revealing the functional domain organization of *Escherichia coli* nucleoid protein H-NS. J Mol Biol *263*, 149-162.

Van Ulsen, P., Hillebrand, M., Zulianello, L., van de Putte, P., and Goosen, N. (1996). Integration host factor alleviates the H-NS-mediated repression of the early promoter of bacteriophage Mu. Mol Microbiol *21*, 567-578.

Wakimoto, B. T. (1998). Beyond the nucleosome: epigenetic aspects of position-effect variegation in *Drosophila*. Cell *93*, 321-324.

Wang, J. C. (1996). DNA topoisomerases. Annu Rev Biochem *65*, 635-692.

White-Ziegler, C. A., Angus Hill, M. L., Braaten, B. A., van der Woude, M. W., and Low, D. A. (1998). Thermoregulation of *Escherichia coli* pap transcription: H-NS is a temperature-dependent DNA methylation blocking factor. Mol Microbiol *28*, 1121-1137.

Whitehouse, C., Chambers, J., Catteau, A., and Solomon, E. (2004). *Brca1* expression is regulated by a bidirectional promoter that is shared by the *Nbr1* gene in mouse. Gene *326*, 87-96.

Workman, J. L., and Kingston, R. E. (1998). Alteration of nucleosome structure as a mechanism of transcriptional regulation. Annu Rev Biochem *67*, 545-579.

Wu, H. Y., Shyy, S. H., Wang, J. C., and Liu, L. F. (1988). Transcription generates positively and negatively supercoiled domains in the template. Cell *53*, 433-440.

Wu, H. Y., Tan, J., and Fang, M. (1995). Long-range interaction between two promoters: activation of the leu-500 promoter by a distant upstream promoter. Cell *82*, 445-451.

Yarmolinsky, M. (2000). Transcriptional silencing in bacteria. Curr Opin Microbiol *3*, 138-143.

Chapter 29

The Biogenesis and Function of MicroRNAs

Yan Zeng and Bryan R. Cullen

Department of Molecular Genetics and Microbiology, Duke University Medical Center, Durham, NC 27710

Key Words: microRNA, RNA interference, post transcriptional gene regulation RNA processing

Summary

MicroRNAs are a large family of approximately 22 nucleotide long, non-coding RNAs processed from stem-loop secondary structures. Current evidence indicates that miRNAs negatively regulate the expression of their target genes in plants and animals.

Introduction and History

In addition to rRNA, mRNA, and tRNA, the three major classes of RNAs responsible for information flow from DNA to protein, there are other types of non-coding RNAs that play important catalytic, structural, and regulatory roles in cells. Examples include RNase P RNA, 7S signal recognition particle RNA, tmRNA involved in bacterial protein degradation, small nucleolar RNAs involved in modification and maturation of rRNAs and other RNAs, spliceosomal RNAs, telomerase RNA, antisense transcripts, and many other RNAs involved in dosage compensation, imprinting, modulating RNA polymerase activity, and stress responses. Some of these RNAs are discussed in other chapters in this volume. In this chapter we will focus on a group of ~22 nucleotide (nt) long RNAs called microRNAs (miRNAs).

The first miRNA was reported in 1993 as the result of an effort to clone the *lin-4* gene, mutation of which caused developmental timing defects in *C. elegans*. More than three years of hard work established that *lin-4* encoded not a protein, but a 21 nt RNA (Lee *et al*, 1993; Lee *et al*., 2004a). Although some regulatory RNAs were known at the time, their mechanisms of action were not well understood, and *lin-4* RNA seemed incredibly small. Fortunately, a target of *lin-4*, *lin-14*, was already known, and it was soon realized that the 3′ untranslated region (3′ UTR) of lin-14 mRNA contained seven sequence elements partially complementary to *lin-4*, giving credence to the notion that the small RNA was indeed functional (Lee *et al*., 1993; Wightman *et al*., 1993; Ruvkun *et al*., 2004). A mechanism was further proposed that through RNA:RNA hybridization, the *lin-4* noncoding RNA inhibited *lin-14* mRNA translation.

The identification of a second miRNA, *let-7*, was reported in 2000, also from *C. elegans* (Pasquinelli, *et al*., 2000; Reinhart *et al*., 2000). Unlike *lin-4*, *let-7* homologs and their potential targets could be readily identified in the genomes of other species, including mammals, which suggested that these small RNAs might not be just developmental oddities in worms after all. Around the same time, RNA interference (RNAi) was drawing great attention to the realm of small RNAs. A gene silencing phenomenon, RNAi was first conclusively demonstrated in 1998, when the introduction of long double-stranded (ds) RNAs into *C. elegans* was shown to cause the degradation of homologous mRNAs (Fire *et al*., 1998). Analogous observations had previously been documented in other organisms, including plants (Baulcombe, 2004). It was then discovered that in plants undergoing gene silencing, ~25 nt long RNAs accumulated that corresponded to the silencing triggers in sequence (Hamilton and Baulcombe, 1999). This result was confirmed and extended in animal systems, as these small RNAs were cloned and turned out to be the processed products of the initial long dsRNAs and the eventual effectors of RNAi, hence the name small interfering RNAs or siRNAs (Yang *et al*., 2000; Zamore

Corresponding Author: Bryan R. Cullen, Duke University Medical Center, Box 3025, Durham, NC USA. Tel: (919) 684-3369, Fax: (919) 681-8979, E-mail: culle002@mc.duke.edu

et al., 2000; Elbashir et al., 2001). Significantly, when cloning methods were applied to naive cells, dozens of endogenous ~22 nt RNAs were identified, revealing a world of tiny RNAs, with *lin-4* and *let-7* being the founding members (Lagos-Quintana et al., 2001; Lau et al., 2002; Lee and Ambros, 2001). There are also virus-encoded miRNAs expressed in virus-infected cells (Pfeffer et al., 2004). Since then, RNA cloning and bioinformatic search efforts have shown that these tiny RNAs, termed miRNAs, are numerous and widespread, and have potentially prevalent functions in multicellular eukaryotes.

How Many miRNAs are There?

Most miRNAs are 20 nt to 25 nt in length. Their sequences are diverse, although a majority have a U residue at their 5' end (Lau et al, 2001). Some miRNAs are expressed relatively broadly, while others have tissue-specific or developmental stage-specific expression patterns. As a result, the abundance of any particular miRNA may vary greatly depending on cell type. The copy number of highly expressed miRNAs can be more than 10^4 per cell, a number on a par with that of U6 snRNA and much higher than those of mRNAs (Lim et al, 2003).

A salient feature of miRNAs is that they are located within stretches of sequences that are predicted by computer folding programs to form stem-loop structures, with the mature miRNA residing within one arm of the stem (Fig.29.1). The miRNA stem-loop structures in plants (Fig.29.1B) tend to be more variable and longer than those in animals (Fig.29.1A). Unlike the long, perfectly complementary dsRNAs that are the precursors of siRNAs, the stem regions of the hairpin structures that eventually give rise to miRNAs (see below) are much shorter and contain bulges and/or internal loops.

The characteristics of such a secondary structure have allowed computational methods to predict new miRNAs on a genome-wide basis (reviewed by Bartel, 2004). Prediction also made use of phylogenetic conservation. Thus, most *C. elegans* stem-loop miRNA features should also be present in the closely related species *C. briggsae*, and a predicted miRNA from *Arabidopsis* is likely a bona fide one if it is also found in rice or maize. Such approaches have estimated that there are hundreds of miRNA genes, constituting ~0.5-1% of the predicted genes in plant and animal genomes (Bartel, 2004). An online miRNA registry (http://www.sanger.ac.uk/Software/Rfam/mirna/) has been established to catalog all the miRNAs in various organisms with links to their genomic DNA information (Ambros et al., 2003a).

So far, no miRNA has been found in unicellular eukaryotes such as budding yeast and fission yeast. Both animals and plants express miRNAs, but no homology is apparent between hundreds of animal miRNAs and plant miRNAs (Bartel, 2004). There is, however, conservation among animal miRNAs or among plant miRNAs. For example, ~30% of *C. elegans* miRNAs have homologs in vertebrates, although their target mRNAs may have diverged. These observations suggest that miRNA genes arose independently and early in the two lineages leading up to animals and plants. Considering that fission yeast has a simplified RNAi apparatus (Volpe et al., 2002), which is also used for the biogenesis and functions of miRNAs in higher eukaryotes, perhaps the expansion of genome size and demand for more complicated developmental and metabolic regulation led to the appearance of miRNAs, processed and used by the more ancient and extant RNAi pathway.

```
    A
              G    UU  U   C   A       U    U
     5' --CCG CCUG  CCC GAGA CUCA GUGUGAG GUA-C  A
         ||| ||||   ||| |||| |||| ||||||| ||| |
     3' --GGC GGAC  GGG CUCU GGGU CACACUU-CGU G  U
              A    CAU  C   C    C          A U


    B
                 UG    U       G   C   C      C  C GAAC    U
     5' ==AAGAA-GA-G AGAG   CGCUGGA GCAG GGUU AUCGAUCU UUC UGU   ACAU  A
         ||||| ||    ||||   ||||||| |||| |||| |||||||| |||  ||    ||||  A
     3' ==UUCUU CU U UCUC   GCGACCU CGUC CCAA UAGCUAGA AAG ACG   UGUA  A
              U  A GU  CGUUU      A   U   A        U  U  AAAA    A
```

Fig.29.1 miRNA-encoding hairpin structures predicted by computer programs. Mature miRNA sequences are indicated in red. (A) *C. elegans lin-4*. (B) *Arabidopsis miR162a*. There are two *miR162* loci, *a* and *b*, and both encode identical miRNAs.

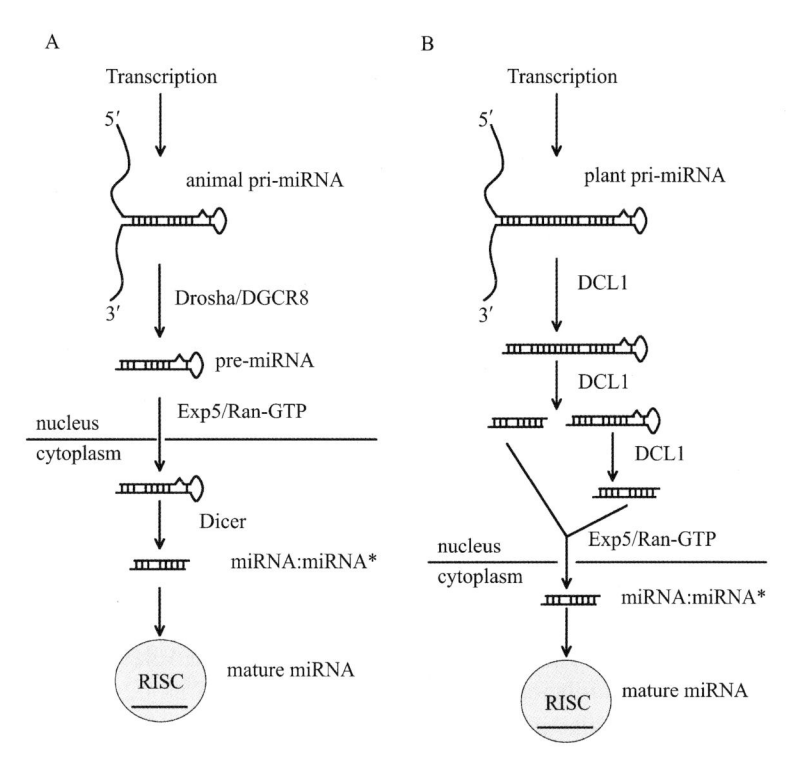

Fig.29.2 miRNA biogenesis pathways in animals (A) and plants (B).

How are miRNAs Made?

The biogenesis of miRNAs follows these stages: transcription, nuclear processing, export, and cytoplasmic processing (Fig.29.2) (Bartel, 2004; Cullen, 2004). There are some differences between the maturation of animal miRNAs (Fig. 29.2A) and that of plant miRNAs (Fig. 29.2B), as will be noted below.

A: Transcription

miRNAs are first transcribed as part of a much longer primary transcript (pri-miRNA, Lee et al., 2002). According to genome sequence information, many miRNAs are located within the intronic regions of annotated protein-coding or non-protein-coding genes (Rodriguez et al., 2004). These miRNAs could therefore use their host gene transcripts as carriers, although the possibility still exists that some are actually transcribed separately. Other miRNAs have their own transcriptional regulatory elements and thus constitute independent transcription units, miRNA transcription is reminiscent of the transcription of another class of RNAs, snoRNAs, snoRNAs are either transcribed independently or encoded by introns of protein-coding host genes (Maxwell and Fournier, 1995).

The primary transcripts of a few miRNAs have been experimentally analyzed (reviewed by Cullen, 2004). Usually their full length versions are quite long (>1 kb) and have a 5′ 7-methyl guanosine cap and a 3′ poly(A) tail. These are the characteristics of RNA polymerase II (Pol II) transcripts. Since the expression of many miRNAs is temporally or spatially regulated, another hallmark of Pol II transcription, those miRNAs are also likely transcribed by Pol II. Available evidence, therefore, indicates that most, if not all, miRNAs are Pol II products, although the involvement of other RNA polymerases (such as Pol III) in the transcription of certain miRNAs cannot be entirely excluded.

B: Cleavage Steps

Like other RNAs in the eukaryotic cells, once transcribed, miRNAs also go through a maturation process. An enzyme called Drosha catalyzes a key reaction whereby a 60-70 nt hairpin RNA (precursor miRNA, or pre-miRNA) containing the mature miRNA is excised from the pri-miRNA in mammals (Fig.29.2A) (Lee et al., 2003). Drosha belongs to a class of RNase III type endonucleases (Fig.29.3A), which generate duplex RNA products containing a 5′ phosphate and a 3′-OH, with usually a 2 nt overhang at the 3′ end. To promote pre-miRNA excision, Drosha requires at least one extra protein partner, DGCR8 in humans, whose gene is often monoallelically deleted in DiGeorge Syndrome (Shiohama et al., 2003; Denli et al., 2004; Gregory et al., 2004; Han et al., 2004). DGCR8 is evolutionarily conserved and contains two double-stranded RNA binding domains (Fig.29.3A). This property likely helps Drosha to recognize the right substrates and to cut at the desired positions within pri-miRNAs, while ignoring a large background of other hairpin RNA structures *in vivo*. In addition, because pri-miRNAs are different in sequence and structural details, some RNAs might be better Drosha substrates than others.

What pri-miRNA features then does the Drosha holoenzyme (Drosha for short) recognize? Both cell culture experiments and *in vitro* Drosha pri-miRNA cleavage assays have shown that, for processing to occur, a few extra paired residues are required outside the eventual pre-miRNA product (Lee et al., 2003; Zeng and Cullen, 2003). The local structure/sequence surrounding the excision sites may also affect the positions where Drosha cleaves (Zeng et al., 2005). Because this region must fit into the active site of Drosha, these requirements perhaps reflect the preference by Drosha for a substrate in a largely helical state before catalysis and the fact that a slight distortion of the RNA helix might alter how individual nucleotides interact with amino acid residues in the catalytic center of the RNase III domains. At the other side of the hairpin, a large terminal loop (usually >10 nt) is also needed for processing (Zeng et al., 2005). Although commonly used computer folding programs tend to predict much smaller terminal loops for pre-miRNAs, mutations that artificially stabilize those small loops compromise Drosha cleavage of pri-miRNAs. Thus, it appears that the structure of the terminal loop region may be inherently flexible, and that Drosha selects the one with a more open conformation. Furthermore, the loop may be an important anchor for Drosha to orient and position its catalytic residues near the bottom of the stem. The region in the middle of the stem-loop structure that includes the mature miRNA is not a major discriminator for Drosha, as long as it maintains a predominantly dsRNA backbone (Zeng and Cullen, 2003).

Recognition by Drosha does not require the 5′ and 3′ end structures of a completely synthesized pri-miRNA, which implies that miRNA processing could take place automatically following transcription. Consistent with this hypothesis, miRNAs can be artificially expressed from irrelevant Pol II and Pol III promoters in animal cells (Zeng et al., 2002, 2005; Chen et al., 2004). It also suggests that differential transcription is likely the main reason for the differential expression of most miRNAs, because if Drosha cleavage or any other downstream event is disrupted in a particular cell type, these cells will make little or no miRNAs at all. Instead, different cell types produce some but not the entire miRNA repertoire encoded by the same genome. On the other hand, just as there is coupling between mRNA transcription and mRNA splicing, a connection between miRNA transcription and processing remains possible.

Drosha is primarily a nuclear protein. For pre-miRNAs produced by Drosha to reach Dicer, which is cytoplasmic in mammalian cells (Billy et al., 2001), pre-miRNAs must exit the nucleus. This step is fulfilled by Exportin 5 (Exp5) and its Ran-GTP cofactor (Yi et al., 2003; Bohnsack et al., 2004; Lund et al., 2004). Exp5 is a member of the karyopherin family that mediates macromolecule transport across the nuclear envelope. Exp5 is specialized at binding to minihelix-containing RNAs with a 3′ overhang (Gwizdek et al, 2003), such as adenovirus VA1 RNA, tRNAs, and pre-miRNAs. Exp5 needs Ran-GTP to bind to its cargo, once in the cytoplasm and stimulated by the RanGTPase-activating protein, Ran-GTP is converted to Ran-GDP, leading to the release of the RNA cargo by Exp5. Dicer would then have access to pre-miRNAs.

Dicer is another RNase III-type enzyme (Fig. 29.3B)

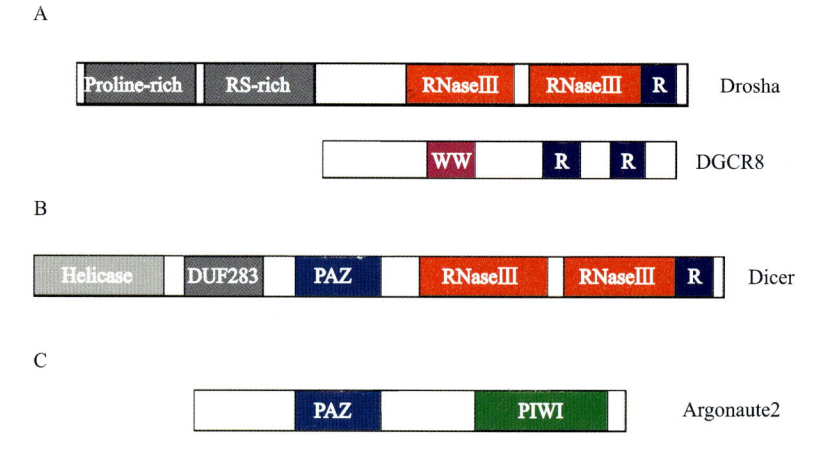

Fig. 29.3 Domain structures of human Drosha and DGCR8 (A), Dicer (B), and Argonaute2 (C). R: double-stranded RNA binding domain.

that excises mature miRNAs from pre-miRNAs (Grishok et al., 2001; Hutvágner et al., 2001; Ketting et al., 2001; Provost et al., 2002; Zhang et al., 2002). Most Dicer proteins possess an ~130-amino-acid-long PAZ domain, which preferentially binds single-stranded 3′ ends of nucleic acids (reviewed by Lingel and Izaurralde, 2004). It has been proposed that human Dicer recognizes a pre-miRNA with an ~2 nt 3′ overhang via its PAZ domain, and then cleaves the double-stranded region ~20 nt away, with a single catalytic center formed intramolecularly by the two RNase III domains (Zhang et al., 2004). The result is a miRNA:miRNA* duplex containing ~ 2 nt overhangs at both 3′ ends (Fig. 29.2). Similarly, human Dicer chews from the free ends of a long dsRNA in a stepwise fashion to generate a series of siRNAs. Interestingly, although mammals and *C. elegans* encode only one Dicer, *Drosophila* has two Dicer proteins, Dcr-1 and Dcr-2, with relatively specialized functions (Liu et al., 2003; Lee et al., 2004c; Pham et al., 2004). Dcr-1 directs miRNA biogenesis, whereas Dcr-2 is the major enzyme that produces siRNAs from long dsRNAs. At the sequence level, Dcr-2 has a functional helicase domain, but lacks a PAZ domain, the opposite is true for Dcr-1. Such an intrinsic difference may explain why Dcr-1 and Dcr-2 prefer different substrates. Furthermore, Dcr-1 and Dcr-2 may recruit different protein partners to aid in their functions *in vivo*.

The plant *Arabidopsis thaliana* encodes four Dicer-like enzymes (DCL1-DCL4), but no Drosha (Bartel, 2004). DCL1 is responsible for generating RNA intermediates from pri-miRNAs, leading to the eventual production of mature miRNAs (Reinhart et al., 2002; Papp et al., 2003; Kurihara and Watanabe, 2004). It has been demonstrated that DCL1 first cuts near the bottom of an extensive stem-loop structure in a pri-miRNA (Fig. 29.2B) (Kurihara and Watanabe, 2004). This releases a long hairpin RNA, which is similar to a long dsRNA. DCL1 then duly trims ~21 nt from the free ends progressively into the stem. As DCL1 is localized in the nucleus, miRNA:miRNA* duplexes are also produced in the nucleus, unlike the situation in mammals. Plants do express Exp5 and Ran orthologs, which likely mediate the nuclear export of miRNA duplexes (Bartel, 2004).

C: Cytoplasmic Selection of the Mature miRNA Strand

When the first batch of endogenous miRNAs was biochemically cloned, it was noted that one precursor almost invariably gives rise to only one strand of mature miRNA, which resides on either the 5′ or 3′ side, but not both sides, of the hairpin (Lagos-Quintana et al., 2001; Lau et al., 2002; Lee and Ambros, 2001). The notation of "miRNA*" has been used to represent the strand that is underrepresented or even "lost" in the final products (Lau et al., 2001; Lim et al., 2003). Such polarity is set after Dicer cleavage but before the incorporation of miRNA into RISC (RNA-induced silencing complex, see below) or a similar functional complex, during which the miRNA:miRNA* duplex is unwound, and the strand (by definition, the miRNA strand) with less stable hydrogen bonding at its 5′ end within the original duplex is retained, while the complementary strand, miRNA*, is released and degraded (Khvorova et al., 2003; Schwarz et al., 2003). The selection procedure thus determines the ratio of final miRNA products by comparing the thermodynamic stability at opposite ends of a ~22 nt long miRNA:miRNA* duplex. This property may partly explain why most miRNAs have a U residue at their 5′ ends, for a U:G base pair is less stable than a U:A pair, which in turn is less stable than a G:C pair. Being incorporated into protein complexes may stabilize mature miRNAs as well. If hydrogen bonds at both ends have similar strength, then miRNA* can also accumulate, which was indeed observed as more miRNAs were cloned and sequenced. The same principle likewise governs the way siRNAs are utilized *in vivo*. In *Drosophila* extracts, the establishment of asymmetry depends on the differential binding to the two individual siRNA strands by Dcr-2 and R2D2, a Dcr-2 interacting protein, and the participation of a helicase(s) (Tomari et al., 2004b). Dicer would then interact with, and therefore transfer, the active siRNA strand to RISC.

By restricting the sequences of the ultimate, functional miRNAs, the miRNA asymmetry rule imposes an additional layer of quality control over miRNA expression. Because the hairpin structure encoding a miRNA has a longer stem than the miRNA:miRNA* intermediate, and because Drosha and/or Dicer may be intrinsically flexible enzymes and may cleave different RNA conformers at slightly different positions, one pri-miRNA sequence can conceivably produce several slightly shifted and therefore distinct miRNA:miRNA* duplexes. Since these duplexes have different 5′ and/or 3′ ends, they would contribute different sequences and strands to the eventual stable products. By eliminating half of the miRNAs in the duplexes, the selection process thus suppresses the "noise" of miRNA expression, yet in rare cases it can also lead to a complete switch of strand bias.

What can miRNAs do?

A: Basic Mechanisms

There are two appreciated modes of action by

miRNAs, both involving basepairing between miRNAs and their target RNAs (mostly, if not exclusively, mRNAs) that guides the relevant effector protein complex(es) to inhibit the expression of target genes (Bartel, 2004)

The first method is to repress the translation of mRNA targets. This mechanism was initially put forwards to explain how *lin-4* negatively regulates its target, *lin-14* mRNA, based on the observation that *lin-4* expression leads to the reduction of LIN-14 protein but not mRNA levels (Lee *et al.*, 1993; Wightman *et al.*, 1993). The 3′ UTR of *lin-14* mRNA confers *lin-4* responsiveness through seven sequence elements (referred to as target sites hereafter) that are partially complementary to *lin-4* (Fig. 29.4A). The presence of multiple sites could allow for a strong and sensitive response to changes in *lin-4* miRNA levels, i.e., cooperativity of *lin-4* regulation.

A

lin-14 mRNA 3′ UTR

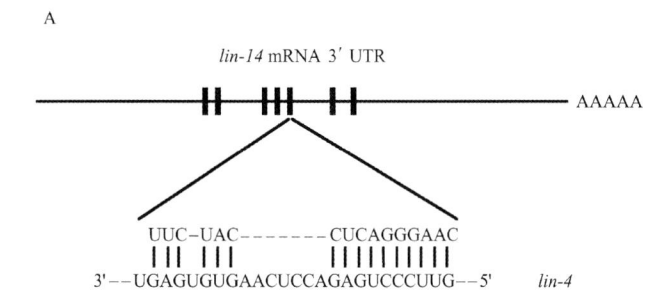

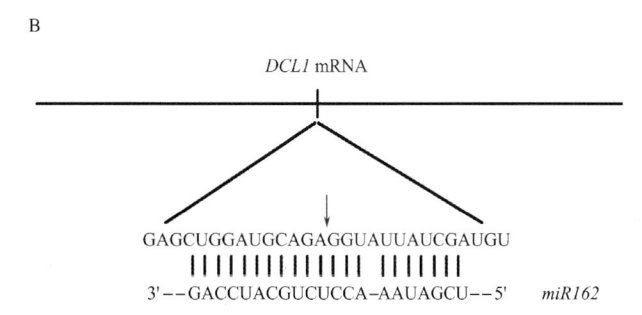

```
                UUC-UAC-------CUCAGGGAAC
                ||| |||       ||||||||||
           3'--UGAGUGUGAACUCCAGAGUCCCUUG--5'    lin-4
```

B

DCL1 mRNA

```
           GAGCUGGAUGCAGAGGUAUUAUCGAUGU
           |||||||||||||| ||||||||
      3'--GACCUACGUCUCCA-AAUAGCU--5'    miR162
```

Fig.29.4 Complementarity between miRNAs and targets. (A) The 3′ UTR of *C. elegans lin-14* mRNA contains seven sequence elements (depicted by black boxes) partial complementary to *lin-4*. Sequence of the fifth element is shown below (Wightman *et al.*, 1993). (B) The protein coding region of *Arabidopsis DCL1* mRNA contains a near perfect match to *miR162*. Cleavage site on the mRNA is indicated by an arrow (Xie *et al.*, 2003).

Following the paradigm of the *lin-4*:*lin-14* pair, subsequent research has indicated that, in the majority of the cases, animal miRNA target sites on mRNAs (experimentally confirmed or computer predicted) contain multiple mismatches, therefore relatively low complementarity, to their respective miRNAs, and that the target sites are located in the 3′ UTR of mRNAs (Bartel, 2004). Reporter assays have indeed demonstrated the ability of animal miRNAs to downregulate the expression of their targets (Zeng *et al.*, 2002; Lewis *et al.*, 2003). How miRNAs inhibit translation, however, has remained obscure. The polysome profile of *lin-14* mRNA does not change significantly whether *lin-4* is present or not (Olsen and Ambros, 1999). Thus, *lin-4* does not block translation initiation of the bulk of *lin-14* mRNA. It has also been observed that the 5′ half of a miRNA contributes more to target selection than the 3′ half does (Lewis *et al.*, 2003; Stark *et al.*, 2003; Doench and Sharp, 2004). This raises an important question as to how the specificity of target recognition by miRNAs is achieved, because a miRNA might be able to target an mRNA after forming a loose duplex with fewer than 10 bp. Such a weak interaction is expected to affect mRNA expression at best modestly, but if several miRNAs could target the same mRNA, then the effect might be significant. Regardless of the details, translation repression is likely the predominant mechanism by which animal miRNAs exert their functions.

The second means by which miRNAs inhibit gene expression is to initiate RNA cleavage and degradation via the canonical RNAi pathway. RNAi and related phenomena are conserved from fungi to higher eukaryotes and triggered by dsRNAs that are either endogenously expressed or exogenously introduced into cells. These triggers may be long, perfectly or near perfectly matched dsRNAs that are virus-encoded or arise due to bi-directional transcription from opposite promoters, cellular ssRNAs that form extended foldback structures, or products of RNA-dependent RNA polymerases. Dicer then converts long dsRNAs into ~21 nt siRNA duplexes. In at least certain organisms siRNAs have been shown to have multifaceted functions (Lippman and Martienssen, 2004): they are involved in genome rearrangement, DNA methylation, heterochromatin formation, and mRNA degradation, the last being the best studied example of RNAi.

The RNA-induced silencing complex (RISC) is the effector for RNAi-mediated mRNA degradation (Hammond *et al.*, 2000). A RISC-bound single stranded siRNA would anneal to a highly homologous mRNA target and guide the protein components of RISC to hydrolyze the linkage in the mRNA at a position corresponding to between nt 10 and 11 of the siRNA as measured from its 5′ end (Elbashir *et al.*, 2001). The cleaved mRNA is no longer functional and is subsequently degraded by other cellular RNases, such as XRN4 in *Arabidopsis* (Souret *et al.*, 2004). Because the biogenesis of miRNAs and that of siRNAs overlap, and because miRNAs are chemically indistinguishable from siRNAs, miRNAs can also enter the RNAi pathway to degrade mRNA targets, provided that the targets have high levels of complementary to the miRNA. In general, plant miRNAs and their target sites are perfectly or nearly perfectly complementary (0-3 mismatches, Fig. 29.4B). Several plant miRNAs have been experimentally confirmed to mediate the cleavage of their respective target mRNAs (Bartel, 2004; Baulcombe, 2004). The cleavage site is not restricted to the 3′ UTR, it can be in the coding region, for instance. RNAi-led degradation permanently changes the fate of a mRNA molecule, therefore if there is any prior effect on translation, it will be masked.

One should note that it is still at least theoretically possible for plant miRNAs to inhibit the translation of mRNAs with weaker homology, as in the case of most animal miRNAs. Conversely, if an animal miRNA is presented with a perfect target, the miRNA is also capable of triggering target cleavage through RNAi (Zeng *et al.*, 2002, 2003; Hutvágner and Zamore, 2002; Doench *et al.*, 2003; Yetka *et al.*, 2004).

B: Protein Cofactors

Genetic and biochemical studies have identified many proteins involved in executing the functions of miRNAs and siRNAs. A family of conserved Argonaute (Ago) proteins turn out to play a pivotal role in this process (Carmell *et al.*, 2002). Present in archaebacteria, unicellular eukaryotes, worms, flies, plants, and mammals, Ago genes encode ~100 kD, highly basic proteins that contain an N-terminal PAZ domain (the same domain as in human Dicer), and a C-terminal ~ 200-amino-acid-long PIWI domain (Fig.29.3C). Ago proteins perform a variety of functions that include stem cell maintenance, DNA elimination, developmental regulation, transcriptional and post-transcriptional gene silencing.

The molecular details of how Ago proteins accomplish their tasks are still being worked out, but a common scheme has emerged that they bind RNA and likely act through RNA-related mechanisms. This revelation has come, in part, from studies of RNAi and miRNAs. At least some Ago family members associate with miRNAs and siRNAs and are required for miRNA accumulation and function *in vivo* (Hutvágner and Zamore, 2002; Mourelatos *et al.*, 2002; Liu *et al.*, 2004; Meister *et al.*, 2004; Okamura *et al.*, 2004; Shi *et al.*, 2004; Vaucheret *et al.*, 2004). Furthermore, they constitute the core component of RISC, and mammalian Ago2 protein serves as the active endonuclease that cleaves target mRNAs (Hammond *et al.*, 2001; Liu *et al.*, 2004). The endonuclease activity resides within the PIWI domain of Ago2 (Liu *et al.*, 2004; Song *et al.*, 2004). Interestingly, although human Ago1 through Ago4 all have the ability to bind siRNAs and miRNAs, only Ago2 has the nuclease activity (Liu *et al.*, 2004; Meister *et al.*, 2004). Perhaps Ago proteins with different primary structures have different biochemical properties and form distinctive RISC-like entities. In addition, they may be differentially expressed in different tissues and at different developmental stages. Related to their functions, one can imagine that they might have different RNA substrates and elicit different biological outcomes, such as RNA degradation, translation repression, and changes in DNA and chromatin structures.

Besides Ago family members, many other proteins also contribute to RNAi and related mechanisms. As mentioned above, fly Dcr-2 and R2D2 help load siRNAs onto RISC (Tomari *et al.*, 2004b). Additional examples include a RNA-dependent RNA polymerase, helicases, and numerous RNA-binding proteins (Sijen *et al.*, 2001; Caudy *et al.*, 2002; Hutvágner and Zamore, 2002; Ishizuka *et al.*, 2002; Tabara *et al.*, 2002; Cook *et al.*, 2004; Tomari *et al.*, 2004a).

C:Examples of miRNA Function

The cellular machinery that generates and utilizes miRNAs is highly conserved in multicellular organisms. In both plant and animal kingdoms, mutations in Argonaute family members, Dicer, Drosha, and DGCR8 tend to cause pleiotropic developmental defects, even embryonic lethality, suggesting that en masse miRNAs and endogenous siRNAs play crucial biological roles (Bohmert *et al.*, 1998; Grishok *et al.*, 2001; Harris and Macdonald, 2001; Kataoka *et al.*, 2001; Ketting *et al.*, 2001; Knight and Bass, 2001; Reinhart *et al.*, 2002; Williams and Rubin, 2002; Bernstein *et al.*, 2003; Denli *et al.*, 2004; Lee *et al.*, 2004c; Liu *et al.*, 2004). Below we describe some of the cases to provide a glimpse into the complex cellular regulatory networks enforced by individual miRNAs.*C. elegans lin-4* as the prototypic miRNA has two known target mRNAs, *lin-14* and *lin-28*, that regulate developmental timing (Fig. 29.5A) (Lee *et al.*, 1993; Wightman *et al.*, 1993; Moss *et al.*, 1997). There are three developmental stages in a worm life cycle: embryonic, larval, and adult stages. The larval (L) stage can be further sub-divided into L1, L2, L3, and L4 stages. Although the exact mechanisms remain unknown, the gene product encoded by *lin-14* activates L1-specific events but inhibits later events, and *lin-28* inhibits L3-specific events. Consequently, progression from L1 to L2 requires downregulating *lin-14* expression, and progression from L2 to L3 requires *lin-28* downregulation. The miRNA encoded by *lin-4* starts to accumulate during the L1 stage and persists afterwards. By binding to target sites present in the 3'UTRs of *lin-14* and *lin-28* mRNAs, *lin-4* miRNA inhibits their expression to clear the way for the appropriate developmental transitions to proceed. Mutations of *lin-4* target sites allow *lin-14* or *lin-28* to bypass the regulatory relationship, so does mutation of *lin-4* itself. There are seven *lin-4* complementary sites in *lin-14* 3'UTR (Fig.29.4A). The other *lin-4* target mRNA, *lin-28*, however, has only one complementary site, so it is not always necessary to have multiple target sites for a single miRNA. An intriguing possibility remains that there is combinatorial regulation by more than one distinct miRNA (Seggerson *et al.*, 2002).

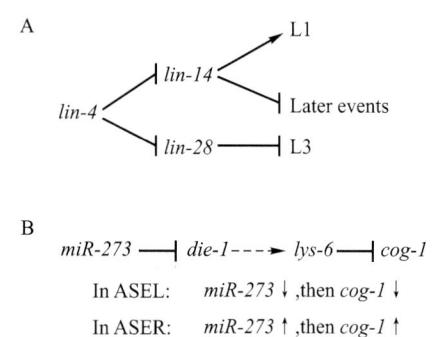

Fig.29.5 Examples of miRNA function. (A) *C. elegans lin-4* inhibits the expression of *lin-14* to promote transition from L1 to L2 stages. In addition, *lin-4* inhibits *lin-28* to promote L2 to L3 transition (Moss *et al.*, 1997). (B) miRNAs control cell fate determination of ASEL and ASER in *C. elegans*. Dashed arrow indicates the relationship between *die-1* and *lys-6* can be direct or indirect.

Another example of miRNA function is neuronal cell fate determination in *C. elegans*. Two worm taste receptor neurons, ASE left (ASEL) and ASE right (ASER), while morphologically similar, express separate chemoreceptors and respond differently to different chemicals. The *cog-1* gene encodes one of the transcription factors specifying ASEL vs ASER: *cog-1* is active in ASER, but not in ASEL (Fig.29.5B) (Hobert, 2004). The 3' UTR of *cog-1* mRNA contains a sequence partially complementary to a miRNA called *lys-6* (Johnston and Hobert, 2003). It has been shown that *lys-6* is both necessary and sufficient for the repression of *cog-1*. Based on a reporter assay, *lys-6* miRNA is expressed in perhaps only 9 neurons in a whole animal, including ASEL, but not ASER. How, then, is *lys-6* miRNA restricted to ASEL? In ASEL, another transcription factor encoded by the *die-1* gene is necessary for *lys-6* expression, whereas in ASER, *die-1*

expression is repressed at least partly due to another miRNA, *miR-273* (Chang *et al.*, 2004). There are two *miR-273* complementary elements in the 3' UTR of *die-1* mRNA, and both are essential for *die-1* downregulation. For some yet-to-be-determined reason, there is less *miR-273* in ASEL, a distinction leading to a cascade that ultimately gives rise to two functionally asymmetric neurons (Fig. 29.5B).

In *Drosophila*, the *bantam* gene encodes a miRNA that enhances cell proliferation and inhibits apoptosis (Brennecke *et al.*, 2003). Thus, reduced *bantam* expression leads to smaller organs and animals. One identified target for *bantam* is the proapoptotic gene *hid*. In the 3' UTR of *hid* mRNA, there are five potential *bantam* binding sites that enable *bantam* miRNA to repress *hid* mRNA translation. Other *bantam* targets likely exist to fully explain the function of *bantam* miRNA.

Identification of animal miRNA targets has been hampered by the generally low homology between miRNAs and their putative targets that is not easily distinguishable from that stemming from chance. The situation is better in plants, as the homology can be much higher (~90%-100%). Quite a few plants miRNAs have firmly established targets (Bartel, 2004; Baulcombe, 2004). For example, the *Arabidopsis JAW* locus yields a miRNA, *miR-JAW*, which negatively regulates the expression of several *TCP* family transcription factors (Palatnik *et al.*, 2003). RNAi-triggered mRNA cleavage seems to be the predominant mechanism, as predicted mRNA cleavage intermediates are stable and can be identified. *JAW* mutations affect leaf morphogenesis, which is rescued by constitutive production of *TCP2* or *TCP4*. Expressing a mutant *TCP4* that supposedly escapes the regulation by *miR-JAW* causes plants to arrest at the seedling stage, indicating that miRNA-mediated suppression of *TCP* expression is important for plant development.

Also in *Arabidopsis*, *DCL1* is necessary for miRNA biogenesis and is itself subject to negative feedback control by a miRNA, as its mRNA is the target of *miR162*-mediated cleavage (Fig. 29.4B) (Xie *et al.*, 2003). Such a feedback mechanism involving miRNAs is likely widespread to help maintaining a dynamic balance of gene expression *in vivo*.

Of hundreds of miRNAs, currently we know only a handful with defined functions. Continuing genetic and bioinformatics analyses will teach us more about the targets and pathways that miRNAs regulate in cells. It is imaginable, however, based on available data, that miRNAs represent a class of transacting regulators of gene expression similar to transcription factors or RNA binding proteins that control the fate of mRNAs transcriptionally or post-transcriptionally. Thus, lessons learned from the studies of transcription factors (Hobert, 2004) and RNA binding proteins (Keene and Tenenbaum, 2002) could easily be transplanted to miRNAs. For example, just as a transcription factor regulates the transcription of multiple genes, and any single gene is controlled by several transcription factors *in vivo*, a single miRNA could potentially influence the expression of numerous mRNAs, and a single mRNA molecule might be under the collective control of several identical or distinct miRNA molecules. The ability of miRNAs to bind to mRNAs with only limited homology, while posing a problem for computational miRNA target prediction, may actually illuminate the complexity *in vivo*. In any given cell, only a subset of miRNAs are made, but they will affect the expression of a much larger number of genes. Even if any individual miRNA only contributes marginally, due to its weak base pairing to its target, the synergistic effort by an assembly of miRNAs could have a profound effect on mRNA expression.

miRNAs, siRNAs, and other Small RNAs

We will end this chapter with a reminder that, besides miRNAs, there are other 20-30 nt long RNA species in an eukaryotic cell. An obvious class is the endogenous siRNAs, processed by Dicer from long dsRNAs. There are siRNAs corresponding to repetitive DNA elements in certain species, and many small antisense RNAs exist that are likely processed from duplexes formed due to bi-directional transcription or that represent RNAi intermediates. Several plant RNAs initially classified as miRNAs are actually siRNAs, for they are derived from long dsRNAs instead of from hairpin-containing transcripts (Vazquez *et al.*, 2004), siRNAs have been shown to participate in processes such as mRNA degradation, DNA methylation, heterochromatin formation, and/or genome rearrangement (Lippman and Martienssen, 2004). Plant siRNAs through RNAi can also protect cells from viruses and transposons (Baulcombe, 2004). It will be interesting to determine if miRNAs can multitask as siRNAs. Cloning and sequencing endeavors have also uncovered other small RNAs of largely unknown origins and functions (Ambros *et al.*, 2003b; Kuwabara *et al.*, 2004). One better-studied example is the ~21 bp NRSE dsRNA (Kuwabara *et al.*, 2004). This dsRNA does not function through the conventional siRNA/miRNA pathways, instead, it appears to interact with the neuronal transcription machinery to induce gene expression. The discovery of small RNAs including miRNAs has undoubtedly enhanced our understanding of the integrated circuitry that regulates gene expression.

References

Ambros, V., Bartel, B., Bartel, D.P., Burge, C.B., Carrington, J.C., Chen, X., Dreyfuss, G., Eddy, S.R., Griffiths-Jones, S., Marshall, M., Matzke, M., Ruvkun, G., and Tuschl, T. (2003a). A uniform system for microRNA annotation. RNA *9*,277-279.

Ambros, V., Lee, R.C., Lavanway, A., Williams, P.T., and Jewell, D. (2003b). MicroRNAs and other tiny endogenous RNAs in *C. elegans*. Curr. Biol. *13*,807-818.

Bartel, D.P. (2004). MicroRNAs: genomics, biogenesis, mechanism, and function. Cell *116*,281-297.

Baulcombe, D. (2004). RNA silencing in plants. Nature *431*, 356-363.

Bernstein, E., Kim, S.Y., Carmell, M.A., Murchison, E.P., Alcorn, H., Li, M.Z., Mills, A.A., Elledge, S.J., Anderson, K.V., and Hannon, G.J. (2003). Dicer is essential for mouse development. Nat. Genet. *35*,215-217.

Billy, E., Brondani, V., Zhang, H., Muller, U., and Filipowicz, W. (2001). Specific interference with gene expression induced by long, double-stranded RNA in mouse embryonal teratocarcinoma cell lines. Proc. Natl. Acad. Sci. U.S.A. *98*,14428-14433.

Bohmert, K., Camus, I., Bellini, C., Bouchez, D., Caboche, M., and Benning, C. (1998). AGO1 defines a novel locus of *Arabidopsis* controlling leaf development. EMBO J. *17*,170-180.

Bohnsack, M.T., Czaplinski, K., and Görlich, D. (2004). Exportin 5 is a RanGTP-dependent dsRNA-binding protein that mediates nuclear export of pre-miRNAs. RNA *10*, 185-191.

Brennecke, J., Hipfner, D.R., Stark, A., Russell, R.B., and Cohen, S.M. (2003). *bantam* encodes a developmentally regulated microRNA that controls cell proliferation and regulates the proapoptotic gene *hid* in *Drosophila*. Cell *113*,25-36.

Carmell, M.A., Xuan, Z., Zhang, M.Q., and Hannon, G.J. (2002). The Argonaute family: tentacles that reach into RNAi, developmental control, stem cell maintenance, and tumorigenesis. Genes Dev. *16*,2733-2742.

Caudy, A.A., Myers, M., Hannon, G.J., and Hammond, S.M. (2002). Fragile X-related protein and VIG associate with the RNA interference machinery. Genes Dev. *16*, 2491–2496.

Chang, S., Johnston, R.J.Jr., Frokjaer-Jensen, C., Lockery, S., and Hobert, O. (2004). MicroRNAs act sequentially and asymmetrically to control chemosensory laterality in the nematode. Nature *430*, 785-789.

Chen, C.Z., Li, L., Lodish, H.F., and Bartel, D.P. (2004). MicroRNAs modulate hematopoietic lineage differentiation. Science *303*, 83–86.

Cook, H.A., Koppetsch, B.S., Wu, J., and Theurkauf, W.E. (2004). The *Drosophila* SDE3 homolog armitage is required for oskar mRNA silencing and embryonic axis specification. Cell *116*,

817-829.

Cullen, B.R. (2004). Transcription and processing of human microRNA precursors. Mol. Cell *16*,861-865.

Denli, A.M., Tops, B., Plasterk, R.H.A., Ketting, R.F., and Hannon, G.J. (2004). Processing of pri-microRNAs by the microprocessor complex. Nature *432*,231-235.

Doench, J.G., Petersen, C.P., and Sharp, P.A. (2003). siRNAs can function as miRNAs. Genes Dev. *17*,438-442.

Doench, J.G., and Sharp, P.A. (2004). Specificity of microRNA target selection in translational repression. Genes Dev. 2004 *18*,504-511.

Elbashir, S.M., Lendeckel, W., and Tuschl, T. (2001). RNA interference is mediated by 21- and 22-nucleotide RNAs. Genes Dev. *15*,188-200.

Fire, A., Xu, S., Montgomery, M.K., Kostas, S.A., Driver, S.E., and Mello, C.C. (1998). Potent and specific genetic interference by double-stranded RNA in *Caenorhabditis elegans*. Nature *391*,806-811.

Gregory, R.I., Yan, K.P., Amuthan, G., Chendrimada, T., Doratotaj, B., Cooch, N., and Shiekhattar, R. (2004). The Microprocessor complex mediates the genesis of microRNAs. Nature *432*,235-240.

Grishok, A., Pasquinelli, A.E., Conte, D., Li, N., Parrish, S., Ha, I., Baillie, D.L., Fire, A., Ruvkun, G., and Mello, C.C. (2001). Genes and mechanisms related to RNA interference regulate expression of the small temporal RNAs that control *C. elegans* developmental timing. Cell *106*, 23–34.

Gwizdek, C., Ossareh-Nazari, B., Brownawell, A.M., Doglio, A., Bertrand, E., Macara, I.G., and Dargemont, C. (2003). Exportin-5 mediates nuclear export of minihelix-containing RNAs. J. Biol. Chem. *278*,5505-5508.

Hamilton, A.J., and Baulcombe, D.C. (1999). A species of small antisense RNA in posttranscriptional gene silencing in plants. Science *286*,950-952.

Hammond, S.M., Bernstein, E., Beach, D., and Hannon, G.J. (2000). An RNA-directed nuclease mediates post-transcriptional gene silencing in *Drosophila* cells. Nature *404*,293-296.

Hammond, S.M., Boettcher, S., Caudy, A.A., Kobayashi, R., and Hannon, G.J. (2001). Argonaute2, a link between genetic and biochemical analyses of RNAi. Science *293*, 1146–1150.

Han, J., Lee, Y., Yeom, K.H., Kim, Y.K., Jin, H., and Kim, V.N. (2004). The Drosha-DGCR8 complex in primary microRNA processing. Genes Dev. *18*,3016-3027.

Harris, A.N., and Macdonald, P.M. (2001). *Aubergine* encodes a *Drosophila* polar granule component required for pole cell formation and related to eIF2C. Development *128*,2823-2832.

Hobert, O. (2004). Common logic of transcription factor and microRNA action. Trends Biochem. Sci. *29*,462-468.

Hutvágner, G., McLachlan, J., Pasquinelli, A.E., Balint, E., Tuschl, T., and Zamore, P.D. (2001). A cellular function for the RNA-interference enzyme Dicer in the maturation of the let-7 small temporal RNA. Science *293*, 834–838.

Hutvágner, G., and Zamore, P.D. (2002). A microRNA in a multiple-turnover RNAi enzyme complex. Science *297*,2056-2060.

Ishizuka, A., Siomi, M.C., and Siomi, H. (2002). A *Drosophila* fragile X protein interacts with components of RNAi and ribosomal proteins. Genes Dev. *16*, 2497–2508.

Johnston, R.J., and Hobert, O. (2003). A microRNA controlling left/right neuronal asymmetry in *Caenorhabditis elegans*. Nature *426*,845-849.

Kataoka, Y., Takeichi, M., and Uemura, T. (2001). Developmental roles and molecular characterization of a *Drosophila* homologue of *Arabidopsis* Argonaute1, the founder of a novel gene superfamily. Genes Cells. *6*,313-325.

Keene, J.D., and Tenenbaum, S.A. (2002). Eukaryotic mRNPs may represent posttranscriptional operons. Mol. Cell *9*,1161-1167.

Ketting, R.F., Haverkamp, T.H., van Luenen, H.G., and Plasterk, R.H.A. (2001). Dicer functions in RNA interference and in synthesis of small RNA involved in developmental timing in *C. elegans*. Genes Dev. *15*, 2654–2659.

Knight, S.W., and Bass, B.L. (2001). A role for the RNase III enzyme DCR-1 in RNA interference and germ line development in *Caenorhabditis elegans*. Science *293*,2269-2271.

Khvorova, A., Reynolds, A., and Jayasena, S.D. (2003). Functional siRNAs and miRNAs exhibit strand bias. Cell *115*, 209–216.

Kurihara, Y., and Watanabe, Y. (2004). *Arabidopsis* micro-RNA biogenesis through Dicer-like 1 protein functions. Proc. Natl. Acad. Sci. U.S.A. *101*,12753-12758.

Kuwabara, T., Hsieh, J., Nakashima, K., Taira, K., and Gage, F.H. (2004). A small modulatory dsRNA specifies the fate of adult neural stem cells. Cell *116*,779-793.

Lagos-Quintana, M., Rauhut, R., Lendeckel, W., and Tuschl, T. (2001). Identification of novel genes coding for small expressed RNAs. Science *294*, 853–858.

Lau, N.C., Lim, L.P., Weinstein, E.G., and Bartel, D.P. (2001). An abundant class of tiny RNAs with probable regulatory roles in *Caenorhabditis elegans*. Science *294*, 858–862.

Lee, R.C., and Ambros, V. (2001). An extensive class of small RNAs in *Caenorhabditis elegans*. Science *294*, 862–864.

Lee. R.C., Feinbaum, R.L., and Ambros, V. (1993). The *C. elegans* heterochronic gene *lin-4* encodes small RNAs with antisense complementarity to *lin-14*. Cell *75*, 843–854.

Lee, R.C., Feinbaum, R.L., and Ambros, V. (2004a). A short history of a short RNA. Cell *116*, S89-S92.

Lee, Y., Jeon, K., Lee, J.T., Kim, S., and Kim, V.N. (2002) .MicroRNA maturation: stepwise processing and subcellular localization. EMBO J. *21*, 4663–4670.

Lee, Y., Ahn, C., Han, J., Choi, H., Kim, J., Yim, J., Lee, J., Provost, P., Radmark, O., Kim, S., and Kim, V.N. (2003). The nuclear RNase III Drosha initiates microRNA processing. Nature *425*, 415–419.

Lee, Y., Kim, M., Han, J., Yeom, K.H., Lee, S., Baek, S.H., and Kim, V.N. (2004b). MicroRNA genes are transcribed by RNA polymerase II. EMBO J. *23*,4051-4060.

Lee, Y.S., Nakahara, K., Pham, J.W., Kim, K., He, Z., Sontheimer, E.J., and Carthew, R.W. (2004c). Distinct roles for *Drosophila* Dicer-1 and Dicer-2 in the siRNA/miRNA silencing pathways. Cell *117*,69-81.

Lewis, B.P., Shih, I.H., Jones-Rhoades, M.W., Bartel, D.P., and Burge, C.B. (2003). Prediction of mammalian microRNA targets. Cell *115*,787-798.

Lim, L.P., Lau, N.C., Weinstein, E.G., Abdelhakim, A., Yekta, S., Rhoades, M.W., Burge, C.B., and Bartel, D.P. (2003). The microRNAs of *Caenorhabditis elegans.* Genes Dev. *17*, 991-1008.

Lingel, A., and Izaurralde, E. (2004). RNAi: finding the elusive endonuclease. RNA *10*,1675-1679.

Lippman, Z., and Martienssen, R. (2004). The role of RNA interference in heterochromatic silencing. Nature *431*, 364-370.

Liu, J., Carmell, M.A., Rivas, F.V., Marsden, C.G., Thomson, J.M., Song, J.J., Hammond, S.M., Joshua-Tor, L, and Hannon, G.J. (2004). Argonaute2 is the catalytic engine of mammalian RNAi. Science *305*,1437-1441.

Liu, Q., Rand, T.A., Kalidas, S., Du, F., Kim, H.E., Smith, D.P., and Wang, X. (2003). R2D2, a bridge between the initiation and effector steps of the *Drosophila* RNAi pathway. Science *301*, 1921-1925.

Lund, E., Güttinger, S., Calado, A., Dahlberg, J.E., and Kutay, U. (2004). Nuclear export of microRNA precursors. Science *303*, 95–98.

Maxwell, E.S., and Fournier, M.J. (1995). The small nucleolar RNAs. Annu. Rev. Biochem. *64*, 897-934.

Meister, G., Landthaler, M., Patkaniowska, A., Dorsett, Y., Teng, G., and Tuschl, T. (2004). Human Argonaute2 mediates RNA cleavage targeted by miRNAs and siRNAs. Mol. Cell *15*, 185-197.

Moss, E.G., Lee, R.C., and Ambros, V. (1997). The cold shock domain protein LIN-28 controls developmental timing in *C. elegans* and is regulated by the lin-4 RNA. Cell *88*, 637-646.

Mourelatos, Z., Dostie, J., Paushkin, S., Sharma, A., Charroux, B., Abel, L., Rappsilber, J., Mann, M., and Dreyfuss, G. (2002). miRNPs: a novel class of ribonucleoproteins containing numerous microRNAs. Genes Dev. *16*, 720-728.

Okamura, K., Ishizuka, A., Siomi, H., and Siomi, M.C. (2004). Distinct roles for Argonaute proteins in small RNA-directed RNA cleavage pathways. Genes Dev. *18*, 1655-1666.

Olsen, P.H., and Ambros, V. (1999). The lin-4 regulatory RNA controls developmental timing in *Camnenorhabditis elegans* by blocking LIN-14 protein synthesis after the initiation of translation. Dev. Biol. *216*, 671-680.

Palatnik, J.F., Allen, E., Wu, X., Schommer, C., Schwab, R., Carrington. J.C., and Weigel, D. (2003). Control of leaf morphogenesis by microRNAs. Nature *425*, 257-263.

Papp, I., Mette, M.F., Aufsatz, W., Daxinger, L., Schauer, S.E., Ray, A., van der Winden, J., Matzke, M., and Matzke, A.J. (2003). Evidence for nuclear processing of plant microRNA and short interfering RNA precursors. Plant Physiol. *132*, 1382-1390.

Pasquinelli, A.E., Reinhart, B.J., Slack, F., Martindale, M.Q., Kuroda, M.I., Maller, B., Hayward, D.C., Ball, E.E., Degnan, B., Muller, P., Spring. J., Srinivasan, A., Fishman, M., Finnerty, J., Corbo, J., Levine, M., Leahy, P., Davidson, E., and Ruvkun, G. (2000). Conservation of the sequence and temporal expression of *let-7* heterochronic regulatory RNA. Nature *408*, 86-89.

Pfeffer, S., Zavolan, M., Grasse, F.A., Chien, M., Russo, J.J., Ju, J., John, B., Enright, A.J., Marks, D., Sander, C., and Tuschl, T. (2004). Identification of virus-encoded microRNAs. Science *304*, 734-736.

Pham, J.W., Pellino, J.L., Lee, Y.S., Carthew, R.W., and Sontheimer, E.J. (2004). A Dicer-2-dependent 80S complex cleaves targeted mRNAs during RNAi in *Drosophila*. Cell *117*, 83-94.

Provost, P., Dishart, D., Doucet, J., Frendewey, D., Samuelsson, B., and Radmark, O. (2002). Ribonuclease activity and RNA binding of recombinant human Dicer. EMBO J. *21*, 5864-5874.

Reinhart, B.J., Slack, F.J., Basson, M., Pasquinelli, A.E., Bettinger, J.C., Rougvie, A.E., Horvitz, H.R., and Ruvkun, G. (2000). The 21-nucleotide *let-7* RNA regulates developmental timing in *Caenorhabditis elegans.* Nature *403*, 901-906.

Reinhart, B.J., Weinstein, E.G., Rhoades, M.W., Bartel, B., and Bartel, D.P. (2002). MicroRNAs in plants. Genes Dev. *16*, 1616-1626.

Rodriguez, A., Griffiths-Jones, S., Ashurst, J.L., and Bradley, A. (2004). Identification of mammalian microRNA host genes and transcription units. Genome Res.*14*, 1902-1910.

Ruvkun, G., Wightman, B., and Ha, I (2004). The 20 years it took to recognize the importance of tiny RNAs. Cell *116*, S93-S96.

Schwarz, D.S., Hutvágner, G., Du, T., Xu, Z., Aronin, N., and Zamore, P.D. (2003). Asymmetry in the assembly of the RNAi enzyme complex. Cell *115*, 199–208.

Seggerson, K., Tang, L., and Moss, E.G. (2002). Two genetic circuits repress the *Caenorhabditis elegans* heterochronic gene *lin-28* after translation initiation. Dev. Biol. *243*, 215-225.

Shi, H., Djikeng, A., Tschudi, C., and Ullu, E. (2004). Argonaute protein in the early divergent eukaryote *Trypanosoma brucei*: control of small interfering RNA accumulation and retroposon transcript abundance. Mol. Cell. Biol. *24*, 420-427.

Shiohama, A., Sasaki, T., Noda, S., Minoshima, S., and Shimizu, N. (2003). Molecular cloning and expression analysis of a novel gene DGCR8 located in the DiGeorge syndrome chromosomal region. Biochem. Biophys. Res. Commun. *304*, 184-190.

Sijen, T., Fleenor, J., Simmer, F., Thijssen, K.L., Parrish, S., Timmons, L., Plasterk, R.H., and Fire, A. (2001). On the role of RNA amplification in dsRNA-triggered gene silencing. Cell *107*, 465-476.

Song, J.J., Smith, S.K., Hannon, G.J., and Joshua-Tor, L. (2004). Crystal structure of Argonaute and its implications for RISC slicer activity. Science *305*, 1434-1437.

Souret, F.F., Kastenmayer, J.P., and Green, P.J. (2004). AtXRN4 degrades mRNA in *Arabidopsis* and its substrates include selected miRNA targets. Mol. Cell *15*, 173-183.

Stark, A., Brennecke, J., Russell, R.B., and Cohen, S.M. (2003). Identification of *Drosophila* MicroRNA targets. PLoS Biol. *1*, 397-409.

Tabara, H., Yigit, E., Siomi, H., and Mello, C.C. (2002). The dsRNA binding protein RDE-4 interacts with RDE-1, DCR-1, and a DExH-box helicase to direct RNAi in *C. elegans*. Cell *109*, 861-871.

Tomari, Y., Du, T., Haley, B., Schwarz, D.S., Bennett, R., Cook, H.A., Koppetsch, B.S., Theurkauf, W.E., and Zamore, P.D. (2004a). RISC assembly defects in the *Drosophila* RNAi mutant *armitage*. Cell 1*16*, 831-841.

Tomari, Y., Matranga, C., Haley, B., Martinez, N., and Zamore, P.D. (2004b). A protein sensor for siRNA asymmetry. Science *306*, 1377-1380.

Vaucheret, H., Vazquez, F., Crete, P., and Bartel, D.P. (2004). The action of *ARGONAUTE1* in the miRNA pathway and its regulation by the miRNA pathway are crucial for plant development. Genes Dev.*18*, 1187-1197.

Vazquez, F., Vaucheret, H., Rajagopalan, R., Lepers, C., Gasciolli, V., Mallory, A.C., Hilbert, J.L., Bartel, D.P., and Crete, P. (2004). Endogenous trans-acting siRNAs regulate the accumulation of *Arabidopsis* mRNAs. Mol. Cell *16*, 69-79.

Volpe, T.A., Kidner, C., Hall, I.M., Teng, G., Grewal, S.I., and Martienssen, R.A. (2002). Regulation of heterochromatic silencing and histone H3 lysine-9 methylation by RNAi. Science *297*,1833-1837.

Wightman, B., Ha, I., and Ruvkun, G. (1993). Posttranscriptional regulation of the heterochronic gene *lin-14* by *lin-4* mediates temporal pattern formation in *C. elegans*. Cell *75*, 855-862.

Williams, R.W., and Rubin, G.M. (2002). *ARGONAUTE1* is required for efficient RNA interference in *Drosophila* embryos. Proc. Natl. Acad. Sci. U.S.A. *99*, 6889-6894.

Xie, Z., Kasschau, K.D., and Carrington, J.C. (2003). Negative feedback regulation of *Dicer-Like1* in *Arabidopsis* by microRNA-guided mRNA degradation. Curr. Biol. *13*,784-789.

Yang, D., Lu, H., and Erickson, J.W. (2000). Evidence that processed small dsRNAs may mediate sequence-specific mRNA degradation during RNAi in *Drosophila* embryos. Curr. Biol. *10*, 191-1200.

Yekta, S., Shih, I.H., and Bartel, D.P. (2004). MicroRNA-directed cleavage of HOXB8 mRNA. Science *304*, 594-596.

Yi, R., Qin, Y., Macara, I.G., and Cullen, B.R. (2003). Exportin-5 mediates the nuclear export of pre-microRNAs and short hairpin RNAs. Genes Dev. *17*, 3011–3016.

Zamore, P.D., Tuschl, T., Sharp, P.A., and Bartel, D.P. (2000). RNAi: double-stranded RNA directs the ATP-dependent cleavage of mRNA at 21 to 23 nucleotide intervals. Cell *101*, 25-33.

Zeng, Y., and Cullen BR (2003). Sequence requirements for microRNA processing and function in human cells. RNA *9*, 112–123.

Zeng, Y., Wagner, E.J., and Cullen, B.R. (2002). Both natural and designed microRNAs can inhibit the expression of cognate mRNAs when expressed in human cells. Mol. Cell *9*, 1327-1333.

Zeng, Y., Yi, R., Cullen, B.R. (2003). MicroRNAs and small interfering RNAs can inhibit mRNA expression by similar mechanisms. Proc. Natl. Acad. Sci. U.S.A. *100*, 9779-9784.

Zeng, Y., Yi, R., and Cullen, B.R. (2005). Recognition and cleavage of primary microRNA precursors by the nuclear processing enzyme Drosha. EMBO J. *24*, 138-148.

Zhang, H., Kolb, F.A., Brondani, V., Billy, E., and Filipowicz, W. (2002). Human Dicer preferentially cleaves dsRNAs at their termini without a requirement for ATP. EMBO J. *21*, 5875-5885.

Zhang, H., Kolb, F.A., Jaskiewicz, L., Westhof, E., and Filipowicz, W. (2004). Single processing center models for human Dicer and bacterial RNase III. Cell *118*, 57-68.

Chapter 30

Transcription Factor Dynamics

Gordon L. Hager and Akhilesh K. Nagaich

Laboratory of Receptor Biology and Gene Expression, Building 41, B602, 41 Library Dr., National Cancer Institute, NIH, Bethesda, MD 20892-5055

Key Words: nuclear receptors, transcription dynamics, chromatin remodeling, laser crosslinking

Summary

The regulation of gene transcription in eukaryotic systems involves a large number of factors. These proteins have been argued to form large, relatively long lived, multi-protein complexes on promoter elements during the process of transcription initiation and elongation. Recent advances in imaging technology have permitted for the first time a dissection of transcription events in real time. Using arrays of gene reporter elements and transcripiton factors tagged with the green fluorescent protein, it is now possible to observe targeting of a regulatory protein to response elements in living cells. Application of photobleaching technology to these systems allows a direct analysis of the rate at which factors are moving in the nucleoplasmic space, and the timing of their interactions with various intranuclear structures, including DNA regulatory sites. These technical breakthroughs have led to the unexpected finding that most transcription factors interact very briefly with promoter elements, and cycle on and off genes at relatively high rates. These findings stand in dramatic contrast to the classic view of large, static initiation complexes, and reveal a level of dynamic action not previously suspected in the process of transcriptional regulation.

Introduction

Transcription factors modulate rates of transcription at target genes both through protein- protein interactions with basal transcription factors and by the recruitment of a variety of factors referred to as coactivators, or corepressors. Some of these interacting proteins serve as bridging factors to other components of the soluble transcription apparatus, while others either harbor intrinsic chromatin modifying activities (such as acetylation or methylation (Berger, 1999; Spencer et al., 1997; Hassig et al., 1997; Chen et al., 1999; Bannister et al., 2002; Selker, 1990; Strahl and Allis, 2000)), or interact with other chromatin remodeling activities (including the Swi/Snf family of nucleosome remodeling proteins (Fryer and Archer, 1998; Becker and Horz, 2002; Kornberg and Lorch, 2002; Francis and Kingston, 2001). The dynamic process by which transcription factors recruit these various activities is poorly understood. The classic view (Fig.30.1) is that a sequence-specific DNA-binding protein binds to a recognition site, and remains at the site for significant periods of time (Becker et al., 1984). Alternatively, the factor may interact only transiently with a response element, recruiting a secondary set of factors that in turn form a stable complex at the regulatory site. This type of mechanism has been referred to as "hit and run", and has been proposed both for steroid receptors (Hager, 2001; Rigaud et al., 1991; Truss et al., 1992; Rigaud et al., 1991), and for enhancer function in general (Suen et al., 1998). The major difficulty in addressing these issues resides in the indirect methods used to detect transcription factor DNA-binding and function. Most techniques currently in use, including the widely utilized chromatin immunoprecipitation (ChIP) approach, would not be sensitive to rapid interactions.

Corresponding Author: Gordon L. Hager, Tel: (301) 496-9867, Fax: (301) 496-4951, E-mail: hagerg@exchange.nih.gov

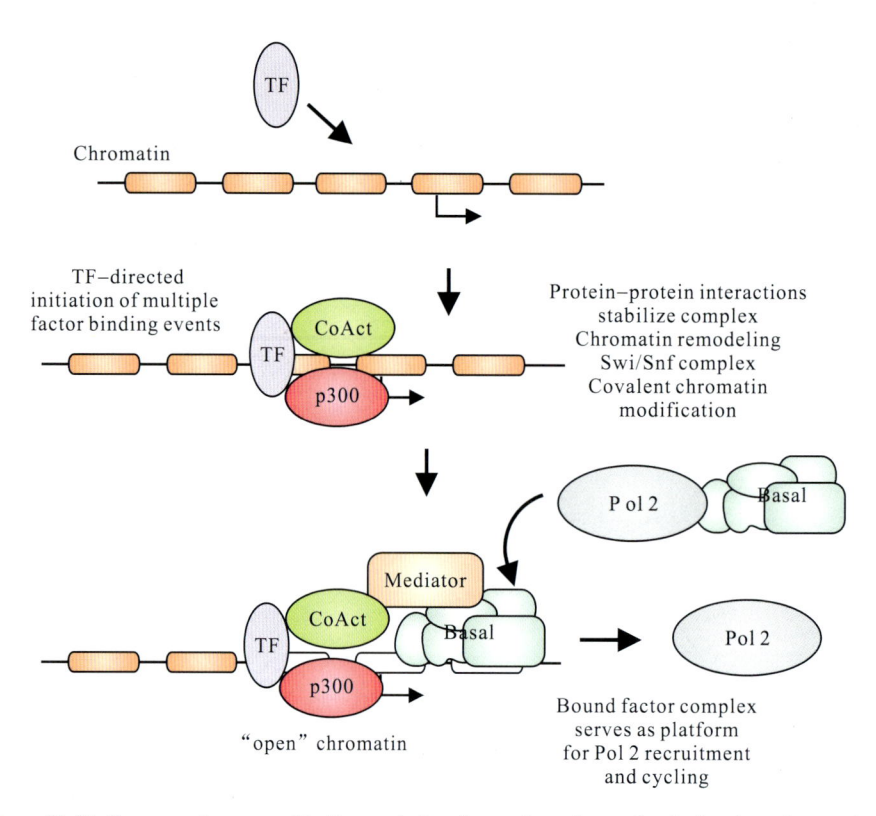

Fig.30.1 "Classic" view of initiation complex assembly. Transcription factors have been classicaly viewed as nucleating the formation of large, stable complexes on the template, involving many coactivators and coregulators. These complexes in turn stimulate the recruitment of members of the general transcription apparatus, and eventually catalyze RNA Pol II initiation events.

Protein Movement and Localization in Living Cells

Use of the green fluorescent protein (GFP) and its derivatives has revolutionized the study of protein mobility in living cells (Prasher, 1995; Cubitt *et al.*, 1995; Heim and Tsien, 1996; Htun *et al.*, 1996; Hager, 1999). Real time characterization of the subcellular localization and inter-compartment movement of factors becomes uniquely possible when the proteins are expressed directly as fusion-chimeras with one of the growing family of fluorescent proteins.

The redistribution of fluorescent tagged proteins through techniques such as time-lapse microscopy becomes an attractive methodology to determine targets and interaction sites for a protein of interest. The technique of greatest utility, however, for studying actual rates of movement in real time is fluorescence recovery after photobleaching, or FRAP (Fig.30.2). By concentrating an intense beam of defined wavelength laser energy on a specific target, the fluorescence associated with that domain can be rapidly extinguished. Subsequent observations on the rate of recovery of the fluorescence provide a unique approach to determine the rates of exchange for the labeled proteins with the target of interest.

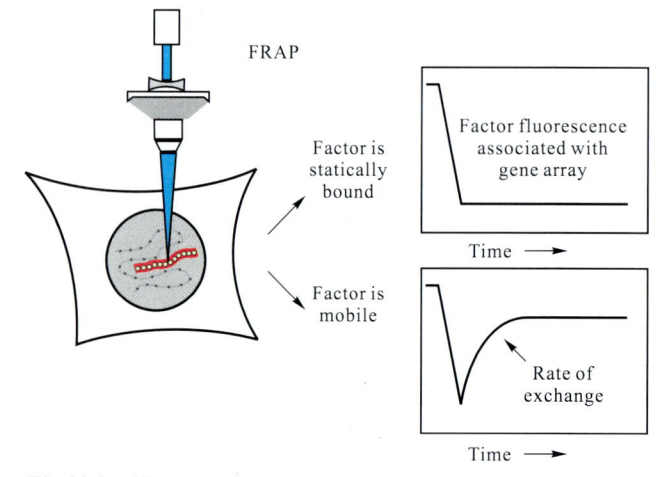

Fig.30.2 FRAP analysis of protein movement. Fluorescence recovery after photobleaching (FRAP) provides a technique to characterize factor interactions with targets in intact cells. If an amplifed copy of a given gene promoter is stably present in one of the cell chromosomes (indicated in the cell nucleus), GFP-labeled factors can be observed to bind to these elements in living cells. Recovery of fluorescence after photobleaching can then be analyzed to determined residence times on the template.

In principle, photobleaching studies carried out with a sequence-specific DNA-binding protein, such as a transcription factor, would provide information concerning the interaction of the factor with its regulatory target. Applying this technique to single copy genes in eucaryotic cells, however, is beyond the sensitivity and resolving power of current imaging systems. Usually many thousands of fluorescent molecules will be present in the nucleus for a given factor, resolving a single complex in this large background cannot be achieved with current techniques. For amplified genes, however, this approach would offer promise if the number of binding sites contained in the amplified set were of sufficient density (Fig.30.2).

Visualizing Transcription in Real Time

Three systems have now been developed, however, that allow visualization of binding events on repeated genetic elements. Two employ artificially developed gene arrays (Kramer *et al.*, 1999; Walker *et al.*, 1999; Belmont *et al.*, 1999), and a third utilizes the naturally occurring repeated ribosomal DNA sequences (Dundr *et al.*, 2002). Andrew Belmont and colleagues have generated artificial arrays containing the lac operator sequence, to which the lac repressor binds with a high affinity. After amplification in a bacterial plasmid to a copy number of 256 sites, the array was co-inserted in chinese hamster ovary cells with the Dhfr gene, and further amplified through methotrexate selection, generating cell lines with very large copy numbers (Belmont *et al.*, 1999). Visualization of GFP-lac repressor targeting to these arrays is dramatic, showing bright concentrations of repressor at the chromosomal integration sites. Protein mobility studies have not been described for the lac repressor itself; however, using fusions of the estrogen receptor (ER) to lac repressor, Stenoien *et al.* (Stenoien *et al.*, 2001a) demonstrated very rapid exchange between a receptor coactivator (SRC-1), or a general coregulator (CBP), and the ER-lac repressor chimera bound to the chromosomal array. These experiments open the way to the study of real time dynamics of protein-protein interactions, although the artificial aspect of the gene targeting element remains an issue for interpretation of the experiments. More recent variations of this system (Janicki *et al.*, 2004) offer reporters with direct read-out of gene activity in living cells.

Steroid Receptors and Cofactors Exchange Rapidly on Response Elements

An alternate gene targeting approach was developed by Hager and colleagues for the glucocorticoid receptor (GR)(Fig. 30.3). This system utilized a cell line (3134) that contains a large tandem array of an MMTV/LTR/v-Ha-ras reporter (Kramer *et al.*, 1999; Walker *et al.*, 1999). The repeat structure arose from the spontaneous chromosomal integration of a 9 kb bovine papilloma virus (BPV) multi-copy episome, creating a head-to-tail array of 1.8×10^6 base pairs (Fig. 30.4). This structure contains approximately 200 copies of the LTR, and thus includes 800-1200 binding sites for GR. Derivatives of this cell line were subsequently developed with a GFP-tagged version of GR expressed from a chromosomal locus under control of the tetracycline repressible promoter (Walker *et al.*, 1999). GFP-GR expressed in these cell lines after removal of tetracycline is resident in the cytoplasm in the absence of ligand, translocation to the nucleus is easily detected by direct live-cell epifluorescence within 10 minutes after addition of hormone. Using this system, direct binding of GR to genomic regulatory elements could be demonstrated (Fig.30.3A)(McNally *et al.*, 2000); unexpectedly, it was found that the receptor exchanges rapidly with the chromosomal regulatory sites, with a residence time of around 10 sec. Although this system also relies on a tandem array, it offers a more physiological target in that the repeated element includes a complete promoter structure with associated regulatory elements.

This approach was extended by Becker *et al.* (Becker *et al.*, 2002) to an analysis of the dynamics for GRIP1, a well known steroid receptor coactivator, and RNA Pol II, using cells containing the amplified MMTV promoter array (Fig.30.3B). The results indicated that upon hormone induction GR and RNA Pol II both rapidly load on to the promoter array. GRIP1, which directly associates with GR, also showed rapid exchange on the promoter, virtually identical to that of GR (t1/2, 5 seconds)(Fig. 30.3B) whereas RNA Pol II showed biphasic recovery kinetics. An initial rapid recovery phase was followed by a much slower complete recovery within 12-15 minutes. The initial rapid recovery of Pol II was ascribed to the abortive initiation events associated with transcription, whereas slower recovery phase of the curve was assigned to elongating polymerase molecules. The results indicated that complete recovery of Pol II is only achieved when elongating polymerases finish transcription and clear the template. The authors further confirmed this interpretation by treating the cells with actinomycin D, a known intercalator of DNA (Becker *et al.*, 2002). Using several variants of the FRAP methodology such as I-FRAP, these investigators showed that the presence of actinomycin D completely immobilized the RNA Pol II on the MMTV array. Therefore, the slower recovery phase of the curve is associated with the block of the RNA Pol II during the

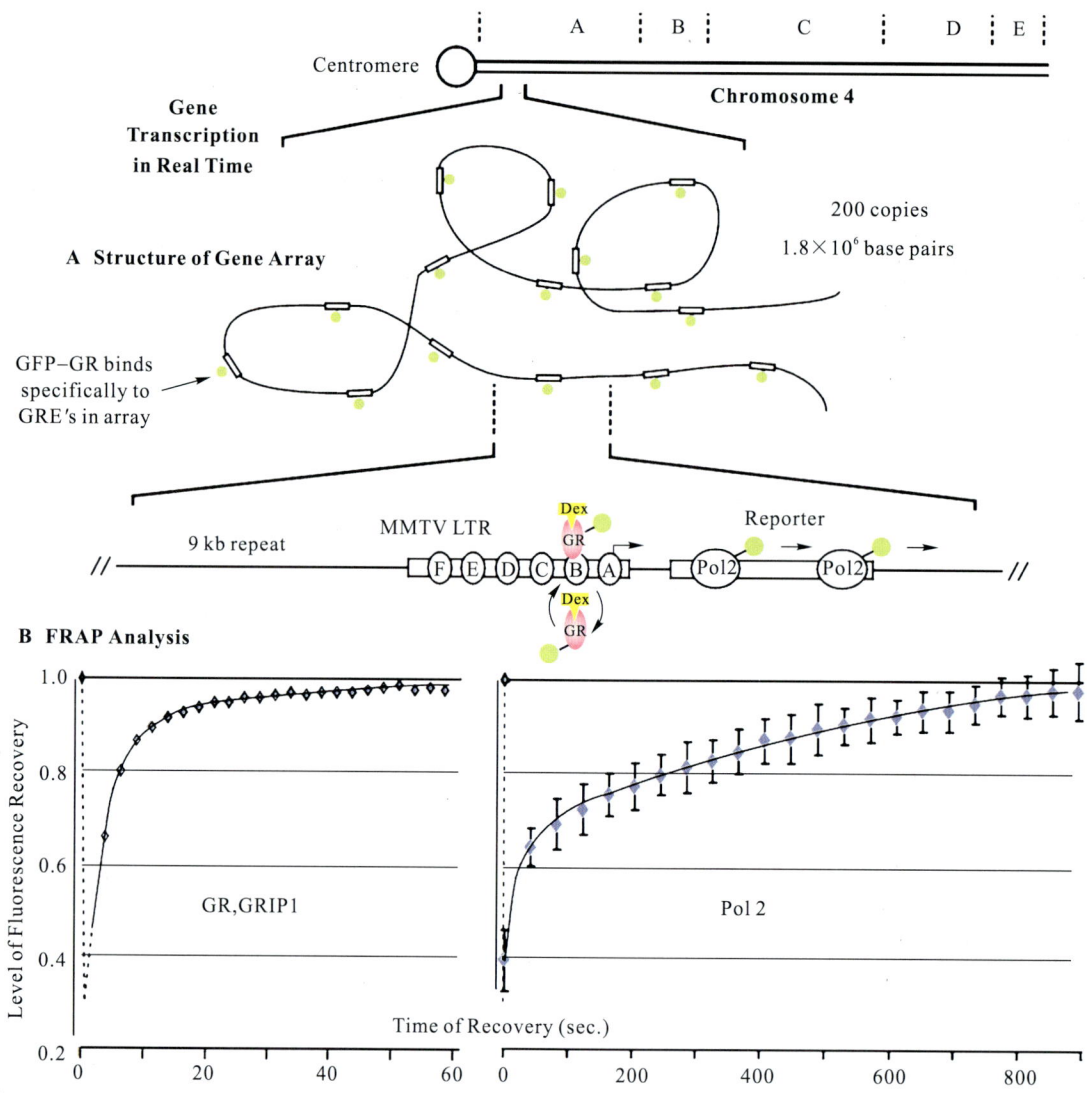

Fig.30.3 A system to study gene transcription in real time. A) In cell line 3134, 200 copies of the MMTV promoter resides as a perfect head-to-tail tandem array near the centromere of chromosome 4. GR can be observed to bind to this structure in living cells. B) Photobleaching analysis measures the mobility of transcription factors on the target gene. Both glucocorticoid (GR) and its co-activator GRIP1 have very short resident times (t1/2 ~ 5 seconds), whereas RNA Pol II requires 13 minutes for complete recovery.

elongation stage. These results confirmed the expected behavior of RNA Pol2 on a transcribing gene, and provided an essential control for the mobility of the transcription factors. It was therefore evident that steroid receptors and cofactors exchange rapidly on response elements in living cells (Hager et al., 2004; Nagaich et al., 2004a; Hager et al., 2002).

In a separate study, Kimura et al. (Kimura et al., 2002), using FRAP methodology and GFP-Pol II showed that, in transcriptionally active CHO cells, RNA Pol II is present in two kinetic pools: one fraction is highly mobile while the other is transiently immobile. As with the gene-specific MMTV observations (Becker et al., 2002), these authors provided evidence that immobilization is due to engagement in transcription, and further showed that the equilibrium between the pools shifts completely toward the mobile fraction after treatment with the transcription inhibitor DRB, likely due to release of RNA Pol II from the template. In addition, after incubation with actinomycin D, RNA Pol II is almost completely immobile due to stalling of the polymerases at sites where the drug has intercalated.

Mancini and coworkers have used fluorescence recovery after photobleaching (FRAP) to examine the intranuclear dynamics of estrogen receptor (ER) (Stenoien et al., 2001b). After bleaching, unliganded ER exhibited high mobility (recovery $t_{1/2}$ < 1 s) while agonist (oestradiol; E2) or partial antagonist (4-hydroxytamoxifen) exhibited slower ER recovery ($t_{1/2}$ 5–6 s). Interestingly, ER liganded with the pure antagonist (ICI 182,780) showed an almost total loss of mobility, indicating that the receptors ability to exchange with nuclear targets was

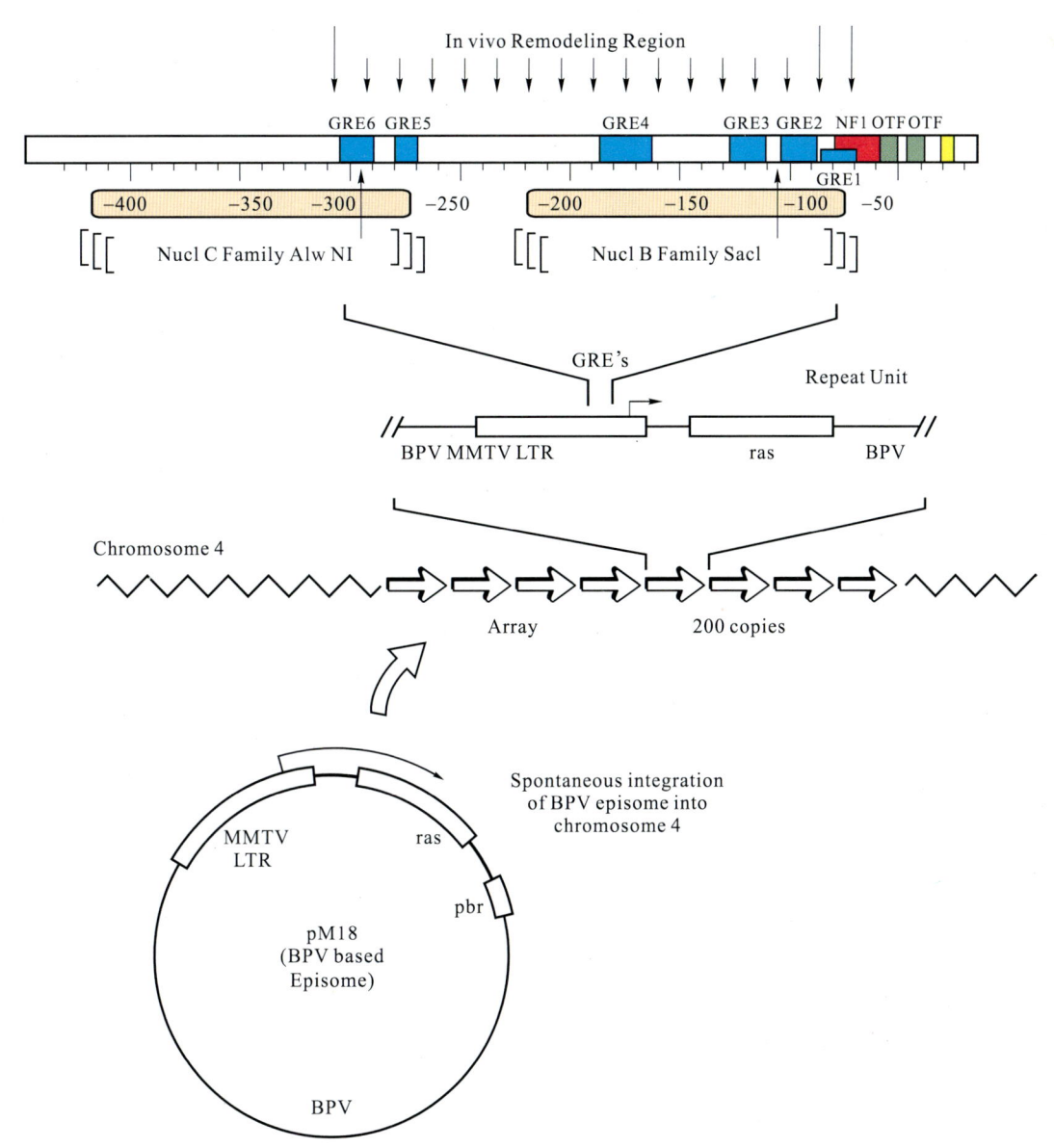

Fig.30.4 **Transcription factor binding sites in the MMTV promoter array.** The gene array was developed by integration and amplification of an episomal MMTV structure present in cell line 904.1. Each copy of the 9 kb MMTV repeat in the chromosome 4 gene array contains a complete promoter structure. Six GR binding sites (shown in blue) are found in this promoter, and hypersensitive access to the Sac I and Alwn I restriction enzyme sites is induced upon treatment of cells harboring the gene array. Location of positioned nucleosomes B & C is shown in relation to the transcription factor binding sites.

compromised when activated with this ligand. Rayasam et. al (Rayasam *et al.*, 2005) studied the dynamics of progesterone receptor (PR) with a natural target promoter in living cells, and also found rapid exchange strongly modulated by ligand specific effects. PR in the presence of the agonist R5020 exhibited rapid exchange with the MMTV promoter. Two PR antagonists, RU486 and ZK98299, showed opposite effects on receptor dynamics *in vivo*. In the presence of RU486, PR showed slower exchange rate than the agonist-activated receptor. In contrast, PR bound to ZK98299 did not localize to the promoter and exhibited higher mobility in the nucleoplasm than the agonist-bound receptor. These experiments suggest that steroid receptors liganded with different agonists and antagonists recruit alternate sets of cofactors and chromatin remodelers to affect transcription and mobility of steroid receptor on target binding sites.

Rapid Dynamics -A Common Feature of Transcription Factor Action

A similarly rapid dynamic behavior has been characterized for RNA Pol I transcripiton factors.

Misteli and colleagues utilized the natural amplification of the rRNA genes to examine the dynamics of Pol I and associated factors (Dundr *et al.*, 2002). The nucleolus can be regarded as the archetype of a "transcription factory". Within this structure, which is dedicated to a high rate of rRNA production, multiple ribosomal genes are located in fibrillar centers containing RNA polymerase I transcription factors. The product rRNA is subsequently assembled into ribosomal precursors. A variety of GFP-tagged RNA Polymerase I components (including preinitiation and assembly factors and different subunits of the RNAP1 complex) were characterized and introduced into mammalian cells. The recruitment to, and residence times at, nucleolar RNAP1-dependent genes was then monitored by several live cell imaging methods including FRAP. The authors observed that each of the RNAP1 factor displayed a different initial rate of fluorescence recovery within the photobleached nucleolus, indicating that the majority of the subunits reach promoters separately and do not reside in preassembled holo-complexes, contrary to the generally accepted view of the Pol I complex as a large, ready-to-use holo-enzyme.

In addition to the rapid recovery of RNAP1 associated factors, FRAP analysis showed secondary slower recovery kinetics of GFP-tagged polymerase subunits but not of the initiation factors. The authors argue that this slower recovery reflects the association of RNAP1 with active rRNA genes. A mathematical model was developed describing nucleolar entry and exit rates, promoter on and off rates, and elongation rates of each analyzed component. This approach presented a global view on the reaction kinetics of RNAP1 transcription within the context of living cells. The residence time of elongating RNAP1 was 2–3 min, which indeed corresponds to previously calculated elongation rates of RNAP1. In general, complex assembly was inefficient as judged from the probability of each factor to associate with the promoter and to actually initiate transcription, and fitted well to a kinetic model based on standard principles of chemical reaction kinetics.

Several labs have now combined kinetic modeling with photobleaching experiments to analyze the binding dynamics of a wide range of proteins in the nucleus of living cells (Rayasam *et al.*, 2005; Elbi *et al.*, 2004; Hager *et al.*, 2004; Nagaich *et al.*, 2004a; Phair *et al.*, 2004; Sprague and McNally, 2005; Carrero *et al.*, 2004). This approach has led to the determination of the basic biophysical properties such as on/off rates and residence time of several chromatin related proteins such as steroid receptors and coactivators, chromatin remodeling complexes, nucleosomal binding protein HMG-17, histone H1, heterochromatin protein (HP1), pre-mRNA splicing factor SF2/ASF and rRNA processing protein fibrillarin. These experiments show that nuclear proteins are highly mobile in the nuclear compartments and engage with binding sites only transiently with residence time in the order of few seconds. Proteins move rapidly through out the entire nucleus, providing the opportunity for factors to explore many regulatory sites in a relatively short time.

Dynamic Behavior of Factors during DNA Repair

To study the nuclear organization and dynamics of nucleotide excision repair (NER), Houtsmuller, used FRAP methodology (Hoogstraten *et al.*, 2002; Houtsmuller *et al.*, 1999) to monitor the mobility of endonuclease ERCC1/XPF and the DNA helix opener TFIIH interactions with chromatin in living Chinese hamster ovary cells. In the absence of DNA damage, the complex moved freely through the nucleus with a diffusion coefficient consistent with its molecular size. DNA damage caused transient dose-dependent immobilization of ERCC1/XPF, likely due to engagement of the complex in the repair events. After 4 minutes of the UV DNA damage, the complex regained mobility. These results suggested that nucleotide excision repair, as for the Pol I and Pol II transcription process, operates by assembly of individual NER factors at the site of DNA damage, rather than by pre-assembly of holo-complexes. The data also suggested that ERCC1/XPF participates in repair of DNA damage in a distributive fashion rather than by processive scanning of genome. The ratio between free and damage bound NER factors was UV dose dependent, in spite of an excessive amount of damage, suggesting that assembly of NER complexes is inefficient, similar to the RNAP 1 and 2 transcription systems. Interestingly, TFIIH appeared to be capable of readily switching between RNAP2 transcription complexes, NER complexes, and nucleolar RNAP1 transcription complexes, showing that stochastic exchange also occurs between common proteins involved in different DNA transactions.

Mechanisms of Factor Movement

Most nuclear proteins reside on a specific chromatin site only for seconds or less. The hit-and-run model of transcriptional control (Hager *et al.*, 2004; McNally *et al.*, 2000) maintains that transcription complexes are assembled in a stochastic fashion from freely diffusible proteins. This is in contrasts to the models involving stepwise assembly of stable holo-complexes. However, the chances of forming a productive complex improve if the binding of one factor promotes the binding of its interacting partners. Two paradigms dominate our thinking: the well-established enhanceosome model and the relatively novel idea that most protein-DNA and protein-protein interactions in the nucleus are very dynamic and highly reversible.

The enhanceosome model is based on the concept of context-dependent interactions among transcription factors, which promote their cooperative assembly on DNA and endow the complex with exceptional stability (Kim and Maniatis, 1997; Thanos and Maniatis, 1995). This is compatible with the high affinity of NF-κB for its DNA binding sites within the interferon-β enhancer (the model enhanceosome) and follows naturally from considerations on the physicochemical equilibria between multiple interacting macromolecules. In fact, *in vitro* measurements confirm the stability of the enhanceosome, and chromatin immunoprecipitation (ChIP) experiments suggest a stepwise recruitment of proteins to chromatin. Although highly successful in explaining the specificity of transcriptional control, and theoretically intuitive, the classical enhanceosome model clashes with the observation that transient and dynamic binding is a common property of all chromatin proteins with the exception of core histones.

In the "hit-and-run" model, transcriptional activation reflects the probability that all components required for activation will meet at a certain chromatin site. The concept of interaction-dependent stabilization of the complex, central to the enhanceosome paradigm, can apply to transient interactions as well. Stability and affinity are thermodynamic concepts that are related to the amount of free energy liberated during complex assembly, and there is no direct constraint on the time frame involved in the interaction. There is however an indirect constraint: affinity is defined as the ratio between the on rate (binding) and the off rate (unbinding), thus, if affinity increases, either the on rate becomes faster or the off rate becomes slower (or both). These rates can be fast (as in a hit-and-run model) or slow (as in a classical enhanceosome model), it is the change in their ratio that brings about a change in stability.

High Resolution Observation of Factor Dynamics

An alternative approach to the study of protein movement involves an *in vitro* analysis of factor interactions with the DNA templates, particularly during the processes of chromatin reorganization and chromatin remodeling. Several lines of evidence suggest that protein mobility in the nucleus and the dynamic exchange of factors with binding sites on chromatin is controlled both by diffusion and by the action of factors that hydrolyze ATP (Rayasam et al., 2005; Fletcher et al., 2002; Fletcher et al., 2000). To dissect the mechanisms involved in the dynamic mobility and exchange of steroid receptors in chromatin, a new approach involving laser UV crosslinking was developed to follow factor interaction with chromatin in real time (Nagaich et al., 2004b; Nagaich and Hager, 2004). A laser UV light source has several advantages over conventional light sources. UV laser mediated crosslinking is highly efficient, proceeds via a biphotonic mechanism, and the crosslinking of proteins to DNA is completed within 1 µs. This approach was applied to study the interaction of glucocorticoid receptor and the chromatin remodeling complex with the MMTV chromatin template during chromatin remodeling. It was found that GR interactions with the template during remodeling process are highly transient and periodic (Fig. 30.5). A sharp peak in laser detected binding is observed 5 minutes after initiation of the reaction, followed by equally rapid loss of receptor (Nagaich et al., 2004b). This cycle repeats periodically, with a cycle time of 5 minutes. A similar cycle of binding was also found for the Swi/Snf complex, although the detailed binding profile was different. There appears to be loss of Swi/Snf interaction as GR binding increases, with a return to the basal level of interactions as GR leaves the template. Laser detected interactions of core histones with the template were also periodic, but more complex. Histones H2A and H2B each manifested a sharp peak during interaction, but these transitions were out of phase with each other. These findings led to the proposal of a dynamic model for GR and chromatin remodeling complex interaction with the template (Fig. 30.6). The model suggests that rapid GR binding results from the initial recruitment of Swi/Snf complex. At this stage nucleosome remodeling opens the structure and increases the number of available GRE response elements. This local perturbed chromatin stage is transient, and the remodeling protein is lost from the template after completion of the remodeling event. This would lead in turn to the collapse of the remodeled nucleosome state. As this ground state is incompatible to with binding of multiple GR homodimers, the receptor would be rapidly ejected from the template. Thus, the receptor is hypothesized to interact dynamically with the template during the remodeling reaction. These observations offer a potential mechanism for ATP dependent mobility in the living cell.

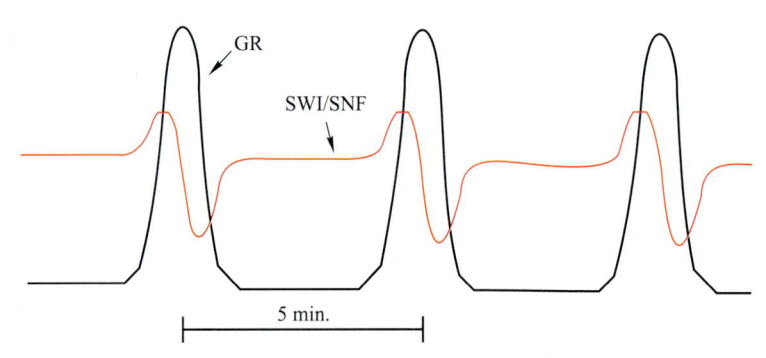

Fig.30.5 Periodic binding of glucocorticoid receptor and the Swi/Snf complex during chromatin remodeling. The profile of laser induced crosslinking during a 15 min. *in vitro* chromatin remodeling reaction is presented schematically for GR (black) and the Swi/Snf remodeling complex (orange). Each complex manifests a transient binding and displacement phase, followed by similar, repetitive events.

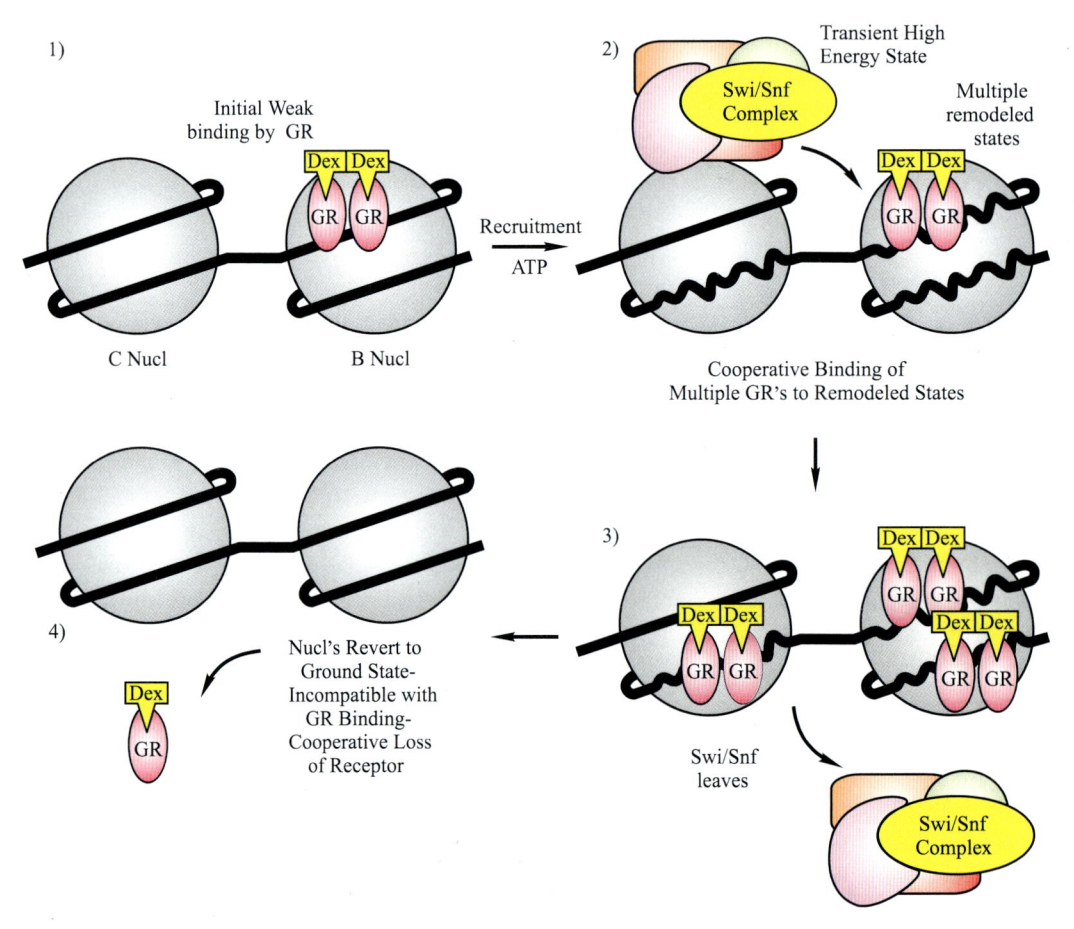

Fig.30.6 Model for the transient, periodic binding behavior of GR and Swi/Snf. Rapid binding of GR results from the initial recruitment of the Swi/Snf complex (step 2). At this stage, nucleosome remodeling "opens" the structure and increases the number of available GR binding sites (steps 2 and 3) (there are a total of six potential binding sites in the B/C region (Fletcher *et al.*, 2000)). We suggest that this local perturbed chromatin state is transient, leading to subsequent loss of the remodeling complex (note in Fig. 30.5 that Swi/Snf binding is significantly reduced after the initial GR loading). Progression of the remodeling process would lead in turn to collapse of the high-energy-state (steps 3 and 4) and return of the local chromatin domain to the ground state (step 4). As this state is incompatible with binding of multiple GR homodimers, GR would be rapidly lost. Thus GR is actively ejected from the chromatin structure as a direct result of the progression of the remodeling process.

Conclusion

A large number of proteins have now been analyzed using live cell imaging techniques such as FRAP. The accumulated body of experimental evidence obtained using real time live cell imaging analysis suggests that the cell nucleus is a highly dynamic environment and the classical approach of describing the nuclear function in terms of a network of interacting proteins is inadequate. An overview of the recent studies favors an alternative image of the nucleus as dynamic integrated system of inter-connected and interdependent metastable molecular organizations realized through stochastic interactions and self organization. Some of these processes rely only on diffusion, but energy dependent mechanisms such as chromatin remodeling are also involved in the observed dynamic movements.

References

Bannister,A.J., Schneider,R., and Kouzarides,T. (2002). Histone methylation: dynamic or static? Cell *109*, 801-806.

Becker,M., Baumann,C.T., John,S., Walker,D., Vigneron,M., McNally,J.G., and Hager,G.L. (2002). Dynamic behavior of transcription factors on a natural promoter in living cells. EMBO Reports *3*, 1188-1194.

Becker,P., Renkawitz,R., and Schutz,G. (1984). Tissue-specific DNaseI hypersensitive sites in the 5′-flanking sequences of the tryptophan oxygenase and the tyrosine aminotransferase genes. EMBO J. *3*, 2015-2020.

Becker,P.B. and Horz,W. (2002). ATP-dependent nucleosome remodeling. Annu. Rev. Biochem. *71*, 247-273.

Belmont,A.S., Li,G., Sudlow,G., and Robinett,C. (1999). Visualization of large-scale chromatin structure and dynamics using the lac operator/lac repressor reporter system. Methods Cell Biol. *58:203-22*, 203-222.

Berger,S.L. (1999). Gene activation by histone and factor acetyltransferases. Curr. Opin. Cell Biol. *11*, 336-341.

Carrero,G., Crawford,E., Th'ng,J., de Vries,G., and Hendzel,M.J. (2004). Quantification of protein-protein and protein-DNA interactions *in vivo*, using fluorescence recovery after photobleaching. Methods Enzymol. *375*, 415-442.

Chen,D., Ma,H., Hong,H., Koh,S.S., Huang,S.M., Schurter,B.T., Aswad,D.W., and Stallcup,M.R. (1999). Regulation of transcription by a protein methyltransferase. Science *284*, 2174-2177.

Cubitt,A.B., Heim,R., Adams,S.R., Boyd,A.E., Gross,L.A., and Tsien,R.Y. (1995). Understanding, improving and using green fluorescent proteins. Trends. Biochem. Sci. *20*, 448-455.

Dundr,M., Hoffmann-Rohrer,U., Hu,Q., Grummt,I., Rothblum,L.I., Phair,R.D., and Misteli,T. (2002). A kinetic framework for a mammalian RNA polymerase *in vivo*. Science *298*, 1623-1626.

Elbi,C., Walker,D.A., Romero,G., Sullivan,W.P., Toft,D.O., Hager,G.L., and DeFranco,D.B. (2004). Molecular chaperones function as steroid receptor nuclear mobility factors. Proc. Natl. Acad. Sci. USA *101*, 2876-2881.

Fletcher,T.M., Ryu,B.-W., Baumann,C.T., Warren,B.S., Fragoso,G., John,S., and Hager,G.L. (2000). Structure and dynamic properties of the glucocorticoid receptor-induced chromatin transition at the MMTV promoter. Mol. Cell. Biol. *20*, 6466-6475.

Fletcher,T.M., Xiao,N., Mautino,G., Baumann,C.T., Wolford,R.G., Warren,B.S., and Hager,G.L. (2002). ATP-dependent mobilization of the glucocorticoid receptor during chromatin remodeling. Mol. Cell. Biol. *22*, 3255-3263.

Francis,N.J. and Kingston,R.E. (2001). Mechanisms of transcriptional memory. Nat. Rev. Mol. Cell Biol. *2*, 409-421.

Fryer,C.J. and Archer,T.K. (1998). Chromatin remodeling by the glucocorticoid receptor requires the BRG1 complex. Nature *393*, 88-91.

Hager,G.L. (2001). Understanding nuclear receptor function: From DNA to chromatin to the interphase nucleus. Prog. Nucleic Acid. Res. Mol. Biol. *66*, 279-305.

Hager,G.L. (1999). Studying nuclear receptors with GFP fusions. Methods Enzymol. *302*, 73-84.

Hager,G.L., Elbi,C.C., and Becker,M. (2002). Protein dynamics in the nuclear compartment. Curr. Opin. Genet. Dev. *12*, 137-141.

Hager,G.L., Nagaich,A.K., Johnson,T.A., Walker,D.A., and John,S. (2004). Dynamics of nuclear receptor movement and transcription. Biochim. Biophys. Acta *1677*, 46-51.

Hassig,C.A., Fleischer,T.C., Billin,A.N., Schreiber,S.L., and Ayer,D.E. (1997). Histone deacetylase activity is required for full transcriptional repression by mSin3A. Cell *89*, 341-347.

Heim,R. and Tsien,R.Y. (1996). Engineering green fluorescent protein for improved brightness, longer wavelengths and fluorescence resonance energy transfer. Curr. Biol. *6*, 178-182.

Hoogstraten,D., Nigg,A.L., Heath,H., Mullenders,L.H., van,D.R., Hoeijmakers,J.H., Vermeulen,W., and Houtsmuller,A.B. (2002). Rapid switching of TFIIH between RNA polymerase I and II transcription and DNA repair *in vivo*. Mol. Cell *10*, 1163-1174.

Houtsmuller,A.B., Rademakers,S., Nigg,A.L., Hoogstraten,D., Hoeijmakers,J.H., and Vermeulen,W. (1999). Action of DNA repair endonuclease ERCC1/XPF in living cells. Science *284*, 958-961.

Htun,H., Barsony,J., Renyi,I., Gould,D.J., and Hager,G.L. (1996). Visualization of glucocorticoid receptor translocation and intranuclear organization in living cells with a green fluorescent protein chimera. Proc. Natl. Acad. Sci. USA *93*, 4845-4850.

Janicki,S.M., Tsukamoto,T., Salghetti,S.E., Tansey,W.P., Sachidanandam,R., Prasanth,K.V., Ried,T., Shav-Tal,Y., Bertrand,E., Singer,R.H., and Spector,D.L. (2004). From silencing to gene expression: real-time analysis in single cells. Cell *116*, 683-698.

Kim,T.K., and Maniatis,T. (1997). The mechanism of

transcriptional synergy of an *in vitro* assembled interferon-beta enhanceosome. Mol. Cell *1*, 119-129.

Kimura,H., Sugaya,K., and Cook,P.R. (2002). The transcription cycle of RNA polymerase II in living cells. J. Cell Biol. *159*, 777-782.

Kornberg,R.D., and Lorch,Y. (2002). Chromatin and transcription: where do we go from here. Curr. Opin. Genet. Dev. *12*, 249-251.

Kramer,P., Fragoso,G., Pennie,W.D., Htun,H., Hager,G.L., and Sinden,R.R. (1999). Transcriptional state of the mouse mammary tumor virus promoter can effect topological domain size *in vivo*. J. Biol. Chem. *274*, 28590-28597.

McNally,J.G., Mueller,W.G., Walker,D., Wolford,R.G., and Hager,G.L. (2000). The glucocorticoid receptor: Rapid exchange with regulatory sites in living cells. Science *287*, 1262-1265.

Nagaich,A.K., and Hager,G.L. (2004). UV laser cross-linking: A real-time assay to study dynamic protein/DNA interactions during chromatin remodeling. Sci. STKE *256*, PL13.

Nagaich,A.K., Rayasam,G.V., Martinez,E.D., Johnson,T.A., Elbi,C., John,S., and Hager,G.L. (2004a). Subnuclear trafficking and gene targeting by nuclear receptors. Ann. N. Y. Acad. Sci. Vol. *1024*, 213-220.

Nagaich,A.K., Walker,D.A., Wolford,R.G., and Hager,G.L. (2004b). Rapid periodic binding and displacement of the glucocorticoid receptor during chromatin remodeling. Mol. Cell *14*, 163-174.

Phair,R.D., Gorski,S.A., and Misteli,T. (2004). Measurement of dynamic protein binding to chromatin *in vivo*, using photobleaching microscopy. Methods Enzymol. *375*, 393-414.

Prasher,D.C. (1995). Using GFP to see the light. Trends Genet. *11*, 320-323.

Rayasam,G.V., Elbi,C., Walker,D.A., Wolford,R.G., Fletcher,T.M., Edwards,D.P., and Hager,G.L. (2005). Ligand specific dynamics of the progesterone receptor in living cells and during chromatin remodeling *in vitro*. Mol Cell Biol *25*, 2406-2418.

Rigaud,G., Roux,J., Pictet,R., and Grange,T. (1991). *In vivo* footprinting of rat TAT gene: dynamic interplay between the glucocorticoid receptor and a liver-specific factor. Cell *67*, 977-986.

Selker,E.U. (1990). DNA methylation and chromatin structure: A view from below. Trends Biochem. Sci. *15*, 103-107.

Spencer,T.E., Jenster,G., Burcin,M.M., Allis,C.D., Zhou,J., Mizzen,C.A., McKenna,N.J., Onate,S.A., Tsai,S.Y., Tsai,M.J., and O'Malley,B.W. (1997). Steroid receptor coactivator-1 is a histone acetyltransferase. Nature *389*, 194-198.

Sprague,B.L. and McNally,J.G. (2005). FRAP analysis of binding: proper and fitting. Trends Cell Biol. *15*, 84-91.

Stenoien,D.L., Nye,A.C., Mancini,M.G., Patel,K., Dutertre,M., O'Malley,B.W., Smith,C.L., Belmont,A.S., and Mancini,M.A. (2001a). Ligand-mediated assembly and real-time cellular dynamics of estrogen receptor alpha-coactivator complexes in living cells. Mol. Cell Biol. *21*, 4404-4412.

Stenoien,D.L., Patel,K., Mancini,M.G., Dutertre,M., Smith,C.L., O'Malley,B.W., and Mancini,M.A. (2001b). FRAP reveals that mobility of oestrogen receptor-alpha is ligand- and proteasome-dependent. Nat. Cell Biol. *3*, 15-23.

Strahl,B.D. and Allis,C.D. (2000). The language of covalent histone modifications. Nature *403*, 41-45.

Suen,C.S., Berrodin,T.J., Mastroeni,R., Cheskis,B.J., Lyttle,C.R., and Frail,D.E. (1998). A transcriptional coactivator, steroid receptor coactivator-3, selectively augments steroid receptor transcriptional activity. J. Biol. Chem. *273*, 27645-27653.

Thanos,D., and Maniatis,T. (1995). Virus induction of human IFN beta gene expression requires the assembly of an enhanceosome. Cell *83*, 1091-1100.

Truss,M., Chalepakis,G., and Beato,M. (1992). Interplay of steroid hormone receptors and transcription factors on the mouse mammary tumor virus promoter. J. Steroid Biochem. Molec. Biol. *43*, 365-378.

Walker,D., Htun,H., and Hager,G.L. (1999). Using inducible vectors to study intracellular trafficking of GFP-tagged steroid/nuclear receptors in living cells. Methods (Companion to Methods in Enzymology) *19*, 386-393.

Chapter 31
Actin, Actin-Related Proteins and Actin-Binding Proteins in Transcriptional Control

Wilma A. Hofmann and Primal de Lanerolle

Department of Physiology and Biophysics, University of Illinois at Chicago, Chicago, IL 60612

Key Words: nuclear actin, nuclear actin-related protein (ARP), nuclear actin-binding protein (ABP), nuclear myosin I (NMI), gelsolin

Summary

Actin is one of the major proteins of the cytoskeleton where, among other functions, it is important for cell movement, defining cell shape, intracellular transport, and muscle contraction. Even though it has been known for many years that conventional actin and actin-related proteins (Arps) are also present in the nucleus, their nuclear functions remained mostly unknown. In recent years, however, it has become clear that actin and Arps are involved in a variety of nuclear processes, especially in transcriptional regulation. It is now apparent that the roles of actin and Arps during transcription are multifaceted and complex ranging from chromatin remodeling to the basic transcription process by RNA polymerases. Furthermore, a number of actin-binding proteins (ABPs) have been identified as transcriptional regulators or co-activators, emphasizing the fundamental role of actin in the nucleus. This chapter specifically addresses the present understanding of actin and Arp function in chromatin remodeling, the role of actin and the actin-binding protein nuclear myosin I in basic transcription, and the role of several other actin-binding proteins as transcriptional co-activators.

Introduction

A: Cytoplasmic Actin

Actin was first purified in Albert Szent-Gyorgyi's lab in Hungary in 1942 (Szent-Gyorgyi, 1945). It was initially purified from skeletal muscle and is still best known as filamentous protein that is involved in muscle contraction. Actin is a highly conserved, widely distributed protein that has been found in all eukaryotic cell types and an ancient form of actin has recently been discovered in bacteria (Jones *et al.*, 2001). It is also one of the most abundant proteins comprising from 1-15% of the total protein in eukaryotic cells. The 43kD protein is expressed in six different isoforms that show >90% amino acid homology. The isoform expression occurs in a tissue and development dependent pattern (McHugh *et al.*, 1991; Vandekerckhove *et al.*, 1986). Four of the isoforms, α-skeletal actin (α-SKA), α-cardiac actin (α-CAA), α-smooth muscle actin (α-SMA), and γ-smooth muscle actin (γ-SMA), are restricted to muscle tissue while β-actin and γ-actin are ubiquitous.

In the cytoplasm actin exists in two forms, as a monomer called G-actin and in a filamentous form called F-actin. Atomic structure determination of the actin molecule shows that it is folded into two large domains each consisting of 2 sub-domains (Otterbein *et al.*, 2001). Monomeric actin has a low ATPase activity and the ATP binding pocket, as well as a high affinity binding site for a divalent cation, is located in a deep cleft, called the actin fold, that is formed by the two large domains (Fig.31.1).

Actin is organized into relatively stable filaments, known as thin filaments, in muscle cells. Actin monomers in the cytoplasm of non-muscle cells also readily polymerize in an ATP dependent process to form polarized actin filaments, known as microfilaments. However, in contrast to thin filaments in muscle cells

---

Corresponding Author: Primal de Lanerolle, Department of Physiology and Biophysics, University of Illinois at Chicago, 835 S. Wolcott, Chicago, IL 60612, Tel: (312) 996-5430, E-mail: primal@uic.edu

that are very stable, microfilaments are double stranded helical structures that are highly dynamic in nature. That is, they have the ability to rapidly form, deform, and reform. Dynamic actin filaments are necessary for defining cell shape in non-muscle cells. They also play crucial roles in cell motility, cell division and intracellular transport. Thus, actin is involved in a host of cellular functions.

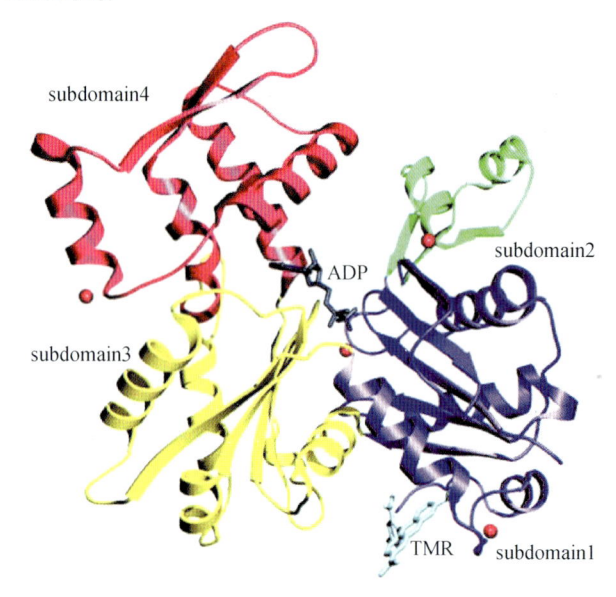

Fig.31.1 The structure of muscle actin. Actin monomer ribbon structure of uncomplexed rabbit skeletal muscle actin in the ADP state at 1.54 Å resolution (Otterbein et al., 2001). Actin is composed of two domains. Each of the major domains is subdivided into two additional domains that are termed subdomain 1-4. ATP/ADP and a divalent cation bind within a deep cleft between subdomain 2 and 4. This cleft is called the ATP-binding pocket or the actin fold. A divalent cation (in this case Ca^{2+}) that is bound adjacent to the ATP/ADP site is shown as a red sphere. Three additional bound calcium ions are also shown. TMR (Tetramethylrhodamine-5-maleimide) is a fluorescently labeled probe that was covalently attached to the actin molecule and used to crystallize monomeric actin.

The numerous forms and functions of actin in the cytoplasm are regulated by a vast array of actin-binding proteins (ABPs). To date, between 60-100 types of ABPs have been identified (dos Remedios et al., 2003). ABPs can be roughly divided into two groups. One group regulates assembly and disassembly of actin filaments as well as the length, stability, and form of the actin filaments. They also regulate interactions between actin filaments and other components of the cytoskeleton. The other group consists of the superfamily of molecular motors called myosins that bind to actin and use the energy of ATP hydrolysis to generate force and move unidirectionally along actin filaments.

B: Nuclear Actin

The presence of actin in the nucleus was first suggested over 30 years ago by Lane (Lane, 1969). Other early studies suggested that actin might be part of the nuclear matrix (Jockusch et al., 1974; Nakayasu and Ueda, 1983). Subsequently, Jokusch and her colleagues reported that a protein that co-purified with RNA polymerase II from Physarum polycephalum was actin (Smith et al., 1979), though no functional role for actin was suggested. These initial studies were followed by numerous others that reported the occurrence of actin within the nucleus (De Boni, 1994). During the last few years there have been quite a few reports of actin in the nucleus. These studies have suggested various and diverse functions for nuclear actin, from a role in transcription to RNA biogenesis and RNA transport to a role in nuclear envelope assembly (Bettinger et al., 2004; Pederson and Aebi, 2002).

Curiously, partly due to the fact that actin is one of the best characterized cytoplasmic proteins, early work on nuclear actin was met by skepticism and the very presence of actin in the nucleus was questioned for many years. One of the main problems was the detection of actin in the nucleus by immunofluorescence. Most of the known functions of actin in the cytoplasm involve polymerization into filaments, which can be stained by fluorescent phalloidin, a specific immunocytochemical probe for actin filaments. However, under normal conditions, nuclei of cells cannot be stained by phalloidin. Therefore, actin detected in isolated nuclei or subnuclear fractions was thought to represent a cytoplasmic contaminant due to the high amount of actin protein in the cytoplasm.

Finally, however, highly pure, hand isolated nuclei from amphibian oocytes that contained none or minimal cytoplasmic contamination clearly revealed the presence of intranuclear actin at concentrations of 3-4 mg/ml (Clark and Merriam, 1977; Clark and Rosenbaum, 1979). Subsequently, actin was detected in nuclei of somatic cells by immunoelectron microscopy studies (Nakayasu and Ueda, 1985). Furthermore, in vitro cross-linking studies showed that actin is associated with DNA (Miller et al., 1991). Recently an anti-actin antibody was developed that, in immunofluorescence microscopy, does not recognize F-actin but recognizes a conformational state of actin that seems to be nucleus-specific (Gonsior et al., 1999). Therefore, it is now generally accepted that actin is present not only in oocyte nuclei but also in interphase nuclei of somatic cells. Moreover, several lines of evidence suggest that in contrast to the cytoplasm, where a significant fraction of the actin is filamentous, most of the nuclear actin is either in monomeric form or organized into very short filaments that are not stained by phalloidin (Ankenbauer et al., 1989; Scheer et al., 1984). Nevertheless, nuclear actin is capable of polymerizing into filaments and usually does so when the nuclear envelope is ruptured.

The entry and exit of actin into and from the nucleus seems to be complex and highly regulated. Even though actin does not contain a known nuclear localization signal, it translocates to the nucleus bound to another small actin-binding protein, cofilin, via the classical importin-β import pathway (Pendleton et al., 2003). Actin is exported from the nucleus by at least two different export pathways. Actin possesses two classical leucin-rich nuclear export signals that are functional and necessary for the export of actin via the export receptor Exportin 1 (Wada et al., 1998). Recently an additional export pathway has been identified. The export receptor Exportin 6 seems to be responsible for the nuclear export of actin that is bound in a complex with the small actin-binding protein profilin (Stuven et al., 2003).

C: The History of Actin in Transcription

Actin first appeared in connection with transcription in 1979 when Smith et al. identified a protein that co-purified with RNA polymerase II from the slime mold *Physarum polycephalum* as actin. Though no functional implications were made at that time, it was suggested that similar proteins found in RNA polymerase II preparations from other species might also be actin. Subsequently, actin was discovered in transcriptionally active extracts from human HeLa cells and from calf thymus (Egly et al., 1984). This study indicated that actin might act as a co-activator for transcription, especially at the level of transcription initiation. The first clear evidence for a direct role of actin in transcription was provided by Scheer et al. (1984) in the same year. They showed that microinjecting antibodies directed against actin, as well as actin-binding proteins like fragmin, into the nuclei of amphibian oocytes, led to a retraction of chromosome loops, sites at which active transcription takes place (Scheer et al., 1984). Even though these experiments gave strong evidence for a function for actin in transcription, no clear role at the molecular level was defined.

This is partly due to the fact that transcription of DNA into RNA is an extraordinarily complex, highly regulated process. Gene transcription depends on a complex molecular machine consisting of more than 100 proteins. It is turned on and off by the exquisitely orchestrated interplay of these proteins with each other and with regulatory DNA elements. Transcription requires chromatin remodeling, the binding of transcription factors to regulatory regions of DNA, the formation of pre-initiation complexes (PICs), and the recruitment of RNA polymerase complexes to the PICs. Ultimately, large transcriptional complexes consisting of a RNA polymerase and other proteins translocate relative to equally large DNA molecules, remodeling chromatin and synthesizing RNA as transcription proceeds. As recent studies have elucidated the role of actin in transcription, it has become obvious that actin is involved in the process of transcription in more than one way. Actin is found in complex with RNA polymerase I, II, and III where it plays specific roles in basic transcription. Furthermore, actin and members of the actin-related proteins (Arps) are functional components of large multiprotein complexes that alter chromatin structure and appear to be involved in the process of chromatin remodeling. In addition, a number of actin-binding proteins (ABPs), such as myosin I, have been identified in the nucleus. It is becoming obvious that these ABPs play an important role in the mechanics of transcription processes and the regulation of individual steps of transcription. In the following sections we will sequentially describe in detail the involvement of actin, Arps and ABPs in transcription.

Nuclear Actin and Basic Transcription

During the basic process of transcription DNA is used as a template by RNA polymerases to generate different types of RNA. Eukaryotic cells contain 3 distinct classes of RNA polymerases that are termed RNA polymerase I, II, and III. Each polymerase synthesizes specific RNAs. RNA polymerase I is located in a specific nuclear compartment, the nucleolus, where it is responsible for ribosomal RNA (rRNA) synthesis (except 5S rRNA). RNA polymerase II is located in the nucleoplasm and responsible for the transcription of all protein genes into messenger RNA (mRNA) and for transcription of most small nuclear RNAs (snRNA). RNA polymerase III is also localized in the nucleoplasm, transcribing 5S rRNA, transfer RNA (tRNA), and some snRNA genes. Each of the RNA polymerase holoenzymes is a multiprotein complex that comprises two large subunits and 12-15 smaller subunits. Mechanistically the process of transcription by all three types of RNA polymerases is similar and occurs roughly in three steps: initiation, elongation, and termination. During transcription initiation a pre-initiation complex consisting of several transcription factors assembles at the promoter region of the gene to be transcribed. After binding of the RNA polymerase the pre-initiation complex is complete and transcription of the gene is initiated. In transcription elongation the RNA polymerase travels along the DNA and assembles ribonucleotides into an RNA strand. When transcription is complete, the RNA is released from the polymerase and the RNA polymerase dissociates from the DNA (transcription termination).

A series of recent studies have now shown that β-actin is associated with all three types of RNA polymerases and directly involved in the basic transcriptional process. The results obtained from studies of each of the polymerases is discussed next.

A: RNA Polymerase II

As mentioned above, early studies on actin in the nucleus suggested a function for actin in transcription by RNA polymerase II. A recent study by Hofmann et al. (2004) investigated the role of actin in transcription by RNA polymerase II in detail. It was shown that β-actin tightly associates with RNA polymerase II in mammalian cells. In fact even highly purified RNA polymerase II contains trace amounts of β-actin, suggesting that β-actin might be a subunit of the RNA polymerase II holoenzyme. The significance of this association has been demonstrated by using a minimal in vitro transcription system that consists of the general transcription factors TFIIB, TFIIF, TBP, and purified RNA polymerase II. This minimal set of factors is able to form a pre-initiation complex at the promoter region of a DNA template that contains the Adenovirus major late promoter (AdMLP), a well studied and typical promoter. This set of factors is also sufficient to promote transcription and RNA production from this DNA template (Kugel and Goodrich, 2000). However, transcription in this system can be inhibited by using antibodies to β-actin. Moreover, accumulation of RNA can be stimulated almost 8-fold by adding exogenous β-actin. This suggests that actin like TFIIB, TFIIF, and TBP is a general transcription factor that constitutively associates with RNA polymerase II (Fig.31.2). The data obtained from in vitro experiments was confirmed in vivo using chromatin immunoprecipitation assays to demonstrate the association of actin with the promoter region of transcribing genes. All of these data indicate that actin might be involved during early stages of transcription. This role for β-actin during transcription initiation was supported by a subsequent study that showed that β-actin interacts with the ribonucleoprotein particle U (hnRNP U) and that both bind to initiation competent RNA polymerase II (Kukalev et al., 2005). The importance of β-actin in pre-initiation complex assembly was conclusively demonstrated when it was found that depleting actin from a nuclear extract prevented the integration of the RNA polymerase II into the developing pre-initiation complex (Hofmann et al., 2004).

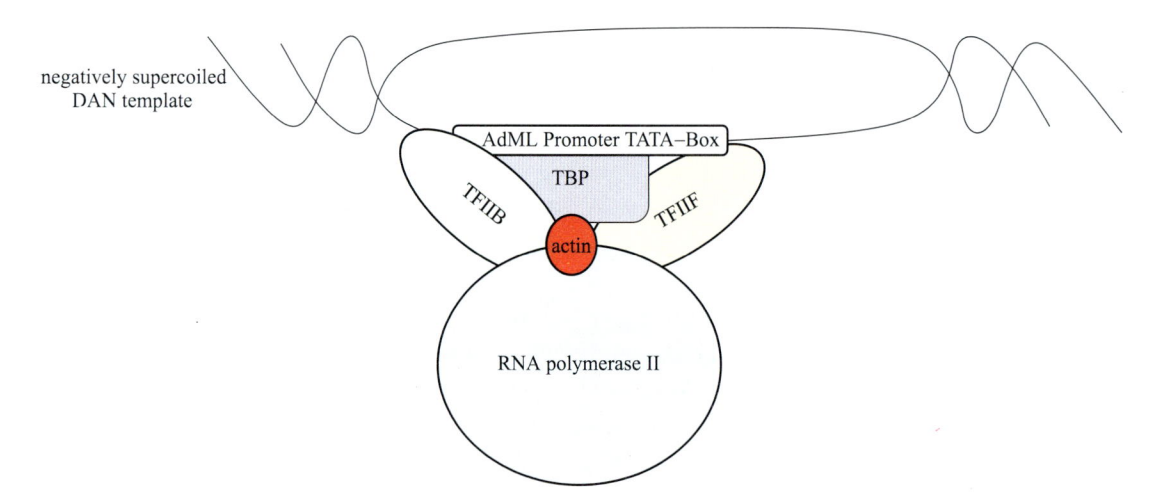

Fig.31.2 Pre-initiation complex assembly at the AdML promoter. This model depicts pre-initiation complex (PIC) formation on the Adenovirus major late (AdML) promoter. In vitro transcription from a negatively supercoiled DNA template containing the AdML promoter requires a minimal set of transcription factors consisting of purified TBP, TFIIF, and TFIIB and purified RNA polymerase II (Kao et al., 1990; Peterson et al., 1990). The first step of transcription is the binding of the TATA-binding protein (TBP) to the TATA box in the promoter region. Next, the transcription factors TFIIB and TFIIF assemble in a complex with TBP around the TATA box. These factors then help to recruit the RNA polymerase II to the forming PIC. Hofmann et al. (2004) have shown that purified RNA polymerase II contains another factor, namely actin, and that actin is necessary for the integration of RNA polymerase II into the assembling PIC. A combination of protein-protein interaction assays and in vitro PIC-formation assays have demonstrated that actin is crucial for the recruitment and binding of the RNA polymerase II to the TBP-TFIIF-TFIIB-complex at the TATA box.

B: RNA Polymerase I

Actin has been shown to localize not only in the nucleoplasm but also in the nucleolus, the place of rRNA transcription by RNA polymerase I, in both mammalian oocytes and somatic cells (Nowak et al., 1997; Fomproix and Percipalle, 2004; Funaki et al., 1995). Philimonenko et al. (2004) demonstrated the physical association of actin with the RNA polymerase I core-enzyme by co-immunoprecipation assays. Microinjection experiments and in vitro transcription assays showed that antibodies to β-actin inhibit transcription by RNA polymerase I in vivo as well as in vitro, demonstrating a functional role for actin in RNA polymerase I transcription. Furthermore, a physical association of actin with the promoter region as well as with the elongation region of rRNA genes in vivo, was shown. These data indicate that actin might play a functional role during the early (initiation) as well as the later (elongation) stages of transcription by RNA polymerase I (Philimonenko et al., 2004).

C: RNA Polymerase III

Hu et al. (2004) have shown that β-actin tightly associates with purified RNA polymerase III. Moreover, in vivo experiments demonstrated that β-actin is present at the promoter region of the actively transcribing U6 snRNA gene, an association that is lost when transcription is inhibited. In addition it was shown that after inhibition of transcription β-actin partly dissociates from the RNA polymerase III complex, which leads to an inactive form of RNA polymerase III. Furthermore, in an in vitro assay, adding exogenous actin to this inactive RNA polymerase III activates transcription, demonstrating a crucial role for actin in transcription by RNA polymerase III (Hu et al., 2004).

Based on these studies on the three eukaryotic RNA polymerases, one can draw the surprising conclusion that actin is a transcription factor that is necessary for transcription by all three RNA polymerases. Moreover, the function of β-actin in transcription shows some striking similarities. In all three cases β-actin seems to be tightly associated with the polymerase core enzyme and this association seems to be crucial for the basal transcriptional activity of the respective polymerase. Interestingly, actin interacts with three RNA polymerase III subunits and two of these subunits are common to all three polymerases (Schramm and Hernandez, 2002). One or both of these subunits could present a common binding site for actin in transcription complexes. Furthermore, actin can be found at the promoter region of transcribing genes from all three RNA polymerases in vivo which also supports the idea of a general role for actin in transcription.

It has also been explicitly shown that actin is necessary for the start of transcription by RNA polymerase II because pre-initiation complexes cannot assemble at the promoter in the absence of actin. Experiments on RNA polymerase I and III, while not as explicit, clearly demonstrate that actin is also necessary either for transcription initiation or transcription elongation or both. Even though detailed studies are necessary to elucidate the concrete role of actin in transcription by RNA polymerase I and III, actin indeed appears to be a fundamental factor for cellular transcription in general. This suggestion is strengthened by studies on viral replication in mammalian cells. Transcription of the negative-strand RNA virus RSV (human respiratory syncytical virus) is carried out by the RNA dependent viral RNA polymerase L and transcription of the viral RNA is dependent on the presence of the cellular host protein actin (Burke et al., 1998; Huang et al., 1993).

Actin and Actin-related Proteins in Chromatin Remodeling

Actin-related proteins (Arps) constitute a group of proteins that share 24-60% homology with conventional actin and are considered an evolutionary conserved ancient class of eukaryotic proteins. Even though all Arps share sequence homology, members of the Arp family are quite diverse among themselves and are divided into 11 classes, termed Arp1-Arp11, in mammals. A common structural feature between Arps and conventional actin is the so-called actin fold that contains the ATP/ADP binding site (Fig.31.1). However, only Arp1 and Arp4 have been shown to bind and hydrolyze ATP. Moreover, several studies have demonstrated substantial differences in the surface structure of individual Arps, suggesting that Arps are functionally distinct (Frankel and Mooseker, 1996).

Of the various Arp classes, Arp1-Arp3 and Arp11 seem to localize exclusively in the cytoplasm while Arp4-Arp10 are also found in the nucleus. In the cytoplasm Arps are mainly involved in nucleation and branching of actin filaments and in microtubule based movement of vesicles. In the nucleus Arps and conventional actin can be found in complexes with a wide variety of chromatin remodeling and modifying complexes (Boyer and Peterson, 2000).

As outlined in detail elsewhere in this book, the structure of chromatin and histones is of extreme importance in regulating gene transcription. A certain group of multiprotein complexes are involved in transcription regulation by modifying histones or altering the chromatin structure. Basically these complexes can be divided into two groups, chromatin remodeling and chromatin modifying complexes. The chromatin remodeling complexes are ATP-dependent complexes that use the energy of ATP hydrolysis to remodel chromatin by locally disrupting or altering the association of histones with DNA. The chromatin modifying complexes, on the other hand, are the histone acetyltransferase (HAT) and histone deacetylase (HDAC) complexes. They regulate the transcriptional activity of genes by altering the level of acetylation of nucleosomal histones associated with the genes.

Actin and Arps were first identified as integral components of the mammalian SWI/SNF chromatin remodeling complex by Zhao *et al.* (1998). To date actin and Arps are found in a wide variety of chromatin remodeling and modifying complexes. Table 31.1 lists complexes from yeast to mammals in which actin and actin-related proteins have been identified. Most of the information we have about the function of Arps in chromatin remodeling have come from studies in yeast because it is relatively easy to manipulate the yeast genome and to produce null mutants or point mutations of proteins. The importance of Arps was demonstrated by studies with Arp null mutants in yeast. Arp null mutants, like null mutants of other components of the chromatin remodeling complex, show a reduced viability and defects in chromatin structure and replication (Cairns *et al.*, 1998; Shen *et al.*, 2003). This indicates that Arps are essential components of specific chromatin remodeling complexes and are necessary for proper function.

Even though the exact role of actin and Arps in these complexes remains to be established, there are several suggested functions for Arps in chromatin remodeling (Fig.31.3). Interestingly, Arps appear to have a structural rather than enzymatic role because point mutations in the ATP-binding pocket had no effect on chromatin remodeling (Cairns *et al.*, 1998; Shen *et al.*, 2003). For instance, Arps could play a role in the assembly of chromatin altering complexes (Fig.31.3A). This is supported by studies on the yeast INO80 complex that showed that Arp4 and Arp5, as well as Arp8, are necessary for proper complex formation (Galarneau *et al.*, 2000; Jonsson *et al.*, 2004; Shen *et al.*, 2003). Another possibility is that Arps could anchor the chromatin altering complexes either to DNA or the nuclear matrix (Fig.31.3B). *In vitro* studies have demonstrated that Arp4 and Arp8 are able to directly bind to core histones, which means that Arp4 and Arp8 could have a role in targeting and/or binding remodeling complexes to certain areas of the chromatin (Galarneau *et al.*, 2000; Harata *et al.*, 1999; Harata *et al.*, 2002; Shen *et al.*, 2003). Furthermore, it was shown that actin and Arps in the mammalian BAF complex are important for nuclear matrix association (Zhao *et al.*, 1998).

Table 31.1

Name	Organism	Actin subfamily	Comments and References
Chromatin remodeler			
SWR1	Yeast	actin, Arp4, Arp6	(Krogan *et al.*, 2003; Mizuguchi *et al.*, 2004)
INO80	Yeast	actin, Arp4, Arp5, Arp8	Arp5 and Arp8 deletion have effect on the complex integrity and show defects in the complex activity, DNA binding and nuclesome mobilization
			Arp8 deletion leads to absence of actin and Arp4 in the complex (Jonsson *et al.*, 2004; Shen *et al.*, 2000; Shen *et al.*, 2003)
SWI/SNF	Yeast	Arp7, Arp9	Deletion of Arp7 and Arp9 leads to defect in structural integrity of RSC
RSC	Yeast	Arp7, Arp9	and SWI/SNF complex (Cairns *et al.*, 1998; Peterson *et al.*, 1998)
Brahma	Drosophila	actin, Arp4	(Papoulas *et al.*, 1998)
BAF	Mammals	actin, Arp4	Actin and Arp4 play a role in signal mediated binding of the BAF and PBAF complex to chromatin and/or matrix (Zhao *et al.*, 1998)
PBAF	Mammals	actin, Arp4	
p400	Mammals	actin, Arp4	(Fuchs *et al.*, 2001)
Chromatin modifier			
NuA4	Yeast	actin, Arp4	Arp4 is required for complex integrity (Galarneau *et al.*, 2000)
Tip60	Mammals	actin, BAF53	(Cai *et al.*, 2003; Ikura *et al.*, 2000)

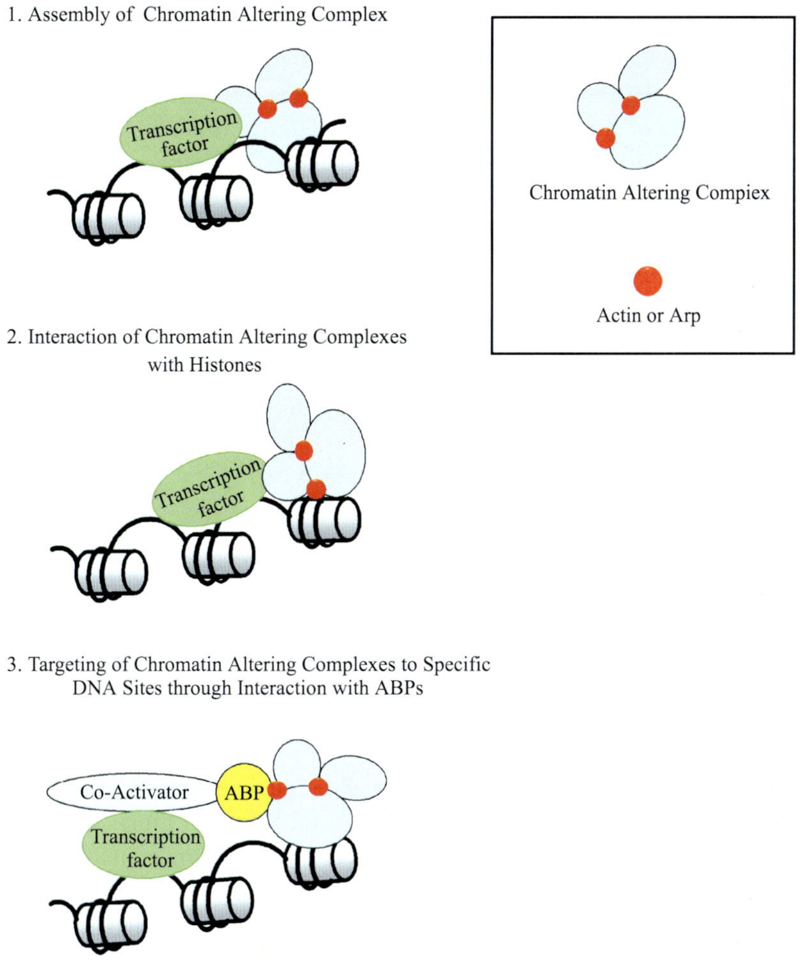

Fig.31.3 Possible functions of actin and actin-related proteins (Arps) in chromatin remodeling. A) Assembly of the chromatin altering complex. Studies in yeast have shown that null and point mutations in certain Arps (namely Arp4, Arp5, and Arp8) can lead to improper or incomplete assembly of chromatin altering complexes. These data suggest that Arps are structural components of chromatin altering complexes that facilitate binding between certain other components of the complex and act to stabilize the complex. B) Interaction of chromatin altering complexes with histones. The main feature of chromatin altering complexes is that they change the structure of the chromatin. *In vitro* as well as *in vivo* experiments have shown that Arps can bind directly to or are associated with core histones or heterochromatin regulating factors. This could indicate that Arps possess histone chaperone functions and/or could recruit or anchor the chromatin altering complexes to the chromatin. C) Targeting of chromatin altering complexes to specific DNA sites through interaction with ABPs. This model clearly depicts the complexity of transcriptional regulation. As described in the text, transcription factors, like hormone receptors, function as regulators by recruiting chromatin altering complexes to certain regions of the DNA. The exact mechanism on how they do this is not known. However, it is known that hormone receptors bind to actin-binding proteins (ABP). As the name implies, actin-binding proteins bind to actin and some also bind to actin-related proteins. Therefore the ABPs could bind to Arps or actin present in chromatin altering complexes. These interactions, in turn, could recruit chromatin altering complexes to regions of the DNA.

Actin-Binding Proteins in Transcription Regulation

Numerous actin-binding proteins have been identified in the nucleus. Some of them are involved in basic transcription and transcriptional regulation. The most important of these are the members of the myosin and gelsolin families and they are discussed below.

A: Nuclear Myosin I

Myosin was first identified in the early part of the last century and, in a classic paper, the husband and wife team of Engelhardt and Ljubimova reported in 1939 that myosin was an ATP hydrolase (Engelhardt and Lyubimova, 1939). It is now known that they purified myosin II and that myosin II is one member of a superfamily of actin based motors. Myosin II is still the best known member of this superfamily and myosin II, itself, represents a sub-family of the myosin superfamily with separate myosin II isoforms found in cardiac, skeletal, smooth, and non-muscle cells. Myosin II is ubiquitously distributed and it is not found only in red blood cells. All forms of myosin II are hexameric proteins that contain two heavy chains and 4 light chains (Fig.31.4). The N-terminals of the heavy chains fold to form globular head regions that contain actin-binding domains and sites of ATP hydrolysis. Light chains, which are associated with regulation, are associated with the head. Much of the heavy chain forms a coiled-coil helix that, through intermolecular interactions, results in the formation of filaments. The ability of myosin II to form filaments is crucial to our understanding of myosin function in cells. Muscles contract because thick (i.e.: myosin II) filaments pull on thin (i.e.: mainly actin) filaments that are eventually attached to the cell membrane. The movement of actin filaments past myosin filaments, called the Sliding-Filament Model, is fundamental to how we visualize the way actin and myosin work inside cells. Moreover, myosin II is known as a "conventional myosin" because it forms filament.

For the better part of 50 years it was thought that there was only one type of myosin (ie: myosin II) and that this was a conventional myosin with two heads and a long coiled-coiled tail that associated into filaments. However, in 1973, Pollard and Korn changed muscle biology by publishing a seminal paper in which they demonstrated the presence of a single headed myosin that had a short tail that did not form filaments. They called this protein myosin I because it has only 1 head (Pollard and Korn, 1973). Because this myosin looked remarkably like a myosin II head, it was initially thought that myosin I was a proteolytic fragment of myosin II that had been hydrolyzed between the head and the tail. However, the cloning of the myosin I gene (Jung *et al.*, 1989; Lee *et al.*, 1999) established without doubt that myosin I is a separate gene product and not a proteolytic fragment. It has become very clear since 1975 that myosin II is but one member of a large superfamily of proteins that are defined as actin-dependent or activated ATPases. It has also become clear that most of the members of the myosin superfamily are "unconventional" in the sense that they do not form filaments. Sequence comparison has shown that the myosin heads are highly conserved among all myosins but that the tail regions are quite diverse and appear to be involved in targeting myosins, and thereby, specifying their functions.

There have been a number of suggestions of myosin or myosin like proteins in the nucleus (Berrios and Fisher, 1986; Berrios *et al.*, 1991; Hagen *et al.*, 1986; Hauser *et al.*, 1975; Milankov and De Boni, 1993). However, the genes for these proteins were not cloned and the proteins themselves were not purified. Consequently, the functional significance of these proteins in the nucleus remained unclear. The first convincing demonstration of a myosin in the nucleus was the discovery of an unique isoform of myosin I, called nuclear myosin I to distinguish it from the myosin I isoform found in the cytoplasm (cytoplasmic myosin I), in the nucleus (Nowak *et al.*, 1997). Nuclear myosin I is a member of the myosin I sub-family of the myosin superfamily. Nuclear myosin I, like other myosin I molecules, consists of a single heavy chain and a globular head that is very similar to the myosin II head because it too contains sites that bind actin and hydrolyze ATP. However, myosin I proteins have a very short tail and they are unable to self-associate into filaments (Fig.31.3).

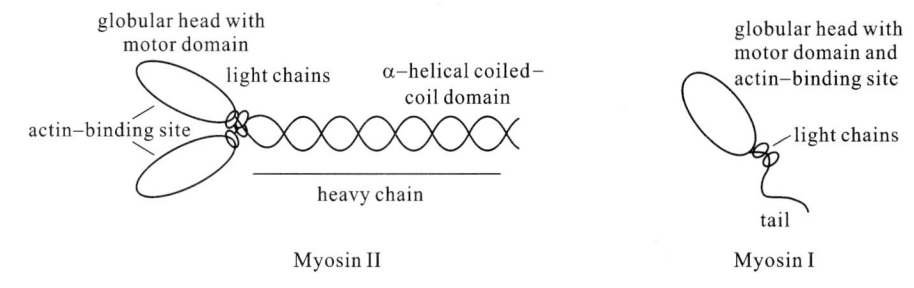

Fig.31.4 Schematic depicting Myosin II and Myosin I structure. Myosin II (left) is a hexamer and consists of two heavy chains and four light chains. The two heavy chains are organized into a globular head at the N-terminus. The head domain contains the actin-binding site and the ATP-binding site and is therefore also considered the motordomain. The C-terminal region of each heavy chain, called the tail region, consists of a α-helix. The tail regions of the two heavy chains coil around each other to form a coiled-coil domain. Myosin II molecules self-assemble into filaments under the appropriate conditions by intermolecular association of tail domains. The four light chains are each wrapped around the neck region of each heavy chain. Myosin I (right) has only one heavy chain with a short tail As with myosin II, the globular head domain is at the N-terminus and contains the actin-binding site and the ATP-binding site. The light chains, made of calmodulin molecules, are also wrapped around the neck region.

Nuclear myosin I contains a unique N-terminal extension that is not found in any other myosin. This extension was discovered by analyzing the myosin I gene and by microsequencing. Analysis of the gene structure revealed the presence of an upstream exon (exon −1) that contains another ATG that is associated with a weak Kozak sequence. The 3′ end of exon −1 and the 5′ end of exon 1 combine to code an additional 48 base pairs that are attached to the 5′ region of the mRNA for cytoplasmic myosin I (Pestic-Dragovich et al., 2000). Microsequencing of nuclear myosin I immunoprecipitated from nuclei confirmed the presence of a unique 16-amino acid N-terminus. This extension does not have sequence homology with any known nuclear localization signal and it does not direct cytoplasmic proteins to the nucleus. Nevertheless, it is required for the nuclear entry or nuclear retention of nuclear myosin I because removing the N-terminal extension results in complete retention of nuclear myosin I in the cytoplasm (Pestic-Dragovich et al., 2000).

Immunofluorescence and immunoelectron microscopy showed that nuclear myosin I co-localizes with RNA polymerase I and II. The localization with RNA polymerase II appears to depend on active transcription because specifically inhibiting transcription by RNA polymerase II results in a loss of the co-localization of nuclear myosin I with RNA polymerase II. The functional association of nuclear myosin I with RNA polymerase II was further supported by the demonstration that nuclear myosin I and RNA polymerase II co-immunoprecipitate. These data suggested that nuclear myosin I might be involved in transcription by RNA polymerase II. An important functional role for nuclear myosin I in transcription was firmly established by the demonstration that antibodies to nuclear myosin I inhibit transcription by RNA polymerase II in an *in vitro* transcription assay using HeLa cell nuclear extract (Pestic-Dragovich et al., 2000).

Nuclear myosin I was also identified in nucleolar structures that transcribe ribosomal genes (Nowak et al., 1997). Subsequently, it was demonstrated that nuclear myosin I associates with RNA polymerase I and transcribing ribosomal genes (Fomproix and Percipalle, 2004; Philimonenko *et al*., 2004). Specifically, Philimonenko *et al*. showed that nuclear myosin I associates with initiation-competent RNA polymerase I complexes through an interaction with the basal transcription factor TIF-IA. Previous studies have shown that only a fraction of the RNA polymerase I is capable of assembling into transcription initiation complexes and supporting rDNA transcription (Miller *et al*., 2001; Schnapp *et al*., 1990; Tower and Sollner-Webb, 1987) and this initiation-competent RNA polymerase I is associated with TIF-IA (Grummt, 2003). TIF-IA interacts with RNA polymerase I and this interaction is required to recruit RNA polymerase I to the rDNA promoter (Miller *et al*., 2001; Yuan *et al*., 2002). Thus the interaction of nuclear myosin I with RNA polymerase I through TIF-IA suggests that nuclear myosin I might be involved during transcription initiation. This is also supported by *in vitro* transcription experiments in which antibodies to nuclear myosin I inhibited the synthesis of the first nucleotides during rRNA transcription.

Because all myosins are actin-activated ATP hydrolases and acto-myosin complexes function as molecular motors to power muscle contraction, cell motility and cell division, it is logical to predict that actin and nuclear myosin I also act together in the nucleus. Transcription, like many other nuclear processes, involves mechanical work. That is, large transcriptional complexes consisting of many different proteins have to move relative to equally large DNA molecules. This form of "motility" is analogous to muscle contraction. In the heart and other muscles, the energy released when actin and myosin hydrolyze ATP is used to move actin filaments relative to myosin filaments. Similarly, the movement of polymerase complexes relative to DNA suggests a role for molecular motors in transcription. However, the polymerase has been shown to have a powerstroke and to generate a great deal of force (Wang *et al*., 1998; Yin *et al*., 1995). Whether the motor function of actin and nuclear myosin I are also required for transcription, when and how they are involved and how they interact with polymerases to power transcription remain major questions that need to be answered.

Even though these initial studies on transcription show that both actin and nuclear myosin I are necessary for the basal transcriptional activity of RNA polymerase I, they also demonstrate that, at least at the initiation stage of transcription, actin and nuclear myosin I have subtly different roles. While actin associates with RNA polymerase I regardless of the transcriptional state, nuclear myosin I only associates with initiation-ompetent RNA polymerase I complexes through an interaction with the basal transcription factor TIF-IA (Philimonenko *et al*., 2004). This is potentially important because actin and myosin work together and the possibility that actin and nuclear myosin I are separately involved in individual steps of transcription is very novel. In addition, whether actin and nuclear myosin I also play roles in the later stages of transcription, such as transcription elongation, remains to be investigated.

B: Gelsolin Family

The gelsolins consist of a class of proteins that are found from lower eukaryotes to mammals. Members of the gelsolin family share 3-6 repeats of a conserved domain and are responsible for capping and/or severing actin filaments in the cytoplasm (Burtnick et al., 2001; Way and Weeds, 1988). Several members of the gelsolin family, namely gelsolin, CapG, supervillin, and FliI have recently been identified in the nucleus where they have roles as transcriptional co-regulators.

Gelsolin is the best studied member of the family and it was first identified in macrophages (Yin and Stossel, 1979; Yin and Stossel, 1980). Gelsolin contains three actin binding sites, two of which can bind actin monomers while the third one binds to F-actin (Bryan, 1988; Pope et al., 1995), and comprises multiple actin binding and regulatory features. Gelsolin binds two actin monomers in the presence of calcium and gelsolin can stimulate actin filament nucleation (Yin et al., 1981). On the other hand, gelsolin also severs F-actin and caps filaments (Way et al., 1992). How gelsolin translocates to and from the nucleus is not known but it may do so in a complex with the androgen receptor. Androgen receptors translocate to the nucleus following androgen stimulation. Studies have shown that gelsolin co-localizes with the androgen receptor in the cytoplasm following androgen stimulation and it is possible that gelsolin enters the nucleus in complex with the androgen receptor (Nishimura et al., 2003).

In contrast to gelsolin, CapG binds and caps actin filaments but does not sever them (Yu et al., 1990). CapG shows the closest similarities to *Dictyostelium discoideum* severin and *Physarum polycephalum* fragminP. CapG, severin and fragminP have been shown to localize to the nucleus (Onoda et al., 1993). However, while severin and fragminP contain a classical leucin-rich nuclear export signal that leads to nuclear localizations only under certain conditions, CapG does not contain a classical nuclear export signal and is found predominantly in the nucleus (Van Impe et al., 2003). How CapG enters the nucleus is not known because it does not contain a known nuclear localization signal. It has been suggested, however, that phosphorylation of the protein might play a role in its nuclear import (Onoda et al., 1993).

Flightless I (FliI) is a highly conserved member of the gelsolin family but its function in the cytoplasm is not well understood. The name flightless I comes from studies in *Drosophila melanogaster* showing that certain point mutations in the FliI gene lead to defects in the flight muscle and to the inability to fly (de Couet et al., 1995). Further studies in mouse and *Drosophila* have shown that it is essential for early development (Campbell et al., 2002; Straub et al., 1996). FliI can be found in the nucleus as well as in cytoplasm (Davy et al., 2000) but the mechanisms of translocation are not known.

Although supervillin was first identified in neutrophils, it is expressed in large amounts in muscle (Pestonjamasp et al., 1997). Supervillin binds directly to myosin II (Chen et al., 2003) and to F-actin and it is found at sites of cell-cell adhesions, suggesting that it might be involved in anchoring the cytoskeleton to the plasma membrane (Pope et al., 1998). In contrast to other members of the gelsolin family, supervillin is the only protein in which a functional nuclear localization signal has been identified so far (Wulfkuhle et al., 1999).

A series of studies has now independently identified CapG, gelsolin, supervillin, and FliI as transcriptional co-activators or repressors for several nuclear hormone receptors. GapG was recently shown to modulate transcriptional activity in an reporter assay (De Corte et al., 2004), supervillin (Ting et al., 2002), and gelsolin, itself (Nishimura et al., 2003) were shown to interact with the androgen receptor and enhance the transcriptional activity of the androgen receptor as well as other nuclear hormone receptors. Another study showed that FliI can associate directly with the estrogen receptor and the thyroid hormone receptor as well as with other co-activators of nuclear hormone receptors thereby enhancing their transactivation (Lee et al., 2004). Activated nuclear hormone receptors translocate to the nucleus and bind to response elements on specific genes. Regulation of gene expression then occurs through the recruitment of chromatin remodeling factors.

How exactly actin-binding proteins enhance or repress the transcriptional activity of the different hormone receptors is not known. The major function of gelsolin family members in the cytoplasm is to regulate actin dynamics by severing and capping actin filament but the gelsolins are also able to bind actin monomers or dimers. Interestingly, besides binding to nuclear receptors or co-activators of nuclear receptors, FliI was also shown to bind to Arp4 (BAF53), the actin-related protein present in several chromatin remodeling complexes (Lee et al., 2004). Therefore, one can speculate that actin-binding proteins regulate transcription by providing a link and connecting the nuclear receptor transcription complex to actin or actin-related proteins in chromatin altering complexes (Fig.35.3C). Depending on the chromatin altering complex that is recruited, the DNA around the nuclear receptor binding site would then be either altered towards a more accessible form which would enhance transcription, or towards a more closed form which would repress transcription.

Concluding Remarks

Recent publications have presented incontrovertible evidence for the presence of actin, Arps and ABPs, especially myosin I, in the nucleus. They have also demonstrated that those proteins play important roles in the mechanics of transcription by RNA polymerase I, II, and III as well as in transcriptional regulation. These papers have changed the landscape and the pertinent question currently is how, exactly, are they involved in transcription?

One issue that has not been addressed is the form of nuclear actin that is required to support transcription. As discussed above, actin can exist as a monomer (G-actin) or it can associate to form a filament (F-actin). Virtually everything we know about actin suggests that polymerization in one form or another is essential for actin to have biological functions (Pollard *et al*., 2000; Pollard *et al*., 2001). However, there is very little evidence for filamentous actin in the nucleus. Gonsoir *et al*. (1999) have used the monoclonal 2G2 antibody to actin to suggest the presence of actin aggregates in the nucleus. Based on the X-ray diffraction studies, the 2G2 recognizes an epitope that is buried in the F-actin structure. Thus, this antibody is unlikely to recognize classical actin filaments. Most importantly, the sine-qua-non for the presence of actin filaments is phalloidin staining, which specifically binds to F-actin with a minimum filament size of 7 monomers (Visegrady *et al*., 2005). But there is no convincing evidence for phalloidin staining in the nucleus and this has been part of the basis for questioning a role for nuclear actin.

If polymeric actin is indeed present in the nucleus, it seems unlikely that its conformation is similar to classical F-actin partly based on the reasons stated above. In addition, β-actin contains 2 nuclear export sequences and the nucleus contains cofilin (Ohta *et al*., 1989) and profilin (Skare *et al*., 2003), proteins that regulate actin polymerization (Pollard *et al*., 2000), and the nuclear import (Pendleton *et al*., 2003) and export (Stuven *et al*., 2003) of actin. Moreover, given steric considerations within the nuclear complexes containing actin (i.e. the chromatin altering complexes or the pre-initiation complexes), it is difficult to visualize a role for long actin filaments in them. When one considers all the available data, there appear to be important conformational differences between nuclear and cytoplasmic actin and the implication is that either monomeric actin (G-actin) or short actin filaments, too short to be stained by phalloidin, are involved in transcription. Whether this is correct remains to be determined experimentally.

One of the most intriguing issue regarding nuclear actin is whether actin, along with its binding partner nuclear myosin I, act as a molecular motor to power transcription and if so, how do they act at the molecular level? All motors have to be anchored to generate force. In the cytoplasm actin and myosin II are polymerized into filaments that are eventually attached to the cell membrane. When cells contract, actin filaments slide past myosin filaments and cells shorten and muscles contract (Kad *et al*., 2003). In fact, it is important to emphasize that the sliding of actin filaments past myosin filaments is central to our understanding of how actin and myosin act as a molecular motor. Here, again, the question whether G- or F-actin is involved in transcription is an important issue. The situation with nuclear myosin I is even more complicated because "unconventional" myosins do not form filaments (Mermall *et al*., 1998).

How, then, are actin and nuclear myosin I involved in transcription? Or, more specifically, what do they bind to in order to generate force? Transcription complexes are composed of a large number of proteins (Lemon and Tjian, 2000) and identifying protein-protein interactions is crucial if we are to understand the details of transcription. Actin binds to all 3 polymerases and nuclear myosin I binds to TIF-IA (Philimonenko *et al*., 2004). The binding of nuclear myosin I to TIF-IA is especially important because it appears to be necessary for the activation of RNA polymerase I. Does myosin have a similar function with respect to RNA polymerase I and III? In addition, Hofmann *et al*. (2004) have shown that actin is necessary for the formation of pre-initiation complexes. What actin binds to, in addition to RNA polymerase II, to stabilize or promote the formation of pre-initiation complexes has not been defined. It is technically feasible to build pre-initiation complexes *in vitro* using an immobilized DNA template, purified transcription factors and purified RNA polymerase II. Therefore, it should be possible, by using antibodies to actin and specific transcription factors and mutant forms of them, to tease out what actin binds to as pre-initiation complexes are formed.

The organization of actin and nuclear myosin I in

elongating complexes also remains to be established. We know that β-actin binds to the large subunit of RNA polymerase II (Hofmann et al., 2004). This appears to be a general characteristic because actin also binds to RNA polymerase I (Philimonenko et al., 2004) and III (Hu et al., 2004). Nuclear myosin I, on the other hand, could bind to DNA. Myosin IC is thought to move cargoes and extend the leading edge of cells by simultaneously binding to negatively charged lipids via a positively-charged domain in the tail and actin filaments via the actin-binding domain on the head (Mermall et al., 1998). The DNA backbone is highly negatively charged and nuclear myosin I could bind to DNA via the same positively charged domain in the tail. It could then generate force during transcription by binding to β-actin, which itself is bound to the polymerase. One issue with such a model is whether the affinity of nuclear myosin I for the DNA is high enough to accommodate force generation. That is, the binding of nuclear myosin I to DNA has to be stronger than the force nuclear myosin I and actin generate during transcription. One possibility is that specific DNA binding proteins stabilize this interaction, another reason to define what proteins actin and nuclear myosin I bind to in the nucleus.

While the possible roles of actin and nuclear myosin I in the mechanics of transcription are intriguing, roles for actin and nuclear myosin I in other aspects of transcription should not be ignored. The demonstration of actin and Arps in chromatin altering complexes and the involvement of several actin-binding proteins in regulating transcription indicates that the forms and functions of actin and actin-related proteins in the nucleus might be as complex as in the cytoplasm. The abundance of actin-related proteins and actin-binding proteins in the nucleus also suggest that they are involved in complex and specific ways in regulating transcription. One important point to remember is that these proteins, along with actin and nuclear myosin I, are involved in increasing the efficiency of transcription and/or regulating gene activation.

In conclusion, we have summarized what is known about actin, actin-related proteins and actin-binding proteins in the nucleus. Important papers that have been published in the last few years have clearly established roles for actin, actin-related proteins and actin-binding proteins in chromatin remodeling, RNA transport and transcription. These papers have opened the door to a new and exciting area of study, namely the role of what have been considered cytoplasmic structural proteins in transcription.

Acknowledgement

Supported, in part, by grants from the National Institutes of Health (GM 596489) and the National Science Foundation (INT 0079298) to PdeL.

References

Ankenbauer, T., Kleinschmidt, J. A., Walsh, M. J., Weiner, O. H., and Franke, W. W. (1989). Identification of a widespread nuclear actin binding protein. Nature 342, 822-825.

Berrios, M., and Fisher, P. A. (1986). A myosin heavy-chain-like polypeptide is associated with the nuclear envelope in higher eukaryotic cells. J Cell Biol 103, 711-724.

Berrios, M., Fisher, P. A., and Matz, E. C. (1991). Localization of a myosin heavy chain-like polypeptide to *Drosophila* nuclear pore complexes. Proc Natl Acad Sci USA 88, 219-223.

Bettinger, B. T., Gilbert, D. M., and Amberg, D. C. (2004). Actin up in the nucleus. Nat Rev Mol Cell Biol 5, 410-415.

Boyer, L. A., and Peterson, C. L. (2000). Actin-related proteins (Arps): conformational switches for chromatin-remodeling machines? Bioessays 22, 666-672.

Bryan, J. (1988). Gelsolin has three actin-binding sites. J Cell Biol 106, 1553-1562.

Burke, E., Dupuy, L., Wall, C., and Barik, S. (1998). Role of cellular actin in the gene expression and morphogenesis of human respiratory syncytial virus. Virology 252, 137-148.

Burtnick, L. D., Robinson, R. C., and Choe, S. (2001). Structure and function of gelsolin. Results Probl Cell Differ 32, 201-211.

Cai, Y., Jin, J., Tomomori-Sato, C., Sato, S., Sorokina, I., Parmely, T. J., Conaway, R. C., and Conaway, J. W. (2003). Identification of new subunits of the multiprotein mammalian TRRAP/TIP60-containing histone acetyltransferase complex. J Biol Chem 278, 42733-42736.

Cairns, B. R., Erdjument-Bromage, H., Tempst, P., Winston, F., and Kornberg, R. D. (1998). Two actin-related proteins are shared functional components of the chromatin-remodeling complexes RSC and SWI/SNF. Mol Cell 2, 639-651.

Campbell, H. D., Fountain, S., McLennan, I. S., Berven, L. A., Crouch, M. F., Davy, D. A., Hooper, J. A., Waterford, K., Chen, K. S., Lupski, J. R., et al. (2002). Fliih, a gelsolin-related cytoskeletal regulator essential for early mammalian embryonic development. Mol Cell Biol 22, 3518-3526.

Chen, Y., Takizawa, N., Crowley, J. L., Oh, S. W., Gatto, C. L., Kambara, T., Sato, O., Li, X. D., Ikebe, M., and Luna, E. J. (2003). F-actin and myosin II binding domains in supervillin. J Biol Chem 278, 46094-46106.

Clark, T. G., and Merriam, R. W. (1977). Diffusible and bound actin nuclei of *Xenopus* laevis oocytes. Cell 12, 883-891.

Clark, T. G., and Rosenbaum, J. L. (1979). An actin filament matrix in hand-isolated nuclei of X. laevis oocytes. Cell 18, 1101-1108.

Davy, D. A., Ball, E. E., Matthaei, K. I., Campbell, H. D., and Crouch, M. F. (2000). The flightless I protein localizes to actin-based structures during embryonic development. Immunol Cell Biol 78, 423-429.

De Boni, U. (1994). The interphase nucleus as a dynamic structure. Int Rev Cytol 150, 149-171.

De Corte, V., Van Impe, K., Bruyneel, E., Boucherie, C., Mareel, M., Vandekerckhove, J., and Gettemans, J. (2004). Increased importin-beta-dependent nuclear import of the actin modulating protein CapG promotes cell invasion. J Cell Sci 117, 5283-5292.

de Couet, H. G., Fong, K. S., Weeds, A. G., McLaughlin, P. J., and Miklos, G. L. (1995). Molecular and mutational analysis of a gelsolin-family member encoded by the flightless I gene of *Drosophila* melanogaster. Genetics 141, 1049-1059.

dos Remedios, C. G., Chhabra, D., Kekic, M., Dedova, I. V., Tsubakihara, M., Berry, D. A., and Nosworthy, N. J. (2003). Actin binding proteins: regulation of cytoskeletal microfilaments. Physiol Rev 83, 433-473.

Egly, J. M., Miyamoto, N. G., Moncollin, V., and Chambon, P. (1984). Is actin a transcription initiation factor for RNA polymerase B? Embo J 3, 2363-2371.

Engelhardt, V. A., and Lyubimova, M. N. (1939). Myosin and adenosinetriphosphatase. Nature 144, 668.

Fomproix, N., and Percipalle, P. (2004). An actin-myosin complex on actively transcribing genes. Exp Cell Res 294, 140-148.

Frankel, S., and Mooseker, M. S. (1996). The actin-related proteins. Curr Opin Cell Biol 8, 30-37.

Fuchs, M., Gerber, J., Drapkin, R., Sif, S., Ikura, T., Ogryzko, V., Lane, W. S., Nakatani, Y., and Livingston, D. M. (2001). The p400 complex is an essential E1A transformation target. Cell 106, 297-307.

Funaki, K., Katsumoto, T., and Iino, A. (1995). Immunocytochemical localization of actin in the nucleolus of rat oocytes. Biol Cell 84, 139-146.

Galarneau, L., Nourani, A., Boudreault, A. A., Zhang, Y., Heliot, L., Allard, S., Savard, J., Lane, W. S., Stillman, D. J., and Cote, J. (2000). Multiple links between the NuA4 histone acetyltransferase complex and epigenetic control of transcription. Mol Cell 5, 927-937.

Gonsior, S. M., Platz, S., Buchmeier, S., Scheer, U., Jockusch, B. M., and Hinssen, H. (1999). Conformational difference between nuclear and cytoplasmic actin as detected by a monoclonal antibody. J Cell Sci 112 (Pt 6), 797-809.

Grummt, I. (2003). Life on a planet of its own: regulation of RNA polymerase I transcription in the nucleolus. Genes Dev 17, 1691-1702.

Hagen, S. J., Kiehart, D. P., Kaiser, D. A., and Pollard, T. D. (1986). Characterization of monoclonal antibodies to Acanthamoeba myosin-I that cross-react with both myosin-II and low molecular mass nuclear proteins. J Cell Biol 103, 2121-2128.

Harata, M., Oma, Y., Mizuno, S., Jiang, Y. W., Stillman, D. J., and Wintersberger, U. (1999). The nuclear actin-related protein of *Saccharomyces cerevisiae*, Act3p/Arp4, interacts with core histones. Mol Biol Cell 10, 2595-2605.

Harata, M., Zhang, Y., Stillman, D. J., Matsui, D., Oma, Y., Nishimori, K., and Mochizuki, R. (2002). Correlation between chromatin association and transcriptional regulation for the Act3p/Arp4 nuclear actin-related protein of *Saccharomyces cerevisiae*. Nucleic Acids Res 30, 1743-1750.

Hauser, M., Beinbrech, G., Groschel-Stewart, U., and Jockusch, B. M. (1975). Localisation by immunological techniques of myosin in nuclei of lower eurkaryotes. Exp Cell Res 95, 127-135.

Hofmann, W. A., Stojiljkovic, L., Fuchsova, B., Vargas, G. M., Mavrommatis, E., Philimonenko, V., Kysela, K., Goodrich, J. A., Lessard, J. L., Hope, T. J., et al. (2004). Actin is part of pre-initiation complexes and is necessary for transcription by RNA polymerase II. Nat Cell Biol 6, 1094-1101.

Hu, P., Wu, S., and Hernandez, N. (2004). A role for beta-actin in RNA polymerase III transcription. Genes Dev 18, 3010-3015.

Huang, Y. Q., Li, J. J., Moscatelli, D., Basilico, C., Nicolaides, A., Zhang, W. G., Poiesz, B. J., and Friedman-Kien, A. E. (1993). Expression of int-2 oncogene in Kaposi's sarcoma lesions. J Clin Invest 91, 1191-1197.

Ikura, T., Ogryzko, V. V., Grigoriev, M., Groisman, R., Wang, J., Horikoshi, M., Scully, R., Qin, J., and Nakatani, Y. (2000). Involvement of the TIP60 histone acetylase complex in DNA repair and apoptosis. Cell 102, 463-473.

Jockusch, B. M., Becker, M., Hindennach, I., and Jockusch, E. (1974). Slime mould actin: homology to vertebrate actin and presence in the nucleus. Exp Cell Res 89, 241-246.

Jones, L. J., Carballido-Lopez, R., and Errington, J. (2001). Control of cell shape in bacteria: helical, actin-like filaments in Bacillus subtilis. Cell 104, 913-922.

Jonsson, Z. O., Jha, S., Wohlschlegel, J. A., and Dutta, A. (2004). Rvb1p/Rvb2p recruit Arp5p and assemble a functional Ino80 chromatin remodeling complex. Mol Cell 16, 465-477.

Jung, G., Schmidt, C. J., and Hammer, J. A., 3rd. (1989). Myosin I heavy-chain genes of Acanthamoeba castellanii: cloning of a second gene and evidence for the existence of a third isoform. Gene 82, 269-280.

Kad, N. M., Rovner, A. S., Fagnant, P. M., Joel, P. B., Kennedy, G. G., Patlak, J. B., Warshaw, D. M., and Trybus, K. M. (2003). A mutant heterodimeric myosin with one inactive head generates maximal displacement. J Cell Biol 162, 481-488.

Kao, C. C., Lieberman, P. M., Schmidt, M. C., Zhou, Q., Pei, R., and Berk, A. J. (1990). Cloning of a transcriptionally active human TATA binding factor. Science 248, 1646-1650.

Krogan, N. J., Keogh, M. C., Datta, N., Sawa, C., Ryan, O. W., Ding, H., Haw, R. A., Pootoolal, J., Tong, A., Canadien, V., et al. (2003). A Snf2 family ATPase complex required for recruitment of the histone H2A variant Htz1. Mol Cell 12, 1565-1576.

Kugel, J. F., and Goodrich, J. A. (2000). A kinetic model for the early steps of RNA synthesis by human RNA polymerase II. J

Biol Chem 275, 40483-40491.

Kukalev, A., Nord, Y., Palmberg, C., Bergman, T., and Percipalle, P. (2005). Actin and hnRNP U cooperate for productive transcription by RNA polymerase II. Nat Struct Mol Biol 12, 238-244.

Lane, N. J. (1969). Intranuclear fibrillar bodies in actinomycin D-treated oocytes. J Cell Biol 40, 286-291.

Lee, W. L., Ostap, E. M., Zot, H. G., and Pollard, T. D. (1999). Organization and ligand binding properties of the tail of Acanthamoeba myosin-IA. Identification of an actin-binding site in the basic (tail homology-1) domain. J Biol Chem 274, 35159-35171.

Lee, Y. H., Campbell, H. D., and Stallcup, M. R. (2004). Developmentally essential protein flightless I is a nuclear receptor coactivator with actin binding activity. Mol Cell Biol 24, 2103-2117.

Lemon, B., and Tjian, R. (2000). Orchestrated response: a symphony of transcription factors for gene control. Genes Dev 14, 2551-2569.

McHugh, K. M., Crawford, K., and Lessard, J. L. (1991). A comprehensive analysis of the developmental and tissue-specific expression of the isoactin multigene family in the rat. Dev Biol 148, 442-458.

Mermall, V., Post, P. L., and Mooseker, M. S. (1998). Unconventional myosins in cell movement, membrane traffic, and signal transduction. Science 279, 527-533.

Milankov, K., and De Boni, U. (1993). Cytochemical localization of actin and myosin aggregates in interphase nuclei in situ. Exp Cell Res 209, 189-199.

Miller, C. A., 3rd, Cohen, M. D., and Costa, M. (1991). Complexing of actin and other nuclear proteins to DNA by cis-diamminedichloroplatinum(II) and chromium compounds. Carcinogenesis 12, 269-276.

Miller, G., Panov, K. I., Friedrich, J. K., Trinkle-Mulcahy, L., Lamond, A. I., and Zomerdijk, J. C. (2001). hRRN3 is essential in the SL1-mediated recruitment of RNA Polymerase I to rRNA gene promoters. Embo J 20, 1373-1382.

Mizuguchi, G., Shen, X., Landry, J., Wu, W. H., Sen, S., and Wu, C. (2004). ATP-driven exchange of histone H2AZ variant catalyzed by SWR1 chromatin remodeling complex. Science 303, 343-348.

Nakayasu, H., and Ueda, K. (1983). Association of actin with the nuclear matrix from bovine lymphocytes. Exp Cell Res 143, 55-62.

Nakayasu, H., and Ueda, K. (1985). Association of rapidly-abelled RNAs with actin in nuclear matrix from mouse L5178Y cells. Exp Cell Res 160, 319-330.

Nishimura, K., Ting, H. J., Harada, Y., Tokizane, T., Nonomura, N., Kang, H. Y., Chang, H. C., Yeh, S., Miyamoto, H., Shin, M., et al. (2003). Modulation of androgen receptor transactivation by gelsolin: a newly identified androgen receptor coregulator. Cancer Res 63, 4888-4894.

Nowak, G., Pestic-Dragovich, L., Hozak, P., Philimonenko, A., Simerly, C., Schatten, G., and de Lanerolle, P. (1997). Evidence for the presence of myosin I in the nucleus. J Biol Chem 272, 17176-17181.

Ohta, Y., Nishida, E., Sakai, H., and Miyamoto, E. (1989). Dephosphorylation of cofilin accompanies heat shock-induced nuclear accumulation of cofilin. J Biol Chem 264, 16143-16148.

Onoda, K., Yu, F. X., and Yin, H. L. (1993). gCap39 is a nuclear and cytoplasmic protein. Cell Motil Cytoskeleton 26, 227-238.

Otterbein, L. R., Graceffa, P., and Dominguez, R. (2001). The crystal structure of uncomplexed actin in the ADP state. Science 293, 708-711.

Papoulas, O., Beek, S. J., Moseley, S. L., McCallum, C. M., Sarte, M., Shearn, A., and Tamkun, J. W. (1998). The *Drosophila* trithorax group proteins BRM, ASH1 and ASH2 are subunits of distinct protein complexes. Development 125, 3955-3966.

Pederson, T., and Aebi, U. (2002). Actin in the nucleus: what form and what for? J Struct Biol 140, 3-9.

Pendleton, A., Pope, B., Weeds, A., and Koffer, A. (2003). Latrunculin B or ATP depletion induces cofilin-dependent translocation of actin into nuclei of mast cells. J Biol Chem 278, 14394-14400.

Pestic-Dragovich, L., Stojiljkovic, L., Philimonenko, A. A., Nowak, G., Ke, Y., Settlage, R. E., Shabanowitz, J., Hunt, D. F., Hozak, P., and de Lanerolle, P. (2000). A myosin I isoform in the nucleus. Science 290, 337-341.

Pestonjamasp, K. N., Pope, R. K., Wulfkuhle, J. D., and Luna, E. J. (1997). Supervillin (p205): A novel membrane-associated, F-actin-binding protein in the villin/gelsolin superfamily. J Cell Biol 139, 1255-1269.

Peterson, C. L., Zhao, Y., and Chait, B. T. (1998). Subunits of the yeast SWI/SNF complex are members of the actin-related protein (ARP) family. J Biol Chem 273, 23641-23644.

Peterson, M. G., Tanese, N., Pugh, B. F., and Tjian, R. (1990). Functional domains and upstream activation properties of cloned human TATA binding protein. Science 248, 1625-1630.

Philimonenko, V. V., Zhao, J., Iben, S., Dingova, H., Kysela, K., Kahle, M., Zentgraf, H., Hofmann, W. A., de Lanerolle, P., Hozak, P., and Grummt, I. (2004). Nuclear actin and myosin I are required for RNA polymerase I transcription. Nat Cell Biol 6, 1165-1172.

Pollard, T. D., Blanchoin, L., and Mullins, R. D. (2000). Molecular mechanisms controlling actin filament dynamics in nonmuscle cells. Annu Rev Biomol Struct 29, 545-576.

Pollard, T. D., Blanchoin, L., and Mullins, R. D. (2001). Actin dynamics. J Cell Sci 114, 3-4.

Pollard, T. D., and Korn, E. D. (1973). Acanthamoeba myosin. I. Isolation from Acanthamoeba castellanii of an enzyme similar to muscle myosin. J Biol Chem 248, 4682-4690.

Pope, B., Maciver, S., and Weeds, A. (1995). Localization of the calcium-sensitive actin monomer binding site in gelsolin to segment 4 and identification of calcium binding sites.

Biochemistry 34, 1583-1588.

Pope, R. K., Pestonjamasp, K. N., Smith, K. P., Wulfkuhle, J. D., Strassel, C. P., Lawrence, J. B., and Luna, E. J. (1998). Cloning, characterization, and chromosomal localization of human supervillin (SVIL). Genomics 52, 342-351.

Scheer, U., Hinssen, H., Franke, W. W., and Jockusch, B. M. (1984). Microinjection of actin-binding proteins and actin antibodies demonstrates involvement of nuclear actin in transcription of lampbrush chromosomes. Cell 39, 111-122.

Schnapp, A., Pfleiderer, C., Rosenbauer, H., and Grummt, I. (1990). A growth-dependent transcription initiation factor (TIF-IA) interacting with RNA polymerase I regulates mouse ribosomal RNA synthesis. Embo J 9, 2857-2863.

Schramm, L., and Hernandez, N. (2002). Recruitment of RNA polymerase III to its target promoters. Genes Dev 16, 2593-2620.

Shen, X., Mizuguchi, G., Hamiche, A., and Wu, C. (2000). A chromatin remodelling complex involved in transcription and DNA processing. Nature 406, 541-544.

Shen, X., Ranallo, R., Choi, E., and Wu, C. (2003). Involvement of actin-related proteins in ATP-dependent chromatin remodeling. Mol Cell 12, 147-155.

Skare, P., Kreivi, J. P., Bergstrom, A., and Karlsson, R. (2003). Profilin I colocalizes with speckles and Cajal bodies: a possible role in pre-mRNA splicing. Exp Cell Res 286, 12-21.

Smith, S. S., Kelly, K. H., and Jockusch, B. M. (1979). Actin co-purifies with RNA polymerase II. Biochem Biophys Res Commun 86, 161-166.

Straub, K. L., Stella, M. C., and Leptin, M. (1996). The gelsolin-related flightless I protein is required for actin distribution during cellularisation in Drosophila. J Cell Sci 109 (Pt 1), 263-270.

Stuven, T., Hartmann, E., and Gorlich, D. (2003). Exportin 6: a novel nuclear export receptor that is specific for profilin.actin complexes. Embo J 22, 5928-5940.

Szent-Gyorgyi, A. (1945). Studies on muscle. Acta Physiol Scandinav 9 (suppl.25).

Ting, H. J., Yeh, S., Nishimura, K., and Chang, C. (2002). Supervillin associates with androgen receptor and modulates its transcriptional activity. Proc Natl Acad Sci USA 99, 661-666.

Tower, J., and Sollner-Webb, B. (1987). Transcription of mouse rDNA is regulated by an activated subform of RNA polymerase I. Cell 50, 873-883.

Van Impe, K., De Corte, V., Eichinger, L., Bruyneel, E., Mareel, M., Vandekerckhove, J., and Gettemans, J. (2003). The Nucleo-cytoplasmic actin-binding protein CapG lacks a nuclear export sequence present in structurally related proteins. J Biol Chem 278, 17945-17952.

Vandekerckhove, J., Bugaisky, G., and Buckingham, M. (1986). Simultaneous expression of skeletal muscle and heart actin proteins in various striated muscle tissues and cells. A quantitative determination of the two actin isoforms. J Biol Chem 261, 1838-1843.

Visegrady, B., Lorinczy, D., Hild, G., Somogyi, B., and Nyitrai, M. (2005). A simple model for the cooperative stabilisation of actin filaments by phalloidin and jasplakinolide. FEBS Lett 579, 6-10.

Wada, A., Fukuda, M., Mishima, M., and Nishida, E. (1998). Nuclear export of actin: a novel mechanism regulating the subcellular localization of a major cytoskeletal protein. Embo J 17, 1635-1641.

Wang, M. D., Schnitzer, M. J., Yin, H., Landick, R., Gelles, J., and Block, S. M. (1998). Force and velocity measured for single molecules of RNA polymerase. Science 282, 902-907.

Way, M., Pope, B., and Weeds, A. G. (1992). Evidence for functional homology in the F-actin binding domains of gelsolin and alpha-actinin: implications for the requirements of severing and capping. J Cell Biol 119, 835-842.

Way, M., and Weeds, A. (1988). Nucleotide sequence of pig plasma gelsolin. Comparison of protein sequence with human gelsolin and other actin-severing proteins shows strong homologies and evidence for large internal repeats. J Mol Biol 203, 1127-1133.

Wulfkuhle, J. D., Donina, I. E., Stark, N. H., Pope, R. K., Pestonjamasp, K. N., Niswonger, M. L., and Luna, E. J. (1999). Domain analysis of supervillin, an F-actin bundling plasma membrane protein with functional nuclear localization signals. J Cell Sci 112 (Pt 13), 2125-2136.

Yin, H., Wang, M. D., Svoboda, K., Landick, R., Block, S. M., and Gelles, J. (1995). Transcription against an applied force. Science 270, 1653-1657.

Yin, H. L., Hartwig, J. H., Maruyama, K., and Stossel, T. P. (1981). Ca2+ control of actin filament length. Effects of macrophage gelsolin on actin polymerization. J Biol Chem 256, 9693-9697.

Yin, H. L., and Stossel, T. P. (1979). Control of cytoplasmic actin gel-sol transformation by gelsolin, a calcium-dependent regulatory protein. Nature 281, 583-586.

Yin, H. L., and Stossel, T. P. (1980). Purification and structural properties of gelsolin, a Ca2+-activated regulatory protein of macrophages. J Biol Chem 255, 9490-9493.

Yu, F. X., Johnston, P. A., Sudhof, T. C., and Yin, H. L. (1990). gCap39, a calcium ion- and polyphosphoinositide-regulated actin capping protein. Science 250, 1413-1415.

Yuan, X., Zhao, J., Zentgraf, H., Hoffmann-Rohrer, U., and Grummt, I. (2002). Multiple interactions between RNA polymerase I, TIF-IA and TAF(I) subunits regulate preinitiation complex assembly at the ribosomal gene promoter. EMBO Rep 3, 1082-1087.

Zhao, K., Wang, W., Rando, O. J., Xue, Y., Swiderek, K., Kuo, A., and Crabtree, G. R. (1998). Rapid and phosphoinositol-dependent binding of the SWI/SNF-like BAF complex to chromatin after T lymphocyte receptor signaling. Cell 95, 625-636.

Chapter 32

Wnt Signaling and Transcriptional Regulation

Xinhua Lin

Division of Developmental Biology, Cincinnati Children's Hospital Medical Center, The University of Cincinnati College of Medicine, Cincinnati, OH 45229, USA

Key Words: Wnt, signaling transduction, beta-catenin, transcriptional regulation

Summary

Cell-cell interactions controlled by the highly conserved Wnt family of secreted proteins play essential roles in embryonic patterning and adult homeostasis. One of the major Wnt signaling pathways, namely Wnt canonical pathway, is mediated by β-catenin whose activity is required for both cell-cell adhesion and for Wnt signaling mediated transcription. Over the past decades, genetic and biochemical analyses in various model systems have uncovered many essential components required for transducing Wnt signaling from the cell surface into transcription events in the nucleus. Multiple extracellular, cytoplasmic, and nuclear regulators are involved in modulating β-catenin levels, its subcellular localizations and its transcriptional activity. Furthermore, de-regulated cell-cell adhesion caused by other signaling pathways can also lead to alteration of β-catenin levels which could subsequently activate transcription of Wnt-target genes. Wnts are required for adult tissue maintenance, and deregulated Wnt signaling events promote both human degenerative diseases and cancers. A full understanding of molecular mechanism(s) of Wnt signaling is likely to shed new lights into rational design of drugs which can cure diseases associated with deregulated Wnt signaling.

Introduction

During development, formation of multi-cellular organisms requires combined actions of several developmental signaling pathways that control patterning of various organs and complex body structures. These signaling pathways, commonly called morphogen signaling pathways, include Wnt/Wingless (Wg), Hedgehog (Hh), transforming growth factor-β (TGF-β), and fibroblast growth factor (FGF). Secreted morphogen molecules relay their signals through their distinct intracellular signaling pathways which ultimately activate transcription of specific genes required for patterning of complex body structures. In the previous chapters, we have discussed transcriptional regulation mediated by TGF-β, Jak/STATs, and NF-κB pathways. Here, I will discuss Wnt signaling pathway and its role in transcription regulation. I will mainly focus on the canonical Wnt signaling pathway mediated by β-catenin.

The Wnt Signal Transduction Pathway: Components and Their Actions

A: Overview of Wnt Signal Transduction Pathways

The Wnt family of secreted signaling molecules are highly conserved and are involved in numerous developmental processes and adult homeostasis in both vertebrates and invertebrates (Logan and Nusse, 2004; Moon, 2005; Wodarz and Nusse, 1998). Wnt-1, the first member of Wnt family protein was initially identified independently as a *Drosophila* segment polarity gene Wingless (Wg) and the murine protooncogene Int-1 (Rijsewijk *et al.*, 1987). The term Wnt was derived from a combination of Wingless and Int-1. Since the discovery of Wnt-1, multiple Wnt members have been found throughout the animal kingdom and the human genome encodes 19 Wnt members (Logan and Nusse, 2004; Moon, 2005).

Intensive studies by *Drosophila* geneticists, vertebrate

Corresponding Author: Tel: (513) 636-2144, Fax: (513) 636-4317, E-mail: linyby@cchmc.org

developmental biologists and molecular biologists have identified essential components of signaling pathways by which Wnt proteins relay their signals into intracellular responses (Logan and Nusse, 2004; Moon, 2005; Wodarz and Nusse, 1998). Wnt proteins can transduce their signaling through distinct intracellular routes which can be mainly divided into two pathways. The most well-studied Wnt signaling pathway is the canonical Wnt pathway which is mediated by β-catenin (Logan and Nusse, 2004; Moon, 2005; Wodarz and Nusse, 1998). Wnts can also relay their signals via the planar polarity pathway or Wnt/Ca^{2+} pathway and these are referred as 'non- anonical' pathways (Moon, 2005; Strutt, 2003; Veeman et al., 2003).

The major canonical Wnt signaling (I will call the Wnt signaling hereafter) components and their actions are illustrated in Fig.32.1. The central player in this pathway is β-catenin whose stability is regulated by a destruction complex which minimally consists of the serine/threonine kinases casein kinase Iα(CK Iα) and glycogen synthase-3β (GSK-3β) bound to a scaffolding complex of the tumour-suppressor gene products Axin and adenomatous polyposis coli (APC). In the absence of Wnt signaling, the cytosolic β-catenin is maintained at low levels since it is phosphorylated at its N-terminal region and targeted for ubiquitination and degradation in the 26S proteosome by the destruction complex. Activation of Wnt signaling through its receptor complex, which consists of a serpentine receptor of the Frizzled (Fz) family and a member of the LDL receptor family related protein (LRP), leads to inhibition of the destruction complex resulting in accumulation of unphosphorylated cytoplasmic β-catenin. As a consequence, accumulated cytoplasmic β-catenin translocates into the nucleus where it binds to the HMG-box transcription factor protein T cell factor (TCF) and activates transcription of Wnt-target genes (Fig.32.1). Thus, the key event in Wnt signaling is β-catenin stabilization and accumulation in the cytoplasm. Like Wnts, the components of the canonical Wnt signaling pathway are highly conserved among animal species. Below, I will discuss the components of Wnt signaling and their actions in transducing Wnt signaling in more details.

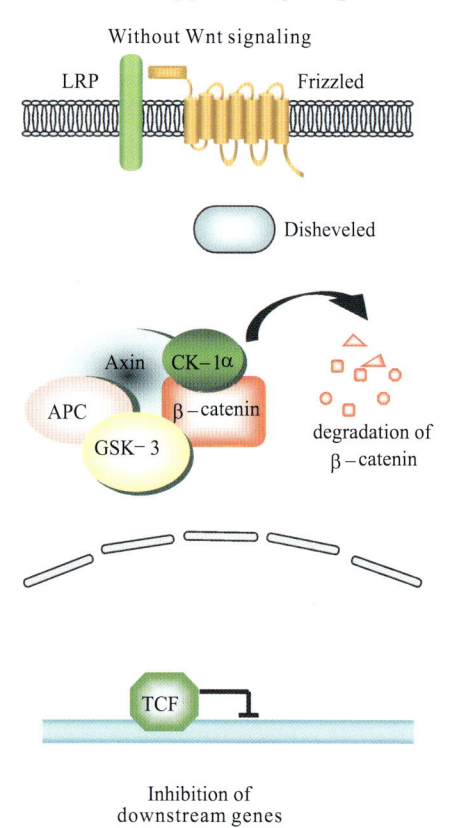

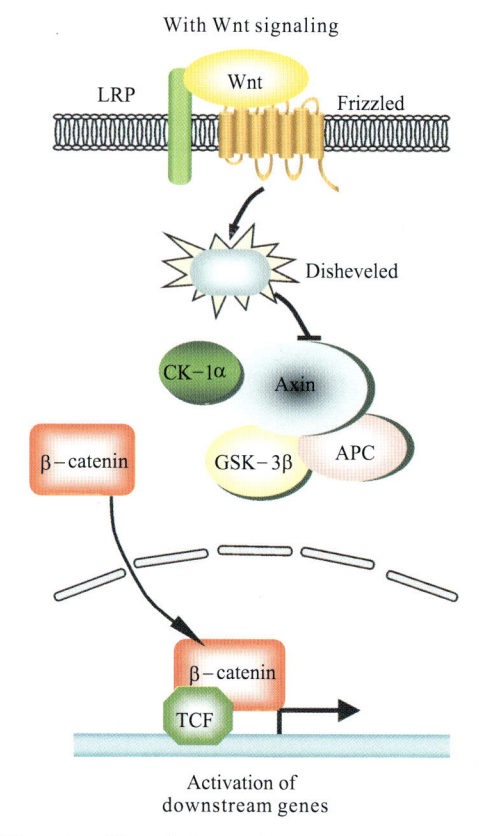

Fig.32.1 The canonical Wnt signaling pathway. In the absence of Wnt signalling (left panel), cytoplasmic β-catenin gets phosphorylated and targeted for degradation through its interaction with the destruction complex consisting of tumor supressors Axin, APC and the protein kinase GSK3-β and CK1α. Wnt proteins (right panel) bind to the Frizzled/LRP receptor complex at the cell surface. These receptors transduce a signal to Dsh which inhibits the activity of the destruction complex. As a consequence, β-catenin is uncoupled from the destruction complex and translocates to the nucleus, where its binds to the TCF transcription factor and activates transcritpion of Wnt-target genes.

B: Cell Surface Receptors and Modulators

The initiation of Wnt canonical signaling pathway begins with the interaction of Wnt proteins with their receptors on the cell surface (Fig.32.1). The first identified Wnt receptors are the Fz family members of seven-pass transmembrane proteins (Bhanot *et al.*, 1996). Fz family members are serpentine receptors closely related to G protein-coupled receptors (Logan and Nusse, 2004; Moon, 2005; Wodarz and Nusse, 1998). Multiple members of Fz proteins are identified in animal kingdom and 10 Fz members are encoded in the human genome. In addition to Fz proteins, the canonical Wnt signaling pathway requires LRP as a co-receptor. LRPs are single-span transmembrane proteins. Both vertebrate Lrp5 and Lrp6 and their *Drosophila* ortholog Arrow have been shown to act co-receptors for Wnt signaling (He *et al.*, 2004; Pinson *et al.*, 2000; Tamai *et al.*, 2000; Wehrli *et al.*, 2000). Studies in *Drosophila* have demonstrated that Fz proteins act as receptors for Wnt proteins by directly interacting with Wnt proteins with high affinity through their extracellular cysteine-rich domain (CRD) (Bhanot *et al.*, 1996; Rulifson *et al.*, 2000; Wu and Nusse, 2002). Experimental results from nematodes, *Xenopus* and mammalian cells also support this view (He *et al.*, 1997; Hsieh *et al.*, 1999; Sawa *et al.*, 1996; Yang-Snyder *et al.*, 1996). However, although co-immunoprecipitation experiments suggest that vertebrate Lrp5 and Lrp6 can bind to Wnt proteins (Kato *et al.*, 2002; Mao *et al.*, 2001; Tamai *et al.*, 2000), a biochemical study showed that *Drosophila* Arrow fails to bind to Wg (Wu and Nusse, 2002). Thus, additional experiments will be needed to further substantiate any Wnt-LRPs interaction.

What's the mechanism for Wnt signal initiation? Current data support a model in which Wnt initiates its signaling event by forming a complex with Fz and LRP proteins. The formation of Wnt-Fz-LRP complex brings the LRP intracellular domain to the intracellular domain(s) of Fz proteins thereby activating Wnt downstream signaling. Consistent with this model, ectopic expression of Dfz2-Arrow fusion protein in which the Arrow intracellular domain is fused with the Dfz2 cytoplasmic tail, can activate Wnt signaling in Wnt ligand-independent manner (Tolwinski *et al.*, 2003).

In addition to Fz and LRP proteins, the cell surface heaparan sulfate proteoglycan can also modulate Wnt signaling by acting as a co-receptor(s) in some developmental contexts (Lin, 2004). Studies in *Drosophila* have shown that glypican members of heaparan sulfate proteoglycans are required for Wnt signaling (Baeg *et al.*, 2001; Han *et al.*, 2005; Lin and Perrimon, 1999; Tao *et al.*, 2005; Tsuda *et al.*, 1999). HSPGs can modulate Wnt signaling by either facilitating Wnt-receptor interaction or preventing Wnt from being degraded on the cell surface (Lin, 2004).

C: Dishevelled (Dsh) and G Proteins: Two Wnt Signaling Components Immediately Downstream of Wnt Receptors

Genetic studies in *Drosophila* have identified Dsh as a Wnt signaling component downstream of Wnt receptor. Dsh proteins are multi-module proteins which contain several essential functional domains including a Dix domain, a PDZ domain, and a DEP domain. The Dix domain is similar to a domain located in Axin (See below) and may promote interaction between these two proteins (Hsu *et al.*, 1999). While *Drosophila* contains only Dsh protein, human genome encodes three Dsh members (DVL1, DVL2, DVL3). Despite intensive studies of the function of Dsh in Wnt signaling, the exact mechanism of action of Dsh is not known. Current data have shown that Dsh may interact with a number of Wnt downstream molecules including Casein Kinase 1 (Peters *et al.*, 1999; Sakanaka *et al.*, 1999) and GBP/Frat1 (Farr *et al.*, 2000; Li *et al.*, 1999; Salic *et al.*, 2000). A current model is that interactions of Dsh with Axin as well as GBP/Frat1 and GSK3β may be required for inactivation of β-catenin destruction complex (Fig. 32.1) thereby transducing Wnt signaling.

Fz proteins are closely related to G protein-coupled receptors which signal to downstream effectors through an associated trimeric G protein complex. Are G proteins involved in Wnt signaling? Several studies have provided evidence for this. Expression of a chimeric rat Fz protein in cultured cells showed that heterotrimeric G proteins may play a role in Fz-stimulated transcriptional response (Liu *et al.*, 2001). Very recently, Katanaev *et al.* further demonstrated that a *Drosophila* Gαo subunit plays a role in Wnt signaling (Katanaev *et al.*, 2005). *Drosophila* genome encodes six Gα genes. Mutations in *Gαo 47A* mimic the phenotypes of mild loss of Wnt signaling while overexpression of Gαo mimics the effects of Wnt gain of function in embryonic patterning. Genetic epistasis experiments further argue that Gαo is an immediate transducer of Fz and acts upstream of *dsh*. Thus, Gαo is likely part of a trimeric G protein complex that directly transduces Fz signals from the membrane to downstream components. These results suggest that Fz receptors act as a guanine nucleotide exchange factors to activate Gαo and promote downstream events (Katanaev *et al.*, 2005).

D: The β-catenin Destruction Complex: Roles of GSK-3β, CKIα, Axin and APC in Regulating Levels of β-catenin

The elevated level of cytoplasmic β-catenin is a hallmark of Wnt canonical pathway activation. In the absence of Wnt signaling, β-catenin directly interacts with Axin and APC (Hart et al., 1998; Kishida et al., 1998) and is recruited into the destruction complex which contains two serine/threonine kinases, GSK-3β (Yost et al., 1996) and CKIα (Amit et al., 2002; Liu et al., 2002; Yanagawa et al., 2002). β-catenin is phosphorylated by CK1α and GSK3β in a sequential manner on four conserved amino (N)-terminal serine and threonine residues (Fig.32.2). The phosphorylated β-catenin is subsequently recognized by β-TrCP, targeted for ubiquitination, and degraded by the proteosome (Aberle et al., 1997; Latres et al., 1999; Liu et al., 1999) (Fig.32.2). Wnt signaling causes elevated β-catenin by disrupting of β-catenin destruction complex through a number of mechanisms.

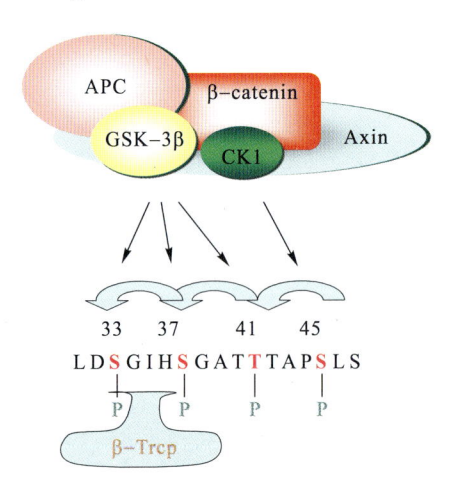

Fig.32.2 The β-catenin destruction complex and the action of β-catenin phosphorylation. In the β-catenin destruction complex consisting of CKIα, GSK-3β, APC and Axin and β-catenin, CKIα and GSK-3β each bind a different domain of Axin. CKIα phosphorylation of S45 allows GSK-3 to phosphorylate T41, then S37 and S33. Phosphorylation of S37 and S33 creates the recognition site for β-Trcp. Wnt signaling disrupts the β-catenin degradation complex and inhibits GSK-3β phosphorylation of T41, S37, and S33. It remains to be determined how CKIα phosphorylation of S45 is regulated.

D1: GSK-3β

GSK-3β is the first identified serine/threonine kinase involved in Wnt signaling. Initial genetics studies in Drosophila isolated GSK-3β (also called shaggy or zeste-white 3 in Drosophila) as a negative regulator in Wg signaling (Siegfried et al., 1992; Siegfried and Perrimon, 1994; Siegfried et al., 1994). Genetic epitasis analyses showed that GSK-3β acts upstream of β-catenin, suggesting its role in regulating the activity of β-catenin (Siegfried et al., 1994). Consistent with this view, studies in Xenopus showed that expression of dominant-negative forms of GSK-3β can lead to constitutive activation of Wnt pathways through β-catenin (Dominguez et al., 1995; He et al., 1995; Pierce and Kimelman, 1995). Together, these data lead to a model where Wnt acts to negatively regulate GSK-3β kinase. The simplest model would suggest that GSK-3β directly phosphorylates β-catenin leading to the degradation of β-catenin (Aberle et al., 1997). There are four potential GSK-3β phosphorylation sites (S33, S37, T41 and S45) in the N-terminal portion of β-catenin that are conserved among different species (Fig. 32.2 and 32.3). Deletions or point mutations in these sites result in stabilization and constitutive activation of β-catenin (Pai et al., 1997; Zecca et al., 1996). These mutant proteins are no longer sensitive to GSK-3β regulation (Pai et al., 1997; Yost et al., 1996). GSK-3β can also phosphorylate β-catenin in vitro in the presence of Axin protein, suggesting further a negative role of GSK-3β in directly phosphorylating β-catenin.

D2: CKIα

In addition to GSK-3β, CKIα is also involved in phosphorylating β-catenin and inducing β-catenin degradation (Amit et al., 2002; Liu et al., 2002; Yanagawa et al., 2002) (Fig 32.2). Depletion of CKIα inhibits β-catenin phosphorylation and degradation and causes abnormal embryogenesis associated with excessive Wnt/β-catenin signaling in Drosophila (Liu et al., 2002). GSK-3β and CKIα bind to different regions of Axin (Fig.32.2). Overexpression of mutant forms of Axin incapable of binding GSK can still stimulate the phosphorylation of S45 at β-catenin. CKIα induce the phosphorylation of S45 at β-catenin, which precedes and is obligatory for subsequent GSK3 phosphorylation at other sites (Liu et al., 2002). A current model is that CKIα phosphorylation of β-catenin at S45 allows GSK-3β to phosphorylate at T41, S37, and S33 (Liu et al., 2002) (Fig.32.2). The phosphorylation of S37 and S33 generates the recognition sites for β-Trcp (Fig.32.2).

D3: Axin

Axin is a negative intracellular regulator in Wnt signaling (Ikeda et al., 1998; Itoh et al., 1998; Sakanaka et al., 1998; Zeng et al., 1997). In vertebrates, there are two Axins, Axin1 and Axin 2, both of which act as negative regulators for Wnt signaling. Axin serves as a scaffolding protein that contains multiple interaction domains capable of directly interacting with β-catenin,

GSK-3β, CKIα, and APC proteins as well as its upstream regulator Dsh and LRPs. Mutations in Axin result in activation of Wnt signaling while overexpression of Axin destabilizes β-catenin and blocks the axis-duplicating activity of *XWnt-8* in *Xenopus* embryos (Itoh *et al*., 1998; Zeng *et al*., 1997). Phosphorylation of β-catenin by GSK3β and CKIα is promoted in the complex between Axin and GSK3 or CKIα (Ikeda *et al*., 1998; Liu *et al*., 2002; Sakanaka *et al*., 1998) (Fig.32.2).

Several studies suggest that Wnt signaling allows Wnt co-receptor LRP to associate with Axin through its cytoplasmic tail, thereby removing Axin from the destruction complex (Mao *et al*., 2001; Tolwinski *et al*., 2003). Consistent with this view, a recent study showed that Axin can bind preferentially to a phosphorylated form of the LRP tail, which is induced by Wnt signaling (Tamai *et al*., 2004).

D4: APC

The *APC* gene was originally discovered in a hereditary cancer syndrome termed familiar adenomatous polyposis (FAP). APC is a large protein that plays many essential cellular functions. In the Wnt signaling pathway, APC acts as a negative regulator by promoting β-catenin degradation (Bienz, 2003; Bienz and Hamada, 2004; Nathke, 2004; Polakis, 1997). APC interacts directly with Axin and β-catenin (Spink *et al*., 2000)(Fig.32.2). The ability of APC to promote degradation of β-catenin depends on APC's interactions with both Axin and β-catenin since APC mutations incapable of interacting with either Axin or β-catenin failed to induce degradation of β-catenin (Kawahara *et al*., 2000). There are two human and two *Drosophila* APC molecules. In human, APC is a tumor suppressor gene. Loss of function mutations in APC1 is associated with colorectal cancer (Bienz, 2003; Bienz and Hamada, 2004; Nathke, 2004). Tumor cell lines producing truncated forms of APC protein have high levels of cytosolic β-catenin (Rubinfeld *et al*., 1996) while expression of wild-type APC in cancer cell lines results in a pronounced reduction of cytoplasmic β-catenin levels (Munemitsu *et al*., 1995). *Drosophila* APC1 and APC2 play redundant role in Wnt signaling both in embryos and in other tissues (Ahmed *et al*., 2002; Akong *et al*., 2002).

E: Wnt Signaling Event in the Nucleus: β-catenin-mediated Transcription Regulation

E1: β-catenin

The key mediator of the canonical Wnt pathway is β-catenin which is the vertebrate orthologue of the *Drosophila* protein Armadillo. β-catenin is composed of 12 imperfect repeats (Arm repeats) flanked by unique N- and C-terminal domains (Peifer *et al*., 1994; Peifer *et al*., 1992). As mentioned above (Fig.32.2), the N-terminal region domain of β-catenin contains four serine/threonine sites which can be phosphorylated by a combined action of CKIα and GSK-3 β (Fig.32.2 and Fig.32.3). Therefore, the N-terminal domain is responsible for its degradation in the absence of Wnt signaling. The center region containing Arm repeats can interact with a number of essential regulators including TCF, APC, E-cadherin, and α-catenin. The C-terminal domain can function as a transcription activation domain (Fig.32.3). β-catenin contains multiple domains that allow it to bind to various transcription factors, co-activators and other adaptor proteins as well as negative regulators, all of which contribute ultimately the transcriptional activity of β-catenin (Fig.32.3).

E2: TCFs: Nuclear Partners for β-catenin Activity in the Wnt Signaling

The main nuclear partners for β-catenin activity are TCF factors. There are four members of TCF factors in human and one member in *Drosophila* (Logan and Nusse, 2004; Moon, 2004; Wodarz and Nusse, 1998). TCFs are HMG box transcription factors and bind to specific DNA sequences with similar specificity. There are usually multiple TCF binding sites located in the promoter/enhancer regions of Wnt-target genes. In addition to β-catenin, TCF also interacts with other nuclear proteins which can regulate the transcriptional activity of TCF. In the absence of the Wnt signal, TCF functions as a repressor of Wnt target genes (Bienz, 1998; Brannon *et al*., 1997; Lin *et al*., 1998; Riese *et al*., 1997) (Fig.32.4). The repression activity of TCFs is mediated by interacting with the transcription co-repressor Groucho, C-terminal binding protein (CtBP), histone deacetylases (Cavallo *et al*., 1998; Chen *et al*., 1999), and Osa-containing Brahma chromatin remodeling complexes (Collins and Treisman, 2000) (Fig. 32.4). In the present of Wnt signaling, Wnt signaling permits accumulated β-catenin to translocate to the nucleus where binding of TCF with β-catenin converts TCF from a transcription repressor into a transcriptional activator, thereby activating the expression of Wnt-target genes (Nusse, 1999) (Fig. 32.4).

E3: Adaptors and Co-activators: Pygopus and Legless/BCL9

Pygopus (Pygo) and Legless (Lgl) are two essential positive players for β-catenin activity and were initially identified from genetic screens in *Drosophila* (Belenkaya *et al*., 2002; Kramps *et al*., 2002; Parker *et*

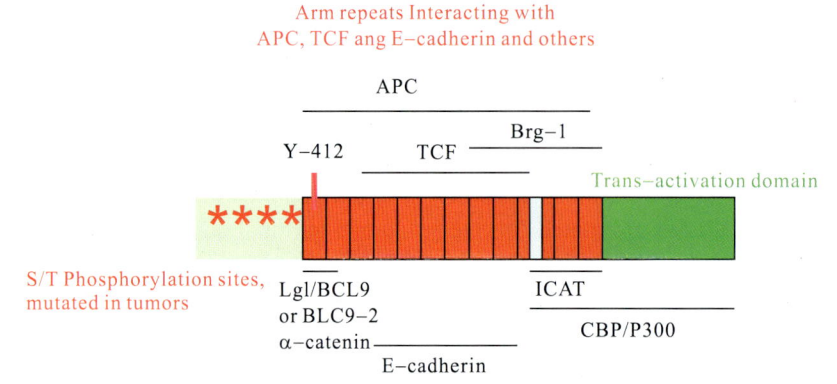

Fig.32.3 **Structure of β-catenin and its interacting domains with various regulators.** β-catenin is composed of 12 imperfect Arm repeats flanked by unique N- and C-terminal domains. The N-terminal region of β-catenin contains serine/threonine sites which can be phosphorylated by the destruction complex and targeted it for degradation. The C-terminal domain can function as a transcription activation domain. β-catenin contains multiple domains that can interact with various factors which regulates the its transcriptional activity.

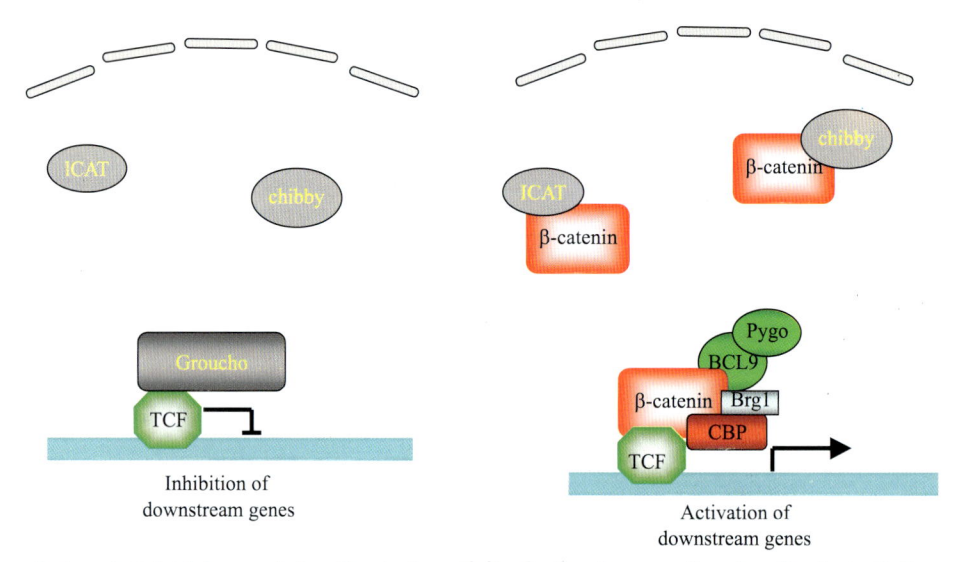

Fig.32.4 **Nuclear factors involved in regulating β-catenin activity.** In the absence of nuclear β-catenin (left panel), TCF binds to DNA elements to repress transcription of Wnt-target genes. This is achieved through its interaction with transcription co-repressors such as Groucho. Wnt signaling leads to elevated nuclear β-catenin which interacts with TCF and converts TCF from a repressor to an activator, thereby activating the expression of Wnt-target gene (right panel). BCL9 (Legless in *Drosophila*) can directly link β-catenin to Pygopus which can also act as a transcriptional co-activator. An alternative model is that BCL9 and Pygopus are involved in β-catenin nuclear import or retention. β-catenin also binds to other positive regulators such as CBP and chromatin remodeling factor Brg1. β-catenin can be negatively regulated by two nuclear factors Chibby and ICAT, both of which bind directly to β-catenin protein and inhibit its interaction with TCF factors in the nuclei.

al., 2002; Thompson et al., 2002). Pygopus is a nuclear protein containing a PHD finger domain which is required for its activity in promoting β-catenin activity. In *Drosophila*, loss of function studies using *pygo* mutants have demonstrated that Pygo is required for Wnt signaling in a variety of developmental processes (Belenkaya et al., 2002; Kramps et al., 2002; Parker et al., 2002; Thompson et al., 2002). There are two Pygo (Pygo1 and Pygo 2) homologues in vertebrates. Loss of Pygo2 in frog leads to reduced Wnt signaling arguing that Pygo-2 is required for Wnt signaling during development (Belenkaya et al., 2002). In colon cancer cells, removal of either Pygo1 or Pygo2 by RNA interference (RNAi) showed reduced Wnt signaling while expression of Pygo1 can restore Wnt signaling defects resulting from reduced levels of Pygo1 or Pygo 2. This data suggests that Pygo1 and Pygo2 may be functionally redundant in Wnt signaling in vertebrate cells (Thompson et al., 2002).

Lgl is a large cytoplasmic protein which regulates

β-catenin signaling activity by linking β-catenin to Pygo. Lgl contains both β-catenin and Pygo interacting domains, mutations of which cause Wnt signaling defects in Drosophila (Kramps et al., 2002). Consistent with a role as an adaptor protein, expression of a deleted form of Lgl protein containing β-catenin and Pygo interacting domains can fully restore Wnt signaling in Drosophila lgl mutants (Kramps et al., 2002). Recently studies showed that Lgl is kept in the nucleus through its interaction with the nuclear protein Pygopus (Townsley et al., 2004).

BCL-9 (Kramps et al., 2002) and BCL-9-2 (Adachi et al., 2004; Brembeck et al., 2004) are two vertebrate homologues of Drosophila Lgl and contain conserved domains that are involved in interacting with β-catenin and Pygo proteins. BCL9 is implicated in human B cell lymphomas (Kramps et al., 2002; Willis et al., 1998). Interestingly, although BCL-9 is a cytoplasmic protein and its nuclear localization depends on its interaction with Pygo (Kramps et al., 2002), BCL-9-2 contains its own nuclear import signal and can translocate into nuclei independent of its interaction with Pygo proteins (Adachi et al., 2004; Brembeck et al., 2004).

While earlier studies have demonstrated that Pygo proteins act as transcription co-activators for β-catenin, recent work suggests that Pygo proteins regulate β-catenin signaling activity by controlling its nuclear localization (Townsley et al., 2004). In Drosophila, β-catenin level is reduced in Pygo mutant embryos (Townsley et al., 2004). These results suggest two models for roles of Pygo and BCL9/BCL-9-2 in β-catenin signaling. The first model is that Pygo targets β-catenin from cytoplasm into nuclei through Lgl/BCL9. This is likely the case for Drosophila as well as for many developmental processes in vertebrates. In a second model, BCL9-2 can interact with cytoplasm β-catenin and directly targets it into nuclei (Adachi et al., 2004; Brembeck et al., 2004). Thus, BCL-9-2 provides additional levels of complexity in β-catenin regulation. The detailed mechanisms of BCL-9 and BCL-9-2 mediated Wnt signaling need to await loss of functional studies in mice.

E4: Other Positive Regulators for β-catenin Signaling Activity in the Nucleus

In addition to the positive regulators mentioned above, β-catenin also interacts with several other transcriptional co-activators through its C-terminal activation domain, which include CBP (Hecht et al., 2000; Takemaru and Moon, 2000; Wolf et al., 2002) and the chromatin remodeling factor Brg-1 (Barker et al., 2001; Chi et al., 2003) (Fig. 32.4).

E5: Negative Regulators for β-catenin Signaling Activity in the Nucleus

β-catenin can also interact with various factors that can negatively regulate its nuclear activity. Two negative regulators for β-catenin were identified in attempts to isolate β-catenin interacting proteins by yeast two-hybrid screen approaches (Tago et al., 2000; Takemaru et al., 2003). Inhibitor of beta-catenin and TCF-4 (ICAT) is a 9-kDa nuclear protein and binds to the Armadillo (Arm) repeats of β-catenin (Tago et al., 2000). Although there is no homology in Drosophila, ICAT is conserved among vertebrates. Overexpression of ICAT can block Wnt signaling in cultured cells and in Xenopus embryos, suggesting that ICAT acts as an inhibitor for β-catenin. Consistent with a role as a negative regulator for β-catenin, knockout of ICAT in mice yields a gain-of-function Wnt signaling phenotype during posterior neural cell fate specification (Satoh et al., 2004). Recent crystal structure of ICAT and β-catenin revealed that ICAT inhibits β-catenin activity by blocking the binding of β-catenin to TCFs (Daniels and Weis, 2002; Graham et al., 2002).

Chibby (Cby) is another nuclear protein that negatively regulates β-catenin signaling activity (Takemaru et al., 2003). Cby is conserved between Drosophila and vertebrates. Cby acts as an antagonist to β-catenin signaling activity by binding to it and preventing it from interacting with TCF. Loss of function of Cby in Drosophila leads to constitutively activation of Wnt signaling pathway, further indicating that Cby functions as a repressor.

Regulation of Wnt Signaling by other Cell Adhesion Molecules

In addition to being an essential transcription co-activator for Wnt-target genes, β-catenin also functions as a cytoplasmic protein critical for adherens junctions (Nelson and Nusse, 2004). In fact, The β-catenin was initially discovered for its role in cell adhesion (Kemler, 1993). As a component of adherens junctions, β-catenin tightly binds to the intracellular domain of the transmembrane protein cadherin, a Ca^{2+}-dependent homotypic adhesion molecule, and bridges it to the actin cytoskeleton through the adaptor protein α-catenin (Fig.32.5).

The adherens junctions provide another level of regulation for Wnt signaling. Cadherin molecules can act as negative regulators for Wnt signaling as they bind β-catenin at the cell surface and thereby can sequester it from being available for Wnt signaling. Indeed, studies in both *Xenopus* and *Drosophila* embryos demonstrated that overexpression of cadherins reduced the availability of β-catenin by sequestering it at the plasma membrane and thereby made it unavailable for Wnt signaling to the nucleus (Heasman *et al.*, 1994; Sanson *et al.*, 1996). On the other hand, alterations of adherens junctions by other signaling pathways or developmental processes can also control β-catenin levels thereby indirectly regulating Wnt signaling. In general, unlike a soluble cytoplasmic β-catenin pool that is highly unstable in the absence of Wnt signaling, the membrane-associated β-catenin is stable. However, Wnt signaling can be coupled to a loosening of adhesion between epithelial cells during epithelial–mesenchymal transitions and other developmental processes (Perez-Moreno *et al.*, 2003) as well as during cancer whose progression depends on inappropriate cell signaling and loss of cadherin-mediated adhesion (Birchmeier *et al.*, 1995).

There are several levels of regulation. First, the structural and functional integrity of the cadherin-catenin complex is regulated by phosphorylation. Serine/threonine phosphorylation of β-catenin (Bek and Kemler, 2002) or epithelial cadherin (E-cadherin) (Lickert *et al.*, 2000) cause increased stability of the cadherin-catenin complex. However, tyrosine phosphorylation of β-catenin by tyrosine kinases can results in a loss of cadherin-mediated cell-cell adhesion and an increase in the level of cytoplasmic β-catenin (Piedra *et al.*, 2001; Roura *et al.*, 1999) (Fig.32.5). Different tyrosine kinases can affect cadherin-catenin complex by different mechanisms. For example, tyrosine phosphorylation of β-catenin at Tyr-142 by cytoplasmic kinase Fer and Fyn disrupts binding of β-catenin to α-catenin (Piedra *et al.*, 2001; Piedra *et al.*, 2003), whereas phosphorylation of β-catenin at Tyr-654 by Src Kinase or the epidermal growth factor (EGF) receptor (Roura *et al.*, 1999) blocks binding of β-catenin to cadherin (Fig. 32.5). In contrast, de-phosphorylation at tyrosine residues by activation of PTPases can stabilize the cadherin-catenin complex and result in increased cadherin-mediated cell-cell adhesion (Balsamo *et al.*, 1998; Hellberg *et al.*, 2002; Nawroth *et al.*, 2002).

Recent data demonstrated that phosphorylation of β-catenin by a tyrosine kinase cMET, the receptor for hepatocyte growth factor, can also directly facilitate its nuclear translocation thereby enhancing its signaling in the nucleus (Bienz, 2005; Brembeck *et al.*, 2004; Danilkovitch-Miagkova *et al.*, 2001). Activated cMET kinase can cause β-catenin phosphorylation at Tyr-142. Interestingly, β-catenin phosphorylation at Tyr-142 can be effectively bound by BCL 9-2 (Bienz, 2005; Brembeck *et al.*, 2004). The interaction of phosphorylated β-catenin with BCL 9-2 allows β-catenin to be effectively translocated into nucleus by BCL 9-2 which contains a strong nuclear importing signal (Brembeck *et al.*, 2004). Thus, β-catenin phosphorylation at Tyr-142 both disrupts its association with α-catenin thereby adhesion junction and induce its nuclear import though BCL 9-2. The coupling of β-catenin phosphorylation and its nuclear import may occur in both developmental processes and during oncogenesis.

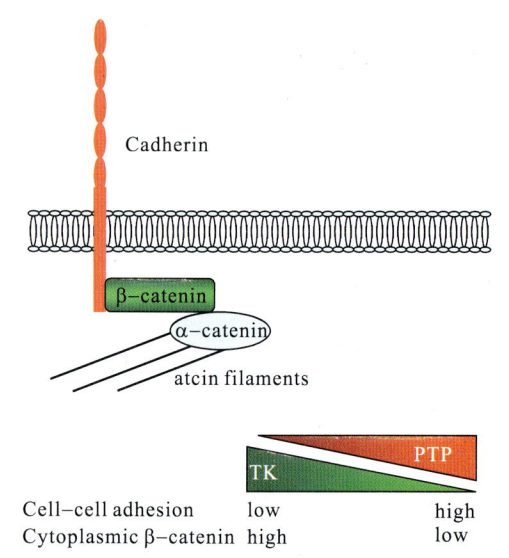

Fig.32.5 Cell adhesion and regulation of β-catenin activity. β-catenin is involved in the formation of cell adhesion by forming the cadherin-catenin complex. The integrity of this complex is regulated by the balance of tyrosine kinase and phosphatase activities. The cadherin-catenin complex is negatively regulated by phosphorylation of β-catenin through tyrosine kinase (TK) including receptor tyrosine kinases (RTKs) and cytoplasmic tyrosine kinases (Fer, Fyn, Yes, and Src). These kinases phosphorylate specific tyrosine residues in β-catenin (Y654, Y142) leading to dissociation of the cadherin-catenin complex, thereby elevating levels of cytoplasmic β-catenin. De-phosphorylation of β-catenin by protein tyrosine phosphatases (PTP) can stabilize the cadherin-catenin complex, leading to reduced levels of cytoplasmic β-catenin.

Wnt Signaling in Cancers and other Human Diseases

Given the essential functions of Wnt signaling in development and in adult homeostasis, it is not surprising that de-regulated Wnt signaling can cause various human diseases and cancers. Constitutive activation of Wnt signaling is implicated in various cancers (Giles *et al.*, 2003), in particular colon cancer (Bienz and Clevers, 2000). The best-known example of this is mutations in tumor suppressor APC. Inactivation mutations in APC, which cause constitutive Wnt signaling activation, are associated with familial adenomatous polyposis (FAP), an autosomal, dominantly inherited disease in which patients display hundreds or thousands of polyps in the colon and rectum (Kinzler et al., 1991; Nishisho et al., 1991). Constitutively activated forms of β-catenin mutations and loss of function APC mutations have also been identified in sporadic colon cancers and a large variety of other tumors (Giles et al., 2003). Furthermore, mutations in human Axin 1 have also been found in hepatocellular carcinomas (Satoh et al., 2000) and mutations in human Axin2 predispose to colon cancer (Lammi et al., 2004). Together, these data demonstrate that uncontrolled β-catenin regulation caused by mutations in the Wnt signaling pathway can lead to many cancers.

De-regulated Wnt signaling is also implicated in a variety of human genetic disorders. Various mutations in human LRP5 are found to be associated with defects in bone formation. While a gain of function mutation in LRP5 causes increased bone density at defined locations (Boyden et al., 2002), loss of function mutations result in reduced bone mass (Gong et al., 2001). These data suggest that Wnt signaling is essential for maintenance of normal bone density. The LRP5 loss of function mutations are also associated with vasculature defects in the eye (osteoperosis-pseudoglioma syndrome or OPPG) (Gong et al., 2001). Furthermore, another hereditary disorder called familial exudative vitreopathy (FEVR), which causes defective vasculogenesis in the peripheral retina, is associated with mutations in both LRP5 and the Fz4 receptor (Robitaille et al., 2002; Toomes et al., 2004).

In summary, it has become increasingly recognized that like other essential signaling pathways, mutations associated de-regulated Wnt signaling can cause various cancers and other human diseases. It is anticipated that further genetic studies will help to uncover other human diseases resulted from mutations or de-regulated Wnt signaling. Understanding Wnt signaling pathway will provide insights into the design of therapeutic drugs which can potentially cure related diseases.

Concluding Remarks

Intensive studies in the past decades have uncovered most of the intracellular signaling component involved in Wnt signaling. Analysis of Wnt signaling components and their functions has also implicated the involvement of Wnt signaling in development and in adult tissue maintenance. It is now well established that de-regulation of Wnt signaling can cause various human diseases, in particular human cancers. However, Wnt signaling is a complex and tightly regulated pathway. The mechanisms of several essential processes during Wnt signaling are still unclear and need to be resolved. These include the processes of how Wnt binding to the Fz/LRP complex transduces a signal to Dsh, how proteins within the β-catenin degradation complex are regulated, and how β-catenin interacts with various positive and negative regulators to integrate Wnt signaling input as well as other signaling inputs to selectively activate the transcription of its target genes. With development of new technology and experimental approaches such as RNAi screens for components of Wnt signaling in various model systems and detailed structural studies of Wnt components (Dasgupta et al., 2005), we will gain a more complete picture of the Wnt signaling pathway. Further, screens for small molecules that can modulate the activity of essential signaling components such as β-catenin will help translate directly our knowledge of Wnt signaling into intervention for curing diseases related with de-regulated Wnt signaling such as cancers (Clevers, 2004; Lepourcelet et al., 2004).

Acknowledgment

I apologize to many investigators whose articles could not be cited due to space constraints. I thank American Heart Association, March of Dime foundation and the National Institutes of Health for support of our research.

References

Aberle, H., Bauer, A., Stappert, J., Kispert, A., and Kemler, R. (1997). beta-catenin is a target for the ubiquitin-proteasome pathway. Embo J *16*, 3797-804.

Adachi, S., Jigami, T., Yasui, T., Nakano, T., Ohwada, S., Omori, Y., Sugano, S., Ohkawara, B., Shibuya, H., Nakamura, T., *et al.* (2004). Role of a BCL9-related beta-catenin-binding protein, B9L, in tumorigenesis induced by aberrant activation of Wnt signaling. Cancer Res *64*, 8496-501.

Ahmed, Y., Nouri, A., and Wieschaus, E. (2002). *Drosophila* Apc1 and Apc2 regulate Wingless transduction throughout development. Development *129*, 1751-62.

Akong, K., McCartney, B. M., and Peifer, M. (2002). *Drosophila* APC2 and APC1 have overlapping roles in the larval brain despite their distinct intracellular localizations. Dev Biol *250*, 71-90.

Amit, S., Hatzubai, A., Birman, Y., Andersen, J. S., Ben-Shushan, E., Mann, M., Ben-Neriah, Y., and Alkalay, I. (2002). Axin-mediated CKI phosphorylation of beta-catenin at Ser 45: a molecular switch for the Wnt pathway. Genes Dev *16*, 1066-76.

Baeg, G. H., Lin, X., Khare, N., Baumgartner, S., and Perrimon, N. (2001). Heparan sulfate proteoglycans are critical for the organization of the extracellular distribution of Wingless. Development *128*, 87-94.

Balsamo, J., Arregui, C., Leung, T., and Lilien, J. (1998). The nonreceptor protein tyrosine phosphatase PTP1B binds to the cytoplasmic domain of N-cadherin and regulates the cadherin-actin linkage. J Cell Biol *143*, 523-32.

Barker, N., Hurlstone, A., Musisi, H., Miles, A., Bienz, M., and Clevers, H. (2001). The chromatin remodelling factor Brg-1 interacts with beta-catenin to promote target gene activation. Embo J *20*, 4935-43.

Bek, S., and Kemler, R. (2002). Protein kinase CKII regulates the interaction of beta-catenin with alpha-catenin and its protein stability. J Cell Sci *115*, 4743-53.

Belenkaya, T. Y., Han, C., Standley, H. J., Lin, X., Houston, D. W., and Heasman, J. (2002). pygopus Encodes a nuclear protein essential for wingless/Wnt signaling. Development *129*, 4089-101.

Bhanot, P., Brink, M., Samos, C. H., Hsieh, J. C., Wang, Y., Macke, J. P., Andrew, D., Nathans, J., and Nusse, R. (1996). A new member of the frizzled family from *Drosophila* functions as a Wingless receptor. Nature *382*, 225-30.

Bienz, M., (1998). TCF: transcriptional activator or repressor? Curr Opin Cell Biol *10*, 366-72.

Bienz, M., (2003). Apc. Curr Biol *13*, R215-6.

Bienz, M. (2005). beta-Catenin: a pivot between cell adhesion and Wnt signalling. Curr Biol *15*, R64-7.

Bienz, M., and Clevers, H. (2000). Linking colorectal cancer to Wnt signaling. Cell *103*, 311-20.

Bienz, M., and Hamada, F. (2004). Adenomatous polyposis coli proteins and cell adhesion. Curr Opin Cell Biol *16*, 528-35.

Birchmeier, W., Hulsken, J., and Behrens, J. (1995). Adherens junction proteins in tumour progression. Cancer Surv *24*, 129-40.

Boyden, L. M., Mao, J., Belsky, J., Mitzner, L., Farhi, A., Mitnick, M. A., Wu, D., Insogna, K., and Lifton, R. P. (2002). High bone density due to a mutation in LDL-receptor-related protein 5. N Engl J Med *346*, 1513-21.

Brannon, M., Gomperts, M., Sumoy, L., Moon, R. T., and Kimelman, D. (1997). A beta-catenin/XTcf-3 complex binds to the siamois promoter to regulate dorsal axis specification in *Xenopus*. Genes Dev *11*, 2359-70.

Brembeck, F. H., Schwarz-Romond, T., Bakkers, J., Wilhelm, S., Hammerschmidt, M., and Birchmeier, W. (2004). Essential role of BCL9-2 in the switch between beta-catenin's adhesive and transcriptional functions. Genes Dev *18*, 2225-30.

Cavallo, R. A., Cox, R. T., Moline, M. M., Roose, J., Polevoy, G. A., Clevers, H., Peifer, M., and Bejsovec, A. (1998). *Drosophila* Tcf and Groucho interact to repress Wingless signalling activity. Nature *395*, 604-8.

Chen, G., Fernandez, J., Mische, S., and Courey, A. J. (1999). A functional interaction between the histone deacetylase Rpd3 and the corepressor groucho in *Drosophila* development. Genes Dev *13*, 2218-30.

Chi, T. H., Wan, M., Lee, P. P., Akashi, K., Metzger, D., Chambon, P., Wilson, C. B,. and Crabtree, G. R. (2003). Sequential roles of Brg, the ATPase subunit of BAF chromatin remodeling complexes, in thymocyte development. Immunity *19*, 169-82.

Clevers, H. (2004). Wnt breakers in colon cancer. Cancer Cell *5*, 5-6.

Collins, R. T., and Treisman, J. E. (2000). Osa-containing Brahma chromatin remodeling complexes are required for the repression of wingless target genes. Genes Dev *14*, 3140-52.

Daniels, D. L., and Weis, W. I. (2002). ICAT inhibits beta-catenin binding to Tcf/Lef-family transcription factors and the general coactivator p300 using independent structural modules. Mol Cell *10*, 573-84.

Danilkovitch-Miagkova, A., Miagkov, A., Skeel, A., Nakaigawa, N., Zbar, B., and Leonard, E. J. (2001). Oncogenic mutants of RON and MET receptor tyrosine kinases cause activation of the beta-catenin pathway. Mol Cell Biol *21*, 5857-68.

Dasgupta, R., Kaykas, A., Moon, R. T., and Perrimon, N. (2005). Functional Genomic Analysis of the Wnt-Wingless Signaling Pathway. Science.

Dominguez, I., Itoh, K., and Sokol, S. Y. (1995). Role of glycogen synthase kinase 3 beta as a negative regulator of dorsoventral axis formation in *Xenopus* embryos. Proc Natl Acad Sci USA *92*, 8498-502.

Farr, G. H., 3rd, Ferkey, D. M., Yost, C., Pierce, S. B., Weaver, C., and Kimelman, D. (2000). Interaction among GSK-3, GBP, axin, and APC in *Xenopus* axis specification. J Cell Biol *148*, 691-702.

Giles, R. H., van Es, J. H., and Clevers, H. (2003). Caught up in a Wnt storm: Wnt signaling in cancer. Biochim Biophys Acta *1653*, 1-24.

Gong, Y., Slee, R. B., Fukai, N., Rawadi, G., Roman-Roman, S., Reginato, A. M., Wang, H., Cundy, T., Glorieux, F. H., Lev, D., *et al.* (2001). LDL receptor-related protein 5 (LRP5) affects bone accrual and eye development. Cell *107*, 513-23.

Graham, T. A., Clements, W. K., Kimelman, D., and Xu, W. (2002). The crystal structure of the beta-catenin/ICAT complex reveals the inhibitory mechanism of ICAT. Mol Cell *10*, 563-71.

Han, C., Yan, D., Belenkaya, T. Y., and Lin, X. (2005). *Drosophila* glypicans Dally and Dally-like shape the extracellular Wingless morphogen gradient in the wing disc. Development *132*, 667-79.

Hart, M. J., de los Santos, R., Albert, I. N., Rubinfeld, B., and Polakis, P. (1998). Downregulation of beta-catenin by human Axin and its association with the APC tumor suppressor, beta-catenin and GSK3 beta. Curr Biol *8*, 573-81.

He, X., Saint-Jeannet, J. P., Wang, Y., Nathans, J., Dawid, I. and Varmus, H., (1997). A member of the Frizzled protein family mediating axis induction by Wnt-5A. Science *275*, 1652-4.

He, X., Saint-Jeannet, J. P., Woodgett, J. R., Varmus, H. E., and Dawid, I. B. (1995). Glycogen synthase kinase-3 and dorsoventral patterning in *Xenopus* embryos. Nature *374*, 617-22.

He, X., Semenov, M., Tamai, K., and Zeng, X. (2004). LDL receptor-related proteins 5 and 6 in Wnt/beta-catenin signaling: arrows point the way. Development *131*, 1663-77.

Heasman, J., Crawford, A., Goldstone, K., Garner-Hamrick, P., Gumbiner, B., McCrea, P., Kintner, C., Noro, C. Y., and Wylie, C. (1994). Overexpression of cadherins and underexpression of beta-catenin inhibit dorsal mesoderm induction in early *Xenopus* embryos. Cell *79*, 791-803.

Hecht, A., Vleminckx, K., Stemmler, M. P., van Roy, F., and Kemler, R. (2000). The p300/CBP acetyltransferases function as transcriptional coactivators of beta-catenin in vertebrates. Embo J *19*, 1839-50.

Hellberg, C. B., Burden-Gulley, S. M., Pietz, G. E., and Brady-Kalnay, S. M. (2002). Expression of the receptor protein-tyrosine phosphatase, PTPmu, restores E-cadherin-dependent adhesion in human prostate carcinoma cells. J Biol Chem *277*, 11165-73.

Hsieh, J. C., Rattner, A., Smallwood, P. M., and Nathans, J. (1999). Biochemical characterization of Wnt-frizzled interactions using a soluble, biologically active vertebrate Wnt protein. Proc Natl Acad Sci USA *96*, 3546-51.

Hsu, W., Zeng, L., and Costantini, F. (1999). Identification of a domain of Axin that binds to the serine/threonine protein phosphatase 2A and a self-binding domain. J Biol Chem *274*, 3439-45.

Ikeda, S., Kishida, S., Yamamoto, H., Murai, H., Koyama, S., and Kikuchi, A. (1998). Axin, a negative regulator of the Wnt signaling pathway, forms a complex with GSK-3beta and beta-catenin and promotes GSK-3beta-dependent phosphorylation of beta-catenin. Embo J *17*, 1371-84.

Itoh, K., Krupnik, V. E., and Sokol, S. Y. (1998). Axis determination in *Xenopus* involves biochemical interactions of axin, glycogen synthase kinase 3 and beta-catenin. Curr Biol *8*, 591-4.

Katanaev, V. L., Ponzielli, R., Semeriva, M., and Tomlinson, A. (2005). Trimeric G protein-dependent frizzled signaling in *Drosophila*. Cell *120*, 111-22.

Kato, M., Patel, M. S., Levasseur, R., Lobov, I., Chang, B. H., Glass, D. A., 2nd, Hartmann, C., Li, L., Hwang, T. H., Brayton, C. F., *et al.* (2002). Cbfa1-independent decrease in osteoblast proliferation, osteopenia, and persistent embryonic eye vascularization in mice deficient in Lrp5, a Wnt coreceptor. J Cell Biol *157*, 303-14.

Kawahara, K., Morishita, T., Nakamura, T., Hamada, F., Toyoshima, K., and Akiyama, T. (2000). Down-regulation of beta-catenin by the colorectal tumor suppressor APC requires association with Axin and beta-catenin. J Biol Chem *275*, 8369-74.

Kemler, R. (1993). From cadherins to catenins: cytoplasmic protein interactions and regulation of cell adhesion. Trends Genet *9*, 317-21.

Kinzler, K. W., Nilbert, M. C., Su, L. K., Vogelstein, B., Bryan, T. M., Levy, D. B., Smith, K. J., Preisinger, A. C., Hedge, P., McKechnie, D., *et al.* (1991). Identification of FAP locus genes from chromosome 5q21. Science *253*, 661-5.

Kishida, S., Yamamoto, H., Ikeda, S., Kishida, M., Sakamoto, I., Koyama, S., and Kikuchi, A. (1998). Axin, a negative regulator of the wnt signaling pathway, directly interacts with adenomatous polyposis coli and regulates the stabilization of beta-catenin. J Biol Chem *273*, 10823-6.

Kramps, T., Peter, O., Brunner, E., Nellen, D., Froesch, B., Chatterjee, S., Murone, M., Zullig, S., and Basler, K. (2002). Wnt/wingless signaling requires BCL9/legless-mediated recruitment of pygopus to the nuclear beta-catenin-TCF complex. Cell *109*, 47-60.

Lammi, L., Arte, S., Somer, M., Jarvinen, H., Lahermo, P., Thesleff, I., Pirinen, S., and Nieminen, P. (2004). Mutations in AXIN2 cause familial tooth agenesis and predispose to colorectal cancer. Am J Hum Genet *74*, 1043-50.

Latres, E., Chiaur, D. S., and Pagano, M. (1999). The human F box protein beta-Trcp associates with the Cul1/Skp1 complex and regulates the stability of beta-catenin. Oncogene *18*, 849-54.

Lepourcelet, M., Chen, Y. N., France, D. S., Wang, H., Crews, P., Petersen, F., Bruseo, C., Wood, A. W., and Shivdasani, R. A. (2004). Small-molecule antagonists of the oncogenic Tcf/beta-catenin protein complex. Cancer Cell *5*, 91-102.

Li, L., Yuan, H., Weaver, C. D., Mao, J., Farr, G. H., 3rd, Sussman, D. J., Jonkers, J., Kimelman, D., and Wu, D. (1999). Axin and Frat1 interact with dvl and GSK, bridging Dvl to GSK in Wnt-mediated regulation of LEF-1. Embo J *18*, 4233-40.

Lickert, H., Bauer, A., Kemler, R., and Stappert, J. (2000). Casein kinase II phosphorylation of E-cadherin increases E-cadherin/beta-catenin interaction and strengthens cell-cell adhesion. J Biol Chem *275*, 5090-5.

Lin, R., Hill, R. J., and Priess, J. R. (1998). POP-1 and anterior-posterior fate decisions in *C. elegans* embryos. Cell *92*, 229-39.

Lin, X. (2004). Functions of heparan sulfate proteoglycans in cell signaling during development. Development *131*, 6009-21.

Lin, X., and Perrimon, N. (1999). Dally cooperates with *Drosophila* Frizzled 2 to transduce Wingless signalling. Nature *400*, 281-4.

Liu, C., Kato, Y., Zhang, Z., Do, V. M., Yankner, B. A., and He, X. (1999). beta-Trcp couples beta-catenin phosphorylation-degradation and regulates *Xenopus* axis formation. Proc Natl Acad Sci USA *96*, 6273-8.

Liu, C., Li, Y., Semenov, M., Han, C., Baeg, G. H., Tan, Y., Zhang, Z., Lin, X., and He, X. (2002). Control of beta-catenin phosphorylation/degradation by a dual-kinase mechanism. Cell *108*, 837-47.

Liu, T., DeCostanzo, A. J., Liu, X., Wang, H., Hallagan, S., Moon,

R. T., and Malbon, C. C. (2001). G protein signaling from activated rat frizzled-1 to the beta-catenin-Lef-Tcf pathway. Science 292, 1718-22.

Logan, C. Y., and Nusse, R. (2004). The Wnt signaling pathway in development and disease. Annu Rev Cell Dev Biol 20, 781-810.

Mao, J., Wang, J., Liu, B., Pan, W., Farr, G. H., 3rd, Flynn, C., Yuan, H., Takada, S., Kimelman, D., Li, L., et al. (2001). Low-density lipoprotein receptor-related protein-5 binds to Axin and regulates the canonical Wnt signaling pathway. Mol Cell 7, 801-9.

Moon, R. T. (2004). Teaching resource. Beta-catenin signaling and axis specification. Sci STKE 2004, tr6.

Moon, R. T. (2005). Wnt/beta-catenin pathway. Sci STKE 2005, cm1.

Munemitsu, S., Albert, I., Souza, B., Rubinfeld, B. and Polakis, P. (1995). Regulation of intracellular beta-catenin levels by the adenomatous polyposis coli (APC) tumor-suppressor protein. Proc Natl Acad Sci USA 92, 3046-50.

Nathke, I. S. (2004). The adenomatous polyposis coli protein: the Achilles heel of the gut epithelium. Annu Rev Cell Dev Biol 20, 337-66.

Nawroth, R., Poell, G., Ranft, A., Kloep, S., Samulowitz, U., Fachinger, G., Golding, M., Shima, D. T., Deutsch, U., and Vestweber, D. (2002). VE-PTP and VE-cadherin ectodomains interact to facilitate regulation of phosphorylation and cell contacts. Embo J 21, 4885-95.

Nelson, W. J., and Nusse, R. (2004). Convergence of Wnt, beta-catenin, and cadherin pathways. Science 303, 1483-7.

Nishisho, I., Nakamura, Y., Miyoshi, Y., Miki, Y., Ando, H., Horii, A., Koyama, K., Utsunomiya, J., Baba, S., and Hedge, P. (1991). Mutations of chromosome 5q21 genes in FAP and colorectal cancer patients. Science 253, 665-9.

Nusse, R. (1999). WNT targets. Repression and activation. Trends Genet 15, 1-3.

Pai, L. M., Orsulic, S., Bejsovec, A., and Peifer, M. (1997). Negative regulation of Armadillo, a Wingless effector in Drosophila. Development 124, 2255-66.

Parker, D. S., Jemison, J,. and Cadigan, K. M. (2002). Pygopus, a nuclear PHD-finger protein required for Wingless signaling in Drosophila. Development 129, 2565-76.

Peifer, M., Berg, S., and Reynolds, A. B. (1994). A repeating amino acid motif shared by proteins with diverse cellular roles. Cell 76, 789-91.

Peifer, M., McCrea, P. D., Green, K. J., Wieschaus, E., and Gumbiner, B. M. (1992). The vertebrate adhesive junction proteins beta-catenin and plakoglobin and the Drosophila segment polarity gene armadillo form a multigene family with similar properties. J Cell Biol 118, 681-91.

Perez-Moreno, M., Jamora, C., and Fuchs, E. (2003). Sticky business: orchestrating cellular signals at adherens junctions. Cell 112, 535-48.

Peters, J. M., McKay, R. M., McKay, J. P., and Graff, J. M. (1999). Casein kinase I transduces Wnt signals. Nature 401, 345-50.

Piedra, J., Martinez, D., Castano, J., Miravet, S., Dunach, M., and de Herreros, A. G. (2001). Regulation of beta-catenin structure and activity by tyrosine phosphorylation. J Biol Chem 276, 20436-43.

Piedra, J., Miravet, S., Castano, J., Palmer, H. G., Heisterkamp, N., Garcia de Herreros, A., and Dunach, M. (2003). p120 Catenin-associated Fer and Fyn tyrosine kinases regulate beta-catenin Tyr-142 phosphorylation and beta- catenin- alpha-catenin Interaction. Mol Cell Biol 23, 2287-97.

Pierce, S. B., and Kimelman, D. (1995). Regulation of Spemann organizer formation by the intracellular kinase Xgsk-3. Development 121, 755-65.

Pinson, K. I., Brennan, J., Monkley, S., Avery, B. J., and Skarnes, W. C. (2000). An LDL-receptor-related protein mediates Wnt signalling in mice. Nature 407, 535-8.

Polakis, P. (1997). The adenomatous polyposis coli (APC) tumor suppressor. Biochim Biophys Acta 1332, F127-47.

Riese, J., Yu, X., Munnerlyn, A., Eresh, S., Hsu, S. C., Grosschedl, R., and Bienz, M. (1997). LEF-1, a nuclear factor coordinating signaling inputs from wingless and decapentaplegic. Cell 88, 777-87.

Rijsewijk, F., Schuermann, M., Wagenaar, E., Parren, P., Weigel, D., and Nusse, R. (1987). The Drosophila homolog of the mouse mammary oncogene int-1 is identical to the segment polarity gene wingless. Cell 50, 649-57.

Robitaille, J., MacDonald, M. L., Kaykas, A., Sheldahl, L. C., Zeisler, J., Dube, M. P., Zhang, L. H., Singaraja, R. R., Guernsey, D. L., Zheng, B., et al. (2002). Mutant frizzled-4 disrupts retinal angiogenesis in familial exudative vitreoretinopathy. Nat Genet 32, 326-30.

Roura, S., Miravet, S., Piedra, J., Garcia de Herreros, A., and Dunach, M. (1999). Regulation of E-cadherin/Catenin association by tyrosine phosphorylation. J Biol Chem 274, 36734-40.

Rubinfeld, B., Albert, I., Porfiri, E., Fiol, C., Munemitsu, S., and Polakis, P. (1996). Binding of GSK3beta to the APC-beta-catenin complex and regulation of complex assembly. Science 272, 1023-6.

Rulifson, E. J., Wu, C. H., and Nusse, R. (2000). Pathway specificity by the bifunctional receptor frizzled is determined by affinity for wingless. Mol Cell 6, 117-26.

Sakanaka, C., Leong, P., Xu, L., Harrison, S. D., and Williams, L. T. (1999). wCasein kinase iepsilon in the wnt pathway: regulation of beta-catenin function. Proc Natl Acad Sci USA 96, 12548-52.

Sakanaka, C., Weiss, J. B., and Williams, L. T. (1998). Bridging of beta-catenin and glycogen synthase kinase-3beta by axin and inhibition of beta-catenin-mediated transcription. Proc Natl Acad Sci USA 95, 3020-3.

Salic, A., Lee, E., Mayer, L., and Kirschner, M. W. (2000). Control of beta-catenin stability: reconstitution of the cytoplasmic steps of the wnt pathway in *Xenopus* egg extracts. Mol Cell *5*, 523-32.

Sanson, B., White, P., and Vincent, J. P. (1996). Uncoupling cadherin-based adhesion from wingless signalling in *Drosophila*. Nature *383*, 627-30.

Satoh, K., Kasai, M., Ishidao, T., Tago, K., Ohwada, S., Hasegawa, Y., Senda, T., Takada, S., Nada, S., Nakamura, T., *et al.* (2004). Anteriorization of neural fate by inhibitor of beta-catenin and T cell factor (ICAT), a negative regulator of Wnt signaling. Proc Natl Acad Sci USA *101*, 8017-21.

Satoh, S., Daigo, Y., Furukawa, Y., Kato, T., Miwa, N., Nishiwaki, T., Kawasoe, T., Ishiguro, H., Fujita, M., Tokino, T., *et al.* (2000). AXIN1 mutations in hepatocellular carcinomas, and growth suppression in cancer cells by virus-mediated transfer of AXIN1. Nat Genet *24*, 245-50.

Sawa, H., Lobel, L., and Horvitz, H. R. (1996). The *Caenorhabditis elegans* gene lin-17, which is required for certain asymmetric cell divisions, encodes a putative seven-transmembrane protein similar to the *Drosophila* frizzled protein. Genes Dev *10*, 2189-97.

Siegfried, E., Chou, T. B., and Perrimon, N. (1992). wingless signaling acts through zeste-white 3, the *Drosophila* homolog of glycogen synthase kinase-3, to regulate engrailed and establish cell fate. Cell *71*, 1167-79.

Siegfried, E., and Perrimon, N. (1994). *Drosophila* wingless: a paradigm for the function and mechanism of Wnt signaling. Bioessays *16*, 395-404.

Siegfried, E., Wilder, E. L., and Perrimon, N. (1994). Components of wingless signalling in *Drosophila*. Nature *367*, 76-80.

Spink, K. E., Polakis, P., and Weis, W. I. (2000). Structural basis of the Axin-adenomatous polyposis coli interaction. Embo J *19*, 2270-9.

Strutt, D. (2003). Frizzled signalling and cell polarisation in *Drosophila* and vertebrates. Development *130*, 4501-13.

Tago, K., Nakamura, T., Nishita, M., Hyodo, J., Nagai, S., Murata, Y., Adachi, S., Ohwada, S., Morishita, Y., Shibuya, H., *et al.* (2000). Inhibition of Wnt signaling by ICAT, a novel beta-catenin-interacting protein. Genes Dev *14*, 1741-9.

Takemaru, K., Yamaguchi, S., Lee, Y. S., Zhang, Y., Carthew, R. W., and Moon, R. T. (2003). Chibby, a nuclear beta-catenin-associated antagonist of the Wnt/Wingless pathway. Nature *422*, 905-9.

Takemaru, K. I., and Moon, R. T. (2000). The transcriptional coactivator CBP interacts with beta-catenin to activate gene expression. J Cell Biol *149*, 249-54.

Tamai, K., Semenov, M., Kato, Y., Spokony, R., Liu, C., Katsuyama, Y., Hess, F., Saint-Jeannet, J. P., and He, X. (2000). LDL-receptor-related proteins in Wnt signal transduction. Nature *407*, 530-5.

Tamai, K., Zeng, X., Liu, C., Zhang, X., Harada, Y., Chang, Z., and He, X. (2004). A mechanism for Wnt coreceptor activation. Mol Cell *13*, 149-56.

Tao, Q., Yokota, C., Puck, H., Kofron, M., Birsoy, B., Yan, D., Asashima, M., Wylie, C. C., Lin, X., and Heasman, J. (2005). Maternal wnt11 activates the canonical wnt signaling pathway required for axis formation in *Xenopus* embryos. Cell *120*, 857-71.

Thompson, B., Townsley, F., Rosin-Arbesfeld, R., Musisi, H., and Bienz, M. (2002). A new nuclear component of the Wnt signalling pathway. Nat Cell Biol *4*, 367-73.

Tolwinski, N. S., Wehrli, M., Rives, A., Erdeniz, N., DiNardo, S., and Wieschaus, E. (2003). Wg/Wnt signal can be transmitted through arrow/LRP5,6 and Axin independently of Zw3/Gsk3beta activity. Dev Cell *4*, 407-18.

Toomes, C., Bottomley, H. M., Jackson, R. M., Towns, K. V., Scott, S., Mackey, D. A., Craig, J. E., Jiang, L., Yang, Z., Trembath, R., *et al.* (2004). Mutations in LRP5 or FZD4 underlie the common familial exudative vitreoretinopathy locus on chromosome 11q. Am J Hum Genet *74*, 721-30.

Townsley, F. M., Cliffe, A., and Bienz, M. (2004). Pygopus and Legless target Armadillo/beta-catenin to the nucleus to enable its transcriptional co-activator function. Nat Cell Biol *6*, 626-33.

Tsuda, M., Kamimura, K., Nakato, H., Archer, M., Staatz, W., Fox, B., Humphrey, M., Olson, S., Futch, T., Kaluza, V., *et al.* (1999). The cell-surface proteoglycan Dally regulates Wingless signalling in *Drosophila*. Nature *400*, 276-80.

Veeman, M. T., Axelrod, J. D., and Moon, R. T. (2003). A second canon. Functions and mechanisms of beta-catenin-independent Wnt signaling. Dev Cell *5*, 367-77.

Wehrli, M., Dougan, S. T., Caldwell, K., O'Keefe, L., Schwartz, S., Vaizel-Ohayon, D., Schejter, E., Tomlinson, A., and DiNardo, S. (2000). Arrow encodes an LDL-receptor-related protein essential for Wingless signalling. Nature *407*, 527-30.

Willis, T. G., Zalcberg, I. R., Coignet, L. J., Wlodarska, I., Stul, M., Jadayel, D. M., Bastard, C., Treleaven, J. G., Catovsky, D., Silva, M. L., *et al.* (1998). Molecular cloning of translocation t(1;14)(q21;q32) defines a novel gene (BCL9) at chromosome 1q21. Blood *91*, 1873-81.

Wodarz, A., and Nusse, R. (1998). Mechanisms of Wnt signaling in development. Annu Rev Cell Dev Biol *14*, 59-88.

Wolf, D., Rodova, M., Miska, E. A., Calvet, J. P., and Kouzarides, T. (2002). Acetylation of beta-catenin by CREB-binding protein (CBP). J Biol Chem *277*, 25562-7.

Wu, C. H., and Nusse, R. (2002). Ligand receptor interactions in the Wnt signaling pathway in *Drosophila*. J Biol Chem *277*, 41762-9.

Yanagawa, S., Matsuda, Y., Lee, J. S., Matsubayashi, H., Sese, S., Kadowaki, T., and Ishimoto, A. (2002). Casein kinase I phosphorylates the Armadillo protein and induces its degradation in *Drosophila*. Embo J *21*, 1733-42.

Yang-Snyder, J., Miller, J. R., Brown, J. D., Lai, C. J., and Moon, R. T. (1996). A frizzled homolog functions in a vertebrate Wnt signaling pathway. Curr Biol *6*, 1302-6.

Yost, C., Torres, M., Miller, J. R., Huang, E., Kimelman, D., and

Moon, R. T. (1996). The axis-inducing activity, stability, and subcellular distribution of beta-catenin is regulated in *Xenopus* embryos by glycogen synthase kinase 3. Genes Dev *10*, 1443-54.

Zecca, M., Basler, K., and Struhl, G. (1996). Direct and long-range action of a wingless morphogen gradient. Cell *87*, 833-44.

Zeng, L., Fagotto, F., Zhang, T., Hsu, W., Vasicek, T. J., Perry, W. L., 3rd, Lee, J. J., Tilghman, S. M., Gumbiner, B. M., and Costantini, F. (1997). The mouse Fused locus encodes Axin, an inhibitor of the Wnt signaling pathway that regulates embryonic axis formation. Cell *90*, 181-92.

Chapter 33
Regulatory Mechanisms for Floral Organ Identity Specification in *Arabidopsis thaliana*

Zhongchi Liu

Dept. of Cell Biology and Molecular Genetics, University of Maryland, College Park, MD 20742

Key Words: ABCE model, flower development, floral organ identity, *LEAFY (LFY), APETALA1 (AP1), APETALA2 (AP2), WUSCHEL (WUS), AGAMOUS (AG), APETALA3 (AP3), PISTALLATA (PI), SEPALLATA (SEP), LEUNIG (LUG), SEUSS (SEU), UNUSUAL FLORAL ORGAN (UFO), miR172* microRNA, MADS-box genes, AP2-domain, floral meristem, shoot apical meristem (SAM), meristem determinancy, homeotic transformation

Summary

In the past decade, a major milestone in plant developmental biology is the elucidation of the molecular genetic basis underlying floral organ identity specification. Based on the genetic characterization of floral homeotic mutants in *Arabidopsis thaliana* and *Antirrhinum majus*, a simple and elegant ABC model was established to explain how the four types of floral organs (sepals, petals, stamens, and carpels) are specified by A, B, and C classes of floral homeotic genes. Reverse genetics later led to the discovery of the E class genes, and the ABC model was renamed the ABCE model. To date, both transcriptional and post-transcriptional mechanisms for the domain-specific expression of A, B, and C class genes are being revealed. In addition to transcriptional activation of the ABC genes by a plant specific regulatory protein *LEAFY* and its co-regulators, positive autoregulation appears to refine and maintain ABCE gene expression. Additional regulatory mechanisms include microRNA-mediated translational block, the recruitment of transcription co-repressors, and the ubiquitin-mediated protein degradation. These discoveries provided molecular insights into how floral organs are evolved from leaves.

Introduction

As indicated by other chapters of this book, pioneering work on regulatory mechanisms of gene expression was performed in yeast, *Drosophila*, and mammalian cells. The establishment of *Arabidopsis thaliana* as a model plant and the development of various molecular and genetic tools such as the availability of the entire genome sequence, easy and rapid transformation protocol, and the large number of Transfer-DNA (T-DNA) and transposon insertion mutations make *Arabidopsis thaliana* an attractive system to study gene regulation. In the past decade, a major milestone in plant developmental biology has been the elucidation of the molecular genetic mechanism underlying floral organ identity specification. The ABCE model for floral organ identity specification provides a framework and molecular genetic tools for uncovering regulatory mechanisms that are unique to plants or common to both plants and animals. In this chapter, I will highlight these regulatory mechanisms. I will focus on research results obtained from studying *Arabidopsis thaliana* and will discuss those findings considered to be most central to the understanding of floral organ identity specification. (Please see Jack, 2004, for a more comprehensive review of flower development). Insights from studying floral organ identity specification should broaden our understanding of regulatory mechanisms in all living organisms.

Corresponding Author: Tel: (301) 405-1586, Fax: (301) 314-9082, E-mail: Zliu@umd.edu

The ABC Model for Flower Development

The *Arabidopsis* flower is typical of many angiosperm flowers. Four types of floral organs are arranged in four concentric circles or whorls (Fig.33.1A and 33.2A). Sepals and petals comprise the outer two whorls (whorls 1, 2), while the reproductive organs stamens and carpels make up the inner two whorls (whorls 3, 4). Despite dramatic variation in the number, color and shape of floral organs in different species, this arrangement of sepal, petal, stamen, and carpel from the outer-most whorl to the inner-most whorl is fixed in the majority of the angiosperm species. This cross-species similarity suggests that the molecular genetic systems responsible for patterning of floral organs are similar in the majority of flowering plants.

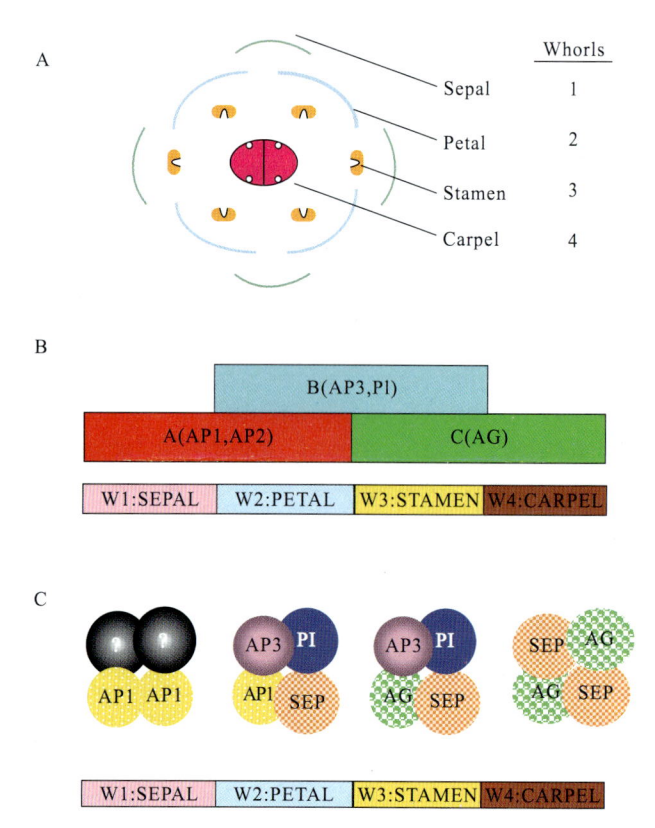

Fig.33.1 The ABCE model for floral organ identity specification. (A) A diagram of an *Arabidopsis* flower showing four sepals in whorl 1, four petals in whorl 2, six stamens in whorl 3, and two fused carpels in whorl 4. (B) The domains of the A, B, and C activities in a wild type flower. Class A activity is provided by *AP1* and *AP2* in whorls 1-2; class B activity is provided by *AP3* and *PI* in whorls 2-3; class C activity is provided by a single gene *AG* in whorls 3-4. W1 to W4 indicate whorl 1 to whorl 4. (C) A revised ABCE model incorporating the SEP protein in tetrameric complexes.

The ABC model for flower development explains how three classes of genes (A, B, C classes) direct the development of these four types of floral organs (Coen and Meyerowitz, 1991; Weigel and Meyerowitz, 1994). The model was established based on genetic characterization of *Arabidopsis thaliana* and *Antirrhinum majus* floral homeotic mutants. Mutations in these "floral homeotic genes" result in the substitutions or replacement of one organ type by another organ type (Fig. 33.2B, 33.2C, and 33.2D). The ABC model places floral homeotic mutants and the corresponding genes into three classes: A, B, or C (Fig. 33.1B). In wild type, class A activity is present in whorls 1-2, class B activity is restricted to whorls 2-3, and class C activity is only present in whorls 3-4. In whorl 1, the class A activity specifies sepal development. In whorl 2, where both A and B class genes are active, petal identity is specified. In whorl 3, B and C together specify stamen identity, and in whorl 4, C activity alone specifies carpel development. In addition, the model predicts that the A and C activities are antagonistic to each other. A function inhibits C function in whorls 1-2, whereas C function inhibits A function in whorls 3-4. Since the primary function of the ABC genes is to specify floral organ identity, the ABC class floral homeotic genes are also termed the "organ identity genes".

APETALA1 (AP1) and *APETALA2 (AP2)* are both class A genes as they are both required to specify sepal and petal identity (Fig. 33.1B) (Bowman et al., 1989; Bowman et al., 1991). However, *ap1* and *ap2* mutants exhibit different phenotypes. In *ap1* mutants, C function is still restricted to whorls 3-4, and *ap1* flowers develop leaf-like organs in whorl 1, no organ formation in whorl 2, but normal stamens and carpels in whorls 3 and 4. In contrast, in *ap2* mutants, C activity is spread to all four whorls resulting in a flower with carpels in whorl 1, stamens in whorls 2 and 3, and carpels in whorl 4 (Fig. 33.2B). This difference in phenotype indicates that *AP2* plays a predominant role in the antagonistic function predicted by the ABC model.

APETALA3 (AP3) and *PISTILLATA (PI)* are the *Arabidopsis* B class genes (Fig. 33.1B). Mutations in either *AP3* or *PI* cause similar homeotic transformations in whorls 2-3 such that whorl 2 organs develop as sepals, and whorl 3 organs develop as carpels (Fig. 33.2C) (Bowman et al., 1989). *AP3* and *PI* both encode MADS domain proteins that have been shown to bind DNA only as AP3/PI heterodimers (Goto and Meyerowitz, 1994; Jack et al., 1992; Riechmann et al., 1996a; Riechmann et al., 1996b). Obligatory heterodimer formation explains why both *AP3* and *PI* are required for the B activity.

Fig.33.2 Phenotypes and expression patterns of ABC genes. (A) A wild-type flower with four sepals in whorl 1, four petals in whorl 2, six stamens in whorl 3, and two fused carpels in whorl 4. (B) A class A floral homeotic mutant, *ap2-2*. All whorl 1 organs develop as carpels. All whorl 2 organs and most of whorl 3 stamens are absent. The whorl 4 carpels are similar to wild-type. (C) A class B floral homeotic mutant, *pi-1*. Organs in the outer two whorls are sepals. Organs in the inner two whorls are carpels. (D) A class C mutant, *ag-1*, consisting of four whorl 1 sepals, four whorl 2 petals, six whorl 3 petals, and a new flower in whorl 4. (E) Class A gene *AP1* expression pattern revealed by RNA *in situ* hybridization. *AP1* mRNA is detected in the entire floral meristem at stage 2. However, *AP1* mRNA is restricted to the emerging sepal primordia and absent from the inner whorl meristem cells at stage 3. Petal primordia are not yet formed at this stage. Number indicates the stage of floral meristems. See Smyth *et al.* (1990) for a description of *Arabidopsis* floral stages. (F) *AP3* expression revealed by RNA *in situ* hybridization. *AP3* is expressed in the region between developing sepals and carpels indicated by S and C respectively. (G) *AG* mRNA expression revealed by RNA *in situ* hybridization. *AG* mRNA is detected in the developing stamen and carpel primordium in a flower. (H) A flower from a transgenic plant containing *35S::AP3* and *35S::PI*. Two outer whorls of petals and two inner whorls of stamens are formed in this flower. (I) A flower from a transgenic plant containing *35S::AP3, 35S::PI* and *35S::AG*. The flower develops stamens in all four whorls. (J) A *sep1 sep2 sep3* triple mutant flower. Sepals or sepaloid organs are formed in whorls 1-3 and a new flower repeating this same pattern is formed in whorl 4. (K) Three-week-old *35S::PI; 35S::AP3; 35S::SEP3* transgenic plant. The two embryonic leaves called cotyledons are normal, true leaves are however transformed into petaloid organs. Numbers indicate the order of leaf development. s, stamens; c, Cotyledons; TF, terminal flower. Photos in J and K are reprinted with permission from Nature Publishing Group.

AGAMOUS (AG) was the first C class gene identified and molecularly isolated (Bowman *et al.*, 1989; Bowman *et al.*, 1991; Yanofsky *et al.*, 1990). *AG* plays a key role both in specifying stamen and carpel identity and in the antagonistic function against class A genes (Fig.33.1B). In *ag* loss-of-function mutants, the class A activity is expanded into whorls 3-4. As a result, stamens are replaced by petals, and carpels are replaced by a new flower (Fig.33.2D). The new flower repeats the same pattern of "sepal, petal, petal", leading to the "flower within a flower" phenotype. This reveals a third role of *AG* in maintaining the determinacy of the floral meristem. In summary, *AG* has at least three functions: repressing class A activity in whorls 3-4, specifying stamen and carpel organ identity, and maintaining the determinacy of floral meristems.

Over 200 years ago, Johann Wolfgang von Goethe proposed that the floral organs are modified leaves. What controls the difference between a floral organ and a leaf? Triple mutants combining mutations in all A, B, and C classes resulted in the formation of flowers with leaf-like organs in all floral whorls (Bowman *et al.*, 1991). This result suggests that leaves are being transformed into floral organs by the action of ABC

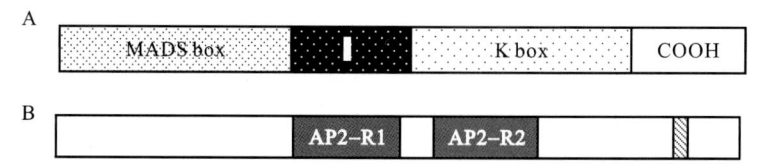

Fig.33.3 Diagrams of MADS box and AP2 domain proteins. (A) The domain structure of MADS box proteins. *AP1, AP3, PI, AG,* and *SEP1, SEP2, SEP3* all encode members of this gene family. The MADS box domain is the most conserved domain required for DNA binding and dimerization. The intervening (I) region between MADS box and K box is important for dimerization specificity. The K-Box domain is important for dimerization, and the C-terminal domain (COOH) is highly divergent among different MADS box proteins and possesses transcription activation function in AP1 and SEP. (B) The domain structure of the AP2 protein. AP2-R1 and AP2-R2 are the two conserved and homologous domains which function in DNA binding. The single hatched box is the binding site of a microRNA, *miR172*.

genes, lending strong supports for von Goethe's theory.

All Floral Homeotic Genes Encode DNA-Binding Transcription Factors

With the exception of *AP2*, all ABC (*AP1, AP3, PI,* and *AG*) genes encode members of a multi-gene family called the MADS-box gene family (Fig.33.3A) (Riechmann and Meyerowitz, 1997b). The name of the MADS-domain was derived from the four founding members: *MCM1*, yeast; *AG*, Arabidopsis; *DEFICIENS* (*DEF*), *Antirrhinum*; and *SRF*, human. The basic domain structure of MADS box proteins is illustrated in Fig.33.3A. The N-terminal half of the MADS domain is essential for DNA binding and the C-terminal half of the MADS domain is required for dimerization (Riechmann *et al.*, 1996b). In the majority of plant MADS domain-containing proteins, a second conserved domain, the K box, was identified because of its similarity to the coiled-coil domain of keratin (Ma *et al.*, 1991). The distinctive feature of the K box is the disposition of hydrophobic residues with a spacing that permits the formation of amphipathic α-helices (Ma *et al.*, 1991; Pnueli *et al.*, 1991). Between the MADS domain and the K box is a less strictly conserved Intervening (I) region. Amino acids in the I region and the K box have been shown to be important for the partner specificity in dimer formation (Riechmann *et al.*, 1996a). Finally, the C-terminal domain of the MADS box proteins is highly divergent. While the C-terminal domain of *AP1* exhibited transcription activation function, the C-terminal domain of *AP3, PI,* and *AG* did not exhibit such an activity (Honma and Goto, 2001).

MADS-domain proteins function as dimers and bind to a core consensus site $CC(A/T)_6GG$ which is known as the CArG-box (Huang *et al.*, 1996; Schwarz-Sommer *et al.*, 1992; Shiraishi *et al.*, 1993; Wynne and Treisman, 1992). Nevertheless, functional specificity (i.e. distinct organ identity activity) of the MADS-box proteins is independent of their DNA-binding specificity.

For example, hybrid genes were generated by swapping the amino terminal half of the MADS domain of the *Arabidopsis* proteins AP1, AP3, PI, and AG with the corresponding portion of human MEF2A or SRF proteins. Such hybrid proteins, having acquired the *in vitro* binding specificity of MEF2A or SRF, are able to perform the specific functions of the corresponding *Arabidopsis* genes in transgenic plants (Krizek and Meyerowitz, 1996b; Riechmann and Meyerowitz, 1997a). Thus, interactions between these MADS proteins with additional cofactors are probably crucial for the specific organ identity function.

AP2 is unique in that it encodes a member of the AP2/EREBP transcription factor family (Jofuku *et al.*, 1994; Riechmann and Meyerowitz, 1998). The AP2/EREBP transcription factors have one or two copies of a conserved 68 amino acid region dubbed the AP2 domain (Fig. 33.3B). Previous studies with EREBPs (ethylene-responsive element binding proteins) with a single AP2 domain demonstrated that the AP2 domain recognizes and binds to DNA specifically in an 11-bp sequence (TAAGAGCCGCC), the GCC box (Ohme-Takagi and Shinshi, 1995). However, *AP2* and another floral regulator *AINTEGUMENTA (ANT)* encode two AP2 domains (Elliott *et al.*, 1996; Klucher *et al.*, 1996; Krizek *et al.*, 2000). Using an *in vitro* selection procedure, the DNA binding specificity of ANT was found to be 5'-gCAC(A/G)N(A/T)TcCC(a/g) ANG(c/t)-3' (Nole-Wilson and Krizek, 2000). Neither single AP2 domain of *ANT* was capable of binding to the selected sequences, suggesting that both AP2 domains make DNA contacts. Therefore, AP2/ANT proteins with two AP2 domains exhibit different DNA-binding properties from the EREBPs.

Most A, B, and C Class Genes are Regulated at Transcription Level

The ABC model gave very specific predictions about where the A, B and C genes are functioning within a flower. Are these ABC genes only expressed in the floral whorls in which their functions are required? RNA *in situ* hybridization revealed that the mRNA expression domains of ABC genes largely coincided with the domains of their function predicted by the ABC model (Fig. 33.2E, 33.2F, and 33.2G) (Drews *et al.*, 1991; Goto and Meyerowitz, 1994; Jack *et al.*, 1992; Mandel *et al.*, 1992b). For example, the class A gene *AP1* is expressed in whorls 1-2 (Fig. 33.2E), the class B gene *AP3* is expressed in whorls 2-3 (Fig. 33.2F), and the class C gene *AG* is expressed in whorls 3-4 (Fig. 33.2G). Furthermore, *AG* mRNA is expanded to all four whorls in *ap2* mutants (Drews *et al.*, 1991). Hence, the spatially restricted function of the ABC genes is regulated either at the transcription level or at the RNA stability level. *AP2* is the only gene whose mRNA expression does not coincide with the domain of its function. *AP2* mRNA is expressed in all floral whorls although its function is limited to whorls 1-2 (Jofuku *et al.*, 1994). *AP2* was later shown to be regulated at translational level mediated by a microRNA (Aukerman and Sakai, 2003; Chen, 2004)

One way to distinguish if the regulation is at transcription level or post-transcriptional (such as RNA stability) level is to utilize reporter genes. *GUS* (*β-glucuronidase*) is a reporter gene commonly used in plant research (Jefferson *et al.*, 1987). The *GUS* gene was fused to the promoter or intron sequences of the ABC genes. The reporter gene was then transformed into *Arabidopsis* to generate stable transgenic plants. The expression pattern of class B genes (such as *AP3*) monitored by the reporter *GUS* was similar to mRNA expression pattern revealed by *in situ* hybridization (Jack *et al.*, 1994; Krizek and Meyerowitz, 1996a). This indicates that the expression of *AP3* is regulated at the transcriptional level rather than at post-transcriptional level. Interestingly, the cis-regulatory element of *AG* resides in the second intron of *AG*. The second intron of *AG* directed *GUS* expression in a pattern similar to the endogenous *AG* mRNA (Busch *et al.*, 1999; Deyholos and Sieburth, 2000; Sieburth and Meyerowitz, 1997). This suggests that the inner whorl-specific activity of *AG* is regulated at transcription level.

If the ABC genes are regulated at the transcription level, one should be able to design and engineer the structure of flowers in a predictable manner simply by altering the mRNA expression domain of the ABC genes. Several studies demonstrated that, indeed, one could ectopically express B and C genes in different genetic backgrounds to generate flowers consisting of, for example, all stamens or all petals. The *35S* promoter from the cauliflower mosaic virus (*CaMV*) is frequently used to drive the constitutive and ectopic expression of plant genes. Ectopic expression of the class C gene *AG* under the 35S promoter caused homeotic conversion from sepals to carpels and petals to stamens (Mandel *et al.*, 1992a; Mizukami and Ma, 1992). Thus, *AG* is both necessary and sufficient for the specification of stamen and carpel identity within a flower. Transgenic plants that constitutively and simultaneously express both class B genes *AP3* and *PI* (i.e. *35S::AP3; 35S::PI*) develop flowers that have petals in whorls 1-2 and stamens in whorls 3-4 (Fig.33.2H) (Krizek and Meyerowitz, 1996a). When the *35S::AP3* and *35S::PI* transgenes were crossed into class A mutant *ap2*, these *35S::AP3, 35S::PI, ap2* plants develop flowers consisting of all stamens (Fig.33.2I) (Krizek and Meyerowitz, 1996a). These experiments not only validated the ABC model but also provided the first example of novel floral varieties through genetic engineering of ABC genes.

The Discovery of E Class Genes Led to a Revised ABC Model

One interesting observation from above ectopic studies is that leaves from the *35S::B* or *35S::C* transgenic plants remain, to a large extent, leaves (Krizek and Meyerowitz, 1996a). Thus, the B and C genes are necessary and sufficient for their function only within the context of a flower. Why can't the ectopic expression of B or C genes change a leaf into a floral organ? One possibility is that another floral-specific factor is required for the floral organ specification and this factor is not expressed in leaves. Alternatively, a floral organ can only be formed after a lateral organ is first turned into a flower. The discovery of class E genes, *SEPALLATA1* (*SEP1*), *SEP2*, and *SEP3* supports the first possibility.

The *SEP* genes were previously named *AGAMOUS-LIKE2* (*AGL2*), *AGL4* and *AGL9* respectively (Krizek and Meyerowitz, 1996a; Ma *et al.*, 1991; Mandel and Yanofsky, 1998), as they all encode highly similar MADS-box proteins. The role of the *SEP* genes in floral organ identity specification was first revealed by reverse genetics (Pelaz *et al.*, 2000). Targeted knockouts or screens for T-DNA insertion mutations in individual *SEP* genes revealed only subtle phenotypes in single *sep* mutants. However, *sep1 sep2 sep3* triple mutants displayed a striking phenotype: all floral organs in the first three whorls were sepals or sepal-like organs and the fourth whorl was converted into a new flower that repeats this same floral pattern (Fig. 33.2J) (Pelaz *et al.*, 2000). The phenotype displayed by the *sep1 sep2 sep3* triple mutant is very similar to "*bc*" double mutants (such as *pi ag* or *ap3 ag*) suggesting that *SEP1, SEP2* and *SEP3* are required for the B and C gene expression or for the B and C activity. A direct physical interaction between the *SEP* gene products and the B and C class gene products (Fan *et al.*, 1997; Honma and Goto, 2001) supports the idea that the SEP proteins may be present in the same protein complex as the B or C proteins. Like AP1, the C-terminal domain of the SEP proteins exhibited transcription activation activity (Honma and Goto, 2001). The formation of higher order protein complexes involving SEP proteins may provide a transcription activation function to these protein complexes. The discovery of *SEP* genes led to a revised ABC model, which is now termed the ABCE model (Goto *et al.*, 2001; Theissen, 2001; Theissen and Saedler, 2001). The ABCE model postulates that sepals are specified by A genes (perhaps with the help of an as yet unidentified factor), petals are specified by A, B, and E genes, stamens are determined by B, C, and E genes, and carpels are determined by C and E (Fig.33.1C).

If SEP proteins are the missing components that are behind a lack of leaf-to-floral transformation in *35S::B* and *35S::C* transgenic plants, one can now directly test this hypothesis by ectopically expressing *SEP* genes together with the *35S::B* or *35S::C*. Indeed, when *SEP3* was expressed ectopically together with *35S::B*, both rosette and cauline leaves were converted to organs that resembled petals (Fig.33.2K) (Honma and Goto, 2001; Pelaz *et al.*, 2001). Because both *AP1* and *SEP* encode a C-terminal domain capable of transcription activation, over expression of *SEP3* by 35S promoter appeared to bypass the requirement for *AP1*. In transgenic plants containing *35S::SEP, 35S::B, 35S::C*, the cauline leaves were converted to organs that resembled stamens (Honma and Goto, 2001). These studies demonstrated that the E class genes together with the ABC genes are sufficient to specify floral organ identity in leaves and that floral organs can form independently of flower formation.

The discovery of E class genes would not be possible in the absence of reverse genetics. Reverse genetics refers to a variety of techniques that can be used to generate mutations in a particular gene whose sequence is known. In the past several years, the *Arabidopsis* research community has benefited tremendously from significant advances in reverse genetic tools such as the T-DNA (Transfer-DNA) or transposon insertion lines, and the Targeted Induced Local Lesions In Genomes (TILLING) facility (Alonso *et al.*, 2003; McCallum *et al.*, 2000; Sessions *et al.*, 2002; Sundaresan *et al.*, 1995; Young *et al.*, 2001), making *Arabidopsis* one of the best systems for reverse genetics. Reverse genetic approaches are now more frequently employed and, as demonstrated here, are crucial to illuminate gene function.

Floral Meristem Identity Genes *LFY* and *AP1* Activate the Floral Program

Now that the ABCE model provides a framework for floral organ formation, one might step back and ask the question how is the decision to flower first made? Insights into this most important developmental switch in higher plants will have profound impact on agriculture, because new strategies in manipulating and controlling plant reproduction are of significant agronomical importance. Regulation of flowering time is the subject of much research. Several regulatory pathways that convey environmental, physiological, and developmental signals control this crucial switch (Hayama and Coupland, 2003; Henderson and Dean, 2004; Jack, 2004; Nilsson and Weigel, 1997; Simpson and Dean, 2002).

Two genes, *LEAFY (LFY)* and *AP1,* are the ultimate targets of various regulatory pathways controlling flowering time and are both necessary and sufficient for this vegetative to reproductive switch (Bowman *et al.*, 1993; Irish and Sussex, 1990; Mandel and Yanofsky, 1995; Weigel *et al.*, 1992; Weigel and Nilsson, 1995). This chapter will focus on how *LFY* and *AP1* switch on the flower program and activate the ABCE genes for floral organ identity specification. Loss-of-function mutations in these two genes cause the conversion (to varying degrees) from flowers to secondary shoots. *lfy ap1* double mutants exhibit synergistic genetic interactions, where all flowers are replaced by shoot-ike structures (Weigel *et al.*, 1992). Conversely, constitutive expression of either *LFY* or *AP1* causes the conversion from shoots to flowers (Mandel and Yanofsky, 1995; Weigel and Nilsson, 1995). Thus *LFY* and *AP1* are referred to as "meristem identity genes" and they play partially redundant roles in floral meristem identity specification.

LFY encodes a plant-specific protein that exhibits no strong sequence similarity to other known families of DNA-binding proteins (Weigel *et al.*, 1992). LFY can bind DNA in a sequence specific manner (Parcy *et al.*, 1998) and may directly or indirectly activate the expression of the ABCE genes. Consistent with its role as an activator of ABCE genes, *LFY* RNA and protein expression precedes the transcriptional activation of ABCE genes (Parcy *et al.* 1998). *AP1*, a MADS box protein described in the previous section as a class A gene, also possesses the function of a meristem identity gene. The dual roles of *AP1* as a meristem identity gene and a class A organ identity gene correlate well with its two phases of expression. *AP1* is initially expressed in the entire floral meristem and later becomes restricted to the first two whorls (Fig. 33.2E) (Bowman *et al.*, 1993; Gustafson-Brown *et al.*, 1994; Mandel *et al.*, 1992b).

Although *LFY* and *AP1* are both meristem identity genes, *AP1* appears to function downstream of *LFY*. Specifically, the flower promoting effect of *35S::LFY* was blocked in *ap1* mutants, but the flower promoting effect of *35S::AP1* was not blocked by *lfy* mutants (Mandel and Yanofsky, 1995; Weigel and Nilsson, 1995). To test if *AP1* is the direct transcriptional target of *LFY*, an inducible form of *LFY* was made (Wagner *et al.*, 1999). This construct (*35S::LFY-GR*) uses the *35S* promoter to express the *LFY* coding sequence that has been fused to a glucocorticoid receptor (GR) hormone binding domain. In the absence of the steroid hormone dexamethasone (DEX), the LFY-GR fusion protein is held in the cytoplasm and is non-functional. In the presence of DEX, the LFY-GR fusion protein moves to the nucleus and is able to perform its function as a transcriptional activator. As the translocation of the LFY-GR protein into the nucleus does not depend on protein synthesis, a direct effect of *LFY* on its target gene transcription can be evaluated in the presence of cyclohexamide (a protein synthesis inhibitor). The LFY-GR protein was able to rescue defects of *AP1* expression at early stages even in the presence of cyclohexamide, indicating that *LFY* directly activates *AP1* at early stages. However the ability of LFY-GR to rescue defects of *AP1* expression during later stages of floral development is dependent on protein synthesis, suggesting different regulatory mechanisms for the later phase of *AP1* activation.

The Floral Program Terminates *WUS* Expression via *AG*

Flowers are formed from lateral meristems which are produced by the shoot apical meristem (SAM). Floral meristem and SAM are homologous stem cell systems that are regulated by a similar set of genes (Sharma and Fletcher, 2002). In *Arabidopsis* floral meristem and SAM, stem cells are specified by signals from an underlying cell group, the organizing center that expresses the *WUSCHEL (WUS)* homeobox gene (Laux *et al.*, 1996; Mayer *et al.*, 1998). Mutations in *WUS* result in premature termination of SAMs as well as floral meristems after forming a few organs. However, floral meristems and SAMs differ fundamentally in that the SAMs are indeterminate, they can produce lateral organs continuously. In contrast, floral meristems are determinate, their meristem activity is terminated after a full set of floral organs is initiated. The determinate nature of floral meristems must be somehow controlled by *LFY* and *AP1*.

The first hint of how the floral meristems become determinate is the pattern of *WUS* expression. During flower development, *WUS* is expressed in the presumptive organizing center of floral meristems until the initiation of fourth whorl organs (Mayer *et al.*, 1998), suggesting that a downregulation of *WUS* expression could terminate floral meristem proliferation. Second, the indeterminate floral phenotype in *ag* loss-of-function mutants indicates a possible role of *AG* in the negative regulation of *WUS*. Indeed, *WUS* expression remained in the center of *ag* mutant flowers (Lenhard *et al.*, 2001; Lohmann *et al.*, 2001). Additionally, *wus* loss of function mutations could suppress the indeterminate floral phenotype of *ag*. These results provided molecular insights into the fundamental difference between floral meristems and SAMs. In SAMs, a lack of *LFY* and *AP1* activity correlates with an absence of *AG* expression. As a result, *WUS* is continuously expressed in the organizing center of SAM and maintains stem cell population. In floral meristems, *LFY* and *AP1* activate *AG* expression, which subsequently leads to a downregulation of *WUS* and a determinate floral meristem. The mechanism of how *AG* negatively regulates *WUS* is still unknown. The repression of *WUS* by *AG* was thought to be indirect (Lenhard *et al.*, 2001; Lohmann *et al.*, 2001).

Many of the most beautiful flowers, including hybrid tea roses, double camellias, and carnations, have many extra whorls of petals. These so-called double flowers were selected from their plain relatives that have only a single whorl of petals. Theophrastus first described double roses more than 2000 years ago, and in the centuries that have followed, numerous descriptions of double flowers occur in the literature (Meyerowitz *et al.*, 1989). By prolonging *WUS* expression or repressing *AG* activity, one may create new double flower varieties without having to depend on naturally occurring mutations.

***WUS* is a Co-regulator of *LFY* for *AG* Expression**

If *LFY* and *AP1* are responsible for activating ABCE genes in floral meristems, what determines the activation of B and C class genes only in a subset of floral meristem cells? Clearly, other factors must act in concert with *LFY* and *AP1* to properly activate B and C class genes in spatially restricted patterns. *WUS* and *UFO* appear to encode such co-regulators of *LFY* and they function to provide domain-specific co-activator activities.

Several lines of evidence indicated a direct and positive regulatory role of *LFY* for *AG*. First, *AG* mRNA expression was delayed and reduced in strong *lfy* mutants (Weigel and Meyerowitz, 1993). Second, two *LFY* binding sites are present near the 3' end of the *AG* second intron. This intron contains cis-regulatory elements both necessary and sufficient for *AG* expression (Busch *et al.*, 1999; Deyholos and Sieburth, 2000; Sieburth and Meyerowitz, 1997). Electrophoretic Mobility Shift Assay (EMSA) confirmed the binding of *LFY* to these two binding sites within the 3' *AG* enhancer (Busch *et al.*, 1999). Mutations in the *LFY* binding sites that abolished *in vitro* binding of LFY largely eliminated the *in vivo* activity of the 3' *AG* enhancer. Finally, *LFY-VP16*, a hyperactive form of *LFY*, can induce ectopic *AG* expression. When *LFY-VP16* was driven by the 35S promoter, *AG* was activated in seedlings even in the absence of flower formation (Parcy *et al.*, 1998). However, wild type *LFY* could not activate *AG* ectopically indicating the need for additional co-activators.

Adjacent to the two *LFY* binding sites in the 3' *AG* enhancer are two consensus binding sites for homeobox proteins. Since *WUS* encodes a homeobox protein and *WUS* is only expressed in a few cells in the center of the floral meristem (beneath the whorls 3-4 precursor cells), *WUS* became an excellent candidate co-activator of *LFY*. To test this possibility, a trimer of a 91 bp fragment that includes the two *LFY* binding sites and two putative *WUS* binding sites was used to drive reporter *LacZ* expression in yeast. While the reporter lacZ could not be activated by *LFY* nor *WUS* alone, coexpression of *LFY* and *WUS* resulted in robust activation of reporter *LacZ* (Lohmann *et al.*, 2001). EMSA experiments further confirmed the binding of WUS protein to these two putative binding sites. LFY and WUS appeared to bind to the *AG* 3' enhancer independently and may each enhance *AG* transcription via independent contacts with transcriptional machineries (Lohmann *et al.*, 2001).

The importance of *WUS* binding sites for 3' *AG* enhancer activity was subsequently verified in transgenic plants (Lohmann *et al.*, 2001). Further, when *WUS* is ectopcally expressed under the control of *LFY* or *AP3* promoters, ectopic *AG* expression as well as homeotic transformation of floral organs was observed (Lohmann, *et al.*, 2001). In flowers, *WUS* was apparently co-opted as a region-specific transcription activator. *WUS*, combined with *LFY* produces a flower- and region-specific pattern of *AG* expression.

UFO is a Co-regulator of *LFY* for *AP3* Expression

Evidence that *LFY* is important for the initial activation of *AP3* comes from the observation that both domain and the level of *AP3* expression are reduced in *lfy* mutants (Weigel and Meyerowitz, 1993). Positive regulation of *AP3* by *LFY* may be direct, because LFY binds *in vitro* to a *LFY* binding site located in an *AP3* promoter element that directs the establishment of *AP3* expression during early floral stages (Hill *et al.*, 1998). However, mutation of this *LFY* binding site does not disrupt *LFY* activation of *AP3* (Lamb *et al.*, 2002) suggesting other redundant *LFY* activation elements in the *AP3* promoter.

Activation of *AP3* by *LFY* apparently relies on additional co-activators as *35S::LFY* or *35S::LFY-VP16* failed to activate *AP3* ectopically (Parcy *et al.*, 1998). A candidate co-activator of *AP3* is *UNUSUAL FLORAL ORGANS (UFO)*. Mutants of *ufo* exhibited a reduced petal and stamen numbers, which correlated with a reduction in the level of *AP3* RNA during early floral stages (Levin and Meyerowitz, 1995; Wilkinson and Haughn, 1995). *UFO* is expressed in second and third whorl primordia during floral stages 3 and 4 (Ingram *et al.*, 1995; Laufs *et al.*, 2003). Ectopic expression of *UFO* (*35S:UFO)* results in the partial conversion of first whorl sepals to petals and fourth whorl carpels to stamens (Lee *et al.*, 1997). Additionally, transgenic plants containing both *35S::UFO* and *35S::LFY* expressed *AP3::GUS* reporters in seedlings, demonstrating that *LFY* and *UFO* together are sufficient to activate *AP3* (Parcy *et al.*, 1998).

UFO encodes an F-box protein (Ingram *et al.*, 1995; Samach *et al.*, 1999). F-box proteins have been shown to be components of a complex, named the SKP1-cullin-F-box (SCF) complex that selects substrates for ubiquitin-mediated protein degradation. *UFO* functions as a component of a SCF complex (Ni *et al.*, 2004; Wang *et al.*, 2003). However, the protein target (or targets) of SCF^{UFO} is unknown. The favored model is that UFO-ediated positive activation of *AP3* occurs as a result of the SCF^{UFO}-mediated degradation of a repressor of *AP3* expression.

B, C, and E Genes Maintain Their Own Expression via Autoregulatory Loops

Careful examination of *in situ* hybridization data indicated that *AP3* and *PI* are not initially expressed in identical domains. *AP3* mRNA is detected in whorls 2-3 plus in a small number of cells at the base of the first whorl (Tilly *et al.*, 1998; Weigel and Meyerowitz, 1993) while *PI* RNA is detected in whorls 2-4 (Goto and Meyerowitz, 1994). At later stages of flower development, the expression of both genes is restricted to petals and stamens. Maintenance of this later expression in petals and stamens requires the activity of both *AP3* and *PI*. In *ap3* and *pi* mutants, both *AP3* and *PI* late phase expression is reduced while the early phase expression is unaffected (Goto and Meyerowitz, 1994; Jack *et al.*, 1994). Therefore, at late stages of flower development, *AP3* and *PI* positively regulate their own expression, leading to similar expression domains.

Three CArG boxes were identified between -90 to -180 of the *AP3* promoter (Hill *et al.*, 1998; Tilly *et al.*, 1998). AP3/PI heterodimers can bind to CArG box 1 and 3 *in vitro* in a sequence specific manner (Hill *et al.*, 1998; Tilly *et al.*, 1998). In addition, *AP3-GR* can induce *AP3* expression in the absence of *de novo* protein synthesis (Honma and Goto, 2000). Thus, direct interaction between *AP3/PI* and *AP3* promoter maintains late phase *AP3* transcription. Interestingly, the promoter or intron sequences of *PI* do not contain any CArG box. Using a similar AP3-GR system, the ability of AP3/PI heterodimer to activate *PI* transcription requires *de novo* protein synthesis (Honma and Goto, 2000), suggesting that autoregulation of *PI* transcription by AP3/PI is indirect.

Recently, the autoregulatory mechanism is emerging as a general theme for B, C, and E gene expression. The *Arabidopsis* ATH1 high-density oligonucleotide array (Affymetrix) in combination with the *35S::AG-GR* was used to identify early targets of *AG* (Gomez-Mena *et al.*, 2005). AG-GR was introduced into the *ap1-1 cal-1* double mutants, which accumulate indeterminate lateral meristems (Kempin *et al.*, 1995). Upon DEX treatment, *AG-GR, ap1-1, cal-1* plants induced stamen and carpel formation in a synchronized fashion. Twelve genes were identified that were activated at multiple time points after a single DEX treatment. Surprisingly, among these twelve genes are *AP3, AG,* and *SEP3*. EMSA and chromatin immunoprecipitation (ChIP) confirmed a direct interaction between AG protein and cis-regulatory elements of *AP3, AG* and *SEP3*. This study revealed a requirement of *AG* in the autoregulation of its own expression as well as its role in the positive regulation of class B and E genes. This finding, however, is consistent with the ABCE model that *AG* is part of the AP3/PI complex involved in the positive autoregulatory loop. The emerging theme is that the expression of ABCE genes is initiated by one mechanism involving *LFY* and other domain-specific regulators such as *WUS* and *UFO* but their later expression is maintained by positive autoregulatory loops (Fig.33.4).

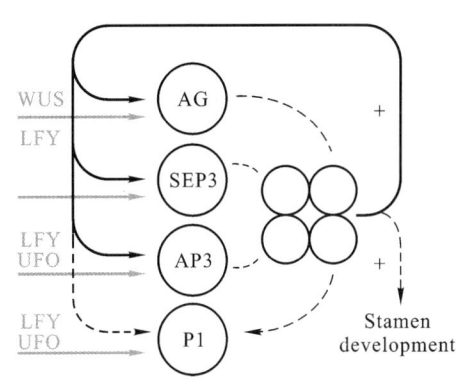

Fig.33.4 Autoregulatory loops maintain B, C, and E gene expression. In stamen, *AG, AP3, PI* and *SEP3* are initially activated independently (grey arrows). *LFY* and *WUS* are responsible for the initial activation of *AG*, while *LFY* and *UFO* are responsible for the initial activation of *AP3* and *PI*. The AG, AP3, PI, and SEP3 proteins (circles) function together in a complex to promote stamen development and to amplify and maintain their own expression. Solid black arrows indicate direct interactions supported by chromatin immunoprecipitation assays. Feedback activation of *PI* may be indirect and is indicated by a dashed arrow. This figure is based on Gomez-Mena *et al.* (2005) and is reprinted with permission from the Company of Biologists Ltd.

AP2 is Regulated by a MicroRNA

AP2 is unique in that it does not encode a MADS box protein but encodes a member of the AP2 domain containing transcription factor family (Jofuku *et al.*, 1994). In addition, *AP2* mRNA is expressed in all floral whorls although its function is only present in whorls 1-2. Two recent reports demonstrated that *AP2* is under post-transcriptional regulation by microRNA (miRNA) (Aukerman and Sakai, 2003; Chen, 2004). miRNAs are ~21-nucleotide noncoding RNAs that have been identified in both animals and plants (Carrington and Ambros, 2003). Complementary pairing of miRNA with their target mRNA either results in specific cleavage or translational inhibition of their target mRNAs (Carrington and Ambros, 2003; Llave *et al.*, 2002; Olsen and Ambros, 1999). Many miRNAs and their putative targets were identified in *Arabidopsis* (Park *et al.*, 2002; Reinhart *et al.*, 2002; Rhoades *et al.*, 2002). One miRNA, *miR172*, was found to be complementary to a single sequence located near the 3' end of the *AP2* open reading frame (Fig. 33.3B).

To test if *miR172* regulates *AP2*, *miR172* was ectopically expressed from the 35S promoter. These *35S::miR172* flowers exhibited an *ap2* phenotype (Fig.33.5B and 33.5C), suggesting that *miR172* down-regulates *AP2* activity (Chen, 2004). Surprisingly, the

AP2 mRNA level was unaffected in the *35S:miR172* plants but the AP2 protein level was reduced, suggesting that *miR172* inhibits *AP2* function by preventing its translation. *In situ* hybridization revealed that *miR172* is expressed at highest levels in whorls 3-4 of wild type flowers, supporting the idea that *AP2* translation is specifically inhibited in whorls 3-4 (Aukerman and Sakai, 2003; Chen, 2004).

To test if the putative *miR172* binding site within *AP2* mRNA subjects *AP2* to *miR172* regulation, mutations were introduced into the putative binding site of *miR172* within *AP2*. These mutations do not alter the amino acid sequence of AP2 protein but render the mutant *AP2* mRNA immune to *miR172* regulation. When wild type *AP2* and a mutant *AP2* with six mismatches to *miR172* (*AP2m1*) were fused to the 35S promoter and introduced into wild type *Arabidopsis*, these two types of transgenic plants gave very different phenotypes. While *35S::AP2* flowers exhibited a wild-type phenotype, *35S::AP2m1* plants exhibited an *ag*-like phenotype including stamen-to-petal transformations and loss of floral determinacy (Fig. 33.5D). Although the *AP2* mRNA levels were comparable between *35S::AP2* normal flower plants and *35S::AP2m1* plants, the levels of AP2 proteins were elevated in *35S::AP2m1* but not in *35S::AP2* (Chen, 2004). These experiments strongly support that *AP2* translation is inhibited by *miR172* present in whorls 3-4 (Chen, 2004).

Transcription Co-repressors Participate in the Class A Antagonistic Function

One important aspect of the ABCE model is that the class A and C genes not only specify sepal/petal or stamen/carpel identities, but also negatively regulate each other's activity (Bowman *et al*., 1991; Coen and Meyerowitz, 1991). What is the molecular mechanism underlying this antagonistic interaction? Genetic screens for floral mutants that exhibit partial or complete homeotic transformation from sepal to carpel-like organs led to the identification of a large number of genes including *AP2, LEUNIG (LUG), SEUSS (SEU), BELLRINGER (BLR), ANT, STERILE APETALA (SAP)*, and *FILAMENTOUS FLOWER (FIL)* (Bao *et al*., 2004; Byzova *et al*., 1999; Elliott *et al*., 1996; Franks *et al*., 2002; Jofuku *et al*., 1994; Klucher *et al*., 1996; Krizek *et al*., 2000; Liu and Meyerowitz, 1995; Sawa *et al*., 1999). Their mutant phenotype and ectopic *AG* expression in outer floral whorls suggest that these genes all participate in the negative regulation of *AG* transcription.

Among the genes mentioned about, *LUG* and *SEU* are the most extensively studied, *lug* and *seu* mutants both exhibit homeotic transformations similar to, but less severe than, *ap2* mutants (Franks *et al*., 2002; Liu and Meyerowitz, 1995). *in situ* hybridization experiments revealed that both ectopic and precocious *AG* RNA was present in *lug* or *seu* single and double mutants. Removing *AG* in *ag lug* or *ag seu* double mutants restored sepal and petal identity suggesting that *LUG* and *SEU* are required for proper repression of *AG* but are not required for the specification of sepal or petal identity. Hence, *LUG* and *SEU* were considered "cadastral" genes analogous to the "gap" genes of *Drosophila* (Liu and Meyerowitz, 1995). In addition, *lug* and *seu* mutants exhibit pleiotropic phenotypes including defects associated with vegetative tissues, indicating that *LUG* and *SEU* may have a more general role in plant development (Franks *et al*., 2002; Liu *et al*., 2000; Liu and Meyerowitz, 1995).

LUG encodes a nuclear protein that has an overall domain structure similar to a class of functionally related transcriptional co-repressors including Tup1 of yeast and Groucho (Gro) of *Drosophila* (Conner and

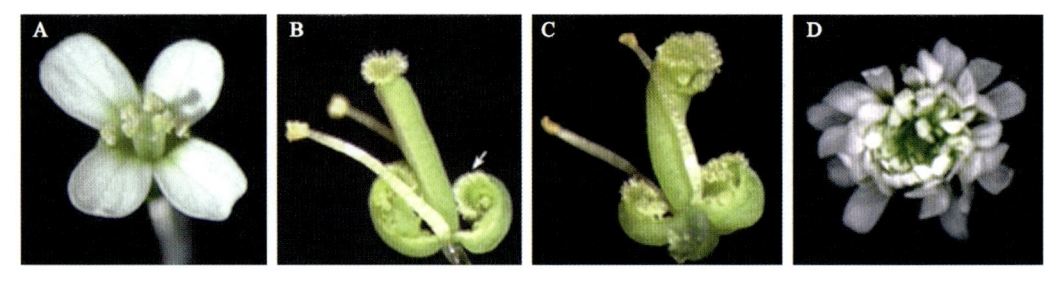

Fig.33.5 **Regulation of *AP2* by *miR172* during *Arabidopsis* flower development.** (A) A wild-type flower. (B) An *ap2-9* mutant flower with first-whorl organs transformed into carpels and a severe reduction of whorl 2 and whorl 3 stamens. (C) A *35S::miR172a-1* flower that closely resembles the *ap2-9* flower. (D) A *35S::AP2m1* flower with numerous petals and loss of floral determinacy, a phenotype resembling *ag* loss-of-function mutants (Chen, 2004). Reprinted with permission from Chen, 2004. Copy right (2004) AAAS.

Liu, 2000; Hartley *et al*., 1988; Williams and Trumbly, 1990). LUG possesses a conserved N-terminal 88 amino acid domain called the LUFS domain, which is both necessary and sufficient for the direct physical interaction with SEU (Sridhar *et al*., 2004). *SEU* encodes a glutamine (Q)-rich protein with a conserved domain that is similar to the dimerization domain of LIM-Domain-inding (Ldb) family of transcriptional co-regulators such as the *Ldb1* in mouse and *Chip* in *Drosophila* (Franks *et al*., 2002). The direct *LUG-SEU* interaction is supported by a parallel study in *Drosophila* and mouse, where the LUFS domain of Single Strand DNA-binding Protein (SSDP) was shown to directly associate with the mouse Ldb1 or *Drosophila* Chip (Chen *et al*., 2002; van Meyel *et al*., 2003).

LUG-GAL4BD or *SEU-GAL4BD* chimeric genes were tethered to the promoters that directed *GUS* or luciferase reporter gene expression in transient expression assays with *Arabidopsis* protoplasts. While *SEU-GAL4BD* did not exhibit any repressor activity in the absence of *LUG*, *LUG-GAL4BD* exhibited strong repressor activity (Sridhar *et al*., 2004). However the repressor activity of *LUG* was eliminated when trichostatin A, a Histone Deacetylase (HDAC) inhibitor, was added to the transient expression assay suggesting that *LUG* represses transcription possibly by recruiting HDACs (Sridhar *et al*., 2004).

The *Tup1* or *Gro* co-repressors are global transcriptional repressors that are recruited by different DNA-binding transcription factors to repress different target genes (Chen and Courey, 2002). Since neither *LUG* nor *SEU* encodes a DNA-binding motif, the *LUG/SEU* co-repressors must depend on other DNA-binding transcription factors that bind to the *AG* cis-elements. Since both *LUG* and *SEU* mRNAs are detected everywhere in a plant (Conner and Liu, 2000; Franks *et al*., 2002), their outer whorl-specific repressor activities may depend on their interaction with other outer whorl-specific factors. The class A genes *AP1* and *AP2* are excellent candidate partners for the *LUG/SEU* co-repressors. First, *AP1* mRNA and possibly AP2 proteins are expressed in whorls 1-2, at stages when the floral organ identities are being specified. Second, although the ability of AP1 or AP2 to bind *AG* intronic sequences has not been demonstrated, several CArG boxes and a putative AP2 domain binding site are present in the *AG* second intron and can serve as the binding sites for AP1 and AP2 (Deyholos and Sieburth, 2000; Nole-Wilson and Krizek, 2000; Hong *et al*., 2003). Finally, while *ap1* single mutants did not exhibit ectopic *AG* expression in flowers, *ap1 lug* and *ap1 seu* double mutants showed much enhanced homeotic transformation of floral organs, indicating enhanced *AG* mis-expression (Liu and Meyerowitz, 1995; V.V. Sridhar and Z. Liu, unpublished data). In addition, *ap2* exhibited dominant genetic interactions with *lug* and *seu* (Liu and Meyerowitz, 1995; Franks *et al*., 2002). Yeast two hybrid assays suggested a direct interaction between *SEU* and *AP1* (V. V. Sridhar and Z. Liu, unpublished). Therefore, *SEU* appears to bridge the interaction between the *LUG* co-repressor and the domain-specific DNA-binding factors encoded by the class A genes (Fig. 33.6). Together, they repress *AG* in the outer two whorls of a flower.

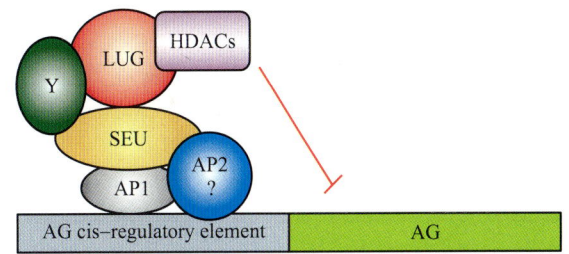

Fig.33.6 A model for the repression of *AG* by transcription co-repressors and class A genes. The LUG co-repressor represses *AG* by recruiting Histone Deacetylases (HDACs). SEU, an adaptor protein, bridges the interaction between LUG and the DNA-binding transcription factors encoded by the class A genes *AP1* and *AP2*. Y represents unidentified component(s) of the co-repressor complex. The question mark indicates a putative role of AP2.

Concluding Remarks

Molecular genetic analyses of homeotic mutations in *Drosophila melanogaster* indicated that the homeotic genes encode master regulatory proteins that switch on or off specific developmental programs in specific segments of the fruit fly (Gehring and Hiromi, 1986). The characterization of floral homeotic mutants and the identification of corresponding genes indicated that the floral homeotic genes also encode master regulatory proteins that act to switch on organ-specific developmental programs. While the *Drosophila* homeotic genes encode the homeobox proteins, plant homeotic genes (the ABCE genes) encode a different type of transcription factor, the MADS box proteins. Flower is an evolutionary novelty that characterizes the most successful group of vascular plants, the angiosperms that first appeared merely 130 million years ago (Crane, 1993). The origin of floral organs was proposed more than 200 years ago by Johann Wolfgang von Goethe to arise from "metamorphosis" of leaves. The radiation and diversification of plant MADS box genes may underlie this floral evolution. The leaf-like floral organs in *abc* triple loss-of-function mutants and the petaloid and staminoid leaves in transgenic plants over-expressing ABCE genes provided answers in molecular terms to the metamorphosis of floral organs.

Acknowledgement

I would like to thank Beth Krizek for Fig 33.2H and 33.2I. Alan Kirschner for proofreading the manuscript. Z.L. is supported by a NSF grant IBN 0212847.

References

Alonso, J. M., Stepanova, A. N., Leisse, T. J., Kim, C. J., Chen, H., Shinn, P., Stevenson, D. K., Zimmerman, J., Barajas, P., Cheuk, R., et al. (2003). Genome-wide insertional mutagenesis of *Arabidopsis thaliana*. Science *301*, 653-657.

Aukerman, M. J., and Sakai, H. (2003). Regulation of flowering time and floral organ identity by a MicroRNA and its APETALA2-like target genes. Plant Cell *15*, 2730-2741.

Bao, X., Franks, R. G., Levin, J. Z., and Liu, Z. (2004). Repression of AGAMOUS by BELLRINGER in floral and inflorescence meristems. Plant Cell *16*, 1478-1489.

Bowman, J. L., Alvarez, J., Weigel, D., Meyerowitz, E. M., and Smyth, D. R. (1993). Control of flower development in *Arabidopsis thaliana* by APETALA1 and interacting genes. Development *119*, 721-743.

Bowman, J. L., Smyth, D. R., and Meyerowitz, E. M. (1989). Genes directing flower development in *Arabidopsis*. Plant Cell *1*, 37-52.

Bowman, J. L., Smyth, D. R., and Meyerowitz, E. M. (1991). Genetic interactions among floral homeotic genes of *Arabidopsis*. Development *112*, 1-20.

Busch, M. A., Bomblies, K., and Weigel, D. (1999). Activation of a floral homeotic gene in *Arabidopsis*. Science *285*, 585-587.

Byzova, M. V., Franken, J., Aarts, M. G., de Almeida-Engler, J., Engler, G., Mariani, C., Van Lookeren Campagne, M. M., and Angenent, G. C. (1999). *Arabidopsis* STERILE APETALA, a multifunctional gene regulating inflorescence, flower, and ovule development. Genes Dev *13*, 1002-1014.

Carrington, J. C., and Ambros, V. (2003). Role of microRNAs in plant and animal development. Science *301*, 336-338.

Chen, G., and Courey, A. J. (2000). Groucho/TLE family proteins and transcriptional repression. Gene *249*, 1-16.

Chen, L., Segal, D., Hukriede, N. A., Podtelejnikov, A. V., Bayarsaihan, D., Kennison, J. A., Ogryzko, V. V., Dawid, I. B., and Westphal, H. (2002). Ssdp proteins interact with the LIM-domain-binding protein Ldb1 to regulate development. Proc Natl Acad Sci USA *99*, 14320-14325.

Chen, X. (2004). A microRNA as a translational repressor of APETALA2 in *Arabidopsis* flower development. Science *303*, 2022-2025.

Coen, E. S., and Meyerowitz, E. M. (1991). The war of the whorls: genetic interactions controlling flower development. Nature *353*, 31-37.

Conner, J., and Liu, Z. (2000). LEUNIG, a putative transcriptional corepressor that regulates AGAMOUS expression during flower development. Proc Natl Acad Sci USA *97*, 12902-12907.

Crane, P.R. (1993). Time for the angiosperms. Nature *336*, 631-632

Deyholos, M. K., and Sieburth, L. E. (2000). Separable whorl-specific expression and negative regulation by enhancer elements within the AGAMOUS second intron. Plant Cell *12*, 1799-1810.

Drews, G. N., Bowman, J. L., and Meyerowitz, E. M. (1991). Negative regulation of the *Arabidopsis* homeotic gene AGAMOUS by the APETALA2 product. Cell *65*, 991-1002.

Elliott, R. C., Betzner, A. S., Huttner, E., Oakes, M. P., Tucker, W. Q., Gerentes, D., Perez, P., and Smyth, D. R. (1996). AINTEGUMENTA, an APETALA2-like gene of *Arabidopsis* with pleiotropic roles in ovule development and floral organ growth. Plant Cell *8*, 155-168.

Fan, H. Y., Hu, Y., Tudor, M., and Ma, H. (1997). Specific interactions between the K domains of AG and AGLs, members of the MADS domain family of DNA binding proteins. Plant J *12*, 999-1010.

Franks, R. G., and Liu, Z. (2001). Floral homeotic gene regulation. Horticultural Reviews *27*, 41-77.

Franks, R. G., Wang, C., Levin, J. Z., and Liu, Z. (2002). SEUSS, a member of a novel family of plant regulatory proteins, represses floral homeotic gene expression with LEUNIG. Development *129*, 253-263.

Gehring, W. J., and Hiromi, Y. (1986). Homeotic genes and the homeobox. Annu Rev Genet *20*, 147-173.

Gomez-Mena, C., de Folter, S., Costa, M. M., Angenent, G. C., and Sablowski, R. (2005). Transcriptional program controlled by the floral homeotic gene AGAMOUS during early organogenesis. Development *132*, 429-438.

Goto, K., Kyozuka, J., and Bowman, J. L. (2001). Turning floral organs into leaves, leaves into floral organs. Curr Opin Genet Dev *11*, 449-456.

Goto, K., and Meyerowitz, E. M. (1994). Function and regulation of the *Arabidopsis* floral homeotic gene PISTILLATA. Genes Dev *8*, 1548-1560.

Gustafson-Brown, C., Savidge, B., and Yanofsky, M. F. (1994). Regulation of the *arabidopsis* floral homeotic gene APETALA1. Cell *76*, 131-143.

Hartley, D. A., Preiss, A., and Artavanis-Tsakonas, S. (1988). A deduced gene product from the *Drosophila* neurogenic locus, enhancer of split, shows homology to mammalian G-protein beta subunit. Cell *55*, 785-795.

Hayama, R., and Coupland, G. (2003). Shedding light on the circadian clock and the photoperiodic control of flowering. Curr

Opin Plant Biol *6*, 13-19.

Henderson, I. R., and Dean, C. (2004). Control of *Arabidopsis* flowering: the chill before the bloom. Development *131*, 3829-3838.

Hill, T. A., Day, C. D., Zondlo, S. C., Thackeray, A. G., and Irish, V. F. (1998). Discrete spatial and temporal cis-acting elements regulate transcription of the *Arabidopsis* floral homeotic gene APETALA3. Development *125*, 1711-1721.

Hong, R. L., Hamaguchi, L., Busch, M. A., and Weigel, D. (2003). Regulatory elements of the floral homeotic gene AGAMOUS identified by phylogenetic footprinting and shadowing. Plant Cell *15*, 1296-1309.

Honma, T., and Goto, K. (2000). The *Arabidopsis* floral homeotic gene PISTILLATA is regulated by discrete cis-elements responsive to induction and maintenance signals. Development *127*, 2021-2030.

Honma, T., and Goto, K. (2001). Complexes of MADS-box proteins are sufficient to convert leaves into floral organs. Nature *409*, 525-529.

Huang, H., Tudor, M., Su, T., Zhang, Y., Hu, Y., and Ma, H. (1996). DNA binding properties of two *Arabidopsis* MADS domain proteins: binding consensus and dimer formation. Plant Cell *8*, 81-94.

Ingram, G. C., Goodrich, J., Wilkinson, M. D., Simon, R., Haughn, G. W., and Coen, E. S. (1995). Parallels between UNUSUAL FLORAL ORGANS and FIMBRIATA, genes controlling flower development in *Arabidopsis* and Antirrhinum. Plant Cell *7*, 1501-1510.

Irish, V. F., and Sussex, I. M. (1990). Function of the apetala-1 gene during *Arabidopsis* floral development. Plant Cell *2*, 741-753.

Jack, T. (2004). Molecular and genetic mechanisms of floral control. Plant Cell *16 Suppl*, S1-17.

Jack, T., Brockman, L. L., and Meyerowitz, E. M. (1992). The homeotic gene APETALA3 of *Arabidopsis thaliana* encodes a MADS box and is expressed in petals and stamens. Cell *68*, 683-697.

Jack, T., Fox, G. L., and Meyerowitz, E. M. (1994). *Arabidopsis* homeotic gene APETALA3 ectopic expression: transcriptional and posttranscriptional regulation determine floral organ identity. Cell *76*, 703-716.

Jefferson, R. A., Kavanagh, T. A., and Bevan, M. W. (1987). GUS fusions: beta-glucuronidase as a sensitive and versatile gene fusion marker in higher plants. Embo J *6*, 3901-3907.

Jofuku, K. D., den Boer, B. G., Van Montagu, M., and Okamuro, J. K. (1994). Control of *Arabidopsis* flower and seed development by the homeotic gene APETALA2. Plant Cell *6*, 1211-1225.

Kempin, S. A., Savidge, B., and Yanofsky, M. F. (1995). Molecular basis of the cauliflower phenotype in *Arabidopsis*. Science *267*, 522-525.

Klucher, K. M., Chow, H., Reiser, L., and Fischer, R. L. (1996). The AINTEGUMENTA gene of *Arabidopsis* required for ovule and female gametophyte development is related to the floral homeotic gene APETALA2. Plant Cell *8*, 137-153.

Krizek, B. A., and Meyerowitz, E. M. (1996a). The *Arabidopsis* homeotic genes APETALA3 and PISTILLATA are sufficient to provide the B class organ identity function. Development *122*, 11-22.

Krizek, B. A., and Meyerowitz, E. M. (1996b). Mapping the protein regions responsible for the functional specificities of the *Arabidopsis* MADS domain organ-identity proteins. Proc Natl Acad Sci USA *93*, 4063-4070.

Krizek, B. A., Prost, V., and Macias, A. (2000). AINTEGUMENTA promotes petal identity and acts as a negative regulator of AGAMOUS. Plant Cell *12*, 1357-1366.

Lamb, R. S., Hill, T. A., Tan, Q. K., and Irish, V. F. (2002). Regulation of APETALA3 floral homeotic gene expression by meristem identity genes. Development *129*, 2079-2086.

Laufs, P., Coen, E., Kronenberger, J., Traas, J., and Doonan, J. (2003). Separable roles of UFO during floral development revealed by conditional restoration of gene function. Development *130*, 785-796.

Laux, T., Mayer, K. F., Berger, J., and Jurgens, G. (1996). The WUSCHEL gene is required for shoot and floral meristem integrity in *Arabidopsis*. Development *122*, 87-96.

Lee, I., Wolfe, D. S., Nilsson, O., and Weigel, D. (1997). A LEAFY co-regulator encoded by UNUSUAL FLORAL ORGANS. Curr Biol *7*, 95-104.

Lenhard, M., Bohnert, A., Jurgens, G., and Laux, T. (2001). Termination of stem cell maintenance in *Arabidopsis* floral meristems by interactions between WUSCHEL and AGAMOUS. Cell *105*, 805-814.

Levin, J. Z., and Meyerowitz, E. M. (1995). UFO: an *Arabidopsis* gene involved in both floral meristem and floral organ development. Plant Cell *7*, 529-548.

Liu, Z., Franks, R. G., and Klink, V. P. (2000). Regulation of gynoecium marginal tissue formation by LEUNIG and AINTEGUMENTA. Plant Cell *12*, 1879-1892.

Liu, Z., and Meyerowitz, E. M. (1995). LEUNIG regulates AGAMOUS expression in *Arabidopsis* flowers. Development *121*, 975-991.

Llave, C., Xie, Z., Kasschau, K. D., and Carrington, J. C. (2002). Cleavage of Scarecrow-like mRNA targets directed by a class of *Arabidopsis* miRNA. Science *297*, 2053-2056.

Lohmann, J. U., Hong, R. L., Hobe, M., Busch, M. A., Parcy, F., Simon, R., and Weigel, D. (2001). A molecular link between stem cell regulation and floral patterning in *Arabidopsis*. Cell *105*, 793-803.

Ma, H., Yanofsky, M. F., and Meyerowitz, E. M. (1991). AGL1-AGL6, an *Arabidopsis* gene family with similarity to floral homeotic and transcription factor genes. Genes Dev *5*, 484-495.

Mandel, M. A., Bowman, J. L., Kempin, S. A., Ma, H., Meyerowitz,

E. M., and Yanofsky, M. F. (1992a). Manipulation of flower structure in transgenic tobacco. Cell 71, 133-143.

Mandel, M. A., Gustafson-Brown, C., Savidge, B., and Yanofsky, M. F. (1992b). Molecular characterization of the *Arabidopsis* floral homeotic gene APETALA1. Nature 360, 273-277.

Mandel, M. A., and Yanofsky, M. F. (1995). A gene triggering flower formation in *Arabidopsis*. Nature 377, 522-524.

Mandel, M. A., and Yanofsky, M. F. (1998). Thbe *Arabidopsis* AGL9 MADS box gene is expressed in young flower primordia. Sexual Plant Reproduction 11, 22-28.

Mayer, K. F., Schoof, H., Haecker, A., Lenhard, M., Jurgens, G., and Laux, T. (1998). Role of WUSCHEL in regulating stem cell fate in the *Arabidopsis* shoot meristem. Cell 95, 805-815.

McCallum, C. M., Comai, L., Greene, E. A., and Henikoff, S. (2000). Targeting induced local lesions IN genomes (TILLING) for plant functional genomics. Plant Physiol 123, 439-442.

Meyerowitz, E. M., Smyth, D. R., and Bowman, J. L. (1989). Abnormal flowers and pattern formation in floral. Development 106, 209-217.

Mizukami, Y., and Ma, H. (1992). Ectopic expression of the floral homeotic gene AGAMOUS in transgenic *Arabidopsis* plants alters floral organ identity. Cell 71, 119-131.

Ni, W., Xie, D., Hobbie, L., Feng, B., Zhao, D., Akkara, J., and Ma, H. (2004). Regulation of flower development in *Arabidopsis* by SCF complexes. Plant Physiol 134, 1574-1585.

Nilsson, O., and Weigel, D. (1997). Modulating the timing of flowering. Curr Opin Biotechnol 8, 195-199.

Nole-Wilson, S., and Krizek, B. A. (2000). DNA binding properties of the *Arabidopsis* floral development protein AINTEGUMENTA. Nucleic Acids Res 28, 4076-4082.

Ohme-Takagi, M., and Shinshi, H. (1995). Ethylene-inducible DNA binding proteins that interact with an ethylene-responsive element. Plant Cell 7, 173-182.

Olsen, P. H., and Ambros, V. (1999). The lin-4 regulatory RNA controls developmental timing in *Caenorhabditis elegans* by blocking LIN-14 protein synthesis after the initiation of translation. Dev Biol 216, 671-680.

Parcy, F., Nilsson, O., Busch, M. A., Lee, I., and Weigel, D. (1998). A genetic framework for floral patterning. Nature 395, 561-566.

Park, W., Li, J., Song, R., Messing, J., and Chen, X. (2002). CARPEL FACTORY, a Dicer homolog, and HEN1, a novel protein, act in microRNA metabolism in *Arabidopsis thaliana*. Curr Biol 12, 1484-1495.

Pelaz, S., Ditta, G. S., Baumann, E., Wisman, E., and Yanofsky, M. F. (2000). B and C floral organ identity functions require SEPALLATA MADS-box genes. Nature 405, 200-203.

Pelaz, S., Tapia-Lopez, R., Alvarez-Buylla, E. R., and Yanofsky, M. F. (2001). Conversion of leaves into petals in *Arabidopsis*. Curr Biol 11, 182-184.

Pnueli, L., Abu-Abeid, M., Zamir, D., Nacken, W., Schwarz-Sommer, Z., and Lifschitz, E. (1991). The MADS box gene family in tomato: temporal expression during floral development, conserved secondary structures and homology with homeotic genes from Antirrhinum and *Arabidopsis*. Plant J 1, 255-266.

Reinhart, B. J., Weinstein, E. G., Rhoades, M. W., Bartel, B., and Bartel, D. P. (2002). MicroRNAs in plants. Genes Dev 16, 1616-1626.

Rhoades, M. W., Reinhart, B. J., Lim, L. P., Burge, C. B., Bartel, B., and Bartel, D. P. (2002). Prediction of plant microRNA targets. Cell 110, 513-520.

Riechmann, J. L., Krizek, B. A., and Meyerowitz, E. M. (1996a). Dimerization specificity of *Arabidopsis* MADS domain homeotic proteins APETALA1, APETALA3, PISTILLATA, and AGAMOUS. Proc Natl Acad Sci USA 93, 4793-4798.

Riechmann, J. L., and Meyerowitz, E. M. (1997a). Determination of floral organ identity by *Arabidopsis* MADS domain homeotic proteins AP1, AP3, PI, and AG is independent of their DNA-binding specificity. Mol Biol Cell 8, 1243-1259.

Riechmann, J. L., and Meyerowitz, E. M. (1997b). MADS domain proteins in plant development. Biol Chem 378, 1079-1101.

Riechmann, J. L., and Meyerowitz, E. M. (1998). The AP2/EREBP family of plant transcription factors. Biol Chem 379, 633-646.

Riechmann, J. L., Wang, M., and Meyerowitz, E. M. (1996b). DNA-binding properties of *Arabidopsis* MADS domain homeotic proteins APETALA1, APETALA3, PISTILLATA and AGAMOUS. Nucleic Acids Res 24, 3134-3141.

Samach, A., Klenz, J. E., Kohalmi, S. E., Risseeuw, E., Haughn, G. W., and Crosby, W. L. (1999). The UNUSUAL FLORAL ORGANS gene of *Arabidopsis thaliana* is an F-box protein required for normal patterning and growth in the floral meristem. Plant J 20, 433-445.

Sawa, S., Watanabe, K., Goto, K., Liu, Y. G., Shibata, D., Kanaya, E., Morita, E. H., and Okada, K. (1999). FILAMENTOUS FLOWER, a meristem and organ identity gene of *Arabidopsis*, encodes a protein with a zinc finger and HMG-related domains. Genes Dev 13, 1079-1088.

Schwarz-Sommer, Z., Hue, I., Huijser, P., Flor, P. J., Hansen, R., Tetens, F., Lonnig, W. E., Saedler, H., and Sommer, H. (1992). Characterization of the Antirrhinum floral homeotic MADS-box gene deficiens: evidence for DNA binding and autoregulation of its persistent expression throughout flower development. Embo J 11, 251-263.

Sessions, A., Burke, E., Presting, G., Aux, G., McElver, J., Patton, D., Dietrich, B., Ho, P., Bacwaden, J., Ko, C., et al. (2002). A high-throughput *Arabidopsis* reverse genetics system. Plant Cell 14, 2985-2994.

Sharma, V. K., and Fletcher, J. C. (2002). Maintenance of shoot and floral meristem cell proliferation and fate. Plant Physiol 129, 31-39.

Shiraishi, H., Okada, K., and Shimura, Y. (1993). Nucleotide sequences recognized by the AGAMOUS MADS domain of *Arabidopsis thaliana in vitro*. Plant J 4, 385-398.

Sieburth, L. E., and Meyerowitz, E. M. (1997). Molecular

dissection of the AGAMOUS control region shows that cis elements for spatial regulation are located intragenically. Plant Cell *9*, 355-365.

Simpson, G. G., and Dean, C. (2002). *Arabidopsis*, the Rosetta stone of flowering time? Science *296*, 285-289.

Smyth, D. R., Bowman, J. L., and Meyerowitz, E. M. (1990). Early flower development in *Arabidopsis*. Plant Cell *2*, 755-767.

Sridhar, V. V., Surendrarao, A., Gonzalez, D., Conlan, R. S., and Liu, Z. (2004). Transcriptional repression of target genes by LEUNIG and SEUSS, two interacting regulatory proteins for *Arabidopsis* flower development. Proc Natl Acad Sci USA *101*, 11494-11499.

Sundaresan, V., Springer, P., Volpe, T., Haward, S., Jones, J. D., Dean, C., Ma, H., and Martienssen, R. (1995). Patterns of gene action in plant development revealed by enhancer trap and gene trap transposable elements. Genes Dev *9*, 1797-1810.

Theissen, G. (2001). Development of floral organ identity: stories from the MADS house. Curr Opin Plant Biol *4*, 75-85.

Theissen, G., and Saedler, H. (2001). Plant biology. Floral quartets. Nature *409*, 469-471.

Tilly, J. J., Allen, D. W., and Jack, T. (1998). The CArG boxes in the promoter of the *Arabidopsis* floral organ identity gene APETALA3 mediate diverse regulatory effects. Development *125*, 1647-1657.

van Meyel, D. J., Thomas, J. B., and Agulnick, A. D. (2003). Ssdp proteins bind to LIM-interacting co-factors and regulate the activity of LIM-homeodomain protein complexes *in vivo*. Development *130*, 1915-1925.

Wagner, D., Sablowski, R. W., and Meyerowitz, E. M. (1999). Transcriptional activation of APETALA1 by LEAFY. Science *285*, 582-584.

Wang, X., Feng, S., Nakayama, N., Crosby, W. L., Irish, V., Deng, X. W., and Wei, N. (2003). The COP9 signalosome interacts with SCF UFO and participates in *Arabidopsis* flower development. Plant Cell *15*, 1071-1082.

Weigel, D., Alvarez, J., Smyth, D. R., Yanofsky, M. F., and Meyerowitz, E. M. (1992). LEAFY controls floral meristem identity in *Arabidopsis*. Cell *69*, 843-859.

Weigel, D., and Meyerowitz, E. M. (1993). Activation of floral homeotic genes in *Arabidopsis*. Science (Washington D C) *261*, 1723-1726.

Weigel, D., and Meyerowitz, E. M. (1994). The ABCs of floral homeotic genes. Cell *78*, 203-209.

Weigel, D., and Nilsson, O. (1995). A developmental switch sufficient for flower initiation in diverse plants. Nature *377*, 495-500.

Wilkinson, M. D., and Haughn, G. W. (1995). UNUSUAL FLORAL ORGANS Controls Meristem Identity and Organ Primordia Fate in *Arabidopsis*. Plant Cell *7*, 1485-1499.

Williams, F. E., and Trumbly, R. J. (1990). Characterization of TUP1, a mediator of glucose repression in *Saccharomyces cerevisiae*. Mol Cell Biol *10*, 6500-6511.

Wynne, J., and Treisman, R. (1992). SRF and MCM1 have related but distinct DNA binding specificities. Nucleic Acids Res *20*, 3297-3303.

Yanofsky, M. F., Ma, H., Bowman, J. L., Drews, G. N., Feldmann, K. A., and Meyerowitz, E. M. (1990). The protein encoded by the *Arabidopsis* homeotic gene agamous resembles transcription factors. Nature *346*, 35-39.

Young, J. C., Krysan, P. J., and Sussman, M. R. (2001). Efficient screening of *Arabidopsis* T-DNA insertion lines using degenerate primers. Plant Physiol *125*, 513-518.

Chapter 34
Transcription Control in Bacteria

Ding Jun Jin[1] and Yan Ning Zhou[2]

[1]*Transcription Control Section, Gene Regulation and Chromosome Biology Laboratory, National Cancer Institute-Frederick, National Institutes of Health, Frederick, MD 21702, USA*
[2]*Laboratory of Molecular Biology, National Cancer Institute, National Institutes of Health, Bethesda, MD 20892, USA*

Key Words: transcription, RNA polymerase, transcription factories, global gene regulation, nucleoid, nutrient starvation response, prokaryotes, bacteria, *E. coli*

Summary

Regulation of transcription is a key step in controlling gene expression in all cells. Many diseases and cancers result from alterations of gene expression caused by defects in transcription machinery. The basic structure and function of RNA polymerase (RNAP) and RNAP-associated proteins are conserved throughout evolution. Sophisticated genetics and advanced biochemistry make single-cell organisms, such as *E. coli,* an ideal model system to study the role of RNAP and transcription factors in gene expression and regulation. Studies of the simple model system have contributed greatly to our knowledge of transcription in principle, which underpin our understanding of gene regulation in much more complex eukaryotic organisms. In addition, studies of *E. coli* and other microorganisms have laid the foundation for the biotechnology industry. This chapter describes transcription machinery including RNAP and RNAP-associated proteins and regulation of transcription cycle in *E. coli,* with a focus on the unique feature of global regulation by RNAP (re)distribution in response to environment cues.

Transcription Machinery
A: RNAP

Escherichia coli RNA polymerase (RNAP) is a multisubunit enzyme (Table 34.1). Unlike in eukaryotes, in which three different RNAPs (Pol I, II, and III) synthesize three different RNA species (rRNA, mRNA, and tRNA/5S rRNA respectively), a single RNAP synthesizes all RNA species in *E. coli* (Burgess *et al.*, 1987). In *E. coli*, RNAP exists in two forms, core RNAP (E) and holoenzyme (Eσ), as shown in Fig. 34.1. The core RNAP consists of two α subunits, and one subunit each of β, β' and ω; and upon binding to a σ factor, it converts to a holoenzyme. The core RNAP is capable of transcription elongation and termination at intrinsic transcription terminators. However, only the holoenzyme can recognize a promoter and engage in transcription initiation (Burgess *et al.*, 1969). Recently, the high-resolution structures of bacterial RNAP have been determined (Murakami *et al.*, 2002; Vassylyev *et al.*, 2002; Zhang *et al.*, 1999). Strikingly, the basic architectures of bacterial RNAP and yeast Pol II are conserved (Ebright, 2000; Fu *et al.*, 1999), which is consistent with the notion that bacterial RNAP shares considerable sequence homology with its eukaryotic counterpart (Allison *et al.*, 1985; Sweetser *et al.*, 1987). These structural studies not only have validated many mutational analyses of RNAP (Jin and Zhou, 1996), but also provided structural basis for the function of RNAP (Borukhov and Nudler, 2003) and the actions of various antibiotics.

As an essential enzyme in the cell, bacterial RNAP has been a target for multiple antibiotics, including rifampicin, streptolydigin, sorangicin A, and Microcin J25 (Adelman *et al.*, 2004; Heisler *et al.*, 1993; Severinov *et al.*, 1995; Yang and Price, 1995). Among

---

Corresponding Author: Ding Jun Jin, Tel: (301) 846-7684, Fax: (301) 846-1456, E-mail: djjin@helix.nih.gov

Table 34.1 RNAP Subunits and Associated Proteins.

A. RNAP Core Subunits

Gene Name	Subunit	M.W. (kDal)	Map Position	Function
rpoA	α	37	74 min.	assembly, contact with DNA and transcriptional activators
rpoB	β	151	90 min.	active site, rifampicin binding
rpoC	β′	155	90 min.	active site, DNA binding
rpoZ	Ω	10	82 min.	unknown

B. RNAP Sigma Factors

Gene Name	Subunit	M.W. (kDal)	Map Position	Function
rpoD	σ^{70}	70	69 min.	housekeeping genes
rpoE	σ^{E}	22	58 min.	genes for periplasmic and envelope proteins
rpoF	σ^{F}	28	43 min.	genes for flagella and chemotaxis
rpoH	σ^{32}	32	78 min.	heat shock stress response genes
rpoN	σ^{N}	54	72 min.	nitrogen limitation genes
rpoS	σ^{S}	38	62 min.	stationary phase genes
fecI	σ^{FecI}	19	97 min.	ferric citrate transport genes

C. RNAP Associated Proteins

Gene Name	M.W. (kDal)	Map Position	Function
dksA	18	3 min.	regulation of rRNA promoters during the stringent response
greA	18	72 min.	cleaves RNA in arrested elongation complexes
greB	19	76 min.	cleaves RNA in arrested elongation complexes
nusA	55	71 min.	termination/antitermination
nusE	12	74 min.	termination/antitermination
nusG	20	90 min.	termination/antitermination
rapA	101	1 min.	RNAP recycling
sspA	24	73 min.	transcription activator; acid tolerance in stationary phase
topA	97	29 min.	introduces (+) supercoils

$$\alpha_2\beta\beta'\omega + \sigma \rightleftharpoons \alpha_2\beta\beta'\omega\sigma$$

core (E)   Holoenzyme (Eσ)

Fig.34.1 The *E. coli* RNAP is a multisubunit enzyme. The association of σ with the core in the holoenzyme differentiates between the two different forms of *E. coli* RNAP.

them, rifampicin and its derivatives have been the most important in clinical use for the treatments of tuberculosis, meningitis and staphylococcal infections (Fisher, 1971; Kapusnik *et al.*, 1984; Leung *et al.*, 1998; Lounis and Roscigno, 2004). Mutations in *E. coli* RNAP conferring rifampicin-resistance (Rif[r]) were reported shortly after the antibiotic was discovered (Ezekiel and Hutchins, 1968; Rabussay and Zillig, 1969). Rif[r] mutations have been located exclusively on the second largest subunit of RNAP, the β subunit encoded by the *rpoB* gene. Most of the Rif[r] mutations are clustered in the middle of the gene defined as the rif region (Jin and Gross, 1988; Severinov *et al.*, 1993). The rif region is well conserved in different bacteria including many pathogens (Aubry-Damon *et al.*, 1998; Ramaswamy and Musser, 1998). Rifampicin inhibits RNAP's function by blocking the transition from transcription initiation to transcription elongation (McClure and Cech, 1978). Cross-linking experiments have indicated that the rifampicin blocks a channel leading a nascent RNA out of the catalytic center of RNAP (Mustaev *et al.*, 1994). The crystal structure of the *T. aquaticus* core RNAP complexed with rifampicin indicates several conserved amino acid residues in the rif region that interact with the antibiotic (Campbell *et al.*, 2001), which adequately account for all known Rif[r] mutants. Rifampicin binds to the rif region of the β subunit, which lies deep within the DNA/RNA channel. Clearly, the critical location of the rif region in RNAP is responsible for the multiple effects of Rif[r] mutations on different aspects of transcription (Jin *et al.*, 1988; Jin and Gross, 1989;

Zhou and Jin, 1998). Interestingly, the antibiotic sorangicin A, which has a different structure than rifampicin, also binds RNAP in the same β subunit pocket, as seen from the structure of the *T. aquaticus* RNAP- sorangicin A complex (Campbell *et al*., 2005). This overlap in binding sites for different antibiotics in RNAP not only explains the reason why Rif^r mutants are cross-resistance to multiple drugs (Xu *et al*., 2005), but also poses a challenge for the development of new generation and antibiotics. Because Rif^r mutations are highly conserved in eubacteria, the well studied set of *E. coli* Rif^r mutant RNAPs could be used to screen for new antibiotics that will inhibit the growth of Rif^r pathogenic bacteria which have emerged as an important clinical issue worldwide.

The size of an *E. coli* cell is about 2-4 μm long and less than 1 μm in width. The *E. coli* genome is close to five million base pairs encoding over 4,000 genes (Blattner *et al*., 1997) and forms loose structures called nucleoids (Drlica, 1987; Hobot *et al*., 1985; Pettijohn, 1996; Robinow and Kellenberger, 1994) without the presence of a membrane separating the genome from the cytoplasm. It is estimated that there are about 2,000 core RNAP molecules in the cell (Ishihama, 2000; Ishihama, 1981). The actual number of genes per cell substantially exceeds that number because a rapidly growing *E. coli* cell contains more than one chromosome (Bremer and Dennis, 1996). Apparently, the number of RNAP is less than the total number of genes in the genome, which suggests that core RNAP is limiting. Thus, control of RNAP distribution in the genome could be critical for global gene regulation in the cell (see below).

There are seven σ factors in *E. coli* (Table 34.1). Among them, σ^{70} encoded by *rpoD* is the major σ factor (Burton *et al*., 1983) and the holoenzyme Eσ^{70} is responsible for the housekeeping functions in the cell. Each of the minor σ factors is required for the expression of a specific set of genes called regulons under different physiological conditions involving various stress responses (Gross *et al*., 1998). For example, the holoenzyme Eσ^S is required for the expression of stationary-phase specific genes (Loewen and Hengge-Aronis, 1994). While Eσ^{32} is responsible for the expression of heat shock genes during the cytoplasmic stress response (Grossman *et al*., 1984), Eσ^E is responsible for the expression of genes encoding periplasmic proteins and envelope components during the extracytoplasmic or envelope stress response (Erickson and Gross, 1989). Eσ^N is important for the expression of genes which are activated during nitrogen limitation (Kustu *et al*., 1989). Eσ^F controls the expression of flagellar and chemotaxis genes (Helmann, 1991), and Eσ^{FecI} is involved in transcription initiation of the *fec* operon (Pressler *et al*., 1988). The expression or the activity of different σ factors is sensitive to the signals induced by different stresses. For example, while unfolding or misfolding of cytoplasmic proteins induced by heat shock stimulates the expression of σ^{32}, heat shock induced unfolding or misfolding of extracytoplasmic or periplasmic proteins activates σ^E (Alba and Gross, 2004; Straus *et al*., 1990). In addition, the cellular levels of some σ factors, such as σ^{32}, σ^E and σ^S are known to be subject to regulated proteolysis (Ades *et al*., 1999; Straus *et al*., 1990; Zhou *et al*., 2001).

B: RNAP-Associated Proteins and Transcription Factors

Many RNAP-associated proteins have been identified in *E. coli*, most of those have regulatory function in transcription (Table 34.1). For example, transcription regulator DksA mediates the transcription of ribosomal operons and the stringent response (Paul *et al*., 2004a). SspA, the stringent starvation protein, is a transcriptional co-activator for the phage P1 late promoters important for the phage development (Hansen *et al*., 2003). In addition, SspA plays a pivotal role in acid tolerance of *E. coli* during stationary phase growth (Hansen *et al*., 2005). Nus factors (NusA, NusB, NusE, and NusG) are involved in transcription elongation and termination/antitermination (Condon *et al*., 1995; Friedman and Court, 1995; Nudler and Gottesman, 2002). TopA is important in modulating DNA topology (Tse-Dinh *et al*., 1997), as an elongating RNAP generates positive supercoils ahead and leaves a negative supercoils behind (Wu *et al*., 1988). While some of these transcription regulators, such as NusA and RapA, bind to RNAP with high affinity, others bind only loosely and their interaction with RNAP cannot be detected during conventional purification of the enzyme (Zhi *et al*., 2003b). Among the RNAP-associated proteins, GreA and GreB are the functional homologs of eukaryotic transcript cleavage factor TFIIS, and are involved in the RNA cleavage reaction of arrested elongation complexes (Borukhov *et al*., 2001). Also, RapA, an ATPase, which is stimulated by the interaction with RNAP, is a bacterial homolog of the SWI/SNF proteins (Sukhodolets and Jin, 1998). Eukaryotic SWI/SNF proteins are important in chromatin/nucleosome remodeling, gene expression and DNA repair (Citterio *et al*., 2000; Muchardt and Yaniv, 1999; Pazin and Kadonaga, 1997; Peterson, 1996), indicating that ATP–mediated chromatin remodeling by the SWI/SNF proteins is important for the regulation of cell growth. RapA plays an important role in stimulating

RNAP recycling in transcription (Sukhodolets *et al.*, 2001), and provides an opportunity to study the detailed mechanism of this homolog of SWI2/SNF2 in transcription and gene regulation.

In addition, there are many transcription factors which modulate transcription at different stages of the transcription cycle (Ishihama, 2000). Numerous repressors and activators control transcription initiation by binding to regulatory regions in or near promoters. A classical example is the Lac repressor and CRP activator, which represses and activates the *lac* operon respectively, depending on the growth condition (Savery *et al.*, 1996; Shuman and Silhavy, 2003). Several elongation factors regulate rate of elongation and/or determine an outcome of termination or antitermination by modulating elongation complexes at regulatory sites in nascent RNA (Friedman and Court, 1995; Weisberg and Gottesman, 1999). One of the major transcriptional termination mechanisms requires the Rho protein which is a RNA-dependent ATPase and DNA:RNA helicase (Platt, 1994).

Another class of proteins, collectively called nucleoid proteins including Fis, Hu, HN-S, and IHF (Azam and Ishihama, 1999; Dorman and Deighan, 2003), is also important for transcription. These small proteins bind to DNA with specific sequences and /or at the bend of DNA. Thus, they exert their role in regulation of transcription either directly by binding to some promoter regions, as a repressor or activator, or indirectly by changing the topology of DNA (Travers *et al.*, 2001).

Transcription Cycle

Transcription is divided into the following stages: initiation, elongation, termination, and RNAP recycling (Fig.34.2). In each stage, however, there are multiple steps involved and each step potentially could be a regulatory step in transcription (von Hippel *et al.*, 1996).

A: Initiation

Initiation of transcription occurs at sites in the DNA called promoters (Fig.34.3). Initiation of transcription is a complex process consisting of several steps: initial binding of RNAP holoenzyme to a promoter, formation of a competent initiation complex, synthesis of the initial phosphodiester bonds, and clearance of RNAP from the promoter (deHaseth *et al.*, 1998), as shown in Fig. 34.4. The basic elements of a promoter recognized by *E. coli* Eσ 70 are the −35 (TTGACA) and −10 (TATAAT) regions relative to the starting site which is defined as +1 (Harley and Reynolds, 1987; Hawley and McClure, 1983). The spacer between the −35 and −10 regions has a consensus length of 17 bp (Ayers *et al.*, 1989; Stefano and Gralla, 1982). At some promoters, there are other regulatory sites either within or near the promoter region where transcription factors, such as repressors or activators, can bind and control transcription initiation. In addition, an A/T rich sequence upstream of the −35 region called the UP element enhances promoter activity by providing additional interaction with RNAP and other transcription factors at some promoters (Blatter *et al.*, 1994; Davis *et al.*, 2005; Ross *et al.*, 1993). In another subset of promoters, there is only the typical −10 region, but lacks the −35 region. At these promoters, an additional activator is required in order for RNAP to initiate transcription. This requirement is overcome at yet another subset of promoters, where there is an extended −10 region (TGN) (Keilty and Rosenberg, 1987; Mitchell *et al.*, 2003), which enables RNAP to recognize these promoters in the absence of the −35 region. Promoter recognition is mainly determined by the σ factor, whose distinct domains bind to and/or act at different elements of promoter (Gross *et al.*, 1998; Young *et al.*, 2002).

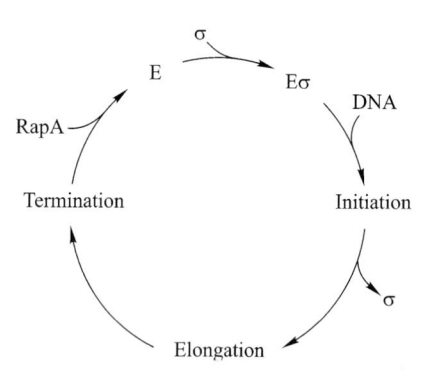

Fig.34.2 The cranscription cycle. Transcription is a cyclical process. Upon the formation of Eσ and the promoter binding, RNAP undergoes the process of initiation, elongation, and termination. The cycle is completed with the help of RapA, which promotes RNAP recycling, and RNAP is ready to start the cycle again.

The first complex to form upon initial binding of RNAP (R) to promoter (P) is a closed complex (RP_c), which transforms into an open complex (RP_o) through several isomerization steps. During this transformation, ~ 10-12 bp segment of the double strand of DNA near the −10 region is melted, and the template DNA strand is positioned into the active center of RNAP making RP_o competent for initiation. There are large scale conformational changes of RNAP in forming these kinetically significant intermediates (Saecker *et al.*, 2002). With the addition of NTPs, RNAP forms initiation complex (RP_{init}). However, before RP_{init} escapes the promoter to become an elongation complex

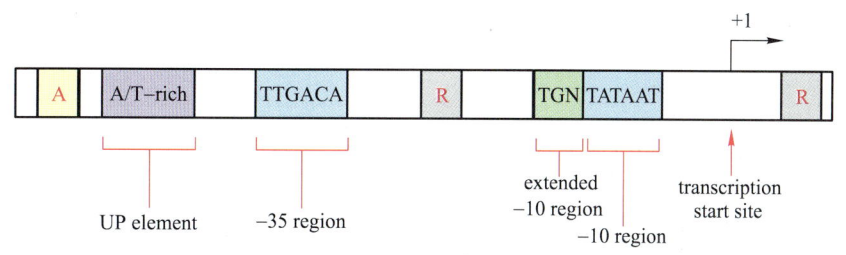

Fig.34.3 Elements in the promoter region. The basic structure of the promoter region includes a −35 region and a −10 region (indicated by the blue boxes) for recognition by RNAP. An A/T-rich UP element upstream (in purple) of the consensus sequence helps enhance transcription. In some cases, the promoter lacks the −35 region, but contains an extended −10 region (highlighted by the green color) or an additional activator region to facilitate the σ recognition. Sites for activator binding (denoted by the red 'A') can enhance transcription initiation. Repressor sites (denoted by the red 'R') found in and/or near the promoter region block transcription through repressor binding.

$$R + P \rightleftarrows RP_{c1} \rightleftarrows RP_{c2} \rightleftarrows RP_{o1} \rightleftarrows RP_{o2} \underset{AP}{\overset{NTP_s}{\rightleftarrows}} RP_{init} \overset{NTP_s}{\longrightarrow} EC$$

binding   isomerization   promoter clearance

Fig.34.4 Intermediates in the process of transcription initiation. The initiation of transcription is a multi-step process that starts with the RNAP holoenzyme (R) recognition and binding to the promoter (P), resulting in a closed complex (RP_c). The closed complex then undergoes several isomerization steps leading to the formation of an open complex (RP_o). Although the open complex of most promoters is stable (irreversible), the formation of the open complex may be reversible in some cases as indicated by the green arrows. Addition of NTPs leads to the formation of the initial phosphodiester bonds in the nascent RNA resulting in the initiation complex (RP_{init}). The initiation complex makes short non-productive initiation or abortive products (AP) at most promoters. The final step of the process involves promoter clearance leading to the transcription elongation complex (EC).

(EC), RNAP generally synthesizes short non-productive RNAs (abortive products), which dissociate from RNAP. There are two kinds of non-productive initiation: abortive and slippage (Hsu et al., 2003; Jin, 1994; Jin and Turnbough, 1994; Liu et al., 1994; McDowell et al., 1994). While abortive initiation is generally found at most promoters, slippage initiation or reiterative RNA synthesis is limited to promoters which have a run of three or more A, T or C sequences at the beginning of the transcript.

Multiple mechanisms are involved in the control of transcription initiation. As mentioned earlier, initiation is carried out by RNAP holoenzyme containing sigma factors. Different holoenzymes containing different σ factors recognize different sets of promoters in the genome. Therefore, operationally, interaction between σ factors and core RNAP is the first step in initiation. Genetically, it has been shown that different σ factors compete for binding to core RNAP (Zhou et al., 1992),

indicating that core RNAP is limiting in the cell. Thus, the amount and/or the affinity of a particular σ factor for core RNAP determines the level of a particular holoenzyme (Maeda et al., 2000), which in turn modulates the transcription profiles in the cell. On the other hand, there are different anti-σ factors (Hughes and Mathee, 1998) in the cell which antagonize either the interaction between σ factors and core RNAP or the activity of holoenzymes containing σ factors. The issue of when σ factor releases from RNAP is not resolved yet (Bar-Nahum and Nudler, 2001; Gill et al., 1991; Greenblatt and Li, 1981; Mukhopadhyay et al., 2001; Shimamoto et al., 1986). However, it is likely that σ factor releases shortly after RNAP enters elongation phase.

Various transcription factors act on promoters with different mechanisms (Ishihama, 2000). For example, while some repressors bind at promoters to prevent the binding of RNAP to the promoters due to steric hindrance, others bind to inhibit the isomerization step(s). In addition, some repressors bind at multiple sites in the promoter regions to form a DNA loop via protein-protein interactions, usually in a concerted action with nucleoid-binding proteins (Adhya et al., 1998). Antibiotics rifampicin and sorangicin A, as well as the nucleoid-binding protein H-NS, prevent RNAP from promoter clearance (Campbell et al., 2005; McClure and Cech, 1978; Xu et al., 2005; Schroder and Wagner, 2000). Activators are able to stimulate transcription at different steps in initiation (Browning and Busby, 2004; Lawson et al., 2004). In general, however, the role of activators is to recruit RNAP to the promoters which usually have weak or no binding to RNAP alone. The two α subunits of RNAP participate in the interaction with various transcriptional activators (Ebright, 1993; McLeod et al., 2002; Ross et al., 1993). The GreA and GreB proteins stimulate promoter escape or promoter clearance (Hsu et al., 1995). In addition,

NTP concentration controls promoter clearance at some promoters (Jin, 1994; Qi and Turnbough, 1995). For example, the non-productive slippage initiation is favored at the *pyrBI* promoter of pyrimidine biosynthetic operons when the concentration of UTP is high, whereas the slippage is minimal when UTP is low (Liu *et al.*, 1994). Thus, the productive initiation of *pyrBI* responds to the concentration of UTP, which is biologically relevant because the end products of the pyrimidine biosynthetic operons are UTP and CTP.

Initiation at the promoters from the ribosomal operons (*rrn*) is an example of multiple controls including several transcription factors and small molecules (Gralla, 2005; Paul *et al.*, 2004b). Regulation of transcription of *rrn* is critical for cell growth and global genome-wide regulation because synthesis of rRNA is the single most important factor that influences the distribution of RNAP inside the cell under different physiological conditions (see next section below). The *rrn* promoters are the most active promoters in the cell grown in rich media, but they have only minimal activity in the cell starved for nutrients. Hence, they are called stringent promoters. The key feature of the *rrn* promoters is that the interaction between wild type RNAP and stringent promoters is intrinsically unstable, presumably because the steps prior to the first phosphodiester bond formation are reversible and the intermediate closed and open complexes are in rapid equilibrium with each other. In contrast, open complexes at non-stringent promoters are generally very stable and the reactions back to closed complexes are negligible (McClure, 1980). Many factors act on this regulatory step. While the DksA protein along with the small molecule ppGpp (the level of ppGpp is minimal in fast growing cells and maximum in nutrient starvation cells) act synergistically to destabilize the open complex (Paul *et al.*, 2004a), the Fis protein stabilizes the complex (Zhi *et al.*, 2003a). In addition, the concentration of the initial NTP to be incorporated into rRNA affects the stability of the open complex (Gaal *et al.*, 1997). The study of the "stringent" mutant RNAPs (Zhou and Jin, 1998), which destabilize the open complex of stringent promoters *in vitro* and reduce the synthesis of rRNA even when cell are grown in rich media, further support the notion that the regulation of the *rrn* operons and the control of the stringent response during nutrient starvation is mainly aimed at the stability of the open complex of the *rrn* promoters (see next section below).

B: Elongation

Escape from the promoter converts RNAP into an elongation complex (Korzheva *et al.*, 2000) where 8-9 nucleotides at the 3' end of the nascent RNA form a hybrid with the template DNA and the 3' end of the RNA in the hybrid is located at the active center of the enzyme (Fig. 34.5). The rate of elongation is not constant because RNAP pauses at some sites called pausing sites (Kingston and Chamberlin, 1981; Landick *et al.*, 1987). Although the formation of an RNA hairpin in the nascent transcript coincides with pausing (Landick *et al.*, 1996), the nature of the pausing at many other sites is not well understood (Artsimovitch and Landick, 2000; Levin and Chamberlin, 1987). However, because pausing in general is sensitive to concentration of NTPs, and some mutant RNAPs with modified elongation rates (pausings) have altered Km for NTPs (Jin and Gross, 1991), there is a kinetic element associated with the pausing: the Km of RNAP for the nucleotide to be incorporated at the pausing site is increased. Transcription factors and/or antitermination factors, such as NusA and N, modulate the rate of elongation by either enhancing pausing or suppressing pausing (Greenblatt *et al.*, 1981; Gusarov and Nudler, 2001; Kingston and Chamberlin, 1981; Rees *et al.*, 1996).

Sometimes the elongation complex becomes arrested at intrinsic arrest sites in DNA *in vitro* and *in vivo* (Komissarova and Kashlev, 1997a; Toulme *et al.*, 2000). Arrest also occurs when RNAP encounters physical obstacles in template such as "roadblock" or chemical lesions in DNA. These arrested complexes cannot be rescued by increasing concentration of NTPs. Detailed biochemical analysis revealed that the arrest is caused by RNAP backtracking along DNA, associated with reverse threading of the RNA through the enzyme (Fig. 34.5). The 3' end of the transcript is extruded from the enzyme during the backtracking, causing its disengagement from the active center of RNAP. Transcription factors GreA/GreB reactivate the arrested complex by stimulating an endonucleolytic cleavage of the nascent RNA at the backtracked position of the active center in the enzyme (Komissarova and Kashlev, 1997b), thus generating a new 3' RNA end at the upstream location of RNAP. This function is conserved: eukaryotic transcript cleavage factor TFIIS reactivates the backtracked complexes by the same mechanism and promotes Pol II transcription through the nucleosomes (Kireeva *et al.*, 2005). In addition, transcription repair coupling factor Mfd (Selby and Sancar, 1993) rescues the arrested complexes by promoting forward translocation (thus reverse backtracking) of RNAP so that the active center of RNAP re-engage the RNA 3' end (Park *et al.*, 2002; Roberts and Park, 2004).

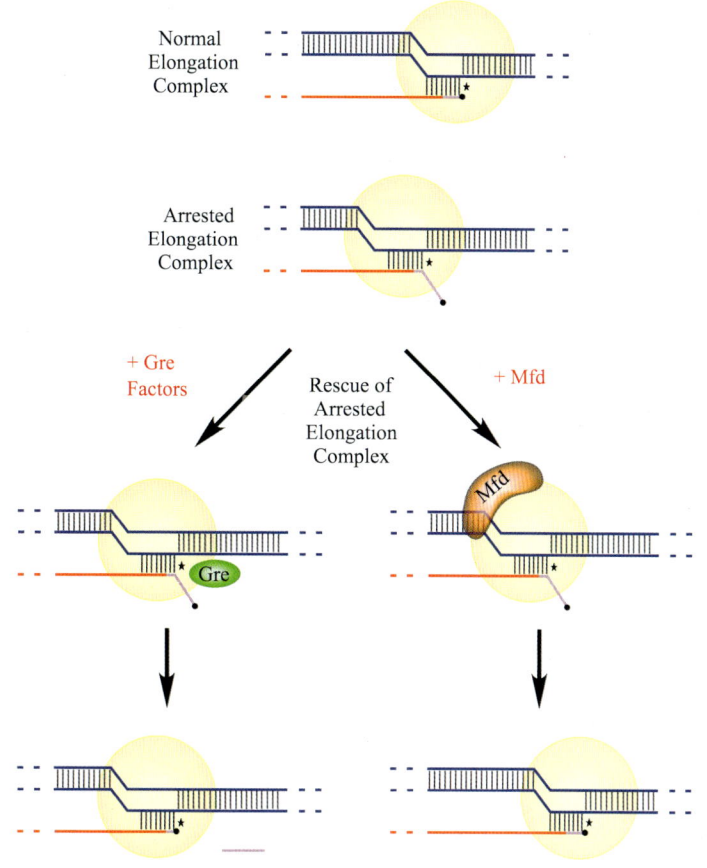

Fig.34.5 **Backtracking of elongation complex and rescuing arrested complex by Gre factors and Mfd.** In the elongation complex, 8-9 nucleotides at the 3' end of the nascent RNA form a hybrid with the template DNA and the 3' end of the RNA (denoted by a black circle) in the hybrid is located at the active center (denoted by the star) of the enzyme. The arrested RNAP complex backtracks along the DNA, simultaneously causing portion of the 3' end of the nascent RNA to extrude out of the active center (indicated by purple color). The arrested complex can be rescued either by Gre factors (GreA/GreB) which cleave the extruded portion of the RNA at the catalytic center, or by Mfd which promotes forward translocation (or reverse backtracking) of the enzyme.

C: Termination and Antitermination

Termination of transcription occurs when elongating RNAP reaches a terminator (Henkin, 2000; Nudler and Gottesman, 2002; von Hippel, 1998). There are two kinds of terminators in *E. coli*. One type is called intrinsic or simple terminator and causes dissociation of elongation complex in the absence of trans-acting protein factors (Brendel *et al.*, 1986; d'Aubenton Carafa *et al.*, 1990). The other type is called a Rho-dependent terminator because termination at those sites requires the transcription factor Rho (Richardson, 2002). However, the kinetic element is important for all termination systems: mutant RNAPs with altered rates of elongation, hence pausing, also have altered properties during termination at both simple and Rho-dependent terminators (Jin *et al.*, 1992; McDowell *et al.*, 1994).

The intrinsic or simple terminators in DNA have two common features: a GC-rich palindromic element, immediately followed by a stretch of T sequence. In RNA, this sequence forms a stable termination hairpin structure followed by 7-9 unpaired U residues (Fig. 34.6). Pausing at the terminators induces the formation of the termination hairpin, which coupled with the intrinsic instability of dA-U pairs (Gusarov and Nudler, 1999; Komissarova *et al.*, 2002; Martin and Tinoco, 1980), facilitates the release of RNA. Thus, formation of the termination hairpin immediately before the stretch of T sequence at the terminators is critical for termination, a step various anti-termination mechanisms act upon (see below).

At Rho-dependent terminators, RNAP does not stop without the assistance of the Rho protein (Platt, 1994; Richardson, 2002). Although there is no common feature for this type of terminator, RNAP generally pauses at the terminator in the absence of Rho. Rho then binds to the upstream part of nascent RNA to be terminated and translocates along the RNA using its ATPase activity (Richardson, 2003). Upon reaching elongation complex, Rho displaces the nascent RNA from the hybrid with DNA in the active center using its DNA:RNA helicase activity powered by ATP hydrolysis (Brennan et al., 1987; Walstrom et al., 1997). NusG stimulates Rho-dependent termination, probably by forming a Rho-NusG-RNAP complex (Burns et al., 1999) since NusG interacts with both RNAP and Rho (Nehrke and Platt, 1994).

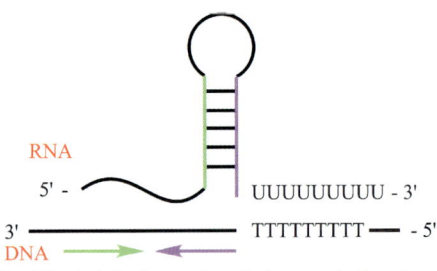

Fig.34.6 The intrinsic or simple transcription terminator. In DNA, there are two components for the intrinsic terminator: a GC-rich palindromic element (indicated by green and purple arrows), immediately followed by a stretch of T sequence. In the RNA to be terminated, this sequence forms a stable termination hairpin structure (shown by the green and purple pairing) followed by 7-9 unpaired U residues.

Elongating RNAP can be modified by transcription or antitermination factors suppressing termination at intrinsic and Rho-dependent terminators in a process called antitermination. The best studied antitermination systems are from λ phage (Friedman and Court, 1995; Gottesman and Weisberg, 2004; Weisberg and Gottesman, 1999). In λN antitermination, the phage N protein binds to a stem-loop structure in RNA called BoxB and modifies elongation complex making it resistant to termination. Although N alone is able to cause antitermination at short distance from BoxB (Gusarov and Nudler, 2001; Rees et al., 1996), formation of a progressive antitermination complex requires binding of general elongation factors NusA, NusE, NusB, and NusG (DeVito and Das, 1994). Mechanisms of how N cause antitermination and the role of Nus factors in N action are not well understood. N may physically interfere with formation of termination hairpins, stabilize the interaction of RNAP with the RNA-DNA hybrid, or block the access of Rho to the RNA-DNA hybrid. In addition, N may reduce the dwell time for the formation of the termination-competent conformation of RNAP at terminators by suppressing pausing of RNAP at the termination sites. Similarly, phage Q protein suppresses termination by directly interfering with hairpin formation, although cellular factors are not required in λQ antitermination (Roberts et al., 1998). Cellular antitermination is important for the transcription of ribosomal RNA (*rrn*) operons (Squires et al., 1993). Like λN antitermination system, this system has BoxB and other *cis* elements; however, factors involved in the *rrn* operons antitermination are less well defined, but they include NusA, NusB, NusE, NusG, and ribosomal protein S4 (Squires et al., 2003).

In some bacterial systems, control of transcription is mediated by alternating termination and antitermination in responding to environmental signals (Henkin and Yanofsky, 2002; Yanofsky, 2000). In these systems, the nascent RNA is able to form two alternating pairings: one pairing results in termination as an intrinsic terminator and the other pairing results in antitermination as it prevents the formation of the termination hairpin (Fig.34.7). Many effectors modulate the alternating pairings in different ways. For example, in *trp* attenuation, the position of a ribosome translating the leader peptide is critical for the regulation (Zurawski et al., 1978). When tryptophan is limiting in the cell, an

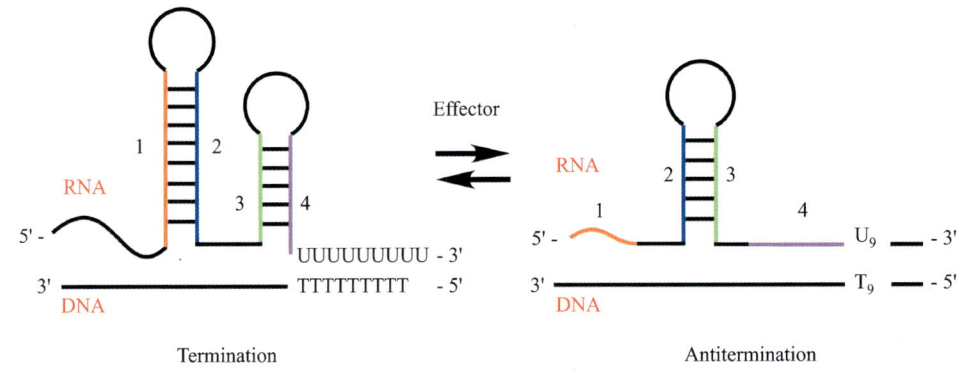

Fig.34.7 Alternating termination and antitermination with different RNA pairings. The nascent RNA sequence is able to form alternative pairings as indicated by regions 1 to 4. One pairing involves the formation of a termination hairpin, represented by the pairing of regions 3 and 4 in green and purple leading to termination. An alternative pairing of regions 2 and 3 (blue and green) forms the alternate hairpin structure, preventing the formation of the termination hairpin, thus favoring antitermination. Many effectors influence the alternating pairings in different ways to control gene expression.

uncharged tRNA^{Trp} binds to the ribosome causing the ribosome to stall at the Trp codons in the leader peptide in such a way that the formation of the antitermination pairing is favored. Conversely, when tryptophan is available, charged tRNA^{Trp} will not stall the ribosome at the leader peptide so that a termination pairing is favored. In other bacterial systems, various effectors, including different tRNAs and some small molecules, are able to modulate the alternating pairings in the absence of translation (Epshtein et al., 2003; Grundy and Henkin, 2003).

D: RNAP Recycling

After the end of one round of transcription, RNAP needs to release from DNA and/or RNA in order for reuse. However, the ability of RNAP to recycle after one round of transcription is limited at least in vitro. RapA activates transcription by stimulating RNAP recycling (Sukhodolets et al., 2001), suggesting that RNAP recycling is potentially a regulatory step in transcription. Probably, RNAP becomes trapped or immobilized in a post transcription or post-termination complex after one round of transcription. As RapA is an ATPase and a member of the SWI/SNF superfamily of helicase-like proteins, it is then able to remodel these complexes so that RNAP is released or becomes mobile for recycling. Also, σ 70 enhances dissociation of RNAP after termination in vitro (Arndt and Chamberlin, 1988); however, the mechanism by which σ 70 promotes RNAP recycling is unknown.

Global Genome-wide Regulation

There are many levels of gene regulation. One is involved in specific operons. The classical example is the *lac* operon, which is specifically induced by the substrate of the operon (Jacob and Monod, 1961). Virtually all carbohydrate catabolic genes or operons can be regulated by substrate-specific induction (Bruckner and Titgemeyer, 2002). Another level of regulation is involved in a limited set of operons which share some common features. One example is carbon catabolite repression that inhibits expression of various alternative carbon utilization pathways when glucose, which is readily usable by the cell, is present (Saier, 1998). When carbon catabolite repression is relaxed, CRP, a global transcription activator for most of carbon catabolic pathways, is activated (Crasnier, 1996). Another example is the expression of different regulons by holoenzymes containing different σ factors as described above.

There is a genome-wide global regulation via RNAP distribution in response to nutrient cues, which affects the overall transcription program in the cell. Under optimal growth conditions such as when the cell is grown in nutrient-rich media, the vast majority of RNAP molecules synthesize rRNA and tRNA (termed stable RNAs) (Bremer and Dennis, 1996). The stable RNAs are encoded by the genes that represent less than 1% of the genome. Thus, the remaining few RNAP molecules not synthesizing stable RNAs are responsible for transcribing the other 99% of genes in the genome. Under suboptimal conditions, such as when cells are growing slowly in nutrient-poor media, few RNAP molecules synthesize stable RNAs. When cells are shifted from nutrient-rich to starvation conditions, such as amino acid starvation, leading to what is termed the stringent response (Cashel et al., 1996), the cellular transcription machinery is dramatically reprogrammed in such a way that the expression of stable RNAs is totally inhibited while that of other genes or operons, such as amino acid biosynthetic operons, is activated. Similarly, global transcriptional programs reveal a carbon source foraging strategy: as the available carbon substrate becomes poorer, cells systematically increase the number of genes expressed and reduce the expression of stable RNAs (Liu et al., 2005).

Until very recently, little has been known about the location and distribution of the RNAP molecule in *E. coli* under different physiological conditions. A functional *rpoC-gfp* gene fusion (the β' subunit of RNAP is fused to the green fluorescent protein (GFP) from marine jellyfish) on the *E. coli* chromosome has been constructed to visualize RNAP in the cell by fluorescence microscopy under different growth conditions (Cabrera and Jin, 2003). The results have demonstrated that the distribution of *E. coli* RNAP is dynamic and sensitive to physiological changes. Several conclusions have been derived from the cell biology study of RNAP.

First, RNAP is located exclusively either within and/or surrounding the nucleoid and there is no RNAP-GFP signal in cytoplasmic space. Second, growth conditions, nutrient deprivation, and transcription activity affect the distribution of RNAP. In fast-growing cells cultured in rich medium, RNAP distribution is heterogeneous, being concentrated in areas of the nucleoid (Fig.34.8). These areas are named transcription foci because they disappear in the presence of antibiotic rifampicin, which inhibits transcription. The transcription foci are likely RNAP molecules actively engaged in stable-RNA synthesis, because in slow-growing cells cultured in nutrient-poor media, the distribution of RNAP is relatively homogeneous and transcription foci are not evident (Fig. 34.8). Also, while the transcription

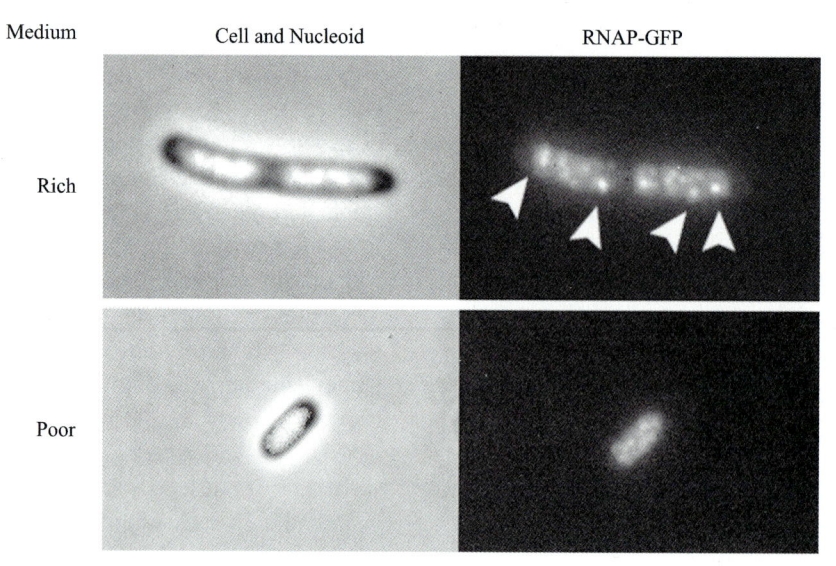

Fig.34.8 RNA polymerase distribution in cells grown in different media. Images of the *rpoC-gfp* cells grown in rich LB and in poor glucose-minimal media are indicated. The larger cell grown in LB has two nucleoids and the smaller cell grown in glucose-minimal medium has one nucleoid. The arrows indicate the transcription foci of the RNAP-GFP in the cell grown in rich medium.

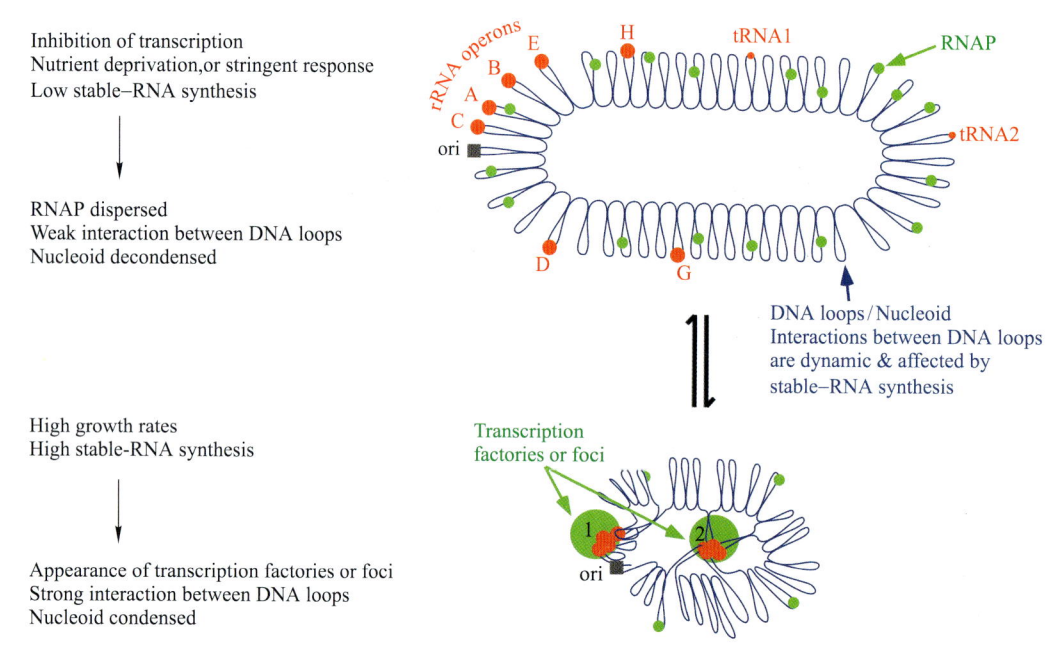

Fig.34.9 Model of stable-RNA synthesis, transcription factories/foci, and chromosome condensation. The *E. coli* chromosome is represented as blue lines folded in loops, the *ori* (origin) of replication as a black square, the seven rRNA operons as large red circles with letters, and the two representative tRNA operons as small red circles. The RNAP molecules are represented as small green circles. For simplicity, only two putative transcription factories/foci, which make the nucleoid more compact by pulling different stable RNA operons into proximity, are indicated here (bottom part of the diagram, large green circles labeled 1 and 2). (Adapted from Cabrera and Jin, 2003). See text for details.

foci disappear rapidly in wild-type cells during amino acid starvation, they remain present in an isogenic *relA* mutant strain in which stable RNAs are still actively synthesized (Metzger *et al.*, 1989). Moreover, the distribution of a "stringent" RNAP, in which stable-RNA synthesis is impaired even in rich media (Zhou and Jin, 1998) resemble that of wild-type RNAP during the stringent response. Thus, the transcription foci are proposed to be transcription "factories" synthesizing stable RNAs, which form structures analogous to the eukaryotic nucleolus (Cook, 1999). Finally, the synthesis of stable RNAs has been suggested to be a driving force in the condensation of the *E. coli* chromosome. The nucleoids become decondensed in wild-type cells when stable-RNA synthesis is preferentially inhibited during the stringent response, whereas they remain condensed in the *relA* mutant cells where stable-RNA synthesis is maintained during amino acid starvation. Moreover, the nucleoids are decondensed in the "stringent" RNAP mutant defective in stable-RNA synthesis even when grown in nutrient-rich media. A working model is postulated to link the synthesis of stable RNAs, RNAP distribution, and chromosome condensation in bacteria (Fig. 34.9).

In summary, systemic studies combining genetics, biochemistry and cell biology approaches not only reveal a link between global gene regulation, such as the stringent (nutrient deprivation) response, and RNAP's (re)distribution in the cell, but also reveal an important role played by RNAP actively engaged in the synthesis of stable RNAs (in particular rRNA), in forming transcription factories or foci and in bringing about chromosome condensation.

Acknowledgment

We thank many colleagues for their contributions in the research. We are grateful for the comments from Mikhail Kashlev, Monica Hui and Julio Cabrera, and for the preparations of figures and table by Monica Hui and Julio Cabrera. This research was supported [in part] by the Intramural Research Program of the NIH, National Cancer Institute, Center for Cancer Research.

References

Adelman, K., Yuzenkova, J., La Porta, A., Zenkin, N., Lee, J., Lis, J. T., Borukhov, S., Wang, M. D., and Severinov, K. (2004). Molecular mechanism of transcription inhibition by peptide antibiotic Microcin J25. Mol Cell *14*, 753-762.

Ades, S. E., Connolly, L. E., Alba, B. M., and Gross, C. A. (1999). The *Escherichia coli* sigma(E)-dependent extracytoplasmic stress response is controlled by the regulated proteolysis of an anti-sigma factor. Genes Dev *13*, 2449-2461.

Adhya, S., Geanacopoulos, M., Lewis, D. E., Roy, S., and Aki, T. (1998). Transcription regulation by repressosome and by RNA polymerase contact. Cold Spring Harb Symp Quant Biol *63*, 1-9.

Alba, B. M., and Gross, C. A. (2004). Regulation of the *Escherichia coli* sigma-dependent envelope stress response. Mol Microbiol *52*, 613-619.

Allison, L. A., Moyle, M., Shales, M., and Ingles, C. J. (1985). Extensive homology among the largest subunits of eukaryotic and prokaryotic RNA polymerases. Cell *42*, 599-610.

Arndt, K. M., and Chamberlin, M. J. (1988). Transcription termination in *Escherichia coli*. Measurement of the rate of enzyme release from Rho-independent terminators. J Mol Biol *202*, 271-285.

Artsimovitch, I., and Landick, R. (2000). Pausing by bacterial RNA polymerase is mediated by mechanistically distinct classes of signals. Proc Natl Acad Sci USA *97*, 7090-7095.

Aubry-Damon, H., Soussy, C. J., and Courvalin, P. (1998). Characterization of mutations in the rpoB gene that confer rifampin resistance in Staphylococcus aureus. Antimicrob Agents Chemother *42*, 2590-2594.

Ayers, D. G., Auble, D. T., and deHaseth, P. L. (1989). Promoter recognition by *Escherichia coli* RNA polymerase. Role of the spacer DNA in functional complex formation. J Mol Biol *207*, 749-756.

Azam, T. A., and Ishihama, A. (1999). Twelve species of the nucleoid-associated protein from *Escherichia coli*. Sequence recognition specificity and DNA binding affinity. J Biol Chem *274*, 33105-33113.

Bar-Nahum, G., and Nudler, E. (2001). Isolation and characterization of sigma(70)-retaining transcription elongation complexes from *Escherichia coli*. Cell *106*, 443-451.

Blatter, E. E., Ross, W., Tang, H., Gourse, R. L., and Ebright, R. H. (1994). Domain organization of RNA polymerase alpha subunit: C-terminal 85 amino acids constitute a domain capable of dimerization and DNA binding. Cell *78*, 889-896.

Blattner, F. R., Plunkett, G., 3rd, Bloch, C. A., Perna, N. T., Burland, V., Riley, M., Collado-Vides, J., Glasner, J. D., Rode, C. K., Mayhew, G. F., *et al.* (1997). The complete genome sequence of *Escherichia coli* K-12. Science *277*, 1453-1474.

Borukhov, S., Laptenko, O., and Lee, J. (2001). *Escherichia coli* transcript cleavage factors GreA and GreB: functions and mechanisms of action. Methods Enzymol *342*, 64-76.

Borukhov, S., and Nudler, E. (2003). RNA polymerase holoenzyme: structure, function and biological implications. Curr Opin Microbiol *6*, 93-100.

Bremer, H., and Dennis, P. (1996). Modulation of Chemical Composition and Other Parameters of the Cell by Growth Rate. In *Escherichia coli* and Salmonella Typhimurium, F. Neidhardt, ed. (Washington, D.C., American Society for Microbiology), pp. 1553.

Brendel, V., Hamm, G. H., and Trifonov, E. N. (1986). Terminators of transcription with RNA polymerase from *Escherichia coli*: what they look like and how to find them. J Biomol Struct Dyn *3*, 705-723.

Brennan, C. A., Dombroski, A. J., and Platt, T. (1987). Transcription termination factor rho is an RNA-DNA helicase. Cell *48*, 945-952.

Browning, D. F., and Busby, S. J. (2004). The regulation of bacterial transcription initiation. Nat Rev Microbiol *2*, 57-65.

Bruckner, R., and Titgemeyer, F. (2002). Carbon catabolite repression in bacteria: choice of the carbon source and autoregulatory limitation of sugar utilization. FEMS Microbiol Lett *209*, 141-148.

Burgess, R. R., Erickson, B., Gentry, D., Gribskov, M., Hager, D., Lesley, S., Strickland, M., and Thompson, N. (1987). Bacterial RNA polymerase subunits and genes. In RNA polymerase and the regulation of transcription, W. S. Reznikoff, ed. (New York, NY, Elsevier Science Publishing), pp. 3-15.

Burgess, R. R., Travers, A. A., Dunn, J. J., and Bautz, E. K. (1969). Factor stimulating transcription by RNA polymerase. Nature *221*, 43-46.

Burns, C. M., Nowatzke, W. L., and Richardson, J. P. (1999). Activation of Rho-dependent transcription termination by NusG. Dependence on terminator location and acceleration of RNA release. J Biol Chem *274*, 5245-5251.

Burton, Z. F., Gross, C. A., Watanabe, K. K., and Burgess, R. R. (1983). The operon that encodes the sigma subunit of RNA polymerase also encodes ribosomal protein S21 and DNA primase in *E. coli* K12. Cell *32*, 335-349.

Cabrera, J. E., and Jin, D. J. (2003). The distribution of RNA polymerase in *Escherichia coli* is dynamic and sensitive to environmental cues. Mol Microbiol *50*, 1493-1505.

Campbell, E. A., Korzheva, N., Mustaev, A., Murakami, K., Nair, S., Goldfarb, A., and Darst, S. A. (2001). Structural mechanism for rifampicin inhibition of bacterial rna polymerase. Cell *104*, 901-912.

Campbell, E. A., Pavlova, O., Zenkin, N., Leon, F., Irschik, H., Jansen, R., Severinov, K., and Darst, S. A. (2005). Structural, functional, and genetic analysis of sorangicin inhibition of bacterial RNA polymerase. EMBO J *24*, 674-682.

Cashel, M., Gentry, D. R., Hernandez, V. J., and Vinella, D. (1996). The stringent response. In *Escherichia coli* and Salmonella typhimurium, F. C. Neidhardt, ed. (Washington, D.C., A.S.M. Press), pp. 1458-1496.

Citterio, E., Van Den Boom, V., Schnitzler, G., Kanaar, R., Bonte, E., Kingston, R. E., Hoeijmakers, J. H., and Vermeulen, W. (2000). ATP-dependent chromatin remodeling by the Cockayne syndrome B DNA repair-transcription-coupling factor. Mol Cell Biol *20*, 7643-7653.

Condon, C., Squires, C., and Squires, C. L. (1995). Control of rRNA transcription in *Escherichia coli*. Microbiol Rev *59*, 623-645.

Cook, P. R. (1999). The organization of replication and transcription. Science *284*, 1790-1795.

Crasnier, M. (1996). Cyclic AMP and catabolite repression. Res Microbiol *147*, 479-482.

d'Aubenton Carafa, Y., Brody, E., and Thermes, C. (1990). Prediction of rho-independent *Escherichia coli* transcription terminators. A statistical analysis of their RNA stem-loop structures. J Mol Biol *216*, 835-858.

Davis, C. A., Capp, M. W., Record, M. T., Jr., and Saecker, R. M. (2005). The effects of upstream DNA on open complex formation by *Escherichia coli* RNA polymerase. Proc Natl Acad Sci USA *102*, 285-290.

deHaseth, P. L., Zupancic, M. L., and Record, M. T., Jr. (1998). RNA polymerase-promoter interactions: the comings and goings of RNA polymerase. J Bacteriol *180*, 3019-3025.

DeVito, J., and Das, A. (1994). Control of transcription processivity in phage lambda: Nus factors strengthen the termination-resistant state of RNA polymerase induced by N antiterminator. Proc Natl Acad Sci USA *91*, 8660-8664.

Dorman, C. J., and Deighan, P. (2003). Regulation of gene expression by histone-like proteins in bacteria. Curr Opin Genet Dev *13*, 179-184.

Drlica, K. (1987). The nucleoid. In *Escherichia coli* and Salmonella typhimurium, F. Neidhardt, ed. (Washington, D.C., American Society for Microbiology), pp. 91-103.

Ebright, R. H. (1993). Transcription activation at Class I CAP-dependent promoters. Mol Microbiol *8*, 797-802.

Ebright, R. H. (2000). RNA polymerase: structural similarities between bacterial RNA polymerase and eukaryotic RNA polymerase II. J Mol Biol *304*, 687-698.

Epshtein, V., Mironov, A. S., and Nudler, E. (2003). The riboswitch-mediated control of sulfur metabolism in bacteria. Proc Natl Acad Sci USA *100*, 5052-5056.

Erickson, J. W., and Gross, C. A. (1989). Identification of the sigma E subunit of *Escherichia coli* RNA polymerase: a second alternate sigma factor involved in high-temperature gene expression. Genes Dev *3*, 1462-1471.

Ezekiel, D. H., and Hutchins, J. E. (1968). Mutations affecting RNA polymerase associated with rifampicin resistance in *Escherichia coli*. Nature *220*, 276-277.

Fisher, L. (1971). Rifampin--new and potent drug for TB treatment. Bull Natl Tuberc Respir Dis Assoc *57*, 11-12.

Friedman, D. I., and Court, D. L. (1995). Transcription antitermination: the lambda paradigm updated. Mol Microbiol *18*, 191-200.

Fu, J., Gnatt, A. L., Bushnell, D. A., Jensen, G. J., Thompson, N. E., Burgess, R. R., David, P. R., and Kornberg, R. D. (1999). Yeast RNA polymerase II at 5 A resolution. Cell *98*, 799-810.

Gaal, T., Bartlett, M. S., Ross, W., Turnbough, C. L., Jr., and Gourse, R. L. (1997). Transcription regulation by initiating NTP concentration: rRNA synthesis in bacteria. Science *278*, 2092-2097.

Gill, S. C., Weitzel, S. E., and von Hippel, P. H. (1991). *Escherichia coli* sigma 70 and NusA proteins. I. Binding interactions with core RNA polymerase in solution and within the transcription complex. J Mol Biol *220*, 307-324.

Gottesman, M. E., and Weisberg, R. A. (2004). Little lambda, who made thee? Microbiol Mol Biol Rev *68*, 796-813.

Gralla, J. D. (2005). *Escherichia coli* ribosomal RNA transcription: regulatory roles for ppGpp, NTPs, architectural proteins and a polymerase-binding protein. Mol Microbiol *55*, 973-977.

Greenblatt, J., and Li, J. (1981). Interaction of the sigma factor and the nusA gene protein of *E. coli* with RNA polymerase in the initiation-termination cycle of transcription. Cell *24*, 421-428.

Greenblatt, J., McLimont, M., and Hanly, S. (1981). Termination of transcription by nusA gene protein of *Escherichia coli*. Nature *292*, 215-220.

Gross, C. A., Chan, C., Dombroski, A., Gruber, T., Sharp, M., Tupy, J., and Young, B. (1998). The functional and regulatory roles of sigma factors in transcription. Cold Spring Harb Symp Quant Biol *63*, 141-155.

Grossman, A. D., Erickson, J. W., and Gross, C. A. (1984). The htpR gene product of *E. coli* is a sigma factor for heat-shock promoters. Cell *38*, 383-390.

Grundy, F. J., and Henkin, T. M. (2003). The T box and S box transcription termination control systems. Front Biosci *8*, d20-31.

Gusarov, I., and Nudler, E. (1999). The mechanism of intrinsic transcription termination. Mol Cell *3*, 495-504.

Gusarov, I., and Nudler, E. (2001). Control of intrinsic transcription termination by N and NusA: the basic mechanisms. Cell *107*, 437-449.

Hansen, A. M., Lehnherr, H., Wang, X., Mobley, V., and Jin, D. J. (2003). *Escherichia coli* SspA is a transcription activator for bacteriophage P1 late genes. Mol Microbiol *48*, 1621-1631.

Hansen, A. M., Qiu, Y., Yeh, N., Blattner, F. R., Durfee, T., and Jin, D. J. (2005). SspA is required for acid resistance in stationary phase by downregulation of H-NS in *Escherichia coli*. Mol Microbiol *56*, 719-734.

Harley, C. B., and Reynolds, R. P. (1987). Analysis of *E. coli* promoter sequences. Nucleic Acids Res *15*, 2343-2361.

Hawley, D. K., and McClure, W. R. (1983). Compilation and analysis of *Escherichia coli* promoter DNA sequences. Nucleic Acids Res *11*, 2237-2255.

Heisler, L. M., Suzuki, H., Landick, R., and Gross, C. A. (1993). Four contiguous amino acids define the target for streptolydigin resistance in the beta subunit of *Escherichia coli* RNA polymerase. J Biol Chem *268*, 25369-25375.

Helmann, J. D. (1991). Alternative sigma factors and the regulation of flagellar gene expression. Mol Microbiol *5*, 2875-2882.

Henkin, T. M. (2000). Transcription termination control in bacteria. Curr Opin Microbiol *3*, 149-153.

Henkin, T. M., and Yanofsky, C. (2002). Regulation by transcription attenuation in bacteria: how RNA provides instructions for transcription termination/antitermination decisions. Bioessays *24*, 700-707.

Hobot, J. A., Villiger, W., Escaig, J., Maeder, M., Ryter, A., and Kellenberger, E. (1985). Shape and fine structure of nucleoids observed on sections of ultrarapidly frozen and cryosubstituted bacteria. J Bacteriol *162*, 960-971.

Hsu, L. M., Vo, N. V., and Chamberlin, M. J. (1995). *Escherichia coli* transcript cleavage factors GreA and GreB stimulate promoter escape and gene expression *in vivo* and *in vitro*. Proc Natl Acad Sci USA *92*, 11588-11592.

Hsu, L. M., Vo, N. V., Kane, C. M., and Chamberlin, M. J. (2003). *In vitro* studies of transcript initiation by *Escherichia coli* RNA polymerase. 1. RNA chain initiation, abortive initiation, and promoter escape at three bacteriophage promoters. Biochemistry *42*, 3777-3786.

Hughes, K. T., and Mathee, K. (1998). The anti-sigma factors. Annu Rev Microbiol *52*, 231-286.

Ishihama, A. (1981). Subunit of assembly of *Escherichia coli* RNA polymerase. Adv Biophys *14*, 1-35.

Ishihama, A. (2000). Functional modulation of *Escherichia coli* RNA polymerase. Annu Rev Microbiol *54*, 499-518.

Jacob, F., and Monod, J. (1961). Genetic regulatory mechanisms in the synthesis of proteins. J Mol Biol *3*, 318-356.

Jin, D. J. (1994). Slippage synthesis at the galP2 promoter of *Escherichia coli* and its regulation by UTP concentration and cAMP.cAMP receptor protein. J Biol Chem *269*, 17221-17227.

Jin, D. J., Burgess, R. R., Richardson, J. P., and Gross, C. A. (1992). Termination efficiency at rho-dependent terminators depends on kinetic coupling between RNA polymerase and rho. Proc Natl Acad Sci USA *89*, 1453-1457.

Jin, D. J., Cashel, M., Friedman, D. I., Nakamura, Y., Walter, W. A., and Gross, C. A. (1988). Effects of rifampicin resistant rpoB mutations on antitermination and interaction with nusA in *Escherichia coli*. J Mol Biol *204*, 247-261.

Jin, D. J., and Gross, C. A. (1988). Mapping and sequencing of mutations in the *Escherichia coli* rpoB gene that lead to rifampicin resistance. J Mol Biol *202*, 45-58.

Jin, D. J., and Gross, C. A. (1989). Characterization of the pleiotropic phenotypes of rifampin-resistant rpoB mutants of *Escherichia coli*. J Bacteriol *171*, 5229-5231.

Jin, D. J., and Gross, C. A. (1991). RpoB8, a rifampicin-resistant termination-proficient RNA polymerase, has an increased Km for purine nucleotides during transcription elongation. J Biol Chem *266*, 14478-14485.

Jin, D. J., and Turnbough, C. L., Jr. (1994). An Escherichia coli RNA polymerase defective in transcription due to its overproduction of abortive initiation products. J Mol Biol *236*, 72-80.

Jin, D. J., and Zhou, Y. N. (1996). Mutational analysis of structure-function relationship of RNA polymerase in Escherichia coli. Methods Enzymol *273*, 300-319.

Kapusnik, J. E., Parenti, F., and Sande, M. A. (1984). The use of rifampicin in staphylococcal infections--a review. J Antimicrob Chemother *13 Suppl C*, 61-66.

Keilty, S., and Rosenberg, M. (1987). Constitutive function of a positively regulated promoter reveals new sequences essential for activity. J Biol Chem *262*, 6389-6395.

Kingston, R. E., and Chamberlin, M. J. (1981). Pausing and attenuation of *in vitro* transcription in the rrnB operon of *E. coli*. Cell *27*, 523-531.

Kireeva, M. L., Hancock, B., Cremona, G. H., Walter, W., Studitsky, V. M., and Kashlev, M. (2005). Nature of the Nucleosomal Barrier to RNA Polymerase II. Mol Cell *18*, 97-108.

Komissarova, N., Becker, J., Solter, S., Kireeva, M., and Kashlev, M. (2002). Shortening of RNA:DNA hybrid in the elongation complex of RNA polymerase is a prerequisite for transcription termination. Mol Cell *10*, 1151-1162.

Komissarova, N., and Kashlev, M. (1997a). RNA polymerase switches between inactivated and activated states By translocating back and forth along the DNA and the RNA. J Biol Chem *272*, 15329-15338.

Komissarova, N., and Kashlev, M. (1997b). Transcriptional arrest: Escherichia coli RNA polymerase translocates backward, leaving the 3' end of the RNA intact and extruded. Proc Natl Acad Sci USA *94*, 1755-1760.

Korzheva, N., Mustaev, A., Kozlov, M., Malhotra, A., Nikiforov, V., Goldfarb, A., and Darst, S. A. (2000). A structural model of transcription elongation. Science *289*, 619-625.

Kustu, S., Santero, E., Keener, J., Popham, D., and Weiss, D. (1989). Expression of sigma 54 (ntrA)-dependent genes is probably united by a common mechanism. Microbiol Rev *53*, 367-376.

Landick, R., Carey, J., and Yanofsky, C. (1987). Detection of transcription-pausing *in vivo* in the trp operon leader region. Proc Natl Acad Sci USA *84*, 1507-1511.

Landick, R., Wang, D., and Chan, C. L. (1996). Quantitative analysis of transcriptional pausing by Escherichia coli RNA polymerase: his leader pause site as paradigm. Methods Enzymol *274*, 334-353.

Lawson, C. L., Swigon, D., Murakami, K. S., Darst, S. A., Berman, H. M., and Ebright, R. H. (2004). Catabolite activator protein: DNA binding and transcription activation. Curr Opin Struct Biol *14*, 10-20.

Leung, M. J., Kell, A. D., and Collignon, P. (1998). Antibiotic guidelines for meningococcal prophylaxis. Med J Aust *169*, 396.

Levin, J. R., and Chamberlin, M. J. (1987). Mapping and characterization of transcriptional pause sites in the early genetic region of bacteriophage T7. J Mol Biol *196*, 61-84.

Liu, C., Heath, L. S., and Turnbough, C. L., Jr. (1994). Regulation of pyrBI operon expression in *Escherichia coli* by UTP-sensitive reiterative RNA synthesis during transcriptional initiation. Genes Dev *8*, 2904-2912.

Liu, M., Durfee, T., Cabrera, J. E., Zhao, K., Jin, D. J., and Blattner, F. R. (2005). Global transcriptional programs reveal a carbon source foraging strategy by *Escherichia coli*. J Biol Chem. *280*:15921-15927.

Loewen, P. C., and Hengge-Aronis, R. (1994). The role of the sigma factor sigma S (KatF) in bacterial global regulation. Annu Rev Microbiol *48*, 53-80.

Lounis, N., and Roscigno, G. (2004). *In vitro* and *in vivo* activities of new rifamycin derivatives against mycobacterial infections. Curr Pharm Des *10*, 3229-3238.

Maeda, H., Fujita, N., and Ishihama, A. (2000). Competition among seven *Escherichia coli* sigma subunits: relative binding affinities to the core RNA polymerase. Nucleic Acids Res *28*, 3497-3503.

Martin, F. H., and Tinoco, I., Jr. (1980). DNA–RNA hybrid duplexes containing oligo(dA:rU) sequences are exceptionally unstable and may facilitate termination of transcription. Nucleic Acids Res *8*, 2295-2299.

McClure, W. R. (1980). Rate-limiting steps in RNA chain initiation. Proc Natl Acad Sci USA *77*, 5634-5638.

McClure, W. R., and Cech, C. L. (1978). On the mechanism of rifampicin inhibition of RNA synthesis. J Biol Chem *253*, 8949-8956.

McDowell, J. C., Roberts, J. W., Jin, D. J., and Gross, C. (1994). Determination of intrinsic transcription termination efficiency by RNA polymerase elongation rate. Science *266*, 822-825.

McLeod, S. M., Aiyar, S. E., Gourse, R. L., and Johnson, R. C. (2002). The C-terminal domains of the RNA polymerase alpha subunits: contact site with Fis and localization during co-activation with CRP at the *Escherichia coli* proP P2 promoter. J Mol Biol *316*, 517-529.

Metzger, S., Schreiber, G., Aizenman, E., Cashel, M., and Glaser, G. (1989). Characterization of the relA1 mutation and a comparison of relA1 with new relA null alleles in *Escherichia coli*. J Biol Chem *264*, 21146-21152.

Mitchell, J. E., Zheng, D., Busby, S. J., and Minchin, S. D. (2003). Identification and analysis of 'extended -10' promoters in *Escherichia coli*. Nucleic Acids Res *31*, 4689-4695.

Muchardt, C., and Yaniv, M. (1999). The mammalian SWI/SNF complex and the control of cell growth. Semin Cell Dev Biol *10*, 189-195.

Mukhopadhyay, J., Kapanidis, A. N., Mekler, V., Kortkhonjia, E., Ebright, Y. W., and Ebright, R. H. (2001). Translocation of sigma(70) with RNA polymerase during transcription: fluorescence resonance energy transfer assay for movement relative to DNA. Cell *106*, 453-463.

Murakami, K. S., Masuda, S., Campbell, E. A., Muzzin, O., and Darst, S. A. (2002). Structural basis of transcription initiation: an RNA polymerase holoenzyme-DNA complex. Science *296*, 1285-1290.

Mustaev, A., Zaychikov, E., Severinov, K., Kashlev, M., Polyakov, A., Nikiforov, V., and Goldfarb, A. (1994). Topology of the RNA polymerase active center probed by chimeric rifampicin-nucleotide compounds. Proc Natl Acad Sci USA *91*, 12036-12040.

Nehrke, K. W., and Platt, T. (1994). A quaternary transcription termination complex. Reciprocal stabilization by Rho factor and NusG protein. J Mol Biol *243*, 830-839.

Nudler, E., and Gottesman, M. E. (2002). Transcription termination and anti-termination in *E. coli*. Genes Cells *7*, 755-768.

Park, J. S., Marr, M. T., and Roberts, J. W. (2002). *E. coli* Transcription repair coupling factor (Mfd protein) rescues arrested complexes by promoting forward translocation. Cell *109*, 757-767.

Paul, B. J., Barker, M. M., Ross, W., Schneider, D. A., Webb, C., Foster, J. W., and Gourse, R. L. (2004a). DksA: a critical component of the transcription initiation machinery that potentiates the regulation of rRNA promoters by ppGpp and the initiating NTP. Cell *118*, 311-322.

Paul, B. J., Ross, W., Gaal, T., and Gourse, R. L. (2004b). rRNA transcription in *Escherichia coli.* Annu Rev Genet *38*, 749-770.

Pazin, M. J., and Kadonaga, J. T. (1997). SWI2/SNF2 and related proteins: ATP-driven motors that disrupt protein-DNA interactions? Cell *88*, 737-740.

Peterson, C. L. (1996). Multiple SWItches to turn on chromatin? Curr Opin Genet Dev *6*, 171-175.

Pettijohn, D. E. (1996). The nucleoid. In *Escherichia coli* and *Salmonella* Typhimurium, F. Neidhardt, ed. (Washington, D. C., American Society for Microbiology), pp. 158.

Platt, T. (1994). Rho and RNA: models for recognition and response. Mol Microbiol *11*, 983-990.

Pressler, U., Staudenmaier, H., Zimmermann, L., and Braun, V. (1988). Genetics of the iron dicitrate transport system of *Escherichia coli.* J Bacteriol *170*, 2716-2724.

Qi, F., and Turnbough, C. L., Jr. (1995). Regulation of codBA operon expression in *Escherichia coli* by UTP-dependent reiterative transcription and UTP-sensitive transcriptional start site switching. J Mol Biol *254*, 552-565.

Rabussay, D., and Zillig, W. (1969). A rifampicin resistent rna-polymerase from *E. coli* altered in the beta-subunit. FEBS Lett *5*, 104-106.

Ramaswamy, S., and Musser, J. M. (1998). Molecular genetic basis of antimicrobial agent resistance in Mycobacterium tuberculosis: 1998 update. Tuber Lung Dis *79*, 3-29.

Rees, W. A., Weitzel, S. E., Yager, T. D., Das, A., and von Hippel, P. H. (1996). Bacteriophage lambda N protein alone can induce transcription antitermination *in vitro.* Proc Natl Acad Sci USA *93*, 342-346.

Richardson, J. P. (2002). Rho-dependent termination and ATPases in transcript termination. Biochim Biophys Acta *1577*, 251-260.

Richardson, J. P. (2003). Loading Rho to terminate transcription. Cell *114*, 157-159.

Roberts, J., and Park, J. S. (2004). Mfd, the bacterial transcription repair coupling factor: translocation, repair and termination. Curr Opin Microbiol *7*, 120-125.

Roberts, J. W., Yarnell, W., Bartlett, E., Guo, J., Marr, M., Ko, D. C., Sun, H., and Roberts, C. W. (1998). Antitermination by bacteriophage lambda Q protein. Cold Spring Harb Symp Quant Biol *63*, 319-325.

Robinow, C., and Kellenberger, E. (1994). The bacterial nucleoid revisited. Microbiol Rev *58*, 211-232.

Ross, W., Gosink, K. K., Salomon, J., Igarashi, K., Zou, C., Ishihama, A., Severinov, K., and Gourse, R. L. (1993). A third recognition element in bacterial promoters: DNA binding by the alpha subunit of RNA polymerase. Science *262*, 1407-1413.

Saecker, R. M., Tsodikov, O. V., McQuade, K. L., Schlax, P. E., Jr., Capp, M. W., and Record, M. T., Jr. (2002). Kinetic studies and structural models of the association of *E. coli* sigma(70) RNA polymerase with the lambdaP(R) promoter: large scale conformational changes in forming the kinetically significant intermediates. J Mol Biol *319*, 649-671.

Saier, M. H., Jr. (1998). Multiple mechanisms controlling carbon metabolism in bacteria. Biotechnol Bioeng *58*, 170-174.

Savery, N., Rhodius, V., and Busby, S. (1996). Protein-protein interactions during transcription activation: the case of the *Escherichia coli* cyclic AMP receptor protein. Philos Trans R Soc Lond B Biol Sci *351*, 543-550.

Schroder, O., and Wagner, R. (2000). The bacterial DNA-binding protein H-NS represses ribosomal RNA transcription by trapping RNA polymerase in the initiation complex. J Mol Biol *298*, 737-748.

Selby, C. P., and Sancar, A. (1993). Molecular mechanism of transcription-repair coupling. Science *260*, 53-58.

Severinov, K., Markov, D., Severinova, E., Nikiforov, V., Landick, R., Darst, S. A., and Goldfarb, A. (1995). Streptolydigin-resistant mutants in an evolutionarily conserved region of the beta' subunit of *Escherichia coli* RNA polymerase. J Biol Chem *270*, 23926-23929.

Severinov, K., Soushko, M., Goldfarb, A., and Nikiforov, V. (1993). Rifampicin region revisited. New rifampicin-resistant and streptolydigin-resistant mutants in the beta subunit of *Escherichia coli* RNA polymerase. J Biol Chem *268*, 14820-14825.

Shimamoto, N., Kamigochi, T., and Utiyama, H. (1986). Release of the sigma subunit of *Escherichia coli* DNA-dependent RNA polymerase depends mainly on time elapsed after the start of initiation, not on length of product RNA. J Biol Chem *261*, 11859-11865.

Shuman, H. A., and Silhavy, T. J. (2003). The art and design of genetic screens: *Escherichia coli.* Nat Rev Genet *4*, 419-431.

Squires, C. L., Condon, C., and Seoh, H. K. (2003). Assay of antitermination of ribosomal RNA transcription. Methods Enzymol *371*, 472-487.

Squires, C. L., Greenblatt, J., Li, J., and Condon, C. (1993). Ribosomal RNA antitermination *in vitro*: requirement for Nus factors and one or more unidentified cellular components. Proc Natl Acad Sci USA *90*, 970-974.

Stefano, J. E., and Gralla, J. D. (1982). Spacer mutations in the lac ps promoter. Proc Natl Acad Sci USA *79*, 1069-1072.

Straus, D., Walter, W., and Gross, C. A. (1990). DnaK, DnaJ, and GrpE heat shock proteins negatively regulate heat shock gene expression by controlling the synthesis and stability of sigma 32. Genes Dev *4*, 2202-2209.

Sukhodolets, M. V., Cabrera, J. E., Zhi, H., and Jin, D. J. (2001). RapA, a bacterial homolog of SWI2/SNF2, stimulates RNA polymerase recycling in transcription. Genes Dev *15*, 3330-3341.

Sukhodolets, M. V., and Jin, D. J. (1998). RapA, a novel RNA polymerase-associated protein, is a bacterial homolog of SWI2/SNF2. J Biol Chem *273*, 7018-7023.

Sweetser, D., Nonet, M., and Young, R. A. (1987). Prokaryotic and eukaryotic RNA polymerases have homologous core subunits.

Proc Natl Acad Sci USA *84*, 1192-1196.

Toulme, F., Mosrin-Huaman, C., Sparkowski, J., Das, A., Leng, M., and Rahmouni, A. R. (2000). GreA and GreB proteins revive backtracked RNA polymerase *in vivo* by promoting transcript trimming. EMBO J *19*, 6853-6859.

Travers, A., Schneider, R., and Muskhelishvili, G. (2001). DNA supercoiling and transcription in *Escherichia coli*: The FIS connection. Biochimie *83*, 213-217.

Tse-Dinh, Y. C., Qi, H., and Menzel, R. (1997). DNA supercoiling and bacterial adaptation: thermotolerance and thermoresistance. Trends Microbiol *5*, 323-326.

Vassylyev, D. G., Sekine, S., Laptenko, O., Lee, J., Vassylyeva, M. N., Borukhov, S., and Yokoyama, S. (2002). Crystal structure of a bacterial RNA polymerase holoenzyme at 2.6 A resolution. Nature *417*, 712-719.

von Hippel, P. H. (1998). An integrated model of the transcription complex in elongation, termination, and editing. Science *281*, 660-665.

von Hippel, P. H., Rees, W. A., Rippe, K., and Wilson, K. S. (1996). Specificity mechanisms in the control of transcription. Biophys Chem *59*, 231-246.

Walstrom, K. M., Dozono, J. M., and von Hippel, P. H. (1997). Kinetics of the RNA-DNA helicase activity of *Escherichia coli* transcription termination factor rho. 2. Processivity, ATP consumption, and RNA binding. Biochemistry *36*, 7993-8004.

Weisberg, R. A., and Gottesman, M. E. (1999). Processive antitermination. J Bacteriol *181*, 359-367.

Wu, H. Y., Shyy, S. H., Wang, J. C., and Liu, L. F. (1988). Transcription generates positively and negatively supercoiled domains in the template. Cell *53*, 433-440.

Xu, M., Zhou, Y. N., Goldstein, B. P., and Jin, D. J. (2005). Cross-resistance of *Escherichia coli* RNA polymerases conferring rifampin resistance to different antibiotics. J Bacteriol *187*, 2783-2792.

Yang, X., and Price, C. W. (1995). Streptolydigin resistance can be conferred by alterations to either the beta or beta' subunits of Bacillus subtilis RNA polymerase. J Biol Chem *270*, 23930-23933.

Yanofsky, C. (2000). Transcription attenuation: once viewed as a novel regulatory strategy. J Bacteriol *182*, 1-8.

Young, B. A., Gruber, T. M., and Gross, C. A. (2002). Views of transcription initiation. Cell *109*, 417-420.

Zhang, G., Campbell, E. A., Minakhin, L., Richter, C., Severinov, K., and Darst, S. A. (1999). Crystal structure of Thermus aquaticus core RNA polymerase at 3.3 A resolution. Cell *98*, 811-824.

Zhi, H., Wang, X., Cabrera, J. E., Johnson, R. C., and Jin, D. J. (2003a). Fis stabilizes the interaction between RNA polymerase and the ribosomal promoter rrnB P1, leading to transcriptional activation. J Biol Chem *278*, 47340-47349.

Zhi, H., Yang, W., and Jin, D. J. (2003b). *Escherichia coli* proteins eluted from mono Q chromatography, a final step during RNA polymerase purification procedure. Methods Enzymol *370*, 291-300.

Zhou, Y., Gottesman, S., Hoskins, J. R., Maurizi, M. R., and Wickner, S. (2001). The RssB response regulator directly targets sigma(S) for degradation by ClpXP. Genes Dev *15*, 627-637.

Zhou, Y. N., and Jin, D. J. (1998). The rpoB mutants destabilizing initiation complexes at stringently controlled promoters behave like "stringent" RNA polymerases in *Escherichia coli*. Proc Natl Acad Sci USA *95*, 2908-2913.

Zhou, Y. N., Walter, W. A., and Gross, C. A. (1992). A mutant sigma 32 with a small deletion in conserved region 3 of sigma has reduced affinity for core RNA polymerase. J Bacteriol *174*, 5005-5012.

Zurawski, G., Elseviers, D., Stauffer, G. V., and Yanofsky, C. (1978). Translational control of transcription termination at the attenuator of the *Escherichia coli* tryptophan operon. Proc Natl Acad Sci USA *75*, 5988-5992.

Chapter 35

Gene Therapy: Back to the Basics

Jim Hu

Programme in Lung Biology Research, Hospital for Sick Children, Departments of Laboratory Medicine and Pathobiology, and Paediatrics, University of Toronto, Toronto, Ontario, Canada

Key Words: gene delivery, transgene expression, airway disease, viral vector, host immune responses, gene therapy, adenoviruses, helper-dependent adenoviral vector, adeno-associated virus, cystic fibrosis, retroviruses, X-linked severe combined immunodeficiency disease, adenosine deaminase-deficient severe combined immunodeficiency, innate (page 559), adaptive, inflammation, macrophages, proinflammatory cytokines, NF-kappaB signaling pathway, DNA regulatory elements

Summary

The concept of gene therapy evolved from the knowledge gained in modern molecular genetics. Following the establishment of the "central dogma" of gene expression—that *DNA begets RNA begets protein*, it was natural to imagine that a human genetic defect could be cured by delivering the correct gene to the affected cells. Therefore, it is fair to state that modern molecular genetics is the forerunner to gene therapy. Over the past few decades biologists have shown that phenotypical changes resulted from mutations in a prokaryotic (*e.g., E. coli*) or eukaryotic (e.g., yeast) gene can be corrected by delivering a copy of the correct gene to the mutant cells. Hence, gene therapy is indeed conceptually sound. The technical difficulty for gene therapy in humans, however, was initially underestimated and successful examples of clinical trials are still rare. As a result, the initial hype of gene therapy led to the great disappointment for the public. This also led to some scientists and journal editors to look for major breakthroughs and ignore the incremental progress in the field. In this chapter, I shall review the major advancements over the past two decades and, using lung gene therapy as an example, discuss the current obstacles and possible solutions to provide a roadmap for future research on gene therapy.

Introduction

The potential of gene therapy in medical applications was recognized soon after the discovery of DNA as genetic material and of genetic information flow (from DNA to RNA to protein). The early history of gene therapy was extensively reviewed by Wolff and Lederberg (1994) and I shall only highlight a few points here prior to the discussion of vectors and approaches used in gene therapy. Actually, attempts at human gene therapy were initiated in the late 1960s and early 1970s when S. Rogers injected the Shope papilloma virus into patients with arginase deficiency, based on his initial observation that the virus induced high levels of arginase activity in rabbit skin tumors and might contain an arginase gene (Rogers and Moore, 1963). Although the attempt eventually turned out to be unsuccessful (Terheggen *et al.*, 1975), it demonstrated the early enthusiasm in gene therapy. Due to the lack of basic understanding of gene expression and effective methods for gene delivery, the early attempts at gene therapy were doomed to fail.

Gene therapy research intensified during the last 15 years following development of various gene delivery methods. Over this period, discoveries were made in recognizing problems associated with *in vivo* gene delivery, including physical barriers and host immune reactions and a lack of sustained expression of transgenes (George, 2003; Koehler *et al.*, 2001). In addition, there were quite a number of successful gene transfer examples performed in animals (Barquinero *et al.*, 2004; Ferrari *et al.*, 2004; George, 2003; Kaplan *et*

Corresponding Author: Tel: (416) 813-6412, Fax: (416) 813-5771, E-mail: jhu@sickkids.ca

al., 1998; Kim *et al.*, 2001; Koehler *et al.*, 2003; Toietta *et al.*, 2003; Wang *et al.*, 2000; Wang *et al.*, 2005) although successful human studies remain scarce (Aiuti *et al.*, 2002; Cavazzana-Calvo *et al.*, 2000; Gaspar *et al.*, 2004). The slow progress in clinical applications brought much disappointment to the public, and a single case of gene therapy-related death in 1999 (NIH Report, 2003, *Hum. Gene Ther.* 13:3-9) (Raper *et al.*, 2003; Raper *et al.*, 2002) further dampened the enthusiasm in the field and raised concerns of safety of gene therapy. Over the past few years, major efforts in gene therapy were directed toward understanding the problems associated with gene delivery, such as acute toxicity associated with adenoviral vectors (Morral *et al.*, 2002), further improving gene therapy vectors (Barquinero *et al.*, 2004; Koehler *et al.*, 2001; Parks, 2000) and exploring experimental conditions for enhancing the efficiency of gene delivery (Glimm and Eaves, 1999; Hennemann *et al.*, 1999; Limberis *et al.*, 2002). In the next part of this chapter, I shall briefly describe vectors commonly used for gene delivery to animal models and to humans, followed by discussing the current problems associated with gene therapy.

Vectors for Gene Therapy

There are two general approaches being employed in gene therapy based on vectors used for gene delivery. One approach employs recombinant viruses to deliver genes and the other uses non-viral vectors. In general, viral vector-mediated gene transfer is more efficient than the non-viral approach, but some vectors elicit stronger immune responses. Among the viral vectors developed for *in vivo* gene delivery, I shall briefly describe the vectors derived from adenovirus (Ad), adeno-associated virus (AAV), and retrovirus since these vectors have been used in clinical trials. Progress has also been made in vectors derived from other viruses, such as herpes simplex virus (Glorioso *et al.*, 1994) and Sendai virus (the murine parainfluenza virus type 1) (Ferrari *et al.*, 2004). Due to the space limitation, however, these vectors will not be covered in this chapter. Among the non-viral vectors, only cationic liposomes were extensively explored and these will be discussed.

A: Viral Vectors
A1: Adenoviral Vectors

Adenoviruses contain a linear double-stranded DNA packaged in an icosahedral, non-enveloped capsid with fiber-like projections from each of the 12 vertices (Kojaoghlanian *et al.*, 2003). In addition to the fiber, the other two major types of capsid proteins are the hexon that forms each geometric face of the capsid and the penton base that anchors the fiber. There are at least 51 human Ad serotypes that are classified into six subgroups (A-F) according to various properties (Kojaoghlanian *et al.*, 2003). Members in subgroup C, such as Ad2 and Ad5, are non-oncogenic and predominantly used as vectors for gene delivery (Cao *et al.*, 2004). The viral genome, which is about 36 kb, is experimentally divided into early and late regions based on whether genes in a region are expressed before or after DNA replication. Early regions, E1a and E1b, contain genes encoding proteins for trans-activating other viral genes or regulating the host's cell cycle, and E2 harbors genes for viral DNA replication, while E3 and E4 genes play roles in modulating host immune responses (Kojaoghlanian *et al.*, 2003) or inhibiting host cell apoptosis (Jornot *et al.*, 2001). The late genes encode proteins for either the viral capsid or gene regulation (Cao *et al.*, 2004).

Cellular receptors are required for efficient transduction by Ad. The coxsackie-adenovirus receptor (CAR) is the primary receptor, and the $\alpha_v\beta_3$ and $\alpha_v\beta_5$ integrins are the secondary receptors for Ad to gain entry into a host cell (Parks, 2000). Infection initiates via attachment of the fiber knob to the CAR and subsequent binding of the penton base proteins to the integrin receptors, which allows a virus to enter the cell via receptor-mediated endocytosis. Ads can also enter cells through CAR-independent transduction, via heparin sulfate glycosoaminoglycans (Smith *et al.*, 2002). Following endocytosis, the virus escapes from the endosome through the lysis of the endosomal membrane and enters the nuclear pore complex via microtubule-mediated translocation. For a wild type Ad, transcription and replication begin upon entering the nucleus. During a lytic life cycle, viral DNA is packaged into virions by self-assembly of the capsid proteins and the viruses are then released following the death of the host cell (Cao *et al.*, 2004). For gene replacement therapy, however, replication-defective Ads are used and these viruses do not go through the lytic life cycle or cause the death of host cells. Ad DNA does not integrate into the genome of the host cell and therefore, poses virtually no risk of insertional mutagenesis to the host cell. Unlike plasmids, the Ad genome is highly stable in transduced cells, making it attractive to be used for gene delivery (Benihoud *et al.*, 1999; Ehrhardt *et al.*, 2003; Hillgenberg *et al.*, 2001).

Adenoviruses have been widely used as tools for gene delivery also because of their ability to infect both dividing and non-dividing cells with high efficiency and to produce high-titre of viral particles in cultured cells

(Cao *et al.*, 2004; Parks, 2000). The early Ad vectors were developed by deleting the E1 region that is required for viral DNA replication, and the E1 function was provided *in trans* by cells used for viral propagation. Ad vectors with E1 deleted can carry only 4 kb foreign DNA. To increase the cloning capacity, other Ad vectors were developed by deleting more than one early region (Parks, 2000). These Ad vectors have been used in a variety of experiments on gene transfer *in vitro* and *in vivo*.

Phase I gene therapy trials with the first generation of Ad vectors in patients with cystic fibrosis (CF) concluded that the level of gene transfer and expression is too low to achieve clinical benefits. CF is the most common monogenic fatal disorder in the Caucasian population and it is caused by recessive mutations in the gene encoding the cystic fibrosis transmembrane conductance regulator (CFTR) (Rommens *et al.*, 1989). CFTR is a cAMP-regulated chloride channel, and defective or absent CFTR in epithelial cells of many internal organs of CF patients, including the lung, pancreas, intestine, gall bladder, and reproductive organs, results in salt and water imbalance across the epithelium (Boucher, 1994; Tsui, 1995; Welsh *et al.*, 1995). Although the disease affects multiple organs, lung failure due to chronic infection and inflammation is currently responsible for most morbidity and mortality (Koehler *et al.*, 2001). Therefore, CF gene therapy studies to date have been aimed at treating the pulmonary manifestations. An early trial (Zabner *et al.*, 1993) showed adenoviral transfer of *CFTR* to human nasal epithelium, with correction of nasal transmembrane potential differences (PD). But this gene transfer was correlated with injury caused by the application device, similar results were reported later by Grubb *et al.* (Grubb *et al.*, 1994). It was confirmed (Knowles *et al.*, 1995) that there is no functional correction of nasal PD in patients resulting from adenoviral *CFTR* transfer in the absence of injury of the epithelium. Several other studies with Ad vectors showed *CFTR* gene transfer, but none demonstrated sustained CFTR expression (Bellon *et al.*, 1997; Crystal *et al.*, 1994; Harvey *et al.*, 1999; Joseph *et al.*, 2001; McElvaney and Crystal, 1995; Perricone *et al.*, 2001).

The human studies mentioned above as well as experiments with animals (Dai *et al.*, 1995; Wilson *et al.*, 1998; Yang *et al.*, 1996; Yang *et al.*, 1995) showed that recombinant Ad vectors delivered to the airways or to the circulation system intravenously induce potent host immune responses that limits both stable transgene expression and the possibility for vector re-administration (Wivel *et al.*, 1999). These strong anti-vector responses are largely attributed to viral particles in the inoculum (Kafri *et al.*, 1998; McCoy *et al.*, 1995) and the expression of viral proteins in transduced cells (Wivel *et al.*, 1999). The transgene product itself, particularly when derived from a species different from the host, can also contribute to host immune responses if it is foreign to the host (Michou *et al.*, 1997; O'Neal *et al.*, 2000; Tripathy *et al.*, 1996). This subject will be discussed further.

To improve Ad vectors, the helper-dependent (or gutted) adenoviral (HD-Ad) vector was developed by deleting all the viral coding sequences, leaving only the viral inverted terminal repeats (ITRs) and packaging signal. The deletion of viral coding sequences indeed reduces host adaptive immune responses (Parks, 2000) and prolongs transgene expression (Kim *et al.*, 2001; Morsy *et al.*, 1998; Toietta *et al.*, 2003) in addition to the expansion of the cloning capacity to ~36 kb. Because HD-Ad vectors have the same capsid proteins, host immune responses to the vectors are still present (Brunetti-Pierri *et al.*, 2004; Morral *et al.*, 2002). However, since non-capsid proteins encoded by Ad (Schaack *et al.*, 2004) can cause inflammation, the innate immune response to HD-Ad vectors is attenuated as demonstrated in gene transfer studies in mice (Kim *et al.*, 2001; Morsy *et al.*, 1998; Toietta *et al.*, 2003). For HD-Ad vector propagation, a helper-virus is required to provide functions for DNA replication and assembly of the virions (Parks, 2000). One past problem in using the HD-Ad vector in clinical studies was the difficulty in large-scale production (Cao *et al.*, 2004). This problem has recently been solved by Ng's group by using a new helper virus and 293 cells capable of growing in liquid suspension (Palmer and Ng, 2003). Although HD-Ad vectors have been shown to be superior to the early generation of Ad vectors, they have not been tested clinically due to the safety concern raised with the conventional Ad vectors (Raper *et al.*, 2003). Since no viral gene expression from the HD-Ad vector, however, transient pharmacological immune modulation and anti-inflammation intervention may be used to control the acute toxicity caused by the capsid proteins.

A2: Adeno-associated Viral Vectors

Recombinant adeno-associated virus (rAAV) vectors are another major type of viral vectors currently used in gene therapy studies (Tal, 2000). AAV is a replication–defective parvovirus that depends on a helper virus, either adenovirus or herpes virus, for its propagation during lytic infection (Berns and Giraud, 1996). AAV has a very small (about 4.7 kb) single-stranded DNA genome (Carter, 2004), including 145 bp

ITRs that are the only sequences required for vector construction. The AAV genome encodes three viral capsid proteins (VP1, VP2, and VP3), Rep68/78 proteins that bind to the ITRs and mediate its integration into the host chromosome (Carter, 2004), and Rep52/40 proteins whose functions are not clear (Tal, 2000). Therefore, rAAV vectors are essentially helper-dependent viral vectors retaining only the ITRs. All the capsid and Rep proteins are provided *in trans* by either a helper virus or sequences integrated into the host genome during vector propagation. The small cloning capacity (4.5 kb) is a disadvantage of AAV vectors because it limits their utilization for therapeutic genes with coding sequences longer than 4.5 kb and with little room for inclusion of DNA regulatory elements even for smaller genes.

AAV uses heparan sulfate proteoglycan as receptor to gain entry into host cells via endocytosis (Summerford and Samulski, 1998) and it can transduce both dividing and non-dividing cells. The viral genome can integrate into a host chromosome or stay episomally in the cell, however the frequency of integration is very low (Hargrove et al., 1997). In nondividing cells, AAV vectors stay episomally as head to tail concatemers (Schnepp et al., 2003). While the wild type AAV integrates specifically at the AAVS1 site on human chromosome 19 (Kotin et al., 1990), rAAV vectors integrate randomly (Kearns et al., 1996) at a much lower frequency than previously believed (Carter, 2004). Although AAV can transduce a broad spectrum of cell types, a very high ratio of viral particles is required to transduce target cells. This may be partially due to the low levels of receptor present on the cell surface, for AAV integration or gene expression, the single-stranded DNA genome has to be converted into double-stranded and this could also be a rate-limiting step especially in nondividing cells or primary cultured cells (Tal, 2000). There are 8 serotypes of AAV identified (AAV1 to AAV8) and some of them show difference in tissue-tropism (Gao et al., 2002). For example, AAV6 (Blankinship et al., 2004) and AAV8 (Gao et al., 2002; Wang et al., 2005) transduce muscle cells very efficiently. In addition, levels of neutralizing antibodies against them are different; Sera from humans show little neutralizing activity to AAV7 and AAV8. Since the viral capsid proteins can be exchanged, a desired serotype can be selected during vector production to maximize the efficiency of gene delivery to a particular tissue. Serotype-switching can also be used to minimize neutralizing antibodies during vector readministration.

Since the cloning capacity of rAAV is limited, inclusion of extra promoter/enhancer elements for better transgene expression would further reduce the capacity.

This problem can be minimized by reducing the size of a therapeutic gene. One example was the construction of an AAV vector for expression of a minedystrophin gene for gene therapy against Duchenne muscular dystrophy (DMD) (Wang et al., 2000). DMD is an X-linked, fatal genetic muscle disease affecting 1 of every 3,500 males born and the progressive muscle degeneration leads to death of patients by their early twenties. The dystrophin gene spans nearly 3 million bp on the X-chromosome (Koenig et al., 1987) and it produces a mRNA of 14 kb. Wang et al. created a 4.2 kb minidystrophin gene and cloned it into an AAV vector containing a muscle-specific creatine kinase (MCK) promoter. They showed that this vector effectively ameliorates muscular dystrophy in the mdx mouse model (Wang et al., 2000). The other example was the development of an AAV vector expressing a mini version of the *CFTR* gene (Sirninger et al., 2004). Since viral promoters are often attenuated or silenced (this subject will be addressed later in this chapter), authors used the beta-actin promoter plus an enhancer from the cytomegalovirus (CMV) to drive a CFTR minigene. In addition to the utilization of minigenes, trans-splicing has been explored to expand the cloning capacity of AAV vectors. The process of trans-splicing was initially discovered in *Trypanosoma brucei* 23 years ago when different surface glycoprotein mRNAs was found to carry a common 39-nucleotide sequence, namely, the spliced leader sequence (Boothroyd and Cross, 1982). In trans-splicing, two primary RNA transcripts are used to produce a mRNA by the RNA splicing machinery and this subject was recently reviewed (Liang et al., 2003). Two AAV vectors can be used to produce two half transcripts for a therapeutic gene and the transcripts can be trans-spliced in cells to produce a functional mRNA (Duan et al., 2001; Pergolizzi and Crystal, 2004; Reich et al., 2003; Yan et al., 2000).

Clinical studies with rAAV vectors for CF gene therapy have yielded mixed results. The major advantage of these vectors is that they are less immunogenic than Ad vectors (Wagner et al., 1999a), but since the AAV vector used in this study contains only the promoter in the ITR sequence to drive *CFTR* expression, limited gene expression was detected when the maxillary sinus of CF patients was used as a delivery test site (Wagner et al., 1999b). The first multi-dose inhalation trial in CF patients demonstrated safety and a transient improvement of lung function (Moss et al., 2004). Clinical studies in other organ systems also showed gene transfer and expression (Tal, 2000); major breakthroughs using AVV vectors in clinical studies remain to be achieved. The recent discovery of AAV8 being able to cross blood vessel barriers efficiently transducing skeletal and heart muscles (Wang *et al.*, 2005) will facilitate the utilization of AAV vectors in clinical applications.

A3: Retroviral Vectors

Retroviral vectors are based on retroviruses that comprise a large class of enveloped viruses that contain two identical single-stranded RNAs (7-11 kb) as the viral genome. The retroviridae family consists of seven genera: alpharetrovirus, betaretrovirus, gammaretrovirus deltar etrovirus, epsilonretrovirus, lentivirus, and spumavirus (Pringle, 1999). The first five genera were also known as oncoretroviruses, and the first retroviral vector was developed based on a gammaretrovirus, Moloney murine leukemia virus (MoMLV). Retroviral vectors were also developed based on lentivirus (Copreni *et al.*, 2004) although traditionally only vectors based on oncoretroviruses were referred as retroviral vectors. The retroviral genome contains three genes (*gag* coding for the group specific antigens or core protein, *pol* for reverse transcriptase and *env* for viral envelope protein), two long terminal repeats (LTRs) and a sequence required for packaging viral RNA during viral propagation. During the infection, the retroviral envelope protein interacts with a cell surface receptor to gain entry into the host cell. Different viruses use different receptors. For example, MoMLV uses a sodium-dependent phosphate transporter as its receptor while lentivirus uses CD4 as its primary receptor (Overbaugh *et al.*, 2001). Upon entering a host cell, the RNA genome is reverse-transcribed into double- stranded DNA that binds to cellular proteins to form a nucleoprotein preintegration complex (PIC) that migrates to the nucleus and integrates into the host genome. The nuclear membrane, however, can be a barrier for some retroviruses. For instance, the PICs of MoMLV cannot cross the membrane and require a mitotic cycle to disrupt the nuclear membrane for the viral genome to reach the nucleus (Barquinero *et al.*, 2004). Therefore MoMLV cannot transduce non- dividing cells. On the other hand, lentiviral vectors based on human immunodeficiency virus-1 (HIV-1) do not have this limitation and thus can transduce nondividing cells (Barquinero *et al.*, 2004).

For development of retroviral vectors, only the 5′ and 3′ LTRs as well as the packaging signal sequence are required for the vector DNA, while the functions of *gag, pol* and *env* are provided by host cells used for viral propagation. Over the years, there were several types of improvements made for the retroviral vectors. The first type was the improvement in transgene expression. This was done by using LTR variants for transgene expression in certain cell types (Smith, 1995) and engineering regulatory regions to enhance transgene expression or reduce transcriptional silencing in specific target cells (Barquinero *et al.*, 2004). Hybrid retroviral vectors with improved expression in hematopoietic cells were developed using sequences from the murine embryonic stem cell virus and the Friend mink cell focus-forming virus or the myeloproliferative sarcoma virus (Baum *et al.*, 1995). Challita *et al.* (1995) increased transgene expression and decreased DNA methylation of retroviral vectors by introducing multiple changes in the LTRs. General principles regarding mechanisms of transcriptional regulation and DNA methylation are extensively reviewed elsewhere in this book and readers are encouraged to visit other chapters.

The second type of improvements was the utilization of alternative envelope proteins, a technique called *pseudotyping*. For example, many hematopoietic cells express higher levels of Glvr1, the receptor for the gibbon ape leukemia virus (GALV), than that of Ram 1 the receptor for amphotropic vectors (Bauer *et al.*, 1995), and GALV-pseudotyped vector particles are more efficient at transducing primate repopulating hematopoietic stem cells than conventional amphotropic vector particles. Another example is the feline endogenous retrovirus (RD114)-pseudotyped vector particles that were shown to be more efficient in transduction of cord blood cells than amphotropic vector particles (van der Loo *et al.*, 2002). In addition, although the vesicular stomatitis virus-G (VSV-G)-pseudotyped oncoretroviral vector particles did not show enhancement in transducing primitive primate hematopoietic cells, (Evans *et al.*, 1999), VSV-G- pseudotyped lentiviral vectors could be useful for gene delivery to other cell types such as airway epithelial cells (Copreni *et al.*, 2004; Sinn *et al.*, 2003). VSV-G is a fusogenic protein that interacts with membrane phospholipids to facilitate transduction, and VSV-G- pseudotyped vector particles are more stable (Yam *et al.*, 1998).

The third type of improvement in retroviral design was the development of self-inactivating (SIN) vectors. A deletion of 229 bp in the 3′ LTR eliminated the enhancer and promoter present in the LTR (Yu *et al.*, 1986). During the reverse transcription, the deletion was transferred to the 5′ LTR, resulting in a vector without viral promoters or enhancers. The absence of the viral promoter and enhancer minimizes the risk of activation of oncogenes as a result of integration and allows transgene expression under the control of desired promoter/enhancer sequences. The most important improvement needed right now is to design a vector that can be safely and specifically integrated at a chromosomal site that allows efficient therapeutic gene expression.

In addition to the improvements made in retroviral vector development, other strategies have also been used to enhance the retroviral gene delivery. Since retroviral vectors have been heavily used for *ex vivo* gene therapy to target hematopoietic stem cells, several techniques were developed to enhance their transduction efficiency. The first strategy was to use cytokines to mobilize the primitive stem cells to peripheral blood and to improve the susceptibility of the cells to retroviral transduction. It was shown that in mice and monkeys, treatment of donor animals with granulocyte colony-stimulating factor (G-CSF) and stem cell factor (SCF) increased the number of $CD34^+$ cells targeted for gene transfer in both peripheral blood and bone marrow, and these cells could be more efficiently transduced (Bodine *et al.*, 1996; Dunbar *et al.*, 1996). The second strategy was to use cytokines to induce the primitive human hematopoietic cells to divide, therefore enhancing the retroviral transduction efficiency while maintaining the hematopoietic potential. Primitive human hematopoietic cells could be stimulated to undergo self-renewal while retaining repopulating potential when SCF and Flt-3 ligand were used in combination with IL-3, IL-6, and G-CSF (Glimm and Eaves, 1999). These stimulated cells could be transduced more efficiently (Veena *et al.*, 1998). The third strategy was to use fibronectin fragments to enhance retroviral transduction efficiency (Hanenberg *et al.*, 1996; Moritz *et al.*, 1994). Fibronectin fragments bind both vector particles and target cells and therefore may facilitate the uptake of vector particles by the cells (Moritz *et al.*, 1996).

Until now, three human gene therapy studies successfully brought clinical benefits to patients and all these trials were conducted by using retroviral vectors. The first trial was reported (Cavazzana-Calvo *et al.*, 2000) in the treatment of children with X-linked severe combined immunodeficiency disease (SCID-X1). The patients have a defective gene encoding the common gamma chain (γc) of receptors for IL-2, -4, -7, -9, -15 and -21, which leads to the absence of functional B, T and NK cells, and they die at very early age. Ten patients were treated by reinfusion of their own CD34+ bone marrow cells transduced with a retroviral vector expressing the wild-type γc gene in the absence of any myelosuppression. Myelosuppression means a decrease in number of blood cells produced from the bone marrow following a pharmacological intervention or under a diseased situation. Nine of ten patients showed almost normal levels of T-cell counts and significantly improved immune function. Despite the insertional mutagenesis resulting in activation of the T-cell proto-oncogene LMO-2 in two patients (Hacein-Bey-Abina *et al.*, 2003), this clinical trial was the first milestone to mark the feasibility of using gene therapy to cure a human disease. The second successful clinical trial was conducted in two children with adenosine deaminase-deficient severe combined immunodeficiency (ADA-SCID) (Aiuti *et al.*, 2002). In ADA-SCID patients, the accumulation of purine metabolites toxic primarily to T cells leads to the immunodeficiency. Autologous CD34+ cells from patients were transduced with a retroviral vector (GIADA1), and four days later, reinfused into the patients who received two doses of busulfan (2 mg/kg/per day) for transient myelosuppression prior to the reinfusion. During the follow-up for more than one year, two patients were in good clinical conditions and did not experience any severe infectious episodes. The third case was reported in UK (Gaspar *et al.*, 2004) on four SCID-X1 patients who received treatment similar to that reported in the first case (Cavazzana-Calvo *et al.*, 2000).

B: Nonviral Vectors

Nonviral methods for gene delivery have also been extensively explored over the past two decades in searching for alternatives safer than viral gene delivery. Liposomes are the most studied types of nonviral vectors. Based on their charge, liposomes can be classified into two classes, positively charged or cationic liposomes and negatively charged or pH-sensitive liposomes (Singhal and Huang, 1994). Cationic liposomes are commonly used for gene delivery and many types of formulations are available. They are made up of a cationic lipid and a neutral lipid, often dioleoylphosphatidylethanolamine (DOPE) or cholestrol. There are many types of cationic lipids that have been tested in gene delivery studies, such as 3β(N(N', N'-dimethylaminoethane) carbamoyl)-cholesterol (DC-chol), 1,2-dioleoyloxy-3-(trimethylammonio) propane (DOTAP), and N-(2,3-(dioleoyloxy)propyl)-N,N,N-trimethyl ammonium chloride (DOTMA). Detailed information regarding various cationic liposomes was covered by Gao and Huang (1995). Negatively charged or pH-sensitive liposomes contain DOPE and another lipid, such as palmitoylhomocysteine or free fatty acids. They fuse with other lipid bilayers at low pH and can be used to deliver molecules to cytoplasm. This type of liposomes was less frequently used for gene delivery (Singhal and Huang, 1994). Although liposomes have been commonly used for delivery of DNA to cultured cells, their potential for *in vivo* gene delivery remains to be shown. Other types of non-viral formulations, such as those using polyethyleneimine (PEI) or polylysine, were also studied for their gene transfer ability (Bragonzi *et al.*, 2000; Kollen *et al.*, 1999). They share with liposomes the same problem of inefficient gene transfer *in vivo*.

Many clinical trials have been conducted with cationic liposomes mostly to assess the potential for treating patients with cystic fibrosis (CF). The first liposome CF trial was carried out in the noses of CF patients (DF508 mutation) using DC-Chol:DOPE complexed to *CFTR* cDNA (Caplen *et al.*, 1995). Later, several nasal trials with various formulations of liposomes were performed (Gill *et al.*, 1997; Porteous *et al.*, 1997; Zabner *et al.*, 1997). Although safe gene transfer and/or a statistically significant partial correction of nasal electrophysiology was reported in these trials, further improvement is needed to achieve the level of efficacy needed for CF gene therapy. A double-blind placebo-controlled nasal trial with p-ethyl- dimyristoylphosphadityl choline (EDMPC) complexed with human *CFTR* cDNA concluded that this lipid-DNA complex was also relatively safe but did not produce evidence of gene transfer to the nasal epithelium by physiological or molecular measures (Noone *et al.*, 2000). The first clinical trial in lung was carried out with *CFTR* cDNA complexed with GL67 liposomes (Alton *et al.*, 1999). One week after nebulizing the DNA/liposome complexes into the lungs of eight subjects, seven of the eight developed mild flu-like symptoms that disappeared within 36 hours. It was found later that DNA produced in bacteria contributes to the inflammatory process. Six of eight patients showed a small change in chloride conductance towards normal values. A trial later demonstrated that even eukaryotic DNA in combination with GL67 elicits immunogenic responses in CF patients (Ruiz *et al.*, 2001). For liposomes to be used in future gene therapy, it is essential to demonstrate their ability to achieve efficient gene delivery and sustained transgene expression in animal models, that remain to be shown.

Obstacles and Solutions

Despite all the advancements in the development of vectors for gene delivery and a few successful gene therapy trials, fundamental problems in gene delivery and transgene expression remain to be solved before gene therapy can be used as a common practice in hospitals. Over the years, many of these problems have been identified. In the following, I shall use gene delivery to the lung airway as an example to point out the obstacles, including physical barriers to gene delivery, host immune responses and maintaining transgene expression, and to propose solutions to overcome these obstacles.

A: Physical Barriers to Gene Delivery

The difficulty in gene delivery was initially underestimated. In fact, the term "gene medicine" may be a little misleading, because most conventional drugs can be delivered orally and/or intravenously or subcutaneously. However, genes delivered in viral vectors or complexed with liposomes have difficulty to reach target cells because of the following reasons. First, viral vectors or DNA/liposome complexes are much larger in size than conventional drugs and therefore they cannot move freely inside a body. For example, most viral vectors once entering a cell cannot get out of the cell. The wild type viruses can spread easily inside a body because they can propagate in an infected cell and cause the cell to burst which allows the progeny to infect other cells and travel to other organs if they can enter the blood circulation. However, for safety reasons, most viral vectors are replication-incompetent. Regarding *in vivo* migration, rAAV is the only known type that can cross the blood vessel barrier because of the relatively small size (Wang *et al.*, 2005). Second, "gene medicine" is biologically labile and vulnerable to host defense system. Viral vectors can be inactivated by the innate immune response or preexisting antibodies, and DNA plasmids are quickly degraded inside and outside cells if they are separated from liposomes. These issues will be discussed below in the context of airway gene delivery.

The lung airway represents an attractive target for gene transfer because vectors can be delivered efficiently to the airway surface. The lung epithelium is a continuous layer of cells lining all the airways and air spaces from the trachea to alveoli, and it is composed of at least eight different cell types that have a range of functions (Spina, 1998). Lung airways provide important defense capabilities to keep infectious or harmful particles from entering the body through the respiratory system (Koehler *et al.*, 2001), in addition to providing a passage for air to go in and out. The surface layer of mucous-containing liquid produced from epithelial goblet cells and submucosal glands can bind foreign particles that are then cleared from airways by the sweeping action of ciliated epithelial cells (Boucher, 1999). The intercellular space of the airway epithelial cells is linked by tight junctions so that the entire epithelium forms a physical barrier keeping infectious particles from penetrating the surface layer of the airway (Boucher, 1999; Koehler *et al.*, 2001). The airway epithelium has also been suggested to play an important role in mucosal immunoglobulin A production, through supplying cytokines responsible for B-cell isotype switch, growth and differentiation into IgA-secreting plasma cells (Salvi and Holgate, 1999). Additionally, lung epithelial cells are important in maintaining ion and water homeostasis (Boucher, 1994; Boucher et al., 1986), as evident from diseases such as cystic fibrosis (Rommens et al., 1989; Tsui, 1995). The alveolar epithelial cells are mainly responsible for air exchange or producing surfactants to keep the airspace open. Macrophages are the major type of non-epithelial cells in the airway and they comprise of more than 90% of the cells present in bronchoalveolar lavage (BAL) fluids (washout fluids from airways and alveoli) (Zsengeller et al., 2000).

When genes in viral or nonviral vectors are delivered to the lung airway, they can be trapped by the layer of mucous and swept out by the mucociliary action. The transmembrane mucin MUC1 as well as other sialoglycoconjugates was reported to inhibit Ad-mediated gene transfer to epithelial cells (Arcasoy et al., 1997). Consistent with these results, glycocalyx on the apical surface of polarized epithelial cells was found to act as a barrier to Ad-mediated gene transfer (Pickles et al., 2000). The mucous layer on the surface of cultured epithelial cells was also found as a barrier to AAV-mediated gene transfer (Bals et al., 1999). One strategy to overcome the problem of mucous layer barrier, as suggested by Koehler et al. (Koehler et al., 2001) is to use mucolytic reagents prior to vector delivery. The second strategy is to neutralize the vector trapping sites on presumably negatively charged mucins, or glycosami noglycans. Interestingly, Kaplan et al. showed that various polycations, such as DEAE-dextran, polylysines, polybrene, protamine, and branched polyethylenimine, could enhance Ad transduction in mouse airways (Kaplan et al., 1998). These polycations are likely able to neutralize the viral vector trapping sites in the airway. The polycations with no side effects in human, such as DEAE-dextran (Pupita and Barone, 1983), may be used to reduce the viral vector particles used in gene transfer, thus decreasing host immune responses in lung gene therapy.

The second physical barrier is the loss of gene therapy vectors (viral or nonviral) to lung macrophages. For example, it was demonstrated that 70% of adenoviral vectors were lost in the mouse airway within 24 h and that depletion of macrophages with liposome/dichloromethylene-biphosphonate (Cl_2MDP or clodro nate) complexes resulted in a 100% increase in vector DNA recovered from the lung (Worgall et al., 1997). The mechanism of vector uptaking by macrophages is likely different from that involved in cell transduction because macrophages are hardly transduced despite they uptake Ad vectors very efficiently (Zsengeller et al., 2000). Since lung macrophages not only destroy gene therapy vectors, but also play a major role in the innate immune response (this will be addressed later), it is important to consider them as a problem in lung gene therapy. One strategy to minimize the problem is to transiently deplete lung macrophages. Gadolinium chloride (Singh and de la Concha-Bermejillo, 1998) or liposome-encapsulated Cl_2MDP (Thepen et al., 1991; Worgall et al., 1997) were used to deplete lung macrophages in animals. But, there is no such a drug identified yet for the depletion of human airway macrophages. The second strategy is to avoid macrophages. For example, when vectors delivered by aerosol, large size aerosol droplets or particles will preferentially deposit in the airway instead of alveoli where macrophages are abundant. Finally, pharmacological intervention may be used to block the macrophage function. It was recently reported that lung macrophages of knockout mice lacking GM-CSF expression or transcription factor PU.1 activity are incapable of uptaking Ad vectors (Berclaz et al., 2002). Therefore, reagents may be developed to block the GM-CSF or PU.1 prior to gene delivery.

The third physical barrier is the lack of viral receptors present on the apical surface of airway epithelial cells for efficient transduction. For example, the major receptor for Ad, CAR, is expressed on the basal lateral side where the virus can not reach unless the tight junction is loosened. One strategy is to modify viral vectors so that they can recognize receptors expressed on the apical surface. This can be done by "pseudotyping" the vectors using different capsid proteins (Ad or AAV) or envelope proteins (Retrovirus) or artificially modifying these proteins so that they can recognize the receptors on the apical side (Barnett et al., 2002). The other strategy is to use reagents to transiently break the tight junctions. For example, application of the Ca^{2+} chelator, EGTA to airways prior to vector delivery can enhance Ad-mediated gene transfer significantly (Chu et al., 2001). In addition, L-α-lysophosphatidylcholine (LPC) can also achieve the same effect in mice (Limberis et al., 2002), and as shown by my group, LPC can be mixed with Ad vectors for aerosol delivery to rabbit airways (Koehler et al., 2005). Finally gene therapy vectors can be lost in other ways such as degradation or inactivation by enzymes secreted to the airway surface fluid. Even delivered inside a cell, plasmid DNA can be quickly degraded by intracellular nucleases. These problems were extensively reviewed by Koehler et al. in 2001 (Koehler et al., 2001).

B: Immunological Barriers

The immunological barriers to gene delivery can be divided into two categories, innate and adaptive immune responses. Although the adaptive immune response to viral vectors was recognized early on and investigated extensively (Dai *et al.*, 1995; Kay *et al.*, 1995; Yang *et al.*, 1995), the innate immune response was greatly underestimated. This was evident from the first incidence of gene therapy related death. In September 1999, a patient who received a high dose of an Ad vector via the hepatic artery succumbed to the acute toxicity cause by the vector and the news shocked the scientific community. Post-mortem analysis confirmed that the patient suffered from systemic inflammation, biochemically detectable disseminated intravascular coagulation, and multiple organ failure within 98 hours (Raper *et al.*, 2003). Clearly this was the result of patient's innate immune system reacting to the high dose of the Ad vector.

B1: Innate Immune Response

The innate immunity is a fast, receptor-mediated, host defense mechanism. The receptors recognize highly conserved structures, called pathogen-associated molecular patterns (PAMPs), present in microorganisms (Medzhitov and Janeway, 2000). These receptors are inherited, therefore limited in numbers, and expressed on many cell types involved in the innate immune system, especially on macrophages, dendritic cells, and B-cells. They can be functionally classified into three categories: secreted, endocytic, and signaling. The secreted receptors, such as the mannan-binding lectin, function as opsonins by interacting with their ligands flagging the microorganisms for recognition by the complement system and phagocytes. The endocytic receptors expressed on the surface of phagocytes, such as the macrophage scavenger receptor, mediate the uptake and delivery of pathogens into lysosomes where they are destroyed (Suzuki *et al.*, 1997). The pathogen-derived peptides are then presented by major-histocompatibility-complex (MHC) molecules on the surface of macrophages. The signaling receptors, such as Toll-like receptors, interact with their ligands and activate signal-transduction pathways for expression of a variety of cytokines.

For airway gene delivery, lung macrophages are the major cell type involved in the initial innate immune response. Macrophages not only uptake gene therapy vectors and destroy them, but also initiate the production of proinflammatory cytokines. For example, TNF-α, IL-6, MIP-2, and MIP-1α were dramatically induced in macrophages, not in other cell types, upon Ad vector delivery to mouse airways within 6 hours (Zsengeller *et al.*, 2000). The proinflammatory cytokines released from macrophages can affect other cells in the lung as well as macrophages themselves leading to a cytokine cascade, which could lead to airway damage, if the host can not shut down the cascade. One of the important pathways involved in the induction of inflammatory cytokines is the NF-κB signaling pathway (Baldwin, 1996). Details of the NF-κB signaling pathway have been described elsewhere in this book. A variety of inflammatory cytokines, such as IL-6 and IL-8, can be induced by the activation of NF-κB, which is inactive when it is associated with its inhibitor IκB in cytoplasm. When cells are induced with microbial products, such as lipopolysaccharide (LPS) or certain cytokines, such as TNF-α and IL-1ß, the signals are channeled inside the cell through a pathway and cause the activation of the IκB kinase (IKK) through phosphorylation. The activated IKK, in turn, phosphoylates IκB leading to its ubiquitination and degradation (Baldwin, 1996), therefore NF-κB is released from inhibition and translocated into nucleus resulting in up-regulation of its target genes. Even non-viral vectors, bacterial DNA can be recognized by Toll-like receptor 9 which activates the NF-κB pathway (Koehler *et al.*, 2004). In humans, a high level of IL-8, a potent neutrophil chemoattractant, can cause neutrophil infiltration that leads to tissue damage (Cao *et al.*, 2005; Matsushima *et al.*, 1988). Roles of other cell types in the initial innate immune response to gene therapy vectors are not clear. Lung epithelial cells may produce IP-10 (Borgland *et al.*, 2000), but very little other cytokines, in response to Ad vectors. The innate immune response not only leads to acute toxicity, but also controls the adaptive immune response (this subject will be discussed below).

Patients receiving Ad vectors often show an acute inflammatory response. This response is characterized by the infiltration of inflammatory cells in tissues of organs targeted in gene therapy, such as liver and lung and the local release of pro-inflammatory cytokines, including TNF-a, IL-1b, IL-6, and IL8 (Cao *et al.*, 2004; George, 2003). Even for nonviral vector-mediated gene transfer, patients show inflammatory response (Ruiz *et al.*, 2001). Therefore, it is very important to reduce the host innate immune response during gene therapy treatment. One of the strategies is to eliminate or inactivate lung macrophages as described earlier. Secondly, anti-inflammatory drugs, such as corticosteroids or Ibuprofen, can be used to reduce inflammation (Koehler *et al.*, 2004). If possible, vectors that do not cause strong host innate immune responses should be used. For example, AAV instead of Ad vectors should be used if they show the same efficiency in a particular gene delivery task.

B2: Adaptive Immune Response

The adaptive immunity relies on T- and B-lymphocytes to produce cellular and humoral responses to infectious agents. During postnatal development, an extremely diverse repertoire of receptors is generated randomly and each type of receptors, that recognizes a unique antigen, is expressed on the surface of one lymphocyte only. The cells bearing useful receptors are subsequently selected from billions of lymphocytes for clonal expansion by interacting with antigens. Antigens are bound to the MHC II molecules on the surface of professional-antigen-presenting cells, normally macrophages or dendritic cells and presented to helper T cells. Activation of helper T cells by antigen presenting cells also requires a co-stimulatory signal, e.g., CD80 or CD86, on the surface of the antigen-presenting cell to bind to CD28 on the surface of the T cell. The expression of co-stimulatory molecules is regulated by innate immunity (Fearon and Locksley, 1996), and therefore, the adaptive immunity is controlled by the innate immunity. After activation, helper T cells control other cells in the adaptive immune system, such as activation of cytotoxic T cells to destroy infected cells and B cells to produce antibodies. Following elimination of an infection, some antigen-specific clones of T and B cells remain as "memory" lymphocytes so that the adaptive immune system remembers the antigens and destroy them quickly when encountered a second time. Compared to the innate immune response, the adaptive immune response is a slow process and it takes three to five days to mobilize enough lymphocytes to take action.

Previous work in several groups (Dai *et al.*, 1995; Kay *et al.*, 1995; Yang *et al.*, 1995) showed clearly that both cellular and humoral responses are involved in Ad vector-mediated gene transfer in mice. Repeated delivery of viral vectors, or primary delivery to individuals with pre-existing immunity, is problematic because of antibodies against the capsid proteins. Various strategies can be used to overcome the problem. First of all, blocking the innate immune response may be used to reduce the adaptive immune response. Reducing the uptake of vectors by macrophages should inhibit antigen presentation. Therefore, all the strategies used to reduce the innate immune response mentioned above can be used to reduce the adaptive immune response. Another strategy for effective repeated delivery of recombinant viruses is "serotype switching" (Mastrangeli *et al.*, 1996). Gene therapy is initiated with one virus serotype, then switched to a second serotype for a subsequent administration, thereby avoiding neutralizing antibodies induced by the first serotype (Mack *et al.*, 1997). However, the level and duration of transgene expression following serotype switching may be limited by cross-reactive cytotoxic T lymphocytes that can also target cells infected by the second serotype virus (Mack *et al.*, 1997; Smith *et al.*, 1998). Furthermore since all the viral gene therapy vectors used in the future will not express any viral coding proteins, transient immune modulation may be used to block the adaptive immune response. Drugs normally used for immunosuppression, such as cyclosporine and cyclophosphamide, may be used to transiently modulate host adaptive immune responses. For example, it was shown that cyclophosphamide alone or in combination with cyclosporine A extended transgene expression mediated by the first generation adenoviral vector (Dai *et al.*, 1995). Finally, blocking co-stimulatory pathways can be used to modulate the host adaptive immune response. Several groups showed that an antibody against CD40 ligand (Scaria *et al.*, 1997; Wilson *et al.*, 1998) or expressing CTLA4Ig, a fusion protein of cytotoxic T lymphocyte-associated protein 4 (CTLA4) and the Fc portion of immunoglobulin G (IgG), by the HD-Ad vector, improved transgene expression in rodents (Jiang *et al.*, 2002; Yamashita *et al.*, 2003),

C: Transgene Expression Barriers

How to control therapeutic gene expression was ignored initially by the gene therapy community since some viral promoter showed high activity in cultured cells. In addition, the initial goal in gene therapy research was to demonstrate gene transfer and gene expression. It was later discovered that viral promoters can be attenuated by host cytokines (Qin *et al.*, 1997; Sung *et al.*, 2001) and some retroviral promoters can be silenced following vector integration (Barquinero *et al.*, 2000; Fearon and Locksley, 1996; Kalberer *et al.*, 2000). As described in the first Chapter by Goodrich and Tjian, we now know that the processes of gene transcription, RNA processing and mRNA transport are coupled. It is important to take into consideration all the DNA elements required for gene expression in the design of a gene expression cassette, such as enhancers for transcription, introns and polyA signals for RNA processing and transport as well as sequences for efficient translation and RNA stability. Some gene introns contain DNA control elements that are required to achieve cell-specific expression (Aronow *et al.*, 1992; Oshima *et al.*, 1990). In addition, for integration vectors, insulators or locus control regions are required to prevent transgene silencing due to the integration. All these regulatory elements are described elsewhere in the book; I shall only discuss their utility here.

C1: Cell-specific Expression

The temporal and spatial expression of genes is determined by the DNA regulatory elements and gene positions on chromosomes. Ideally, for expression of a therapeutic gene, its own DNA regulatory elements should be used to drive the expression, since non-specific expression of a gene could result in adverse effects. Unfortunately, for many genes, their regulatory elements are not characterized. Initially, most gene therapy studies were done with promoters from viruses, such as Simian virus 40 (SV40), CMV and respiratory syncytial virus (RSV). As mentioned above, the viral promoters can be shut down in mammals by host cytokines although these promoters are quite active in cultured cells. Another problem associated with the viral promoters is that they are not cell or tissue- specific.

One strategy to solve this problem is to use DNA regulatory elements from a gene that shows similar cell-specificity to that of a therapeutic gene. My group applied this strategy to our gene therapy research at the very beginning. For CF gene transfer, we developed a gene expression cassette using DNA control elements from the human cytokeratin 18 gene which displays a very similar expression pattern as the *CFTR* gene (Chow *et al.*, 1997; Chow *et al.*, 2000). We showed that this gene expression cassette can be used to efficiently express transgenes in HD-Ad vectors in mice (Toietta *et al.*, 2003). We further demonstrated that CF knockout mice treated with a HD-Ad vector expressing the *CFTR* gene from our K18 expression cassette became resistant, like the wild-type mice, to acute lung infection by a clinically relevant *Burkholdera cepacia* strain (Koehler *et al.*, 2003). Recently, we showed that our vectors efficiently transduced airway epithelia of rabbits (Koehler *et al.*, 2005). The other strategy is to use hybrid promoters with regulatory elements from different promoters. As mentioned early, an AAV vector using the beta-actin promoter plus an enhancer from the CMV promoter exhibited stronger expression of a *CFTR* minigene gene than that from vectors using viral promoters (Sirninger *et al.*, 2004).

C2: Sustained and Regulated Expression

Sustained therapeutic gene expression is critical for gene therapy. One major problem in CF lung gene therapy trials is the lack of sustained transgene expression. For liposome-mediated gene transfer, this can be attributed, at least partially, to the instability of the plasmid DNA in the transfected cells. For Ad vectors, host immune responses are the major problem. For AAV vectors, the small capacity for carrying DNA does not allow it to include enough DNA regulatory sequences. Since all the lung gene therapy trials conducted so far have not demonstrated efficient gene transfer, the lack of sustained transgene expression would further diminish any hope for showing clinical benefits. Lentiviral vectors have not been used for lung gene therapy trials, it is not clear whether silencing of transgene expression will be a problem. Therefore, solutions to the lack of sustained transgene expression may be different for different vectors. For Ad vectors, simply switching to HD-Ad vectors can greatly improve the length of transgene expression in the lung (Toietta *et al.*, 2003) and in other organs as well (Kim *et al.*, 2001; Maione *et al.*, 2001; Morsy *et al.*, 1998). In addition, reducing the innate and adaptive immune responses will also enhance the sustained transgene expression if the therapeutic gene product were not previously present in patients. However, even for HD-Ad vectors to be used in lung gene therapy, there is an additional obstacle, epithelial cell turnover. This is also problematic for rAAV vectors because their frequency of integration is very low (Schnepp *et al.*, 2003). This requires readministration of vectors, which may not be as difficult as many think since the mucosal antibody response is short-lived (Ahmed and Gray, 1996). For lentiviral vectors, unless the integration at a specific, benign chromosomal site can be established, they are too risky to be used for lung gene therapy, because of so many cells that need to be targeted. If integration vectors are used in gene therapy, all silencing elements should be removed and insulators or locus control regions may be added to prevent silencing caused by integration. I do not think that the current nonviral vectors can be used for efficient gene delivery to the lung and it is no use at this point to propose strategies to enhance sustained transgene expression for them.

Regulated transgene expression is a subject that did not attract much attention of the gene therapy community. As described in other chapters of this book, the level as well as the temporal and spatial expression of each gene in eukaryotic organisms is carefully regulated. For current gene therapy studies, most therapeutic genes are expressed from heterologous promoters. Modification of these promoters may be needed to make the therapeutic gene expression match the normal expression pattern and/or level. This type of fine-tuning of design will require comprehensive knowledge of gene regulation. This will still be a daunting task for years to come. It is no exaggeration to say that improvement of human gene therapy is still dependent on the advancement of our knowledge of gene regulation.

Concluding Remarks

Gene therapy evolved largely from modern molecular genetics in anticipation of inventing novel treatments for human diseases. Our knowledge of gene expression and regulation plays a critical role in the design of gene therapy studies. Despite the initial underestimation of the difficulty in safety and efficiency of gene delivery, a few successful clinical studies have shown the feasibility of using gene therapy to cure fatal human diseases. Previous work in the field not only enhanced our understanding of the problems involved in gene therapy, but also will allow us to continue improving our methodology. Although the obstacles discussed here still impede our current progress in bringing clinical benefits of gene therapy to patients, all these problems can be solved theoretically. The complexity of human biology is expected to dictate different approaches to be used for different diseases, such as an *ex vivo* approach for blood diseases and a direct gene transfer method for airway diseases. Because of the differences in organ anatomy and in drug tolerance between rodents and humans, more large animal studies will be needed to check the safety and efficiency of gene therapy vectors before being clinical tested. As we know more about genes involved in diseases, more therapeutic "drugs" will be available. One can predict that more clinical benefits of gene therapy will be brought to society by a combined effort of the persistent basic and clinical researchers in the field.

Acknowledgment

I thank Drs. Carl A. Price and Jun Ma, and Mr. Quinn Hu for reviewing the manuscript. Research in my laboratory was supported by Operating Grants from the Canadian Institutes of Health Research, the Canadian Cystic Fibrosis Foundation, the Foundation Fighting Blindness-Canada and NIH, USA. J.H. is a CCFF Scholar, a recipient of the CCFF Zellers Senior Scientist Award, and holds a Premier's Research Excellence Award of Ontario, Canada.

References

Ahmed, R., and Gray, D. (1996). Immunological memory and protective immunity: understanding their relation, Science *272*, 54-60.

Aiuti, A., Slavin, S., Aker, M., Ficara, F., Deola, S., Mortellaro, A., Morecki, S., Andolfi, G., Tabucchi, A., Carlucci, F., *et al.* (2002). Correction of ADA-SCID by stem cell gene therapy combined with nonmyeloablative conditioning, Science *296*, 2410-3.

Alton, E. W., Stern, M., Farley, R., Jaffe, A., Chadwick, S. L., Phillips, J., Davies, J., Smith, S. N., Browning, J., Davies, M. G., *et al.* (1999). Cationic lipid-mediated CFTR gene transfer to the lungs and nose of patients with cystic fibrosis: a double-blind placebo-controlled trial, Lancet *353*, 947-54.

Arcasoy, S. M., Latoche, J., Gondor, M., Watkins, S. C., Henderson, R. A., Hughey, R., Finn, O. J., and Pilewski, J. M. (1997). MUC1 and other sialoglycoconjugates inhibit adenovirus-mediated gene transfer to epithelial cells, Am J Respir Cell Mol Biol *17*, 422-35.

Aronow, B. J., Silbiger, R. N., Dusing, M. R., Stock, J. L., Yager, K. L., Potter, S. S., Hutton, J. J., and Wiginton, D. A. (1992). Functional analysis of the human adenosine deaminase gene thymic regulatory region and its ability to generate position-independent transgene expression, Mol Cell Biol *12*, 4170-85.

Baldwin, A. S., Jr. (1996). The NF-kappa B and I kappa B proteins: new discoveries and insights, Annu Rev Immunol *14*, 649-83.

Bals, R., Xiao, W., Sang, N., Weiner, D. J., Meegalla, R. L., and Wilson, J. M. (1999). Transduction of well-differentiated airway epithelium by recombinant adeno-associated virus is limited by vector entry, J Virol *73*, 6085-8.

Barnett, B. G., Crews, C. J., and Douglas, J. T. (2002). Targeted adenoviral vectors, Biochim Biophys Acta *1575*, 1-14.

Barquinero, J., Eixarch, H., and Perez-Melgosa, M. (2004). Retroviral vectors: new applications for an old tool, Gene Ther *11 Suppl 1*, S3-9.

Barquinero, J., Segovia, J. C., Ramirez, M., Limon, A., Guenechea, G., Puig, T., Briones, J., Garcia, J., and Bueren, J. A. (2000). Efficient transduction of human hematopoietic repopulating cells generating stable engraftment of transgene-expressing cells in NOD/SCID mice, Blood *95*, 3085-93.

Bauer, T. R., Jr., Miller, A. D., and Hickstein, D. D. (1995). Improved transfer of the leukocyte integrin CD18 subunit into hematopoietic cell lines by using retroviral vectors having a gibbon ape leukemia virus envelope, Blood *86*, 2379-87.

Baum, C., Hegewisch-Becker, S., Eckert, H. G., Stocking, C., and Ostertag, W. (1995). Novel retroviral vectors for efficient expression of the multidrug resistance (mdr-1) gene in early hematopoietic cells, J Virol *69*, 7541-7.

Bellon, G., Michel-Calemard, L., Thouvenot, D., Jagneaux, V., Poitevin, F., Malcus, C., Accart, N., Layani, M. P., Aymard, M., Bernon, H., *et al.* (1997). Aerosol administration of a recombinant adenovirus expressing CFTR to cystic fibrosis patients: a phase I clinical trial, Hum Gene Ther *8*, 15-25.

Benihoud, K., Yeh, P., and Perricaudet, M. (1999). Adenovirus vectors for gene delivery, Curr Opin Biotechnol *10*, 440-7.

Berclaz, P. Y., Zsengeller, Z., Shibata, Y., Otake, K., Strasbaugh, S., Whitsett, J. A., and Trapnell, B. C. (2002). Endocytic internalization of adenovirus, nonspecific phagocytosis, and cytoskeletal organization are coordinately regulated in alveolar macrophages by GM-CSF and PU.1, J Immunol *169*, 6332-42.

Berns, K. I., and Giraud, C. (1996). Biology of adeno-associated virus, Curr Top Micriobiol Immunol *218*, 1-23.

Blankinship, M. J., Gregorevic, P., Allen, J. M., Harper, S. Q., Harper, H., Halbert, C. L., Miller, D. A., and Chamberlain, J. S. (2004). Efficient transduction of skeletal muscle using vectors based on adeno-associated virus serotype 6, Mol Ther *10*, 671-8.

Bodine, D. M., Seidel, N. E., and Orlic, D. (1996). Bone marrow collected 14 days after *in vivo* administration of granulocyte colony-stimulating factor and stem cell factor to mice has 10-fold more repopulating ability than untreated bone marrow, Blood *88*, 89-97.

Boothroyd, J. C., and Cross, G. A. (1982). Transcripts coding for variant surface glycoproteins of Trypanosoma brucei have a short, identical exon at their 5′ end, Gene *20*, 281-289.

Borgland, S. L., Bowen, G. P., Wong, N. C., Libermann, T. A., and Muruve, D. A. (2000). Adenovirus vector-induced expression of the C-X-C chemokine IP-10 is mediated through capsid-dependent activation of NF-kappaB, J Virol *74*, 3941-7.

Boucher, R. C. (1994). Human airway ion transport. Part 1., Am J Respir Crit Care Med *150*, 271-281.

Boucher, R. C. (1999). Status of gene therapy for cystic fibrosis lung disease. (Letter; Comment). (Review) (22 refs), J Clin Invest *103*, 441-5.

Boucher, R. C., Stutts, M. J., Knowles, M. R., Cantley, L., and Gatzy, J. T. (1986). Na+ transport in cystic fibrosis respiratory epithelia. Abnormal basal rate and response to adenylate cyclase activation, J Clin Invest *78*, 1245-52.

Bragonzi, A., Dina, G., Villa, A., Calori, G., Biffi, A., Bordignon, C., Assael, B. M., and Conese, M. (2000). Biodistribution and transgene expression with nonviral cationic vector/DNA complexes in the lungs, Gene Ther *7*, 1753-1760.

Brunetti-Pierri, N., Palmer, D. J., Beaudet, A. L., Carey, K. D., Finegold, M., and Ng, P. (2004). Acute toxicity after high-dose systemic injection of helper-dependent adenoviral vectors into nonhuman primates, Hum Gene Ther *15*, 35-46.

Cao, H., Koehler, D. R., and Hu, J. (2004). Adenoviral vectors for gene replacement therapy, Viral Immunol *17*, 327-33.

Cao, H. B., Wang, A., Martin, B., Koehler, D. R., Zeitlin, P. L., Tanawell, A. K., and Hu, J. (2005). Down-regulation of IL-8 expression in human airway epithelial cells through helper-dependent adenoviral-mediated RNA interference, Cell Res *15*, 111-119.

Caplen, N. J., Alton, E. W., Middleton, P. G., Dorin, J. R., Stevenson, B. J., Gao, X., Durham, S. R., Jeffery, P. K., Hodson, M. E., Coutelle, C., *et al.* (1995). Liposome-mediated CFTR gene transfer to the nasal epithelium of patients with cystic fibrosis [see comments], Nat Med *1*, 39-46.

Carter, B. J. (2004). Adeno-associated virus and the development of adeno-associated virus vectors: A historical perspective, Mol Ther *10*, 981-989.

Cavazzana-Calvo, M., Hacein-Bey, S., de Saint Basile, G., Gross, F., Yvon, E., Nusbaum, P., Selz, F., Hue, C., Certain, S., Casanova, J. L., *et al.* (2000). Gene therapy of human severe combined immunodeficiency (SCID)-X1 disease, Science *288*, 669-72.

Challita, P. M., Skelton, D., el-Khoueiry, A., Yu, X. J., Weinberg, K., and Kohn, D. B. (1995). Multiple modifications in cis elements of the long terminal repeat of retroviral vectors lead to increased expression and decreased DNA methylation in embryonic carcinoma cells, J Virol *69*, 748-55.

Chow, Y. H., O'Brodovich, H., Plumb, J., Wen, Y., Sohn, K. J., Lu, Z., Zhang, F., Lukacs, G. L., Tanswell, A. K., Hui, C. C., *et al.* (1997). Development of an epithelium-specific expression cassette with human DNA regulatory elements for transgene expression in lung airways, Proc Natl Acad Sci USA *94*, 14695-14700.

Chow, Y. H., Plumb, J., Wen, Y., Steer, B. M., Lu, Z., Buchwald, M., and Hu, J. (2000). Targeting Transgene Expression to Airway Epithelia and Submucosal Glands, Prominent Sites of Human CFTR Expression., Mol Ther *2*, 359-367.

Chu, Q., St George, J. A., Lukason, M., Cheng, S. H., Scheule, R. K., and Eastman, S. J. (2001). EGTA enhancement of adenovirus-mediated gene transfer to mouse tracheal epithelium *in vivo*, Hum Gene Ther *12*, 455-67.

Copreni, E., Penzo, M., Carrabino, S., and Conese, M. (2004). Lentivirus-mediated gene transfer to the respiratory epithelium: a promising approach to gene therapy of cystic fibrosis, Gene Ther *11 Suppl 1*, S67-75.

Crystal, R. G., McElvaney, N. G., Rosenfeld, M. A., Chu, C. S., Mastrangeli, A., Hay, J. G., Brody, S. L., Jaffe, H. A., Eissa, N. T., and Danel, C. (1994). Administration of an adenovirus containing the human CFTR cDNA to the respiratory tract of individuals with cystic fibrosis, Nat Genet *8*, 42-51.

Dai, Y., Schwarz, E. M., Gu, D., Zhang, W. W., Sarvetnick, N., and Verma, I. M. (1995). Cellular and humoral immune responses to adenoviral vectors containing factor IX gene: tolerization of factor IX and vector antigens allows for long-term expression, Proc Natl Acad Sci USA *92*, 1401-5.

Duan, D., Yue, Y., Engelhardt, J. F., and Evans, D. (2001). Expanding AAV packaging capacity with trans-splicing or overlapping vectors: a quantitative comparison
A complex containing CstF-64 and the SL2 snRNP connects mRNA 3′ end formation and trans-splicing in *C. elegans* operons, Mol Ther *4*, 383-91.

Dunbar, C. E., Seidel, N. E., Doren, S., Sellers, S., Cline, A. P., Metzger, M. E., Agricola, B. A., Donahue, R. E., and Bodine, D. M. (1996). Improved retroviral gene transfer into murine and Rhesus peripheral blood or bone marrow repopulating cells primed *in vivo* with stem cell factor and granulocyte colony-stimulating factor, Proc Natl Acad Sci USA 93, 11871-6.

Ehrhardt, A., Xu, H., and Kay, M. A. (2003). Episomal persistence of recombinant adenoviral vector genomes during the cell cycle in vivo, J Virol 77, 7689-95.

Evans, J. T., Kelly, P. F., O'Neill, E., and Garcia, J. V. (1999). Human cord blood CD34+CD38- cell transduction via lentivirus-based gene transfer vectors, Hum Gene Ther 10, 1479-89.

Fearon, D. T., and Locksley, R. M. (1996). The instructive role of innate immunity in the acquired immune response, Science 272, 50-3.

Ferrari, S., Griesenbach, U., Shiraki-Iida, T., Shu, T., Hironaka, T., Hou, X., Williams, J., Zhu, J., Jeffery, P. K., Geddes, D. M., et al. (2004). A defective nontransmissible recombinant Sendai virus mediates efficient gene transfer to airway epithelium in vivo, Gene Ther 11, 1659-64.

Gao, G. P., Alvira, M. R., Wang, L., Calcedo, R., Johnston, J., and Wilson, J. M. (2002). Novel adeno-associated viruses from rhesus monkeys as vectors for human gene therapy, Proc Natl Acad Sci USA 99, 11854-9.

Gao, X., and Huang, L. (1995). Cationic liposome-mediated gene transfer, Gene Ther 2, 710-22.

Gaspar, H. B., Parsley, K. L., Howe, S., King, D., Gilmour, K. C., Sinclair, J., Brouns, G., Schmidt, M., Von Kalle, C., Barington, T., et al. (2004). Gene therapy of X-linked severe combined immunodeficiency by use of a pseudotyped gammaretroviral vector, Lancet 364, 2181-2187.

George, J. S. (2003). Gene Therapy Progress and Prospects: Adenoviral Vector, Gene Ther 10, 1135-1141.

Gill, D. R., Southern, K. W., Mofford, K. A., Seddon, T., Huang, L., Sorgi, F., Thomson, A., MacVinish, L. J., Ratcliff, R., Bilton, D., et al. (1997). A placebo-controlled study of liposome-mediated gene transfer to the nasal epithelium of patients with cystic fibrosis, Gene Ther 4, 199-209.

Glimm, H., and Eaves, C. J. (1999). Direct evidence for multiple self-renewal divisions of human in vivo repopulating hematopoietic cells in short-term culture, 94, 2161-2168.

Glorioso, J. C., DeLuca, N. A., Goins, W. F., and Fink, D. J. (1994). Development of herpers simplex virus vectors for gene transfer to central nervous system. In Gene Therapeutics, J. A. Wolff, ed. (Boston, Birkhauser), pp. 281-302.

Grubb, B. R., Pickles, R. J., Ye, H., Yankaskas, J. R., Vick, R. N., Engelhardt, J. F., Wilson, J. M., Johnson, L. G., and Boucher, R. C. (1994). Inefficient gene transfer by adenovirus vector to cystic fibrosis airway epithelia of mice and humans, Nature 371, 802-6.

Hacein-Bey-Abina, S., Von Kalle, C., Schmidt, M., McCormack, M. P., Wulffraat, N., Leboulch, P., Lim, A., Osborne, C. S., Pawliuk, R., Morillon, E., et al. (2003). LMO2-associated clonal T cell proliferation in two patients after gene therapy for SCID-X1., Science 302, 415-419.

Hanenberg, H., Xiao, X. L., Dilloo, D., Hashino, K., Kato, I., and Williams, D. A. (1996). Colocalization of retrovirus and target cells on specific fibronectin fragments increases genetic transduction of mammalian cells, Nat Med 2, 876-82.

Hargrove, P. W., Vanin, E. F., Kurtzman, G. J., and Nienhuis, A. W. (1997). High-level globin gene expression mediated by a recombinant adeno-associated virus genome that contains the 3' gamma globin gene regulatory element and integrates as tandem copies in erythroid cells, Blood 89, 2167-75.

Harvey, B. G., Hackett, N. R., El-Sawy, T., Rosengart, T. K., Hirschowitz, E. A., Lieberman, M. D., Lesser, M. L., and Crystal, R. G. (1999). Variability of human systemic humoral immune responses to adenovirus gene transfer vectors administered to different organs, J Virol 73, 6729-42.

Hennemann, B., Conneally, E., Pawliuk, R., Leboulch, P., Rose-John, S., Reid, D., Chuo, J. Y., Humphries, R. K., and Eaves, C. J. (1999). Optimization of retroviral-mediated gene transfer to human NOD/SCID mouse repopulating cord blood cells through a systematic analysis of protocol variables, Exp Hematol 27, 817-25.

Hillgenberg, M., Tonnies, H., and Strauss, M. (2001). Chromosomal integration pattern of a helper-dependent minimal adenovirus vector with a selectable marker inserted into a 27.4-kilobase genomic stuffer, J Virol 75, 9896-908.

Jiang, Z., Feingold, E., Kochanek, S., and Clemens, P. R. (2002). Systemic delivery of a high-capacity adenoviral vector expressing mouse CTLA4Ig improves skeletal muscle gene therapy, Mol Ther 6, 369-76.

Jornot, L., Petersen, H., Lusky, M., Pavirani, A., Moix, I., Morris, and Rochat, T. (2001). Effects of first generation E1E3-deleted and second generation E1E3E4-deleted/modified adenovirus vectors on human endothelial cell death, Endothelium 8, 167-79.

Joseph, P. M., O'Sullivan, B. P., Lapey, A., Dorkin, H., Oren, J., Balfour, R., Perricone, M. A., Rosenberg, M., Wadsworth, S. C., Smith, A. E., et al. (2001). Aerosol and lobar administration of a recombinant adenovirus to individuals with cystic fibrosis. I. Methods, safety, and clinical implications, Hum Gene Ther 12, 1369-1382.

Kafri, T., Morgan, D., Krahl, T., Sarvetnick, N., Sherman, L., and Verma, I. (1998). Cellular immune response to adenoviral vector infected cells does not require de novo viral gene expression: implications for gene therapy, Proc Natl Acad Sci USA 95, 11377-82.

Kalberer, A., Zimmerman-Phillips, S., Barker, M. J., Geier, L., and Kalberer, C. P. (2000). Preselection of retrovirally transduced bone marrow avoids subsequent stem cell gene silencing and age-dependent extinction of expression of human beta-globin in engrafted mice.(comment), Annals of Otol Rhinol Laryngol - Supplement 185, 75-7.

Kaplan, J. M., Pennington, S. E., St., George, J. A., Woodworth, L. A., Fasbender, A., Marshall, J., Cheng, S. H., Wadsworth, S. C., Gregory, R. J., and Smith, A. E. (1998). Potentiation of gene transfer to the mouse lung by complexes of adenovirus vector and polycations improves therapeutic potential, Hum Gene Ther 9, 1469-79.

Kay, M. A., Holterman, A. X., Meuse, L., Gown, A., Ochs, H. D., Linsley, P. S., and Wilson, C. B. (1995). Long-term hepatic adenovirus-mediated gene expression in mice following CTLA4Ig administration, Nat Genet *11*, 191-7.

Kearns, W. G., Afione, S. A., Fulmer, S. B., Pang, M. C., Erikson, D., Egan, M., Landrum, M. J., Flotte, T. R., and Cutting, G. R. (1996). Recombinant adeno-associated virus (AAV-CFTR) vectors do not integrate in a site-specific fashion in an immortalized epithelial cell line, Gene Ther *3*, 748-55.

Kim, I. H., Jozkowicz, A., Piedra, P. A., Oka, K., and Chan, L. (2001). Lifetime correction of genetic deficiency in mice with a single injection of helper-dependent adenoviral vector, Proc Natl Acad Sci USA *98*, 13282-7.

Knowles, M. R., Hohneker, K. W., Zhou, Z., Olsen, J. C., Noah, T. L., Hu, P. C., Leigh, M. W., Engelhardt, J. F., Edwards, L. J., Jones, K. R., et al. (1995). A controlled study of adenoviral-vector-mediated gene transfer in the nasal epithelium of patients with cystic fibrosis (see comments), Nat Genet *333*, 823-31.

Koehler, D. R., Downey, G. P., Sweezey, N. B., Tanswell, A. K., and Hu, J. (2004). Lung inflammation as a therapeutic target in cystic fibrosis, Am J Respir Cell Mol Biol *31*, 377-81.

Koehler, D. R., Frndova, H., Leung, K., Louca, E., Palmer, D., Ng, P., McKerlie, C., Cox, P., Coates, A. L., and Hu, J. (2005). Aerosol delivery of an enhanced helper-dependent adenovirus formulation to rabbit lung using an intratracheal catheter, J Gene Med, In press.

Koehler, D. R., Hitt, M. M., and Hu, J. (2001). Challenges and strategies for cystic fibrosis lung gene therapy., Mol Ther *4*, 84-91.

Koehler, D. R., Sajjan, U., Chow, Y.-H., Martin, B., Kent, G., Tanswell, A. K., McKerlie, C., Forstner, J. F., and Hu, J. (2003). Protection of Cftr knockout mice from acute lung infection by a helper-dependent adenoviral vector expressing Cftr in airway epithelia, Proc Natl Acad Sci USA *100*, 15364-15369.

Koenig, M., Hoffman, E. P., Bertelson, C. J., Monaco, A. P., Feener, C., and Kunkel, L. M. (1987). Complete cloning of the Duchenne muscular dystrophy (DMD) cDNA and preliminary genomic organization of the DMD gene in normal and affected individuals, Cell *50*, 509-17.

Kojaoghlanian, T., Flomenberg, P., and Horwitz, M. S. (2003). The Impact of Adenovirus Infection on the Immunocompromised Host, Rev Med Virol *13*, 155-171.

Kollen, W. J., Mulberg, A. E., Wei, X., Sugita, M., Raghuram, V., Wang, J., Foskett, J. K., Glick, M. C., and Scanlin, T. F. (1999). High-efficiency transfer of cystic fibrosis transmembrane conductance regulator cDNA into cystic fibrosis airway cells in culture using lactosylated polylysine as a vector, Hum Gene Ther *10*, 615-22.

Kotin, R. M., Siniscalco, M., Samulski, R. J., Zhu, X. D., Hunter, L., Laughlin, C. A., McLaughlin, S., Muzyczka, N., Rocchi, M., and Berns, K. I. (1990). Site-specific integration by adeno-associated virus, Proc Natl Acad Sci USA *87*, 2211-5.

Liang, X. H., Haritan, A., Uliel, S., and Michaeli, S. (2003). trans and cis splicing in trypanosomatids: mechanism, factors, and regulation, Eukaryotic Cell *2*, 830-40.

Limberis, M., Anson, D. S., Fuller, M., and Parsons, D. W. (2002). Recovery of airway cystic fibrosis transmembrane conductance regulator function in mice with cystic fibrosis after single-dose lentivirus-mediated gene transfer.[erratum appears in Hum Gene Ther. 2002 Nov 20;13(17)2112, Hum Gene Ther *13*, 1961-70.

Mack, C. A., Song, W. R., Carpenter, H., Wickham, T. J., Kovesdi, I., Harvey, B. G., Magovern, C. J., Isom, O. W., Rosengart, T., Falck-Pedersen, E., et al. (1997). Circumvention of anti-adenovirus neutralizing immunity by administration of an adenoviral vector of an alternate serotype, Hum Gene Ther *8*, 99-109.

Maione, D., Della Rocca, C., Giannetti, P., D'Arrigo, R., Liberatoscioli, L., Franlin, L. L., Sandig, V., Ciliberto, G., La Monica, N., and Savino, R. (2001). An improved helper-dependent adenoviral vector allows persistent gene expression after intramuscular delivery and overcomes preexisting immunity to adenovirus, Proc Natl Acad Sci USA *98*, 5986-91.

Mastrangeli, A., Harvey, B. G., Yao, J., Wolff, G., Kovesdi, I., Crystal, R. G., and Falck-Pedersen, E. (1996). "Sero-switch" adenovirus-mediated *in vivo* gene transfer: circumvention of anti-adenovirus humoral immune defenses against repeat adenovirus vector administration by changing the adenovirus serotype, Hum Gene Ther *7*, 79-87.

Matsushima, K., Morishita, K., Yoshimura, T., Lavu, S., Kobayashi, Y., Lew, W., Appella, E., Kung, H. F., Leonard, E. J., and Oppenheim, J. J. (1988). Molecular cloning of a human monocyte-derived neutrophil chemotactic factor (MDNCF) and the induction of MDNCF mRNA by interleukin 1 and tumor necrosis factor, J Exp Med *167*, 1883-93.

McCoy, R. D., Davidson, B. L., Roessler, B. J., Huffnagle, G. B., Janich, S. L., Laing, T. J., and Simon, R. H. (1995). Pulmonary inflammation induced by incomplete or inactivated adenoviral particles, Hum Gene Ther *6*, 1553-60.

McElvaney, N. G., and Crystal, R. G. (1995). IL-6 release and airway administration of human CFR cDNA adenovirus vector, Nat Med *1*, 182-4.

Medzhitov, R., and Janeway, C., Jr. (2000). Innate immunity, N Engl J Med *343*, 338-44.

Michou, A. I., Santoro, L., Christ, M., Julliard, V., Pavirani, A., and Mehtali, M. (1997). Adenovirus-mediated gene transfer: influence of transgene, mouse strain and type of immune response on persistence of transgene expression, Gene Ther *4*, 473-82.

Moritz, T., Dutt, P., Xiao, X., Carstanjen, D., Vik, T., Hanenberg, H., and Williams, D. A. (1996). Fibronectin improves transduction of reconstituting hematopoietic stem cells by retroviral vectors: evidence of direct viral binding to chymotryptic carboxy-terminal fragments, Blood *88*, 855-62.

Moritz, T., Patel, V. P., and Williams, D. A. (1994). Bone marrow extracellular matrix molecules improve gene transfer into human hematopoietic cells via retroviral vectors, J Clin Invest 93, 1451-7.

Morral, N., O'Neal, W. K., Rice, K., Leland, M. M., Piedra, P. A., Aguilar-Cordova, E., Carey, K. D., Beaudet, A. L., and Langston, C. (2002). Lethal toxicity, severe endothelial injury, and a threshold effect with high doses of an adenoviral vector in baboons, Hum Gene Ther 13, 143-54.

Morsy, M. A., Gu, M., Motzel, S., Zhao, J., Lin, J., Su, Q., Allen, H., Franlin, L., Parks, R. J., Graham, F. L., et al. (1998). An adenoviral vector deleted for all viral coding sequences results in enhanced safety and extended expression of a leptin transgene, Proc Natl Acad Sci USA 95, 7866-71.

Moss, R. B., Rodman, D., Spencer, L. T., Aitken, M. L., Zeitlin, P. L., Waltz, D., Milla, C., Brody, A. S., Clancy, J. P., Ramsey, B., et al. (2004). Repeated adeno-associated virus serotype 2 aerosol-mediated cystic fibrosis transmembrane regulator gene transfer to the lungs of patients with cystic fibrosis: a multicenter, double-blind, placebo-controlled trial, Chest 125, 509-21.

Noone, P. G., Hohneker, K. W., Zhou, Z., Johnson, L. G., Foy, C., Gipson, C., Jones, K., Noah, T. L., Leigh, M. W., Schwartzbach, C., et al. (2000). Safety and biological efficacy of a lipid-CFTR complex for gene transfer in the nasal epithelium of adult patients with cystic fibrosis, Mol Ther 1, 105-14.

O'Neal, W. K., Rose, E., Zhou, H., Langston, C., Rice, K., Carey, D., and Beaudet, A. L. (2000). Multiple advantages of alpha-fetoprotein as a marker for in vivo gene transfer, Mol Ther 2, 640-8.

Oshima, R. G., Abrams, L., and Kulesh, D. (1990). Activation of an intron enhancer within the keratin 18 gene by expression of c-fos and c-jun in undifferentiated F9 embryonal carcinoma cells, Genes Dev 4, 835-848.

Overbaugh, J., Miller, A. D., and Eiden, M. V. (2001). Receptors and entry cofactors for retroviruses include single and multiple transmembrane-spanning proteins as well as newly described glycophosphatidylinositol-anchored and secreted proteins, Microbiol Mol Biol Rev 65, 371-89.

Palmer, D., and Ng, P. (2003). Improved system for helper-dependent adenoviral vector production, Mol Ther 8, 846-52.

Parks, R. J. (2000). Improvements in adenoviral vector technology: overcoming barriers for gene therapy, Clin Genet 58, 1-11.

Pergolizzi, R. G., and Crystal, R. G. (2004). Genetic medicine at the RNA level: modifications of the genetic repertoire for therapeutic purposes by pre-mRNA trans-splicing, Comptes Rendus Biologies 327, 695-709.

Perricone, M. A., Morris, J. E., Pavelka, K., Plog, M. S., O'Sullivan, B. P., Joseph, P. M., Dorkin, H., Lapey, A., Balfour, R., Meeker, D. P., et al. (2001). Aerosol and lobar administration of a recombinant adenovirus to individuals with cystic fibrosis. ii. transfection efficiency in airway epithelium, Hum Gene Ther 12, 1383-94.

Pickles, R. J., Fahrner, J. A., Petrella, J. M., Boucher, R. C., and Bergelson, J. M. (2000). Retargeting the coxsackievirus and adenovirus receptor to the apical surface of polarized epithelial cells reveals the glycocalyx as a barrier to adenovirus-mediated gene transfer, J Virol 74, 6050-7.

Porteous, D. J., Dorin, J. R., McLachlan, G., Davidson-Smith, H., Davidson, H., Stevenson, B. J., Carothers, A. D., Wallace, W. A., Moralee, S., Hoenes, C., et al. (1997). Evidence for safety and efficacy of DOTAP cationic liposome mediated CFTR gene transfer to the nasal epithelium of patients with cystic fibrosis, Gene Ther 4, 210-8.

Pringle, C. R. (1999). Virus taxonomy--1999. The universal system of virus taxonomy, updated to include the new proposals ratified by the International Committee on Taxonomy of Viruses during 1998, Arch Virol 144, 421-9.

Pupita, F., and Barone, A. (1983). Clinical pharmacology of DEAE-dextran for long-term administration (one year), Int J Clin Pharmacol Res 3, 287-93.

Qin, L., Ding, Y., Pahud, D. R., Chang, E., Imperiale, M. J., and Bromberg, J. S. (1997). Promoter attenuation in gene therapy: Interferon-gamma and tumor necrosis factor-alpha inhibit transgene expression, Hum Gene Ther 8, 2019-2029.

Raper, S. E., Chirmule, N., Lee, F. S., Wivel, N. A., Bagg, A., Gao, G. P., Wilson, J. M., and Batshaw, M. L. (2003). Fatal systemic inflammatory response syndrome in a ornithine transcarbamylase deficient patient following adenoviral gene transfer, Mol Genet Metab 80, 148-58.

Raper, S. E., Yudkoff, M., Chirmule, N., Gao, G. P., Nunes, F., Haskal, Z. J., Furth, E. E., Propert, K. J., Robinson, M. B., Magosin, S., et al. (2002). A pilot study of in vivo liver-directed gene transfer with an adenoviral vector in partial ornithine transcarbamylase deficiency, Hum Gene Ther 13, 163-75.

Reich, S. J., Auricchio, A., Hildinger, M., Glover, E., Maguire, A. M., Wilson, J. M., and Bennett, J. (2003). Efficient trans-splicing in the retina expands the utility of adeno-associated virus as a vector for gene therapy, Hum Gene Ther 14, 37-44.

Rogers, S., and Moore, M. (1963). Studies of the mechanism of action of the shope rabbit papilloma virus. I. Concerning the nature of the induction of arginase in the infected cells, J Exp Med 117, 521-542.

Rommens, J. M., Iannuzzi, M. C., Kerem, B., Drumm, M. L., Melmer, G., Dean, M., Rozmahel, R., Cole, J. L., Kennedy, D., Hidaka, N., et al. (1989). Identification of the cystic fibrosis gene: chromosome walking and jumping, Science 245, 1059-65.

Ruiz, F. E., Clancy, J. P., Perricone, M. A., Bebok, Z., Hong, J. S., Cheng, S. H., Meeker, D. P., Young, K. R., Schoumacher, R. A., Weatherly, M. R., et al. (2001). A clinical inflammatory syndrome attributable to aerosolized lipid-DNA administration in cystic fibrosis, Hum Gene Ther 12, 751-61.

Salvi, S., and Holgate, S. T. (1999). Could the airway epithelium play an important role in mucosal immunoglobulin A production?, Clin Exp Allergy 29, 1597-605.

Scaria, A., St George, J. A., Gregory, R. J., Noelle, R. J., Wadsworth, S. C., Smith, A. E., and Kaplan, J. M. (1997). Antibody to CD40 ligand inhibits both humoral and cellular immune responses to adenoviral vectors and facilitates repeated administration to mouse airway, Gene Ther 4, 611-7.

Schaack, J., Bennett, M. L., Colbert, J. D., Torres, A. V., Clayton, G. H., Ornelles, D., and Moorhead, J. (2004). E1A and E1B proteins inhibit inflammation induced by adenovirus, Proc Natl Acad Sci USA 101, 3124-9.

Schnepp, B. C., Clark, K. R., Klemanski, D. L., Pacak, C. A., and Johnson, P. R. (2003). Genetic fate of recombinant adeno-associated virus vector genomes in muscle, J Virol 77, 3495-504.

Singh, B., and de la Concha-Bermejillo, A. (1998). Gadolinium chloride removes pulmonary intravascular macrophages and curtails the degree of ovine lentivirus-induced lymphoid interstitial pneumonia, Int J Exp Pathol 79, 151-62.

Singhal, A., and Huang, L. (1994). Gene transfer in mammalian cells uisng liposomes as carriers. In Gene Therapeutics, J. A. Wolff, ed. (Boston, Birkhauser), pp. 118-142.

Sinn, P. L., Hickey, M. A., Staber, P. D., Dylla, D. E., Jeffers, S. A., Davidson, B. L., Sanders, D. A., and McCray, P. B., Jr. (2003). Lentivirus vectors pseudotyped with filoviral envelope glycoproteins transduce airway epithelia from the apical surface independently of folate receptor alpha, J Virol 77, 5902-10.

Sirninger, J., Muller, C., Braag, S., Tang, Q., Yue, H., Detrisac, C., Ferkol, T., Guggino, W. B., and Flotte, T. R. (2004). Functional characterization of a recombinant adeno-associated virus 5-pseudotyped cystic fibrosis transmembrane conductance regulator vector, Hum Gene Ther 15, 832-41.

Smith, A. E. (1995). Viral vectors in gene therapy, Annu Rev Microbiol 49, 807-38.

Smith, C. A., Woodruff, L. S., Rooney, C., and Kitchingman, G. R. (1998). Extensive cross-reactivity of adenovirus-specific cytotoxic T cells, Hum Gene Ther 9, 1419-27.

Smith, T., Idamakanti, N., Kylefjord, H., Rollence, M., King, L., Kaloss, M., Kaleko, M., and Stevenson, S. C. (2002). In vivo hepatic adenoviral gene delivery occurs independently of the coxsackievirus-adenovirus receptor, Mol Ther 5, 770-9.

Spina, D. (1998). Epithelium smooth muscle regulation and interactions, Am J Respir Crit Care Med 158, S141-5.

Summerford, C., and Samulski, R. J. (1998). Membrane-associated heparan sulfate proteoglycan is a receptor for adeno-associated virus type 2 virions, J Virol 72, 1438-45.

Sung, R. S., Qin, L., and Bromberg, J. S. (2001). TNF-alpha and IFN-gamma induced by innate anti-adenoviral immune responses inhibit adenovirus-mediated transgene expression, Mol Ther 3, 757-67.

Suzuki, H., Kurihara, Y., Takeya, M., Kamada, N., Kataoka, M., Jishage, K., Ueda, O., Sakaguchi, H., Higashi, T., Suzuki, T., et al. (1997). A role for macrophage scavenger receptors in atherosclerosis and susceptibility to infection, Nature 386, 292-6.

Tal, J. (2000). Adeno-associated virus-based vectors in gene therapy, J Biomed Sci 7, 279-91.

Terheggen, H. G., Lowenthal, A., Lavinha, F., Colombo, J. P., and Rogers, S. (1975). Unsuccessful trial of gene replacement in arginase deficiency, J Exp Med 119, 1-3.

Thepen, T., McMenamin, C., Oliver, J., Kraal, G., and Holt, P. G. (1991). Alveolar macrophage elimination in vivo is associated with an increase in pulmonary immune response in mice, Eur J Immunol 21, 2845-50.

Toietta, G., Koehler, D. R., Finegold, M., Lee, B., Hu, J., and Beaudet, A. L. (2003). Reduced inflammation and improved airway expression using helper-dependent adenoviral vectors with a K18 promoter, Mol Ther 7, 649-658.

Tripathy, S. K., Black, H. B., Goldwasser, E., and Leiden, J. M. (1996). Immune responses to transgene-encoded proteins limit the stability of gene expression after injection of replication-defective adenovirus vectors, Nat Med 2, 545-50.

Tsui, L. C. (1995). The Cystic Fibrosis Transmembrane conductance Regulator gene, Am J Respir Crit Care Med 151, S47-S53.

van der Loo, J. C., Liu, B. L., Goldman, A. I., Buckley, S. M., and Chrudimsky, K. S. (2002). Optimization of gene transfer into primitive human hematopoietic cells of granulocyte-colony stimulating factor-mobilized peripheral blood using low-dose cytokines and comparison of a gibbon ape leukemia virus versus an RD114-pseudotyped retroviral vector, Hum Gene Ther 13, 1317-30.

Veena, P., Traycoff, C. M., Williams, D. A., McMahel, J., Rice, S., Cornetta, K., and Srour, E. F. (1998). Delayed targeting of cytokine-nonresponsive human bone marrow CD34(+) cells with retrovirus-mediated gene transfer enhances transduction efficiency and long-term expression of transduced genes, Blood 91, 3693-701.

Wagner, J. A., Messner, A. H., Moran, M. L., Daifuku, R., Kouyama, K., Desch, J. K., Manley, S., Norbash, A. M., Conrad, C. K., Friborg, S., et al. (1999a). Safety and biological efficacy of an adeno-associated virus vector- cystic fibrosis transmembrane regulator (AAV-CFTR) in the cystic fibrosis maxillary sinus, Laryngoscope 109, 266-74.

Wagner, J. A., Nepomuceno, I. B., Shah, N., Messner, A. H., Moran, M. L., Norbash, A. M., Moss, R. B., Wine, J. J., and Gardner, P. (1999b). Maxillary sinusitis as a surrogate model for CF gene therapy clinical trials in patients with antrostomies, J Gene Med 1, 13-21.

Wang, B., Li, J., and Xiao, X. (2000). Adeno-associated virus vector carrying human minidystrophin genes effectively ameliorates muscular dystrophy in mdx mouse model.[see comment], Proc Natl Acad Sci USA 97, 13714-9.

Wang, Z., Zhu, T., Qiao, C., Zhou, L., Wang, B., Zhang, J., Chen, C., Li, J., and Xiao, X. (2005). Adeno-associated virus serotype 8 efficiently delivers genes to muscle and heart, Nat Biotechnol 23, 321-8.

Welsh, M. J., Tsui, L. C., Boat, T. F., and Beaudet, A. L. (1995). Cystic Fibrosis. In The Metabolic and Molecular Basis of Inherited Disease, C. R. Scriver, A. L. Beaudet, W. S. Sly, and D.

Valle, eds. (New York, McGraw-Hill), pp. 3799-3876.

Wilson, C. B., Embree, L. J., Schowalter, D., Albert, R., Aruffo, A., Hollenbaugh, D., Linsley, P., and Kay, M. A. (1998). Transient inhibition of CD28 and CD40 ligand interactions prolongs adenovirus-mediated transgene expression in the lung and facilitates expression after secondary vector administration, J Virol 72, 7542-50.

Wivel, N. A., Gao, G. P., and Wilson, J. M. (1999). Adenovirus vectors. In The Development of Human Gene Therapy, T. Friedmann, ed. (San Diego, CA, Cold Spring Harbor Laboratory Press), pp. 87-110.

Wolff, J. A., and Lederberg, J. (1994). A history of gene transfer and therapy. In Gene Therapeutics, J. A. Wolff, ed. (Boston, Birkhauser), pp. 3-25.

Worgall, S., Leopold, P. L., Wolff, G., Ferris, B., Van Roijen, N., and Crystal, R. G. (1997). Role of alveolar macrophages in rapid elimination of adenovirus vectors administered to the epithelial surface of the respiratory tract, Hum Gene Ther 8, 1675-84.

Yam, P. Y., Yee, J. K., Ito, J. I., Sniecinski, I., Doroshow, J. H., Forman, S. J., and Zaia, J. A. (1998). Comparison of amphotropic and pseudotyped VSV-G retroviral transduction in human CD34+ peripheral blood progenitor cells from adult donors with HIV-1 infection or cancer, Exp Hematol 26, 962-8.

Yamashita, K., Masunaga, T., Yanagida, N., Takehara, M., Hashimoto, T., Kobayashi, T., Echizenya, H., Hua, N., Fujita, M., Murakami, M., *et al.* (2003). Long-term acceptance of rat cardiac allografts on the basis of adenovirus mediated CD40Ig plus CTLA4Ig gene therapies, Transplantation 76, 1089-96.

Yan, Z., Zhang, Y., Duan, D., and Engelhardt, J. F. (2000). Trans-splicing vectors expand the utility of adeno-associated virus for gene therapy.(see comment), Proc Natl Acad Sci USA 97, 6716-21.

Yang, Y., Jooss, K. U., Su, Q., Ertl, H. C., and Wilson, J. M. (1996). Immune responses to viral antigens versus transgene product in the elimination of recombinant adenovirus-infected hepatocytes *in vivo*, Gene Ther 3, 137-144.

Yang, Y., Li, Q., Ertl, H. C., and Wilson, J. M. (1995). Cellular and humoral immune responses to viral antigens create barriers to lung-directed gene therapy with recombinant adenoviruses, J Virol 69, 2004-15.

Yu, S. F., von Ruden, T., Kantoff, P. W., Garber, C., Seiberg, M., Ruther, U., Anderson, W. F., Wagner, E. F., and Gilboa, E. (1986). Self-inactivating retroviral vectors designed for transfer of whole genes into mammalian cells, Proc Natl Acad Sci USA 83, 3194-8.

Zabner, J., Cheng, S. H., Meeker, D., Launspach, J., Balfour, R., Perricone, M. A., Morris, J. E., Marshall, J., Fasbender, A., Smith, A. E., and Welsh, M. J. (1997). Comparison of DNA-lipid complexes and DNA alone for gene transfer to cystic fibrosis airway epithelia *in vivo*, J Clin Invest 100, 1529-37.

Zabner, J., Couture, L. A., Gregory, R. J., Graham, S. M., Smith, A. E., and Welsh, M. J. (1993). Adenovirus-mediated gene transfer transiently corrects the chloride transport defect in nasal epithelia of patients with cystic fibrosis, Cell 75, 207-16.

Zsengeller, Z., Otake, K., Hossain, S. A., Berclaz, P. Y., and Trapnell, B. C. (2000). Internalization of adenovirus by alveolar macrophages initiates early proinflammatory signaling during acute respiratory tract infection, J Virol 74, 9655-67.